KIRK-OTHMER

ENCYCLOPEDIA OF CHEMICAL TECHNOLOGY

Third Edition

VOLUME 23

Thyroid and Antithyroid Preparations
to
Vinyl Polymers

KIRK-OTHMER

ENCYCLOPEDIA OF CHEMICAL TECHNOLOGY

THIRD EDITION

VOLUME 23

THYROID AND ANTITHYROID PREPARATIONS
TO
VINYL POLYMERS

A WILEY-INTERSCIENCE PUBLICATION

John Wiley & Sons

NEW YORK · CHICHESTER · BRISBANE · TORONTO · SINGAPORE

Library of Congress Cataloging in Publication Data:

Main entry under title:
    Encyclopedia of chemical technology.

    At head of title: Kirk-Othmer.
    "A Wiley-Interscience publication."
    Includes bibliographies.
    1. Chemistry, Technical—Dictionaries. I. Kirk, Raymond
Eller, 1890–1957. II. Othmer, Donald Frederick, 1904—
    III. Grayson, Martin. IV. Eckroth, David. V. Title:
Kirk-Othmer encyclopedia of chemical technology.

TP9.E685   1978   660'.03   77-15820
ISBN 0-471-02076-1

CONTENTS

Thyroid and antithyroid
    preparations, 1
Tin and tin alloys, 18
Tin compounds, 42
Tire cords, 78
Titanium and titanium alloys, 98
Titanium compounds, 131
Toluene, 246
Tool materials, 273
Trace and residue analysis, 310
Trademarks and copyrights, 348
Transportation, 375
Triphenylmethane and related
    dyes, 399
Tungsten and tungsten alloys, 413
Tungsten compounds, 426
Ultrafiltration, 439
Ultrasonics, 462
Units and conversion factors, 491

Uranium and uranium compounds,
    502
Urea, 548
Urethane polymers, 576
Uric acid, 608
Uv stabilizers, 615
Vaccine technology, 628
Vacuum technology, 644
Vanadium and vanadium alloys,
    673
Vanadium compounds, 688
Vanillin, 704
Vegetable oils, 717
Veterinary drugs, 742
Vinegar, 753
Vinylidene chloride and
    poly(vinylidene chloride),
    764
Vinyl polymers, 798

EDITORIAL STAFF
FOR VOLUME 23

Executive Editor: **Martin Grayson**
Associate Editor: **David Eckroth**
Production Supervisor: **Michalina Bickford**
Editors: **Joyce Brown**   **Caroline L. Eastman**   **Carolyn Golojuch**
             Anna Klingsberg   **Mimi Wainwright**

CONTRIBUTORS
TO VOLUME 23

V. E. Archer, *Vulcan Materials Co., Birmingham, Alabama,* Detinning under Tin and tin alloys

Derek Bannister, *CIBA-GEIGY Corporation, Toms River, New Jersey,* Triphenylmethane and related dyes

Edmund F. Baroch, *International Titanium, Inc., Moses Lake, Washington,* Vanadium and vanadium alloys

Jon A. Baumgarten, *Paskus, Gordon & Hyman, Washington, D. C.,* Trademarks and copyrights

David L. Cincera, *Air Products and Chemicals, Inc., Allentown, Pennsylvania,* Poly(vinyl alcohol) under Vinyl polymers

J. A. Cowfer, *BFGoodrich Co., Brecksville, Ohio,* Vinyl chloride under Vinyl polymers

Wiley Daniels, *Air Products and Chemicals, Inc., Allentown, Pennsylvania,* Poly(vinyl acetate) under Vinyl polymers

John A. Davidson, *BFGoodrich Co., Avon Lake, Ohio,* Poly(vinyl chloride) under Vinyl polymers

J. D. Desai, *General Electric Company, Schenectady, New York,* Tool materials

Martin Dexter, *CIBA-GEIGY Corporation, Ardsley, New York,* Uv stabilizers

R. B. Dougherty, *American Cyanamid, Princeton, New Jersey,* Veterinary drugs

John Elliott, *CIBA-GEIGY Corporation, Toms River, New Jersey,* Triphenylmethane and related dyes

Keith L. Gardner, *BFGoodrich Co., Avon Lake, Ohio,* Poly(vinyl chloride) under Vinyl polymers

R. C. Gasman, *GAF Corporation, West Milford, New Jersey,* Vinyl ether monomers and polymers under Vinyl polymers

William Germain, *Vulcan Materials Co., Birmingham, Alabama,* Detinning under Tin and tin alloys

Dale S. Gibbs, *Dow Chemical U.S.A., Midland, Michigan,* Vinylidene chloride and poly(vinylidene chloride)

Melvin H. Gitlitz, *M&T Chemicals, Inc., Rahway, New Jersey,* Tin compounds

W. Hamm, *Unilever Research, Colworth Laboratory, London, United Kingdom,* Vegetable oils

M. C. Hoff, *Amoco Chemicals Corporation, Naperville, Illinois,* Toluene

Stanley Hoffman, *Union Carbide Corporation, Danbury, Connecticut,* Transportation

Eugene V. Hort, *GAF Corporation, Wayne, New Jersey,* Vinyl ether monomers and polymers; N-Vinyl monomers and polymers both under Vinyl polymers

V. A. Jagede, *Lederle Laboratories, American Cyanamid Co., Pearl River, New York,* Vaccine technology

T. K. Kim, *GTE Products Corporation, Towanda, Pennsylvania,* Tungsten compounds

P. Klinkowski, *Dorr-Oliver, Inc., Stamford, Connecticut,* Ultrafiltration

Donald Knittel, *Cabot Corporation, Kokomo, Indiana,* Titanium and titanium alloys

R. Komanduri, *General Electric Company, Schenectady, New York,* Tool materials

K. J. Kowal, *Lederle Laboratories, American Cyanamid Co., Pearl River, New York,* Vaccine technology

Edward Lavin, *Monsanto Company, Indian Orchard, Massachusetts,* Poly(vinyl acetal)s under Vinyl polymers

Pedro A. Lehmann F., *Centro de Investigación y de Estudios Avanzados del I.P.N. México 0700 D. F.,* Thyroid and antithyroid preparations

Charles H. Lieb, *Paskus, Gordon & Hyman, New York, New York,* Trademarks and copyrights

W. Lin, *Lederle Laboratories, American Cyanamid Co., Pearl River, New York,* Vaccine technology

Robert P. Lukens, *American Society for Testing and Materials, Philadelphia, Pennsylvania,* Units and conversion factors

M. B. MacInnis, *GTE Products Corporation, Towanda, Pennsylvania,* Tungsten compounds

A. J. Magistro, *BFGoodrich Co., Brecksville, Ohio,* Vinyl chloride under Vinyl polymers

Ivo Mavrovic, *Consultant, New York, New York,* Urea

D. J. Maykuth, *Tin Research Institute, Inc., Columbus, Ohio,* Tin and tin alloys

Bruce McDuffie, *State University of New York at Binghamton, Binghamton, New York,* Trace and residue analysis

Ron E. McKeighen, *Krautkramer-Branson, Inc., Lewiston, Pennsylvania,* Ultrasonics, low power

Craig S. Miller, *Krautkramer-Branson, Inc., Lewiston, Pennsylvania,* Ultrasonics, low power

Norman Milleron, *EMR Photoelectric, Princeton, New Jersey,* Vacuum technology

Marguerite K. Moran, *M&T Chemicals, Inc., Rahway, New Jersey,* Tin compounds

James A. Mullendore, *GTE Product Corp., Towanda, Pennsylvania,* Tungsten and tungsten alloys

David M. Petrick, *American Cyanamid Co., Princeton, New Jersey,* Veterinary drugs

M. Ritchey, *Lederle Laboratories, American Cyanamid Co., Pearl River, New York,* Vaccine technology

Christian S. Rondestvedt, Jr., *E. I. du Pont de Nemours & Co., Inc., Wilmington, Delaware,* Titanium compounds, organic

Joseph Rose, *Drexel University, Philadelphia, Pennsylvania,* Ultrasonics, low power

Joe B. Rosenbaum, *Consultant, Salt Lake City, Utah,* Vanadium compounds

A. Ray Shirley, Jr., *Applied Chemical Technology, Muscle Shoals, Alabama,* Urea

Andrew Shoh, *Branson Sonic Power Company, Danbury, Connecticut,* Ultrasonics, high power

Leonard Skolnik, *BFGoodrich Co., University Heights, Ohio,* Tire cords

James A. Snelgrove, *Monsanto Company, Indian Orchard, Massachusetts,* Poly(vinyl acetal)s under Vinyl polymers

L. G. Sylvester, *CIBA-GEIGY Corporation, Ardsley, New York,* Uric acid

Henri Ulrich, *The Upjohn Company, North Haven, Connecticut,* Urethane polymers

J. H. Van Ness, *Monsanto Company, St. Louis, Missouri,* Vanillin

B. H. Waxman, *GAF Corporation, Wayne, New Jersey,* N-Vinyl monomers and polymers under Vinyl polymers

A. D. Webb, *University of California, Davis; Davis, California,* Vinegar

Fritz Weigel, *University of Munich, Munich, Federal Republic of Germany,* Uranium and uranium compounds

R. A. Wessling, *Dow Chemical U.S.A., Midland, Michigan,* Vinylidene chloride and poly(vinylidene) chloride

J. Whitehead, *Tioxide Group PLC, Stockton on Tees, Cleveland, United Kingdon,* Titanium compounds, inorganic

H. P. Wilson, *Vulcan Materials Co., Birmingham, Alabama,* Detinning under Tin and tin alloys

NOTE ON CHEMICAL ABSTRACTS SERVICE REGISTRY NUMBERS AND NOMENCLATURE

Chemical Abstracts Service (CAS) Registry Numbers are unique numerical identifiers assigned to substances recorded in the CAS Registry System. They appear in brackets in the *Chemical Abstracts* (CA) substance and formula indexes following the names of compounds. A single compound may have many synonyms in the chemical literature. A simple compound like phenethylamine can be named β-phenylethylamine or, as in *Chemical Abstracts*, benzeneethanamine. The usefulness of the *Encyclopedia* depends on accessibility through the most common correct name of a substance. Because of this diversity in nomenclature careful attention has been given the problem in order to assist the reader as much as possible, especially in locating the systematic CA index name by means of the Registry Number. For this purpose, the reader may refer to the CAS Registry Handbook-Number Section which lists in numerical order the Registry Number with the *Chemical Abstracts* index name and the molecular formula; eg, **458-88-8,** Piperidine, 2-propyl-, (*S*)-, $C_8H_{17}N$; in the *Encyclopedia* this compound would be found under its common name, coniine [*458-88-8*]. The Registry Number is a valuable link for the reader in retrieving additional published information on substances and also as a point of access for such on-line data bases as Chemline, Medline, and Toxline.

In all cases, the CAS Registry Numbers have been given for title compounds in articles and for all compounds in the index. All specific substances indexed in *Chemical Abstracts* since 1965 are included in the CAS Registry System as are a large number of substances derived from a variety of reference works. The CAS Registry System identifies a substance on the basis of an unambiguous computer-language description of its molecular structure including stereochemical detail. The Registry Number is a machine-checkable number (like a Social Security number) assigned in sequential order to each substance as it enters the registry system. The value of the number lies in the fact that it is a concise and unique means of substance identification, which is independent of, and therefore bridges, many systems of chemical nomenclature. For polymers, one Registry Number is used for the entire family; eg, polyoxyethylene (20) sorbitan monolaurate has the same number as all of its polyoxyethylene homologues.

Registry numbers for each substance will be provided in the third edition cumulative index and appear as well in the annual indexes (eg, Alkaloids shows the Registry Number of all alkaloids (title compounds) in a table in the article as well, but the intermediates have their Registry Numbers shown only in the index). Articles such as Analytical methods, Batteries and electric cells, Chemurgy, Distillation, Economic evaluation, and Fluid mechanics have no Registry Numbers in the text.

Cross-references are inserted in the index for many common names and for some systematic names. Trademark names appear in the index. Names that are incorrect, misleading or ambiguous are avoided. Formulas are given very frequently in the text to help in identifying compounds. The spelling and form used, even for industrial names, follow American chemical usage, but not always the usage of *Chemical Abstracts* (eg, *coniine* is used instead of *(S)-2-propylpiperidine*, *aniline* instead of *benzenamine*, and *acrylic acid* instead of *2-propenoic acid*).

There are variations in representation of rings in different disciplines. The dye industry does not designate aromaticity or double bonds in rings. All double bonds and aromaticity are shown in the *Encyclopedia* as a matter of course. For example, tetralin has an aromatic ring and a saturated ring and its structure appears in the

Encyclopedia with its common name, Registry Number enclosed in brackets, and parenthetical CA index name, ie, tetralin, [*119-64-2*] (1,2,3,4-tetrahydronaphthalene). With names and structural formulas, and especially with CAS Registry Numbers the aim is to help the reader have a concise means of substance identification.

CONVERSION FACTORS, ABBREVIATIONS, AND UNIT SYMBOLS

SI Units (Adopted 1960)

A new system of measurement, the International System of Units (abbreviated SI), is being implemented throughout the world. This system is a modernized version of the MKSA (meter, kilogram, second, ampere) system, and its details are published and controlled by an international treaty organization (The International Bureau of Weights and Measures) (1).

SI units are divided into three classes:

BASE UNITS

length	meter[†] (m)
mass[‡]	kilogram (kg)
time	second (s)
electric current	ampere (A)
thermodynamic temperature[§]	kelvin (K)
amount of substance	mole (mol)
luminous intensity	candela (cd)

SUPPLEMENTARY UNITS

plane angle	radian (rad)
solid angle	steradian (sr)

[†] The spellings "metre" and "litre" are preferred by ASTM; however "-er" are used in the Encyclopedia.

[‡] "Weight" is the commonly used term for "mass."

[§] Wide use is made of "Celsius temperature" (t) defined by

$$t = T - T_0$$

where T is the thermodynamic temperature, expressed in kelvins, and $T_0 = 273.15$ K by definition. A temperature interval may be expressed in degrees Celsius as well as in kelvins.

SUPPLEMENTARY UNITS

plane angle   radian (rad)
solid angle   steradian (sr)

DERIVED UNITS AND OTHER ACCEPTABLE UNITS

These units are formed by combining base units, supplementary units, and other derived units (2–4). Those derived units having special names and symbols are marked with an asterisk in the list below:

Quantity	Unit	Symbol	Acceptable equivalent
*absorbed dose	gray	Gy	J/kg
acceleration	meter per second squared	m/s^2	
*activity (of ionizing radiation source)	becquerel	Bq	1/s
area	square kilometer	km^2	
	square hectometer	hm^2	ha (hectare)
	square meter	m^2	
*capacitance	farad	F	C/V
concentration (of amount of substance)	mole per cubic meter	mol/m^3	
*conductance	siemens	S	A/V
current density	ampere per square meter	A/m^2	
density, mass density	kilogram per cubic meter	kg/m^3	g/L; mg/cm^3
dipole moment (quantity)	coulomb meter	C·m	
*electric charge, quantity of electricity	coulomb	C	A·s
electric charge density	coulomb per cubic meter	C/m^3	
electric field strength	volt per meter	V/m	
electric flux density	coulomb per square meter	C/m^2	
*electric potential, potential difference, electromotive force	volt	V	W/A
*electric resistance	ohm	Ω	V/A
*energy, work, quantity of heat	megajoule	MJ	
	kilojoule	kJ	
	joule	J	N·m
	electron volt[†]	eV[†]	
	kilowatt hour[†]	kW·h[†]	
energy density	joule per cubic meter	J/m^3	

[†] This non-SI unit is recognized by the CIPM as having to be retained because of practical importance or use in specialized fields (1).

Quantity	Unit	Symbol	Acceptable equivalent
*force	kilonewton	kN	
	newton	N	kg·m/s²
*frequency	megahertz	MHz	
	hertz	Hz	1/s
heat capacity, entropy	joule per kelvin	J/K	
heat capacity (specific), specific entropy	joule per kilogram kelvin	J/(kg·K)	
heat transfer coefficient	watt per square meter kelvin	W/(m²·K)	
*illuminance	lux	lx	lm/m²
*inductance	henry	H	Wb/A
linear density	kilogram per meter	kg/m	
luminance	candela per square meter	cd/m²	
*luminous flux	lumen	lm	cd·sr
magnetic field strength	ampere per meter	A/m	
*magnetic flux	weber	Wb	V·s
*magnetic flux density	tesla	T	Wb/m²
molar energy	joule per mole	J/mol	
molar entropy, molar heat capacity	joule per mole kelvin	J/(mol·K)	
moment of force, torque	newton meter	N·m	
momentum	kilogram meter per second	kg·m/s	
permeability	henry per meter	H/m	
permittivity	farad per meter	F/m	
*power, heat flow rate, radiant flux	kilowatt	kW	
	watt	W	J/s
power density, heat flux density, irradiance	watt per square meter	W/m²	
*pressure, stress	megapascal	MPa	
	kilopascal	kPa	
	pascal	Pa	N/m²
sound level	decibel	dB	
specific energy	joule per kilogram	J/kg	
specific volume	cubic meter per kilogram	m³/kg	
surface tension	newton per meter	N/m	
thermal conductivity	watt per meter kelvin	W/(m·K)	
velocity	meter per second	m/s	
	kilometer per hour	km/h	
viscosity, dynamic	pascal second	Pa·s	
	millipascal second	mPa·s	
viscosity, kinematic	square meter per second	m²/s	
	square millimeter per second	mm²/s	

Quantity	Unit	Symbol	Acceptable equivalent
volume	cubic meter	m^3	
	cubic decimeter	dm^3	L(liter) (5)
	cubic centimeter	cm^3	mL
wave number	1 per meter	m^{-1}	
	1 per centimeter	cm^{-1}	

In addition, there are 16 prefixes used to indicate order of magnitude, as follows:

Multiplication factor	Prefix	Symbol	Note
10^{18}	exa	E	
10^{15}	peta	P	
10^{12}	tera	T	
10^9	giga	G	
10^6	mega	M	
10^3	kilo	k	
10^2	hecto	h[a]	
10	deka	da[a]	
10^{-1}	deci	d[a]	
10^{-2}	centi	c[a]	
10^{-3}	milli	m	
10^{-6}	micro	μ	
10^{-9}	nano	n	
10^{-12}	pico	p	
10^{-15}	femto	f	
10^{-18}	atto	a	

[a] Although hecto, deka, deci, and centi are SI prefixes, their use should be avoided except for SI unit-multiples for area and volume and nontechnical use of centimeter, as for body and clothing measurement.

For a complete description of SI and its use the reader is referred to ASTM E 380 (4) and the article Units and Conversion Factors which will appear in a later volume of the *Encyclopedia*.

A representative list of conversion factors from non-SI to SI units is presented herewith. Factors are given to four significant figures. Exact relationships are followed by a dagger. A more complete list is given in ASTM E 380-79(4) and ANSI Z210.1-1976 (6).

Conversion Factors to SI Units

To convert from	To	Multiply by
acre	square meter (m²)	4.047×10^3
angstrom	meter (m)	1.0×10^{-10}†
are	square meter (m²)	1.0×10^2†
astronomical unit	meter (m)	1.496×10^{11}
atmosphere	pascal (Pa)	1.013×10^5
bar	pascal (Pa)	1.0×10^5†
barn	square meter (m²)	1.0×10^{-28}†
barrel (42 U.S. liquid gallons)	cubic meter (m³)	0.1590

† Exact.

To convert from	*To*	*Multiply by*
Bohr magneton (μ_β)	J/T	9.274×10^{-24}
Btu (International Table)	joule (J)	1.055×10^3
Btu (mean)	joule (J)	1.056×10^3
Btu (thermochemical)	joule (J)	1.054×10^3
bushel	cubic meter (m^3)	3.524×10^{-2}
calorie (International Table)	joule (J)	4.187
calorie (mean)	joule (J)	4.190
calorie (thermochemical)	joule (J)	$4.184^\dagger$
centipoise	pascal second (Pa·s)	$1.0 \times 10^{-3\dagger}$
centistokes	square millimeter per second (mm^2/s)	$1.0^\dagger$
cfm (cubic foot per minute)	cubic meter per second (m^3/s)	4.72×10^{-4}
cubic inch	cubic meter (m^3)	1.639×10^{-5}
cubic foot	cubic meter (m^3)	2.832×10^{-2}
cubic yard	cubic meter (m^3)	0.7646
curie	becquerel (Bq)	$3.70 \times 10^{10\dagger}$
debye	coulomb·meter (C·m)	3.336×10^{-30}
degree (angle)	radian (rad)	1.745×10^{-2}
denier (international)	kilogram per meter (kg/m)	1.111×10^{-7}
	tex‡	0.1111
dram (apothecaries')	kilogram (kg)	3.888×10^{-3}
dram (avoirdupois)	kilogram (kg)	1.772×10^{-3}
dram (U.S. fluid)	cubic meter (m^3)	3.697×10^{-6}
dyne	newton (N)	$1.0 \times 10^{-5\dagger}$
dyne/cm	newton per meter (N/m)	$1.0 \times 10^{-3\dagger}$
electron volt	joule (J)	1.602×10^{-19}
erg	joule (J)	$1.0 \times 10^{-7\dagger}$
fathom	meter (m)	1.829
fluid ounce (U.S.)	cubic meter (m^3)	2.957×10^{-5}
foot	meter (m)	$0.3048^\dagger$
footcandle	lux (lx)	10.76
furlong	meter (m)	2.012×10^{-2}
gal	meter per second squared (m/s^2)	$1.0 \times 10^{-2\dagger}$
gallon (U.S. dry)	cubic meter (m^3)	4.405×10^{-3}
gallon (U.S. liquid)	cubic meter (m^3)	3.785×10^{-3}
gallon per minute (gpm)	cubic meter per second (m^3/s)	6.308×10^{-5}
	cubic meter per hour (m^3/h)	0.2271
gauss	tesla (T)	1.0×10^{-4}
gilbert	ampere (A)	0.7958
gill (U.S.)	cubic meter (m^3)	1.183×10^{-4}
grad	radian	1.571×10^{-2}
grain	kilogram (kg)	6.480×10^{-5}
gram-force per denier	newton per tex (N/tex)	8.826×10^{-2}
hectare	square meter (m^2)	$1.0 \times 10^{4\dagger}$

† Exact.
‡ See footnote on p. xiv.

To convert from	To	Multiply by
horsepower (550 ft·lbf/s)	watt (W)	7.457×10^2
horsepower (boiler)	watt (W)	9.810×10^3
horsepower (electric)	watt (W)	$7.46 \times 10^{2\dagger}$
hundredweight (long)	kilogram (kg)	50.80
hundredweight (short)	kilogram (kg)	45.36
inch	meter (m)	$2.54 \times 10^{-2\dagger}$
inch of mercury (32°F)	pascal (Pa)	3.386×10^3
inch of water (39.2°F)	pascal (Pa)	2.491×10^2
kilogram-force	newton (N)	9.807
kilowatt hour	megajoule (MJ)	$3.6^\dagger$
kip	newton (N)	4.48×10^3
knot (international)	meter per second (m/s)	0.5144
lambert	candela per square meter (cd/m²)	3.183×10^3
league (British nautical)	meter (m)	5.559×10^3
league (statute)	meter (m)	4.828×10^3
light year	meter (m)	9.461×10^{15}
liter (for fluids only)	cubic meter (m³)	$1.0 \times 10^{-3\dagger}$
maxwell	weber (Wb)	$1.0 \times 10^{-8\dagger}$
micron	meter (m)	$1.0 \times 10^{-6\dagger}$
mil	meter (m)	$2.54 \times 10^{-5\dagger}$
mile (statute)	meter (m)	1.609×10^3
mile (U.S. nautical)	meter (m)	$1.852 \times 10^{3\dagger}$
mile per hour	meter per second (m/s)	0.4470
millibar	pascal (Pa)	1.0×10^2
millimeter of mercury (0°C)	pascal (Pa)	$1.333 \times 10^{2\dagger}$
minute (angular)	radian	2.909×10^{-4}
myriagram	kilogram (kg)	10
myriameter	kilometer (km)	10
oersted	ampere per meter (A/m)	79.58
ounce (avoirdupois)	kilogram (kg)	2.835×10^{-2}
ounce (troy)	kilogram (kg)	3.110×10^{-2}
ounce (U.S. fluid)	cubic meter (m³)	2.957×10^{-5}
ounce-force	newton (N)	0.2780
peck (U.S.)	cubic meter (m³)	8.810×10^{-3}
pennyweight	kilogram (kg)	1.555×10^{-3}
pint (U.S. dry)	cubic meter (m³)	5.506×10^{-4}
pint (U.S. liquid)	cubic meter (m³)	4.732×10^{-4}
poise (absolute viscosity)	pascal second (Pa·s)	$0.10^\dagger$
pound (avoirdupois)	kilogram (kg)	0.4536
pound (troy)	kilogram (kg)	0.3732
poundal	newton (N)	0.1383
pound-force	newton (N)	4.448
pound-per square inch (psi)	pascal (Pa)	6.895×10^3
quart (U.S. dry)	cubic meter (m³)	1.101×10^{-3}
quart (U.S. liquid)	cubic meter (m³)	9.464×10^{-4}
quintal	kilogram (kg)	$1.0 \times 10^{2\dagger}$

† Exact.

To convert from	*To*	*Multiply by*
rad	gray (Gy)	1.0×10^{-2}†
rod	meter (m)	5.029
roentgen	coulomb per kilogram (C/kg)	2.58×10^{-4}
second (angle)	radian (rad)	4.848×10^{-6}
section	square meter (m²)	2.590×10^{6}
slug	kilogram (kg)	14.59
spherical candle power	lumen (lm)	12.57
square inch	square meter (m²)	6.452×10^{-4}
square foot	square meter (m²)	9.290×10^{-2}
square mile	square meter (m²)	2.590×10^{6}
square yard	square meter (m²)	0.8361
stere	cubic meter (m³)	1.0†
stokes (kinematic viscosity)	square meter per second (m²/s)	1.0×10^{-4}†
tex	kilogram per meter (kg/m)	1.0×10^{-6}†
ton (long, 2240 pounds)	kilogram (kg)	1.016×10^{3}
ton (metric)	kilogram (kg)	1.0×10^{3}†
ton (short, 2000 pounds)	kilogram (kg)	9.072×10^{2}
torr	pascal (Pa)	1.333×10^{2}
unit pole	weber (Wb)	1.257×10^{-7}
yard	meter (m)	0.9144†

Abbreviations and Unit Symbols

Following is a list of commonly used abbreviations and unit symbols appropriate for use in the *Encyclopedia*. In general they agree with those listed in *American National Standard Abbreviations for Use on Drawings and in Text (ANSI Y1.1)* (6) and *American National Standard Letter Symbols for Units in Science and Technology (ANSI Y10)* (6). Also included is a list of acronyms for a number of private and government organizations as well as common industrial solvents, polymers, and other chemicals.

Rules for Writing Unit Symbols (4):

1. Unit symbols should be printed in upright letters (roman) regardless of the type style used in the surrounding text.

2. Unit symbols are unaltered in the plural.

3. Unit symbols are not followed by a period except when used as the end of a sentence.

4. Letter unit symbols are generally written in lower-case (eg, cd for candela) unless the unit name has been derived from a proper name, in which case the first letter of the symbol is capitalized (W,Pa). Prefix and unit symbols retain their prescribed form regardless of the surrounding typography.

5. In the complete expression for a quantity, a space should be left between the numerical value and the unit symbol. For example, write 2.37 lm, *not* 2.37lm, and 35 mm, *not* 35mm. When the quantity is used in an adjectival sense, a hyphen is often used, for example, 35-mm film. *Exception:* No space is left between the numerical value and the symbols for degree, minute, and second of plane angle, and degree Celsius.

6. No space is used between the prefix and unit symbols (eg, kg).

7. Symbols, not abbreviations, should be used for units. For example, use "A," not "amp," for ampere.

8. When multiplying unit symbols, use a raised dot:

$$\text{N·m for newton meter}$$

In the case of W·h, the dot may be omitted, thus:

$$\text{Wh}$$

An exception to this practice is made for computer printouts, automatic typewriter work, etc, where the raised dot is not possible, and a dot on the line may be used.

9. When dividing unit symbols use one of the following forms:

$$\text{m/s } or \text{ m·s}^{-1} or \frac{\text{m}}{\text{s}}$$

In no case should more than one slash be used in the same expression unless parentheses are inserted to avoid ambiguity. For example, write:

$$\text{J/(mol·K) } or \text{ J·mol}^{-1} \cdot \text{K}^{-1} or \text{ (J/mol)/K}$$

but *not*

$$\text{J/mol/K}$$

10. Do not mix symbols and unit names in the same expression. Write:

$$\text{joules per kilogram } or \text{ J/kg } or \text{ J·kg}^{-1}$$

but *not*

$$\text{joules/kilogram } nor \text{ joules/kg } nor \text{ joules·kg}^{-1}$$

ABBREVIATIONS AND UNITS

A	ampere
A	anion (eg, H*A*); mass number
a	atto (prefix for 10^{-18})
AATCC	American Association of Textile Chemists and Colorists
ABS	acrylonitrile–butadiene–styrene
abs	absolute
ac	alternating current, *n.*
a-c	alternating current, *adj.*
ac-	alicyclic
acac	acetylacetonate
ACGIH	American Conference of Governmental Industrial Hygienists
ACS	American Chemical Society
AGA	American Gas Association
Ah	ampere hour
AIChE	American Institute of Chemical Engineers
AIME	American Institute of Mining, Metallurgical, and Petroleum Engineers
AIP	American Institute of Physics
AISI	American Iron and Steel Institute
alc	alcohol(ic)
Alk	alkyl
alk	alkaline (not alkali)
amt	amount
amu	atomic mass unit
ANSI	American National Standards Institute
AO	atomic orbital
AOAC	Association of Official Analytical Chemists
AOCS	American Oil Chemists' Society
APHA	American Public Health Association
API	American Petroleum Institute
aq	aqueous
Ar	aryl
ar-	aromatic
as-	asymmetric(al)

ASH-RAE	American Society of Heating, Refrigerating, and Air Conditioning Engineers	coml	commercial(ly)
		cp	chemically pure
		cph	close-packed hexagonal
ASM	American Society for Metals	CPSC	Consumer Product Safety Commission
ASME	American Society of Mechanical Engineers		
		cryst	crystalline
ASTM	American Society for Testing and Materials	cub	cubic
		D	debye
at no.	atomic number	D-	denoting configurational relationship
at wt	atomic weight		
av(g)	average	$\mathbf{d}$	differential operator
AWS	American Welding Society	d-	*dextro-*, dextrorotatory
b	bonding orbital	da	deka (prefix for 10^1)
bbl	barrel	dB	decibel
bcc	body-centered cubic	dc	direct current, *n.*
BCT	body-centered tetragonal	d-c	direct current, *adj.*
Bé	Baumé	dec	decompose
BET	Brunauer-Emmett-Teller (adsorption equation)	detd	determined
		detn	determination
bid	twice daily	Di	didymium, a mixture of all lanthanons
Boc	*t*-butyloxycarbonyl		
BOD	biochemical (biological) oxygen demand	dia	diameter
		dil	dilute
bp	boiling point	DIN	Deutsche Industrie Normen
Bq	becquerel	dl-; DL-	racemic
C	coulomb	DMA	dimethylacetamide
°C	degree Celsius	DMF	dimethylformamide
C-	denoting attachment to carbon	DMG	dimethyl glyoxime
		DMSO	dimethyl sulfoxide
c	centi (prefix for 10^{-2})	DOD	Department of Defense
c	critical	DOE	Department of Energy
ca	circa (approximately)	DOT	Department of Transportation
cd	candela; current density; circular dichroism		
		DP	degree of polymerization
CFR	Code of Federal Regulations	dp	dew point
		DPH	diamond pyramid hardness
cgs	centimeter–gram–second	dstl(d)	distill(ed)
CI	Color Index	dta	differential thermal analysis
cis-	isomer in which substituted groups are on same side of double bond between C atoms		
		(E)-	entgegen; opposed
		ϵ	dielectric constant (unitless number)
cl	carload	e	electron
cm	centimeter	ECU	electrochemical unit
cmil	circular mil	ed.	edited, edition, editor
cmpd	compound	ED	effective dose
CNS	central nervous system	EDTA	ethylenediaminetetraacetic acid
CoA	coenzyme A		
COD	chemical oxygen demand	emf	electromotive force

emu	electromagnetic unit		grd	ground
en	ethylene diamine		Gy	gray
eng	engineering		H	henry
EPA	Environmental Protection Agency		h	hour; hecto (prefix for 10^2)
epr	electron paramagnetic resonance		ha	hectare
			HB	Brinell hardness number
eq.	equation		Hb	hemoglobin
esca	electron-spectroscopy for chemical analysis		hcp	hexagonal close-packed
			hex	hexagonal
esp	especially		HK	Knoop hardness number
esr	electron-spin resonance		hplc	high pressure liquid chromatography
est(d)	estimate(d)			
estn	estimation		HRC	Rockwell hardness (C scale)
esu	electrostatic unit		HV	Vickers hardness number
exp	experiment, experimental		hyd	hydrated, hydrous
ext(d)	extract(ed)		hyg	hygroscopic
F	farad (capacitance)		Hz	hertz
F	faraday (96,487 C)		^{i}(eg, Pr^i)	iso (eg, isopropyl)
f	femto (prefix for 10^{-15})		i-	inactive (eg, i-methionine)
FAO	Food and Agriculture Organization (United Nations)		IACS	International Annealed Copper Standard
			ibp	initial boiling point
fcc	face-centered cubic		IC	inhibitory concentration
FDA	Food and Drug Administration		ICC	Interstate Commerce Commission
FEA	Federal Energy Administration		ICT	International Critical Table
			ID	inside diameter; infective dose
FHSA	Federal Hazardous Substances Act		ip	intraperitoneal
			IPS	iron pipe size
fob	free on board		IPTS	International Practical Temperature Scale (NBS)
fp	freezing point			
FPC	Federal Power Commission		ir	infrared
FRB	Federal Reserve Board		IRLG	Interagency Regulatory Liaison Group
frz	freezing			
G	giga (prefix for 10^9)		ISO	International Organization for Standardization
G	gravitational constant = 6.67×10^{11} N·m^2/kg^2			
			IU	International Unit
g	gram		IUPAC	International Union of Pure and Applied Chemistry
(g)	gas, only as in H_2O(g)			
g	gravitational acceleration		IV	iodine value
gc	gas chromatography		iv	intravenous
gem-	geminal		J	joule
glc	gas-liquid chromatography		K	kelvin
g-mol wt; gmw	gram-molecular weight		k	kilo (prefix for 10^3)
			kg	kilogram
GNP	gross national product		L	denoting configurational relationship
gpc	gel-permeation chromatography			
			L	liter (for fluids only)(5)
GRAS	Generally Recognized as Safe		l-	$levo$-, levorotatory

(l)	liquid, only as in $NH_3(l)$	ms	mass spectrum
LC_{50}	conc lethal to 50% of the animals tested	mxt	mixture
		μ	micro (prefix for 10^{-6})
LCAO	linear combination of atomic orbitals	N	newton (force)
		N	normal (concentration); neutron number
LCD	liquid crystal display		
lcl	less than carload lots	N-	denoting attachment to nitrogen
LD_{50}	dose lethal to 50% of the animals tested		
		n (as n_D^{20})	index of refraction (for 20°C and sodium light)
LED	light-emitting diode		
liq	liquid	^{n} (as Bu^n), n-	normal (straight-chain structure)
lm	lumen		
ln	logarithm (natural)	n	neutron
LNG	liquefied natural gas	n	nano (prefix for 10^9)
log	logarithm (common)	na	not available
LPG	liquefied petroleum gas	NAS	National Academy of Sciences
ltl	less than truckload lots		
lx	lux	NASA	National Aeronautics and Space Administration
M	mega (prefix for 10^6); metal (as in MA)		
		nat	natural
M	molar; actual mass	NBS	National Bureau of Standards
$\overline{M}_w$	weight-average mol wt		
$\overline{M}_n$	number-average mol wt	neg	negative
m	meter; milli (prefix for 10^{-3})	NF	*National Formulary*
		NIH	National Institutes of Health
m	molal		
m-	meta	NIOSH	National Institute of Occupational Safety and Health
max	maximum		
MCA	Chemical Manufacturers' Association (was Manufacturing Chemists Association)		
		nmr	nuclear magnetic resonance
		NND	New and Nonofficial Drugs (AMA)
MEK	methyl ethyl ketone		
meq	milliequivalent	no.	number
mfd	manufactured	NOI-(BN)	not otherwise indexed (by name)
mfg	manufacturing		
mfr	manufacturer	NOS	not otherwise specified
MIBC	methyl isobutyl carbinol	nqr	nuclear quadruple resonance
MIBK	methyl isobutyl ketone	NRC	Nuclear Regulatory Commission; National Research Council
MIC	minimum inhibiting concentration		
		NRI	New Ring Index
min	minute; minimum	NSF	National Science Foundation
mL	milliliter	NTA	nitrilotriacetic acid
MLD	minimum lethal dose	NTP	normal temperature and pressure (25°C and 101.3 kPa or 1 atm)
MO	molecular orbital		
mo	month		
mol	mole		
mol wt	molecular weight	NTSB	National Transportation Safety Board
mp	melting point		
MR	molar refraction	O-	denoting attachment to

	oxygen	ref.	reference
o-	ortho	rf	radio frequency, *n.*
OD	outside diameter	r-f	radio frequency, *adj.*
OPEC	Organization of	rh	relative humidity
	Petroleum Exporting	RI	Ring Index
	Countries	rms	root-mean square
o-phen	*o*-phenanthridine	rpm	rotations per minute
OSHA	Occupational Safety and	rps	revolutions per second
	Health Administration	RT	room temperature
owf	on weight of fiber	^{s} (eg,	secondary (eg, secondary
Ω	ohm	Bu^s);	butyl)
P	peta (prefix for 10^{15})	*sec*-	
p	pico (prefix for 10^{-12})	S	siemens
p-	para	(*S*)-	sinister (counterclockwise
p	proton		configuration)
p.	page	*S*-	denoting attachment to
Pa	pascal (pressure)		sulfur
pd	potential difference	*s*-	symmetric(al)
pH	negative logarithm of the	s	second
	effective hydrogen ion	(s)	solid, only as in $H_2O(s)$
	concentration	SAE	Society of Automotive
phr	parts per hundred of resin		Engineers
	(rubber)	SAN	styrene–acrylonitrile
p-i-n	positive-intrinsic-negative	sat(d)	saturate(d)
pmr	proton magnetic resonance	satn	saturation
p-n	positive-negative	SBS	styrene–butadiene–styrene
po	per os (oral)	sc	subcutaneous
POP	polyoxypropylene	SCF	self-consistent field;
pos	positive		standard cubic feet
pp.	pages	Sch	Schultz number
ppb	parts per billion (10^9)	SFs	Saybolt Furol seconds
ppm	parts per million (10^6)	SI	Le Système International
ppmv	parts per million by volume		d'Unités (International
ppmwt	parts per million by weight		System of Units)
PPO	poly(phenyl oxide)	sl sol	slightly soluble
ppt(d)	precipitate(d)	sol	soluble
pptn	precipitation	soln	solution
Pr (no.)	foreign prototype (number)	soly	solubility
pt	point; part	sp	specific; species
PVC	poly(vinyl chloride)	sp gr	specific gravity
pwd	powder	sr	steradian
py	pyridine	std	standard
qv	quod vide (which see)	STP	standard temperature and
R	univalent hydrocarbon		pressure (0°C and 101.3
	radical		kPa)
(*R*)-	rectus (clockwise	sub	sublime(s)
	configuration)	SUs	Saybolt Universal
r	precision of data		seconds
rad	radian; radius	syn	synthetic
rds	rate determining step	^{t} (eg,	tertiary (eg, tertiary butyl)

Bu^t),	
t-,	
tert-	
T	tera (prefix for 10^{12}); tesla (magnetic flux density)
t	metric ton (tonne); temperature
TAPPI	Technical Association of the Pulp and Paper Industry
tex	tex (linear density)
T_g	glass-transition temperature
tga	thermogravimetric analysis
THF	tetrahydrofuran
tlc	thin layer chromatography
TLV	threshold limit value
trans-	isomer in which substituted groups are on opposite sides of double bond between C atoms
TSCA	Toxic Substance Control Act
TWA	time-weighted average
Twad	Twaddell

UL	Underwriters' Laboratory
USDA	United States Department of Agriculture
USP	*United States Pharmacopeia*
uv	ultraviolet
V	volt (emf)
var	variable
vic-	vicinal
vol	volume (not volatile)
vs	versus
v sol	very soluble
W	watt
Wb	weber
Wh	watt hour
WHO	World Health Organization (United Nations)
wk	week
yr	year
(*Z*)-	zusammen; together; atomic number

Non-SI (Unacceptable and Obsolete) Units		*Use*
Å	angstrom	nm
at	atmosphere, technical	Pa
atm	atmosphere, standard	Pa
b	barn	cm^2
bar†	bar	Pa
bbl	barrel	m^3
bhp	brake horsepower	W
Btu	British thermal unit	J
bu	bushel	m^3; L
cal	calorie	J
cfm	cubic foot per minute	m^3/s
Ci	curie	Bq
cSt	centistokes	mm^2/s
c/s	cycle per second	Hz
cu	cubic	exponential form
D	debye	C·m
den	denier	tex
dr	dram	kg
dyn	dyne	N
dyn/cm	dyne per centimeter	mN/m
erg	erg	J
eu	entropy unit	J/K
°F	degree Fahrenheit	°C; K
fc	footcandle	lx
fl	footlambert	lx
fl oz	fluid ounce	m^3; L
ft	foot	m

† Do not use bar (10^5Pa) or millibar (10^2Pa) because they are not SI units, and are accepted internationally only for a limited time in special fields because of existing usage.

Non-SI (Unacceptable and Obsolete) Units		*Use*
ft·lbf	foot pound-force	J
gf/den	gram-force per denier	N/tex
G	gauss	T
Gal	gal	m/s²
gal	gallon	m³; L
Gb	gilbert	A
gpm	gallon per minute	(m³/s); (m³/h)
gr	grain	kg
hp	horsepower	W
ihp	indicated horsepower	W
in.	inch	m
in. Hg	inch of mercury	Pa
in. H₂O	inch of water	Pa
in.-lbf	inch pound-force	J
kcal	kilogram-calorie	J
kgf	kilogram-force	N
kilo	for kilogram	kg
L	lambert	lx
lb	pound	kg
lbf	pound-force	N
mho	mho	S
mi	mile	m
MM	million	M
mm Hg	millimeter of mercury	Pa
mμ	millimicron	nm
mph	mile per hour	km/h
μ	micron	μm
Oe	oersted	A/m
oz	ounce	kg
ozf	ounce-force	N
η	poise	Pa·s
P	poise	Pa·s
ph	phot	lx
psi	pound-force per square inch	Pa
psia	pound-force per square inch absolute	Pa
psig	pound-force per square inch gauge	Pa
qt	quart	m³; L
°R	degree Rankine	K
rd	rad	Gy
sb	stilb	lx
SCF	standard cubic foot	m³
sq	square	exponential form
thm	therm	J
yd	yard	m

BIBLIOGRAPHY

1. The International Bureau of Weights and Measures, BIPM (Parc de Saint-Cloud, France) is described on page 22 of Ref. 4. This bureau operates under the exclusive supervision of the International Committee of Weights and Measures (CIPM).

2. *Metric Editorial Guide (ANMC-78-1)* 3rd ed., American National Metric Council, 5410 Grosvenor Lane, Bethesda, Md. 20814, 1981.

3. *SI Units and Recommendations for the Use of Their Multiples and of Certain Other Units (ISO 1000-1981)*, American National Standards Institute, 1430 Broadway, New York, N. Y. 10018, 1981.

4. Based on *ASTM E 380-82 (Standard for Metric Practice)*, American Society for Testing and Materials, 1916 Race Street, Philadelphia, Pa. 19103, 1982.

5. *Fed. Regist.*, Dec. 10, 1976 (41 FR 36414).

6. For ANSI address, see Ref. 3.

R. P. LUKENS

American Society for Testing and Materials

THYROID AND ANTITHYROID PREPARATIONS

The main role of the thyroid gland is the production of the thyroid hormones (iodinated amino acids) which are essential for adequate growth, development, and energy metabolism (1–6). Thyroid underfunction is a frequent occurrence that can be treated successfully with thyroid preparations. In addition, the thyroid secretes calcitonin (also known as thyrocalcitonin), a polypeptide that lowers excessively high calcium blood levels. Thyroid hyperfunction is another important clinical entity that can be corrected by treatment with a variety of substances known as antithyroid drugs.

Related substances include thyroid-stimulating hormone [9002- 71-5] (thyrotropin or TSH), which is secreted by the pituitary gland; thyrotropin-releasing hormone [24305-27-9] or thyroliberin (TRH), a hypothalamic tripeptide; D-thyroxine [51-49-0] (dextrothyroxine), the synthetic unnatural enantiomer of one of the thyroid hormones with blood-cholesterol lowering activity; various radioactive iodine-containing preparations that are used to destroy excessive thyroid tissue or to measure thyroid function; and long-acting thyroid stimulator [9034-48-4] (LATS), an immunoglobulin.

Perhaps the most important development in thyroid hormone research during the last 20 yr is the medicinal chemistry aspect. Detailed descriptions of the structural requirements for thyromimetic activity and of the molecular interactions of the thyroid hormones with their receptors and carriers are available. In this sense, the thyroid hormones are perhaps the best understood of all hormones.

1

Thyroid Function and Malfunction

Human life without thyroid hormones is possible but of minimal quality since these hormones have profound biological effects. In the fetus, they affect growth and differentiation; in the mature human, they regulate metabolism. The two principal thyroid hormones, L-thyroxine [51-48-9] (T_4) (**1**) and L-triiodothyronine [6893-02-3] (T_3) (**2**) (see Fig. 1) are produced by the thyroid gland and secreted into the blood stream. The minute amounts secreted are regulated by a complex system (Fig. 2) that originates in the CNS and is amplified by both the hypothalamus and the anterior pituitary; it can, however, be diminished by feedback loops in which circulating levels of free T_3 and T_4 repress production of the pituitary TSH (and perhaps of this hormone itself by inhibiting release in the hypothalamus of its liberating hormone, TRH). In addition, the amounts of the hormones reaching the cells to preserve an optimal (euthyroid) condition are regulated by two plasma proteins, ie, thyroid hormone-binding globulin [9010-34-8] (TBG) and thyroid hormone-binding prealbumin [632-79-1] (TBPA); only a small fraction (<0.3%) of the total hormones in circulation is free. Finally, tissue deiodinases convert T_4 (possibly a prohormone) into the fivefold more active T_3.

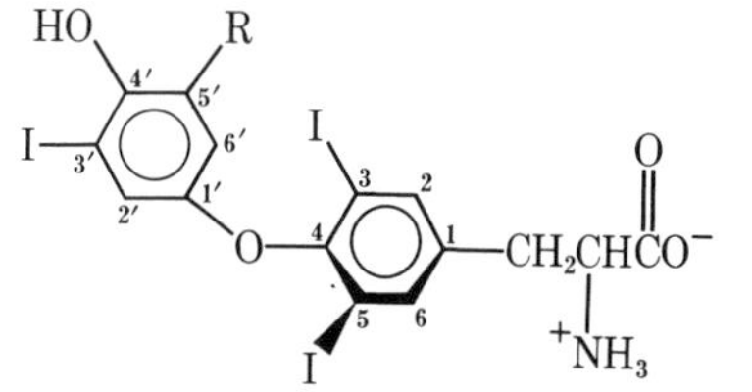

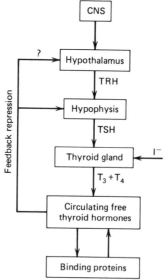

(**1**) R = I: L-thyroxine, 3,5,3′,5′-tetraiodo-L-thyronine, T_4
(**2**) R = H: 3,5,3′-triiodo-L-thyronine, T_3

Figure 1.   The principal thyroid hormones.

Figure 2.   Mechanisms controlling free thyroid-hormone levels.

Thyroid hormones affect growth and development by stimulating protein synthesis. It is thought that a specific receptor protein that strongly binds the hormones is present in cell nuclei (7) (see Fig. 3). This protein is closely associated with nuclear deoxyribonucleic acid (DNA), a complex involved in DNA transcription. The binding of the hormones is a specific signal to a DNA template that, when activated, stimulates the synthesis and release of a specific messenger ribonucleic acid (mRNA). The latter stimulates the synthesis of structural and functional proteins, eg, enzymes and other hormones, which then bring about growth and development.

In mature animals, the main action of the thyroid hormones is their calorigenic effect which is caused by an increase in the basal metabolic rate (BMR). Although many theories have been advanced (8), the mechanism of action of this effect at the molecular level is not understood. Conceivably it might just be another expression of the above mentioned stimulation of protein synthesis.

Given the importance of the thyroid hormones in bringing about and then maintaining a normal metabolic state, it is not surprising that malfunctions of the thyroid gland have grave consequences (2,9).

Thyroid underfunction results in a series of hypothyroid states clinically known as cretinism if present in a fetus or an infant, and myxedema in an adult. If the hypothyroidism is owing to insufficient iodine intake, it is known as simple goiter, a state characterized by an enlarged but functionally underactive thyroid gland. Today, endemic goiter can be avoided by adding iodine to the diet in a convenient form, eg, iodate. In the United States, iodized table salt contains 100 μg of iodate per gram of NaCl (2) (see also Sodium compounds, sodium chloride). Even so, endemic goiter is still an important health problem in many areas of the world, especially in those where underdevelopment coincides with remoteness from oceans.

Myxedema and goiter are the main conditions for which thyroid preparations are indicated. The treatment of cretinism is very difficult because it is recognized only at or after birth. Even if they could be diagnosed *in utero*, thyroid hormones do not readily cross the placental barrier. In addition, the fetus, as well as premature infants, rapidly deactivates the thyroid hormones. A possible cure is on the horizon in the form

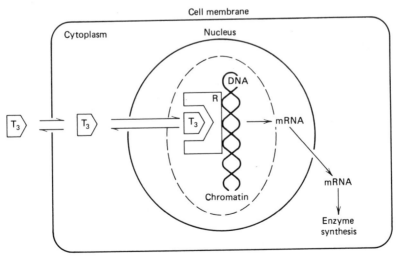

Figure 3. Early events in thyroid-hormone action. Interaction of T$_3$ with cell nuclear receptors (6).

of the halogen-free analogue DIMIT [26384-44-7] (5), which is resistant to fetal deiodinases.

Thyroid hyperfunction occurs as diffuse toxic goiter, also known as Graves' disease, seen mainly in young adults and considered to be caused by an immune disorder. It is characterized by protruding eyeballs (exophthalmos). Another form of thyroid hyperfunction is thyrotoxicosis, ie, a collection of symptoms caused by excessive production of thyroid hormones, including hyperthermia, rapid heart rate, increased appetite and loss of weight, insomnia, anxiety, etc. Toxic nodular goiter (Plummer's disease) is less common. Severe cases are treated by partial surgical removal of the thyroid gland or its partial destruction with radioactive iodine. Milder cases are controlled with antithyroid drugs.

Thyromimetic Compounds

Thyroidal Amino Acids. Toward the end of the 19th century, it was discovered that the consumption of fresh sheep thyroid glands was beneficial in hypothyroidism. In an attempt to isolate the active principle, an extract was prepared (10) and commercialized (11). In the course of this work, it was discovered that thyroid glands were rich in iodine (qv). In 1914, a biologically active pure compound was isolated from thyroid extracts and was called thyroxin on the mistaken assumption that it had an oxyindole structure. Some years later, its correct structure (1) was established by degradation and synthesis. About 25 yr later two groups simultaneously identified another biologically active compound that is now recognized as the main thyroid hormone. It is the 5′–desiodo analogue of thyroxine, known as triiodothyronine or T_3 (2). More recently, two more iodinated thyronines have been found in the thyroid (see Fig. 4). They are 3,3′,5′-triiodothyronine [5817-39-0] (reverse-T_3 (3)) and 3,3′-diiodothyronine [4604-41-5] (T_2 (4)). These compounds are hormonally inactive and are secreted by the thyroid or arise by partial deiodination of T_3 and T_4. Their physiological significance is not clear. Some properties of these compounds are listed in Table 1.

For many years it was believed that iodine, or at least some halogen, had to be present in these compounds to endow them with thyromimetic activity. This was shown to be incorrect when a halogen-free analogue (DIMIT (5)) was found to have 20% of the potency of T_4 in a variety of *in vivo* tests (12).

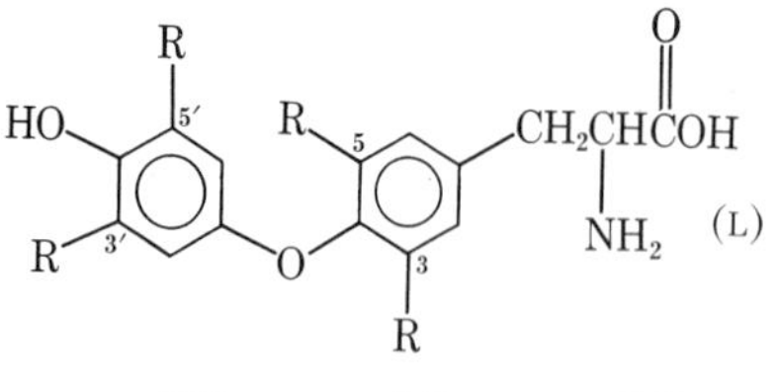

Compound	3-R	5-R	3′-R	5′-R
(1) T$_4$	I	I	I	I
(2) T$_3$	I	I	I	H
(3) r-T$_3$	I	H	I	I
(4) 3,3′-T$_2$	I	H	I	H
(5) DIMIT	CH$_3$	CH$_3$	Pr^i	H

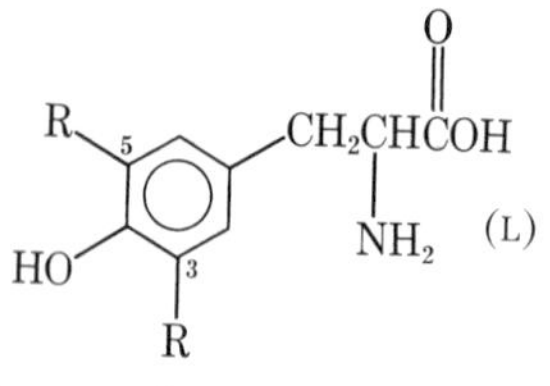

Compound	3-R	5-R
(6) TYR	H	H
(7) MIT	I	H
(8) DIT	I	I

Figure 4.   Structures of the thyroidal iodinated amino acids and the halogen-free analogue DIMIT. Compound (3) is reverse-T$_3$.

Table 1.   Thyroidal Iodinated Amino Acids[a]

Name	CAS Reg. No.	Abbreviation	Compound	Mol wt	I, %	pK_a (OH)	$[\alpha]_D$
3-iodo-L-tyrosine	[70-78-0]	MIT	(7)	307.1	41.3	8.70	−4.4[b]
3,5-diiodo-L-tyrosine	[66-02-4]	DIT	(8)	433.0	58.6	6.48, 6.36	2.75[b]
3,5-diiodo-L-thyronine	[1041-01-6]	3,5-T_2		525.1	48.3	9.29	26.0[c]
3,3′-diiodo-L-thyronine	[4604-41-5]	3,3′-T_2	(4)	525.1	48.3		18.8[c]
3,5,3′-triiodo-L-thyronine	[6893-02-3][d]	T_3	(2)	650.9	58.5	8.45	21.5[c]
3,3′,5′-triiodo-L-thyronine	[5817-39-0]	r-T_3	(3)	650.9	58.5	6.5 (estd)	16.7[c]
3,5,3′,5′-tetraiodo-L-thyronine	[51-48-9][e]	T_4	(1)	776.8	65.3	6.73, 6.45	17.5[c]

[a] Data mainly from refs. 6 and 10. These compounds decompose; their melting points are indistinct.
[b] In 4.8% HCl.
[c] In 1 N HCl-C_2H_5OH.
[d] Anhydrous Na salt [55-06-7].
[e] Anhydrous Na salt [55-03-8]. Na salt. 5 H_2O [25416-65-3].

Structure-Activity Relationships. In spite of the considerable synthetic and bioassay effort involved in establishing the thyromimetic potency of thyroid-hormone analogues, several groups of researchers, mainly E. C. Jorgensen and co-workers, have studied more than 100 compounds over the years (see Table 2). The main structural requirements for thyromimetic activity can be summarized as follows (see also Table 2) (6,12–16):

1. Two aromatic rings insulated electronically from each other by connecting oxygen, sulfur, or carbon bridges, forming a central lipophilic core in which the two rings are angled 120°.

2. Substitution at the 3 and 5 positions with alkyl groups or with halogens large enough to force the diphenyl ether nucleus to adopt a minimum energy conformation in which the two rings are approximately in mutually perpendicular planes.

Table 2.   Thyromimetic Compounds: Relative Binding Affinities (BA) to Solubilized Rat Hepatic Nuclear Protein Receptor and Antigoiter Potencies (AG)[a]

3′-R	5′-R	CAS Reg. No.	BA	AG	3′-R	5′-R	CAS Reg. No.	BA	AG
I	H	[6893-02-3]	1	1	I	I	[51-48-9]	0.14	0.18
Br	H	[58437-19-7]	0.16	0.24	Br	Br	[2500-09-6]	0.05	0.02
Cl	H	[4299-63-2]	0.04	0.05	Cl	Cl	[4299-64-3]	0.04	0.04
F	H	[348-94-7]	0.02	ca 0.01	Pr^i	I	[3458-12-6]	0.12	
Pr^i	H	[51-23-0]	0.89	1.42	Pr^i	Br	[75628-30-7]	0.22	
Bu^s	H	[3415-06-3]	0.78	0.80	Pr^i	Cl	[75628-29-4]	0.53	
Pr^n	H	[72468-99-6]	0.24	0.40	Pr^i	Pr^i	[30804-63-8]	0.01	
Bu^t	H	[857-98-7]	0.08	0.22					
CH$_3$	H	[2378-96-3]	0.03	0.14					

[a] Affinities and potencies are relative to L-T_3 taken as 1 (6).

3. At position 1, there should be present an acidic side chain two or three carbons long. The natural L-alanyl side chain reduces receptor binding but enhances *in vivo* activity by increasing access to the receptor and by retarding metabolism and excretion. The enantiomeric D-analogues retain considerable activity in contrast to other bioactive substances (17).

4. The presence of a small substituent capable of forming hydrogen bonds in the 4′ position. Isosteric groups such as NH_2 reduce activity, whereas any other group that cannot be converted metabolically to a 4′-OH group results in inactive compounds.

5. The minimal activity residing in the core structure so far described is greatly enhanced by one lipophilic substituent ortho to the 4′-OH group. High activity is imparted by iodine or alkyl groups of similar size, eg, isopropyl. Inspection of Figure 1 shows that for T_3 two atropisomers (preferred conformers due to restricted rotation about single bonds) are possible. In the one shown, the 3′-I is distal to the other ring, whereas a second one is obtained by rotating the phenolic ring 180° about the C-1′—O bond, in which the lone iodine is proximal to the other ring. Both atropisomers have been detected in a variety of compounds by x-ray crystallography (18). In their interaction with biomacromolecules, however, affinity is increased when the 3′-substituent is distal. A second lipophilic group at the other position ortho to the 4′-OH (ie, at C-5′) always reduces activity because of steric hindrance at the binding site. In a variety of *in vivo* test systems, one synthetic analogue (3,5-diiodo-3′-isopropyl-L-thyronine, Table 2) has been shown to be more potent than T_3.

Quantitative structure–activity relationships have been established using the Hansch multiparameter approach (14). For rat antigoiter activities it was found that

$$\log \text{AG} = 1.354 \ \pi 35 + 1.344 \ \pi 3' - 1.324 \ [(\text{size-}3') > \text{I}]$$

$$-0.359 \ \pi 5' - 0.658 \ \sigma 3'5' - 0.890 \ (\text{OCH}_3\text{-}4') - 2.836$$

$$n = 36, \qquad r = 0.938, \qquad s = 0.304$$

In statistical regression equations (see any text), n = number of compounds, r = regression coefficient, and s = standard deviation of the dependent variable. In this equation AG is the relative antigoiter potency, π and σ are lipophilic and electronic substituent parameters, $(\text{OCH}_3\text{-}4')$ is a dummy parameter indicating the presence (1) or absence (0) of this substituent, and (size-3′ > I) is a computed estimate of the size of this substituent beyond that of an iodine atom. This equation does not include a π^2 term (optimal lipophilicity), and therefore the substituent pattern is unimportant in the overall partitioning behavior. Analogous equations have been derived for the interactions of thyroid hormone analogues with other systems (TBG, TBPA, or nuclear receptors) (14). Slight but significant differences in the regression constants point to differences in the shape of the recognition site. Nevertheless, good correlations exist between them and *in vivo* activities (see Table 2), and they can be used with confidence to predict intact animal activity.

Although in a variety of *in vivo* tests DIMIT has shown only 20% the potency of T_4, it is resistant to fetal deiodinase activity and crosses the placental barrier readily. It may be clinically useful in treating fetal hypothyroidism (cretinism) and in the prevention of respiratory distress syndrome in premature birth (19).

Biosynthesis, Distribution, and Metabolism.   Although iodine is a trace element in the environment (0.006% in ocean water) and in the diet, the thyroid gland avidly extracts it from the blood as iodide ion via an active transport system. In the thyroid cells it is converted by a peroxidase to a form that is capable of iodinating tyrosyl residues present in a large glycoprotein called thyroglobulin [9010-34-8]. It is now believed that the resulting mono- and diiodinated residues react with each other in the protein matrix (possibly by a free-radical coupling mechanism) to form all the possible di-, tri-, and tetraiodothyronines (20–22). Thyroglobulin is a glycoprotein with several subunits. Its mol wt is 660,000 (19S), and it contains ca 10% carbohydrate (corresponding to 300 residues) and ca 5500 amino acid residues of which only 2–5 are thyroxine. Its amino acid composition, and further details of its structure and physical properties are found in ref. 23.

The iodinated thyroglobulin is stored as a colloid in thyroidal follicular cells, and T_3 and T_4 are liberated from it by proteolysis as required. It is estimated that ca 90 μg of T_4 and 6 μg of T_3 are secreted daily by the thyroid gland, giving mean plasma concentrations of 80 and 2 μg/L, respectively, of which only 0.03% and 0.3% are in the free form, ie, not protein bound (21). The half-life of T_4 in the body is very long (6–7 d) (2); that of T_3 is somewhat shorter (2 d).

The biosynthesis and release of the hormones can be interfered with in various ways, which is the basis of action of certain antithyroid preparations.

Only the small amounts of T_4 and T_3 that are free in the circulation can be metabolized. The main route is deiodination of T_4 to T_3 and rT_3, and from these to other inactive thyronines (21). Most of the liberated iodide is reabsorbed in the kidney. Another route is the formation of glucuronide and sulfate conjugates at the 4'-OH in the liver. These are then secreted in the bile and excreted in the feces as free phenols after hydrolysis in the lower gut.

Synthesis.   In the syntheses of T_4 and its congeners, formation of the sterically hindered diaryl ether core is difficult, as is the introduction of the alanyl side chain (or the preservation of its L (S) absolute configuration) and iodination to the desired degree (T_3 or T_4).

The most widely employed route is the so-called Glaxo method (24) (see Fig. 5). The starting material is tyrosine, which is readily available in the L-form and accessible in the D-form. Nitration and protection of the side chain give the key intermediate, N-acetyl-3,5-dinitrotyrosine, ethyl ester [29358-99-4]. The activating effect of the two ortho nitro groups allows the phenolic OH to be displaced readily by pyridine. The resulting quaternary pyridinium adduct, in turn, displays a very high reactivity toward nucleophilic displacement by phenoxides, which is the key factor in the successful formation of the ether link in spite of the formidable steric hindrance presented by the nitro groups. In the original procedure the pyridinium tosylate was isolated and purified (24). A modification, in which methanesulfonyl chloride is used, obviates this isolation and results in a faster reaction with higher yields (25). The dinitro ether is then subjected to a Sandmeyer procedure in which the nitro groups are converted to iodines, followed by removal of the blocking groups to give 3,5-diiodothyronine. Finally, one or two additional iodines are introduced to give T_3 and T_4, respectively. In the original procedure, L-T_4 was obtained in an overall yield of 26% based on L-tyrosine (20). By starting with D-tyrosine, D-T_4, which is used clinically to reduce high cholesterol blood levels, can be obtained (26).

This versatile synthetic route has been used extensively with a great variety of

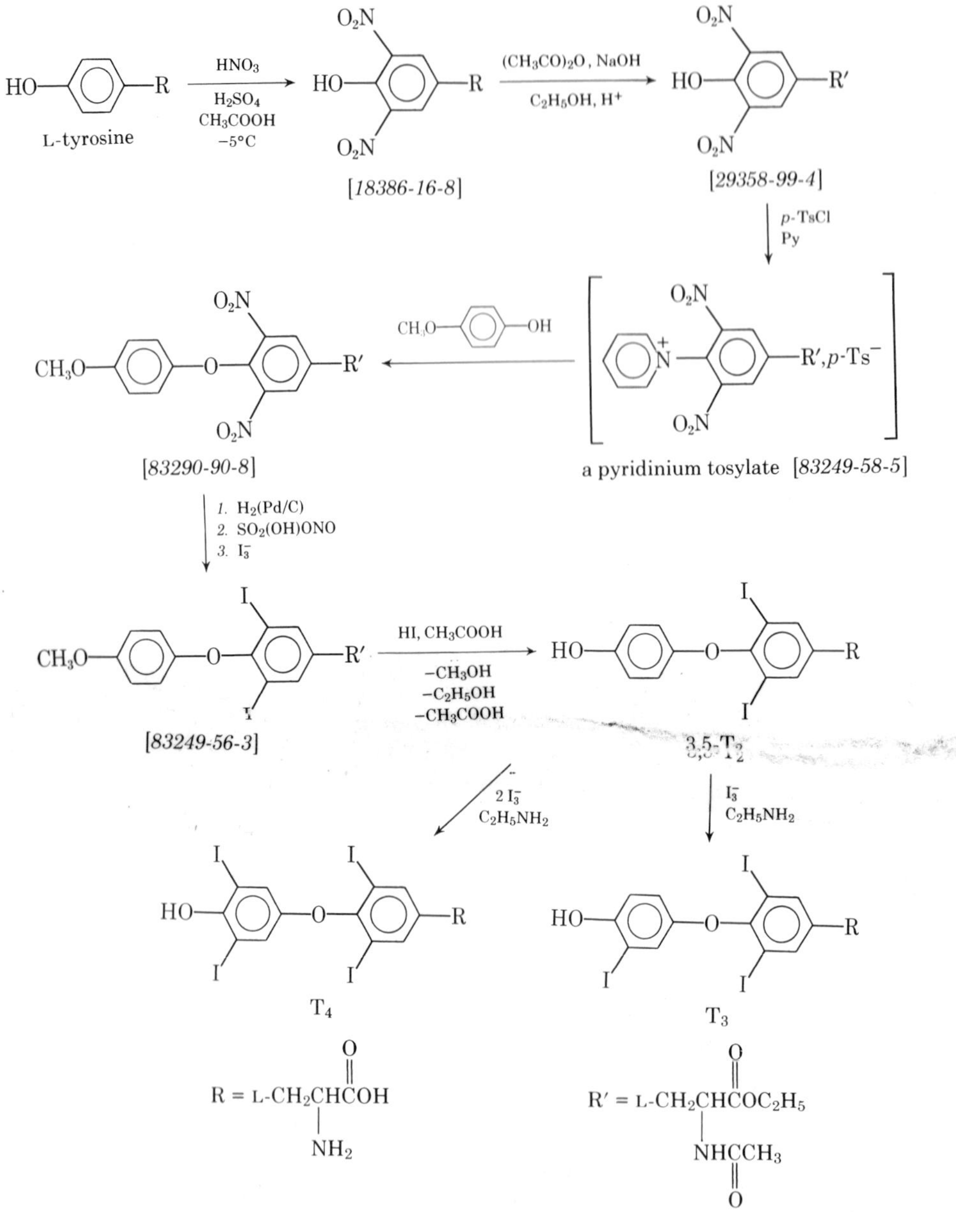

Figure 5. Glaxo synthesis of T_3 and T_4 (p-TsCl = p-toluenesulfonyl chloride; Py = pyridine) (24).

phenols and thiophenols to establish structure–activity relationships for thyromimetic activity.

Other routes can be summarized as follows (13):

In the first synthesis of T_4, the diphenyl ether was formed from p-methoxyphenol and 3,4,5-triiodonitrobenzene. The nitro group was replaced by a nitrile which was then built up into the alanyl side chain by a series of steps (10).

In a biomimetic synthesis, two DIT (**8**) molecules have been coupled to give T_4 (27).

In a procedure known as the iodonium condensation, substituted diaryliodonium compounds react with tyrosines. Thyronines with substituents other than NO_2 or I at the 3 and 5 positions are obtained.

Many other routes have been used to prepare analogues with structures that differ more from T_4, eg, a methylene, carbonyl, or no bridge in place of the ether linkage (13). The halogen-free analogue DIMIT was synthesized by an ingenious route involving the replacement of the iodines at the 3 and 5 positions with cyano groups and their reduction to methyl groups (28).

Chemical Assay. In view of the similarity of their chemical and physical properties (see Table 1) (29), the main problem in the chemical analysis of the thyroid hormones is their separation. A USP procedure gives the details of a paper chromatographic separation in which T_3 is examined for contamination by T_4 and 3,5-diiodothyronine (30). Other systems are also employed (29).

When the purity of the preparation has been ascertained, both T_3 and T_4 are assayed on the basis of their iodine content after combustion in an oxygen flask (29–30).

Body fluids are analyzed for T_3 and T_4 by a variety of radioimmunoassay procedures (31). The important clinical parameter for estimating thyroid function, the protein-bound iodine (PBI), is measured as described in treatises of clinical chemistry.

Recently high performance liquid chromatographic methods have replaced tlc (32–33).

Bioassay. Although the chemical assays described above have replaced bioassays for the determination of T_3 and T_4, several *in vivo* and *in vitro* bioassays are used to determine the potency of thyroglobulin preparations and to establish the thyromimetic or antithyroid potency of new compounds.

In Vivo Tests. The rat antigoiter assay is the most common test for thyromimetic activity. Rats are fed an antithyroid compound, eg, propylthiouracil, for 10 d. At the end of this period, they have developed a goiter of such size that the thyroid weighs ca six times that of control rats. A group of rats is injected daily with standard doses of T_4 (2.5 µg/100 g body wt) or T_3 (0.5 µg/100 g body wt) which are sufficient to prevent goiter formation. Other groups of rats are treated with the appropriate amounts of the thyromimetic compound. Comparison of the equiactive dose with that of T_3 or T_4 establishes its relative potency. Putative antagonists are administered concurrently with T_3 or T_4 and their activity is assessed from the weight of the goiter formed. Strictly speaking, this assay is based on the relative efficacy of the analogues in their interaction with a thyroid-hormone receptor found in anterior pituitary cells which modulates the secretion of TSH (see Fig. 2).

The mouse anoxia or oxygen-consumption test is based on the stimulation of the basal metabolic rate by thyromimetic compounds. Mice or other small animals are placed in airtight containers of known volume and their survival time is determined (34).

The amphibian metamorphosis test is based on the ability of the thyroid hormones to induce precocious transformation of a tadpole into a frog or of the axolotl into a salamander. At present, it is little used because of solubility problems and the difficulty of applying the results to man.

In Vitro Biological Tests.   The inherent complexities, vagaries, and high cost of whole animal assays spurred the recent development of a series of *in vitro* binding assays to various macromolecules that avidly bind thyroid hormones. In general, and allowing for differences in metabolism, excellent agreement with *in vivo* assays was found, and studies of thyroid-hormone structure–activity relationships (SAR) have been greatly simplified.

Using any of the carrier proteins that are available in highly purified form (eg, TBG or TBPA), a convenient and accurate quantitative determination of T_3 and T_4 is possible by displacement of radioiodinated T_3 or T_4. This procedure enables their quick determination at very low concentrations even in the presence of countless other substances that occur in body fluids (31).

In a similar fashion, intact cell nuclei or solubilized proteins from rat liver cell nuclei, which display very high affinities for thyroid hormones, especially T_3, have been used to establish the relative binding affinities of many thyromimetic compounds (7).

Antithyroid Substances

In principle, antithyroid effects (35–39) can be produced by destroying excess thyroid gland tissue surgically or by treatment with radioiodine; blocking synthesis of thyroid hormones with goitrogens such as certain thionamides; inhibition of thyroid-hormone release with lithium; inhibition of the peripheral deiodination of T_4 to the more active T_3 with thiouracils; increasing excretion of thyroid hormone (*n*-butyl 3,5-diiodo-4-hydroxybenzoate [*51-38-7*]) as a result of displacing T_3 and T_4 from serum proteins; and competitive antagonism at the receptor level (r-T_3).

Intrathyroidal Inhibitors.   *Iodide and Other Inorganic Anions.*   When large doses of iodide ion are administered, a transient inhibition of synthesis and release of the thyroid hormones is brought about by the so-called Wolff-Chaikoff effect.

The selective uptake of iodide ion by the thyroid gland is the basis of radioiodine treatment in hyperthyroidism, mainly with ^{131}I, although various other radioactive isotopes are also used (40–41). With a half-life of 8 d, the decay of this isotope produces high energy β particles which cause selective destruction within a 2 mm sphere of their origin. The γ rays also emitted are not absorbed by the thyroid tissue and are employed for external scanning.

Certain inorganic monovalent anions, similar in size to I, are also taken up by the thyroid gland and competitively inhibit active iodide transport with the following decreasing potencies:

$$TcO_4^- \gg ClO_4^- > ReO_4^- > BF_4^- > I^-$$

Clinical use of perchlorate salts (Na or K) is very limited because of side effects.

Thiocyanate ion (SCN) inhibits formation of thyroid hormones by inhibiting the iodination of tyrosine residues in thyroglobulin by thyroid peroxidase. This ion is also responsible for the goitrogenic effect of cassava (manioc, tapioca), since CN is liberated by hydrolysis from the cyanogenic glucoside linamarin it contains, which in turn is biodetoxified to SCN.

Thionamides.   A large group of compounds incorporating thionamide

$$\left(-\overset{\overset{\textstyle S}{\|}}{C}N \diagup \right)$$

or thiourea

moieties are potent antithyroid agents. They inhibit the peroxidases which catalyze the iodination of tyrosine residues in thyroglobulin and their coupling. Although several hundred such compounds are known (42), only four (Fig. 6 and Table 3) are used clinically and only two are accepted by the USP XX.

The imidazoles (**12**) [*60-56-0*] (methimazole, MMI) and (**13**) [*22232-54-8*] (Carbimazole) act by inhibiting intrathyroidal hormone synthesis, whereas the thiouracils (**10**) [*51-52-5*] and (**11**) [*56-04-2*] also inhibit the peripheral deiodination of T_4 to T_3. Thus, the latter are preferred in the treatment of thyroid storm (thyrotoxic crisis) where a quick drop in circulating T_3 is desired (2,9). In general, the imidazoles are 10 times as active as the thiouracils.

The synthesis of these compounds is shown in Figure 6. Extensive compilations of the chemical, chromatographic, and spectral properties of compounds (**10**) and (**12**) are given in refs. 43 and 44, respectively.

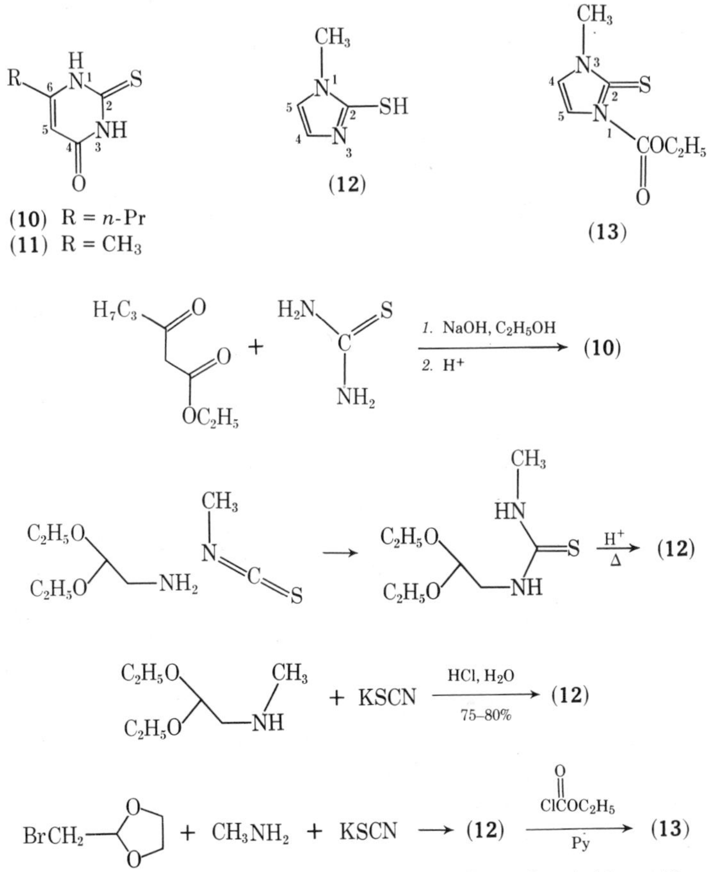

Figure 6. Structures and syntheses of the clinically employed thionamides.

Table 3.   Antithyroidal Thionamides

Name	CAS Reg. No.	Struc-ture	Abbrevi-ation	Composition	Mol wt	S, %	Mp, °C
6-propyl-2-thiouracil (propylthiouracil)	[51-52-5]	(10)	PTU	$C_7H_{10}N_2OS$	170.23	18.84	219–221
6-methyl-2-thiouracil	[56-04-2]	(11)		$C_5H_6N_2OS$	142.18	22.55	325 dec
1-methyl-2-mercapto-imidazole (methimazole)	[60-56-0]	(12)	MMI	$C_4H_6N_2S$	114.16	28.09	146–148
3-methyl-1-carbethoxy-2-thioimidazoline (carbimazole)	[22232-54-8]	(13)		$C_7H_{10}N_2O_2S$	186.23	17.22	122–125

Although several metabolites of propylthiouracil have been found (36,44), it is mainly excreted in the urine as the glucuronide. Its relatively short plasma half-life requires that it be administered 4 times daily.

Extensive studies have been carried out on the metabolic fate of compounds (12) and (13) (36). After initial accumulation in the thyroid gland, the unchanged drugs and various metabolites appear in the urine. The carbethoxy group in carbimazole, which was introduced to mask the bitter taste of methimazole, is metabolically removed, and therefore carbimazole can be considered a prodrug of methimazole.

Recommended daily maintenance doses are 50–200 mg for PTU, and 5–20 mg (three times per day) for the imidazoles. The incidence of side effects is low.

It has been known for a long time that some foodstuffs (eg, turnips, rutabaga) were goitrogenic because of the presence of progoitrin. This substance is hydrolyzed to goitrin, or (S)-5-vinyl-2-oxazolidinethione [500-12-9] (14), which is goitrogenic when iodine intake is low.

$$S = \overset{\displaystyle O}{\underset{\displaystyle NH}{\diagdown}} \!\!- CH\!\!=\!\!CH_2$$

(14)

Aromatic Amines and Phenols.   Approximately 40 yr ago, it was serendipitously discovered that sulfaguanidine [57-67-0] was goitrogenic to rats. Many related compounds were examined, and the aniline moiety was usually present (2,6). Such compounds, as well as resorcinol-like phenols, may act as goitrogens by inhibiting thyroid peroxidases. They are not used clinically.

Lithium.   In the lithium carbonate treatment of certain psychotic states, a low incidence (3.6%) of hypothyroidism and goiter production have been observed as side effects (6,36) (see Psychopharmacological agents). It has been proposed that the mechanism of this action is the inhibition of adenyl cyclase. Lithium salts have not found general acceptance in the treatment of hyperthyroidism (see also Lithium and lithium compounds).

Peripheral Antagonists. The relatively long duration of action of the thyroid hormones makes it desirable to have compounds capable of blocking them competitively at their site of action. This is desirable in the treatment of thyroid storm where the reduction of circulating hormone levels brought about by the inhibition of their synthesis is too slow.

A large number of thyroid hormone analogues have been tested for this effect (6). Among others, r-T_3 (**3**) and 3,3'-T_2 (**4**) and their propionic acid side-chain analogues decrease oxygen consumption at molar ratios of 50–200:1 of T_4. Nevertheless, no potent or clinically useful peripheral antagonists have been found.

The level of circulating hormones is lowered indirectly by n-butyl 3,5-diiodo-4-hydroxybenzoate which displaces them from their carriers (TBG and TBPA) and thus accelerates their metabolism and excretion (6).

Calcitonin

Several years ago, it was discovered that the thyroid gland was also the source of a hypocalcemic hormone whose effects in general oppose those of the parathyroid hormone. This hormone is produced in mammals by the parafollicular C cells and in other vertebrates by the ultimobrachial bodies (45). Originally called thyrocalcitonin, it is now referred to as calcitonin (CT).

The calcitonins from several species have been characterized and synthesized. They are all single-chain 32-residue polypeptides (ca 3600 mol wt), although a disulfide link between the first and seventh cysteine residues results in a cyclic structure that is indispensable for activity (Fig. 7).

Calcitonin is secreted when abnormally high calcium levels occur in plasma. Although plasma concentrations are normally minute (<100 pg/mL), they increase two- to threefold after calcium infusion. Calcitonin has a very short plasma half-life (ca 10 min). Certain thyroid tumors are the result of CT concentrations 50–500 times normal.

The mechanism of action is a direct inhibition of bone resorption. Calcitonin is used clinically in various diseases in which hypercalcemia is present, eg, Paget's disease (46).

Commercial Preparations

Thyroid Preparations. At present, four basic preparations are used in the treatment of hypothyroidism: thyroxine, triiodothyronine, thyroglobulin, and desiccated thyroid extracts. Liotrix is just a 4:1 mixture of T_4 and T_3. Iodinated casein and albumin are

(H)
$$
\begin{cases}
\text{H}_2\text{N-Cys-Gly-Asn-Leu-Ser-Thr-Cys-Met-Leu-Gly-Thr-Tyr-Thr-Gln-Asp-Phe-} \\
\qquad\qquad\quad 5\qquad\qquad\qquad\qquad 10\qquad\qquad\qquad\qquad 15 \\
\text{Asn-Lys-Phe-His-Thr-Phe-Pro-Gln-Thr-Ala-Ile-Gly-Val-Gly-Ala-Pro-CONH}_2 \\
\qquad 20\qquad\qquad\qquad\quad 25\qquad\qquad\qquad\qquad 30\quad\ 32
\end{cases}
$$

(P)
$$
\begin{cases}
\text{H}_2\text{N-Cys-Ser-Asn-Leu-Ser-Thr-Cys-Val-Leu-Ser-Ala-Tyr-Trp-Arg-Asn-Leu-} \\
\qquad\qquad\quad 5\qquad\qquad\qquad\qquad 10\qquad\qquad\qquad\qquad 15 \\
\text{Asn-Asn-Phe-His-Arg-Phe-Ser-Gly-Met-Gly-Phe-Gly-Pro-Glu-Thr-Pro-CONH}_2 \\
\qquad 20\qquad\qquad\qquad\quad 25\qquad\qquad\qquad\qquad 30\quad\ 32
\end{cases}
$$

Figure 7. Amino acid sequence of human (H) and porcine (P) calcitonins.

used now only as animal-feed supplements (see Pet and livestock feeds). Equivalent daily maintenance doses are given in Table 4. Prices are given in Table 5.

Sodium Levothyroxine. As one of the active principles of the thyroid gland, sodium levothyroxine [55-03-8] (levothyroxine sodium) can be obtained either from the thyroid glands of domesticated animals (10) or synthetically. It should contain 61.6–65.5% iodine, corresponding to $100 \pm 3\%$ of the pure salt calculated on an anhydrous basis. Its chiral purity must also be ascertained since partial racemization may occur during synthesis and because dl-T_4 is available commercially. Sodium levothyroxine melts with decomposition at ca 235°C. It is prepared as pentahydrate [6106-07-6] from L-thyroxine and sodium carbonate.

Sodium L-thyroxine is a light yellow or buff-colored odorless, tasteless, hygroscopic powder that is stable when dry and protected from light. It is slightly soluble in water (1 g/700 mL) and ethanol (1 g/300 mL) and insoluble in most organic solvents; it is soluble in aqueous alkaline solutions (48).

The sodium salt is reported to be better absorbed than the free acid although its bioavailability is still low (50%). Its plasma half-life is 5 d.

Table 4.   Daily Maintenance Doses of Thyroid Preparations

Preparation	Daily dose, μg
thyroxine	100
triiodothyronine	25
thyroglobulin[a]	65,000
liotrix[b]	
T_4	50
T_3	12.5

[a] Or thyroid powder.
[b] Mixture of T_4 and T_3.

Table 5.   Current (1982) Approximate Retail Prices of Some Thyroid and Antithyroid Preparations[a]

Preparation	$/g
L-T_4, sodium salt pentahydrate	16
D-T_4, sodium salt pentahydrate	25
L-T_3, sodium salt	23
thyroglobulin	
bovine	84
porcine	64
thyroid acetone powder	
bovine	18
porcine (0.67% I)	0.12
thyrocalcitonin	
powder, bovine	2,130
synthetic, human	245,000
synthetic, salmon	245,000
PTU	0.23
MMI	0.44
methylthiouracil	0.08

[a] Ref. 47.

An extensive compilation of its chemical, spectroscopic, and chromatographic characteristics is given in ref. 29.

Sodium levothyroxine is available in tablets containing 25, 50, 100, 150, 200, 300, or 500 μg under the trade names Letter (Armour), Synthroid (Flint), Levothyroxine-Natrium (Glaxo), L-Thyroxin (Henning), Euthyrox, Cytolen, and Levoid. Nonproprietary preparations are also available. In a 4:1 combination with T_3 it is available in tablets of various strengths as Euthroid (Warner-Chilcott), Thyrolar (Armour), Novothyral, and Thyroxin-T_3 (Henning).

Sodium Liothyronine. Sodium liothyronine [55-06-1] is the sodium salt of L-3,5,3'-triiodothyronine. It is made by the controlled iodination of L-3,5-diiodothyronine. It may be contaminated by starting material or L-T_4. The USP assay (49) describes a chromatographic separation specifying 3,5-T_2 2% max and T_4 5% max. Iodine content is specified at 95–101%. Chiral purity must also be ascertained. Detailed information on its chromatographic behavior is given in ref. 29.

Sodium liothyronine is available in tablets containing 5, 25, or 50 μg as Cytomel (SK & F) or Thybon.

Thyroglobulin. Thyroglobulin is obtained by fractionating hog thyroid glands until a preparation is obtained containing not less than 0.7% of organically bound iodine (50). It is a cream- to tan-colored powder with a characteristic odor and taste. It is stable in air but sensitive to light and is insoluble in water, alcohol, and other organic solvents (51). It is standardized by chemical and biological assay to contain a T_4:T_3 ratio of 2.5:1. It is available in tablets containing varying amounts of iodine (32–650 μg) as Proloid (Warner-Chilcott).

Thyroid. *Glandulae Thyroideae siccatae* is the cleaned, dried, and powdered thyroid gland previously deprived of connective tissue and fat from domesticated animals used for food by man (52). It contains 0.20 ± 0.03% iodine in organically bound form and is free of inorganic iodine. Batches of high or low iodine content should be adjusted to the specified concentration by blending or with a suitable diluent. It is dispensed in tablets of various strengths (15–300 mg). Nonproprietary as well as proprietary preparations are available, eg, Thyroidin (Merck), Thyrar, Thyrocrine, and Thyreoid-Dispert.

Calcitonin. Calcitonin is available commercially from pork and salmon extracts (Calcimar, Armour) as well as by synthesis. Preparations are bioassayed on the basis of their calcium-lowering activity in comparison to the potency of pure pork calcitonin of which ca 4 μg is equivalent to 1 MRC unit (Medical Research Council, UK). For clinical use, vials containing 400 units in 4 mL are available. The recommended daily dosage is 100 units to be administered subcutaneously or intramuscularly since its plasma half-life is very short (4–12 min).

Antithyroid Drugs. *Propylthiouracil.* This compound is a white, powdery crystalline substance of starchlike appearance with a bitter taste. It is slightly soluble in water, chloroform, and ethyl ether, sparingly soluble in ethanol, and soluble in aqueous alkaline solutions (53).

An extensive compilation of its chemical, spectral, and chromatographic properties is given in ref. 43. It is assayed titrimetrically with NaOH (53).

It is available in 5- and 10-mg tablets from various manufacturers (Propacil, Propycil, Prothyran, Procasil, Propylthyracil, Thyreostat II).

Methimazole. This compound is a white to pale buff crystalline powder with a faint characteristic odor. It is soluble in water, ethanol, and chloroform (1 g/5 mL) and only slightly soluble in other organic solvents. A detailed chemical, analytical, spectral, and chromatographic description is given in ref. 44. It is assayed titrimetrically with NaOH (54). It is available in 5- and 10-mg tablets as Tapazole, Thiamethazole, Bazolan, Danantizol, Favistan, Frentorox, Mercazole, Metazole, Thacapazol, Strumazole, and Metothyrine.

BIBLIOGRAPHY

"Thyroid and Antithyroid Preparations (Antithyroid Substances)" in *ECT* 1st ed., Vol. 14, pp. 132–135, by R. G. Jones, Eli Lilly and Co.; "Thyroid and Antithyroid Preparations" in *ECT* 2nd ed., Vol. 20, pp. 260–272, by Reuben G. Jones, Eli Lilly and Company and James B. Lesh (Armour Pharmaceutical Co.).

1. S. C. Werner and S. H. Ingbar, eds., *The Thyroid*, 4th ed., Harper & Row Publishers Inc., New York, 1978.
2. R. C. Haynes, Jr. and F. Murad in A. Goodman Gilman and co-eds., *The Pharmacological Basis of Therapeutics*, 6th ed., Macmillan Publishing Co., New York, 1980, pp. 1397–1419.
3. F. Neuman and B. Schenk in W. Forth and co-eds., *Allgemeine und Spezielle Pharmakologie und Toxikologie*, 2nd ed., Wissenschaftsverlag, Mannheim, FRG, 1977, pp. 349–359.
4. M. A. Greer and D. H. Solomon, *Thyroid*, Vol. III, Section 7 of *Handbook of Physiology*, American Physiological Society, Washington, D.C., 1974.
5. J. Robbins and L. E. Braverman, eds., *Thyroid Research: 7th International Thyroid Conference*, Excerpta Medica, Amsterdam, The Netherlands, 1976.
6. E. C. Jorgensen, in M. E. Wolff, ed., *Burger's Medicinal Chemistry*, Pt. III, John Wiley & Sons, New York, 1981, pp. 103–145.
7. J. D. Baxter and co-workers, *Recent Prog. Horm. Res.* **35,** 97 (1979).
8. P. A. Lehmann F., *J. Med. Chem.* **15,** 404 (1972).
9. H. F. Conn, ed., *Current Therapy 1978*, W. B. Saunders Co., Philadelphia, Pa., 1978, pp. 479–501.
10. R. Pitt-Rivers and J. R. Tata, *The Thyroid Hormones*, Pergamon Press, New York, 1959.
11. C. L. Lautenschläger, *50 Jahre Arzneimittelforschung*, Thieme, Stuttgart, FRG, 1955, pp. 229–238.
12. E. C. Jorgensen, W. J. Murray, and P. Block, Jr., *J. Med. Chem.* **17,** 434 (1974); somewhat earlier the tetramethyl analogue had been shown to have 1–5% of the activity of T_4 by J. A. Pittman and co-workers, as mentioned in *J. Med. Chem.* **16,** 306 (1973).
13. E. C. Jorgensen, *Thyroid Hormones and Analogs*, Vol. VI of C. H. Li, ed., *Hormonal Proteins and Peptides*, Academic Press, Inc., New York, 1978, pp. 57–105, 107–204.
14. S. W. Dietrich and co-workers, *J. Med. Chem.* **20,** 863 (1977).
15. E. C. Jorgensen, *Pharmacol. Ther. B* **2,** 661 (1976).
16. M. Bolger and E. C. Jorgensen, *J. Biol. Chem.* **255,** 10271 (1980).
17. P. A. Lehmann F., *Trends Pharmacol. Sci.* **3,** 103 (1982).
18. V. Cody, *Recent Prog. Horm. Res.* **34,** 437 (1978).
19. P. L. Ballard and co-workers, *J. Clin. Invest.* **65,** 1407 (1980).
20. H. L. Schwartz and J. H. Oppenheimer, *Pharmacol. Ther. B* **3,** 349 (1978).
21. J. J. DiStefano III and D. A. Fisher, *Pharmacol. Ther. B* **2,** 539 (1976).
22. J. Robbins and co-workers, *Recent Prog. Horm. Res.* **34,** 477 (1978).
23. S. Lissitzky, *Pharmacol. Ther. B* **2,** 219 (1976).
24. J. R. Chalmers and co-workers, *J. Chem. Soc.*, 3424 (1949); U.S. Pat. 2,823,164 (Feb. 11, 1958), R. Pitt-Rivers and J. Gross (to National Research Development Corp.); Brit. Pat. 671,070 (Apr. 30, 1952), G. T. Dickson (to Glaxo Laboratories, Ltd.).
25. R. I. Meltzer and co-workers, *J. Org. Chem.* **22,** 1577 (1957); **26,** 1977 (1961).
26. Ref. 2, p. 844.
27. U.S. Pat. 2,889,363 (June 2, 1959), L. G. Ginger and P. Z. Anthony (to Baxter Laboratories, Inc.).
28. P. J. Block, Jr., and D. H. Coy, *J. Chem. Soc. Perkin Trans. 1*, 633 (1972).
29. A. Post and R. J. Wagner in K. Florey, ed., *Analytical Profiles on Drug Substances*, Vol. 5, Academic Press, Inc., New York, 1976, pp. 225–281.

30. *The United States Pharmacopeia XX (USP XX–NF XV)*, The United States Pharmacopeial Convention, Inc., Rockville, Md., 1980, p. 412.
31. B. M. Jaffe and H. R. Behrman, eds., *Methods of Hormone Radioimmunoassay*, 2nd ed., Academic Press, Inc., New York, 1979.
32. B. Hepler and co-workers, *Anal. Chim. Acta* **113,** 269 (1980).
33. B. v. D. Walt and H. J. Cahmann, *Proc. Natl. Acad. Sci. USA* **79,** 1492 (1982).
34. A. Osol and co-eds., *Remington's Pharmaceutical Sciences*, 16th ed., Mack Publishing Co., 1980, p. 130.
35. P. Langer and M. A. Greer, *Antithyroid Substances and Naturally Occurring Goitrogens*, S. Karger, Basel, Switz., 1977.
36. B. Marchant, J. F. H. Lees, and W. D. Alexander, *Pharmacol. Ther. B* **3,** 305 (1978).
37. D. H. Solomon in ref. 1, p. 814.
38. T. Yamada and co-workers in ref. 4, pp. 345–357.
39. W. L. Green in ref. 1, p. 77.
40. Ref. 30, pp. 407–411.
41. Ref. 34, pp. 480–481.
42. G. W. Anderson in *Medicinal Chemistry*, Vol. 1, John Wiley & Sons, Inc., New York, 1951, pp. 1–150.
43. H. Y. Aboul-Enein in ref. 29, Vol. 6, 1977, pp. 457–486.
44. H. Y. Aboul-Enein and A. A. Al-Badr in ref. 29, Vol. 8, 1979, pp. 351–370.
45. H. Rasmussen and M. Pechet in H. Rasmussen, ed., *Parathyroid Hormone, Thyrocalcitonin and Related Drugs*, Vol. 1, Section 51, International Encyclopedia of Pharmacology and Therapeutics, Pergamon Press, Oxford, England, 1970, pp. 237–260.
46. Ref. 2, pp. 1536–1538.
47. Sigma Chemical Co., St. Louis, Mo., price list, Feb. 1982.
48. Ref. 30, p. 446.
49. Ref. 30, p. 452.
50. Ref. 10, p. 189.
51. Ref. 30, p. 799.
52. Ref. 30, p. 800.
53. Ref. 30, p. 686.
54. Ref. 30, p. 505.

PEDRO A. LEHMANN F.
Centro de Investigación y de Estudios Avanzados del I.P.N.

TIN AND TIN ALLOYS

Tin and tin alloys,   18
Detinning,   35

TIN AND TIN ALLOYS

Tin [7440-31-5] is one of the world's most ancient metals. When and where it was discovered is uncertain, but evidence points to tin being used in 3200–3500 BC. Ancient bronze weapons and tools found in Ur contained 10–15 wt % tin. In 79 AD, Pliny described an alloy of tin and lead now commonly called solder (see Solders and brazing alloys). The Romans used tinned copper vessels, but tinned iron vessels did not appear until the fourteenth century in Bohemia. Tinned sheet for metal containers and tole (painted) ware made its appearance in England and Saxony about the middle of the seventeenth century. Although tinplate was not manufactured in the United States until the early nineteenth century, production increased rapidly and soon outstripped that in all other countries (1).

In most cases, tin is used on or in a manufactured material in small amounts, much out of proportion to the purpose it serves. Nevertheless, tin in some form has been associated with the economic and cultural growth of civilization. Food preservation and canning developed rapidly with the invention of tin-coated steel; transportation and high speed machinery became a reality with the invention of tin-base bearing metals; the casting of type metal was an important advance in printing technology; bronze alloys became weapons, tools, and architectural objects; tin alloys are used in organ pipes and bells; and telecommunications and electronic equipment depend upon the tin–lead soldered joint. In modern technology, new uses of tin include the plating of protective coatings, nuclear energy, plastics and other polymers, agriculture, biochemistry, electronic packaging, and glassmaking.

Of the nine different tin-bearing minerals found in the earth's crust, only cassiterite [1317-45-9], SnO_2, is of importance. Over 80% of the world's tin ore occurs in low grade alluvial or eluvial placer deposits where the tin content of the ore can be as low as 0.015%. Complex tin sulfide minerals such as stannite [12019-29-3], Cu_2S.$FeS.SnS_2$; teallite [12294-02-9], $PbSnS_2$; cylinderite [59858-98-9], $PbSn_4FeSb_2S_{14}$; and canfieldite [12250-27-0], Ag_8SnS_6, are found in the lode deposits of Bolivia and Cornwall associated with cassiterite and granitic rock. In the lode mines, the ores often contain 0.8–1 wt % of tin metal. No workable tin deposits have been found in the United States.

Tin-mining methods depend on the character of the deposit. Primary deposits are embedded in underground granitic rock and recovery methods are complex. The more important secondary deposits are in the form of an alluvial mud in the stream beds and placers and the recovery is simpler than lode mining. Cassiterite is recovered from alluvial deposits by dredging, hydraulicking where a head of water permits it, jets and gravel pumps on level ground, or open-pit mining.

Gravel-pump mining is widely used in Southeast Asia and probably accounts for 40% of the world's tin production. Powerful jets of water are directed onto the mine face to break down the tin-bearing soil which is allowed to collect in a sump. A gravel pump in the sump elevates the watery mud to a wooden trough termed a palong in the trade. The palong has a gentle slope and as the ore flows down the slope, the tin oxide particles, which are 2.5 times heavier than sand, are trapped behind wooden slats or riffles. Periodically, the preliminary concentrates are collected and transferred to the dressing shed for final concentration. Hydraulicking and open-pit mining methods also involve gravity separation with water in palongs.

Dredging is mining with a floating dredge on an artificial pond in a placer. Chain buckets or suction cutters, digging at depths of 46 m, transfer the tin-bearing mud to revolving screens, hydrocyclons, jigs, shaking tables, classifiers, and similar equipment. The sand and dirt are removed in these preliminary roughing steps. The mineral is further beneficiated in dressing sheds on shore with modern techniques such as flotation (qv), heavy-media separation, and electrostatic, magnetic, and spiral separators for the removal of associated minerals (see Gravity concentration; Magnetic separation). Final concentrates ready for direct smelting contain 70–77 wt % tin, which is almost pure cassiterite.

Underground-lode deposits in Bolivia are located at very high altitudes, 4000–5000 m above sea level, whereas the lode deposits in Cornwall are ca 430 m below sea level. Access to the lodes is by shaft sinking or by adits (passages driven into the side of a mountain), depending on the terrain. The ore is broken from the working face by blasting and drilling. Further crushing and grinding above ground is necessary to produce the finely divided ore capable of being concentrated by the various gravity-concentration methods and mechanical separations commonly used for alluvial deposits.

Tin concentrates from the lode deposits are 40–60 wt % tin and must be further upgraded before smelting. Roasting the ore removes sulfur and arsenic; the sulfides of iron, copper, bismuth, and zinc are converted to oxides and lead sulfide is oxidized to sulfate. When the concentrates contain considerable quantities of sulfides, the impurities are sometimes removed by a chloridizing roast, followed by leaching. Advances in froth-flotation methods offer another alternative for the removal of unwanted sulfide minerals. When the concentrates are roasted with 1–5 wt % salt (NaCl) in an oxidizing atmosphere, sodium sulfate and the chlorides of the metals are formed without attack on the tin oxide. Many chlorides are volatile; bismuth, lead, arsenic, antimony, and silver may be partially removed in the form of fume; leaching with water removes the remaining chlorides which are readily soluble.

Physical Properties

Physical, mechanical, and thermal constants of tin are shown in Table 1.

Although the pure metal has a silvery-white color, in the cast condition it may have a yellowish tinge caused by a thin film of protective oxide on the surface. When highly polished, it has high light reflectivity. It retains its brightness well during exposure, both outdoors and indoors.

The melting point (232°C) is low compared with those of the common structural metals, whereas the boiling point (2625°C) exceeds that of most metals except tungsten and the platinum group. Loss by volatilization during melting and alloying with other metals is insignificant. Tin is a soft, pliable metal easily adaptable to cold working by rolling, extrusion, and spinning. It readily forms alloys with other metals, imparting hardness and strength. Only small quantities of some metals can be dissolved in pure liquid tin near its melting point. Intermetallic compounds are freely formed, partic-

Table 1.   Physical Properties of Tin[a]

Property	Value
mp, °C	231.9
bp, °C	2625
sp gr	
α-form (gray tin)	5.77
β-form (white tin)	7.29
liquid at mp	6.97
transformation temp $\beta \rightleftharpoons \alpha$, °C	13.2
vapor pressure at K, Pa[b]	
1000	986×10^{-6}
1300	1.1
1500	22.6
2000	4.08×10^3
2550	91×10^3
surface tension at mp, mN/m (= dyn/cm)	544
viscosity at mp, mPa·s (= cP)	1.85
specific heat at 20°C, J/(kg·K)[c]	222
latent heat of fusion, kJ/(g·atom)[c]	7.08
thermal conductivity at 20°C, W/(m·K)	65
coefficient of linear expansion, $\times 10^{-6}$	
at 0°C	19.9
at 100°C	23.8
shrinkage on solidification, %	2.8
resistivity of white tin, $\mu\Omega$·cm	
at 0°C	11.0
at 100°C	15.5
at 200°C	20.0
at mp (solid)	22.0
at mp (liquid)	45.0
volume conductivity, % IACS	15
Brinell hardness, 10 kg, 5 mm, 180 s	
at 20°C	3.9
at 220°C	0.7
tensile strength, as cast, MPa[d]	
at 15°C	14.5
at 200°C	4.5
at −40°C	20.0
at −120°C	87.5
latent heat of vaporization, kJ/mol[c]	296.4

[a] Ref. 2.
[b] To convert Pa to mm Hg, multiply by 0.0075.
[c] To convert J to cal, divide by 4.184.
[d] To convert MPa to psi, multiply by 145.

ularly with metals of high melting point, and some of these compounds are of metallurgical importance. Copper, nickel, silver, and gold are appreciably soluble in liquid tin. A small amount of tin oxide dispersed in tin has a hardening effect.

Molten tin wets and adheres readily to clean iron, steel, copper, and copper-base alloys, and the coating is bright. It provides protection against oxidation of the coated metal and aids in subsequent fabrication because it is ductile and solderable. Tin coatings can be applied to most metals by electrodeposition (see Electroplating).

Tin exists in two allotropic forms: white tin (β) and gray tin (α). White tin, the form which is most familiar, crystallizes in the body-centered tetragonal system. Gray tin has a diamond cubic structure and may be formed when very high purity tin is exposed to temperatures well below zero. The allotropic transformation is retarded if the tin contains small amounts of bismuth, antimony, or lead. The spontaneous appearance of gray tin is a rare occurrence because the initiation of transformation requires, in some cases, years of exposure at $-40°C$. Inoculation with α-tin particles accelerates the transformation.

Chemical Properties

Tin, at wt 118.69, falls between germanium and lead in Group IV A of the periodic table. It has ten naturally occurring isotopes which, in the order of abundance, are 120, 118, 116, 119, 117, 124, 122, 112, 114, and 115 (see Isotopes).

Tin is amphoteric and reacts with strong acids and strong bases, but is relatively resistant to nearly neutral solutions. Distilled water has no effect on tin. Oxygen greatly accelerates corrosion in aqueous solutions. In the absence of oxygen, the high overpotential of tin (0.75 V) causes a film of hydrogen to be retained on the surface which retards acid attack. The metal is normally covered with a thin protective oxide film which thickens with increasing temperature.

A reversal of potential of the tin–iron couple occurs when tin-coated steel (tinplate) is in contact with acid solutions in the absence of air. The tin coating acts as an anode; it is the tin that is slowly attacked and not the steel. This unique property is the keystone of the canning industry because dissolved iron affects the flavor and appearance of the product. Thus, presence of tin protects the appearance and flavor of the product.

Tin does not react directly with nitrogen, hydrogen, carbon dioxide, or gaseous ammonia. Sulfur dioxide, when moist, attacks tin. Chlorine, bromine, and iodine readily react with tin; with fluorine, the action is slow at room temperature. The halogen acids attack tin, particularly when hot and concentrated. Hot sulfuric acid dissolves tin, especially in the presence of oxidizers. Although cold nitric acid attacks tin only slowly, hot concentrated nitric acid converts it to an insoluble hydrated stannic oxide. Sulfurous, chlorosulfuric, and pyrosulfuric acids react rapidly with tin. Phosphoric acid dissolves tin less readily than the other mineral acids. Organic acids such as lactic, citric, tartaric, and oxalic attack tin slowly in the presence of air or oxidizing substances.

Dilute solutions of ammonium hydroxide and sodium carbonate have little effect on tin, but strong alkaline solutions of sodium or potassium hydroxide, cold and dilute, dissolve tin to form stannates.

Neutral aqueous salt solutions react slowly with tin when oxygen is present but oxidizing salt solutions, such as potassium peroxysulfate, ferric chloride and sulfate, and aluminum and stannic chlorides dissolve tin. Nonaqueous organic solvents, lubricating oils, and gasoline have little effect.

Processing

Smelting. Although the metallurgy of tin is comparatively simple, several complicating factors must be dealt with in tin smelting: The temperature necessary for the reduction of tin dioxide with carbon is high enough to reduce the oxides of other metals which may be present. Thus, reduced iron forms troublesome, high-melting compounds with tin, the so-called hard head of the tin smelter. Tin at smelting temperatures is more fluid than mercury at room temperature. It escapes into the most minute openings and soaks into porous refractories. Furthermore, tin reacts with either acid or basic linings, and the slags produced contain appreciable quantities of tin and silica and must be retreated.

Because of the high tin content of the slag, a primary smelting is used to effect a first separation, followed by a second stage to process the slag and hardhead from the first smelting plus refinery drosses.

In primary smelting, carbon (in the form of coal or fuel oil) is the reducing agent. During heat-up, carbon monoxide is formed by reaction with carbon dioxide of the furnace atmosphere. The carbon monoxide reacts with the solid cassiterite particles to produce tin and carbon dioxide:

$$2\,CO + SnO_2 \rightarrow Sn + 2\,CO_2 \tag{1}$$

As the temperature rises, silica (which is present in nearly all concentrates) also reacts with cassiterite under reducing conditions to give stannous silicate:

$$SnO_2 + CO + SiO_2 \rightarrow SnSiO_3 + CO_2 \tag{2}$$

Iron, also present in all concentrates, reacts with silica to form ferrous silicate:

$$Fe_2O_3 + CO + 2\,SiO_2 \rightarrow 2\,FeSiO_3 + CO_2 \tag{3}$$

The silicates formed in reactions 2 and 3 fuse with the added fluxes to form a liquid slag at which point carbon monoxide loses its effectiveness as a reducing agent. Unreacted carbon from the fuel then becomes the predominant reductant in reducing both stannous silicate to tin and ferrous silicate to iron. The metallic iron, in turn, reduces tin from stannous silicate:

$$SnSiO_3 + Fe \rightleftarrows FeSiO_3 + Sn \tag{4}$$

This is the equilibrium established at the end of each smelting cycle. By this time, a considerable proportion of the tin produced in the heat-up stages has been drained from the furnace and only the metal remaining in the furnace at the final tapping time comes into equilibrium with the slag (3).

Primary smelting can be carried out in a reverberatory, rotary, or electric furnace. The choice depends more on economic circumstances than on technical considerations (3). Thus, in the Far East, reverberatory furnaces fired with anthracite coal as the reductant were and still are widely used. Indonesia and Singapore use slow-speed rotary furnaces. Both Malaysia and Thailand have added new electric-furnace smelting capacity in order to improve smelting efficiencies. Reverberatory and rotary furnaces are also used in Indonesia. On the other hand, the smelters in Central Africa, including those in Zaire and Rwanda as well as those in South Africa, which are far away from coal sources, use electric furnaces because of the availability of electric power.

The development of satisfactory processes for the fuming of tin from slags has been one of the greatest contributions to tin smelting in recent years. This work, stimulated by the need for better metal recoveries, relies on the formation and volatilization of tin as SnO_2 in a type of blast furnace (4). The process requires the addition of pyrites (FeS_2) to the tin-rich slag where it reacts to produce $FeSiO_3$ and SnS. The tin sulfide vapor oxidizes to SnO_2 and is carried out in the furnace exhaust gases where it is collected and recycled.

Fuming is also an alternative to roasting in the processing of low grade concentrates (5–25 wt % tin). This procedure yields a tin oxide dust, free of iron, which is again fed back to a conventional smelting furnace.

Other variations of the fuming process under development by the Commonwealth Scientific and Industrial Research Organization in Clayton, Australia, promise even greater efficiencies of metal recovery (5).

Refining. The crude tin obtained from slags and by smelting ore concentrates is refined by further heat treatment, or sometimes electrolytic processes.

The conventional heat-treatment refining includes liquidation or sweating and boiling, or tossing.

In liquidation, tin is heated on the sloping hearth of a small reverberatory furnace to just above its melting point. The tin runs into a so-called poling kettle, and metals that melt sufficiently higher than tin remain in the dross. Most of the iron is removed in this manner. Lead and bismuth remain, but arsenic, antimony, and copper are partly removed as dross.

In the final refining step, the molten tin is agitated in the poling kettles with steam, compressed air, or poles of green wood which produce steam. This process is referred to as boiling. The remaining traces of impurities form a scum which is removed and recirculated through the smelting cycle. The pure tin is cast in iron molds in the form of 45-kg ingots. Purity is guaranteed to exceed 99.8%.

Iron, copper, arsenic, and antimony can be readily removed by the above pyrometallurgical processes or variations on these (3). However, for the removal of large quantities of lead or bismuth, either separately or together, conventional electrolysis or a newly developed vacuum-refining process is used. The latter is now in use in Australia, Bolivia, Mexico, and the USSR (5).

Electrolytic refining is more efficient in regard to both the purity of the product and the ratio of tin to impurities in by-products. However, a large stock of the crude-tin anodes is tied up in the cells, requiring a high capital investment for equipment. An electrolytic plant working with an acid stannous salt at ca 108 A/m^2 requires ca 25 metric tons of working anodes for every ton of refined tin produced per day. Because of these high costs, fire refining should be used as much as possible. The by-products containing high lead, bismuth, and other metal impurities can then be treated in a modest electrolytic plant (6).

An electrorefining plant may operate with either an acid or an alkaline bath. The acid bath contains stannous sulfate, cresolsulfonic or phenolsulfonic acids (to retard the oxidation of the stannous tin in the solution), and free sulfuric acid with β-naphthol and glue as addition agents to prevent treelike deposits on the cathode which may short-circuit the cells. The concentration of these addition agents must be carefully controlled. The acid electrolyte operates at room temperature with a current density of ca 86–108 A/m^2, cell voltage of 0.3 V, and an efficiency of 85%. Anodes (95 wt % tin) have a life of 21 d, whereas the cathode sheets have a life of 7 d. Anode slimes may be a problem if the lead content of the anodes is high; the anodes are removed at frequent intervals and scrubbed with revolving brushes to remove the slime (7).

The alkaline bath contains potassium or sodium stannate and free alkali and operates without addition agents. The solution must, however, be heated to 82°C. Stannous tin must be absent because it passivates the anodes. The tin dissolves as stannate (SnO_3^{2-}), but only if the anodes are initially coated with a film of yellow-green hydrated oxide ($SnO_2.2H_2O$). This so-called filming is accomplished by passing a high current through the anode after insertion in the cell; the current density is reduced to normal once the film is formed. Slow insertion of the anode with the current turned on gives the same result. The advantages of the alkaline electrolyte are ease of operation and the capability of using a lower grade of anode. The disadvantages are that the solution must be heated and that the current-carrying species is Sn^{4+}, giving an electrochemical equivalent half that of the acid electrolyte. At equilibrium, the lead plumbite is almost completely precipitated as hydroxide in the slime, but some antimony remains in solution. The spent anode is returned to the fire-refining process for recovery of lead, antimony, etc. The plated cathode sheets, weighing ca 90 kg, are melted in a holding pot and cast into ingots. Part of the metal from the holding pots is used to form the starter cathode sheets weighing ca 7.2 kg each.

Secondary Tin. In 1980, >14,700 metric tons of tin were recovered in the United States from scrap (8). Sources include bronze rejects and used parts, solder in the form of dross or sweepings, dross from tinning pots, sludges from tinning lines, babbitt from discarded bearings, type-metal scrap, and clean tinplate clippings from container manufacturers. High purity tin is recovered by detinning clean tinplate (see Recycling).

Alloy scrap containing tin is handled by secondary smelters as part of their production of primary metals and alloys; lead refineries accept solder, tin drosses, babbitt, and type metal. This type of scrap is remelted, impurities such as iron, copper, antimony, and zinc are removed, and the scrap is returned to the market as binary or ternary alloy. The dross obtained by cleaning up the scrap metal is returned to the primary refining process.

Economic Aspects

Tin has long been regarded as a strategic metal because of its importance in canning, electrical, and transportation applications. Accordingly, it is stockpiled by the General Services Administration (GSA) at various locations in the country. On December 31, 1979, the U.S. Government stocks of pig tin totaled 203,691 metric tons which included 170,670 metric tons above the goal of 33,021 t (9). On May 2, 1980, the Federal Emergency Agency set the new National Defense Stockpile Goal for tin at 42,700 metric tons. On January 2, 1980, the Strategic and Critical Materials Transaction Authorization Act became effective. This authorizes the President to dispose of materials determined to be excessive to the current needs of the stockpile. This act provides for the sale of up to 35,600 metric tons of tin including a contribution of up to 5100 metric tons of tin to the International Tin Council (ITC) buffer stock (see below). The GSA set up a schedule to offer about 500 metric tons of Grade A tin, for domestic sales and consumption only, every other Tuesday beginning July 1, 1980. On December 14, 1981, the restrictions on exporting the GSA tin sold were lifted; sales increased immediately. Thus, from July 1, 1980, through December 11, 1981, the total GSA sales were 3170 metric tons. An additional 1815 metric tons were sold soon thereafter, mostly to traders (10).

Since the mid 1950s, the price of tin has been subject to an international agreement between producing and consuming nations. Under the International Tin Agreement (ITA), the ITC seeks to deal effectively with situations where a shortage or surplus of tin arises and prevent excessive price fluctuations (11). It attempts to ensure that a fair price is paid for tin on the world market and tries to stimulate export earnings from tin, especially for developing producing countries. To these ends, the ITC establishes floor and ceiling prices for tin. Prices within this range are subject to buffer-stock management. The floor price is maintained by ITC buffer stock purchases, or by applying export controls on member producing nations. The ceiling price is maintained through buffer-stock sales. The United States was a member of the Fifth ITA which expired on June 30, 1982. On October 17, 1981, the ITC price range was the equivalent of $11.92/kg for the floor price and $15.51/kg for the ceiling price (12).

World trading in tin occurs mostly at Penang, London, and New York. As shown in Table 2, most of the world's tin is produced in Southeast Asia, and generally, the marketing at Penang establishes the world tin price. The Penang price is determined daily at two Malaysian smelters by comparing bids from dealers and consumers with the available metal supply. The London Metal Exchange (LME) and New York market offer cash and forward metal prices. The LME price is based on the tin smelted in the UK, whereas the New York price is an average of dealer-quoted prices. New York prices for the past 20 years are given in Table 3.

The United States is by far the largest consumer of tin, followed by Japan and the FRG. The 1980 uses distribution in the six principal consuming countries is given in Table 4. A more detailed breakdown of U.S. tin uses in 1979 and 1980 is given in Table 5. The United States uses ca 4800 t/yr in the form of tin compounds.

Tinplate provides an outlet for over one third of the primary tin used in the United States. In 1980, ca 3.7×10^6 t of tinplate were produced in U.S. steel mills. Total world production in 1980 was 13.6×10^6 t. In the United States in 1980, ca 56×10^6 base boxes

Table 2. World Tin Production, Metric Tons [a]

Year	World	Malaysia	Thailand	Indonesia	Bolivia	Brazil	Zaire	UK
1925[b]	165,600	57,746	8,335	33,831	37,763		982	2,700
1950	164,800	58,694	10,530	32,617	31,714	183	11,947	904
1960	138,700	52,813	12,275	22,958	20,543	1,581	9,350	1,218
1965	154,400	64,692	19,353	14,935	23,407	1,220	6,311	1,334
1970	185,800	73,794	21,779	19,092	30,100	3,610	6,458	1,722
1975	177,700	64,364	16,406	25,346	28,324	5,000	4,562	3,330
1978	196,900	62,650	30,186	27,410	30,881	6,320	3,450	2,802
1979	200,700	62,995	33,962	29,440	27,781	6,645	3,300	2,374
1980	199,300	61,404	33,685	32,527	27,271	6,756	3,159	3,028

[a] Tin-in-concentrates (tin content) as compiled by the International Tin Council (13).
[b] From ref. 1.

Table 3. New York Average Annual Tin Prices [a]

Year	¢/kg	Based on constant 1978 dollars, ¢/kg
1960	223.08	493.97
1966	360.84	714.82
1972	390.43	593.65
1974	871.79	1142.57
1978	1385.08	1385.08
1980	1861.20	

[a] *Metals Week* New York market price, 1958–1975; thereafter, *Metals Week* composite price (9).

Table 4.   Distribution of 1980 Tin Consumption, Metric Tons[a]

Product	United States	Japan[b]	FRG	France[b]	UK	Combined
tinplate	16,346	11,997	4,329	4,885	3,155	40,712
solder	15,618	10,878	2,446	2,678	892	32,512
bronze	7,478	1,521	375	493	1,245	11,112
babbitt	2,380	855	376	332	1,472	5,415
tinning	2,577	1,396	549	262	844	5,628
chemicals	4,800[c]	1,649	2,050[c]	510	1,282	10,291
other	7,163	2,583	4,146	899	1,010	15,801
Total	*56,362*	*30,879*	*14,271*	*10,059*	*9,900*	*121,471*

[a] Combined primary and secondary tin consumed, except where otherwise stated. As compiled by International Tin Council, except as modified by the indicated estimates.
[b] Primary only.
[c] Estimated.

Table 5.   Distribution of U.S. Tin Consumption, Metric Tons[a]

Product	1979		1980	
	Primary	Secondary	Primary	Secondary
tinplate	17,929		16,346	
solder	13,249	4,773	11,653	3,965
babbitt	1,830	413	1,537	843
bronze and brass	2,709	5,981	2,147	5,331
collapsible tubes and foil	686		526	
tinning	2,498	86	2,531	46
bar tin and anodes	567		486	
tin powder	1,435		1,098	
alloys (misc)	2,248	180		134
white metal	1,258		914	
chemicals	4,797			
other	290	1,536	7,104	1,701
Total	*49,496*	*12,969*	*44,342*	*12,020*
Grand Total		*62,465*		*56,362*

[a] Ref. 14.

was used to make food containers (one base box comprises an area of 202,000 cm^2 or ca 31 in.^{2}).

Specifications and Analytical Methods

The ASTM Classification of Pig Tin B 339 lists seven grades as shown in Tables 6–7 (15).

In the field, cassiterite ore is usually recognized by its high density (7.04 g/cm^3), low solubility in acid and alkaline solutions, and extreme hardness. Tin in solution is detected by the white precipitate formed with mercuric chloride. Stannous tin in solution gives a red precipitate with toluene-3,4-dithiol.

The tin content of ores, concentrates, ingot metal, and other products is determined by fire assay, fusion method, and volumetric wet analysis.

In a fire-assay method used at the smelters, a weighed quantity of concentrate

Table 6. ASTM B 339-72 Classification of Pig Tin[a]

| Grade designation | | Description[b] | | General applications |
ASTM	Commercial	Class	Tin, % min	
AAA	electrolytic	extra high purity	99.98	analytical standards and research
AA	electrolytic	high purity	99.95	research, pharmaceuticals, fine chemicals
A	A	standard	99.80	food containers, foil, collapsible tubes, unalloyed (block) tin products, electrotinning, tin-alloyed cast iron, high grade solders
B	B	general purpose	99.80	less exacting than A; general purpose
C	C	intermediate grade	99.65	general purpose; alloys
D	D	lower intermediate grade	99.50	general purpose; alloys
E	E	common	99.00	cast bronze, bearing metal, general-purpose solders, lead-base alloys

[a] Ref. 15.
[b] A more complete description of these grades is given in Table 7 and in the full ASTM Standard B 339.

Table 7. Chemical Composition and Impurity Contents[a]

| Element, % max | ASTM Classification[b] | | |
	AAA	AA	A
tin, min	99.98	99.95	99.80
antimony	0.008	0.02	0.04
arsenic	0.0005	0.01	0.05
bismuth	0.001	0.01	0.015
cadmium	0.001	0.001	0.001
copper	0.002	0.02	0.04
iron	0.005	0.01	0.015
lead	0.010	0.02	0.05
nickel + cobalt	0.005	0.01	0.01
sulfur	0.002	0.01	0.01
zinc	0.001	0.005	0.005

[a] Ref. 15.
[b] The only figures available for the chemical compositions and impurity contents of other ASTM classifications are: B, tin 99.80% min; arsenic 0.05% max; C, tin 99.65% min; D, tin 99.50% min; E, tin 99.00% min.

is mixed with sodium cyanide in a clay or porcelain crucible and heated in a muffle furnace at red heat for 20–25 min. The tin oxide is reduced to metal which is cleaned and weighed. Preliminary digestion of the concentrate with hydrochloric and nitric acids to remove impurities normally precedes the sodium cyanide fusion.

Tin ores and concentrates can be brought into solution by fusing at red heat in a nickel crucible with sodium carbonate and sodium peroxide, leaching in water, acidifying with hydrochloric acid, and digesting with nickel sheet. The solution is cooled in carbon dioxide, and titrated with a standard potassium iodate–iodide solution using starch as an indicator.

The determination of tin in metals containing over 75 wt % tin (eg, ingot tin) requires a special procedure (16). A 5 g sample is dissolved in hydrochloric acid, reduced with nickel, and cooled in CO_2. A calculated weight of pure potassium iodate (dried at 100°C) and an excess of potassium iodide (1:3) are dissolved in water and added to the reduced solution to oxidize 96–98 wt % of the stannous chloride present. The reaction is completed by titration with 0.1 N KIO$_3$–KI solution to a blue color using starch as the indicator.

Several ASTM methods are available for the determination of tin in tin-containing alloys such as solder, babbitt, and bronze (17).

The purity of commercial tin is under strict control at the smelters. Photometric, chemical, atomic absorption, fluorimetric, and spectrographic methods are available for the determination of impurities (16).

Health and Safety Factors

Tests have shown that considerable quantities of tin can be consumed without any effect on the human system. Small amounts of tin are present in most liquid canned products; the permitted limit of tin content in foods is 300 mg/kg in the United States and 250 ppm in the UK, which far exceed the amount in canned products of good quality (18) (see also Tin compounds).

Uses

Tin is used in various industrial applications as cast and wrought forms obtained by rolling, drawing, extrusion, atomizing, and casting; tinplate, ie, low carbon steel sheet or strip rolled to 0.15–0.25 mm thick and thinly coated with pure tin; tin coatings and tin alloy coatings applied to fabricated articles (as opposed to sheet or strip) of steel, cast iron, copper, copper-base alloys, and aluminum; tin alloys; and tin compounds.

Cast and Wrought Forms. Thousands of tons of tin ingots are cast into anodes for plating processes. Tin foil is used for electrical condensers, bottle-cap liners, gun charges, and wrappings for food. Tin wire is used for fuses and safety plugs. Extruded tin pipe and tin-lined brass pipe are the first choice for conveying distilled water and carbonated beverages. Sheet tin is used to line storage tanks for distilled water. Tin powder is used in powder metallurgy, the largest use going to tin powder mixtures with copper to form bronze parts. It is also used for coating paper and for solder pastes. In the float-glass process, adopted by all leading plate-glass manufacturers, the molten glass is allowed to float and solidify on the surface of a pool of molten tin which provides an ideally flat surface. The endless glass ribbon has a surface so smooth that costly grinding and polishing are unnecessary.

Tinplate. The development of tinplate was associated with the need for a reliable packaging material for preserving foods. It comprises in one inexpensive material the strength and formability of steel and the corrosion resistance, solderability, absence of toxicity, and good appearance of tin. The tin coating is applied by electroplating in a continuous process or by passing cut sheet through a bath of molten tin. In the United States, tinplate is now made mainly by the electrolytic process with less than 1% of production from hot-tinning machines.

The electrolytic process is flexible and capable of applying tin coatings from 250 nm to 2.5 μm on each face. A thick coating can be applied to one side of the sheet and a thinner coating to the other (differential tinplate) providing a cost savings to the can manufacturer if less protection is needed on the outside of the can. The thinner coatings usually require a baked enamel coating over the tin, except when packaging dry foods and nonfood products.

In addition to the packaging of foods and beverages in regular containers, a large quantity of tinplate is used in the form of aerosol containers for cosmetics, paint, insecticides, polishes, and other products. Decorative trays, lithographed boxes, and containers of unusual shape are additional outlets for tinplate.

A tinplate container has an energy advantage over some of the competitive container materials, as shown in Table 8. By virtue of its magnetic properties, tinplated material is more readily separated by other forms of industrial or domestic waste.

Tin Coatings. The coating may be applied by hot-dipping the fabricated article in liquid tin, by electroplating using either acid or alkaline electrolytes, and by immersion tinning (see below). The hot-tinned coating is bright. Electrodeposited coatings are normally dull as plated, but may be flow-brightened by heating momentarily to the melting point by induction or in hot air or oil. Proprietary bright tin-plating processes have been developed and are in commercial use. These incorporate long-chain organic molecules as brightening and leveling agents in a stannous sulfate electrolyte. Brighteners for other electrolytes have not been developed.

The coating thickness may range from 0.0025 to 0.05 mm, depending on the type of protection required. Pure tin coatings are used on food-processing equipment, milk cans, kitchen implements, electronic and electrical components, fasteners, steel and copper wire, pins, automotive bearings, and pistons.

For articles that require only a very thin film of tin, seldom exceeding 0.8 μm immersion tin coatings are applied. The process is based on chemical displacement by immersion in a solution of tin salts. Recently, a new autocatalytic tin-deposition process was developed at the research level. It promises to be useful to coat any base material including plastics, in addition to providing coatings of any thickness desired (20).

Tin Alloys. *Coatings.* Tin-alloy coatings provide harder, brighter, and more corrosion-resistant coatings than tin alone. Tin–copper electrodeposited coatings (12 wt % tin) have the appearance of 24-carat gold and provide a bronze finish for furniture hardware, trophies, and ornaments (see also Coatings). They also provide a stop-off coating (resist) for nitriding.

Table 8. Energy Consumption in Container Manufacturing [a]

	Energy consumed in producing raw material for 1 t, GJ [b]	Number of containers per ton	Energy consumed per container, kJ [b]
tinplate can	49	16,500	3,010
aluminum can	395	44,500	8,660
bimetallic can	77	18,400	4,210
glass bottle			
returnable	54.8	2,000	27,540
nonreturnable	54.8	4,000	13,770

[a] Ref. 19.
[b] To convert J to Btu, divide by 1054.

Tin–lead coatings (10–60 wt % tin) can be applied by hot-dipping or electrodeposition to steel and copper fabricated articles and sheet. A special product is terne plate used for roofing and flashings, automobile fuel tanks and fittings, air filters, mufflers, and general uses such as covers, lids, drawers, cabinets, consoles for instruments, and for radio and television equipment. Terne plate is low carbon steel, coated by a hot-dip process with an alloy of tin and lead, commonly about 7–25 wt % tin, remainder lead. Electroplating is another possibility.

Because of the ease with which they can be soldered, electroplated tin–lead coatings of near eutectic composition (62 wt % tin) are extensively used in the electronics industry for coating printed circuit boards and electrical connectors, lead wires, capacitor and condenser cases, and chassis.
for

Tin–nickel electrodeposited coatings (65 wt % tin) provide a bright decorative finish with a corrosion resistance that exceeds that of nickel and copper–nickel–chromium coatings. A new and expanding use for tin–nickel is for printed circuit boards. Such coatings deposited on the copper-clad boards are etch resistant and provide protection for the conducting path. The well-established decorative and functional uses are for watch parts, drawing instruments, scientific apparatus, refrigeration equipment, musical instruments, and handbag frames.

Tin–zinc coatings (75 wt % tin) have application as a solderable coating for radio, television, and electronic components. They also provide galvanic protection for steel in contact with aluminum.

A ternary tin–copper–lead coating, 2 Cu–8 Sn–90 Pb, is a standard overplate for steel-backed copper–lead automotive bearings.

Tin–cadmium coatings are particularly resistant to marine atmospheres and have applications in the aviation industry.

Solder. Tin and lead combine easily to form a group of alloys known generally as soft solders. The joining of metals with tin-containing solders can be attributed to several properties. Their low melting point allows simple equipment to be used for melting and joining, the alloys are unsurpassed in wetting and adhering to clean metal surfaces and flowing into small spaces, and they are relatively cheap. The tin–lead solders have no serious competitors in the field of low temperature joining.

Tin is the important constituent in solders because it is the element that wets the base metal, such as copper and steel by alloying with it. Solders are used mainly in auto radiators, air conditioners, heat exchangers, plumbing and sheet-metal joining, container seaming, electrical connections in radio and television, generating equipment, telephone wiring, electronic equipment and computers, and aerospace equipment (see Table 9). Lead-free solder alloys, where the tin is alloyed individually with antimony, silver, gold, zinc, or indium, are available for special joining applications where properties such as high strength, absence of toxicity, and special corrosion resistance are required.

Low melting or fusible alloys (mp, 20–176°C) may be loosely described as solders and are employed for sealing and joining materials which may be damaged in ordinary soldering practice. They have applications in automatic safety devices, foundry patterns, electroforming, tube bending, tempering baths, molds for plastics, and denture models. Fusible alloys, mostly eutectic alloys, are usually two, three, four, or even five-component mixtures of bismuth, tin, lead, cadmium, indium, and gallium.

Table 9. Solders and Their Uses

Alloy content, wt %	Melting range, °C	Typical tensile strength of cast solder, MPa[a]	Uses
general engineering			
60 Sn, 40 Pb	183–188	53	electronics and instruments
50 Sn, 50 Pb	183–212	45	sheet-metal work and light engineering
40 Sn, 60 Pb	183–234	43	general engineering and capillary fittings
30 Sn, 70 Pb	183–255	43	plumber's solder, cable joining, automobile radiators
20 Sn, 80 Pb	183–276	42	automobile radiators
40 Sn, 57.8 Pb, 2.2 Sb	185–227	51	similar to 40 Sn, 60 Pb solder
special purpose			
2 Sn, 98 Pb	315–322	28	tinplate can side seams
10 Sn, 90 Pb	267–301	37 ⎫	cryogenics
5 Sn, 93.5 Pb, 1.5 Ag	296–301	39 ⎭	
62 Sn, 36 Pb, 2 Ag	178	43 ⎫	
95 Sn, 5 Sb	236–243	40 ⎬	creep resistance
95 Sn, 5 Ag	221–225	59 ⎭	
98 Sn, 2 Ag	221–235	26	food and beverage containers
100 Sn	232	14	
52 Sn, 30 Pb, 18 Cd	145	43	low melting solder
80 Sn, 20 Zn	200–265	70	soldering aluminum

[a] To convert MPa to psi, multiply by 145.

Bronze. Copper-tin alloys, with or without other modifying elements, are classed under the general name of bronzes. They can be wrought, sand cast, or continuously cast into shapes. Binary tin–copper alloys are difficult to cast because they are prone to gassing, which can be alleviated by additions of phosphorus and zinc. The phosphor bronzes (5–10 wt % tin) are preferred because they have superior elastic properties, excellent resistance to alternating stress and corrosion fatigue and to corrosive attack by the atmosphere and water, and superior bearing properties. So-called gun metals containing 1–6 wt % zinc and 5–10 wt % tin are gas-free, pressure-tight alloys used for valves and fittings for water and steam lines. These alloy types may be further modified with lead to improve machinability and, in the case of the phosphor bronzes, to obtain a more conformable bearing alloy. The 85 Cu–5 Sn–5 Zn–5 Pb alloy is a popular composition. Bronzes are especially applicable to marine and railway engineering pumps, valves and pipe fittings, bearings and bushings, gears and springs, and ship propellers. Included in special bronze alloys is bell metal, known for its excellent tonal quality, containing 20 wt % tin, and statuary bronze.

Bearing Metals. Metals used for casting or lining bearing shells are classed as white bearing alloys, but are known commercially as babbitt (see also Bearing materials). The term white metal was used by Isaac Babbitt in 1839 in his description of tin-base bearing metals supported by a stronger shell. Although white metal is a general term for many white-colored alloys of relatively low melting point, the white-metal product mentioned in Table 5 is made from pewter, britannia, or jeweler's metal (alloys containing approx. 90–95 wt % tin, 1–8 wt % antimony and 0.5–3 wt % copper).

The term babbitt includes high tin alloys (substantially lead free) containing >80 wt % tin, and high lead alloys containing ≥70 wt % lead and ≤12 wt % tin. Both have the characteristic structure of hard compounds in a soft matrix, and although they contain the same or similar types of compounds, they differ in composition and properties of the matrix.

The common high tin babbitts are all based upon the tin–antimony–copper system. Compositions and properties of the more widely used tin-base bearing alloys are given in Table 10. Antimony up to 8.0% strengthens the bearing alloy matrix by dissolving in the tin. Above 8.0%, hard particles of tin–antimony compound (SnSb) are formed which tend to float. Additions of copper secure a uniform distribution of these hard particles in a soft but rigid bearing matrix.

The lead-base babbitts are based upon the lead–antimony–tin system, and, like the tin-base, have a structure of hard crystals in a relatively soft matrix. The lead-base alloys are however, more prone to segregation, have a lower thermal conductivity than the tin-base babbitts, and are employed generally as an inexpensive substitute for the tin-base alloys. Properly lined, however, they function satisfactorily as bearings under moderate conditions of load and speed.

Both types of babbitt are easily cast and can be bonded rigidly to cast iron, steel, and bronze backings. They perform satisfactorily when lubricated against a soft steel shaft, and occasional corrosion problems with lead babbitt can be corrected by increasing the tin content or shifting to high tin babbitt.

Babbitt alloys are suitable for hundreds of types of installations involving the movement of machinery, eg, the main, crankshaft, connecting rod big end, camshaft, and journal bearings associated with marine propulsion, railroad and automotive transportation, compressors, motors, generators, blowers, fans, rolling-mill equipment, etc.

The field of bearing metals also includes bronze and aluminum alloys. The aluminum–tin alloys have fatigue properties comparable with white-metal alloys at ordinary temperatures but at the temperatures encountered in automobile engines (up to 250°C) they are much stronger. An aluminum–tin alloy containing 6.5 wt % Sn, 1 wt % Cu, and 1 wt % Ni is used in applications such as bushings and solid bearings in aircraft landing gear assemblies subject to shock loads of 48 MPa (ca 7000 psi), as tracks for vertical boring mills, and as floating bearings in gas-turbine engines, diesel locomotive and tractor engines, and cold-rolling mills. A modification of this alloy contains

Table 10.   Composition and Properties of ASTM B 23 Bearing Alloys[a]

Alloy no.	Nominal composition, wt %	Compressive strength, MPa[b]		Brinell hardness[e]
		Yield[c]	Ultimate[d]	
1	91 Sn, 4.5 Sb, 4.5 Cu	30	89	17.0
2	89 Sn, 7.5 Sb, 3.5 Cu	42	103	24.5
3	84 Sn, 8 Sb, 8 Cu	46	121	27.0
4	75 Sn, 12 Sb, 10 Pb, 3 Cu	38	111	24.5
5	65 Sn, 15 Sb, 18 Pb, 2 Cu	35	104	22.5

[a] Ref. 21.

[b] To convert MPa to psi, multiply by 145.

[c] Determined at a total deformation of 0.125% reduction in gauge length.

[d] Determined from unit load required to produce a 25% deformation of specimen length.

[e] Average of three values using a 10 mm ball and applying a 500 kg load for 30 s.

also 2.5 wt % silicon. Increasing the tin content to 20 wt % gives a good compromise between fatigue strength and softness, and such steel-backed aluminum–tin alloys are used for connecting rod and main bearings for passenger cars, automatic transmission bushings, camshaft bearings, and thrust washers.

Pewter. Modern pewter may have a composition of 90–95 wt % tin, 1–8 wt % antimony, and 0.5–3 wt % copper. Lead should be avoided by contemporary craftsman because it causes the metal surface to blacken with age. Pewter metal can be compressed, bent, spun, and formed into any shape, as well as being easily cast. A wide variety of consumer articles are available from domestic and foreign manufacturers. Reproductions of pewter objects from colonial times, some cast from the original molds, are popular. The annual U.S. production of pewter exceeds 1100 t.

Type Metals. The printing trade still requires significant amounts of lead-based alloys containing 10–25 wt % antimony and 3–13 wt % tin. By varying the tin and antimony content, type suitable for each printing process can be obtained. Linotype machines demand a fluid and mobile metal with a short freezing range. Casting stereotype plates requires cool metal and a hot box with progressive solidification from the bottom up. Monotype, with the higher percentages of tin and antimony, provides a type with a fine face and superior wear resistance. Foundry type is extra hard wearing and has extra long life which reduces the need for recasting for duplicating jobs.

Alloyed Iron. Tin-alloyed flake and nodular cast irons are now widely used throughout the world, although 20 years ago, world usage of tin as an alloying element for cast iron was probably <1 t/yr. Estimated 1980 consumption was ca 1,200 t, largely due to research carried out by the International Tin Research Institute in the 1950's. As little as 0.1% tin when added to flake and spheroidal graphite cast irons in the pouring ladle gives the iron a structure that is completely pearlitic. Tin-inoculated iron has a uniformity of hardness, improved machinability, wear resistance, and better retention of shape on heating. Where pearlitic and heat-resistant cast irons are required, such as for engine blocks, transmissions, and automotive parts, tin additions may provide a suitable material.

Special Alloys. Alloys of tin with the rarer metals, such as niobium, titanium, and zirconium have recently been developed. The single-phase alloy Nb_3Sn [12035-04-0] has the highest transition temperature of any known superconductor (18 K) and appears to keep its superconductivity in magnetic fields up to at least 17 T (170 kG) (see Superconducting materials). Niobium–tin ribbon, therefore, is of practical importance for the construction of high field superconducting solenoid magnets (see Superconducting materials).

Tin is an important addition to titanium. As a nominal addition (2–4% Sn), tin is a solid-state strengthener, retards interstitial diffusion, and promotes plasticity and free-scaling. Some of the more widely used commercial alloys include 92.5 Ti–5 Al–2.5 Sn, 86 Ti–6 Al–6 V–2 Sn, and 86 Ti–6 Al–2 Sn–4 Zn–2 Mo.

Because of its low neutron absorption, zirconium is an attractive structural material and fuel cladding for nuclear power reactors, but it has low strength and highly variable corrosion behavior. However, Zircalloy-2, with a nominal composition of 1.5 wt % tin, 0.12 wt % iron, 0.05 wt % nickel, 0.10 wt % chromium, and the remainder zirconium, can be used in all nuclear power reactors that employ pressurized water as coolant and moderator (see Nuclear reactors).

Dental amalgams, mainly silver–tin–mercury alloys, have been used as fillings for many years (see Dental materials). The most common alloy contains 12 wt % tin.

Other Uses.   There have been various developments in recent years with modest usage of tin today but good potential for the future. The production of finished shapes from iron powder by compacting and sintering utilizes about 100,000 t of iron powder annually; copper powder (2–10 wt %) is normally added as a sintering aid. Addition of 2% tin powder or equal amounts of tin and copper powder considerably lowers the sintering temperature and time of sintering at a cost saving. The tin addition also improves dimensional control. Iron powder plus 10 wt % powdered lead–tin metal is pressed and sintered to make pistons for use in automotive hydraulic brake cylinders.

The electronics and aerospace industries have for a number of years used gold-plated printed circuit boards and component leads where highest reliability is desired. Problems in the use of gold coatings have plagued the industry and the trend is toward the substitution of tin–lead or tin coatings for the gold coatings. Tin–nickel coatings with a thin flash of tin or gold are also used as a substitute for heavy gold coatings (see also Electrical connectors).

BIBLIOGRAPHY

"Tin and Tin Alloys" in *ECT* 2nd ed., Vol. 20, pp. 273–293, by R. M. MacIntosh, Tin Research Institute, Inc.

1. C. L. Mantell, *Tin: Its Mining, Production, Technology, and Applications*, 2nd ed., Reinhold Publishing Corp., New York, 1949.
2. C. J. Faulkner, *The Properties of Tin*, Publication 218, International Tin Research Institute, London, England, 1965, 55 pp.
3. S. C. Pearce in J. Cigan, T. S. Mackey, and T. O'Keefe, eds., *Proceedings of a World Symposium on Metallurgy and Environmental Control at 109th AIME Annual Meeting*, The Metallurgical Society of AIME, Warrendale, Pa., 1980, pp. 754–770.
4. S. M. Kolodin, *Vetoritznoe Olova (Secondary Tin)*, Moscow, 1964, p. 207.
5. T. S. Mackey, *J. Met.* **34,** 72 (April 1982).
6. P. A. Wright, *Extractive Metallurgy of Tin*, American Elsevier Publishing Co., New York, 1966.
7. C. L. Mantell, *J. Met.* **15,** 152 (1963).
8. U.S. Bureau of Mines, *Metal Statistics 1981*, American Metal Market Fairchild Publications, New York, 1981, p. 219.
9. J. F. Carlin, Jr., *Tin, Bulletin 671*, a chapter from *Mineral Facts and Problems*, 1980 ed., Superintendent of Documents, Washington, D.C., 1980.
10. *Am. Met. Mark.* **89**(248), 2 (Dec. 24, 1981).
11. P. M. Dinsdale, *A Guide to Tin*, Publication No. 540, International Tin Research Institute, London, England.
12. J. F. Carlin, Jr., *Tin in October 1981*, *Tin Industry Monthly*, Mineral Industry Surveys, U.S. Bureau of Mines, Washington, D.C., Dec. 31, 1981.
13. *Monthly Statistical Bulletin*, International Tin Council, London, UK.
14. U.S. Bureau of Mines, *Metal Statistics 1982*, American Metal Market Fairchild Publications, New York, 1982, p. 211.
15. *Nonferrous Metals; Electrodeposition Coatings; Metal Powders; Surgical Implants, Part 7, 1972 Annual Book of ASTM Standards*, American Society for Testing Materials, Philadelphia, Pa., 1972.
16. J. W. Price, and R. Smith in W. Fresenius, ed., *Handbook of Analytical Chemistry*, Part III, Vol. 4a, Springer-Verlag, Berlin-Heidelberg, New York, 1978.
17. *Nonferrous Metals; Book of ASTM Standards*, Parts 7 and 32, American Society for Testing Materials, Philadelphia, Pa., 1980.
18. G. W. Monier-Williams, *Trace Elements in Food*, John Wiley & Sons, Inc., New York, 1950, p. 138.
19. *Environmental Data Handbook*, American Can Company, Technical Information Center, Barrington, Ill.

20. M. E. Warwick and B. J. Shirley, *Trans. Inst. Met. Finish.* **58,** 9 (1980).
21. *Metals Handbook, Vol. 1, Properties and Selection of Metals*, 8th ed., American Society for Metals, Metals Park, Ohio, 1961, pp. 843–864.

D. J. MAYKUTH
Tin Research Institute, Inc.

DETINNING

Detinning refers to the mechanical and chemical processing of tinplate scrap for the recovery of the tin coating as a tin chemical or the metal and the reclamation of the base metal, which is normally steel. This scrap is generated by fabricators of tin cans and other miscellaneous articles down to electronic-component stampings. Used tin cans, which are a potentially large source of tin-bearing scrap, have not been utilized except occasionally in small volumes because of the currently unfavorable economics caused by collection, delivery, and processing problems.

An important secondary supply of detinning scrap is tin-plated copper alloys, particularly brass, such as small brass stampings and wire. This material is subjected to special alkaline detinning processes.

In the past, an important market for detinned steel scrap was its use for copper cementation (precipitation iron) for copper recovery from waste liquors. However, this market was dependent upon the production of the copper industry. Scrap utilized for this type of operation is usually shredded and sold as unbaled scrap.

A problem is the removal of the alloy that is formed between the tin and the steel substrate, commonly called hardhead, $FeSn_2$ [*12023-01-7*], during the elevation of temperature caused by the reflowing operation of tinplate production (see Steel). In the older detinning processes, such as alkaline electrolytic and anhydrous chlorination, this alloy is attacked and removed very slowly. However, the alkaline chemical process removes this alloy more efficiently, partly because of better solution contact obtained by the mechanical action of modern detinning systems. The alloy is seldom completely removed, but to meet current specifications for tin residues on scrap, most of this alloy must be removed.

The average thickness of tinplate is ca 0.15–0.76 μm. The average tin coating amounts to ca 0.404%, which makes the steel unsuitable for direct melting. In addition to the recovery of the tin value, detinning lowers the residual tin below the minimum allowable for melting in steel furnaces. Excessive tin results in embrittlement and loss of ductility.

The development of better lacquers for tinplate coatings has led to the reduction of average tinplate thickness, as shown below:

Year	Available tin, kg per metric ton tinplate
1933	18
1960	5
1981	4.04

Processes

The patent literature of detinning, starting in the middle of the nineteenth century, includes claims on melting, volatilization, chemical attack, amalgamation, and electrolytic stripping.

Alkaline Electrolytic Process. In the 1880s, the availability of adequate electric power enabled the Goldschmidt Detinning Company in Germany to develop the alkaline electrolytic process. The tinplate scrap was the anode, steel plate or steel-tank sides the cathodes, and a hot alkaline solution the electrolyte. The tin dissolved from the scrap was deposited on the cathode as a spongy metal which was scraped or melted off and cast as pig tin (1).

In this process, banded baskets of wide-mesh steel construction, with the bands hooked to positive bus bars, were suspended in order to submerse the scrap in the solution. The baskets contained 48–56 kg/m^3. The cathodes were suspended with sufficient clearance to prevent short circuits. This process permitted the direct detinning of relatively small batches of tinplate. However, good process control was difficult because of constantly changing current conditions. Detinning times were 2–7 h, depending upon tin content, electrolytic temperature, electrolyte concentration, the number of baskets per cell, and the quantity of scrap per basket. The anode-current efficiency could be 80%, whereas generally lower cathode-current efficiency caused a build-up of tin concentration in the electrolyte. Tin-bearing slimes settled at the bottom because of particles falling from the electrodes plus an accumulation of impurities. This process produced detinned scrap containing <0.1% tin (1).

Chlorination Process. The chlorine detinning process was developed in the middle of the nineteenth century both in the UK and the United States. Plants were established in Switzerland and Belgium (1).

Detinning by chlorination became a full-scale industry to satisfy the greatly increased demand by the silk-weighting industry for anhydrous stannic chloride [7646-78-8]. In the chlorination process, dry chlorine reacts with tin at <50°C to produce anhydrous stannic chloride ($SnCl_4$, bp 114°C), without substantial attack on the steel substrate. Both the chlorine and scrap had to be very dry; even traces of moisture limited the efficiency of the process. Absence of dirt, grease, or oil improved efficiency, and the corrosion was limited to the steel reaction vessels at <38°C. Removal of ferric chloride film from the detinned scrap prevented rusting (1).

This process was capable of handling large quantities of tinplate scrap per batch, which reduced operating costs. Furthermore, anhydrous stannic chloride is produced directly and can be converted immediately to the oxide.

Some of the operational difficulties were poor contact of chlorine to scrap, explosive accumulations of hydrogen, corrosion problems, and losses caused by the drying out of scrap; a distillation step was required for a grade A product.

The anhydrous stannic chloride was used for pink salt, which is ammonium chlorostannate [16960-53-5], used for silk weighting. Principal secondary markets were for stannic oxide for the ceramic industry and tin chemicals.

Increased use of lacquer on tinplate plus the decline of the silk-weighting industry in the 1930s led to the displacement of the chlorine detinning process by the alkaline chemical process which is preferred today.

Alkaline Chemical Process. In the alkaline chemical process, the scrap is treated with a hot alkaline solution, normally caustic soda, containing an oxidizing agent to dissolve the tin as sodium stannate. The steel, relatively unaffected by this treatment, is rinsed and pressed into bales for shipping.

The tin may be recovered from the stannate solution as sodium stannate crystals, as metal by electrolysis, or as tin oxide by neutralization of the solution with sulfuric acid or carbon dioxide:

$$Na_2Sn(OH)_6 + H_2SO_4 \rightarrow Sn(OH)_4 + Na_2SO_4 + 2\ H_2O$$
$$[12027\text{-}70\text{-}2] \qquad\qquad [12054\text{-}72\text{-}7]$$

$$Na_2Sn(OH)_6 + CO_2 \rightarrow Sn(OH)_4 + Na_2CO_3 + H_2O$$

Potassium hydroxide in place of caustic soda gives potassium stannate [12027-61-1]. The tin is recovered from the stannate solution by electrolysis. Electrolytic recovery offers certain cost advantages over the other processes and enhances product purity. A flow sheet of the alkaline detinning process is shown in Figure 1.

Tinplate scrap typically consisting of punch-press skeleton sheets, punchings, trimmings, shredded cans, and defective sheets, is loaded directly into detinning baskets by magnet, ca 10–15 t of scrap per basket. The baskets consist of large perforated steel-plate cylinders, trunnion-supported in a horizontal position in a structural steel frame. Access is gained through a removable section. The baskets are placed in the detinning tanks which are fitted with individual drives to revolve the baskets for maximum contact of the solution with scrap. They are moved about by an electrically operated bridge crane.

The general material of construction is steel. The solution temperature is maintained at 96–99°C. Following the detinning cycle, the baskets containing the scrap are removed from the detinning solution by crane and moved through a series of countercurrent tanks where the scrap is washed as free as possible of detinning solution and soluble tin. The scrap is then dumped from the baskets and moved to the baling presses where it is hydraulically pressed into bales of suitable dimensions for marketing. The bales weigh an average of 360 ± 90 kg. The residual tin remaining on the surface of the detinned steel scrap is <0.03%.

Detinned and shredded scrap which is to be sold to the copper industry is not baled, but utilized by the copper industry for copper cementation (see Copper and copper alloys).

Some processes are specially designed to utilize certain secondary supplies of tin-bearing scrap including meters, white-metal alloys, and aluminum–tin bearings. In general, the procedure described above is followed. The detinned scrap from these sources is usually handled as separate streams.

Hot sodium hydroxide solutions, containing sodium nitrite as the most effective oxidizing agent to dissolve the tin metal from steel, are generally used. The sodium nitrite reacts with the tin to form soluble sodium stannate at a rate high enough to meet production requirements. Without an oxidizing agent, the sodium hydroxide reacts

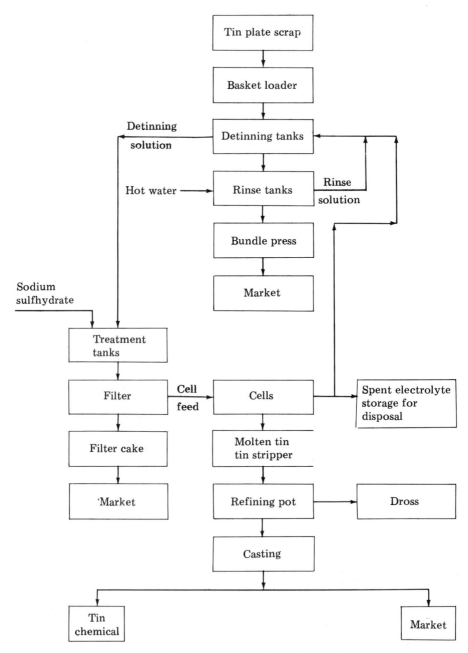

Figure 1. Alkaline detinning process, Vulcan Materials Co., Metals Division.

very slowly to form sodium stannite [*12214-41-4*] and hydrogen. The hydrogen tends to passivate the metal and slow down the reaction, whereas the stannite disproportionates slowly to sodium stannate and finely divided tin which can be lost to the insoluble residues which settle out as a mud. Sodium nitrate can be used as a substitute for sodium nitrite or even mixed with it, but nitrite is more efficient as well as less detrimental to electrowinning (see Extractive metallurgy).

The tumbling action of the scrap in the baskets plus solution circulation promote the detinning reaction, whereas a solution temperature of 96–99°C promotes both detinning and delacquering. Detinning also is aided by galvanic action between tin metal and the steel base metal.

Sodium nitrite or nitrate reacts with tin metal as follows:

$$3\,Sn + 3\,NaOH + 3\,NaNO_2 + 9\,H_2O \rightarrow 3\,Na_2Sn(OH)_6 + N_2 + NH_3$$

$$2\,Sn + 3\,NaOH + NaNO_3 + 6\,H_2O \rightarrow 2\,Na_2Sn(OH)_6 + NH_3$$

A fairly high concentration of sodium hydroxide sustains the tin-dissolved reaction and removes various lacquers and resin coatings, including pigments and extenders. High alkalinity also removes greases, oils, waxes, and dirt. Occasionally, various solvents and delacquering agents are added to remove stubborn lacquers, usually highly cross-linked epoxies, vinyls, and urethanes (2–3).

A small amount of aluminum and aluminum alloy scrap mixed in with tinplate scrap presents a perennial problem. Aluminum metal reacts quickly with generation of hydrogen in the hot alkaline detinning solution to increase tin loss to mud. The reaction consumes caustic and sodium nitrate and galvanically counters the detinning reaction. The sodium aluminate also depresses the current efficiency in the electrowinning cells. However, the aluminate combines with any silicate present to form an insoluble precipitate.

Generally, the detinning solutions are made up with low tin solution recycled from the cells, fresh 50% caustic solution, and strong rinse solution (ca 70 g NaOH/L). The concentration of sodium nitrite for an adequate detinning rate is maintained by additions of sodium nitrite solution. When the soluble tin concentration has built up to about 30 g/L, the solution is pumped to storage tanks. Mud settled out on the detinning tank bottom is removed for tin recovery. It also contains an appreciable amount of small steel particles.

Lead and zinc are precipitated with sodium sulfhydrate:

$$Na_2PbO_2 + NaSH + H_2O \rightarrow PbS + 3\,NaOH$$

$$Na_2ZnO_2 + NaSH + H_2O \rightarrow ZnS + 3\,NaOH$$

Some of the sulfides settle out and the detinning solution is filtered before passing to the electrowinning cells. The ideal equation for the electrochemical recovery of tin is

$$Na_2Sn(OH)_6 \xrightarrow{\text{electrolysis}} Sn + 2\,NaOH + 2\,H_2O + O_2$$

$$2\,H_2O \xrightarrow[\text{NaOH}]{\text{electrolysis}} O_2 + 2\,H_2$$

The second reaction expresses the relative inefficiency of the electrolytic recovery process for tin.

The electrowinning cells are steel tanks connected in series in rows; a temperature of 82–93°C is maintained. The tanks are fitted with thin steel or stainless steel anode plates and cathodes with rows of steel chains hanging from frames or steel plates. The tanks are connected in series for dc and each tank has a potential of ca 2.2 V at a current density of 0.86–1.29 A/dm^2. The tin metal is electroplated in solid form with considerable organic and inorganic occlusion on the cathode chains or plates with some evolution of hydrogen and release of free caustic. Oxygen is evolved at the anodes. The average cathode current efficiency is ca 45%. When the cathodes have accumulated sufficient tin, they are immersed in molten tin in the stripping tank to melt off the tin deposit. Some molten tin overflows to the refining pot where the tin is given a reduction treatment with the aid of hardwood sawdust and the relatively small amount of hardhead dross containing 85–90% tin is skimmed off. Some zinc can be recovered with this dross which is sold to tin smelters. Since these are surface reactions, an agitator in the refining pot is required. Impurities such as hardhead can also be removed from the molten tin by filtration through high temperature media such as porous alumina or asbestos.

Ammonium chloride can be used to separate lead or zinc, and anhydrous stannous chloride has been used as a scavenger for iron, zinc, or lead residues. Arsenic, antimony, copper, and iron residue can be removed by adding aluminum, magnesium, or both by forming alloys with the impurities; these alloys can then be separated. Care must be taken to prevent the additional dross from reacting with water to form extremely poisonous arsine and stibine. Any excess of these alloying metals can be removed by air oxidation such as sparging or intensive agitation followed by a skimming operation.

The general purity of the refined electrolytic tin is >99.9%. Special refining methods are available for obtaining very high purity. These methods, such as secondary electrolysis in acid electrolyte (for ultrapure grade) and zone refining (qv) (for spectrographic grade), are limited to small batches.

The refined electrolytic tin is generally cast in 45-kg pigs. However, the tin also is cast in special forms such as anodes.

A substantial amount of the tin is consumed in the manufacture of tin chemicals. Formerly, sodium and potassium stannates, in particular, were made as adjuncts to the detinning processes. Now, these stannates, for the most part, are produced directly from the metal by chemical processes, mainly alkaline oxidation methods or electrolytic processes (4). High purity solder and white-metal production consume a smaller amount of tin.

Economic Aspects

The largest single use of tin is in making tin-plated steel with usage for solders (qv) a very close second. Traditionally, 10–15% of the tin-plated steel results in scrap. However, with the advent of the two-piece drawn can, the scrap rate has increased to 15–20%. Although during World War II fairly substantial quantities of used cans were collected for detinning, that practice essentially stopped in the mid 1950s. In the last ten years, there has been a resurgence of can recovery at solid-waste processing facilities designed for resource recovery. These types of facilities are currently recovering in the United States in excess of 10^5 t/yr of mixed ferrous metals, most of which was used cans (5) (see Recycling). Production figures for the U.S. detinning industry are given in Table 1.

Table 1. Detinning Operations in the United States, Metric Tons

Year	Scrap treated	Tin recovered
1940[a]	273,232	4394
1950[a]	486,235	3938
1960[a]	684,247	3328
1970[b]	744,728	2497
1979[b]	836,168	2427
1980[b]	773,176	2157
1981[b]	681,736	1855

[a] Ref. 6.
[b] Author's estimate.

The total volume of scrap processed increased more than threefold between 1940 and 1979. The decline from 1979 to 1981 is due to the impact of the use of aluminum for beverage cans instead of tin-plated steel. The use of thinner coatings of tin on the steel has reduced the total yield of tin from detinning operations.

Detinning and other tin-recovery processes represent the only domestic U.S. sources of tin. Although tin-plated steel as a packaging medium is in decline, the growth in the recovery of used cans from resource recovery facilities could represent an important source for the detinning industry if economically practical methods for detinning used cans are developed.

Health and Safety Factors; Environmental Considerations

There are no health and safety measures that are exclusive to the detinning industry. Detinning is similar to the steel or scrap industries, using large, heavy equipment for material handling. Obvious risks are associated with this type of equipment. In addition, there is the risk of burns from hot, weak caustic solutions.

Environmentally, detinning is a relatively clean industry. There is no air pollution associated with detinning. The detinning process generates a solid residue which is a mixture of tin compounds, steel fines, lacquer flakes, and dirt. At present, the tin contained in the residue has sufficient value for tin smelters to purchase the residue for processing. Spent electrolyte is considered a hazardous waste under present EPA regulations and must be disposed of accordingly. Certain industries are willing to purchase the spent electrolyte for its alkali content. The spent electrolyte is the only liquid waste generated in the detinning process.

BIBLIOGRAPHY

"Tin and Tin Alloys (Detinning)" in *ECT* 1st ed., Vol. 14, pp. 151–156, by L. E. Swanson and R. H. Taylor, The Vulcan Detinning Company; "Tin and Tin Alloys (Detinning)" in *ECT* 2nd ed., Vol. 20, pp. 294–304, by C. Kenneth Banks, M&T Chemicals, Inc.

1. C. L. Mantel, *Tin*, 2nd ed., Reprint Hafner Publishing, New York, 1970, pp. 519–529.
2. U.S. Pat. 3,022,161 (Feb. 20, 1962), F. Finkener and co-workers (to Th. Goldschmidt A.G.).
3. U.S. Pat. 3,168,477 (Feb. 2, 1965), L. Swanson and H. Wilson (to Vulcan Materials Company).
4. U.S. Pat. 4,066,518 (Jan. 3, 1978), R. E. Horn (to Pitt Metals and Chemicals, Inc.).
5. *Solid Waste Processing Facilities*, American Iron and Steel Institute, Washington, D.C., Sept. 1, 1981.

6. *Minerals Yearbook*, *U.S. Bureau of Mines*, U.S. Government Printing Office, Washington, D.C., yearly editions.

WILLIAM GERMAIN
H. P. WILSON
V. E. ARCHER
Vulcan Materials Co.

TIN COMPOUNDS

Tin is between germanium and lead in Group IVA of the periodic table. As a bronze component, tin was used as early as 3500 BC and the pure metal was not used until 600 BC. The history of tin compounds dates to the Copts of Egypt, who reportedly used basic tin citrate [59178-29-9] in dye preparation, and to the preparation in 1605 of stannic chloride; the alchemists' symbol for tin is ♃. Tin occurs in the earth's crust to the extent of 40 grams per metric ton and is present in the form of nine different minerals from two types of deposits, ie, the most commercially significant ore cassiterite [1317-45-9] (tinstone), SnO_2, and the complex sulfidic ores which are combinations with the sulfides of base metals and pyrites. The more economically valuable deposits of cassiterite, which are heavily concentrated in bands and layers of varying thickness, are in Malaysia, Thailand, Indonesia, and the People's Republic of China. The complex sulfidic ores, which are in lode deposits and economically significant only in Bolivia, are stannite [12019-29-3], $SnS_2.Cu_2S.FeS$; herzenbergite [14752-27-3], SnS; teallite [12294-02-9], $SnS.PbS$; franckeite [12294-04-1], $2SnS_2.Sb_2S_3.5PbS$; cylindrite [12294-05-0, 59858-98-9], $Sn_6Pb_6Sb_2S_{11}$; plumbostannite $2SnS_2.2PbS.2(Fe,Zn)S.-Sb_2S_3$; and canfieldite [12250-27-0], $4Ag_2S.SnS_2$. The important tin-producing countries are Malaysia, Bolivia, Indonesia, Nigeria, Thailand, Zaire, and the People's Republic of China, with smaller quantities produced in the UK, Burma, Japan, Canada, Portugal, Spain, and Australia. Tin is also normally present in natural waters, in soil, in marine organisms and animals, in the milk of lactating animals, in meteorites, in the tissues of animals, and in minor amounts in human organs.

Tin with a valence of +2 and +4 forms stannous, ie, tin(II), compounds and stannic, ie, tin(IV), compounds. Types of tin compounds include ones containing tin(II) and tin(IV) compounds, complex stannites ($MSnX_3$) and stannates (M_2SnX_6), coordination complexes, organic tin salts where the tin is not bonded through carbon, and organotin compounds, which contain one-to-four carbon atoms bonded directly to tin.

Of the large volume of tin compounds reported in the literature, possibly only ca 100 are commercially important. The most commercially significant tin compounds include stannic chloride, stannic oxide, potassium stannate, sodium stannate, stannous chloride, stannous fluoride, stannous fluoroborate, stannous oxide, stannous pyro-phosphate, stannous sulfate, stannous 2-ethylhexanoate, stannous oxalate, and or-ganotins of the dimethyltin, dibutyltin, tributyltin, dioctyltin, triphenyltin, and tri-cyclohexyltin families (see Inorganic Tin Compounds and Organotin Compounds for CAS Registry Numbers).

Commercially available tin compounds whose annual production or gross sales exceeds 2.3 metric tons or $5000.00 are listed with their producers in refs. 1–2. Principal U.S. producers of inorganic tin compounds include M&T Chemicals, Inc., Vulcan Materials Co., and Allied Corp. M&T Chemicals, Inc., is the largest U.S. producer of organotin compounds followed by Carstab Corp., Witco Chemical Corp., and Cardinal Chemical Co.; minor producers are Interstab, Synthetic Products Co., Tenneco Chemicals Co., and Ferro Chemical Co.

Tin compounds are used for a wide variety of purposes, eg, catalysts, and stabi-lizers for many materials including polymers (see Heat stabilizers); biocidal agents, eg, bactericides, insecticides, fungicides, wood preservatives, acaricides, and antifouling paints; ceramic opacifiers; textile additives; in metal finishing operations; as food additives; and in electroconductive coatings (see Catalysis; Coatings, marine; Insect control technology; Fungicides; Dentifrices; Industrial antimicrobial agents).

In 1975, the estimated world annual production of tin chemicals represented the consumption of 12,000–14,000 t of tin metal or 5% of total tin consumption (3). In 1978, ca 20,000 t/yr was consumed worldwide with equal amounts represented by inorganic tin and organotin compounds (4). It is established that worldwide production of or-ganotins rose from ca 50 t in 1950 to a possible 30,000–35,000 t in 1980.

Consumption of primary tin for chemical applications by the five reporting countries from 1975 to 1980 is shown in Table 1. The use of primary tin in chemical applications has increased sharply in recent years in the United States and the FRG but has declined in the UK and Italy. The U.S. demand for tin for tin-chemicals pro-duction has been: 3196 t in 1970, 4970 t in 1974, 5800 t in 1977, and 6120 t in 1979. In 1979, ca 11% of the total U.S. demand for tin was accounted for by tin chemicals. In 1977, the tin consumption in tin-chemicals production was based on ca 80% primary tin usage and 20% secondary tin usage. Since 1979, these statistics have been withheld by the U.S. Bureau of Mines (BOM). The BOM estimates that the use of tin in tin chemicals in the year 2000 will be 11,000 t. The approximate use of tin chemicals in the western world has been estimated as follows for inorganic tins: ca 6000 t including

Table 1. Consumption of Primary Tin for Chemical Application, t[a]

Year	France	FRG	Italy[b]	UK[c]	United States	
					Primary	Secondary
1975	345	492	650	1210	2735	1263
1976	330	998	700	1463	4718	903
1977	420	1137	650	1468	4655	1072
1978	540	1414	600	1374	4557	
1979	590	1894	500	1305	4797	
1980	700[d]	2050[d]	450[d]	1282	4800[d]	

[a] Ref. 5.
[b] Includes secondary tin.
[c] Includes tin powder.
[d] Estimates from ref. 6.

ca 3000 t stannic oxide; 1000 t stannic chloride; 500 t stannous sulfate and sodium and potassium stannates; 200 t stannous fluoride, fluoroborate, and pyrophosphate; and 1000 t stannous chloride, oxide, and octanoate (7). The approximate use of organotins is ca 30,000 t, including ca 21,000 t for mono- and diorganotins (representing 20,000 t for PVC stabilizers and 1000 t for homogeneous catalysts) and 9000 t for triorganotins (representing 1000 t for agricultural fungicides and 8000 t in other uses). More detailed approximate world consumption data of inorganic tin compounds are reported in ref. 8. Other sources estimate the consumption of stannous octoate as a catalyst in flexible urethane foams at 1000 t/yr, stannic chloride as a perfume stabilizer in soaps at 250 t/yr, and stannic oxide as an opacifier for ceramic glazes and vitreous enamels at 2500 t/yr (5).

The most rapidly increasing use for tin, particularly organotin compounds, is in chemicals. In the FRG, the amount of tin consumed in organotin production rose from 691 t in 1973 to 1760 t in 1979 (5). In 1979, organotins accounted for ca 93% of tin consumption in chemicals in the FRG. In 1980, the estimated world production of organotin compounds was 30,000–35,000 t, with 75% used in the manufacture of PVC stabilizers. The market for PVC stabilizers is large and, depending on the country, organotins represent 10–25% of the market. At the beginning of 1980, the organotin share of the market for PVC stabilizers was ca 20% in the United States, 15% in Japan, 25% in the FRG, and 12% in the rest of Western Europe (5). The estimated 1981 U.S. consumption of organotins as PVC stabilizers was 10,650 t or 27% of the market (9). The estimated 1981 U.S. consumption of tin compounds as urethane catalysts was 808 t or 34% of the market (10). The 1978 U.S. market for stannous fluoride in dental preparations was 186 t valued at $3.72 × 10⁶ (11). An important earlier report with market data on organotins was issued in early 1976 (12).

The price of tin chemicals depends to a large extent on the fluctuating price of tin. The 1982 prices and the CAS Registry Numbers for selected inorganic tin compounds and organotins are listed in Table 2.

Table 2.   Prices of Selected Inorganic Tin and Organotin Compounds, 1982[a]

Compound	CAS Registry No.	Price, $/kg
inorganic[b]		
potassium stannate	[12142-33-5]	9.66
stannic chloride, anhydrous	[7646-78-8]	9.66
stannic oxide	[18282-10-5]	26.40
stannous chloride, anhydrous	[7772-99-8]	14.10
stannous fluoroborate	[13814-97-6]	5.39
stannous oxide	[21651-19-4]	20.30
stannous sulfate	[7488-55-3]	16.13
organic[c]		
bis(tributyltin) oxide	[56-35-9]	17.51
dibutyltin dichloride	[683-18-1]	11.55
tributyltin acetate	[56-36-0]	19.93
tributyltin chloride	[1461-22-9]	17.12
tributyltin fluoride	[1983-10-4]	18.44

[a] Based on largest quantity price.
[b] Ref. 13.
[c] Ref. 14.

Inorganic Tin Compounds

Because of its amphoteric nature, tin reacts with strong acids and strong bases but remains relatively resistant to neutral solutions. A thin oxide film forms on tin exposed to oxygen or dry air at ordinary temperatures; heat accelerates this oxide formation. Tin is not attacked by gaseous ammonia even when heated. Chlorine, bromine, and iodine react with tin at normal temperatures, and fluorine reacts at 100°C forming the appropriate stannic halides. Tin is easily attacked by hydrogen iodide and hydrogen bromide, but less readily by hydrogen chloride; it is weakly attacked by gaseous hydrogen fluoride, and it slowly dissolves in aqueous hydrochloric acid. Hot concentrated sulfuric acid reacts with tin forming stannous sulfate, whereas dilute sulfuric acid reacts only slowly with tin at room temperature. Reaction of tin with dilute nitric acid yields soluble tin nitrates; in concentrated nitric acid, tin is oxidized to insoluble hydrated stannic oxide. No reaction occurs upon direct union of tin with hydrogen, nitrogen, or carbon dioxide.

If tin and sulfur are heated, a vigorous reaction takes place with the formation of tin sulfides. At 100–400°C, hydrogen sulfide reacts with tin forming stannous sulfide; however, at ordinary temperatures no reaction occurs. Stannous sulfide also forms from the reaction of tin with an aqueous solution of sulfur dioxide. Molten tin reacts with phosphorus forming a phosphide. Aqueous solutions of the hydroxides and carbonates of sodium and potassium, especially when warm, attack tin. Stannates are produced by the action of strong sodium hydroxide and potassium hydroxide solutions on tin. Oxidizing agents, eg, sodium or potassium nitrate or nitrite, are used to prevent the formation of stannites and to promote the reactions.

Stannic and stannous chloride are best prepared by the reaction of chlorine with tin metal. Stannous salts are generally prepared by double decomposition reactions of stannous chloride, stannous oxide, or stannous hydroxide with the appropriate reagents. Metallic stannates are prepared either by direct double decomposition or by fusion of stannic oxide with the desired metal hydroxide or carbonate. Approximately 80% of inorganic tin chemicals consumption is accounted for by tin chlorides and tin oxides.

Halides. The tin halides of the greatest commercial importance are stannous chloride, stannic chloride, and stannous fluoride. Tin halides of less commercial importance are stannic bromide [7789-67-5], stannic iodide [7790-47-8], stannous bromide [10031-24-0], and stannous iodide [10294-70-9] (15).

Stannous Chloride. Stannous chloride is available in two forms: anhydrous stannous chloride ($SnCl_2$) and stannous chloride dihydrate ($SnCl_2.2H_2O$), also called tin crystals or tin salts. Both forms are sometimes used interchangeably; however, where stability, concentration and adaptability are important, anhydrous stannous chloride is the preferred material. Even after long storage, changes in the stannous tin content of anhydrous stannous chloride are extremely low. Physical properties of the tin chlorides are listed in Table 3.

Anhydrous stannous chloride, a water-soluble white solid, is the most economical source of stannous tin and is especially important in redox and plating reactions. Preparation of the anhydrous salt may be by direct reaction of chlorine and molten tin, heating tin in hydrogen chloride gas, or reducing stannic chloride solution with tin metal followed by dehydration. It is soluble in a number of organic solvents (g/100 g solvent at 23°C): acetone 42.7, ethyl alcohol 54.4, methyl isobutyl carbinol 10.45,

Table 3. Physical Properties of Tin Chlorides

Property	$SnCl_2$	$SnCl_2.2H_2O$ [10025-69-1]	$SnCl_4$	$SnCl_4.5H_2O$ [10026-06-9]
mol wt	189.60	225.63	260.50	350.58
mp, °C	246.8	37.7	−33	ca 56 (dec)
bp, °C	623		114	
density (at 25°C), g/cm³	3.95	2.63	2.23[a]	2.04

[a] At 20°C.

isopropyl alcohol 9.61, methyl ethyl ketone 9.43; isoamyl acetate 3.76, diethyl ether 0.49, and mineral spirits 0.03; it is insoluble in petroleum naphtha and xylene (16).

Solutions of anhydrous stannous chloride are strongly reducing and thus are widely used as reducing agents. Dilute aqueous solutions tend to hydrolyze and oxidize in air, but addition of dilute hydrochloric acid prevents this hydrolysis; concentrated solutions resist both hydrolysis and oxidation. Neutralization of tin(II) chloride solutions with caustic causes the precipitation of stannous oxide or its metastable hydrate. Excess addition of caustic causes the formation of stannites. Numerous complex salts of stannous chloride, known as chlorostannites, have been reported (17). They are generally prepared by the evaporation of a solution containing the complexing salts.

Anhydrous stannous chloride is used extensively in the plating industry, eg, in the high speed electrotinning of continuous strip steel by the halogen process involving an aqueous solution of stannous chloride and alkali-metal fluorides, in a variety of formulations in immersion tinning processes, and in tin-alloy plating (18–19) (see Electroplating). The strongly reducing nature of this chloride has contributed to its commercial development and success. Established applications of this property include its use as an analytical reducing reagent, a reducing agent in inorganic and organic chemicals manufacture and in the photoleaching of dyes, and as a sensitizing agent for nonconductive surfaces before silver coating or other metallization processes. Originally, surface sensitization with stannous chloride solutions was used only on glass prior to silvering (20). It is now also used to sensitize plastics prior to their electroless coating with metals, eg, nickel or copper (21–23) (see Electroless plating).

Stannous chloride is also used as a food additive, for which use it has FDA GRAS approval (24). Other approvals include its use as a preservative for canned soda water, a color-retention agent in canned asparagus, and a component in food-packaging materials (24–28). It catalyzes a variety of organic reactions, eg, condensation, curing of resins and rubbers, esterification, halogenation, hydrogenation, oxidation, polymerization, hydrocarbon conversion, etc. Minor applications include the use of stannous chloride as an additive to drilling muds, an antisludge agent for oils, in tin coating of sensitized paper, and to improve the dyeing fastness of synthetic fibers, eg, polyamides (see Petroleum, drilling fluids; Polyamides, polyamide fibers).

Stannous Chloride Dihydrate. A white crystalline solid, stannous chloride dihydrate is prepared either by treatment of granulated tin with hydrochloric acid followed by evaporation and crystallization or by reduction of a stannic chloride solution with a cathode or tin metal followed by crystallization. It is soluble in methanol, ethyl acetate, glacial acetic acid, sodium hydroxide solution, and dilute or concentrated hydrochloric acid. It is soluble in less than its own weight of water, but with much water it forms an insoluble basic salt.

Stannic Chloride. Stannic chloride is available commercially as anhydrous stannic chloride, $SnCl_4$ (tin(IV) chloride); stannic chloride pentahydrate, $SnCl_4.5H_2O$; and in proprietary solutions for special applications. Anhydrous stannic chloride, a colorless fuming liquid, fumes only in moist air, with the subsequent hydrolysis producing finely divided hydrated tin oxide or basic chloride. It is soluble in water, carbon tetrachloride, benzene, toluene, kerosene, gasoline, methanol, and many other organic solvents. With water, it forms a number of hydrates, of which the most important is the pentahydrate. Although stannic chloride is an almost perfect electrical insulator, traces of water make it a weak conductor (see Table 3 for physical properties).

Stannic chloride is made by the direct chlorination of tin at 110–115°C. Any stannous chloride formed in the process is separated from the stannic chloride by volatilization and subsequently chlorinated to stannic chloride. The latter is inert to steel in the absence of moisture and is shipped in plain steel drums of special design. Since prolonged contact with the skin causes burns, goggles and protective clothing should be used in the handling of stannic chloride. Stannic chloride, like stannous chloride, also forms many complexes (17).

The main uses of stannic chloride are as a raw material for the manufacture of other tin compounds, especially organotins, and in the surface treatment of glass (qv) and other nonconductive materials, whereby deposition of stannic oxide from stannic chloride solutions onto the nonconductive substrate gives it strength, abrasion-resistance, and conductivity (29). Very thin stannic oxide films (less than 100 nm thick) are thus used to strengthen glassware for returnable and nonreturnable foodstuff bottles and jars and for restaurant and catering glasses, which are subject to rigorous use (30). Glass treated in this way can also be made considerably lighter, which is advantageous to packing, shipping, and handling. The process involves passing freshly formed glassware through an oven maintained in an atmosphere containing stannic chloride vapor. The chloride breaks down, leaving at the glass temperature a stannic oxide deposit on the glass surface (31–32).

Where electrical conductivity and optical transparency are required, thicker (greater than 1 μm) films of stannic oxide are necessary. Such treated glasses are used in low intensity lighting panels and display signs, fluorescent lights, electron-beam control in cathode-ray tubes, and deicing windshields in aircraft. The deposition of stannic oxide films on glass surfaces is accomplished by the decomposition of stannic chloride vapor at 500–600°C and depositing it onto the glass surface or by spraying it from aqueous or mixed-organic solutions.

Stannic chloride is also used widely as a catalyst in Friedel-Crafts acylation, alkylation and cyclization reactions, esterifications, halogenations, and curing and other polymerization reactions. Minor uses are as a stabilizer for colors in soap (33), as a mordant in the dyeing of silks, in the manufacture of blueprint and other sensitized paper, and as an antistatic agent for synthetic fibers (see Dyes, application and evaluation—application; Antistatic agents).

Stannic Chloride Pentahydrate. Stannic chloride pentahydrate is a white, crystalline, deliquescent solid that is soluble in water or methanol and stable at 19–56°C. It is used in place of the anhydrous chloride where anhydrous conditions are not mandatory. It is easier to handle than the fuming anhydrous liquid form. The pentahydrate is prepared by dissolving stannic chloride in hot water, thereby forming the pentahydrate at a temperature above the melting point and crystallizing by cooling. The cake is broken into small lumps for packaging.

A stannic chloride pentahydrate–ammonium bifluoride formulation for fire-proofing wool is commercially available and used in New Zealand and Australia (34) (see Flame retardants in textiles).

Stannous Fluoride.   Stannous fluoride [7783-47-3] (tin(II) fluoride, mol wt 156.7, mp 219.5°C) occurs as opaque, white, lustrous crystals, which are soluble in potassium hydroxide, fluorides, and water (31 g/100 g H_2O at 0°C, 78.5 g/100 g H_2O at 106°C) and practically insoluble in methanol, ether, and chloroform. Dilute aqueous solutions hydrolyze unless stabilized with excess acid. The specific gravity of a saturated aqueous solution at 25°C is 1.51. Commercially, stannous fluoride is produced by the reaction of stannous oxide and aqueous hydrofluoric acid or by dissolving tin in anhydrous or aqueous hydrofluoric acid.

The principal commercial use of stannous fluoride is in toothpaste formulations and other dental preparations, eg, topical solutions, mouthwash, chewing gum, etc, for preventing demineralization of teeth (35–37).

Oxides.   *Stannous Oxide.*   Stannous oxide (SnO, tin(II) oxide, mol wt 134.70, sp gr 6.5) is a stable, blue-black, crystalline product that decomposes at above 385°C. It is insoluble in water or methanol, but is readily soluble in acids and concentrated alkalies. It is generally prepared from the precipitation of a stannous oxide hydrate from a solution of stannous chloride with alkali. Treatment at controlled pH in water near the boiling point converts the hydrate to the oxide. Stannous oxide reacts readily with organic acids and mineral acids, which accounts and for its primary use as an intermediate in the manufacture of other tin compounds. Minor uses of stannous oxide are in the preparation of gold–tin and copper–tin ruby glass.

Stannous Oxide Hydrate.   Stannous oxide hydrate [12026-24-3], $SnO.H_2O$ (sometimes erroneously called stannous hydroxide or stannous acid), mol wt 152.7, is obtained as a white amorphous crystalline product on treatment of stannous chloride solutions with alkali. It dissolves in alkali solutions forming stannites. The stannite solutions, which decompose readily to alkali-metal stannates and tin, have been used industrially for immersion tinning.

Stannic Oxide.   Stannic oxide (tin(IV) oxide, white crystals, mol wt 150.69, mp >1600°C, sp gr 6.9) is insoluble in water, methanol, or acids but slowly dissolves in hot, concentrated alkali solutions. In nature, it occurs as the mineral cassiterite. It is prepared industrially by blowing hot air over molten tin, by atomizing tin with high pressure steam and burning the finely divided metal, or by calcination of the hydrated oxide. Other methods of preparation include treating stannic chloride at high temperature with steam, treatment of granular tin at room temperature with nitric acid, or neutralization of stannic chloride with a base.

In the ceramics and glass industries, stannic oxide is used for the production of opaque glasses; as an opacifier for glazes and, to a lesser extent, enamels for metals (eg, cast iron) as used in bathtubs, sinks, tile, and other sanitary ware; as a base for certain ceramic colors, eg, chrome–tin pink, vanadium–tin yellow, and antimony–tin blue; and as a component of ceramic capacitor dielectrics. More than 15 large glass-melting furnaces in the world use stannic oxide electrodes in the electromelting of lead glass (38).

Other important uses of stannic oxide are as a putty powder for polishing marble, granite, glass, and plastic lenses and as a catalyst. The most widely used heterogeneous tin catalysts are those based on binary oxide systems with stannic oxide for use in organic oxidation reactions. The tin–antimony oxide system is particularly selective in the oxidation and ammoxidation of propylene to acrolein, acrylic acid, and acrylonitrile. Research has been conducted for many years on the catalytic properties of stannic oxide and its effectiveness in catalyzing the oxidation of carbon monoxide at below 150°C has been described (39).

Transparent electroconductive coatings of stannic oxide are deposited on nonconductive substrates for electrical and strengthening applications. However, the agents used to deposit the oxide film are actually stannic chloride and, more recently, some organotin compounds.

Hydrated Stannic Oxide. Hydrated stannic oxide of variable water content is obtained by the hydrolysis of stannates. Acidification of a sodium stannate solution precipitates the hydrate as a flocculent white mass. The colloidal solution, which is obtained by washing the mass free of water-soluble ions and peptization with potassium hydroxide, is stable below 50°C and forms the basis for the patented Tin Sol process for replenishing tin in stannate tin-plating baths (see Soluble Stannates). A similar type of solution (Stannasol A and B) is prepared by the direct electrolysis of concentrated potassium stannate solutions (40).

Metal Stannates. Soluble Stannates. Many metal stannates of formula $M_n Sn(OH)_6$ are known. The two main commercial products are the soluble sodium and potassium salts, which are usually obtained by recovery from the alkaline detinning process. They are also produced by the fusion of stannic oxide with sodium hydroxide or potassium carbonate, respectively, followed by leaching and by direct electrolysis of tin metal in the respective caustic solutions in cells using cation-exchange membranes (41). Another route is the recovery from plating sludges.

Potassium stannate, $K_2 Sn(OH)_6$ (mol wt 298.93) and sodium stannate [12058-66-1], $Na_2 Sn(OH)_6$ (mol wt 266.71) are colorless crystals and are soluble in water. The solubility of potassium stannate in water is 110.5 g/100 mL water at 15°C and that of sodium stannate is 61.5 g/100 mL water at 15°C. The solubility of sodium stannate decreases with increasing temperature, whereas the solubility of potassium stannate increases with increasing temperature. The solubility of either sodium or potassium stannate decreases as the concentration of the respective free caustic increases. Hydrolysis of stannates yields hydrated stannic oxides and is the basis of the Tin Sol solution, which is used to replenish tin in stannate tin-plating baths (42–43).

Although sodium stannate formed the basis for the first successful alkaline tin-electroplating bath, both stannates are used in these baths with potassium stannate being favored in the United States. Potassium stannate is used for an alkaline tin-electroplating bath yielding higher cathode efficiencies and higher conductivities than any other alkaline bath. The stannates are also used in immersion tinning, particularly the immersion plating of aluminum pistons and other parts for the automotive industry, and in the electroplating of alloy coatings, especially tin–zinc and tin–copper alloys from mixed stannate–cyanide baths. Reviews of the use of stannates in tin plating are given in refs. 18 and 44–45.

Other. Insoluble alkaline-earth metal and heavy metal stannates are prepared by the metathetic reaction of a soluble salt of the metal with a soluble alkali-metal stannate. They are used as additives to ceramic dielectric bodies (46). The use of bismuth stannate [12777-45-6], $Bi_2(SnO_3)_3 \cdot 5H_2O$, with barium titanate produces a ceramic capacitor body of uniform dielectric constant over a substantial temperature range (47). Ceramic and dielectric properties of individual stannates are given in ref. 48. Other typical commercially available stannates are: barium stannate [12009-18-6],

$BaSnO_3$; calcium stannate [12013-46-6], $CaSnO_3$; magnesium stannate [12032-29-0], $MgSnO_3$; and strontium stannate [12143-34-9], $SrSnO_3$.

Certain anhydrous stannates are effective as smoke suppressants in glass-reinforced polyester, especially $Na_2Sn(OH)_6$ [12058-66-1] and $ZnSnO_3$ [12036-37-2]. This use has not yet been commercialized (49).

Salts. *Stannous Sulfate.* Stannous sulfate ($SnSO_4$, tin(II) sulfate, mol wt 214.75), is a white crystalline powder which decomposes above 360°C. Because of internal redox reactions and a residue of acid moisture, the commercial product tends to discolor and degrade at ca 60°C. It is soluble in concentrated sulfuric acid and in water (330 g/L at 25°C). The solubility in sulfuric acid solutions decreases as the concentration of free sulfuric acid increases. Stannous sulfate can be prepared from the reaction of excess sulfuric acid (specific gravity 1.53) with granulated tin for several days at 100°C until the reaction has ceased. Stannous sulfate is extracted with water and the aqueous solution evaporates *in vacuo*. Methanol is used to remove excess acid. It is also prepared by reaction of stannous oxide and sulfuric acid and by the direct electrolysis of high grade tin metal in sulfuric acid solutions of moderate strength in cells with anion-exchange membranes (50).

The main use for stannous sulfate is in tin plating. The sulfate bath is widely used commercially for general plating, especially barrel plating. Significant tin-plating processes involving stannous sulfate baths include flow melting, ie, momentarily melting the coating to attain a bright finish on tin or tin–lead alloy deposits, which is used primarily in the production of printed circuit boards; bright-acid tin plating, which is used in finishing electrical contacts, radio chassis, domestic articles, and kitchen utensils; electrotinning steel strip by the vertical acid process, and liquor finishing, ie, immersion plating of steel wire with tin or copper–tin alloy prior to drawing. All tin-plating and tin-alloy-plating processes are reviewed in ref. 18.

Stannous Fluoroborate. Stannous fluoroborate ($Sn(BF_4)_2$) is available only in solution, as the solid form has not been isolated. It is prepared by dissolving stannous oxide in fluoroboric acid or by direct electrolysis of tin metal in fluoroboric acid in cells with anion-exchange membranes (50). The commercially available 47 wt % solution is widely used in tin and tin–lead-alloy plating, especially in the deposition of tin–lead solder alloys for the electronics industry (51–52). Stannous fluoroborate solutions are important in plating because of their good throwing and covering power and high solubility, which promote high rates of deposition. They are used in the tin plating of copper wire, backing of electrotypes, and barrel tin plating of components for subsequent soldering (18).

Stannous Pyrophosphate. Stannous pyrophosphate [15578-26-4], $Sn_2P_2O_7$ (mol wt 411.32, sp gr 4.009 at 16°C) is an amorphous white powder, decomposes at above 400°C, and is insoluble in water and soluble in concentrated mineral acids and sodium pyrophosphate. It is prepared from stannous chloride and sodium pyrophosphate, and it is used as a caries preventative in toothpaste and as a diagnostic aid in radioactive bone scanning and red-blood-cell labeling (37,53–55).

Other Inorganic Tin Compounds. Other inorganic tin compounds, which have been used industrially in the past and are not now used or which have limited use, include stannic sulfide [1315-01-1] as a pigment and bronzing agent, stannous sulfide [1314-95-0] as a pigment, stannic vanadate [66188-22-5] as an oxidation catalyst and in ceramic pigments, stannic molybdate [34782-17-7] as a source of gamma rays in Mössbauer spectroscopy of tin compounds, stannic arsenate [35568-59-3] as an anthelmintic, and tin naphthenate. Stannic phosphate [15142-98-0] gels are effective ion exchangers.

Toxicology. Inorganic tin and its compounds are generally of a low order of toxicity, largely because of the poor absorption and rapid excretion from the tissues of the metal (56–63). The acidity and alkalinity of their solutions make assessment of their parenteral toxicity difficult. The oral LD_{50} values for selected inorganic tin compounds are listed in Table 4. It is estimated that the average U.S. daily intake of tin, which is mostly from processed foods, is 4 mg (see Food processing).

Tin is normally present in animals, including man. In the human body, it is present in small amounts in nearly all organs (68). Tin is eliminated almost completely by the alimentary tract and is scantily absorbed by the alimentary tissues. The output and intake of tin balances during adult life (69). In human subjects fed packaged C (canned) rations for 22 successive days, all tin ingested was accounted for in fecal excretion (70). A recent study suggests that tin is an essential trace element for the growth of the mammalian organism (see Mineral nutrients). Many inorganic tin compounds have been approved for human contact or use by the FDA (see Uses) (71).

The inorganic tin compound that has received the most study from a toxicological viewpoint is stannic oxide. Autopsies performed on workers in the tin mining and refining industry, who inhaled tin oxide dust for as long as 20 yrs, disclosed no pulmonary fibrosis (72). Inhalation for long periods produces a benign, symptomless pneumoconiosis with no toxic systemic effects (73).

Stannous chloride, an FDA-approved direct food additive with GRAS status, has also been extensively studied (74–77). In three FDA-sponsored studies, it was determined that stannous chloride is nonmutagenic in rats; when administered orally up to 50 mg/kg to pregnant mice for ten consecutive days, stannous chloride has no discernible effect on nidation or on maternal or fetal survival; and, when administered orally at 41.5 mg/kg to pregnant rabbits for 13 consecutive days, it produced no discernible effect on nidation or on maternal or fetal survival (78–80).

Other studies of the toxicity of stannous fluoride, sodium pentafluorostannite, sodium pentachlorostannite, sodium chlorostannate, stannous sulfide [1314-95-0], stannous and stannic oxides, stannous pyrophosphate [15578-26-4], stannous tartrate [815-85-0], and other inorganic tin compounds are reviewed in refs. 81–87. The OSHA

Table 4. Acute Oral Toxicity of Selected Inorganic Tin Compounds

Compound	CAS Registry No.	LD_{50}, mg/kg	Test animal	Ref.
stannous chloride	[7772-99-8]	700	rat	64
		1,200	mouse	64
stannous ethylene glycoxide	[68921-71-1]	>10,000	rat	65
stannous 2-ethylhexanoate	[301-10-0]	5,870	rat	65
stannous fluoride	[7783-47-3]	128.4	mouse	66
		188.2	rat	66
stannous oxalate	[814-94-8]	3,400	rat	65
stannous oxide	[21651-19-4]	>10,000	rat	65
sodium pentafluorostannite	[22578-17-2]	595	male mouse	66
		221	male rat	66
		227	female rat	66

TLV standard for inorganic tin compounds is two milligrams of inorganic tin compounds as tin per cubic meter of air averaged over an eight-hour work shift (61).

Organotin Compounds

In an organotin compound, there is at least one tin–carbon bond. The oxidation state of tin in most organotin compounds is +4, although organotin compounds with bulky groups bonded to divalent tin have been reported (88). Five classes of organotin compounds are known: R_4Sn (tetraorganotins), R_3SnX (triorganotins), R_2SnX_2 (diorganotins), $RSnX_3$ (monoorganotins), and R_6Sn_2 (hexaorganoditins) (see Organometallics). Of commercial importance are those organotin where R is methyl, butyl, octyl, cyclohexyl, phenyl, or β,β-dimethylphenethyl (neophyl). The noncarbon-bonded anionic group is commonly halide, oxide, hydroxide, carboxylate or mercaptide.

It was not until the 1940s that the commercial potential of organotins was realized. Organotins first were used as stabilizers for poly(vinyl chloride), which is normally processed just below its decomposition temperature (see Heat stabilizers). The high biocidal activity of the triorganotins is one of the most applied areas of their usefulness. In addition, organotins are widely used as catalysts and curing agents and in the treatment of glass. A number of organotin subjects, including structural organotin chemistry and industrial applications, are discussed in refs. 89–93.

Properties. As a member of Group IVA of the periodic table, tin has four valence electrons available for bonding. In its usual tetravalent state, tin assumes a typical sp^3 hydrization and the configuration of its covalent bonds is tetrahedral. In the tin atom, d orbitals are available and are utilized in the formation of pentacoordinate and hexacoordinate complexes by Lewis bases with organotin halides. These complexes are frequently trigonal bipyramidal or octahedral. Tin forms predominately covalent bonds to other elements, but these bonds exhibit a high degree of ionic character, with tin usually acting as the electropositive member.

Although the mean dissociation energy of tin–carbon bonds is less than that normally associated with carbon–carbon bonds ($\overline{D}_{Sn-C}$ = 188–230 kJ/mol (45–55 kcal/mol), $\overline{D}_{C-C}$ = 335–380 kJ/mol (80–90 kcal/mol) (90)), the difference is not great enough to render the tin–carbon bond very reactive. The bond is stable to water and atmospheric oxygen at normal temperatures. The tin–carbon bond is also quite stable to heat and many organotins can be distilled under reduced pressure with little decomposition. Strong acids, halogens, and other electrophilic reagents readily cleave the tin–carbon bond, although other reactions that are common with other organometallics, eg, Grignard and organolithium reagents, do not occur with organotins. For example, the tin–carbon bond does not add to the carbonyl group, nor does it react with alcohols.

The ionicity of organotins leads to dissimilar chemical properties. For example, triorganotin hydroxides behave not as alcohols, but more like inorganic bases, although strong bases remove the proton in certain triorganotin hydroxides since tin is amphoteric. The bis(triorganotin) oxides, $(R_3Sn)_2O$, are strong bases and react with inorganic and organic acids forming normal saltlike but nonconducting and water-insoluble compounds. They do not in the least resemble organic ethers, though they can occasionally form peroxides. Tin doubly bonded to oxygen, which is analogous to an organic ketone, does not exist and diorganotin oxides, R_2SnO, are polymers, ie,

$+Sn(R_2)O+_n$, and usually are highly cross-linked via intermolecular tin–oxygen bonds. Unlike the halocarbons, organotin halides are reactive compounds and, because of their ionic character, readily enter into metathetical substitution reactions resembling the inorganic metal halides. Tin–hydrogen bonds are unlike carbon–hydrogen bonds and, although essentially covalent, their partial ionicity makes them true hydrides with hydrogen as the formal electronegative partner. Organotin hydrides are strong reducing agents and are similar to lithium aluminum hydride. Many are organic-soluble and easily distilled and are used increasingly in organic syntheses. Unlike carbon, tin shows much less tendency to catenate, ie, form chains of Sn atoms bonded to each other. Although tin–tin-bonded compounds are known (see Hexaorganoditins), the tin–tin bond is easily cleaved by oxygen, halogens, and acids.

Tetraorganotins. *Physical Properties.* Physical properties of typical tetraorganotin compounds are shown in Table 5. All tetraorganotin compounds are insoluble in water but are soluble in many organic solvents.

Chemical Properties. The most important reactions which tetraorganotins undergo are heterolytic, ie, electrophilic and nucleophilic, cleavage and Kocheshkov redistribution (96–99). The tin–carbon bond in tetraorganotins is easily cleaved by halogens, hydrogen halides, and mineral acids:

$$R_4Sn + Br_2 \rightarrow R_3SnBr + RBr$$

$$R_3SnBr + Br_2 \rightarrow R_2SnBr_2 + RBr$$

$$R_4Sn + HCl \rightarrow R_3SnCl + RH$$

$$R_3SnCl + HCl \rightarrow R_2SnCl_2 + RH$$

With tetraaryltin compounds, the reaction can proceed further to the aryltin trihalides:

$$\underset{[4713\text{-}59\text{-}1]}{(C_6H_5)_2SnBr_2} + Br_2 \rightarrow \underset{[7727\text{-}17\text{-}5]}{C_6H_5SnBr_3} + C_6H_5Br$$

$$\underset{[1135\text{-}99\text{-}5]}{(C_6H_5)_2SnCl_2} + HCl \rightarrow \underset{[1124\text{-}19\text{-}2]}{C_6H_5SnCl_3} + C_6H_6$$

In practice, these cleavage reactions are difficult to control, and usually mixtures of

Table 5. Physical Properties of Typical Tetraorganotin Compounds [a]

Compound	CAS Registry No.	Mp, °C	Bp, °C	n_D^{20}	d^{20}, g/cm³
$(CH_3)_4Sn$	[594-27-4]	−54	78	1.4415	1.2905[b]
$(C_4H_9)_4Sn$	[1461-25-2]	−97	127$_{1.3\ kPa}$[c,d]	1.4727	1.0541
$(C_8H_{17})_4Sn$	[3590-84-9]			1.4677[d]	0.9609[d]
$(C_6H_5)_4Sn$	[595-90-4]	228			1.521
$(C_6H_{11})_4Sn$	[1449-55-4]	261	160–163		
$(CH_2{=}CH)_4Sn$	[1112-56-7]		70$_{0.59\ kPa}$[c]	1.4914[b]	1.257
$(CH_3)_2(C_4H_9)_2Sn$	[1528-00-3]		73–75$_{0.5\ kPa}$[c]	1.4640[b]	1.124[b]
$(C_2H_5)_3(C_4H_9)Sn$	[17582-53-5]			1.4736	1.1457

[a] Ref. 94, except where noted.
[b] At 25°C.
[c] To convert kPa to mm Hg, multiply by 7.5.
[d] Ref. 95.

products form, even with stoichiometric quantities of reagents. Selectivity improves at lower temperatures, higher dilutions, and in the presence of polar solvents, eg, pyridine. This method is not used to prepare the lower alkylated–arylated organotins outside the laboratory.

The most widely utilized reaction of tetraorganotins is the Kocheshkov redistribution reaction, by which the tri-, di-, and in some cases the monoorganotin halides can be readily prepared:

$$R_4Sn + SnCl_4 \rightarrow 2\,R_2SnCl_2$$

$$R_2SnCl_2 + R_4Sn \rightarrow 2\,R_3SnCl$$

$$3\,R_4Sn + SnCl_4 \rightarrow 4\,R_3SnCl$$

$$R_4Sn + 3\,SnCl_4 \rightarrow 4\,RSnCl_3$$

These reactions proceed rapidly and in good yield with primary alkyl and phenyl organotin compounds at ca 200°C. The reactions proceed at lower temperatures if anhydrous aluminum chloride is used as a catalyst.

If the reaction temperature is controlled through the use of a low boiling solvent or other means, it is possible to isolate equimolar quantities of monoalkyltin trichloride and trialkyltin chloride using a 1:1 ratio of tetraorganotin and tin tetrachloride:

$$R_4Sn + SnCl_4 \xrightarrow{<100°C} R_3SnCl + RSnCl_3$$

When R is a lower alkyl, the organotin trichloride can be easily separated from the reaction mixture by extraction with dilute aqueous hydrochloric acid, in which it is soluble. This reaction also works well with unsymmetrical tetraorganotins and has been practiced commercially (100).

With tetraaryltins, the redistribution reaction can be made to proceed to the monoorganotin stage with the proper stoichiometry of reactants:

$$Ar_4Sn + 3\,SnCl_4 \rightarrow 4\,ArSnCl_3$$

Preparation. The tetraorganotins, although of little commercial utility by themselves, are important compounds since they are the starting materials for many of the industrially important mono-, di-, and triorganotins. Among the most widely used preparations of tetraalkyl- and tetraaryltin compounds is the reaction of stannic chloride with tetrahydrofuran-based Grignard reagents or organoaluminum compounds:

$$4\,RMgX + SnCl_4 \xrightarrow[\text{or R}_2O]{\text{THF}} R_4Sn + 4\,MgXCl$$

$$4\,R_3Al + 3\,SnCl_4 \xrightarrow[\text{or R}_2O]{R_3N} 3\,R_4Sn + 4\,AlCl_3$$

Excess alkylating reagent is required if the tetraorganotin is desired as the exclusive product. In commercial practice, the stoichiometry is kept at or below 4:1, since the crude product is usually redistributed to lower organotin chlorides in a subsequent step and an ether is used as the solvent (101). The use of diethyl ether in the Grignard reaction has been generally replaced with tetrahydrofuran.

Organolithium and organosodium reagents can also be used to prepare tetraorganotins:

$$4\,RLi + SnCl_4 \rightarrow R_4Sn + 4\,LiCl$$

$$4\,RNa + SnCl_4 \rightarrow R_4Sn + 4\,NaCl$$

The Wurtz reaction, which relies on *in situ* formation of an active organosodium species, is also useful for preparing tetraorganotin compounds and is practiced commercially. Yields are usually only fair and a variety of by-products, including ditins, also form:

$$SnCl_4 + 8\,Na + 4\,RCl \rightarrow R_4Sn + 8\,NaCl$$

A variant of the Wurtz reaction is the preparation of tetrabutyltin from activated magnesium chips, butyl chloride, and stannic chloride in a hydrocarbon mixture. Only a small amount of tetrahydrofuran is required for the reaction to proceed in high yield (101).

The use limitations of an active metal organometallic, eg, Grignard or organolithium reagents, allow preparation of only tetraorganotins, which have no functional groups reactive to the organometallic reagent on the molecule. The preparation of tetraorganotins with functional groups, eg, hydroxyl, amino, nitrile, etc, bonded to the organic group requires special measures, eg, blocking the functional group with an inert function then deblocking, usually mildly, after the formation of the tin–carbon bonds. The nitrile derivative, tetrakis(cyanoethyl)tin [*15961-16-7*], is prepared in good yield via a unique electrochemical reaction of tin metal with acrylonitrile (102). Unsymmetrical tetraorganotins can be prepared from the mono, di-, or triorganotin halides and the appropriate organometallic reagent of magnesium, lithium, sodium, or aluminum:

$$RSnCl_3 + 3\,R'MgX \rightarrow RR'_3Sn + 3\,MgXCl$$

$$R_2SnCl_2 + 2\,R'MgX \rightarrow R_2R'_2Sn + 2\,MgXCl$$

$$R_3SnCl + R'MgX \rightarrow R_3R'Sn + MgXCl$$

$$R_2SnCl_2 + 2\,R'Li \rightarrow R_2R'_2Sn + 2\,LiCl$$

Unsymmetrical functional tetraorganotins are generally prepared by tin hydride addition (hydrostannation) to functional unsaturated organic compounds (103) (see also Hydroboration). The realization that organotin hydrides readily add to aliphatic carbon–carbon double and triple bonds forming tin–carbon bonds led to a synthetic method which does not rely on reactive organometallic reagents for tin–carbon bond formation and, thus, allows the synthesis of organofunctional tetraorganotins containing a wide variety of functional groups. Typical compounds which undergo such a reaction include tributyltin hydride and triphenyltin hydride, which can be prepared by the reaction of the chlorides with lithium aluminum hydride or sodium borohydride (104–105). Representative organic substrates include acrylonitrile, acrylate and methacrylate esters, allyl alcohol, vinyl ethers, styrene, and other olefins:

$$R_3SnH \; + \; {\textstyle >}C{=}CHX \; \longrightarrow \; R_3Sn\overset{\displaystyle |}{\underset{\displaystyle |}{C}}CH_2X$$

Compounds with active halogens, eg, allyl chloride, also undergo reduction. Diorganotin dihydrides, monoorganotin trihydrides, and even stannane [*2406-52-2*], SnH_4, undergo analogous reactions, but the stability of the organotin hydrides decreases with increasing number of hydride groups, so these hydrostannation reactions generally proceed in poorer yield with more by-products.

Other methods for preparing tetraorganotin compounds include the use of diorganozinc compounds, halomethylzinc halides, electrolysis of organoaluminum reagents with a tin anode, and the electrolysis of diethyl sulfate with a zinc cathode and a tin anode (106–109). The latter method probably involves the *in situ* generation of an organozinc intermediate.

The reaction of an organotin–lithium, organotin–sodium, or organotin–magnesium reagent is occasionally useful for the preparation of tetraorganotins in the laboratory (93). These reagents or organostannylanionoids are air- and moisture-sensitive and can be prepared from most triorganotin halides and some tetraorganotins:

$$(CH_3CH_2CH_2)_3SnCl \xrightarrow[\text{or Li in THF}]{\text{Na in liquid } NH_3} (CH_3CH_2CH_2)_3SnNa \text{ or } \quad (Li)$$
$$[84474-09-9] \qquad [84474-08-8]$$

$$(CH_3CH_2CH_2)_3SnNa(Li) + RX \rightarrow (CH_3CH_2CH_2)_3SnR + Na(Li)X$$

$$(CH_3)_4Sn \xrightarrow{\text{Na in liquid } NH_3} (CH_3)_3SnNa + NaNH_2 + CH_4$$
$$[16643-09-7]$$

Primary and secondary alkyl halides react well, but *tert*-alkyl halides are preferentially dehydrohalogenated by the tin reagents.

Uses. The main use for tetraorganotin compounds is as (usually captive) intermediates for the tri-, di-, and monoorganotins. Although there have been reports in the patent literature of the use of tetraorganotins as components of Ziegler-Natta-type catalysts for the polymerization of olefins, there is no evidence that such catalysts are used commercially.

Triorganotins. Triorganotins and diorganotins constitute by far the most important classes of organotins.

Physical Properties. Physical properties of some typical triorganotin halides are listed in Table 6 and those of commercially important triorganotin compounds are listed in Table 7. The triorganotin halides are insoluble in water, except for $(CH_3)_3SnCl$ which is completely water soluble, but are soluble in most organic solvents. The fluorides are insoluble in most organic solvents because of their highly associated structure resulting from strong SnF---Sn interactions.

Table 6. Physical Properties of Typical Triorganotin Halides [a]

Compound	CAS Registry No.	Mp, °C	Bp, °C	n_D^{20}	d^{20}, g/cm^3
$(CH_3)_3SnCl$	[1066-45-1]	37.5	154–156		
$(CH_3)_3SnBr$	[1066-44-0]	26–27	163–165		
$(C_4H_9)_3SnCl$	[1461-22-9]		152–156$_{1.9 \text{ kPa}}$[b]	1.4930	1.2105
$(C_4H_9)_3SnF$	[1983-10-4]	218–219 (dec)			1.27[c]
$(C_6H_5)_3SnCl$	[639-58-7]	106			
$(C_6H_5)_3SnF$	[379-52-2]	357 (dec)			
$(C_6H_{11})_3SnCl$	[3091-32-5]	129–130			

[a] Ref. 110.
[b] To convert kPa to mm Hg, multiply by 7.5.
[c] At 25°C.

Table 7. Physical Properties of Commercially Important Triorganotin Compounds

Compound	CAS Registry No.	Mp, °C	Bp, °C	n_D^{20}	d^{20}, g/cm³
$[(C_4H_9)_3Sn]_2O$	[56-35-9]	<−45	$210-214_{1.3 \text{ kPa}}{}^a$	1.488	1.17
$(C_4H_9)_3SnF$	[1983-10-4]	218–219 (dec)			1.27^b
$(C_4H_9)_3SnOCOC_6H_5$	[4342-36-3]		$166-168_{0.13 \text{ kPa}}{}^a$	1.5157	1.1926
$(C_4H_9)_3SnOCOCH_3$	[56-36-0]	80–85			1.27
$(C_6H_5)_3SnOH$	[76-87-9]	118–120 (dec)			1.552^b
$(C_6H_5)_3SnF$	[379-52-2]	357 (dec)			1.53
$(C_6H_5)_3SnOCOCH_3$	[900-95-8]	119–120			
$(C_6H_{11})_3SnOH$	[13121-70-5]	c			
$(C_6H_{11})_3SnN_3C_2H_2{}^d$	[41083-11-8]	218.8			
$(Neoph_3Sn)_2O{}^e$	[13356-08-6]	$138-139^f$			

^{a} To convert kPa to mm Hg, multiply by 7.5.
^{b} At 25°C.
^{c} No true melting point, converts to bis-oxide at above 120°C.
^{d} $N_3C_2H_2$ = 1,2,4-triazole.
^{e} Neoph = neophyl = β,β-dimethylphenethyl.
^{f} Technical material.

Reactions. The utility of triorganotin chlorides and their application as starting materials for most other triorganotin compounds results from the ease of nucleophile displacement, as indicated in Figure 1. The commercially important triorganotin compounds are most frequently the oxides or hydroxides, the fluorides, and the carboxylates.

The basic hydrolysis of trialkyltin halides and other salts forms bis(oxide)s since, except for trimethyltin, hydroxides are unstable towards dehydration at room temperature. With tin aryl, aralkyl, and cycloalkyltin compounds, the hydroxides can be isolated. Although quite stable, they exist in mobile equilibrium with the bisoxide and water and are easily dehydrated. Trimethyltin hydroxide is exceptionally stable towards dehydration.

Triorganotin oxides and hydroxides are moderately strong bases and react readily with a wide variety of acidic compounds:

$$R_3SnOH + HX \rightarrow R_3SnX + H_2O$$
$$(R_3Sn)_2O + 2 HX \rightarrow 2 R_3SnX + H_2O$$

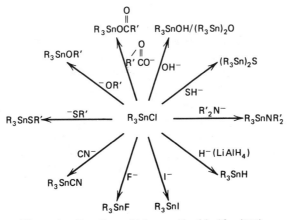

Figure 1. Reactions of triorganotin chlorides (110).

This reaction is useful in the preparation of anionic derivatives from the chlorides when the nucleophilic displacement route is unsatisfactory. Even weak acids, eg, phenols, mercaptans, and cyclic nitrogen compounds, can be made to undergo reaction with triorganotin hydroxides or bisoxides if the water of reaction is removed azeotropically as it forms.

Triorganotin compounds of strong acids are generally quite stable to hydrolysis under neutral conditions. Under basic conditions, the hydroxide or bisoxide forms. Strong acids, halogens, and other electrophiles can cause cleavage of tin–carbon bonds with the formation of diorganotins. The triorganotin oxides of lower alkyl groups (C_1–C_4) are sufficiently basic to react with carbon dioxide in air, resulting in the precipitation of triorganotin carbonates.

Preparation. Triorganotin chlorides of the general formula R_3SnX are the basic starting materials for other triorganotins. They are generally prepared by Kocheshkov redistribution from the crude tetraorganotin:

$$4\,R_4Sn + SnCl_4 \rightarrow 4\,R_3SnCl$$

The stoichiometric reaction of Grignard or alkylaluminum reagents with stannic chloride to give the trialkyltin chloride usually gives a mixture of products. Only in a very few cases is it possible to alkylate tin tetrachloride directly to the triorganotin chloride in good yield with few by-products using a Grignard reagent. In such cases, the formation of the triorganotin is favored because of steric hindrance (111):

$$3\,C_6H_{11}MgCl + SnCl_4 \rightarrow (C_6H_{11})_3SnCl + 3\,MgCl_2$$

Acid, hydrogen halide, or halogen cleavage of tetraorganotins is not used except on a laboratory scale because they are wasteful of tin–carbon bonds and uneconomical on a commercial scale.

Tribenzyltin chloride [*3151-41-5*] is a unique example of a triorganotin chloride that can be prepared directly from the organic halide and tin metal:

$$3\,C_6H_5CH_2Cl + 2\,Sn \xrightarrow[\text{reflux}]{H_2O} (C_6H_5CH_2)_3SnCl + SnCl_2$$

This reaction only proceeds in water. In a solventless system, only organic condensation products of benzyl chloride form, including dibenzyl. In toluene, dibenzyltin dichloride [*3002-01-5*] is the principal reaction product (112).

The production of triphenyltin hydroxide [*76-87-9*] and triphenyltin acetate [*900-95-8*] start with triphenyltin chloride, which is prepared by the Kocheshkov redistribution reaction from tetraphenyltin and tin tetrachloride. The hydroxide is prepared from the chloride by hydrolysis with aqueous sodium hydroxide. The acetate can be made directly from the chloride using sodium acetate or from the hydroxide by neutralization with a stoichiometric quantity of acetic acid.

For the preparation of tricyclohexyltin chloride, the Kocheshkov redistribution reaction is not suitable, since tetracyclohexyltin decomposes in the presence of stannic chloride at the normal redistribution temperatures. Two alternative routes are practiced for the manufacture of tricyclohexyltin chloride. The closely controlled reaction of cyclohexylmagnesium chloride and stannic chloride in a three-to-one molar ratio can be made to give the desired product in a good yield (111). Another method involves two steps for the preparation of tricyclohexyltin chloride (113). In the first step, butyltin trichloride [*1118-46-3*] reacts with three moles of cyclohexylmagnesium chloride forming butyltricyclohexyltin [*7067-44-9*]. This tetraorganotin then reacts with stannic chloride under mild conditions in an inert solvent, cleaving a butyl group and yielding tricyclohexyltin chloride and butyltin trichloride. The latter is recovered and recycled. The reactions are shown below:

$$C_4H_9SnCl_3 + 3\ C_6H_{11}MgCl \rightarrow (C_6H_{11})_3SnC_4H_9$$

$$(C_6H_{11})_3SnC_4H_9 + SnCl_4 \rightarrow (C_6H_{11})_3SnCl + C_4H_9SnCl_3$$

Tricyclohexyltin chloride is converted to the hydroxide with sodium hydroxide. The triazole can be prepared from the chloride with sodium or potassium hydroxide and 1,2,4-triazole.

Bis(trineophyltin) oxide [60268-17-4] is prepared from the chloride in the normal manner. The chloride can either be prepared directly from the reaction of three moles of neophylmagnesium chloride and stannic chloride or by the butyl transfer reaction between butyltrineophyltin and stannic chloride. The hydroxide derivative initially formed on hydrolysis of the chloride is readily dehydrated to the bis(oxide) at ca 100°C.

Uses. Triorganotin compounds are widely used as industrial biocides, agricultural chemicals, wood preservatives, and marine antifoulants. Although the *in vitro* fungicidal biological activity of the triorganotins was recognized in the mid-1950s, commercial development was not seriously undertaken until the early 1960s (114–115). The triorganotins that are most useful as biological control agents, in general, are the tributyltins, triphenyltins, and tricyclohexyltins.

The lower trialkyltins from trimethyl to tri-*n*-pentyl show high biological activity. The trimethyltins are highly insecticidal and the tripropyl-, tributyl-, and tripentyltin compounds have a high degree of fungicide and bactericide activity. Dialkyltin compounds are less active than the analogous trialkyltins. The maximum activity towards bacteria and fungi is exhibited by the tripropyl and tributyltin compounds, with the tributyltins providing the optimum balance between fungicidal and bactericidal activity and mammalian toxicity. Tributyltin compounds, especially the oxide and benzoate, are used as antimicrobials and slimicides for cooling-water treatment and as hard-surface disinfectants. These and similar compounds have been used as laundry sanitizers and mildewcides to prevent mildew formation in the dried film of water-based emulsion paints. In most microbiocide applications, the tributyltin compound is used in conjunction with another biocide, usually a quaternary ammonium compound, to complement the activity of the organotin which is most effective against gram-positive bacteria.

Although the lower trialkyltins show high fungicide activity, they are unlikely candidates for agricultural fungicide use because of their high phytotoxicity to the host plant. Various attempts have been made to moderate the phytotoxicity of the lower trialkyltins by changing the anion portion of the molecule. These have not been successful since the nature of the anionic group has little influence on the spectrum of biological activity, provided that the anion is not biologically active and it confers a sufficient minimal solubility on the compound.

In the early 1960s, the first organotin-based agricultural fungicide, triphenyltin acetate, was introduced in Europe commercially by Farbwerke Hoechst A.G. as Brestan. Brestan, which is a protectant foliar fungicide, was recommended for the control of *phytophthora* (late blight) on potatoes and *cercospora* on sugar beets at application rates of a few ounces per acre (116). Shortly thereafter, triphenyltin hydroxide was introduced as Du-Ter by Philips-Duphar, N.V., with about the same activity and spectrum of disease control as Brestan. Du-Ter is registered with the EPA in the United States as a fungicide for potatoes, sugar beets, pecans, and peanuts. Both compounds also exhibit a strong antifeedant effect on some insects and are fly sterilants at sublethal concentrations. Triphenyltin hydroxide formulations are also supplied by Griffin Corp. in the United States.

Tricyclohexyltin hydroxide was introduced into the U.S. market by the Dow Chemical Company as Plictran. Plictran was originally recommended for the control of phytophagous (plant-feeding) mites on apples and pears. It is also registered in the United States and in many European and Asian countries for this use as well as for mite control on citrus, stone fruits, and hops. This product has since been joined in the market by similar-acting, competitive products marketed by Shell and Bayer. Other triorganotin compounds with significant agricultural uses are tricyclohexyltin hydroxide (Plictran, Dow Chemical), and hexaneophyldistannoxane (Vendex, Shell U.S.A.; Torque, Shell International Chemical).

Bis(tributyltin) oxide [56-35-9] is widely used in Europe for the preservation of timber, millwork, and wood joinery, eg, window sashes and door frames. It is applied from organic solution by dipping or vacuum impregnation. It imparts resistance to attack by fungi and insects but is not suitable for underground use. An advantage of bis(tributyltin) oxide is that it does not interfere with subsequent painting or decorative staining and does not change the natural color of the wood. Tributyltin phosphate, $[(C_4H_9)_3Sn]_3PO_4$, has also been suggested as a wood preservative.

Most surfaces in prolonged contact with seawater and freshwater are susceptible to the attachment of marine growths, eg, algae and barnacles.

The most common method for preventing marine fouling has been to paint the underwater structure of the vessel with an antifouling paint containing a toxicant. For many years, the antifouling agent of choice was cuprous oxide, but there has been a strong trend towards the use of triorganotin compounds, both alone and in combination with cuprous oxide (117). Preferred compounds for use in this application are tributyltin fluoride, triphenyltin hydroxide, and triphenyltin fluoride since they are highly active against a wide range of fouling species. Bis(tributyltin) oxide, tributyltin acetate, and other tributyltin carboxylates have also been successfully used as antifoulants. Triorganotin compounds offer many advantages over cuprous oxide. Since they are colorless, they can be used in the preparation of paints of a variety of colors. Unlike cuprous oxide, they do not contribute to galvanic corrosion on steel or aluminum hulls. The triorganotins are rapidly degraded into lower alkylated species and then to nontoxic inorganic tin once released from the coating. Inorganic copper, on the other hand, is toxic in all its forms.

There has been much interest in eroding antifouling paints that are based on tributyltin acrylate [13331-52-7] or methacrylate [2155-70-6] copolymers with various organic acrylate esters as the combined toxicant and paint binder resin (117). Such paints erode in moving seawater because the triorganotin portion slowly hydrolyzes from the acrylic backbone in normally basic seawater, releasing the active species tributyltin chloride and bis(tributyltin) oxide. The depleted surface layer of the paint film, containing hydrophilic-free carboxylic acid groups, becomes water-swollen and is easily eroded by moving seawater. A fresh surface of triorganotin acrylate polymer is thereby exposed and the process repeats. Coatings based on organotin polymers can be formulated to release the toxicant at a rate which is linear with time. Such coatings are claimed to reduce fuel costs over and above the savings resulting from a clean hull by providing a surface which becomes smoother with time. M&T Chemicals, Inc., is a worldwide supplier of a variety of tributyltin methacrylate copolymers with different hydrolysis and erosion rates (bioMeT 300 series antifoulant polymers). Paints based on organotin copolymers are offered by the principal marine paint companies, including Hempel's Marine Paints (Nautic Modules), International Paint Co., Ltd. (Intersmooth SPC), Nippon Oil & Fats Co., Ltd., and Jotun Marine Coatings (Takata LLL) (see Coatings, marine).

The advantages claimed for organotin polymer-based antifouling paints include constant toxicant delivery versus time, erosion rate and toxicant delivery are controllable, no depleted paint residue to remove and dispose, 100% utilization of toxicant, polishing at high erosion rates, surface is self-cleaning, and function is continuously reactivated.

Triorganotin compounds have also been used experimentally in controlled-release formulations to control the infective snail vector in the debilitating tropical disease schistosomiasis (bilharzia) and to control mosquitoes in stagnant ponds (118). As yet, the large-scale use of such methods has little support in the host third world countries where these problems are most severe. Tributyltin chloride has been used to confer rodent-repellent properties on wire and cable coatings (119) (see Repellents, Supplement Volume).

Diorganotins. *Physical Properties.* Physical properties of some typical diorganotin compounds are shown in Table 8. The diorganotin chlorides, bromides, and iodides are soluble in many organic solvents and, except for dimethyltin dichloride, are insoluble in water.

Commercial grades of diorganotin carboxylates frequently have wider melting ranges because of the use of less pure grades of carboxylic acids in their manufacture which, for many applications, permits more facile handling of the liquids.

Reactions. Although there are few industrial applications for the diorganotin halides, these compounds are the basic intermediates for the preparation of all the commercially important diorganotin derivatives. They are prepared by nucleophilic displacement similar to that used for triorganotin derivatives (see Fig. 1). Basic hydrolysis of the diorganotin halides gives the diorganotin oxides in high yield. Except in rare cases, dihydroxide derivatives are unknown. As with the triorganotins, diorganotin oxides are sometimes used as intermediates from which other derivatives can be obtained by neutralization with strong or weak acids:

$$R_2SnCl_2 \xrightarrow{\text{OH}^-} R_2SnO$$

$$R_2SnO + 2\,HY \rightarrow R_2SnY_2 + H_2O$$

Diorganotin dihalides are moderately strong Lewis acids and form stable complexes with ammonia and amines. The commercially important diorganotin compounds are most frequently the oxides, carboxylates, and mercaptocarboxylic acid esters. The oxides are amorphous or polycrystalline, highly polymeric, infusible, and insoluble solids. They are moderately strong bases and react readily with a wide variety of strongly and weakly acidic compounds. Their insolubility in all nonreactive solvents makes the choice of proper reaction conditions for such a neutralization reaction an important consideration for optimum yields.

Diorganotin esters of strong acids are relatively stable to hydrolysis under neutral conditions, but, generally, diorganotin compounds are more reactive chemically than

Table 8. Physical Properties of Diorganotin Compounds[a]

Compound	CAS Registry No.	Mp, °C	Bp, °C	n_D^{20}	d^{20}, g/cm³
$(CH_3)_2SnCl_2$	[753-73-1]	107–108	185–190		
$(C_4H_9)_2SnCl_2$	[683-18-1]	41–42	140–143$_{1.3\ kPa}$[b]		
$(C_4H_9)_2SnBr_2$	[996-08-7]	21–22	90–92$_{0.04\ kPa}$[b]	1.5400	1.3913[c]
$(C_4H_9)_2SnI_2$	[2865-19-2]		145$_{0.8\ kPa}$[b]	1.6042	1.996[c]
$(C_6H_5)_2SnCl_2$	[1135-99-5]	42–44	180–185$_{0.7\ kPa}$[b]		
$(CH_3OC(O)CH_2CH_2)_2SnCl_2$	[10175-01-6]	132			
$(CH_3)_2Sn(SC_4H_9)_2$	[1000-40-4]		81$_{0.013\ kPa}$[b]	1.5400	1.280
$(C_4H_9)_2Sn(OCH_3)_2$	[1067-55-6]		126–128$_{7\ Pa}$[b]	1.4880	
$[(C_4H_9)_2SnS]_3$	[15220-82-3]	63–69			
$(C_4H_9)_2Sn(O\!\!\!\overset{\displaystyle O}{\overset{\|}{C}}\!CH_3)_2$	[1067-33-0]	8.5–10	142–145$_{1.3\ kPa}$[b]	1.4706	
$(C_4H_9)_2Sn(C_{11}H_{23}\overset{\displaystyle O}{\overset{\|}{C}}O)_2$	[77-58-7]	22–24		1.4683	1.05
$(C_4H_9\underset{\displaystyle C_2H_5}{CH}\overset{\displaystyle O}{\overset{\|}{C}}O)_2Sn(C_4H_9)_2$	[2781-10-4]	54–60	215–220$_{0.3\ kPa}$[b]	1.4653	1.070[c]
$[(C_2H_5O)_2P\overset{\displaystyle S}{\overset{\|}{}}S]_2Sn(C_6H_5)_2$	[74097-03-3]	149.5			

[a] Refs. 110 and 120.

[b] To convert kPa to mm Hg, multiply by 7.5.

[c] At 25°C.

the triorganotins. Diorganotin esters of weak acids are somewhat susceptible to hydrolysis, even under neutral conditions, but this reactivity is somewhat moderated by their hydrophobicity.

On partial hydrolysis, diorganotin halides and carboxylates may form basic salts of rather complicated structure:

$$2\ R_2SnY_2\ +\ 2\ OH^-\ \longrightarrow\ \underset{\overset{\displaystyle |}{Y}\ \overset{\displaystyle |}{Y}}{R_2SnOSnR_2}\ +\ 2\ Y^-\ +\ H_2O$$

Diorganotin sulfides can be prepared from the chlorides or oxides by the exchange of a reactive substituent for sulfur:

$$R_2SnCl_2 + S^{2-} \rightarrow R_2SnS + 2\ Cl^-$$

$$R_2SnO + CS_2 \rightarrow R_2SnS + COS$$

The sulfides are associated like the oxides, but to a lesser degree. They are crystalline, sharp-melting, soluble in many organic solvents, and resistant to hydrolysis. Most are cyclic trimers (121).

Some diorganotin compounds, eg, the alkoxides, add to hetero-unsaturated systems, eg, isocyanates. This reaction is believed to occur in stages (122).

$$Bu_2Sn(OCH_3)_2 \xrightarrow{RN=C=O} Bu_2SnOCH_3 \xrightarrow{RN=C=O} Bu_2Sn(NCOCH_3)_2$$

Preparation. Diorganotin dichlorides are the usual precursors for all other diorganotin compounds; three primary methods of manufacture are practiced. Dibutyltin dichloride is manufactured by Kocheshkov redistribution from crude tetrabutyltin and stannic chloride and usually is catalyzed with a few tenths of a percent aluminum trichloride:

$$(C_4H_9)_4Sn + SnCl_4 \rightarrow 2\ (C_4H_9)_2SnCl_2$$

Yields are almost quantitative and product purity is good with formation of only minute amounts of mono- and tributyltin by-products.

Many organic halides, especially alkyl bromides and iodides, react directly with tin metal at elevated temperatures (>150°C). Methyl chloride reacts with molten tin metal giving good yields of dimethyltin dichloride, which is an important intermediate in the manufacture of dimethyltin-based PVC stabilizers (see Uses). The presence of catalytic metallic impurities, eg, copper and zinc, is necessary to achieve optimum yields (123):

$$2\ CH_3Cl + Sn \xrightarrow[Cu]{235°C} (CH_3)_2SnCl_2$$

The reaction of higher alkyl chlorides with tin metal at 235°C is not practical because of the thermal decomposition which occurs before the products can be removed from the reaction zone. The reaction temperature necessary for the formation of dimethyltin dichloride can be lowered considerably by the use of certain catalysts. Quaternary ammonium and phosphonium iodides allow the reaction to proceed in good yield at 150–160°C (124). An improvement in the process involves the use of amine–stannic chloride complexes or mixtures of stannic chloride and a quaternary ammonium or phosphonium compound (125). Use of these catalysts is claimed to yield dimethyltin dichloride containing less than 0.1 wt % trimethyltin chloride. Catalyzed direct reactions under pressure are used commercially to manufacture dimethyltin dichloride.

The direct reaction of tin metal with higher haloalkanes is less satisfactory even when catalysts are used, except with alkyl iodides. The reaction of butyl iodide with tin metal is used commercially in Japan to prepare dibutyltin diiodide, from which dibutyltin oxide is obtained on hydrolysis with base:

$$2\ C_4H_9I + Sn \rightarrow (C_4H_9)_2SnI_2 \xrightarrow{OH^-} (C_4H_9)_2SnO + 2\ I^-$$

The economics of this process depend on near-quantitative recovery and recycle of the iodine to prepare butyl iodide.

Tin metal also reacts directly with a number of activated organic halides, including allyl bromide, benzyl chloride, chloromethyl methyl ether, and β-halocarboxylic esters and nitriles giving fair-to-good yields of diorganotin dihalides (112,126–129).

The facile reaction of metallic tin in the presence of hydrogen chloride with acrylic esters to give high yields of bis(β-alkoxycarbonylethyl)tin dichlorides is reported in refs. 130–131. This reaction proceeds at atmospheric pressure and room temperature and has been practiced commercially. Halogenostannanes have been postulated as intermediates (120).

Uses. Poly(vinyl chloride) stabilizers. The largest single industrial application for organotin compounds is in the stabilization of PVC. Of the estimated 30,000-t world production of organotins, it is believed that 20,000 t or two thirds of production, is accounted for by PVC stabilization (7). The estimated 1981 U.S. consumption of organotins as PVC stabilizers was 10,650 t, representing 27% of the market. Organotins are added to PVC to prevent its degradation by heat (180–200°C) during processing and by long-term exposure to sunlight (132–154).

Dialkyltin compounds are the best general-purpose stabilizers for PVC, especially if colorlessness and transparency are required. Commercial organotin stabilizers include the carboxylates, especially the maleates, laurates, and substituted maleates; the mercaptide, the mercaptoacid, and mercaptoalcohol ester derivatives (see Monoorganotin compounds); and the estertins, 2-carboalkoxyethyltin derivatives. The common industrial organotin stabilizers are listed in Table 9. U.S. producers of organotin stabilizers and the trade names of their products are: Argus (Witco), Mark; Cardinal, Cardinal Clear; Thiokol, Carstab; Ferro, Thermchek and Polychek; Interstab (Akzo), Interstab and Stanclere; M&T Chemicals, Thermolite; Tenneco, Nuostabe; and Synthetic Products (Dart), Synpron.

Sulfur-containing organotins impart excellent heat stability to PVC, but non-sulfur-containing organotins are used when resistance to light and weathering are required. The two main markets for organotin stabilizers are in the packaging and building industries. In the packaging industry, certain organotin stabilizers are used in PVC food packaging and drink containers. In the United States and the FRG, dioctyltin maleate [*16091-18-2*], dioctyltin bis(isooctylmercaptoacetate), and butylthiostannoic acid [*26410-42-4*] are approved for use in PVC food packaging; in the FRG,

Table 9.   Typical Commercially Significant Organotin PVC Stabilizers

Compound	CAS Registry No.	Structure
dibutyltin bis(isooctyl mercaptoacetate)	[*25168-24-5*]	$(C_4H_9)_2Sn(SCH_2CO_2C_8H_{17}\text{-i})_2$
dioctyltin bis(isooctyl mercaptoacetate)	[*26401-97-8*]	$(C_8H_{17})_2Sn(SCH_2CO_2C_8H_{17}\text{-i})_2$
dimethyltin bis(isooctyl mercaptoacetate)	[*26636-01-1*]	$(CH_3)_2Sn(SCH_2CO_2C_8H_{17}\text{-i})_2$
bis(2-carbobutoxyethyltin) bis(isooctyl mercaptoacetate)	[*63397-60-4*]	$(C_4H_9OCOCH_2CH_2)_2Sn(SCH_2CO_2C_8H_{17}\text{-i})_2$
dibutyltin sulfide	[*4253-22-9*]	$(C_4H_9)_2SnS$
dibutyltin bis(lauryl mercaptide)	[*1185-81-5*]	$(C_4H_9)_2Sn(SC_{12}H_{25})_2$
dibutyltin β-mercaptopropionate	[*27380-35-4*]	$\left[(C_4H_9)_2SnSCH_2CH_2COO\right]_n$ (n = 1 to 3)
dibutyltin bis(mercaptoethyldecanoate) (also other esters)	[*28570-24-3*]	$(C_4H_9)_2Sn(SCH_2CH_2OC(O)C_{11}H_{25})_2$
butylthiostannoic acid anhydride	[*15666-29-2*]	$(C_4H_9\overset{\displaystyle S}{\overset{\displaystyle \|}{Sn}})_2S$
butyltin tris(isooctylmercaptoacetate)	[*25852-70-4*]	$C_4H_9Sn(SCH_2CO_2C_8H_{17}\text{-i})_3$
dibutyltin dilaurate	[*77-58-7*]	$(C_4H_9)_2Sn(OOCC_{11}H_{23})_2$
dibutyltin maleate (dioctyltin derivative)	[*32076-99-6*] [*16091-18-2*]	$\left[(C_4H_9)_2SnOOCCH{=}CHCOO\right]_n$ (n = 1 to 3)
dibutyltin bis(monoisooctylmaleate) (also other alkyl maleate esters)	[*25168-21-2*]	$(C_4H_9)_2Sn(OOCCH{=}CHCOOC_8H_{17}\text{-i})_2$

dimethyltin bis(isooctylmercaptoacetate), 2-carbobutoxyethyltin tris(isooctylmer-captoacetate) [63438-80-2], and bis(2-carbobutoxyethyltin) bis(isooctylmercaptoa-cetate) are also approved. These uses reflect the low toxicity of these organotin sta-bilizers.

In the building industry, rigid PVC is stabilized with diorganotin carboxylates, especially dibutyltin maleate, for use in floorings and light fixture glazing and with diorganotin mercaptides and mercaptoacid esters for use in sidings, profiles, roofing, fencing, window frames, and piping. The dibutyltin, dimethyltin, and estertin sul-fur-containing derivatives are used for these nonfood applications as well as in PVC potable-water piping.

Polyurethane foam catalysts. Early production of polyurethane foams involved a two-step reaction in which a polyether glycol reacted with toluene diisocyanate forming a urethane prepolymer having reactive isocyanate end groups. Water was then added to condense the neighboring isocyanate groups to urethane linkages. In the process, carbon dioxide formed, which acted on the gelling polymer to produce a rigid or elastomeric foam. Inorganic tin compounds and diorganotin compounds, eg, di-butyltin diacetate [1067-33-0], dilaurate [77-58-7], and di(2-ethylhexanoate) [2781-10-4], catalyze the glycol–isocyanate reaction as well as the urethane conden-sation step and enable the preparation of foams in one step in a semicontinuous process (155–156). In the United States, dibutyltin compounds are used mostly in the catalysis of rigid foams and the laurate has been the catalyst of choice (157) (see Urethane polymers).

Diorganotin compounds have been used increasingly as catalysts for high resil-iency foam in automotive seating. In high resiliency foam, diorganotin mercaptocar-boxylates and mercaptides as catalysts improve some key physical properties (158). Some diorganotins, eg, the mercaptocarboxylates and mercaptides, are stable enough to be used in the preparation of masterbatches containing premixed polyol, water surfactant, amine, and organotin catalyst which are stored for up to six months (159). The principal suppliers of organotin-based polyurethane catalysts in the United States are M&T Chemicals, Inc., and Witco Chemical Corp.

Esterification catalysts. Dibutyltin compounds as well as monobutyltins are used increasingly as esterification (qv) catalysts for the manufacture of organic esters used in plasticizers (qv), lubricants, and heat-transfer fluids (see Lubrication and lubricants; Heat-exchange technology). Although esterification reactions catalyzed by organotins require higher temperatures (200–230°C) than those involving strong acid catalysts, eg, *p*-toluenesulfonic acid, side reactions are minimized and the products need no extensive refining to remove acidic ionic catalyst residues. Additionally, equipment corrosion is eliminated and the products have better color and odor properties because fewer by-products form (160). Usual catalyst levels are 0.05–0.3 wt % based on the total reactants charged. Dibutyltin compounds are also useful in catalyzing the transesterification and polycondensation of dimethyl terephthalate to poly(ethylene terephthalate) for packaging applications and in the manufacture of polyester-based alkyd resins (161). Both solid and liquid and insoluble and soluble organotin-based esterification catalysts are marketed by M&T Chemicals, Inc., as Fascat (see Polyesters).

Other. Dibutyltin dilaurate [77-58-7] has been successfully used for many years as a coccidiostat in the treatment of intestinal worm infections in chickens and turkeys (see Chemotherapeutics, antiprotozoal).

In Japan and Europe, dimethyltin dichloride that has been purified to remove all traces of trimethyltin chloride is used to provide a thin coating of stannic oxide on glass upon thermal decomposition at 500–600°C. Thin deposits of stannic oxide improve the abrasion resistance and bursting strength of glass bottles. Dimethyltin dichloride is manufactured and marketed as Glahard by Chugoku Toryo Co., Ltd., Shiga, Japan. Electroconductive films can be formed with thick coatings of tin oxide that is deposited in this manner.

Dibutyltin and dioctyltin diacetate, dilaurate, and di-(2-ethylhexanoate) are used as catalysts for the curing of room-temperature-vulcanized (RTV) silicone elastomers to produce flexible silicone rubbers used as sealing compounds, insulators, and a wide variety of other uses. Diorganotin carboxylates also catalyze the curing of thermosetting silicone resins, which are widely used in paper-release coatings.

In addition, diorganotin compounds are used as transesterification catalysts for the curing of cathodic, electrocoated paints (162). The biological activity and toxicity of diorganotins are much less than of analogous triorganotins with the same carbon-bonded organic groups.

Monoorganotins. *Physical Properties.* Properties of some monoorganotin trihalides are listed in Table 10. The monoorganotin trihalides are hygroscopic, low melting solids or liquids which are to varying extents hydrolyzed in water or moist air, liberating the hydrogen halides. They are soluble in most organic solvents and in water that contains enough acid to retard hydrolysis.

Chemical Properties. The monoorganotin trihalides are strong Lewis acids and form complexes with ammonia, amines, and many other oxygenated organic compounds, eg, ethers. In many ways, they resemble acid chlorides. As with the diorganotin dichlorides, the halogens on the molecule are easily replaced by a wide variety of nucleophilic reagents, making these trihalides useful intermediates for other monoorganotins. Typical compounds, which are easily formed by displacement reactions, include tris(alkoxides), tris(carboxylates), tris(mercaptides), and tris(mercaptocarboxylate esters). These compounds are generally more easily hydrolyzed than the analogous diorganotins.

The oxide monobutyltin oxide [51590-67-1], is a sesquioxide, $C_4H_9SnO_{1.5}$, from which it is difficult to remove the last traces of water. It is an infusible, insoluble, amorphous white powder that forms when butyltin trichloride is hydrolyzed with base. The partially dehydrated material, butylstannoic acid [2273-43-0], is slightly acidic and forms alkali-metal salts. These salts, ie, alkali-metal alkylstannonates, form when excess alkali is used to hydrolyze the organotin trichloride:

$$RSnCl_3 + 4\ NaOH \rightarrow RSnO_2Na + 3\ NaCl + 2\ H_2O$$

Partially hydrolyzed products of the form $RSn(OH)_2Cl$ are believed to be mixtures in most cases.

Table 10. Physical Properties of Typical Organotin Trihalides

Compound	CAS Registry No.	Mp, °C	Bp, °C	n_D^{20}
CH_3SnCl_3	[993-16-8]	45–46		
CH_3SnBr_3	[993-15-7]	53	211	
$C_4H_9SnCl_3$	[1118-46-3]		$102–103_{1.6\ kPa}{}^a$	1.5233
$C_6H_5SnCl_3$	[1124-19-2]		$142–143_{3.3\ kPa}{}^a$	1.5871

^{a} To convert kPa to mm Hg, multiply by 7.5.

When organotin trihalides are treated with alkali-metal sulfide, the sesquisulfides form (163):

$$2 \, RSnCl_3 + 3 \, Na_2S \rightarrow 2 \, RSnS_{1.5} + 6 \, NaCl$$

At least one, the monobutyl compound, is a tetramer in benzene (163).

Preparation and Manufacture. Monoorganotin halides are the basic raw materials for all other triorganotin compounds and are generally prepared by Kocheshkov redistribution from the tetraorganotin, eg, tetrabutyltin or the higher organotin halides:

$$R_4Sn + 3 \, SnCl_4 \rightarrow 4 \, RSnCl_3$$

$$R_2SnCl_2 + SnCl_4 \rightarrow 2 \, RSnCl_3$$

The oxidative addition of aliphatic organic halides to stannous chloride has long been of interest for the preparation of monoorganotin trihalides:

$$SnCl_2 + RCl \rightarrow RSnCl_3$$

This reaction gives fair-to-good yields of monoorganotin tribromides and trichlorides when quaternary ammonium or phosphonium catalysts are used (164). Better yields are obtained with organic bromides and stannous bromide than with the chlorides. This reaction is also catalyzed by trialkylantimony compounds at 100–160°C, bromides are more reactive than chlorides in this preparation (165–166). α,ω-Dihaloalkanes also react in good yield giving ω-haloalkyltin trihalides when catalyzed by organoantimony compounds (167).

A significant advance in the synthesis of monoorganotin trihalides was the preparation of β-substituted ethyltin trihalides in good yield from the reaction of stannous chloride, hydrogen halides, and α,β-unsaturated carbonyl compounds, eg, acrylic esters, in common solvents at room temperature and atmospheric pressure (168–169). The reaction is believed to proceed through a solvated trichlorostannane intermediate (170):

$$SnCl_2 + HCl \xrightarrow{(C_2H_5)_2O} HSnCl_3 \cdot 2(C_2H_5)_2O \xrightarrow{\quad CH_2=\!\!\!\!\!\overset{\overset{\displaystyle O}{\|}}{C}\!\!-\!OR \quad} Cl_3SnCH_2CH_2\overset{\overset{\displaystyle O}{\|}}{C}OR$$

This reaction can be extended to unsaturated nitriles, eg, acrylonitrile, which can give trihalostannyl-functional carboxylic acids, esters, and amides by the proper choice of solvents and reaction conditions (171).

Uses. *Poly(vinyl chloride) stabilizers.* Although generally less effective as PVC stabilizers than dialkyltin derivatives, monoalkyltin compounds added to the dialkyltin compounds in amounts of 5–20 wt % exert a synergistic effect on stabilizer effectiveness, preventing early yellowing. They supposedly function by reacting more quickly and at lower processing temperatures than the dialkyltin species, thus preventing the early onset of yellowing; conversely, diorganotins are more effective in retarding the long-term degradation of the polymer. Butylthiostannoic acid anhydride is used as a sole stabilizer for certain grades of PVC in the FRG but elsewhere is rarely used alone (172). It is approved in the FRG and in the United States for food packaging (173). In the FRG, the following monoorganotins alone or in mixtures are also approved for this use: butylthiostannoic acid anhydride with either dioctyltin compounds or 2-

carbobutoxyethyltin compounds, 2-carbobutoxyethyltin tris(isooctylmercaptoacetate) alone or mixed with its dicounterpart, monomethyltin tris(isooctylmercaptoacetate) [56225-49-1] plus its dicounterpart in a 24:76 wt % ratio, and monooctyltin tris[alkyl (C_{10}–C_{16}, isooctyl) mercaptoacid esters] with their dicounterparts.

Treatment of glass. The use of monobutyltin trichloride in the hot-end coating of glass to improve the abrasion resistance and bursting strength of glass bottles has been patented, and the deposition process variables and product advantages have been described (174–176). Highly efficient utilization of tin is one of the main benefits.

Compounds with Tin–Tin Bonds. The most important class of catenated tin compounds is the hexaorganoditins. The ditin compounds are usually prepared by reductive coupling of an triorganotin halide with sodium in liquid ammonia:

$$2\,R_3SnCl + 2\,Na \xrightarrow{NH_3(l)} R_3SnSnR_3 + 2\,NaCl$$

This reaction proceeds in stages via an organostannylsodium compound:

$$R_3SnCl + 2\,Na \rightarrow R_3SnNa + NaCl$$

$$R_3SnNa + R_3SnCl \rightarrow R_3SnSnR_3 + NaCl$$

Lithium metal in tetrahydrofuran can also be used as the coupling reagent, and unsymmetrical ditins can be prepared when the reaction is conducted in stages (177–178).

Hexaorganoditins with short-chain aliphatic groups are colorless liquids, distillable under vacuum, soluble in organic solvents other than the lower alcohols, and insoluble in water. They are generally unstable in air, undergoing ready oxidation to a mixture of organotin compounds. Hexaarylditins are usually crystallline solids and are much more stable towards oxidation.

The ditins as of yet are insignificant commercially, although there has been interest in hexamethylditin [661-69-8] (Pennwalt TD-5032) as an insecticide (179–180).

Salts. Organic tin salts are tin compounds containing an organic radical in which the tin is bonded with an element other than carbon. The most common of these are the tin carboxylates, especially the stannous carboxylates. The latter are manufactured by reaction of stannous oxide or chloride with the appropriate acid. The most commercially significant of the stannous carboxylates is stannous 2-ethylhexanoate [301-10-0]. It is estimated that in 1979, worldwide annual consumption of tin-based catalysts for polyurethanes was ca 2500 t with stannous 2-ethylhexanoate accounting for 95% of this usage (155). The second most important industrial organic tin salt is stannous oxalate [814-94-8]. Other commercially available, organic tin salts that are of minor commercial importance are listed in Table 11.

Stannous 2-Ethylhexanoate. Stannous 2-ethylhexanoate, $Sn(C_8H_{15}O_2)_2$ (sometimes referred to as stannous octanoate, mol wt 405.1, sp gr 1.26), is a clear, very light yellow, and somewhat viscous liquid that is soluble in most organic solvents and in silicone oils (181). It is prepared by the reaction of stannous chloride or oxide with 2-ethylhexanoic acid.

The primary use for stannous 2-ethylhexanoate is as a catalyst with certain amines for the manufacture of one-shot polyether urethane foams (182). Resulting foams exhibit good dry-heat stability over a wide range of catalyst concentrations. Food-grade

Table 11. Physical Properties of Organic Tin Salts of Minor Commercial Importance

Salt	CAS Registry No.	Mp, °C	Density, g/cm^3	Use
stannous acetate	[638-39-1]	182.5–183	2.31	promotes dye uptake by fabrics
stannous ethylene glycoxide	[68921-71-1]	dec >300	2.87	esterification catalyst
stannous formate	[2879-85-8]	dec >100		catalyst for hydrogenation of liquid fuels
stannous gluconate	[35984-19-1]		1.35	silicone catalyst
stannous oleate	[1912-84-1]		1.06	silicone catalyst
stannous stearate	[6994-59-8]	90; dec 340	1.05	catalyst
stannous tartrate	[815-85-0]	dec 280	2.6	dyeing and printing of textiles

stannous 2-ethylhexanoate is approved by the FDA for use in polymers and resins used in food packaging (183). Other industrial applications include its use as a catalyst in silicones, including room-temperature-vulcanizing (RTV) silicone rubbers and silicone–oil emulsions; in epoxy formulations; and in various urethane coatings and sealants (qv) (184–185). Proprietary catalyst formulations based on stannous 2-ethylhexanoate are also available.

Stannous Oxalate. Stannous oxalate, $Sn(C_2O_4)$ (mol wt 206.71, dec 280°C, sp gr 3.56 at 18°C), is a white crystalline powder, is soluble in hot concentrated hydrochloric acid and mixtures of oxalic acid and ammonium oxalate, and is insoluble in water, toluene, ethyl acetate, dioctyl phthalate, THF, isomeric heptanes, and acetone (186). It is prepared by precipitation from a solution of stannous chloride and oxalic acid and is stable indefinitely.

Stannous oxalate is used as an esterification and transesterification catalyst for the preparation of alkyds, esters, and polyesters (187–188). In esterification reactions, it limits the undesirable side reactions responsible for the degradation of esters at preparation temperatures. The U.S. Bureau of Mines conducted research on the use of stannous oxalate as a catalyst in the hydrogenation of coal (189) (see Coal).

Toxicology. The toxicological properties of organotin compounds are reviewed in refs. 57, 89–92, 190. The toxicity of organotin compounds is a reflection of their biological activity. Thus, the most toxic to mammals, including man, are the lower trialkyltin compounds, ie, trimethyltin and triethyltin. As with the fungicidal activity, the toxicity seems little affected by the nature of the anionic group bonded to the trialkyltin moiety. There is some evidence that triorganotin compounds that are five-coordinate and intramolecularly chelated are less toxic than similar unchelated four-coordinate compounds (191–192). As a general rule, the toxicity of the trialkyltins decreases with increasing chain length of the alkyl group.

The acute oral mammalian toxicities of typical triorganotin compounds, including some which are not used commercially, are listed in Table 12. In some cases, two or more substantially different values are reported in the literature for the same test animal. In these cases, both high and low values are tabulated. The toxicity of triorganotins is strongly dependent on the nature of the organic groups bonded to tin. The toxicity varies from the highly toxic lower alkyl trimethyl and triethyltins, which are not used in any commercial applications, to the substantially less toxic trineophyl and trioctyl derivatives. The widely used tributyl-, triphenyl-, and tricyclohexyltin de-

Table 12.   Acute Oral Toxicities of Triorganotin Compounds

Compound	CAS Registry No.	LD_{50}, mg/kg	Test animal	Ref.
$(CH_3)_3SnOCCH_3$ (with O double bond)	[1118-14-5]	9	rat	200
$(C_2H_5)_3SnOCCH_3$ (with O double bond)	[1907-13-7]	4	rat	200
$(C_3H_7)_3SnOCCH_3$ (with O double bond)	[3267-78-5]	118	rat	200
$(C_4H_9)_3SnOCCH_3$ (with O double bond)		133	rat	190
		380	rat	200
$[(C_4H_9)_3Sn]_2O$	[56-35-9]	ca 200	rat	190
$(C_4H_9)_3SnF$		200	rat	190
$(C_6H_{13})_3SnOCCH_3$ (with O double bond)	[2897-46-3]	1,000	rat	200
$(C_8H_{17})_3SnOCCH_3$ (with O double bond)	[919-28-8]	>1,000	rat	200
$(C_6H_5)_3SnOCCH_3$ (with O double bond)		136	rat	190
		491	rat	190
$(C_6H_5)_3SnOH$		108	rat	208
		209	mouse	208
$(C_6H_{11})_3SnOH$		540	rat	208
		780	guinea pig	208
$(Neoph_3Sn)_2O^a$		1,450	mouse	208
		>1,500	dog	208
		2,630	rat	208

^{a} Neoph = neophyl = β,β-dimethylphenethyl.

rivatives are intermediate in mammalian oral toxicity. The highest trialkyl and triaryltins are less toxic when given orally than when given parentally because of their poor absorption from the gastrointestinal tract (193). Uncoupling of oxidative phosphorylation in cellular mitochondria has been suggested as one of the mechanisms of lower trialkyltin toxicity (194).

Most triorganotins that have been studied and all commercial ones are eye and skin irritants. Animal studies have shown that, particularly with tributyl and triphenyltin compounds, untreated eye contact can result in permanent corneal damage. If allowed to remain in contact with the skin for prolonged periods, these compounds can produce severe irritation and, in some cases, severe chemical burns (195). Thus, eye and skin protection must be worn when handling triorganotin compounds. Sometimes the irritant effect is delayed and may not be apparent for several hours. In the event of acute local dermal contact episodes with tributyltin compounds, pruritis, minor edema, and follicular pustules in hirsute areas occur. Systemic effects are observed in percutaneous tests of tributyltin iodide, bromide, chloride and bis-oxide in tests with rabbits, so it could be assumed that these compounds are absorbed through the skin (196). Bis(tributyltin) oxide produces a typical lower trialkyltin response dermally, characterized by redness, swelling, and skin discoloration in test animals. Its effects on the eyes are serious with damage to the cornea (196).

Among the most widely studied triorganotin compounds are triphenyltin hydroxide, triphenyltin acetate, and tricyclohexyltin hydroxide because of their use as agricultural chemicals. Triphenyltin hydroxide is a severe eye irritant in rabbits but is nonirritating to dry rabbit skin (197). In contrast, triphenyltin chloride on rabbits produces erythema and edema with tissue damage. The injuries are worsened by washing with organic solvent (196). In feeding tests on rats and mice, triphenyltin hydroxide shows no evidence of carcinogenicity (198).

Diorganotin compounds as a class are substantially less toxic than the analogous triorganotins. Some compounds of this class are used as additives in plastics intended to be in contact with food or potable water or used as PVC stabilizers. The acute oral toxicities of common commercial diorganotin compounds are given in Table 13. The dialkyltin chlorides and oxides generally show decreasing oral toxicity with increasing length of the alkyl chain. The toxicity of the lower dialkyltins is believed to be related to their ability to combine with enzymes containing two thiol groups in a suitable stereochemical conformation and thereby inhibiting the oxidation of α-ketoacids in the cell (199). 2,3-Dimercapto-1-propanol, $HSCH_2CH(SH)CH_2OH$, has been reported as an effective antidote for lower dialkyltin poisoning (83). The lower dialkyltin halides are somewhat less irritating to the skin than the analogous triorganotins, but skin contact should be avoided. Other studies of the toxicities of specific diorganotin compounds are reported in refs. 200–202.

Table 13. Acute Oral Toxicities of Diorganotin Compounds

Compound	CAS Registry No.	LD$_{50}$ (rat), mg/kg	Ref.
$(CH_3)_2SnCl_2$		74	190
$(CH_3)_2Sn(SCH_2\overset{O}{\overset{\|}{C}}OC_8H_{17}\text{-}i)_2$	[26636-01-1]	1380	190
$(C_4H_9)_2SnCl_2$		126	190
$(C_4H_9)_2SnO$	[818-08-6]	600–800	204
$(C_4H_9)_2Sn(O\overset{O}{\overset{\|}{C}}C_{11}H_{23})_2$		175	190
$(C_4H_9)_2Sn(SCH_2\overset{O}{\overset{\|}{C}}OC_8H_{17}\text{-}i)_2$	[25168-24-5]	500	190
$(C_8H_{17})_2SnCl_2$	[3542-36-7]	5500	190
$(C_8H_{17})_2SnO$	[870-08-6]	2500	190
$(C_8H_{17})_2Sn(O\overset{O}{\overset{\|}{C}}C_{11}H_{23})_2$	[3648-18-8]	6450	190
$(C_8H_{17})_2Sn(SCH_2\overset{O}{\overset{\|}{C}}OC_8H_{17}\text{-}i)_2$	[26401-97-8]	2000	190
$(RO\overset{O}{\overset{\|}{C}}CH_2CH_2)_2SnCl_2{}^a$		2350	203
$(RO\overset{O}{\overset{\|}{C}}CH_2CH_2)_2Sn(SCH_2\overset{O}{\overset{\|}{C}}OC_8H_{17}\text{-}i)_2{}^a$		1430	203

^{a} R is undefined; it is probably C_2H_5.

Monoorganotin compounds present no special toxicological problems. In general, they show the familiar trend of decreasing toxicity with increasing alkyl chain length, but of a lower order of toxicity than the diorganotins. As with most organotin compound classes, there are conflicting toxicity data and exceptions to general rules. Monobutyltin sulfide [15666-29-2] (butylthiostannoic anhydride, BTSA, poly[(1,3-dibutyldistannthiondiylidene)-1,3-dithio] is allowed as a stabilizer in semirigid or rigid PVC used in food packaging. Typical LD_{50} values for monoorganotin compounds are

$$CH_3Sn(SCH_2\overset{\overset{\displaystyle O}{\|}}{C}OC_8H_{17}\text{-}i)_3$$

920 mg/kg (rats) (203);

$$C_4H_9Sn(SCH_2\overset{\overset{\displaystyle O}{\|}}{C}OC_8H_{17}\text{-}i)_3$$

1063 mg/kg (rats) (203); $(C_4H_9SnS_{1.5})n$, >20,000 mg/kg (rats) (204);

$$C_8H_{17}Sn(SCH_2\overset{\overset{\displaystyle O}{\|}}{C}OC_8H_{17}\text{-}i)_3$$

3400 mg/kg (rats) (203). The lower monoorganotin trihalides can present special problems, however, because of their facile reaction with moisture, resulting in the liberation of hydrochloric acid.

The toxicity of the tetraorganotins has been little studied. Available literature indicates that tetrabutyltin and the higher tetraalkyltins are substantially less toxic than triorganotins to mammals if taken orally (190). The high toxicity reported for tetraethyltin (LD_{50} = 9–16 mg/kg) appears to be caused by its rapid conversion in the liver to a triethyltin species.

The inhalation toxicities (50% fatality in rats) of dimethyltin dichloride, monomethyltin trichloride, and dibutyltin dichloride are 1070, 600, and 73 mg/(L·h) (205).

The current OSHA TLV standard for exposure to all organotin compounds is 0.1 mg of organotin compounds (as tin)/m³ air averaged over an 8-h work shift (206). NIOSH has recommended a permissible exposure limit of 0.1 mg/m³ of tin averaged over a work shift of up to 10 h/d, 40 h/wk; ref. 207 should be consulted for more detailed information. Additional information on the health effects of organotin compounds is given in ref. 62.

BIBLIOGRAPHY

"Tin Compounds" in *ECT* 1st ed., Vol. 14, pp. 157–165, by H. Richter, Metal & Thermit Corp.; "Tin Compounds" in *ECT* 2nd ed., Vol. 20, pp. 304–327, by C. Kenneth Banks, M&T Chemicals, Inc.

1. *SRI International, 1981 Directory of Chemical Producers—United States of America*, SRI International, Palo Alto, Calif., 1981, pp. 459, 471, 493, 798, 855, 892, 897, 960–962.
2. *SRI International, 1981 Directory of Chemical Producers—Western Europe*, 4th ed., Vol. 2, SRI International, Palo Alto, Calif., 1981, pp. 1555, 1567, 1636, 1690, 1760–1763.

3. M. J. Fuller, *Industrial Uses of Inorganic Tin Chemicals*, ITRI Publication 499, International Tin Research Institute, Middlesex, UK, 1975.
4. *Annual Report 1978*, International Tin Research Institute, Middlesex, UK, p. 20.
5. *The Economics of Tin*, 3rd ed., Roskill Information Services, Ltd., London, UK, July 1981, pp. 110–117, 129–130, 349–354.
6. *Annual Review of the World Tin Industry 1981–1982*, Hargreaves & Williamson, London, UK, 1981, pp. 11–12, 15.
7. *The World Tin Industry—Supply and Demand*, Australian Mineral Economics Pty., Ltd., Sydney Australia, 1980, pp. 176–180, 202–206.
8. P. A. Cusack and P. J. Smith in R. Thomson, ed., *Speciality Inorganic Chemicals*, Royal Society of Chemistry, London, 1981, pp. 285–310.
9. *Mod. Plast.*, 70 (Sept. 1981).
10. *Mod. Plast.*, 82 (Sept. 1981).
11. *Cosmetic and Toiletry Raw Materials*, Charles H. Kline & Co., Inc., Fairfield, N.J., 1979, pp. 275–278.
12. T.W. Lapp, *The Manufacture and Use of Selected Alkyltin Compounds—Final Report*, EPA 5601-6-76-011, EPA, Washington, D.C., 1976, 123 pp.
13. *Chem. Mark. Rep.* 39 (Jan. 11, 1982).
14. Price lists, M&T Chemicals, Inc., Rahway, N.J.; Bis(tributyltin) oxide, Jan. 4, 1982; Dibutyltin dichloride (solid), Mar. 1, 1981; Tributyltin acetate, Jan. 4, 1982; Tributyltin chloride, Jan. 4, 1982; Tributyltin fluoride, Jan. 4, 1982.
15. *Toxic Substances Control Act Chemical Substances Inventory*, Vol. 3, EPA, Washington, D.C., May 1979, Cumulative Suppl., July 1980.
16. *Stannochlor Compound*, Technical Bulletin 161, M&T Chemicals Inc., Rahway, N.J., 1976.
17. *Gmelin Handbuch der Organischen Chemie*, Band 46, Parts C5 and C6, Springer-Verlag, New York, 1977, and 1978.
18. F. A. Lowenheim, ed., *Modern Electroplating*, 3rd ed., John Wiley & Sons, Inc., New York, 1974, pp. 377–417.
19. U.S. Pat. 2,407,579 (Sept. 10, 1946), E. W. Schweikher (to E. I. du Pont de Nemours & Co., Inc.).
20. S. Wein, *Glass Ind.*, 367 (1954).
21. C. J. Evans, *Tin Its Uses* **98**, 7 (1973).
22. J. I. Duffy, ed., *Electroless and Other Nonelectrolytic Plating Techniques*, Noyes Data Corp., Park Ridge, N.J., 1980.
23. F. A. Domino, *Plating on Plastics—Recent Developments*, Noyes Data Corp., Park Ridge, N.J., 1979, p. 108.
24. CFR Title 21, 182.3845, pp. 617, 631 (1977); *Fed. Reg.* **33**, 5619 (April 11, 1968).
25. *Fed. Reg.* **34**, 12087 (July 18, 1969).
26. *Fed. Reg.* **33**, 3375 (Feb. 27, 1968); **35**, 15372 (Oct. 27, 1970).
27. *Fed. Reg.* **35**, 8552 (June 3, 1970).
28. U.S. Pat. 2,785,076 (March 12, 1957), G. Felton (to Hawaiian Pineapple Co.).
29. U.S. Pat. 3,414,429 (Dec. 3, 1968), H. G. Bruss, W. J. Schlientz, and B. E. Wiens (to Owens-Illinois Co.); U.S. Pat. 3,554,787 (Jan. 12, 1971), C. E. Plymale (to Owens-Illinois Co.).
30. T. Williamson, *Tin Its Uses* **129**, 14 (1981).
31. U.S. Pat. 3,623,854 (Nov. 20, 1971), C. A. Frank (to Owens-Illinois Co., Inc.).
32. U.S. Pat. 3,952,118 (April 20, 1976), M. A. Novice (to Dart Industries, Inc.).
33. U.S. Pat. 2,162,255 (June 13, 1939), R. F. Heald (to Colgate-Palmolive-Peet Co.).
34. P. A. Cusack, P. J. Smith, J. S. Brooks, and R. Smith, *J. Text. Inst.* **7**, 308 (1979).
35. J. C. Muhler, W. H. Nebergall, and H. G. Day, *J. Dent. Res.* **33**, 33 (1954).
36. U.S. Pat. 2,876,166 (March 3, 1959), W. H. Nebergall (to Indiana University Foundation).
37. U.S. Pat. 2,946,725 (July 26, 1960), P. E. Norris and H. C. Schweizer (to Procter & Gamble Co.).
38. W. B. Hampshire and C. J. Evans, *Tin Its Uses* **118**, 3 (1978).
39. M. J. Fuller and M. E. Warwick, *J. Catal.* **42**, 418 (1976).
40. U.S. Pat. 3,723,273 (March 27, 1973), H. P. Wilson (to Vulcan Materials Co.).
41. U.S. Pat. 4,066,518 (Jan. 3, 1978), R. E. Horn (to Pitt Metals and Chemicals, Inc.).
42. U.S. Pat. 3,346,468 (Oct. 10, 1967), J. C. Jongkind (to M&T Chemicals, Inc.).
43. U.S. Pat. 3,462,373 (Aug. 19, 1969), J. C. Jongkind (to M&T Chemicals, Inc.).
44. S. Karpel, *Tin Its Uses* **124**, 3 (1980).

45. J. C. Jongkind, *Metal Finishing Guidebook & Directory*, Metals and Plastics Pub. Inc., Hackensack, N.J., 1982, p. 346.
46. W. W. Coffeen, *J. Am. Ceram. Soc.* **36**, 207, 215 (1953); **37**, 480 (1954).
47. U.S. Pat. 2,658,833 (Nov. 10, 1953), W. W. Coffeen and H. W. Richter (to M&T Chemicals, Inc.).
48. *Electronic Ceramic Stannates*, Technical Data Sheet CER-322, M&T Chemicals, Inc., Rahway, N.J., 1969.
49. P. A. Cusack, P. J. Smith, and L. T. Arthur, *J. Fire Retardant Chem.* **7**, 9 (1980).
50. U.S. Pat. 3,795,595 (March 5, 1974), H. P. Wilson (to Vulcan Materials Co.).
51. H. Silman, *Prod. Finish. (London)* **34**(1), 15 (1981).
52. N. J. Spiliotis, *Metal Finishing Guidebook & Directory*, Metals and Plastics Pub. Inc., Hackensack, N.J., 1982, pp. 333, 338.
53. U.S. Pat. 4,075,314 (Feb. 21, 1978), R. G. Wolfangel and H. A. Anderson (to Mallinckrodt, Inc.).
54. A. M. Zimmer, *Am. J. Hosp. Pharm.* **34**(3), 264 (1977).
55. D. G. Pavel, A. M. Zimmer, and V. N. Patterson, *J. Nucl. Med.* **18**, 305 (1977).
56. R. A. Hiles, *Toxicol. Appl. Pharmacol.* **27**, 366 (1974).
57. J. M. Barnes and H. B. Stoner, *Pharmacol. Rev.* **11**, 211 (1959).
58. *Toxicants Occurring Naturally in Foods*, 2nd ed., Natural Academy of Sciences, Washington, D.C., 1973, pp. 63–64.
59. E. Browning, in *Toxicity of Industrial Metals*, 2nd ed., Butterworths, London, 1969, pp. 323–330.
60. H. Cheftel, *Tin in Food*, Joint FAO/WHO Food Standards Program, 4th Meeting of the Codex Committee on Food Additives, PEPT, 1967, Joint FAO/WHO Food Standards Branch (Codex Alimentarius), FAO, Rome.
61. *Occupational Health Guidelines for Inorganic Tin Compounds (as Tin)*, U.S. Department of Labor, Occupational Safety and Health Administration, Washington, D.C., Sept. 1978.
62. *Environmental Health Criteria, Vol. 15: Tin and Organotin Compounds: A Preliminary Review*, WHO, Geneva, 1980, *Chem. Abstr.* 93:232412.
63. H. E. Stokinger in G. D. Clayton and F. F. Clayton, eds., *Patty's Industrial Hygiene and Toxicology*, 3rd rev. ed., Vol. 2A, Wiley-Interscience, New York, 1981, p. 1945.
64. H. O. Calvery, *Food Res.* **7**, 313 (1942).
65. Unpublished data, M&T Chemicals, Inc., Rahway, N.J., 1976, 1978.
66. J. K. Lim, G. J. Renaldo, and P. Chapman, *Caries Res.* **12**(3), 177 (1978).
67. D. L. Conine, J. C. Muhler, and R. B. Forney, *Toxicol. Appl. Pharmacol.* **22**, 303 (1972).
68. R. A. Kehoe and co-workers, *J. Nutr.* **19**, 597 (1940); **20**, 85 (1940).
69. H. A. Schroeder, J. J. Balassa, and I. H. Tipton, *J. Chronic Dis.* **17**, 483 (1964).
70. D. H. Calloway and J. J. McMullen, *Am. J. Clin. Nutr.* **18**, 1 (1966).
71. CFR Title 21, 172.180, 175.105, 175.300, 176.130, 176.170, 177.120, 177.1210, 177.2600, 178.3910, 181.29, 182.3845, 1981.
72. C. C. Dundon and J. P. Hughes, *Am. J. Roentgenol.* **63**, 797 (1950).
73. G. E. Spencer and W. C. Wycoff, *Arch. Ind. Hyg. Occup. Med.* **10**, 295 (1954).
74. H. A. Schroeder and co-workers, *J. Nutr.* **96**, 37 (1968).
75. A. P. DeGroot, *Voeding* **37**(2), 87 (1976).
76. M. Kanisawa and H. A. Schroeder, *Cancer Res.* **29**, 892 (1969).
77. O. J. Stone and C. J. Willis, *Toxicol. Appl. Pharmacol.* **13**, 332 (1968).
78. *Mutagenic Evaluation of Compound FDA 71-33, Stannous Chloride*, PB 245,461, prepared by Litton Bionetics Inc. for the FDA, National Technical Information Service, Springfield, Va., Dec. 31, 1974.
79. *Teratologic Evaluation of FDA 71-33, Stannous Chloride in Mice*, PB 221,780, prepared by Food & Drug Research Labs. for FDA, National Technical Information Service, Springfield, Va., Oct. 1972.
80. *Teratologic Evaluation of Compound FDA 71-33, Stannous Chloride in Rabbits*, PB 267,192, prepared by Food & Drug Research Labs. for FDA, National Technical Information Service, Springfield, Va., Sept. 17, 1974.
81. J. K. J. Lim, G. K. Hensen, and O. H. King, Jr., *J. Dent. Res.* **54**, 615 (1975).
82. R. C. Theuer, A. W. Mahoney, and H. P. Sarett, *J. Nutr.* **101**, 525 (1971).
83. H. G. Stoner, J. M. Barnes, and J. I. Duff, *Br. J. Pharm. Chemother.* **10**(1), 16 (1955).
84. F. J. C. Roe, E. Boyland, and K. Millican, *Food Cosmet. Toxicol.* **3**, 277 (1965).
85. M. Walters and F. J. C. Roe, *Food Cosmet. Toxicol.* **3**, 271 (1965).
86. A. P. DeGroot, V. J. Feron, and H. P. Til, *Food Cosmet. Toxicol.* **11**, 19 (1973).

87. D. M. Smith and co-workers, *A Preliminary Toxicological Study of Silastic 386 Catalyst*, Report LA-7367-MS, Los Alamos Scientific Laboratory, Los Alamos, N.M., June 1978.

88. M. P. Bigwood, P. J. Corvan, and J. J. Zuckerman, *J. Am. Chem. Soc.* **103,** 7643 (1981).

89. A. Sawyer, ed., *Organotin Compounds*, Marcel Dekker, New York, 1971.

90. R. C. Poller, *The Chemistry of Organotin Compounds*, Academic Press, New York, 1970.

91. W. Neumann, *The Organic Chemistry of Tin*, John Wiley & Sons, Inc., New York, 1970.

92. A. G. Davies and P. J. Smith, *Adv. Inorg. Chem. Radiochem.* **23,** 1 (1980).

93. H. G. Kuivila in J. J. Zuckerman, ed., *Organotin Compounds—New Chemistry and Applications*, *Advances in Chemistry Series No. 157*, American Chemical Society, Washington, D.C., 1976, p. 41.

94. Ref. 90, pp. 19–20.

95. Ref. 89, p. 666.

96. C. Eaborn, *J. Organomet. Chem.* **100,** 43 (1975).

97. O. A. Reutov, *J. Organomet. Chem.* **100,** 219 (1975).

98. K. A. Kocheshkov, *Ber.* **62,** 996 (1929).

99. K. A. Kocheshkov and co-workers, *Ber.* **67,** 717, 1348 (1934).

100. M. H. Gitlitz, ref. 93, p. 169.

101. U.S. Pat. 2,675,398 (April 13, 1954), H. E. Ramsden and H. Davidson (to Metal & Thermit Corp.).

102. A. P. Tomilov and I. N. Brago in A. N. Frumkin and A. B. Ershler, eds., *Progress in Electrochemistry of Organic Compounds*, Plenum Press, London, 1971.

103. G. J. M. van der Kerk, J. G. A. Luijten, and J. Noltes, *J. Appl. Chem.* 356 (1957); *Angew. Chem.* **70,** 298 (1958).

104. A. F. Finholt, A. C. Bond, Jr., K. E. Wilzbach, and H. J. Slesinger, *J. Am. Chem. Soc.* **69,** 2692 (1947).

105. E. R. Birnbaum and P. H. Javora, *Inorg. Synth.* **12,** 45 (1970).

106. R. F. Chambers and P. C. Scherer, *J. Am. Chem. Soc.* **48,** 1054 (1926).

107. D. Seyferth and S. B. Andrews, *J. Organomet. Chem.* **18,** P21 (1969).

108. U.S. Pat. 3,028,320 (April 3, 1962), P. Kobetz and R. C. Pinkerton (to Ethyl Corp.).

109. G. Mengoli and S. Daolio, *J. Chem. Soc. Chem. Commun.*, 96 (1976).

110. Ref. 90, p. 59.

111. U.S. Pat. 3,355,468 (Nov. 28, 1967), J. L. Hirshman and J. G. Natoli (to M&T Chemicals, Inc.).

112. K. Sisido, Y. Takeda, and Z. Kinugawa, *J. Am. Chem. Soc.* **83,** 538 (1961).

113. U.S. Pat. 3,607,891 (Sept. 21, 1971), B. G. Kushlefsky, W. J. Considine, G. H. Reifenberg, and J. L. Hirshman (to M&T Chemicals, Inc.).

114. G. J. M. van der Kerk and J. G. A. Luijten, *J. Appl. Chem.* **4,** 314 (1954).

115. *Ibid.*, **6,** 56 (1956).

116. K. Hartel, *Agric. Vet. Chem.* **3,** 19 (1962).

117. M. H. Gitlitz, *J. Coat. Technol.* **53**(678), 46 (1981).

118. N. F. Cardarelli, *Controlled Release Pesticide Formulations*, CRC Press Inc., Boca Raton, Fla., 1976.

119. C. J. Anthony, Jr., and J. R. Tigner, *Wire Wire Prod.* **43**(2), 72 (1968).

120. J. Burley, P. Hope, R. E. Hutton, and C. J. Groenenboom, *J. Organomet. Chem.* **170,** 21 (1979).

121. H. Schumann and M. Schmidt, *Angew. Chem.* **77,** 1049 (1965).

122. A. G. Davies and P. G. Harrison, *J. Chem. Soc. C*, 298, 1313 (1967).

123. U.S. Pat. 2,679,506 (May 25, 1954), E. R. Rochow (to M&T Chemicals, Inc.).

124. U.S. Pat. 3,519,665 (July 7, 1970), K. R. Molt and I. Hechenbleikner (to Carlisle Chemical Works, Inc.).

125. U.S. Pat. 3,901,824 (Aug. 26, 1975), V. Knezevic, M. W. Pollock, K.-L. Liauu, and G. Spiegelman (to Witco Chemical Corp.).

126. V. V. Pozdaev and V. E. Gel'fan, *J. Gen. Chem. USSR* **43**(5), 1201 (1973).

127. V. I. Shiryaev, E. I. M. Stepina, and V. F. Mironov, *Zh. Prikl. Khim.* (*Leningrad*), **45**(9), 2124 (1972).

128. S. Matsuda, S. Kikkawa, and I. Omae, *J. Organomet. Chem.* **18,** 95 (1969).

129. U.S. Pat. 3,440,255 (April 22, 1969), S. Matsuda and S. Kikkawa.

130. R. E. Hutton and V. Oakes, ref. 93, p. 123.

131. U.S. Pat. 4,130,573 (Dec. 19, 1978), R. E. Hutton and J. Burley (to Akzo, N.V.).

132. B. B. Cooray and G. Scott, *Dev. Polym. Stab.* **2,** 53 (1980).

133. G. Ayrey and R. C. Poller, *Dev. Polym. Stab.* **2,** 1 (1980).

134. H. O. Wirth and H. Andreas, *Pure Appl. Chem.* **49,** 627 (1977).
135. L. I. Nass, *Encyclopedia of PVC*, Vol. 1, Marcel Dekker, New York, 1976, pp. 313–384.
136. P. Klimsch, *Plaste Kautsch.* **24,** 380 (1977).
137. U.S. Pat. 4,028,337 (June 7, 1977), W. H. Starnes, Jr. (to Bell Telephone Laboratories, Inc.).
138. K. Figge and W. Findeiss, *Angew. Makromol. Chem.* **47,** 141 (1975).
139. W. H. Starnes, Jr., *Dev. Polym. Degrad.* **3,** 135 (1981).
140. D. Braun, *Dev. Polym. Degrad.* **3,** 101 (1981).
141. U.S. Pats. 2,219,463 (Oct. 29, 1940), 2,267,777 (Dec. 30, 1941), 2,267,779 (Dec. 20, 1941), and 2,307,092 (Jan. 5, 1943), V. Yngve (to Carbide & Carbon Chemicals Corp.).
142. U.S. Pat. 2,307,157 (Jan. 5, 1943), W. M. Quattlebaum, Jr., and C. A. Noffsinger (to Carbide and Carbon Chemicals Corp.).
143. U.S. Pat. 2,731,484 (Jan. 17, 1956), C. E. Best (to Firestone Tire & Rubber Co.).
144. U.S. Pat. 2,648,650 (Aug. 11, 1953), E. L. Weinberg and E. W. Johnson (to Metal & Thermit Corp.).
145. J. G. A. Luijten and S. Pezarro, *Br. Plast.* **30,** 183 (1957).
146. *Thermolite 813 PVC Stabilizer*, Technical Information Bulletin 273, M&T Chemicals, Inc., Rahway, N.J., Feb. 1979.
147. *Thermolite 831 PVC Stabilizer*, Technical Information Bulletin 316, M&T Chemicals, Inc., Rahway, N.J., Feb. 1979.
148. *Fed. Regist.* **33,** 16334 (Nov. 7, 1968); **37,** 5019 (March 9, 1972); **39,** 28899 (Aug. 12, 1974); **40,** 2798 (Jan. 16, 1975).
149. U.S. Pat. 3,424,717 (Jan. 28, 1969), J. B. Gottlieb and W. E. Mayo (to M&T Chemicals, Inc.).
150. U.S. Pat. 3,769,263 (Oct. 30, 1973), W. E. Mayo and J. B. Gottlieb (to M&T Chemicals, Inc.).
151. U.S. Pat. 3,810,868 (May 14, 1974), L. B. Weisfeld and R. C. Witman (to Cincinnati Milacron Chemicals, Inc.).
152. J. W. Burley, *Tin Its Uses* **111,** 10 (1977).
153. D. Lanigan and E. L. Weinberg, ref. 93, p. 134.
154. U.S. Pat. 4,080,363 (March 21, 1978), R. E. Hutton and J. W. Burley (to Akzo, N.V.).
155. S. Karpel, *Tin Its Uses* **125,** 1 (1980).
156. F. G. Willeboordse, F. E. Critchfield, and R. L. Mecker, *J. Cell. Plast.* **1,** 3 (1965).
157. L. R. Brecker, *Plast. Eng.*, 39 (March 1977).
158. R. V. Russo, *J. Cell. Plast.* **12,** 203 (1976).
159. J. Kenney, Sr., *Plast. Eng.*, 32 (May 1978).
160. *Fascat Esterification Catalysts from M&T*, Bulletin Quick Facts No. 45, M&T Chemicals, Inc., Rahway, N.J., 1978.
161. *M&T Fascat 4201 Organotin Esterification Catalyst*, Technical Data Sheet No. 345, M&T Chemicals, Inc., Rahway, N.J., 1978.
162. U.S. Pat. 4,170,579 (Oct. 9, 1979), J. F. Bosso and M. Wismer (to PPG Industries, Inc.).
163. M. Komura and R. Okawara, *Inorg. Nucl. Chem. Lett.* **2,** 93 (1966).
164. Brit. Pat. 1,146,435 (Oct. 7, 1966), P. A. Hoye (to Albright & Wilson, Ltd.).
165. U.S. Pat. 3,824,264 (July 16, 1974), E. J. Bulten (to Cosan Chemical Corp.).
166. E. J. Bulten, *J. Organomet. Chem.* **97,** 167 (1975).
167. E. J. Bulten, H. F. M. Gruter, and H. F. Martens, *J. Organomet. Chem.* **117,** 329 (1976).
168. J. W. Burley, R. E. Hutton, and V. Oakes, *J. Chem. Soc. Chem. Commun.*, 803 (1976).
169. U.S. Pat. 4,105,684 (Aug. 8, 1978), R. E. Hutton, V. Oakes, and J. Burley (to Akzo, N.V.).
170. R. E. Hutton, J. W. Burley, and V. Oakes, *J. Organomet. Chem.* **156,** 369 (1978).
171. U.S. Pat. 4,195,029 (March 25, 1980), E. Otto, W. Wehner, and H. O. Wirth (to CIBA-GEIGY Corp.).
172. U.S. Pat. 3,021,302 (Feb. 13, 1962), H. H. Frey and C. Dorfelt (to Farbwerke Hoechst A.G.).
173. *Fed. Regist.* **35,** 13124 (Aug. 18, 1970); **37,** 4077 (Feb. 26, 1972).
174. U.S. Pat. 4,130,673 (Dec. 19, 1978), W. Larkin (to M&T Chemicals, Inc.).
175. U.S. Pat. 4,144,362 (March 13, 1979), W. Larkin (to M&T Chemicals, Inc.).
176. G. H. Lindner, *Verres Refract.* **35,** 262 (1981).
177. C. Tamborski, F. E. Ford, and E. J. Soloski, *J. Org. Chem.* **28,** 181 (1963).
178. Ref. 177, p. 237.
179. K. Harrendorf and R. E. Klutts, *J. Econ. Entomol.* **60,** 1471 (1967).
180. K. R. S. Ascher and J. Moscowitz, *Int. Pest Control*, 17 (Jan.–Feb. 1969).

181. *Stannous Octanoate (Stannous 2-Ethylhexanoate)*, Technical Data Sheet 176, M&T Chemicals, Inc., Rahway, N.J., Dec. 1981.
182. S. L. Axelrood, C. W. Hamilton, and K. C. Frisch, *Ind. Eng. Chem.* **53,** 889 (1961).
183. *Fed. Regist.* **39,** 43390 (Dec. 13, 1974).
184. C. J. Evans, *Tin Its Uses* **89,** 5 (1971).
185. U.S. Pat. 3,578,616 (May 11, 1971), L. D. Harry and G. A. Sweeney (to the Dow Chemical Co.).
186. *Stannous Oxalate*, Technical Data Sheet 227, M&T Chemicals, Inc., Rahway, N.J., 1975.
187. U.S. Pat. 3,153,010 (Oct. 13, 1964), L. L. Jenkins and N. R. Congiundi (to Monsanto Co.).
188. U.S. Pat. 3,194,791 (July 13, 1965), E. W. Wilson and J. E. Hutchins (to Eastman Kodak Co.).
189. Hydrogenation of Coal in the Batch Autoclave, U.S. Bureau of Mines Bulletin 622, U.S. Bureau of Mines, Washington, D.C., 1965.
190. P. J. Smith, *Toxicological Data on Organotin Compounds*, ITRI Publication No. 538, International Tin Research Institute, Perivale, UK, 1977.
191. Ger. Pat. 63,490 (June 10, 1968), A. Tzschach, E. Reiss, P. Held, and W. Bollmann.
192. W. N. Aldridge, J. E. Casida, R. H. Fish, E. C. Kimmel, and B. W. Street, *Biochem. Pharmacol.* **26,** 1997 (1977).
193. R. D. Kimbrough, *Environ. Health Perspectives* **14,** 51 (1976).
194. H. G. Verschuuren, E. J. Ruitenberg, F. Peetoom, P. W. Helleman, and G. J. Van Esch, *Toxicol. Appl. Pharmacol.* **16,** 400 (1970).
195. Ref. 63, p. 1964.
196. Ref. 63, p. 1962.
197. Ref. 63, p. 1959.
198. *Fed. Regist.* **43,** 49575 (1978).
199. W. N. Aldridge, ref. 93, p. 189.
200. J. M. Barnes and H. B. Stoner, *Br. J. Ind. Med.* **15,** 15 (1958).
201. J. M. Barnes and L. Magos, *Organometal. Chem. Rev.* **3,** 137 (1968).
202. Ref. 63, p. 1956.
203. D. Lanigan and E. W. Weinberg, ref. 93, p. 135.
204. Ref. 91, p. 233.
205. L. B. Weisfeld, *Kunststoffe* **65,** 298 (1975).
206. *Occupational Health Guidelines for Organic Tin Compounds (as Tin)*, U.S. Department of Labor, Occupational Safety and Health Administration, Washington, D.C., 1978.
207. *Occupational Exposure to Organotin Compounds*, DHEW (NIOSH) Publication No. 77-115, U.S. Department of Health, Education, and Welfare, National Institute for Occupational Safety and Health, Washington, D.C., Nov. 1976.
208. C. R. Worthington, ed., *The Pesticide Manual*, 6th ed., British Crop Protection Council, Croydon, UK, 1979, pp. 142, 259, 266, 267.

General References

J. D. Donaldson, *A Review of the Chemistry of Tin(II) Compounds*, ITRI Publication 348, 1964.
Gmelin Handbook of Inorganic Chemistry, System No. 46, Pts. A–D (1971–1978); Pts. 1–8 (1975–1981), Springer-Verlag, New York.
C. L. Mantell, *Tin—Its Mining, Production, Technology and Applications*, 2nd ed., Hafner Publishing Co., New York, 1970, pp. 407–449.
J. W. Mellor, *A Comprehensive Treatise on Inorganic and Theoretical Chemistry*, Vol. 7, Longmans, Green & Co., New York, 1937.
Tin Times, International Tin Research Institute, Middlesex, England, Nos. 1–6, 1981.
A. F. Trotman-Dickenson, ed., *Comprehensive Inorganic Chemistry*, Vol. 2, Pergamon Press, Oxford, 1973, pp. 43–104.
J. W. Price and R. Smith in W. Fresenius, ed., *Tin, Handbook of Analytical Chemistry*, Vol. 4a, Pt. 3, Springer-Verlag, New York, 1978.

MELVIN H. GITLITZ
MARGUERITE K. MORAN
M&T Chemicals, Inc.

TIRE CORDS

A pneumatic tire is a remarkable engineering achievement; many of its properties are derived from the high strength cords it contains. A tire cord gives the tire its size and shape, bruise and fatigue resistance, and load-carrying capacity. It also affects the wear, comfort, and maneuverability of the tire. Yet tire cords comprise only ca 10 wt % of tires.

When R. W. Thompson patented the first pneumatic tire in the UK in 1845, he used textile fabric as the strength member. The term tire cord came into usage in 1893 when tire cord was patented in the UK and the United States.

After World War I, with the ascendancy of the automobile in the United States, square woven cotton tire fabric was replaced by several bias plies of warp fabric made with cotton cords and held together by a few cotton picks and carcass rubber.

In the 1930s, rayon (qv) was introduced for tire application. However, it lacked the adhesion of cotton to the carcass rubber. This problem was solved with the development of the resorcinol–formaldehyde–latex (RFL) adhesive system (1) (see also Phenolic resins).

Although nylon-6,6 had been developed by DuPont in the 1930s, it was not until World War II that it was tried in airplane tires where cost was less important than toughness and light weight. Again adhesion was a problem, but RFL with poly(2-vinylpyridine) latex as part of the latex component improved adhesion (2). Both nylon-6,6 and nylon-6 bias tires had a drawback called flatspotting, which was unacceptable to automobile manufacturers, and rayon-bias tires were preferred during the 1950s and 1960s.

Partially because of the competition between nylon and rayon, properties of tire cords were continually improved (3–4); for example, two-ply rather than four-ply rayon tires were successfully commercialized in 1961. They remained standard original equipment on most passenger cars through 1969.

The flatspot resistance of nylon was improved in a mixture of 70 parts nylon-6 and 30 parts polyester, trade named EF-121 (5), or a melt blend of 70 parts nylon-6,6 and 30 parts poly(hexamethyleneisophthalamide), trade named N-44 (6). However, these yarns were rejected for original-equipment passenger tires because of flatspotting and poor thermal stability during processing (see Polyamides; Polyesters).

Although polyester appeared to be the best replacement for nylon in tires, it proved difficult to obtain good cord-to-rubber adhesion. For mass production, water-based adhesives were needed, and several two-step and one-step dip systems were developed (7). In the early 1960s, polyester passenger tires were introduced. They gave excellent field results, but attempts to extend their use to airplanes or trucks did not succeed because of the poor dynamic performance of polyester at temperatures above those normally found in passenger tires, eg, 82°C, which is the glass temperature T_g of polyester. However, various process modifications have permitted the use of polyester as the carcass reinforcement material for light truck tires and radial truck tires.

The introduction of the radial and belted-bias constructions had a tremendous influence on tire-cord reinforcement. In the radial tire, one or more plies of carcass cords lie radially. In addition, several plies of cords, in the form of a belt, are placed under the tread in such a manner that the belt cords lie in an angle close to the circumferential direction (see Fig. 1). The Michelin Company of France was the first to commercialize this type of tire construction in 1947 (8). Their passenger tires had rayon cords in the carcass with a special three-ply wire belt, and later they introduced all-wire radial truck tires. Radial tires have several advantages over bias tires, eg, up to 100% more tread wear, 7–10% less rolling resistance, excellent bruise resistance in the tread region, greater cornering ability, good control during nibbling (eg, driving on or off the edge of the highway), and generally greater endurance because of lower operating temperatures. Their disadvantages are mainly lower bruise resistance of sidewalls, more difficult retreadability, poorer high speed performance, and poor ride characteristics with soft automobile suspensions. Although rayon belts offer better ride and high speed endurance than steel belts, the latter have longer tread life and better bruise resistance and cornering performance.

The radial constructions became very popular in Europe. The preferred reinforcing materials for passenger tires were rayon in the carcass and rayon or steel in the belts. All steel was preferred in heavy-duty tires.

Although U.S. manufacturers recognized the advantages of the radial construction and BF Goodrich introduced an all-rayon radial passenger tire to the U.S. market, no rapid changeover from bias to radial tires took place because of the tremendous capital investment required. Instead, belted-bias tires, which could be manufactured on existing equipment, were developed.

Belted-bias construction is more efficient than simple bias construction. It allows the use of high modulus cords such as glass to reinforce the crown region of the tire, which increases tread life up to 150% of that of conventional bias tires. The conversion to the belted-bias construction was facilitated by the development of a new high modulus glass tire cord. Owens-Corning introduced glass as a tire cord in a composite impregnated with 15–30% RFL (9), which coated and protected the individual glass filaments. The first commercial glass belted-bias tires had a nylon carcass and a glass belt. Other high modulus materials like rayon, vinal (vinyl alcohol polymer fibers), and steel were also used. Because of the success of polyester as a flatspot-resistant carcass material for bias tires, it became the preferred cord for belted-bias carcasses.

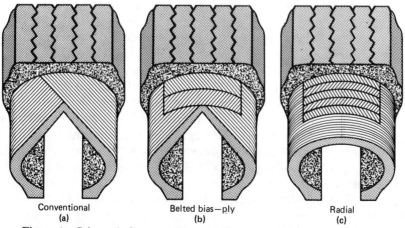

| Conventional | Belted bias—ply | Radial |
| (a) | (b) | (c) |

Figure 1.   Schematic diagram of tire cords in commercial tire constructions.

After the 1973 oil embargo, new motor vehicles were equipped with radial tires in order to increase fuel efficiency, and today radial tires dominate the market.

The aramid fiber, Kevlar, developed in the 1970s, was a breakthrough in fiber technology (10) (see Aramid fibers). This aramid has a tensile strength of 1.98 N/tex (22 gf/den), over twice that of nylon, and a tensile modulus approaching that of steel. It is the only organic textile fiber developed for reinforcement. It is used in the belts of a few premium radial passenger tires and appears to offer advantages in the carcass of radial heavy-duty tires. If its high strength could be fully utilized in a tire cord, it should permit lighter tires, save materials, reduce processing energy, and improve tire performance.

New forms of vinal, a textile fiber developed and used in Japan since the 1950s, have tensile strengths and moduli considerably higher than nylon, rayon, and polyester. Although vinal has been described as a super rayon, it is more sensitive than rayon to moisture at high temperatures. Vinal tire cord acts as reinforcement in both radial carcass and radial-belt tires.

Many methods have been suggested for making radial tires by a one-step process in order to use the conventional, bias-building and curing equipment. For radial passenger tires, a slip technique was developed for sliding the green belts and tread over the carcass during the tire lift into the curing press. For large tires, this technique was impractical. However, the invention of stretch cord permitted BF Goodrich to make large radial farm tires with circumferential rayon stretch-cord belts on their conventional tire-building equipment (11).

Economic Aspects

During 1981, ca 3×10^5 metric tons of tire cords (almost one third of which was steel) was consumed in tire manufacture in the United States; world consumption exceeded 4.5×10^5 t.

Cotton was replaced by rayon which, in turn, was replaced by nylon (see Fig. 2). Actually, polyester replaced rayon, whereas use of glass and steel increased as the materials for the belted-bias and radial tires. In the United States, rayon still retains a small share of the market, although a larger market exists in Europe. Both nylon and polyester seem to have reached a plateau at 9×10^4 t/yr. Glass consumption remained steady at 16,000 t/yr; it is expected to grow slowly as it replaces some steel and perhaps more of the rayon in the belts of radial passenger tires. Steel consumption continues to grow in both the carcass and belts of truck and heavy-duty radial tires at the expense of nylon. The auspicious entry of aramid into the tire-cord market has so far been cut short, but prospects are good if the price becomes competitive.

In order to compare the economic value of one fiber to another, the cost of building equivalent tires from each fiber should be compared. However, there is no single property for all fibers that limits the amount needed in any equivalent tire. For example, in comparing rayon with nylon, the room-temperature tenacity indicates that 0.6 kg nylon is equivalent to 1 kg rayon; however, when the two kinds of cords are tested at high speeds and elevated temperatures (110°C), 0.8 kg nylon is equivalent to 1 kg rayon. Therefore, room-temperature tenacity alone cannot be used for accurate cost comparison of these tire cords. Yet, tenacity is one of the main performance criteria, and it is often used to compare cost. The 1982 U.S. tire-yarns price information is given in Table 1.

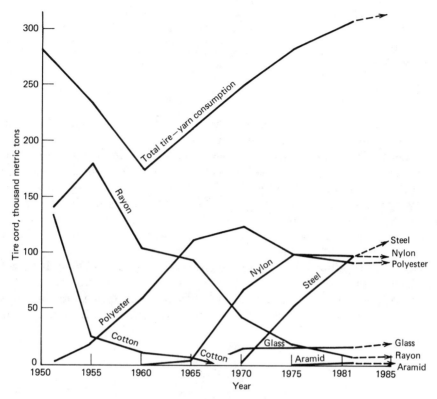

Figure 2. Tire-yarn consumption in the United States. The arrows indicate the estimated direction on the curves in the next few years (12).

Table 1. Cost per Tenacity

Type of yarn	Yarn cost, $/kg	Twist and dip cost[a], ¢/kg	Tenacity of cord N/tex[b]	Dipped cost to tenacity ratio, $/(N/tex)[b]
rayon	3.26	26.4	0.36	9.77
nylon-6,6	3.72	33	0.65	6.23
polyester	2.97	44	0.59	5.79
glass (dipped)	2.97		0.56	5.3
steel wire				
passenger tires	2.40		0.30	7.99
truck tires	2.64–4.09		0.30	8.80–13.60
aramid	13.31	44	1.50	9.17

[a] Estd.
[b] To convert N/tex to gf/den, multiply by 11.33.

The cost of the fiber reinforcement in a tire and its performance in a particular design are the principal criteria of choice. Other marketing considerations are availability and the public's perception of a tough and strong material. Since cost per unit tenacity is important, glass is utilized wherever possible. Although the cost–tenacity ratio of low-twist nylon is below that of polyester, development costs have so far prevented commercialization of the former in heavy-duty radial tires, where it has the best chance to succeed. Steel wire offers many advantages in radial tires which explains its growing usage despite its very high cost–tenacity ratio. The slow growth of aramid is due to its high cost and poor strength utilization at the higher twists needed for adequate fatigue resistance.

Processing

Most tire companies produce tire cord from high tenacity, continuous-filament yarn supplied by various fiber manufacturers.

Twisting. Typically, nylon yarn is received on small beams (rolls) with ca 200 ends of 186.7 tex (1680 den) yarn. These yarns are twisted to 394 turns per meter (tpm) in the Z direction (spiraling counterclockwise). Then, two spools of twisted yarn are back-twisted together to 354 tpm in the S direction (spiraling clockwise) to make 186.7/2 (1680/2) nylon tire cord (the number in the denominator indicates the number of plies) (see Fig. 3). If 186.7/3 cord is desired, 3 spools of Z-twisted yarn are back-twisted together in the S direction.

Weaving. Twisting is followed by loosely weaving by 1000–1500 cords (warp) with $1/3$–1 pick (or weft) per centimeter of light cotton, rayon, polyester–cotton, or nylon yarn threads. A type of fabric is obtained in which the cords in one direction, the warp, are far stronger than the threads in the other direction, the weft. In fact, the weft only

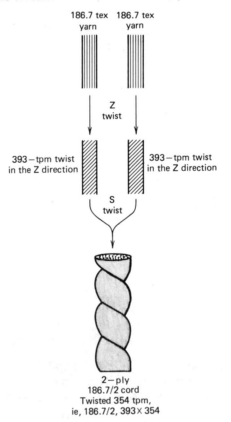

Figure 3. A schematic example of how tire yarn is twisted into tire cord (tpm = turns per meter; 186.7 tex = 1680 den).

serves to give the construction the minimum coherence necessary for the subsequent operations. Rolls of this fabric are then dipped in adhesive, baked, and heat set.

For radial-tire production a stretch pick (Olboplast 25) has been commercialized by Olbo Textilewerke GmbH in the FRG. This material consists of partially oriented, nylon-filament yarn twisted together with cotton fibers for stability during processing. When the so-called green tire plies are lifted into the curing press, enough extension is generated to break the cotton portion of these picks first, and then the pick plus the rubber plies continue smoothly to extend into the mold. The cords separate uniformly and the spread-cord fault is avoided.

Dipping, Heat Setting, and Calendering. A simplified processing train for tire cord is shown in Figure 4. Flexible controls of heat, tension, and speed are required in each processing zone. The fabric, taken off from a roll of 1825 m length, is spliced onto the tail received from the previous roll either with adhesive or with a high speed sewing machine. Continuous operation is maintained during splicing by the fabric stored in the festoon. Then, under controlled tension or controlled length change, the fabric is dipped in an aqueous adhesive RFL dip. After dipping, the fabric passes vacuum suction lines or rotating beater bars to remove the excess dip before entering a drying oven where the water is removed. The dried fabric is baked and heat-set under tension at 175°C for rayon, 230°C for nylon, 245°C for polyester, and 260°C for aramid. After stretching and heat setting, the viscoelastic fabrics (nylon or polyester) are usually relaxed at 175°C under a tension lower than the heat-setting tension. The relaxed fabric is covered on both sides with a precise amount of rubber by a four-roll calender. Tension can also be controlled on viscoelastic fabrics as they are cooled after calendering and before reaching the take-up festoon. At the take-up, a removable liner is inserted to prevent the rubberized fabric from sticking to itself. Usually the dipped, heat-set, and baked fabric is not calendered immediately, but taken up into large rolls, wrapped to protect it from light, ozone, and moisture, and shipped to tire plants for calendering. Before actually calendering the precise amount of rubber onto the fabric, it is dried by reheating to provide better final adhesion.

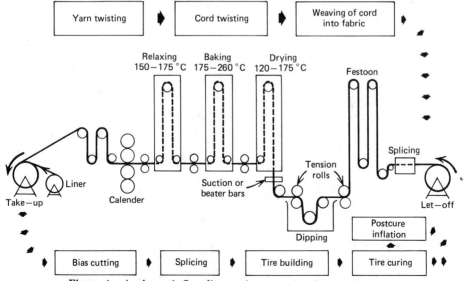

Figure 4. A schematic flow diagram for processing tire yarn into tires.

Typical formulations of aqueous adhesive dips for the organic tire cords are shown in Table 2. For rayon, vinal, and nylon, similar RFL dips are employed in a single-step treatment, whereas polyester and aramid are usually treated with an epoxy-blocked isocyanate dip followed by an RFL dip. Adhesion-activated polyesters reduce the need for the first-step treatment (15), but the strength of adhesion is generally lower than that obtained from the two-step treatment. Several other one-step adhesive systems develop adequate but usually weaker adhesion (16–17). For aramids, a simple first-step dip of epoxy has been recommended by DuPont (18); it avoids the high baking temperatures required to unblock the isocyanates. The serious problem of adhesion degradation caused by exposure of the RFL coating to ozone and humidity has been greatly reduced by the partial substitution of acrylic resins for the RF resin in the basic dip (19). Waxes that tend to bleed to the surface and protect the RFL coating have also been proposed as an RFL modification (20); they are usually contained in glass cords. Further discussion of adhesive dips for tire cords is given in refs. 13, 15, 21–22.

Cutting and Splicing; Tire Building. The calendered fabric is cut to the required angle and width, and the pieces are butt-spliced together automatically into a continuous sheet.

The sheet of parallel cords arranged at the proper angle is cut or torn to the proper length and built into the so-called green tire on a cylindrical drum; the other tire components are included here, eg, beads, chafer, sidewalls, belt plies, cushion gums, undertread, and tread. The green bias or belted-bias tires are then ready for cure or vulcanization.

Table 2. Typical Tire-Cord Adhesive Dips

Dip and composition	Dry, parts	Wet, parts
D-417[a]		
Hylene MP[b] (40% dispersion)	3.60	9.00
NER-010A[c]	1.36	1.36
gum tragacanth (2%)	0.04	2.00
water, soft		87.64
Total	*5.00*	*100.00*
D-5 ammoniated[d]		
resorcinol	11	11
formaldehyde (37%)	6	16.2
sodium hydroxide	0.3	0.3
water, soft		238.5
Total	*17.3*	*266.0*
Final nylon dip		
Gen-Tac (41%)	100	244
resin master	17.3	266
water, soft		60
ammonium hydroxide (28%)		11.3
Total	*117.3*	*581.3*

[a] DuPont. For first-step polyester or aramid (13). Total solids = 5.0%; pH min = 10.3.

[b] Hylene MP is phenol blocked methylenebis(4-phenyl isocyanate) (DuPont).

[c] NER-010A is glycerol polyglycidyl ether, water-soluble polyepoxide (Nagase and Co., Ltd.).

[d] General Tire & Rubber Co. For nylon, rayon, and second step polyester or aramid (14). Total solids = 5.4%; pH min = 10.

Radial tires usually require a second step after the radial carcass has been built onto the cylindrical drum. The carcass is lifted into a toroidal shape before applying the belt plies, cushion gums, undertread, and tread to the green tire. During this process, the carcass cords must separate evenly in order to avoid a blowout and form a uniform tire. This spread-cord fault is minimized in a carcass rubber with high green strength and when stretch picks are used in the fabric.

Curing and Postcure Inflation. The green tires are cured in automatic press molds at 165–180°C and under 1.48–2.86 MPa (200–400 psig). After a minimal curing time of ca 12 min for passenger tires, the tire is automatically ejected and moved to the postcure inflators. Upon release from the press molds, the tire cords are almost completely free to shrink. The tire is reinflated to 0.31–0.51 MPa (30–60 psig), which reloads and stretches the cords. At the same time, cure of the compounded rubber in the tire continues. Control of the postcure inflation process is very important for viscoelastic cords like nylon or polyester because the growth, groove-crack resistance, size, shape, and uniformity of the tire depend on the time, temperature, and load on the cords during postcure inflation. In bias and belted-bias tires, postcure inflation is necessary for polyester tires in order to reduce initial heat generation and increase endurance life. Radial tires, on the other hand, do not need postcure inflation since the stiff belt prevents excessive distortion which may be caused by shrinkage of the viscoelastic cords in the carcass. In fact, the quality of the radial tire sometimes suffers slightly from postcure inflation since it tends to emphasize undulations in the sidewall at splices and uneven cord spacings.

Storage and Shipping. Proper storage and shipping of tire cord is vital to the quality of the fabric and therefore the quality of the tire. The fabric rolls must be packaged to withstand both mechanical and environmental damage. Polyethylene bags, kraft paper, and heavy cardboard are generally employed for packaging. Silica gel reduces the moisture content in the packages and prevents shrinkage of exposed edges of the fabric (so-called tight edges). Tight edges in a roll of greige, dipped, or calendered fabric result in handling difficulties, scrap, and nonuniform tires. For overseas shipment, the fabric rolls are boxed.

Glass and Steel Tire Cords. The processing of glass and steel tire cords is similar to the processing of polymeric tire cords once they are prepared for calendering. However, preparation for adhesion to the rubber compound is different.

Large spools of glass tire cords completely impregnated with a modified RFL are obtained from the supplier. The RFL solids content is >15 wt % in order to thoroughly coat each glass filament. The cord constructions for the belts of bias-belted and radial tires are shown in Table 3.

Steel tire cord is supplied to the tire industry in various constructions. The cords are made from finely drawn high carbon (0.67 wt %) steel and brass plated just before the last draw. The brass coating acts as both a drawing lubricant and an adhesive to the rubber. These wires are twisted together in a closing operation by either cabling or bunching and thus form the appropriate lay length (twist in millimeters per turn) of the strands or cords. Lay lengths of the strands and cords are chosen to hold the wires together and yet obtain the smallest degree of fretting during tire operation. A spiral wrap of a single wire (transfil) is frequently applied to the steel cords in order to increase the cord's compression resistance and bending rigidity. However, this wrap decreases the elasticity of the cord.

For radial passenger tire belts, the 4×0.28 cord has evolved into the preferred

Table 3.   Glass[a] Cord Constructions Used in Belted-Bias and Radial Tires

Year introduced	Tire design	Cord construction[b]		Impregnant[c]
1966	belted bias	G 150	10/0	063T
1968	belted bias	G 75	5/0	065T
1970	belted bias	G 15	1/0	075T
1970 (unsuccessful)	radial	G 15	3/0	075T
1976	belted bias	H 15	1/0	086T
1976	radial	H 15	3/0	086T

[a] Type E glass, a high strength glass with a low alkali content and a high alumina content.

[b] G and H designate filament diameters, where G = 8.89 μm dia with 204 filaments per dia, and H = 10.12 μm dia with 158 filaments per dia. The twist of all the above strands is 59 twists per meter. The remaining figures indicate cord length per weight of strand and the number of strands per ply. For example, the first entry describes a cord with G filament dia (8.89 μm dia with 204 filaments), 150 yds/lb (302.4 m/kg) of strand, and 10 strands per ply (10/0).

[c] All impregnants are modifications of RFL with code numbers of Owens-Corning Fiberglass Corporation. Code 086T has a latex with a significantly lower T_g. This improved its low temperature properties so that radial glass belts are operable in cold weather.

construction from the 5×0.25 construction. As shown in Table 4, the 5×0.25 cords tend to form into a pentagonal cross section with the five wires touching each other and sealing off a large hole in the center. The rubber is thus prevented from penetrating into the central hole, but moisture can collect there during cure and reduce the initial adhesion. Furthermore, upon exposure of the tire to high humidity, this hole provides easy access for moisture, which then degrades the adhesive bond. When the tire is accidentally cut, corrosion from salts and water spreads from this hole. The 4×0.28 construction costs less and tends to collapse into a close-packed parallelogram which permits more rubber to flow into the interstices. This flow, in turn, permits more complete surface contact for adhesion and blocks the migration of moisture from adversely affecting adhesion. A more expensive construction that allows better penetration of the rubber is the 2+7×0.22+1×0.15 construction popularized by Firestone. Radial truck tires use larger constructions such as 3×0.20+6×0.35 or 3+9×0.28+1×0.15 in the belt and 3+9+15×0.175+1×0.15 in the carcass.

Steel-wire tire cords have been coated with a thin layer of brass to promote adhesion to the rubber compounds for many years. Yet, experts still speculate as to the exact mechanism of adhesion and adhesion degradation in the presence of moisture.

Such speculation should explain the following phenomena:

1. Maximum adhesion is obtained when free sulfur is present in the initial rubber compound.

2. The maximum initial adhesion is obtained when the rate of cure of the rubber compound is optimized to the rate of adhesion development.

3. Heat aging degrades adhesion.

4. Moisture tends to lower initial adhesion and increase the rate of adhesion degradation.

5. Higher zinc to copper ratios in the brass tend to retard the rate of adhesion development and the rate of adhesion degradation owing to heat aging or moisture.

6. Thin brass coatings tend to resist adhesion degradation better than thick coatings.

Table 4. Nomenclature of Popular Wire Construction[a,b]

construction	5 × 0.25	4 × 0.28	2 + 7 × 0.22 + 1 × 0.15	3 × 0.20 + 6 × 0.35	3 + 9 + 15 × 0.175 + 1 × 0.15	3 × 7 × 0.22 (high elongation, H.E.)
cross section						
lay length, mm/turn	10S	12.5S	6.3S/12.5S/3.5Z	10S/18Z	5S/10S/16Z/3.5S	4S/7S
breaking load, N	620	620	880	1,620	1,730	1,710
tensile modulus, MPa[c]	196,200	198,200	203,000	185,500	193,500	93,000
compression modulus, MPa[c]	47,100	78,100	99,000	93,000	68,000	7,800
principal use	passenger belt	passenger belt	passenger belt	truck belt	truck carcass	off-the-road (OTR) belt

[a] Ref. 23.

[b] ASTM has adopted a wire nomenclature system. The objective of this system is to describe the wire cord construction in the same manner as it is manufactured. The general format is as follows:

core (S × F × D) + intermediate part (S × F × D) . . . + outermost part (S × F × D) + spiral wrap (S × F × D)

Where S = number of strands (when S = 1, it is omitted); F = number of filaments; and D = nominal diameter of filaments. Various parts of the cord are separated by a "+" sign. If the filament diameter is the same for two or more parts (with the exception of the wrap), omit the diameter except for the last part before the change. If the filament diameter of the wrap is different than the filament parts (with the exception of the wrap), omit the diameter except for the last part before the change. If the filament diameter of the wrap is different than the filament in the cord, the diameter is given twice. The general rules are to work from the center of the cord to the outside. Various parts to the nearest 0.01 mm.

[c] To convert MPa to psi, multiply by 145.

7. Cobalt ions in the rubber compound tend to retain the maximum obtainable level of adhesion for the system more consistantly than without it. Cobalt also improves adhesion to zinc.

8. Copper tends to migrate into the rubber, especially in the presence of moisture.

Two separate articles have been written that try to explain the above adhesion phenomena in light of new data obtained by such modern analytical techniques as esca (24–25).

The principal problem of adhesion degradation, especially in the presence of moisture, has been overcome through improvements in rubber compounds; by avoidance of exposure to moisture of the uncured tire in the tire plants; by development of wire construction that increase rubber penetration; and by development of modified brass coatings, eg, an alloy of 70 wt % Cu–4 wt % Co–26 wt % Zn (25).

Tire Performance Related to Tire Cords

Tire performance characteristics are affected by the physical properties of the tire cords. Although a few characteristics such as burst strength, tire size, and shape are amenable to mathematical analysis based on the physical properties of tire cords (26–30), most relationships are determined empirically and correlated statistically. Tire performance is difficult to define. For example, a tire must safely support a specified load under dynamic conditions with a minimum of power loss, overcome minor obstacles and provide a safe endurance life even beyond its normal life expectancy. And a tire must provide maximum wear, a smooth and quiet ride with good cornering, traction, and skid resistance under various road conditions. A tire must also have a pleasing appearance to complement the vehicle.

Tire performance is therefore measured in a variety of ways and under a variety of conditions in the laboratory, on indoor roadwheels, on outdoor tracks, and on special test fleets. Before commercialization in the United States, a tire has to pass certain qualification tests set up by the tire companies, the Rubber Manufacturers Association (RMA), the DOT, and the original equipment manufacturers, eg, the automobile companies. Competition among tire manufacturers has continually raised the standards of tire performance.

The tire performance most difficult to measure is the endurance life. In order to measure it quantitatively and empirically relate it to the physical properties of the cord, the test tires have to be run to failure in a reasonable length of time. Testing conditions are much more severe than those any tire would normally undergo. Therefore, the accelerated endurance life of a tire usually is measured in combinations of overloads up to 150% of the recommended loads, inflation pressures as low as 75% of the recommended pressure, and speeds >161 km/h. Tests are usually taken on indoor roadwheels where conditions can be most readily controlled. However, prolonged outdoor-track tests are run before an improvement of tire performance is accepted.

Burst Strength. The tensile strength is related to the burst strength by the following relationship (31–32):

$$\text{burst strength} = N t_u K = \frac{\pi P_B (r_c^2 - r_{max}^2)}{\sin \alpha}$$

where N = total number of cords in a tire; t_u = average ultimate tensile strength of the cords tested at the same rate of elongation as the tire; P_B = burst pressure; r_c = radius from the center of rotation to the crown of tire; r_{max} = radius from the center of rotation to the maximum section width of a tire; α = crown angle between the cord path and the circumferential plane through the crown of a tire; and K = efficiency factor which depends on the distribution of ultimate cord strengths, always <1 in an assembly of cords.

The actual burst strength is always somewhat lower than the calculated strength because the tensile strength and stress distribution among all the cords are never exactly the same. The safety factor used in designing most tires is >10. Thus, a tire normally inflated cold to 274 kPa (25 psig) resists more than 1.82 MPa (250 psig) before burst. When extreme operating temperatures reach 135°C, most organic fibers temporarily lose almost half of their tensile strength, still leaving a safety factor of 5 which is more than adequate to allow for minor weaknesses and normal degradation from oxidation, fatigue, etc.

Bruise Resistance. Years ago, a tire would blow out when the car was driven over a 15-cm curb at 25–30 km/h. Today, the steel rim may bend, but the tire remains intact even at speeds up to 100 km/h. Thus, bruise resistance has been increased by improved tire cords and tire design and is maintained by quality-control tests.

Bruise resistance is usually tested by measuring the energy required to penetrate a 19-mm cylindrical plunger at a rate of 51 mm/min crosshead speed into the crown of an inflated passenger tire at room temperature. The area under the load–elongation curve is interpreted as the slow penetration bruise or bruise energy of a tire. Since no rigorous mathematical analysis of this test has as yet been made, the correlations with the tensile strength or the tensile product (tensile strength × ultimate percent elongation) of the cord are not exact.

Bruise resistance should be measured at different operating conditions for different materials (33). For example, rupture energy tests at high speeds and elevated temperatures show that rayon resembles nylon at ca 135°C and 6000% elongation per second (34–35). Apparently, rayon dries out, which strengthens it and reduces the normal strength loss expected from increased temperatures. Furthermore, the strengthening effect of high speed straining generally is greater with the high modulus materials. Both of these effects favor rayon over nylon.

A number of high speed bruise tests have been developed which correlate better with field experience than the low speed test (36–38). The DOT has recognized the difference in nylon and rayon tires and specifies that rayon tires with a low speed bruise energy of 186.4 J (137.5 ft·lbf) are equivalent to nylon tires with a low speed bruise energy of 293.8 J (217 ft·lbf). Although high speed tests are more realistic, low speed tests are simpler and faster and are used for the quality control of tires.

Tire Endurance (Separations). Several different types of separations related to tire cords can occur. Most separations are due to poor adhesion between the cord and the carcass rubber. Initial adhesion between cord and rubber are important, as well as uniformity and adhesion retention. Adhesion uniformity is more important than a high average adhesion since unevenness causes stress concentrations which tend to promote separations.

Degradation of the adhesive system tends to occur because the system is subject to elevated temperatures, moisture, oxygen, and stress–strain cycling. High temperatures weaken all the materials in the system and increase their rate of degradation from oxygen or moisture. Oxygen tends mainly to degrade the rubber portion of the system. Moisture tends to hydrolyze the polyester cords, degrade the brass–rubber bonds of steel wire cords, or swell rayon cords which then tends to increase the stress at the cord–rubber interface and reduce the hydrogen bonds between the adhesive and the cord's surface.

Another separation phenomenon is fatigue failure. At speeds >145 km/h or under loads of ≥20% over those recommended, or inflation pressures of ≥34.5 kPa (5 psi) below those recommended, fatigue failure can occur in the sidewall near the tire-tread shoulder or near the bead where the cord plies end. Both are regions of relatively high stress concentrations and heat generation. When a tire is operated under severe conditions for ≥80 km, it can reach temperatures over 121°C. The combination of heat, oxygen (mainly from the internal air inflation), and stress–strain cycling causes the cord, adhesive, and especially the rubber matrix to degrade and weaken. Continued operation of the tire under such conditions initiates small internal flaws, which then grow in the weakened material and ultimately lead to cord breakage and loss of inflation pressure. Fatigue separations result from interrelated factors that depend on tire-cord construction and design, adhesion, temperature, rubber compound, oxygen concentration, stress concentrations, and severity of the tire's operating conditions (39–41).

High Speed Endurance.   During operation, a tire continually stores and releases mechanical energy. Any loss of useful energy develops heat. At very high test speeds (up to 193 km/h), a tire sometimes generates enough heat to cause a tread separation or chunk-out, often starting at a small internal flaw in the hottest and thickest part of the tire, usually the tread shoulder. The heat generated by the tire cords in a bias tire is sufficient to significantly affect the maximum operating speed of the tire. Thus, at high speeds nylon bias tires usually run cooler than rayon bias tires before tread separations occur, whereas rayon radial tires run cooler than rayon bias tires of the same size. The radial tire usually tolerates slightly higher loads than bias tires of the same size but not higher speeds.

Separations are also caused by stress. At high speeds, a tire can go into resonance called a standing wave which violently distorts the tire, causing large stresses and high temperatures to be generated in the regions that distort the most. For passenger bias tires, the onset of resonance generally begins in the crown at speeds >225 km/h. With special designs, onset speed can be increased to >400 km/h. The radial tire is more sensitive to speed. The onset of resonance for a passenger radial tire generally begins in the sidewall at speeds >113 km/h; however, the distortions in the sidewall are mild at first and develop slowly into severe distortions at speeds >137 km/h. Again the onset speed can be raised to much higher values by changing the tire design. For instance, by increasing the resonant frequency of the sidewall (shorter and stiffer sidewall) or reducing the centrifugal force (lower tread or belt weight), the resonant-onset speed can be significantly increased.

Power Loss.   In a bias tire, 30–40% of the total power loss, a significant portion of the heat generated, is because of the carcass cords (42). Measurable temperature differences have been found among tires made with different kinds of cords (43–45).

Since power loss is synonymous with heat-generation rate, similar tires made from different cords should show equilibrium operating temperatures corresponding to the heat-generation rates of the cords. However, it is almost impossible to make truly equivalent tires from different cords. A good comparative test demonstrated that under severe conditions, nylon-6-bias truck tires operate 10–25°C cooler than those made of nylon-6,6. These differences apparently correspond mainly to the relative rates of heat generation of the cords.

Tread Wear. Tread wear partially depends upon load distribution and movement (scuffing) of the tread elements in the tire print. Since the cord modulus contributes to the stiffness and load distribution of the tire print, a higher cord modulus implies less tread wear and longer tread life. Belts improve tread-wear resistance. Ranking the relative wear-rating properties based on the bias tire as 100, the steel-belted radial is ca 200, and the textile-belted radial ca 160 (30,46). Glass-belted radials should rank between the steel and textile belts.

Tire Size and Shape. The size and shape of an inflated tire depends upon the design of the tire and the modulus of the cord. In addition, some tire growth occurs while the tire is inflated over a long period of time, depending on the creep properties of the cord as affected by the inflation load, tire temperature, and moisture regain (plasticization). Control of moisture regain is especially important for moisture-sensitive rayon and nylon. Because cord processing affects modulus and creep, careful process control must be maintained in order to obtain uniform size and shape.

Tire-Groove Cracking. When the tread grooves of a tire are subjected to excessive strain, small cracks appear in the rubber at the bottom of the grooves, possibly from ozone attack. Cords with less creep produce less groove cracking. Postcure inflation of nylon tires reduces tire growth and the tendency of the grooves to crack.

Flatspotting and Tire Nonuniformity. Thermoelastic tire cords such as nylon and polyester tend to shrink readily when they are heated above their T_g. A running tire generates heat, which increases the temperature above the T_g of its cord (usual maximum operating temperature of a passenger tire is 77°C). When the tire is stopped, the cords in the tire-print region are essentially unloaded and free to shrink. The cords in the remainder of the tire are prevented from shrinking by the inflation pressure. When the tire cools below the cord T_g, the cord lengths become frozen, and a flatspot or out-of-roundness remains where the cords shrank. It remains until the tire is reheated to the temperature at which the flatspot was originally introduced. Above that temperature, cord lengths are again equal (47–48).

In a similar manner, a nonuniformity develops in a tire that is postcure inflated hot, and cooled nonuniformly, followed by release of postcure inflation pressure above the T_g of the cord. The hotter cords shrink more than the cooler cords, and the tire is distorted and nonuniform (47). In this simplified analysis, the effect of long-term creep is not considered.

Tire-Cornering Force. Generally, the higher the cord modulus, the stronger the cornering force developed at any particular steering angle. However, tire design changes, eg, in the cord-crown angle, can often overshadow the modulus effect.

Spring Rate. Spring rate is a measure of the tire deflection under load. The cord property that most affects spring rate is the modulus.

Noise. Several types of noise are generated by a running tire. The most irritating noise is a high-pitched whine produced by the interaction of certain tread patterns and tire speeds. A low-pitched rumble or boom generated by irregularities in the road can also be heard. Tire squeal, which develops while accelerating or decelerating the tires, is another type of noise. Different noise levels have been found for tires made from different types of cord. However, the role that tire cord plays in developing tire noise and in transmitting this noise is not well understood and needs further investigation (see also Noise pollution).

Test Methods

Tire cord is tested for quality control and to predict performance. Some tests determine the physical properties, whereas others are intended to simulate processing or use. Properties tested include ultimate tensile strength, ultimate elongation, total work to break, and cord modulus in tension. They are obtained together by recording the load–elongation curve from a tensile-testing machine. The environment is standardized for control purposes but changed for evaluation and design to simulate use conditions. Thus the following testing conditions are varied: temperature, humidity, test speed, and type of loading rate (either constant elongation or constant load).

Yarn-to-Cord Conversion Efficiency. This property is usually measured by relating the ultimate tensile strength of the untwisted yarn to the ultimate tensile strength of the cord (49). Higher cord twist and/or larger diameter of the individual yarns result in lower cord efficiency. This cording loss or loss in efficiency has several causes but is mainly due to the reorientation of the filaments from the axially loaded tensile direction to the twisted angle of the yarn or cord. Higher twist corresponds to higher twist angle (or disorientation) and lower tensile strength of the yarn or cord. Any unevenness in the alignment of these filaments or damage that might have occurred during the twisting operations lowers tensile strength and efficiency. Another cause of tensile-efficiency loss is the inability of the filaments to adjust to the load. Therefore, a lubricating finish increases tensile efficiencies. Because the high modulus yarns are more difficult to adjust to the twisted structure, they tend to have lower twist efficiencies. In addition, it is more difficult to uniformly align the filaments in large twisted structures than those in small ones. Thus, large cords are less efficient.

Length Stability to Moisture. The length of stability of a cord in the presence of moisture is measured by the amount of shrinkage, usually as percent free shrinkage (under minimal load) in hot water. Although swelling and shrinkage in the presence of moisture can aid in the processing when controlled, length stability to moisture promotes and maintains uniformity. Both rayon and nylon are readily plasticized by moisture which tends to release any built-in shrinkage forces. Polyester is not plasticized by moisture and remains stable.

Length Stability to Heat. During cord processing and tire operation, temperatures above the T_g are frequently reached. It is difficult to control length in thermoelastic fibers like nylon and polyester which tend to shrink considerably when heated above their T_g. The amount of shrinkage depends upon the force built into the fibers during their thermal and load history (50). The free shrinkage of a cord is measured by heating the cord under a minimal load and noting its length change. A better evaluation can be obtained by closely simulating cord-processing and tire-operating conditions (47).

Length Stability to Load. Simple creep tests under various loads and temperatures can be related to tire growth, groove cracking, and possible nonuniformity. A low creep rate is desirable.

Second-Order Glass Transition Temperature (T_g). The glass temperature of many polymeric fibers are usually close to operating and processing temperatures. Hence, the cords change from the glassy to the rubbery state as they are heated through their T_g. This affects properties such as dynamic-storage and loss modulus, shrinkage, and even chemical reactivity. There are several methods to measure T_g, and the actual value depends on the measurement. The torsion-pendulum method is useful, but the best method is to test at various temperatures physical changes that are of direct interest to the user. Comparative T_g values have been correlated with tire uniformity and flatspotting (47–48). Table 5 gives the T_g of various tire cords measured by simulating the loads and temperatures that a cord is exposed to during postcure inflation of a tire. Thus, a processed cord is heated to 166°C and loaded to 10 mN/tex (0.9 gf/ den). By cooling under load and then releasing this load at various temperatures, the length of the unloaded cord can be measured for each release temperature. A plot of the percent elongation vs release temperature in degrees Celsius develops a curve that can be approximated by straight lines. The slope of these lines is the percent elongation per degree Celsius, or coefficient of retraction (CR). The intersection of these straight lines represents the T_g of the polymer. Additional transitions, T_c, were noted in the cords made from two different polyamides.

Dynamic Properties. Tire cords operate under dynamic conditions. The mode of cycling, construction, internal structure, and temperature affect the rate of heat generation. Although the exact modes of cycling are not known, reasonable stress and strain values can be applied to simulate the cycling modes. Thus, dynamic properties can be measured that indicate how much heat one cord may generate relative to another cord under the same conditions. From the surface temperatures of rotating tires and the material properties and standard heat-transfer equations, the internal tire temperatures and heat generation of the various parts of the tire have been calculated (43,51–52). In bias tires, the heat generation of the lower sidewall carcass correlates with the dynamic properties of the cords in the constant stress-cycling mode. The heat

Table 5. Coefficient of Retraction (CR)[a] and T_g of Tire-Cord Material

Tire-cord material, 186.7[b]/2	CR per °C, % length change		Transition temperatures	
	Value	Range, °C	T_g, °C	T_c, °C[c]
nylon-6	0.029	27–53	54	
	0.050	55–165		
nylon-6,6	0.022	27–47	48	
	0.041	49–165		
nylon-6,6 plus aromatic polymer, 70/30	0.022	27–80	81	120
	0.085	82–119		
	0.033	121–165		
nylon-6,6 plus aromatic copolymer, 70/30	0.022	27–48	49	140
	0.043	49–139		
	0.017	141–165		
nylon-6 plus PET[d], 70/30	0.022	27–59	60	
	0.054	61–165		
nylon-6,6 plus PET[d], 70/30	0.018	27–58	59	
	0.043	60–165		
polyester (PET[d])	0.011	27–84	85	
	0.050	86–165		

[a] After release from a load of 10 mN/tex (0.9 gf/den).
[b] To convert tex to den, divide by 0.1111.
[c] Additional transition.
[d] PET = poly(ethylene terephthalate).

generation of the carcass in the crown and shoulder region of bias tires correlates best with the dynamic properties of the cords in the constant strain-cycling mode. Radial tires perform less work and generate much less heat (44).

Fatigue. During the lifetime of a tire, the cords undergo innumerable stress-and-strain cycles which sometimes lead to degradation to the point of failure. Many tests have been devised in attempts to simulate fatigue failure, but none have been completely successful because of the complexity of true simulation. Fatigue tests are most meaningful when performed under controlled operating conditions. However, a number of laboratory tests have been developed, classified as bare-cord tests and in-rubber tests (53). The bare-cord tests cycle cords between various tensions in controlled atmospheres and temperatures and measure oxidative and mechanical damage (interfilamentary abrasion). They are sensitive to the yarn lubricant, filament placement, cord twist, uniformity, and oxidative resistance at elevated temperatures (54–55).

The in-rubber fatigue tests more closely simulate the fatigue mechanism under use conditions. Such in-depth studies show that the failure mechanism begins with a localized separation of the adhesive system, usually in the degraded rubber next to the cord. This leads to more violent distortions of the cord and its subsequent failure (39–40).

Thermal and Chemical Degradation. Most materials lose strength at elevated temperatures, and some textile materials are degraded by moisture, oxygen, or rubber chemicals. The loss in tensile strength is measured after aging treatments in rubber compounds as well as in other controlled environments. These values are compared with those obtained from tire cords. Minimal strength loss is sought, but compromises are usually made with other properties such as adhesion.

Cord-to-Rubber Adhesion. Adhesion tests are either peel tests or pull-out tests. In the peel test, one layer of cord fabric is peeled from another after they have been cured in rubber. The force required to separate a strip 2.5-cm wide that is pulled at 180° is the peel-adhesion value. It depends not only on the adhesive strength of the system but also on the force required to bend each layer during testing. Since the bending stiffness of each layer is difficult to control or measure separately, more reliance is placed on a subjective appearance rating than the peel force. Pull-out tests measure the force required to pull a cord out of a rubber block. The size of the cord and its embedment length must be standardized for purposes of comparison. Pull-out tests can be very precise and independent of the rubber modulus (56–57). A specially molded pull-out test that measures the energy of adhesion at the surface of the cord has been proposed to ASTM (58). Both the pull-out force and the modulus of the rubber are used to calculate the energy of adhesion. In all adhesive testing, it is important to know where the failure occurs as well as measure the adhesion strength. Adhesion testing should be standardized near the highest normal operating temperature of the tire (100–150°C). With increasing test temperature, most rubber compounds lose strength much more rapidly than either the adhesive or the tire cord. Therefore, at some elevated temperature the rubber compound becomes the weakest link, and the rubber will tear during the test. The limit of good adhesion values is usually obtained at a low temperature when the test breaks cord filaments. Adhesion is improved by strengthening the weakest link in the system.

Compaction and Dip Penetration. When the adhesive dip penetrates into the filament bundle, it ties the outside filaments together and blocks some of the void space between the filaments. Uniform dip penetration maintains the integrity of each individual yarn bundle making up the cord during flexing. Compaction promotes uniformity and reduces the void spaces, which provide both a reservoir and a conducting tube for air and moisture. The oxygen in the air tends to degrade the cord and surrounding materials especially when hot, whereas the moisture can cause separation problems during cure. In steel cords it degrades the adhesion and increases corrosion.

Compaction and dip penetration are measured by examination of cord cross sections under the microscope. In another method, the time is measured for a standard amount of air to flow through a standard length of cord embedded in a rubber matrix while under a standard air pressure differential (41,59).

Flatspotting and Evaluation of Uniformity. The flatspot test for tire cord is based on a simulation procedure of what a tire cord "sees" during flatspot formation and elimination (47–48,60–61). A simulation test can be used to predict both the relative tendency for a tire cord to flatspot as well as its tendency to introduce nonuniformity in tires during postcure inflation. The coefficients of retraction or relative length change per degree Celsius to be expected from various cord materials above and below its T_g are shown in Table 5. The higher the CR, the greater is the tendency for nonuniformity and flatspotting. Flatspotting is no problem for radial tires because radials have stiff nonshrinkable belts.

Bending Stiffness. During tire building the calendered fabric plies are folded around the bead and these ply turnups are held in place by the tackiness of the rubber. Stiff cord is difficult to fold, and it is even more difficult to keep the plies from separating while the green tire is awaiting cure. The stiffness of the processed cord can be measured by the angle of recovery after folding under a standard load.

Tire-Cord Status

Different types of tires require tire cords with different properties. Such tire designs as the radial tire or the belted-bias tire take advantage of these different requirements by placing different types of cords in the belt and in the carcass. The carcass of a bias or belted-bias tire requires a high strength, tough, flexible cord like rayon, nylon, or polyester. For radial tires, high modulus cords like steel or aramid can be used. Successful experiments have been conducted with radial glass carcasses (62). The belt requires a high strength high modulus cord like steel, glass, or high modulus rayon. Airplane tires require lightweight, high strength, tough, flexible cords with low heat generation; nylon seems to fill these requirements best. As airplanes increase in size and speed, new cords will be needed to withstand the high loads and high temperatures. Truck tires, off-the-road tires, high speed tires, and passenger tires all require different cord properties for optimum safe and economical performance. Chemical and fiber producers have increased the number of possible tire cords such as aramids. In turn, new and improved methods of evaluation have been developed.

Nylon and polyester cords are very strong, but rayon is superior to nylon at high speed impact. Polyester resists high speed impact, but loses modulus and strength faster than rayon at 150°C. Nylon and polyester retain tensile strength better than rayon when highly twisted. High modulus nonadjusted fibers, eg, aramid, glass, and steel, lose even more tensile strength than rayon when highly twisted. Polyester is dimensionally stable to moisture, and rayon is dimensionally stable to heat. Polyester creeps the least under load. At 93°C, rayon has a high storage modulus and a low loss tangent; ie, less heat is generated while cycling under constant stress limits. Aramid is even better under stress cycling. Under constant strain limits, nylon generates the least heat. The fatigue life of rayon is good, that of nylon is better, and that of polyester varies from longer than rayon below 93°C to shorter than rayon above 93°C. Under high flex conditions, glass and steel tend to fatigue readily. Aramid needs a relatively high twist for flex-fatigue resistance which lowers its tensile strength. Rayon retains its strength best at 150°C, whereas polyester is least affected by dry air at 150°C. Both nylon and rayon resist hydrolysis and aminolysis at 150°C, whereas polyester is much less resistant. In general, it is easier to process rayon than nylon, polyester, or aramid, except with respect to dimensional stability to moisture, where both rayon and nylon show some shrinkage but polyester and aramid are unaffected. Each tire cord, whether old or new, represents a different combination of properties which have to be optimized. Since the changing demands on tires emphasize different properties, the search for the best tire cord is never-ending.

BIBLIOGRAPHY

"Tire Cords" in *ECT* 2nd ed., Vol. 20, pp. 328–346, by Leonard Skolnik, The B. F. Goodrich Company.

1. U.S. Pat. 2,128,635 (Aug. 30, 1938), W. H. Charch and D. G. Maney (to DuPont).
2. U.S. Pat. 2,561,215 (July 17, 1951), C. J. Mighton (to DuPont).
3. U.S. Pat. 3,282,039 (Nov. 1, 1966), C. J. Geyer and J. B. Curley (to FMC).
4. J. B. Curley, *Rubber Age* **98,** 124 (Apr. 1966).
5. *Chem. Eng. News* **44,** 59 (Apr. 25, 1966).
6. Brit. Pat. 918,637 (Feb. 13, 1963), (to DuPont).
7. G. W. Rye, R. S. Bhakuni, and D. M. Callahan, *Polyester Adhesive Systems*, Apr. 24, 1968, American Chemical Society, Rubber Chemicals Division, Cleveland, Ohio.
8. U.S. Pat. 2,493,614 (Jan. 3, 1950), Bourdon (to Mfg. de Caoutchouc Michelin).
9. U.S. Pat. 3,869,306 (Mar. 4, 1975), A. Marzocchi (to Owens Corning).
10. U.S. Pats. 3,671,542 (June 20, 1972); 3,819,587 (June 25, 1974); and 3,888,965 (June 10, 1975), S. L. Kwolek (to DuPont).
11. U.S. Pat. 3,455,100 (July 15, 1969); U.S. Pat. 3,486,546 (Dec. 30, 1969), J. Sidles, D. P. Skala, and L. Skolnik (to BF Goodrich).
12. *Textile Organon* and *Tire Construction and Reinforcement Forecasts*, Monsanto Chemical Co., St. Louis, Mo., 1971–1981.
13. U.S. Pat. 3,307,966 (Mar. 7, 1967), C. J. Shoaf (to DuPont).
14. *Gen-Tac (Vinylpyridine Latex)*, Chemical/Plastics Division bulletin, The General Tire and Rubber Company, Akron, Ohio, 1964.
15. A. L. Promislow in *Akron Rubber Group Technical Symposiums*, Akron, Ohio, 1977–1978, p. 98.
16. W. D. Timmons, *Adhes. Age* **10**(10), 27 (1967).
17. Belg. Pat. 688,424 (1967); Fr. Pat. 1,496,951 (1967); and Neth. Pat. 6,614,669 (Apr. 19, 1967), (to Imperial Chemical Industries, Ltd.).
18. Y. Iyen in ref. 15, p. 110.
19. U.S. Pat. 3,968,295 (July 6, 1976), T. S. Solomon (to BF Goodrich).
20. R. E. Hartz and H. T. Adams, *J. Appl. Polym. Sci.* **21,** 525 (1977).
21. R. G. Aitken, R. L. Griffith, J. S. Little, and J. W. McClellan, *Rubber World* **151**(5), 58 (1965).
22. T. Takeyama and J. Matsui, *Rubber Chem. Technol.* **42,** 159 (1969).
23. Bekaert Steel Wire Corp., Pittsburgh, Pa., steel cord catalogue, Mar. 1982.
24. W. J. van Ooij, *Rubber Chem. Technol.* **52,** 605 (1979).
25. G. Haemers, *Rubber World*, 26 (Sept. 1980).

26. R. S. Rivlin, *Arch. Rat. Mech. Anal.* **2,** 447 (1959).
27. W. Hofferberth, *Kautsch. Gummi* **9,** 225 (1956).
28. H. G. Lauterbach and F. W. Ames, *Tex. Res. J.* **29,** 890 (1959).
29. S. K. Clark, *Tex. Res. J.* **33,** 295 (Apr. 1963).
30. J. D. Walters, *Rubber Chem. Technol.* **51,** 565 (1978).
31. C. B. Budd, private communications, BF Goodrich, Brecksville, Ohio, Aug. 9, 1957 and Mar. 30, 1964.
32. B. D. Coleman, *J. Mech. Phys. Solids* **7,** 60 (1958).
33. K. B. O'Neil, M. F. Dague, and J. E. Kimmel, *International Symposium on High Speed Testing*, Boston, Mass., Mar. 7, 1967.
34. J. B. Curley, *Rubber World*, (Sept. 1967).
35. E. W. Lothrop, *Appl. Polym. Symp.* **1,** 111 (1965).
36. F. S. Vukan and T. P. Kuebler, *Determination of Passenger Tire Performance Levels—Tire Strength and Endurance*, Society of Automotive Engineers, Chicago, Ill., May 23, 1969.
37. C. Z. Draves, Jr., T. P. Kuebler, and S. F. Vukan, *ASTM Mat. Res. Stand.* **10**(6), 26 (1970).
38. R. L. Guslitser, *Tr. Nauchno-Issled. Inst. Shinnoi Promsti.* **3,** 154 (1957); transl. by R. J. Moseley, Research Association of British Rubber Manufacturers, translation 817.
39. S. Eccher, *Rubber Chem. Technol.* **40,** 1014 (1967).
40. G. Butterworth, *Chem. Eng. News*, 45 (Sept. 1967).
41. C. Z. Draves, Jr. and L. Skolnik, *Processing Rayon Cord for Optimum Tire Performance*, 2nd Dissolving Pulp Conference TAPPI, New Orleans, La., June 1968.
42. J. M. Collins, W. L. Jackson, and P. S. Oubridge, *Trans. Inst. Rub. Ind.* **40,** T239 (1964).
43. N. M. Trivisonno, *Thermal Analysis of a Rolling Tire*, paper 700474, Society of Automotive Engineers meeting, Detroit, Mich., May 1970.
44. D. J. Schuring, *Rubber Chem. Technol.* **53,** 600 (1980).
45. F. S. Conant, *Rubber Chem. Technol.* **44,** 397 (1971).
46. V. E. Gough, *Rubber Chem. Technol.* **41,** 988 (1968); *Kautsch. Gummi Kunstst.* **20,** 469 (1967).
47. C. Z. Draves, Jr. and L. Skolnik, *Tire Cord Process Simulation and Evaluation*, DKG Rubber Conference, Berlin, FRG, May 1968; *Kautsch. Gummi Kunstst.* **22,** 561, Oct. 1969.
48. P. V. Papero, R. C. Wincklhofer, and H. J. Oswald, *Rubber Chem. Technol.* **38**(4), 999 (1965).
49. J. K. Van Wijngaarden in *Recent Development in Tire Yarns*, American Enka Corp. bulletin 9-6, May 8, 1963.
50. G. Heidemann, G. Stein, and R. Kuhn in B. Von Falkai, ed., *Synthesefasern*, Verlag Chemie, Weinheim, FRG, 1981, pp. 379–423.
51. U.S. Pat. 3,893,331 (July 8, 1975), D. C. Prevorsek, Y. D. Kwon, R. H. Butler, and R. K. Sharma (to Allied Chemical Corporation).
52. Y. D. Kwon, R. K. Sharma, and D. C. Prevorsek, *Tire Reinforcement and Tire Performance*, Symposium at Montrose, Ohio, Oct. 23–25, 1978, American Society for Testing and Materials, Philadelphia, Pa., 1979, pp. 239–262.
53. *ASTM D 885-64*, American Society for Testing and Materials, Philadelphia, Pa., 1967.
54. W. H. Bradshaw, *ASTM Bulletin*, American Society for Testing and Materials, Philadelphia, Pa., Oct. 1945, pp. 13–17.
55. C. B. Budd, *Text. Res. J.* **21,** 174 (Mar. 1951).
56. L. Skolnik, *Rubber Chem. Technol.* **47,** 434 (1974).
57. L. C. Coates and C. Lauer, *Rubber Chem. Technol.* **45,** 16 (1972).
58. G. S. Fielding-Russell, D. I. Livingston, and D. W. Nicholson, *Rubber Chem. Technol.* **53,** 950 (1980).
59. J. B. Curley, *A New Development in Rayon Tire Cord Technology*, Buffalo Rubber Group technical meeting, Mar. 8, 1966.
60. G. W. Rye and J. E. Martin, *Rubber World* **149,** 75 (Oct. 1963).
61. W. E. Claxton, M. J. Forester, J. J. Robertson, and G. R. Thurman, *Text. Res. J.* **35,** 903 (1965).
62. J. A. Gooch, *Akron Rubber Group Technical Symposiums*, Akron, Ohio, 1978–1979, pp. 56–58.

LEONARD SKOLNIK
BF Goodrich

TITANIUM AND TITANIUM ALLOYS

Titanium [7440-32-6] is a metal element of group IVB and at wt 47.90. Titanium metal has become known as a space-age metal because of its high strength-to-weight ratio and inertness to many corrosive environments. Its principal use, however, is as TiO_2 as paint filler (see also Paint; Pigments). The whiteness and high refractive index of TiO_2 is unequaled for whitening paints, paper, rubber, plastics, and other materials. A small amount of mineral-grade TiO_2 is used in fluxes and ceramics.

Titanium is the ninth most abundant element in the earth's crust and the fourth most abundant structural element. Its elemental abundance is about five times less than iron and 100 times greater than copper, yet for structural applications titanium's annual use is ca 200 times less than copper and 2000 times less than iron. When it is considered that commercial titanium metal production began in 1948, its lack of widespread application is not surprising (1).

The principal titanium mineral, ilmenite, $FeTiO_3$, is found in either alluvial sands or hard-rock deposits. After concentrating, the titanium ore color is black. This is the black sand often found concentrated in bands along sandy beaches. The density of this concentrate is ca 4–5 g/cm^3. The concentrate is processed to either pigment-grade TiO_2 or metal. The titanium metal is won from the ore in a physical form called sponge. The sponge is consolidated to an ingot which is further processed to mill products. In metallic form, titanium has a dull silver luster and its appearance is similar to stainless steel.

Titanium was first identified as a constituent of the earth's crust in the late 1700s. In 1790, William Gregor, an English clergyman and mineralogist, discovered a black magnetic sand (ilmenite) which he called menaccanite after his local parish. In 1795, a German chemist found that a Hungarian mineral, rutile, was the oxide of a new element he called titan, after the mythical Titans of ancient Greece. In the early 1900s, a sulfate purification process was developed to commercially obtain high purity TiO_2 for the pigment industry, and titanium pigment became available both in the United States and Europe. During this period, titanium was also used as an alloying element in irons and steels. In 1910, 99.5% pure titanium metal was produced at General Electric from titanium tetrachloride and sodium in an evacuated steel container. Since the metal did not have the desired properties, further work was discouraged. However, this reaction formed the basis for the commercial sodium reduction process. In the 1920s, ductile titanium was prepared with an iodide dissociation method combined with Hunter's sodium reduction process.

In the early 1930s, a magnesium vacuum reduction process was developed for reduction of titanium tetrachloride to metal. Based on this process, the U.S. Bureau of Mines (BOM) initiated a program in 1940 to develop commercial production. Some years later, the BOM publicized its work on titanium and made samples available to the industrial community. By 1948, the BOM produced batch sizes of 104 kg. In the same year, E. I. du Pont de Nemours & Co., Inc., announced commercial availability of titanium, and the modern titanium metals industry began (2).

By the mid-1950s, this new metals industry had become well established, with six producers, two other companies with tentative production plans, and more than 25 institutions engaged in research projects. Titanium, termed the wonder metal, was billed as the successor to aluminum and stainless steels. When in the 1950s, the DOD (titanium's most staunch supporter) shifted emphasis from aircraft to missiles, the demand for titanium sharply declined. Only two of the original titanium metal plants are still in use, the Titanium Metals Corporation of America's (TMCA) plant in Henderson, Nevada, and National Distillers & Chemical Corporation's two-stage sodium reduction plant built in the late 1950s at Ashtabula, Ohio, which now houses the sponge production facility for RMI Corporation (formerly Reactor Metals, Inc.).

Overoptimism followed by disappointment has characterized the titanium-metals industry. In the late 1960s, the future again appeared bright. Supersonic transports and desalination plants were intended to use large amounts of titanium (see Water, supply and desalination). Oregon Metallurgical Corporation, a titanium melter, decided at that time to become a fully integrated producer (ie, from raw material to mill products). However, the supersonic transports and the desalination industry did not grow as expected. Nevertheless, in the late 1970s and early 1980s, the titanium-metal demand again exceeded capacity and both the United States and Japan expanded capacities. This growth was stimulated by greater acceptance of titanium in the chemical process industry, power-industry requirements for seawater cooling, and commercial and military aircraft demands. However, with the economic recession of 1981–1983, the demand dropped well below capacity and the industry was again faced with hard times.

Titanium mineral occurs in nature as ilmenite ($FeTiO_3$), rutile (tetragonal TiO_2), anatase (tetragonal TiO_2), brookite (rhombic TiO_2), perovskite ($CaTiO_3$), sphene ($CaTiSiO_5$), and geikielite ($MgTiO_3$). Ilmenite is by far the most common, although rutile has been an important source of raw material. Some deposits of anatase and perovskite are rich enough to be of commercial interest, but the plentiful availability of high grade deposits of ilmenite and rutile have postponed their development.

Titanium ore bodies are uniformly distributed throughout the continents of the world. They occur either as hard-rock deposits, magnetic in origin, or as secondary placer.deposits (see Tables 1 and 2). The titanium contained in placer deposits is close to 10^8 metric tons as ilmenite and 15×10^6 t as rutile. Hard-rock deposits contain over 2×10^8 metric tons, excluding the oil sands of Alberta, which contain over 2×10^9 metric tons of titanium (4–7).

In 1972, ca 45% of the 2×10^6 metric tons of ilmenite mining capacity was concentrated in four hard-rock mineral deposits exploited by QIT Fer et Titane, Inc. (Allard Lake, Canada) and NL Industries (Tellnes, Norway; Tahawas, New York; and Otanmaki, Finland). Titanium ore is gravity-concentrated and magnetically separated wet to yield a concentrate of 45% TiO_2 in all the hard-rock mining operations except at Allard Lake; QIT Fer et Titane, Inc. ships concentrate from the mine at 35% TiO_2 to a smelter where the ore is reduced to produce pig iron and a 71% TiO_2 slag (Sorel slag). This slag is called titaniferous slag. The remaining 55% of ilmenite mining capacity is at placer deposits in Australia (42%), Republic of South Africa (18%), India (18%), United States (13%), and Sri Lanka and Malaysia (9%). All rutile mining is in placer deposits, 58% in Australia, 20% in Sierra Leone, and 12% in the Republic of South Africa. Titanium recovery from placer deposits is shown in Figure 1. Titanium ore concentrated from ilmenite ranges from 45% TiO_2 in hard-rock and some placer deposits to 64% TiO_2 in some Florida placer deposits. Rutile is ca 95% TiO_2. Chemical analysis of some selected titanium mineral concentrations is given in Table 3 (6,8).

Table 1.   Hard Rock Titanium Ore Deposits [a]

Location	Company	Ore grade, % TiO$_2$	Resources, 10^6 t Ti content
Brazil [b]			
Minas Gerais		23–27	85
Canada			
Allard Lake	QIT Fer et Titane, Inc. [c]	33.8	26
St. Urbain		38	5
Terrebonne	Canadian Nickel Co. [d]	20	6
Alberta	Sun Corp. [e]	0.35	very large
Finland			
Otarmaki	Rautaruukki	9–10	3
Norway			
Tellnes	Titania A/S	18	33
United States			
Tahawas, N.Y.	NL Industries, Inc.	17	6
Colorado	Buttes Oil & Gas Co. [f]	12	27
Minnesota		10	12
Wyoming		20	5
USSR			
Ukraine	Verkhnedneprosk Ishansk		4

[a] Refs. 3–7.

[b] Anatase deposits.

[c] 75% high grade (33.8% TiO$_2$) ore the remainder low grade (6% TiO$_2$) ore.

[d] 30% high grade (20% TiO$_2$) ore the remainder low grade (6% TiO$_2$) ore.

[e] Oil sands of Alberta. There are an estimated 1.5 × 10^{12} metric tons of oil sands containing 0.17% Ti content as rutile, anatase, and ilmenite. This deposit also contains about 5 × 10^8 t Zr.

[f] Perovskite deposit.

Alloys

Titanium alloy systems have been extensively studied. A single company evaluated over 3000 compositions in 8 yr. Alloy development has been aimed at elevated temperature aerospace applications, strength for structural applications, and aqueous corrosion resistance. The principal effort has been in aerospace applications to replace nickel- and cobalt-base alloys in the 500–900°C ranges. To date, titanium alloys have replaced steel in the 200–500°C ranges. The useful strength and corrosion-resistance temperature limit is ca 550°C.

The addition of alloying elements alters the α–β transformation temperature. Elements that raise the transformation temperature are called α stabilizers, elements that depress the transformation temperature β stabilizers; the latter are divided into β-isomorphous and β-eutectoid types. The β-isomorphous elements have limited α solubility and increasing additions of these elements progressively depresses the transformation temperature. The β-eutectoid elements have restricted beta solubility and form intermetallic compounds by eutectoid decomposition of the β phase. The binary phase diagram illustrating these three types of alloy systems is shown in Figure 2.

The important α-stabilizing alloying elements include aluminum, tin, zirconium, and the interstitial alloying elements (ie, elements that do not occupy lattice positions)

Table 2.　Placer Titanium Ore Deposits[a]

| Location | Company | Reserves, 10^6 t Ti and Zr content | | |
		Ilmenite	Rutile	Zircon
Australia				
east coast	Associated Minerals Consolidated, Ltd.		0.9	0.2
	Mineral Deposits, Ltd.	0.4	0.5	0.2
	Consolidated Rutile, Ltd.		0.5	0.4
	RZ Mines (Newcastle), Ltd.	0.2	0.2	0.3
	Minsands Exploration Pty., Ltd.		0.2	
west coast	Associated Minerals Consolidated, Ltd.	5.2	1.3	2.6
	Westralian Sand–Tioxide Group	3.6	0.2	0.6
	Cable Sands, Ltd.	1.1		0.2
	Allied Encabba, Ltd.	1.6	0.5	1.1
	Western Mining, Ltd.	0.3		
	Metals Exploration, Ltd.	0.2		0.3
Brazil				
Paraiba	State	0.9	0.1	
India				
Chavara	Kerala Minerals & Metals Co.	12.6	1.7	
Orissa	Indian Rare Earths, Ltd.	11.2	1.5	
other	Indian Rare Earths, Ltd.	22	2.3	5
Sierra Leone				
	Sierra Rutile, Ltd.		1.7	
South Africa				
Richards Bay	Richards Bay Minerals Organization	7.3	1.4	1.8
other		23	1.8	1.8
Sri Lanka				
Pulmoddai	Sri Lanka Mineral Sands Corp.	0.9	0.2	0.1
United States				
Florida, Georgia	E. I. du Pont de Nemours & Co., Inc.	5.4		
Florida	Associated Minerals Consolidated, Ltd.	0.3	0.2	
New Jersey	Asarco, Inc.	1.5		
Total		*97.7*	*15.2*	*14.6*

[a] Refs. 4, 6–7.

oxygen, nitrogen, and carbon. Small quantities of interstitial alloying elements, generally considered to be impurities, have a very great effect on strength and ultimately embrittle the titanium at room temperature (9). The effects of oxygen, nitrogen, and carbon on the ultimate tensile properties and elongation are shown in Table 4. These elements are always present and difficult to control. Nitrogen has the greatest effect and commercial alloys specify its limit to be less than 0.05 wt %. It may also be present as nitride inclusions (TiN) which are detrimental to critical aerospace structural applications. Oxygen additions increase strength and serve to identify several commercial grades. This strengthening effect diminishes at elevated temperatures and under creep conditions at room temperature. For cryogenic service, low oxygen content is specified (<1300 ppm) because high concentrations of interstitial impurities increase sensitivity to cracking, cold brittleness, and fracture temperatures. Alloys with low interstitial content are identified as ELI (extra-low interstitials) after the alloy name. Carbon does not affect strength at concentration above 0.25 wt % because carbides (TiC) are formed. Carbon content is usually specified at 0.08 wt % max (10).

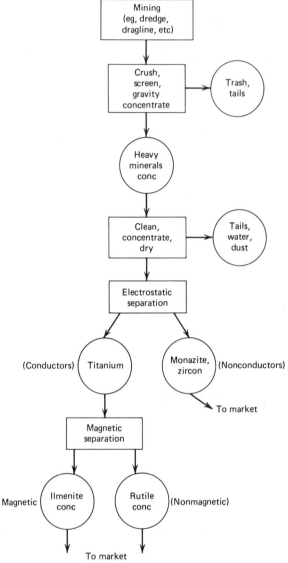

Figure 1.   Typical processes and products of a titanium beach-sand mining and beneficiating operation (8).

The most important alloying element is aluminum, an α stabilizer. It is not expensive, and its atomic weight is less than that of titanium; hence, aluminum additions lower the density. The mechanical strength of titanium can be increased considerably by aluminum additions. Even though the solubility range of aluminum extends to 27 wt %, above 7.5 wt % the alloy becomes to difficult to fabricate and embrittles. The embrittlement is caused by a coherently ordered phase based on Ti$_3$Al [12635-69-7]. Other α-stabilizing elements also cause phase ordering. An empirical relationship below which ordering does not occur is given below (11):

$$\text{wt \% Al} + \frac{\text{wt \% Sn}}{3} + \frac{\text{wt \% Zr}}{6} + 10 \times \text{wt \% O} \leq 9$$

Table 3. Analyses of Selected Titanium Mineral Concentrates, Wt % [a]

| Constituent | Ilmenite | | | | Sorel Slag Canada | Rutile, eastern Australia |
| | Placer | | | Hard Rock | | |
	Florida	Western Australia	Sri Lanka	New York		
TiO_2	64.10	55.30	53.45	44.40	71.00	96.40
ZrO_2		0.10	0.16	0.01		0.30
FeO	4.70	26.70	20.45	36.70	13.00	
Fe_2O_3	25.60	15.40	22.18	4.40		0.25
P_2O_5	0.21	0.04	0.21	0.07	0.33	0.02
SiO_2	0.30	0.20	0.52	3.20		0.56
Cr_2O_3	0.10	0.03	0.09	trace	0.20	0.15
Al_2O_3	1.50	0.38	0.58	0.19	5.70	0.17
V_2O_5	0.13	0.08		0.24	0.59	0.61
MnO	1.35	1.64	0.93	0.35	0.22	trace
CaO	0.13	0.17		1.00	1.00	0.05
MgO	0.35	0.29	1.46	0.80	5.00	0.04

[a] Ref. 8.

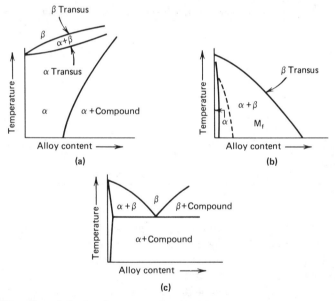

Figure 2.   The effect of alloying elements on the phase diagram of titanium. (**a**) α-stabilized system, (**b**) β-isomorphous system, and (**c**) β-eutectoid system.

The important β-stabilizing alloying elements are the bcc elements vanadium, molybdenum, tantalum, and niobium of the β-isomorphous type and manganese, iron, chromium, cobalt, nickel, copper and silicon of the β-eutectoid type. The β-eutectoid elements arranged in order of increasing tendency to form compounds are shown in Table 5. The elements copper, silicon, nickel, and cobalt are termed active eutectoid forms because of a rapid decomposition of β to α and a compound. The other elements in Table 5 are sluggish in their eutectoid reactions.

Alloys of the β type respond to heat treatment, are characterized by higher density than pure titanium, and are easily fabricated. The purpose of β alloying is to form an

Table 4. Effects of O, N, and C on the Ultimate Tensile Strength [a,b]

Concentration of impurity, wt %	Oxygen [c,d]		Nitrogen [c,d]		Carbon [c,d]	
	UT, MPa [e]	Elong., %	UT, MPa [e]	Elong., %	UT, MPa [e]	Elong., %
0.025	330	37	380	35	310	40
0.05	365	35	460	28	330	39
0.1	440	30	550	20	370	36
0.15	490	27	630	15	415	32
0.2	545	25	700	13	450	26
0.3	640	23	embrittles		500	21
0.5	790	18			520	18
0.7	930	8			525	17

[a] Ref. 9.
[b] Tests were conducted using titanium produced by the iodide process.
[c] UT = ultimate tensile stress.
[d] Elongation on 2.54 cm.
[e] To convert MPa to psi, multiply by 145.

Table 5. β-Eutectoid Elements in Order of Increasing Tendency to Form Compounds [a]

Element	Eutectoid composition, wt %	Eutectoid temperature, °C	Composition for β retention on quenching, wt %
manganese	20	550	6.5
iron	15	600	4.0
chromium	15	675	8.0
cobalt	9	685	7.0
nickel	7	770	8.0
copper	7	790	13.0
silicon	0.9	860	

[a] Ref. 12.

all β-phase alloy with commercially useful qualities, form alloys with duplex α and β structure to enhance heat-treatment response (ie, changing the α and β volume ratio), or use β-eutectoid elements for intermetallic hardening. The most important commercial β-alloying element is vanadium.

Aerospace Alloys. The alloys of titanium for aerospace use can be divided into three categories: an all-α structure, a mixed α–β structure, and an all-β structure. The α–β structure alloys are further divided into near-α alloys (<2% β stabilizers). Most of the ca 100 commercially available alloys (ca 30 in the United States, 40 in the USSR, and 10 in Europe and Japan) are of the α–β structure type (8). Some of these, produced in the United States, are given in Table 6 along with some wrought properties (13–15). The most important commercial alloy is Ti–6 Al–4 V, an α–β alloy with a good combination of strength and ductility. It can be age-hardened and has moderate ductility, and an excellent record of successful applications. It is mostly used for compressor blades and disks in aircraft gas-turbine engines, and also in lower temperature engine applications such as rotating disks and fans. It is also used for rocket-motor cases, structural forgings, steam-turbine blades, and cryogenic parts for which ELI grades are usually specified.

Other commercially important α–β alloys are Ti–3 Al–2.5 V, Ti–6 Al–6 V–2 Sn, and Ti–10 V–2 Fe–3 Al (see Table 6). As a group, these alloys have good strength, moderate ductility, and can be age-hardened (15–16). Weldability becomes more difficult with increasing β constituents, and fabrication of strip, foil, sheet, and tubing may be difficult. Temperature tolerances are lower than those of the α or near-α alloys. The alloy Ti–3 Al–2.5 V (called one half Ti–6 Al–4 V) is easier to fabricate than Ti–6 Al–4 V and is used primarily as seamless aircraft-hydraulic tubing. The alloy Ti–6 Al–6 V–2 Sn is used for some aircraft forgings because it has a higher strength than Ti–6 Al–4 V. The recent alloy Ti–10 V–2 Fe–3 Al is easier to forge at lower temperatures than Ti–6 Al–4 V because it contains more β-alloying constituents and has good fracture toughness. This alloy can be hardened to high strengths (1.24–1.38 GPa or $(1.8–2) \times 10^5$ psi) and is expected to be used as forgings for airframe structures to re-place steel below 300°C (17).

The only α alloy of commercial importance is Ti–5 Al–2.5 Sn. It is weldable, has good elevated temperature stability, and good oxidation resistance to about 600°C. It is used for forgings and sheet-metal parts such as aircraft-engine compressor cases because of weldability.

The commercially important near-α alloys are Ti–8 Al–1 Mo–1 V and Ti–6 Al–2 Sn–4 Zr–2 Mo. They exhibit good creep resistance and the excellent weldability and high strength of α alloys; temperature limit is ca 500°C. Alloy Ti–8 Al–1 Mo–1 V is used for compressor blades because of its high elastic modules and creep resistance; however, it may suffer from ordering embrittlement. Alloy Ti–6 Al–2 Sn–4 Zr–2 Mo is also used for blades and disks in aircraft engines. The service temperature limit of 470°C is ca 70°C higher than that of Ti–8 Al–1 Mo–1 V (11).

Commercialization of β alloys has not been very successful. Even though alloys with high strength (up to 1.5 GPa (217,500 psi)) were made, they suffered from in-termetallic and ω-phase embrittlement. These alloys are metallurgically unstable and have little practical use above 250°C. They are fabricable but welds are not ductile. This alloy type is used in the cold-drawn or cold-rolled condition and finds application in spring manufacture (alloy Ti–13 V–11 Cr–3 Al) (18). There is one commercially available alloy of the β-eutectoid type (Ti–2.5 Cu [37270-40-9]) that uses a true pre-cipitation-hardening mechanism to increase strength. The precipitate is Ti$_2$Cu [12054-13-6]. This alloy is only slightly heat-treatable; it is used in engine castings and flanges (11) (see High temperature alloys).

Nonaerospace Alloys. The nonaerospace alloys are used primarily in industrial applications. The four grades (ASTM grade 1 through grade 4) differ primarily in oxygen and iron content (see Table 7). ASTM grade 1 has the highest purity and the lowest strength (strength is controlled by impurities). The two other alloys of this group are ASTM grade 7, Ti–0.2 Pd and ASTM grade 12, Ti–0.8 Ni–0.3 Mo [12793-98-5]. The alloys in this group are distinguished by excellent weldability, formability, and corrosion resistance. The strength, however, is not maintained at elevated temperatures (see Table 6). The primary use of alloys in this group is in industrial processing equipment (ie, tanks, heat exchangers, pumps, electrodes, etc), even though there is some use in airframes and aircraft engines. The ASTM grade 1 is used where higher purity is desired, eg, as weld wire for grade 2 fabrication and as sheet for explosive bonding to steel. Grade 1 is manufactured from high purity sponge. The ASTM grade 2 is the most commonly used grade of commercially pure titanium. The chemistry for this grade is easy to meet with most sponge. The ASTM grades 3 and 4 are higher

Table 6.   Properties, Specifications, and Applications of Wrought Titanium Alloys[a]

Nominal composition, wt %	CAS Registry No.	ASTM B-265	CLTE[b], μm/(m·K) 21–100°C	21–538°C	Average physical properties				
					Modulus of elasticity[c], GPa[d]	Modulus of rigidity[c], GPa[d]	Poisson's[c] ratio	Density, g/cm³	Condition
commercially pure									
99.5 Ti		grade 1	8.7	9.8	102	39	0.34	4.5	annealed
99.2 Ti		grade 2	8.7	9.8	102	39	0.34	4.5	annealed
99.1 Ti		grade 3	8.7	9.8	103	39	0.34	4.5	annealed
99.0 Ti		grade 4	8.7	9.8	104	39	0.34	4.5	annealed
99.2 Ti[h]		grade 7	8.7	9.8	102	39	0.34	4.5	annealed
98.9 Ti[i]								4.5	annealed
Ti–5 Al–2.5 Sn[j]	[11109-19-6]	grade 6	9.4	9.6	110			4.5	annealed
Ti–8 Al–1 Mo, 1 V[j]	[39303-55-4]		8.5	10.1	124	47	0.32	4.4	duplex annealed
Ti–6 Al–2 Sn, 4 Zr–2 Mo[j]	[11109-15-2]		7.8	8.1	114			4.5	annealed
Ti–3 Al–2.5 V[j]	[11109-23-2]		9.6	9.9	107			4.5	annealed
Ti–6 Al–4 V[j]	[12743-70-3]	grade 5	8.7	9.6	114	42	0.342	4.4	annealed
Ti–6 Al–6 V, 2 Sn[j]	[12606-77-8]		9.0	9.6	110			4.5	annealed
Ti–10 V–2 Fe, 3 Al[j]	[51809-47-3]				112			4.6	solution and age

[a] Refs. 14–15.
[b] CLTE = coefficient of linear thermal expansion.
[c] Room temperature.
[d] To convert GPa to psi, multiply by 145,000.
[e] To convert MPa to psi, multiply by 145.
[f] To convert J/m to ft·lb/in., divide by 53.38.
[g] HB = Brinell, HRC = Rockwell (C-scale).
[h] Also contains 0.2 Pd.
[i] Also contains 0.8 Ni and 0.3 Mo.
[j] Numerical designations = wt % of element.

strength versions of grade 2; grades 7 and 12 have better corrosion resistance than grade 2 in reducing acids and acid chlorides. However, grade 7 is expensive and grade 12 is not readily available.

Other Alloys.   Other alloying ranges include the aluminides (TiAl [12003-96-2] and Ti₃Al), the superconducting alloys (Ti–Nb type), the shape-memory alloys (Ni–Ti type), and the hydrogen-storage alloys (Fe–Ti) (see also Superconducting materials; Shape-memory alloys). The aluminides TiAl and Ti₃Al have excellent high temperature strengths, comparable to those of nickel- and cobalt-base alloys, with less than half the density. These alloys exhibit ultimate strengths of 1 GPa (145,000 psi), and 800 MPa (116,000 psi) yield, respectively, 4–5% elongation, and 7% reduction in area. Strengths are maintained to 800–900°C. The modulus of elasticity is high (125–165 GPa (18–24) × 10⁶ psi), and oxidation resistance is good (13). The aluminides are intended for both static and rotating parts in the turbine section of gas-turbine aircraft engines.

Titanium alloyed with niobium exhibits superconductivity, and a lack of electrical resistance below 10 K. Composition ranges from 25 to 50 wt % Ti. These alloys are β-phase alloys with superconducting transitional temperatures at ca 10 K. Their use is of interest for power generation, propulsion devices, fusion research, and electronic devices (19).

Titanium alloyed with nickel exhibits a memory effect, that is, the metal form switches from one specific shape to another in response to temperature changes. The group of Ti–Ni alloys (nitinol) was developed by the Navy in the early 1960s for F-14

Table 6 (continued)

\multicolumn{10}{c}{Average mechanical properties}	Charpy	Hardness[g]

Room temperature				Test temperature, °C	Extreme temperatures				Charpy impact strength, J/m[f]	Hardness[g]
Tensile strength, MPa[e]	Yield strength, MPa[e]	Elonga-tion, %	Reduction in area, %		Tensile strength, MPa[e]	Yield strength, MPa[e]	Elonga-tion, %	Reduction in area, %		
331	241	30	55	315	152	97	32	80		HB 120
434	346	28	50	315	193	117	35	75	43	HB 200
517	448	25	45	315	234	138	34	75	38	HB 225
662	586	20	40	315	310	172	25	70	20	HB 265
434	346	28	50	315	186	110	37	75	43	HB 200
517	448	25	42	315	324	207	32			
862	807	16	40	315	565	448	18	45	26	HRC 36
1000	952	15	28	540	621	517	25	55	33	HRC 35
979	896	15	35	540	648	490	26	60		HRC 32
690	586	20		315	483	345	25			
993	924	14	30	540	531	427	35	50	19	HRC 36
1069	1000	14	30	315	931	807	18	42	18	HRC 38
1276	1200	10	19	315	1103	979	13	42		

fighter jets. The compositions are typically Ti with 55 wt % Ni. The transition temperature ranges from −100°C to >100°C and is controlled by additional alloying elements. These alloys are of interest for thermostats, recapture of waste heat, pipe joining, etc. The nitinols have not been extensively used because of high price and fabrication difficulties (20).

Titanium alloyed with iron is a leading candidate for solid-hydride energy storage material for automotive fuel. The hydride, $FeTiH_2$, absorbs and releases hydrogen at low temperatures. This hydride stores 0.9 kW·h/kg. To provide the energy equivalent to a tank of gasoline would require about 800 kg $FeTiH_2$ (13).

Table 7.   ASTM Requirements for Different Titanium Grades, % [a]

Element	Grade 1	Grade 2	Grade 3	Grade 4	Grade 7	Grade 12
nitrogen, max	0.03	0.03	0.05	0.05	0.03	0.03
carbon, max	0.10	0.10	0.10	0.10	0.10	0.08
hydrogen, max	0.015	0.015	0.015	0.015	0.015	0.015
iron, max	0.20	0.30	0.30	0.50	0.30	0.30
oxygen, max	0.18	0.25	0.35	0.40	0.25	0.25
palladium					0.12–0.25	
molybdenum						0.2–0.4
nickel						0.6–0.9
residuals, max						
each	0.1	0.1	0.1	0.1	0.1	0.1
total	0.4	0.4	0.4	0.4	0.4	0.4
titanium	remainder	remainder	remainder	remainder	remainder	remainder

[a] Ref. 10.

Physical Properties

The physical properties of titanium are given in Table 8. The most important physical property of titanium from a commercial viewpoint is the ratio of its strength (ultimate strength > 690 MPa (100,000 psi)) at a density of 4.507 g/cm^3. Titanium alloys have a higher yield strength-to-density rating between −200 and 540°C than either aluminum alloys or steel (12,21). Titanium alloys can be made with strength equivalent to high strength steel, yet with density ca 60% that of iron alloys. At ambient temperatures, titanium's strength-to-weight ratio is equal to that of magnesium, 1.5 times greater than that of aluminum, two times greater than that of stainless steel, and three times greater than that of nickel. Alloys of titanium have much higher strength:weight ratios than alloys of nickel, aluminum, or magnesium, and stainless steel. Because of its high melting point, titanium can be alloyed to maintain strength well above the useful limits of magnesium and aluminum alloys. This property gives titanium a unique position in applications between 150–550°C where the strength:weight ratio is the sole criterion.

Solid titanium exists in two allotropic crystalline forms. The α phase, stable below 882.5°C, is a hexagonal closed-packed structure, whereas the β phase, a bcc crystalline

Table 8.   Physical Properties of Titanium

Property	Value
melting point, °C	1668 ± 5
boiling point, °C	3260
density, g/cm^3	
$\quad \alpha$ phase at 20°C	4.507
$\quad \beta$ phase at 885°C	4.35
allotropic transformation, °C	882.5
latent heat of fusion, kJ/kg[a]	440
latent heat of transition, kJ/kg[a]	91.8
latent heat of vaporization, MJ/kg[a]	9.83
entropy at 25°C, J/mol[a]	30.3
thermal expansion coefficient at 20°C per °C	8.41×10^{-6}
thermal conductivity at 25°C, W/(m·K)	21.9
emissivity	9.43
electrical resistivity at 20°C, nΩ·m	420
magnetic susceptibility, mks	180×10^{-6}
modulus of elasticity, GPa[b]	
$\quad$ tension	ca 101
$\quad$ compression	103
$\quad$ shear	44
Poisson's ratio	∼0.41
lattice constants, nm	
$\quad \alpha$, 25°C	$a_0 = 0.29503$
	$c_0 = 0.46531$
$\quad \beta$, 900°C	$a_0 = 0.332$
vapor pressure, kPa[c]	$\log P_{kPa} = 5.7904 - 24644/T - 0.000227\ T$
specific heat, J/(kg·K)[d]	$C_p = 669.0 - 0.037188\ T - 1.080 \times 10^7/T^2$

[a] To convert J to cal, divide by 4.184.
[b] To convert GPa to psi, multiply by 145,000.
[c] To convert $\log P_{kPa}$ to $\log P_{atm}$, add 2.0056 to the constant.
[d] $T > 298$ K.

structure, is stable between 882.5°C and the melting point of 1668°C. The high temperature β phase can be found at room temperature when β-stabilizing elements are present as impurities or additions (see above under Alloys). Alpha and beta phases can be distinguished by examining an unetched polished mount with polarized light. Alpha is optically active and changes from light to dark as the microscope stage is rotated. The microstructure of titanium is difficult to interpret without knowledge of the alloy content, working temperature, and thermal treatment (12,22–23).

The heat-transfer qualities of titanium are characterized by the coefficient of thermal conductivity. Even though this is low, heat transfer in service approaches that of admiralty brass (thermal conductivity seven times greater) because titanium's greater strength permits thinner-walled equipment, relative absence of corrosion scale, erosion–corrosion resistance permitting higher operating velocities, and the inherently passive film.

Corrosion Resistance

Titanium is immune to corrosion in all naturally occurring environments. It does not corrode in air, even if polluted or moist with ocean spray. It does not corrode in soil and even the deep salt-mine-type environments where nuclear waste might be buried. It does not corrode in any naturally occurring water and most industrial wastewater streams. For these reasons, titanium has been termed the metal for the earth, and 20–30% of consumption is used in corrosion-resistance applications (see also Corrosion and corrosion inhibitors).

Even though titanium is an active metal,

$$\mathrm{Ti^{2+} + 2\,e \rightarrow Ti} \qquad E° = -1.63\ \mathrm{V}$$

$$\mathrm{TiO_2 + 4\,H^+ + 4\,e \rightarrow Ti + 2\,H_2O} \qquad E° = -0.86\ \mathrm{V}$$

it resists decomposition because of a tenacious protective oxide film. This film is insoluble, repairable, and nonporous in many chemical media and provides excellent corrosion resistance. However, where this oxide film is broken, the corrosion rate is very rapid. However, usually the presence of a small amount of water is sufficient to repair the damaged oxide film. In a seawater solution, this film is maintained in the passive region from ca -0.2 to 10 V versus the saturated calomel electrode (24–25).

Titanium is resistant to corrosion attack in oxidizing, neutral, and inhibited reducing conditions. Examples of oxidizing environments are nitric acid, oxidizing chloride ($FeCl_3$ and $CuCl_2$) solutions, and wet chlorine gas. Neutral conditions include all neutral waters (fresh, salt, and brackish), neutral salt solutions, and natural soil environments. Examples of inhibited reducing conditions are in hydrochloric or sulfuric acids with oxidizing inhibitors and in organic acids inhibited with small amounts of water. Corrosion resistances to a variety of media are given in Table 9 (26). Titanium resistance to aqueous chloride solutions and chlorine account for most of its use in corrosion-resistant applications.

Titanium corrodes very rapidly in acid fluoride environments. It is attacked in boiling HCl or H_2SO_4 at acid concentrations >1% in or ca 10 wt % acid concentration at room temperature. Titanium is also attacked by hot caustic solutions, phosphoric acid solutions (concentrations above 25 wt %), boiling $AlCl_3$ (concentrations >10 wt %), dry chlorine gas, anhydrous ammonia above 150°C, and dry hydrogen–dihydrogen sulfide above 150°C.

Table 9. Corrosion Data for ASTM Grade 2 Titanium[a]

Media	Conc, wt %	Temperature, °C	Corrosion rate, mm/yr
acetaldehyde	100	149	0.0
acetic acid	5–99.7	124	0.0
adipic acid	67	232	0.0
aluminum chloride, aerated	10	100	0.002
	10	150	0.03
	20	149	16
	25	20	0.001
	25	100	6.6
	40	121	109
ammonia + 28% urea + 20.5% H_2O + 19% CO_2 + 0.3% inerts + air	32.2	182	0.08
ammonium carbamate	50	100	0.0
ammonium perchlorate aerated	20	88	0.0
aniline hydrochloride	20	100	0.0
aqua regia	3:1	RT	0.0
	3:1	79	0.9
barium chloride, aerated	5–20	100	<0.003
bromine–water solution	.	RT	0.0
calcium chloride		RT	0.0
	5	100	0.005
	10	100	0.007
	20	100	0.02
	55	104	0.0005
	60	149	<0.003
	62	154	0.05–0.4
	73	177	2.1
calcium hypochlorite	6	100	0.001
chlorine gas, wet	>0.7 H_2O	RT	0.0
	>1.5 H_2O	200	0.0
chlorine gas, dry	<0.5 H_2O	RT	may react
chlorine dioxide in steam	5	99	0.0
chloracetic acid	100	189	<0.1
chromic acid	50	24	0.01
citric acid	25	100	0.0009
copper sulfate + 2% H_2SO_4	saturated	RT	0.02
cupric chloride, aerated	1–20	100	<0.01
cyclohexane (plus traces of formic acid)		150	0.003
ethylene dichloride	100	boiling	0.005–0.1
ferric chloride	10–30	100	<0.1
formic acid, nonaerated	10	100	2.4
hydrochloric acid, aerated	5	35	0.04
	20	35	4.4
HCl, chlorine saturated	5	190	<0.03
HCl + 10% HNO_3	5	38	0.0
HCl + 1% CrO_3	5	93	0.03
hydrofluoric acid	1–48	RT	rapid
hydrogen peroxide	3	RT	<0.1
hydrogen sulfide, steam and 0.077% mercaptans	7.65	93–110	0.0
hypochlorous acid + Cl_2O and Cl_2	17	38	0.00003
lactic acid	10	boiling	<0.1
manganous chloride, aerated	5–20	100	0.0

Table 9 (continued)

Media	Conc, wt %	Temperature, °C	Corrosion rate, mm/yr
magnesium chloride	5–40	boiling	0.0
mercuric chloride, aerated	1	100	0.0003
	5	100	0.01
	10	100	· 0.001
	55	102	0.0
mercury	100	RT	0.0
nickel chloride, aerated	5–20	100	0.0004
nitric acid	17	boiling	0.08–0.1
	70	boiling	0.05–0.9
nitric acid, red fuming	<about 2% H_2O	RT	ignition sensitive
	>about 2% H_2O	RT	nonignition sensitive
oxalic acid	1	37	0.3
oxygen, pure			ignition sensitive
phenol	saturated	21	0.1
phosphoric acid	10–30	RT	0.02–0.05
	10	boiling	10
potassium chloride	saturated	60	<0.0002
potassium dichromate			0.0
potassium hydroxide	50	27	0.01
	50	boiling	2.7
seawater, ten year test			0.0
sodium chlorate	saturated	boiling	0.0
sodium chloride	saturated	boiling	0.0
sodium chloride, titanium in contact with Teflon	23	boiling	crevice attack
sodium dichromate	saturated	RT	0.0
sodium hypochlorite + 12–15% sodium chloride + 1% sodium hydroxide + 1–2% sodium carbonate	1.5–4	66–93	0.03
stannic chloride	5	100	0.003
	24	boiling	0.04
sulfuric acid	1	boiling	2.5
sulfuric acid + 0.25% $CuSO_4$	5	93	0.0
terephthalic acid	77	218	0.0
urea–ammonia reaction mass		elevated temperature and pressure	no attack
zinc chloride	20	104	0.0
	50	150	0.0
	75	200	0.5
	80	200	203

[a] Refs. 3, 21.

Titanium is susceptible to pitting and crevice corrosion in aqueous chloride environments. The area of susceptibility is shown in Figure 3 as a function of temperature and sodium chloride content (26). The susceptibility also depends on pH. The susceptibility temperature increases parabolically from 65°C as pH is increased from zero. With ASTM grades 7 or 12, crevice corrosion attack is not observed above pH 2 until ca 270°C. Noble alloying elements shift the equilibrium potential into the passive region where a protective film is formed and maintained.

Titanium does not stress-crack in environments that cause stress-cracking of other

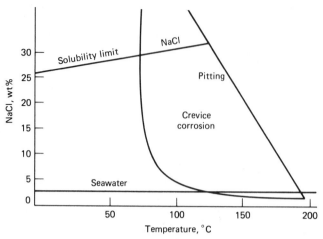

Figure 3.   Corrosion characteristics of titanium in aqueous NaCl solution (27).

metal alloys (ie, boiling 42% $MgCl_2$, NaOH, sulfides, etc). Some of the alloys are susceptible to hot-salt stress-cracking; however, this is a laboratory observation and has not been confirmed in service. Titanium stress-cracks in methanol containing acid chlorides or sulfates, red fuming nitric acid, nitrogen tetroxide, and trichloroethylene.

Titanium is susceptible to failure by hydrogen embrittlement. Hydrogen attack initiates at sites of surface iron contamination or when titanium is galvanically coupled with iron (27). In hydrogen-containing environments, titanium absorbs hydrogen above 80°C or in areas of high stress. If the surface oxide is removed by vacuum annealing or abrasion, pure dry hydrogen reacts at lower temperatures. Small amounts of oxygen or water vapor repair the oxide film and prevent this occurrence. Molybdenum alloys are less susceptible to hydrogen attack. Titanium resists oxidation in air up to 650°C. Noticeable scale forms and embrittlement occurs at higher temperatures. Surface contaminants accelerate oxidation. In the presence of oxygen, the metal does not react significantly with nitrogen. Spontaneous ignition occurs in gas mixtures containing more than 40% oxygen under impact loading or abrasion. Ignition occurs in dry halogen gases.

Titanium resists erosion–corrosion by fast-moving sand-laden water. In a high velocity, sand-laden seawater test (8.2 m/s) for a 60-d period, titanium performed more than 100 times better than 18 Cr–8 Ni stainless steel, Monel, or 70 Cu–30 Ni. Resistance to cavitation (ie, corrosion on surfaces exposed to high velocity liquids) is better than by most other structural metals (3,26).

In galvanic coupling, titanium is usually the cathode metal and consequently not attacked. The galvanic potential in flowing seawater in relation to other metals is shown in Table 10 (3). Since titanium is a cathode metal, hydrogen attack may be of concern, as it occurs with titanium complexed to iron.

Manufacture

Ore-Concentrate Refining.   The TiO_2 content of ore concentrates determines further processing steps. High grade ore (>85% TiO_2) is refined to pigment-grade TiO_2 via chlorination (DuPont chlorinates 70% TiO_2 concentrate). Lower grade ore is pro-

Table 10.   Galvanic Series in Flowing Seawater; 4 m/s at 24°C[a]

Metal	Potential, V[b]
T304 stainless steel, passive	0.08
Monel alloy	0.08
Hastelloy alloy C	0.08
unalloyed titanium	0.10
silver	0.13
T410 stainless steel, passive	0.15
nickel	0.20
T430 stainless steel, passive	0.22
70–30 copper–nickel	0.25
90–10 copper–nickel	0.28
admiralty brass	0.29
G bronze	0.31
aluminum brass	0.32
copper	0.36
naval brass	0.40
T410 stainless steel, active	0.52
T304 stainless steel, active	0.53
T430 stainless steel, active	0.57
carbon steel	0.61
cast iron	0.61
aluminum	0.79
zinc	1.03

[a] Ref. 27.
[b] Steady-state potential, negative to saturated calomel half-cell.

cessed via the sulfate route. The chlorination process, commercialized by DuPont in the early 1960s, produces a better quality pigment, requires less processing energy (1800 kW·h/t compared to 2500 kW·h/t) (28–30), and has less waste discharge (ie, discharge of sulfuric acid is more than twice the product weight). However, high grade ore is required, ie, TiO_2 content >70% with <1% MgO and 0.2% CaO, since ores with high MgO and CaO clog the chlorinator. Environmental problems have forced the industry to either shut down sulfate plants or install expensive pollution-control equipment (ca $300/t of capacity in 1980 dollars) (28). Because of the shortage of high grade TiO_2 reserves, the pigment industry must adapt the ore to the chloride process. To date, the trend is toward ore beneficiation. Ore with 50–60% TiO_2 content is beneficiated by partial reduction, then leached with sulfuric or hydrochloric acid to yield a concentrate containing >90% TiO_2, so-called synthetic rutile (28,31). Beneficiation plants have been built by Ishihara Sangy Kaisha, Associated Minerals Consolidated, Ltd., Kerr-McGee Chemical, Dhrangadhra Chemical, and Taiwan Alkali. Capacities are 25,000–100,000 t/yr.

Sulfate Process. In the sulfate process (see Fig. 4), ilmenite ore is treated with sulfuric acid at 150–180°C:

$$5 \, H_2O + FeTiO_3 + 2 \, H_2SO_4 \rightarrow FeSO_4 \cdot 7H_2O + TiOSO_4$$

The undissolved solids are removed and the liquid is evaporated under vacuum and cooled. The precipitated $FeSO_4 \cdot 7H_2O$ is filtered and the filtrate concentrated to ca 230 g/L. Heating to 90°C hydrolyzes titanyl sulfate to insoluble titanyl hydroxide:

$$TiOSO_4 + 2 \, H_2O \rightarrow TiO(OH)_2 \downarrow + H_2SO_4$$

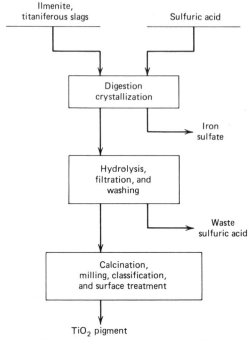

Figure 4.   The sulfate process (28).

To ensure the rutile crystal form, seed crystals are added, otherwise anatase is obtained. The precipitate is throughly washed with water and sulfuric acid to remove all traces of discoloring elements (iron, chromium, vanadium, and manganese). The $TiO(OH)_2$ is finally calcined at 1000°C to TiO_2 (29).

Chloride Process.   In the chloride process (see Fig. 5), a high grade titanium oxide ore is chlorinated in a fluidized-bed reactor in the presence of coke at 925–1010°C:

$$TiO_2 + 2\,C + 2\,Cl_2 \rightarrow 2\,CO + TiCl_4 \qquad \Delta G_{1300°C} = -125 \text{ kJ (30 kcal)}$$

The volatile chlorides are collected and the unreacted solids and nonvolatile chlorides are discarded. Titanium tetrachloride is separated from the other chlorides by double distillation (32). Vanadium oxychloride ($VOCl_3$), which has a boiling point close to $TiCl_4$, is separated by complexing with mineral oil, reducing with H_2S to $VOCl_2$, or complexing with copper. The $TiCl_4$ is finally oxidized at 985°C to TiO_2 and the chlorine gas is recycled (29,31,33) (see also Titanium compounds; Pigments, inorganic).

Tetrachloride Reduction.   Titanium tetrachloride for metal production must be of very high purity. The required purity of technical grade $TiCl_4$ for pigment production is compared with that for metal production in Table 11. Titanium tetrachloride for metal production is prepared by the same process as described above, except that a greater effort is made to remove impurities, especially oxygen and carbon containing compounds.

Sodium Reduction (Hunter) Process.   The sodium reduction process employed by RMI Company is a two-step process. In the first step, $TiCl_4$ is reduced to dichloride ($TiCl_2$) and trichloride ($TiCl_3$) at ca 230°C. In the second stage, the lower chlorides are reduced to metal by adding more sodium to the salt bath (NaCl) at ca 1000°C:

$$TiCl_4(g) + 2\,Na(l) \rightarrow TiCl_2(s) + 2\,NaCl(s) \qquad \Delta G_{230°C} = -460 \text{ kJ } (-110 \text{ kcal})$$

$$TiCl_2(s) + 2\,Na(g) \rightarrow Ti(s) + 2\,NaCl(l) \qquad \Delta G_{1000°C} = -267 \text{ kJ } (-64 \text{ kcal})$$

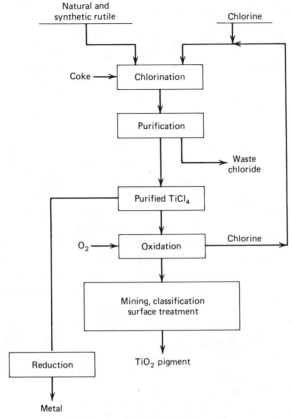

Figure 5. The chloride process.

The sponge forms in the center of the reaction vessel encased by sodium chloride. The titanium and salt mixture (spalt) is chipped from the pot, crushed, leached with dilute acid, washed, dried, and stored to await consolidation. The energy required to make Hunter sponge ($TiCl_4$ to sponge ready to melt) is ca 31 kW·h/kg (30). Ninety-seven percent of the energy is required for production of the reducing metal, sodium.

Table 11. Chemical Composition of Titanium Tetrachloride Grades, Wt %[a]

Impurity	Technical[b]	Purified[c]
$VOCl_3$	0.33	0.0034
$AlCl_3$	0.02	0.05
$SiCl_4$	0.4	0.006
Si_2OCl_6	0.04	0.003
$FeCl_3$	0.012	0.0029
CCl_3COCl	0.005	0.0002
CS_2	0.01	0.00002
COS	0.009	0.00002
$Si_3O_2Cl_8$	0.007	0.002
$COCl_2$	0.5	0.00002
other[d]	0.175	0.001

[a] Refs. 8, 31.
[b] Pigment grade.
[c] Sponge grade.
[d] Includes oxychlorides, CO_2, Cl_2, CCl_4, and C_6Cl_6.

Magnesium Reduction (Kroll) Process. In the magnesium reduction process (see Fig. 6):

$$\text{TiCl}_4(\text{g}) + 2\,\text{Mg}(\text{l}) \rightarrow \text{Ti}(\text{s}) + 2\,\text{MgCl}_2(\text{l}) \qquad \Delta G_{900°C} = -301 \text{ kJ } (-72 \text{ kcal})$$

$\text{TiCl}_4(\text{g})$ is metered into a carbon-steel or 304 stainless-steel reaction vessel which contains liquid magnesium. An excess of 25% magnesium over the stoichiometric amount ensures that the lower chlorides of titanium (TiCl_2 and TiCl_3) are reduced to metal. The highly exothermic reaction ($\Delta H_{900°C} = -420$ kJ/mol (-100 kcal/mol)) is controlled by the feed rate of TiCl_4 at ca 900°C. The reaction atmosphere is helium or argon. Molten magnesium chloride is tapped from the reactor bottom and recycled using conventional magnesium-reduction methods, including I. G. Farben, Alcan, and USSR VAMI cells. The production is in batches of 1.40–6.35 t (31,33–35). The product, so-called sponge (see Fig. 7), is further processed to remove the unreacted titanium chlorides, magnesium, and residual magnesium chlorides. These impurities, which can be as much as 30 wt %, are removed by acid leaching in dilute nitric and hydrochloric acids with low energy requirement of ca 0.3 kW·h/kg of sponge but effluent production of 8 L/kg of sponge; vacuum distillation at 960–1020°C for as much as 60 h; or the argon sweep at 1000°C used by the Oregon Metallurgical Plant. After purification, the sponge is crushed, screened, dried, and placed in air-tight 23-kg drums to await consolidation. The energy required to convert TiCl_4 to sponge, which is ready for further processing by the leaching routes (TMCA's process) is ca 37 kW·h/kg of sponge (30) of which ca 97% is required for magnesium production. The Japanese claim an energy consumption of 30 kW·h/kg of sponge using vacuum distillation instead of acid leaching for purification. They also use more efficient magnesium-electrolysis cells than those used by U.S. producers (35).

Comparison of purity of sponge produced by magnesium reduction and acid-leached, magnesium reduction and vacuum-distilled, and sodium reduction is given in Table 12 (8,36). Hardness, indicating the degree of purity, is affected by the interstitial impurities, oxygen, nitrogen, and carbon, and the noninterstitial impurity, iron.

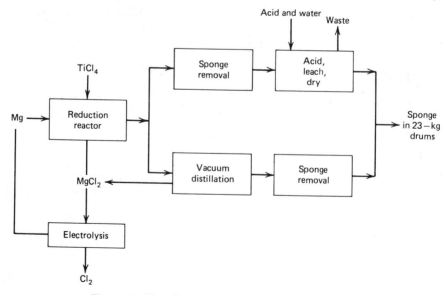

Figure 6. Flow diagram for titanium-sponge production.

Figure 7. Vacuum-distilled titanium sponge produced by magnesium reduction at Teledyne Wah Chang Albany.

Hardness numbers range from 80 to 150 HB units; typical commercial sponge is characterized by 110–120 HB units. Some processes (electrolysis reduction) produce sponge with 60–90 HB units (37). Iron impurities in Kroll sponge are difficult to control because of diffusion into the sponge from the reactor wall. In the sodium-reduction process, the sponge is protected from the wall by sodium chloride. The other impurities originate from tetrachloride, residual gases in the reactor, helium or argon impurities, or magnesium or sodium residues.

Table 12.   Comparison of ASTM Specifications for Titanium Sponge, Wt % on a Dry Basis[a]

| Property | ASTM B 299 69 | | | Electrolytic |
	MD 120 type A[b]	ML 120 type B[c]	SL 120 type C[d]	
nitrogen, max	0.015	0.015	0.010	0.003
carbon, max	0.020	0.025	0.020	0.011
sodium, total max			0.190	
magnesium, max	0.08	0.50		
chlorine, max	0.12	0.20	0.20	0.035
iron, max	0.12	0.10	0.05	0.02
silicon, max	0.04	0.04	0.04	
hydrogen, max	0.005	0.03	0.05	0.005
oxygen, max	0.10	0.10	0.10	0.065
all other impurities	0.05	0.05	0.05	
titanium balance, nominal	99.3	99.1	99.3	
HB, max	120	120	120	60–90

[a] Refs. 36–37.
[b] Type A magnesium reduced and finished by vacuum distillation.
[c] Type B magnesium reduced and finished by acid leaching on inert gas sweep distillation.
[d] Type C sodium reduced and finished by acid leaching.

Other Reduction Processes. Other methods of producing titanium have been studied with the objectives of finding another reduction route, collecting the metal as an ingot instead of sponge, and designing a continuous process. The most successful and most studied noncommercialized process is the electrolytic reduction of $TiCl_4$. Primary electrical energy reduces $TiCl_4$ to titanium metal at the cathode and chlorine gas at the anode. This process was conceived and developed at the Bureau of Mines in the mid-1950s. Since that time both TMCA and Dow-Howmet Titanium Company (D-H Titanium) have built large-scale pilot plants to study the commercial feasability. In both processes, $TiCl_4$ is fed into a molten-salt electrolyte, which for TMCA is NaCl operated at 900°C and for Dow-Howmet is KCl–LiCl eutectic operated at 520°C. The feed, which is insoluble in the electrolyte, is immediately reduced to soluble $TiCl_2$ at a feed electrode. In both processes, the anode is isolated from the bulk of the electrolyte by a diaphram which minimizes reaction of titanium ions with chlorine to liberate $TiCl_4$. The main problem is suppression of reactions of titanium ions with chlorine. Each cell has a capacity of ca 1–2 t per run which requires ca 20 kAh (about two weeks per run). The titanium electrodeposit is transferred from the salt bath into an argon chamber, removed from the cathode (after cooling by TMCA and hot by D-H Titanium), and finally acid leached. The energy requirement is considerably lower than that of other reduction processes at ca 18 kW·h/kg of leached sponge equivalent (37). The TMCA electrolytic sponge is shown in Figure 8.

Other methods include hydrogen reduction of $TiCl_4$ to $TiCl_3$ and $TiCl_2$, reduction above the melting point of titanium metal with sodium which presents a container problem, plasma reduction, in which titanium is collected as a powder however ionized and vaporized titanium combine with chlorine gas to reform $TiCl_2$ on cool-down, and aluminum reduction which reduces $TiCl_4$ to lower chlorides (38–39).

Methods that do not utilize $TiCl_4$ include reduction of TiO_2 with Al, Ca, or C, to name a few. The problems are the purity of the TiO_2, the amount of reductant remaining in the metal, and the interstitial elements remaining in the metal. Ductile metal has not been produced by direct TiO_2 reduction (40–42) (see also Electrochemical processing).

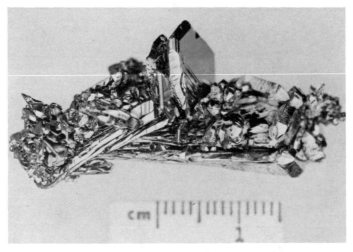

Figure 8.   Titanium metal produced by electrolytic reduction of $TiCl_4$ at Titanium Metals Corporation of America.

Sponge Consolidation. The next step is the consolidation of the sponge into ingot. The crushed sponge is blended with alloying elements or other sponge, consumable electrodes are produced by welding 90-kg sponge compactions (electrode compacts) in an inert atmosphere, and then double-vacuum-arc-remelted (VAR). The ingots are ca 71–81 cm dia and long enough to weigh 4.5 t. The double melt, included in aerospace specifications, is required for high yields because vaporization of volatiles during the first melt leaves a rough, porous surface. The double melt removes residual volatiles like Mg, $MgCl_2$, Cl_2, and H_2. Sometimes triple melts are specified for quality.

A two-station VAR furnace for double melting has an annual production capacity for ca 1400 t. The energy requirement is ca 1.1 kW·h/kg per single melt. Although now used primarily to consolidate scrap, plasma melting, electron-beam melting, and nonconsumable-electrode melting are also being investigated for sponge consolidation. In all these processes, the electrode-production steps are eliminated. Each has an energy requirement of ca 2.2 kW·h/kg of sponge.

Casting. Consolidated titanium is cast either by precision castings or investment casting. The metal is melted in a protected atmosphere, usually in a water-cooled copper crucible. In precision casting, rammed graphite molds are used; in investment casting, ceramic molds. Hot and cold isostatic pressing promotes property optimization and porosity closure. Casting companies include Titec International, Inc., Tiline Corp., Oregon Metallurgical Corp., Howmet Turbine Components Corp., Precision Cast Parts Corp., and Kobe Steel, Ltd.

Powder. In the 1940s, powder metallurgy (qv) was investigated before the advent of VAR consumable-electrode melting. In the early 1960s, DuPont had an extensive program studying powder consolidation. However, the company lost interest in 1964, partly because of technical difficulties associated with sodium and chlorine residues in sponge powder used for welding. The powder fines used were a product of the sodium-reduction process and were termed elemental powder.

Other powder-making processes include the hydride–dehydride process giving blocky powder particles that are often cracked; the electron-beam rotating-disk method (43); the Crucible Research Center Colt Titanium process which uses hydrogen charged into the metal to disintegrate it when melted; the Battelle Columbus Laboratory's process called pendant drop; and the Nuclear Metals, Inc., rotating-electrode process. Characteristics of powder made by these different processes are shown in Table 13 (44).

Table 13. Comparison of Titanium Powder-Making Processes[a]

Process	Particle size, μm Range[b]	50%[c]	Theoretical density, %	Flow rate per ASTM B 213, s	Shape
hydride–dehydride	175–400	200	50	45	blocky
electron-beam rotating disk	250–500	350	60–65	45	spherical
Colt titanium	50–600	200	60–65	40	spherical
pendant drop	100–175	125	45	no flow	fibrous
rotating electrode	150–275	175	60–65	28	spherical

[a] Ref. 44.

[b] 80% of the powder balance lies in this size range.

[c] 50% of the powder balance is below this size.

Contamination limits the amount of handling and atmospheric exposure the powder can be subjected to before consolidation. The consolidation method most widely used is cold compaction and sintering.

A principal motive for developing consolidation techniques is to reduce cost by improving yields. For aircraft parts, the so-called Buy-to-fly ratio is ca 6:1. Product yields based on sponge are shown in Figure 9. The potential improvements in yield using powder are obvious. However, low cost reliable powder-production techniques are not yet fully developed.

Metal Working. The ingots are further processed by the conventional methods of forging, hot-rolling, cold-rolling, extrusion, etc. The mill product forms include billet, bar, plate, sheet, strip, foil, extrusion, wire, and tubing. Mill practices differ somewhat from those of other metal products. Minimum heating time at the lowest practical temperatures attain the best mechanical properties, minimize contamination and oxidation, and avoid excess grain growth. A protected atmosphere of inert gas or a vacuum is preferred. Hydrogen must be excluded because it absorbs and forms hydrides, rendering the metal brittle. An oxidizing atmosphere minimizes absorption. The furnace must be kept clean to avoid contact with other metal oxides which titanium can reduce, thereby absorbing oxygen and releasing metal into the furnace. Primary working generally takes place above the phase-transition temperature, and secondary working below. Typical hot-working temperatures for some common titanium alloys are given in Table 14.

The surface is conditioned by lathe turning, grit blasting, belt grinding, centerless grinding, and caustic or acid pickling. Although other metals are similarly treated, with titanium lathe speeds are usually slower, grinding fines could present a fire hazard, and acid pickling is in 2–4 wt % hydrofluoric acid, 15–30 wt % nitric acid, and the remainder water. The amount of HF and the temperature determine the pickling rate. Hydrofluoric acid etches titanium by reacting with the oxide surface to form titanium

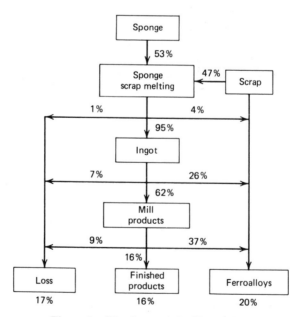

Figure 9. Titanium-metal utilization (8).

Table 14. Typical Hot-Working Temperature Ranges[a]

Alloy	β transus, °C	Forging temperature, °C Ingot	Rolling temperature, °C Bar	Plate	Sheet	Extrusion, °C
commercially pure	799–954	950–980	760–820	760–790	704–760	700–900
Ti–5 Al–2.5 Sn	1032	1120–1180	1010–1070	980–1040	980–1010	840–1040
Ti–6 Al–2 Sn–4 Zr–2 Mo	993	1090–1150	950–1010	950–980	930–982	500–950
Ti–8 Al–1 Mo–1 V	1038	1120–1170	1010–1070	980–1040	982–1040	950–1010
Ti–6 Al–4 V	993	1090–1150	950–1010	930–980	900–930	980–1150
Ti–6 Al–6 V–2 Sn	946	1040–1090	900–950	870–930	870–900	800–1040

[a] Ref. 45.

fluoride complex ions. Nitric acid is added to minimize hydrogen pickup (see Metal surface treatments).

Sheet, thin plate, welded tubing, and small-diameter bar are manufactured by conventional cold-working techniques. The formability of titanium and its alloys, when worked at room temperature, is like that of cold-rolled stainless steel. At 650°C, the formability compares with stainless steel annealed at room temperature. Cold-working may be difficult for some titanium alloys and heat may be required, especially for severe forming operations. Generally, titanium and its alloys are worked between 200 and 300°C. Lubricants reduce friction and galling. Slow forming speeds at controlled rates improve workability and are recommended for more difficult operations. After forming, the parts usually require a stress-relief annealing at ca 550°C for commercial grades (21).

Fabrication. Fabrication of titanium into useful parts like tanks, heat exchangers, pressure vessels, etc, is comparable with the fabrication of austenitic steel in method, degree of difficulty, and cost. Commercial-grade titanium can be bent 105° without cracking around a radius of 2–2.5 times the sheet thickness. The bend radius for alloys is as high as five times the sheet thickness. A loss of 15–25° in the included bend angles is normal because of springback at room temperature. Heat is required to form most titanium metal parts.

Welding (qv) of titanium requires a protected atmosphere of inert gas. Furthermore, parts and filler wire are cleaned with acetone (trichloroethylene is not recommended); the pieces to be welded are clamped, not tacked, unless tacks are shielded with inert gas; a test sample should be welded; coated electrodes are excluded; and higher purity metal is preferred as filler. Titanium cannot be fusion-welded to other metals because of formation of brittle intermetallic phases in the weld zone.

In some applications, a titanium cladding, especially Detaclad, is desirable for cost reduction. In this process, ca 2 mm of ASTM grade-1 titanium sheet is explosively bonded to steel plate (see Metallic coatings, explosively clad metals). The cladding is cost effective for wall thickness >1.5 cm. Another common cladding method is loose lining. Experimental methods include vacuum rolling and resistance welding.

Economic Aspects

Titanium raw-material utilization can be broken down as illustrated in Figure 10. About 5% of the titanium mined is used as metal, 93% is used as pigment-grade TiO_2, and 2% as ore-grade rutile for fluxes and ceramics. In 1979, the U.S. estimated

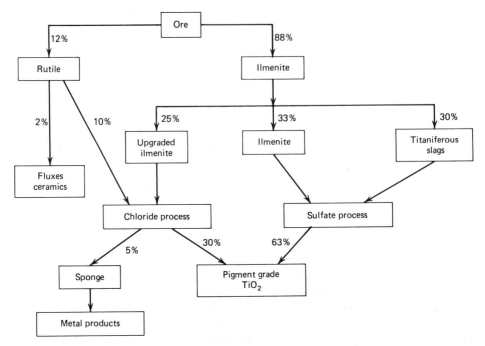

Figure 10.   U.S. titanium raw-material utilization (4,6–7).

demand of titanium ore (TiO_2 content) for these applications was as follows: pigments, 882,000 metric tons; titanium metal, 35,000 t TiO_2 (= 20,000 t metal); and fluxes and ceramics, 40,000 t. In the United States, only 4% of the processed raw material goes to metal production. In 1979, ca 1.8×10^6 t of titanium content was produced in the world. At \$2.20/kg for contained titanium, the pigment industry's sales amounted to ca 3.7×10^9. The metal industry's sales were valued at 2.5×10^9 at an average of \$27.60/kg for mill products, including USSR metal production which is ca 50% of the industry's capacity (6,46).

The principal world producers of pigment-grade TiO_2 are the United States, Western Europe, and Japan (see Table 15); consumption is shown in Fig. 11. The growth rate from 1960 to 1973 was ca 8% annually. Consumption decreased sharply after the 1973 oil crisis, from 2×10^6 to 1.5×10^6 t. The demand has recovered to about the peak level before the oil crisis. The U.S. TiO_2 price history in constant 1980 dollars is also shown in Figure 11 (see Pigments, inorganic).

The history of titanium sponge production in the United States and Japan is shown in Figure 12. The U.S. metal demand has been greater than sponge production, which has been supplemented by imports primarily from Japan, with the USSR participating heavily in the mid-1970s. Imports supply ca 10–20% of the U.S. demand. The United States does not supply its own demands because the producers have been reluctant to add capacity for peak demand periods since capital requirement is large (\$22/kg of annual capacity); producers prefer to melt sponge to ingots and consequently do not supply sponge to nonproducer melters; and the quality of U.S.-produced sponge is inferior to foreign vacuum-distilled sponge. Companies engaged in sponge melting and their estimated capacities are given in Table 16. The price history of titanium sponge is given in Table 17.

Table 15.   TiO₂ Pigment Production Plants [a]

Country	Company	Capacity, (10^3 t/yr) Sulfate process	Capacity, (10^3 t/yr) Chloride process	Percent of world capacity
United States	E. I. du Pont de Nemours & Co., Inc.		657	24
UK	BTP Tioxide, Ltd.	353	30	14
United States	N.L. Industries, Inc.	329	16	13
FRG	Bayer A.G.	118	35	6
UK	LaPorte Ind.	84	40	5
USSR	Tecmasluimport	124		5
United States	American Cyanamid Co.	77	36	4
United States	SCM Corp.	48	65	4
Japan	Ishihara Sangyo Kaisha	92	12	4
France	Thann et Mulhouse S.A.	84		3
Italy	Montedison SpA	81		3
United States	Gulf & Western Natural Resources		72	2
	others [b]	311	47	13
Total		*1701*	*1010*	

[a] Refs. 4, 6, 28, 47.

[b] Companies with capacity less than 60,000 t/yr and include companies in the United States, Finland, Japan, Poland, the FRG, Spain, Czechoslovakia, and the Republic of Korea.

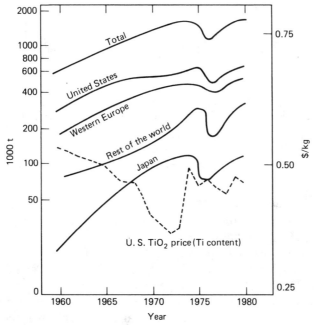

Figure 11.   World TiO₂ consumption with U.S. selling price (Ti content) in 1980 dollars (---) (4,7).

The U.S. titanium market distribution is shown in Table 18 (28,48). Before 1970, more than 90% of the titanium produced was used for aerospace, which fell to ca 70–80% by 1982. World consumption is given in Table 19. In contrast to the United States, aerospace uses in Western Europe and Japan account for only 40–50% of the demand (49). The USSR consumption of titanium metal is about one half of the world

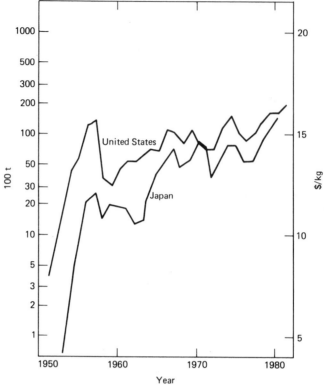

Figure 12.   Titanium-sponge production in the United States and Japan (6,46).

consumption. Although the USSR distribution is not known, it is believed that in 1980 considerable amounts were used for deep-submersible construction. It is also believed that their use in nonaerospace applications is greater than in the United States. The world production facilities for titanium metal extraction, including capacity and process type, are given in Table 20 (48,50). The distribution of corrosion-resistant nonaerospace applications is given in Table 21.

Specifications, Standards, and Quality Control

The alloys of titanium have compositional specifications tabulated by ASTM. The ASTM specification number is given in Table 6 for the commercially important alloys. Military specifications are found under MIL-T-9046 and MIL-T-9047, and aerospace material specifications for bar, sheet, tubing, and wire under specification numbers 4900–4980. Each aircraft company has its own set of alloy specifications.

The alloy name in the United States usually includes a company name or trademark in conjunction with the composition for alloyed titanium or the strength (ultimate tensile strength for TMCA and yield strength for other U.S. producers) for unalloyed titanium. The common alloys and company designations are shown in Table 22.

Since titanium alloys are used in a variety of applications, several different material and quality standards are specified. Among them are ASTM, ASME, AMS, U.S. military, and a number of proprietary sources. The correct chemistry is basic to obtaining mechanical and other properties required for a given application. Minor elements controlled by specification include carbon, iron, hydrogen, nitrogen, and oxygen.

Table 16. Titanium Melters and Capacities[a]

Company	Capacity, metric tons
Western Europe	
IMI Titanium, UK	6,000
Krupp Stahl AG, FRG	1,200
Sézus, France	
Contimet, FRG	2,800
W. C. Heraeus, FRG	
Total	*10,000*
Japan	
Kobe Steel, Ltd.	6,000
Nippon Mining Co., Ltd.	1,200
Osaka Titanium	3,000
Toho Titanium	
Total	*10,200*
United States	
Crucible Steel	900
Lawrence Aviation	900
Howmet Turbine Components Corp.	4,500
Martin-Marietta	3,600
Oregon Metallurgical	4,000
RMI Company	9,000
Teledyne Allvac, Inc.	2,700
Titanium Metals Corporation of America	13,600
Teledyne Wah Chang Albany	1,400
Total	*40,600*
Grand total	*60,800*

[a] Refs. 6, 37.

Table 17. Price History of US Titanium Sponge

Year	$/kg
1948–1955	11.02
1964–1975	2.91
early 1982	16.86
mid-1982	12.20

For more stringent applications, yttrium may also be specified. In addition, control of thermomechanical processing and subsequent heat treatment is vital to obtaining desired properties. For extremely critical applications, such as rotating parts in aircraft gas turbines, raw materials, melting parameters, chemistry, thermomechanical processing, heat treatment, test, and finishing operations must be carefully and closely controlled at each step to assure that required characteristics are present in the products supplied.

Health and Safety Factors

Titanium and its corrosion products are nontoxic. A safety problem does exist with titanium grindings, turnings, and some corrosion products which are pyrophoric. Grindings and turnings should be stored in a closed container and not left on the floor. Smoking must be prohibited in areas where titanium is ground or turned and, if a fire

Table 18.   U.S. Titanium Market Distribution [a]

Market	1955	1961	1966	1975	1979
aerospace, %					
military	94	72	74	54	45
civilian	3	20	19	20	35
industrial, %	3	6	7	26	20
mill products, t	910	2940	6580	7030	9750

[a] Refs. 13, 48.

Table 19.   Titanium Sponge and Scrap Consumption in the World, 1000 t [a]

Year	United States	Western Europe	Japan	USSR
1968	17.2	10.6	2.0	13.2
1970	21.5	14.0	4.3	12.0
1972	18.9	12.7	2.4	28.3
1974	34.0	18.4	5.8	26.9
1976	20.4	19.0	4.3	35.8
1978	29.2	22.3	6.2	45.8
1980	38.4			
1981	42.1			

[a] Refs. 7, 46.

occurs, it must be extinguished with a class D extinguisher (for use against metal fires). The larger the surface area, the more pyrophoric the titanium fines. When titanium equipment is being worked on, all flammable products and corrosive products must be removed, and the area well ventilated. A pyrophoric corrosion product has been observed in environments of dry Cl_2 gas and in dry red fuming nitric acids.

Uses

Titanium is primarily used in the form of high purity titanium oxide. Although the principal application of high purity (pigment-grade) TiO_2 is in paint pigments, other important uses are in plastics (for color in floor-covering products and to help protect plastic products and foodstuffs contained in plastic bags from ultraviolet radiation deterioration), paper (as a filler and whitener), and in rubber (see Table 23). Future applications areas include TiO_2 single-crystal electrodes for water decomposition for the production of hydrogen fuel, flue-gas denitrification catalysts, and high purity TiO_2 to make barium titanate thermistors.

Titanium metal was first established as a material for aerospace, "metal-for-air" applications. In the late 1970s, it was developed as "metal-for-sea" uses. The metal-for-air and metal-for-sea designations characterize Japanese market development goals. In terms of volume, the U.S. titanium industry is still in the metal-for-air development stage; the statements about metal-for-earth and -sea reflect an optimistic outlook.

In the United States, the high strength:weight ratio of titanium accounts for ca 70% of its uses. Before 1970, the high strength:weight ratio was the basis of over 90% of applications, such as engines, where the advantage of light weight is translated to

Table 20. Production Capacity of Titanium Sponge, Metric Tons/Yr[a]

Location	Producer	Process	1981–1982	Future prospect, 1985
Japan				
Amagasaki	OTC	Mg-distillation	18,000	25,000
Chigasaki	TTC	Mg-distillation	12,000	18,000
Nihongi	NMI	Na-leach	2,000	4,800
Akita	Mitsubishi			6,000
Total			*32,000*	*53,800*
United States				
Henderson, Nevada	Timet	Mg-leach	13,600	14,500
Ashtabula, Ohio	RMI	Na-leach	8,600	8,600
Albany, Oregon	Ormet	Mg–Ar distillation	3,000	3,400
Free Port, Texas	D-H Titanium	electrolysis	500	900
Albany, Oregon	TWCA	Mg-distillation	1,500	1,500
Moses Lake, Washington	International Titanium	Mg-distillation	2,000	4,500
Total			*29,200*	*33,400*
Europe				
UK	ICI	Na-leach	3,000	
	Deeside Titanium	Na-leach		5,000
People's Republic of China	5 plants	Mg/Na distillation/leach	2,000	5,000
USSR	4 plants	Mg-distillation	45,000	60,000
World Total			*111,200*	*157,200*

[a] Refs. 34, 48, 50.

higher flying, faster planes. Aerospace applications have shaped and controlled the titanium-metal industry.

The use of titanium in aircraft is divided about equally between engines and airframes. For engine components, titanium is limited because of temperature constraints at the compressor area where it is used as blades, casings, and disks. In the frame, it is used in bulkheads, firewall, flap tracks, landing-gear parts, wiring pivot structures, fasteners, rotor hubs, and hot-area skins. In the F-15, titanium accounts for about 32% of the structural weight. Design changes and weight savings owing to the use of titanium in Pratt and Whitney's JT3D engine, employed to power Boeing 707 and Douglas DC 8 aircraft, resulted in 42% more takeoff thrust, 13% lower specific fuel consumption, and 18% less weight than the prior JT3C engine (9).

The other outstanding property of titanium metal is its corrosion resistance, although its use in corrosion-resistance applications in 1980 in the United States was a mere 5000 t or ca 0.001% of the metal used in corrosion-resistance markets. The largest application was heat-exchanger pipes and tubing (ca 800 μm or 22 gauge welded) for the power industry, and marine and desalination applications, where titanium provides protection against corrosion by seawater, brackish water, and other estuary waters containing high concentrations of chlorides and industrial wastes (see Heat-exchange technology).

Titanium metal is especially utilized in environments of wet chlorine gas and bleaching solutions, ie, in the chlor-alkali industry and the pulp and paper industries.

Table 21. Distribution of Titanium in Nonaerospace Corrosion-Resistant Applications, 1980[a]

Application	1000 metric tons[b]
power industry[b]	2.3
chlor-alkali industry	1.6
tanks, mixers, and heat exchangers for chemical production	1.6
bleaching and chemical generation equipment for pulp and paper	1.1
metal coating and recovery	1.4
oil and gas production	0.5
marine	0.5
desalinization	0.5
other[d]	1.4
Total	10.9

[a] Ref. 51.
[b] United States, Western Europe, and Japan.
[c] Especially heat-exchanger tubing.
[d] Prosthetic, environmental, etc.

Table 22. Company Names of Common Titanium Alloys[a]

Alloys	ASTM	Cabot	IMI[b]	RMI	Timet	USSR
99.5 Ti	grade 2	CABOT Ti 40	IMI-125	RMI 40	Ti–50 A	VT1-0
99.2 Ti	grade 3	CABOT Ti 55	IMI-130	RMI 55	Ti–65 A	VT1
99.0 Ti	grade 4	CABOT Ti 70	IMI-155	RMI 70	Ti–75 A	VT1-1
Ti–5 Al–2.5 Sn	grade 6		IMI-315	RMI 5 Al–2.5 Sn	Ti–5 Al–2.5 Sn	VT5-1
Ti–6 Al–4 V	grade 5	CABOT Ti–6 Al–4 V	IMI-317	RMI 6 Al–4 V	Ti–6 Al–4 V	VT6

[a] Refs. 8, 14.
[b] IMI = IMI Limited, Witton, Birmingham, UK.

Table 23. Distribution of Titanium Dioxide Consumption, 1978, %[a]

Property	Japan	UK	Western Europe	North America	World
paint	56	70	66	52	63
plastics	17	13	15	12	12
paper	10	5	10	22	14
others	27	12	9	14	11

[a] Ref. 7.

Here, titanium is used as anodes for chlorine production, chlorine–caustic scrubbers, pulp washers, and Cl_2, ClO_2, and $HClO_4$ storage and piping equipment (see also Alkali and chlorine products; Paper; Pulp).

In the chemical industry, titanium is used in heat-exchanger tubing for salt production, in the production of ethylene glycol, ethylene oxide, propylene oxide, and terephthalic acid, and in industrial wastewater treatment. Titanium is used in environments of aqueous chloride salts ($ZnCl_2$, NH_4Cl, $CaCl_2$, $MgCl_2$, etc), chlorine gas, chlorinated hydrocarbons, and nitric acid.

In metal recovery, titanium is used for ore-leaching solutions and as racks for metal plating. The leaching solutions contain HCl or H_2SO_4 with enough ferric or cupric ions to inhibit the corrosion of titanium. In metal-plating applications, titanium is anodically protected against H_2SO_4 and chrome-plating solution corrosion.

In oil and gas refinery applications, titanium is used as protection in environments of H_2S, SO_2, CO_2, NH_3, caustic solutions, steam, and cooling water. It is used in heat-exchanger condensers for the fractional condensation of crude hydrocarbons, NH_3, propane and desulfurization products using seawater or brackish water for cooling.

Other application areas include nuclear-waste storage canisters, pacemaker castings, implantations, geothermal equipment, automotive connection rods, ordnance, etc.

BIBLIOGRAPHY

"Titanium and Titanium Alloys" in *ECT* 1st ed., Vol. 14, pp. 190–213, by C. H. Winter, Jr., and E. A. Gee, E. I. du Pont de Nemours & Co., Inc.; "Titanium and Titanium Alloys" in *ECT* 1st ed., 2nd Suppl., pp. 866–873, by H. R. Ogden, Battelle Memorial Institute, Columbus, Ohio; "Titanium and Titanium Alloys" in *ECT* 2nd ed., Vol. 20, pp. 347–379, by H. B. Bomberger, Reactive Metals, Inc.

1. M. J. Donachie, Jr., ed., *Titanium and Titanium Alloys*, American Society for Metals, Metals Park, Ohio, 1982.
2. S. C. Williams, *Report on Titanium*, J. W. Edwards, Inc., Ann Arbor, Mich., 1965.
3. L. C. Covington, R. W. Schultz, and I. A. Fronson, *Chem. Eng. Prog.* **74,** 67 (1978).
4. L. E. Lynd in *Mineral Facts and Problems*, Bulletin 671, U.S. Department of the Interior, Bureau of Mines, Washington, D.C., 1980.
5. L. W. Trevoy in R. F. Meyer and C. T. Steele, eds., *International Conference on The Future of Heavy Crude Oils and Tar Sands*, Mining Informational Services, McGraw-Hill, New York, 1981, p. 698.
6. L. E. Lynd in *Mineral Commodity Profiles MCP-18*, U.S. Department of the Interior, Bureau of Mines, Washington, D.C., 1978.
7. H. Schmidt and P. Eggert, *Unterschungen üder Angebot und Nachfrage Mineral ischer Roh Stoffe Titan*, ISSN 0343-8120, Deutsches Institut Fur Wirtschaftsforschary, Berlin, FRG, 1980.
8. R. A. Wood, *The Titanium Industry in the Mid-1970's*, Battelle Report MCIC-75-26, Battelle Memorial Institute, Columbus, Ohio, June 1975.
9. A. D. McQuillan and M. K. McQuillan in H. M. Finniston, ed., *Metallurgy of the Rarer Metals*, Academic Press, Inc., New York, 1956, p. 335.
10. *ASTM Standard Specification for Titanium and Titanium Alloy Strip, Sheet, and Plate*, ANSI-ASTM B265-79, American Society for Testing and Materials, Philadelphia, Penn., Oct. 1980.
11. R. M. Duncan, P. A. Blenkinsop, and R. E. Goosey in G. W. Meethan, ed., *The Development of Gas Turbine Materials*, John Wiley & Sons, Inc., New York, 1981, p. 63.
12. *Facts About the Metallography of Titanium*, RMI Company, Niles, Ohio, 1975.
13. R. I. Jaffee in H. Kimura and O. Izumi, eds., *Titanium '80 Science and Technology*, The Metallurgical Society/American Institute of Mining, Metallurgical and Petroleum Engineers, Warrendale, Penna., 1980, p. 53.
14. H. Hucek and M. Wahll, *Handbook of International Alloy Compositions and Designations*, Vol. 1, Battelle Report MCIC-HB-09, Battelle Memorial Institute, Columbus, Ohio, Nov. 1976.
15. *Metals Prog. Databook* **110,** 94 (June 1976).
16. S. G. Glazunov in N. P. Sazhin and co-workers, eds., *Titanium Alloys for Modern Technology*, NASA TT F-596, National Aeronautics and Space Administration, Washington, D.C., March 1970, p. 11.
17. C. C. Chen and R. R. Boyer, *J. Met.* **31,** 33 (1979).
18. E. L. Hayman, D. W. Greenwood, and B. G. Martin, *Exp. Mech.* **17,** 161 (May 1977).
19. E. M. Savitskiy, M. I. Bychkova, and V. V. Baron in ref. 13, p. 735.
20. C. M. Wayman, *J. Met.* **32,** 129 (1980).
21. *How to Use Titanium Properties and Fabrication of Titanium Mill Products*, Titanium Metals Corporation of America, Pittsburgh, Penn., 1975.

22. H. R. Ogden and F. C. Holden, *Metallography of Titanium Alloys*, TML Report No. 103, Battelle Memorial Institute, Columbus, Ohio, May 29, 1958.
23. *Metals Handbook*, Vol. 7, American Society for Metals, Metals Park, Ohio, 1972.
24. T. R. Beck in R. W. Staehle, B. F. Brown, J. Kruger, and A. Agarwal, eds., *Localized Corrosion*, Vol. Nace-3, National Association of Corrosion Engineers, Houston, Texas, 1974, p. 644.
25. E. E. Millaway, *Mater. Prot.* **4,** 16 (1965).
26. *Titanium for Industrial Brine and Sea Water Service*, Titanium Metal Corporation of America, Pittsburgh, Penn., 1968.
27. L. C. Covington and R. W. Schultz in E. W. Kleefisch ed., *Industrial Applications of Titanium and Zirconium*, STP 728, American Society for Testing and Materials, Philadelphia, Penn., 1981, p. 163.
28. N. Ohta, *Chem. Eco. Eng. Rev.* **13,** 22 (1981).
29. G. E. Haddeland and S. Morikawa, *Titanium Dioxide Pigment*, Process Economics Program Report No. 117, Stanford Research Institute International, Menlo Park, Calif., April 1978.
30. *Energy Use Patterns in Metallurgical and Nonmetallic Mineral Processing*, Report No. PB-246 357, Battelle Columbus Laboratories, U.S. Department of Commerce, Washington, D.C., Sept. 16, 1975.
31. W. W. Minkler and E. F. Baroch, *The Production of Titanium, Zirconium and Hafnium*, 1981.
32. R. C. Weast, ed., *CRC Handbook of Chemistry*, 62nd ed., CRC Press, Inc., Boca Raton, Fla., 1982.
33. W. W. Minkler in ref. 13, p. 217.
34. W. W. Minkler, *J. Met.* **33,** 41 (1981).
35. K. Takahashi in ref. 13, p. 15.
36. *ASTM Standard Specification for Titanium Sponge*, ANSI-ASTM B 265-74, American Society for Testing and Materials, Philadelphia, Penn., Oct. 1980.
37. G. Cobel, J. Fisher, and L. E. Snyder in ref. 13, p. 1969.
38. Brit. Pat. 1,355,433 (June 5, 1974), P. D. Johnston, J. Lawton, and I. M. Parker (to the Electricity Council).
39. Fr. Demande 2,002,771 (Oct. 31, 1969), A. G. Halomet.
40. U.S. Pat. 3,429,691 (Feb. 25, 1969), W. J. McLauglin (to Aerojet-General Corp.).
41. U.S. Pat. 3,794,482 (Feb. 26, 1974), R. N. Anderson and N. A. D. Parlee (to Parlee Anderson Corp.).
42. G. V. Samsonov and V. S. Sinelinikova in A. T. Dogvinenko, ed., *Metalloterm. Protsessy Khim. Met. Maters. Konf.*, Nauka. Sib Otd., Novosibirsk, USSR, 1971, p. 32.
43. R. Ruthardt, H. Stephan, and W. Dietrich in ref. 13, p. 2289.
44. R. E. Peebles and C. A. Kelto in F. H. Froes and J. E. Smugeresky, eds., *Powder Metallurgy of Titanium Alloys*, The Metallurgical Society/American Institute of Mining Engineers, Warrendale, Penn., 1980, p. 47.
45. K. Lane and H. Stenger, *Extrusion*, American Society for Metals, Metals Park, Ohio, 1981, p. 180.
46. *Mineral Industry Surveys*, *Titanium*, quarterly report, U.S. Department of the Interior, Bureau of Mines, Washington, D.C.
47. *Chem. Week*, 67 (Oct. 13, 1982).
48. W. W. Minkler in *Assessment of Selected Materials Issues*, National Materials Advisory Board-National Academy of Sciences, Washington, D.C., 1981.
49. L. Kovisars, *Titanium Metal Winning and Market Outlook*, SRI Project ECC-6356, private report prepared by Stanford Research Institute International, Menlo Park, Calif., Aug. 1980.
50. K. Kitaoka, Kobe Steel, Japan, private communication, Aug. 6, 1981.
51. *Chem. Week*, 42 (Aug. 26, 1981).

Donald Knittel
Cabot Corporation

TITANIUM COMPOUNDS

Inorganic,   131
Organic,   176

INORGANIC

Titanium (at no. 22; ionization potentials: first 6.83 eV, second 13.57 eV, third 27.47 eV, fourth 43.24 eV) is the first member of Group IVB of the periodic chart. It has four valence electrons, and Ti(IV) is the most stable valence state. The lower valence states Ti(II) and Ti(III) exist, but these are readily oxidized to the tetravalent state by air, water, and other oxidizing agents. The ionization potentials indicate that the Ti^{4+} ion would not be expected to exist and, indeed, Ti(IV) compounds are generally covalent. Titanium is able to expand its outer group of electrons and can form a large number of addition compounds by coordinating other substances having a donor atom, eg, oxygen or sulfur. The most important commercial forms are titanium dioxide, over 2×10^6 metric tons of which were sold in 1981, and titanium metal. Titanium dioxide is amphoteric, giving rise to a series of titanates as well as salts of Ti(IV), which are readily hydrolyzed in aqueous solution. This readiness to form hydrated titanium oxides is the basis of the first commercial process for the manufacture of titanium dioxide pigments (see Pigments, inorganic; Paint). Titanium properties and chemistry are reviewed in refs. 1–10.

Thermochemical Data

Data relating to changes of state of selected titanium compounds are listed in Table 1. Table 2 gives values for heats of formation, free energy of formation, and entropy of a number of titanium compounds at two temperatures, 298 K and 1300 K.

Titanium–Hydrogen System

The absorption of hydrogen by titanium is a reversible process, proceeding rapidly at temperatures above 400°C to a maximum composition of $TiH_{1.7}$. The vapor pressure of the hydride at this composition is high, and this effectively prevents further hydrogen absorption. At each temperature and composition of the solid, there is an equilibrium pressure of hydrogen. The phase diagram has been investigated by pressure–temperature–composition and x-ray studies (11). Three phases can exist in the solid; they are, in order of increasing hydrogen content, an alpha phase which has an hcp structure containing hydrogen up to a maximum of 7.8 at % ($TiH_{0.085}$), a beta phase which has a body-centered cubic structure, and a gamma phase which is face-centered cubic. Change to the latter phase is complete at a hydrogen composition of 48–60 at % ($TiH_{0.9}$–$TiH_{1.5}$). The gamma phase is stable at room temperature and has sometimes been identified with the compound TiH_2 [7704-98-5] (66.7 at % H). A compound with the stoichiometric formula TiH_2 has been claimed to have been prepared by the reduction of pure titanium dioxide with calcium hydride, ie, starting with uncontaminated titanium. The existence of TiH_2 is not well substantiated.

Table 1.   Thermal Data for Changes of State of Titanium Compounds

Compound	CAS Registry No.	Change	Temperature, K	ΔH, kJ/mol[a]
$TiCl_4$	[7550-45-0]	mp	249.05	9.966
		bp	409	35.77
$TiCl_3$	[7705-07-9]	sublimation temp	1104.1	166.15
$TiCl_2$	[10049-06-6]	sublimation temp	1581.5	248.5
TiI_4	[7720-83-4]	mp	428	19.83 ± 0.63
		bp	652.6	56.48 ± 2.09
TiF_4	[7783-63-3]	sublimation temp	558.6	97.78 ± 0.42
$TiBr_4$	[7789-68-6]	mp	311.4	12.89
		bp	504.1	45.19
TiO_2	[13463-67-7]	phase change (anatase to rutile)		ca −12.6

[a] To convert kJ/mol to kcal/mol, divide by 4.184.

Table 2.   Thermochemical Data for Formation of Titanium Compounds

Compound	CAS Registry No.	State	Heat of formation ($\Delta H_f^{\circ a}$), kJ/mol[b] at 298 K	at 1300 K	Free energy of formation (ΔG_f°), kJ/mol[b] at 298 K	at 1300 K	Entropy S, J/(mol·K)[b] at 298 K	at 1300 K
TiO	[12137-20-1]	crystal	−519.6	−515.9	−495.1	−417.1	50.2	127.0
TiO_2								
anatase		crystal	−933.0	−930.0	−877.6	−697.4	49.9	150.6
rutile		crystal	−944.7	−942.4	−889.5	−707.9	50.3	149.0
TiC	[12070-08-05]	crystal	−184.1	−188.2	−180.5	−169.1	24.2	92.25
TiN	[25583-20-4]	crystal	−337.6	−337.6	−309.0	−215.0	30.2	100.8
$TiCl_2$		crystal	−515.5	−504.2	−465.5	−314.8	87.4	204.2
$TiCl_3$		crystal	−721.7	−698.2	−654.5	−445.9	139.7	290.9
$TiCl_4$		liquid	−804.2	−771.4	−737.3	−604.6	252.4	431.2
$TiCl_4$		gas	−763.2	−765.8	−726.8	−607.0	354.8	507.4
$TiBr_4$		crystal	−618.0	−654.5	−593.3	−347.2	243.6	440.6
$TiBr_3$	[13135-31-4]	crystal	−550.2	−546.3	−525.6	−341.1	176.4	368.7
TiF_4		crystal	−1649.3	−1611.1	−1559.2	−1290.3	134.0	338.7
TiI_4		crystal	−413.4	−453.0	−370.7	−164.6	246.1	486.6
TiI_3	[13783-08-9]	crystal	−322.2	−385.7	−318.5	−149.2	192.5	368.5
TiI_2	[13783-07-8]	crystal	−266.1	−310.1	−258.9	−129.6	122.6	253.7

[a] H_f and G_f refer to the formation of the named substances in the named states at 298 K and 1300 K from their elements in the standard states of those elements at these temperatures.
[b] To convert J to cal, divide by 4.184.

Titanium hydride is a light gray powder of metallic appearance, contains ca 4 wt % hydrogen, is stable in air, and has a density of 3.9 g/cm³. When heated to ca 600°C, it decomposes and the liberated hydrogen is a very convenient source of pure hydrogen. Experimental details have been given on the use of titanium for purifying cylinder hydrogen by alternate formation and decomposition of titanium hydride (12).

Titanium hydride has been used as the starting point for the preparation of titanium borides, nitrides, and silicides, the evolved hydrogen providing a convenient reducing atmosphere.

It has been used for forming glass- or ceramic-to-metal seals and for producing titanium coatings on copper or copper-plated metals. It acts as a catalyst in the hydrogenation of certain unsaturated organic compounds. A recent review on research and development into the titanium–hydrogen system includes a discussion of the possibility of its use for thermal storage (13) (see Hydrogen energy).

Titanium–Boron System

The equilibrium diagram of the titanium–boron system shows the presence of five phases, ie, TiB_2 [12405-63-5], Ti_2B [12305-68-9], TiB [12007-08-8], Ti_2B_5 [12447-59-5], and TiB_{12} [51311-04-7] (14). Of these, only the diboride, TiB_2, is well known. It has a hexagonal crystal structure with parameters $a = 0.3028$ nm and $c = 0.3228$ nm and the following properties: density 4.52 g/cm³, hardness (qv) 9 on Mohs scale and 33.3 GPa (3400 kgf/mm²) HV, mp 2920°C, and electrical conductivity 28.4 $\mu\Omega$·cm at 20°C and becoming superconducting at 1.26 K.

Titanium diboride is a gray crystalline solid that is not attacked by cold concentrated hydrochloric or sulfuric acids, but dissolves slowly at boiling temperatures. It dissolves readily in mixtures of nitric acid and hydrogen peroxide or sulfuric acid. It also decomposes upon fusion with alkali hydroxides, carbonates, or bisulfates. Titanium diboride may be prepared by a number of methods (15): direct combination of titanium metal or its hydride with boron at 2000°C; reduction of oxides of titanium and boron with carbon, alkali metals, magnesium, and aluminum (a recent method involves heating the mixed oxides in a graphite crucible with an argon-plasma burner) (16); reaction of titanium metal, its oxides, or carbide with boron carbide or a mixture of boron carbide and boric anhydride; fused-salt electrolysis of titanium oxides and boron and double fluorides; or vapor-phase reactions of titanium tetrachloride and boron trichloride, eg, with hot plasma gas (17–18).

Because of its hardness and resistance to oxidation, titanium boride has become increasingly important in the aircraft industry and as a replacement for diamond dust in cutting and grinding applications. Titanium diboride is also useful as a protective coating for metals, eg, tungsten, molybdenum, and tantalum, in high temperature applications. Titanium diboride is an important base material for cermet production (see High temperature composites).

Titanium–Carbon System

Only one compound of titanium and carbon has been isolated, ie, TiC. Titanium carbide has an fcc structure similar to NaCl with a lattice parameter of 0.4328 nm and a density of 4.939 g/cm³. It is, however, very difficult to prepare TiC that is free from impurities, eg, oxygen and nitrogen, and some variations in the lattice parameter and density values have been reported. The titanium–carbon system has a wide composition range (19–20). The fcc structure is stable from $TiC_{1.0}$ to $TiC_{0.47}$; with reduction in the carbon content, the lattice parameter and density decrease. From $TiC_{0.47}$ to $TiC_{0.08}$, the phases α, β-Ti and TiC are observed, and below $Ti_{0.08}$ only α Ti is present (see Carbides).

Titanium carbide is manufactured by the reduction of titanium dioxide with carbon.

$$TiO_2 + 3\,C \rightarrow TiC + 2\,CO$$

An intimate mixture of titanium dioxide and carbon is heated to 1900–2100°C in an electric arc furnace or a graphite tube. The product is a central mass of titanium carbide surrounded by partially reacted products, in which lower oxides of titanium can be identified. The reaction proceeds in stages where titanium-reduced oxides are intermediates.

$$TiO_2 \longrightarrow \underset{[12065\text{-}65\text{-}5]}{Ti_3O_5} \longrightarrow \underset{[1344\text{-}54\text{-}3]}{Ti_2O_3} \longrightarrow TiO \longrightarrow TiC$$

Alternatively, the mixture of titanium dioxide and graphite can be heated in a tungsten tube under an atmosphere of hydrogen, in which case hydrocarbons are possible intermediate compounds, eg,

$$TiO + C_2H_2 \rightarrow TiC + CO + H_2$$

In another process, a metal or metallic oxide is mixed with excess carbon and is heated to ca 3000°C. Reduction and solution of the metals and carbon in the menstruum occurs, and either immediately or on cooling, stoichiometric carbides form. This process is suitable for the manufacture of mixed carbides for use in cermets (21).

Titanium carbide may also be made by the reaction at high temperature of titanium with carbon; titanium tetrachloride with organic compounds, eg, methane, chloroform, or poly(vinyl chloride); titanium disulfide [12039-13-3] with carbon; and titanium tetrachloride with hydrogen and carbon monoxide.

Titanium carbide has been used as an intermediate in the manufacture of titanium tetrachloride from ores, eg, ilmenite [12168-52-4] and perovskite [12194-71-7]. A process in which a titanium-bearing ore is mixed with an alkali metal chloride and carbonaceous material and heated to 2000°C is described in ref. 22. After removal of volatile impurities, cooling, grinding, and acid-leaching, highly pure TiC is obtained.

Work carried out on the production of titanium carbide from perovskite (essentially $CaTiO_3$) and ilmenite concentrate by the Bureau of Mines of the U.S. Department of the Interior has been summarized (23). In the case of perovskite, a mixture with carbon is heated in a single-phase arc-melting furnace at 2400–1800°C. The cooled mass is ground and leached with water, which decomposes calcium carbide with the formation of acetylene. The TiC is then separated from the aqueous slurry by elutriation. Approximately 72% of the titanium is recovered as the purified product: analysis 63.0 wt % Ti and 23.2 wt % C.

In the case of ilmenite, in order to reduce the impurity content of the final product, it is necessary to reduce the ilmenite carbothermically in the presence of lime at ca 1600°C. Molten iron is separated, and the remaining $CaTiO_3$ slag can then be processed as perovskite.

Compact titanium carbide is light gray when fractured but can be polished to a silver gray. The maximum melting point in the Ti–C system is 3067°C for $TiC_{0.8}$. The melting point of a mixture with an overall carbon:titanium ratio of 1.0 is slightly lower, and in such a mixture there is evidence of two phases, ie, carbon and titanium carbide, the structure of the latter phase showing carbon vacancies. The boiling point of titanium carbide is ca 4800°C.

Titanium carbide is one of the hardest pure carbides known (9–10 Mohs). The diamond pyramid hardness is ca 31.4 GPa (3200 kgf/mm^2), but the hardness is influenced by the presence of oxygen which also causes embrittlement, so that the recorded values disagree but still are very high.

Titanium carbide conducts electricity, and values for the specific resistance of $(50–200) \times 10^{-6}$ Ω·cm have been recorded. The temperature coefficient of resistance is positive, and the carbide becomes superconducting at 1.1 K (see Superconducting materials). It is possible for titanium carbide to be a semiconductor, notably when deposited in thin films on silica (24) (see Semiconductors).

Titanium carbide is not attacked by acids, hot or cold, except nitric acid, aqua regia, mixtures of nitric and hydrofluoric acids, and mixtures of nitric and sulfuric acids. Similarly, it is resistant to aqueous alkali except in the presence of oxidizing agents. It is stable to hydrogen up to 2400°C, but it almost completely reacts with nitrogen at 1000–1300°C.

In oxygen at 450°C, a nonadhering nonprotective coating of anatase forms; at 700–1000°C, the coating changes to rutile and becomes protective. Carbon monoxide reacts with titanium carbide at ca 1500°C, giving an oxide of titanium, carbon, and oxygen-saturated carbide. Carbon dioxide reacts readily at 1200°C:

$$3\,CO_2 + TiC \rightarrow 4\,CO + TiO_2$$

Solid solutions form between titanium carbide and a number of other metallic carbides. These, particularly with tungsten carbide, are highly significant in the technology of hard metals.

Titanium carbide is used extensively either alone or in admixture with other metallic carbides as a component for cutting tools. Its refractory nature and corrosion resistance make it valuable as a constituent of temperature-resistant alloys for use in turbine blades for high temperature applications and in components for jet and rocket engines. Manufacture of these may involve infiltration of a preformed and sintered body by a transition metal followed by further heat treatment. The metal, usually cobalt, nickel, or iron, acts as a binder for the carbide particles, with which it does not react, although it wets the carbide particles and readily penetrates the interstices between them.

Thermodynamic data for titanium carbide are given in Table 2.

Titanium–Nitrogen System

The titanium–nitrogen system has been studied (14,25). Titanium dissolves nitrogen up to a nitrogen content of <20 at % ($TiN_{0.23}$). The cubic titanium nitride phase exists over a wide range of compositions starting at $TiN_{0.42}$ (30 at % N).

Titanium Nitride. Titanium nitride melts at 2950°C, but there is some evidence of decomposition at this temperature. It is a better conductor of electricity than titanium metal; the resistivity is 21.7×10^{-6} Ω·cm at 20°C, and it increases to 340×10^{-6} Ω·cm at 2930°C. It becomes superconductive at 1.2–1.6 K; thin layers deposited on silica show semiconductivity (4,24). Thermodynamic data for titanium nitride are given in Table 2.

Titanium nitride is not attacked by acids except boiling aqua regia, and it is decomposed by boiling alkalies with the evolution of ammonia. It is stable to heat under vacuum; when heated in an atmosphere of oxygen, nitric oxide, or carbon dioxide to

1200°C, it rapidly oxidizes. Hydrogen, nitrogen, and carbon monoxide do not react with it. Titanium nitride is brittle and best used infiltrated with cobalt or nickel or as a hard and corrosion-resistant thin layer deposited on other metals.

Titanium nitride can be prepared by direct synthesis from finely divided titanium and nitrogen at 1000–1400°C. A stoichiometric product is obtained by grinding the product and repeating the process many times. A very pure product is obtained by passing a mixture of titanium tetrachloride with nitrogen and hydrogen over a tungsten wire heated to 1450°C (26). The TiN is deposited on the wire in polycrystalline form, or single crystals can be grown:

$$2\,TiCl_4 + N_2 + 4\,H_2 \rightarrow 2\,TiN + 8\,HCl$$

Alternatively, titanium tetrachloride and nitrogen can be passed over a heated iron wire (27):

$$2\,TiCl_4 + N_2 + 4\,Fe \rightarrow 2\,TiN + 4\,FeCl_2$$

Titanium dioxide can be heated at 1600°C for 2.5 h with nitrogen in a reducing atmosphere to give 98–99% TiN (28):

$$2\,TiO_2 + 4\,C + N_2 \rightarrow 2\,TiN + 4\,CO$$

Ammonia has been used as the source of nitrogen under the following conditions:
1. Reaction of titanium dioxide with ammonia at 1100–1500°C.
2. Reaction of titanium tetrachloride with ammonia at 1400°C.
3. Reaction of titanium hydride with ammonia at 1000°C for 100 h.
4. Heating of finely divided titanium dioxide with calcium hydride to form titanium. Nitrogen or ammonia is then admitted to the hot reaction zone to react with the titanium forming titanium nitride. The presence of calcium oxide, which forms during the reduction process, prevents the titanium nitride particles from sticking together (29).

A recent preparation involves the use of a plasma jet at 2000–3600°C (30). Titanium dioxide is heated in an atmosphere of nitrogen with a d-c transport-type plasma torch that has a water-cooled copper crucible as an outer anode. A deeper penetration of the nitride layer is obtained, and the reaction can be accelerated by addition of carbon, eg,

$$2\,TiO_2 + 4\,C + 2\,N \rightarrow 2\,TiN + 4\,CO$$

Titanium nitride is a brown-yellow powder which, after pressing and sintering, can be polished to a golden-yellow mirror. It crystallizes in the cubic system with the sodium chloride structure. The lattice spacing a and density vary with composition

	a, nm	Density, g/cm^3
TiN$_{0.42}$	0.421	4.870
TiN$_{1.0}$	0.4235	5.213

a decreases as the nitrogen content exceeds 50 at %; eg, TiN$_{1.16}$ has a value of 0.4213 nm. The hardness is rated at 8–9 on the Mohs scale and the diamond-pyramid hardness is 16.7 GPa (1700 kgf/mm^2).

Titanium Nitrate. Prolonged reaction of liquid nitrogen dioxide on titanium tetrachloride at −60° to −20°C produces an intermediate compound 2 TiCl$_4$.3N$_2$O$_4$, which decomposes in the presence of excess nitrogen dioxide to form nitrosyl chloride and titanium nitrate [13860-02-1], Ti(NO$_3$)$_4$ (31–32). Titanium nitrate is a clear yellow powder that is stable in a sealed tube at room temperature. It is less sensitive to moisture than titanium tetrachloride and does not fume in air. It reacts with water vigorously, thereby liberating nitrogen oxides.

Aqueous solutions of titanium nitrate can readily be prepared by dissolving hydrated titanium oxide in nitric acid. Miscellaneous titanium–nitrogen compounds that have been described are titanous amide [15190-25-9] $Ti(NH_2)_3$; titanic amide [15792-80-0], $Ti(NH)_2$; and various products in which the amide group has replaced chlorine in titanium tetrachloride.

Titanium–Oxygen System

The phase diagram of the titanium–oxygen system is discussed in ref. 4. Metallic α-titanium can dissolve oxygen up to a composition $TiO_{0.042}$. The hexagonal structure of α-titanium is maintained up to this concentration, although the a and c parameters both increase with the latter being greater. As the oxygen content increases, the transition temperature from α to β increases. At 900–950°C, a new phase forms between α-titanium and TiO_2 and is tetragonal ($a = 0.533$ nm, $c = 0.6645$ nm, $c/a = 1.247$); this includes the composition Ti_3O_2. The phase TiO exists at 52–54 at % oxygen. It is fcc and can exist with either oxygen or titanium vacancies; the parameter $a = 0.417$ nm decreases with increasing oxygen content. Titanium sesquioxide, Ti_2O_3, exists at 59.4–60.8 at % oxygen and has a corundum-type structure ($a = 0.5155$ nm, $c = 1.361$ nm, $c/a = 2.64$) (33). Trititanium pentoxide, Ti_3O_5, exists at 62.3–64.3 at % oxygen and has an orthorhombic structure ($a = 0.3747$ nm, $b = 0.9465$ nm, $c = 0.9715$ nm). Seven phases exist at 63.6–65.5 at % oxygen with the general formula TiO_{2n-1} (34). As n increases from 4 to 10, the structure of the phases approaches more and more closely that of rutile TiO_2, the oxygen deficit being accommodated by the formation of crystallographic shear planes in an otherwise normal rutile lattice. Recent work has shown that shear planes can result in a wide variety of stable, slightly reduced forms of rutile TiO_2 (35). The rutile structure is stable at 66.5–66.7 at % oxygen, ie, from $TiO_{1.988}$ to $TiO_{2.003}$, and has a tetragonal structure with parameters $a = 0.4584$ nm, $c = 0.2953$ nm, and $c/a = 0.6442$. Thus, there is a wide range of possible compounds and structures in the titanium–oxygen system. Although the range over which the rutile structure can exist is narrow, the possibility that irregularities can exist in the crystal lattice is of profound importance to the behavior of titanium dioxide as a pigment. For example, TiO_2 is white whereas $TiO_{1.995}$ is blue. The enthalpies of the titanium oxygen system have been discussed (36).

The two other crystalline forms of titanium dioxide, brookite [12188-41-9] and anatase [1317-70-0], do not appear in the phase diagram, whereas rutile is the thermally stable form at all temperatures. The conversion of these forms to rutile may, however, be extremely slow as shown by their occurrence in nature.

Titanium Monoxide and Titanium Sesquioxide. Titanium monoxide and titanium sesquioxide (Ti_2O_3) are best made by heating a compressed stoichiometric mixture of titanium powder and titanium dioxide at 1600°C under vacuum in an aluminum or molybdenum capsule.

$$Ti + TiO_2 \rightarrow 2\,TiO$$

$$Ti + 3\,TiO_2 \rightarrow 2\,Ti_2O_3$$

Alternative methods are reduction of TiO_2 with hydrogen, which yields a product of variable composition depending on the experimental conditions:

Conditions	Composition
1300°C	Ti_3O_5
12–1500°C	Ti_2O_3 + some TiO
2000°C and 15.4 kPa (2.2 psi)	TiO only

Reduction with magnesium at 1000°C yields TiO only. When titanium monoxide is heated in air at 150–200°C, titanium sesquioxide forms and, at 250–350°C, it changes to Ti_3O_5.

Trititanium Pentoxide. Trititanium pentoxide can also be made by heating titanium metal with titanium dioxide or by reduction of titanium dioxide with hydrogen or carbon monoxide at 700–1100°C. It is a bluish-black solid with a metallic luster, has a density of 4.29 g/cm^3, and crystallizes in the orthorhombic system. The properties of titanium monoxide and titanium sesquioxide are given in Table 3.

There are no commercial uses for titanium monoxide or titanium sesquioxide as yet, but they are used in the synthesis of various titanate semiconductors. The formation of an oxide film on the surface of titanium metal is important in improving the ability of the metal to withstand corrosion (see Corrosion and corrosion inhibitors).

Hydrated Titanium Oxides. If an alkali-metal hydroxide is added to a solution of a salt of Ti(II) or Ti(III), precipitates of the hydroxides of Ti(II) (black) and Ti(III) (brown) form. They are unstable and readily oxidize in air forming hydrated titanium dioxide. The precipitates are powerful reducing agents and are difficult to make in the pure state (37).

The precipitate obtained when an alkali metal hydroxide is added to a solution of a Ti(IV) salt consists essentially of hydrated titanium oxide. Its exact composition depends on the conditions under which it precipitates. If precipitation is carried out at room temperature, a voluminous and gelatinous precipitate forms which, because of its large surface area, is extremely adsorbent. It is known as orthotitanic acid [20338-08-3] and has the formula $TiO_2.2H_2O$ ($Ti(OH)_4$), but it may contain more water than is indicated by the formula. Its adsorbent nature makes it difficult to obtain as a pure compound, and reprecipitation may be necessary to eliminate all of the impurities. It is readily soluble in dilute hydrochloric acid.

Table 3.   **Properties of Titanium Monoxide and Titanium Sesquioxide**

Property	TiO	Ti_2O_3
color	golden yellow	violet
density, g/cm^3	4.888	4.486
melting point, °C	1750	1900
structure	fcc	hexagonal
solubility		
hot 40 wt % HF	dissolves rapidly	dissolves
hot concentrated HCl	slow attack	no action
hot concentrated H_2SO_4	slow attack	slow attack
hot concentrated HNO_8	slow surface attack	no action
hot concentrated NaOH	slow attack	no action
30 wt % H_2O_2	yellow solution	weak yellow color after two days

If the suspension is boiled or if precipitation is from hot solutions, then a less hydrated oxide forms and is known as metatitanic acid [*12026-28-7*], $TiO_2.H_2O$ ($TiO(OH)_2$). This form is more difficult to dissolve in acid and is only soluble in hydrofluoric or hot concentrated sulfuric acid. Metatitanic acid may not precipitate and, in the presence of acid electrolytes, colloidal solutions in concentrations as high as 600 g/L are possible. The flocculation of hydrated titanium dioxide from such colloidal suspensions is an important stage in the manufacture of titanium dioxide pigments. The kinetics of precipitation of hydrated titanium dioxide from titanyl sulfate solutions are described in ref. 38.

In view of its ready interchange with other compounds, hydrated titanium dioxide is used as an ion-exchange medium (see Ion exchange). The preparation, properties, and performance of hydrated titanium oxide as an ion-exchange agent have been studied (39–42). Separations include those of the alkali and alkaline-earth metals; zinc, copper, and cobalt; cesium, strontium, and barium. The kinetics of ion exchange of calcium on hydrated titanium dioxide are discussed in ref. 43. The possibility of using hydrated titanium dioxide as an ion-exchange agent for the treatment of liquid radioactive wastes from nuclear-reactor installations and for the separation of uranium from seawater have also been proposed (44–45). Finally, hydrated titanium dioxide is a useful starting material for the manufacture of other titanium compounds, as it is readily available in large quantities and in a highly pure state.

Titanium Dioxide. *Physical and Chemical Properties.* Titanium dioxide occurs naturally in three crystalline forms: anatase, brookite, and rutile. These crystals are substantially pure titanium dioxide but usually contain small amounts of impurities, eg, iron, chromium, or vanadium, which darken them. A summary of the crystallographic properties of the three varieties is given in Table 4.

Although anatase and rutile are both tetragonal, they are not isomorphous. Anatase occurs usually in near-regular octahedra, and rutile forms slender prismatic crystals which are frequently twinned. Rutile is the thermally stable form and is one of the two most important ores of titanium.

The three allotropic forms of titanium dioxide have been prepared artificially but only rutile, the thermally stable form, has been obtained in the form of transparent large single crystals. The transformation from anatase to rutile is accompanied by the evolution of ca 12.6 kJ/mol (3.01 kcal/mol), but the rate of transformation is greatly affected by temperature and by the presence of other substances which may either catalyze or inhibit the reaction. The lowest temperature at which conversion of anatase

Table 4. Crystallographic Properties of Anatase, Brookite, and Rutile

Property	Anatase	Brookite	Rutile
crystal structure	tetragonal	orthorhombic	tetragonal
optical	uniaxial, negative	biaxial, positive	uniaxial, positive
density, g/cm^3	3.9	4.0	4.23
hardness, Mohs scale	5½–6	5½–6	7–7½
unit cell dimensions, nm	$D_4a^{19}4.TiO_2$	$D_2h^{15}.8TiO_2$	$D_4h^{12}.?TiO_2$
a	0.3758	0.9166	0.4584
b		0.5436	
c	0.9514	0.5135	2.953

to rutile takes place at a measurable rate is ca 700°C, but this is not a transition temperature. The change is not reversible; ΔG for the change from anatase to rutile is always negative (see Tables 1 and 2 for thermodynamic data).

Brookite has been produced by heating amorphous titanium dioxide, prepared from an alkyl titanate or sodium titanate [12034-34-3] with sodium or potassium hydroxide in an autoclave at 200–600°C for several days. The important commercial forms of titanium dioxide are anatase and rutile, and these can readily be distinguished by x-ray diffraction spectrometry.

Since both anatase and rutile are tetragonal, they are both anisotropic, and their physical properties, eg, refractive index, vary according to the direction relative to the crystal axes. In most applications of these substances, the distinction between crystallographic directions is lost because of the random orientation of large numbers of small particles, and it is the mean value of the property that is significant.

Measurements of physical properties, in which the crystallographic directions are taken into account, may be made of both natural and synthetic rutile, natural anatase crystals, and natural brookite crystals. Measurements of the refractive index of titanium dioxide must be made by using a crystal that is suitably orientated with respect to the crystallographic axis as a prism in a spectrometer. Crystals of suitable size of all three modifications occur naturally and have been studied (46). However, rutile is the only form that can be obtained in large artificial crystals from melts (47). The refractive index of rutile is 2.75 (48). The dielectric constant of rutile varies with direction in the crystal and with any variation from the stoichiometric formula, TiO_2; an average value for rutile in powder form is 114. The dielectric constant of anatase powder is 48 (49–50).

Titanium dioxide is used extensively in the electronics, plastics, and ceramics industry because of its electrical properties. Rutile is a semiconductor whose specific conductivity rises rapidly with increasing temperature and is also very sensitive to any oxygen deficiency. For example, it is an insulator at 20°C, but at 420°C the conductivity increases 10^7 times. Stoichiometric titanium dioxide has a specific conductivity of $<10^{-10}$ S/cm, whereas $TiO_{1.9995}$ yields a value of 10^{-1} S/cm. The electrical properties of rutile are summarized in Table 5 (51).

Titanium dioxide is thermally stable (mp 1855°C) and very resistant to chemical attack. When it is heated strongly under vacuum, there is a slight loss of oxygen corresponding to a change in composition to $TiO_{1.97}$. The product is dark blue but reverts to the original white color when it is heated in air.

Hydrogen and carbon monoxide reduce it only partially at high temperatures, yielding lower oxides or mixtures of carbide and lower oxides. At ca 2000°C and under vacuum, carbon reduces it to titanium carbide. Reduction by metals, eg, Na, K, Ca, and Mg, is not complete. Chlorination is only possible if a reducing agent is present; the position of equilibrium in the system is

$$TiO_2 + 2\,Cl_2 \rightleftharpoons TiCl_4 + O_2$$

The reactivity of titanium dioxide towards acids is very dependent on the temperature to which it has been heated. For example, titanium dioxide that has been prepared by precipitation from a titanium(IV) solution and gently heated to remove water is soluble in concentrated hydrochloric acid. If the titanium dioxide is heated to ca 900°C, then its solubility in acids is considerably reduced. It is slowly dissolved by hot concentrated sulfuric acid, the rate of solvation being increased by the addition

Table 5.   Electrical Properties of Rutile

Property	Value
dielectric constant	
powder	114
single crystal	
a direction	170
c direction	86
loss tangent	
a direction	0.0110–0.0002
c direction	0.35–0.0016
electrical conductivity, S/cm	
single crystal	
a direction	
30°C	10^{-10}
227°C	10^{-7}
c direction	
30°C	10^{-13}
227°C	10^{-6}
breakdown voltage, mV/m	15.2–17.8
dipole moment, C·m × 10^{-30} [b]	9.3–11.0
magnetic susceptibility	$(0.078-0.089) \times 10^{-6}$

[a] Ref. 51.
[b] To convert C·m to debye, divide by 3.336×10^{-30}.

of ammonium sulfate which raises the boiling point of the acid. The only other acid in which it is soluble is hydrofluoric acid, which is used extensively in the analysis of titanium dioxide for trace elements. Aqueous alkalies have virtually no effect, but molten sodium and potassium hydroxides, carbonates, and borates dissolve titanium dioxide readily. An equimolar molten mixture of sodium carbonate and sodium borate is particularly effective as is molten potassium pyrosulfate.

Preparation. Chemically pure titanium dioxide is best prepared from titanium tetrachloride that has been purified by repeated distillation. Titanium tetrachloride is hydrolyzed in aqueous solution resulting in a precipitate of hydrated titanium dioxide which, after washing and drying at 110°C, can be calcined at 800°C to remove combined water and chloride. Preparation of high purity titanium dioxide with silica, magnesium, and iron contents of ca 10^{-5} wt % has been described (52). The method used involves precipitation from titanium tetrachloride as hydrated titanium dioxide, conversion of the precipitate to the double oxalate [10580-02-6] $(NH_4)_2[TiO(C_2O_4)_2]$, recrystallization of this from methanol, and subsequent calcination.

Titanium dioxide can also be prepared by precipitation with ammonium hydroxide from potassium hexachlorotitanate solution; the salt must have been purified previously by recrystallization. Isopropyl titanate [546-68-9], $Ti[OCH(CH_3)_2]_4$, may also be used as a starting point because it can be obtained in a highly pure state by repeated distillation (53). It is hydrolyzed by boiling with water, and the precipitated hydrated titanium dioxide is then washed, dried, and calcined.

Production. Australian and South African sources of rutile consist of coastal or inland beaches where the rutile has been weathered and classified by the action of tidal forces. Small deposits of massive rutile exist and there is a large alluvial deposit in Sierra Leone (54) (see Table 6).

Table 6.   Sources of Titanium Concentrates, 1000 Metric Tons

Country	1969	1974	1977	1980[a]
Rutile				
Australia	363	319	326	294
India	3	6	6	7
Sierra Leone	28			50
South Africa	0.5		5	48
Sri Lanka	3	3	1	15
USSR[a]		27	9	9
United States		6	[b]	[b]
Total	*397.5*	*361*	*347*	*423*
Titaniferous slag				
Canada	681	846	694	876
Japan	6	5	1	
South Africa				344
Total	*687*	*851*	*695*	*1220*

[a] Estimate.
[b] Withheld to avoid disclosing individual company information.

Brookite is a very rare mineral and has no importance commercially. Extensive deposits of anatase have recently been located in Brazil, but impurities associated with the deposit make it unsuitable for direct use in titanium pigment manufacture (see Titanium Dioxide Pigments).

Nonpigment Uses.   Large tonnages of titanium dioxide are used in the vitreous enamel industry (see Enamels). The titanium dioxide improves the acid resistance of the enamel and its excellent opacifying properties make it possible to use thinner coats, which result in improved resistance to flexure as well as savings in cost and weight. For improved ease of handling and decreased dusting, a coarse grade of titanium dioxide is required. It is important also to specify carefully the concentration of elements, eg, tungsten and niobium, which can produce pronounced color effects in the enamel. The opacity of the enamel does not result from the properties of the titanium dioxide used, since a recrystallization process is involved in producing the enamel.

A second important use for titanium dioxide is in the production of components for electronic equipment. Its high dielectric constant and high resistance make it ideal for use in the miniaturization of capacitors. These are made by pressing high purity rutile into the desired shape and then heating the pressing until the particles sinter at ca 1400–1450°C. Additions of materials, eg, clay, assist in producing the mold and reduce the firing temperature. These additions must be kept small to avoid adverse effects on the electrical properties of the product.

Many substances have been added to titanium to improve or modify its electrical characteristics. Some of these, eg, zirconium dioxide, simply act as diluents but others, eg, barium oxide, form mixed oxides or titanates which can also be manufactured and used independently (see Ferroelectrics).

Synthetic gems have been produced from rutile and strontium titanate [12060-59-2] by a process similar to the Verneuil process used for the manufacture of synthetic sapphires and rubies (55–56) (see Gems, synthetic). The high refractive index of these materials results in gems of high brilliance. The stones also provide a fine display of colors because of their high dispersion, which is superior to that of diamond; however, they are not very hard.

Mineral rutile is used as an ingredient of welding-rod coatings, for which impurities such as iron are acceptable. In 1980, 6251 t was used for this purpose in the United States.

Titanium dioxide is used increasingly as a catalyst, either as an active agent or an inert support. The catalytic oxidation of o-xylene to phthalic anhydride using a mixture of titanium dioxide and vanadium pentoxide as catalyst on an inert support is a well-established commercial process (57). The same agents have also been used for the reduction of nitrogen oxides in exhaust gases from internal combustion engines (58). Titanium dioxide has also been used as an oxygen sensor to monitor automobile engine performance; the feedback from the detector controls the air–fuel ratio, giving optimum low pollution performance (59).

Titanium dioxide impregnated with precious metals (eg, platinum, rhodium, or ruthenium) or nickel is used in the Fischer-Tropsch synthesis for the production of hydrocarbons from carbon monoxide and hydrogen (60) (see Fuels, synthetic). There is much research interest in the photocatalytic splitting of water to give hydrogen, which is used as a fuel. Titanium dioxide impregnated with platinum is used as the catalyst (61–62).

Numerous other reactions, which as yet are only of academic interest, have been studied in which titanium dioxide is used as a catalyst. These include the oxidation of hydrogen sulfide to sulfur dioxide, the dehydration of alcohols, ammonoxidation, methylation, isomerization, and alkylation (63–68) (see Catalysis).

Titanium Dioxide Pigments. *Properties.* The high refractive index, lack of absorption of visible light, ability to be produced in the correct size range, and the stability and nontoxicity of titanium dioxide are the reasons why it has become the predominant white pigment in the world. Titanium dioxide pigments are produced in two forms, anatase and rutile. These names indicate that the pigments have the same crystal structure as the minerals of those names but do not necessarily indicate that they have been made from those ores (see Pigments, inorganic).

Preparation. *Raw materials.* The raw materials in the manufacture of titanium pigments are ilmenite and rutile (see Tables 7 and 8). Titanium dioxide pigments are produced by two processes, ie, the classical sulfate process and the chloride process. In the first, by which the bulk of the world's pigment is produced, the essential step is hydrolysis under carefully controlled conditions of an acid solution of titanyl sulfate [13825-74-6], followed by calcination of the hydrous precipitate. In the second, the essential step is burning titanium tetrachloride in oxygen to yield titanium dioxide and chlorine. These two routes make use of the two feasible methods in which titanium ores react by solution in sulfuric acid and by chlorination in the presence of a reducing agent.

Ilmenite is, ideally, ferrous titanate, $FeO.TiO_2$ or $FeTiO_3$, but it usually contains some ferric iron. It occurs massive, as in Norway, Finland, and New York state, or as a constituent of beach sands, as in Florida, southern India, Western Australia, and Queensland and New South Wales in eastern Australia (see Table 9). Ancient beaches may be inland and not necessarily on the present shoreline. Ilmenite may also be associated with other ores; in a large deposit near Lake Allard in Quebec, Canada, ilmenite occurs closely associated with hematite and the two cannot be separated mechanically. The mixture can, however, be smelted to give iron and slag, which is rich

Table 7.   Typical Analyses of Ilmenite, wt %

Component	Norway, massive	United States New York, massive	United States Florida, sand	W. Australia, beach sand	Malaysia, sand	Finland, massive	India, beach sand
TiO_2	43.8	44.5	64.1	54.3	52.6	44.0	59.8
FeO	34.6	36.7	4.7	23.6	34.3	39.4	10.1
Fe_2O_3	12.7	6.5	25.6	18.5	6.5	9.1	24.8
Cr_2O_3	0.02	0.01	0.001	0.06	0.02	<0.01	0.13
P_2O_5	0.04	0.06	0.21	0.07	0.10	0.08	0.18
Al_2O_3	1.2	2.0	1.5	0.4	1.3	0.1	1.1
SiO_2	1.9	3.6	0.3	0.5	0.7	2.0	1.0
ZrO_2	0.01	0.02	0.10	0.04	0.1	<0.01	0.5
CaO	0.5	0.5	0.13	0.01	0.1	0.85	0.15
MgO	3.2	2.4	0.35	0.1	0.1	1.8	0.8
V_2O_5	0.2	0.18	0.24	0.03	0.04	0.3	0.25
MnO	0.33	0.4	1.35	1.7	2.6	0.8	0.48
Nb_2O_5	0.01	0.01		0.16	0.27	<0.01	0.17

Table 8.   Typical Analysis of Rutile and Beneficiated Ilmenite, wt %

Component	Rutile E. Australia, beach sand	Rutile S. Africa, beach sand	Rutile Sierra Leone, alluvial	Beneficiated ilmenites QIT^a, slag	Beneficiated ilmenites Richards Bay, slag	Beneficiated ilmenites Benelite, Kerr McGee	Beneficiated ilmenites Rupaque	Beneficiated ilmenites W. Australia
TiO_2	96.0	95.4	95.7	71.5	85.5	92.4	96.1	92.1
trivalent Ti as wt % TiO_2				9.7	31.4			6.7
Fe metallic				0.5	0.2			0.2
FeO				11.5	10.8			3.0
Fe_2O_3	0.7	0.7	0.9			2.5	1.7	
Cr_2O_3	0.27	0.1	0.23	0.18	0.22	0.07	0.5	0.1
P_2O_5	0.02	0.02	0.04	0.01	<0.01	0.07	0.17	0.03
Al_2O_3	0.15	0.65	0.2	4.4	1.2	0.7	0.46	1.4
SiO_2	1.0	1.75	0.7	4.0	1.6	1.0	0.15	0.85
ZrO_2	0.5	0.65	1.0	0.06	0.1	0.65	0.15	0.15
CaO	0.02	0.05	<0.01	0.7	0.07	0.01	0.01	0.06
MgO	0.02	0.03	<0.01	5.3	0.9	0.02	0.07	0.02
V_2O_5	0.5	0.46	0.67	0.57	0.5	0.01	0.2	0.19
MnO	0.02	<0.01	<0.01	0.2	1.6	0.04	0.03	1.0
Nb_2O_5	0.45	0.32	0.21	0.01	0.13	0.51	0.25	0.41

^{a} QIT Fer et Titane, Inc.

in titanium and suitable for pigment production. When ilmenite occurs as a constituent of beach sands, a considerable degree of concentration has occurred by wave action; concentration may also occur as the result of wind action, giving aeolian deposits, such as in South Africa. Ilmenite in sands is often much altered by oxidation and leaching. Consequently, the ratio of ferric to ferrous iron increases, and this ratio indicates the extent to which weathering has occurred. At the same time, the total amount of iron tends to fall and the amount of titanium dioxide tends to increase correspondingly.

Table 9.   Sources of Titanium Concentrates (Ilmenite), Thousand Metric Tons

Country	1969	1974	1977	1980[a]
Australia				
ilmenite	722[b]	803	1035	1312
leucoxene		15	11	27
Brazil	20	7	14	20
Finland	138	153	125	150
India	52	77	137	180
Malaysia	133	154	154	160
Norway	492	850	830	830
Sri Lanka	83	81	34	58
USSR	na	na	400	418
United States[c]	846	677	581	499
Total	*2486[d]*	*2820[d]*	*3321*	*3654*

[a] Estimate.
[b] Includes leucoxene.
[c] Includes a mixed product containing ilmenite, leucoxene, and rutile.
[d] Excludes USSR.

The form of titanium dioxide also changes, passing through leucoxene [1358-95-8] and pseudorutile [1310-39-0] stages to rutile. Leucoxene, pseudorutile, and rutile are not attacked by sulfuric acid under the conditions normally used for pigment manufacture; the upper limit for titanium dioxide in an ilmenite suitable for attack by sulfuric acid is ca 60 wt %. Leucoxene and rutile can be chlorinated, but the iron content of leucoxene is too high for economic chlorination unless chlorine is recovered from the iron chlorides or a commercial use can be found for the latter, eg, as a water treatment agent.

Sulfate process.   Since 1970, there has been virtually no change in sulfate pigment production and an increase of 250% for chloride pigment production. The capacity for sulfate pigment is still twice that for chloride pigment, and it remains an important process for producing titanium pigments. Both processes are presented in simplified block form in Figure 1.

Chloride process.   Pigments produced by the chloride process, which involves reaction of titanium tetrachloride with oxygen, were first introduced commercially by DuPont in the United States in 1958 and in Europe by British Titan Products Co., Ltd. (now the Tioxide Group) in 1965. There has been a rapid growth in production capacity of chloride pigment since 1970, particularly in the United States where it amounts to 75% of the total titanium dioxide pigment capacity (see Table 10). The reasons for the changeover to the chloride process from the sulfate process are that it is a continuous operation, it is a more compact process, recovery and recycle of chlorine can be achieved, by-products are considerably less, and a purer product with superior color is obtained because the titanium tetrachloride can be purified by distillation. However, there are certain disadvantages: raw materials of high titanium content, eg, rutile, are expensive; it is technologically an extremely difficult process; there are severe high temperature and corrosion problems; there are toxicity hazards caused by large amounts of chlorine, titanium tetrachloride, and the waste products of chlorination; it has not yet been possible to produce anatase pigment on a large scale; and in certain applications the erosive nature of the pigment can be a cause for concern.

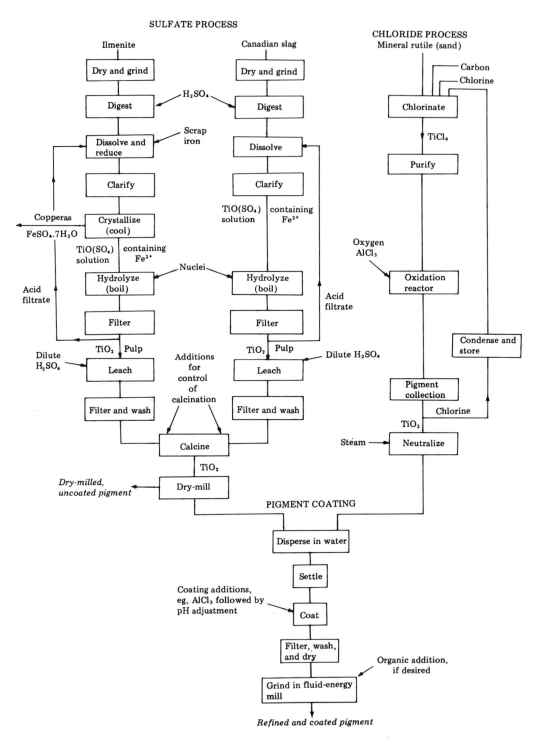

Figure 1. Flow diagram for the manufacture of titanium dioxide pigments.

Table 10.   Annual World Capacity for Titanium Dioxide Pigment Production, 1000 t

Country	1971 Sulfate	1971 Chloride	1975 Sulfate	1975 Chloride	1981 Sulfate	1981 Chloride
Americas						
Brazil	20		25		34	
Canada	59	9	69		69	
Mexico	14		25.3			35
United States	437.5	218.2	354	498	237	705
Europe						
Belgium	39		65		65	
Czechoslovakia	24		24		22	
Finland	55		85		80	
France	105	3.6	166	3	165	
FRG	259.5	39	251	25	280	56
Italy	96		95		54	
Netherlands	30		31		35	
Norway	17.5		20		25	
Poland					36	
Spain	18		21		71	
United Kingdom	171	60	179	62	100	65
Yugoslavia			20		25	
Asia						
People's Republic of China	12		20		15	
India	6.5		18		24.5	
Japan	194.7		203.6	12	205.8	24
Republic of China			2.4		11.8	
Republic of Korea	8.6		8.6		15.6	
USSR	100 (estd)		80		101?	
Africa						
South Africa	22		26		28	
Oceania—Australia	43		53		68	
Total	*1732.3*	*329.8*	*1841.9*	*600*	*1767.7*	*885*

The position of equilibrium in the system $TiCl_4(g) + O_2(g) \rightarrow TiO_2$ (rutile, solid) $+ 2 Cl_2(g)$ is favorable to the production of TiO_2. Typical values of energy changes are given in Table 11. The rate of reaction is negligible below 600°C but increases rapidly above this temperature. It is necessary therefore to bring the gases together at a suitable temperature for reaction to occur and then to chill the mixture to ensure completion of the reaction, ensure the provision of suitable nucleating agents to control flocculation

Table 11.   Energy Changes in the Reaction

$$TiCl_4(g) + O_2(g) \rightarrow TiO_2(\text{rutile, solid}) + 2 Cl_2(g)$$

Temperature, °C	ΔH, kJ/mol[a]	ΔG, kJ/mol[a]	$\log K_p$ [b]
827	−174.7	−116.0	5.51
1027	−174.7	−105.4	
1327	−170.9	−90.0	

[a] To convert kJ to kcal, divide by 4.184.
[b] $K_p = p[Cl_2]^2/p[TiCl_4]p[O_2]$.

and crystalline form, prevent excessive growth of the pigment particles, and prevent massive growths of titanium dioxide on the reactor walls.

Coating. Surface coating involves the formation of a coating over the surface of the pigment particles. The coating agent must be a white hydrous oxide; silica and alumina are commonly used, but others including titania and zirconia are used for particular purposes. The original purpose of coating was to improve durability and to lessen the yellowing which occurred in certain types of paints. The coating improves durability by preventing or reducing the breakdown of the paint medium by the action of uv radiation. It was subsequently realized, however, that variation in surface treatment can be used to improve the dispersibility of pigments in different media, which include systems where there is appreciable pigment–medium reaction, systems in which there is little pigment–medium interaction, and systems in which there is a phase change involving the pigment. The first system includes most paints, both air-drying and stoving. These media are often spoken of as polar, owing to a certain amount of ionization of the medium and adsorption of molecules from the medium on the pigment surface. The second includes most plastics where the molecules of the medium are large, relatively inert, and nonpolar. And the third includes, for instance, emulsion paints where the pigment is dispersed in an aqueous phase which evaporates during drying (see Paint).

The coating process involves dispersion of the pigment from the calciner or from the oxidation stage in the chloride process in water. This may be done in a ballmill or sand mill; sodium silicate is frequently used as the dispersing agent. The dispersion obtained may be allowed to settle, resulting in the removal of oversized particles. A source of, for example, hydrous alumina in the form of aluminum sulfate or sodium aluminate is added to the dispersion; the solution is stirred constantly during the addition. Adjustment of pH follows to cause precipitation of hydrous alumina onto the surface of the pigment particles. The form of this precipitate may vary according to the conditions of precipitation. Coatings of two or more hydrous oxides may be applied to a pigment, either separately by successive operations or simultaneously in one operation. Once the coating is formed, the pigment is washed and dried before being ground in a micronizer or other fluid-energy mill. In some cases the coated pigment may be subjected to heat treatment to achieve particular properties.

Some pigments, particularly those designed for incorporation into plastics, are given an organic surface treatment in addition to the inorganic treatments. This is usually carried out during the final milling stage, and a variety of organic compounds has been used, the most effective being dimethyl siloxane, pentaerythritol, and triethanolamine. There are many variables in the coating process, and detailed procedures are well-documented in the patent literature. The important feature is that the alumina, silica, etc, should form a coating on the pigment particles and not be present as a simple admixture.

A procedure has been patented for the application of a dense silica coating to titanium pigments (69–70). Such coatings, which consist of hydrated amorphous silica in concentrations of 5–10%, are extremely compact and form a dense protective skin around the particle of titanium dioxide. Pigments prepared in this way are extremely durable and have outstanding ease of dispersibility in paint systems.

Rutile pigments produced by the chloride route are basically similar to those from sulfate and require coating in the same way and for the same reason. Coating is usually done by the wet process, but another possibility is that of depositing a pyrogenic coating on these pigments during manufacture.

Manufacturers produce many grades of titanium dioxide pigments, and their literature should be consulted for recommendations as to which grade should be used for any particular purpose. A comprehensive account of the properties of titanium dioxide and of the use of titanium dioxide pigments has been published (71).

Vesiculated bead systems. The principle of using air bubbles as a means of opacifying paints has been known for many years, but poor opacity in the wet paint film and a tendency for the bubbles to collapse under slight pressure hindered successful development. In 1969, a method for encapsulating clusters of water vesicles in rigid polymer beads was developed (72). With drying, the water evaporated leaving vesicles of air. Further work involved the development of polymer beads containing air and titanium dioxide pigment. Typically the beads are made from a copolymer of styrene and an unsaturated polyester and are generally fairly uniform in size, the maximum diameter being controllable during manufacture. The vesicles occupy a large proportion of the total bead volume and, though they vary in size, the mean diameter is 0.7–1.0 μm. The particles of pigment within the vesicles in contact with air give the vesiculated beads enhanced light-scattering power, as light scattering depends on the difference in refractive index at the interphase (pigment:air ratio is 1.75:compared with a pigment:medium ratio of ca 1.2:). In practice, extra pigment is usually needed to produce the correct balance of properties in the paint film. The product is marketed as Spindrift and its use in paint preparation has been described (73). The advantages of this type of product over conventional aqueous paint pigments are better utilization of titanium dioxide pigment; good film integrity; films with improved stain resistance and ease of stain removal, high resistance to burnishing, and good weathering resistance; easier incorporation of the predispersed system into paints; prevention of flocculation and resultant decreased hiding power when the paint film dries; and spherical beads, which give good brushing properties in paints.

Colored pigments. Traces of certain elements may affect the color of titanium dioxide pigments. An example is chromium, the presence of which imparts a yellow tone, making ilmenite ores containing this element in substantial quantity unsuitable for the manufacture of pigment. The color of titanium dioxide pigments can be modified by the incorporation of guest ions into the rutile host lattice. Insertion of trivalent metal ions, for example, should be accompanied by addition of an equal number of pentavalent metal ions. The cation-to-anion ratio of 1:2 of the host lattice should be maintained, and the dark color resulting from oxygen deficiency must be avoided. The preparation of mixed-phase pigments and some of the colors obtainable by different additions are described in ref. 74. The general method of preparation is by heating an intimate mixture of the oxides or compounds which yields the oxides on heating. As a class, these colored pigments are fast to light and chemical reagents, and they are not marked by the difficulties resulting from the separation of components or normal tints during paint manufacture. Further, deterioration of the paint or plastic by weathering does not cause a change in color.

The best known of these pigments is yellow and contains nickel and an equivalent amount of antimony. It is sold as Titanium Nickel Yellow, Titanate Yellow, and Sun Yellow. It is produced by adding a nickel salt and an antimony compound to the pulp after hydrolysis (in the sulfate process for the manufacture of titanium dioxide pigment), mixing thoroughly, and then calcining at ca 1000°C. This pigment is very stable to acids, alkalies, and oxidizing and reducing agents, is insoluble in all solvents, and has excellent heat stability; it is, however, tinctorially weak compared with organic yellows (75) (see Pigments, organic).

Production and Shipment.   A list of the manufacturers of titanium dioxide pigments and the trade names of their respective grades of pigments is given in Table 12 (76). The annual distribution of titanium dioxide pigment shipments among industries in the United States from 1969 to 1980 is shown in Table 13.

Packaging of titanium dioxide pigments with particular reference to its effect on energy consumption in the coatings industry has been discussed (78). During the

Table 12.   Manufacturers of Titanium Dioxide Pigments, 1981

Trade name	Manufacturer	Country
Ajantox	Travancore Titanium Products Ltd.	India
Bayertitan	Farbenfabriken Bayer A.G.	FRG
	Bayer SA/NV	Belgium
	TIBRAS (Titanio de Brasil SA)	Brazil
Dia White	Tohoku Kagaku KK	Japan
Finntitan	Kemira Oy	Finland
Fujititan	Fuji Titan Kogyo KK	Japan
Furukawa	Furukawa Kogyo KK	Japan
Hombitan	Sachtleben Chemie GmbH	FRG
Horse Head	Gulf & Western	United States
Kotiox	Hankook Titanium Ind. Co., Ltd.	Republic of Korea
Kronos	Kronos SA/NV	Belgium
	Kronos-Titan A/S	Norway
	Kronos-Titan GmbH	FRG
	Titan Kogyo KK	Japan
Montedison	Soc. Italiana Biossido Di Titanio	Italy
Pai Tai	China Metal & Chemical Co.	Taiwan
Pretiox	Prerovske Chemicke Zavody	Czechoslovakia
Suntiox	Titan Kogyo KK	Japan
Teika	Teikoko Kako KK	Japan
Ti-chlor	Pigmentos y Productos Quimicos SA de CV	Mexico
Ticon	Tam Ceramics Inc.	United States
Tiona	Laporte Australia Ltd.	Australia
	Laporte Industries Ltd.	UK
Tiofine	TDF Tiofine NV	Netherlands
Tioxide	BTP Tioxide Ltd.	UK
	SA Tioxide (Pty) Ltd.	South Africa
	Tioxide Australia Pty. Ltd.	Australia
	Tioxide Canada Inc.	Canada
	Tioxide S.A.	France
	Titanio S.A.	Spain
Tipaque	Ishihara Sangyo KK	Japan
Ti-pure	E. I. du Pont de Nemours & Co., Inc.	United States
	Pigmentos y Productos Quimicos SA de CV	Mexico
Tiso-Lite	Gulf & Western	United States
Titafrance	Thann et Mulhouse SA	France
Titanox	NL Chem Canada Inc.	Canada
	NL Chemicals	United States
Titone	Sakai Kagaku Kogyo KK	Japan
Tronox	Kerr-McGee Chemical Corp.	United States
Tytanpol	Zjednoczenie Przemyslu Nieorganicznego	Poland
Unitane	American Cyanamid Co.	United States
Unkitox	Dow Quimica Iberica S.A.	Spain
Zopaque	Glidden Pigments Group	United States

Table 13. Distribution of Titanium Dioxide Pigment Shipments in the United States by Industries and Total U.S. Production[a]

Industry distribution, wt %	1969	1973	1977[b]	1980[b]
paints, varnishes, and lacquers	58.5	52.7	52.0	44.1
paper	17.0	19.6	20.7	24.3
floor coverings	2.3	1.3	c	c
rubber	2.6	3.2	3.1	c
coated fabrics and textiles	1.3	1.3	c	c
printing ink	2.3	2.0	2.0	c
roofing granules	0.9	0.6	c	c
ceramics	2.0	2.5	1.9	c
plastics (excluding floor coverings, vinyl-coated fabrics, and textiles)	6.2	9.8	11.7	10.6
other (including exports)	6.9	7.0	8.6	21.0
Total U.S. production, 1000 t	604[b]	715[b]	624[d]	605[b]

[a] Ref. 77.
[b] Based on TiO_2 content.
[c] Included in "other" category.
[d] Gross weight of pigment.

1970s, a move away from the packing of titanium dioxide in paper sacks containing 25 kg began, and much emphasis has been placed on the development of bulk methods of delivery. For large-scale users, pigment may be delivered in tanks and stored in silos. This method, however, involves large capital (equipment) costs, and an alternative is the use of semibulk containers. Most commercially produced containers are unsuitable, as they are difficult to empty completely and they tend to compact the pigment which seriously impairs its flow properties. A special form of container has been designed to overcome these problems (79). The 500-kg and 100-kg containers are designed so that the rate of discharge of pigment can either be controlled at a slow rate or be extremely rapid; eg, the larger container can be emptied in less than one minute. The materials from which the bag is made have antistatic properties to eliminate the risk of sparks being produced during discharge of pigment from the bag; this could result in an explosion if the pigment is being discharged into a vessel containing a flammable vapor. The bags can be folded flat to facilitate their return shipment when empty.

The delivery of titanium dioxide pigments in aqueous slurry form has many advantages (80–81), eg, removal of bag handling and disposal problems and costs; elimination of pigment and dust wastage; no requirement to mill the slurry; and higher productivity and greater automation. On the other hand, there are certain disadvantages: high capital investment; necessity of agitating the slurry and not allowing it to dry; protection of slurries from frost; and increased transport charges.

Economic Aspects. Sales of slurries in the United States increased rapidly during the 1970s until, in 1980, it was estimated that they amounted to 20% of TiO_2 consumption. Consumption of slurries also increased in Canada. Slurries are sold (1982) for $0.044/kg.

In the rest of the world, there has been very little movement in the slurry market, with less than five grades of pigment available from titanium pigment manufacturers outside North America. This compares with ca 40 grades from eight producers in the

United States. Prices for anatase and rutile titanium dioxide pigments are listed in Table 14.

Specifications. Pigment manufacturers control the composition and properties of their products within close limits and have their own specifications for their various grades of pigments. There are also national and international specifications available, eg, ASTM D 476-66 Titanium Dioxide Pigments; ASTM D 13911-63 Chemical Analysis of White Titanium Pigments; British Standard 239-1967 (which includes BS 1851) White Pigments for Paints; ISO 591-1977 Titanium Dioxide Pigments for Paints; Cosmetic, Toiletry, and Fragrancy Association Inc. O119 and SP-1R/115 SP, 1971; and the *American Industrial Hygiene Association Journal*, 1966.

Peroxidic Compounds. When hydrogen peroxide is added to a solution of titanium(IV) compounds, an intense, stable, yellow solution is obtained and forms the basis of a sensitive method for determining small amounts of titanium. The color is discharged by the addition of soluble fluorides, which distinguishes it from vanadium which also forms a yellow complex, whose color is not discharged.

The action of hydrogen peroxide on freshly precipitated hydrated titanium(IV) oxide or the hydrolysis of a peroxide compound, eg, $K_2[TiO_2(SO_4)_2]$ gives a solid which, on dehydration with acetone, gives a yellow solid that is stable below 0°C. This solid has the formula $TiO_3.2H_2O$ and behaves as a true peroxide of structure $(OH)_3TiOOH$ aq or

$$O=Ti \underset{O}{\overset{O}{\diagup\!\!\!\diagdown}} \cdot 2H_2O$$

It loses oxygen and water when heated and liberates chlorine from hydrochloric acid. When freshly prepared, it is stable in acids or alkalies giving peroxy salts. The constitution of aqueous solutions has been investigated by physical measurements of dilute solutions, and it appears that the ion $[Ti(OH)_2]^{2-}$ forms a complex with hydrogen peroxide; this complex is a peracid, possibly containing the complex ion $[(HO)_5\text{-}TiOOH]^{2-}$ in which the titanium remains tetravalent with the coordination number 6 (see Peroxides and peroxy compounds, inorganic).

Inorganic Titanates. Titanium forms a series of compounds with other metals in which it is present in the basic form. There is, however, no evidence from x-ray crystallographic studies for the existence of a titanate anion, and the compounds must

Table 14.   Prices for 20-t Lots of Titanium Dioxide Pigment (Jan. 1982), $/kg

Country	Anatase	Rutile
United States	1.52 (slurry 1.56)	1.65 (slurry 1.69)
UK	1.47	1.58
Belgium	1.21	1.23
France	1.19	1.25
FRG	1.28	1.32
Netherlands	1.28	1.32
Italy	1.28	1.34
Japan	1.69	1.83

be considered as multiple oxides. Three types can be distinguished:

Type	General formula	Remarks
meta	$M(II)TiO_3$	amorphous, ilmenite or perovskite type
ortho	$M(II)_2TiO_4$	cubic spinel type
poly	$M(I)_2Ti_2O_5$	$TiO_2/M(I)O > 1$

In addition to these three types, the alkaline earth metals form titanates with the general formula $3M(II)OTiO_2$ or $M_3(II)TiO_5$.

The titanates are prepared by heating mixtures of titanium dioxide or hydrated titanium dioxide with the oxide, hydroxide, or carbonate of the metal. The correct proportion of the two components and the correct temperature selected for the particular titanate must be used. For example, lithium titanates are prepared by heating lithium carbonate and titanium dioxide to ca 950°C, whereas potassium titanate can be made by heating hydrated titanium dioxide and potassium hydroxide to 160–170°C. Certain titanates can also be formed in the liquid phase; eg, barium and potassium titanates can be formed by mixing the hydroxides of barium and potassium with a solution of a Ti(IV) salt.

Titanates are in general very stable compounds and are insoluble in water, but they are decomposed by acid. The alkali-metal titanates tend to be more reactive and are less stable than the other titanates. This is evident in their tendency to hydrolyze in water and their easy dissolution by dilute acids.

Alkali Metal Titanates. A few properties of alkali metal titanates are listed in Table 15. All the compounds listed in Table 15 are white. Commercially, sodium titanate has been used as an intermediate in the manufacture of titanium pigments.

The polytitanates of potassum are of considerable commercial interest, particularly in Japan, and they can be manufactured in fibrous form. The composition of the titanate may be $K_2Ti_6O_{13}$ [12056-51-8] or $K_2Ti_4O_9$ [12056-49-4], and the fibers have an average diameter of ca 1 μm (82). They may be several millimeters long, each fiber being made up of several fibrils which have a definite crystal structure. They are chemically stable and melt sharply at 1370°C. The earliest method of preparation appears to have been a hydrothermal process in which titanium dioxide reacted with aqueous potassium hydroxide at high pressure (ca 20 MPa or 200 atm) and temperature (600–700°C). A later patent refers to recrystallization of potassium titanate from a saturated solution of titanium dioxide or nonfibrous potassium titanate in molten

Table 15. Properties of Alkali-Metal Titanates

Compound	Formula	CAS Registry No.	Sp gr	Mp, °C	Structure
lithium metatitanate	Li_2TiO_3	[12031-82-2]	3.42		cubic
lithium dititanate	$Li_2Ti_2O_5$	[12600-48-5]	3.50		cubic
lithium orthotitanate	Li_4TiO_4	[12768-28-4]			
sodium metatitanate	Na_2TiO_3	[12034-34-3]	2.19	1030	
sodium dititanate	$Na_2Ti_2O_5$	[12164-19-1]		985	
sodium trititanate	$Na_2Ti_3O_7$	[12034-36-5]		1128	
sodium pentatitanate	$4Na_2O.5TiO_2$	[12034-52-5]			
sodium trititanate	$2Na_2O.3TiO_2$	[12503-05-8]			
potassium metatitanate	K_2TiO_3	[12030-97-6]		806	
potassium dititanate	$K_2Ti_2O_5$	[12056-46-1]		980	

alkali chloride (83). Preparation by reaction of titanium dioxide and potassium carbonate in a melt of potassium chloride and potassium fluoride has also been described (84). The material is produced in lumps, which can be broken down under high shear to give a water-dispersed pulp. This may be treated in much the same way as paper pulp to give papers, felts, mats, and blocks. The blocks have good mechanical strength and recover well from deformation under heavy loads.

Potassium titanate fiber has a low thermal conductivity and a high refractive index and crystallite size, making it opaque to infrared radiation (0.9–2.4 μm). Fibrous potassium titanate has been discussed in detail in refs. 85–89. Commercial applications include its use as a reinforcement material for organic plastics, a filtration medium, a component of friction brakes, and an insulating and reflective material.

Pigmentary potassium titanate was manufactured by one company only, in the United States, from 1965 to 1972. It was used principally in the paper industry, but a small amount was used in vinyl plastics. Its manufacture, properties, and application are described in ref. 90.

Alkaline Earth Metal Titanates. In general, alkaline earth metal titanates exhibit unique electrical characteristics which make them well-suited for electrical applications. Few of these compounds are naturally occurring, and those that do occur in nature are too impure for electrical use.

Barium titanate. Barium titanate [12047-27-7], $BaTiO_3$ (mp ca 1625°C) is a crystalline solid having five crystalline modifications: hexagonal, cubic, tetragonal, orthorhombic, and trigonal. The tetragonal, which is the important form electrically, has a specific gravity of ca 6 and a very high dielectric constant of ca 4000 in the a direction; that in the c direction is much smaller.

The ceramic form of barium titanate is prepared by heating a mixture of the correct proportions of barium carbonate and titanium dioxide at 1300°C until the reaction is complete. When commercial ceramic bodies are produced, the sinter is ground to a small particle size with small amounts of other constituents which are added to facilitate manufacture or to modify the characteristics of the finished product. The mass is then pressed into the finished shape and is fired. Considerable care is necessary in the firing cycle and in controlling the furnace atmosphere to obtain satisfactory crystal growth and maximum density in the product. These factors have an appreciable effect on the electrical properties.

Barium titanate becomes permanently polarized when subjected to a high d-c field at temperatures below its Curie point of 120°C. At higher temperatures, it is transformed from the tetragonal to the cubic form and the effect is destroyed.

Polarized barium titanate exhibits two unusual properties: it is ferroelectric and piezoelectric. The latter property permits the interconversion of electrical and mechanical oscillations; application of pressure to the crystal generates electrical potential differences, and application of electrical potential differences causes expansion or contraction of the crystal. Because of this reversible effect, barium titanate is used in phonograph pickups, underwater detection equipment, and equipment for the generation of ultrasonic vibrations. The high dielectric constant of barium titanate and its change with temperature also make it a useful material for small, high value capacitors and for temperature-compensating elements. The technical applications of barium titanate are described in ref. 91. A review of the crystal structure, electrical properties, and preparation of semiconducting barium titanate is given in ref. 92 (see Ferroelectrics).

Strontium titanate. Strontium titanate [*12060-59-2*], $SrTiO_3$ (mp 2080°C) has a cubic structure and a density of 5.12 g/cm^3. It is used as an additive to barium titanate in the manufacture of electronic components, since it lowers the Curie point and changes the temperature coefficient.

A particularly pure preparation obtained by calcining the double strontium titanium oxalate precipitate from titanium tetrachloride solution has been used for the preparation of boules by the Verneuil process (56). The refractive index of the boule is close to that of diamond but the dispersion is greater, so that gems cut from the boule are of good brilliance and color (see Table 16). Absorption of uv radiation is complete with a sharp threshold at ca 395 nm.

Calcium titanate. Calcium titanate [*12049-50-2*], $CaTiO_3$, occurs naturally as the mineral perovskite. The pure compound can be synthesized by heating equimolar amounts of the oxides to 1350°C. It has a density of 4.02 g/cm^3 and a melting point of 1980°C; like strontium titanate, it is used as an additive to barium titanate. Other calcium titanates include $3CaO.2TiO_2$ [*12013-80-8*] and $3CaO.TiO$ [*12013-70-6*].

Magnesium titanate. There are three magnesium titanates: the metatitanate [*1312-99-8*], $MgTiO_3$; the orthotitanate [*12032-52-9*], Mg_2TiO_4; and the dititanate [*12032-35-8*], $MgTi_2O_5$ (see Table 17). The metatitanate is the most commonly used form with its principal use as an additive to ceramic dielectric components. It is also used as a gemstone and as a pigment in uv-cured systems (93).

Iron Titanates. Ferrous metatitanate [*12168-52-4*], $FeTiO_3$ (mp ca 1470°C, sp gr 4.72) is rhombohedral and opaque black with a submetallic luster. It occurs in nature as the mineral ilmenite, which is used extensively in the manufacture of titanium pigments. It has been made artificially by heating a mixture of ferrous oxide and titanium oxide in a sealed silica tube for several hours at 1200°C and by reduction of a mixture of ferric oxide and titanium dioxide at 450°C.

Ferrous orthotitanate [*12160-20-2*], Fe_2TiO_4 (mp ca 1470°C) is orthorhombic and opaque. It has been prepared by fusing a mixture of ferrous oxide and titanium

Table 16. Comparison of Strontium Titanate and Diamond

Property	Strontium titanate	Diamond
refractive index		
n_C	2.380	2.4103
n_D	2.409	2.4175
n_F	2.488	2.4354
dispersion, $n_F - n_C$	0.108	0.025
optical transmission, nm	395–1700	
hardness, Mohs	6–6½	10

Table 17. Properties of Magnesium Titanates

Property	Meta	Ortho	Dititanate (pyro)
sp gr	3.84	3.53	
mp, °C	1565	1840	1645
crystal habit	rhombohedral	cubic	orthorhombic

dioxide. Ferrous dititanate [12160-10-0], Fe_2TiO_5, is orthorhombic and has been formed by heating ilmenite with carbon at 1000°C. The metallic iron formed in the reaction is removed, leaving a composition that is essentially the dititanate.

Ferric titanate (pseudo brookite) [1310-39-0], Fe_2TiO_5, is orthorhombic and occurs to a limited extent in nature. It has been prepared by heating a mixture of ferric oxide and titanium dioxide in a sealed quartz tube at $\geq 1000°C$.

Lead Titanate. Lead titanate [12060-00-3], $PbTiO_3$, is a yellow solid (density 7.3 g/cm^3). It can be made by heating the calculated amounts of the two oxides together at 400°C, and it is the only compound formed from the two oxides. It had been used as a pigment in paint systems, but it has been superseded by high durability rutile pigments (94). Lead titanate zirconate [12626-81-2] has also been prepared, and its use in ferroelectric ceramics has been proposed (95–96).

Zinc Titanate. Zinc orthotitanate [12036-69-0], Zn_2TiO_4 (sp gr 5.12) is a white solid having the spinel structure and is obtained by heating the correct mixture of the two oxides at 1000°C. Zinc orthotitanate forms a series of solid solutions with titanium dioxide extending to the composition $Zn_2TiO_4.1\frac{1}{2}TiO_2$. These solid solutions are not stable at elevated temperatures and begin to dissociate at 775°C with the formation of rutile titanium dioxide. Although there are many references in the literature to other compounds of zinc and titanium oxides, none has been established.

Nickel Titanate. Nickel titanate [12035-39-1], $NiTiO_3$ (sp gr 5.08) is a canary-yellow solid with a rhombohedral structure. When antimony oxide is incorporated in a mixture of nickel carbonate and titanium dioxide and heated to 980°C, nickel antimony titanate [8007-18-9] forms and is used as a yellow pigment (97).

Other Titanates. Other titanates that have been made are listed in Table 18. Rare-earth-element titanates are discussed in refs. 99 and 100.

Titanium–Halogen Compounds

The known halides and oxyhalides are shown below with their oxidation states.

Oxidation state	Fluoride	Chloride	Bromide	Iodide
II	TiF_2 [13814-20-5]	$TiCl_2$	$TiBr_2$ [13873-04-5]	TiI_2
III	TiF_3 [13470-08-1], TiOF [17497-75-5]	$TiCl_3$, TiOCl [15605-36-4]	$TiBr_3$ [13135-31-4]	TiI_3
IV	TiF_4 [7783-63-3]	$TiCl_4$, $TiOCl_2$ [13780-39-8]	$TiBr_4$, $TiOBr_2$ [13596-01-5]	TiI_4 [7720-83-4], $TiOI_2$ [14899-61-7]

The chemistry of the halides in general is predictable with the tetravalent compounds showing the maximum stability. The ease of preparation of the lower valent halides increases with increasing atomic mass of the halide. There is considerable doubt as to the existence of TiF_2. Titanium tetrachloride is manufactured in large quantities and is used for the manufacture of titanium pigments and titanium metal. The chemistry of titanium–halogen compounds has been reviewed (6,101). Thermodynamic data for the halogen compounds of titanium are given in Tables 1 and 2.

Table 18. Properties of Miscellaneous Titanates

Compound	Formula	CAS Registry No.	Crystal structure	Mp, °C	Sp gr
manganese metatitanate	$MnTiO_3$	[12032-74-5]	hexagonal	1360	4.54
manganese orthotitanate	Mn_2TiO_4	[12032-93-8]	cubic	1455	4.49
aluminum titanate[a]	Al_2TiO_5	[12004-39-6]		1800	
cadmium titanate	$CdTiO_3$	[12014-14-1]	orthorhombic		6.5
cobalt metatitanate	$CoTiO_3$	[12017-01-5]	rhombohedral		5.00
cobalt orthotitanate	Co_2TiO_4	[12017-38-8]	cubic		5.07

[a] Ref. 98.

Titanium Fluorides. *Titanium Trifluoride.* Titanium trifluoride is a blue crystalline solid of sp gr 2.98. It is stable in air at ordinary temperatures but decomposes to titanium dioxide when heated to 100°C. It is insoluble in water, dilute acids, and alkalies, but it is decomposed by hot concentrated acids. It sublimes under vacuum at ca 900°C but decomposes to form titanium and titanium tetrafluoride at higher temperatures.

Titanium trifluoride may be made by the action of gaseous hydrogen fluoride on titanium or titanium hydride. It is necessary to control the conditions carefully to stop the reaction at the trifluoride stage:

$$2\,Ti + 6\,HF \rightarrow 2\,TiF_3 + 3\,H_2$$

Excess hydrogen fluoride converts the trifluoride to the tetrafluoride or the trifluoride can decompose at >900°C as follows:

$$4\,TiF_3 \rightarrow Ti + 3\,TiF_4$$

This reaction can be reversed if titanium and titanium tetrafluoride are heated in a sealed tube at 900°C.

An alternative method of preparation is to pass a mixture of hydrogen and hydrogen fluoride in a 1:4 ratio over titanium hydride at 700°C for several hours. At temperatures >700°C, the tetrafluoride forms.

Titanium trifluoride can also be made by an exchange reaction between hydrogen fluoride and titanium trichloride:

$$TiCl_3 + 3\,HF \rightarrow TiF_3 + 3\,HCl$$

The temperature of the mixture is held at 450°C for several hours and then raised rapidly to 700°C. The reaction product is then quickly cooled.

Titanium Tetrafluoride. The principal method for making titanium tetrafluoride is the reaction of anhydrous hydrogen fluoride with titanium tetrachloride. The mixture is distilled in the absence of all traces of moisture and titanium tetrafluoride sublimes at 284°C as a white solid of sp gr 2.798. It can also be made from the reaction of gaseous hydrogen fluoride with titanium tetrachloride at 100–200°C, boron trifluoride with titanium dioxide at 500°C, direct synthesis of the elements, or by heating barium titanium fluoride [31252-79-6], $BaTiF_6$.

Fluorotitanates. Titanium dioxide is soluble in hydrofluoric acid and forms fluorotitanic acid [17439-11-1]:

$$TiO_2 + 6\,HF \rightarrow H_2TiF_6 + 2\,H_2O$$

The acid is very stable in solution as it resists hydrolysis and decomposition, but the compound has not been isolated. A well-defined series of salts exists corresponding to this acid, perhaps the most important of which is potassium fluorotitanate [16919-27-0], K_2TiF_6. This can be made by dissolving titanium dioxide in hydrofluoric acid, adding potassium fluoride, and allowing the solution to crystallize forming $K_2TiF_6 \cdot H_2O$ [23969-67-7]. Alternatively, titanium dioxide can be fused with potassium fluoride and the melt extracted in water. The anhydrous salt forms brilliant transparent crystals (density 3.022 g/cm^3, mp 780°C). When heated in air to 500°C, it decomposes forming titanium dioxide and potassium fluoride. It can be reduced by sodium metal above 500°C to form titanium metal. It has been used as a constituent of molten electrolytes, for the formation of titanium metal, in abrasive grinding wheels, as a gelling agent for rubber, and as a grain-refining agent for aluminum.

Titanium Chlorides.   *Titanium Dichloride.* Titanium dichloride (mp 1035 ± 10°C, bp 1500 ± 40°C, density 3.1 g/cm^3) is a black solid crystallizing in the hexagonal system with the cadmium iodide-type structure. It reacts vigorously with water, thereby liberating hydrogen and leaving a solution of titanium trichloride.

Titanium dichloride may be prepared by one of the following methods:

1. The thermal disproportionation of titanium trichloride at 475°C in a vacuum:

$$2 \ TiCl_3 \rightarrow TiCl_2 + TiCl_4$$

The product is difficult to obtain in a pure state because the dichloride slowly decomposes according to the following equation:

$$2 \ TiCl_2 \rightleftharpoons TiCl_4 + Ti$$

2. The reduction of titanium tetrachloride with hydrogen, sodium, or titanium. Reduction with titanium is carried out in a sealed tube at 1100°C. A recent patent has described the production of gaseous titanium dichloride from the reaction of scrap titanium with titanium tetrachloride at >800°C and in the presence of an inert gas (102). The titanium dichloride is collected in a molten salt system and reacts with magnesium to produce pure titanium metal. In the industrial process for making titanium, sodium reacts with titanium tetrachloride; the formation of the dichloride is the first stage of the reaction.

In both methods of preparation, it is necessary to remove the product from the reaction zone as quickly as possible and to protect it from the atmosphere as it is very reactive.

Titanium Trichloride. Titanium trichloride is a dull purple, crystalline solid having a density of 2.66 g/cm^3. It reacts readily with moisture and is easily oxidized on exposure to the atmosphere. Various forms exist and reduction of titanium tetrachloride with hydrogen yields α-TiCl$_3$ in a practically pure form. When titanium tetrachloride is mixed with an aluminum alkyl, eg, triethylaluminum, to form a Ziegler catalyst which is used in polymerization reactions, a brown precipitate consisting essentially of titanium trichloride forms. Variations in the activity of this catalyst were examined and the existence of four modifications of titanium trichloride have been described (108). The modification produced depends critically on the method of preparation. The β and γ forms are made from the reaction of TiCl$_4$ with aluminium alkyls under various conditions, eg, β-TiCl$_3$ at low temperature (80°C); γ-TiCl$_3$ at 150–200°C; δ-TiCl$_3$ is made by prolonged dry grinding the α- or γ-forms. The β-form is brown, and the other three are violet.

The β-form has a fiberlike structure and a hexagonal unit cell of dimensions a = 0.627 nm and c = 0.582 nm. The position of the titanium atoms is such that the whole appears to be made up of bundles of fibers of $TiCl_3$ molecules, which may be regarded as one-dimensional linear polymers. The α- and γ-forms consist of planar sheets of titanium and chlorine atoms forming two-dimensional polymers of titanium trichloride. The x-ray powder-diffraction data are

	a, nm	c, nm
α-TiCl$_3$	0.356	0.587
γ-TiCl$_3$	0.614	1.740

In α-TiCl$_3$, the sheets are so placed as to give hexagonal close packing of the chlorine atoms and, in γ-TiCl$_4$, the same sheets are so placed as to give cubic close packing of the chlorine atoms. There is a close resemblance between the two structures, which results in a close similarity of properties. The x-ray diagram of the δ-form shows characteristics of both the α- and γ-forms, indicating that a certain amount of gliding of the sheets of atoms common to both structures has occurred during grinding.

Titanium trichloride is prepared by the reduction of titanium tetrachloride with hydrogen at 700°C. It is necessary to cool the titanium trichloride quickly to <450°C as temperatures above this cause it to disproportionate rapidly to the di- and tetrachlorides. Other reducing agents may be used, and recent examples include carbon or titanium, a mixture of carbon and hydrogen, and aluminum (109–112). The plasma arc has also been used in the production of titanium trichloride (113–116).

Titanium trichloride is used extensively as a catalyst for the polymerization of hydrocarbons (117–125). The trichloride is usually prepared from the tetrachloride by reduction with hydrogen, aluminum alkyls, or aluminum powder. Principal suppliers of titanium trichloride are, in the United States: Apache Chemicals, Inc., Seward, Ill.; Aran Isles Chemicals, Inc., Rockport, Mass.; Davos Chemical Corporation, Fort Lee, N.J.; Mitsubishi International Corporation, New York; Purechem Co., Catalyst Division of Dart Industries, Houston, Texas; Stauffer Chemical Co., Specialty Chemical Division, Westport, Conn.; and Texstar Chemical Corporation, Kearny, N.J. European manufacturers are Angler SpA (Italy), Brenntag (FRG), Kronos Titan GmbH (FRG), E. Merck (FRG), and BDH Chemicals Ltd. (UK). Japanese manufacturers are Toho Titanium Co., Ltd. and Toyo Stauffer Chemical Co., Ltd. The cost of titanium trichloride depends on the amount bought ($3500–5000/t).

Titanium Trichloride Hexahydrate. Titanium trichloride hexahydrate [19114-57-9] can be prepared by dissolving anhydrous titanium trichloride in water or by reducing a solution of titanium tetrachloride electrolytically or with sodium or zinc amalgam. Evaporation and crystallization of the solution yield violet crystals of the hexahydrate. The hydrated salt has had some commercial application because of its reducing properties as a stripping or bleaching agent in the dyeing industry, particularly where chlorine must be avoided.

Titanium Tetrachloride. Titanium tetrachloride is the most important of the halides of titanium and is manufactured commercially on a large scale. Production capacity in 1980 was ca 2.5×10^6 t. Its main uses are in the manufacture of titanium dioxide pigments and titanium metal, as a catalyst in many organic syntheses, and as a starting material in the manufacture of a range of titanium organic and inorganic compounds. It is also used by the military for smoke-screen purposes (see Chemicals in war).

Properties.   Physical properties of titanium tetrachloride are given in Table 19. Chlorine is completely miscible with titanium tetrachloride. The system shows a eutectic, which solidifies at $-108°C$ and contains 87.5 at % Cl. Above ca $-34°C$ (the normal boiling point of chlorine), complete miscibility only occurs in a closed system. The partial pressure of chlorine over solutions in titanium tetrachloride is proportional to the mole fraction of chlorine in the mixture, ie, Henry's law is obeyed. The variation in Henry's law constant H with temperature is given by $\log H = -941/T + 4.00$, and the heat of solution of chlorine in titanium tetrachloride is calculated from the variation of solubility with temperature as 16.7 kJ/mol (4 kcal/mol).

The apparent maximum solubilities of chlorine at 15.45 kPa (116 mm Hg) total pressure are as follows:

	Solubility	
Temp, °C	*Mole fraction of Cl_2*	*Cl_2 in 100 g of $TiCl_4$*
-10	0.38	23.4
0	0.28	14.3
20	0.155	6.85

Titanium tetrachloride is very susceptible to hydrolysis, and the liquid fumes strongly when in contact with moist air; the fumes are dense and consist of finely divided oxychlorides. Titanium tetrachloride has such an affinity for water that it is a very good desiccating agent. It is miscible with all common liquids, but reaction occurs with those containing hydroxyl, carboxyl, or diketone groups (in the enol form). Substitution products are formed with the elimination of hydrogen chloride. For example, water forms clear mixtures with titanium tetrachloride at room temperatures, but these are acidic because of hydrolysis. A series of oxychlorides and, ultimately, hydrous titanium dioxide, according to the conditions of temperature, acidity, and concentration, may be obtained from these solutions.

Table 19.   Physical Properties of Titanium Tetrachloride

color	colorless
density (at 20°C), g/cm^3	1.70
freezing point, °C	-24
heat of fusion, kJ/mol[a]	9.37
boiling point, °C	135.8
vapor pressure, kPa[b]	
at 20°C	1.33
at 50°C	5.52
at 100°C	35.47
heat of vaporization, kJ/mol[a]	
at 25°C	38.1
at 135.8°C	35.1
specific heat (at 20°C), J/g[a] (J/mol[a])	0.81 (153.1)
critical temperature, °C	358
heat of formation of liquid (at 25°C), kJ/mol[a]	-804.6 ± 3.8
viscosity, mPa·s (= cP)	0.079
refractive index, n_D^{20}	1.6985
magnetic susceptibility	-0.287×10^{-6}
dielectric constant (at 20°C)	2.79

[a] To convert J to cal, divide by 4.184.
[b] To convert kPa to mm Hg, multiply by 7.5.

A similar reaction occurs with alcohols, leading to the formation of alkoxides, which are also called esters of titanic acid (see Alkoxides, metal).

$$TiCl_4 \xrightarrow{\text{ROH}} TiCl_3OR \rightarrow TiCl_2(OR)_2 \rightarrow TiCl(OR)_3 \rightarrow Ti(OR)_4$$

In order that the reaction go to completion, ammonia or pyridine must be present to react with the liberated hydrogen chloride.

Titanium tetrachloride is soluble in simple organic solvents, eg, hydrocarbons, carbon tetrachloride, and chlorinated hydrocarbons, but addition compounds form with those that contain a donor atom, eg, ketonic oxygen, nitrogen, and sulfur. Addition compounds also form with a number of inorganic substances; their existence is deduced from phase-rule studies, examples of which are given in ref. 4. A summary of the addition compounds is given in ref. 126.

The reaction of titanium tetrachloride with anhydrous ammonia is complex and yields amidochlorides of titanium, $Ti(NH_2)_x Cl_{4-x}$ (127). Reduction of titanium tetrachloride by sodium, calcium, and magnesium yields the metal. Titanium tetrachloride is readily reduced by hydrogen yielding, at 700°C, titanium trichloride, which can be collected as a crystalline solid in a receiver held at a temperature above the boiling point of the tetrachloride. Complete reduction to the metal by hydrogen is only possible at very high temperatures, and the resultant metal sponge tends to take up hydrogen on cooling.

Manufacture. Detailed accounts of the manufacture of titanium tetrachloride are given in refs. 2, 128–132. A paper has been published recently on the chlorination of mineral rutile in fluidized beds (133). The fluidized-bed chlorination of titanium slags and ores has also been the subject of a report by the U.S. Bureau of Mines (134).

Titanium tetrachloride is manufactured by the chlorination of titanium compounds. These must contain as high a titanium content as possible, and the compounds used by industry are mineral rutile, beneficiated ilmenite, and leucoxene. As these are all oxygen-containing compounds, it is necessary to add carbon during the chlorination process to act as a reducing agent. Attention has recently been directed towards the possibility of using titanium carbide as the starting material, as the conditions required for chlorination are much less severe and titanium carbide contains a very high proportion of titanium (23).

Mineral rutile has been the preferred starting material for many years and coke, made from either coal or crude oil, has been the source of carbon. The reactions are as follows:

$$TiO_2 + 2\,Cl_2(g) + C(s) \rightarrow TiCl_4(g) + CO_2(g)$$
rutile

At 1100 K, $\Delta H = -217.6$ kJ/mol (-52.01 kcal/mol) and $\Delta G = -280.0$ kJ/mol (-66.92 kcal/mol). At 1300 K, $\Delta H = -221.7$ kJ/mol (-52.99 kcal/mol) and $\Delta G = -290.8$ kJ/mol (-69.50 kcal/mol).

$$TiO_2 + 2\,Cl_2(g) + 2\,C(s) \rightarrow TiCl_4(g) + 2\,CO(g)$$
rutile

At 1100 K, $\Delta H = -50.1$ kJ/mol (-12 kcal/mol) and $\Delta G = -302.0$ kJ/mol (-72.2 kcal/mol). At 1300 K, $\Delta H = -54.2$ kJ/mol (-13.0 kcal/mol) and $\Delta G = -305.7$ kJ/mol

(−73.06 kcal/mol). The chlorination process was first carried out as a batch process, which has been superseded by a continuous process in a fluid-bed reactor. The bed consists of a mixture of mineral rutile and coke, which are usually introduced at the top of the bed and which are fluidized by a stream of chlorine introduced at the base. The reaction conditions are chosen so that the bed is fluidized but so that fine reactants are not carried over by the exit gases; also, the first reaction forming carbon dioxide must occur to enough of an extent to ensure thermal balance. Chlorination occurs readily at 800–1000°C; it is necessary to establish the correct conditions experimentally for the use of every ore supply; therefore, continuity of one supply is very desirable.

The reaction products are condensed and impurities are removed by a sequential process involving a solids separator and a liquid scrubbing system. Impurities not readily removed by this process are vanadium as vanadium(V) oxychloride and tin as stannic chloride; both have boiling points close to that of titanium tetrachloride. In order to remove vanadium, it is necessary to add a reducing agent to reduce the vanadium to a chloride of appreciably lower volatility and to carry out a simple distillation. A very large number of compounds has been patented. Those most frequently used are lubricating oils and soaps. A more efficient distillation is required to separate tin.

The increasing cost of and demand for mineral rutile has led manufacturers to consider alternative materials, eg, ilmenite, leucoxene, and beneficiated ilmenite. It is necessary to consider the impurities that are likely to be present in these materials: impurities present in the chlorinator feed which chlorinate and pass over into the product stream and those which, when chlorinated, remain and make the fluid bed sticky, eg, sodium, calcium, and magnesium. In general, the purification system described above is efficient and the purity of the final distilled titanium tetrachloride is excellent. If the titanium tetrachloride is to be used for pigment manufacture, then particular attention must be paid to the vanadium content since only a few ppm can have an adverse effect on the color of the pigment.

Elements that, when chlorinated, remain in the fluid bed are particularly troublesome because they can cause the bed particles to sinter together, so that the bed is no longer fluidized. It is necessary to lay down stringent specifications for the concentrations of these elements.

The use of ilmenite as a starting material leads to problems in disposing of the large amounts of impurity chlorides (principally those of iron) that are produced. The chlorides may be dumped or decomposed to regenerate chlorine, depending on environmental and economic factors. Another alternative is to process the chlorides to produce a reasonably pure form of ferric chloride, which may be used in water treatment. The use of leucoxene or blends of ilmenite and mineral rutile are attractive in that they reduce the amount of iron chloride produced.

Much work has been done to upgrade ilmenite to provide a suitable feedstock for chlorination and sulfation. The success of a beneficiation process depends on many factors, eg, the current price of mineral rutile, environmental considerations, energy costs, and the ability to produce by-products in a salable form. The product must have a size range similar to rutile if it is to be an acceptable substitute in a fluid-bed process and it must chlorinate at a similar rate.

Alternative processes to the fluidized-bed method include the chlorination of titanium slags in chloride melts, chlorination with hydrogen chloride, flash chlorination which is claimed to be particularly advantageous for titanium minerals with high impurity contents, and chlorination with hydrogen chloride using titanium carbide as starting material (135–138).

Producers and Economic Aspects. The main producers of titanium tetrachloride throughout the world are, in the United States: American Cyanamid Company, E. I. du Pont de Nemours & Co., Inc. (Chemicals, Dyes, and Pigments Department), Gulf & Western, Kerr-McGee Chemical Corporation, and NL Industries; in Europe: Bayer A.G. (FRG), Dynamit Nobel A.G. (FRG), Kronos Titan GmbH (FRG), Thann et Mulhouse (France), and Tioxide UK Ltd.; and in Japan: Osaka Titanium Co. Ltd. and Toho Titanium Co. Ltd. Other suppliers who are probably not large-scale producers are Angler SpA (Italy), BDH Chemicals Ltd., Ferak Berlin OHG (FRG), and Steetley Chemicals Ltd. (UK). Current prices for titanium tetrachloride are: U.S. bulk deliveries, $0.66/kg (Dec. 1981); UK 10-t lots, $736/metric ton (Jan. 1982); UK 50-t lots, $726/t (Jan. 1982); and UK 10-kg lots, $3220/t (Jan. 1982).

Hexachlorotitanates. The complex acid H_2TiCl_6 forms when dry hydrogen chloride dissolves in titanium tetrachloride; the solution becomes yellow and is decolorized by the addition of water. The existence of this acid has been verified by a number of physical measurements, eg, conductivity and freezing-point measurements. This acid and its salts are less stable to hydrolysis than the corresponding fluorine derivatives. Ammonium [*21439-26-9*], potassium [*16918-46-0*], rubidium [*16902-24-2*], and cesium [*16918-47-1*] hexachlorotitanates are known. These salts are yellow and can undergo hydrolysis.

Oxychlorides. An oxychloride TiOCl can be prepared by the action of TiO_2, Fe_2O_3, or oxygen on $TiCl_3$ (139). For example, an excess of $TiCl_3$ is heated with TiO_2 in an evacuated sealed tube at 650°C.

$$2\ TiCl_3 + TiO_2 \rightarrow 2\ TiOCl + TiCl_4$$

The $TiCl_4$ can be separated by freezing and the excess $TiCl_3$ removed by solution in dimethylformamide. Titanous oxychloride forms yellow tablets that are rhombic. It is stable in air and inert in water and mineral acids. When heated in air, it yields $TiCl_4$ and TiO_2.

Hydrolysis of $TiCl_4$ yields a number of products, the compositions of which vary according to the conditions of hydrolysis. A definite compound $TiOCl_2$ can be prepared by the reaction of chlorine monoxide or, better, ozone with titanium tetrachloride (140). It is a pale yellow powder, insoluble in nonpolar solvents, but soluble in ether–HCl. Water hydrolyzes it to a white powder. Its ir spectrum shows a band characteristic of the grouping O—Ti—O, which indicates a polymeric structure.

Titanium Bromides. *Titanium Dibromide.* Titanium dibromide is a black powder of sp gr 4.31. It is a powerful reducing agent, dissolving in water to liberate hydrogen and form a solution containing trivalent titanium. It is spontaneously flammable in air. It reacts with titanium tetrabromide at 300°C to form the tribromide.

Titanium dibromide can be made by one of the following methods. The first is the direct synthesis from the elements; the bromine should be cooled until it solidifies to avoid a violent reaction. The reaction then takes place as the bromine melts. The other method is disproportionation of titanium tribromide at 400°C:

$$2\ TiBr_3 \rightarrow TiBr_2 + TiBr_4$$

At high temperatures (500–600°C), the dibromide decomposes to form titanium tetrabromide:

$$2 \text{ TiBr}_2 \rightarrow \text{TiBr}_4 + \text{Ti}$$

Titanium Tribromide. Titanium tribromide is a blue-black crystalline powder, which exists in two crystalline forms, ie, hexagonal plates or needles. It dissolves in water to give a dark violet solution which on evaporation yields crystals of the hexahydrate $\text{TiBr}_3 \cdot 6\text{H}_2\text{O}$ [15162-99-9]. It disproportionates at 400°C to give the di- and tetrabromides.

Titanium tribromide can be prepared either by a reduction of the tetrabromide with hydrogen:

$$2 \text{ TiBr}_4 + \text{H}_2 \rightarrow 2 \text{ TiBr}_3 + 2 \text{ HBr}$$

or by a reaction of the tetrabromide with titanium:

$$3 \text{ TiBr}_4 + \text{Ti} \rightarrow 4 \text{ TiBr}_3$$

Titanium Tetrabromide. Titanium tetrabromide is a lemon-yellow solid of density 3.25 g/cm^3, and it crystallizes in the cubic system. It has a melting point of 38°C and boils at 233°C. It dissolves in water and is readily hydrolyzed. Addition compounds are formed with NH_3, PH_3, SO_2, H_2S, and some organic compounds.

Titanium tetrabromide is the best known of the titanium–bromine compounds. It is prepared either by double decomposition of titanium tetrachloride and hydrogen bromide (141):

$$\text{TiCl}_4 + 4 \text{ HBr} \rightarrow \text{TiBr}_4 + 4 \text{ HCl}$$

or by direct combination of the elements (142):

$$\text{Ti} + 2 \text{ Br}_2 \rightarrow \text{TiBr}_4$$

Care must be taken in the first reaction because it proceeds vigorously. The temperature is increased gradually until titanium tetrabromide distills from the mixture. In the second reaction, solidified bromine and titanium in the stoichiometric quantities are placed in an evacuated, sealed, thick-walled tube. The reaction proceeds rapidly as the bromine melts. The product can be purified by distillation in the absence of moisture.

Titanium tetrabromide can also be prepared by bromination of an intimate mixture of carbon and titanium dioxide at 650–700°C.

Many patents have been issued for the use of titanium tetrabromide as a catalyst. Recent examples include the catalytic conversion of high boiling to low boiling hydrocarbons, the polymerization of epoxides, the polymerization of isobutene, the low temperature polymerization of ethylene, the production of methacrylate esters, and the production of terpenes (143–149).

Titanium Iodides. **Titanium Diiodide.** Titanium diiodide is a black powder, crystallizing in the hexagonal system. It reacts rapidly with water, liberating hydrogen and producing a solution of titanous iodide, TiI_3.

Titanium diiodide can be made by direct combination of the elements, with the reaction mixture heated to 440°C to remove the tri- and tetraiodides (149). It can also be made from the reaction of solid potassium iodide with titanium tetrachloride or by reduction of TiI_4 with silver or mercury (150).

Titanium Triiodide. Titanium triiodide can be made by direct combination of the elements. It is stable under vacuum up to 300°C, but it disproportionates at 350°C forming a mixture of the di- and tetraiodides (151). It is a violet, crystalline compound, reacting slowly with water without liberation of hydrogen. The hexahydrate $\text{TiI}_3 \cdot 6\text{H}_2\text{O}$

[*84282-52-0*] can be prepared by electrolytic reduction of an aqueous solution of titanium tetraiodide followed by crystallization in the presence of hydriodic acid (152). The crystals oxidize rapidly in air.

Titanium Tetraiodide. Titanium tetraiodide forms reddish brown, octahedral crystals that dissolve in water and undergo hydrolysis. The crystals have a density of 4.40 g/cm^3, a melting point of 150°C, and a boiling point of 377°C. It does not form addition compounds, such as are formed by the tetrachloride and tetrabromide.

Titanium tetraiodide can be prepared by direct combination of the elements under carefully controlled conditions to prevent decomposition of the product. It can also be made by the reaction of gaseous hydrogen iodide with titanium tetrachloride and by gradually raising the temperature until titanium tetrachloride distills from the mixture. A hydrogen atmosphere is maintained during the distillation.

Titanium tetraiodide is mentioned extensively in the literature as a catalyst in organic reactions. A recent example of its application is in the production of ethylene glycol from acetylene (153). It is also an important intermediate in the production of pure titanium metal. The tetraiodide forms from the reaction of iodine with the impure metal under vacuum at ca 200°C. The sublimed titanium tetraiodide then decomposes on a tungsten filament at 1300°C. Very pure titanium metal can be made by this process (154).

Titanium–Silicon System

The titanium–silicon system includes three silicides: $TiSi_2$, $TiSi$, and Ti_5Si_3 (155). A fourth silicide, Ti_5Si_4, has been isolated from the ternary system Cu–Ti–Si and characterized by electron microscopy (156). The compound was also detected in the reaction products of titanium and silicon tetrachlorides on graphite at 900–1300°C (157). The crystal structures and physical constants of the silicides are given in Table 20.

In general, the various titanium silicides can be made by direct synthesis of the elements. Special methods for their manufacture involve the use of fused salt baths, in which reduction is brought about by electrolysis or by the use of alkali or alkaline earth elements (157–158). By these means, $TiSi_2$ and Ti_5Si_3 have been produced. Mixtures of titanium and silicon have been melted in an argon atmosphere in an arc at a temperature of 1100°C to produce $TiSi_2$ (159). If titanium hydride is used instead of titanium, then a reducing atmosphere is generated simultaneously (160).

Table 20.   Structure Parameters and Physical Constants of Titanium Silicides

Property	TiSi [*12039-70-2*]	TiSi$_2$ [*12039-83-7*]	Ti$_5$Si$_3$ [*12067-57-1*]	Ti$_5$Si$_4$ [*12413-10-4*]
structure	rhombic	orthorhombic	hexagonal	
dimensions, nm				
a	0.3611	0.8236	0.7465	0.6173
b	0.4960	0.4773		
c	0.6479	0.8523	0.5162	1.2171
sp gr	4.34	4.39	4.31	
mp, °C	1760 (incongruent)	ca 1540	2120	
hardness, 100-g load	1039	870	986	
hardness, Mohs		4–5		
resistivity, $\mu\Omega\cdot$cm		123		

Titanium tetrachloride and silicon tetrachloride or silicon have also been used as starting materials. By the reaction of a mixture of $TiCl_4$, $SiCl_4$, and H_2 at ca 1150°C and by selecting the correct reactant quantities, TiSi or $TiSi_2$ can be prepared (161). A process has been described in which $TiCl_3$, $AlCl_3$, Si, and Al are heated together in an argon atmosphere at 1000°C; 99.8% pure TiSi is obtained (162). Argon or helium plasma units have been used as the source of energy with silicon, silicon dioxide, titanium, or titanium dioxide as the starting materials (163–164).

The compound Ti_5Si_3 can be prepared by heating a mixture of titanium dioxide and calcium silicide to 900°C and then to 1200–1300°C. The overall reaction is

$$4\,Ca_2Si + 5\,TiO_2 \rightarrow Ti_5Si_3 + 6\,CaO + 2\,CaO{\cdot}SiO_2$$

The lime and calcium silicate are dissolved with acetic acid, leaving a residue of pure Ti_5Si_3.

Titanium disilicide is a silvery gray, crystalline material that oxidizes slowly in air when heated to red heat (700–800°C). It is resistant to mineral acids and 50 wt % aq solns of sodium hydroxide and potassium hydroxide, but dissolves in hydrofluoric acid. It is decomposed by molten borax, sodium hydroxide, and potassium hydroxide. It reacts explosively with chlorine at high temperatures. Like the pentatitanium trisilicide, it is very friable at room temperature.

The titanium silicides are used in the preparation of abrasion and heat-resistant refractories (qv). A mixture of Ti_5Si_3 with diamond and titanium carbide, hot pressed at 1450°C and 6 GPa (9×10^5 psi), forms a cutting tip with a much longer life than conventional tools (165). It also is used as an electrical resistance material and in electrically conducting ceramics (166) (see Ceramics as electrical materials). An interesting application is its use in pressure-sensitive elastic resistors, the electrical resistance of which varies with external pressure (167).

Titanium–Phosphorus Compounds

Titanium Phosphides. Titanium phosphide is a substance of variable composition; among the phases that have been recognized are TiP_2 [12037-76-2], TiP [12037-65-9], $TiP_{0.95}$ [12037-65-9], Ti_3P_2 [12202-46-9], and Ti_3P [12037-66-0]. Titanium monophosphide can be prepared by heating phosphine with titanium tetrachloride or titanium sponge; alternatively, titanium metal powder may be heated with phosphorus in a sealed tube. The final compound is slightly deficient in phosphorus and, in this, it resembles the phosphides of the other Group IVB elements. It is a gray metallic powder of hexagonal structure and has a density of 4.08 g/cm³. The density decreases as the phosphorus content increases:

TiP_n, n	0.26	0.68	0.84	0.95
Density, g/cm³	4.66	4.27	4.18	4.08

It has unusually high thermal stability, and it does not decompose when heated to 1100°C in vacuum or a protective atmosphere. It is not attacked by any of the common dilute or concentrated acids, is only slightly attacked by aqua regia, but is oxidized on heating in air. It is classed as a hard substance having a diamond pyramid hardness of 718 (titanium carbide's hardness is 1700). Titanium diphosphide can be prepared by heating TiP and red P in an evacuated sealed tube in a two-zone furnace. The temperature of the TiP zone is held at 700°C and the P zone at 547°C (168).

The only commercial use for titanium phosphide reported to date is as a catalyst in polycondensation reactions.

Titanium(III) Phosphate, Titanous Phosphate. Titanium(III) phosphate [24704-65-2], $Ti(PO_4)_3$, is soluble in dilute acids, giving solutions that are relatively stable and resistant to hydrolysis. This compound is made by adding a soluble phosphate to a solution of titanous chloride or sulfate, decreasing the acidity of the solution until precipitation occurs, and filtering and washing the product, which is purple.

Titanium(IV) Phosphate. The addition of soluble phosphates to a solution of titanium(IV) sulfate [13693-11-3] or chloride yields a precipitate of titanium(IV) phosphate [17017-60-6], which is a white compound that is insoluble in dilute sulfuric acid. Titanium phosphate prepared in this way is used in the dyeing and leather tanning industries (see Dyes; Leather). Recently, attention has been directed towards the production of titanium phosphate for use as a white pigment, and a process for the recovery of titanium phosphate pigment from wastes from the sulfuric acid production of TiO_2 has been described (169–170). It has also been used as a catalyst in the oxidation of ethane and reduction of ethane to ethylene (171).

Titanium(IV) Bis(Hydrogen Phosphate) Dihydrate. There has been considerable interest recently in the use of titanium(IV) bis(hydrogen phosphate) dihydrate [14635-14-4], $Ti(HPO_4)_2 \cdot 2H_2O$, as an ion-exchange reagent (172). Processes involving the compound either alone or in admixture with a carbonaceous material and phosphoric acid have been described. These include the removal of K^+ from seawater, $^{42}K^+$ and $^{137}Cs^+$ from strong mineral acid media, and NH_4^+ and NH_3 from waste solutions (173–175). A complex phosphate, titanium vanadophosphate, has been shown to have a high selectivity for univalent metal ions, and nine representative separations have been reported (176).

Titanium Pyrophosphate. Titanium pyrophosphate [13470-09-2], TiP_2O, is a white powder that crystallizes in the cubic system and has a density, as calculated from x-ray data, of 3.106 g/cm^3. It is insoluble in water and can be prepared by heating a mixture of 1 mol TiO_2 as hydrous TiO_2 and 1 mol P_2O_5 as phosphoric acid to 900°C. Some interest has been shown in titanium pyrophosphate as an ultraviolet reflecting pigment.

Titanium–Sulfur Compounds

Titanium Sulfides. The titanium–sulfur system is summarized in refs. 4 and 177. Nonstoichiometry occurs extensively in the system, and the limits of composition for each sulfide are wide. Five sulfides have been reported.

Titanium Subsulfide. Titanium subsulfide [12039-08-6], Ti_2S, forms as a gray solid of density 4.60 g/cm^3 when titanium monosulfide [12039-07-5] is heated at 1000°C with titanium in a sealed tube. It can also be formed by carefully heating a mixture of the two elements at 800–1000°C. The sulfide is soluble in concentrated hydrochloric and sulfuric acids but is insoluble in alkalies.

Titanium Monosulfide. Titanium monosulfide, TiS, is a dark brown solid which can be prepared either by direct combination of the elements or by reduction of the disulfide [12039-13-3] with hydrogen at a high temperature:

$$TiS_2 + H_2 \rightarrow TiS + H_2S$$

It is dimorphous, one form being unstable and readily converted by heating to a stable form, which has a hexagonal structure and a density of 4.05 g/cm^3; this change is irreversible. Titanium monosulfide is attacked by concentrated hydrochloric and nitric acids but not by alkalies.

Titanium Sesquisulfide. Titanium sesquisulfide [*12039-16-6*], Ti_2S_3, is a black, crystalline solid (density 3.52 g/cm^3) and has a hexagonal structure. It is prepared by direct combination of the elements at 800°C, by reduction of titanium disulfide with hydrogen, or by pyrolysis of titanium disulfide *in vacuo* at 1000°C. It has similar chemical properties to the other sulfides.

Titanium Disulfide. Titanium disulfide, TiS_2, can be made by the action of H_2S on titanium tetrachloride at 600°C, by heating titanium sponge with sulfur in a sealed tube, or by pyrolysis of titanium trisulfide at 550°C (178–179). It has a hexagonal structure and a consistency similar to that of graphite, a bronze color, and a density of 3.22 g/cm^3. When heated, titanium disulfide loses sulfur at 1000°C *in vacuo* forming a lower sulfide. In moist air, it forms H_2S and TiO_2 on heating. It is not attacked by hydrochloric acid but is soluble in cold and hot sulfuric acid. Unlike the other titanium sulfides, it is decomposed by hot sodium hydroxide solution.

Titanium Trisulfide. Titanium trisulfide [*12423-80-2*], TiS_3, is a black, crystalline substance that can be prepared by passing a mixture of titanium tetrachloride vapor and H_2S through a pyrex tube heated to 480–540°C. The reaction product is mixed with sulfur and heated to 600°C in a sealed tube. The product is contaminated with other sulfides and the pure trisulfide can be separated by sublimation (TiS_3 at 390°C and TiS_2 at 500°C). Alternatively, a mixture of titanium sponge and sulfur can be heated in an evacuated sealed container at 600°C for about two weeks, yielding monoclinic crystals of the trisulfide which can be further purified by sublimation.

Titanium trisulfide has a density of 3.21 g/cm^3 and is insoluble in hydrochloric acid but soluble in hot and cold sulfuric acid. It is attacked by concentrated nitric acid forming hydrated titanium dioxide.

There are no significant commercial applications for the sulfides of titanium, although a patent was issued covering the use of titanium disulfide as a solid lubricant (180). Recently, there has been considerable investigation of the possibility of using titanium disulfide as cathode material in high efficiency batteries (see Batteries). The titanium disulfide, which replaces graphite, acts as a host material for various intercalation reactions involving alkali- or alkaline-earth elements (181).

Titanium Sulfates.   *Titanous Sulfate, Titanium(III) Sulfate.* Solutions of titanous sulfate [*10343-61-0*] (titanium(III) sulfate) are readily made by reduction of titanium(IV) sulfate in sulfuric acid solution by chemical or electrolytical means (182). A solution of $Ti(SO_4)_2$ containing the equivalent of 100 g/L of TiO_2 in a divided cell with a lead anode and a rotating amalgamated lead cathode can be used. Zinc, zinc amalgam, cadmium, aluminum, and lead have been used as reducing agents, and the reaction is the basis of the most used titrimetric procedure for the determination of titanium. Titanous sulfate solutions are violet and slowly oxidize in contact with the atmosphere. Storing them in an oxygen-free gas or covering the solution surfaces with a layer of oil prevents their contact with oxygen. If all the titanium has been reduced to the trivalent form and the solution is evaporated, crystals of an acid sulfate $3Ti_2(SO_4)_3.H_2SO_4.25H_2O$ [*10343-61-0*] are produced. This is a purple salt, stable in air at ordinary temperatures, and soluble in water, giving a stable violet solution. It is insoluble in 60 wt % sulfuric acid, alcohol, and ether. Heated in air, it decomposes to TiO_2, water, sulfuric acid, and sulfur dioxide. The hydrogen in the sulfuric acid can be replaced by metals to form double sulfates, of which $NH_4Ti_3(SO_4)_5.9H_2O$ [*55231-27-1*] and $RbTi_3(SO_4)_5.12H_2O$ [*35064-61-0*] are typical.

If the acid sulfate is dissolved in dilute sulfuric acid and evaporated at 200°C, crystals of the anhydrous neutral salt form and can be purified by washing with dilute sulfuric acid, alcohol, and ether, followed by drying under a nitrogen atmosphere at 140°C. The anhydrous salt is green, is insoluble in water, alcohol, and concentrated sulfuric acid, but is soluble in dilute sulfuric or hydrochloric acid solutions yielding a violet solution. It is hexagonal and isomorphous with chromic sulfate, and it is stable up to 450°C. It is used commercially as a reducing agent.

Titanium(IV) Sulfate. Titanium(IV) sulfate is extremely sensitive to moisture, which decomposes it into hydrolysis products. If heated, it changes to titanyl sulfate [13825-74-6], $TiOSO_4$, and then TiO_2. It is a white hygroscopic solid. Several hydrates have been reported, but the existence of these and of the anhydrous salt in a neutral condition is questionable.

The preparation and conservation of titanium(IV) sulfate are extremely difficult. It can be prepared by reaction of titanium tetrachloride with sulfur trioxide dissolved in sulfuryl chloride.

$$TiCl_4 + 6\,SO_3 \rightarrow Ti(SO_4)_2 + 2\,S_2O_5Cl_2$$

The precipitate can be purified by boiling under reflux, filtering, and washing with sulfuryl chloride. It is then dried by pressure and finally under vacuum.

Titanyl Sulfate. Titanyl sulfate is the best known of the basic sulfates of titanium; it can be prepared as a white, crystalline, hygroscopic powder by the controlled reaction of sulfuric acid with titanium dioxide or hydrated titanium dioxide. Alternatively, anhydrous $Ti(SO_4)_2$ can be heated under controlled conditions. It is soluble in water with hydrolysis, and a precipitate of hydrated TiO_2 forms on warming. The oxide also forms if titanyl sulfate is heated strongly.

A stable dihydrate $TiOSO_4.2H_2O$ [58428-64-1] forms upon evaporation of a solution of hydrated TiO_2 in sulfuric acid until crystals are formed. The crystals are separated, washed with dilute sulfuric acid and then acetone, and then dried. The dihydrate is a white, needlelike crystal powder, which dissolves slowly in cold water. These solutions are unstable on warming and readily hydrolyze forming a precipitate of hydrated TiO_2. When heated in air, the dihydrate loses water and then SO_3 at a starting temperature of ca 450°C forming TiO_2.

Titanyl sulfate is used in the dying industry for the preparation of titanous sulfate, which is useful in dye stripping and as a mordant. Its main use is as an intermediate in the manufacture of titanium pigments by the sulfuric acid process. It is also used as a starting material for the preparation of other titanium compounds. Titanyl sulfate has been reported to be of use in the tanning of leather to produce an acid-resistant book-binding leather and flexible sole leathers with high abrasion resistance (183–184).

Various double sulfates of the type $M_2[Ti(SO_4)_3]$ and $M_2[TiO(SO_4)_2]$ have been described, where M is a monovalent element. Titanyl ammonium disulfate [19468-87-2], $(NH_4)_2TiO(SO_4)_2.H_2O$, is used increasingly in the leather industry as a tanning agent (185–186). It can be prepared from sulfate liquors of titanium by precipitation with ammonium sulfate and sulfuric acid (see Leather).

Analytical Methods

The analytical chemistry of titanium has been reviewed in refs. 187–188. Titanium ores can be dissolved by fusion with potassium pyrosulfate; the fusion is best carried out in borosilicate glass flasks that are covered to avoid losses by splattering of the mixture. The cooled melt can be dissolved in dilute sulfuric acid. Any residue should be treated to remove silica, fused in a sodium carbonate–borate mixture, dissolved, and added to the original solution. Alternatively, the latter fusion mixture can be used by itself. Titanium can then be determined directly in the solution using the titrimetric procedure outlined below. Of the elements present, only chromium, vanadium, and niobium interfere.

Titanium dioxide pigment samples can be dissolved in a boiling mixture of concentrated sulfuric and ammonium sulfate. The solution is cooled and diluted, and the titanium is determined titrimetrically. Alternatively, the sample may be dissolved in hydrofluoric acid, which is the recommended procedure if trace elements are to be determined in the pigment sample.

Titanium can be determined in solution by reduction methods, eg, with the Jones or Nakazono reductors, or by heating the solution in a flask with the reducing agent. Zinc, amalgamated zinc, cadmium, lead, aluminum, fusible alloys, and bismuth have been used as reducing agents. The solutions must be protected from reoxidation by atmospheric oxygen and they can be titrated with standard ferric iron solutions, which must be standardized against a sample of known titanium content. Potassium thiocyanate is used as an indicator, or the titration can be carried out potentiometrically.

Small amounts of titanium can be determined by adding hydrogen peroxide solution to the acidic sample solution. A yellow solution containing the peroxy–titanium complex is obtained, and the intensity is determined spectrophotometrically. Of the more common elements, only vanadium interferes and this may be distinguished as the color of the peroxy–titanium complex is discharged by the addition of hydrofluoric acid. This method is reliable but has limited sensitivity. A more sensitive method involves Tiron (disodium 1,2-dihydroxybenzene-3,5-disulfonate), which is used to detect 10 ppb in solution. The solution to be analyzed is neutralized, and a sodium acetate–acetic acid buffer is added. Tiron solution is then added followed by the addition of a small amount of solid sodium dithionite, and the optical density of the solution is determined. Titanium can also be determined by any of the standard instrumental methods, eg, polarography, x-ray spectrometry, atomic absorption, and plasma-emission spectrometry.

Health and Safety Aspects, Toxicology

Much work has been carried out on the toxicology of titanium during the last four or five decades, largely as a result of the widespread use of titanium dioxide pigments (189–191). The following discussion is concerned only with the behavior of titanium as it is present in inorganic compounds and not with the effects of the compound itself. For example, titanium tetrachloride must be treated with care because of the toxic effects of hydrochloric acid and heat produced when it reacts with water, not because of the possible toxicity of titanium.

There is no evidence that titanium is an essential element for human life; nor is there any evidence that it is a toxic element. The human body appears to have a wide range of tolerance to titanium. Animal studies have confirmed that TiO_2 is not carcinogenic (192), mutagenic, or teratogenic. The metal is frequently used as a surgical implant, and several titanium compounds have been used in medical and food products without any ill effects. Titanium dioxide has also been used as a control in the study of rejection of other substances from the lungs, where the highest concentration in the human body occurs (193–195). Titanium accumulates in the lungs throughout life. Persons exposed to titanium compounds in their working environment have higher titanium concentrations in their bodies; eg, a mean level of 119 μg/g dry weight was determined to be in the lungs of mine workers compared with 19 μg/g for persons not so exposed (196). Significantly higher levels of titanium occurred in the lungs of three workers who had been engaged in the manufacture of titanium dioxide pigments for nine years when compared with normal subjects (197). Deposits in the pulmonary interstices occurred in association with cell destruction and slight fibrosis. The presence of titanium in the lymphatic system led to the conclusion that this system was responsible for the elimination of titanium dioxide from the lungs. In light of this evidence, titanium dioxide first was classified as a slight irritant on the pulmonary interstices, but it was shown later that the adverse effects resulted from substances such as quartz or silicates and not titanium dioxide (198).

Titanium dioxide pigment, because of its small particle size, is classified as a nuisance dust and a TLV rating of 10 mg/m³ total dust or 5 mg/m³ respirable dust has been specified by the ACGIH (1980) (199) and the UK Health and Safety Executive (1980) (200). The short-term exposure limit is 30 mg/m³ (199). In the FRG, a MAK-value of 8 mg/m³ has been specified. It is not possible to predict what proportion of titanium dioxide pigment is in the respirable range, as this depends on the storage and handling history of the material. During production at the packing stage, respirable dust amounts to ca 5–15% of the total dust. If it is likely that the concentration exceeds the specified levels, workers must wear suitable respiratory protection and dustproof goggles. Because of their fine particle size, titanium dioxide pigments absorb water and natural oils from the skin thereby causing dryness. Prolonged exposure should therefore be avoided and any dryness remedied by washing with soap and water.

The concentration of titanium in blood samples taken from average persons has been determined to be 30–160 μg/L blood by a number of investigators (194,201–202). The concentration in urine for normal persons is ca 10 μg/L (203).

BIBLIOGRAPHY

"Titanium Compounds, Inorganic" in *ECT* 1st ed., Vol. 14, pp. 213–237, by L. R. Blair, H. H. Beacham, and W. K. Nelson, Titanium Div., National Lead Company; "Titanium Compounds, Inorganic" in *ECT* 2nd ed., Vol. 20, pp. 380–424, by G. H. J. Neville, British Titan Products Co., Ltd.

1. *Gmelins Handbuch der Anorganischen Chemie*, 8th ed., System-Nummer 41, Springer-Verlag, Berlin, FRG, 1979.
2. J. Barksdale, *Titanium, Its Occurrence, Chemistry and Technology*, 2nd ed., The Ronald Press Co., New York, 1966.
3. G. Skinner, H. L. Johnson, and C. Beckett, *Titanium and Its Compounds*, Herrick L. Johnston Enterprises, Columbus, Ohio, 1954.
4. P. Pascal, *Nouveau Traité de Chimie Minérals*, Tome IX, Marson et Cie, Paris, France, 1963.
5. R. Feld and P. L. Crowe, *The Organic Chemistry of Titanium*, Butterworth & Co., Ltd., London, 1965.
6. R. J. H. Clark, *The Chemistry of Titanium and Vanadium*, Elsevier Publishing Co., Amsterdam, The Netherlands, and New York, 1968.
7. G. P. Luchinskii, *Chemistry of Titanium*, Khimiya, Moscow, USSR, 1971.
8. Ya. G. Goroshchenko, *Chemistry of Titanium*, Nankova Dumka, Kiev, USSR, 1970.
9. F. D. Rossini and co-workers, *Properties of Titanium and Related Compounds*, U.S. Office of Naval Research, Washington, D.C., 1956.

10. *JANAF Thermochemical Tables*, 2nd ed., U.S. National Bureau of Standards, June 1971, NSRDS-NBS37.
11. W. M. Mueller, J. P. Blackledge, and G. G. Libowitz, *Metal Hydrides*, Academic Press, Inc., New York and London, 1968, pp. 336–383.
12. G. Brauer, *Handbook of Preparative Inorganic Chemistry*, 2nd ed., Vol. 1, Academic Press, Inc., New York, 1963, p. 114.
13. Y. Osumi and H. Suzuki, *Chitaniumi Jirukoniumi* **27,** 142 (1979).
14. A. E. Palty, H. Margolin, and J. P. Nielsen, *Trans. Am. Soc. Metals* **46,** 312 (1954).
15. A. K. Ganesan and co-workers, *High Temp. Mater., Proc. Symp. Mater. Sci. Res. 3rd,* 281 (1972).
16. F. Marx and co-workers, *Keram Z.* **26**(7), 382 (1974).
17. Ger. Pat. 2,523,325 (Aug. 19, 1976), R. S. Sheppard (to PPG Industries, Inc.).
18. U.S. Pat. 4,022,872 (May 10, 1977), D. R. Carson and C. B. Calvin (to PPG Industries, Inc.).
19. E. K. Storms, *The Refractory Carbides*, Academic Press, Inc., New York and London, 1967, pp. 1–17.
20. B. W. Davis and R. G. Varsinik, *J. Colloid Interface Sci.* **37,** 870 (1971).
21. J. R. Tinklepaugh and W. B. Crandall, *Cermets*, Reinhold Publishing Corporation, New York, 1960.
22. U.S. Pat. 3,786,133 (Jan. 15, 1974), S-T. Chiu (to Quebec Iron & Titanium Corp.).
23. G. W. Elger, *Preparation and Chlorination of Titanium Carbide from Domestic Titaniferous Ores*, Report of Investigations 8497, U.S. Department of the Interior, Bureau of Mines, Washington, D.C., 1980.
24. A. Münster, *Angew. Chem.* **69,** 281 (1957).
25. R. I. Jaffee, H. R. Ogden, and D. J. Maykuth, *J. Metals, Trans.* **188,** 1261 (1950).
26. A. E. Van Arkel and J. H. Boer, *Z. Anorg. Chem.* **148,** 345 (1925).
27. A. Münster and W. Ruppert, *Naturwissenschaften* **39,** 349 (1952).
28. A. I. Karasev and co-workers, *Vopr. Khim. Khim. Tekhnol.* **31,** 153 (1973).
29. U.S. Pat. 2,461,018 (Feb. 8, 1949), P. P. Alexander (to Metal Hydrides Inc.).
30. O. Matsumoto, *Kinzoku* **38,** 69 (1968).
31. A. Reihlen and A. Hake, *Liebigs Ann.* **452,** 47 (1927).
32. J. R. Partington and A. L. Whynes, *J. Chem. Soc.*, 3135 (1949).
33. C. E. Rice and W. R. Robinson, *Acta Crystallogr. Sect. B* **B33**(5), 1342 (1977).
34. M. Marezio and P. D. Dernier, *J. Solid State Chem.* **3,** 340 (1971).
35. R. J. D. Tilley in M. W. Roberts and J. M. Thomas, eds., *Chemical Physics of Solids and their Surfaces*, Vol. 8, Royal Society of Chemistry Specialist Publications, London, 1981, Chapt. 6, p. 121.
36. T. V. Charlu, O. J. Kleppa, and T. B. Reed, *J. Chem. Thermodyn.* **6,** 1065 (1974).
37. Gutbier and co-workers, *Z. Anorg. Chem.* **162,** 87 (1927).
38. O. Söhnel, *Collect. Czech. Chem. Commun.* **44,** 2560 (1974).
39. Y. Ionue and M. Tsuji, *J. Nucl. Sci. Technol.* **13**(2), 85 (1976).
40. Y. Ionue and M. Tsuji, *Bull. Chem. Soc. Jpn.* **49,** 111 (1976).
41. Y. Ionue and M. Tsuji, *Bull. Chem. Soc. Jpn.* **51,** 479 (1978).
42. *Ibid.*, p. 794.
43. K. V. Chumutov and co-workers, *Russ. J. Phys. Chem.* **46**(5), 660 (1972).
44. G. R. Doshi and V. N. Sastry, *Indian J. Chem.* **15A,** 904 (1977).
45. A. M. Andrianov and co-workers, *J. Appl. Chem. USSR* **51,** 1789 (1978).
46. A. Schroder, *Z. Kristallographie* **67,** 485 (1928).
47. W. L. Bond, *J. Appl. Phys.* **36,** 1676 (1965).
48. J. R. De Vore and A. H. J. Pfund, *J. Opt. Soc. Am.* **37,** 826 (1947).
49. D. C. Cronemeyer, *Phys. Rev.* **87,** 876 (1952).
50. F. A. Grant, *Rev. Mod. Phys.* **31,** 646 (1959).
51. W. A. Kampfer in T. C. Patton, ed., *Pigment Handbook*, Vol. 1, Wiley-Interscience, New York and London, 1973, p. 6.
52. W. Piekarczyk, *Int. Symp. Reinst. Wiss. Tech. Tagungsber.* **1,** 213 (1966); *Chem. Abstr.* **70,** 21376 (1969).
53. E. M. Gladrow and H. G. Ellert, *J. Chem. Eng. Data* **6**(2), 318 (1961).
54. *Ind. Miner.* (*London*), 14 (July 1979).
55. C. H. Moore, *Trans. Am. Inst. Min. Metall. Pet. Eng.* **184,** 194 (1949).
56. L. Merker, *Min. Eng.* (*N.Y.*) **7,** 645 (1955).
57. M. S. Wainwright and N. R. Foster, *Catal. Rev. Sci. Eng.* **19**(2), 211 (1979).

58. T. Ohtsuka and Y. Ishihara, *J. Inst. Fuel* **51,** 82 (1977).
59. Ger. Pat. 2,833,993 (Mar. 22, 1979), T. Y. Tien, D. J. Romine, and D. E. Davis (to Bendix Autolite Corp.).
60. M. A. Vannice and R. L. Garten, *J. Catal.* **56,** 236 (1979).
61. A. J. Bard, *Science* **207,** 139 (1980).
62. E. Borgarello and co-workers, *J. Am. Chem. Soc.* **104**(11), 2996 (1982).
63. Jpn. Pat. 78 94,299 (Jan. 28, 1978), A. Kato and co-workers (to Babcock Hitachi K.K.).
64. Jpn. Pat. 80 49,322 (Apr. 9, 1980), T. Veno and co-workers (to Mitsubishi Gas Chemical Co., Inc.).
65. U.S. Pat. 4,044,042 (Aug. 23, 1977), H. P. Angstad (to Suntech Inc.).
66. Jpn. Pat. 79 135,733 (Oct. 28, 1979), H. Hayami and S. Yoda (to Nippon Kayaku Co., Ltd.).
67. K. Arata and K. Tanabe, *Chem. Lett.*, (8), 1017 (1979).
68. Jpn. Pat. 79 95,524 (July 28, 1979), N. Cho and co-workers (to Mitsui Petrochemical Industries Ltd.).
69. U.S. Pat. 2,885,366 (May 12, 1959), R. K. Iles (to DuPont).
70. U.S. Pat. 3,437,502 (Apr. 8, 1969), A. J. Werner (to DuPont).
71. *The Kronos Guide* (*Kronos Leitfaden*), The Kronos Titanium Companies (Titangesellschaft m.G.H.), Leverkusen, FRG, 1968.
72. R. W. Kershaw, *Aust. OCCA Proc. News* **E8**(4), 4 (1970).
73. K. Goldsbrough, L. A. Simpson, and D. F. Tunstall, *Prog. Org. Coat.* **10**(1), 35 (1982).
74. F. Hund, *Angew. Chem. Int. Ed.* **1,** 41 (1962).
75. E. Herrmann, *Farbe Lacke* **68,** 174 (1962).
76. *Chem. Week*, 67 (Oct. 13, 1982).
77. *Metals and Minerals*, Vol. 1 of *Minerals Yearbook*, Bureau of Mines, U.S. Department of the Interior, Washington, D.C., 1980.
78. J. M. Rackham, *Proc. XVth Fatipec Congr.* **1,** 279 (1980).
79. Brit. Pat. 1,575,297 (Sept. 17, 1980), F. Massey (to Tioxide Group PLC.).
80. W. W. Greer, *Am. Paint. Coat. J.* **63**(46), 64 (1979).
81. R. W. Porter, *Mod. Paint Coat.* **66**(10), 51 (1976).
82. T. Shimizu, *Kagaku Kogyo* **31**(5), 503 (1980).
83. K. L. Berry and co-workers, *J. Inorg. Nucl. Chem.* **14,** 231 (1960).
84. E. K. Ovechkin and co-workers, *Inorg. Mater.* (*English Transl.*) **7,** 1000 (1968).
85. S. Chijiwa, *Purasuchikkusu* **31**(8), 106 (1980).
86. Y. Fujiki, *Jidosha Gijutsu* **34**(8), 874 (1980).
87. T. Shimizu, *Kagaku Kogyo* **31**(7), 752 (1980).
88. S. Taki, *Baruka Rebyu* **23**(1), 8 (1979).
89. Y. Fujiki, *Kinzoku* **45**(12 Suppl.), 61 (1975).
90. W. W. Riches in ref. 51, pp. 105–108.
91. B. E. Waye, *Introduction to Technical Ceramics*, MacLaren & Sons Ltd., London, 1967.
92. J. C. Moure, *Bol. Soc. Esp. Ceram. Vidrio* **18**(6), 389 (1979).
93. I. Shindo, *Kagaku Kogyo* **32**(2), 181 (1981).
94. F. H. W. Wachholtz, *Chim. Peint.* **16,** 141 (1953).
95. G. Helke and G. Roeder, *Silikattechnik* **30**(8), 245 (1979).
96. H. Wierzba, *Electronika* **21**(12), 15 (1980).
97. J. R. Hackman in ref. 51, pp. 419–427.
98. P. Nuetzenadel, *Freiberg. Forschungsh. A* **A604,** 7 (1979).
99. L. Shcherbakova et alia, *Usp. Khim.* **48**(3), 423 (1979).
100. V. A. Reznichenko and G. A. Menyailova, *Synthetic Titanates Nauka*, Moscow, USSR, 1977, 136 pp.
101. R. Colton and J. H. Canterford, *Halides of the Transition Elements, Halides of the First Row Transition Metals*, Wiley-Interscience, London, 1969.
102. Jpn. Pat. 76 28,599 (Aug. 20, 1976), K. Egi.
103. Ger. Pat. 1,292,385 (Apr. 10, 1954), G. Pieper, H. F. Rickert, and E. Stein.
104. V. A. Khodzhemirov and co-workers, *Vysokomol. Soedin., Ser. B* **13**(6), 402 (1971).
105. G. D. Bukatov and co-workers, *Kinet. Katal.* **16**(3), 645 (1975).
106. Ger. Pat. 1,420,390 (Sept. 7, 1972), H. F. Rickert, E. Stein, and G. Pieper.
107. Ger. Pat. 2,206,571 (Aug. 24, 1972), C. A. Finch and G. Barker (to Croda International Ltd.).
108. G. Natta, P. Corrodini, and G. Allegra, *J. Polym. Sci.* **51,** 399 (1961).
109. Jpn. Pat. 80 60,028 (May 6, 1980), K. Egi.

110. Jpn. Pat. 78 106,691 (Mar. 1, 1978), K. Egi.

111. Ital. Pat. 802,297 (Feb. 15, 1968), (to "Montecatini" Societa Generale per l'Industria Mineraria e Chimica).

112. Jpn. Pat. 70 28,285 (Sept. 16, 1970), N. Okudaira (to Toho Titanium Co., Ltd.).

113. V. V. Krapukhin and E. A. Korolev, *Izv. Vyssh. Ucheb. Zaved. Tsvet. Met.* **11**(6), 66 (1968).

114. Ger. Pat. 1,961,339 (June 25, 1970), T. Kugler and J. Silbiger (to Lonza Ltd.).

115. Ger. Pat. 1,962,989 (July 9, 1970), T. Kugler and J. Silbiger (to Lonza Ltd.).

116. Ger. Pat. 2,158,956 (May 30, 1973), K. Wisseroth and R. Scholl (to Badische Anilin- und Soda-Fabrik A.-G.).

117. Ger. Pat. 2,110,380 (Oct. 7, 1971), J. P. Hermans and P. Henriqulie (to Solvay et Cie).

118. U.S. Pat. 3,681,256 (Feb. 16, 1972), H. W. Blunt (to Hercules, Inc.).

119. Jpn. Pat. 74 20,476 (May 24, 1974), S. Okudaira (to Toho Titanium Co., Ltd.).

120. Ger. Pat. 2,533,511 (Feb. 19, 1976), K. Yamaguchi, G. Kakogawa, and Y. Maruyawa (to Mitsubishi Chemical Industries, Ltd.).

121. Ger. Pat. 2,600,593 (July 15, 1976), N. Kuroda, T. Shiraishi, and A. Itoh (to Nippon Oil Co.).

122. Jpn. Pat. 77 82,987 (July 11, 1977), G. Kakogawa, M. Hasuo, and Y. Uehara (to Mitsubishi Chemical Industries, Co., Ltd.).

123. Jpn. Pat. 76 115,797 (Sept. 28, 1977), K. Yamaguchi and co-workers (to Mitsubishi Chemical Industries Co., Ltd.).

124. Jpn. Pat. 77 120,284 (Oct. 8, 1977), T. Miyake and co-workers (to Asahi Chemical Industry Co., Ltd.).

125. U.S. Pat. 4,124,532 (Nov. 7, 1978), U. Giannini and co-workers (to Montedison S.p.A.).

126. A. Slawisch, *Chem. Ztg.* **92**, 311 (1968).

127. G. W. A. Fowles and F. H. Pollard, *J. Chem. Soc.*, 2588 (1955).

128. D. J. Jones, *The Production of Titanium Tetrachloride*, R. H. Chandler Ltd., London, 1969.

129. R. Powell, *Titanium Dioxide and Titanium Tetrachloride*, Noyes Development Corporation, Park Ridge, N.J., 1968.

130. *Titanium Dioxide Production, 1965–1975*, R. H. Chandler Ltd., London, 1976.

131. N. A. Baitenev and co-workers, *Production of Titanium Tetrachloride and Dioxide* (Nauka: Alma-Ata, Kaz. SSR) (1974).

132. M. K. Baikekov, V. D. Popov, and I. M. Cheprasov, *Production of Titanium Tetrachloride* (*Metallurgiya* Moscow), (1980).

133. P. L. Vijay, C. Subramanian, and Ch. S. Rao, *Trans. Indian Inst. Met.* **29**(5), 355 (1976).

134. E. C. Perkins and co-workers, *Fluidised Bed Chlorination of Ores and Slags*, U.S. Bureau of Mines Rept. Invest. 6317, Department of the Interior, Washington, D.C., 1963.

135. A. Z. Bezukladnikov, *J. Appl. Chem. USSR* (*English Transl.*) **40**, 25 (1967).

136. Ger. Pat. 1,043,290 (Nov. 13, 1958), R. H. Walsh (to Columbia Southern Chem. Corp.).

137. Norw. Pat. 92,999 (Dec. 8, 1958), A. G. Oppegaard, A. Helge, and H. Barth (to Titan Co. A/S).

138. U.S. Pat. 2,962,353 (Nov. 29, 1960), J. N. Haimsohn (to Stauffer Chem. Co.).

139. H. Schäfer, F. Wartenpfuhl, and E. Weise, *Z. Anorg. Allgem. Chem.* **295**, 268 (1958).

140. K. Kehnicke, *Angew. Chem.* **75**, 419 (1963).

141. G. Brauer, *Präparativen Anorg. Chem. Stuttgart*, 899 (1954).

142. W. Klemm, W. Tilk, and S. von Müllenheim, *Z. Anorg. Chem.* **176**, 1 (1928).

143. U.S. Pat. 3,668,109 (June 6, 1972), E. Thomas and M. M. Wald (to Shell Oil Co.).

144. Jpn. Pat. 74 40,638 (Nov. 2, 1974), A. Oda, K. Ura, N. Kibayashi, and S. Tokuda (to Osaka Soda Co.).

145. J. Pecka and co-workers, *Makromol Chem.* **176**(6), 1725 (1975).

146. G. Henlein and co-workers, *Z. Chem.* **15**(4), 150 (1975).

147. Jpn. Pat. 78 144,524 (Dec. 15, 1978), J. Nishikido, N. Tamura, and Y. Fukuoka (to Asahi Chemical Industry Co., Ltd.).

148. H. Morikawa and S. Kitazume, *Ind. Eng. Chem. Prod. Res. Dev.* **18**(4), 254 (1979).

149. W. Klemm and L. Grimun, *Z. Anorg. Chem.* **249**, 198 (1942).

150. L. Hock and W. Knauf, *Z. Anorg. Chem.* **228**, 204 (1936).

151. J. D. Fast, *Rec. Trav. Chim.* **58**, 174 (1939).

152. A. Stähler, *Ber.* **37**, 4405 (1904).

153. Jpn. Pat. 78 21,107 (Feb. 27, 1978), T. Okano, N. Wada, and Y. Kobayashi (to Mitsubishi Chemical Industries Ltd.).

154. I. E. Campbell and co-workers, *Trans. Electrochem. Soc.* **93**, 271 (1948).

155. Hansen, Kessler, and McPherson, *Trans. Am. Soc. Met.* **44,** 518 (1952).

156. H. Springer and J. Nickl, *Naturwissenschhaften* **54**(24), 645 (1967).

157. J. Beaudouin, *C. R. Acad. Sci. Paris, Ser. C* **263**(17), 993 (1966).

158. Brit. Pat. 901,402 (July 18, 1962), (to du Pont de Nemours Ltd.).

159. K. Tamara, *Nippon Kinzoku Gakkaishi* **24,** 707 (1960).

160. W. Rutkowski and T. Gibas, *Szklo Ceram.* **13**(11), 341 (1962).

161. I. V. Petrusevich, L. A. Nisel'sen, and A. I. Belyaev, *Izv. Akad. Nauk SSSR, Metally*, (6), 52 (1965).

162. Jpn. Pat. 78 58,998 (May 27, 1978), T. Mitamura, T. Mori, and N. Kawakami (to Toyo Soda Mfg. Co., Ltd.).

163. A. N. Rubtsov and co-workers, *Izv. Akad. Nauk SSSR, Neorg. Mater.* **12**(6), 1125 (1976).

164. Y. A. De Vynck, *Silicates Ind.* **39**(4), 109 (1974).

165. Jpn. Pat. 80 62,850 (May 12, 1980), (to Mitsubishi Metal Corp.).

166. Ger. Pat. 2,261,523 (June 28, 1973), H. Carbonnel and L. Hamon (to Groupement Atomique Alsacienne Atlantique).

167. U.S. Pat. 4,028,276 (June 7, 1977), J. C. Harden and S. V. R. Mastrangelo (to E. I. du Pont de Nemours & Co.).

168. A. A. Illarianov, *Mater. Vses. Vauchn. Stud. Konf. Novosib. Gos. Univ. Khim.* **12,** 12 (1974); *Chem. Abstr.* **84,** 53282j (1976).

169. Russ. Pat. 815,012 (Mar. 23, 1981), N. Z. Yaramenko and I. P. Dobrovolskii.

170. V. P. Titov and co-workers, *Izv. Vyssh. Uchebn. Zaved. Khim. Khim. Tekhnol.* **23**(1), 64 (1980).

171. V. A. Obrubov and co-workers, *Geterog. Katal. Protsessy Vzveshennom Filtruyushchem Sloe*, 109 (1978); *Chem. Abstr.* **94,** 30091p (1981).

172. E. Kobayashi, *Kagaku Ryoiki* **33**(10), 815 (1979).

173. Jpn. Pat. 80 51,442 (Apr. 15, 1980), E. Kobayashi and T. Kanayama (to Agency of Industrial Sciences and Technology).

174. A. Ludmany, G. Torok, and L. G. Nagy, *Radiochem. Radioanal. Lett.* **45**(6), 387 (1980).

175. G. Albert and co-workers, *J. Inorg. Nucl. Chem.* **42**(11), 1637 (1980).

176. N. J. Singh and S. N. Tandon, *Indian J. Chem. Sect. A* **19A**(5), 502 (1980).

177. Biltz and co-workers, *Z. Anorg. Chem.* **234,** 97 (1937).

178. M. S. Whittingham and J. A. Panella, *Mat. Res. Bull.* **16,** 37 (1981).

179. T. Moeller, *Inorganic Syntheses*, Vol. V, McGraw-Hill, Inc., New York, 1957.

180. Aust. Pat. 205, 568 (Nov. 16, 1956), F. K. McTaggart (to Commonwealth Industrial & Scientific Research Organisation).

181. M. S. Whittingham, *Prog. Solid State Chem.* **12,** 41 (1978).

182. K. C. Narasimkam and co-workers, *Trans. Soc. Adv. Electrochem. Sci. Technol.* **15**(2), 147 (1980).

183. U. Manivel, S. Bangaruswamy, and J. B. Rao, *Leather Sci. (Madras)* **27**(8), 257 (1980).

184. *Ibid.*, p. 260.

185. Brit. Pat. 2,062,596A (May 12, 1980), D. L. Motov and co-workers.

186. A. I. Metelkin, *Kozh. Obuvna Promst.* **21**(4), 30 (1979); *Chem. Abstr.* **91,** 6415q (1979).

187. E. R. Scheffer in I. M. Kolthoff and P. J. Elving, eds., *Treatise on Analytical Chemistry*, Vol. 5, Pt. II, Interscience Publishers, New York and London, 1961, pp. 1–60.

188. W. T. Elwell and J. Whitehead in C. L. and D. W. Wilson, eds., *Comprehensive Analytical Chemistry*, Vol. 1c, Elsevier Publishing Co., Amsterdam, The Netherlands, 1962, pp. 497–506.

189. M. Berlin and C. Nordman in L. Friberg and co-eds., *Handbook on the Toxicology of Metals*, Elsevier/North Holland Biomedical Press, Amsterdam, The Netherlands, 1979, Chapt. 38, pp. 627–636.

190. H. Valentin and K. H. Scholler, *Titanium*, Vol. 3 of *Human Biological Monitoring of Chemicals*, Directorate-General Employment and Social Affairs, Commission of the European Communities, Brussels, 1980, EUR 6609 EN.

191. W. K. Poole and co-workers, *Estimating Population Exposure to Selected Metals—Titanium*, Research Triangle Institute, Research Triangle Park, N.C., 1969.

192. *Bioassay of Titanium Dioxide for Possible Carcinogenicity*, National Cancer Institute technical report series no. 97, U.S. Department of Health, Education, and Welfare, Washington, D.C., 1979.

193. J. Ferin, "Inhaled Particles-3," *Proceedings of an International Symposium, 1970*, 3rd ed., Unwin Bros. Ltd., Old Woking, Surrey, England, 1971, pp. 283–292.

194. I. H. Tipton and M. J. Cook, *Health Phy.* **9,** 103 (1963).

195. E. I. Hamilton, M. J. Minski, and J. J. Cleary, *Sci. Total Environ.* **1,** 341 (1972).

196. J. V. Crabb and co-workers, *Am. Ind. Hyg. Assoc.* **28**, 8 (1967); **29**, 106 (1968).
197. R. Elo and co-workers, *Arch. Path.* **94**, 417 (1972).
198. K. Määttä and A. V. Arstila, *Lab. Invest.* **33**, 342 (1975).
199. *Threshold Limit Values for Chemical Substances and Physical Agents in the Workroom Environment*, American Conference of Governmental Hygienists, Cincinnati, Ohio, 1980.
200. Guidance Note EH 15/80, The Health and Safety Executive HMSO, London (1980).
201. L. C. Maillard and J. E. Hori, *Bull. Acad. Med. Paris* **115**, 631 (1936).
202. N. P. Timakan and co-workers, *Vap. Teor. Khim. Tomsk. Med. Inst.*, 114 (1967).
203. H. M. Perry, Jr., and E. F. Perry, *J. Clin. Invest.* **38**, 1452 (1959).

J. WHITEHEAD
Tioxide Group PLC

ORGANIC

Organic titanium compounds are compounds with a covalent bond between titanium and another atom that is bonded to a carbon-containing group. Titanium tetrachloride, which is manufactured from titaniferous ores like ilmenite or rutile, is the basic raw material for organic titanium compounds. It is readily converted to titanate esters, in particular tetraisopropyl titanate [546-68-7] (TPT), by the Nelles process. This ester can be converted by alkoxy exchange to a wide variety of pure tetraalkoxy titaniums, $(RO)_4Ti$, which are sold commercially worldwide. Tetraisopropyl titanate reacts with chelating reagents, eg, acetylacetone, lactic acid, and triethanolamine, which are important items of commerce.

True organometallic compounds with a titanium–carbon bond are prepared from titanium tetrachloride by reaction with main-group organometallics such as organomagnesium, sodium, or lithium reagents. Most simple C–Ti compounds are very unstable; the prominent exceptions are bis(cyclopentadienyl)titanium dichloride [1271-19-8] (titanocene dichloride), Cp_2TiCl_2, and its analogues (see Organometallics, metal Ti-complexes).

Many organometallics and complexes with trivalent titanium are stable at room temperature, but most are attacked by oxygen and moisture. Organic derivatives of divalent titanium are less common. Titanium(II) and titanium(O) compounds are potent reducing agents. Hydridotitanium derivatives are involved in olefin and acetylene isomerization and hydrogenation as well as in some types of Ziegler-Natta dimerizations and polymerizations (see Olefin polymers; Polymerization mechanisms and processes).

In industry, titanium alkoxides are superb catalysts for esterification, transesterification, and cross-linking of ester-containing resins and epoxides (see Catalysis). Water-soluble titanium chelates are widely used in cross-linking (gelling) of dilute polysaccharide solutions; cross-linked guar is used to carry sand in hydraulic fracturing of oil and gas wells to increase their flow (see Petroleum, drilling fluids). Titanium's great affinity for oxygen atoms is reflected in its bonding to oxide surfaces. Thus, titanates react with glass surfaces yielding a scratch-resistant oxide coating and with masonry to improve its properties (see Glass; Cement). Titanates cross-link many silicone resins. They disperse pigments and often help bond resins to fillers (qv) and reinforcements.

Alkoxides (Titanate Esters)

The most useful titanium alkoxide is tetraisopropoxytitanium (TPT) made from titanium tetrachloride and isopropyl alcohol; hydrogen chloride is formed reversibly at each stage:

$$TiCl_4 + ROH \rightleftharpoons ROTiCl_3 + HCl$$

$$ROTiCl_3 + ROH \rightleftharpoons (RO)_2TiCl_2 + HCl$$

$$(RO)_2TiCl_2 + ROH \rightleftharpoons (RO)_3TiCl + HCl$$

$$(RO)_3TiCl + ROH \rightleftharpoons (RO)_4Ti + HCl$$

Alkylene oxides react stepwise with $TiCl_4$ forming 2-chloroalkoxides; one example is $Ti(OCH_2CH_2Cl)_4$ [*19600-96-5*] (1) (see Alkoxides, metal).

Higher alkoxides can be prepared by alcohol interchange in a solvent such as benzene or cyclohexane, to form a volatile azeotrope with the displaced isopropyl alcohol. The affinity of an alcohol for titanium decreases in the order: primary > secondary > tertiary, and unbranched > branched. The process is more convenient than the direct synthesis from $TiCl_4$, an alcohol, and a base because a metal chloride need not be handled. Silanols, eg, R_3SiOH and $(CH_3)_3SiCH_2OH$, behave similarly:

$$(i\text{-}C_3H_7O)_4Ti + 4\,ROH \rightleftharpoons (RO)_4Ti + 4\,i\text{-}C_3H_7OH\uparrow$$
$$(\text{TPT})$$

Phenols react readily with TPT and the equilibrium lies far to the right, especially with diphenols, such as substituted catechol and *vic*-dihydroxycoumarins (2). These aromatic heterocyclic products are yellow or orange.

Mixed esters can be prepared from $Cl_n Ti(OR)_{4-n}$ and the second alcohol in the presence of a base or by mixing a tetraalkoxytitanium and the second alcohol in the desired proportion and flash-evaporating the mixed alcohols. However, there is little interest in such compounds because of their tendency to redistribute.

A rarely used preparation involves the reaction of an amidotitanium with an alcohol. The reaction goes to completion because of the much greater affinity of titanium for oxygen than for nitrogen. However, amides are more difficult to prepare and handle than the esters:

$$\geqslant\!TiNR_2 + R'OH \rightarrow\, \geqslant\!TiOR' + HNR_2$$

Properties and Reactions. *Association.* A dominant property of organic titanates is their tendency to associate (3). A titanium(IV) atom achieves coordination number 6 by sharing electron pairs from nearby ester molecules. This tendency is opposed to steric crowding. Thus, *tert*-alkyl titanates are monomeric (degree of complexity = 1.0). The degree of complexity increases to 1.4 in TPT (*sec*-alkyl), 2 in primary unbranched alkyl titanates, 3 in tetraethoxytitanium [*3087-36-3*] in solution, and 4 in tetramethoxytitanium [*992-92-7*] and crystalline tetraethoxytitanium, as measured by cryoscopy or ebullioscopy in benzene or other solvents and by x-ray crystallography. The aggregate may dissociate somewhat in dilute solutions of donor solvents.

Association may involve single or double bridges. As shown in x-ray studies, in the crystalline tetramers, each titanium atom is octahedrally coordinated to six oxygens; edge sharing of four TiO_6 octahedra gives the tetramer, eg:

$$(RO)_3Ti \overset{\overset{R}{\underset{\displaystyle O}{}}}{\diagdown} Ti(OR)_4 \qquad\qquad (RO)_3Ti \overset{\overset{R}{\underset{\displaystyle O}{}}}{\underset{\underset{R}{O}}{\diamond}} Ti(OR)_3$$

Increased molecular association increases viscosity. Tetra-t-butoxytitanium and tetraisopropoxytitanium are mobile liquids; tetra-n-butoxytitanium and n-propoxy-titanium are thick and syrupy. The boiling point also reflects this association (see Table 1).

Hydrolysis and Condensation. The lower titanium alkoxides, except the methoxide, are rapidly hydrolyzed by moist air or water giving a series of condensed titanoxanes (—TiOTiO—). Titanium methoxides, aryloxides, and higher $(RO)_4Ti$ (R = C_{10} and longer) are hydrolyzed much more slowly. Complete hydrolysis to TiO_2 with cleavage of all the alkoxy groups is difficult to achieve.

Most simply, titanoxane formation may follow the path:

$$(RO)_3TiOR + H_2O \xrightarrow{-ROH} (RO)_3TiOH \xrightarrow[-ROH]{(RO)Ti(OR)_3} (RO)_3TiOTi(OR)_3$$

$$\xrightarrow{H_2O} (RO)_3TiOTi(OR)_2OTi(OR)_3$$
$$\text{trimer}$$

The trimer may continue to undergo condensation at the less-hindered terminals, forming linear oligomers. The slower reaction at the central titanium yields a branched structure.

The hydrolysis of tetraethoxytitanium has also been considered in terms of its trimeric form. Thus:

$$2\ Ti_3(OC_2H_5)_{12} \xrightarrow{4\ H_2O} Ti_3(OC_2H_5)_8(O)_4Ti_3(OC_2H_5)_{11}$$

Further condensation gives $[Ti_3O_4(C_2H_5)_4]$ (4).

Quantitative studies by various workers have yielded conflicting results. The longer the chain, the more hindered the access of water to the titanium atom. The less branched, ie, more associated, isomers are more slowly hydrolyzed because titanium atoms are more fully coordinated. Extensive studies of the condensation of (n-$C_4H_9O)_4Ti$ were aimed at producing oligomers useful in heat-resistant protective coatings and involved the measurement of weight loss, apparent molecular weight, and Ti content as a function of water added (5–7). Oligomers have also been obtained by pyrolysis at 200–250°C:

$$2\ (n\text{-}C_4H_9O)_4Ti \rightarrow (n\text{-}C_4H_9O)_3TiOTi(OC_4H_9\text{-}n)_3 + n\text{-}C_4H_9OH + C_4H_8, \text{ etc}$$
$$[7393\text{-}46\text{-}6]$$

Titanoxanes can also be prepared from the reaction of alkoxides with acid anhydrides (8):

$$n\ (RO)_4Ti + \frac{n-1}{2}(R'\overset{\overset{\displaystyle O}{\|}}{C})_2O \rightarrow RO[Ti(OR)_2O]_nR + (n-1)\ R'C\overset{\overset{\displaystyle O}{\|}}{O}R$$

Pure dimers can be made from the oxidation of Ti(III) alkoxides (9):

$$4\ (RO)_3Ti + O_2 \rightarrow 2\ (RO)_3TiOTi(OR)_3$$

Oligomers can be prepared directly from titanium tetrachloride and various alcohols containing enough water to yield the desired degree of polymerization; the reaction is driven to completion by the addition of ammonia. The progress of the reaction is followed by viscosity increase.

The alkoxide oligomers remain at least somewhat soluble in organic solvents until the TiO_2 content reaches ca 50 wt %. As with most reactions of titanium compounds, the tendency to associate by expanding the coordination shell facilitates disproportionation, ie, redistribution, of alkoxy groups on heating to yield lower oligomers and even monomeric $(RO)_4Ti$.

Reactions with Alcohols. The tendency of titanium(IV) to reach coordination number 6 accounts for the rapid exchange of alkoxy groups with alcohols. Departure of an alkoxy group with the proton is the first step in the ultimate exchange of all four alkoxyls:

$$(RO)_3Ti \overset{R.O.}{\underset{O.R}{\diamond}} Ti(OR)_3 + R'OH \rightleftharpoons (RO)_3Ti \overset{R'OH \quad R.O.}{\underset{O.R}{\diamond}} Ti(OR)_3 \rightleftharpoons ROH + (RO_2)Ti \overset{R'O \quad R.O.}{\underset{O.R}{\diamond}} Ti(OR)_3$$

five-coordinated dimer   six coordinate   five coordinate

The four-coordinated monomer is expected to react more rapidly than the dimer. The depicted process is very fast, as evidenced by nmr studies at room temperature, which show no difference in chemical shift between free ROH and combined ROTi; at sufficiently lower temperatures, both kinds of alkoxyl can be seen (10).

For preparative purposes, the equilibrium must be shifted either with an excess of the exchanging alcohol or by distilling the more volatile lower alcohol. Tetraisopropoxytitanium is the preferred starting ester because of the low boiling point of 2-propanol. Monohydric alcohols of varying complexity, diols and polyols, phenols, alkanolamines, and the enolic forms of β-diketones and β-ketoesters react readily with TPT; often all four alkoxy groups can be interchanged.

Reactions with Esters. Ester interchange catalyzed by titanates is an important industrial reaction (11):

$$\underset{RCOR'}{\overset{O}{\parallel}} + R''OH \rightleftharpoons \underset{RCOR''}{\overset{O}{\parallel}} + R'OH\uparrow$$

One commercial example is the formation of basic methacrylates from dialkylaminoethanols and methyl methacrylate:

$$R_2NCH_2CH_2OH + \underset{\underset{CH_3}{|}}{\overset{O}{\overset{\parallel}{CH_3OCC}}}{=}CH_2 \rightleftharpoons R_2NCH_2CH_2O\underset{\underset{CH_3}{|}}{\overset{O}{\overset{\parallel}{CC}}}{=}CH_2 + CH_3OH\uparrow$$

The same reaction can be used to convert one alkoxide to another by distillation of a lower boiling ester:

$$Ti(OCH(CH_3)_2)_4 + 4\ \underset{CH_3COC_4H_9}{\overset{O}{\parallel}} \rightleftharpoons Ti(OC_4H_9)_4 + 4\ \underset{CH_3COCH(CH_3)_2}{\overset{O}{\parallel}}\uparrow$$

Table 1. Titanium(IV) Tetraalkoxides and Tetraaryloxides

Compound	CAS Registry No.	Mp, °C	Bp, °C$_{Pa}$[a]	Other properties
methoxide, $Ti(OCH_3)_4$	[992-92-7]	210	$170_{1.3}$ sublimes	dipole moment, 5.37×10^{-30} C·m[b]
ethoxide, $Ti(OC_2H_5)_4$	[3087-36-3]	<−40	103_{13}	n_D^{35}, 1.5051; d_4^{35}, 1.107 g/cm^3; η_{25}, 44.45 mPa·s (= cP)
allyloxide, $Ti(OCH_2CH{=}CH_2)_4$	[5128-21-2]		$141–142_{133}$	n_D^{35}, 1.5381; d_4^{35}, 0.9970 g/cm^3
n-propoxide, $Ti(OCH_2CH_2CH_3)_4$	[3087-37-4]		124_{13}	n_D^{35}, 1.4803; η_{25}, 161.35 mPa·s (= cP)
isopropoxide, $Ti[OCH(CH_3)_2]_4$	[546-68-9]	18.5	49_{13}	n_D^{35}, 1.4568; η_{25}, 4.5 mPa·s (= cP); d_4^{20}, 0.9711 g/cm^3
n-butoxide, $Ti(OC_4H_9)_4$	[5593-70-4]	ca −50	142_{13}	d_4^{35}, 0.9927 g/cm^3; n_D^{35}, 1.4863; η_{25}, 67 mPa·s (= cP); dipole moment, 3.84×10^{-30} C·m[b]
isobutoxide, $Ti[OCH_2CH(CH_3)_2]_4$	[7425-80-1]		141_{133}	n_D^{54}, 1.4749; d_4^{50}, 0.9601 g/cm^3; η_{50}, 97.40 mP·s (= cP)
sec-butoxide, $Ti[O(CH_3)CHC_2H_5]_4$	[3374-12-7]	ca −25	81_{13}	n_D^{35}, 1.4550; d_4^{35}, 0.9196 g/cm^3
tert-butoxide, $Ti[OC(CH_3)_3]_4$	[3087-39-6]		$59–60_{40}$	n_D^{20}, 1.4436; d_4^{20}, 0.8893 g/cm^3
n-pentoxide, $Ti(OC_5H_{11})_4$	[10585-24-7]		158_{13}	η_{25}, 79.24 mPa·s (= cP); d_4^{25}, 0.9735 g/cm^3; n_D^{35}, 1.4813
cyclopentyloxide, $Ti(OC_5H_9)_4$	[1517-19-7]	45	$200–201_{667}$	

Compound	CAS	mp, °C	bp, °C (Pa)[a]	Properties
n-hexyloxide, Ti(OC$_6$H$_{13}$)$_4$	[7360-52-3]		176_{13}	η_{25}, 64.90 mPa·s (= cP); d_4^{25}, 0.9499 g/cm^3; n_D^{20}, 1.4830
cyclohexyloxide, Ti(OC$_6$H$_{11}$)$_4$	[6426-39-7]		$190.5-192_{133}$	dipole moment, 5.54 × 10^{-30} C·m[b]; d_4^{25}, 1.0589 g/cm^3; n_D^{35}, 1.5155
benzyloxide, Ti(OCH$_2$C$_6$H$_5$)$_4$	[103-50-4]		250_{267} dec	n_D^{20}, 1.4810
n-octyloxide, Ti(OC$_8$H$_{17}$)$_4$	[3061-42-5]		214_{13}	d_4^{20}, 0.9339 g/cm^3; dipole moment, 5.68 × 10^{-30} C·m[b]
2-ethylhexyloxide, Ti[OCH$_2$(C$_2$H$_5$)CHC$_4$H$_9$]$_4$	[1070-10-6]	<−25	$248-249_{1467}$	n_D^{35}, 1.4750
nonyloxide, Ti(OC$_9$H$_{19}$)$_4$	[6167-42-6]		$264-265_{200}$	n_D^{20}, 1.4785; d_0^{20}, 0.9241 g/cm^3; dipole moment, 5.60 × 10^{-30} C·m[b]
n-decyloxide, Ti(OC$_{10}$H$_{21}$)$_4$		68	265_{27}	d_4^{25}, 0.87 g/cm^3; η_{25}, 76.0 mPa·s (= cP)
isobornyloxide, Ti(OC$_{10}$H$_{17}$)$_4$	[84215-64-5]			
benzhydryloxide, Ti[OCH(C$_6$H$_5$)$_2$]$_4$	[84215-65-6]			
oleyloxide, Ti(OC$_{18}$H$_{35}$)$_4$	[26291-85-0]	57–59		d_4^{20}, 0.87 g/cm^3
phenoxide, Ti(OC$_6$H$_5$)$_4$	[2892-89-9]	153	288_{13}	orange–red solid
o-chlorophenoxide, Ti(OC$_6$H$_4$Cl)$_4$	[22922-75-4]	145.5–147	267_{40}	red solid
p-chlorophenoxide, Ti(OC$_6$H$_4$Cl)$_4$	[13438-75-0]	84–86		
o-nitrophenoxide, Ti(OC$_6$H$_4$NO$_2$)$_4$	[55535-59-6]	154–158		shiny black solid
p-nitrophenoxide, Ti(OC$_6$H$_4$NO$_2$)$_4$	[22922-76-5]	97		grayish-black solid
o-methylphenoxide, Ti(OC$_6$H$_4$CH$_3$)$_4$	[22922-73-2]	48–51		
m-methylphenoxide, Ti(OC$_6$H$_4$CH$_3$)$_4$	[22949-88-0]		$323-325_{40}$	red crystalline solid
1-naphthyloxide, Ti(OC$_{10}$H$_7$)$_4$	[84215-66-7]		does not distill	
2-naphthyloxide, Ti(OC$_{10}$H$_7$)$_4$	[36452-22-9]	60–64	does not distill	black solid
resorcinyloxide, Ti(O$_2$C$_6$H$_4$)$_2$	[34075-40-6]			dark-brown solid
2,4,6-trinitrophenoxide, Ti[OC$_6$H$_2$(NO$_2$)$_3$]$_4$	[84215-67-8]			

[a] To convert Pa to mm Hg, divide by 133.
[b] To convert C·m to D, divide by 3.336 × 10^{-30}.

181

Titanium-catalyzed ester interchange can be used to prepare polyesters from diesters and diols and from diacids and diols at considerably higher temperatures. Polymer chains bearing pendant ester and hydroxy functions can be cross-linked with titanates.

Reactions with Acids. Organic acids form acylates when heated with alkoxy titaniums. Best results are obtained with only one or two moles of acid, as attempts to force the reaction with 3 or 4 mol of acid yields polymers (see Titanium acylates):

$$n \text{ RCOOH} + \text{Ti(OR')}_4 \longrightarrow (\text{RCOO})_n\text{Ti(OR')}_{4-n} + n \text{ R'OH}$$

$$n \text{ (RCO)}_2\text{O} + \text{Ti(OR')}_4 \longrightarrow (\text{RCOO})_n\text{Ti(OR')}_{4-n} + n \text{ RCOOR'}$$

With acid anhydrides, the exothermic reaction yields similar products, subject to the limitation on n. With higher anhydride ratios, condensed acylates form:

$$3 \text{ mol anhydride/mol alkoxide} \longrightarrow [\text{R'OTi(OOCR)}_2]_2\text{O}$$

$$4 \text{ mol anhydride/mol alkoxide} \longrightarrow [(\text{RCOO})_3\text{Ti}]_2\text{O}$$

Phthalic anhydride does not give a cyclic product:

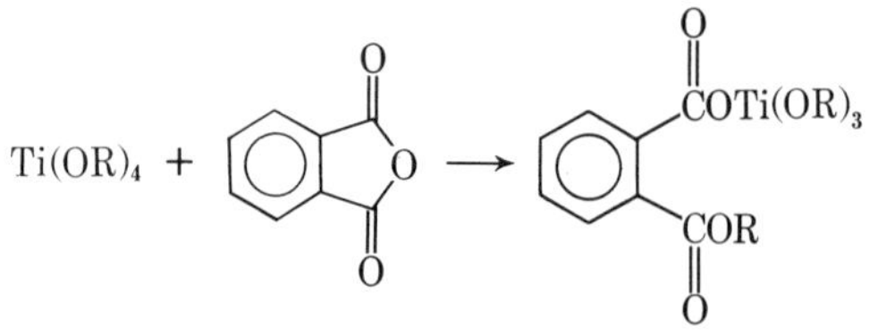

Thermolysis. Lower alkoxides are reasonably stable and can be distilled quickly at atmospheric pressure. Protracted heating forms condensation polymers plus, usually, alcohol and alkene. Longer or more branched chains are less stable. Thus, tetra-n-pentoxytitanium can be distilled at 314°C/101.3 kPa (= 1 atm), whereas tetra-n-hexoxytitanium must be distilled at below 18.7 kPa (140 mm Hg) and tetra-n-hexadecyloxytitanium [34729-16-3] cannot be distilled at all even under high vacuum. Unsaturated groups, as in the allyloxide, also decrease thermal stability.

Thermolysis is used in the coating of glass and other surfaces with a film of titanium dioxide. When a lower alkoxide, eg, TPT, vaporizes in a stream of dry air and is blown onto hot glass bottles above ca 500°C, a thin transparent protective coating of TiO_2 is deposited.

Production and Economic Aspects.   The use of titanium alkoxides has grown considerably since the mid-1960s. The estimated worldwide production of titanium alkoxides in 1967 was 1250 metric tons. By early 1982, production had grown to ca 4500 t. However, the estimates are subject to uncertainty because much of the primary production is for captive use. Principal producers are, in the United States: E. I. du Pont de Nemours & Co., Inc., Kay-Fries, Kenrich Petrochemicals, Inc., and Stauffer Chemical Co.; in the UK: Titanium Intermediates, Ltd.; in the FRG: Dynamit-Nobel and Kronos-Titan; in France: Rhone-Poulenc (Thann et Mulhouse); and in Japan: Nippon Soda Co., Ltd., Matsumoto Trading Co., and Mitsubishi Gas Chemicals Co.

The alkoxides offered commercially in the United States include a variety of titanium tetraalkoxides, chelates, and simple and complex acylates. Prices of some of these are listed in Table 2.

The alkoxides are manufactured by a modification of the classical Nelles process, in which titanium tetrachloride in an inert solvent, eg, heptane, is treated with a monohydric lower alcohol (12). If the hydrogen chloride by-product is expelled by heating or by sweeping with dry nitrogen, the product is the dialkoxydichlorotitanate (13–14). Expulsion of HCl is facilitated by adding the required alcohol gradually with continuous sweeping (15). An acid acceptor is required to attain the tetraalkoxide stage; ammonia is preferred because of its low cost and because the product ammonium chloride is mostly insoluble and can be removed by filtration or centrifugation. The solvent is then distilled and the product is purified by distillation, if required. Because of the facile hydrolysis of the products, all reagents and equipment must be dry and all operations must be performed in a water-free atmosphere. The chemical reactions involved are shown:

$$TiCl_4 + 2\,ROH \rightarrow (RO)_2TiCl_2 + 2\,HCl$$

$$(RO)_2TiCl_2 + ROH \rightarrow (RO)_2TiCl_2.HOR \text{ (isolable as alcoholate)}$$

$$(RO)_2TiCl_2 + 2\,ROH + 2\,NH_3 \rightarrow (RO)_4Ti + 2\,NH_4Cl$$

In a variation, titanium tetrachloride in an inert solvent is converted to the ammine $TiCl_4.8NH_3$. This ammine reacts with almost any alcohol yielding the tetraalkoxide. This operation can be performed as a continuous process (16–17).

Alkoxides can also be prepared from titanium disulfide (18). The problems of manufacturing the disulfide and disposing of hydrogen sulfide make it unlikely that this route will become a commercial process:

$$4\,ROH + TiS_2 \rightarrow (RO)_4Ti + 2\,H_2S$$

Table 2.   Prices of Titanium Alkoxides Sold in the United States, $/kg

Alkoxide	CAS Registry No.	1971	1976	1981
titanium isopropoxide (TPT)		2.23	3.00	4.63
titanium-n-butoxide (TBT)		2.89	3.92	6.04
titanium-2-ethylhexyloxide		2.34	3.17	4.96
triethanolamine titanate	[37481-13-3]	2.12	3.79	5.32
diisopropoxy bis(acetylacetone)titanate (AA)	[17927-72-9]	4.41	6.00	8.00
diisopropoxy bis(ethyl acetoacetate)titanate	[27858-32-8]			13.28
dihydroxydilactatotitanate	[65104-06-5]	4.41	6.00	8.00

Higher alkoxides are readily prepared from tetraisopropyl titanate (TPT) or tetrabutyl titanate (TBT) by alcohol interchange, with removal of 2-propanol or n-butanol by distillation. Tetraisopropyl titanate is most commonly used because of its convenient properties, and TBT can be made from it.

The chelates are made simply by mixing the chelating agent with TPT or another tetraalkoxide. The liberated isopropyl alcohol is usually left in the product to maintain the product's fluidity, but it may be distilled. In the latter case, the product may oligomerize as the alcohol is removed.

In the United States, titanates are shipped in 208-L steel drums. Tank trailer delivery is also available for large quantities. Since most titanates are moisture-sensitive, they must be handled with care, preferably under dry nitrogen.

Alkoxy Halides

Titanium alkoxy halides have the formula $Ti(OR)_n X_{4-n}$, where R may be alkyl, alkenyl, or aryl and X is F, Cl, or Br but not I.

Properties. Alkoxytitanium fluorides and chlorides are colorless or pale-yellow solids or viscous liquids which darken on standing, especially in the light. Bromides are yellow crystalline solids. Aryloxytitanium halides are orange-to-red solids. The alkoxy halides are hygroscopic. Most dissolve in water without immediate decomposition, but they hydrolyze slowly to a hydrogen halide, alcohol, alkyl halide, and hydrous titanium dioxide. With less than one mole of water, poly(alkoxytitanium) compounds can be prepared if the liberated hydrogen halide is neutralized by ammonia. Primary alkoxy halides are thermally stable, though they disproportionate on heating. Secondary and tertiary alkoxy halides decompose gradually on standing (more rapidly on heating) yielding alkyl halides, polymers, and titanium oxychloride. Physical properties of some alkoxytitanium halides have been tabulated and are shown in Table 3.

Synthesis. Titanium alkoxy halides are intermediates in the preparation of alkoxides from a titanium tetrahalide (except the fluoride) and an alcohol or phenol. If $TiCl_4$ is heated with excess primary alcohol, only two chlorine atoms can be replaced and the product is dialkoxydichlorotitanium alcoholate $(RO)_2TiCl_2 \cdot ROH$ and 2 HCl. The yields are poor, and some alcohols such as allyl, benzyl, and t-butyl alcohols are converted to chlorides (19). With excess $TiCl_4$ at 0°C, the trichloride $(RO)TiCl_3$ is obtained nearly quantitatively, even from sec- and $tert$-alcohols (20–21).

With phenols, the number of chlorines replaced depends in part on the acidity of the phenol. Under mild conditions, phenol or m-nitrophenol displace only one chlorine from $TiCl_4$, whereas p-chlorophenol, o-nitrophenol, p-nitrophenol, picric acid, and 2-naphthol give $(ArO)_2TiCl_2$. If the mixtures are heated sufficiently, eg, in refluxing phenol, all four chlorines can be displaced (22–25).

Tetraalkoxides can be cleaved by hydrogen chloride or bromide in an inert solvent. Dialkoxytitanium dichloride is obtained as an alcoholate (26):

$$Ti(OR)_4 + 2\,HX \rightarrow Ti(OR)_2X_2 \cdot ROH + ROH$$

The same products may be made from primary alkoxides by the violent reaction with elementary chlorine or bromine; a radical mechanism has been proposed to account for the oxidation of some of the alkoxy groups (27).

$$Ti(OCH_2R)_4 + X_2 \xrightarrow{[O]} Ti(OCH_2R)_2X_2 \cdot RCH_2OH + 1/2\ \overset{\displaystyle O}{\overset{\|}{RCOCH_2R}}$$

Table 3. Titanium(IV) Alkoxyhalides and Aryloxyhalides

Compound	CAS Registry No.	Mp, °C	Bp, °C$_{Pa}$[a]	Other properties
ethoxytrifluoride, Ti(OC$_2$H$_5$)F$_3$	[1524-67-0]	220		
ethoxytrichloride, Ti(OC$_2$H$_5$)Cl$_3$	[3112-67-2]	decomposes 80–81	185–186	
ethoxytribromide, Ti(OC$_2$H$_5$)Br$_3$	[2489-72-8]	indefinite		
diethoxydifluoride, Ti(OC$_2$H$_5$)$_2$F$_2$	[650-27-1]	115		
diethoxydichloride, Ti(OC$_2$H$_5$)$_2$Cl$_2$	[3582-00-1]	40–50	142$_{240}$	
diethoxydibromide, Ti(OC$_2$H$_5$)$_2$Br$_2$	[3981-88-2]	47–50	95–105$_{67}$	red solid
triethoxyfluoride, Ti(OC$_2$H$_5$)$_3$F	[1868-77-5]	75–78	162–163$_{27}$	
triethoxychloride, Ti(OC$_2$H$_5$)$_3$Cl	[3712-48-9]		176$_{240}$	dipole moment, 9.57 × 10^{-30} C·m[b]
propoxytrichloride, Ti(OC$_3$H$_7$)Cl$_3$	[7569-98-4]	65–66	83–85$_{147}$	
isopropoxytrichloride, Ti(OC$_3$H$_7$-i)Cl$_3$	[3981-83-7]	78–79	65$_{13}$	
dipropoxydichloride, Ti(OC$_3$H$_7$)$_2$Cl$_2$	[1790-23-4]		159$_{240}$	
diisopropoxydibromide, Ti(OC$_3$H$_7$-i)$_2$Br$_2$	[37943-35-4]		100–102$_{27}$	yellow solid
triisopropoxyfluoride, Ti(OC$_3$H$_7$-i)$_3$F	[757-61-9]	83–85	140–150$_{80}$	
tripropoxychloride, Ti(OC$_3$H$_7$)$_3$Cl	[24287-11-4]		168$_{160}$	d$_4^{25}$, 1.1348 g/cm^3
butoxytrichloride, Ti(OC$_4$H$_9$)Cl$_3$	[3112-68-3]	67.5–70	124–127$_{540-567}$	
isobutoxytrichloride, Ti(OC$_4$H$_9$-i)Cl$_3$	[17754-61-9]	81–83	92–94$_{120}$	
dibutoxydichloride, Ti(OC$_4$H$_9$)$_2$Cl$_2$	[1790-25-6]		146–147$_{133}$	
diisobutoxydichloride, Ti(OC$_4$H$_9$-i)$_2$Cl$_2$	[1790-25-6]		184$_{213}$	
tributoxyfluoride, Ti(OC$_4$H$_9$)$_3$F	[84215-68-9]	45–48	175–180$_{27}$	
tributoxychloride, Ti(OC$_4$H$_9$)$_3$Cl	[4200-76-4]		154–155$_{27}$	n_D^{20}, 1.5169 d$_4^{20}$, 1.0985 g/cm^3
triisobutoxychloride, Ti(OC$_4$H$_9$-i)$_3$Cl	[52027-15-3]		125$_{27}$	n_D^{20}, 1.5158 d$_4^{20}$, 1.1043 g/cm^3
phenoxytrichloride, Ti(OC$_6$H$_5$)Cl$_3$	[4403-68-3]			dark-red crystalline solid; dipole moment, 9.91 × 10^{-30} C·m[b]
diphenoxydichloride, Ti(OC$_6$H$_5$)$_2$Cl$_2$	[2234-06-2]	116		reddish-brown solid
triphenoxychloride, Ti(OC$_6$H$_5$)$_3$Cl	[4401-43-8]		265$_{53}$	
tri-o-xylenoxychloride, Ti[(CH$_3$)$_2$C$_6$H$_5$O]Cl	[84501-82-6]		480$_{27}$	

[a] To convert Pa to mm Hg, divide by 133.
[b] To convert C·m to D, divide by 3.336 × 10^{-30}.

More useful than the preceding methods is cleavage of alkoxides by acetyl chloride or bromide. One, two, three, or four alkoxyls can be replaced by chloride or bromide. Benzoyl chloride gives poor yields, however. The tri- and tetrachlorides, which are stronger Lewis acids than mono- and dichlorides, coordinate with the alkyl acetate formed yielding distillable complexes (19,28–29):

$$Ti(OR)_4 + n\ CH_3\overset{O}{\overset{\|}{C}}X \longrightarrow Ti(OR)_{4-n}X_n + n\ CH_3\overset{O}{\overset{\|}{C}}OR$$

A very useful method is the proportionation of alkoxides with a stoichiometric quantity of titanium tetrachloride or bromide, preferably in an inert hydrocarbon solvent (28,30):

$$n\ Ti(OR)_4 + (4 - n)\ TiCl_4 \rightarrow 4\ Ti(OR)_nCl_{4-n} \qquad n = 1, 2, \text{or } 3$$

Alkoxy fluorides are prepared with acetyl fluoride. Alternatively, antimony trifluoride replaces one alkoxyl by fluorine (31):

$$3\ Ti(OR)_4 + SbF_3 \rightarrow 3\ Ti(OR)_3F + Sb(OR)_3$$

Titanium tetrafluoride reacts with $Ti(OR)_4$ in a manner similar to $TiCl_4$, especially if pyridine is present.

A principal use for the alkoxy halides is their reaction with organo-lithium or -magnesium compounds, R′M, in a Wurtz-type reaction to form compounds with carbon–titanium bonds. For this purpose, it is inconvenient that proportionation is reversible, and the products $Cl_2Ti(OR)_2$ and $ClTi(OR)_3$ disproportionate on vacuum distillation. This occurs because the alkoxyhalides tend to satisfy their coordination needs through bridge structures, in which the halogen and alkoxyl readily interchange:

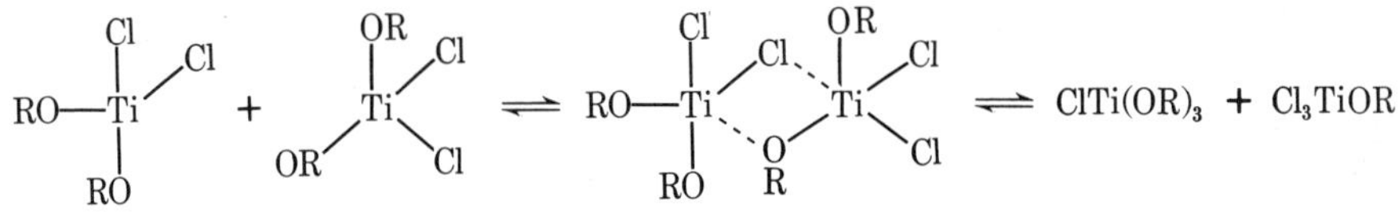

This can be prevented by supplying external electron pairs to satisfy the coordination needs of titanium; the strong base piperidine serves well, and a wide variety of compounds $Ti(OR)_nCl_{4-n}\cdot C_5H_{10}NH$ can be prepared (32). Alcohols and pyridine are also effective.

In a given $(RO)_nTiCl_{4-n}$, the alkoxy group can be exchanged by a higher alcohol if the resulting lower alcohol is removed by distillation (33). In the intermediate, HOR departs much more easily than HCl:

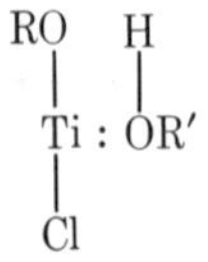

Chelates

Titanium chelates are formed from tetravalent titanium alkoxides or halides with bi- or polydentate ligands. One of the functional groups is usually alcoholic or enolic hydroxyl which interchanges with an alkoxy group, RO, on titanium to liberate ROH. If the second function is hydroxyl or carboxyl, it may react similarly. Diols and polyols, α-hydroxy acids, and oxalic acid are examples of this type. A carbonyl or amino group will donate an electron pair to increase the coordination number without liberating more ROH. β-Keto esters, β-diketones, and alkanolamines illustrate this second class of ligand.

Glycol Titanates. Primary diols (HOGOH), eg, ethylene glycol and 3-methyl-pentane-1,5-diol, react by alkoxide interchange at both ends yielding insoluble white solids, which are cross-linked polymers (5,34–36):

$$(RO)_4Ti \ + \ HOGOH \ \longrightarrow \ (RO)_3TiOGOTi(OR)_3 \ \longrightarrow$$

$$(RO)_3TiOGO\overset{\overset{\displaystyle (OR)_2}{|}}{Ti}OGOTi(OR)_3 \ \longrightarrow \ (RO)_3TiOGO\overset{\overset{\displaystyle OR}{|}}{\underset{\underset{\displaystyle OGOTi(OR)_3}{|}}{Ti}}OGOTi(OR)_3 \quad etc$$

Where the glycol contains one or two secondary or tertiary hydroxyls, the products are more soluble and some are even monomeric cyclic chelates (37–38). Three compounds are obtained from 2-methylpentane-2,4-diol, depending on the mole ratio (39–42):

(1)	(2)	(3)
	[62444-92-2]	[83562-90-7]

where G =

Structure (3) represents an isolable but labile alcoholate of (2) (41). An alternative structure (1) has both hydroxyls of one HOGOH molecule involved in covalent bonds instead of one hydroxyl from each of two glycol molecules. The solvating glycol molecule can be driven off by heating.

Silanediols, eg, $(C_6H_5)_2Si(OH)_2$ and $HOSi(C_6H_5)_2OSi(C_6H_5)_2OH$, yield four- and six-membered rings with titanium alkoxides. Pinacols and cyclic 1,2-diols form chelates rather than polymers. The more branched the diol molecule, the more likely that its titanate derivatives are soluble and even monomeric. "2,4-Diorgano-1,3-diols," eg, 2-ethylhexane-1,3-diol and other reduced aldols, furnish soluble products with lower titanates (43).

Titanate α-Hydroxy Acid Complexes. Titanate α-hydroxy acid complexes are frequently prepared in aqueous solutions and are stable in water over a wide pH range. Adding TPT to one or two moles of aqueous lactic acid yields solutions that do not precipitate hydrated TiO_2 (titanic acid) on standing. The strongly acidic solutions can be neutralized fully or partially with bases, eg, ammonia, amines and alkanol-amines, or sodium hydroxide. The structures of the products are uncertain and probably depend on pH and concentration; either the hydroxyl or carboxyl or both may be bonded covalently to titanium. Thus, the following structures show, respectively, hydroxyl coordination and monodentate carboxylate covalent bonding, hydroxyl covalent bonding and monodentate carboxylate coordination, and full coordination of two moles of lactic acid with bidentate carboxylate giving titanium a coordination number of 6:

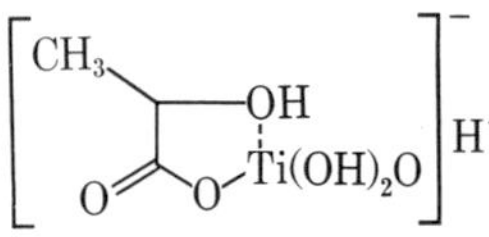

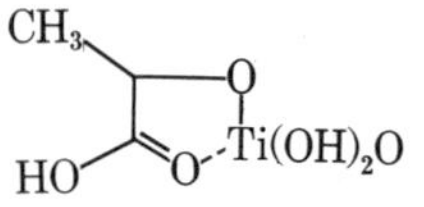

[83562-91-8]

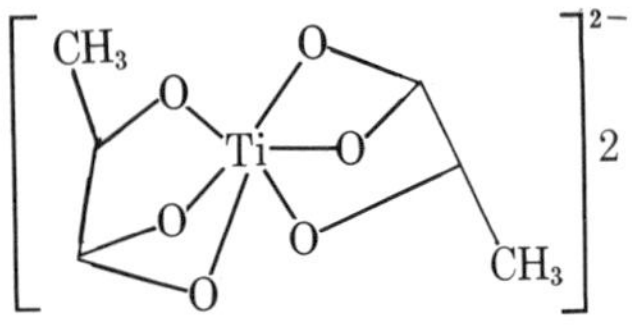

[83562-92-9]

The last structure explains best the stability of Tyzor LA titanium chelate [65104-06-5], the ammonium salt of the 2 lactic acid:1 titanate product. It is likely, however, that this product contains oligomers with Ti–O–Ti titanoxane bonds, and ligand molecules can bridge two titanium atoms (44).

A wide variety of other α-hydroxy acids coordinate similarly. A typical preparation involves adding an alkoxytitanate solution in, eg, acetone, tetrahydrofuran (THF), dioxane, etc, to the α-hydroxy acid in the same solvent. Alternatively, titanic acid can be prepared by hydrolysis of a lower alkyl titanate and is dissolved in aqueous hydroxy acid. By a third approach, aqueous titanium(IV) chloride, sulfate, or nitrate is treated with a hydroxy acid; chelate formation is detected spectroscopically. The colors of complexes with salicylic acid are like other phenolic titanates.

These procedures have been used to prepare water-soluble complexes of Ti(IV) with glycolic, lactic, α-hydroxybutyric, malic, tartaric, and citric acids, and numerous sugar-derived acids from glyceric to saccharic and gluconic acids. Many have been prepared only in solution and rarely are characterized by other than an ultraviolet spectrum. They are reported to be polymeric in concentrated solution and in the solid state (44). Ternary complexes in which copper or another metal is added have been studied (45).

Oxalic acid behaves as an α-hydroxy acid, and it yields crystalline ammonium or potassium salts from either aqueous titanium(IV) solutions or tetraalkoxytitaniums. These are written as

$$[O{=}Ti(O\overset{O}{\overset{\|}{C}}C\overset{O}{\overset{\|}{C}}OM)_2]\cdot n\,H_2O$$

(46). Dicarboxylic acids, eg, succinic or adipic, do not dissolve titanic acid. A phthalate has been prepared by adding acidic titanium sulfate solution to sodium phthalate solution.

β-Dicarbonyl Chelates. β-Diketones, reacting as enols, readily form chelates with titanium alkoxide and liberate one mole of an alcohol. As prepared, the 1:1 and 2:1 adducts are monomeric. Tyzor AA titanium chelate [*17927-72-9*] is the product mixture from TPT and 2 mol of acetylacetone reacting in the enol form; the structure is commonly written as:

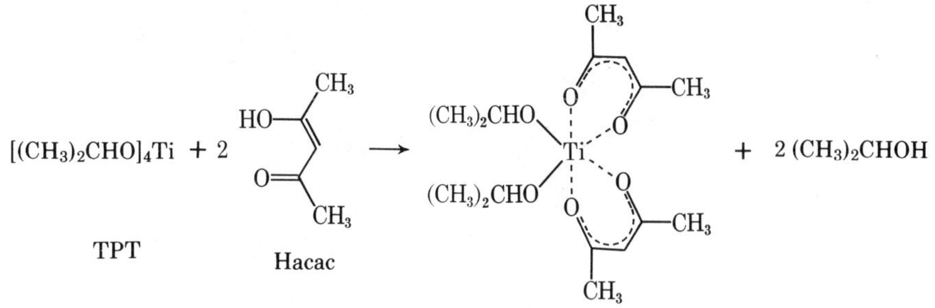

$[(CH_3)_2CHO]_4Ti \; + \; 2$ [Hacac enol structure] $\longrightarrow$ [(i-PrO)$_2$Ti(acac)$_2$ structure] $+ \; 2 \,(CH_3)_2CHOH$

     TPT   Hacac

(i-PrO)$_2$ Ti(acac)$_2$ [*17927-72-9*]

The dotted bonds indicate electron delocalization, since acac chelates are symmetrical. The chelate may be isolated by careful vacuum stripping of 2 $(CH_3)_2CHOH$, but if the distillation is pushed, an oligomeric titanoxane forms:

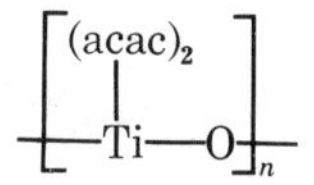

$$\left[\begin{array}{c} (acac)_2 \\ | \\ -Ti-O- \end{array} \right]_n$$

The orange-red chelates are soluble in common solvents. Since they are coordinately saturated (coordination number = 6 or CN 6); they are much more resistant to hydrolysis than the parent alkoxides (coordination number = 4 or CN 4). The isopropoxyls are the groups removed by hydrolysis. Hydrolysis is slowest at ca pH 4.5 (47). The titandiol self-condenses on drying to the above oligomeric titanoxane.

Chelates can be prepared from titanium tetrachloride and β-diketones. The product from Hacac is yellow (mp 192–194°C) when recrystallized from hydrocarbons; it is monomeric in solution (42,48):

$TiCl_4 \; + \; 2 \; Hacac \; \longrightarrow \; 2 \; HCl \; + \; Cl_2Ti\,(acac)_2 \; \equiv$ [Cl$_2$Ti(acac)$_2$ structure]

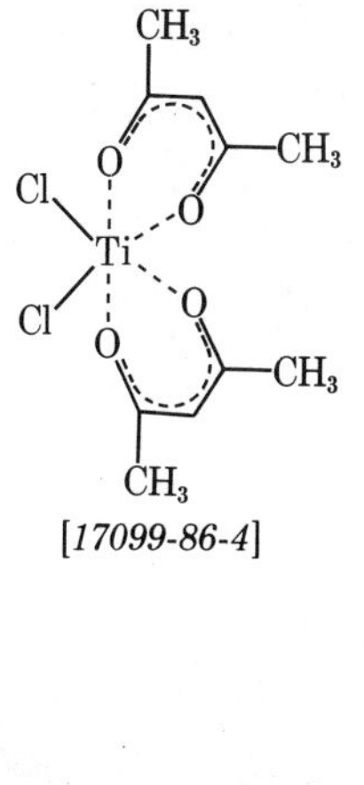

[*17099-86-4*]

It should be noted that the two chlorines are cis. This is the preferred configuration for practically all X_2TiL_2, where X is monodentate (usually RO or a halogen) and L is bidentate (often unsaturated like acac) (49). Rapid assessment of configuration can be made by nmr studies. Hydrolysis of $Cl_2Ti(acac)_2$ yields the same products formed from $(RO)_2Ti(acac)_2$.

β-Ketoesters yield similar chelates. Tyzor DC titanium chelate [27858-32-8] is the light-yellow liquid from TPT and two moles of ethyl acetoacetate after removal of the isopropyl alcohol.

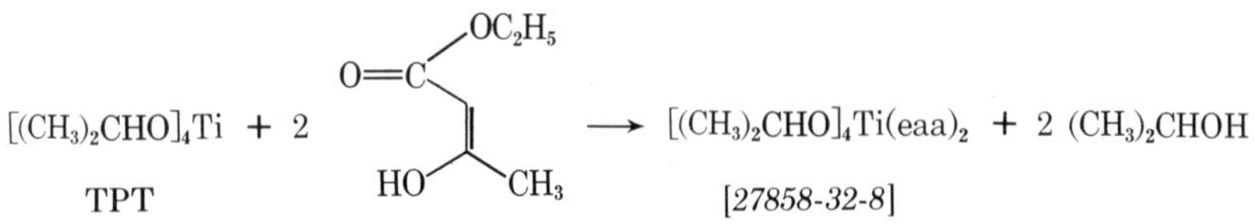

Heaa
(enol form of ethyl acetoacetate)

These products are also solvent-soluble monomers and hydrolyze slowly.

Alkanolamine Chelates. Alkanolamine chelates are industrially important and are used primarily in cross-linking water-soluble polymers. The products are used in thixotropic emulsion paints, in hydraulic fracturing and drilling of oil and gas wells, and in many other fields. Their preparation was originally described in ref. 50:

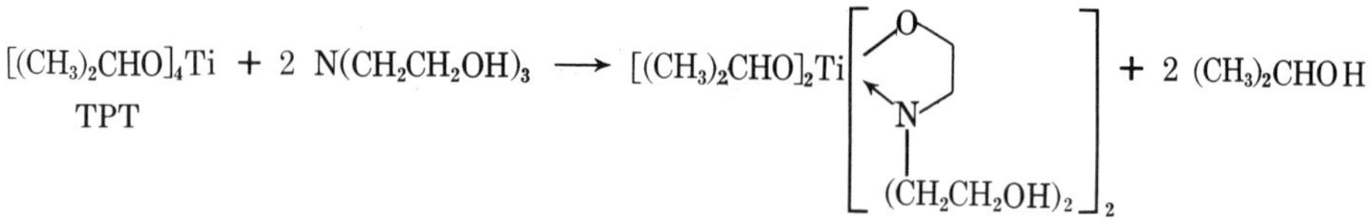

[36673-16-2]

The structure of the product indicates that titanium has received electron pairs from both nitrogens to complete the coordination shell, but it ascribes no role to the four free hydroxyls. If the liberated isopropyl alcohol is left in place, the product is a mobile liquid, but if the isopropyl alcohol is removed, the product is viscous and sticky and doubtless oligomeric.

At a 1:1 ratio, a solid product forms, for which the following structure has been assigned but not established (51). Note the eight-membered rings and that the coordination number is 4 and that no role is assigned to the nitrogen:

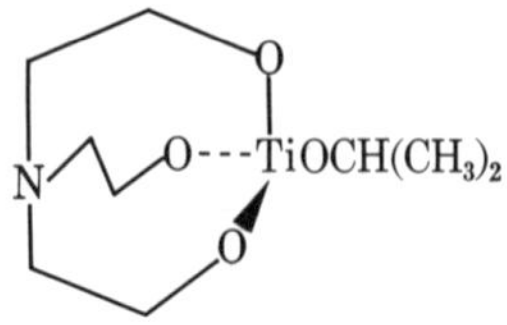

[55231-30-6]

Even if the nitrogen were coordinated within the structure or between molecules, only a coordination number of 5 could be attained.

At a 4:3 amine-to-alkoxide ratio, a pale-yellow solid resin is obtained by exhaustive stripping of the isopropyl alcohol (52). Its composition corresponds to the following structure:

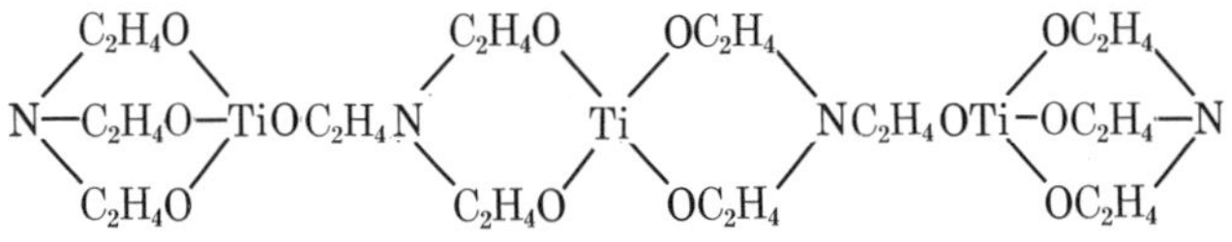

[83562-92-9]

Other ligands of this type include 1,1',1''-nitrilotris(2-propanol),

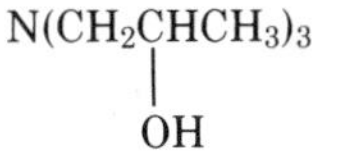

$$N(CH_2CHCH_3)_3$$
$$|$$
$$OH$$

and the general class

$$HOCH_2CH_2NRNR'R''$$
          R'
          /
     (with R' above N)

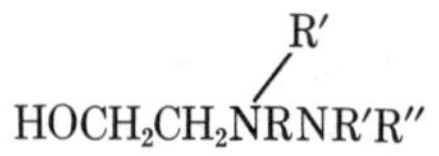

where the R groups are alkyl, hydroxyalkyl, and aminoalkyl (51,53).

The alkanolamine titanates from these polyfunctional ligands are water-soluble and slowly hydrolyzed at ca pH 9. Lowering the pH increases the rate of hydrolysis, which is shown by the development of turbidity. Turbidity also occurs above ca pH 11. Addition of secondary chelating agents, eg, polyols such as sorbitol or mannitol or the strongly chelating α-hydroxy acids such as citric or oxalic, prevents development of turbidity outside the pH 9–11 range (54).

Alkanolamine titanates react rapidly with aqueous solutions of polyhydric polymers forming gels. Soluble cellulose derivatives, guar gum, and poly(vinyl alcohol) are thus cross-linked. The strength of the gel depends on the molecular weight of the base polymer, the number of cross-links per chain, ie, the titanium-to-polymer ratio, the polymer structure, and the pH. The strongest gels are obtained in the more alkaline solutions; as the mixture is made more acidic, the gels become weaker and are eventually destroyed. Adding powerful ligands, eg, citric or lactic acids, breaks down the gels rapidly.

The reactions of simpler alkanolamines with titanium alkoxides are not completely understood. Ethanolamine reacts with the lower titanium alkoxides giving insoluble white solids. N,N-Dialkylethanolamines, $R_2NCH_2CH_2OH$, react with TPT to yield, depending on the mol ratio, all members of the family $[(CH_3)_2CHO]_{4-n}$-$Ti(OCH_2CH_2NR_2)_n$, where R = CH_3 or C_2H_5 (55–56). When R = CH_3, all products are distillable: bp$_{27\ Pa\ (0.2\ mm\ Hg)}$ 81°C (n = 1, [40881-31-0]), 95–105°C (n = 2, [52406-69-6]), 140°C (n = 3, [52406-70-9]), 148°C (n = 4, [55235-61-5]) (55). When R = C_2H_5, the products disproportionate on attempted distillation (56). All of these products are monomeric (ebullioscopy in benzene), suggesting that electron donation from nitrogen completes the coordination sphere of Ti. The compounds derived from $CH_3NHCH_2CH_2OH$ are dimeric crystalline solids.

The results reported in ref. 57 do not agree with those reported in refs. 55–56. Whereas Bharara and co-workers (55–56) prepared their products from aminoalcohols and TPT by azeotroping the isopropyl alcohol with benzene, Alyea and Merrell (57) used the lithium salt of the ligand (derived from C_4H_9Li) with $(RO)_3TiCl$ in hexane, because they could not displace isopropyl alcohol from TPT with the neutral ligands. From trimethylethylenediamine, dimethylethanolamine, and dimethylisopropanolamine with TPT, they obtained, respectively, (4), (5), and (6):

$[(CH_3)_2CHO]_3Ti$ ⟨structure with N-CH₃ top, N(CH₃)₂ bottom⟩

$[(CH_3)_2CHO]_3Ti$ ⟨structure with O top, N(CH₃)₂ bottom⟩

$[(CH_3)_2CHO]_3Ti$ ⟨structure with O-CH₃ top, N(CH₃)₂ bottom⟩

[40506-23-8]   [40881-31-0]   [40506-24-9]
(4) (orange)   (5) (yellow)   (6) (pale green)

Each of these compounds is five-coordinated because: the three compounds are monomeric (cryoscopy in cyclohexane); they exhibit no ir bands near 2820 or 2760 cm^{-1}, whereas uncoordinated $(CH_3)_2NR$ exhibits a pair of bands (C–H stretch) at these frequencies; and in ^1H nmr, the CH_3N protons move downfield when coordinated, compared to the free ligand, and all three compounds show $\delta N(CH_3)_2$ of 0.20–0.25 ppm (as low as −50°C, the peaks remain sharp and do not shift) implying no exchange between free and coordinated $N(CH_3)_2$ (58–59). However, Alyea and Merrell (57) were unable to distill compound (5) because it disproportionated to TPT and a chelate containing more than one ligand per titanium. The triethyl analogue of (5) was distillable (bp 160°C$_{13\ Pa\ (0.1\ mm\ Hg)}$). However, they could not displace alcohol from TPT, because the transition state for disproportionation must be at least as crowded as that for alcohol displacement. Moreover, the boiling point of the triethyl analogue of (5) is much higher than that recorded by Bharara (55–56) for the triisopropyl product of supposed structure (5).

Acylates. Titanium acylates are prepared either from titanium tetrachloride or tetraalkoxide. Since it is difficult to obtain titanium tetraacylates, most compounds reported are either chloro- or alkoxyacylates. Under most conditions, $TiCl_4$ and acetic acid give dichlorotitanium diacetate [4644-35-3]. The best method involves passing preheated (136–170°C) $TiCl_4$ and acetic acid simultaneously into a heated chamber; the product separates as an HCl-free white powder (60):

$$TiCl_4\ +\ 2\ HOCCH_3\ \longrightarrow\ Cl_2Ti(OCCH_3)_2\ +\ 2\ HCl$$

Alternatively, $TiCl_4$ reacts with a cold mixture of acetic acid and acetic anhydride. If the mixture is heated, condensation occurs with elimination of acetyl chloride to yield hexaacetoxydititanoxane [4861-18-1] (61–62).

$$ClTi(OCCH_3)_3\ +\ Cl_2Ti(OCCH_3)_2\ \longrightarrow\ O[TiCl(OCCH_3)_2]_2\ +\ CH_3CCl$$

[84215-69-0]   [84215-70-3]

$$\downarrow\ 2\ (CH_3C)_2O$$

$$O[Ti(OCCH_3)_3]_2\ +\ 2\ CH_3CCl$$

Trichlorotitanium monoacylates form during thermal decomposition of $TiCl_4$–ester complexes (63):

$$TiCl_4 \cdot R\overset{\overset{O}{\|}}{C}OR' \longrightarrow TiCl_3(O\overset{\overset{O}{\|}}{C}R) + R'Cl$$

Tetraacylates are prepared from titanium tetrabromide and excess carboxylic acid in an inert solvent. After solvent removal, the residue is heated to remove hydrogen bromide:

$$TiBr_4 + 4 \, R\overset{\overset{O}{\|}}{C}OH \longrightarrow Ti(O\overset{\overset{O}{\|}}{C}R)_4 + 4 \, HBr$$

Tetraacylates have been prepared in this way from stearic, benzoic, cinnamic, and other acids, as well as from succinic and adipic acids (64):

In some cases, titanium tetrachloride may be used.

The usual products from alkoxides and acids are dialkoxytitanium diacylates (65–66). The third acyl group, but not the fourth, can often be introduced by azeotroping the lower alcohol with benzene (67). With acetic anhydride, the same hexaacetoxydititanoxane as prepared from the chloride forms.

An acylate group is potentially a bidentate ligand. It may bond once or twice to one titanium or may bridge two titanium atoms:

Dimeric (ebullioscopy in benzene)

$$[(CH_3)_2CHO]Ti(O\overset{\overset{O}{\|}}{C}R)_3$$

is represented with eight-coordinate Ti. If structure (7)

(7)

is correct, isopropoxy is a better bridging ligand than acylate (67). This saturation of titanium coordination may explain the difficulty in preparing titanium tetraacylates by mild methods. Dimeric dichloro- and dibromotitanium diacylates exhibit three different carbonyl frequencies in the ir region, ie, at 1650, 1540, and 1400 cm^{-1}, associated with three different types of bonding (68).

Polymeric acyltitanoxanes have been prepared in several ways. From $(RO)_4Ti$ and $R'COOH$ in 1:2, 1:2.5 and 1:3 mole ratios, the respective products

$$
\begin{bmatrix} OR \\ | \\ -TiO- \\ | \\ OCR' \\ \| \\ O \end{bmatrix}_n , \quad
\begin{bmatrix} OR\ (OCR')_2 \\ | \\ -TiOTiO- \\ | \\ OCR' \\ \| \\ O \end{bmatrix}_n , \quad \text{and} \quad
\begin{bmatrix} (OCR')_2 \\ | \\ -TiO- \end{bmatrix}_n
$$

have been prepared (69). As described in another report, equimolar amounts of $TiCl_4$ and $RCOOH$ react in the presence of an amine, followed by 1–2 mol addition of H_2O (70). The product is:

$$
\begin{bmatrix} OH \\ | \\ -TiO- \\ | \\ OCR \\ \| \\ O \end{bmatrix}_n
$$

In another study, acid and alkoxytitanium are heated together. Addition of water to the clear solution followed by stripping of ROH yields a material with the preceding formula as a solvent-soluble, brown wax (71). In these structures, the degree of polymerization is low, but extensive association makes molecular weight determination difficult.

Alkoxytitanium acylates

$$
(RO)_n Ti(O\overset{\overset{\displaystyle O}{\|}}{C}R')_{4-n}
$$

react with alcohols resulting in the exchange of alkoxy groups. The acylate group is unaffected.

Titanium(IV) Complexes with Other Ligands

The d^0-titanium(IV) atom is hard, ie, not very polarizable, and would be expected to form its most stable complexes with hard ligands, eg, fluoride, chloride, oxygen, and nitrogen. Soft or relatively polarizable ligands containing second- and third-row elements or multiple bonds should give less stable complexes. The stability depends on the coordination number of titanium, on whether the ligand is mono- or polydentate, and on the mechanism of the reaction used to measure stability.

A partial list of ligands that bond to titanium(IV) includes —OSOR (sulfinate), —OSO$_2$R (sulfonate), O$_2^{2-}$ (peroxido), O$_2^-$ (superoxido), —NO$_2$ (nitro), —ONO (nitrito), —ONO$_2$ (nitrato), CO$_3^{2-}$ (carbonato), PO$_4^{3-}$ (phosphato), —NR$_2$ (amido), —N(R)- (COX) (acylamido), —N=CR$_2$ (Schiff base), N in a heterocyclic ring, —N=NR (azo), —N$_3$ (azido), —N=C=O (isocyanato), —N=C=S (isothiocyanato), —N=C=Se (isoselenocyanato), —SR (alkanethio), —SAr (arenethiol),

$$\underset{\text{—SCOR}}{\overset{\displaystyle \text{S} \atop \|}{}}$$

(xanthato),

$$\underset{\text{—SCNR}_2}{\overset{\displaystyle \text{S} \atop \|}{}}$$

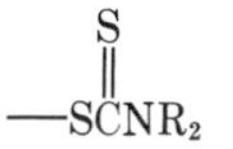

(dialkyldithiocarbamato), BH$_4$ (tetrahydridoborato or borohydride), —SiR$_3$ (trialkylsilyl), and Ge and Sn analogues. Phosphine, arsine, and stibine ligands, as well as CO and hydrido, associate only with lower valent titanium. Carbon ligands, eg, alkyl, aryl, dienyl, and cyclopentadienyl give organometallic compounds. The early literature on many of these complexes has been reviewed (72).

Peroxides. Titanates may influence reactions of organic peroxides; thus, t-butyl hydroperoxide epoxidizes olefins:

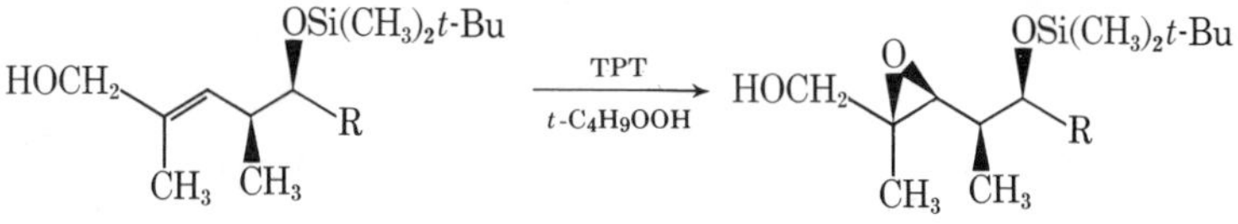

The ratio of syn-epoxide (shown above) to $anti$-epoxide is 10–25:1 with TPT catalysis, whereas vanadylacetylacetonate is less selective and m-chloroperoxybenzoic acid gives the reverse 1:25 ratio. It is supposed that TPT esterifies the free hydroxyl, then coordinates with the peroxide to favor syn-epoxidation (73). This procedure is related to that for enantioselective epoxidation of other allylic alcohols in 90–95% enantiomeric excess (73).

Titanates trigger peroxide-initiated curing of unsaturated polyesters to give products of superior color, compared to conventional cobalt-initiated curing (74–75) (see Initiators). Titanium coordinated to a porphyrin bonds hydrogen peroxide as (76):

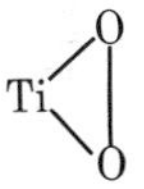

Hydrogen peroxide produces an intense yellow color with Ti(IV) in aqueous solution and has long been used as a qualitative test for Ti. Solid inorganic peroxides have been reported.

Amides. Titanium amides, $Ti(NR_2)_4$, are typically prepared from $TiCl_4$ or another TiCl compound and $LiNR_2$. Alternatively, $TiCl_4$ and 8 mol amine react to form $Ti(NR_2)_4$ and $4 R_2NH \cdot HCl$. Amides have been prepared from both primary and secondary amines by these procedures and by their reaction with titanium disulfide TiS_2 (77–79).

The properties of the tetrakisdialkylamides resemble those of the alkoxides, and the ir and nmr spectra have been studied (80–81). They are hydrolyzed rapidly by water and are sensitive to oxygen. They undergo reversible amine exchange with other secondary amines. With alcohols, the exchange is irreversible and tetraalkoxides are the products, reflecting the fact that $:\ddot{O}\!<$ is a harder ligand than $:N\!\leqslant$; ie, $TiNR_2$ is more reactive than TiOR as a titanating agent. However, $Ti(NHR)_4$ compounds do not exchange with alcohols, probably because of hydrogen bonding. A recent example is shown in the following equation (82–83):

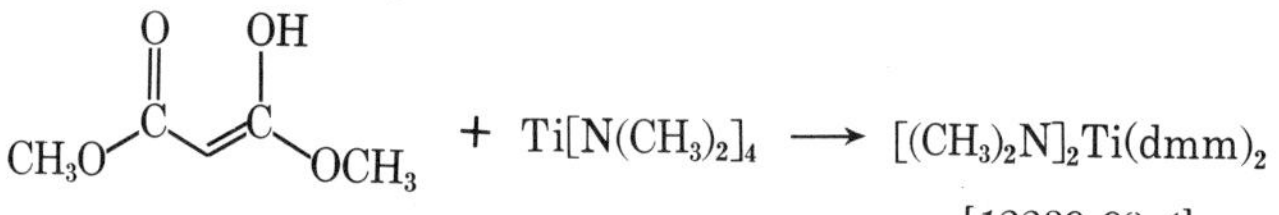

enol form of dimethyl
malonate, Hdmm

$[12239\text{-}98\text{-}4]$

The unstable compounds formed from β-diketones decompose to enamines, which are further aminated (83):

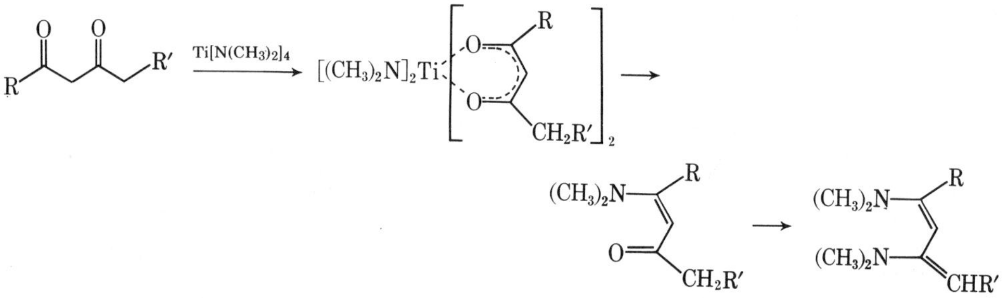

The amides also react with other HA compounds, which include cyclopentadiene, decaborane, and mercaptans, as well as the ROH and R_2NH compounds mentioned above, forming Ti–A bonds. Haloamides $Cl_n Ti(NRR')_{4-n}$ and alkoxyamides $(R''O)_n$-$Ti(NRR')_{4-n}$ are discussed in ref. 84.

An interesting type of amide is derived from hexamethyldisilazane, $(CH_3)_3SiNSi(CH_3)_2$ salts and TiCl compounds (85):

$$\geqslant\!TiCl + LiN[Si(CH_3)_3]_2 \rightarrow \ \geqslant\!TiN[Si(CH_3)_3]_2$$

Amides of formula $\geqslant\!TiN(R)COX$ can be prepared by insertion of a titanium alkoxide into an isocyanate (55–56,86–87):

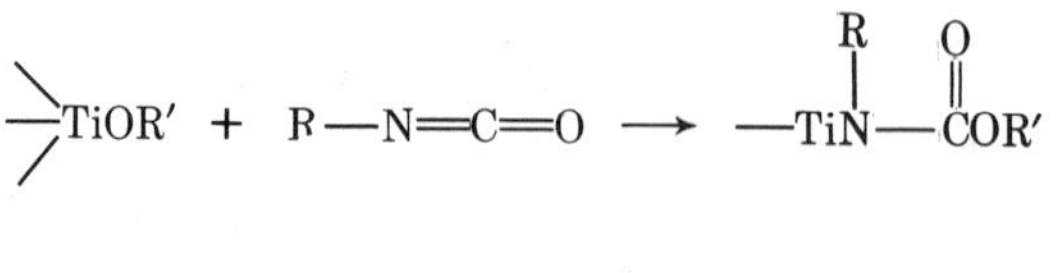

This occurs also with the cyclopentadienyl (Cp) alkoxides $CpTi(OR')_3$ once, twice, or three times (88). Thus, titanates catalyze conversion of isocyanate to urethane (86). Sometimes the isocyanate is trimerized to isocyanurate or polymerized to low mol wt nylon-1 (86). No reaction occurs with the heterocumulenes CS_2 and ArNCS. With carbodiimides, a 2:1 adduct forms regardless of the molar proportions of the reagents (86):

$$Ti[OCH(CH_3)_2]_4 + 2\,ArN{=}C{=}NAr \longrightarrow [(CH_3)_2CHO]_2Ti[\underset{\overset{|}{Ar}}{N}C[OCH(CH_3)_2]{=}NAr]_2$$

In compounds, eg, $\geqslant TiOCH_2CH_2NHCH_3$, the insertion reaction of $\geqslant TiOR'$ is faster than the expected reaction at the NH group. The same titanium amides can be prepared from $RCONH_2$:

$$Ti[OCH(CH_3)_2]_4 + n\,R\overset{\overset{\text{O}}{\|}}{C}NH_2 \longrightarrow [(CH_3)_2CHO]_{4-n}Ti(NHC\overset{\overset{\text{O}}{\|}}{R})_n + n\,(CH_3)_2CHOH$$

$$n = 1{-}4$$

Replacement of the fourth isopropoxyl requires 16 h of refluxing in benzene. The compounds are N-bonded, not O-bonded, as shown by ir studies. When the solvent and $(CH_3)_2CHOH$ are boiled out, the residue polymerizes through hydrogen bonds and becomes insoluble even in dimethylformamide (DMF). The reaction is not reversed by boiling with $(CH_3)_2CHOH$, yet the product is hydrolyzed by moist air (89).

Schiff Base ($>$C$=$N—). The nitrogen of a Schiff base unit in a polydentate ligand coordinates readily. One example involving a tridentate ONS ligand is salicylaldehyde dithiocarbazate, which is dimeric through double isopropoxy bridges (39).

Though this compound is readily hydrolyzed, the related TiL_2, where L = ligand, is monomeric and stable in hot water because the coordination number of Ti is 6. The related ONO ligands derived from enol ketones or aldehydes, eg, acetylacetone or salicylaldehyde, and hydroxyamines, react similarly with TPT:

Other products are analogous, pentacoordinate, dimeric, 1:1 water-sensitive complexes and hexacoordinate, monomeric, 2:1 water-stable complexes. Both 1:1 complexes react with pyridine becoming six-coordinate and react with

$$\text{HO—CH}_2\text{—C(CH}_3)_2\text{—CH}_2\text{—CH(OH)—CH}_3$$

by displacing isopropoxy forming pentacoordinate chelates

which also add pyridine (90). Several similar ligands and their five- and six-coordinate complexes with Ti have been reported (91–93). Complexes with 8-hydroxyquinoline are well known (94–95).

Azo ligands are sometimes encountered in metallized azo dyes (qv) when other groups, eg, phenol, thiophenol, or amine functions, provide the primary bonds. They are more common with lower-valent titanium, as they form sometimes upon efforts to fix nitrogen (see Nitrogen fixation).

Ambident ligands, eg, —N=C=O, —N=C=S, and —N=C=Se, have been studied by several workers; the azido group is included because of its formal similarity to the heterocumulenes. All of these are N-bonded to titanium, but —OC≡N used to be written for cyanato on the basis of some spectral evidence and because O is harder than N. A good illustration of the effect of soft and hard groups is in the complex where soft Ni bonds to soft S and hard Ti to hard N (96):

$$(\text{NCS})_4\text{Ni} \underset{\text{SCN}}{\overset{\text{SCN}}{\cdots\cdots}} \text{Ti(NCS)}_4$$

Most examples are derived from cyclopentadienyltitanium compounds.

Alkanethio. Simple sulfur ligands bond to titanium. Mercaptoacetic (thioglycolic) esters displace alcohol from TPT when the alcohol is azeotroped with benzene. The incoming ligand is the bidentate enol

$$\text{HSCH}=\underset{\overset{|}{\text{OH}}}{\text{C}}\text{—OC}_2\text{H}_5$$

and it displaces two moles of i-C_3H_7OH. The bidentate structure has been supported by ir spectra. Predictably, the 1:1 adducts $\{[(CH_3)_2CHO]_2TiL\}$ are dimeric with isopropoxy bridges and the 2:1 adducts (TiL_2) are monomeric. Attempts to add a third mole to give six-coordinate TiL_3 have failed. The reaction:

$$[(CH_3)_2CHO]_2TiL + 2\,t\text{-BuOH} \rightarrow (t\text{-BuO})_2TiL + 2\,(CH_3)_2CHOH$$

proceeds readily when the isopropyl alcohol is removed by distillation; curiously, no transesterification of $R = CH_3$ or C_2H_5 occurs in the ligand $HSCH_2CO_2R$, indicating that interchange in the enol form is not possible (97).

Complexes of $\geqslant$TiSR can also be prepared from $\geqslant$TiNR$_2$ (81). The ease of displacement of Z in TiZ by A of HA is $NR_2 > SR > Cp \gg OR$ (80).

Bidentate ligands with two coordinating sulfurs are represented by xanthates and dithiocarbamates. The reaction of $TiCl_4$ in CH_2Cl_2 with

$$\overset{\overset{\displaystyle S}{\|}}{NaSCNR_2}$$

forms (98):

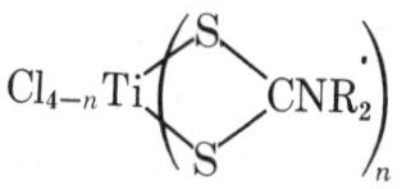

Even if $n = 4$, the ligand is bidentate; therefore, the product is eight-coordinate (99–100). X-ray studies show that the structure is dodecahedral. The products in which $R = CH_3$ are quite insoluble in typical laboratory solvents, but those in which $R = C_2H_5$, i-C_3H_7, or i-C_4H_9 are reasonably soluble (98).

Bis(thioethers) can coordinate without displacing another ligand. Methyltitanium trichloride [2747-38-8], CH_3TiCl_3, coordinates with a series of ligands YCH_2CH_2Z, where Y and Z = CH_3O, CH_3S, $(CH_3)_2N$, PH_2 or PO in various combinations, yielding the following:

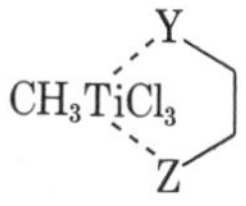

These adopt a meridional (mer) configuration in which the harder atom is trans to the methyl group. Facile air oxidation yields:

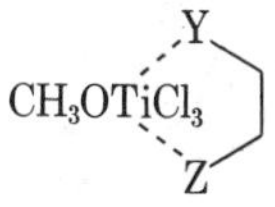

Dithiocatechol reacts with TPT in the presence of tertiary amines yielding the 3:1 adduct as a dianion (101):

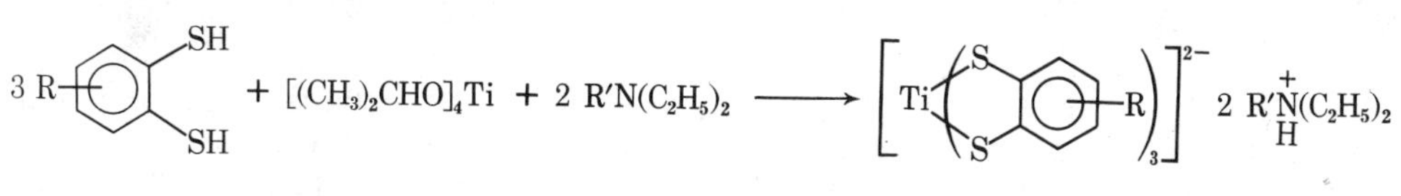

Seleninate. Seleninate bonds to titanium through both oxygens yielding water-stable compounds (102):

$$TiCl_4 + 4\ RSeO_2Na \xrightarrow{\ CH_3CN\ } Ti(O\overset{\overset{\textstyle O}{\|}}{S}eR)_4$$

[71560-14-0] R = CH_3
[71648-91-4] R = C_6H_5

The same compounds can be prepared from $TiCl_3$, which is oxidized by seleninate.

Borohydrides. Titanium(IV) borohydrides are unknown because BH_4^- immediately reduces Ti(IV); however, $Ti(BH_4)_3$ [62387-90-0] has been reported (103). It is not known whether hypothetical $Ti(BH_4)_4$ includes the same unusual triply-bridged BH_4^- that occurs in $Zr(BH_4)_4$ and $Hf(BH_4)_4$ (104).

Lower Valent Titanates

Titanium tetraethoxide is reduced by sodium and ethanol to a dark-blue compound (105). Use of potassium as the reducing agent in the alcohol permits the isolation and identification of $Ti(OC_3H_7)_3$ [22922-82-3] and $Ti(OC_4H_9)_3$ [5058-41-3] (106–107). The products precipitate as solid alcoholates, $(RO)_3Ti.2ROH$, which can be dried to solvent-free material. Air oxidation in a radical reaction yields solvent-soluble materials formulated as $(C_3H_7O)_2Ti{=}O$ [20644-85-3], mp 100–113°C and $(C_4H_9O)_2Ti{=}O$ [30860-71-0], mp 112–115°C, plus C_3H_7OH or C_4H_9OH and the corresponding aldehydes. These materials may be oligomers. With potassium *in ether*, triethoxytitanium [19726-75-1] can be prepared as a pink–lilac compound which, on exposure to oxygen, yields $(C_2H_5O)_3TiOTi(OC_2H_5)_3$ [84215-71-4] (9). The ethoxy groups in the titanoxane exchange with C_3H_7OH or C_3H_7OH without breaking the Ti–O–Ti bond. Table 4 is a list of a few of the organotitanium compounds of valency lower than four.

A family of Ti(III) derivatives roughly parallels those of Ti(IV). Titanium(III) chelates are known, eg, the trisacetylacetonate prepared in benzene from titanium trichloride, acetylacetone, and ammonia (108). This deep-blue compound is soluble in benzene but insoluble in water:

$$TiCl_3 + 3\ Hacac + 3\ NH_3 \rightarrow Ti(acac)_3 + 3\ NH_4Cl$$
[14284-96-9]

It is oxidized by air to orange–red crystals, which are possibly $O{=}Ti(acac)_2$ or an oligomer. If, however, the mixture is refluxed in the absence of ammonia, a red dimer is formed (mp 214°C), to which the doubly-bridged structure is assigned (109). The color is not consistent, however:

$$(acac)_2Ti\underset{Cl}{\overset{Cl}{\diagdown\diagup}}Ti(acac)_2$$
[61436-17-7]

Titanium(III) β-diketonates can also be prepared by reduction of the Ti(IV) chelates (110).

Alpha-hydroxy acids are represented by oxalic acid. A titanous oxalate is prepared in water from $TiCl_3$ and oxalic acid, and it precipitates upon addition of ethanol as a yellow solid (111). It forms double salts with metal oxalates, $MTi(C_2O_4)_2.2H_2O$. A

Table 4.   Organotitanium Compounds of Lower Valence

Compound	CAS Registry No.	Type	Appearance	Mp, °C	Other properties
Ti(CH$_3$)$_3$	[32835-60-2]	Ti(III) trialkyl	not isolated; green in THF solution		solutions give positive Gilman test; decompose above −20°C
(C$_5$H$_5$)$_2$TiCl	[60955-54-6]	Ti(III) Cp$_2$ halide	green crystals	279–281	
C$_5$H$_5$TiCl$_2$	[31781-62-1]	Ti(III) Cp halide	violet	sublimes in vacuo, 150	insoluble in hydrocarbons; very sensitive to oxygen; blue solution in acetonitrile
(CH$_2$=CH)$_2$TiCl		Ti(III) (divinyl) halide	infusible powder		
(C$_9$H$_7$)$_2$TiCl	[12113-02-9]	Ti(III) (indenyl) halide	yellowish red		
(C$_5$H$_5$)$_2$TiCH$_2$-CH=CH$_2$	[12110-59-7]	Ti(III) Cp$_2$ allyl	purple–blue		monomeric; extremely air sensitive
(C$_5$H$_5$)$_2$TiOCCH$_3$ (O)	[12248-00-9; 56260-60-7]	Ti(III) Cp$_2$ carboxylate	blue	110	air sensitive
(C$_5$H$_5$)$_2$TiOCC$_9$H$_{19}$ (O)	[12248-77-0]		blue	ca −5	air sensitive
(C$_5$H$_5$)$_2$TiBH$_4$	[12772-20-2]	Ti(III) Cp$_2$ hydroborate	black–violet needles		solution very sensitive to air
(C$_5$H$_5$)$_2$TiBF$_4$	[83562-93-0]	Ti(III) fluoroborate	light blue		very sensitive to air; aqueous alkali → Ti$_2$O$_3$ (hydrate)
(C$_5$H$_5$)$_3$Ti	[39333-58-9]	Ti(III) tricyclopentadienyl	green	125 sublimes	extremely air sensitive; gives (C$_5$H$_5$)$_2$Ti(CO)$_2$ with CO under pressure
Ti(C$_6$H$_5$)$_2$	[14724-88-0]	Ti(II) diphenyl	black solid		pyrophoric; gives phenylmercury chloride with HgCl$_2$
C$_5$H$_5$TiC$_6$H$_5$	[12109-06-7]	Ti(II) cyclopentadienyl phenyl	black solid		sensitive to air and moisture; thermally stable to 170°C
(C$_5$H$_5$)$_2$Ti	[1271-29-0]	Ti(II) dicyclopentadienyl	dark green	200	pyrophoric; catalyst for polymerization of olefins and acetylenes

group of violet titanium(III) acylates has been prepared from TiCl$_3$ and alkali carboxylates. All of the acylates are strong reducing agents similar to TiCl$_3$ (112).

Recent studies of Ti(III) compounds include the following reaction (113):

$$\text{TiCl}_3 + (\text{RO})_3\text{P}{=}\text{O} \longrightarrow \left[\text{TiCl}_2[\overset{\overset{\text{O}}{\|}}{\text{OP}}(\text{OR})_2] \right] + \text{RCl} \longrightarrow \text{Ti}[\overset{\overset{\text{O}}{\|}}{\text{OP}}(\text{OR})_2]_3$$

R = C$_2$H$_5$ [56170-51-5]
R = Pr [56170-52-6]
R = CH$_2$CH=CH$_2$ [56170-53-7]
R = Bu [56170-54-8]

The tris compounds are highly bridged three-dimensional polymers. Photoreduction of aqueous Ti(IV)-containing alcohols or glycols, but not ethylene glycol, yields Ti(III) and the aldehyde or ketone corresponding to the alcohol (114–115). A possible mechanism is

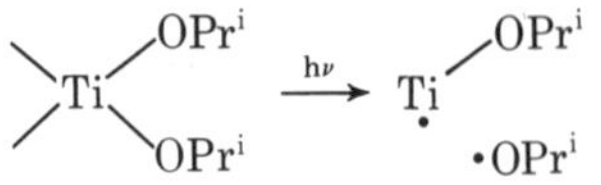

$$2 \cdot OPr^i \longrightarrow (CH_3)_2CO + HOPr^i$$

A broad selection of Ti(III) compounds coordinated to α-hydroxy acids, diboric acids, and 8-hydroxyquinoline has been prepared by the reaction:

$$TiCl_3 + n \, HL \xrightarrow[DMF]{(C_2H_5)_3N} TiCl_{3-n}L_n$$

The esr has been measured at 77 K and 298 K. The compounds are dimers, but the Ti–Ti distances vary with the ligand (116–117).

Exposure of $TiCl_4$ in ethyl, propyl, and butyl alcohols for four weeks results in the precipitation of green octahedral Ti(III) complexes. Similar products form on irradiating Ti(OR)$_4$ in ROH containing an equivalent of a lithium halide (115).

Electron-spin resonance measurements have been made on 0.05 M aqueous solutions of Ti(III) complexed with a variety of α-hydroxy acids and poly(acrylic acid). The complexes appear to be polynuclear with different multiplets, s $> \frac{1}{2}$. The mandelate complex is isolated as an air-stable yellow powder (117). Titanium(III) mandelate and lactate complexes have also been studied by nuclear magnetic relaxation. The manner of bonding depends markedly on the pH of the solution. As the acidity is reduced, TiL complexes shift to TiL$_2$ complexes. The mandelate shown below yields polymeric complexes, and it is suggested that the π-system of the phenyl groups coordinates with titanium; however, this proposal can not be evaluated from the evidence given (118):

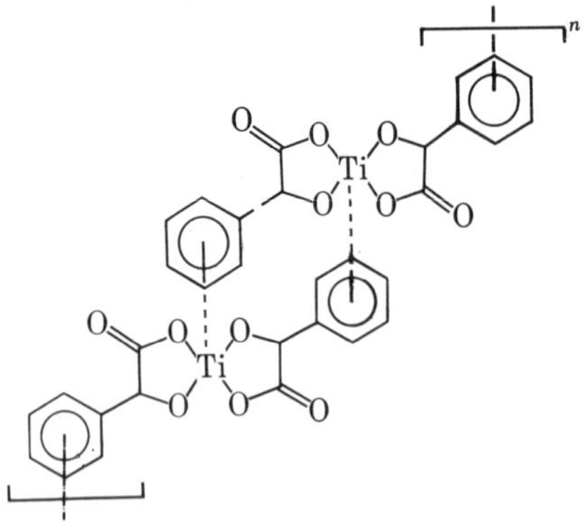

Poly(titanium(IV) dimandelate)

[83548-01-0]

Titanium(IV) citrate [52215-05-1] can be reduced polarographically to Ti(III) citrate [53971-27-0], and the procedure has been advocated as a method for determining citric acid (119).

The Ti(III) silylamide is obtained as blue monomeric crystals from pentane. This material is very sensitive to air and moisture:

$$TiCl_3.2[N(CH_3)_3] + 3\ LiN[Si(CH_3)_3]_2 \rightarrow Ti(N[Si(CH_3)_3]_2)_3$$
$$[23064\text{-}71\text{-}3] \qquad\qquad\qquad [37512\text{-}29\text{-}1]$$

The compound's esr, ms, ir, uv, and magnetic properties were compared to those of the silyl amides from Sc, V, Cr, and Fe. The esr curves result from a large trigonal crystal field on d^1-Ti (120).

Ti(acac)$_3$ fragments in a mass spectrometer giving M$^+$ and (M-acac)$^+$. The respective appearance potentials are 7.1 and 11.8 ± 0.1 eV. Other fragments correspond to oxygen atoms remaining on Ti (121).

Titanium dichloride has yielded uncharacterized colored materials with alcohols and acids; however, these have not been investigated with modern techniques (113–114).

Organometallics

Titanium(IV) Organometallics. In classical organometallic chemistry, Grignard reagents or organolithium compounds reacted with halides of less active metals forming new C–M bonds. These failed uniformly with titanium halides until it was realized that many simple titanium alkyls are extraordinarily unstable thermally and to moisture and air. This thermal instability is derived from the presence of unfilled, low lying $3d$ orbitals (see Organometallics). In titanium metal, the electron configuration is $3s^23p^63d^24s^2$, and in simple tetralkyltitaniums it is $3s^23p^63d^64s^2$ or includes hybrid $3d$–$4s$ orbitals. Supplying two extra electron pairs by coordination with, for example, pyridine, diamines, or strong donor ethers, gives stable molecules with the configuration $3s^23p^63d^{10}4s^2$ (59,122–123). Another source of instability is the availability of facile decomposition mechanisms, eg, β-elimination:

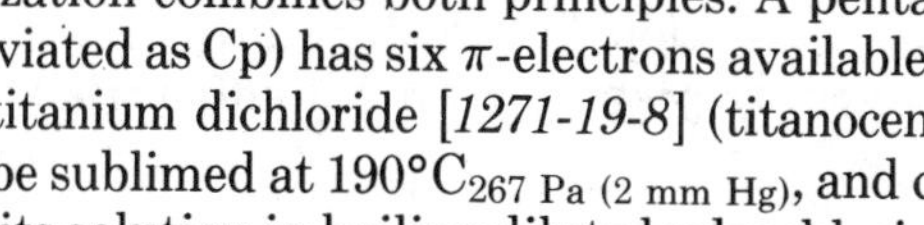

This can be circumvented by choosing alkyl groups with no β-H, eg, methyl, neopentyl, trimethylsilylmethyl, phenyl and other aryl groups, and benzyl. The linear transition state for β-elimination can also be made sterically impossible. The most successful technique for stabilization combines both principles. A pentahaptocyclopentadienyl ring (η^5-C$_5$H$_5$, abbreviated as Cp) has six π-electrons available to share with titanium. Biscyclopentadienyltitanium dichloride [1271-19-8] (titanocene dichloride, Cp$_2$TiCl$_2$) melts at 289°C, can be sublimed at 190°C$_{267\ Pa\ (2\ mm\ Hg)}$, and can be recovered almost quantitatively from its solution in boiling dilute hydrochloric acid (pH < 1). The Cp ligand and its substitution products (abbreviated Cp′) can also stabilize otherwise labile Ti–R compounds; thus, Cp$_2$Ti(C$_6$H$_5$)$_2$ [1273-09-2] is unchanged for several days at room temperature.

Thermal stability is enhanced in chelates; thus

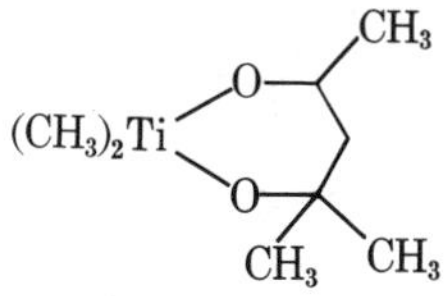

[23916-35-0]

is much more stable than $(CH_3)_3Ti[OCH(CH_3)_2]_2$ (40). The structure of the former has been shown by x-ray diffraction to be dimeric, five-coordinate through oxygen bridges. The more highly substituted the six-membered ring, the more thermally stable the compound.

Covalent Noncyclopentadienyl Compounds. The general synthesis of covalent non-Cp compounds, R_nTiX_{4-n} (where R = alkyl or aryl and X = halogen, alkoxyl, or amido) involves a lithium, sodium, or magnesium organometallic with a titanium–halogen compound in an inert atmosphere; solvents are usually ethers, eg, $(C_2H_5)_2O$, THF, or glyme, or hydrocarbons, eg, hexane or benzene, and a low temperature is required to compensate for the low thermal stability. Schlenk-tube techniques are commonly used. Grignard reagents and alkylaluminum compounds are reducing agents; organoalkalies generally give less reduction. The halides may be TiX_4 or a readily prepared alkoxytitanium halide $X_nTi(OR)_{4-n}$. Since the R–Ti bond is generally broken by protic reagents (HA), an alkoxyl cannot be introduced by alcoholysis after the R–Ti bond forms.

An unusual reaction leading to a Ti–C bond is unrelated to those just discussed. Diphenylketene adds $Ti(OR)_4$ (124):

$$Ti(OR)_4 + n\ (C_6H_5)_2C{=}C{=}O \longrightarrow (RO)_{4-n}Ti(C(C_6H_5)_2\overset{\overset{O}{\|}}{C}OR)_n$$

$n = 1, R = C_2H_5$ [78319-02-5]
$n = 1, R = CH(CH_3)_2$ [78319-03-6]
$n = 2, R = C_2H_5$ [78319-05-8]
$n = 2, R = CH(CH_3)_2$ [78319-06-9]
$n = 1, R = C_6H_5$ [78319-04-7]

Other heterocumulenes, eg, $C_6H_5N{=}C{=}O$, react similarly to yield carbamic esters (55–56).

There are numerous alkyltitaniums, and many of their reactions resemble those of alkyllithiums and alkylmagnesium halides. They are protolyzed by water and alcohols, $R{-}Ti\lessgtr + HA \rightarrow RH + A{-}Ti\lessgtr$, they insert oxygen, $R{-}Ti\lessgtr + O_2 \rightarrow ROTi\lessgtr$, and they add to a carbonyl group:

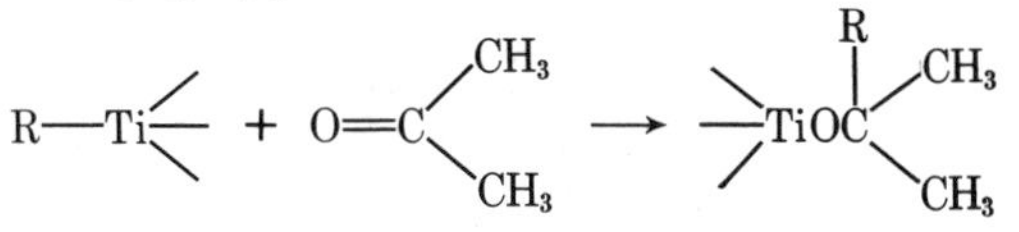

One reaction, disproportionation, had made it difficult to prepare di- and triaryltitanium alkoxides. It is easy to prepare phenyltitanium trialkoxide, $C_6H_5Ti(OR)_3$. However, the reaction

$$2\ C_6H_5Li + Cl_2Ti[OCH(CH_3)_2] \rightarrow (C_6H_5)_2Ti[OCH(CH_3)_2]$$
$$[762\text{-}99\text{-}2] \qquad\qquad [84215\text{-}72\text{-}5]$$

is followed by disproportionation, doubtless promoted by the intermolecular bridging of alkoxyl groups, which creates a molecule differing only slightly from a reasonable transition state for phenyl migration. Thermodynamics takes over and the products are $C_6H_5Ti(OR)_3$ and $(C_6H_5)_4Ti$:

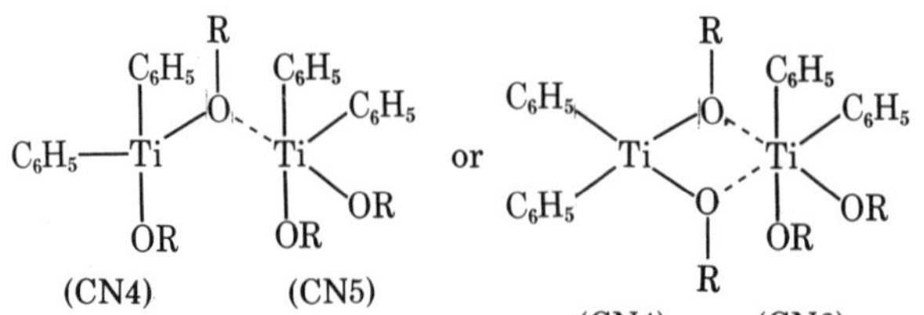

(CN4)   (CN5)   or   (CN4)   (CN6)

The unstable CH_3TiCl_3 [2747-38-8] from $(CH_3)_2Zn + TiCl_4$ forms stable complexes, with donors, eg, $(CH_3)_2NCH_2CH_2N(CH_3)_2$, THF, and sparteine, which methylate carbonyl groups stereoselectively. They give 80% of the isomer shown and 20% of the diastereomer; this is considerably more selective than the more active CH_3MgBr (125). Such complexes or $CH_3Ti(OC_3H_7\text{-}i)_3$ methylate tertiary halides or ethers (126):

$$R_3CCl \rightarrow R_3CCH_3$$
$$R_2CCl_2 \rightarrow R_2C(CH_3)_2$$
$$(CH_3)_3COCH_3 \rightarrow (CH_3)_3CCH_3$$

Such reactions can even be performed with $(CH_3)_2Zn$ and catalytic amounts of $TiCl_4$ at $-78°C$. Gem dialkylation of a ketone has been achieved in a one-pot reaction (127–128):

$$R_2CO + R'MgX \xrightarrow{(HCl)} R_2CR'Cl \xrightarrow[TiCl_4,\ -30°C]{(CH_3)_2Zn} R_2CR'CH_3$$

Grignard reagents and lithium alkyls add to ester groups, but the CH_3Ti reagents do not; this selectivity has synthetic value (129). Titanium alkyls and aryls discriminate between aldehydes and ketones (130). Titanium alkyls minimize side reactions, eg, elimination, rearrangement, and enolization, which often occur with aluminum alkyls and other active organometallics (128).

The solvent is important, as illustrated with mesityl (2,4,6-trimethylphenyl or Mes):

$$MesLi + TiCl_4 \xrightarrow{THF} TiCl_3\cdot3THF + Mes\text{-}H + Mes\text{-}Mes$$

$$\downarrow \text{excess MesLi}$$

$$LiTiMes_4\cdot4THF$$
$$[63916\text{-}94\text{-}1]$$

but

$$Mes_2Zn + 2\ TiCl_4 \xrightarrow{hydrocarbon} 2\ MesTiCl_3 + ZnCl_2$$
$$[77801\text{-}18\text{-}4]$$

$MesTiCl_3$ reacts with THF yielding $TiCl_3$ (131).

Other $RTiX_3$ (X-Hal or OR') compounds are selective at $-20°C$. A chiral R'OH, eg, $S(-)C_2H_5(CH_3)CHCH_2OH$, permits asymmetric synthesis; 98:2 or 99:1 selectivity for aldehyde over ketone has been reported (132–134). The activation energy difference is ca 42 kJ/mol (10 kcal/mol) for RTi, and only 4 kJ/mol (1 kcal/mol) for RLi or RMgX. The kinetic reactivity of $RTi(OR')_3$ decreases in the order:

(for R'O): $n\text{-}C_3H_7O > (CH_3)_2CHO > CH_3CH_2CH(CH_3)O > CH_3CH_2(CH_3)CHCH_2O$

(for R):

The titanium alkyls were prepared simply:

$$TiCl_4 + 3\ TPT \rightarrow 4\ ClTi[OCH(CH_3)_2]_3 \text{ (distillable)}$$
$$[20717\text{-}86\text{-}6]$$

or

$$CH_3\overset{\overset{\displaystyle O}{\|}}{C}Cl + TPT \longrightarrow ClTi[OCH(CH_3)_2]_3 + CH_3\overset{\overset{\displaystyle O}{\|}}{C}OCH(CH_3)_2$$

followed by

$$RLi + ClTi[OCH(CH_3)_2]_3 \rightarrow RTi[OCH(CH_3)_2]_3 + LiCl$$

Titanium alkyls are "tamed Grignard reagents." They do not add to esters, nitriles, epoxides, or nitroalkanes at low temperatures. They add exclusively in a 1,2 fashion to unsaturated aldehydes (132–134).

Tetraneopentyltitanium [36945-13-8], Np_4Ti, forms from the reaction of $TiCl_4$ and neopentyllithium in hexane at $-80°C$ in modest yield only because of extensive reduction of Ti(IV). Tetranorbornyltitanium [36333-76-3] can be prepared similarly. When exposed to oxygen, $(NpO)_4Ti$ forms. If it is boiled in benzene, it decomposes to neopentane. When dissolved in monomers, eg, α-olefins or dienes, styrene, or methyl methacrylate, it initiates a slow polymerization (135–136). Results from copolymerization studies indicate a radical mechanism (136). Ultraviolet light increases the rate of dissociation to radicals. The titanocycle is planar and it may be an intermediate in some forms of olefin metathesis (137).

$$Cl_2Ti\diamondsuit$$
$$[79953\text{-}32\text{-}5]$$

Tetraneopentyltitanium and $[(CH_3)_3SiCH_2]_4Ti$ [33948-28-6] react with nitric oxide yielding compounds with R groups bonded in two different ways, as shown by nmr (138). The reactions of NO with R_3TiCl and R_2TiCl_2 yield similar products with Cl instead of the —ONRNO substituent.

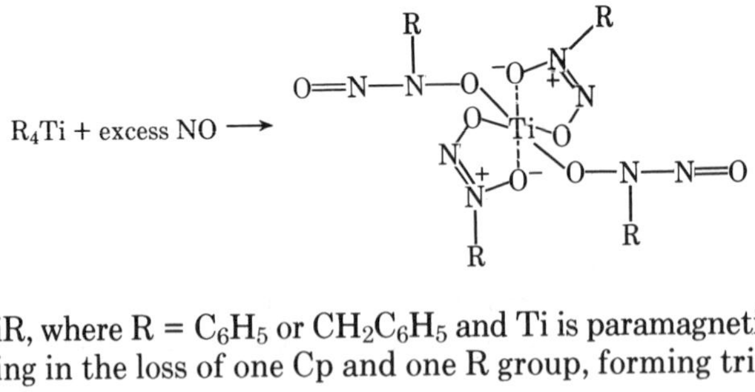

R₄Ti + excess NO ⟶

However, Cp_2TiR, where $R = C_6H_5$ or $CH_2C_6H_5$ and Ti is paramagnetic Ti(III), reacts with NO resulting in the loss of one Cp and one R group, forming trinuclear:

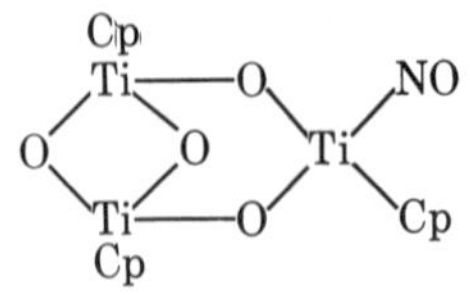

$$[76722\text{-}03\text{-}7]$$

Carbometalation is an important reaction of RTi(IV) compounds; RTi$\leqslant$ adds to a C=C or C≡C multiple bond and results in a net R–H addition. Carbometalation is involved in Ziegler-Natta polymerization:

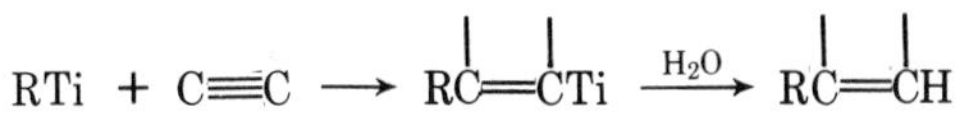

$$RTi + C\equiv C \longrightarrow R\overset{|}{C}=\overset{|}{\underset{|}{C}}Ti \xrightarrow{H_2O} R\overset{|}{C}=\overset{|}{\underset{|}{C}}H$$

In the following cases, only those reactions in which there is no chain growth, or at most dimerization, are considered (see Olefin polymers).

Alkyltitanium halides can be prepared from alkylaluminum derivatives. The ring structure imparts regiospecificity to the ensuing carbometalation (139):

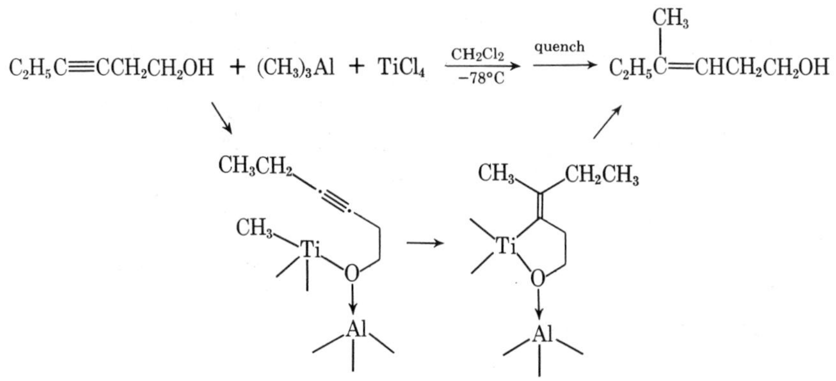

The metallocycle (8) undergoes an apparent β-elimination to a carbenelike reagent, which adds regiospecifically to terminal acetylenes (140):

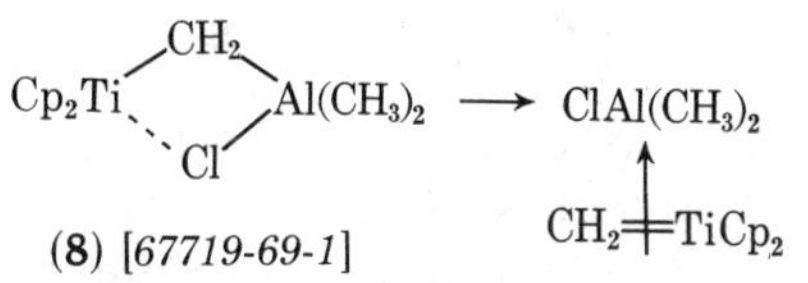

(8) [67719-69-1]

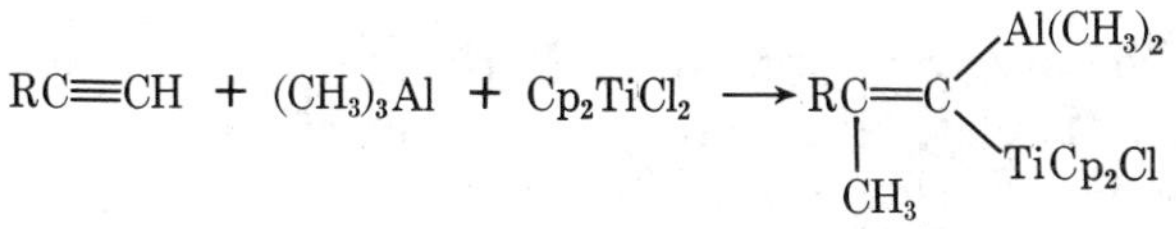

$$RC\equiv CH + (CH_3)_3Al + Cp_2TiCl_2 \longrightarrow RC=C\begin{matrix} Al(CH_3)_2 \\ \\ TiCp_2Cl \end{matrix}$$

(9)

Intermediate (9) reacts with ketones (R$_2'$CO) forming cumulenes RC(CH$_3$)=C=CR$_2'$. The indenyl derivative reacts similarly.

Olefin isomerization is often catalyzed by titanium. An example is the conversion of vinylnorbornene to the comonomer ethylidenenorbornene (141). The catalyst is a mixture of a sodium suspension, AlCl$_3$, and (RO)$_4$Ti or Cp$_2$TiCl$_2$. Although isomerization is slow, the yield is high. The active reagent is doubtless a Ti(III) compound.

$$\text{(structures)} \qquad \overset{\text{CH}=\text{CH}_2}{} \qquad\qquad \overset{\text{CHCH}_3}{}$$

Cyclopentadienyltitanium Compounds (Tetravalent). *Properties.* The structure of Cp_2TiCl_2 has been shown by x-ray diffraction to be a distorted tetrahedron: Ti–Cl = 0.2364 nm, Ti–centroid = 0.2058 nm, C–C = 0.1339–0.1419 nm (142–143). It has also been studied by electron diffraction (144). Changes in the structure are imposed by bridging the Cp rings with —$(CH_2)_n$— or other groups (145). In $Cp_2Ti(C_6H_5)_2$, the C–C bonds in the Cp rings have different lengths, and the same is true of those in the phenyl rings (146). Chemical ionization mass spectrometry of Cp_2TiCl_2 gives mainly $M^+(-Cl)$ (147–148); ^{13}C-nmr results are discussed in ref. 149. The rings of Cp_2TiCl_2 spin rapidly, so the ^1H-nmr spectrum shows no split; bridging the rings prevents spinning and the splitting can be detected (150–151). Electron spectroscopy for Cp_2TiX_2, where X = halogen, and other CpTi(IV) compounds has been conducted (152–154). Infrared and Raman spectra are discussed in refs. 155–158, and the photoelectron spectrum has been determined (159). Molecular orbital calculations have been reported for many Cp–Ti compounds (159–167). Vapor-pressure equations for Cp_2TiCl_2 and $Ti(OC_4H_9)_4$ have been given, and other thermodynamic quantities have been reported (168–169). A selected group of titanium(IV) cyclopentadienyls and biscyclopentadienyls is shown in Tables 5 and 6, respectively.

Titanium tetrachloride is a Lewis acid with many useful properties for the organic chemist (170). Replacing Cl by OR weakens the acid. Acid-base complexes with dimethyl sulfoxide (DMSO), DMF,

$$\overset{\displaystyle O}{\underset{\displaystyle CH_3\overset{\|}{C}NHCH_3}{}}$$

(all O-bonded), pyridine, and ethylenediamine have been reported (171). Although $CpTiCl_3$ forms stable sublimable complexes with ditertiary amines or arsines, eg, $(CH_3)_2ECH_2CH_2E(CH_3)_2$, where E = N or As, the less-basic analogues where E = O, S, or P, do not form stable complexes. The monodentates pyridine, $(CH_3)_2S$, $(CH_3)_3As$, and $(C_6H_5)_3P$ also do not form complexes. The bromide and iodide behave similarly (172). Biscyclopentadienyltitanium dichloride is virtually devoid of Lewis acidity. The properties of ring-substituted Cp derivatives differ from those of the parent Cp derivatives in a manner predictable from the electronic properties of the substituent.

Synthesis. The discovery of stable cyclopentadienyltitanium compounds, in particular Cp_2TiCl_2, stimulated the synthesis and study of a host of related compounds. These include both $CpTiX_3$ and Cp_2TiX_2.

The basic laboratory synthesis involves a salt of cyclopentadiene, eg, lithium, sodium, potassium, or magnesium, and a titanium(IV) halide, usually in an ether solvent (173–175). However, this is probably too expensive for industrial use. Different cyclopentadienyl groups can be introduced in two separate steps (176). A USSR patent shows that merely heating CpH and $TiCl_4$ in dioxane at 60–80°C in the presence of diethylamine yields Cp_2TiCl_2 (177). Titanium alkoxides react with CpMgBr yielding $CpTiBr(OR)_2$ or Cp_2TiBr_2, depending on the mole ratio (178). Another patent describes the reaction (179):

$$CpH + Cl_2Ti(OR)_2 + (C_2H_5)_3N \text{ (or piperidine)} \rightarrow CpTi(OR)_2Cl$$

(80% yield, distillable)

However, no additional amine is necessary for the following reaction (80):

$$n \text{ CpH} + \text{Ti(NR}_2)_4 \rightarrow \text{Cp}_n\text{Ti(NR}_2)_{4-n} \qquad n = 1, 2$$

Should a real demand develop for Cp_2TiCl_2, new manufacturing processes will be required.

For laboratory use, cyclopentadienylthallium [34822-90-7] reacts cleanly with TiX compounds (X = halogens) (180–186). The cost and toxicity of thallium compounds are drawbacks to its large-scale use. Cp_2Pb may, in certain cases, be useful (187). Monocyclopentadienyl compounds can be prepared by the above techniques with appropriate control of the stoichiometric proportion of reagents or by use of reagents like $ClTi(OR)_3$ (188). One technique involves redistribution, which may not be convenient nor produce high yields (175–176,188):

$$\text{Cp}_2\text{TiCl}_2 + \text{TiCl}_4 \rightarrow 2 \text{ CpTiCl}_3$$
$$[1270\text{-}98\text{-}0]$$

Biscyclopentadienyltitanium dichloride can be carefully chlorinolyzed with Cl_2 or with SO_2Cl_2 in refluxing $SOCl_2$ (176,189). It is inert to Br_2 (176). An alternative method has been reported to give high purity $CpTiCl_3$ (172):

$$\text{CpNa} + \text{ClSi(CH}_3)_3 \rightarrow \text{CpSi(CH}_3)_3 \xrightarrow{\text{TiCl}_4} \text{CpTiCl}_3 + \text{ClSi(CH}_3)_3$$
$$(92\%)$$

A $(CH_3)_3SiCpTi$ derivative can be made as follows:

$$\text{CpSi(CH}_3)_3 \xrightarrow[\text{(2) ClSi(CH}_3)_3]{\text{(1) Na}} \text{C}_5\text{H}_4[\text{Si(CH}_3)_3]_2 \xrightarrow{\text{TiCl}_4} (\text{CH}_3)_3\text{SiCpTiCl}_3 \ (82\%)$$
$$[7173\text{-}68\text{-}3]$$

The corresponding bromides and iodides are available by the same chemistry with $TiBr_4$ or TiI_4 (172).

A wide variety of ring-substituted cyclopentadienes have been converted to Cp'–$Ti\lessgtr$ compounds. Methyl and other alkyl and aryl groups and $(CH_3)_3Si$— are most common; pentamethylcyclopentadiene is a popular, although expensive, ligand (190–191). Bridged bis(cyclopentadiene)s, eg, $Cp(CH_2)_nCp$ and $CpSi(CH_3)_2Cp$, have received recent attention (150,192). The properties of derivatives have been compared with various length bridges (193–194). Indene and, to a very limited extent, fluorene function as ligands.

Cyclopentadienyltitanium Compounds with Other Carbon–Titanium Links. Cyclopentadienyltitanium trichloride and, particularly, Cp_2TiCl_2 react with RLi or with $RAl\lessgtr$ compounds to form one or more R–Ti bonds. As noted, the Cp groups stabilize the Ti–R bond considerably against thermal decomposition, although the sensitivity to air and moisture remains. Depending on the temperature, mol ratio, and structure of R, reduction of Ti(IV) may be a serious side reaction which often has preparative value for $Cp_n Ti(III)$ compounds (190,195–196).

Methyl and aryl groups are most commonly used; higher alkyltitaniums tend to decompose by β-elimination except where R = $CH_2Si(CH_3)_3$ or $CH_2C(CH_3)_3$ (197). Ring compounds or titanocycles have been made from Cp_2TiCl_2 and 1,n-dilithioalkanes ($Li\text{+}CH_2\text{+}_n Li$) at $-78°C$; they are quite thermolabile, though more stable than

Table 5. Organotitanium(IV) Compounds—Sandwich Structures—Cyclopentadienyls

Compound	CAS Registry No.	X in $C_5H_5TiX_3$	Appearance	Mp, °C	Bp, °C$_{Pa}^{a}$	Other properties
$C_5H_5TiCl_3$	[1270-98-0]	halide	orange	140–142		on hydrolysis gives $(C_5H_5TiClO)_n$
$C_5H_5TiBr_3$	[12240-42-5]		orange	163–165		
$C_5H_5TiI_3$	[12240-43-6]		deep red	185–190		
$C_5H_5TiBrCl_2$	[70568-74-0]		orange	165–170		
$C_5H_5Ti(OCH_3)_3$		alkoxy		50–52	88_{200}	readily hydrolyzes; darkens on storage
$C_5H_5Ti(OC_2H_5)_3$	[1282-41-3]				$106–107_{400}$	very sensitive to moisture; monomeric in benzene
$C_5H_5Ti(OC_3H_7)_3$	[12242-59-0]				$106–107_{67}$	very sensitive to moisture; monomeric in benzene
$C_5H_5Ti(OC_4H_9)_3$	[84215-74-7]				$124.5–125.5_{67–133}$	n_D^{20}, 1.5224
$C_5H_5Ti[OC(CH_3)_3]_3$	[12148-33-3]				102_{133}	n_D^{20}, 1.5065
$C_5H_5Ti(OC_6H_{13})_3$	[12290-79-8]				$177–181_{133}$	n_D^{20}, 1.5082
$C_5H_5Ti(OCH_3)Cl_2$	[12192-52-8]	alkoxyhalide	yellow crystals	93–96		
$C_5H_5Ti(OC_2H_5)Cl_2$	[1282-32-2]			49		
$C_5H_5Ti(OC_3H_7)Cl_2$	[70046-23-0]		yellow–green		$159–161_{267}$ dec	
$C_5H_5Ti(OCH_3)_2Cl$	[84215-75-8]					
$C_5H_5Ti(OC_2H_5)_2Cl$	[1282-38-8]		yellow–green		$109–111_{133}$	
$C_5H_5Ti(OC_3H_7)_2Cl$	[84215-76-9]				$132–145_{133}$ dec	
$C_5H_5Ti(OC_4H_9)_2Cl$	[84215-77-0]				$145–150_{267–400}$	n_D^{20}, 1.5818
$C_5H_5Ti(OC_4H_9)_2Br$	[84215-78-1]				$36–45_{107}$	
$C_5H_5Ti(OC_6H_5)Cl_2$	[12288-59-4]	aryloxyhalide				very sensitive to moisture

Compound	CAS No.	Type	m.p./b.p.	Color	Physical constants	Remarks
$C_5H_5Ti(OC_2H_5)_2O\overset{O}{\overset{\|}{C}}CH_3$	[84215-79-2]	alkoxyacetate	$106-108_{267}$			on heating, $C_5H_5Ti(OC_2H_5)_3$ is formed
$C_5H_5Ti_5(O\overset{O}{\overset{\|}{C}}CH_3)_3$	[1282-42-4]	acetate	$115-117$			hydrolytically and thermally unstable
$C_5H_5Ti(SCH_3)_3$	[84215-80-5]	mercapto				
$C_5H_5Ti(SCH_3)_2Cl$	[84215-81-6]	mercaptohalide				
$C_5H_5Ti[OSi(CH_3)_3]_3$	[57665-25-5]	siloxy	$138-139_{133}$		n_D^{20}, 1.4582 d_4^{20}, 0.9436 g/cm^3	
$CH_3C_5H_4TiCl_3$	[1282-31-1]	halide (substituted cyclopentadiene)	$98-99$			
$C_2H_5C_5H_4TiCl_3$	[1282-33-3]		136_{133} sub	orange solid		crystals liquefy in air
$(CH_3)_5C_5TiCl_3$	[12129-06-5]		$225-227$	red solid green		
$CH_3C_5H_4Ti(OC_2H_5)Cl_2$	[1282-39-9]	alkoxyhalide (substituted cyclopentadiene)	n_D^{20}, 1.5730			viscous mass
$CH_3C_5H_4Ti(OC_2H_5)_2Cl$	[1282-36-6]	alkoxy (substituted cyclopentadiene)	$143-145_{267}$			easily hydrolyzed in air
$CH_3C_5H_4Ti(OC_2H_5)_3$	[1282-47-9]		$80-81_{133}$		n_D^{20}, 1.5401 d^{20}, 1.0780 g/cm^3	
$C_2H_5C_5H_4Ti((OC_2H_5)_3$	[1282-46-8]		$101-102_{267}$		n_D^{20}, 1.5359 d^{20}, 1.0717 g/cm^3	easily hydrolyzed; darkens on storage even at −50°C

[a] To convert Pa to mm Hg, divide by 133.

Table 6. Organotitanium(IV) Compounds—Sandwich Structures—Biscyclopentadienyls

Compound	CAS Registry No.	X in $(C_5H_5)_2TiX_2$	Appearance	Mp, °C	Other properties
$(C_5H_5)_2TiCl_2$	[1271-19-8]	halide	vivid red crystals	289	slightly soluble in water; forms salts with acids
$(C_5H_5)_2TiBr_2$	[1293-73-8]		vivid red crystals	314	diamagnetic, soluble in nonpolar solvents
$(C_5H_5)_2TiI_2$	[12152-92-0]		purple crystals	319	
$(C_5H_5)_2TiF_2$	[309-89-7]		yellow		
$(C_5H_5)_2Ti(OC_4H_9)_2$	[12303-65-0]	alkoxy			
$(C_5H_5)_2Ti(OC_6H_5)_2$	[12246-19-4]	aryloxy	yellow	142	thermally stable; hydrolyzed only by heating with conc NaOH
$(C_5H_5)_2Ti(o\text{-}ClC_6H_4O)_2$	[12309-06-7]		orange–yellow	145–147	thermally stable; hydrolyzed only by heating with conc NaOH
$(C_5H_5)_2Ti(o\text{-}CH_3C_6H_4O)_2$	[12309-37-4]		yellow	162	thermally stable; hydrolyzed only by heating with conc NaOH
$(C_5H_5)_2Ti(o\text{-}NO_2C_6H_4O)_2$	[12309-11-4]		red	122–124	thermally stable; hydrolyzed only by heating with conc NaOH
$(C_5H_5)_2Ti(p\text{-}ClC_6H_4O)_2$	[12309-07-8]		yellow	125–127	thermally stable; hydrolyzed only by heating with conc NaOH
$(C_5H_5)_2Ti(OC_2H_5)Cl$	[12129-76-9]	alkoxyhalide		91–92	
$(C_5H_5)_2Ti(OC_3H_7)Cl$	[12715-66-1]			57–58	
$(C_5H_5)_2Ti(OC_6H_5)Cl$	[62652-01-1]	aryloxyhalide		71–73	
$(C_5H_5)_2Ti(SH)_2$	[12170-34-2]	mercapto		150–160 (dec)	stable to air and water
$(C_5H_5)_2Ti(SCH_3)_2$	[12089-78-0]	alkylmercapto	deep red	193–197	
$(C_5H_5)_2Ti(SC_2H_5)_2$	[1291-79-8]			107–110	
$(C_5H_5)_2Ti(SC_3H_7)_2$	[1292-07-5]			88–93	
$(C_5H_5)_2Ti(SC_6H_5)_2$	[1292-72-4]	arylmercapto		199–201	
$(C_5H_5)_2Ti(SCH_2C_6H_5)_2$	[1292-61-1]			172–174	
$(C_5H_5)_2Ti(SCH_2CH_2C_6H_5)_2$	[1292-47-3]			92–94	
$(C_5H_5)_2Ti(\overset{O}{\overset{\|}{C}}CH_3)_2$	[1282-51-5]	acyl	orange	126–128	
$(C_5H_5)_2Ti(\overset{O}{\overset{\|}{C}}C_3H_7)_2$	[12290-20-9]		red–orange	114–116	readily hydrolyzed and thermally unstable
$(C_5H_5)_2Ti(\overset{O}{\overset{\|}{C}}CH_2Cl)_2$	[1282-44-6]		red–orange	98–99	readily hydrolyzed and thermally unstable

Formula	CAS No.	Type	Color	M.p. (°C)	Remarks
$(C_5H_5)_2Ti(O\overset{O}{C}CCl_3)_2$	[12212-37-2]		red–orange	192–194	readily hydrolyzed and thermally unstable
$(C_5H_5)_2Ti(O\overset{O}{C}CF_3)_2$	[1282-45-7]		orange	178–180	soluble in benzene, ethyl acetate, ethanol; moderately sol in chloroform; thermally stable
$(C_5H_5)_2Ti(O\overset{O}{C}C_6H_5)_2$	[51178-00-8]	benzoyl	yellow	188	reacts with aqueous alkali to give benzoic acid; n_D^{20} 1.4582; d_4^{20}, 0.9436 g/cm^3
$(C_5H_5)_2Ti[OSi(CH_3)_3]_2$	[12319-01-6]	siloxy			stable in air but not in acidic or basic media
$(C_5H_5)_2Ti(OSi(C_6H_5)_3)_2$	[12321-33-4]				
$(C_5H_5)_2Ti(OSi(CH_3)_3)Cl$	[12319-01-6]	siloxyhalide	orange	137.5	stable in air but not in acidic or basic media
$(C_5H_5)_2Ti(OSi(C_6H_5)_3)Cl$	[12320-99-9]			210–212	
$(C_5H_5)_2Ti(CO)_2$	[12129-51-0]	carbonyl	dark reddish-brown crystals	dec ca 90	extremely air sensitive
$(C_5H_5)_2Ti(CH_3)_2$	[1271-66-5]	alkyl	yellow–orange	dec ca 100	
$(C_5H_5)_2Ti(CH_3)Cl$	[1278-83-7]	alkylhalide	orange–red	168–170 dec	
$(C_5H_5)_2Ti(C_3H_7)Cl$	[12715-66-1]		orange	160	
$(C_5H_5)_2Ti(C_6H_5)_2$	[1273-09-2]	aryl	orange–yellow	146–148 dec	
$(C_5H_5)_2Ti(m\text{-}CH_3C_6H_4)_2$	[12156-57-9]		red	137–139	
$(C_5H_5)_2Ti(p\text{-}CH_3C_6H_4)_2$	[12156-58-0]		yellow–orange	133–134	
$(C_5H_5)_2Ti[p\text{-}[(CH_3)_2NC_6H_4]_2$	[12156-86-4]		maroon	137–139 dec	
$(C_5H_5)_2Ti(p\text{-}FC_6H_4)_2$	[12155-98-5]		orange	120 dec	
$(C_5H_5)_2Ti(p\text{-}ClC_6H_4)_2$	[12155-97-4]		orange	130 dec	
$(C_5H_5)_2Ti(p\text{-}BrC_6H_4)_2$	[12155-95-2]		orange	130 dec	
$(C_5H_5)_2Ti[m\text{-}CF_3C_6H_4]_2$	[12156-38-6]		orange–yellow	145–146	
$(C_5H_5)_2Ti(p\text{-}CF_3C_6H_4)_2$	[12156-39-7]			142–143	
$(C_5H_5)_2Ti(C{\equiv}CC_6H_5)_2$	[12303-93-4]		orange–brown crystals	141	
$(C_5H_5)_2Ti(C_6F_5)_2$	[12155-89-4]	fluorophenyl	orange needles	228–230	thermally stable in vacuo at 110°C
$(C_5H_5)_2TiCl(C_6F_5)$	[50648-18-5]		pale–orange needles	201–203	
$(C_5H_5)_2Ti(OC_2H_5)(C_6F_5)$	[84215-82-7]		yellow solid	117	stable in dry air; soluble in organic solvents
$(C_5H_5)_2Ti(OH)(C_6F_5)$	[84215-83-8]		yellow solid	183–185 dec	
$(C_5H_5)_2Ti(C_6F_5)F$	[84501-83-7]		yellow solid	240 dec	soluble in organic solvents

213

$Cp_2Ti(C_4H_9)_2$ [52124-69-3; 71297-31-9]. In the compound where $n = 4$, β-elimination is suppressed because the Ti–C–C–H dihedral angle is far from 0°. The compound inserts carbon monoxide yielding a titanoketone, which expels Ti at 25°C. These reactions do not occur when $n = 5$:

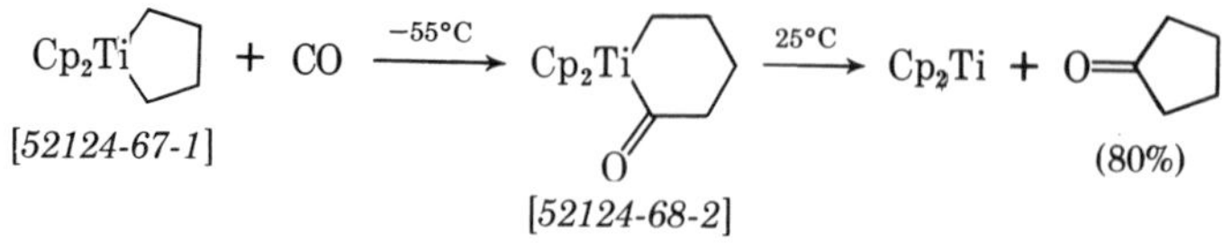

[52124-67-1]   [52124-68-2]   (80%)

Other titanocene synthons may undergo similar reactions with CO (198–199). 2,2'-Dilithiobiphenyl closes a ring with Cp_2TiCl_2, but the crowded, noncoplanar 2,3,6,2',3',6'-hexachloro-3,3'-dilithio-4,4'-bipyridyl reacts at only one Li atom (200):

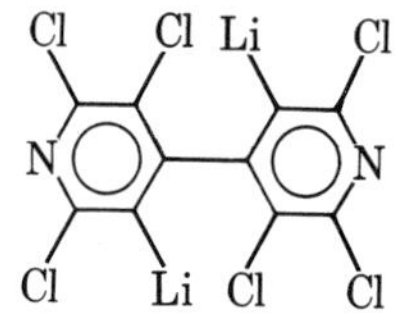

The ring can be closed with SCl_2, however.

Both vinylic and ethynylic groups can be attached to the Cp_2Ti framework; they tend to be stable thermally and to air and moisture (201).

The following rather air-stable titanocycle has been prepared in meso form only (202):

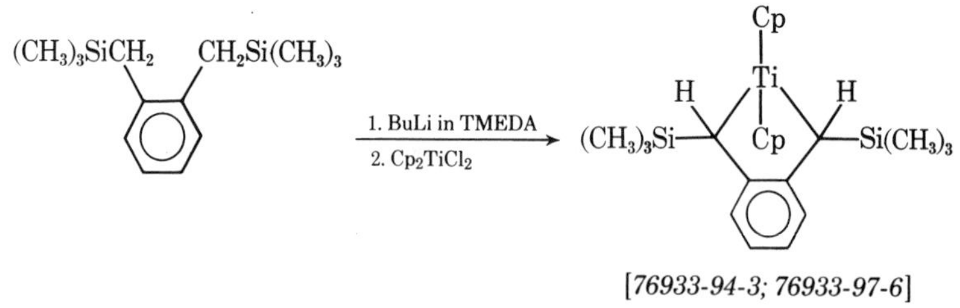

[76933-94-3; 76933-97-6]

Related examples of silicon-containing titanocycles are prepared as shown below (203):

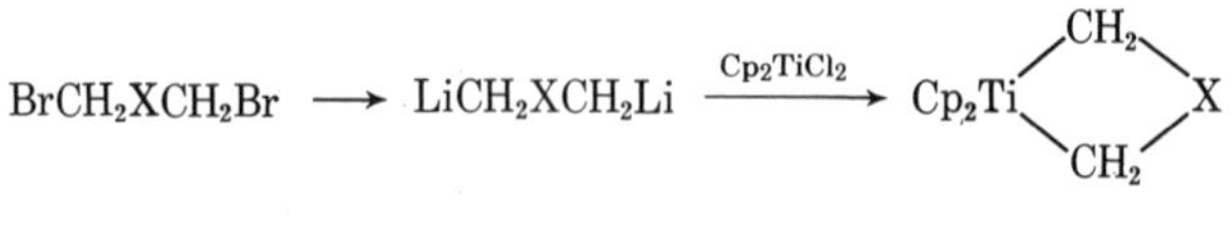

In the above titanium product:

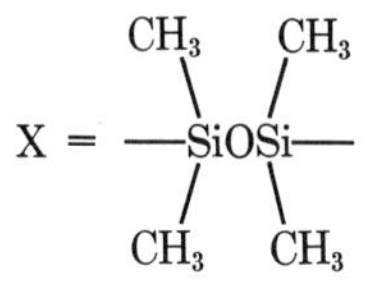

$$X = \begin{array}{c} CH_3 \quad CH_3 \\ | \quad | \\ -SiOSi- \\ | \quad | \\ CH_3 \quad CH_3 \end{array}$$

[65585-81-1; 71297-30-8]

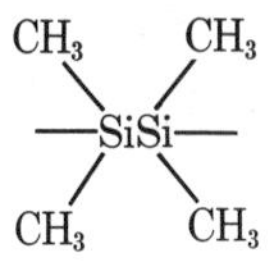

$$\begin{array}{c} CH_3 \quad CH_3 \\ | \quad | \\ -SiSi- \\ | \quad | \\ CH_3 \quad CH_3 \end{array}$$

[65585-83-3]

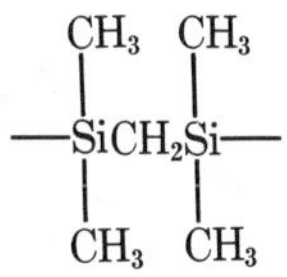

$$\begin{array}{c} CH_3 \quad CH_3 \\ | \quad | \\ -SiCH_2Si- \\ | \quad | \\ CH_3 \quad CH_3 \end{array}$$

[65585-80-0]

$$\begin{array}{c} CH_3 \quad CH_3 \quad CH_3 \\ | \quad | \quad | \\ -SiSiSi- \\ | \quad | \quad | \\ CH_3 \quad CH_3 \quad CH_3 \end{array}$$

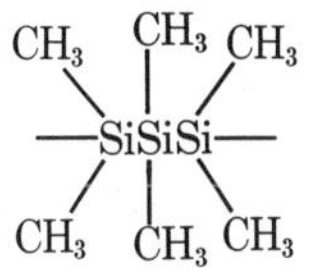

[65585-82-2]

These compounds are air-stable and are not decomposed by methanol.

When CpCp′TiClOAr reacts with Grignard reagents, the aryloxy group is replaced ·by R, not Cl (205). Compounds of the formula $(Cp_2TiR)_2O$ prepared from $(Cp_2TiCl)_2O$ + RLi (R = CH_3, C_2H_5, C_6H_5, p-tolyl, $C_6H_5C{\equiv}C$, $CH_2{=}CH$) are thermally stable and quite stable to air (205–206).

Alkyl and aryl groups are cleaved by iodine, but Cp groups are not affected. The following reaction

$$Cp_2TiR_2 + 2\,I_2 \rightarrow Cp_2TiI_2 + 2\,RI$$
[12152-92-0]

has been proposed for quantitative analysis of such compounds (207).

Carbometalation of olefins and acetylenes is a useful reaction; an example follows:

$$Cp_2TiCl_2 \;+\; R_nAlCl_{3-n} \longrightarrow Cp_2TiRCl.AlR_{n-1}Cl_{4-n} \longrightarrow Cp_2TiR^+$$

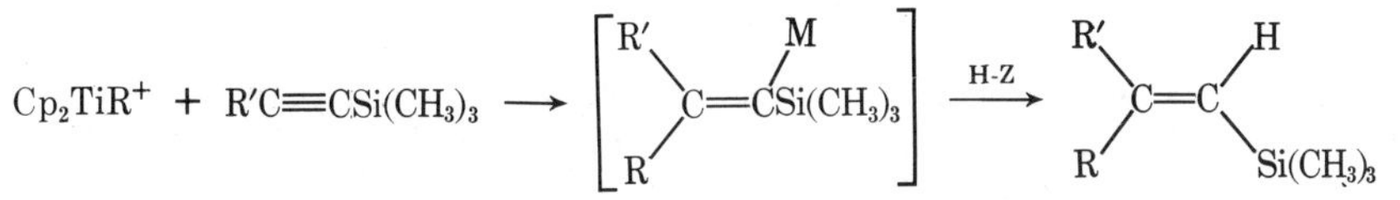

$$Cp_2TiR^+ \;+\; R'C{\equiv}CSi(CH_3)_3 \longrightarrow \left[\begin{array}{c} R' \qquad M \\ \diagdown \quad \diagup \\ C{=}CSi(CH_3)_3 \\ \diagup \\ R \end{array}\right] \xrightarrow{\text{H-Z}} \begin{array}{c} R' \qquad H \\ \diagdown \quad \diagup \\ C{=}C \\ \diagup \qquad \diagdown \\ R \qquad Si(CH_3)_3 \end{array}$$

up to 90% trans addition where M is a complex of Ti and Al. The hydrogen in the product is acquired from the solvent or from elsewhere in the reaction mixture, because if the mixture is quenched with $(C_2H_5)_3N$ followed by D_2O, no deuterium occurs in the olefin. If R′ in the acetylene is phenyl or 1-cyclohexenyl, equal amounts of cis and trans products are obtained. If the unsilylated parent R′C≡CH is exposed to this catalyst system, it is cyclotrimerized to the 1,3,5-$C_6H_3R'_3$. However, this system does not polymerize propylene (208–209).

Other authors disagree with the results of the previous carbometalation sequence. They find that the initial product results strictly from cis-carbometalation; but with a trace of base, H is abstracted from the medium homolytically with loss of stereochemistry (210). They show that M in the above structure must be Ti, not Al. If the reaction mixture is quenched with D_2O containing NaOD, 95 mol % D is incorporated in the olefin.

Alkynols are ethylated by $(C_2H_5)_2AlCl$ catalyzed by Cp_2TiCl_2 (211); $(CH_3Cp)_2TiCl_2$ [1282-40-2] is sometimes preferred because it is more soluble in nonpolar solvents. Ten-to-fifty percent of the titanium compound is required, because many alkynols rapidly deactivate the titanium. In one example, when a prereacted mixture of $(C_2H_5)_2AlCl$ and $HOCH_2CH_2C{\equiv}CH$ is treated with Cp_2TiCl_2, a 1:1 mixture of

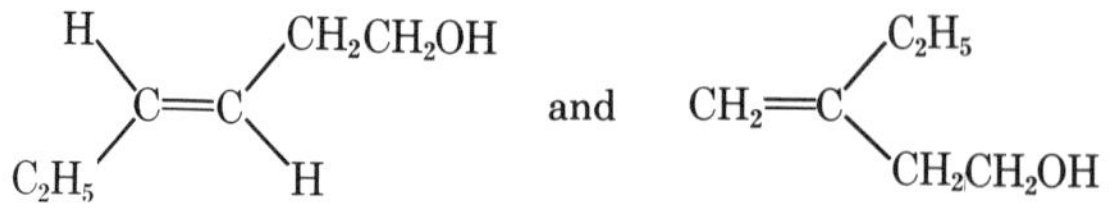

forms in only 55% yield. Reducing the amount of Cp_2TiCl_2 to 10 mol % raises the yield to 80–90%.

The corresponding olefin also is carbometalated by ethylaluminum–titanium combinations. There are marked differences among the titanate esters of the unsaturated alcohol $HOCH_2CH_2CH{=}CH_2$ prepared with $TiCl_4$, $Cl_2Ti(acac)_2$, or Cp_2TiCl_2 (212):

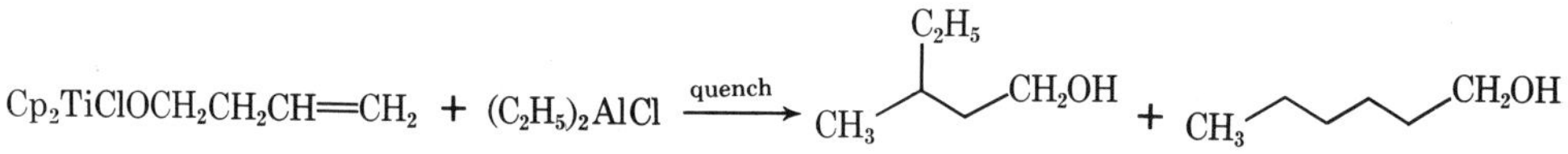

The other two derivatives give chiefly the terminally alkylated product. The mechanism involves a coordination of the double bond to the ethyltitanium species; the shift of the ethyl group to one side or the other of the double bond is influenced by the substituents on Ti.

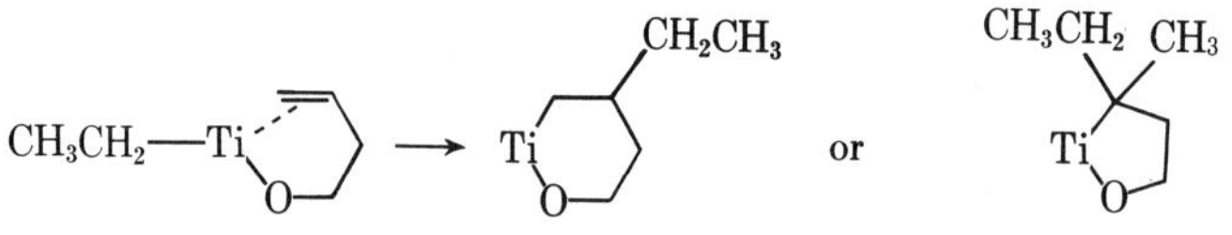

Olefin dimerization involves carbometalation. High selectivity (98%) for the ethylene-to-1-butene dimerization is achieved by adding 10 mol % of $Ti(OC_4H_9)_4$ and 0.5 mol % of Cp_2TiCl_2 to the $(C_2H_5)_3Al$ catalyst (213).

Photolysis has been intensively studied. For example, $Cp_2Ti(C_6H_5)_2$ yields a green polymer $(Cp_2TiH)_x$ [11136-22-4]. At low temperature in benzene or THF, a dark-green transient Cp_2Ti-solvent species forms and quickly dimerizes. The phenyl groups appear as benzene and biphenyl. In the presence of CO or $C_6H_5C{\equiv}CC_6H_5$, $Cp_2Ti(CO)_2$ [12159-51-0] or

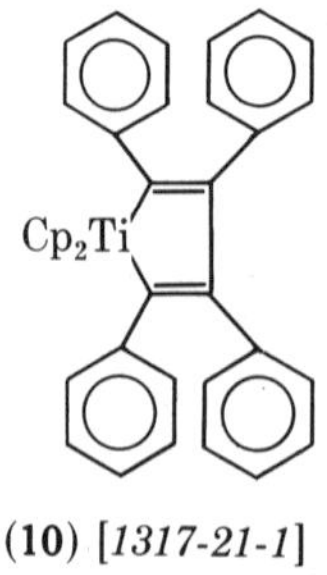

(10) [1317-21-1]

form (214–215). However, in the presence of carbon dioxide, the following compound did not form:

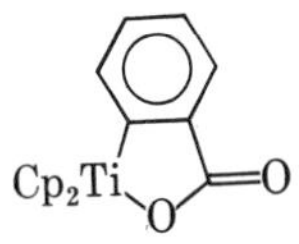

[*11105-82-1*]

Photolysis of Cp_2TiAr_2 in benzene solution yields titanocene and a variety of aryl products derived both intra- and intermolecularly (215). Aryl radicals attack the solvent (215–219):

$$Cp_2Ti(C_6H_5)_2 \xrightarrow{h\nu,\ C_6D_6} (C_6H_5)_2 + C_6H_5C_6D_5$$

[*1273-09-2*]   (yields are in 1:1 ratio)

$$Cp_2TiAr_2 \xrightarrow{h\nu,\ C_6H_6} ArC_6H_5 + ArH + ArAr$$

Dimethyltitanocene photolyzed in hydrocarbons yields methane, but the hydrogen is derived from the other methyl group and from the cyclopentadienyl rings, as demonstrated by deuteration. Photolysis in the presence of diphenylacetylene yields the dimeric titanocycle (**10**) and a titanomethylation product:

$$Cp_2Ti(CH_3)_2 \ + \ C_6H_5C\equiv CC_6H_5 \ \xrightarrow{h\nu} \ (\mathbf{10}) \ + \ \ \ \underset{[65090\text{-}11\text{-}1]}{\ }$$

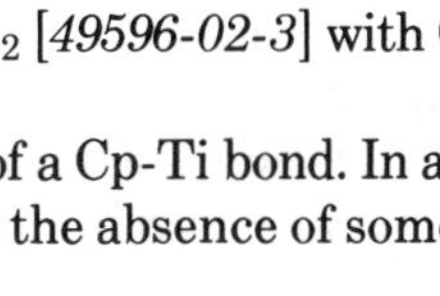

[*65090-11-1*]

The fluorinated titanocycle related to (**10**) is not obtained from $C_6H_5C\equiv CC_6F_5$.

Photolysis of the indenyl analogue $(Ind)_2Ti(CH_3)_2$ [*49596-02-3*] with CO similarly yields $(Ind)_2Ti(CO)_2$ [*56770-60-6*] (220).

Photolysis of Cp_2TiX_2 always gives first scission of a Cp-Ti bond. In a chlorinated solvent, the place vacated by Cp is assumed by Cl. In the absence of some donor, the radical dimerizes (221).

$$Cp_2TiX_2 \xrightarrow[CCl_4]{h\nu} CpTiX_2Cl \qquad Cp = C_5H_5 \text{ or } C_5(CH_3)_5; \ X = Cl, Br, OR, \text{ or } CH_3$$

$$Cp_2TiX_2 \xrightarrow{h\nu} CpTiX_2 \cdot \rightarrow (CpTiX_2)_2$$

When $Cp_2Ti(CH_3)_2$ is photolyzed in toluene, 2 mol CH_4 is produced; but in monomers only 1 mol (1.08 mol in styrene, 0.90 mol in methyl methacrylate) CH_4 is liberated as the monomers polymerize (222). $Cp_2Ti(CD_3)_2$ [*65554-67-8*] photolyzed in $C_6D_5CD_3$ gives CD_3H but not CD_4, thus ruling out free CD_3. In methyl methacrylate, $Cp_2Ti(^{14}CH_3)_2$ yields a polymer containing 0.8–1.1 ^{14}C per polymer chain, but no Ti. If tritiated $(Cp\text{-}T)_2Ti(CH_3)_2$ is used in methyl methacrylate, the polymer contains only traces of tritium. Clearly, the hydrogen for the methane comes from the Cp groups.

Pyrolysis of solid $Cp_2Ti(CD_3)_2$ yields CD_3H but not CD_4. Pyrolysis of $(C_5D_5)_2Ti(CH_3)_2$ yields CH_3D. These results show that the radical attacks the Cp rings (223–224). Pyrolysis of $Cp_2Ti(C_6H_5)_2$ proceeds via a benzyne intermediate, as shown by trapping experiments involving cycloadditions (215,225–228):

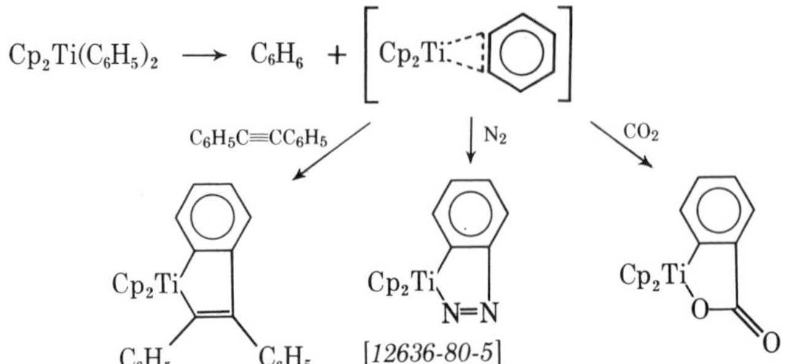

The detailed mechanism of pyrolysis of Cp_2TiR_2 compounds has been studied recently (229–235).

A useful titanocycle is formed from Cp_2TiCl_2 and trimethylaluminum; triethylaluminum gives a different product (236). The titanocycle adds to terminal olefins in the presence of 4-dimethylaminopyridine; the adduct expels olefin above 0°C to yield a bistitanocyclobutane in 85.9% yield (237):

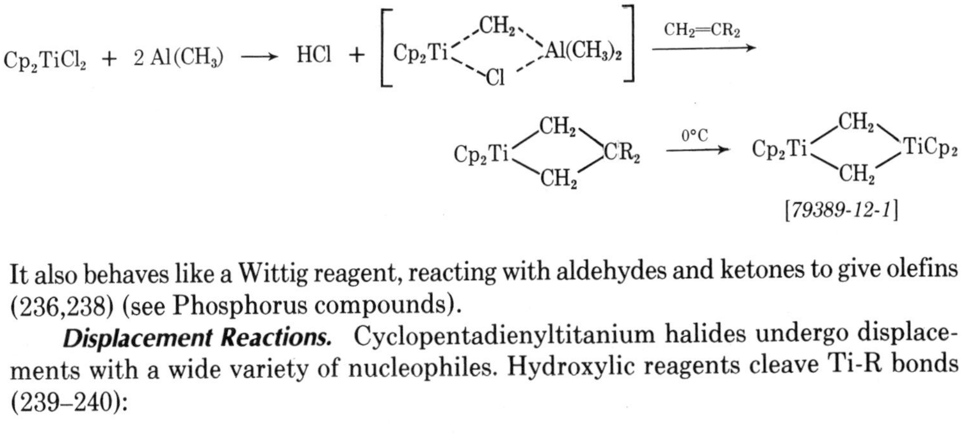

It also behaves like a Wittig reagent, reacting with aldehydes and ketones to give olefins (236,238) (see Phosphorus compounds).

Displacement Reactions. Cyclopentadienyltitanium halides undergo displacements with a wide variety of nucleophiles. Hydroxylic reagents cleave Ti-R bonds (239–240):

$$Cp_2TiCH_3Cl + H_2O \rightarrow CH_4 + (Cp_2TiCl)_2O$$
$$[12118\text{-}17\text{-}1]$$

Amides are formed with amines, often with strong base assistance (241–242). Occasionally Cp groups are lost (243):

$$Cp_2TiCl_2 + NaNRAr \rightarrow Cp_2Ti(Cl)NRAr \rightarrow Cp_2Ti\,(NRAr)_2$$
$$R = H, ArC_6H_5$$
$$[56778\text{-}28\text{-}0]$$
$$Cp_2TiCl_2 + LiN(CH_3)_2 \text{ (excess)} \rightarrow CpTi[N(CH_3)_2]_3$$
$$[58057\text{-}99\text{-}1]$$

Although pyrrole is isoelectronic with cyclopentadiene, the compound $Cp_2Ti(NC_4H_4)_2$ [11077-90-0, 12636-83-8] displays only η^1-bonding of the pyrrole ligands (244).

The ambident pseudohalides displace one or two chlorides (245):

$$Cp_2TCl_2 + 1 \text{ or } 2 \text{ NaA} \rightarrow Cp_2TiClA \rightarrow Cp_2TiA_2$$

A = —NCO [85204-57-5]	A = —NCO [12109-61-4]
—NCS [70149-18-7]	—NCS [12109-64-7]
—NCSe [85204-58-6]	—NCSe [69567-79-9]
—N$_3$ [71737-48-9]	—N$_3$ [1298-37-9]
—N(CN)$_2$ [85204-59-7]	—N(CN)$_2$ [51831-59-5]

These products all contain Ti-N bonds (246–252). An isocyanate or isothiocyanate ligand can be displaced by a harder ion (253–254):

$$Cp_2Ti(NCO)_2 + \text{NaOR} \rightarrow Cp_2Ti(OR)NCO$$

In titanium acylates, the carboxylate ligands are unidentate, not bidentate, as shown by ir studies (255–256). They are generally prepared from the halide and silver acylate (257). The benzoate is available also from a curious oxidative addition with benzoyl peroxide (257–260):

$$Cp_2TiCl_2 + \text{AgO}\overset{O}{\overset{\|}{C}}R \rightarrow Cp_2Ti(O\overset{O}{\overset{\|}{C}}R)_2$$

$$Cp_2Ti(C_6H_5)_2 \;+\; (C_6H_5\overset{O}{\overset{\|}{C}}O)_2 \;+\; 2\,(CH_3)_2CHOH \longrightarrow$$

$$Cp_2Ti(O\overset{O}{\overset{\|}{C}}C_6H_5)_2 \;+\; 2\,C_6H_6 \;+\; 2\,(CH_3)_2C{=}O$$
$$[12156\text{-}48\text{-}8]$$

The acylates undergo facile hydrolysis or alcoholysis with loss of one or both Cp groups (257).

A silyl group can be attached to titanium by using Al[Si(CH$_3$)$_3$]$_3$ or KSi(C$_6$H$_5$)$_3$; LiSi(CH$_3$)$_3$ causes reduction (261–262). Germyl and stannyl groups can also be attached to titanium (263–264).

Organometallic ligands also can be attached to titanium by displacement reactions; both chlorines are displaceable (265–268):

$$Cp_2TiCl_2 + \text{LiOC[Co}_3(CO)_9] \rightarrow Cp_2ClTi\text{-OCCo}_3(CO)_9$$
$$[60383\text{-}58\text{-}6]$$
$$Cp_2TiCl_2 + \text{LiOC(R)Cr(CO)}_5 \rightarrow Cp_2ClTi\text{-OC(R)Cr(CO)}_5$$

Unusual titanocycles are formed from sulfur and phosphorus reagents (163–164,184,269–274):

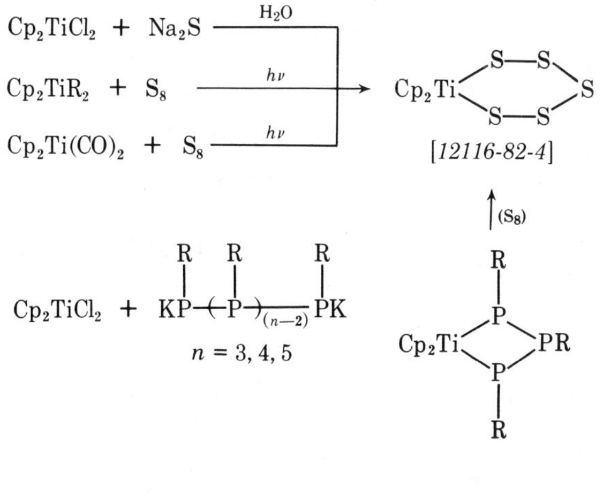

$$Cp_2TiCl_2 + \text{Na}_2S \xrightarrow{\;H_2O\;}$$
$$Cp_2TiR_2 + S_8 \xrightarrow{\;h\nu\;}$$
$$Cp_2Ti(CO)_2 + S_8 \xrightarrow{\;h\nu\;}$$

$$Cp_2Ti\begin{smallmatrix} S{-}S \\ \diagup \qquad \diagdown \\ \diagdown \qquad S \\ S{-}S \end{smallmatrix}$$
$$[12116\text{-}82\text{-}4]$$

↑ (S$_8$)

$$Cp_2TiCl_2 + \text{KP}\overset{R}{\underset{}{|}}{(}\overset{R}{\underset{}{|}}P{)}_{(n-2)}\overset{R}{\underset{}{|}}\text{PK}$$
$$n = 3, 4, 5$$

$$Cp_2Ti\begin{smallmatrix} R \\ | \\ P \\ \diagup \quad \diagdown \\ \quad \qquad PR \\ \diagdown \quad \diagup \\ P \\ | \\ R \end{smallmatrix}$$

R = CH$_3$ [79375-38-5]
= C$_2$H$_5$ [79375-38-5]
= t-C$_4$H$_9$ [79375-39-6]
= C$_6$H$_5$ [37299-20-0]

Reaction of cyclopentadienyltitanium halides with oxygen-bonding reagents is confusing. On the one hand, alcohols cleave one Cp group readily from Cp_2TiCl_2, the second more slowly. Sodium alkoxides or aryloxides in aprotic solvents give less cleavage (275–286). Silanols and ambient nitrite and nitrate do not cleave Cp groups (287–288). Several studies have attempted to explain the often facile cleavage by O-bonding reagents, which contrasts with normal displacement by N- or S-bonding reagents (272,289–304). No satisfactory explanation has yet emerged, and it is not yet possible to predict whether a desired Cp_2TiXY (X, Y = O-bonding ligand) can be prepared by direct displacement (305–310).

On the other hand, Cp_2TiCl_2 and $CpTiCl_3$ dissolve in boiling dilute hydrochloric acid yielding aquo cations which retain the Cp groups; they can be isolated as salts (183,311).

$$Cp_2TiCl_2 + 2\,H_2O \xrightarrow[\text{pH} <4]{100°C} [Cp_2Ti(OH).H_2O]^+Cl^-$$

[12116-84-6]
hydrate

$$CpTiCl_3 + 3\,H_2O \xrightarrow[\text{pH} <1]{100°C} [CpTi.3H_2O]^{3+}\,3\,Cl^-$$

[1270-90-0]
trihydrate

The ions shown are in equilibrium with other aquo ions (274,313–317). These ions undergo displacements with various nucleophiles to give the same Cp_2TiA_2 products prepared in nonaqueous solution (273,298–302,315,318–320). These ions are probably involved in the preparation of copolymers containing —Cp_2Ti— groups by interfacial polycondensation (312–314,321–328). Some of these polymers have promising high temperature properties.

Insertion into the CpTi-R Bond.   Sulfur dioxide yields sulfones and ultimately sulfinates; the latter are available also from RSO_2Na:

$$X\,Cp_2TiR + SO_2 \rightarrow X\,Cp_2Ti\overset{O}{\underset{O}{\overset{\|}{\underset{\|}{S}}}}R \rightarrow X\,Cp_2TiO\overset{O}{\overset{\|}{S}}R$$

$$R = CH_3, C_2H_5, C_4H_9, C_6H_5;\ X = F, Cl$$

Such titanium sulfinates are reported to increase crop yields (329–331).

Isocyanides insert to yield imines (332–333):

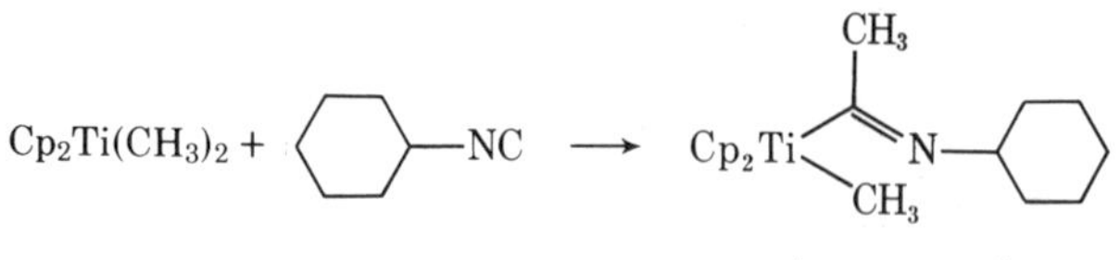

[58699-89-1]

Organic isocyanates and isothiocyanates, and nitric oxide, insert similarly (333). Carbon monoxide inserts to yield very stable acyltitaniums (334–335):

$$ClCp_2TiR + CO \longrightarrow ClCp_2Ti\overset{O}{\overset{\|}{C}}R$$

Miscellaneous Reactions of Cp$_2$Ti Derivatives. Coupling of fluxional pentadienide ion with allyl bromide is regiospecifically catalyzed by Cp$_2$TiCl$_2$ (337–338). In contrast, cuprous chloride gives the linear triene:

$$(C_5H_7)_2Mg \ + \ BrCH_2CH{=}\!{=}CH_2 \ \xrightarrow{\text{Cp}_2\text{TiCl}_2} \ \text{[structure]}$$

Reactions of titanium alkyls with aldehydes and ketones are generally more stereospecific and selective than the corresponding Grignard reactions (339).

Transition metal-catalyzed polymerizations of β-propiolactone and CH$_2$=CH-OCH$_2$CH$_2$Cl are markedly accelerated by Cp$_2$TiCl$_2$ (340–341).

Formation of dicyclopentadienylmagnesium is promoted by Cp$_2$TiCl$_2$, but not by TiCl$_4$ nor CpTiCl$_3$ (342):

$$3 \ CpH + Mg \rightarrow Cp_2Mg + \text{cyclopentene}$$

In olefin metathesis, the potent WCl$_6$ and WOCl$_4$ catalysts are tamed sufficiently by CpTiCl$_2$ to permit metathesis with unsaturated esters (343).

Cyclopentadienyl Derivatives of Titanium(III) and Titanium(II). The following references have been reported to reactions of lower-valent titanium compounds.

Properties and Preparation. Refs. 147, 166, 181, 196, 344–363.

Displacement Reactions of CpTiCl$_2$ and Cp$_2$TiCl. Refs. 294, 364–378.

Titanium(III) Hydrides. Refs. 167, 191, 224, 262, 379–405.

Dehalogenation and Deoxygenation of Organic Compounds. Refs. 397–409.

Hydrometalations. Refs. 410–414.

Reduction and Reductive Dimerization of Carbon-Carbon and Carbon-Oxygen Multiple Bonds. Refs. 339, 415–441.

Oxidation of Ti(III) Compounds. Refs. 442–445.

Carbometalations. Refs. 437, 446–450.

Coordination with Nitrogen (N$_2$). Refs. 377, 451–455.

Stereochemistry. Refs. 280, 339, 456–464.

Titanocene Dicarbonyl. Properties and Preparation. Refs. 220, 335–336, 391, 465–473.

Reactions. Refs. 164, 290, 344, 354–356, 370, 468, 474–486.

Titanocene. Refs. 215–219, 243, 487–496.

Other Ti(II) Sandwiches. Refs. 160, 466, 497–514.

Tricyclopentadienyl- and Tetracyclopentadienyl-Titanium. Refs. 375, 459, 515–519.

Health and Safety Factors

Commercial titanates should be handled according to good industrial practice. The tetraalkoxides have a low acute oral toxicity (LD$_{50}$ 7,500–11,000 mg/kg in rats). Because of their rapid hydrolysis, they can cause severe eye damage. They cause mild-to-moderate irritation of guinea pig skin, but they are not sensitizers. The titanium chelates possess the added toxicity of the chelating agent; for example, the LD$_{50}$

(rat, oral) of acetylacetonate AA is 5000 mg/kg. The chelates containing isopropyl alcohol have flash points of 12–27°C. The toxicology of titanium compounds has recently been reviewed (520).

Uses

Titanates with a carbon–titanium bond are extensively involved in Ziegler-Natta polymerizations of olefins. Nitrogen fixation by low valent titanium compounds has achieved no commercial importance.

Cross-linking of aqueous guar gum solutions to give gels capable of suspending sand is being used in hydraulic fracturing of oil and gas wells to stimulate their production. Other cross-linked gels are components of drilling muds. Titanates are being used extensively to coat particles of titanium dioxide, silica, alumina, and a variety of siliceous fillers used in plastics fabrication or metal powders in electrical and magnetic applications (see Fillers). Titanocene dichloride is active against cancer, although titanium compounds in general are devoid of reported physiological activity (520–521).

Many applications for titanates are based on the great affinity of tetravalent titanium atoms for oxygen atoms and on their tendency to form ···TiOTiO··· chains. Branched structures and networks are common, since titanium can bond covalently to four oxygens. Six oxygens can be attached by coordination. Moreover, structures, eg, $\geq$Ti–O–Si$\leq$ or $\geq$Ti–O–Al$<$, form readily so that titanates can bond to a siliceous substrate and simultaneously to something else, ie, they can be used as adhesives (qv).

Titanate Oligomer or Titania Formation. When titanium alkoxides are exposed to atmospheric moisture, they are rapidly hydrolyzed, and the resulting TiOH compounds condense to TiOTiO chains. Although hydrolysis rarely goes to completion, most of the organic residues disappear leaving a polymeric residue which approximates $(TiO_2)_x$. Pyrolysis of ROTi compounds gives olefin and $(TiO_2)_x$ (see Titanium compounds, inorganic).

Glass-Surface Coating. A thin (<100 nm) film of $(TiO_2)_x$ is virtually transparent. On glass, these films are bonded by Ti–O–Si bridges. After application of a lubricant, they impart considerable scratch resistance to glass and consequently greatly reduce its fragility (522–523). The lower alkoxides, particularly the tetraisopropoxide (TPT), are preferred for glass treatment. They can be applied undiluted, ie, in a hot process, or in a solvent, which may be hot or cold. Chelates, usually with triethanolamine (TE), may be used in water solution and applied hot or cold. Mixtures of TPT and chelates, eg, TE and the acetylacetone adduct, are said to give more uniform coatings (12).

The $(TiO_2)_x$ films are also applied to glass or vitreous enamel for decorative purposes. Thin films enhance brilliance; thicker films impart a silver-gray luster. Milk glass can be produced by mixing the titanate with a low melting enamel, which sinters when the coating is baked (524).

When the $(TiO_2)_x$ film is ca 150 nm thick, ie, one-quarter wavelength of average visible light, it is antireflective toward visible light, yet reflective toward heat-producing infrared radiation. Precisely coated window glass is used in hot countries to reduce solar heating of houses (525). Thicker coatings pass infrared light for solar cells and are valuable as antireflective coatings for lightwave guides, photo diodes, or semiconductors (526–528). In some of these applications, mixtures of titanate with a silicate ester are valuable. These protective coatings have been used in liquid-crystal devices and electronic products and sometimes as mixtures with silicates or zirconates (529–532) (see Liquid crystals). The coatings adhere to polytetrafluoroethylene (PTFE) (530). In an unusual application, titanoxane–siloxane mixtures, which may contain a borate, are applied to glass or glass particles which are then fired in an oxidizing atmosphere to yield a crystalline glass (pyroceram) of superior thermal properties (533) (see Glass-ceramics).

The bonding properties of $(TiO_2)_x$ have been used for size reinforcing of glass fibers so that they adhere to asphalt or to a PTFE–polysulfide mixture to impart enhanced flex endurance (534–536). Poly(vinyl alcohol) (PVA) solutions mixed with sucrose can be cross-linked (see below) with the lactic acid chelate and used generally for glass-fiber sizing (537).

Nonemulsion Paints. Heat-resistant paints (up to 500–600°C) must yield films containing little or no organic residues since most C—C or C–H bonds are pyrolyzed below those temperatures (see Paints). Pyrolysis of oligomeric titanates, obtained by controlled hydrolysis of $(C_4H_9O)_4Ti$, furnishes adherent films of nearly inorganic $(TiO_2)_x$. Since these oligomers will suspend pigments, particularly aluminum, alumina, and silica, paints were formulated from oligomers, such pigments, ethylcellulose to prevent pigment settling, and mineral–spirit solvent. Encouraging results were reported for these paints on rocket launchers, smokestacks, motor exhaust systems, and fire doors. Incorporation of zinc dust imparts some corrosion resistance as well (538–539). There is scant recent patent literature on such applications, which implies that the needs for heat-resistant paints are met by silicones, polyurethanes, epoxies, and other formulations.

Titanates are valuable in other paint applications. Corrosion-resistant coatings have been described for tinplate, steel, and aluminum (540–544). Incorporation of phosphoric acid or polyphosphates enhances the corrosion resistance. Since titanates promote hardening of epoxy resins, they are often used in epoxy-based paints (545). Silicones (polysiloxanes) are often cured by titanates. Pigments, eg, TiO_2, SiO_2, Al_2O_3, ZrO_2, etc, are frequently pretreated with titanates before incorporation into paints (541,546). In these applications, the $Ti(OR)_4$ compounds are often mixed with $Si(OR)_4$, $Al(OR)_3$, $Zr(OR)_4$, and other metal alkoxides (3).

Titanates react with ester groups in paint vehicles, eg, linseed oil, tung oil, and alkyds, and with hydroxy groups, eg, in castor oil and some alkyds. They prevent wrinkling of paint films (50,547).

Adhesives. Tetrafunctional titanates react with hydroxyl, ester, amide, imide, and other functions, and with oxide groups on metals and nonmetal oxides. Titanates bond such materials together (51). For unexplained reasons, they bond to polyethylene and fluorinated polymers (548). Packaging films, such as Mylar polyester or aluminized films, are coated with a titanate, then warmed in moist air to hydrolyze and oligomerize the titanate and evaporate volatile organics (549). The resulting film is not tacky and can be rolled and stored. Subsequently, polyethylene is extruded onto the surface and bonded by calendering. Printing inks and decorations can also be bonded.

Recent developments include the bonding of fluorinated resins to packaging films, poly(hydantoin)–polyester to polyester wire enamel, polysulfide sealant to polyurethane (here, a phosphated titanate is recommended), titanated polyethylenimine for bonding polyethylene to cellophane, and silicone rubber sealant to metal or plastic support using polysilane (Si–H) plus polysiloxane (Si–OR) and titanate as the adhesive ingredients (550–554). Polyester film coated first with a titanium alkoxide, then with a poly(vinyl alcohol)–polyethylenimine blend, becomes impermeable to gases (555).

Water Repellents. Titanate–wax compositions have been used for the reproofing of textiles that have been dry cleaned. Typically, a slowly hydrolyzable titanate ester from octylene glycol or 3,5,5-trimethyl-1-hexanol is dissolved in a dry-cleaning solvent with wax and applied to the garment before the final drying stage. Hydrolysis and bonding to the cellulosic textile occurs during steam pressing. In a variation of this process, a silicone is applied with the titanate; the former furnishes repellency, the latter bonds it for durability. Leather waterproofed with silicones possesses improved properties when the resin is bonded with tetrabutoxytitanium (556) (see Waterproofing and water/oil repellency).

Catalysts. Titanates accelerate many organic reactions and frequently provide significant advantages in product purity and yield over conventional catalysts (see Catalysis). Their polyfunctionality permits assembling oxygen-containing reactants at one location in a geometrical, usually octahedral, arrangement, which permits facile shuffling of groups to yield products.

Esterification. The reaction

$$RCOOH + R'OH \rightleftharpoons R\overset{\overset{\displaystyle O}{\|}}{C}OR' + H_2O$$

can be catalyzed by small quantities of titanium alkoxides·(see Esterification). Although the water which forms can hydrolyze and inactivate the titanate, titanoxane oligomers are cleaved by carboxylic acids to bicoordinated monomeric acylates. A simplified mechanism is illustrated:

$$[(CH_3)_2CHO]_4Ti + R'OH \longrightarrow [(CH_3)_2CHO]_4Ti\text{-}\text{-}\overset{\displaystyle |}{\underset{\displaystyle H}{O}}R' \longrightarrow (CH_3)_2CHOH + R'OTi[OCH(CH_3)_2]_3$$

In plasticizer manufacture, ie, of phthalates, sebacates, etc, with sulfuric or *p*-toluenesulfonic acid catalysts, the temperature (140–150°C) required for rapid reaction and high conversion may dehydrate or oxidize the alcohol and may yield a dark or foul-smelling product; neutral titanates do not cause such side reactions. Although a temperature of 200°C is required, esterifications can easily be forced to over 99% conversion without the formation of odors or discoloration. Recent preparations include long-chain esters from neopentyl glycol, trimethylolpropane, and pentaerythritol for synthetic lubricants, and triglycerides from mixed long-chain acids for suppositories (557–559).

In polyester manufacture from dibasic acids and diols, color is particularly important for fiber and film products. Moreover, destruction of diol by strong acids upsets the stoichiometric balance required for high molecular weight and forms by-products, eg, diethylene glycol or toxic dioxane from ethylene glycol or tetrahydrofuran from tetramethylene glycol. Titanate catalysts, eg, $MgTi(OR)_6$ and $(RO)_4Ti$, are devoid of these problems (560–563). With hydroxyl-terminated prepolymers, chain extension by diisocyanates is also accelerated by titanates (560).

Ester Interchange (Alcoholysis, Transesterification). Ester interchange is exemplified by the following reactions, all of which are strongly promoted by titanates:

$$RCOR' + R''OH \rightleftharpoons RCOR'' + R'OH$$

$$RCOR' + R''COH \rightleftharpoons R''COR' + RCOH$$

$$RCOR' + R''COR''' \rightleftharpoons RCOR''' + R''COR'$$

The first is the most common synthetically, whereas the third is important in cross-linking polyesters. Classical catalysts, ie, sulfuric acid and NaOR, react with other functional groups to lower yield and product purity (564). In transesterification, the methyl ester of sebacic or other acid is heated with an equivalent amount of a long-chain alcohol and titanate (0.1–2.0 mol % of TPT) at atmospheric or reduced pressure to distill methanol and drive the reaction to completion. A solvent, eg, benzene or cyclohexane, which forms a low boiling azeotrope with methanol, may be added. Esters prepared recently from methyl esters include diaryl carbonates from phenols, long-chain carbamates (urethanes), and diethylaminoethyl methacrylate (565–567). In the first example, the catalyst is the solid prepared by calcining the precipitate from cohydrolysis of $TiCl_4$ and $SiCl_4$. The third example is typical of the preparation of a host of methacrylate or acrylate esters from low cost methyl methacrylate (MMA) or ethyl acrylate and other alcohols under nonpolymerizing conditions. The product methacrylate is often pure enough for subsequent polymerization. To remove the titanate catalyst, if required, one need only add two moles of water per mole of titanate and remove the $+Ti(OCH(CH_3)_2)O+_x$ [66593-86-0] by filtration through diatomaceous earth.

Polyesters. Transesterification is the classical preparative method for polyesters (qv). Titanates are much more active than traditional catalysts, eg, Pb_3O_4; $(TiO_2)_x$ is also active, especially at higher temperatures. Thus, condensation of dimethyl terephthalate (DMT) with p-$C_6H_4(COOCH_2CH_2OH)_2$ begins with titanium lactate and is completed with titanium dioxide (568). Oligomers from titanium lactate with ethylene glycol or $[(CH_3)_2CHO]_2Ti(acac)_2$ with polyester prepolymers yield high molecular weight polyesters (551,569–570). Kinetics of the DMT–tetramethylene glycol polycondensation catalyzed by tetrabutoxytitanium has been reported (571).

Polyesters bearing pendant ester groups react with such groups on adjacent chains resulting in alkyd resin formation. These reactions are catalyzed by titanates. Acrylic polymers are cross-linked with glycols or to hydroxyl-containing alkyds (572–573) (see Alkyd resins). Oligoester titanates cure epoxies and the oligoester acts as internal plasticizer (574). Polyurethanes react with titanium acylates forming artificial leather (575) (see Leatherlike materials). An unsaturated polyester filled with crushed marble is cross-linked with triethanolamine titanate [30177-87-8] (TE) and cumene hydroperoxide; the product is a moldable artificial marble (74).

Epoxy Resins. Titanates react with free hydroxy groups in the resin or with the epoxy group itself:

$$\text{RCH}\underset{\underset{O}{\diagdown\diagup}}{\text{—CH}_2} + \text{Ti(OR')}_4 \longrightarrow \text{RCH}\underset{\underset{O}{\diagdown\diagup}}{\text{—CH}_2} \longrightarrow \underset{\underset{\text{OTi(OR')}_3}{|}}{\text{RCHCH}_2\text{OR'}}$$
$$\text{Ti(OR')}_4$$

With sufficient titanate, the polymer is fully hardened. With only enough titanate to react with free hydroxyls, the resin may subsequently be cured at lower cost with conventional cross-linking agents. The titanated epoxy resin has a low power factor, which is important in electrical applications, eg, potting components and insulation (see Embedding). Titanates improve adhesion of metals to epoxies.

Epoxy cross-linking is catalyzed by TPT and TBT, alone or with piperidine, and by triethanolamine titanate. The solid condensation product from 3 TPT:4 TEA (triethanolamine) has also been applied to epoxy curing (52). Titanate curing is accelerated by selected phenolic ethers and esters at 150°C; the mixtures have a long pot life at 50°C (576) (see Epoxy resins).

Copper powder suspended in titanate-cured epoxy resin shields electronic apparatus against electromagnetic interference (577). Magnetic particles, ie, iron, iron oxides, and magnetic alloys, suspended in resin containing a titanate yield a superior recording tape (578–579) (see Magnetic tape). Titanates are very effective in dispersing and suspending particles. Such compositions can be formed into larger magnets (580). Other titanate-curable resins useful for these purposes include silicones, polyesters, phenolics, polyurethanes, polyamides, and acrylics (580). The combination $(C_4H_9O)_4Ti$ and $(C_4H_9O)_3B$ provides a fast-curing system suitable for metal coating (555).

Peroxide Activation. Organic peroxides coordinate with titanates, then dissociate to radicals capable of initiating vinyl polymerization (581). Cumene hydroperoxide with TE cures unsaturated polyesters (74). Methyl ethyl ketone peroxide with cobalt naphthenate and a titanium chelate gives a light-colored, cured, unsaturated polyester; use of cobalt naphthenate alone gives a discolored product (74–75). A peroxide–titanyl sulfate complex is used in tanning leather (582).

Cross-Linking of Polyols. Polyols such as natural polysaccharides (cellulose, starch, guar gum, and their derivatives) and poly(vinyl alcohol) and its derivatives are cross-linked by titanates.

Paints. Water-based emulsion paints are large consumers of water-soluble, hydrolysis-resistant titanates, eg, TE. Titanates cross-link cellulose nitrate and acetate-based lacquers.

Pigments suspended in an aqueous polyacrylate latex tend to settle, so thickening agents are incorporated. Cellulose ethers and carboxymethyl cellulose are protective colloids. They are more effective when cross-linked with a titanate. Such paints are thixotropic; the very thick paint thins when sheared by brushing, rolling, or passing through a spray nozzle. Paint on the brush does not drip; when applied to the surface, it thins enough to brush out easily and long enough to be self-leveling. Yet it rethickens so rapidly that an excessively heavy application does not run down the wall or drip off the ceiling. Titanium chelates in thixotropic emulsion paints were first claimed in a UK patent (583). They not only improve the rheology of the paint but also minimize syneresis and give a more water-repellent, durable paint film. The ratio of titanate to polysaccharide greatly influences paint properties and the properties of the dried film.

The exact chemical nature of polyol cross-links has been inferred from the behavior of simple alcohols and glycols. First, primary alcohol functions on each glucose unit of (modified) cellulose are coupled by monomeric or oligomeric titanate. The primary 6-CH_2OH is probably the most active site, but the secondary 2-CHOH of the pyranose ring is also active, since it is labilized toward etherification by the adjacent acetal function; these conclusions are extrapolated from studies of the reactivity of polysaccharides toward ethylene and propylene oxide and halides, eg, methyl iodide and chloroacetic acid. Second, reaction of a monomeric or oligomeric titanate may occur at two neighboring hydroxyls of one chain and two or one of a nearby chain if the hydroxyls are favorably oriented. In polyglucosides, eg, starch or cellulose, the 2,3-diol is trans, but in galactomannans, eg, guar, cis-hydroxyl pairs are prevalent. In poly(vinyl alcohol), secondary 1,3-diol pairs lie on a flexible open chain, which may readily assume an orientation capable of forming 6-membered titanate rings. Third, cross-linking can involve hydrogen bonds, which are postulated to explain shear sensitivity of titanate-cross-linked gels; the location and shape of these hydrogen bonds to titanates has not been established. They probably form more readily, break more readily, and reform more readily than the covalent bonds thought to be responsible for forming rigid gels which are irreversibly destroyed by shear. Links formed from oligomeric titanates should be more shear-resistant and reform more readily when broken than rigid, single-titanium links.

Organic aluminates, silicates, or silicas may be added to the titanates. Two patents describe a paint used for recoating old chalked films (584–585). An acrylic hydrosol containing TiO_2, $CaCO_3$, and crushed marble is thickened with methylhydroxypropyl cellulose cross-linked with a mixture of isopropyl tristearoyl titanate [58766-22-6] and isopropyl tris(octyl pyrophosphate) titanate. Poly(vinyl butyral) hydrolyzed with aqueous phosphoric acid is mixed with ethyl silicate and TBT-diol chelate and baked on steel to give a transparent, pigment-free, rustproof coating (240). Another rust-proofing composition is formulated from a complex acrylic latex, silica, and an alkyd resin plus TE or a diol chelate (513). Paints containing a polyphosphate are advantageously cured with a mixed titanate–aluminate–glycol oligomer (586). A solution of AA in water–isopropyl alcohol–methyl ethyl ketone is recommended as a cleaner-primer for metal, wood, plastic, or wallboard (587).

Cellulose. Cellulose (qv) or starch xanthate cross-linked by titanates adsorbs uranium from seawater (588). Carboxymethylcellulose cross-linked with isopropyl tristearoyl titanate is the bonding agent for clay, talc, wax, and pigments to make colored pencil leads of unusual strength (589). Long-chain titanium acylates are added to drilling muds to aid in barite suspension and to prevent emulsion flipping on contact with subterranean deposits of carnallite ($KCl.MgCl_2.6H_2O$) (590).

Titanate-Cross-linked Guar Gels. Water under high pressure (69–172 MPa or 10,000–25,000 psi) is forced down well bores into the oil-bearing formation to fracture the rock so as to open channels through which oil can flow more rapidly. The cracks are prevented from closing under the lithostatic pressure of the overburden by passing sand down with water. Because the sand must be suspended until it reaches the cracks it is to prop open, the water is thickened with a soluble polymer, eg, PVA, or a natural polysaccharide or derivative. Up to 10% of the polyol may be required to attain sand-suspending viscosity. Cross-linking the polymer with titanates permits dramatic reductions of the polymer concentration to 0.3–1 wt % of the water (see Petroleum, drilling fluids).

Guar gum, a natural galactomannan, has cis-2,3-diol units in the mannan backbone and cis-3,4-diol units on each galactose moiety (see Gums). Cis-diols react with borate ion yielding cyclic borates. Borate-cross-linked guar is useful up to ca 79°C; however, wells may have bottom-hole temperatures up to 204°C. The discovery that guar solutions could be cross-linked with titanates or other transition metal ions to yield firm gels that are stable at quite high temperatures provoked rapid development of hydraulic fracturing into an important tool for oil- and gas-well stimulation (591). Titanium and zirconium are effective and environmentally innocuous cross-linking agents (590–595).

Paper Sizing. Various materials are added to paper to improve its wet strength and ink acceptance, to make possible clay coating, etc (see Papermaking additives). Titanic acid precipitation on paper fibers can be controlled (55). Aqueous triethanolamine titanate, when neutralized, rapidly deposits titanic acid. Addition of a monosaccharide or derivative delays precipitation for several hours or months, depending on the stereochemistry of the sugar. Glucose, lacking cis-hydroxyl pairs, is less effective than mannose (2,3-cis-diol). Lactose is also more effective than glucose. Yet fructose is better than mannose, indicating that the hemiketal structure, ie, the furanose ring, furnishes a cis-diol structure. Alditols are better than the corresponding aldose, and mannitol surpasses sorbitol. Handsheets prepared from semibleached kraft pulp treated with sugar-stabilized TE solutions exhibit a considerably higher wet strength than those from untitanated pulp.

Poly(vinyl Alcohol). Poly(vinyl alcohol) (PVA) is used extensively as a paper size, alone or in combination with dyes or pigments; it is rendered insoluble by cross-linking with titanates. A rayon-based paper has been made resistant to boiling water (596–597). A size of PVA and hydroxyethyl starch cross-linked with titanium citrate or lactate prevents linting in the sizing press (598). A mixed chelate (diol + hydroxyacid) has been recommended (599). A mixture of PVA and a silicone with TE provides a release treatment for papers, though silicones alone or with an epoxyamine also provide release coatings or waterproofed paper (566,599–603).

The patent literature includes many other examples of titanate-cross-linked PVA. A smokeable sausage casing has been prepared from PVA, hydrolyzed ethylene–vinyl acetate copolymer, and glycerol on paper (604). Poly(vinyl alcohol) and poly(oxyethylene)sorbitan laurate suspend kaolin in cosmetic mud packs (605). Cross-linked PVA is foamed and then dried to give porous material for an unspecified medicinal use (606). Cotton textiles sized with cross-linked PVA are wrinkle-resistant (607). Clay or humus soils are stabilized with PVA–titanate products (608–609). Poly(vinyl alcohol) titanates are used as capsule walls for microencapsulated dyes for copy paper (610). A PVA solution reinforced with lignin has been recommended as a temporary plug for leaks in equipment containing $TiCl_4$ (611).

Amino Polymers and Leather. Although the Ti–N bond is not as strong as the Ti–O bond, several applications are based upon it. Polyurethanes cross-linked with titanium chelates have been claimed as artificial leather (612). Titanium solutions with or without PVA or with a silicone are used in tanning or waterproofing leather because titanium bonds to collagen (556,582,613). TE and its homologues have been used in shrink-resistant treatments for wool, and, when combined with poly(vinyl chloride) and glycerol–ethylene oxide adducts, they impart flame resistance as well (50,544,614).

Polyamides, polyhydantoins, polyimides, and polyethylenimine have been cross-linked for various applications (551,553,615–616). Gelatin that is cross-linked first with glutaraldehyde, then with titanium chelates, bonds enzymes; these immobilized enzymes retain their activity. In one example, glucoamylase on this support has been used to prepare high-dextrose equivalent syrups (617). More simply, gelatinous titanic acid itself immobilizes enzymes (618) (see Enzymes, immobilized).

Particle Dispersants. Complex titanates suspend particles of metals, metal and nonmetal oxides, minerals like clays and talc, carbon black, etc. When titanated particles are blended with polymers, the mixtures are frequently much less viscous and better adapted to processes such as extrusion, molding, or coating on substrates.

Kenrich Petrochemicals Company furnishes a series of couplers, which are titanates of nominal formula $(RO)(R'O)(R''O)(R'''O)Ti$ in which each RO–R'''O is selected from lower alkoxyl (i-C_3H_7O), long-chain acyl (isostearato), long-chain sulfo (alkylbenzenesulfonato), phosphato (dioctylpyrophosphato), unsaturated (methacrylato), and difunctional analogues. One such coupler is (i-PrO)Ti(isostearato)$_3$ [68443-53-8], which is used to disperse fillers in oils, rubbers, and plastics (619). The lower alkoxy group provides reactivity with oxide-type particles and the long chain compatibility with the binding polymer. The variety of available functionality permits tailoring of the titanate to the characteristics of the polymer. Some examples include coating MgO granules to make waterproof insulating coatings for radiator tubes; pigments for acrylic antichalk exterior paints; Al(OH)$_3$ in intumescent coatings; talc in polypropylene; SiO$_2$ in nylon-6, epoxies, polyurethane, and polypropylene; pigments in polyester-molding compositions; and glass fibers in poly(phenylene sulfide) and silicones as electrical potting compounds (535,584–585,620–625). Dramatic reductions in viscosity of these blends are frequently claimed. Reviews of their use in corrosion-resistant paints and in mineral-filled RIM (reaction injection-molded) urethanes are given in ref. 626. A chelated long-chain acylate can suspend a powdered fluorinated resin in toothpaste to prevent buildup of calculus on teeth (627) (see Dentifrices).

Metal powders are efficiently suspended in polymeric binders with titanates (628–629). Carbon black, when suspended in a silicone containing TE and coated on a substrate, cannot be rubbed off after curing in moist air (630).

Titanium Dioxide. The literature shows numerous examples of treating TiO$_2$ particles with organic reagents which are the same chelating agents for titanium as discussed in this article. Perhaps these are not organic titanates, but they illustrate the chemistry of organic titanates. Titania has been treated with hydroxy acids, alkanolamines and their esters, or polyols to yield a more dispersible pigment which does not agglomerate on storage (547,550,631–642). Frequently, the titania contains alumina or silica throughout the inside of the particle or as a surface coating. Titanic acid dried at a low temperature is a powerful adsorbent for basic materials, including amino acids, peptides, proteins, and antibiotics (643). In all of these cases, Ti–O-organic bonds or Ti–O–H-organic bonds are probably responsible for the observed effects.

Miscellaneous. Epoxidation by oxygen of propylene or butylenes is favored by the use of a silver catalyst treated with $(C_2H_5)_2TiCl_2$ (644). Titanium dioxide, alone or in mixtures, is a vapor-phase oxidation catalyst. Titanium compounds also catalyze photoxidation.

Dyes for polystyrene are prepared from titanated o-hydroxyazobenzene derivatives (645). Phthalein dyes with free phenolic and sulfo groups can be complexed to —OCp$_2$(F)O— by interfacial copolymerization (646). Tinplate coated with tetraoctyl titanate and uv-cured provides good adhesion for inks. A titanium–catechol chelate develops color when applied as an ink to alkaline paper (647).

The reaction product from long-chain

$$\text{Ti(OR)}_4 \; + \; \text{HS}\overset{\overset{\displaystyle S}{\|}}{\text{P}}(\text{OR}')_2$$

is used in lubricating oils (648). Olefins are dimerized and isomerized by RTi compounds (141,649).

Organic-soluble vanadium compounds containing titanium, eg, OV[O-Ti(OC$_4$H$_9$)$_3$]$_2$ [78552-17-7] are claimed as superior catalysts for vinyl or epoxide polymerization (650). Titanium tetraalkoxides and tetraacylates are claimed to polymerize moist formaldehyde to high molecular weight polyformals (651).

Insecticides (and other agricultural chemicals) normally contain functional groups that can react with titanates. When dispersed in a titanate-cross-linked silicone and applied to plant leaves, the insecticidal activity is retained for some time despite rains (652) (see Insect control technology).

A silicone–titanate coating on a substrate promotes dropwise water condensation. The ice produced on freezing is easily released (653).

Titanium(IV) compounds on exposure to ultraviolet radiation dissociate to Ti(III) with an unpaired electron, which can react with oxygen:

$$\geq\!\text{TiX} \xrightarrow{\text{light}} \geq\!\text{Ti}\cdot + \cdot\text{X} \xrightarrow{\text{O}_2} \geq\!\text{TiOO}\cdot$$

where X = general substituent. Thus, titanates are catalysts in photooxidation of alcohols to aldehydes or ketones and alanine to lactic acid (via hydrolysis of imine) and oxalic acid (654–655). The rate of photodissociation depends on the ligands; effects have been noted with pyridine, borate plus hydroxy acid or amino acid, and borate plus PVA (656–657). When a Ti(IV) cross-link is photolyzed to Ti(III), the cross-link is weakened or destroyed. This phenomenon has been turned to advantage in the preparation of photoresists (657–660). Polyethylene containing Cp$_2$TiCl$_2$ is degraded in sunlight far more rapidly than regular polyethylene (661). This property should be valuable in preparing environmentally acceptable plastics (see Plastics, environmentally degradable, Supplement Volume).

The properties of titanium ions in solution are greatly modified by chelation. Chemists in the USSR and in India have been particularly active in this area since the mid-1960s. Chelation has been used to mask titanium to prevent its interfering in the determination of other elements, eg, iron, germanium, or aluminum, spectroscopically, polarographically, or titrimetrically (see Chelating agents). Photometric determination of titanium is facilitated by chelation, which shifts and intensifies absorption maxima. Hydroxy acids are most commonly used for such analyses, especially oxalic, tartaric, citric, malic, mandelic, and salicylic acid. The extractability of titanium from aqueous solution by organic solvents is enhanced by ligands, eg, 8-hydroxyquinoline and other amines. Selective dialysis and electrophoresis have also been discussed. The use of organotitaniums in analytical chemistry is reviewed in refs. 311, 662–682.

BIBLIOGRAPHY

"Titanium Compounds" in *ECT* 1st ed., Vol. 14, pp. 213–241, by L. R. Blair, H. H. Beacham, and W. K. Nelson, National Lead Company; "Titanium Compounds (Organic)" in *ECT* 2nd ed., Vol. 20, pp. 424–503, by R. H. Stanley, Titanium Intermediates, Ltd.

1. U.S. Pat. 2,709,174 (May 24, 1955), J. B. Rust and L. Spialter (to Montclair Research and Ellis-Foster Co.).
2. B. D. Jain and R. Kumar, *Proc. Indian Acad. Sci. Sect. A* **60,** 265 (1964); *Indian J. Chem.* **1,** 317 (1963); V. Patrovsky, *Coll. Czech. Chem. Commun.* **27,** 1824 (1962).
3. D. C. Bradley, R. C. Mehrotra, and D. P. Gaur, *Metal Alkoxides*, Academic Press, New York, 1978.
4. Ref. 3, p. 150.
5. I. Kraitzer, F. K. McTaggart, and G. Winter, *J. Counc. Sci. Ind. Res.* **21,** 328 (1948).
6. T. Boyd, *J. Polym. Sci.* **7,** 591 (1951).
7. T. Ishino and S. Minami, *Tech. Rep. Osaka Univ.* **3,** 357 (1953).
8. Can. Pat. 583,036 (Sept. 8, 1959), J. H. Haslam (to E. I. du Pont de Nemours & Co., Inc.).
9. A. N. Nesmeyanov, O. V. Nogina, and R. K. Freidlina, *Bull. Acad. Sci. USSR Div. Chem. Sci.*, 355 (1956).
10. Ref. 3, p. 27ff.
11. D. Seebach and co-workers, *Synthesis*, 138 (1982).
12. Brit. Pat. 1,510,587 (May 10, 1978), (to St. Gobain Industries).
13. Ger. Pat. 934,352 (Oct. 20, 1955), D. F. Herman (to Titangesellschaft).
14. U.S. Pats. 3,091,625 (May 28, 1963) and 3,119,852 (Jan. 28, 1964), R. T. Gilsdorf (to E. I. du Pont de Nemours & Co., Inc.).
15. Brit. Pat. 997,892 (July 14, 1965), L. J. Lawrence (to British Titan).
16. U.S. Pat. 2,684,972 (July 27, 1954), J. H. Haslam (to E. I. du Pont de Nemours & Co., Inc.).
17. Brit. Pat. 787,180 (Dec. 4, 1957), J. H. Haslam (to E. I. du Pont de Nemours & Co., Inc.).
18. Brit. Pat. 719,346 (1954), (to Monsanto Chemical Co.).
19. J. S. Jennings, W. Wardlaw, and W. J. Way, *J. Chem. Soc.*, 637 (1936).
20. A. N. Nesmeyanov and co-workers, *Izv. Akad. Nauk SSSR Otd. Khim. Nauk*, 1037 (1952).
21. G. A. Razuvaev and co-workers, *Dokl. Akad. Nauk SSSR* **122,** 618 (1958).
22. S. Prasad and J. B. Tripathi, *J. Indian Chem. Soc.* **35,** 177 (1958).
23. G. P. Luchinskii, *Zh. Obshch. Khim.* **7,** 2044 (1937).
24. O. C. Dermer and W. C. Fernelius, *Z. Anorg. Allg. Chem.* **221,** 83 (1934).
25. H. Funk and E. Rogler, *Z. Anorg. Allg. Chem.* **252,** 323 (1944).
26. R. C. Mehrotra, *J. Indian Chem. Soc.* **30,** 731 (1953).
27. A. N. Nesmeyanov, O. V. Nogina, and R. Kh. Freidlina, *Izv. Akad. Nauk SSSR Otd. Khim. Nauk*, 518 (1951).
28. D. C. Bradley, D. C. Hancock, and W. Wardlaw, *J. Chem. Soc.*, 2773 (1952).
29. I. D. Verma and R. C. Mehrotra, *J. Less Common. Met.* **1,** 263 (1959).
30. A. N. Nesmeyanov, E. M. Brainina, and R. Kh. Freidlina, *Dokl. Akad. Nauk SSSR* **94,** 249 (1954).
31. A. B. Bruker, R. I. Frenkel, and L. Z. Soborovskii, *Zh. Obshch. Khim.* **28,** 2413 (1958).
32. A. N. Nesmeyanov, E. M. Brainina, and R. Kh. Freidlina, *Izv. Akad. Nauk SSSR Otd. Khim. Nauk*, 987 (1954).
33. O. V. Nogina, R. Kh. Freidlina, and A. N. Nesmeyanov, *Izv. Akad. Nauk SSSR Otd. Khim. Nauk*, 74 (1952).
34. D. M. Puri and R. C. Mehrotra, *Indian J. Chem.* **5,** 448 (1967).
35. R. E. Reeves and L. W. Mazzeno, *J. Am. Chem. Soc.* **76,** 2533 (1954).
36. Brit. Pat. 1,586,671 (March 25, 1981), M. S. Howarth and B. W. H. Terry (to Imperial Chemical Industries).
37. N. Baggett, D. S. P. Poolton, and W. B. Jennings, *J. Chem. Soc. Dalton Trans.*, 1128 (1979); *J. Chem. Soc. Chem. Commun.*, 239 (1975).
38. R. C. Fay and A. F. Lindmark, *J. Am. Chem. Soc.* **97,** 5928 (1975); P. Finocchiaro, *J. Am. Chem. Soc.* **97,** 4443 (1975).
39. R. V. Singh, R. V. Sharma, and J. P. Tandon, *Synth. React. Inorg. Met. Org. Chem.* **11,** 139 (1981).
40. H. Sugahara and Y. Shuto, *J. Organomet. Chem.* **24,** 709 (1970).
41. A. Yamamoto and S. Kambara, *J. Am. Chem. Soc.* **81,** 2663 (1959).
42. *Ibid.*, **79,** 4344 (1957).

43. U.S. Pat. 2,643,262 (June 23, 1953), C. O. Bostwick (to E. I. du Pont de Nemours & Co., Inc.).

44. Brit. Pat. 757,190 (Sept. 12, 1956), (to E. I. du Pont de Nemours & Co., Inc.).

45. S. P. Biswas and co-workers, *J. Indian Chem.* **16A,** 972 (1978), and previous papers.

46. U.S. Pat. 3,699,137 (Oct. 17, 1972), E. Termin and O. Bleh (to Dynamit-Nobel A.G.); V. Zatka and O. Hoffman, *Analyst (London)* **95,** 200 (1970).

47. S. Minami, H. Takano, and T. Ishino, *J. Chem. Soc. Jpn.* **60,** 1406 (1957).

48. D. M. Puri and R. C. Mehrotra, *J. Less Common. Met.* **3,** 247 (1961).

49. D. C. Bradley and C. E. Holloway, *J. Chem. Soc. A,* 282 (1969).

50. Brit. Pat. 755,728 (Aug. 27, 1956), (to National Lead Co.).

51. U.S. Pat. 2,935,522 (May 3, 1960), C. M. Samour (to Kendall Co.).

52. Brit. Pat. 994,717 (June 10, 1965), (to Dr. Beck and Co.).

53. Brit. Pat. 786,388 (Nov. 20, 1957), (to National Lead Co.).

54. P. Lagally and H. Lagally, *TAPPI* **39,** 747 (1956).

55. P. C. Bharara, V. D. Gupta, and R. C. Mehrotra, *Z. Anorg. Allg. Chem.* **403,** 337 (1974).

56. P. C. Bharara, V. D. Gupta, and R. C. Mehrotra, *J. Indian Chem. Soc.* **51,** 859 (1974).

57. E. C. Alyea and P. H. Merrell, *Inorg. Nucl. Chem. Lett.* **9,** 69 (1973).

58. D. A. Baldwin and G. H. Leigh, *J. Chem. Soc. A,* 1432 (1968).

59. R. J. H. Clark and A. J. McAlees, *J. Chem. Soc. A,* 2026 (1970).

60. U.S. Pat. 2,670,363 (Feb. 23, 1954), J. P. Wadington (to National Lead Co.).

61. K. C. Pande and R. C. Mehrotra, *J. Prakt. Chem.* **5,** 101 (1957).

62. K. C. Pande and R. C. Mehrotra, *Chem. Ind.,* 114 (1957).

63. Y. A. Lysenko and O. A. Osipov, *Zh. Obshch. Khim.* **28,** 1724 (1958).

64. S. Prasad and R. C. Srivastava, *J. Indian Chem. Soc.* **39,** 9 (1962).

65. Brit. Pat. 787,180 (Dec. 4, 1957), J. H. Haslam (to E. I. du Pont de Nemours & Co., Inc.).

66. I. D. Varma and R. C. Mehrotra, *J. Prakt. Chem.* **8,** 235 (1959).

67. A. N. Solanki, K. R. Nahar, and A. M. Bhandari, *Synth. React. Inorg. Met. Org. Chem.* **8,** 335 (1978).

68. J. Amaudrut, *Bull. Soc. Chim. Fr.,* 624 (1977); J. Amaudrut, B. Viard, and R. Mercier, *J. Chem. Res. Synop.,* 138 (1979).

69. U.S. Pat. 2,621,193 (Dec. 9, 1952), C. Langkammerer (to E. I. du Pont de Nemours & Co., Inc.).

70. U.S. Pat. 2,621,194 (Dec. 9, 1952), J. H. Balthis (to E. I. du Pont de Nemours & Co., Inc.).

71. U.S. Pat. 2,621,195 (Dec. 9, 1952), J. H. Haslam (to E. I. du Pont de Nemours & Co., Inc.).

72. R. Feld and P. L. Cowe, *The Organic Chemistry of Titanium,* Butterworths, London, 1965.

73. M. Isobe and co-workers, *Tetrahedron Lett.* **23,** 221 (1982); cf. B. E. Rossiter, T. Katsuki, and K. B. Sharpless, *J. Am. Chem. Soc.* **102,** 5976 (1980); **103,** 464 (1981).

74. USSR Pat. 771,056 (Oct. 15, 1980), V. K. Skubin and co-workers.

75. Brit. Pat. 2,051,093 (Jan. 14, 1981) and Ger. Offen. 3,017,887 (May 9, 1979), H. Kamio, Y. Ogina, and K. Nakamura (to Nippon Mining Co.).

76. J. M. Latour, B. Galland, and J. C. Marchon, *J. Chem. Soc. Chem. Commun.,* 570 (1979).

77. D. C. Bradley and I. M. Thomas, *J. Chem. Soc.,* 3854 (1960).

78. D. C. Bradley, *Adv. Inorg. Chem. Radiochem.* **15,** 259 (1972).

79. U.S. Pat. 2,579,413 (Dec. 18, 1951), T. Boyd (to Monsanto).

80. G. Chandra and M. F. Lappert, *Inorg. Nucl. Chem. Lett.* **1,** 83 (1965); *J. Chem. Soc. A,* 1940 (1968).

81. D. C. Bradley and M. H. Gitlitz, *J. Chem. Soc. A,* 980 (1969).

82. H. Weingarten, M. G. Miles, and N. K. Edelmann, *Inorg. Chem.* **7,** 879 (1968).

83. H. Weingarten and M. G. Miles, *J. Org. Chem.* **33,** 1506 (1968).

84. E. Benzing and W. Kornicker, *Chem. Ber.* **94,** 2263 (1961).

85. H. Burger and U. Wannagat, *Monatsh.* **94,** 761 (1963), and subsequent papers.

86. O. Meth-Cohn, D. Thorpe, and H. J. Twitchett, *J. Chem. Soc. C,* 132 (1970).

87. H. Burger, *Monatsh.* **95,** 671 (1964).

88. G. K. Parashar, P. C. Bharara, and R. C. Mehrotra, *Z. Naturforsch. Teil B* **34,** 109 (1979).

89. K. R. Nahar, A. K. Solanki, and A. M. Bhandari, *Z. Anorg. Allg. Chem.* **449,** 187 (1979).

90. R. K. Sharma, R. V. Singh, and J. P. Tandon, *Synth. React. Inorg. Met. Org. Chem.* **9,** 519 (1979).

91. E. C. Alyea, A. Malek, and P. H. Merrell, *Transition Met. Chem.* **4,** 172 (1979).

92. E. C. Alyea and A. Malek, *Inorg. Nucl. Chem. Lett.* **13,** 587 (1977).

93. E. C. Alyea, A. Malek, and P. H. Merrell, *J. Coord. Chem.* **4,** 55 (1974).

94. I. R. Unny, S. Gopinathan, and C. Gopinathan, *Indian J. Chem.* **19A,** 598 (1980).

95. J. F. Harrod and K. R. Taylor, *Inorg. Chem.* **14,** 1541 (1975).

96. P. P. Singh and S. B. Sharma, *Can. J. Chem.* **54,** 1563 (1976).
97. D. Bhatia and co-workers, *Synth. React. Inorg. Met. Org. Chem.* **9,** 95 (1979).
98. E. C. Alyea and co-workers, *Inorg. Nucl. Chem. Lett.* **9,** 399 (1973).
99. M. Colapietro, A. Vaciago, and D. C. Bradley, *J. Chem. Soc. Chem. Commun.,* 743 (1970); *J. Chem. Soc. Dalton Trans.,* 1052 (1972).
100. D. C. Bradley and M. H. Gitlitz, *J. Chem. Soc. A,* 1152 (1969).
101. J. Jones and J. Douek, *J. Inorg. Nucl. Chem.* **43,** 406 (1981).
102. I. P. Lorenz, *Inorg. Nucl. Chem. Lett.* **15,** 127 (1979).
103. T. J. Marks and J. R. Kolb, *Chem. Rev.* **77,** 263 (1977); P. A. Wegner in E. L. Muetterties, ed., *Boron Hydride Chemistry,* Academic Press, New York, 1973, Chapt. 12.
104. A. P. Hitchcock and co-workers, *Inorg. Chem.* **21,** 793 (1982).
105. D. W. MacCorquodale and H. Adkins, *J. Am. Chem. Soc.* **50,** 1938 (1928).
106. A. N. Nesmeyanov, O. V. Nogina, and R. Kh. Freidlina, *Dokl. Akad. Nauk SSSR* **95,** 813 (1954).
107. A. N. Nesmeyanov and co-workers, *Bull. Acad. Sci. USSR Div. Chem. Sci.,* 1117 (1960).
108. B. N. Chakravarti, *Naturwissenschaften* **45,** 286 (1958).
109. A. Pflugmacher and co-workers, *Naturwissenschaften* **45,** 490 (1958).
110. Ger. Pat. 1,091,105 (Oct. 20, 1960), A. Gumboldt and W. Herwig (to Hoechst).
111. A. Stahler and H. Wirthwein, *Chem. Ber.* **38,** 2619 (1905).
112. A. Monnier, *Ann. Chim. Anal.* **20,** 1 (1915).
113. C. M. Mikulski and co-workers, *J. Inorg. Nucl. Chem.* **41,** 1671 (1979).
114. F. E. McFarlane and G. W. Tindall, *Inorg. Nucl. Chem. Lett.* **9,** 907 (1973).
115. M. R. Hunt and G. Winter, *Inorg. Nucl. Chem. Lett.* **6,** 529 (1970).
116. T. D. Smith, T. Lund, and J. R. Pilbrow, *J. Chem. Soc. A,* 2786 (1971).
117. A. N. Glebov, P. A. Vasil'ev, and Yu. I. Sal'nikov, *Russ. J. Inorg. Chem.* **26,** 140 (1981).
118. A. A. Popel' and co-workers, *Russ. J. Inorg. Chem.* **24,** 1336 (1979).
119. Ya. I. Turyan and E. V. Saksin, *Russ. J. Anal. Chem.* **25,** 860 (1970).
120. E. C. Alyea, D. C. Bradley, and R. G. Copperthwaite, *J. Chem. Soc. Dalton Trans.,* 185, 191 (1973).
121. G. M. Bancroft, C. Reichert, and J. B. Westmore, *Inorg. Chem.* **7,** 870 (1968).
122. R. J. H. Clark and A. J. McAlees, *Inorg. Chem.* **11,** 342·(1972).
123. R. J. H. Clark and A. J. McAlees, *J. Chem. Soc. Dalton Trans.,* 640 (1972).
124. C. Blandy and D. Gervais, *Inorg. Chim. Acta* **47,** 197 (1981).
125. M. T. Reetz and J. Westermann, *Synth. Commun.* **11,** 647 (1981).
126. M. T. Reetz, R. Steinbach, and B. Wenderoth, *Synth. Commun.* **11,** 261 (1981).
127. M. T. Reetz, J. Westermann, and R. Steinbach, *Angew. Chem. Int. Ed. Engl.* **19,** 900 (1980).
128. M. T. Reetz, J. Westermann, and R. Steinbach, *J. Chem. Soc. Chem. Commun.,* 237 (1981).
129. M. T. Reetz, J. Westermann, and R. Steinbach, *Angew. Chem. Int. Ed. Engl.* **19,** 901 (1980).
130. M. T. Reetz and co-workers, *Angew. Chem. Int. Ed. Engl.* **19,** 1011 (1980).
131. W. Seidel and E. Riesenberg, *Z. Chem.* **20,** 450 (1980).
132. B. Weidmann and co-workers, *Helv. Chim. Acta* **64,** 357 (1981).
133. A. G. Olivero, B. Weidmann, and D. Seebach, *Helv. Chim. Acta* **64,** 2485 (1981).
134. B. Weidmann and D. Seebach, *Helv. Chim. Acta* **63,** 2451 (1980).
135. P. J. Davidson, M. F. Lappert, and R. Pearce, *J. Organomet. Chem.* **57,** 269 (1973).
136. J. C. W. Chien, J. Wu, and M. D. Rausch, *J. Am. Chem. Soc.* **103,** 1180 (1981).
137. A. K. Rappe and W. A. Goddard, *J. Am. Chem. Soc.* **104,** 297 (1982).
138. A. R. Middleton and G. Wilkinson, *J. Chem. Soc. Dalton Trans.,* 1888 (1980).
139. M. D. Schiavelli, J. J. Plunkett, and D. W. Thompson, *J. Org. Chem.* **46,** 807 (1981).
140. T. Yoshida and E. Negishi, *J. Am. Chem. Soc.* **103,** 1276 (1981).
141. U.S. Pat. 3,694,517 (Sept. 26, 1972), W. Schneider (to B. F. Goodrich); Jpn. Kokai Tokkyo Koho 75 88,059 (to Montedison); Ital. Pat. 932,191 (Nov. 15, 1972), (to Montedison).
142. A. Clearfield, I. Bernal, and co-workers, *Can. J. Chem.* **53,** 1622 (1975).
143. V. V. Tkachev and L. O. Atomyan, *Zh. Strukt. Khim.* **13,** 287 (1972).
144. I. A. Ronova and N. V. Alekseev, *Zh. Strukt. Khim.* **18,** 212 (1977).
145. E. F. Epstein and I. Bernal, *Inorg. Chim. Acta* **7,** 211 (1973).
146. V. Kocman and co-workers, *J. Chem. Soc. Chem. Commun.,* 1340 (1971).
147. D. F. Hunt, J. W. Russell, and R. L. Torian, *J. Organomet. Chem.* **43,** 175 (1972).
148. A. N. Nesmeyanov and co-workers, *J. Organomet. Chem.* **61,** 225 (1973).
149. A. N. Nesmeyanov and co-workers, *Zh. Strukt. Khim.* **13,** 1033 (1972); **16,** 759 (1975).
150. M. Hillman and A. J. Weiss, *J. Organomet. Chem.* **42,** 123 (1972).
151. A. N. Nesmeyanov and co-workers, *Dokl. Chem.* **163,** 704 (1965).

152. C. Cauletti, J. C. Green, and co-workers, *J. Electron Spectrosc. Relat. Phenom.* **18**, 61 (1980).
153. A. A. MacDowell, J. C. Green, and co-workers, *J. Chem. Soc. Chem. Commun.*, 427 (1979).
154. O. S. Roshchupkina and co-workers, *Soviet J. Coord. Chem.* **1**, 1052 (1975).
155. E. Samuel, R. Ferner, and M. Bigorgne, *Inorg. Chem.* **12**, 881 (1973).
156. G. Balducci and co-workers, *J. Mol. Struct.* **64**, 163 (1980).
157. M. Spoliti and co-workers, *J. Mol. Struct.* **65**, 105 (1980).
158. N. N. Vyshinskii and co-workers, *Tr. Khim. Khim. Tekhnol.*, 64 (1973); 119, 123 (1974).
159. G. Condorelli and co-workers, *J. Organomet. Chem.* **87**, 311 (1975).
160. D. W. Clack and K. D. Warren, *Theor. Chim. Acta* **46**, 313 (1977); *Inorg. Chim. Acta* **24**, 35 (1977); **30**, 251 (1978); *J. Organomet. Chem.* **162**, 83 (1978).
161. V. E. L'vovskii, *Soviet J. Coord. Chem.* **4**, 1266 (1978).
162. V. E. L'vovskii and G. B. Erusalimskii, *Soviet J. Coord. Chem.* **2**, 934, 1221 (1976).
163. E. G. Muller, S. F. Watkins, and L. F. Dahl, *J. Organomet. Chem.* **111**, 73 (1976).
164. E. G. Muller, J. L. Petersen, and L. F. Dahl, *J. Organomet. Chem.* **111**, 91 (1976).
165. J. L. Petersen, D. L. Lichtenberger, R. F. Fenske, and L. F. Dahl, *J. Am. Chem. Soc.* **97**, 6433 (1975).
166. A. M. McPherson, G. D. Stucky, and co-workers, *J. Am. Chem. Soc.* **101**, 3425 (1979).
167. V. E. Lvovsky, E. A. Fushman, and F. S. Dyachkovsky, *J. Mol. Catal.* **10**, 43 (1981).
168. T. D. Grabik and co-workers, *Russ. J. Phys. Chem.* **52**, 894 (1978).
169. V. I. Tel'noi and I. B. Rabinovich, *Conf. Int. Thermodyn. Chim.* [*C.R.*], *4th*, 1, 98 (1975).
170. T. Mukaiyama, *Angew. Chem. Int. Ed. Engl.* **16**, 817 (1977); J. Lange, J. M. Kanabus-Kaminska, and A. Kral, *Synth. Commun.* **10**, 473 (1980); M. T. Reetz, *Angew. Chem. Int. Ed. Engl.* **21**, 96 (1982).
171. R. C. Paul and co-workers, *J. Less Common Met.* **17**, 437 (1969).
172. A. M. Cardoso, R. J. H. Clark, and S. Moorhouse, *J. Chem. Soc. Dalton Trans.*, 1156 (1980).
173. G. Wilkinson and J. M. Birmingham, *J. Am. Chem. Soc.* **76**, 4281 (1954).
174. L. Summers and R. H. Uloth, *J. Am. Chem. Soc.* **76**, 2278 (1954); **77**, 3604 (1955).
175. R. B. King, *Organometallic Syntheses*, Vol. 1, Academic Press, New York, 1965, pp. 75–78.
176. R. D. Gorsich, *J. Am. Chem. Soc.* **82**, 4211 (1960).
177. USSR Pat. 825,534 (May 7, 1981), Y. A. Sorokin and co-workers.
178. Brit. Pat. 793,354 (April 16, 1958), (to National Lead Co.).
179. Brit. Pat. 798,001 (July 9, 1958), (to National Lead Co.).
180. J. A. Marsella, K. G. Moloy, and K. G. Caulton, *J. Organomet. Chem.* **201**, 389 (1980).
181. L. E. Manzer, *J. Organomet. Chem.* **110**, 291 (1976).
182. F. H. Kohler and D. Cozek, *Z. Naturforsch. Teil B* **33**, 1274 (1978).
183. D. Nath, R. K. Sharma, and A. N. Bhat, *Inorg. Chim. Acta* **20**, 109 (1976).
184. K. Chandra, P. Soni, B. S. Garg, and R. P. Singh, *J. Indian Chem. Soc.* **58**, 10 (1981).
185. R. K. Tuli, B. S. Garg, and co-workers, *Trans. Met. Chem.* **5**, 49 (1980).
186. K. Chandra, B. S. Garg, and co-workers, *Inorg. Chim. Acta* **37**, 125 (1979).
187. A. K. Holliday, P. H. Makin, R. J. Puddephatt, and J. D. Wilkins, *J. Organomet. Chem.* **57**, C45 (1973).
188. A. N. Nesmeyanov and co-workers, *Bull. Acad. Sci. USSR Div. Chem. Sci.*, 777 (1967).
189. K. Chandra, B. S. Garg, and co-workers, *Chem. Industry* (London), 288 (1980).
190. M. F. Lappert and co-workers, *J. Chem. Soc. Dalton Trans.*, 805 (1981).
191. J. E. Bercaw, *J. Am. Chem. Soc.* **96**, 5087 (1974).
192. T. J. Katz, N. Acton, and G. Martin, *J. Am. Chem. Soc.* **95**, 2934 (1973).
193. J. A. Smith, J. v. Seyerl, G. Huttner, and H. H. Brintzinger, *J. Organomet. Chem.* **173**, 175 (1979).
194. J. A. Smith and H. H. Brintzinger, *J. Organomet. Chem.* **218**, 159 (1981).
195. J. Jeffery, M. F. Lappert, J. L. Atwood, and co-workers, *J. Chem. Soc. Dalton Trans.*, 1593 (1981).
196. G. K. Barker and M. F. Lappert, *J. Organomet. Chem.* **76**, C45 (1974).
197. B. Wozniak, J. D. Ruddick, and G. Wilkinson, *J. Chem. Soc. A*, 3116 (1971).
198. J. X. McDermott, M. E. Wilson, and G. M. Whitesides, *J. Am. Chem. Soc.* **98**, 6529 (1976).
199. J. X. McDermott and G. M. Whitesides, *J. Am. Chem. Soc.* **96**, 947 (1974).
200. N. J. Foulger and B. J. Wakefield, *J. Organomet. Chem.* **69**, 161 (1974).
201. R. Jimenez and co-workers, *J. Organomet. Chem.* **174**, 281 (1979).
202. M. F. Lappert and C. L. Raston, *J. Chem. Soc. Chem. Commun.*, 1284 (1980).
203. H. Sakurai and H. Umino, *J. Organomet. Chem.* **142**, C49 (1977); Jpn. Pat. 81 6437 (Feb. 10, 1981), (to Mitsubishi Chemical Industries KK).
204. T. Bounthakna, J. C. LeBlanc, and C. Moise, *C. R. Acad. Sci. Ser. C* **280**, 1431 (1975).

205. S. A. Giddings, *Inorg. Chem.* **3,** 684 (1964).
206. H. Surer, S. Claude, and A. Jacot-Guillarmod, *Helv. Chim. Acta* **61,** 2956 (1978).
207. T. P. Bryukhanova and co-workers, *Tr. Khim. Khim. Tekhnol.*, 101 (1974); *Chem. Abstr.* **83,** 141516.
208. J. J. Eisch and R. J. Manfre, *Fundam. Res. Homog. Catal.* **3,** 397 (1978).
209. J. J. Eisch, R. J. Manfrę, and D. A. Komar, *J. Organomet. Chem.* **159,** C13 (1978).
210. B. B. Snider and M. Karras, *J. Organomet. Chem.* **179,** C37 (1979).
211. D. C. Brown, S. A. Nichols, H. B. Gilpin, and D. W. Thompson, *J. Org. Chem.* **44,** 3457 (1979).
212. H. E. Tweedy, D. W. Thompson, and co-workers, *J. Mol. Catal.* **3,** 239 (1977–1978).
213. U.S. Pat. 3,969,429 (July 13, 1976), G. P. Belov and co-workers.
214. M. Peng and C. H. Brubaker, *Inorg. Chim. Acta* **26,** 231 (1978).
215. M. D. Rausch, W. H. Boon, and E. A. Mintz, *J. Organomet. Chem.* **160,** 81 (1978).
216. M. D. Rausch, W. H. Boon, and H. G. Alt, *J. Organomet. Chem.* **141,** 229 (1977).
217. W. H. Boon and M. D. Rausch, *J. Chem. Soc. Chem. Commun.*, 397 (1977).
218. J. L. Atwood, W. E. Hunter, H. Alt, and M. D. Rausch, *J. Am. Chem. Soc.* **98,** 1454 (1976).
219. H. Alt and M. D. Rausch, *J. Am. Chem. Soc.* **96,** 5936 (1974).
220. H. G. Alt and M. D. Rausch, *Z. Naturforsch. Teil B* **30,** 813 (1975).
221. R. W. Harrigan, G. S. Hammond, and H. B. Gray, *J. Organomet. Chem.* **81,** 79 (1974).
222. C. H. Bamford, R. J. Puddephatt, and D. M. Slater, *J. Organomet. Chem.* **159,** C31 (1978).
223. H. G. Alt, F. P. Di Sanzo, M. D. Rausch, and P. C. Uden, *J. Organomet. Chem.* **107,** 257 (1976).
224. G. A. Razuvaev, V. P. Mar'in, and Yu. A. Andrianov, *J. Organomet. Chem.* **174,** 67 (1979); G. A. Razuvaev and co-workers, *J. Organomet. Chem.* **164,** 41 (1979).
225. J. Mattia, M. B. Humphrey, R. D. Rogers, J. L. Atwood, and M. D. Rausch, *Inorg. Chem.* **17,** 3257 (1978).
226. E. G. Berkovich and co-workers, *Chem. Ber.* **113,** 70 (1980).
227. V. B. Shur and co-workers, *J. Organomet. Chem.* **78,** 127 (1974).
228. M. Kh. Grigoryan and co-workers, *Bull. Acad. Sci. USSR, Div. Chem. Sci.* 1024 (1978).
229. M. D. Rausch and H. B. Gordon, *J. Organomet. Chem.* **74,** 85 (1974).
230. G. J. Erskine, J. Hartgerink, E. L. Weinberg, and J. D. McCowan, *J. Organomet. Chem.* **170,** 51 (1979).
231. G. J. Erskine, D. A. Wilson, and J. D. McCowan, *J. Organomet. Chem.* **114,** 119 (1976).
232. J. A. Waters, V. V. Vickroy, and G. A. Mortimer, *J. Organomet. Chem.* **33,** 41 (1971).
233. C. P. Boekel, J. H. Teuben, and H. J. de Liefde Meijer, *J. Organomet. Chem.* **102,** 161 (1975).
234. *Ibid.*, p. 317.
235. C. P. Boekel, J. H. Teuben, and H. J. de Liefde Meijer, *J. Organomet. Chem.* **81,** 371 (1974).
236. F. W. Hartner and J. Schwartz, *J. Am. Chem. Soc.* **103,** 4974 (1981).
237. K. G. Ott and R. H. Grubbs, *J. Am. Chem. Soc.* **103,** 5922 (1981); (for a very recent discussion) *Chem. Eng. News*, 34 (April 19, 1982).
238. T. Yoshida and E. I. Negishi, *J. Am. Chem. Soc.* **103,** 1276 (1981).
239. Y. Le Page, J. D. McCowan, B. K. Hunter, and R. D. Heyding, *J. Organomet. Chem.* **193,** 201 (1980).
240. A. Glivicky and J. D. McCowan, *Can. J. Chem.* **51,** 2609 (1973).
241. L. J. Baye, *Synth. React. Inorg. Met. Org. Chem.* **5,** 95 (1975).
242. *Ibid.*, **2,** 47 (1972).
243. C. R. Bennett and D. C. Bradley, *J. Chem. Soc. Chem. Commun.*, 29 (1974).
244. R. V. Bynum, W. E. Hunter, R. D. Rogers, and J. L. Atwood, *Inorg. Chem.* **19,** 2368 (1980).
245. P. L. Maxfield and E. Lima, *Proceedings of the 16th International Conference on Coord. Chem.*, R58, 1974.
246. A. Jensen, J. L. Burmeister, and co-workers, *Acta Chem. Scand.* **26,** 2898 (1972).
247. J. L. Burmeister, A. Jensen, and co-workers, *Inorg. Chem.* **9,** 58 (1970).
248. S. J. Anderson, D. S. Brown, and K. J. Finney, *J. Chem. Soc. Dalton Trans.*, 152 (1979).
249. S. J. Anderson, D. S. Brown, and A. H. Norbury, *J. Chem. Soc. Chem. Commun.*, 996 (1974).
250. J. Besançon and D. Camboli, *Compt. Rend.* **288C,** 121 (1979).
251. A. Chiesi-Villa, A. G. Manfredotti, and C. Guastini, *Acta Cryst.* **B32,** 909 (1976).
252. K. Issleib, H. Köhler, and G. Wille, *Z. Chem.* **13,** 347 (1973).
253. D. Camboli, J. Besançon, and B. Trimaille, *Compt. Rend.* **290C,** 365 (1980).
254. J. Besançon, J. Tirouflet, and co-workers, *Compt. Rend.* **287C,** 573 (1978).
255. L. Saunders and L. Spirer, *Polymer* **6,** 635 (1965).

256. R. B. King and R. N. Kapoor, *J. Organomet. Chem.* **15**, 457 (1968).
257. G. A. Razuvaev, V. N. Latyaeva, and L. I. Vyshinskaya, *Dokl. Chem.* **138**, 592 (1961).
258. T. S. Kuntsevich and co-workers, *Kristallografiya* **21**, 80 (1976).
259. V. P. Nistratov and co-workers, *Tr. Khim. Khim. Tekhnol.*, 54 (1975).
260. J. C. G. Calado and co-workers, *J. Chem. Soc. Dalton Trans.*, 1174 (1981).
261. L. Rösch and co-workers, *J. Organomet. Chem.* **197**, 51 (1980).
262. E. Hengge and W. Zimmermann, *Angew. Chem. Intl. Ed.* **7**, 142 (1968).
263. G. A. Razuvaev and co-workers, *Bull. Acad. Sci. USSR Div. Chem. Sci.*, 2310 (1978).
264. G. A. Razuvaev and co-workers, *J. Organomet. Chem.* **87**, 93 (1975).
265. B. Stütte, V. Bätzel, R. Boese, and G. Schmid, *Chem. Ber.* **111**, 1603 (1978).
266. G. Schmid, V. Bätzel, and B. Stütte, *J. Organomet. Chem.* **113**, 67 (1976).
267. H. G. Raubenheimer and E. O. Fischer, *J. Organomet. Chem.* **91**, C23 (1975).
268. E. O. Fischer and S. Fontana, *J. Organomet. Chem.* **40**, 159 (1972).
269. H. Köpf and R. Voigtländer, *Chem. Ber.* **114**, 2731 (1981).
270. K. Issleib, F. Krech, and E. Lapp, *Synth. React. Inorg. Met. Org. Chem.* **7**, 253 (1977).
271. K. Issleib, G. Wille, and F. Krech, *Angew. Chem. Intl. Ed.* **11**, 527 (1972).
272. J. M. McCall and A. Shaver, *J. Organomet. Chem.* **193**, C37 (1980).
273. E. Samuel, *Bull. Soc. Chim. France*, 3548 (1966).
274. J. L. Petersen and L. F. Dahl, *J. Am. Chem. Soc.* **96**, 2248 (1974).
275. J. C. LeBlanc, C. Moise, and T. Bounthakna, *Compt. Rend.* **278C**, 973 (1974).
276. M. B. Bert and D. Gervais, *J. Organomet. Chem.* **165**, 209 (1979).
277. G. V. Drozdov, A. L. Klebanskii, and V. A. Bartashev, *J. Gen. Chem. USSR* **33**, 2362 (1963).
278. M. A. Chaudhari and F. G. A. Stone, *J. Chem. Soc. A*, 838 (1966); M. A. Chaudhari, P. M. Treichel, and F. G. A. Stone, *J. Organomet. Chem.* **2**, 206 (1964).
279. K. Andrä, *J. Organomet. Chem.* **11**, 567 (1968).
280. J. C. LeBlanc, C. Moise, and J. Tirouflet, *Nouv. J. Chim.* **1**, 211 (1977).
281. T. Marey, J. Besançon, and co-workers, *Compt. Rend.* **284C**, 967 (1977).
282. J. Besançon, S. Top, and J. Tirouflet, *Compt. Rend.* **281C**, 135 (1975).
283. J. Besançon, F. Hug, and M. Colette, *J. Organomet. Chem.* **96**, 63 (1975).
284. R. Sharan, G. Gupta, and R. N. Kapoor, *Indian J. Chem.* **20A**, 94 (1981).
285. R. Sharan, G. Gupta, and R. N. Kapoor, *J. Less Com. Met.* **60**, 171 (1978).
286. A. R. Dias, M. S. Salema, and J. A. Martinho Simões, *J. Organomet. Chem.* **222**, 69 (1981).
287. H. Suzuki and T. Takiguchi, *Bull. Chem. Soc. Jpn.* **48**, 2460 (1975).
288. R. S. Arora, M. B. Bhalla, and R. K. Multani, *Indian J. Chem.* **16A**, 169 (1978).
289. R. S. P. Coutts, J. R. Surtees, J. M. Swan, and P. C. Wailes, *Austr. J. Chem.* **19**, 1377 (1966).
290. M. Sato and T. Yoshida, *J. Organomet. Chem.* **67**, 395 (1974).
291. H. Köpf and M. Schmid, *Angew. Chem. Intl. Ed.* **4**, 953 (1965).
292. A. Kutoğlu, *Acta Cryst.* **B29**, 2891 (1973).
293. A. Kutoğlu, *Z. Anorg. Allgem. Chem.* **390**, 195 (1972).
294. D. Sen and U. N. Kantak, *Indian J. Chem.* **13**, 72 (1975).
295. S. K. Sengupta, *Indian J. Chem.* **20A**, 515 (1981).
296. R. S. Arora and co-workers, *J. Chinese Chem. Soc. (Taipei)* **27**, 65 (1980).
297. W. L. Steffen, H. K. Chun, and R. C. Fay, *Inorg. Chem.* **17**, 3498 (1978).
298. K. Chandra, R. K. Sharma, B. S. Garg, and R. P. Singh, *J. Inorg. Nucl. Chem.* **43**, 663 (1981).
299. K. Chandra, R. K. Tuli, B. S. Garg, and R. P. Singh, *J. Inorg. Nucl. Chem.* **43**, 29 (1981).
300. K. Chandra, R. K. Sharma, B. S. Garg, and R. P. Singh, *Trans. Met. Chem.* **5**, 209 (1980).
301. N. K. Kaushik, B. Bhushan, and G. R. Chhatwal, *Z. Naturforsch.* **34b**, 949 (1979).
302. B. Bhushan, I. P. Mittal, G. R. Chhatwal, and N. K. Kaushik, *J. Inorg. Nucl. Chem.* **41**, 159 (1979).
303. N. K. Kaushik, B. Bhushan, and G. R. Chhatwal, *Trans. Met. Chem.* **3**, 215 (1978).
304. N. K. Kaushik, B. Bhushan, and G. R. Chhatwal, *Synth. React. Inorg. Met. Org. Chem.* **8**, 467 (1978).
305. P. C. Bharara, *J. Organomet. Chem.* **121**, 199 (1976).
306. O. N. Suvorova, V. V. Sharutin, and G. A. Domrachev, *3rd Mater. Vses Semin. 1977*, 132 (1978); *Chem. Abstr.* **91**, 157839 (1979).
307. G. Fachinetti and co-workers, *J. Am. Chem. Soc.* **100**, 1921 (1978).
308. G. G. Dvoryantseva and co-workers, *Dokl. Chem.* **161**, 303 (1965).

309. A. N. Nesmeyanov, O. V. Nogina, and V. A. Dubovitskii, *Bull. Acad. Sci. USSR Div. Chem. Sci.*, 1395 (1962).
310. A. N. Nesmeyanov, O. V. Nogina, and A. M. Berlin, *Bull. Acad. Sci. USSR Div. Chem. Sci.*, 743 (1961).
311. G. Wilkinson and J. M. Birmingham, *J. Am. Chem. Soc.* **76**, 4281 (1954).
312. K. Döppert and R. Sanchez, *J. Organomet. Chem.* **210**, C9 (1980).
313. K. Döppert, *Makromol. Chem. Rapid Commun.* **1**, 519 (1980).
314. K. Döppert, *J. Organomet. Chem.* **178**, C3 (1979).
315. K. Chandra, R. K. Tuli, N. K. Bhatia, and B. S. Garg, *J. Indian Chem. Soc.* **58**, 122 (1981).
316. D. Pacheco and co-workers, *Inorg. Chim. Acta* **18**, L24 (1976).
317. H. Klein and U. Thewalt, *Z. Anorg. Allgem. Chem.* **476**, 62 (1981).
318. P. M. Druce, M. F. Lappert, and co-workers, *J. Chem. Soc. A*, 2106 (1969).
319. V. K. Jain, N. K. Bhatia, K. C. Sharma, and B. S. Garg, *J. Indian Chem. Soc.* **57**, 6 (1980).
320. D. Nath and A. N. Bhat, *Indian J. Chem.* **14**, 281 (1976).
321. C. E. Carraher and J. D. Piersma, *Makromol. Chem.* **152**, 49 (1972).
322. C. E. Carraher and co-workers, *J. Macromol. Sci. Chem.* **A16**, 195 (1981).
323. C. E. Carraher and J. L. Lee, *J. Macromol. Sci. Chem.* **A9**, 191 (1975).
324. C. E. Carraher and J. L. Lee, *Am. Chem. Soc. Div. Org. Coat. Plast. Chem. Pap.* **34**, 478 (1974).
325. C. E. Carraher and R. Frary, *Br. Polym. J.* **6**, 255 (1974).
326. C. E. Carraher and R. A. Frary, *Makromol. Chem.* **175**, 2307 (1974).
327. C. E. Carraher and J. D. Piersma, *J. Macromol. Sci. Chem.* **A7**, 913 (1973).
328. C. E. Carraher and L. S. Wang, *Angew. Makromol. Chem.* **25**, 121 (1972).
329. P. C. Wailes, H. Weigold, and A. P. Bell, *J. Organomet. Chem.* **33**, 181 (1971).
330. U.S. Pat. 3,782,917 (Jan. 1, 1974), J. J. Mrowca (to DuPont).
331. U.S. Pat. 3,728,365 (April 17, 1973), J. J. Mrowca (to DuPont).
332. K. W. Chiu, G. Wilkinson, and co-workers, *J. Chem. Soc. Dalton Trans.*, 2088 (1981).
333. R. J. H. Clark, J. A. Stockwell, and J. D. Wilkins, *J. Chem. Soc. Dalton Trans.*, 120 (1976).
334. G. Fachinetti and C. Floriani, *J. Organomet. Chem.* **71**, C5 (1974).
335. G. Fachinetti and C. Floriani, *J. Chem. Soc. Chem. Commun.* 654 (1972).
336. H. Masai, K. Sonogashira, and N. Hagihara, *Bull. Soc. Chem. Jpn.* **41**, 750 (1968).
337. H. Yasuda and co-workers, *Bull. Chem. Soc. Jpn.* **53**, 1089 (1980).
338. S. Akutagawa and S. Otsuka, *J. Am. Chem. Soc.* **97**, 6870 (1975).
339. F. Sato and co-workers, *J. Chem. Soc., Chem. Commun.*, 1140 (1981).
340. K. Kaeriyama, *Makromol. Chem.* **175**, 2285 (1974).
341. K. Kaeriyama, *Makromol. Chem.* **153**, 229 (1972).
342. T. Saito, *J. Chem. Soc. Chem. Commun.*, 1422 (1971).
343. J. Tsuji and S. Hashiguchi, *Tetrahedron Lett.* **21**, 2955 (1980).
344. D. R. Corbin, G. D. Stucky, and co-workers, *J. Am. Chem. Soc.* **102**, 5969 (1980).
345. D. Sekutowski, R. Jungst, and G. D. Stucky, *Inorg. Chem.* **17**, 1848 (1978).
346. R. S. P. Coutts, R. L. Martin, and P. C. Wailes, *Austr. J. Chem.* **24**, 2533 (1971).
347. C. R. Lucas and M. L. H. Green, *Inorg. Synth.* **16**, 237 (1976).
348. T. Chivers and E. D. Ibrahim, *J. Organomet. Chem.* **77**, 241 (1974).
349. M. F. Lappert, P. I. Riley, and P. I. W. Yarrow, *J. Chem. Soc. Chem. Commun.*, 305 (1979); *J. Chem. Soc. Dalton Trans.*, 805 (1981).
350. A. Chaloyard, A. Dormond, J. Tirouflet and N. El Murr, *J. Chem. Soc. Chem. Commun.*, 214 (1980).
351. N. El Murr, A. Chaloyard, and J. Tirouflet, *J. Chem. Soc. Chem. Commun.*, 446 (1980).
352. E. Laviron, J. Besançon, and F. Hug, *J. Organomet. Chem.* **159**, 279 (1978).
353. R. S. P. Coutts, R. L. Martin, and P. C. Wailes, *Austr. J. Chem.* **26**, 2101 (1973).
354. L. C. Francesconi, G. D. Stucky, and co-workers, *Inorg. Chem.* **20**, 2059 (1981).
355. L. S. Kramer, G. D. Stucky, and co-workers, *Inorg. Chem.* **20**, 2070 (1981).
356. L. C. Francesconi, G. D. Stucky, and co-workers, *Inorg. Chem.* **20**, 2078 (1981).
357. D. R. Corbin, G. D. Stucky, and co-workers, *Inorg. Chem.* **20**, 2084 (1981).
358. J. C. Huffman, K. G. Moloy, J. A. Marsella, and K. G. Caulton, *J. Am. Chem. Soc.* **102**, 3009 (1980).
359. B. Çetinkaya, M. F. Lappert, J. L. Atwood, and co-workers, *J. Organomet. Chem.* **188**, C31 (1980).
360. M. C. Vanderveer and J. M. Burlitch, *J. Organomet. Chem.* **197**, 357 (1980).
361. D. G. Sekutowski and G. D. Stucky, *Inorg. Chem.* **14**, 2192 (1975).

362. R. S. P. Coutts, R. L. Martin, and P. C. Wailes, *Austr. J. Chem.* **26**, 47 (1973).
363. R. S. P. Coutts, *Inorg. Nucl. Chem. Lett.* **13**, 41 (1977).
364. D. R. Corbin, G. D. Stucky, and co-workers, *Inorg. Chem.* **18**, 3069 (1979).
365. L. C. Francesconi, G. D. Stucky, and co-workers, *Inorg. Chem.* **18**, 3074 (1979).
366. R. S. P. Coutts and P. C. Wailes, *Austr. J. Chem.* **27**, 2483 (1974).
367. A. A. Pasynskii and co-workers, *Soviet J. Coord. Chem.* **3**, 1177 (1977).
368. R. S. P. Coutts, R. L. Martin, and P. C. Wailes, *Austr. J. Chem.* **26**, 941 (1973).
369. F. Bottomley, I. J. B. Lin, and P. S. White, *J. Organomet. Chem.*, 341 (1981).
370. G. Fachinetti and C. Floriani, *J. Chem. Soc. Dalton Trans.*, 2433 (1974).
371. J. G. Kenworthy, J. Myatt, and M. C. R. Symons, *J. Chem. Soc. A*, 3428 (1971).
372. K. M. Melmed, D. Coucouvanis, and S. J. Lippard, *Inorg. Chem.* **12**, 232 (1973).
373. T. J. Marks and co-workers, *Inorg. Chem.* **11**, 2540 (1972).
374. G. L. Soloveichik and co-workers, *Tezisy Dokl. Vses. Chugaevskoe Soveshch Khim. Komplesksn. Soedin* **3**, 466 (1975); *Chem. Abstr.* **86**, 5574 (1977).
375. M. L. H. Green and C. R. Lucas, *J. Organomet. Chem.* **73**, 259 (1974).
376. M. L. H. Green and C. R. Lucas, *J. Chem. Soc. Dalton Trans.*, 1000 (1972).
377. J. H. Teuben and H. J. de Liefde Meijer, *J. Organomet. Chem.* **46**, 313 (1972).
378. D. Ytsma, J. G. Hartsuiker, and J. H. Teuben, *J. Organomet. Chem.* **74**, 239 (1974).
379. R. H. Grubbs, C. P. Lau, R. Cukier, and C. H. Brubaker, *J. Am. Chem. Soc.* **99**, 4517 (1977).
380. W. D. Bonds, C. H. Brubaker, and co-workers, *J. Am. Chem. Soc.* **97**, 2128 (1975).
381. D. E. Bergbreiter and G. L. Parsons, *J. Organomet. Chem.* **208**, 47 (1981).
382. F. Sato and co-workers, *Tetrahedron Lett.*, 3745 (1979).
383. M. Koide, E. Tsuchida, and Y. Kurimura, *Makromol. Chem.* **182**, 749 (1981).
384. M. Koide and co-workers, *Polymer J.* **12**, 793 (1980).
385. D. W. Macomber and co-workers, *J. Am. Chem. Soc.* **104**, 884 (1982).
386. D. Slotfeldt-Ellingson, I. M. Dahl, and O. H. Ellestad, *J. Mol. Catal.* **9**, 423 (1980).
387. W. Skupinski, I. Cieslowska, and S. Malinowski, *J. Organomet. Chem.* **182**, C33 (1979).
388. V. V. Saraev and co-workers, *Soviet J. Coord. Chem.* **3**, 1064 (1977).
389. F. A. Shmidt and co-workers, *React. Kinet. Catal. Lett.* **5**, 101 (1976).
390. H. A. Martin, M. Van Gorkom, and R. O. De Jongh, *J. Organomet. Chem.* **36**, 93 (1972).
391. J. E. Bercaw, H. H. Brintzinger, and co-workers, *J. Am. Chem. Soc.* **94**, 1219 (1972).
392. F. N. Tebbe and L. J. Guggenberger, *J. Chem. Soc. Chem. Commun.*, 227 (1973).
393. F. N. Tebbe, G. W. Parshall, and G. S. Reddy, *J. Am. Chem. Soc.* **100**, 3611 (1978).
394. D. G. H. Ballard and R. Pearce, *J. Chem. Soc. Chem. Commun.*, 621 (1975).
395. J. Holton, M. F. Lappert, J. L. Atwood, and co-workers, *J. Chem. Soc. Dalton Trans.*, 45 (1979).
396. L. N. Sosnovskaya and co-workers, *J. Mol. Catal.* **9**, 411 (1980).
397. L. E. Manzer, *Inorg. Chem.* **15**, 2567 (1976).
398. H. Schmidbaur, W. Scharf, and H. J. Füller, *Z. Naturforsch.* **32b**, 858 (1977).
399. G. Hencken and E. Weiss, *Chem. Ber.* **106**, 1747 (1973).
400. H. Antropiusová and co-workers, *Trans. Met. Chem.* **6**, 90 (1981).
401. K. Mach and co-workers, *Trans. Met. Chem.* **5**, 5 (1980).
402. K. Mach, H. Antropiusová, and J. Poláček, *J. Organomet. Chem.* **194**, 285 (1980).
403. F. Tureček and co-workers, *Tetrahedron Lett.* **21**, 637 (1980).
404. H. Antropiusová and co-workers, *React. Kinet. Catal. Lett.* **10**, 297 (1979).
405. H. Antropiusová, V. Hanuš, and K. Mach, *Trans. Met. Chem.* **3**, 121 (1978).
406. B. Meunier, *J. Organomet. Chem.* **204**, 345 (1981).
407. E. Colomer and R. Corriu, *J. Organomet. Chem.* **82**, 367 (1974).
408. R. J. P. Corriu and B. Meunier, *J. Organomet. Chem.* **65**, 187 (1974).
409. F. Sato and co-workers, *Chem. Lett.*, 103 (1980).
410. H. Lehmkuhl and S. Fustero, *Liebigs Ann. Chem.*, 1353, 1361, 1371 (1980).
411. K. Tamao and co-workers, *J. Organomet. Chem.* **226**, C9 (1982).
412. L. G. Cannell, *J. Am. Chem. Soc.* **94**, 6867 (1972).
413. A. B. Amerik, V. M. Vdovin, and V. A. Poletaev, *Bull. Acad. Sci. USSR Div. Chem. Sci.*, 144 (1977).
414. A. B. Amerik and V. M. Vdovin, *Bull. Acad. Sci. USSR Div. Chem. Sci.*, 851 (1979).
415. F. Sato, H. Ishikawa, and M. Sato, *Tetrahedron Lett.* **22**, 85 (1981); **21**, 365 (1980).
416. U.S. Pat. 4,200,716 (April 29, 1980), G. Pez (to Allied Chemical Corp.).
417. E. Samuel, *J. Organomet. Chem.* **198**, C65 (1981).

418. P. C. Wailes, H. Weigold, and A. P. Bell, *J. Organomet. Chem.* **43**, C32 (1972).

419. Ger. Offen. 2,925,626 (Jan. 22, 1981), P. Haenssle and R. Streck (to Hüls).

420. K. Isagawa and co-workers, *Chem. Lett.*, 1155 (1978).

421. K. Isagawa, K. Tatsumi, and Y. Otsuji, *Chem. Lett.*, 1117 (1977).

422. K. Isagawa and co-workers, *Chem. Lett.*, 1017 (1977).

423. K. Isagawa, K. Tatsumi, and Y. Otsuji, *Chem. Lett.*, 1145 (1976).

424. E. V. Evdokimova, B. M. Bulychev, and G. L. Soloveichik, *Kinet. Catal.* **22**, 144 (1981).

425. B. M. Bulychev and co-workers, *J. Organomet. Chem.* **179**, 263 (1979).

426. B. M. Bulychev, *Bull. Acad. Sci. USSR Inorg. Mater.* **14**, 1350 (1978).

427. S. E. Tokaeva and co-workers, *Bull. Acad. Sci. USSR Inorg. Mater.* **14**, 1355 (1978).

428. G. L. Soloveichik and co-workers, *Soviet J. Coord. Chem.* **4**, 909 (1978).

429. G. L. Soloveichik, B. M. Bulychev, and K. N. Semenenko, *Soviet J. Coord. Chem.* **4**, 913 (1978).

430. N. A. Yakovleva, G. L. Soloveichik, and B. M. Bulychev, *Bull. Acad. Sci. USSR Div. Chem. Sci.*, 1228 (1978).

431. K. Isagawa and co-workers, *Chem. Lett.*, 1069 (1979).

432. Jpn. Kokai Tokkyo Koho 80 162,726 (Dec. 18, 1980), (to Grelan Pharmaceutical Co.).

433. J. Tsuji and T. Mandai, *Chem. Lett.*, 975 (1977); Jpn. Kokai Tokkyo Koho 78 149,907 (Dec. 27, 1978), (to Chisso Corp.).

434. C. R. Lucas, *Inorg. Synth.* **17**, 91 (1977).

435. V. Kadlec, H. Kadlecová, and O. Štrouf, *J. Organomet. Chem.* **82**, 113 (1974).

436. F. Sato, T. Jimbo, and M. Sato, *Synthesis*, 871 (1981).

437. E. Klei and J. H. Teuben, *J. Organomet. Chem.* **222**, 79 (1981).

438. R. S. P. Coutts, P. C. Wailes, and R. L. Martin, *J. Organomet. Chem.* **50**, 145 (1973).

439. R. S. P. Coutts, R. L. Martin, and P. C. Wailes, *Inorg. Nucl. Chem. Lett.* **9**, 49 (1973).

440. R. Dams, M. Malinowski, I. Westdorp, and H. J. Geise, *J. Org. Chem.* **46**, 2407 (1981); **47**, 248 (1982).

441. E. J. M. de Boer and J. H. Teuben, *J. Organomet. Chem.* **153**, 53 (1978).

442. F. Bottomley and I. J. B. Lin, *J. Chem. Soc. Dalton Trans.*, 271 (1981).

443. F. Bottomley, I. J. B. Lin, and M. Mukaida, *J. Am. Chem. Soc.* **102**, 5238 (1980).

444. F. Bottomley and H. H. Brintzinger, *J. Chem. Soc. Chem. Commun.*, 234 (1978).

445. R. S. P. Coutts and P. C. Wailes, *J. Organomet. Chem.* **73**, C5 (1974).

446. E. Klei and J. H. Teuben, *J. Organomet. Chem.* **214**, 53 (1981).

447. E. Klei, J. H. Telgen, and J. H. Teuben, *J. Organomet. Chem.* **209**, 297 (1981).

448. E. Klei and J. H. Teuben, *J. Organomet. Chem.* **188**, 97 (1980).

449. J. Blenkers, H. J. de Liefde Meijer, and J. H. Teuben, *Rec. Trav. Chim.* **99**, 216 (1980).

450. E. J. M. de Boer, J. H. Teuben, and co-workers, *J. Organomet. Chem.* **181**, 61 (1979).

451. J. H. Teuben, *J. Organomet. Chem.* **57**, 159 (1973).

452. J. H. Teuben and H. J. de Liefde Meijer, *Rec. Trav. Chim.* **90**, 360 (1971).

453. G. Henrici-Olivé and S. Olivé, *Angew. Chem. Intl. Ed.* **8**, 650 (1969); E. E. van Tamelen, *Acc. Chem. Res.* **3**, 361 (1970); J. Chatt, J. R. Dilworth, and R. L. Richards, *Chem. Rev.* **78**, 589 (1978).

454. Yu. G. Borodko and co-workers, *J. Chem. Soc. Chem. Commun.*, 1178 (1972).

455. A. E. Shilov and co-workers, *J. Chem. Soc. Chem. Commun.*, 1590 (1971).

456. M. L. Martin, J. Tirouflet, and B. Gautheron, *J. Organomet. Chem.* **97**, 261–73 (1975).

457. A. Dormond, Ou-Khan, and J. Tirouflet, *J. Organomet. Chem.* **110**, 321 (1976).

458. A. Dormond, T. Kolavudh, and J. Tirouflet, *Compt. Rend.* **282C**, 551 (1976).

459. A. Dormond, T. Kolavudh, and J. Tirouflet, *J. Organomet. Chem.* **165**, 319 (1979).

460. A. Dormond and T. Kolavudh, *J. Organomet. Chem.* **125**, 63 (1977).

461. J. LeBlanc and C. Moise, *J. Organomet. Chem.* **120**, 65 (1976).

462. A. Dormond, T. Kolavudh, and J. Tirouflet, *J. Organomet. Chem.* **164**, 317 (1979).

463. J. Besançon, J. Tirouflet, and co-workers, *Bull. Soc. Chim. France II*, 465 (1978).

464. D. LeClerc, J. Tirouflet, and co-workers, *J. Mol. Struct.* **38**, 203 (1977).

465. J. L. Thomas and K. T. Brown, *J. Organomet. Chem.* **111**, 297 (1976).

466. B. Demerseman, G. Bouquet, and M. Bigorgne, *J. Organomet. Chem.* **101**, C24 (1975).

467. G. Fachinetti, G. Fochi, and C. Floriani, *J. Chem. Soc. Chem. Commun.*, 230 (1976).

468. D. J. Sikora, M. D. Rausch, R. D. Rogers, and J. L. Atwood, *J. Am. Chem. Soc.* **103**, 1265 (1981).

469. B. Demerseman, G. Bouquet, and M. Bigorgne, *J. Organomet. Chem.* **93**, 199 (1975).

470. B. Demerseman, G. Bouquet, and M. Bigorgne, *J. Organomet. Chem.* **145**, 41 (1978).

471. J. L. Atwood and co-workers, *J. Organomet. Chem.* **96**, C4 (1975).

472. I. Fragolà, E. Ciliberto, and J. L. Thomas, *J. Organomet. Chem.* **175**, C25 (1979).
473. B. Demerseman and P. H. Dixneuf, *J. Chem. Soc. Chem. Commun.*, 665 (1981).
474. J. C. Huffman, J. G. Stone, W. C. Krusell, and K. G. Caulton, *J. Am. Chem. Soc.* **99**, 5829 (1977).
475. D. J. Sikora, M. D. Rausch, R. D. Rogers, and J. L. Atwood, *J. Am. Chem. Soc.* **103**, 982 (1981).
476. G. Fachinetti, C. Floriani, and H. Stoeckli-Evans, *J. Chem. Soc. Dalton Trans.*, 2297 (1977).
477. C. Floriani and G. Fachinetti, *J. Chem. Soc. Chem. Commun.*, 790 (1972).
478. M. Moran and V. Fernandez, *J. Organomet. Chem.* **153**, C4 (1978).
479. M. Pasquali, C. Floriani, A. Chiesi-Villa, and C. Guastini, *Inorg. Chem.* **20**, 349 (1981).
480. M. Pasquali, C. Floriani, A. Chiesi-Villa, and C. Guastini, *J. Am. Chem. Soc.* **101**, 4740 (1979).
481. G. Fachinetti, G. Fochi, and C. Floriani, *J. Organomet. Chem.* **57**, C51 (1973).
482. G. Fachinetti and co-workers, *Inorg. Chem.* **17**, 2995 (1978).
483. G. Fachinetti, C. Biran, C. Floriani, A. Chiesi-Villa, and C. Guastini, *J. Chem. Soc. Dalton Trans.*, 792 (1979).
484. B. Demerseman and co-workers, *J. Organomet. Chem.* **117**, C10 (1976).
485. G. Fachinetti and C. Floriani, *J. Chem. Soc. Chem. Commun.*, 66 (1974).
486. M. D. Rausch and co-workers, *Inorg. Chem.* **19**, 3817 (1980).
487. G. P. Pez and S. C. Kwan, *Ann. N.Y. Acad. Sci.* **295**, 174 (1976).
488. Ger. Offen. 2,243,664 (March 22, 1973), G. P. Pez (to Allied Chemical Corp.).
489. J. N. Armor, *Inorg. Chem.* **17**, 203 (1978).
490. *Ibid.*, p. 213.
491. E. E. Van Tamelen, W. Cretney, N. Klaentschi, and J. S. Miller, *J. Chem. Soc. Chem. Commun.*, 481 (1972).
492. T. R. Nelsen and J. J. Tufariello, *J. Org. Chem.* **40**, 3159 (1975).
493. A. Merijanian, T. Mayer, J. F. Helling, and F. Klemick, *J. Org. Chem.* **37**, 3945 (1972).
494. E. J. Corey, R. L. Danheiser, and S. Chandrasekaran, *J. Org. Chem.* **41**, 260 (1976).
495. E. E. Van Tamelen and J. A. Gladysz, *J. Am. Chem. Soc.* **96**, 5290 (1974).
496. S. Tyrlik, I. Wolochowicz, and H. Stepowska, *J. Organomet. Chem.* **93**, 353 (1975).
497. H. T. Verkouw and H. O. Van Oven, *J. Organomet. Chem.* **59**, 259 (1973).
498. C. J. Groenenboom, H. J. de Liefde Meijer, and F. Jellinek, *Rec. Trav. Chim.* **93**, 6 (1974).
499. J. D. Zeinstra and J. L. DeBoer, *J. Organomet. Chem.* **54**, 207 (1973).
500. S. Evans, J. C. Green, and co-workers, *J. Chem. Soc. Dalton Trans.*, 304 (1974).
501. J. D. Zeinstra, and W. C. Nieuwpoort, *Inorg. Chim. Acta* **30**, 103 (1978).
502. C. J. Groenenboom, G. Sawatzky, H. J. de Liefde Meijer, and F. Jellinek, *J. Organomet. Chem.* **76**, C4 (1974).
503. C. J. Groenenboom and F. Jellinek, *J. Organomet. Chem.* **80**, 229 (1974).
504. C. J. Groenenboom, H. J. de Liefde Meijer, and F. Jellinek, *J. Organomet. Chem.* **69**, 235 (1974).
505. H. T. Verkouw, H. J. de Liefde Meijer, and co-workers, *J. Organomet. Chem.* **102**, 49 (1975).
506. J. L. Thomas and R. G. Hayes, *Inorg. Chem.* **11**, 348 (1972).
507. E. Samuel, G. Labauze, and D. Vivien, *J. Chem. Soc. Dalton Trans.*, 2353 (1981).
508. G. Labauze, J. B. Raynor, and E. Samuel, *J. Chem. Soc. Dalton Trans.*, 2425 (1980).
509. M. Vliek, C. J. Groenenboom, H. J. de Liefde Meijer, and F. Jellinek, *J. Organomet. Chem.* **97**, 67 (1975).
510. G. Knol, A. Westerhof, H. O. van Oven, and H. J. de Liefde Meijer, *J. Organomet. Chem.* **96**, 257 (1975).
511. M. E. E. Veldman, J. H. de Liefde Meijer, and co-workers, *J. Organomet. Chem.* **197**, 59 (1980).
512. C. G. Salentine and M. F. Hawthorne, *J. Chem. Soc. Chem. Commun.*, 848 (1975); *J. Am. Chem. Soc.* **97**, 426, 428 (1975); *Inorg. Chem.* **15**, 2872 (1976); A. I. Kovredov and co-workers, *J. Gen. Chem. USSR* **51**, 708 (1981).
513. A. Zwijnenburg, H. J. de Liefde Meijer, and co-workers, *J. Organomet. Chem.* **94**, 23 (1975).
514. K. H. Thiele, A. Röder, and W. Mörke, *Z. Anorg. Allgem. Chem.* **441**, 13 (1978).
515. R. A. Forder and K. Prout, *Acta Cryst.* **B30**, 491 (1974).
516. A. Dormond, Ou-Khan, and J. Tirouflet, *Compt. Rend.* **280C**, 389 (1975).
517. A. Dormond, Ou-Khan, and J. Tirouflet, *Compt. Rend.* **278C**, 1207 (1974).
518. J. L. Calderon, F. A. Cotton, and J. Takats, *J. Am. Chem. Soc.* **93**, 3587, 3592 (1971).
519. R. J. Daroda, G. Wilkinson, and co-workers, *J. Chem. Soc. Dalton Trans.*, 2315 (1980).
520. D. F. Williams, *Syst. Aspects Biocompat.* **1**, 169 (1981); *Chem. Abstr.* **95**, 198364.
521. P. Köpf-Maier, W. Wagner, and H. Köpf, *Naturwissenschaften* **68**, 272, 273 (1981); **67**, 415 (1980); *J. Cancer Res. Clin. Oncol.* **96**, 31, 43 (1980).

522. U.S. Pat. 4,272,588 (June 9, 1981), B. E. Yoldas, A. M. Filippi, and R. W. Buckman (to Westinghouse Electric Corp.).

523. Brit. Pat. 2,067,540 (July 30, 1981), J. H. Novak and G. L. Smay (to American Glass Research).

524. Jpn. Kokai Tokkyo Koho 81 88,843 (July 18, 1981), (to Matusushita Electric Works).

525. U.S. Pat. 3,094,436 (June 18, 1963), H. Schröder (to Jenaer Glaswerk Schott and Gen.).

526. Jpn. Kokai Tokkyo Koho 81 60,068 (May 23, 1981), (to Tokyo Shibaura Electric, Ltd.).

527. Jpn. Kokai Tokkyo Koho 81 114,904 (Sept. 9, 1981), (to Nippon Telegraph and Telephone).

528. Jpn. Kokai Tokkyo Koho 81 37,173 (Aug. 29, 1981), (to Sharp KK).

529. Jpn. Kokai Tokkyo Koho 81 116,015 (Sept. 11, 1981), (to Suwa Seikosha KK).

530. Jpn. Kokai Tokkyo Koho 81 116,016 (Sept. 11, 1981), (to Suwa Seikosha KK).

531. Jpn. Kokai Tokkyo Koho 81 63,846 (May 30, 1981), (tò Suwa Seikosha KK).

532. Jpn. Kokai Tokkyo Koho 81 94,651 (July 31, 1981), (to Tokyo Denshi Kagaku).

533. U.S. Pat. 4,279,654 (July 21, 1981), S. Yajima and co-workers (to Research Institute for Special Inorganic Materials); S. Yajima and co-workers, *J. Mater. Sci.* **16,** 1349 (1981).

534. U.S. Pat. 4,246,314 (Jan. 20, 1981), A. Marzocchi, M. G. Roberts, and C. E. Bolen (to Owens Corning Fiberglas Corp.).

535. U.S. Pat. 4,269,756 (May 26, 1981), T. Y. Su (to Union Carbide).

536. Ger. Offen. 3,026,987 (Feb. 12, 1981), R. G. Adams and S. J. Milletari (to J. P. Stevens and Co.).

537. USSR Pat. 235,907 (Jan. 24, 1969), L. V. Golosova and K. S. Zatsepin.

538. G. Winter, *J. Oil Color Chem. Assoc.* **36**(402), 689 (1953); **34**(367), 30 (1951); A. Hancock and R. Sidlow, *J. Oil Color Chem. Assoc.* **35**(379), 28 (1952).

539. G. Pagliara, *Pitture Vernici* **40,** 279 (1964).

540. Jpn. Kokai Tokkyo Koho 80 11,147 (March 22, 1980), (to Kansai Paint Co.).

541. U.S. Pat. 4,224,213 (June 9, 1978), S. D. Johnson (to Cook Paint & Varnish Co.).

542. Jpn. Kokai Tokkyo Koho 80 152,759 (Nov. 28, 1980), (to Dainippon Toryo Co.).

543. U.S. Pat. 3,524,799 (Aug. 18, 1970), K. H. Dale (to Reynolds Metals Co.).

544. Jpn. Kokai Tokkyo Koho 80 141,573 (Nov 5, 1980), (to Showa Keikinzoku KK).

545. USSR Pat. 744,014 (June 2, 1980), (to Enam Chem. EQP Res.).

546. Jpn. Kokai Tokkyo Koho 81 104,973 (Aug. 21, 1981), (to Toray Industries).

547. Brit. Pat. 786,388 (Nov. 20, 1957), (to National Lead Co.).

548. U.S. Pat. 2,888,367 (May 26, 1959), W. L. Greyson (to Hitemp Wires, Inc.).

549. U.S. Pat. 3,862,099 (Jan. 21, 1975), N. S. Marans (to W. R. Grace).

550. Jpn. Kokai Tokkyo Koho 77 10,322 (Jan. 26, 1977), T. Yoshimura and co-workers (to Kaikin Kogyo Co.).

551. Jpn. Kokai Tokkyo Koho 81 30,472 (March 27, 1981), (to Furukawa Electric Co.).

552. A. M. Usmani and co-workers, *Rubber Chem. Technol.* **54,** 1081 (1981).

553. Jpn. Kokai Tokkyo Koho 77 132,082 (Nov. 5, 1977), I. Sugiyama, H. Doi, and Y. Takaoka (to Matsumoto Seiyaku Kogyo Co.).

554. Belg. Pat. 887,145 (July 20, 1981), (to Toray Silicone Co.).

555. Jpn. Pat. 71 22,878 (June 30, 1971), I. Honda (to Kuraray Co.).

556. N. V. Vakrameeva and co-workers, *Kozh. Obuvn. Promst.* **23,** 37, 43 (1981).

557. Brit. Pat. 1,374,263 (Nov. 20, 1974), T. Keating (to Imperial Chemical Industries).

558. U.S. Pat. 4,234,497 (Nov. 18, 1980), M. Honig (to Standard Lubricants).

559. Ger. Offen. 2,004,098 (Aug. 12, 1971), R. Tuma and R. Lebender (to Dynamit Nobel A.G.).

560. A. G. Okuneva and co-workers, *Plast. Massy*, 10 (1981).

561. U.S. Pat. 4,260,735 (April 7, 1981), J. A. Bander, S. D. Lazarus, and I. C. Twilley (to Allied Corp.).

562. Jpn. Kokai Tokkyo Koho 80 125,120 (Sept. 26, 1980), (to Toray Industries, Inc.).

563. Ger. Offen. 2,751,385 (May 24, 1978), S. P. Elliot (to E. I. du Pont de Nemours & Co., Inc.).

564. *Organic Titanium Compounds as Acrylic Ester Alcoholysis Catalysts*, Titanium Intermediates, Ltd., 1967.

565. Jpn. Kokai Tokkyo Koho 79 125,617 (Sept. 29, 1979), T. Onoda, K. Tano, and Y. Hara (to Mitsubishi Chemical Industries).

566. Ger. Offen. 2,922,343 (Dec. 4, 1980), B. Luthingshauser and C. Lindzus (to Dynamit Nobel A.G.).

567. Jpn. Kokai Tokkyo Koho 74 95,918 (Sept. 11, 1974), K. Kimura and H. Ito (to Toa Gosei Chemical Industry Co.).

568. Jpn. Kokai Tokkyo Koho 78 98,393 (Aug. 28, 1978), W. Funakoshi, K. Nawata, and K. Tsunawaki (to Teijin, Ltd.).

569. Jpn. Kokai Tokkyo Koho 78 106,792 (Sept. 18, 1978), W. Funakoshi and K. Nawata (to Teijin, Ltd.).
570. USSR Pat. 821,452 (April 15, 1981), A. L. Suvorov and co-workers.
571. E. Tuček and H. D. Dinse, *Acta Polym.* **31,** 429 (1980).
572. Eur. Pat. 32,587 (July 29, 1981), (to Union Carbide Corp.).
573. Jpn. Kokai Tokkyo Koho 79 77,635 (June 21, 1979), M. Yoshiaki and co-workers (to Kansai Paint Co.).
574. A. L. Suvorov and co-workers, *Vysokomol. Soedin. Ser. A* **20,** 2592 (1978).
575. Jpn. Pat. 81 5869 (Feb. 7, 1981), (to Dainippon Ink Chem KK).
576. U.S. Pats. 4,297,447–4,297,449 (Oct. 27, 1981), C. J. Stark (to General Electric Co.).
577. Ger. Offen. 3,028,114 (Feb. 12, 1981), (to Acheson Industries, Inc.).
578. Jpn. Kokai Tokkyo Koho 81 88,471 (July 17, 1981), (to Mitsui Toatsu Chem.).
579. Ger. Offen. 3,038,646 (April 23, 1981), K. Kawasumi, H. Watanabe, and J. Seto (to Sony Corp.).
580. Jpn. Kokai Tokkyo Koho 81 75,544 (June 22, 1981), (to Suwa Seikosha KK).
581. A. V. Zholnin, V. N. Podchainova, and A. I. Salova, *Sb. Nauch. Tr. Chelyabinsk. Politekh. Inst.*, (91), 86 (1971); *Chem. Abstr.* **76,** 80670.
582. Ger. Offen. 3,016,508 (May 21, 1981), D. L. Motov and co-workers.
583. Brit. Pat. 922,456 (April 3, 1963), G. E. Westwood (to Berger, Jensen, and Nicholson, Ltd.).
584. Jpn. Kokai Tokkyo Koho 81 99,266 (Aug. 10, 1981), (to Dainippon Toryo Co.).
585. Jpn. Kokai Tokkyo Koho 81 112,973 (Sept. 5, 1981), (to Dainippon Toryo KK).
586. U.S. Pat. 4,159,209 (June 26, 1979), P. Womersley (to Manchem., Ltd.).
587. U.S. Pat. 4,281,037 (July 28, 1981), H. R. Choung (to DAP, Inc.).
588. T. Sakaguchi, A. Nakajima, and T. Horikoshi, *Nippon Kagaku Kaishi*, 788 (1979).
589. Jpn. Kokai Tokkyo Koho 81 109,266 (Aug. 29, 1981), (to Mitsubishi Pencil Co.).
590. U.S. Pat. 3,878,111 (April 15, 1975), R. E. McGlothin and T. E. Cox (to Dresser Industries).
591. U.S. Pat. 3,301,723 (Jan. 31, 1967), J. Chrisp (to E. I. du Pont de Nemours & Co., Inc.).
592. U.S. Pat. 3,888,312 (June 10, 1975), R. L. Tiner, M. D. Holtmeyer, B. J. King, and R. A. Gatlin (to Halliburton Co.).
593. M. W. Conway, R. W. Pauls, and L. E. Harris, *Soc. Pet. Eng. of AIME*, paper SPE 9333 (1980).
594. M. W. Conway, S. W. Almond, J. E. Briscoe, and L. E. Harris, *Soc. Pet. Eng. of AIME*, paper SPE 9334 (1980).
595. Ger. Offen. 3,034,721 (April 2, 1981), J. M. Dees and E. A. Elphingstone (to Halliburton Co.).
596. Jpn. Pat. 71 12,085 (Mar. 27, 1971), K. Hamahiro, Y. Yoshioka, and H. Sakurai (to Kuraray Co.); Jpn. Pat. 71 12,089 (March 27, 1971), Y. Yoshioka, S. Kurokawa, and K. Hashita (to Kuraray Co.); Jpn. Pat. 71 405 (Jan. 7, 1971), T. Ashikaga and U. Maeda (to Kuraray Co.).
597. Jpn. Pat. 71 38,410 (Nov. 12, 1971), T. Asikaga and S. Higashimori (to Kuraray Co.); Jpn. Pat. 71 38,411 (Nov. 12, 1971), T. Ashikaga and U. Maeda (to Kuraray Co.).
598. U.S. Pats. 3,941,728–3,941,730 (March 2, 1976), J. C. Solenberger (to E. I. du Pont de Nemours & Co., Inc.).
599. U.S. Pat. 4,113,757 (Sept. 12, 1978), P. D. Kay (to Tioxide Group).
600. USSR Pat. 380,774 (May 15, 1973), A. F. Tishchenko and co-workers.
601. Ger. Offen. 2,326,828 (Nov. 29, 1973), M. Camp and J. Dumoulin (to Rhône-Poulenc S.A.).
602. Czech. Pat. 181,452 (Jan. 15, 1980), E. Wurstová.
603. U.S. Pat. 4,288,496 (Sept. 8, 1981), R. E. Reusser and B. E. Jones (to Phillips Petroleum Co.).
604. Jpn. Kokai Tokkyo Koho 78 24,408 (March 7, 1978), T. Noguchi and co-workers (to Kureha Chemical Industry Co.).
605. Jpn. Kokai Tokkyo Koho 80 167,211 (to Matsumoto Seiyaku Kogyo Co.).
606. USSR Pat. 443,859 (Sept. 25, 1974), T. P. Osipova and A. A. Kas'yanova.
607. Jpn. Kokai Tokkyo Koho 74 36,998 (April 5, 1974), M. Komeyama, S. Murakami, and H. Ohnishi (to Kuraray Co.).
608. S. D. Voronkevich, L. K. Zgadzai, and M. T. Kuleev, *Plast. Massy*, 64 (1973).
609. A. A. Panasevich, S. P. Nichiporenko, and G. M. Nikitina, *Tr. Mezhvuz. Konf. Primen. Plastmass Stroit.*, 3rd, 174 (1970); *Chem. Abstr.* **82,** 63610.
610. Ger. Offen. 2,310,820 (Sept. 13, 1973), A. E. Vassiliades, D. N. Vincent, and M. P. Powell (to Champion Paper Co.).
611. *Research Disclosure 20402*, E. I. du Pont de Nemours & Co., Inc., Wilmington, Del., April 1981.
612. Ger. Offen. 2,442,686 (March 27, 1975), (to Dainippon Ink Chem. KK).
613. A. S. Roman and co-workers, *Izv. Vyssh. Uchebn. Zaved. Tekhnol. Legk. Promsti.*, 31, 61.

614. Eur. Pat. 32,637 (July 29, 1981), (to Wool Development International).
615. E. Santi, S. G. Babe, and J. Fontan, *Rev. Plast. Mod.* **22,** 1521 (1971).
616. Jpn. Kokai Tokkyo Koho 80 58,227 (April 30, 1980), (to Showa Electric Wire and Cable Co.).
617. J. F. Kennedy and B. Kalogerakis, *Biochimie* **62,** 549 (1980).
618. J. F. Kennedy, S. A. Barker, and J. D. Humphreys, *J. Chem. Soc. Perkin Trans. 1*, 962 (1976).
619. Ger. Offen. 2,515,863 (Oct. 23, 1975), S. J. Monte and P. F. Bruins (to Kenrich Petrochemicals).
620. Ger. Offen. 3,007,655 (Sept. 17, 1981), (to Licentia Patent GmbH).
621. S. J. Monte and G. Sugerman, *Fire Retard., Proc. Int. Symp. Flammability Fire Retard.*, 240 (1978); *Chem. Abstr.* **93,** 27149.
622. Eur. Pat. 7,748 (Feb. 6, 1980), S. J. Monte and G. Sugerman (to Kenrich Petrochemicals).
623. Ger. Offen. 2,623,472 (April 7, 1977), S. J. Monte and G. Sugerman (to Kenrich Petrochemicals); see also S. J. Monte and G. Sugerman, *Polym. Plast. Technol. Eng.* **17,** 95 (1981).
624. U.S. Pat. 4,096,110 (June 20, 1978), S. J. Monte and G. Sugerman (to Kenrich Petrochemicals).
625. Brit. Pat. 1,592,260 (July 1, 1981), (to BASF Wyandotte).
626. S. J. Monte, G. Sugerman, and S. Spindel, *Mod. Paint Coatings*, 14 (April 1980).
627. U.S. Pat. 3,317,396 (May 2, 1967), T. Istvan.
628. Jpn. Kokai Tokkyo Koho 81 111,129 (Sept. 2, 1981), (to Sony Corp.).
629. Ger. Offen. 2,927,379 (Jan. 8, 1981), G. Buxbaum and co-workers (to Bayer A.G.).
630. U.S. Pat. 4,221,693 (Sept. 9, 1980), J. C. Getson and C. G. Neuroth.
631. U.S. Pat. 4,165,239 (Aug. 21, 1979), H. Linden and H. Bornmann (to Henkel).
632. U.S. Pat. 4,042,557 (Aug. 16, 1977), W. L. Dills (to E. I. du Pont de Nemours & Co., Inc.).
633. Fr. Demande 2,260,610 (Sept. 5, 1975), C. F. Carter and co-workers (to Plessy Handel and Investments A.G.).
634. Jpn. Kokai Tokkyo Koho 81 62,538 (May 28, 1981), (to Kao Soap Co.).
635. Jpn. Kokai Tokkyo Koho 81 62,537 (May 28, 1981), (to Kao Soap Co.).
636. Fr. Demande 2,045,426 (Feb. 26, 1971), (to British Titan Products Co.).
637. Ger. Pat. 1,717,026 (Sept. 4, 1969), W. R. Whately and G. M. Sheehan (to American Cyanamid Co.).
638. L. A. Aleinikova and A. V. Yudin, *Izv. Vyssh. Uchebn. Zaved., Tekhnol. Legk. Promsti.*, 39 (1969).
639. Ger. Offen. 2,834,941 (Feb. 22, 1979), G. A. Philpot and co-workers (to Eastman Kodak).
640. Can. Pat. 980,955 (Jan. 6, 1976), Y. Gladu (to Canadian Titanium Pigments).
641. A. A. Guzairova, T. P. Repich, and N. P. Ogorodova, *Lakokras. Mater. Ikh Primen.*, 21 (1981).
642. Ger. Offen. 2,924,849 (Jan. 22, 1981), K. Koehler and co-workers (to Bayer A.G.).
643. Jpn. Kokai Tokkyo Koho 80 165,143 (Dec. 23, 1980), (to Kureha Chemical Industries KK).
644. Jpn. Pat. 78 12,489 (May 1, 1978), Y. Oda, T. Otoma, and K. Uchida (to Asahi Glass Co.).
645. T. Misono and co-workers, *Shikizai Kyokaishi* **54,** 15 (1981).
646. C. E. Carraher and co-workers, *J. Macromol. Sci. Chem.* **15,** 773 (1981).
647. U.S. Pat. 4,043,820 (Aug. 23, 1977), R. Landau (to Ozalid Group Holdings).
648. U.S. Pat. 4,137,183 (Jan. 30, 1979), G. Caspari (to Standard Oil of Indiana).
649. U.S. Pats. 3,879,485 (April 22, 1975), 3,911,042 (Oct. 7, 1975), and 3,969,429 (July 13, 1976), G. P. Belov and co-workers.
650. Ger. Offen. 2,934,277 (March 26, 1981), W. Josten and H. J. Vahlensieck (to Dynamit Nobel A.G.).
651. U.S. Pat. 3,647,754 (March 7, 1972), J. Bemesma and E. M. J. Pijpers (to Stamicarbon NV).
652. U.S. Pats. 4,200,664 (April 29, 1980) and 4,283,387 (Aug. 11, 1981), R. W. Young, S. Prussin, and N. G. Gaylord (to Young Prussin, MGK, JV).
653. U.S. Pat. 4,271,215 (June 2, 1981), D. L. Coon (to Dow Corning).
654. F. E. McFarlane and G. W. Tindall, *Inorg. Nucl. Chem. Lett.* **9,** 907 (1973).
655. R. S. Natarajan and G. D. Kalyankar, *Bull. Indian Natl. Sci. Acad.* **49,** 473 (1974); *Chem. Abstr.* **84,** 180585; *Proc. Indian Natl. Sci. Acad.* **39B,** 473 (1973); *Chem. Abstr.* **82,** 86590.
656. Jpn. Kokai Tokkyo Koho 80 139,392 (Oct. 31, 1980), I. Kiijima.
657. U.S. Pat. 4,002,574 (Jan. 11, 1977), R. C. Wade (to Ventron Corp.).
658. Jpn. Kokai Tokkyo Koho 78 103,732 (Sept. 9, 1978), K. Shirakawa and E. Watahiki (to Toppan Printing Co.).
659. Y. Shirai and co-workers, *Nippon Shashin Gakkaishi* **40,** 97 (1977); *Chem. Abstr.* **88,** 30321.
660. Y. Shirai and G. Miyamato, *Nippon Shashin Gakkaishi* **42,** 161 (1979).
661. Australian Pat. 481,536 (April 10, 1975), P. J. Reed, E. H. Brooks, and J. R. Jennings (to ICI Australia Ltd.).
662. T-C Chai and C-M. Wei, *Fen Hsi Hua Hsueh* **7,** 327 (1979); *Chem. Abstr.* **92,** 103744.

663. V. D. Bakalov, V. V. Dunina, and V. M. Potapov, *Zh. Anal. Khim.* **34**, 2138 (1979).

664. G. D. Brykina and T. A. Belyavskaya, *Vestn. Mosk. Univ. Khim.* **13**, 608 (1972).

665. C. G. Macarovici and E. Motiu, *Stud. Univ. Babes-Bolyai Ser. Chem.* **16**, 39 (1971).

666. R. M. Dranitskaya, A. I. Gavril'chenko, and L. A. Okhitina, *Zh. Anal. Khim.* **25**, 1740 (1970).

667. Ya. I. Tur'yan and co-workers, *Zh. Neorg. Khim.* **23**, 2061 (1978).

668. Yu. K. Tselinskii and V. K. Gadzhun, *Soviet J. Coord. Chem.* **4**, 1028 (1978); *Chem. Abstr.* **89**, 186769.

669. E. I. Stepanovskii, G. M. Fofanov, and G. A. Kitaev, *Zh. Neorg. Khim.* **24**, 941 (1979).

670. F. G. Banica and L. Carlea, *Rev. Chim. (Bucharest)* **30**, 640 (1979); *Chem. Abstr.* **91**, 217716.

671. A. Y. Nazarenko and I. V. Pyatnitskii, *Zh. Neorg. Khim.* **23**, 2655 (1978).

672. A. A. Popel, A. N. Glebov, and Y. I. Sal'nikov, *Zh. Neorg. Khim.* **24**, 2409 (1979).

673. I. V. Pyatnitskii and R. S. Kharchenko, *Ukr. Khim. Zh.* **33**, 734 (1967).

674. V. M. Savostina, F. I. Lobanov, and V. M. Peshkova, *Zh. Neorg. Khim.* **12**, 2162 (1967).

675. V. K. Zolotukhin, O. M. Gnatishin, and E. I. Senchishin, *Visn. L'viv. Derzh. Univ. Ser. Khim.* **17**, 40 (1975).

676. I. V. Pyatnitskii and A. Yu. Nazarenko, *Zh. Anal. Khim.* **32**, 853 (1977); *Zh. Neorg. Khim.* **22**, 1816 (1977).

677. Yu. K. Tselinskii, L. Y. Kvyatokovskaya, and V. K. Gadzhun, *Zh. Fiz. Khim.* **50**, 3002 (1976).

678. R. S. Ramakrishna and D. T. A. Seneratyapa, *J. Inorg. Nucl. Chem.* **39**, 333 (1977).

679. Ger. Offen. 1,811,502 (June 26, 1969), R. H. Stanley, D. W. Brook, and L. L. Lawrence (to British Titan Products Co.).

680. C. G. Macarovici and L. Czegledi, *Rev. Roum. Chim.* **14**, 57 (1969); *Stud. Univ. Babes-Bolyai Ser. Chem.* **22**, 25 (1977).

681. F. I. Lobanov and co-workers, *Zh. Neorg. Khim.* **14**, 1077 (1969); *Vestn. Mosk. Univ. Khim.* **24**, 121 (1969).

682. Jpn. Pat. 67 26,628 (Dec. 16, 1967), I. Sugiyama, K. Takahashi, and N. Takahashi (to Matsumoto Pharmaceutical Industry Co.).

General References

R. J. H. Clark, *The Chemistry of Titanium and Vanadium*, Elsevier, Amsterdam, 1968.

R. J. H. Clark in J. C. Bailar, H. J. Emeleus, R. S. Nyholm, and A. F. Trotman-Dickenson, eds., *Comprehensive Inorganic Chemistry*, Vol. 3, Pergamon Press, London, 1973, Chapt. 32, pp. 355–417.

P. C. Wailes, R. S. P. Coutts, and H. Weigold, *Organometallic Chemistry of Titanium, Zirconium, and Hafnium*, Academic Press, New York, 1974.

R. J. H. Clark, D. C. Bradley, and P. Thornton, *Chemistry of Titanium, Zirconium, and Hafnium*, Pergamon Press, New York, 1975.

R. J. H. Clark, S. Moorhouse, and J. A. Stockwell, *J. Organometal. Chem. Library* **3**, 223 (1977) (literature covered through 1975).

E. Müller, ed., *Houben-Weyl's Methods in Organic Chemistry*, 4th ed., Vol. 13, part 7, Georg Thieme, Stuttgart, 1975.

Gmelin, Handbuch der Anorganischen Chemie, 8th ed., Syst. No. 41, Springer Verlag, Berlin, 1977.

Annual Surveys

K. S. Mazdiyasni in K. Niedenzu and H. Zimmer, eds., *Annual Reports in Inorganic and General Synthesis*, Academic Press, New York, 1972, pp. 73–81; 1973, pp. 137–147.

J. J. Alexander in K. Niedenzu and H. Zimmer, eds., *Annual Reports in Inorganic and General Synthesis*, Academic Press, New York, 1974, pp. 130–141; 1975, pp. 138–151; 1976, pp. 151–166.

F. Calderazzo, *Organomet. Chem. Rev. B* **4**, 12 (1968); **5**, 547 (1969); **6**, 1001 (1970); **9**, 137 (1972); *J. Organomet. Chem.* **53**, 179 (1973); **89**, 193 (1975). Surveys covering the years 1967–1972.

P. C. Wailes, *J. Organomet. Chem.* **79**, 201 (1974); **103**, 475 (1975); **126**, 361 (1977). Surveys covering the years 1973–1975.

J. A. Labinger, *J. Organomet. Chem.* **138**, 185 (1977); **167**, 19 (1979); **180**, 187 (1979); **196**, 37 (1980); **227**, 341 (1981). Surveys covering the years 1976–1980.

R. C. Fay, *Coord. Chem. Rev.* **37**, 9 (1981); this review covers the literature on titanium for 1979; reviews for later years may appear subsequently.

Specialist Periodical Reports of the Chemical Society (*London*), *Organometallic Compounds*, contain numerous references on titanium compounds which can be located through the volume indexes.

Titanium Alkoxides and Amides; Polyalkoxides

D. C. Bradley in F. G. A. Stone and W. A. Graham, eds., *Inorganic Polymers*, Academic Press, New York, 1962, pp. 410–446.
D. C. Bradley, *Prep. Inorg. React.* **2,** 169 (1965).
D. C. Bradley, *Coord. Chem. Rev.* **2,** 299 (1967).
D. C. Bradley, *Inorg. Macromol. Rev.* **1,** 141 (1970).

Ziegler-Natta Polymerization

G. Henrici-Olivé and S. Olivé, *Chemtech*, 746 (1981); this is a very readable overview of this complex topic.
J. Boor, *Ziegler-Natta Catalysts and Polymerization*, Academic Press, New York, 1979.
C. E. Schildknecht and I. Skeist, "Polymerization Processes," in *High Polymer Series*, Vol. 29, John Wiley and Sons, Inc., New York, 1977.
J. C. W. Chien, *Coordination Polymerization*, Academic Press, New York, 1975.
H. Sinn and W. Kaminsky, *Adv. Organomet. Chem.* **18,** 99 (1980).

Lower Valent Titanium Compounds

R. S. P. Coutts and P. C. Wailes, *Adv. Organometal. Chem.* **9,** 136 (1970).

CHRISTIAN S. RONDESTVEDT, JR.
E. I. du Pont de Nemours & Co., Inc.

TOBIAS ACID. See Naphthalene derivatives.

TOCOPHEROLS. See Vitamins, Vitamin E.

TOILET PREPARATIONS. See Cosmetics.

TOLIDINES. See Benzidine and related biphenyldiamines.

TOLU BALSAM. See Perfumes.

TOLUENE

Toluene [*108-88-3*], C_7H_8, is a colorless, mobile liquid with a distinctive aromatic odor somewhat milder than that of benzene. The name toluene derives from a natural resin, balsam of Tolu, named for a small town in Colombia, South America. Toluene was discovered among the degradation products obtained by heating this resin.

Prior to World War I, the main source of toluene was coke ovens (see Coal, coal conversion process, carbonization). At that time, trinitrotoluene (TNT) was the preferred high explosive, and large quantities of toluene were required for its manufacture (see Explosives and propellants). To augment the supply, toluene was for the first time obtained from petroleum sources by subjecting narrow-cut naphthas containing relatively small amounts of toluene to thermal cracking. The toluene concentrate so produced was then purified and used for the manufacture of TNT. Production from petroleum was discontinued shortly after World War I. Petroleum again became the source for toluene with the advent of catalytic reforming and the need for large quantities of toluene for use in aviation fuel during World War II. Since then, manufacture of toluene from petroleum sources has continued to increase, and manufacture from coke ovens and coal-tar products has continued to decrease (see BTX processing; Petroleum).

Toluene is generally produced along with benzene (qv), xylenes (qv), and C_9 aromatics by the catalytic reforming of C_6–C_9 naphthas. The resulting crude reformate is extracted, most frequently with sulfolane, to yield a mixture of benzene, toluene, xylenes, and C_9 aromatics, which are then separated by fractionation. About 90–95% of the nearly 31×10^6 metric tons (9.4×10^9 gal) of toluene produced annually in the United States is not isolated but is blended directly into the gasoline pool as a component of reformate and of pyrolysis gasoline. Capacity exists to isolate ca 5.3×10^6 t (1.6×10^9 gal) per year of which about half is used for chemicals and solvents. The remainder is blended into gasoline to increase octane number (see Gasoline).

Physical Properties

The physical and thermodynamic properties of toluene are summarized in Table 1 (1–2). Vapor-pressure data for toluene are summarized in Table 2 (3). Toluene forms azeotropes with many hydrocarbons and most alcohols that boil in a similar range. All are minimum-boiling azeotropes. Composition and boiling data are summarized in Table 3 (4). Toluene, water, and alcohols frequently form ternary azeotropes.

As in benzene, the carbon–carbon bond lengths of the ring carbons in toluene are all the same length, 0.1397 nm, intermediate between normal single and double carbon–carbon bond lengths. Quantum mechanical studies show that the bonding electrons in the benzene ring of toluene occupy three sp^2 hybrid orbitals and one $2p_x$ orbital per carbon atom and one $1s$ electron per hydrogen atom. One sp^2 electron of each carbon forms a bond with a $1s$ hydrogen atom and the other two form bonds with the adjacent carbon atoms to form a planar ring. The remaining six $2p_x$ electrons combine to produce a π bond with a doughnut-shaped probability distribution on each side of the ring. These electrons are delocalized over the entire ring. This delocalization accounts for the stabilization energy of 163.2 kJ/mol (39 kcal/mol) calculated from

Table 1.   Physical and Thermodynamic Properties of Toluene[a]

Property	Value	
mol wt	92.14	
freezing pt, °C	−94.965	
boiling pt, °C	110.629	
density, g/cm^3		
at 25°C	0.8623	
at 20°C	0.8667	
critical properties		
temperature, °C	318.64	
pressure, MPa (atm)	4.109 (40.55)	
volume, L/mol	0.316	
heat of combustion, at 25°C constant pressure, kJ/mol (kcal/mol)	3910.3 (934.5)	
heat of vaporization, kJ/mol (kcal/mol)		
at 25°C	37.99 (9.080)	
at bp	33.18 (7.931)	
heat capacity, J/(g·K) (cal/(g·K))		
ideal gas	1.125 (0.2688)	
liquid at 101.3 kPa (1 atm)	1.970 (0.4709)	
surface tension at 25°C, mN/m (= dyn/cm)	27.92	
	Gas	*Liquid*
heat of formation, ΔH_f°, kJ/mol (kcal/mol)	50.00 (11.950)	12.00 (2.867)
entropy, S°, kJ/K (kcal/K)	319.7 (76.42)	219.6 (52.48)
free energy of formation, ΔF_f°, kJ/K (kcal/K)	93.00 (22.228)	114.1 (27.282)

[a] Refs. 1–2.

Table 2.   Vapor Pressure of Toluene[a]

Temperature, °C	Pressure, kPa (mm Hg)	Temperature, °C	Pressure, kPa (atm)
0	0.91 (6.8)	160	343.0 (3.385)
20	2.92 (21.9)	180	516.8 (5.100)
40	7.91 (59.3)	200	749.3 (7.395)
60	18.56 (139.2)	220	1053 (10.39)
80	38.86 (291.5)	240	1441 (14.22)
100	74.19 (556.5)	260	1927 (19.02)
120	131.2 (1.295)	280	2530 (24.97)
140	217.8 (2.150)	300	3273 (32.30)

[a] Ref. 3.

the observed heats of combustion compared to calculated heats of combustion for a cyclohexatriene-type structure with alternate and fixed single and double carbon–carbon bonds. The same value for stabilization energy is obtained from heats of hydrogenation.

Table 3. Azeotropes of Toluene [a]

Component	Bp, °C	Azeotrope	
		Bp, °C	Toluene, wt %
Paraffins			
n-heptane	98.4	(nonazeotrope)	
2,5-dimethylhexane	109.4	107.0	35
2,3,4-trimethylpentane	113.5	109.5	60
2-methylheptane	117.6	110.3	82
n-octane	125.4	(nonazeotrope)	
Cycloparaffins			
methylcyclohexane	100.85	(nonazeotrope)	
ethylcyclopentane	103.5	103.0	7
1,1,3-trimethylcyclopentane	104.9	103.8	16
cis,trans,cis-1,2,4-trimethylcyclopentane	109.3	107.0	39
cis,trans,cis-1,2,3-trimethylcyclopentane	110.4	108.0	39
cis-1,3-dimethylcyclohexane	120.1	110.6	96
1,3-dimethylcyclohexane	120.7	(nonazeotrope)	
Hydroxyl compounds			
methanol	64.7	63.8	31
ethanol	78.3	76.7	32
2-propanol	82.4	81.5	23
1-propanol	97.2	92.6	50
water	100.0	84.1	86.5
2-butanol	99.5	95.3	45
1-butanol	117.8	105.5	72
2-pentanol	119.8	107	72
3-pentanol	116.0	106	65

[a] Ref. 4.

Chemical Properties

Because of the high electron density in the aromatic ring, toluene behaves as a base both in formation of charge-transfer π complexes and in the formation of complexes with super acids. In this regard, toluene is intermediate between benzene and the xylenes as shown by the data in Table 4. In the formation of π complexes with electrophiles such as silver ion, hydrogen chloride, and tetracyanoethylene, toluene differs from either benzene or the xylenes by less than a factor of two in relative basicity. The difference is small because the complex is formed almost entirely with the π electrons of the aromatic ring; the inductive effect of the methyl group provides only minor enhancement. In contrast, with HF and BF_3 which form a sigma-type complex, or in the case of reaction as with nitronium ion or chlorine where formation of sigma bonds and complexes plays a predominant role, the methyl group participates by hyperconjugation and the relative reactivity of toluene is enhanced by several orders of magnitude compared to that of benzene. Reactivity of xylenes is enhanced again by several magnitudes over that of toluene. Thus, when only the π electrons are involved, toluene behaves much like benzene and the xylenes.

When sigma bonds are involved, toluene is a much stronger base than benzene and a much weaker base than the xylenes. The reasons for this difference are readily shown by contrasting the complexes of toluene with hydrogen chloride in the absence and presence of aluminum chloride. In the absence of aluminum chloride, hydrogen

Table 4. Relative Basicity and Reactivity toward Electrophiles

Electrophile	Benzene	Toluene	Xylene		
			Ortho	Meta	Para
Ag$^+$ [a]	0.90	1.00	1.08	1.13	0.98
HCl [b]	0.66	1.00	1.23	1.37	1.09
TCE [c]	0.54	1.00	1.89	1.62	2.05
HF—BF$_3$ [d]		1	200	2000	100
NO$_2^+$ [e]	0.045	1			
Cl$_2$ [f]	0.003	1	13.1	1250	6.3

[a] Solubility in aqueous Ag$^+$ (5).
[b] K for Ar + HCl $\rightleftharpoons$ Ar.HCl in n-heptane at $-78°$C (6).
[c] K for association with tetracyanoethylene in CH$_2$Cl$_2$ (7).
[d] Basicity by competitive protonation (8–9).
[e] CH$_3$CONO$_2$ in (CH$_3$C)$_2$O at 24°C (10).
[f] Cl$_2$ in CH$_3$COH at 24°C (11).

chloride is loosely attracted to the π cloud of electrons above and below the plane of the ring. With aluminum chloride present, the electrophilicity is greatly enhanced and a sigma bond is formed with a specific electron pair; structures involving the methyl group contribute to the stabilization.

For attack at either of the two ortho positions or the para position, three such structures can be written.

Chemical derivatives of toluene are formed by substitution of the hydrogen atoms of the methyl group, by substitution of the hydrogen atoms of the ring, and by addition to the double bonds of the ring. Toluene can also undergo a disproportionation reaction in which two molecules react to yield one molecule of benzene and one molecule of xylene.

Substitutions on the Methyl Group. The reactions that give substitution on the methyl group are generally high temperature and free-radical reactions. Thus, chlorination at ca 100°C, or in the presence of ultraviolet light or other free-radical initiators, successively gives benzyl chloride, benzal chloride, and benzotrichloride (see Benzaldehyde; Benzoic acid).

With oxygen in the liquid phase and particularly in the presence of catalysts, eg, bromine-promoted cobalt and manganese, very good yields of benzoic acid are obtained.

CH_3 → COOH (O_2, Br, Co, Mn / 50 °C)

In the presence of alkali metals such as potassium and sodium, toluene is alkylated on the methyl group to yield, successively, normal propylbenzene, 3-phenylpentane, and 3-ethyl-3-phenylpentane (12).

CH_3 →(C_2H_4, Na / 195 °C) CH_2CH_2CH_3 → CH(CH_2CH_3)_2 → C(CH_2CH_3)_3

In the presence of a potassium catalyst dispersed on calcium oxide, toluene reacts with 1,3-butadiene to yield 5-phenyl-2-pentene (13).

CH_3 →(C_4H_6, K, CaO / 100 °C) CH_2CH_2CH=CHCH_3

When lithium is used as a catalyst in conjunction with a chelating compound such as tetramethylethylenediamine (TMEDA), telomers are generally obtained from toluene and ethylene (14).

CH_3 →(C_2H_4, Li, TMEDA / 110 °C) CH_2$-$(C_2H_4)$_n$$-$CH_2CH_3

$n = 0$–10

The intermediates in these base-catalyzed reactions are believed to be of the nature of a benzyl cation because the reaction product from toluene and propylene is isobutylbenzene, not n-butylbenzene, and the reaction rate is slower than with ethylene (15).

Addition to the Ring. Additions to the double bonds in the aromatic ring of toluene result from both free-radical and catalytic reactions. Chlorination using free-radical initiators at temperatures $<0°C$ saturates the ring. However, this reaction is not entirely selective, for in addition to saturating the ring to yield hexachlorohexane derivatives, the reaction also effects substitution on the methyl group (16). Hydrogenation with typical hydrogenation catalysts readily yields methylcyclohexane. However, rates for hydrogenation of toluene are only 60–70% of that for benzene (17). The commercial technology used for hydrogenating benzene to cyclohexane (18) can be directly applied to manufacture of methylcyclohexane. Both of these ring-saturating reactions probably proceed stepwise, but since the initial reaction must overcome the high resonance energy of the aromatic ring, saturation of the second and the third double bond is much more rapid with the result that partially saturated intermediates are not normally detected (19).

Substitution on the Ring. Substitution of the ring hydrogen atoms by electrophilic attack takes place with all of the same reagents that react with benzene. Some of the common groups with which toluene can be substituted directly are

$$-\text{Cl}, \ -\text{Br}, \ -\overset{\overset{\text{O}}{\|}}{\text{C}}\text{CH}_3, \ -\text{SO}_3\text{H}, \ -\text{NO}_2, \ -(\text{C}_n\text{H}_{2n+1}), \ \text{and} \ -\text{CH}_2\text{Cl}.$$

and —CH$_2$Cl. Typical electrophilic substitutions are summarized in Tables 5 and 6. The reactivity ratios in Table 5 show that under the same conditions toluene reacts more rapidly than benzene and that those reactions that exhibit the highest selectivity to the ortho and para positions also show the most greatly enhanced reactivity relative to benzene.

Table 5. Isomer Distribution and Reactivity Ratio for Selected Reactions[a]

Reaction	Conditions	Isomer Distribution			Reactivity ratio
		Ortho	Meta	Para	
chlorination	Cl$_2$ in HO$\overset{\overset{\text{O}}{\|}}{\text{C}}CH_3$ at 24°C	58	<1	42	353
chloromethylation	CH$_2$O in HO$\overset{\overset{\text{O}}{\|}}{\text{C}}CH_3$ at 60°C with HCl and ZnCl$_2$	34.7	1.3	64.0	112
nitration	90% HO$\overset{\overset{\text{O}}{\|}}{\text{C}}CH_3$ at 45°C	56.5	3.5	40.0	24.5
mercuration	Hg(O$\overset{\overset{\text{O}}{\|}}{\text{C}}CH_3$)$_2$ in HO$\overset{\overset{\text{O}}{\|}}{\text{C}}CH_3$ with HClO$_4$ at 25°C	21.0	9.5	69.5	7.9
sulfonylation	CH$_3$SO$_2$Cl with AlCl$_3$ at 100°C	49	15	36	
isopropylation	C$_3$H$_6$ at 40°C with AlCl$_3$	37.0	28.5	33.9	2.1

[a] Ref. 20.

Table 6. Isomer Distributions in the Monoalkylations of Toluene[a], %

Entering group	Ortho	Meta	Para
methyl	53.8	17.3	28.8
ethyl	45	30	25
isopropyl	37.5	29.8	32.7
t-butyl	0	7	93

[a] Ref. 20.

Generally, these increased reactivities of toluene and the related selectivity to the ortho and para positions can be explained in terms of the inductive effect of the methyl group, which increases the electron density in the ring, and by the ability of the methyl group to hyperconjugate (as shown below) and thereby to stabilize the reaction intermediates.

In addition to these effects, there is a steric effect at the ortho position as shown by the data in Table 6. These data clearly demonstrate that bulky groups cannot enter easily into the position adjacent to the methyl group and therefore attack selectively at the para position.

Toluene itself does not undergo substitution by nucleophilic attack of anions but requires substitution by strongly electronegative groups, such as nitro groups, before the ring becomes sufficiently electrophilic to react with anions. Detailed treatment of aromatic electrophilic substitutions can be found in references 21 and 22.

Miscellaneous Reactions.   Several other types of reactions of toluene are also used commercially or are of potential commercial interest and are discussed in more detail in Utilization and Potential Uses of Toluene. These are thermal hydrogenolysis to yield benzene, methane, and biphenyl;

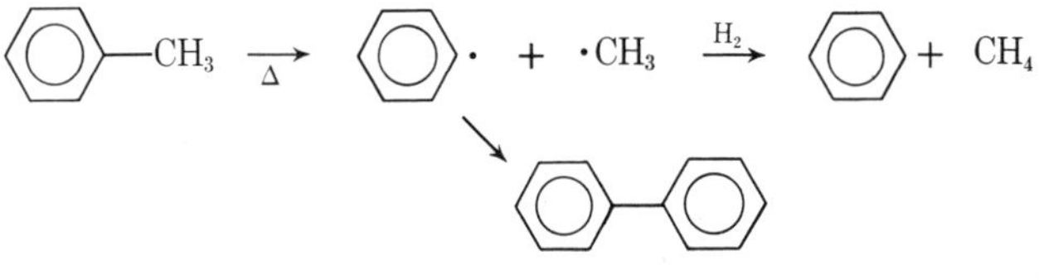

partial oxidation to yield stilbene;

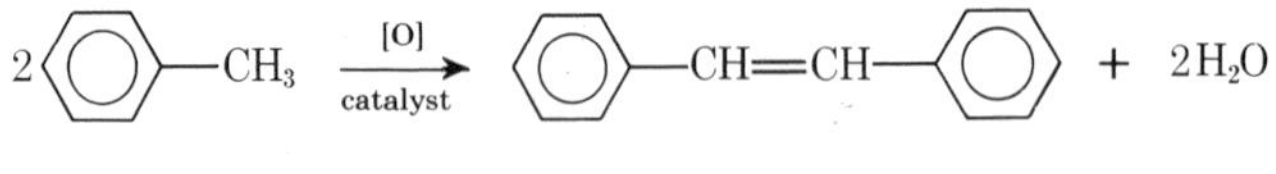

and the disproportionation to yield benzene and xylenes.

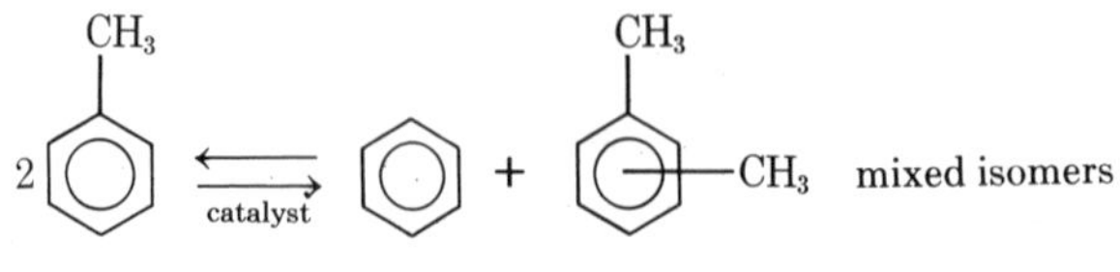

Manufacture of Toluene.   The principal source of toluene is catalytic reforming of refinery streams. This source accounts for ca 87% of the total toluene produced. An additional 9% is separated from pyrolysis gasoline produced in steam crackers during the manufacture of ethylene and propylene. Other sources are an additional 2% recovered as a by-product from styrene manufacture and 1–2% entering the market from separation from coal tars. The reactions taking place in catalytic reforming to yield aromatics are dehydrogenation or aromatization of cyclohexanes, dehydroisomerization of substituted cyclopentanes, and the cyclodehydrogenation of paraffins. The formation of toluene by these reactions is shown below.

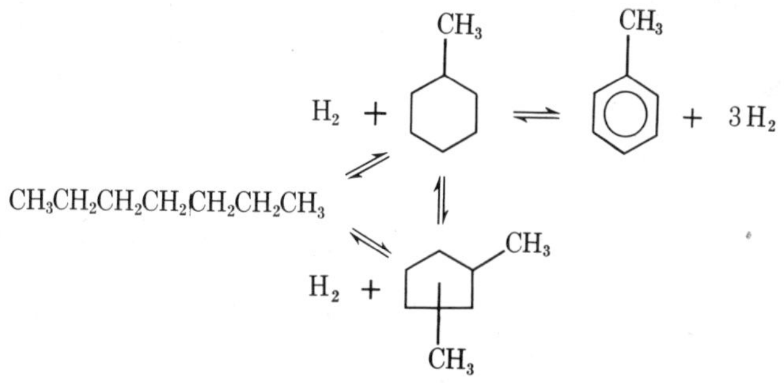

$$CH_3CH_2CH_2CH_2CH_2CH_2CH_3$$

Of the main reactions, aromatization takes place most readily and proceeds ca 7 times as fast as the dehydroisomerization reaction and ca 20 times as fast as the dehydrocyclization. Hence, feeds richest in cycloparaffins are most easily reformed. Hydrocracking to yield paraffins lower boiling than feedstock proceeds at about the same rate as dehydrocyclization.

In order to obtain pure aromatics, crude reformate is extracted to separate the aromatics from unreacted paraffins and cycloparaffins. The aromatics are, in turn, separated by simple fractional distillation to yield high purity benzene, toluene, xylenes, and C_9 aromatics.

Catalytic reforming, which was introduced primarily to increase octane values for both aviation and automotive fuels, has since become the main source of benzene and xylenes as well as of toluene (see Feedstocks). Before 1940, both fixed-bed and fluidized-bed units, typically using a 10–15% $Mo–Al_2O_3$ catalyst or similar catalysts promoted with 0.5–2% cobalt, predominated. Improved operation was obtained in 1940 by the introduction of a 0.3–0.6% $Pt–Al_2O_3$ catalyst. Since ca 1970, further improvement has been obtained by promoting the $Pt–Al_2O_3$ catalyst with up to 1% chloride, by using bimetallic catalysts containing 0.3–0.6% of both platinum and rhenium to retard deactivation, and by using molecular sieves (qv) as part of the catalyst base to gain activity. Continuous catalytic reforming was introduced ca 1971.

Because catalytic reforming is an endothermic reaction, most reforming units comprise about three reactors with reheat furnaces in between to minimize kinetic and thermodynamic limitations caused by decreasing temperature. There are three basic types of operations, ie, semiregenerative, cyclic, and continuous. In the semiregenerative operation, feedstocks and operating conditions are controlled so that the unit can be maintained on-stream from 6 mo to 2 yr before shutdown and catalyst regeneration. In cyclic operation, a swing reactor is employed so that one reactor can be regenerated while the other three are in operation. Regeneration, which may be as frequent as every 24 h, permits continuous operation at high severity. Since ca 1970, continuous units have been used commercially. In this type of operation, the catalyst is continuously withdrawn, regenerated, and fed back to the system. Flow sheets for representatives of each of the three types of processes, ie, Rheniforming (23), Ultraforming (24), and Platforming (25), are shown in Figures 1, 2, and 3.

The predominant feeds for reforming are straight-run naphthas from crude stills. Naphthas from catalytic crackers and naphthas from coke stills are also used. Typical compositions are summarized in Table 7. Typical operating conditions for catalytic reforming are 1.135–3.548 MPa (150–500 psig), 455–549°C, 0.356–1.069 m³ H_2/L (2000–6000 ft³/bbl) of liquid feed, and a space velocity (wt feed per wt catalyst) of 1–5 h. Operation of reformers at low pressure, high temperature, and low hydrogen recycle rates favors the kinetics and the thermodynamics for aromatics production and reduces operating costs. However, all three of these factors, which tend to increase coking,

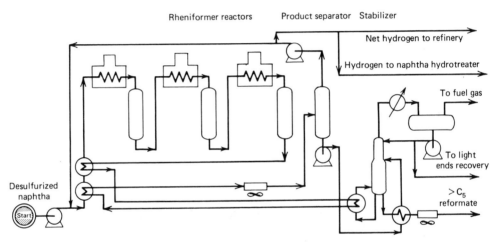

Figure 1. Chevron Research Co. Rheniforming process. Courtesy of Gulf Publishing Co. (23).

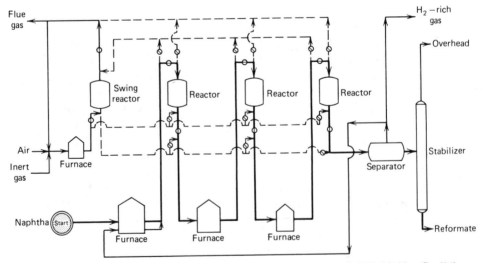

Figure 2. Standard Oil (IN) Co. Ultraforming process. Courtesy of Gulf Publishing Co. (24).

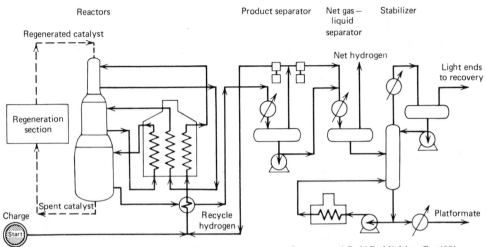

Figure 3. Universal Oil Products Platforming process. Courtesy of Gulf Publishing Co. (25).

Table 7. Composition of Typical 93–204°C Reformer Feeds, Vol %

Source	Paraffins	Cycloparaffins	Aromatics
crude still	40–55	40–30	10–20
catalytic cracker	30–40	15–25	40–50
coking still	50–55	30–35	10–15

increase the deactivation rate of the catalyst; therefore, operating conditions are a compromise. More detailed treatment of the catalysis and chemistry of catalytic reforming is available in refs. 26–28. Typical reformate compositions are shown in Table 8.

The composition of aromatics centers on the C_7 and C_8 or C_9 fraction, depending somewhat on the boiling range of the feedstock used. Most catalytic reformate is used directly in gasoline. That part which is converted to benzene, toluene, and xylenes for commercial sale is separated from the unreacted paraffins and cycloparaffins by extraction. The two processes primarily in use now are both liquid–liquid extractions. The first, developed by Shell and licensed by UOP, uses sulfolane as the extraction agent (29). A flow sheet for this process is shown in Figure 4. The second process developed and licensed by Union Carbide (30), use tetraethylene glycol (Fig. 5). These

Table 8. Composition of Typical Reformate, Vol %

paraffins	20–30
cycloparaffins	2–3
aromatics	67–77
C_6	2–3
C_7	15–20
C_8	20–28
C_9	15–25
C_{10}	1–10

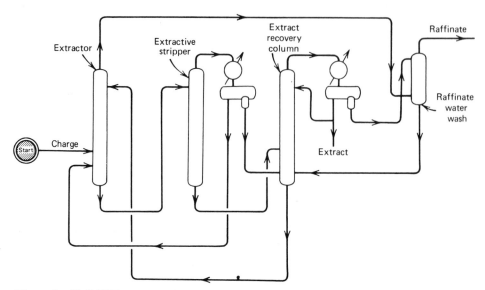

Figure 4. Shell-UOP process, sulfolane-extraction process. Courtesy of Gulf Publishing Co. (28).

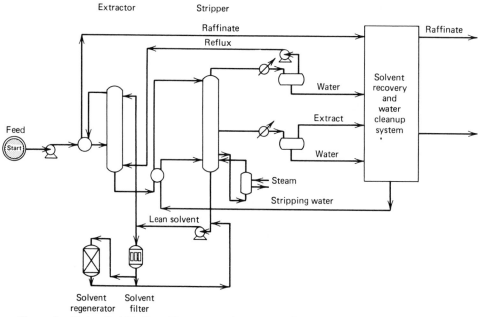

Figure 5. Union Carbide Corp. Tetra-extraction process. Courtesy of Gulf Publishing Co. (29).

processes both supplant the Udex process, which utilized diethylene glycol. *N*-Methylpyrrolidinone is also used as extractant but has not achieved the preeminence of the sulfolane or tetraethylene glycol extraction. Once the nonaromatic paraffins and cycloparaffins are removed, fractionation to separate the C_6–C_9 aromatics is relatively simple.

Proper choice of feedstocks and use of relatively severe operating conditions in the reformers produce streams high enough in toluene to be directly usable for hydrodemethylation to benzene without the need for extraction.

Toluene is recovered from pyrolysis gasoline usually by mixing the pyrolysis gasoline with reformate and processing the mixture in a typical aromatics extraction unit. Yields of pyrolysis gasoline and the toluene content depend on the feedstock to the steam-cracking unit as shown in Table 9 (see Ethylene). Pyrolysis gasoline is hydrotreated to eliminate dienes and styrene before processing to recover aromatics.

Economic Aspects

Production and sales of toluene are summarized in Table 10 and in Figure 6. These data show that the contribution of coal-derived toluene to the total production has diminished rapidly over the last 30 yr and now contributes probably <1% of the total. Aberrations in the continued smooth growth of the use of toluene result from the OPEC oil embargos in 1973 and from the current adjustment of the market to rapidly changing needs and values for use in chemicals and automotive fuels. Reforming capacity in the United States is capable of producing ca 31×10^6 t (9.4×10^9 gal) of toluene per year. Of this amount, ca 17% is actually isolated as toluene, and in 1980, about half of this material was used for chemicals manufacture, the remainder being returned to the gasoline pool for octane-number improvement. The 32 U.S. corporations that isolate toluene are listed in Table 11 with their annual capacities. In the cases where the primary use of the toluene is for dealkylation to benzene, actual production is very difficult to predict because the amount of benzene manufactured by hydrodealkylation units (see subsequent discussion) generally supplies the swing capacity to meet benzene demands that are not filled by other sources.

Table 9.   Toluene Content of Pyrolysis Gasoline, C_5 to 200°C[a]

Feedstock	Wt % to pyrolysis gasoline	Wt % toluene in pyrolysis gasoline
C_2–C_4 paraffin	0.5–10	7–15
naphthas	15–21	11–22
gas oils	17–20	13–19

[a] Ref. 31.

Table 10.   U.S. Production and Sales of Toluene[a], 1000 t (10^6 gal)

Year	Production	Sales
1950	277 (84)	231 (70)
1955	613 (186)	454 (138)
1960	902 (274)	659 (200)
1965	1808 (549)	1067 (324)
1970	2733 (830)	1416 (430)
1973	3155 (958)	1689 (513)
1975	2322 (705)	1452 (441)
1976	3290 (999)	2035 (618)
1977	3349 (1017)	1505 (457)
1978	3471 (1054)	2569 (780)
1979	3326 (1010)	2858 (868)
1980	5104 (1550)	2157 (655)

[a] Ref. 32.

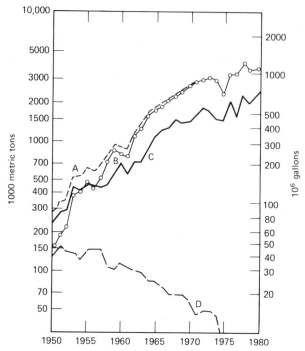

Figure 6.   Production and sales of toluene (33). A, Total production; B, production from petroleum operations; C, sales; D, production from coal-related operations.

The future production of toluene is also very difficult to predict because a number of factors will affect it (35). Octane requirements for gasoline are expected to increase and will increasingly require the use of high octane components such as toluene as lead-free gasolines become dominant in the market. On the other hand, the use of gasoline has generally decreased and is predicted to remain fairly constant for several years. The amounts of toluene and benzene produced from steam-cracking operations are currently diminishing because of increasing use of feedstocks containing ethane, propane, and butanes, which produce low yields of these aromatics. Benzene and toluene from this source will then need to be replaced by toluene derived from catalytic reforming. However, as more steam crackers are built or if the light feedstocks become less available, heavier feedstocks will be steam-cracked and they will in turn produce more benzene and toluene. Additionally, the use of toluene as a solvent is expected to decrease. However, because only ca 7–9% of the total toluene produced in the United States is now used for chemicals manufacture, there is a very broad base to draw from should the chemical needs increase even very significantly.

Prices for nitration-grade toluene are shown in Table 12. During 1960–1970, the price of toluene was relatively stable at ca $60/t (ca 20¢/gal). Owing to increased prices of imported crude, a need for high octane components in gasoline resulting from the phasing out of lead, and most recently, deregulation of the price of crude in the United States, the price of toluene has risen considerably to the current $450/t (ca $1.50/gal). The price actually paid for toluene varies ±10% from values shown depending upon point of manufacture and point of delivery.

The minimum price for the toluene used in chemicals is set by its value in unleaded gasoline, which is the principal use. The ceiling price is set by the relative values

Table 11. U.S. Producers of Toluene from Petroleum and Their Annual Capacities, Jan. 1, 1981[a], 1000 t (10[6] gal)

Continental United States	
American Petrofina Company of Texas/Union Oil Company of California	125 (38)
ARCO Chemical Company	280 (85)
Ashland Chemical Company (Petrochemicals Division)	181 (55)
Bethlehem Steel Corporation	3 (1)
Champlin Petroleum Company	99 (30)
Charter Chemicals–Charter International Oil Company	40 (12)
Conoco Chemicals (now part of DuPont)	165 (50)
Coastal States Marketing Inc.	56 (17)
Cosden Oil & Chemical Company	165 (50)
Crown Central Petroleum Corporation	36 (11)
Dow Chemical U.S.A.	13 (4)
Exxon Company U.S.A	412 (125)
Getty Refining and Marketing Company	132 (40)
Gulf Oil Chemicals Company (Aromatics and Derivatives)	461 (140)
Jones and Laughlin Steel Corporation	7 (2)
Koch Industries	138 (42)
Marathon Oil Company	72 (22)
Mobil Chemical Company (Petrochemicals Division)	296 (90)
Nueces Petrochemical Company	86 (26)
Pennzoil Products Company (Manufacturing Division)	86 (26)
Phillips Chemical Company	33 (10)
Shell Chemical Company	198 (60)
Southwestern Refining Company, Inc.	148 (45)
Sun Company, Inc.	464 (141)
Tenneco Oil Processing and Marketing	115 (35)
Texaco Chemical Company	224 (68)
Union Carbide Corporation (Engineering and Hydrocarbons Division)	60 (20)
Union Oil Company of California (Union Chemicals Division)	56 (17)
USS Chemicals	30 (9)
Caribbean	
Commonwealth Petrochemicals, Inc.	445 (135)
Hess Oil Virgin Islands Corporation	46 (140)
Phillips Puerto Rico Core, Inc.	336 (102)
Total	5427 *(1648)*

[a] Ref. 34.

of benzene and toluene. When the value of benzene is such that the differential between benzene and toluene exceeds the cost of converting toluene to benzene, then the price of toluene is set by its value for the conversion to benzene. A differential of $91/t (ca 30¢/gal) is generally required to make conversion of toluene to benzene economically attractive.

U.S. imports and exports of toluene are summarized in Table 13. From 1966 to 1978, ca 90% of the imports were from Canada and Japan. Toluene was exported to about a dozen countries, Italy and the Netherlands receiving ca 60–70% of the total; most of this material was for chemicals use. World production capacities for toluene are summarized in Table 14 for geographical areas along with the main producer in each area and that country's percentage of the total production.

Table 12.   U.S. Toluene Prices[a]

Year	Price, $/t (¢/gal)
1960	58 (19)
1965	49 (16)
1970	55 (18)
1975	140 (46)
1976	164 (54)
1977	161 (53)
1978	167 (55)
1979	331 (109)
1980	389 (128)
1981	440 (145)

[a] Midyear prices. Depending on source and delivery point, prices vary by ca ±10% (32,36).

Table 13.   U.S. Toluene Trade[a]**, 1000 t (10⁶ gal)**

Year	Imports	Exports	Balance
1955		38.2 (11.6)	+38.2 (+11.6)
1960	9.9 (3.0)	167.0 (50.7)	+157.1 (+47.7)
1965	44.5 (13.5)	154.4 (46.9)	+109.9 (+33.4)
1968	106.7 (32.4)	118.9 (36.1)	+12.2 (+3.7)
1970	306.9 (93.2)	74.1 (22.6)	−235.5 (−70.6)
1972	472.6 (143.5)	84.6 (25.7)	−388.0 (−117.8)
1975	68.5 (20.8)	383.3 (116.4)	+314.8 (+95.6)
1977	216.0 (65.6)	525.9 (159.7)	+309.9 (+94.1)
1978	192.3 (58.4)	362.2 (110)	+169.9 (+51.6)
1979	163.0 (49.5)	337.9 (102.6)	+174.9 (+53.1)
1980	252.3 (76.6)	130.4 (39.6)	−121.9 (+37.0)

[a] Ref. 33.

Specifications, Test Methods, and Analysis

Toluene is marketed mostly as nitration and industrial grades. The general accepted quality standards are given by ASTM D 841 and D 362, which are summarized in Tables 15 and 16 with the appropriate ASTM test method specified for determining the specification properties (38). Although the actual concentration of toluene in samples is not stipulated by these specifications, the purity is in fact controlled by the specific gravity and the boiling-range requirements of the method.

Purity of toluene samples as well as the number, concentration, and identity of other components can be readily determined using standard gas-chromatography techniques (39–41). Toluene content of high purity samples can also be accurately measured by freezing point as outlined in ASTM D 1016. Toluene exhibits characteristic uv, ir, nmr, and mass spectra which are useful in many specific control and analytical problems (42–45).

Table 14. Worldwide Annual Toluene Capacities, Jan. 1, 1981[a], 1000 t (10⁶ gal)

North America	6,234 (1,893)
U.S., 87%	
South America	382 (116)
Brazil, 59%	
Western Europe	1,666 (506)
Italy, 31%	
Eastern Europe[b]	>1,179 (>358)
USSR, 39%	
Middle East	63 (19)
Israel, 94%	
Far East	>2,193 (>666)
Japan, 76%	
Oceania	46 (14)
Africa	43 (13)
Total	*>11,805 (>3,585)*

[a] Ref. 37.
[b] Includes capacity data for three USSR toluene plants which may not as yet have been completed.

Table 15. Specifications For Nitration Grade Toluene, ASTM D 841-80[a]

Property	Specification	ASTM test method
sp gr, 20°/20°C	0.8690–0.8730	D 891
color	not darker than 20 max on the Pt–Co scale	D 1209
distillation range at 101.3 kPa (1 atm)	not more than 1°C including 110.6°C for any one sample	D 850
paraffins	not more than 1.5 wt %	D 851
acid-wash color	not darker than no. 2 color standard	D 848
acidity	no free acid, no evidence of acidity	D 847
sulfur compound	free of H_2S and SO_2	D 853
copper corrosion	copper strip shall not show iridescence nor gray or black deposit or discoloration	D 849

[a] Ref. 38.

Safety and Handling

The Manufacturing Chemists' Association Inc. (MCA) has published the Chemical Safety Data Sheet SD 63 which describes in detail procedures for safe handling of use of toluene (46). The Interstate Commerce Commission classifies toluene as a flammable liquid. Accordingly, it must be packaged in authorized containers, and shipping must comply with ICC regulations. Properties related to safe handling are

explosive limits	1.27–7.0 vol % in air
flash point	4.4°C, closed cup

Current permissible exposure limits established by the U.S. Department of Health and Human Services and the U.S. Department of Labor are summarized below with

Table 16. Specifications for Industrial-Grade Toluene, ASTM D 362-80[a]

Property	Specification	ASTM test method
sp gr, 20°/20°C	0.860–0.874	D 891
color	not darker than 20 max on the Pt–Co scale	D 1209
distillation range at 101.3 kPa (1 atm)	not more than 2°C from initial boiling point to dry point, including 110.6°C	D 850, D 1078
odor	characteristic aromatic hydrocarbon odor as agreed on by buyer and seller	D 1296
water	not sufficient to show turbidity at 20°C	
acidity	not more than 0.005 wt % (free acid calculated as acetic acid) equivalent to 0.047 mg KOH (0.033 mg NaOH) per gram of sample or no free acid; that is, no evidence of acidity	D 847
acid-wash color	not darker than no. 4 color standard	D 848
sulfur compounds	free of H_2 and SO_2	D 853
corrosion ½ h at 100°C	copper strip shall not show greater discoloration than Class 2 in Method D 1616	D 1616
solvent power	100 min kauri-butanol value	D 1113

[a] Ref. 38.

the more restrictive levels proposed by NIOSH (47):

	OSHA, mg/m^3 (ppm)	NIOSH, mg/m^3 (ppm)
average during 8-h shift (TWA)	752 (200)	376 (100)
not to exceed	1129 (300)	
except for 10-min average (TLV)	1881 (500)	752 (200)

Toluene generally resembles benzene closely in its toxicological properties; however, it is devoid of benzene's chronic negative effects on blood formation (48). General effects of inhalation are summarized in Table 17. A detailed discussion of physiological response may be found in ref. 48. The odor threshold for toluene has been determined to be ca 9.5 mg/m^3 (2.5 ppm) (49). In the human system, toluene is oxidized to benzoic acid which in turn reacts with glycine to form hippuric acid (N-benzoylglycine) which is excreted in the urine.

Utilization of Toluene

The overall pattern for utilization of toluene in 1978 and 1980 in the United States is summarized in Table 18. The actual production capacity can only be estimated since

Table 17. Physiological Response to Inhaled Toluene[a]

Level, mg/L (ppm)	Result
0.38 (100)	transient irritation, psychological effects
0.76 (200)	transitory mild upper respiratory-tract irritation
1.52 (400)	mild eye irritations, lacrimation, hilarity
2.28 (600)	lassitude, hilarity, slight nausea
3.03 (800)	rapid irritation, nasal secretion, metallic taste, drowsiness, and impaired balance

[a] Ref. 48.

Table 18. U.S. Utilization of Toluene, 1000 t (10^6 gal)

Use	1978	1980
Gasoline[a]		
est contained	24,732 (7,510)	24,765 (7,520)
est isolated	2,444 (742)	2,829 (859)
Total	*27,176 (8,252)*	*27,594 (8,379)*
Chemicals[b]		
benzene	1,680 (510)	1,581 (480)
solvent	395 (120)	379 (115)
TDI	201 (61)	194 (59)
benzoic acid	66 (20)	53 (16)
benzyl chloride	36 (11)	33 (10)
other	161 (49)	165 (50)
Total	*2,539 (771)*	*2,405 (730)*
Grand Total	*29,715 (9,023)*	*29,999 (9,109)*

[a] Gasoline utilization figures are industry estimate.
[b] Chemical utilization figures are based on ref. 33.

it is dependent on the feedstocks used, the number of units operated, and the operating conditions of the units. Figures used in Table 18 are based on typical general conditions. About 17% of the total production capacity for toluene is isolated in extraction units; about half of the isolated material is used in chemicals production, and half is returned to the gasoline pool for blending to increase the octane number of premium fuels. During the past three years, the use in fuels has increased and the use in chemicals has shown a slight decline. By far the greatest use of toluene for chemicals (about two thirds of the total) is conversion by demethylation into benzene. The use of toluene as a solvent in surface coatings and other formulations is expected gradually to diminish because of substitutions required by environmental considerations. Because gasoline demand is estimated to remain fairly constant, or even to decline, in the early 1980s it is unlikely that new reforming capacity, which is capital intensive, will be built. Therefore, the capacity to produce toluene will remain relatively constant and will be increased only if the severity of reforming operations is increased or catalyst improvements permit greater production of aromatics from existing units.

Table 19. Blending Octane Number, (R + M)/2[a], for Selected Components in Unleaded Gasoline

Component	(R + M)/2	References
methanol	120, 117	50–51
ethanol	119, 113, 117	50–52
methyl *tert*-butyl ether	108, 106, 111	50–52
tert-butyl alcohol	97.5, 94.5, 96	50–52
toluene	106, 103.5, 102.9	51–53
C$_8$ aromatics	105.5	53
unleaded regular	88.0	54
unleaded premium	93.0	54

[a] R = Research-method octane rating, ASTM D 2699. M = Motor-method octane rating, ASTM D 2700.

Automotive Fuels.   About 90% of the toluene generated by catalytic reforming is blended into gasoline as a component of >C_5 reformate. The octane number (R + M/2) of such reformates is typically in the range of 88.9–94.5 depending on severity of the reforming operation. Toluene itself has a blending octane number of 103–106, which, as shown in Table 19, is exceeded only by oxygenated compounds such as methyl *tert*-butyl ether, ethanol, and methanol.

Toluene is, therefore, a valuable blending component, particularly in unleaded premium gasolines. Although reformates are not extracted solely for the purpose of generating a high octane blending stock, the toluene that is coproduced when xylenes and benzene are extracted for use in chemicals, and that exceeds demands for use in chemicals, has a ready market as a blending component for gasoline.

As a blending component in automotive fuels, toluene has several advantages. First, as shown in Table 19, it has a high octane number compared to regular and premium unleaded gasoline. Secondly, its relatively low volatility permits incorporation into gasoline blends of other less expensive and available materials, eg, *n*-butane, with relatively high volatility. Since the main use of toluene is in gasoline, with only ca 9% or less used in chemicals, there will always be an available supply for chemicals manufacture at a price essentially fixed by the value of toluene as a blending component in gasoline.

Manufacture of Benzene.   Toluene is converted to benzene by hydrodemethylation either under thermal or catalytic conditions. Benzene produced from this source generally supplies 25–30% of the total benzene demand. Reaction conditions generally range from 600–800°C at 3.55–7.00 MPa (500–1000 psig), and the reaction is exothermic. Conversion per pass is 60–90% with selectivities to benzene >95%. With catalysts, typically supported Cr_2O_3, Mo_2O_3, and CoO, operating temperatures are lower than in the thermal process, and selectivities are higher. These gains, however, are offset by the need to decoke the catalyst periodically. Losses to by-product formation, particularly biphenyls, are controlled by recycle of these materials to the reaction zone (55–56).

A flow scheme for a typical catalytic process (56) is shown in Figure 7. The feedstock is usually extracted toluene, but some reformers are operated under sufficiently severe conditions or with selected feedstocks to provide toluene pure enough to be fed directly to the dealkylation unit without extraction. In addition to toluene, xylenes

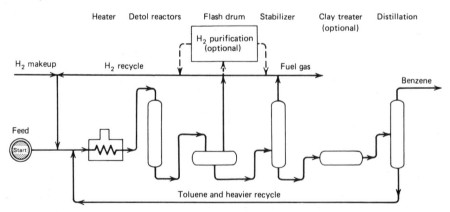

Figure 7.   Air Products and Chemicals toluene-dealkylation (Detol) process. Courtesy of Gulf Publishing Co. (56).

can also be fed to a dealkylation unit to produce benzene. Table 20 lists the producers and their capacities for manufacture of benzene by hydrodealkylation of toluene. Additional information on hydrodealkylation is available in references 55 and 57.

Use as Solvent. Toluene is more important as a solvent than either benzene or xylene. Solvent use accounts for ca 14% of the total U.S. toluene demand for chemicals. About two thirds of the solvent use is in paints and coatings; the remainder is in adhesives, inks, pharmaceuticals, and other formulated products utilizing a solvent carrier. Use of toluene as solvent in surface coatings is declining and is expected to continue to decline primarily because of various environmental and health regulations. It is being replaced by other solvents such as esters and ketones and by changing of product formulation to use either fully solid systems or water-based emulsion systems (see Solvents, industrial).

Toluene Diisocyanate. Toluene diisocyanate is the basic raw material for production of flexible polyurethane foams. It is produced by the reaction sequence shown below in which toluene is dinitrated, the dinitrotoluene hydrogenated to yield 2,4-diaminotoluene, and this diamine in turn treated with phosgene to yield toluene 2,4-diisocyanate.

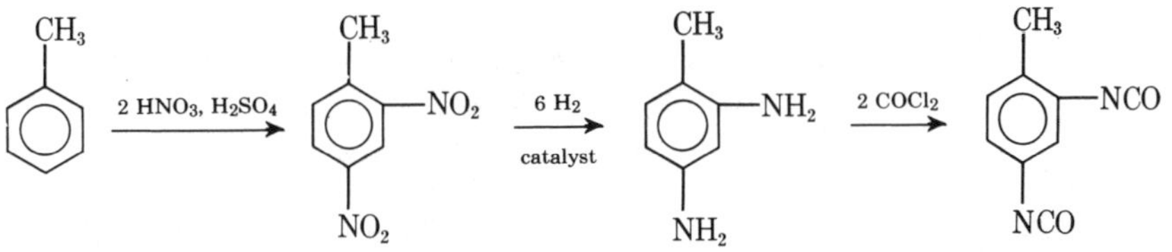

The nitration step produces two isomers, 2,4-dinitrotoluene and 2,6-dinitrotoluene, the former predominating. Mixtures of the two isomers are frequently used, but if single isomers are desired, particularly the 2,4-dinitrotoluene, nitration is stopped at the mono stage and pure *para*-nitrotoluene is obtained by crystallization. Subsequent nitration of this material yields only 2,4-dinitrotoluene for conversion to the diisocyanate.

Table 20. U.S. Producers of Benzene by Hydrodealkylation of Toluene and Their Annual Capacities, Jan. 1982[a], 1000 t (10^6 gal)

American Petrofina Company of Texas	76 (23)
ARCO Chemical Company	33 (10)
Ashland Oil Company	66 (20)
Coastal States Petroleum	148 (45)
Commonwealth Petrochemicals, Inc.	379 (115)
Cosden Oil and Chemical Company	132 (40)
Crown Central Petroleum Corporation	59 (18)
Dow Chemical U.S.A.	214 (65)
Gulf Oil Chemicals Company (Aromatics and Derivatives)	356 (108)
Koch Industries	66 (20)
Nueces Petrochemical Company	287 (87)
Phillips Puerto Rico Core, Inc.	310 (94)
Shell Chemical Company	23 (7)
Sun Company, Inc.	293 (89)
Total	*2442 (741)*

[a] Ref. 37.

Polyurethane foams are formed by reaction with glycerol; with poly(propylene oxide), sometimes capped with poly(ethylene oxide) groups; with a reaction product of trimethylolpropane and propylene oxide; or with other appropriate polyols. A typical reaction sequence is shown below in which HO—R—OH represents the diol. If a triol is used, a cross-linked product is obtained.

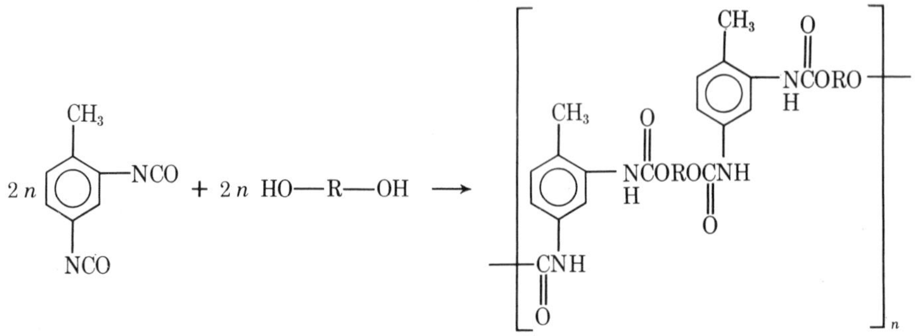

Water, in small amount, reacts with the diisocyanate to generate carbon dioxide and amine and is used most frequently as the foaming agent. Polyurethanes are treated in detail in references 58–60 (see Foamed plastics; Urethane polymers).

Benzoic Acid. Benzoic acid (qv) is manufactured from toluene by oxidation in the liquid phase using air and a cobalt catalyst. Typical conditions are 308–790 kPa (30–100 psig) and 130–160°C. The crude product is purified by distillation, crystallization, or both. Yields are generally >90 mol %, and product purity is generally >99%. Kalama Chemical Company, the largest producer, converts about half of its production to phenol, but most producers consider the most economic process for phenol (qv) to be peroxidation of cumene (qv). Other uses of benzoic acid are for the manufacture of benzoyl chloride, of plasticizers such as butyl benzoate, and of sodium benzoate for use in preservatives. In Italy, Snia Viscosa uses benzoic acid as raw material for the production of caprolactam, and subsequently nylon-6, by the sequence shown below.

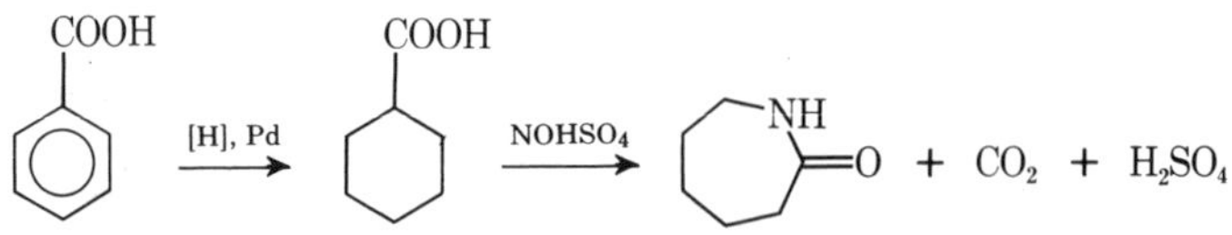

The Henkel process employing potassium benzoate has been used in Japan for the manufacture of terephthalic acid by the following scheme:

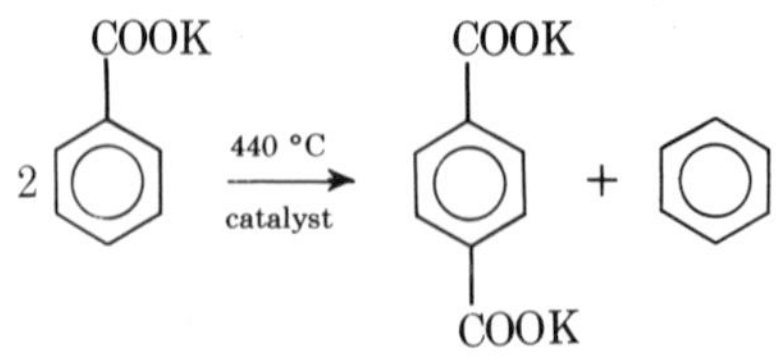

The Henkel process provides a means to convert toluene to benzene and at the same time makes use of the methyl group. Neither of these two processes is economically attractive for use in the United States.

Benzyl Chloride. Benzyl chloride is manufactured by high temperature free-radical chlorination of toluene. The yield of benzyl chloride is maximized by use of excess toluene in the feed. More than half of the benzyl chloride produced is converted to butyl benzyl phthalate by reaction with monosodium butyl phthalate. The remainder is hydrolyzed to benzyl alcohol (qv), which is converted to aliphatic esters for use in soaps, perfume, and flavors. Benzyl salicylate is used as a sunscreen in lotions and creams (see Cosmetics). By-product benzal chloride can be converted to benzaldehyde, which is also produced directly by oxidation of toluene and as a by-product during formation of benzoic acid. By-product benzotrichloride is not hydrolyzed to make benzoic acid but is allowed to react with benzoic acid to yield benzoyl chloride.

$$\text{PhCCl}_3 + \text{PhCOOH} \longrightarrow 2\,\text{PhCOCl} + \text{HCl}$$

Disproportionation to Benzene and Xylenes. With acidic catalysts, toluene can transfer a methyl group to a second molecule of toluene to yield one molecule of benzene and one molecule of mixed isomers of xylene.

$$2\,\text{PhCH}_3 \rightleftharpoons \text{PhH} + \text{(CH}_3)\text{C}_6\text{H}_4\text{—CH}_3$$

mixed
isomers

This disproportionation is an equilibrium reaction for which typical distributions are shown in Table 21. Disproportionation generates benzene from toluene and at the same

Table 21. Equilibrium Distribution for Toluene Disproportionation[a], Mol %

Temperature, K	Benzene	Toluene	Xylenes
300	31.3	37.4	31.3
400	28.8	42.4	28.8
500	26.5	47.0	26.5
600	23.9	52.2	23.9
700	21.4	57.2	21.4
800	18.7	62.6	18.7
900	16.2	67.6	16.2
1000	13.6	72.8	13.6

[a] Ref. 61.

time takes full advantage of the methyl group to generate a valuable product, ie, xylene. Economic utility of the process is strongly dependent on the relative values of toluene, benzene, and the xylenes. This xylene, which contains little or no ethylbenzene, at one time would have commanded a premium as a feed for *para*-xylene units. However, ethylbenzene-free feeds have little advantage because the zeolite-based isomerization catalysts, eg, the Mobil ZSM-5 and the Amoco AMS-1B molecular sieves (qv), very selectively hydrodeethylate and disproportionate ethylbenzene as they isomerize xylenes. Accordingly, toluene disproportionation processes have little advantage over catalytic reforming for production of xylenes. Two companies, Atlantic Richfield Co. and Sun Co., operate disproportionation plants in the United States. Operation of such plants can be justified only where there is an excess of toluene and where both xylenes and benzene are a desired product. The disproportionation plant can then replace a toluene demethylation unit. By proper selection of catalysts, the xylene production can be controlled to give high selectivity to the para isomer; however, in order to accomplish this, catalyst reactivity is greatly diminished (62).

Vinyltoluene. Vinyltoluene is produced by Dow Chemical Co. and is used as a resin modifier in unsaturated polyester resins. Its manufacture is similar to that of styrene (qv); toluene is alkylated with ethylene, and the resulting ethyltoluene is dehydrogenated to yield vinyltoluene. Annual production is in the range of 18,000–23,000 t/yr requiring 20,000–25,000 t (6–7.5 × 10^6 gal) of toluene.

Toluenesulfonic Acid. Toluene reacts readily with fuming sulfuric acid to yield toluenesulfonic acid. By proper control of conditions, *para*-toluenesulfonic acid is obtained. The primary use is for conversion, by fusion with NaOH, to *para*-cresol. The resulting high purity *para*-cresol is then alkylated with isobutylene to produce 2,6-di-*tert*-butyl-*para*-cresol (BHT), which is used as an antioxidant in foods, gasoline, and rubber. Mixed cresols can be obtained by alkylation of phenol and by isolation from certain petroleum and coal-tar process streams.

The toluenesulfonic acid prepared as an intermediate in the preparation of *para*-cresol also has a modest use as a catalyst for various esterifications and condensations. Sodium salts of the toluenesulfonic acids are also used in surfactant formulations. Annual use of toluene for sulfonation is ca 100,000–150,000 t (30–45 × 10^6 gal).

Benzaldehyde. Annual production of benzaldehyde (qv) requires ca 6,500–10,000 t (2–3 × 10^6 gal) of toluene. It is produced mainly as by-product during oxidation of toluene to benzoic acid, but some is produced by hydrolysis of benzal chloride. The main use of benzaldehyde is as a chemical intermediate for production of fine chemicals used for food flavoring, pharmaceuticals, herbicides, and dyestuffs.

Toluenesulfonyl Chloride. Toluene reacts with chlorosulfonic acid to yield both *ortho*- and *para*-toluenesulfonyl chlorides, of which Monsanto is the only producer. The ortho isomer is converted to saccharin (see Sweeteners).

The para isomer is used for preparation of specialty chemicals. Annual toluene requirements are ca 6500 t (2×10^6 gal).

Miscellaneous Derivatives. Other derivatives of toluene, none of which is estimated to consume more than ca 3000 t (10^6 gal) of toluene annually, are mono- and dinitro-toluene hydrogenated to amines; benzotrichloride, and chlorotoluene, all used as dye intermediates; *tert*-butylbenzoic acid from *tert*-butyltoluene, used as a resin modifier; dodecyltoluene converted to a benzyl quaternary ammonium salt for use as a germicide; and biphenyl, obtained as by-product during demethylation, used in specialty chemicals. Toluene is also used as a denaturant in specially denatured alcohol (SDA) formulas 2-B and 12-A.

Potential Uses of Toluene

Because much toluene is demethylated for use as benzene, considerable effort has been expended on developing processes in which toluene can be used in place of benzene to make directly from toluene the same products that are derived from benzene. Such processes both save the cost of demethylation and utilize the methyl group already on toluene. Most of this effort has been directed toward manufacture of styrene. An alternative approach, currently near commercialization, is the manufacture of *para*-methylstyrene by selective ethylation of toluene followed by dehydrogenation. Resins from this monomer are expected to displace polystyrene because of price and performance advantages. Another approach to developing large-scale uses of toluene is to find a reagent that reacts selectively in the para position to yield a derivative readily converted to a carboxylic acid. Such a process would provide a feedstock for manufacture of terephthalic acid and eliminate the need for separation of *para*-xylene, the current feedstock, from mixed C_8 aromatics.

Styrene from Toluene. Processes for forming styrene by reaction of methanol with toluene have been reported in both the Japanese and USSR literature, and in the United States a patent has been issued to Monsanto (63–66). In the latter case, an X-type faujasite aluminosilicate, exchanged with cesium and promoted with either boron or phosphorus, was used as the catalyst. Toluene and methanol at a 5:1 mol ratio react at 400–475°C. About half the methanol is converted to an ethylbenzene–styrene mixture and about half is converted to carbon monoxide and hydrogen. Yields on toluene are very high. Provision must still be made for dehydrogenation of the ethyl-benzene in the mixture. The product stream is quite dilute, ca 5%, necessitating large recycles. Because of the generation of carbon monoxide and hydrogen, such a plant would need to operate in conjunction with a methanol-synthesis plant, a significant process disadvantage which may possibly be overcome by catalyst development.

Chem Systems Inc. has proposed a process in which benzyl alcohol obtained by an undisclosed direct oxidation of toluene is homologated with synthesis gas to yield 2-phenylethyl alcohol which is then readily dehydrated to styrene (67). This process eliminates the intermediate formation of methanol from synthesis gas but does require the independent production of benzyl alcohol. This technology is in a very early stage of development.

A different approach, taken by both Monsanto (68) and Gulf Research and Development Co. (69), involves the oxidative coupling of two molecules of toluene to yield stilbene. The stilbene is then subjected to a metathesis reaction with ethylene to yield two molecules of styrene.

$$2 \bigcirc\!\!-CH_3 \xrightarrow[\text{catalyst}]{\text{air}} \bigcirc\!\!-CH{=}CH{-}\!\!\bigcirc + 2\,H_2O$$

$$\bigcirc\!\!-CH{=}CH{-}\!\!\bigcirc + CH_2{=}CH_2 \longrightarrow 2 \bigcirc\!\!-CH{=}CH_2$$

A principal problem appears to be the dehydrocoupling reaction which proceeds only at low yields per pass and is accompanied by rapid deactivation of the catalyst. The metathesis step, although chemically feasible, requires that polar contaminants resulting from partial oxidation be removed so that they will not deactivate the metathesis catalyst. In addition, apparently both *cis-* and *trans-*stilbenes are obtained; consequently, a means to convert the unreactive *cis-*stilbene to the more reactive trans isomer must also be provided, complicating the process.

None of these potential toluene processes appears to be simple enough to compete with existing technology.

para-**Methylstyrene.**   Mobil Chemical has recently announced plans to commercialize manufacture of *para*-methylstyrene from toluene (70–71). This monomer will be produced by alkylating toluene with ethylene using the Mobil ZSM-5 zeolite catalyst. The alkylation is highly selective, reportedly producing the para isomer with 97% selectivity. Conventional technology, employing a special catalyst to minimize by-products and optimize conversion, will be used to dehydrogenate the *para*-ethyltoluene to *para*-methylstyrene.

Vinyltoluene, comprising a mixture of ca 33% *para-* and 67% *meta*-methylstyrene, has been marketed for ca 30 yr by Dow Chemical Co. and also by Cosden. However, the performance properties of the polymers prepared from the para isomer are not only superior to those of the polymer prepared from the typical mixed isomers, but are generally superior to those of polystyrene (70). This advantage, coupled with a raw-material cost advantage over styrene, suggests that *para*-methylstyrene may displace significant amounts of styrene, currently a 3.2×10^6 t/yr domestic market (see Styrene polymers).

Terephthalic Acid from Toluene.   Both carbon monoxide and methanol can react with toluene to yield intermediates that can be oxidized to terephthalic acid. In work conducted mainly by Mitsubishi Gas Chemical Co. (72–73), toluene reacts with carbon monoxide and molar excesses of HF and BF_3 to yield a *para*-tolualdehyde–HF–BF_3 complex. Decomposition of this complex under carefully controlled conditions recovers HF and BF_3 for recycle and *para*-tolualdehyde which can be oxidized in place of *para*-xylene to yield terephthalic acid. A detriment to the process is the energy-intensive, and therefore high cost, decomplexing step. The need for corrosion-resistant materials for construction and the need for extra design features to handle the relatively hazardous HF and BF_3 also add to the cost. Currently, the process is not being used. It might, however, be advantageous where toluene is available and xylenes are in short supply (see Polyesters; Phthalic acid and other benzene polycarboxylic acids).

A second approach is the selective alkylation of toluene with methanol to yield C_8 aromatic mixtures containing 70–90% *para*-xylene and generally <1% of ethylbenzene (74). Such C_8 aromatic mixtures are excellent feedstocks for recovery of high purity *para*-xylene. The high selectivity to *para*-xylene is achieved by modifying typical HZMS-5 silica aluminate zeolites (75) with phosphorus and boron. To date this process is not used commercially, probably because current feedstock needs for manufacture of terephthalic acid are met by *para*-xylene from typical reformate, and because conversions of toluene in the process are relatively low (ca 20% per pass) and significant amounts of methanol are converted to by-products such as HCHO, CO, CO_2, CH_4, C_2H_4 and C_3H_6 when stoichiometric quantities of methanol are used. Best results are obtained at 4:1 and higher mole ratios of toluene to methanol. Improved selectivity of the catalyst to permit better utilization of the methanol would enhance the economics for this process (see Xylenes and ethylbenzene).

BIBLIOGRAPHY

"Toluene" in *ECT* 1st ed., Vol. 14, pp. 262–273, by Maury Lapeyrouse, Esso Research and Engineering Co.; "Toluene" in *ECT* 2nd ed., Vol. 20, pp. 527–565, by Harry E. Cier, Esso Research and Engineering Co.

1. *Selected Values of Physical and Thermodynamic Properties of Hydrocarbons and Related Compounds*, API research project 44, Carnegie Press, Pittsburgh, Pa., 1953.
2. *Physical Constants of Hydrocarbons C_1 to C_{10}*, ASTM Data Series Pub. DS4A, American Society for Testing and Materials, Philadelphia, Pa., 1971.
3. *Washington University Thermodynamics Research Laboratory Recommended Values*, Washington University, St. Louis, Mo.
4. L. H. Horsley, "Azeotropic Data-III," *Adv. Chem. Ser.* **116**, 373 (1973).
5. L. J. Andrews and R. M. Keefer, *J. Am. Chem. Soc.* **71**, 3644 (1949); **72**, 3113 (1950).
6. H. C. Brown and J. D. Brady, *J. Am. Chem. Soc.* **74**, 3570 (1952).
7. R. E. Merrifield and W. D. Phillips, *J. Am. Chem. Soc.* **80**, 2778 (1958).
8. D. A. McCaulay and A. P. Lien, *J. Am. Chem. Soc.* **73**, 2013 (1951).
9. D. A. McCaulay and co-workers, *Ind. Eng. Chem.* **42**, 2103 (1950).
10. C. K. Ingold and co-workers, *J. Chem. Soc.*, 1959 (1931).
11. F. E. Condon, *J. Am. Chem. Soc.* **70**, 1963 (1948).
12. H. Pines and co-workers, *J. Chem. Soc.* **77**, 554 (1955).
13. G. G. Eberhardt and H. J. Peterson, *J. Org. Chem.* **30**, 82 (1965).
14. G. G. Eberhardt and W. A. Butte, *J. Org. Chem.* **29**, 2928 (1964).
15. H. Pines and L. A. Schaap in *Advances in Catalysis*, Vol. XII, Academic Press, New York, 1960, p. 117.
16. M. S. Kharasch and M. J. Berkman, *J. Org. Chem.* **6**, 810 (1941).
17. H. Pines, *The Chemistry of Catalysis Hydrocarbon Conversions*, Academic Press, Inc., New York, 1981, pp. 173–174.
18. *Hydrocarbon Process.* **60**(11), 147 (1981).
19. R. T. Morrison and R. N. Boyd, *Organic Chemistry*, 3rd ed., Allyn & Bacon, Inc., Boston, Mass., 1973, p. 323.
20. B. T. Brooks and co-workers, *The Chemistry of Petroleum Hydrocarbons*, Van Nostrand Reinhold Company, New York, 1955, Chapt. 56.
21. J. March, *Advanced Organic Chemistry*, McGraw-Hill Book Company, New York, 1977, Chapt. 11.
22. G. A. Olah, *Friedel-Crafts Reactions*, Vols. 2 and 3, Interscience Publishers, a division of John Wiley & Sons, Inc., New York, 1964.
23. *Hydrocarbon Process.* **55**(5), 75 (May 1976).
24. R. Coates and co-workers, *Proc. API Div. Refin.* **53**, 251 (1973).
25. E. A. Sutton, A. R. Greenwood, and F. H. Adams, *Oil Gas. J.* **70**(21), 52 (May 22, 1972).
26. Ref. 17, pp. 101–110.
27. J. H. Gary and G. E. Handwark, *Petroleum Refining Technology and Economics*, Marcel Dekker, Inc., New York, 1975, pp. 65–85.
28. J. E. Germain, *Catalytic Conversion of Hydrocarbons*, Academic Press, Inc., New York, 1969.
29. D. B. Broughton and G. F. Asselin, *Proc. Second World Pet. Congr.* **4**, 65 (1967).
30. *Hydrocarbon Process.* (9), 204 (Sept. 1980).
31. L. Kniel, O. Winter, and Chung-Hu Tsai, "Ethylene" in *ECT* 3rd ed., Vol. 9, p. 404.
32. U.S. International Trade Commission, *Synthetic Organic Chemicals, U.S. Production and Sales*, U.S. Government Printing Office, Washington, D.C., annual publication.

33. *Toluene* in *Chemical Economics Handbook*, SRI International, Menlo Park, Calif., July 1979; update, Jan. 1982.
34. *World Petrochemicals—Toluene Report*, SRI International, Menlo Park, Calif., 1982, p. 427.
35. *Chem. Week*, 42 (Mar. 18, 1981).
36. *Chem. Mark. Rep.*, Schnell Publishing Company, New York.
37. Ref. 34, p. 405.
38. *1980 Annual Book of ASTM Standards*, Pt. 29, American Society for Testing and Materials, Philadelphia, Pa., 1980.
39. *1972 Annual Book of ASTM Standards*, Pt. 29, American Society for Testing and Materials, D 2360-68 and D 2306-67.
40. C. L. Stucky, *J. Chromatogr. Sci.* **7**, 177 (1969).
41. H. M. McNaire and E. J. Bonelli, *Basic Gas Chromatography*, Varian Aerograph, Walnut Creek, Calif.; R. R. Freeman, ed., *High Resolution Gas Chromatography*, 2nd ed., Hewlett-Packard Company, Palo Alto, Calif., 1981.
42. *American Petroleum Institute Project 44 data sheets*, API Data Distribution Office, A & M Press, College Station, Texas.
43. *The Sadtler Standard Spectra*, Sadtler Research Laboratories, Philadelphia, Pa., 1971.
44. C. J. Pouchert, *The Aldrich Library of Infrared Spectra*, 2nd ed., Aldrich Chemical Co., 1975.
45. J. G. Grasselli, *Atlas of Spectral Data and Physical Constants for Organic Compounds*, CRC Press, Cleveland, Ohio, 1973.
46. *Chemical Safety Data Sheet SD-63, Toluene*, Manufacturing Chemists' Association, Washington, D.C., 1956.
47. U.S. Department of Health and Human Services, *Occupational Health Guidelines for Chemial Hazards*, U.S. Government Printing Office, Washington, D.C., Jan. 1981, DHHS (NIOSH) Pub. No. 81-123.
48. G. D. Clayton and F. E. Clayton, eds., Patty's Industrial Hygiene and Toxicology, 3rd ed., Vol. 2B, John Wiley & Sons, Inc., New York, 1981.
49. C. P. Carpenter and co-workers, *Toxicol. Appl. Pharmacol.* **26**, 473 (1976).
50. H. L. Hoffman, *Hydrocarbon Process.* (2), 56 (Feb. 1980).
51. G. H. Unzelman, *46th Refining Midyear Meeting*, API, Chicago, Ill., May 12, 1981.
52. T. O. Wagner and co-workers, *Congress of Society of Automotive Engineers*, Detroit, Mich., Feb. 26, 1979.
53. W. E. Morris, *National Petroleum Refiners Association Annual Meeting*, New Orleans, La., Mar. 23–25, 1980.
54. J. R. Dosher, *Hydrocarbon Process.* (5), 123 (May 1979).
55. K. Weissermel and H. Arpe, *Industrial Organic Chemistry*, Verlag Chemie, New York, 1978, pp. 288–289.
56. *Hydrocarbon Process.* (11), 138 (Nov. 1981).
57. A. L. Waddams, *Chemicals from Petroleum*, 4th ed., John Murray, London, 1978, Chapt. 13.
58. H. J. Saunders and co-workers, *Polyurethanes: Chemistry and Technology*, Pt. 1, John Wiley & Sons, Inc., 1962, pp. 273–314.
59. P. F. Bruins, *Polyurethane Technology*, Interscience Publishers, a division of John Wiley & Sons, Inc., New York, 1969, pp. 1–37.
60. E. N. Doyle, *The Development and Use of Polyurethane Products*, McGraw-Hill Book Company, New York, 1971, pp. 233–255.
61. D. R. Stull and co-workers, *Chemical Thermodynamics of Hydrocarbon Compounds*, John Wiley & Sons, Inc., New York, 1969, p. 368.
62. W. W. Kaeding and co-workers, *J. Catal.* **69**, 392 (1981).
63. T. Yashima and co-workers, *J. Catal.* **26**, 303 (1972).
64. H. Itoh and co-workers, *J. Catal.* **64**, 284 (1980).
65. USSR Pat. 188,958 (Oct. 11, 1965), (to USSR).
66. U.S. Pat. 4,115,424 (Dec. 22, 1976), M. L. Unland and G. E. Barker (to Monsanto Company).
67. A. P. Gelbein, *Toluene-Synthesis Gas Based Routes to Styrene, An Assessment*, ACS Petro Chemical Division, New York, Aug. 23–28, 1981.
68. U.S. Pat. 3,965,206 (Dec. 12, 1974), P. D. Montgomery, R. N. Moore, and W. R. Knox (to Monsanto Company).
69. R. A. Innes and H. E. Swift, *Chemtech*, 244 (Apr. 1981).
70. *Chem. Week*, 42 (Feb. 17, 1982).
71. *Chem. Eng. News*, 20 (May 31, 1982).
72. S. Fujiyama and T. Kashara, *Hydrocarbon Process.* **11**, 147 (1978).

73. A. Mitsutani, *Terephthalic Acid from Toluene*, R&D Review Report No. 8, Nippon Chemtec Consulting, Inc., Feb. 1978.
74. W. W. Kaeding and co-workers, *J. Catal.* **67,** 159 (1981).
75. U.S. Pat. 3,702,886 (Oct. 19, 1969), R. J. Arganer and R. G. Landolt (to Mobil Oil Corporation).

M. C. HOFF
Amoco Chemicals Corporation

TOLUENEDIAMINES. See Amines, aromatic, diaminotoluenes.

TOOL MATERIALS

Machining of materials with a cutting tool harder than the work material is a common operation in the production of a variety of parts. The unwanted material is removed from the workpiece in the form of chips, using a single- or multiple-point cutting tool on a machine tool. The machining process involves extensive plastic deformation (shear strain γ ca 2–8) of the work material ahead of the tool in a narrow region; high temperature (ca 1000°C); freshly generated, chemically active surfaces (underside of the chip and the machined surface) that can interact extensively with the tool material; and high mechanical and thermal stresses on the tool. The exact nature and extent of these effects depend upon the cutting conditions and the combination of tool and work material. A successful cutting tool material must resist these severe conditions and provide a sufficiently long tool life. Currently, >100×10^9/yr is spent on labor and overhead costs for machining in the United States alone (1) (see also Ceramics; Carbon, diamond; Boron compounds, refractory boron compounds; High temperature alloys).

The cutting tool is a very important component of the machining system. Consequently, the tool material significantly affects the productivity of the machining operation. Other important elements include cutting conditions, tool geometry, and the characteristics of the work material (chemical and metallurgical state), nature of the parts produced (geometry, accuracy, finish, and surface-integrity requirements), machine tool (adequate rigidity with high horsepower, and wide speed and feed ranges), and support system (operator's ability, sensors, controls, method of lubrication, and chip control).

A tool material must meet several stringent requirements dictated by the cutting process. Both the deformation energy, generated by extensive plastic deformation, and the frictional energy, generated by severe frictional interactions at the tool-chip and tool-machined surface interfaces, are converted into heat. Although most of this heat (ca 80%) is dissipated with the chip, half of the remaining portion concentrates at an extremely small area near the tool tip; consequently, tool temperatures are very high (ca 1000°C). The other half is absorbed by the machined surface. Also, the mechanical stresses on the tool are high. Consequently, tool materials must meet the following requirements: high hardness and wear resistance; high hardness at elevated temperature; toughness; deformation resistance (of the body and at the tool tip); chemical stability; chemical inertness (or negligible affinity) with the work material; adequate thermal properties; high stiffness; easy fabricability; availability; and low cost.

Some of the above requirements are conflicting, and their relative importance will depend upon several factors including: work material; type of machining operation—continuous vs intermittent cutting or roughing vs finishing; condition of the machine tool; volume of production—one piece, small batch, or mass production; geometry, finish, accuracy, and surface integrity requirements; adequate tool life; and other constraints—tool delivery schedule, production schedule demands and bottlenecks.

A wide range of cutting-tool materials is available with a variety of properties, performance capabilities, and cost. These include high carbon steels and low/medium alloy steels, high speed steels (HSS), cast cobalt alloys, cemented carbides, cast carbides, coated carbides, ceramics, sintered polycrystalline cubic boron nitride (CBN), sintered polycrystalline diamond and single-crystal natural diamond. Most tool materials used today were developed during this century (see also Carbides, cemented carbides).

High speed steels (HSS) and cemented carbides (coated and uncoated) are currently the most extensively used tool materials. Diamond and cubic boron nitride (CBN) are used for special applications where despite their cost, their use is justified. Cast-cobalt alloys are being phased out because of the high cost of raw materials and the increasing availability of alternative materials. New ceramics will have significant impact on future manufacturing productivity.

Properties and salient features of various cutting-tool materials are shown in Table 1. In many categories, a wide range of compositions, grades, and properties is offered. The methodology for the selection of the tools is illustrated in Figure 1 (2–3). In general, the selection of a particular class of tool material for a given application is easy. However, selection of a precise grade, shape, geometry, and size is more difficult and depends on the tool and work materials and overall economics.

General guidelines for the selection of tool materials are given in Tables 2 and 3.

The variation of hot hardness (indentation hardness measured at various temperatures) and recovery hardness (hardness measured at room temperature following exposure at elevated temperatures as indicated) with temperature is shown in Figures 2(a) and 2(b) (4).

Suppliers of tool steels, cemented carbides, ceramics, and diamond and CBN are given in references 5, 6, and 7, respectively.

Carbon Steels and Low–Medium Alloy Steels

Plain carbon tool steels, the most common cutting-tool materials in the early nineteenth century, were replaced by low–medium alloy steels because of increased machining productivity in many applications. Today, low–medium carbon steels have been largely superseded by other tool materials, except for some low speed applications.

Low–medium alloy steels contain elements such as Mo and Cr for hardenability, and W and Mo for wear resistance (see Table 4) (4,8–9). These alloy steels, however, lose their hardness rapidly when heated above 150–340°C (see Fig. 2). Furthermore, because of the low volume fraction of hard, refractory carbide phase present in these alloys, their abrasion resistance is limited. Hence, low–medium alloy steels are used in relatively inexpensive tools, eg, drills, taps, dies, reamers, broaches, and chasers, for certain low speed cutting applications where the heat generated is not high enough to reduce hardness (qv).

Table 1. Properties and Salient Features of Various Cutting Tool Materials[a]

Property and salient features	Carbon and low–medium alloy steels	High speed steels	Cast-cobalt alloys	Cemented carbides	Coated carbides	Ceramics	Polycrystalline CBN[b]	Diamond
hot hardness			increasing					→
toughness			increasing					↓
impact strength			increasing					↓
wear resistance			increasing					→
chipping resistance			increasing					↓
cutting speed			increasing					→
depth of cut	light to medium	light to heavy	light to heavy	light to heavy	light to heavy	light to heavy	light to heavy	very light for single-crystal diamond
finish obtainable	rough	rough	rough	good	good	very good	very good	excellent
method of processing	wrought	wrought, cast, HIP[c] sintering	cast and HIP[c] sintering	cold pressing and sintering	CVD[d]	cold pressing and sintering or HIP sintering	high pressure–high temperature sintering	high pressure–high temperature sintering
fabrication	machining and grinding	machining and grinding	grinding	grinding		grinding	grinding and polishing	grinding and polishing
thermal shock resistance			increasing →					
cost			increasing →					

[a] Overlapping characteristics exist in many cases. Exceptions to the rule are very common, and, in many classes, a wide range of compositions and properties is available.
[b] Cubic boron nitride.
[c] Hot isostatic pressure.
[d] Chemical vapor deposition.

275

Table 2. Guidelines for Tool Materials

Tool materials	Work materials	Machining operation and cutting-speed range	Modes of tool wear or failure	Limitations
carbon steels	low strength, softer materials, nonferrous alloys, plastics	tapping, drilling, reaming; low speed	buildup, plastic deformation, abrasive wear, microchipping	low hot hardness, limited hardenability and wear resistance, low cutting speed, low-strength materials
low–medium alloy steels	low strength–soft materials, nonferrous alloys, plastics	tapping, drilling, reaming; low speed	buildup, plastic deformation, abrasive wear, microchipping	low hot hardness, limited hardenability and wear resistance, low cutting speed, low-strength materials
HSS	all materials of low–medium strength and hardness	turning, drilling, milling, broaching; medium speed	flank wear, crater wear	low hot hardness, limited hardenability and wear resistance, low to medium cutting speed, low- to medium-strength materials
cemented carbide	all materials up to medium strength and hardness	turning, drilling, milling, broaching; medium speed	flank wear, crater wear	not for low speed because of cold welding of chips and microchipping, not suitable for low speed application
coated carbides	cast iron, alloy steels, stainless steels, superalloys	turning; medium to high speed	flank wear, crater wear	not for low speed because of cold welding of chips and microchipping, not for titanium alloys, not for nonferrous alloys since the coated grades do not offer additional benefits over uncoated

Tool material	Workpiece material	Operation; cutting speed	Failure modes	Characteristics and limitations
ceramics	cast iron, Ni-base superalloys, nonferrous alloys, plastics	turning; high speed to very high speed	DCL[a] notching, microchipping, gross fracture	low strength and thermomechanical fatigue strength, not for low speed operations or interrupted cutting, not for machining Al, Ti alloys
CBN	hardened alloy steels, HSS, Ni-base superalloys, hardened chill-cast iron, commercially pure nickel	turning, milling; medium to high speed	DCL notching, chipping, oxidation, graphitization	low strength and chemical stability at higher temperature, but high strength, hard materials otherwise
diamond	pure copper, pure aluminum, aluminum–Si alloys, cold-pressed cemented carbides, rock, cement, plastics, glass–epoxy composites, nonferrous alloys, hardened high carbon alloy steels (for burnishing only), fibrous composites	turning, milling; high to very high speed	chipping, oxidation, graphitization	low strength and chemical stability at higher temperature, not for machining low carbon steels, Co, Ni, Ti, Zr

[a] Depth of cut line.

277

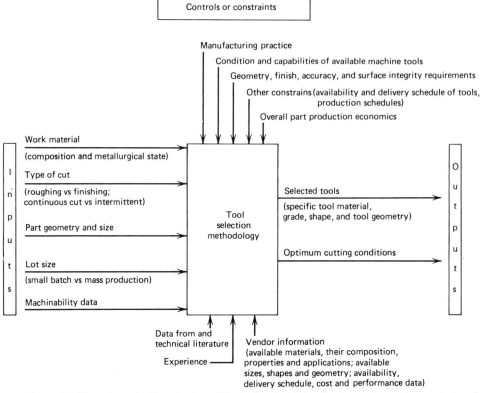

Figure 1. Procedures for the selection of the tool material (grade, shape, size, and geometry) and cutting conditions for a given application. Courtesy of Pergamon Press (2).

Low–medium alloy steels are relatively inexpensive and readily available on short notice or for a short run of parts, and can be heat-treated by simple hardening and tempering with relatively inexpensive equipment; easily formed and ground; and processed in many job shops fabricating their own tools. However, they have the following limitations: low hot hardness (Fig. 2); low wear resistance; poor hardenability; susceptibility to forming quench cracks and grinding cracks; and poor dimensional stability. Choice of a given grade depends upon the tool requirement, availability, cost, and other factors.

High Speed Steels

In the United States toward the latter part of the nineteenth century, a new heat-treatment technique for tool steels that enabled increased metal-removal rates and cutting speeds was developed (4,10) by Taylor and White. This new tool material was appropriately termed high speed steel (HSS). Today, however, cemented carbides and ceramics surpass the cutting-speed capabilities of HSS by three to six times.

High speed steels contain significant amounts of W, Mo, Co, V, and Cr, in addition to Fe and C. The presence of these alloying elements strengthens the matrix beyond the tempering temperature and increases the hot hardness and wear resistance.

High speed steels are easily available at reasonable cost and exhibit the following: through hardenability; higher hardness than carbon steel and low–medium alloy steels; good wear resistance; toughness (a feature especially desirable in intermittent cutting);

Table 3. Tool Materials for Different Cutting Operations[a]

Operation	Tool materials	Speed range[b]
single-point turning	low–medium alloy steels, HSS, cemented carbide, coated carbide, CBN, diamond	low to very high
drilling	low–medium alloy steels, HSS, solid cemented carbide <2.54 cm	low
tapping	carbon steels, low–medium alloy steels, cemented carbides[c]	low
reaming	HSS, cemented carbides, diamond	low to medium
broaching	HSS, cemented carbide	low to medium
end milling	HSS, solid cemented carbide <2.54 cm, brazed carbides >2.54 cm	low to medium
face milling	HSS, brazed carbides, cemented-carbide inserts, diamond, CBN	medium to very high

[a] Refs. 1, 3.
[b] Low: 30 m/min; medium: 30–150 m/min; high: 150–300 m/min; very high: 300 m/min.
[c] Limited application.

Table 4. Compositions of Carbon Steels and Low–Medium Alloy Steels[a] **Wt %**[b]

Type	C	Mn	Si	Cr	V	W	Mo
Carbon steels[c]							
W1	0.6–1.4						
W2	0.6–1.4				0.25		
W3	0.6–1.4			0.5			
Low–medium alloy steels[d]							
O1	0.9	1.00		0.5		0.5	
O2	0.9	1.60					
O6	1.45		1.00				0.25
O7	1.20			0.75		1.75	0.25

[a] Refs. 4, 8–9.
[b] Remainder Fe in all cases.
[c] Available in ranges of 0.1 wt % of carbon content; W = water-hardening grade.
[d] Cold work; O = oil-hardening grade.

and ability to alter hardness appropriately by suitable heat treatment, which facilitates manufacturing of complex tools and grinding of tools and cutters to final shape. Associated with these advantages are the following limitations: hardness decreases sharply beyond 540°C (see Figs. 2(**a**) and 2(**b**)), limiting the use of these tools to low speed cutting operations (<30 m/min); wear resistance, chemical stability, and propensity to interact chemically with the chip and the machined surface are limited; and the chips tend to adhere to the tool.

Tool steels are broadly classified as the T-type and the M-type depending on whether tungsten or molybdenum is the principal alloying element (see Table 5) (4,8–9). The two types can be used interchangeably because they possess the same properties and have comparable cutting performance. However, M-type steels tend to decarburize more during heat treatment, for which the temperature range is narrow, and hence, care should be exercised during this treatment. In general, M-type tool steels are more popular (ca 85% of all tool steels) as they are less expensive (ca 30%)

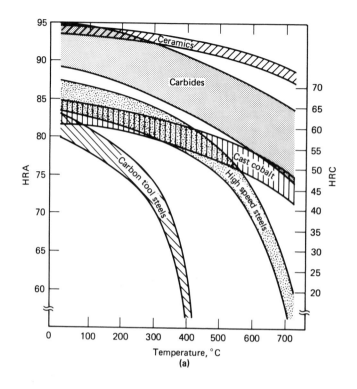

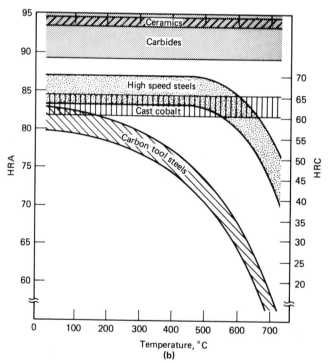

Figure 2. (a) Variation of hot hardness (Rockwell A and C scales, HRA and HRC, respectively) with temperature for various tool materials (4). (b) Variation of recovery hardness with temperature for various tool materials (4).

Table 5. Chemical Compositions of Different Grades of High Speed Steels[a], Wt %[b]

AISI HSS tool type	C	Cr	V	W	Mo	Co	Cb	W_{eq}[c]
Tungsten high speed steel								
T1[d]	0.70	4.0	1.0	18.0				18.0
T2[d]	0.85	4.0	2.0	18.0				18.0
T3	1.00	4.0	3.0	18.0	0.60			18.8
T4	0.75	4.0	1.0	18.0	0.60	5.0		19.2
T5	0.80	4.25	1.0	18.0	0.90	8.0		19.8
T6	0.80	4.25	1.5	2.0	0.90	12.0		21.8
T7	0.80	4.0	2.0	14.0				14.0
T8	0.80	4.0	2.0	14.0	0.90	5.0		15.8
T9	1.20	4.0	4.0	18.0				18.0
T15	1.55	4.50	5.0	12.0	0.60	5.0		13.2
Molybdenum high speed steels								
M1[d]	0.80	4.0	1.00	1.5	8.0			17.5
M2[d]	0.85	4.0	2.00	6.0	5.0			16.0
M3	1.00	4.0	2.75	6.0	5.0			16.0
M4	1.30	4.0	4.00	5.5	4.5			14.5
M6	0.80	4.0	1.50	4.0	5.0	12.0		14.0
M7[d]	1.00	4.0	2.00	1.75	8.75			19.25
M8	0.80	4.0	1.50	5.0	5.0		1.25	15.0
M10	0.85	4.0	2.00		8.0			16.0
High hardness (molybdenum base) cobalt high speed steels								
M30	0.85	4.0	1.25	2.0	8.0	5.0		18.0
M34	0.85	4.0	2.00	2.0	8.0	8.0		18.0
M35	0.85	4.0	2.00	6.0	5.0	5.0		16.0
M36	0.85	4.0	2.00	6.0	5.0	8.0		16.0
M41	1.10	4.25	2.00	6.75	3.75	5.0		14.25
M42	1.10	3.75	1.15	1.50	9.50	8.25		20.5
M43	1.20	3.75	1.60	2.75	8.00	8.25		18.75
M44	1.15	4.25	2.00	5.25	6.50	12.00		18.25
M45	1.25	4.25	1.60	8.25	5.0	5.50		18.25
M46	1.25	4.00	3.20	2.00	8.25	8.25		18.0

[a] Courtesy of American Society of Metals (4,8–9).

[b] Normal ranges of manganese, silicon, phosphorus, and sulfur are assumed; the remainder is Fe in all cases.

[c] $\% W_{eq} = 2 \, (wt \% \, Mo) + wt \% \, W$.

[d] Widely available.

than the corresponding T-type steels. The latter is due to the greater availability and lower cost of molybdenum and that only about one half as much in weight of molybdenum as tungsten is needed to yield the same properties and performance.

High speed steel tools are available in cast, wrought, and sintered forms. Improper processing of both cast and wrought products can lead to undesirable microstructure carbide segregation, formation of large carbide particles, significant variation of carbide size, and nonuniform distribution of carbides in the matrix. Such a material is difficult to grind to shape and causes wide fluctuations of properties, inconsistent performance, distortion, and cracking.

A new processing technique involves atomization of the prealloyed molten tool steel alloy into fine powder, followed by consolidation under hot isostatic pressure (HIP) (3,11–12). This technique combined with suitable hardening and tempering provides a microstructure consisting of a uniform and fine dispersion of carbides in a tempered martensite matrix (see Figs. 3(**a**) and 3(**b**)). Tool steels made in this manner grind more easily, exhibit more uniform properties, and perform more consistently (13) (see also Powder metallurgy).

The heat-treatment procedure generally consists of first preheating the HSS tool steel to 730–840°C and then rapidly to 1177–1220°C for 2–5 min to fully austenitize the steel, followed by quenching (initially in a suitable molten salt bath to a certain intermediate temperature (ca 600°C)), and then cooling in air (8–10,14). This treatment is followed by single or double tempering, where the steel is heated to 540–590°C for ca 1 h and then air cooled to produce a tempered martensite structure and to relieve residual stresses.

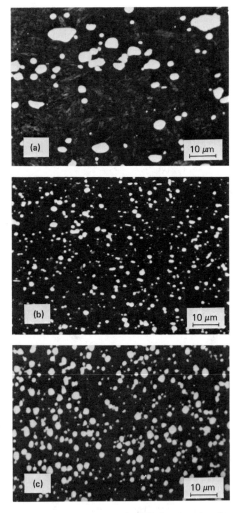

Figure 3. Micrographs of the microstructure of a fully hardened and tempered AISI M-42 tool steel, (**a**) produced by cast/wrought technique showing inhomogeneous distribution and large carbide particles in the matrix; (**b**) produced by powder metallurgy technique showing uniform distribution and fine grain size carbide particles in the matrix; (**c**) produced by powder metallurgy technique of a cobalt-free tool steel showing higher concentration of fine grain size carbide particles in the matrix.

Recent shortages and escalating costs of cobalt have prompted tool-steel producers to seek an appropriate substitute for cobalt. It was found that hot hardness can be maintained without cobalt by an appropriate increase of Mo–W or V content, or both (15). The higher concentrations of these elements in the matrix provide equivalent solid-solution strengthening at elevated temperatures. The compositions of two grades with and without cobalt, yielding similar performances, are given in Table 6. Figures 3(**b**) and 3(**c**) are micrographs of a heat-treated (quenched and tempered) AISI M-42 tool steel with and without cobalt, showing a significantly higher volume fraction of carbides in the cobalt-free steel. Despite heavy competition from other tool materials such as cemented carbide, coated carbides, and ceramics, HSS still accounts for the largest tonnage of tool materials used today because of its unique properties (chiefly the toughness and the fracture resistance), flexibility in fabrication, and the fact that many cutting operations are still conducted at a speed range low enough for HSS to perform efficiently and economically. Characteristics of high carbon steels, low-medium alloy steels, and HSS are given in Table 7 (16).

Cast-Cobalt Alloys

Cast-cobalt alloys were introduced for cutting-tool applications about the same time as HSS. Popularly known as Stellite tools, these materials are cobalt-rich chromium–tungsten–carbon cast alloys with properties and applications in the intermediate range between HSS and cemented carbides. Though comparable in room-temperature hardness to HSS tools, cast-cobalt alloy tools retain their hardness to a much higher temperature, and can be used at higher (25%) cutting speeds than HSS tools. Unlike the high speed steels that can be heat treated to obtain desired hardness, cast-cobalt alloys are hard as cast, and cannot be softened or hardened by heat treatment (see also High temperature alloys).

Cast-cobalt alloys contain a primary phase of Co-rich solid solution strengthened by Cr and W, and dispersion hardened by complex, hard, refractory carbides of W and Cr (5–7). Nominal compositions and properties of cast-cobalt alloy grades are given in Table 8 (4,8–9,17–18). Other alloying elements include V, B, Ni, and Ta. The casting provides a tough core and elongated grains normal to the surface. However, the structure is inhomogeneous. By atomizing the preformed molten alloy into powder and subsequent consolidation by HIP, a more homogeneous structure results with fine carbide particles uniformly dispersed in the matrix.

Tools of cast-cobalt alloys are generally cast to shape and finished to size by

Table 6. Compositions of Equivalent-Grade HSS with and without Cobalt[a], Wt %[b]

HSS type	C	Cr	V	W	Mo	Co	W_{eq}[c]
T15	1.55	4	5	12.25		5	12.25
Co-free T15	1.08	4	5	12.5	6.5		25.5
M42	1.1	3.75	1.1	1.5	9.5	8	20.5
Co-free M42	1.3	3.75	2.0	6.25	10.5		27.25

[a] Ref. 15.
[b] Remainder Fe in all cases.
[c] $\% W_{eq} = 2 (\% Mo) + \% W$.

Table 7. Characteristics of Tool Steels [a]

AISI-SAE designation	Wear resistance[b]	Tough-ness[b]	Hardness retention at elevated temp[b]	Depth of hardening[c]	Nondeforming properties[b]	Safety in hardening[b]	Machin-ability[d]	Grind-ability[e]
Water hardening[f]								
W1	F–G	G[g]	P	S	P	F	B	B
W5	F–G	G[g]	P	S	P	F	B	B
S1	F	VG	F	M	F	G	F	B
S7	F	VG	VG	D	A, P; OQ, P	VG	F	B
Cold work[h]								
O1	G	F	P	M	VG	VG	G	B
O7	G	F	P	M	OQ, VG; WQ, P	OQ, VG; WQ, P	G	B
A2	VG	F	F	D	B	B	F	G
A10	VG	F	G	D	B	B	G	F
D2	VG	P	G	D	B	B	P	F
D7	B	P	G	D	B	B	P	P
Hot work								
H10	F	G	G	D	VG	B	F	G
H19	F–G	G	G	D	A, G; OQ, F	G	F	G
H25	F	G	VG	D	AC, G; OQ, F	G	F	G
H43	G	P	VG	D	S, AC, G; OQ, F	F	F	G
HSS[i]								
T1, T2	VG	P	VG	D	G	G	F	F
T6	VG	P	B	D	G	F	F	F
T15	B	P	B	D	G	F	F	P
M1	VG	P	VG	D	G	F	F	F
M10	VG	P	VG	D	G	F	F	F
M30	VG	P	B	D	G	F	F	F
M41	VG	P	B	D	G	F	F	F
M47	VG	P	B	D	G	F	F	F
Special purpose[j]								
L2	F	G[g]	P	M	OQ, F; WQ, P	OQ, F; WQ, P	G	B
L6	F	G	P	M	G	G	F	B
F1	G	P	P	S	P	P	G	P
P2	G	G	P	S	G	G	G[k]	B
P20	F–G	G	P	S	G	G	G[k]	B
P21	G	G	G	D	B	B	G	G

[a] Courtesy of McGraw-Hill Book Company (16).

[b] Ratings: P = poor; F = fair; G = good; VG = very good; B = best. Other symbols: WQ = water quench; OQ = oil quench; S = salt quench; AC = air cool.

[c] Depth of hardening: S = shallow; M = medium; D = deep.

[d] Machinability: Same ratings as given in footnote[b]. Approximate speeds: best = 40% of B1112; poor = 15% of B1112.

[e] Grindability: same ratings as given in footnote[b]. In terms of grindability index (metal removed/wheel wear): best = 40–90; good = 10–30; fair = 2–10; poor = 0.4–2.

[f] In the W group, chromium increases the depth of hardening whereas vanadium retains a fine grain size to higher hardening temperatures, thereby permitting a range of case depths in one steel.

[g] For W1 to W7 and L2, the toughness decreases with increasing carbon content and depth of hardened case.

[h] Types O, A, and D.

[i] Types T and M.

[j] Types L, F, and P.

[k] Annealed stock before carburizing.

Table 8. Composition and Properties of Cast-Cobalt Alloys[a]

Grade	Cr	W	Mo	C	Mn	Si	Ni	HRC[c]	Tensile strength, MPa[d]	Compressive strength, MPa[d]	Impact strength, J[e]	Density, g/cm^3
roughing	30	4.5	1.5	1.1	1.0	1.5	3.0	46	834	1517	14.9	8.4
general purpose	31	10.5		1.7	1.0	1.0	3.0	55	758	2137	15.6	8.4
finishing	32	17.0		2.5	1.0	1.0	2.5	62	524	2309	4.7	8.7

[a] Refs. 4, 8–9, 17–18.
[b] Remainder cobalt in all grades.
[c] Rockwell hardness C scale.
[d] To convert MPa to psi, multiply by 145.
[e] To convert J to ft·lbf, divide by 1.356.

grinding. They are available only in simple shapes, such as single-point tools and saw blades, because of limitations in the casting process and expense involved in the final shaping (grinding). The high cost of fabrication is primarily due to the hardness of the material in the as-cast condition. Cast-cobalt alloy tools are stiffer (Young's modulus E = 276 GPa (4×10^7 psi)) than high speed steel tools (E = 207 GPa (3×10^7 psi)). They can be used with a positive rake geometry for both roughing and interrupted cutting operations. Their higher hot hardness, eg, 350 HV (Vickers hardness) instead of 75 HV for HSS at 900°C (see Figs. 2(**a**) and 2(**b**)), permits them to be used at higher (20%) cutting speeds than HSS tools. At equal cutting speed, the cast-cobalt alloy tool provides a longer tool life. Work materials that are machined with this tool material include plain carbon steels, alloy steels, nonferrous alloys, and cast iron.

Cast-cobalt alloys are currently being phased out for cutting-tool applications as a result of rapidly increasing costs and a potential shortage of the strategic raw materials Co, W, and Cr used in these alloys.

Cemented Carbides

Cemented carbides are a class of tool material containing a large volume fraction ($\geq$90%) of fine-grain, refractory carbides in a metal binder produced by cold pressing, followed by liquid-phase sintering (19–21). Cemented carbides differ from HSS in many important respects. They are much harder, chemically more stable, and superior in hot hardness but are generally lower in toughness than HSS. They can be used at cutting speeds 3–6 times higher than HSS. Carbide is the predominant phase in cemented carbides as the metallic phase in HSS. Consequently, the Young's modulus E of cemented carbide is two to three times that of HSS (414–689 GPa (6–10×10^7 psi)). As a result, the carbides are 2–3 times stiffer than HSS. Furthermore, a specific grade of cemented carbide can be used to machine a specific work material, thus minimizing chemical interaction between the tool and the work material. This is possible in cemented carbides because the chemistry of the primary phase (carbide phase) can be altered to provide the needed stability. Cemented carbides have less tendency for adhesion but are more brittle and more expensive to fabricate and shape. Hard refractory coatings can be deposited without adversely affecting the substrate characteristics. Unfortunately, strategic materials such as tungsten, cobalt, and tantalum are used more extensively in cemented carbides.

Most cemented-carbide tools are WC-based (either straight WC or multicarbides of W–Ti or W–Ti–Ta, depending upon the work material to be machined) with cobalt as the binder. Other materials based on TiC were developed primarily for automobile industry applications (predominantly with a Ni–Mo binder). They are used for higher speed (>300 m/min) finish machining of steels and gray cast irons.

The first cemented-carbide tool material, introduced in Germany in the 1920s, was a straight WC in a Co binder. In the machining of cast iron, it had a long tool life even at three to six times the cutting speeds used with HSS but developed a deep crater on the tool face when machining steels, thus leading to rapid wear. However, the development of multicarbides of W–Ti or W–Ti–Ta provided considerable resistance to crater wear. Different grades of cemented carbides were developed by varying the Co content, the amount of different carbides, and the carbide grain size (see Table 9). The higher cobalt grades or coarser carbide size grades are tougher but less hard, and the more complex carbides are harder, and chemically more resistant (especially to steels), but weaker than WC–Co alloys.

Figures 4(a)–(c) are micrographs of representative nonsteel machining grades of cemented carbides (roughing, general-purpose, and finish-machining grades, respectively) containing straight WC with decreasing grain size or Co content (3). Figures 5(a)–(c) are micrographs of similar grades for steels containing different amounts of complex multicarbides in a cobalt binder (3). Progressing from a roughing to a finishing grade, the hardness increases, toughness decreases, and resistance to high temperature deformation and wear resistance increases. Figures 6(a)–(d) show the variation of hardness, transverse rupture strength, impact strength, and elastic modulus with percent cobalt binder content for straight cemented-WC grades. Hardness and elastic modulus decrease with an increase in cobalt content while the impact strength and transverse rupture strength increases.

There are at least four different classifications of cemented carbides (6). The U.S. system is based on relative performance, the UK system on properties, and the USSR system on composition; the fourth system, which is used widely in Europe and supported by the ISO, is based on application and chip form.

The C-classification (C-1 to C-8) for cemented carbides, used unofficially in the United States for machining applications, was originally developed by the automobile industry to obtain a relative performance index of tools made by different tool producers. It is by far the simplest system. The grades are broadly divided into two classes (C-1 to C-4 and C-5 to C-8), according to the type of work material to be machined. Grades C-1 to C-4 are recommended for machining cast iron, nonferrous alloys, and nonmetallics (nonsteel work materials), whereas C-5 to C-8 are recommended for machining carbon steels and alloy steels. Although the grades to be used for machining other difficult to machine materials (eg, the titanium alloys and nickel-base and cobalt-base superalloys) have not been specified explicitly in this classification, the

Table 9. Composition and Properties of Some Representative Grades of Cemented-Carbide Tools[a]

Grade	Composition, wt % WC	TiC	TaC	Co	Grain size	Density, g/cm³	HRA[b]	TRS[c], MPa[d]	Elastic modulus E, GPa[e]	Impact strength, J[f]	Compressive strength, MPa[d]	Tensile strength, MPa[d]	Relative abrasion resistance, vol loss/cm³	Thermal conductivity, W/(m·K)	Thermal expansion, per (°C × 10⁻⁶)
Nonsteel grades[g]															
roughing	94			6	coarse	15.0	91	2210	640	16	5170	1520	15	120	4.3
general purpose	94			6	medium	15.0	92	2000	650	16	5450	1950	35	100	4.5
finishing	97			3	fine		92.8	1790	610	12	5930	1790	60		4.3
Steel grades[h]															
roughing	72	8	11.5	8.5	coarse	12.6	91.1	1720	560	11	5170		8	50	5.8
general purpose	71	12.5	12	4.5	medium	12.0	92.4	690	570	9	5790		7	35	5.2
finishing	64	25.5	4.5	6	medium	9.9	93.0	130	460	5	4900	480	5		5.9

Note: the TRS, Elastic modulus, Impact strength, Compressive strength, Tensile strength, Relative abrasion resistance, Thermal conductivity, and Thermal expansion columns are grouped under the heading *Properties*.

[a] Ref. 2.
[b] Rockwell hardness A scale.
[c] Transverse rupture strength.
[d] To convert MPa to psi, multiply by 145.
[e] To convert GPa to psi, multiply by 145,000.
[f] To convert J to ft·lbf, divide by 1.356.
[g] C-1 to C-4.
[h] C-5 to C-8.

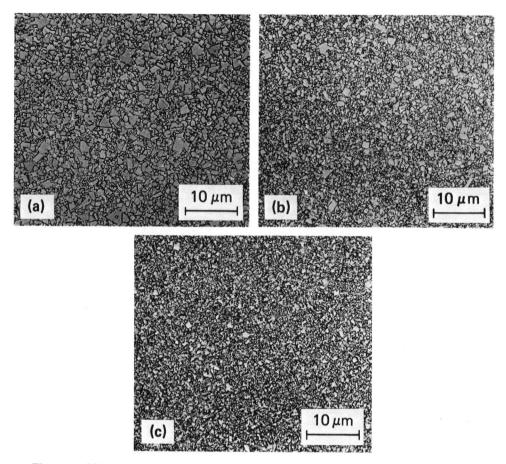

Figure 4.   Micrographs of three representative grades of cemented, straight WC in a cobalt binder for (**a**) roughing, (**b**) general purpose, and (**c**) finish machining of materials other than steels (3).

nonsteel grades C-1 to C-4 are applicable for machining such nonferrous alloys. Many users of this classification system are not familiar with this, as they consider grades C-1 to C-4 as cast-iron grades, and not as grades for machining materials other than steels (or nonsteels). In general, the nonsteel grades are straight WC in a Co binder, whereas the steel grades are multicarbides in a Co binder. Individual cemented-carbide producers, however, may deviate from this rule slightly to gain a competitive edge.

Within each class (ie, C-1 to C-4 and C-5 to C-8), each grade is distinguished by the type of machining operation: C-1 and C-5 for roughing, C-2 and C-6 for general purpose, C-3 and C-7 for finishing, and C-4 and C-8 for precision-finishing operations. In general, from grades C-1 to C-4 or C-5 to C-8 within each class, the shock resistance decreases, hardness increases, high temperature deformation resistance and wear resistance increase, and the cobalt content and carbide grain size decrease. Roughing and general-purpose grades require more toughness to withstand heavy loads, whereas finishing and semifinishing grades require a high temperature deformation-resistant and a wear-resistant sharp edge. At one time, each tool producer associated one or more carbide grades with the eight grades in the C-classification. However, this comparison involves competition only of grades identified within each class (eg, C-1 or C-5 by

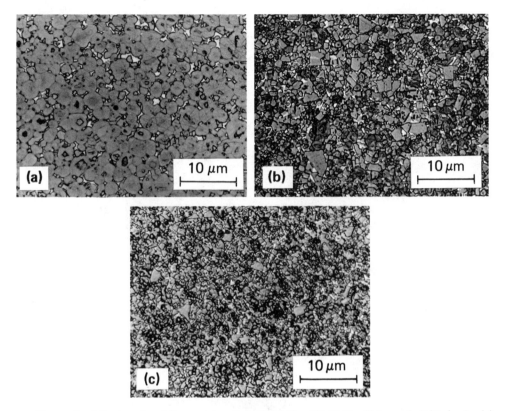

Figure 5.   Micrographs of three grades of multicarbides (W–Ti–Ta–C) in a cobalt binder for (a) roughing, (b) general purpose, and (c) finish machining of steels (3). The carbide particles with angular morphology constitute the WC phase, whereas the rounded morphology constitutes complex multicarbides.

different manufacturers), and hence there is a trend to disassociate with this classification.

The cemented-carbide classification used in the UK is based on the following three tool requirements as reflected by an appropriate property or means to achieve such a property: wear resistance, as measured by relative hardness; toughness or shock resistance, as measured by transverse rupture strength (TRS); and crater resistance, as indicated by the percentage TiC and TaC present in the cemented tungsten carbide. This classification was developed by the British Hard Metal Association (BHMA) and consists of three digits. The first digit (1 through 9) is an indicator of wear resistance. Number 1 represents low hardness up to 1300 HV, no. 2 up to 1450 HV, and no. 3 up to 1550 HV. From nos. 4 to 8, the hardness increases by 200 HV; eg, no. 5 represents hardness of 1600–1650 HV, and no. 9 represents hardness >1800 HV. The second digit (1 through 9) is an indicator of toughness. Number 1 represents low toughness (TRS ca 1.1 GPa (ca 1.6×10^5 psi)). Within this group, TRS increases by 138 MPa (2 $\times 10^4$ psi) for each number up to 8. For example, no. 5 represents TRS of 1.52–1.65 GPa (($2.2–2.4) \times 10^5$ psi). The third digit is an indicator of crater resistance and is given graphically by the relative percentage of TiC and TaC present in the cemented tungsten carbide. Within this group, the lower number indicates lower TiC–TaC content and hence lower crater resistance. This system has the disadvantage that performance requirements do not match one-to-one with the properties selected.

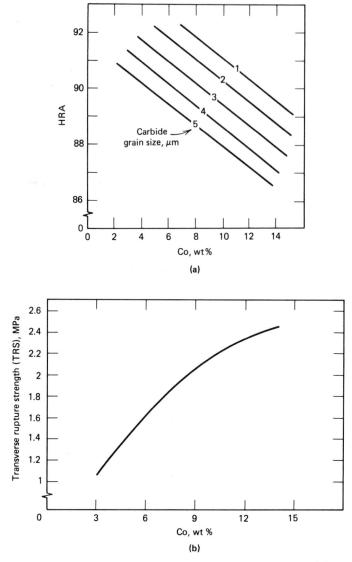

Figure 6 (a) to **(d)**.   Variation of hardness, transverse rupture strength (TRS), impact strength and elastic modulus with percent cobalt binder content of cemented WC cutting tool grades (22).

The cemented-carbide classification used in the USSR is based on composition. The relationship between various properties and performance of different compositions is available in general terms for customer use. Although this system is used in the USSR, it would be unacceptable in Western countries for proprietary reasons.

In view of different classification systems practiced by various manufacturers in different countries, an attempt was initiated in 1953 under the auspices of the ISO to arrive at a universally acceptable classification. The first draft was approved in 1964 by 27 member countries, excluding the United States and Canada. Because of the proprietary considerations, this classification was confined to applications. It is broadly divided into three categories and color coded for convenience: P-grades (blue) are highly alloyed multicarbides used mainly for machining hard steels and steel castings; M-

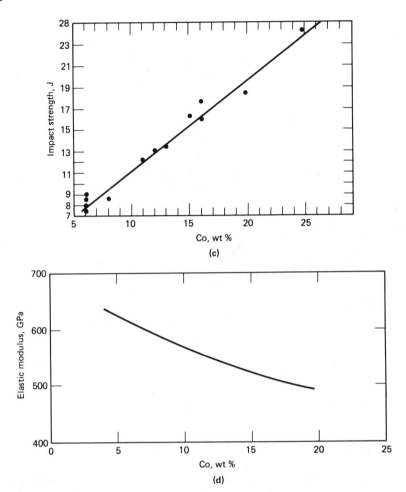

Figure 6. (*continued*)

grades (yellow) are low alloy multicarbide alloys which are multipurpose nonsteel grades used for machining high temperature alloys, low strength steels, gray cast iron, free machining steels, and nonferrous metals and their alloys; and K-grades (red) are straight WC grades for machining very hard gray cast iron, chilled castings, nonferrous metals and their alloys, and nonmetallics such as plastics, glass, glass–epoxy composites, hard rubber, and cardboard. Details of the work material to be machined, the type of cutting operation (continuous vs intermittent, roughing vs finishing), and the type of chip formed are included in this classification. There are six basic categories in the P-type (P 01, P 10, P 20, P 30, P 40, and P 50), four in the M-type (M 10, M 20, M 30, and M 40), and five in K-type (K 01, K 10, K 20, K 30, and K 40). Carbide grades differing from these basic categories can be designated by appropriate in-between numbers within each class (eg, between P 01 and P 10). Thus, the coated grades that do not fall under the basic categories can be placed in these categories. The lower numbers (ie, P 01, M 10, or K 10) are for higher speed, finishing (lighter cut) applications (harder with low cobalt–finer carbide grain size), and the higher numbers (P 50, M 40, or K 40) are for lower speed, roughing (heavier cut) applications (tougher with higher cobalt–coarser carbide grain size).

The ISO classification for cemented-carbide cutting tools is given in Table 10 (6). Some prior knowledge of machining is expected when consulting this table.

Table 10. ISO Classification of Cemented-Carbide Tools According to Use [a]

Main groups of chip removal		Groups of application			Increase or decrease in characteristic	
Symbol (color)	Categories of material to be machined	Designation	Material to be machined [b]	Use and working conditions	Of cut	Of carbide
P (blue)	ferrous metals with long chips	P 01	steel, steel castings	finish turning and boring; high cutting speeds; small chip section; accuracy of dimensions and fine finish; vibration-free operation		
		P 10	steel, steel castings	turning, copying, threading, milling; high cutting speeds; small or medium chip sections		
		P 20	steel, steel castings, malleable cast iron with long chips	turning, copying, milling; medium cutting speeds; medium chip sections; planing with small chip sections		
		P 30	steel, steel castings, malleable cast iron with long chips	turning, milling, planing; medium or low cutting speeds; medium or large chip sections; machining in unfavorable conditions [b]		
		P 40	steel, steel castings with sand inclusion and cavities	turning, planing, slotting; low cutting speeds; large chip sections; possibility of large cutting angles for machining in unfavorable conditions [b] and work on automatic machines		
		P 50	steel, steel castings of medium or low tensile strength, with sand inclusion and cavities	for operations demanding very tough carbides; turning, planing slotting; low cutting speeds; large chip sections; possibility of large cutting angles for machining in unfavorable conditions [b] and work on automatic machines		
M (yellow)	ferrous metals with long or short chips and nonferrous metals	M 10	steel, steel castings, manganese steel, gray cast iron, alloy cast iron	turning; medium or high cutting speeds; small or medium chip sections		
		M 20	steel, steel castings, austenitic or manganese steel, gray cast iron	turning, milling; medium cutting speeds; medium chip sections		
		M 30	steel, steel castings, austenitic steel, gray cast iron, high temperature-resistant alloys	turning, milling, planing; medium cutting speeds; medium or large chip sections		
		M 40	mild free cutting steel, low tensile steel, nonferrous metals and light alloys	turning, parting off, particularly on automatic machines		

Increase or decrease in characteristic — Of cut: ← increasing speed — decreasing feed →; Of carbide: ← wear resistance — decreasing toughness →

K (red) — ferrous metals with short chips, nonferrous metals, and nonmetallic materials

increasing speed →
decreasing feed →
wear resistance →
decreasing toughness →

Grade	Material	Operation
K 01	very hard gray cast iron, chilled castings of over 85 Shore, high silicon–aluminum alloys, hardened steel, highly abrasive plastics, hard cardboard, ceramics	turning, finish turning, boring, milling, scraping
K 10	gray cast iron over 220 Brinell, malleable cast iron with short chips, hardened steel, silicon aluminium alloys, copper alloys, plastics, glass, hard rubber, hard cardboard, porcelain, stone	turning, milling, drilling, boring, broaching, scraping
K 20	gray cast iron up to 220 Brinell, nonferrous metals (copper, brass, aluminium)	turning, milling, planing, boring, broaching, demanding very tough carbide
K 30	low hardness gray cast iron, low tensile steel, compressed wood	turning, milling, planing, slotting, for machining in unfavorable conditions[b] and with the possibility of large cutting angles
K 40	soft wood or hard wood, nonferrous metals	turning, milling, planing, slotting, for machining in unfavorable conditions[b] and with the possibility of large cutting angles

[a] Ref. 6.
[b] Raw material or components in shapes which are awkward to machine: casting or forging skins, variable hardness, variable depth of cut, interrupted cut, work subject to vibrations, etc.

The original objective of the ISO classification was to issue detailed standards for cemented carbides in terms of microstructure, composition, and properties for quality control and performance reliability. This objective, however, is yet to be realized. When and if achieved, such standardization would considerably simplify the choice of carbide grade for a given application.

In selecting a carbide grade for a given application, the following general guidelines should be followed: the grade with the lowest cobalt content and the finest grain size consistent with adequate strength to eliminate chipping should be chosen; straight WC grades can be employed if cratering, seizure, or galling is not experienced and for work materials other than steels; to reduce cratering and abrasive wear when machining steels, TiC grades are preferred; and for heavy cuts in steel where high temperature and high pressure deform the cutting edge plastically, a multicarbide grade with low binder content and containing W–Ti–Ta should be used.

Composition, microstructure, and performance properties depend on cobalt binder content, carbide grain size, and type and composition of various carbides. With increasing cobalt content, toughness as reflected by the transverse rupture strength (TRS) and impact strength increase, whereas hardness, Young's modulus, and thermal conductivity decrease (Figs. 6(a) to (d)). Finer grain size gives a harder product than coarser grain size carbides (Fig. 6(a)). Multicarbides increase chemical stability and both room-temperature and hot hardness, whereas TiC addition to WC controls crater wear, especially for machining steels. The proper grade of cemented carbide for a given work material should provide adequate crater-wear resistance, abrasion resistance, and toughness to prevent microchipping of the cutting edge.

Another type of carbide tool material, developed for high speed (>300 m/min) finish machining of steels, is based on titanium carbide cemented with a nickel–molybdenum binder (23). Because of the popularity of cemented WC tools and the initial reputation gained by cemented TiC for being relatively brittle and easy to chip, there was a natural reluctance to change from the cemented WC tools to cemented TiC tools. This trend may, however, change in the future for some steel machining applications because of the escalating costs of W and Co in cemented WC, the ready availability of TiC at a reasonable price, and significant technological progress in the processing of this material. Not all tool manufacturers offer cemented TiC tools; about half of the cemented TiC compositions currently produced are based primarily on the original Ford patent (24). Compositions and properties of various grades (C-5 to C-8 steel grades only) of cemented TiC are given in Table 11. The TRS of cemented TiC is higher than that of most ceramics, but lower than that of cemented WC. Furthermore, the Young's modulus of cemented TiC, although double that of HSS, is ca 25% less than that of cemented WC. Addition of Ni to Mo wets the TiC and hence is used as a binder. Small amounts of Ti and Mo enter the binder phase during liquid-phase sintering (25). This reaction strengthens the tool material and increases tool life. The percentage of C is also critical for strength. Similarly, small additions of TiN and V increase the resistance to hot deformation and thermal cracking. Like the cemented WC tools, cemented TiC is processed by cold pressing followed by liquid-phase vacuum sintering. Although this material was originally used for high speed finish turning and boring of steels, the new improved grades are used in the high speed milling of steels and malleable cast irons.

Cemented-carbide tools are available in insert form in squares, triangles, diamonds, and rounds. They can be either brazed or preferably clamped on to the tool

Table 11. Composition and Properties of Steel Grades of Cemented Titanium Carbide[a]

Grade[b]	Composition, wt %			HRA[c]	Transverse rupture strength, MPa[d]	Young's modulus, GPa[e]	Density, g/cm³
	TiC	Ni	Mo				
roughing	67–69	22	9–11	91	1900	413	5.8
general purpose	72–74	17	9–11	92	1620	431	5.6
finishing	77–79	12	9–11	92.8	1380	440	5.5

[a] Ref. 3.
[b] C-5 to C-8.
[c] Rockwell hardness A scale.
[d] To convert MPa to psi, multiply by 145.
[e] To convert GPa to psi, multiply by 145,000.

shank. Since cemented-carbide tools are relatively brittle, they are used with a tool geometry of −5° rake and 5° clearance angle to provide a 90° included angle at the cutting edge. This will enable a square insert to cut with all its eight corners successively. To strengthen the edge, it is generally rounded off by honing or an appropriate chamfer, or a negative land on the rake face is provided. A molded-in chip groove with a positive rake angle just at the tool tip reduces cutting forces without reducing the overall strength of the insert significantly and facilitates disposal of the chips.

Cemented carbides are not recommended for low speed cutting operations because the chips tend to weld to the tool face and cause microchipping and there is no economic incentive to use them at lower speeds. They are effective at higher speed, generally in the 45–180 m/min range. This speed can be much higher (>300 m/min) with materials that are easier to machine, eg, aluminum alloys, and much lower (ca 30 m/min) with materials that are more difficult to machine, eg, titanium alloys. In interrupted cutting applications, edge chipping is prevented by appropriate choice of cutter geometry and cutter position with respect to the workpiece in such a way as to transfer the point of application of the load away from the tool tip. Finer grain size and higher cobalt content improve toughness in straight-WC–Co grades and are considered desirable in materials used for interrupted cutting. Because of the high hardness of cemented carbides, they can be finished only by diamond grinding. Abusive grinding can lead to thermal cracks and poor performance.

To conserve the strategic materials (W, Co, and Ta) and to reduce costs, cemented-carbide inserts (so-called throwaway inserts) are generally recycled after use by separating Ta, WC, and Co for reuse. Alternatively, these tools can be reground for application where the actual size of the insert is not of critical concern. Several commercial fabricators provide regrinding service on a regular basis.

Coated Tools

Since the 1960s, numerous materials have been introduced with high strength, abrasive constituents, chemical reactivity, poor thermal characteristics, and high temperature deformation resistance. The difficulty of machining many of these materials presented new challenges to the cutting-tool industry, leading to the introduction of coated cemented carbides in the mid-1960s (26–28) and coated HSS in the late 1970s. Although rapid advances in coated cemented-carbide technology took place during the last decade, coating technology for HSS is lagging. The technique commonly used for coated cemented carbides, ie, chemical vapor deposition (CVD), requires that the tools be heated to ca 1000°C. At this temperature, the metallurgical structure of HSS is altered significantly. Thus, only coatings that require heating below the HSS transformation temperature can be applied to HSS. Some success has been reported with a coating of TiN on HSS applied by physical vapor deposition (PVD) at lower temperatures (see Coated Tools; Refractory coatings).

An analysis of the cutting process indicates that the material requirements at or near the surface of the tool are different from those of the body. The surface, to be abrasion resistant, has to be hard and to prevent chemical interaction has to be chemically inert. A thin, chemically stable, hard, refractory binderless coating often satisfies these requirements. The tool body, by contrast, should be tough enough to withstand high temperature plastic deformation of the nose and the body of the cutting tool under the conditions of cutting. Since these requirements conflict, tool-material design should be considered in terms of composite materials. Consequently, a thin coating (ca 5 μm thick) of TiC was developed for cemented-carbide tools in the late 1960s.

An effective coating should be hard; refractory; chemically stable; chemically inert to shield the constituents of the tool and the workpiece from interacting chemically under cutting conditions; binder free; of fine grain size with no porosity; metallurgically bonded to the substrate with a graded interface to match the properties of the coating and the substrate; thick enough to prolong tool life, but thin enough to prevent brittleness; free from the tendency of the clip to adhere or seize to the tool face; easy to deposit in bulk quantities; and inexpensive. In addition, coatings should have low friction and exhibit no detrimental effects on the substrate or bulk properties of the tool.

Several refractory coatings have been developed over the last decade, including single coatings of TiC, TiN, Al_2O_3, HfN, or HfC, and multiple coatings of Al_2O_3 or TiN on top of TiC, generally deposited by CVD. The tools are initially heated to ca 1000°C for a coating of either TiC or Al_2O_3, or their combination; this initial temperature is <1000°C for TiN. For a TiC coating, $TiCl_4$ and methane react in a hydrogen atmosphere at a pressure of ca 101 kPa (1 atm) or less. The TiC deposits on the tool face and establishes a metallurgical bond with the substrate at high temperatures. Alternatively, $TiCl_4$ and carbon (from the cemented carbide substrate) react in a hydrogen atmosphere. Depletion of carbon from the substrate, however, results in the formation of a brittle η-phase, $Co_xW_yC_z$. A coating of ca 5 μm thickness is deposited for optimum performance, and the process requires ca 8 h. Since the coating is extremely thin, the edge of a coated tool should be prepared (eg, by honing a radius or providing a small negative rake land) prior to coating and should not be altered subsequently.

Multiple coatings prolong tool life, create a strong metallurgical bond between the coating and the substrate, and provide protection for different work materials. Figures 7(**a**) and 7(**b**) are micrographs of single (TiC) and multiple (TiN over TiC) coatings on cemented-carbide tools (3). A very thin coating (ca 5 μm) effectively reduces crater formation on the tool face by one or two orders of magnitude relative to uncoated tools.

To take full advantage of the coating potential, substrates are carefully matched or appropriately altered to optimize properties, resulting in significant gains in pro-

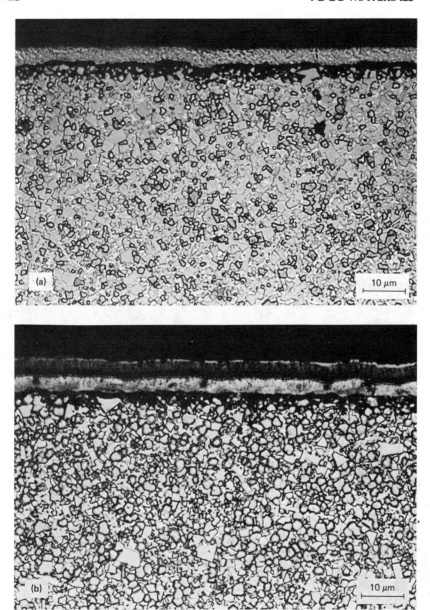

Figure 7. (a) Micrograph of the cross section of a cemented-carbide tool coated with a single layer of TiC (3). (b) Micrograph of the cross section of a cemented-WC tool coated with TiN over TiC (3).

ductivity. For example, various coatings of Al_2O_3 or TiC, or TiN–TiC or Al_2O_3–TiC, combined with different cemented-carbide substrate materials, provide a range of combination properties (substrate–coating) tailored for a given application. Coated tools with steel-grade substrates (multiple carbides) are still recommended for machining steels, whereas tools with nonsteel-grade substrates are recommended for other materials. However, the coated tools can be used at higher speeds or higher removal rates, or for longer life at the prevailing speeds (6).

Since the mode of wear is different on the rake face than on the clearance face, coated-tool technology is advancing in the direction of selective compositions and modifications of the substrate in these areas to prolong the tool life even further and to make the coated tool more versatile (29). An example of the approach in this direction is the recent development of a multiphase (Ti–C–N) coated tool with a straight WC–cobalt-enriched layer (ca 25 μm thick) on the rake face. This layer provides superior edge strength and a multicarbide–Co nonenriched flank face for combined high abrasion resistance and high temperature deformation resistance. Figure 8(a) shows an optical micrograph of this tool, and Figures 8(b) and 8(c) are micrographs of the areas around the rake face and flank face, respectively.

Although coated tools have demonstrated significant performance gains over comparable uncoated tools, several factors have contributed to making the acceptance somewhat less than feasible, such as inadequate machine-tool systems (the significant performance gains possible with coated tools are accomplished at higher speeds, at higher removal rates, and with more rigid, high power machine tools; most older machine tools are somewhat limited in this respect); nonuniform and inconsistent performance of some earlier coated tools owing to quality-control problems; limited user knowledge, partly because many small-scale users have less knowledge of the perfor-

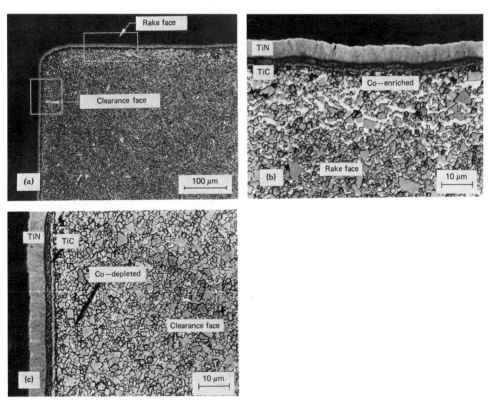

Figure 8. Micrographs of a multicoating (TiC–Ti(C,N)–TiN) over an engineered cemented tungsten carbide substrate. (**a**) A low magnification micrograph showing the areas around the rake face and the clearance face of the tool; (**b**) micrograph at higher magnification of an area around the rake face showing a thin layer of cobalt-enriched straight-WC layer near the rake face for increased toughness and edge strength; (**c**) micrograph at higher magnification of an area around the clearance face showing a thin layer of cobalt-depleted multicarbide near the clearance face for increased wear resistance and high temperature deformation resistance.

mance and application of these coated tools, and partly because the technology is advancing rather rapidly; slightly higher cost, which should not be a consideration since the cost of coated tools is only fractionally higher than that of uncoated tools; larger inventory of different tool grades; and slightly lower toughness reported with some coated tools.

Of the carbide tools used today, 30–50% are coated. This percentage is expected to increase considerably (as high as ca 80%) in the next five years with advances in coating technology and the need to improve productivity.

Ceramics

Ceramics, the newest class of tool materials, are used for a wide range of high speed finishing operations and at high removal-rate machining of difficult to machine materials, eg, superalloys, hard chill-cast iron, and high strength steels. Ceramics are predominantly alumina based, although silicon nitride-based materials (also called nitrogen ceramics) are just beginning to appear with attractive features for certain applications. Ceramics, in general, are harder; more wear-resistant, highly refractory, and chemically more stable than cemented carbides and high speed steels. Notching at the depth of cut line (known as DCL notching), microchipping of the tool edge, and gross fracture of the tool are the predominant modes of wear experienced with ceramic tools. Because of poor thermal and mechanical shock resistance, interrupted cutting is especially severe on ceramic tools owing to repeated entry and exit of the cut. Both the tool and the part to be machined must be fully supported, and to prolong life, the machine tool must be extremely rigid. Ceramics are machined either dry or with a heavy stream of coolant, since intermittent application of coolant can cause thermal shock leading to fracture.

Although ceramic tools were considered for certain machining applications as early as 1905, strength under the conditions of cutting was inadequate and the performance inconsistent (30–31).

In the mid-1950s, ceramic tools were re-introduced for high speed machining of gray cast iron for the automobile industry, and slow speed, high removal-rate machining of extremely hard (and difficult to machine) cast or forged steel rolls used in the steel industry. These materials were basically fine grain (<5 μm) alumina-based materials, either almost pure or alloyed with suboxides of titanium or chromium to form solid solutions, and contained small amounts of magnesia as a sintering aid (30). The General Electric Carboloy Systems Department developed an alumina–TiO ceramic (grade 0-30) that is characterized by an extremely fine (<3 μm) grain and uniform microstructure; the TiO constitutes ca 10%. A hardness of ca 93–94 HRA and a TRS >550 MPa (ca 80,000 psi) were achieved with cold-pressing and sintering techniques. The Carborundum Company developed a nearly pure alumina (Stupalox or CCT 707) with minor additions of MgO as a sintering aid and grain-growth inhibitor. The Vascoloy Ramet/Wesson Corporation manufactures a similar material (VR 97), originally developed by the Norton Company. Extremely rigid, high powered (up to 450 kW or 600 hp) and/or high speed (up to 5000 rpm) machine tools were specially built for some applications to take full advantage of the potential of this tool material.

Several factors have recently generated interest in the development and application of ceramic cutting tools (31–32), such as advances in the ceramic-processing technology; rapidly rising manufacturing costs; the need to use materials that are increasingly more difficult to machine; rapidly increasing costs and decreasing availability of tungsten, tantalum, and cobalt, which are the principal and strategic raw materials in the manufacture of cemented-carbide tools; and advances in machining science and technology (see Ceramics).

A comparison of the physical properties of ceramic tools and carbide tools is given in Table 12 (3). Ceramics are harder (hence, are more abrasion resistant), have a higher melting temperature (hence, are more refractory), and are chemically more stable up to their melting temperatures. They are, however, less dense and less tough (lower TRS), and have lower thermal conductivity and lower thermal expansion than cemented carbides. Toughness of ceramics with smaller grain size can be improved by the introduction of a more ductile second phase. Because of lower TRS and high refractory characteristics, ceramics are generally recommended for high cutting speed ($\geq$300 m/min), a lower rate of material removal (ie, high speed finish machining), and continuous-cutting applications.

Recently, pure alumina and alumina–TiC dispersion-strengthened ceramics have attracted attention (32). Alumina–TiC-based ceramics contain ca 30 wt % TiC and small amounts of yttria as a sintering agent, resulting in a density close to 99.50% theoretical. High purity, fine grain size, and elimination of porosity are the principal reasons for the high TRS (700–900 MPa or ca 1.0–1.3 $\times$ 10^5 psi). The hot-pressed materials are slightly more expensive, but they usually have higher TRS and more consistent performance than cold-pressed materials. A micrograph of a hot-pressed alumina–TiC ceramic, showing a white TiC phase and a dark alumina phase with some fine porosity, is given in Figure 9. Other recent alumina-based ceramic tools include alumina–titanium diboride, alumina–zirconia–tungsten compound, and silicon–aluminum–oxygen–nitrogen (Si–Al–O–N) complex compound. The latter two materials are slightly less hard than alumina or alumina–TiC, but significantly tougher. When machining superalloys, hard chill-cast irons, and high strength steels in the medium speed range (ca 150 m/min), longer tool life results because of lower flank wear and, more important, lower DCL notching.

Table 12. Physical Properties of Ceramic and Cemented-Carbide Cutting Tools[a]

Property[b]	Ceramics	Cemented carbide[c]
hardness, HRA[d]	91–95	90–93
TRS[e] for alumina-based ceramics, MPa[f]	690–930	1590–2760
melting range, °C	ca 2000	ca 1350
density, g/cm^3	3.9–4.5	12.0–15.3
modulus of elasticity E, GPa[g]	410	70–648
grain size, μm	1–3	0.1–6
compressive strength, MPa[f]	2760	3720–5860
tensile strength, MPa[f]	240	1100–1860
thermal conductivity, W/(m·K)		41.8–125.5
thermal expansion coefficient, 10^{-6}/°C	7.8	4–6.5

[a] Ref. 3.

[b] The exact properties depend upon the materials used, grain size, binder content, volume fraction of each constituent, and processing method.

[c] Coated carbides are not included.

[d] Rockwell hardness A scale.

[e] Transverse rupture strength.

[f] To convert MPa to psi, multiply by 145.

[g] To convert GPa to psi, multiply by 145,000.

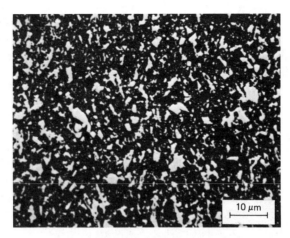

Figure 9.   Optical micrograph of a hot-pressed alumina–TiC ceramic tool material (3).

An alumina–zirconia–tungsten alloy (Cer Max, grades 440 and 460) ceramic has recently been introduced by General Electric Carboloy Systems Dept. A similar material performs exceptionally well in the grinding industry as an abrasive in heavy-stock grinding operations, such as cutoff and snagging (see Abrasives). The high toughness of this alloy is due to rapid freezing from the melt, which results in a dendritic freezing structure and superior grinding performance. The three popular compositions contain 10, 25, and 40% ZrO_2; the remainder is alumina. The 40% ZrO_2 composition is close to the eutectic. The higher ZrO_2 compositions are less hard, but tougher.

The micrograph of a fracture surface of an alumina–zirconia–tungsten alloy is shown in Figure 10 (3). The zirconia particles are concentrated predominantly at the alumina–grain boundaries. Although the fracture is intergranular, the presence of these particles is believed to provide additional toughness before failure can occur by fracture. In some recent tests on a tough chill-cast iron (HRC 42–44) used for steel rolls,

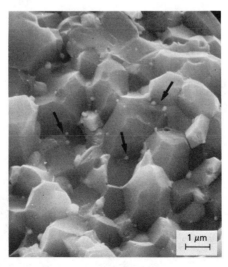

Figure 10.   Micrograph of a fracture surface of an alumina–zirconia–tungsten alloy ceramic showing the concentration of zirconia at the alumina grain boundaries (arrows) (3).

this material performed exceptionally well with very little wear in plunge cuts at 150 m/min cutting speed, a feed rate of 0.4 mm per revolution, and a width of cut of 2.54 cm over a straight-alumina tool. Chipping was the predominant mode of wear in the latter case. Similarly, when machining solution-treated and aged Inconel 718, a nickel–iron base superalloy (HRC 42–44), at a cutting speed of 150 m/min, a feed rate of 0.3 mm per revolution, and a depth of cut of 0.3175 mm, this material (2.54-cm round) gave a tool life of >8 min and yielded an excellent finish (1–2.5 μm) on the machined surface. These tests were conducted dry. Based on other successful high speed machining tests (ca 300 m/min) where a coolant–lubricant was used, it appears feasible to extend the tool life at this speed or increase the cutting speed for the same tool life.

The second interesting class of ceramic tool material currently under development is based on silicon nitride with various additions of aluminum oxide, yttrium oxide, and titanium carbide. It is a spin-off of the high temperature-structural ceramics technology developed in the 1970s for automotive gas turbines and other high temperature applications. Ford Motor Co. developed a ceramic tool of silicon nitride with additions of up to ca 12% yttria (grade S 8). GTE Products Corp. developed a ceramic tool of silicon nitride with additions of yttria (ca 6%), alumina (ca 2%), and titanium carbide (ca 30%) (grade Quantum 5000). Joseph Lucas Industries Ltd. of the UK developed a ceramic tool of a complex compound of Si–Al–O–N (SiAlON), and additions of yttria and marketed under the trademark SYALON (in the U.S., this material is marketed as KYON 2000 by Kennametal Inc. under license). Some properties of the SiAlON material are given in Table 13 (33).

SiAlON tools are currently produced by sintering. The powder charge consists of a mixture of Si_3N_4, AlN, Al_2O_3, and Y_2O_3; the latter is added as a sintering aid for full densification. The powder mix is first ball-milled, then preformed by cold isostatic pressing, and subsequently sintered at a maximum temperature of ca 1800°C under isothermal conditions for ca 1 h before it is allowed to cool slowly. The resulting material has a β-Si_3N_4 glassy intergranular phase. Because of its high toughness and good thermal shock resistance, test results indicate the possibility of using square-, triangular-, and diamond-shaped tools of this material for machining superalloys in the intermediate speed range (ca 150 m/min) where only round tools are used currently

Table 13.   Composition and Properties of SiAlON Material [a]

Composition, wt %	
Si_3N_4	77
Al_2O_3	13
Y_2O_3	10
Properties	
density, g/cm³	3.2–3.4
hardness, GPa (kgf/mm²)	17.65 (1800)
Young's modulus, GPa [b]	300
compressive strength, MPa [c]	>3500
thermal conductivity, W/(m·K)	20–25
thermal expansion coefficient, per °C	32×10^{-6}

[a] Ref. 33.

[b] To convert GPa to psi, multiply by 145,000.

[c] To convert MPa to psi, multiply by 145.

with other ceramics. Further developments in microstructure and composition may yield an even tougher material consisting of β'-Si_3N_4 and an intergranular phase of yttrium aluminum garnet (YAG) without an intergranular glassy phase. Similar to the alumina–zirconia ceramic, this material offers significant improvements in tool life, consistency in tool performance (more reliability), and higher removal rates possible at reasonable cutting speed (90–125 m/min). With the increasing trend toward computer-controlled machining, consistency and reliability of tool performance are crucial. Furthermore, the trend toward more than one machine tool per operator is resulting in lower and more manageable cutting speeds. SiAlON, alumina–TiC, alumina–zirconia, and straight alumina are some of the tool materials that might meet the needs of these trends.

Ceramic tools are inherently more brittle than cemented carbides, and a tool geometry of −10° rake and +10° clearance is recommended instead of −5° rake and +5° clearance (for cemented carbides). In interrupted cutting, attempts should be made to shift the point of application of the load away from the cutting edge to minimize chipping. Suitable edge preparation involving honing a small radius or a small negative land on the rake face is also recommended. The work materials recommended include hardened steel, chill-cast iron, and superalloys (Ni-base and Co-base). Improved design (dynamic characteristics) and performance (speed and power) of the machine tools will further improve the efficiency of machining with ceramics.

Certain ceramic tools, especially those based on alumina, are not suitable for machining aluminum, titanium, and similar materials because of a strong tendency to react chemically. They are also not generally suited for low speed and intermittent cutting operations because of failure by chipping. Poor thermal shock resistance forbids intermittent application of cutting fluids. Either heavy flooding or no coolant at all is recommended for machining with ceramic tools for this reason.

Diamond

Diamond is the hardest (Knoop hardness ca 78.5 GPa (ca 8000 kgf/mm^2)) of all known materials. Both the natural (single-crystal) and synthetic (polycrystalline sintered body) form can be used for cutting-tool applications. Diamond tools exhibit good thermal conductivity; ability to form a sharp edge by cleavage (single-crystal natural diamond); low friction; nonadherence to most materials; ability to maintain a sharp edge for a long period of time, especially when machining soft materials like copper and aluminum; and high wear resistance. Disadvantages include extensive chemical interaction with elements of Group IVB to Group VIII of the periodic table (diamond wears rapidly when machining or grinding mild steel; it wears less rapidly with high carbon alloy steels than with low carbon steel and is occasionally employed to machine gray cast iron (high carbon content) with long life); a tendency to revert at higher temperatures (ca 700°C) to graphite and oxidize in air; extreme brittleness (single-crystal diamond cleaves easily); difficulty in shaping and reshaping after use; and high cost (see Carbon, diamond, natural).

Natural-diamond tools are used mainly for special applications where it outperforms other tool materials and for applications involving light cuts and high speed, where long life can be expected to justify the cost. Examples of the former include finish machining of copper-front surface mirrors and use of a single-crystal diamond as microtome knives. High quality, single-crystal industrial diamonds are the tools of choice for these applications because of their long life and ability to machine accurately (with an extremely sharp edge formed by cleavage). Lower-quality industrial diamonds are extensively used in high speed machining of aluminum–silicon alloys in the automobile industry; in polymers and glass–epoxy composites in the aircraft industry; in copper commutators in the electrical industry; for machining nonferrous (brass, bronze) and nonmetallic materials; for cold-pressed sintered-carbide preforms for the metal-cutting and metal-forming industries; to shape and cut stone, concrete; and as dressing tools for alumina grinding wheels. Natural diamond gives unreliable performance caused by easy cleavage and unknown amounts of impurities and imperfections. Regrinding of these tools is difficult and expensive. Limited supply, increasing demand, and high cost have led to an intense search for an alternative, dependable source of diamond. This search led to the ultrahigh pressure (ca 5 GPa (50,000 atm)), high temperature (ca 1500°C) synthesis of diamond from graphite in the mid 1950s (34–35) and the subsequent development of polycrystalline sintered diamond tools in the late 1960s (36) (see Carbon, diamond, synthetic).

The polycrystalline diamond tools consist of a thin layer (ca 0.5–1.5 mm) of fine grain size particles sintered together and metallurgically bonded to a cemented-carbide base. The cemented carbide provides the necessary elastic support for the hard and brittle diamond layer above it (see Fig. 11(**a**)). These tools are formed by a high temperature–high pressure process at conditions close to those used for the synthesis of diamond from graphite. Fine diamond powder (1–30 μm) is first packed on a support base of cemented carbide in the press. At the appropriate sintering conditions of pressure and temperature (in the diamond-stable region), complete consolidation and extensive diamond-to-diamond bonding takes place (see Fig. 11(**b**)). Stress concentration at the sharp corners of the diamond crystals during sintering subjects these areas to local stresses of perhaps an order of magnitude higher than nominal (ca 5 GPa (50,000 atm)). As a result, individual diamond crystals are work hardened, resulting in a sintered diamond compact which is probably much harder than an undeformed diamond, consequently the abrasion resistance of the tool is increased. In addition

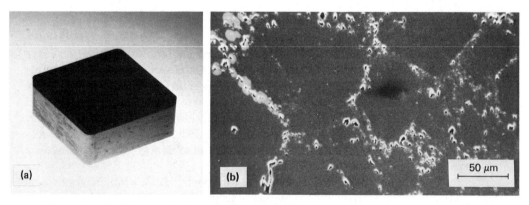

Figure 11. (**a**) Photograph of a sintered polycrystalline diamond tool showing a dark, thin diamond layer (ca 0.5 mm) metallurgically bonded to a lighter colored cemented WC substrate. Sintered polycrystalline-CBN tools also look similar to this tool (3). (**b**) Micrograph of a polished polycrystalline sintered diamond tool showing diamond-to-diamond bonding (3).

to diamond-to-diamond bonding, good metallurgical bonding is established between the diamond layer and the cemented-carbide support base in this process. The tools are then finished to shape, size, and accuracy by laser cutting, electrodischarge machining, grinding, and polishing or lapping.

Sintered polycrystalline diamond tools are fabricated in an assortment of shapes (squares, rounds, triangles, and sectors of a circle of different included angles) and sizes from round blanks. The main advantages of sintered polycrystalline tools over natural single-crystal tools are better control over inclusions and imperfections, higher quality, and greater toughness and wear resistance (resulting from the random orientation of the diamond grains and the corresponding lack of simple cleavage planes). In addition, the availability of sintered diamond tools is not dictated by nature or some artificial control; thus, such tools can be manufactured to meet strategic needs.

Sintered polycrystalline diamond tools are much more expensive than conventional cemented-carbide or ceramic tools because of the high cost of the processing technique, which involves high temperature and high pressure sintering, and the finishing methods. Diamond tools, however, are economical on an overall-cost-per-part basis for certain applications because of long life and increased productivity.

Sintered diamond tools are used for applications similar to the lower quality industrial diamonds. Because of their high reactivity, they are not recommended for machining soft low-carbon steels, titanium, nickel, cobalt, or zirconium. Since they are inherently brittle, they are used with a negative rake $(-5°)$ geometry with suitable edge preparation on materials that are difficult to machine, such as pressed and sintered cemented tungsten carbide stone and concrete. For softer materials, eg, aluminum–silicon alloys, aluminum- or copper-front surface mirrors, and motor commutators, in contrast a high positive rake $(+15°)$ geometry is used. Positive rake inserts with polycrystalline-diamond tips are among the most commonly used tools for this application; the tips can be resharpened and are available in cartridges.

Cubic Boron Nitride (CBN)

Cubic boron nitride (CBN), next only to diamond in hardness (Knoop hardness 46.1 GPa (ca 4700 kgf/mm^2)), was developed in the late 1960s (34–36). It is a remarkable material in that it does not exist in nature and is produced by high temperature–high pressure synthesis in a process similar to that used to produce diamond from graphite (see Boron compounds, refractory; Nitrides). Cubic boron nitride, although not as hard as diamond, is less reactive with ferrous materials like hardened steels, hard chill-cast iron, and nickel-base and cobalt-base superalloys. It can be used efficiently and economically at higher speed (ca 5 times), with a higher removal rate (ca 5 times) than cemented carbide, and with superior accuracy, finish, and surface integrity. Sintered CBN tools are fabricated in the same manner as sintered diamond tools and are available in the same sizes and shapes. Their costs are higher than those of either cemented-carbide or ceramic tools because of higher processing and shaping costs. Like the sintered polycrystalline diamond tools, CBN tools are held on standard tool holders.

To compete with low cost tools such as ceramics, coated carbides, new fabrication technology is currently under development. For example, instead of fabricating sintered CBN tools on a cemented-carbide base, tools of ca 1.5 mm thickness are fabricated without the base with the tool faces ground on either side (37). Such a tool can be used on both sides, which roughly doubles its life. The tool, however, has to be properly supported and clamped during use to prevent premature failure. Another example is the fabrication of sintered CBN tools by the same high temperature-high pressure, with the use of binder phase (either metallic or nonmetallic) to increase toughness (38). In respect to phase distribution, these CBN tools resemble cemented-carbide or alumina–TiC ceramic tools, but they are tougher and have greater chemical stability.

The two predominant wear modes of CBN tools are DCL notching and microchipping. Polycrystalline CBN tools exhibit flank wear where alumina ceramic tools fail catastrophically. These tools have been used successfully for heavy interrupted cutting and for milling white cast iron and hardened steels. Negative lands and honed cutting edges were used. Like diamond, CBN is thermally unstable at elevated temperatures. The reaction products, however, when machining materials like steel- or nickel-base alloys, are generally not damaging to the process. At present, CBN tools are not recommended for very low or very high speed cutting applications. Nevertheless, these tools are capable of very high removal rates when used with machine tools of adequate power and stiffness. In fact, they perform better with heavier cuts than with lighter cuts. Since they are inherently brittle, CBN tools are used with a negative rake ($-5°$) geometry. Suitable edge preparation, consisting of honing a small radius or a small negative land on the rake face, is also recommended.

Since diamond and CBN tools provide significantly higher performance capability, new demands are being placed on the machine tools and manufacturing practice in order to take full advantage of the potential of these materials. Since they are extremely hard but brittle, rigid machine tools must be used with gentle entry and exit of the cut in order to prevent microchipping by cleavage. High precision machine tools offer the advantage of producing high finish and accuracy. Use of machine tools with higher power and rigidity enable higher removal rates.

Economic Aspects

The current U.S. estimate of $>\$100 \times 10^9$/yr as the cost of labor and overhead for machining is based on an estimated number of total metal-cutting machine tools in various metal-cutting industries (1). It does not take into account the cost of raw stock (work material), cutting tools, and many other support facilities. An estimated breakdown of cutting tool costs is given in Table 14. Because of the competitive nature, most industries prefer to keep cost information proprietary.

The estimated tool costs per year in the United States are based on ref. 39. In 1981, the value of disposable metal-cutting tools shipped to various U.S. manufacturing plants was estimated at $\$2.13 \times 10^9$, which is only ca 2% of the total estimated U.S.

Table 14. Estimated Breakdown of Cutting Tool Costs

Material	Cost, $
high carbon, low alloy, HSS	$(0.8–1) \times 10^9$
cemented carbides	400×10^6 [a]
ceramics	10×10^6
diamond, CBN	2.5×10^6

[a] About half is for coated grades.

manufacturing costs. In other words, the cost of cutting tools per se is only a small fraction of the total costs, although the tooling costs may be significant in a large manufacturing facility. Optimization of tool grade, geometry, and cutting conditions is highly recommended to improve the efficiency of cutting and to reduce manufacturing costs.

Health and Safety Factors

Threshold-limit values for the components of cemented carbides and tool steels are given in Table 15 (40). There is generally no fire or explosion hazard involved with tool steels, cemented carbides, or other tool materials. Fires can be handled as metal fires (eg, with Type D fire extinguishers) (see Plant safety; Fire-extinguishing agents, Supplement Volume). Most constituents of tool materials do not polymerize.

During machining operations, eye protection is recommended; during grinding operations, NIOSH-approved respirators for metal fumes and dust are recommended (41). Some dust of cobalt is reported to cause dermatitis and pulmonary changes in susceptible individuals. Most manufacturers supply safety information with their products (42).

Safety is of particular concern in metal-cutting and metal-forming operations (42). Precautions should be taken to ensure protection of personnel and equipment from potential flying fragments and sharp edges. Safety devices and protective shields or screens must be installed on metal-cutting machines. Chips should be handled with some mechanical device, never by hand. Tool overhang must be as short as possible to avoid deflection resulting in breakage or chatter.

Some cutting fluids, eg, oils, may present a fire hazard. Some work materials, eg, magnesium, aluminum, titanium (under certain conditions), and uranium, in finely divided form, also present fire hazards. Very small metal chips or dust may ignite.

Adequate ventilation of grinding operations should be established to comply with existing government regulations, and management should remain alert to the possibility of symptoms, even in grinders working within established government standards.

Table 15. Threshold Limit Values (TLV)[a]

Constituent	TLV, mg/m^3
tungsten carbide	5
titanium carbide	na
tantalum carbide	5
chromium carbide	0.5
cobalt	0.1
nickel	1
iron	na
tungsten	5
copper	1

[a] Ref. 40.

Outlook

With the introduction of new grades of ceramics of improved properties and performance and the need to machine many difficult-to-machine materials at higher cutting speeds and higher removal rates for reducing the rapidly escalating manufacturing costs, the ceramic market may grow rather rapidly. Similarly, the improved performance of coated grades over uncoated grades in many applications may also expand its market. The high speed steel tool market may remain at present levels or grow steadily, especially with the introduction of coated HSS. Diamond and cubic boron nitride superabrasive cutting tools are used today in applications where their use is justified, despite their higher cost compared to other tool materials. With increasing use of aluminum alloys containing hard, abrasive constituents such as SiO_2 for many automotive applications, the use of glass-reinforced epoxy composite materials for the aircraft industry applications, and the need to machine many alloy steels in their hardened condition to reduce manufacturing costs, the market for superabrasives will continue to grow. The specific percentage of market shared by any tool material may change, but the total market of each tool material is expected to grow steadily.

BIBLIOGRAPHY

"Tool Materials for Machining" in *ECT* 1st ed., Suppl. 2, pp. 873–882, by Roland B. Fischer, Battelle Memorial Institute; "Tool Materials for Machining" in *ECT* 2nd ed., Vol. 20, pp. 566–578, by Roland B. Fischer, The Dow Chemical Company.

1. *Machinability Data Handbook*, 3rd ed., Vols. 1 and 2, Machinability Data Center, Metcut Research Associates, Cincinnati, Ohio, 1980.
2. R. Komanduri, "Cutting Tool Materials" in M. B. Bever, ed., *Encyclopedia of Materials Science and Engineering*, Pergamon Press, Oxford, UK, 1983; *General Electric TIS report no. 82CRD176*, Schenectady, N.Y., June 1982.
3. R. Komanduri and J. D. Desai, *Tool Materials for Machining*, General Electric TIS report no. 82CRD220, Schenectady, N.Y., Aug. 1982.
4. H. G. Swinehart, ed., *Cutting Tool Materials Selection*, American Society of Tool and Manufacturing Engineers (now the Society of Manufacturing Engineers (SME)), Dearborn, Mich., 1968.
5. *Manuf. Eng.*, 49 (Oct. 1977).
6. K. J. A. Brooks, *World Directory and Handbook of Hard Metals*, An Engineer's Digest Ltd., UK, 1976.
7. R. L. Hatschek, *Am. Mach.* (733), 165 (May 1981).
8. *Properties and Selection of Tool Materials*, American Society for Metals, Metals Park, Ohio, 1975.
9. *Properties and Selection: Stainless Steels, Tool Materials and Special-Purpose Metals*, Vol. 3 of *Metals Handbook*, 9th ed., American Society for Metals, Metals Park, Ohio, 1979, pp. 421–488.
10. G. A. Roberts, J. C. Hamaker, and A. R. Johnson, *Tool Steels*, American Society for Metals, Metals Park, Ohio, 1962.
11. E. J. Dulis and T. A. Neumeyer, *Materials for Metal Cutting*, ISI P216, The Iron and Steel Institute, UK, 1970, pp. 112–118.
12. M. G. H. Wells and L. W. Lherbier, eds., *Processing and Properties of High Speed Tool Steels*, Metallurgical Society of AIME, Warrendale, Pa., 1980.
13. R. Komanduri and M. C. Shaw, *Proceedings of the Third North American Metal Working Research Conference (NAMRC-111), May 5–7, 1975*, Carnegie Press, Pittsburgh, Pa.
14. R. Wilson, *Metallurgy and Heat Treatment of Tool Steels*, McGraw-Hill Book Company, New York, 1975.
15. W. T. Haswell, W. Stasko, and F. R. Dax in ref. 12, pp. 147–158.
16. D. B. Dallas, ed., *Tool and Manufacturing Engineering Handbook*, McGraw-Hill Book Company, New York, 1976, pp. 1-10–1-12.

17. *Cobalt Monograph*, prepared by Battelle Memorial Institute, Columbus, Ohio for the Centre D'information du Cobalt, Brussels, Belgium, 1960.
18. Technical information brochures, Stellite Division of Cabot Corporation.
19. W. Dawihl, *Handbook of Hard Metals*, English transl., Her Majesty's Stationery Office, London, 1955.
20. P. Schwarzkopf and R. Keiffer, *Cemented Carbides*, The MacMillan Company, New York, 1960.
21. H. E. Exner, *Int. Met. Rev.* (243), 149 (1979).
22. J. Gurland and P. Bardzil, *Trans. AIME, J. Met.*, 311 (Feb. 1955).
23. J. E. Mayer, D. Moskowicz, and M. Humenik, *Materials for Metal Cutting*, ISI Special Report No. P126, The Iron and Steel Institute, UK, 1970, pp. 143–151.
24. U.S. Pat. 2,967,349 (Jan. 10, 1961), M. Humenik and D. Moskowicz (to Ford Motor Co.).
25. D. Moskowitz and M. Humenik, Jr., *Proceedings of the 1975 International Powder Metallurgy Conference*, New York.
26. S. J. Whalen, *Vapor Deposition of Titanium Carbide*, ASTME (now SME) Tech. Paper No. 690, Society of Manufacturing Engineers, Dearborn, Mich., 1965.
27. *Proceedings from the 1st International Cemented Carbide Conference, Chicago, Ill., February 1–3, 1971*, Vols. 1 and 2, Society of Manufacturing Engineers, Dearborn, Mich.
28. R. Komanduri, *Advances in Hard Materials Tool Technology*, Carnegie Press, Pittsburgh, Pa., 1976.
29. B. J. Nemeth, A. T. Santhanam, and G. P. Grab in H. M. Ortner, ed., *Proceedings of the 10th Plansee Seminar on "Trends in Refractory Metals and Special Materials and Their Technology,"* Metallwerk Plansee, Ruette, Austria, 1981, pp. 613–627.
30. A. G. King and W. M. Wheildon, *Ceramics in Machining Processes*, Academic Press, Inc., New York, 1966.
31. F. W. Wilson, ed., *Machining with Carbides and Oxides*, McGraw-Hill Book Company, 1962.
32. E. Dow Whitney, *SAE International Congress and Exposition*, SAE Tech. Paper Ser. 80319, Detroit, Mich., Feb. 23–27, 1981.
33. *Engineering*, 1009 (Sept. 1980).
34. F. P. Bundy, *Sci. Am.* **231,** 62 (Aug. 1974).
35. R. H. Wentorf, Jr., R. C. DeVries, and F. P. Bundy, *Science* **208,** 873 (May 23, 1980).
36. L. E. Hibbs, Jr., and R. H. Wentorf, Jr., *High Temp. High Pressure* **6,** 409 (1974).
37. K. S. Reckling, *Tool. Prod.*, 74 (Dec. 1981).
38. N. Tabuchi, A. Hara, S. Yazu, Y. Kono, K. Asai, K. Tsuji, S. Nakatani, T. Uchida, and Y. Mori, *Sumitomo Elec. Tech. Rev.* (18), 57 (Dec. 1978).
39. P. M. Klutznick, *1981 U.S. Industrial Outlook for 200 Industries with Projections for 1985*, U.S. Department of Commerce, Washington, D.C., 1981.
40. Material safety data sheets, Carboloy Systems Department, General Electric Company, Detroit, Mich.
41. M. E. Lichtenstein, F. Bartl, and R. T. Pierce, *Am. Ind. Hyg. Assoc. J.*, 879 (Dec. 1975).
42. *Turning Handbook of High Efficiency Cutting*, GT9-262, Carboloy Systems Business Department, General Electric Company, Detroit, Mich., 1980.

General Reference

E. M. Trent, *Metal Cutting*, Butterworth and Co., Ltd., London, 1977.

R. KOMANDURI
J. D. DESAI
General Electric Company

TOOTHPASTE. See Dentifrices.

TORPEX.   See Explosives and propellants.

TOXAPHENE.   See Insect control technology.

TOXICOLOGY.   See Supplement Volume; also Industrial hygiene and toxicology.

TOXICOLOGY, CARCINOGENS.   See Supplement Volume.

TRACE AND RESIDUE ANALYSIS

Trace analysis, the determination of substances comprising <0.01 wt % of a sample
(1–2), is an area of analytical chemistry that has flourished in recent years (3). Three
fourths of the abstracted articles in *Analytical Abstracts* for 1981 describe trace
analysis or methodology—totaling about 10,000 per year, a fourfold increase since 1967
(4). This increase is partly a consequence of the development of ever more sensitive
methods, but it is also the result of discoveries concerning the importance of trace
substances in living organisms and in various materials of commerce, from semiconductors to foods (5–6). Some of these discoveries have led to industrial specifications
for optimum trace impurity levels in a material, others have led to government regulations as to the maximum permissible concentrations of various contaminants in air,
water, foods, sewage sludge, etc, by Federal agencies in the United States such as the
FDA, EPA, and OSHA, as well as by state and local agencies.

Environmental toxicology, in particular, is an area where determinations of
residues of pesticides and other toxic substances have become commonplace. Gas
chromatographic (gc) analysis of a mixture of chloroorganic insecticides, after sample
cleanup, typifies the methods developed during the last 15 years to analyze samples
containing many different trace substances. Acronyms for instrumental methods are
given in Table 1.

As the need has increased for monitoring lower concentrations, chemists have
search for (and generally have found) methods to meet the need. There are now many
methods for routine determinations of concentrations at the ppm level, and some at
the ppb or ultratrace levels. Nevertheless, the search for simpler, more convenient,
more sensitive, and more comprehensive methods remains a continuing challenge.

Some general principles of sampling, preconcentration, and measurement—
common to many different trace analytical procedures—are presented here, along
with highlights (over the past decade) of methods development particularly relevant

to trace and residue analysis. Also, selected applications in a number of current areas are reviewed, with emphasis on those fields in which regulatory requirements as to methods of analysis have been established. Finally, a brief literature review is included.

Sampling and Sample Preservation

The noun sample has two meanings in chemical analysis, depending on the context: (1) it refers to the entire quantity of a substance collected at a given time and place, ie, the total sample, and (2) it refers to that portion of the substance actually analyzed, ie, the analytical sample which is measured out and treated according to a particular analytical procedure (see Sampling). The terms sample collection and sample handling refer to meaning (1); the term sample size refers to meaning (2), which also is the meaning when replicate samples are analyzed. Sometimes the total sample is the analytical sample (eg, an impinger-bottle solution collected for a specific air pollutant).

Sample Collection. The representativeness of the sample must be considered in trace analysis as in an analysis for major or minor constituents (see Analytical methods). The fact that the concentration of a substance sought (symbolized here as X) is at the trace level does not guarantee homogeneity, especially when contamination is a principal source of X in the sample. However, if the required precision for a trace analysis is only ±10% (less stringent than in assay work, for example), then the criterion for representativeness is correspondingly less stringent.

The method of collecting a sample, and the container used, should preserve essential sample parameters (eg, the dissolved oxygen concentration in a water sample), should introduce a minimum of contamination, and should not cause the loss (eg, by adsorption) of any substances of interest. Of practical importance, sampling should be easy to do safely, by only semiskilled personnel.

There is interest today in knowing the precise chemical form or speciation of a substance in a sample (22). Obviously, processes such as precipitation, sorption, hydrolysis, oxidation–reduction, or microbial action (eg, biomethylation of Hg(II) to methylmercury, CH_3Hg^+) can change the speciation pattern of a sample. A procedure for preserving some particular sample constituents may require a reagent that unavoidably alters the speciation of other constituents.

A grab sample may be sufficient for a particular trace analysis, but often a more systematic sampling protocol is necessary, such as: to characterize a given quantity of material (eg, a batch, a tankcar, a lake), a number of samples must be taken with a specified spatial distribution, depending on available information and assumptions about the homogeneity of the bulk material and on the questions that are to be answered by the analysis; or if the material to be sampled is a flowing stream of air, liquid, or even of solid particles (eg, a pharmaceutical production line), automatic samplers are available for collecting composite samples (by time fraction or by flow fraction) or for collecting separate samples sequentially according to a programmed schedule. Thus, sampling can be a four-dimensional problem in which time is added to the three dimensions of space.

Preliminary Operations and Sample Preservation. Some operations and precautions used in handling many types of samples are listed in Table 2.

Table 1. Instrumental Methods of Trace Determination

Nature of method	Acronym	Name	Characterization	Refs.
Spectrometric atomic (outer *e* excited):	aes	atomic emission spectrometry	multielement; flame or inductively coupled plasma (icp)	
	afs	atomic fluorescence spectrometry	flame (solutions; hydride generators)	
	aa	atomic absorption:	flameless (graphite furnace; Hg vapor)	
	zaa	Zeeman effect aa	magnetic-field background correction	
	es	emission spectrometry	arc or spark excitation; semiquantitative trace multielement	
x-ray and surface (inner and next-to-outer *e* excited):	ssms	spark source–mass spectrometry		
	xrf	x-ray fluorescence	multielement; wavelength- and energy-dispersive detectors	
	pixe	particle-induced x-ray emission	uses proton accelerator spectra	
	esca and Auger	electron spectroscopy-chemical analysis	characteristic of chemical environments	
	sims	secondary-ion mass spectrometry	multielement	
	sem–xrf	scanning electron microscope–xrf	surface mapping	
	cm	chemical microscopy	eg, identification of forensic samples	
	emp–xrf	electron microprobe–xrf	small spot analysis; depth profiling	
	imp	ion microprobe		
	im	ion microscope	multielement surface pattern	
radiochemical (excited nuclei): sample activation:	naa	neutron activation analysis	thermal neutrons; most elements except Be, Cd, Pb, P, S, Tl	7–8
	faa	fast neutron activation	for O, Cl, Si	
	paa	photon activation	for As, Br, Ca, Ce, Cl, Cr, I, Na, Ni, Sb, Ti, Zn, Zr	
	id	isotopic dilution		
	sid	substoichiometric id		
radiotracers:	ria	radioimmunoassay method	addition of chelating agent	7

312

molecular (species-specific) excitation:				
absorption:	absorption spectrophotometry	uv-vis	LD ca 10^{-6} M	
	absorption spectrophotometry	ir	trace method in gc-ir detector	
absorption (magnetic field nuclear-spin states):	nuclear magnetic resonance	nmr	seldom a trace method	
emission:	fluorescence spectrometry	fs	excitation and emission monochromators; sensitive and selective; fluorescent derivatization	9
	phosphorescence spectrometry	ps	increased sensitivity at room temperature by heavy atom effect or by sorption on filter paper	9
scattering:	Raman spectrometry	rs	complements ir; lasers lower LD	
	turbidimetric spectrometry	ts	eg, water analysis	10–11
	nephelometric spectrometry	ns	eg, particulates in air	12
Electrometric potentiometric:	ion-selective electrodes	ise	cations, anions, gases, enzyme–substrates	
	null-point potentiometry	npp		
	potentiometric titrations	pt	microtitrations end-point determinations	
voltammetric steady state (slow sweep):	d-c polarography	dcp	current-voltage (i-E) curve at dropping mercury electrode (dme); LD 3×10^{-6} M	
	voltammetry at rotated electrode	dcv	rpe = rot. Pt; rde = rot. disk; rrde = rot. ring-disk using dme or rotated electrode	
	amperometric titration	amp		
modified steady state:	pulse polarography	pp	linearly increasing pulse applied to dme prior to drop fall; i (faradaic) measured after decay of i (capacity); LD 5×10^{-7} M	
	differential pp	dpp	small pulse (50 mV) added to polarograph voltage at dme prior to drop fall; i-increase measurement after decay of i (capacity); LD 10^{-7} M	
	differential pulse anodic stripping voltammetry	dpasv	dpp after electrodeposition at Hg drop (hmde) or thin Hg-film electrode (mfe); LD 10^{-9} M	
	a-c polarography	ac	small a-c added to d-c polarograph voltage; LD 10^{-6} M if fast e transfer	

313

Table 1 (*continued*)

Nature of method	Acronym	Name	Characterization	Refs.
fast d-c sweep:	lsv	linear sweep voltammetry	stationary electrode (Hg, Pt, graphite, glassy carbon); LD 10^{-7} M	
constant current:	cv	cyclic voltammetry	mechanism and e transfer rate studies	
	cp	chronopotentiometry	potential–time (E–t) curve at small stationary electrode; anodic stripping at constant-i (eg, Sn plate thickness)	
			$i \times t$ = coulombs; sensitive; automated	
electrolysis (exhaustive):	ct	coulometric titn.	add carrier to collect traces	
	ed	electrodeposition	Hg cathode electrolysis; distillation of Hg and polarograph residue	13
	ep	electrolytic-polarographic method		
	cpe	controlled-E electrolysis	i-t curve (chronoamperometry), or coulomb–time (Q–t) curve (chronocoulometry)	
Separation-chromatographic				
elution chromatographic:				
relative volatilities:	gc	gas chromatography	selected stationary phases; packed or capillary columns; temperature-programmed column oven; detectors: thermal conductivity (tc); flame-ionization (fid); N and P-fid; electron-capture (ecd); Hall coulometric; P and S flame-photometric (fpd); photoionization (pid); ms; ir	14
	gsc	solid stationary phase		
	glc	liquid stationary phase		
relative sorptivities:	lc	liquid chromatography	gravity flow, low pressure, low resolution chromatography compared to hplc	15–16
	hplc	high performance liquid chromatography	selected stationary and mobile phases; gradient elution; detectors: refractive index (ri); uv; fluorescence, photoconductivity, amperometry, ms, ir, aa	

314

	hplc–np	normal phase hplc	polar column, nonpolar mobile phase	
	hplc–rp	reverse phase hplc	nonpolar column, polar mobile phase (includes ion pairing)	
relative sizes:	hplc–sec	size exclusion or	mol wt distribution	
	gpc	gel permeation chromatography		
relative selectivities:	ixc	ion-exchange chromatography	cations; anions; Chelex-100*	17
reaction tendencies + ix:	riex	reactive ion exchange	in ligand exchange, amino acid analyzers	
	ic	ion chromatography	suppressor column; conductance detector	18
mass: charge ratios:	ms	mass spectrometry	direct insertion of sample: (g), (l), (s); ion sources: electron impact (ei), field desorption, chemical ionization (ci), or negative-ion sources; double-focusing magnetic sectors, quadrupole or time-of-flight mass filter; low ($1/1000$) or high ($1/20{,}000$) mass resolution)	19–20
	ms–ms	double quadrupole ms	analyzes crude mixtures: separation in first section, ci in middle chamber, final separation in second section	
	gc–ms	gc–mass spectrometry	effluent from gc into ms; interface design	20
	lc–ms	lc–mass spectrometry	effluent from hplc into ms; interface design	21

Table 2. Common Operations in Sample Handling and Preservation

Operation	Method
separation of dissolved gases and volatiles from solution	purge with inert gas; if desired, collect vapors in cold trap, on solid adsorbent, or in trapping solution (see Table 3)
separation of solids from solution	filter to remove particulate matter (use 0.45-μm membrane filter and suction to remove bacteria and fine particles); centrifuge, eg, at 8000 rpm for 20 min to remove clay particles >1 μm
separation of solids into size fractions	use sieves (stainless steel or plastic often preferred) to isolate large particles (eg, down to 62 μm dia (230 mesh)); use pipet-sedimentation method[a] to separate finer particles (eg, silt, 62–4 μm, from clay, <4 μm); use continuous-flow ultracentrifuge to size fractionate suspended particles in a large volume of sample solution[b]; use zonal centrifugation to fractionate particles (within a size fraction) by density gradient[c]
precautions in storage of samples (see Sampling, Tables 3 and 4, Vol. 20, p. 540)	
atmosphere	use inert gas as necessary to prevent oxidation
humidity	store samples moist (unless unstable) in sealed containers, or dry in desiccators; oven-dry if necessary; air-dry (eg, soils) or freeze-dry (eg, foods) to preserve organic matter
light	store in the dark, or in amber or brown containers, to prevent photolysis or autooxidation reactions
pH adjustment of solutions	acidify to prevent loss of ammonia and other volatile bases, to minimize "plating" of heavy-metal hydrolysis products on container surfaces, and to minimize bacterial action on C–, N–, and S– compounds; make basic to prevent loss of volatile acids (eg, HCN, H_2S)
biocide addition	add trace toxic substance (eg, $HgCl_2$) to prevent bacterial growth
temperature	refrigerate at 4°C to minimize biological and chemical changes; freeze to prevent biodegradation, minimize volatilization, and slow chemical reactions

[a] Ref. 23.　[b] Ref. 24.　[c] Ref. 25.

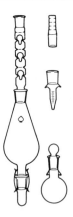

Figure 1. Kuderna-Danish evaporator, with receiving flasks. Courtesy of Ace Glass, Inc., Vineland, N.J.

Table 3. Separation–Preconcentration Methods in Trace Analysis, Classified on the Basis of Phases and Phase Changes

Phase change[a]	Method or process (comment or example)	Ref.
Volatilization		26
(l) → less (l)	solvent evaporation at <100% reflux (Kuderna-Danish apparatus (Fig. 1) or rotary-evaporator; stream of N_2 or Ar)	27
	adsorptive bubble techniques (gas–liquid interface; foam fractionation)	28
(l) → (g) → (l)	fractional distillation (rare for trace analysis; loss of inorganic halides by distillation)	26
	(preparative gc: fractional volatilization and sorption)	
(l) → (g) → (s)	purge and trap (volatile organics; Fig. 2)	29–30
	Gutzeit method: As (and Sb, Se, Sn) hydrides	38
(s) → (g)	ms of volatile species: direct insertion probe	
(s) → (g) → (s)	fractional sublimation (purification of solids)	
Extraction		
(g) or (l) → (l)	transport through semipermeable membrane (dialysis; gas and enzyme–substrate electrodes)	12, 31
	reverse osmosis: isolation of trace organics	32
(l)$_A$ → (l)$_B$	liquid–liquid extraction: batch, multistage, countercurrent, or continuous (Fig. 3)	33–35
	solubility: trace organics (solvents: Table 4)	15
	organic acids and bases (adjust pH)	36
	derivatizations: inorganic ions (chelate and ion-association reagents for metals: Table 4)	26, 33–35, 37
(s) → (l)	extraction (Soxhlet: fats and pesticide residues in foods)	38
Sorption–Coprecipitation		
(g) → (s)	condensation, cold traps (air pollutants)	
	physical sorption (air pollutants: Table 5, A)	39
	reactive sorption (air pollutants: Table 5, B; Fig. 4)	40
(g) → (l)	reactive sorption impingers (Table 5, B; Fig. 5)	39
(l) → (s)	physical sorption (water pollutants: Table 5, C)	41
	lc column separations (cleanup; fractional elution)	29, 42–43
	hplc pre-enrichment columns and cartridges (Table 5, C)	
	preparatory hplc: normal and reverse phase (Table 5, C)	15
	extraction chromatography	26
	size exclusion: gel filtration; (preparatory hplc: mol wt fractions)	
	reactive uptake: from solution (Table 5, D)	
	ion exchange (ix): batch; column; preparatory hplc	17
	chelating resins (heavy metals from seawater)	44
	reactive ix (riex: chemical reaction + ix)	45
	(elution chromatography and ion chromatography: Table 1)	
	precipitation and coprecipitation ($Cu_2Fe(CN)_6$ carrier for scavenging Cs radionuclides from seawater)	26, 46–48
	electrodeposition (usually as metal or amalgam)	10
	mercury cathode (ca 40 metals)	13
	(anodic stripping: Table 1)	
(l) → (s)	zone melting (ultrapure materials; organics)	49

[a] (g) = gas or vapor, (l) = liquid, and (s) = solid.

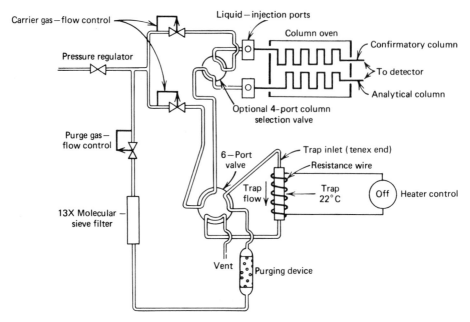

Figure 2. Purge and trap system (purge–sorb mode) for trapping volatile organics prior to desorption and gc detection. Note: all lines between trap and gc should be heated to 80°C (50).

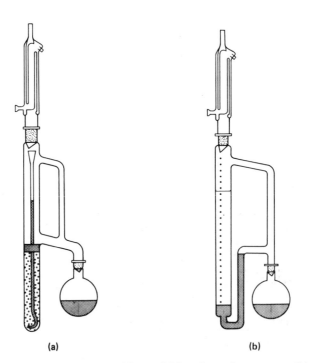

(a)   (b)

Figure 3. Continuous extractors for solvents (**a**) less dense than water or (**b**) denser than water (51).

Table 4. Extraction Solvents and Reagents

Compound[a]	Dielectric constant	Liquid range mp to bp, °C	Density, g/mL (20°C)	P' [b]
Some common solvents and their properties				
water	78.0	0–100	0.998	10.2
acetonitrile[c]	36.0	−45–82	0.777	5.8
methanol[c]	32.6	−98–65	0.792	5.1
acetone[c]	20.7	−95–56	0.791	5.1
methyl isobutyl ketone (MIBK)	13.1	−85–118	0.801	ca 4.5
ethanol[c]	24.3	−117–78	0.789	4.3
chloroform[d]	4.8	−64–61	1.489	4.1
methylene chloride	8.9	−97–40	1.336	3.1
diethyl ether	4.3	−116–35	0.708	2.8
benzene[d]	2.3	5.5–80	0.879	2.7
toluene	2.38	−95–111	0.868	2.4
carbon tetrachloride[d]	2.24	−23–77	1.594	1.6
n-hexane	1.91	−94–69	0.659	0.1
isooctane	1.94	−107–99	0.692	0.1

Type of reagent	Name of reagent (acronym)	Typical solvent
Some chelate and ion-association reagents for metals[e]		
Chelates		
β-diketone	acetylacetone (Acac)	benzene
	thenoyltrifluoroacetylacetone (TTA)	MIBK, hexane
8-quinolinol	8-hydroxyquinoline (Oxine)	CHCl$_3$
nitrosoaryl hydroxylamine	cupferron	CHCl$_3$
hydroxamic acid	N-benzoyl-N-phenyl-hydrox-amine (BPHA)	CHCl$_3$
thiocarbazone	dithizone (H$_2$Dz)	CHCl$_3$, CCl$_4$
dithiocarbamate	sodium diethyldithiocarbamate (NaDDTC)	CHCl$_3$, CCl$_4$
dithiocarbamate	ammonium pyrrolidine dithiocarbamate (APDC)	MIBK
Ion-association (from acid media)		
alkyl esters of H$_3$PO$_4$		
$\overset{\displaystyle O}{\overset{\displaystyle \|}{ROP}}(OH)_2$	mono-2-ethylhexyl ester	toluene
$\overset{\displaystyle O}{\overset{\displaystyle \|}{(RO)_2P}}OH$	dibutyl ester	toluene
(RO)$_3$PO	tributyl phosphate (TBP)	many solvents
trialkyl phosphine oxide		
R$_3$PO	tri-*n*-octylphosphine oxide (TOPO)	cyclohexane
alkyl amines (high mol wt)	methyldioctylamine (MDOA)	xylene, trichloroethylene
quaternary ions	R$_4$N$^+$, R$_4$P$^+$, (C$_6$H$_5$)$_4$As$^+$ for complex anions; (C$_6$H$_5$)$_4$B$^-$ for complex cations	polar solvents
halides and pseudohalides	Cl$^-$, I$^-$, SCN$^-$	CCl$_4$ and many polar solvents

[a] Liquid CO$_2$ (triple pt −56°C, critical pt 31°C at 7.4 MPa (73 atm), density 0.914 g/cm^3 at 0°C) is an inexpensive, nontoxic, nonflammable solvent used under moderate pressure as an extractant for organic compounds; also, if used as a mobile phase in lc, fractions are self-freezing as they are collected, then the CO$_2$ sublimes leaving pure solutes (perhaps useful for lc–ms interface) (52).

[b] P' = solvent polarity, from liquid chromatography theory (15); decreasing polarity represents increasing power to extract hydrophobic organics.

[c] Miscible with water; used to extract solids or modify aqueous phase.

[d] Carcinogen; use with caution.　　　[e] Refs. 26, 33–35, 37.

Table 5. Selected Sorbents

A. Some physical sorbents for air contaminants[a]

cold traps: miscellaneous vapors
activated charcoal: solvent vapors
XAD-2 macroreticular resin (polystyrene-divinyl benzene): neutral organic compounds
Chromosorb: bis(chloromethyl) ether
Thermosorb/N: for nitrosamines
Tenax-GC (poly(diphenyl-p-phenylene oxide)): neutral or slightly polar organic compounds
polyurethane foam: polychlorinated biphenyls (PCBs)
Florisil (PR grade)
silica gel: aromatic amines
alumina (deactivated): polar organic compounds
solvents in impinger tubes or bubblers: water can trap many soluble gases (including HCl, HNO_3, SO_2, SO_3 and H_2SO_4 mist, NH_3); glycols and cellosolves may be preferred, being less volatile than water

B. Some reactive sorbents for air contaminants

Solids

sequential sorption of Hg species[b]
    $HgCl_2$, by 3% methylsilicone on Chromosorb-W
    CH_3HgCl, by 0.05 M on Chromosorb-W
    elemental Hg, by Ag-coated glass beads
    $(CH_3)_2Hg$, by Au-coated glass beads (Fig. 4)
colored indicator tubes[c]
    mine safety and military applications
SO_2 "candle"
    cylinder of PbO paste → insoluble $PbSO_4$

Solutions for trapping gases in impinger tubes (Fig. 5)[a]

SO_2
    0.04 M potassium tetrachloromercurate, plus trace of EDTA; forms stable dichlorosulfitomercurate complex
ozone
    0.1 M boric acid and 1 wt % KI (calibration procedure); O_3 forms equivalent amount of I_2
formaldehyde
    1 wt % $NaHSO_3$ soln; forms sulfite addition product

C. Some physical sorbents for water contaminants[d]

lc collector–cleanup columns: solid sorbents from part A above, plus the following:
    XAD-4 (similar to XAD-2; more highly crosslinked, smaller pores)[e]
    XAD-7 (acrylic ester groups; more polar than XAD-4)
hplc, bonded phase microparticulate packings
    normal phase—amino, nitrile, diol groups
    reverse phase—C_{18}, C_8, and phenyl groups (Waters Sep-Pak collector cartridges contain disposable sorbent for organics)
    (prepacked reverse-phase, polar, and ix columns are available from many vendors)

D. Some reactive sorbents for water contaminants[f]

ion-exchange resins and zeolites (including resin-loaded papers[g], Chelex for trace heavy metals[h], and reactive ion exchange (riex)[i]
coated silica beads: bonded functional-group layer[j]
clathrate compounds; crown ethers; valinomycin for K^+
W-wire: adsorbs part per trillion (10^{12}) Cd, Cu, Pb, and Zn from water; determined by gf-aa[k]

---

[a] Ref. 39.   [g] Ref. 17.
[b] Ref. 40.   [h] Ref. 44.
[c] Ref. 53.   [i] Ref. 45.
[d] Ref. 41.   [j] Ref. 56.
[e] Ref. 54.   [k] Ref. 57.
[f] Ref. 55.

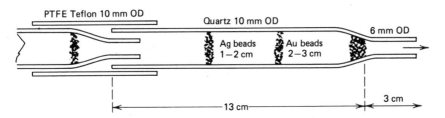

Figure 4.   Stackable tube for trapping Hg species from air: Ag-coated beads for Hg, Au-coated beads for $(CH_3)_2Hg$ (58). Courtesy of American Chemical Society.

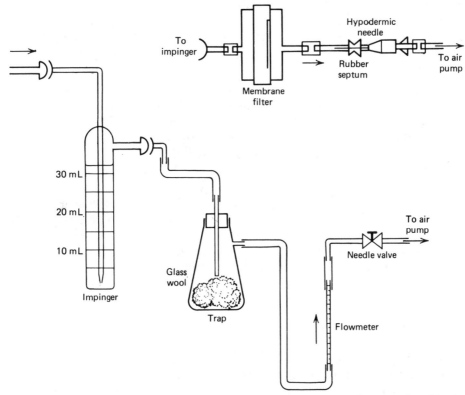

Figure 5.   Sampling train for gaseous pollutant, showing filter, orifice flow control, and impinger tube (59).

The Analytical Sample.   Portions taken for analysis must be representative of the total sample: solids can be mixed and split by quartering. and may be ground if necessary to render the material more homogeneous; slurries subject to size fractionation by gravity can be mixed and sampled using a thief (ie, a hollow tube inserted to the bottom of the container) for withdrawing a vertical cross section; liquids can be stirred or shaken to homogenize particulate matter.

As a general rule, relatively large portions are required for trace analysis—0.1 g to 1 kg of solids, 1 mL to 2 L of liquids, 1 cm^3 to 2000 m^3 of gases. With a large sample, the procedure includes a preconcentration step so that X, the substance (or substances) sought, is brought to a final concentration, C_f, which is high enough to be measured precisely. In such a case, the following equation applies:

$$C_f = \frac{S_i}{S_f} \cdot C_i \cdot f_r$$

where S_i and S_f are the initial and preconcentrated (ie, final) sample sizes (in g or mL) (the fraction S_i/S_f is the preconcentration factor), C_i is the concentration of X in the initial sample, and f_r is the fraction of X recovered in the total procedure (reflecting primarily the efficiency of the preconcentration step). Rearranging this equation, one can calculate the sample size S_i if the other terms are known or can be estimated:

$$S_i = \frac{C_f}{C_i} \cdot \frac{S_f}{f_r}$$

The final preconcentrated sample size S_f is generally fixed (within narrow limits) by the procedure chosen. Ideally, f_r will approach unity and C_f should be 5–10 times the limit of detection (LD) or the minimum detectable quantity (MDQ) of X in the final sample by the particular method of determination used. (The LD is defined as the concentration equivalent to three times the noise or standard deviation of the measurement (1).) It follows that with more sensitive methods (ie, lower LD values), lower C_i values can be measured or smaller sample sizes can be taken.

In some cases, methods are applied directly to the bulk sample material: eg, gc for trace organics in air, aa for metals in water samples or dilute solutions, neutron activation of solids, xrf of alloys. The direct approach is feasible if a method is quite selective and sensitive for X, provided the sample matrix (eg, air or water) does not cause significant, irreproducible interference with the X-signal.

Pretreatment and Preconcentration Methods

A concise, general scheme for trace analysis is shown in Table 6 (26). This scheme begins after the sample has been collected, and includes many operations and factors that are of vital importance to a successful trace analysis. Too often, the inexperienced chemist focuses entirely on the measurement operation, and overlooks the care and analytical skill that must precede the final step.

General Precautions. *Contamination.* The addition of contaminants from the laboratory atmosphere, hood superstructures, crucibles, tongs, filter media, glassware, and reagents (including boiling chips) must be kept to a minimum. Some operations are best carried out in glove boxes and, in some ultratrace determinations, a clean room with laminar-flow hood or positive-pressure ventilation may be necessary (60). All glassware must be washed carefully with its specific use in mind: for trace organics, a common practice is to wash items with detergent, rinse thoroughly with tap water and distilled water, then rinse with a volatile solvent like acetone and oven-dry; finally, cap with aluminum foil and store in a drawer.

It is essential that blanks be carried through the entire procedure to correct for unavoidable contamination. Water must be highly purified for trace analysis:

Table 6.   General Outline of Trace Analysis[a]

Factor affecting result of analysis	Analytical operation	Method of determination
homogeneity ⎫ storage ⎬	material to be analyzed ⟶	nondestructive
tools ⎫ atmosphere ⎪ reagents ⎬ changes in composition ⎪ of analytical sample ⎭	⎧ sampling ⎪ grinding ⎨ mixing ⎩ surface cleaning	
	laboratory sample ⎫ ⎬ ⟶ ashing ⎭	direct, for solids
	dissolved sample ⟶	direct, for solutes
vessels ⎫ reagents ⎪ atmosphere ⎬ volatility ⎭	separation of separation of matrix traces preconcentration	
	dissolved traces ⟶	simultaneous determination of many constituents
	separation individual trace constituents ⟶	determination of individual constituents

[a] Ref. 26.

mixed-bed ion exchange and distillation from borosilicate or (preferably) quartz removes most inorganics; carbon-adsorption and double distillation from glass removes most organics. Reagents must be of known purity (61). Ultrapure reagents are available from chemical supply houses (eg, one such lot of HNO_3, analyzed for 22 metals and five nonmetals, contained <0.005 ppm for 17 of them). Even higher grades of sealed-in-quartz acids are available. However, HCl and NH_3 can be purified conveniently in the laboratory by isothermal (ie, "isopiestic") distillation (62–63). Specially purified solvents, low in pesticides or in uv-absorbance, are available (64) (see also Fine chemicals).

Losses. The loss of sample constituents or products from volatilization, from incomplete separations, and from failure to redissolve sample constituents completely after evaporation to dryness should be minimized. It is essential that fortified samples be analyzed periodically to check on the overall recovery of a trace analysis procedure.

Quality Control. As in all analytical operations, a program of quality assurance is necessary (65). This involves the use (and development) of reference or standard methods, the certification of equivalent methods, a reliance on standard reference materials, control charts, and round-robin analyses with other laboratories to monitor accuracy (66).

Preparing the Sample Solution. Unless a direct method of determination is available, the trace procedure (for solid materials and sorbed samples), starts with preparation of a sample solution. This may involve extraction with water or a non-aqueous solvent, but often a more drastic chemical attack is used. The approach depends on the objective, which may be to isolate either a single trace substance or a group of related substances, eg, heavy metals or chloroorganic pesticides (see Insect-control technology). The following steps are considered:

Selective removal of sorbed or dissolved substances (organics, organometallics, some inorganics) by thermal volatilization (eg, moisture, trace solvent residue) or selective solubility; efficiency of dissolution depends on the relative volumes of sample and solvent, and on the procedure for mixing or shaking: sonication facilitates contact between solvent and inner pores of the solid, speeding up equilibration (67); extraction is commonly used to separate fats and fat-soluble substances (eg, pesticide residues) from solid samples.

Removal of available surface inorganics by selective leaching is based on solubility, ion exchange, acid–base reactions, formation of complexes, and redox reactions (Table 7).

Complete dissolution or digestion of sample. Inorganics are dissolved using appropriate chemical attack, with fusion if necessary (eg, refractory oxides, silicates); and organics are dissolved after dry ashing (ignition) or are wet ashed (digested) in acids (eg, conc HNO_3, aqua regia, 30% H_2O_2, $HNO_3 + H_2SO_4$, $HNO_3 + HClO_4$). In many procedures, excess acid is evaporated before preparation of sample solution; use special hood for $HClO_4$ evaporations (71–72).

Other pretreatment operations (*that change the sample*) include degradation of organic matter by uv irradiation, ozonolysis, or microbial action; radiochemical conversion by neutron, proton, or photon activation (7); derivatization of inorganics (73) or organics (74) for gc, hplc, or voltammetric determinations; adjustment of pH and oxidation state; masking (especially for inorganics) (75).

Table 7. Typical Reagents for Selective Leaching of Inorganics

Reagents	Substances leached
water, or other polar solvent	soluble salts
electrolyte soln ($CH_3CO_2NH_4$ or $MgCl_2$)	exchangeable ions[a]
complexing agents: Cl^-, CN^-, citrate, tartrate, EDTA, DPTA[b]	heavy metals (Groups IIIB–VIIB, VIII, IB, and IIB of periodic table)
weak acids: CH_3CO_2H	carbonates; some oxides
HF	silicates ($SiF_4\uparrow$)
strong mineral acids: dil solns of HCl, HNO_3, H_2SO_4, $HClO_4$	most oxides; surficial heavy metals;[c] some anions
oxidizing or reducing agents: H_2O_2, HNO_3; dithionite	remove organic C; remove Fe- and Mn-oxide coatings

[a] Ref. 68.
[b] Ref. 69.
[c] Ref. 70.

Separation and Preconcentration Operations. Trace substances generally are separated or preconcentrated prior to the final determination (26,76–77). With many samples, it is necessary to sidetrack major constituents by a prefractionation or cleanup procedure before the preconcentration step. Two approaches are used: removal of the matrix constituents from the trace substances to be determined, or (typically) removal of the trace substances from the matrix (1).

Separation and preconcentration methods can be classified based on the phases and phase changes involved. The methods so classified in Table 3 include volatilization, extraction, and sorption–precipitation. Some methods use the batch equilibrium approach: the rate of equilibration between phases is generally fast, but can be slow when particles of solid phase or a chemical reaction are involved. Other methods use a continuous process, the system finally approaching a steady-state condition. (A few separation methods, such as centrifugation and electrophoresis, are not based on phase changes, but depend on external forces, ie, gravity or electric fields.)

Detailed procedures reflect practical considerations: volatilization efficiency depends on temperature control; batch extraction efficiency is improved if solvent is added in several small portions; emulsions can be broken by adding salts, by using mixed solvents to create larger density differences, by using wetting agents (de-emulsifiers), or by filtering through glass wool; sorption efficiency increases with decreasing size of sorbent particles, and also can depend on the linear flow rate of mobile phase (gas or solution) through column; and short backup columns are used to detect breakthrough.

Methods of Determination

Most methods for analytical determinations can be used in trace analysis. Even the classical methods of gravimetry (using a sensitive, fifth-place balance), titrimetry (using $0.01 \, M$ or $0.001 \, M$ standard solutions), and, of course, colorimetry (or the equivalent, uv–vis spectrophotometry) have a place in current practices (78). With these older methods, amounts as low as 10–$100 \, \mu g$ can be determined with fair precision, but instrumental methods developed in the years 1940–1960 (eg, naa, ssms, aa, fs, dcp, and gc) lowered the limits for quantitative determination to the nanogram (ng) range. This refers to the absolute limit of determination, which in many cases is equal to the relative (or concentration) limit multiplied by S_f, the smallest practical final weight or volume prepared for the measurement step.

Highlights of Recent Developments. Developments in trace analysis since 1960 have lowered these absolute limits to the subnanogram range for many techniques, at the same time making the 10–100 ng range more accessible for routine automated determinations. Some of the developments are listed below.

Automated analyzers: for routine analysis, especially of clinical samples (blood serum) by vis methods (79).

New detectors for x- and gamma-ray spectrometry: Li-drifted Ge and thin Si (low energy photons); 100 times better resolution (see Fig. 6) and hence vastly improved sensitivity for xrf and naa determinations (81); also, improved multichannel analyzers (and computers) for storage and display of data (a general improvement, affecting all types of instruments).

New sample-introduction methods for aa and aes: graphite furnace requires only a few μL of solution, determining 0.1 ng of many elements; cold vapor method for Hg lowered limit to a few ng; hydride method for As, Se, and Sn improved sensitivity a hundredfold; icp source (see Fig. 7) for aes gives much steadier and stronger signals than flame emission and thus lower limits of detn applicable to routine, multielement analysis of sample solutions or slurries (55).

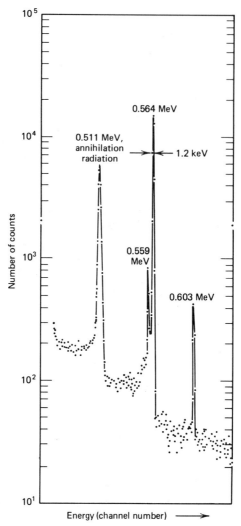

Figure 6.   Gamma spectrum taken with Ge(Li) detector, showing 1.2 keV resolution (80). Courtesy of Wadsworth Publishing Co., Belmont, Calif.

High intensity, stable excitation sources for spectrofluorimetry: high pressure xenon arc lamps (and presently lasers) give subnanogram limits of detection (9).

Pp and dpp: from advances in controlled potential and current measurement circuitry (based on op. amps. and electronic switching); increased sensitivity a hundredfold (10–11,83).

Dpasv: a combination of preconcentration (at hmde or mfe electrode) of metals such as Cu, Cd, Pb, and Zn with the sensitive dp anodic stripping step, lowering the limit to the ng range (see Fig. 8) (84).

Ion-selective electrodes (joining the time-tested Hg–Hg$_2$Cl$_2$, Ag–AgCl, and glass or pH electrodes): made potentiometry more useful for trace analysis, down to 10^{-7} M in many cases (12) (see Ion-selective electrodes).

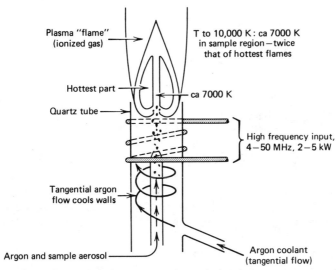

Figure 7. Inductively coupled plasma source for optical emission spectrometry (82). Courtesy of Wadsworth Publishing Co., Belmont, Calif.

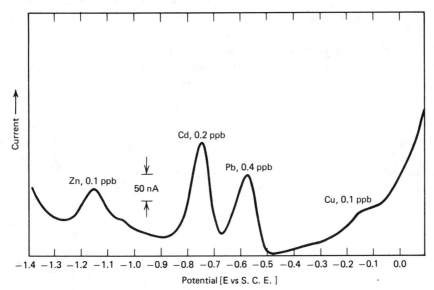

Figure 8. Differential-pulse-anodic stripping voltammetry of $2 \times 10^{-9}\,M$ Zn, Cd, Pb, Cu, at Hg-coated electrode (84). S. C. E = standard calomel electrode. Courtesy of American Chemical Society.

Gc developments: new detectors—more sensitive and more selective (eg, fid, ecd, fpd, N&P detector (halide bead))—pushed the determination limits down to subnanogram amounts (85); temperature programming made gc a much more versatile tool for analyzing a diverse mixture of trace substances (eg, petroleum products, foods, flavors, and pesticides); and open tubular capillary columns gave vastly improved resolution (see Fig. 9), with somewhat smaller sample size requirements (87).

Computerized gc–ms methods, with data storage, retrieval, presentation, and comparison with spectral files (spectral stripping for mixtures), have revolutionalized the identification of trace organics (eg, EPA Priority Pollutants (88)) and other sub-

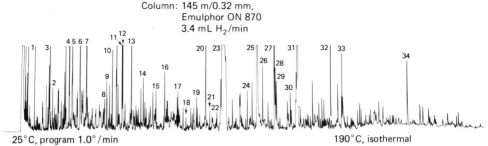

Column: 145 m/0.32 mm,
Emulphor ON 870
3.4 mL H_2/min

25°C, program 1.0°/min   190°C, isothermal

Figure 9. Capillary gc of middle fraction of cigarette smoke (86). Courtesy of K. Grob and *Chromatographia*, Pergamon Press, Elmsford, N.Y.

stances, from pesticide residues to research samples (19); very important in forensic chemistry (qv). The design of a quadrupole mass separator, used in popular commercial gc–ms instruments and well suited for the chemical ionization (ci) method, is shown in Figure 10.

High Performance Liquid Chromatography (hplc): both normal-phase and reverse-phase, with high resolution microparticulate (5–10 μm dia), bonded-phase packed columns, developed into excellent separation–determination methods at the 1–100 ng level; gradient elution has extended the versatility, and new detectors (eg, fluorescence, photoconductivity, voltammetry, variable wavelength uv and ir) have improved selectivity and lowered limits of determination (15).

Classification of Instrumental Trace Methods. An outline of instrumental methods for trace determination, based on the physical principles employed, is presented in Table 1, with an acronym for each method. Mass spectrometry (except for ssms) is grouped with the separation–chromatographic methods.

Another classification is based on the distinction between methods for determining elements and those for determining molecular or ionic species. Among the former, single-element and multielement methods have different features, but an evaluation of lunar sample analyses shows that multielement methods such as naa and ssms, in the hands of skilled operators, can have great utility for trace element determinations without appreciable sacrifice in precision (90).

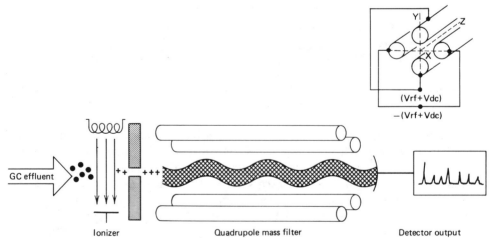

Figure 10. Schematic of quadrupole mass spectrometer; inset at upper right shows spatial arrangement of rods and ac ± dc voltages applied to quadrupole mass filter (89). Courtesy of Marcel Dekker, Inc.

Standard Reference Materials

Standard materials, standard methods, and standards-setting organizations play a vital role in the development of trace analytical chemistry (66).

About 450 standard reference materials (SRMs), certified for the concentrations of trace substances, are available from the National Bureau of Standards (Table 8) (91). Also, the EPA makes available small samples of standard pesticides and organic pollutants to workers in the environmental field (92). In addition, other governmental agencies such as the FDA and the International Atomic Energy Agency, and many professional associations (eg, The American Association of Clinical Chemistry, API, ANSI, and ASTM) prepare and distribute standard materials, often of a specialized nature (93). Examples of other standards, available from private vendors (94), are EPA Priority Organic Pollutants (volatiles, base–neutrals, acids, pesticides, PCBs); metal solutions for aa (aq and nonaqueous); classes of pure organic compounds (eg, amines, carbonyls, chlorinated phenols, mercaptans, phthalate esters, polynuclear aromatic hydrocarbons); water and wastewater contaminants; air pollution particulates; food additives; carcinogens and mutagens (including mycotoxins), teratogens, and other biochemicals.

Applications

Applications of trace analysis are so extensive that only a few selected aspects can be treated here, to serve as an introduction.

Agriculture, Foods, Drugs, and Consumer Products. Methods dealing with the composition of agricultural materials, foods of all kinds, drugs, and household products are found in ref. 38, published by the AOAC, which was founded by government regulatory chemists. This society approves methods only after rigorous interlaboratory testing (95–96). Included are trace methods for metals and pesticide residues in foods; for drugs in feeds and in animal tissues; for food additives (qv) and natural toxins (see Food toxicants, naturally occurring). The "Official Methods" are used by the FDA to monitor compliance with government regulations in the food and drug industry (97).

Gc surpassed spectrophotometry in number of new methods approved during the period 1974–1978 (38).

A manual of procedures for determining multiple and single pesticide residues is published by the FDA (43). The methods for organochlorine and organophosphate pesticides generally involve extraction with a nonpolar organic solvent, cleanup by lc through a Florisil column, then evaporative concentration and gc–ecd determination.

Tolerances for over 300 pesticides in or on agricultural commodities, foods, and feeds have been established by the EPA (98); the tolerances are 0–110 ppm (based on raw weight), with the acceptable methods of analysis being those in ref. 43. Each year, the FDA analyzes a Market Basket sampling of 12 classes of foods, selected from stores in 20 different urban regions of the country; composites for each food class in each region are analyzed. Results from the 1975–1976 survey (published five years later), for metal and pesticide residues and for two industrial pollutants, are summarized in Table 9 (99) (see Mineral nutrients; Insect-control technology). In general, the pesticide residues found were quite low.

Table 8. NBS Standard Reference Materials for Trace Analysis [a,b]

Samples certified for multielements[c]

Metals, ferrous—analyzed for up to 39 different elements
steels (chip form): plain carbon (16), low alloy (20), high alloy (4), stainless (8), tool (4)
steels (granular form): low alloy, stainless
steels (solid form): ingot iron and low alloy (31), stainless (16), specialty (2), high temperature alloy (9), tool (8)
steelmaking alloys (10)
cast irons (chip form) (13)
cast steels, white cast irons, ductile irons, blast furnace irons (7)

Metals, nonferrous—analyzed for 8–24 different elements
chip form: alloys (including some unalloyed) based on Al (4), Cu (15), Pb (2), Mg, Ni (5), Sn, Ti (3), Zn (2), Zr
solid form: alloys (including some unalloyed) based on Al (4), Cu (42), Pb (2), Ni (2), Ti (7), Zn (7), Zr

High purity metals—analyzed for 5–12 different elements
Au (2), Pt (2), Zn (2)

Solid inorganic materials—analyzed for 5–30 different elements
minerals (8), refractories (6), glasses (5), cements (7), trace elements in glass (4 concentration levels: 0.2, 1, 50, and 500 ppm)

Environmental standards—analyzed for up to 16 different trace elements
coal (bituminous and subbituminous), coal fly ash, fuel oil, water, river sediment, estuarine sediment, urban particulates

Environmental standards—certified for organics
shale oil, urban dust, oil (for PCBs); priority pollutant polynuclear aromatic hydrocarbons (in acetonitrile)

Biological standards (8)—analyzed for up to 45 elements
oyster tissue, wheat flour, rice flour, brewer's yeast, spinach, citrus leaves, tomato leaves, pine needles, bovine liver

River sediment—for radioactive measurements

Research materials
albacore tuna, for Hg, Se, Pb, As, Zn, etc; homogeneous river sediment

Samples certified for one or two elements

Gases in metals
nitrogen in steels (10), cast irons (3), Ti-base alloy (3), Zr-base alloy
oxygen and nitrogen in steels (10)
hydrogen or oxygen in Ti-base alloys (4)

Gases in N$_2$
CO, CO$_2$, NO, or SO$_2$

Gases in air
CO or methane and propane

Trace elements in feldspar
  Sr, 65 ppm; Rb, 524 ppm
Environmental standards (analyzed liquids and solids)
  Hg in water, 1 ng/mL; trace Hg in coal, 0.1 µg/g; Pb in reference fuel (4), 8–528 mg Pb/L (0.03–2 g Pb/gal); S in distillate fuel oil, 0.1%;
    S in residual fuel oil (5), 0.2–5%
Urine, freeze-dried (2)
  low and high Hg and F
Special reference material (distributed by NBS)
  residual oil—for Ni (93 ppm) and V (79 ppm)

Special materials used in trace analysis

Clinical laboratory standards
  human serum, freeze-dried: normal and containing anticonvulsant drugs (2)
  certified inorganic compounds (5) and organic compounds (10)
Hydrocarbon blends (to calibrate mass spectrometry and gc instruments) (4)
  for analysis of gasoline, naphthas, and blending stocks
Industrial hygiene standards (known weights of specified organic solvent added to charcoal tubes)
  benzene, m-xylene, p-dioxane, 1,2-dichloroethane, $CHCl_3$, trichloroethylene, CCl_4
Ion-activity standards
  for pH, pD, pNa, pK, pCl, pF
Isotopic reference standards (for mass spectrometry)
  12 nuclides: B, Cl, Cu, Br, Ag, Cr, Mg, Pb (three 206/208 ratios), Rb, Sr, Rh, Si
Materials on filters
  Be; Pb; metals (Cd, Pb, Mn, Zn); sulfate and nitrate; quartz and clay
Metallo-organic compounds (to make solutions of metals in oils)
  24 compounds (of 21 metals and of B, P, and Si)
Nuclear materials
  Pu and U isotopic ratio standards; alpha, beta, and gamma standards
Permeation tubes (for air pollution calibration) (see Fig. 15)
  SO_2 (3), 0.6–3.0 µg/min; NO_2, 0.5–1.5 µg/min at 25°C
Spectrophotometer calibration standards
  neutral density filters; metal-on-quartz uv-vis absorbance standards
  $K_2Cr_2O_7$—for uv; quinine sulfate—for fluorescence

---

[a] Ref. 91.
[b] If more than one standard reference material of a given type is available, the number is shown in parentheses.
[c] The chip form of metals and alloys is used for checking chemical and instrumental methods involving solutions, the solid form for checking optical emission and x-ray
  fluorescence methods.

Other countries, of course, have agencies comparable to the FDA and, on the international level, the UN affiliate agencies, FAO and WHO, are concerned with methods and standards related to foods and drugs, as well as with environmental pollutants and other concerns. In addition, states and some local governments have regulations necessitating trace analyses. California and Florida have elaborate residue-monitoring programs. New York, spurred by several well-documented pollution episodes, has done extensive testing of fish from Lake Erie, Lake Ontario, and other waters for mercury and chlorinated organic compounds; on the basis of such studies, the New York State Department of Health issued with fishing licenses in 1981–1982 a "Health Advisory" recommending very restricted consumption of fish from state waters, based mainly on mercury, PCBs, and mirex concentrations (see Water, water pollution).

In the United States, the CPSC issues regulations or advisories concerning trace hazardous substances such as an organobromine compound (Tris) in children's sleepwear (100), and formaldehyde in foam insulation (101):

$$\text{BrCH}_2\text{-CHBr-CH}_2\text{-O-P(=O)(-OCH}_2\text{-CHBr-CH}_2\text{Br)-O-CH}_2\text{-CHBr-CH}_2\text{Br}$$

Tris

Research in the areas of soil science, agronomy, and animal husbandry often requires analyses for trace inorganics, organics, and biochemicals in soils, nutrient media, feeds, tissues, seeds, and crops (102).

Some journals in this applied area are *J. Assoc. Off. Anal. Chem.; J. Agric. Food Chem.; J. Environ. Qual.; J. Soil Sci. Soc. Am.; Pestic. Monit. J.;* see also journals listed in other sections.

Clinical Chemistry, and Biomedical Research.   Trace analysis is commonplace today in hospital laboratories and clinics (79). The American Association of Clinical Chemistry publishes selected methods, in an effort to standardize laboratory procedures and practices (103). Using automated analyzers (involving the operations of sampling, mixing, dialyzing, heating, color development, and measurement), qualified technicians determine many substances in blood serum at the ppm level (see Biomedical automated instrumentation). Some typical automated trace blood analyses are shown in section A of Table 10. Trace metals are done by flame emission (Na, K), atomic absorption (Cu, Hg, Mg, Zn), spectrophotometry (Mn, Pb), or by anodic stripping (Pb). Hormones, steroids, and various other trace biochemical substances are determined by the very sensitive competitive protein-binding assay procedures, in particular, by the radioimmunoassay (ria) method (see B in Table 10) (see Medical diagnostic reagents).

Two other developments in clinical analysis illustrate new aspects of trace analysis: determining the concentrations of the creatine phosphokinase (CPK) isomers (ie, the isoenzyme patterns, used to diagnose cardiac muscle damage), determined by electrophoresis with fluorescent development of the gel matrix, or by an immuno-inhibition method (105); measuring the concentration of high density lipoproteins (related to the tendency to develop atherosclerosis), determined by a precipitation–colorimetric method (106).

Table 9. Residues in Foods of U.S. Adults, 1975–1976[a]

Food class[b]	Average concentration, ppm, in prepared food composites[c]											
	I	II	III	IV	V	VI	VII	VIII	IX	X	XI	XII
Metals												
zinc	4.9	32.0	9.0	5.2	2.7	7.6	2.3	2.1	2.4	4.1	3.0	0.5
cadmium	0.002	0.01	0.03	0.05	0.04	0.01	0.03	0.02	0.003	0.02	0.01	0.002
lead	nd	0.01	0.05	0.03	0.007	0.26	0.04	0.08	0.04	0.03	0.02	0.004
selenium	0.004	0.20	0.19	0.006	tr	0.008	0.002	nd	nd	nd	nd	nd
arsenic	0.004	0.19	0.02	nd	nd	tr	0.004	tr	nd	nd	nd	0.008
mercury	nd	0.02	nd	nd	nd	nd	nd	nd	tr	tr	tr	0.001

Pesticides[d]

Average concentration in ppb per food class, if >0.5 ppb and if more than one composite per food class contained >1 ppb of the pesticide.

    organochlorines: dieldrin 7 (II), 2 (VIII); DDE 2 (I), 10 (II), 3 (V); DDT 2 (II); endosulfan sulfate 4 (V); endosulfan I 1 (V); dichloran 2 (V), 9 (IX); 1,2,4,5-tetrachloro-3-nitrobenzene 7 (IV); dicofol 3 (IX); pentachloroaniline 2 (X); perthane 3 (V)

    organophosphates: malathion 20 (III), 3 (X)

    carbamates, etc: chlorpropham 40 (IV); carbaryl 2 (VIII); captan 3 (IX)

[a] Ref. 99.

[b] I, dairy products; II, meat, fish, poultry; III, grain and cereal products; IV, potatoes; V, leafy vegetables; VI, legume vegetables; VII, root vegetables; VIII, garden fruits; IX, fruits; X, oils, fats, shortening; XI, sugar and adjuncts; XII, beverages.

[c] tr = trace; nd = not detected.

[d] Structures of the pesticides are shown below. Trace concentrations (<1 ppb) of PCBs were found in three out of the 240 composites (one dairy, two meat-fish-poultry); another industrial chemical, pentachlorobenzene, was found at 4 and 5 ppb in two fat composites.

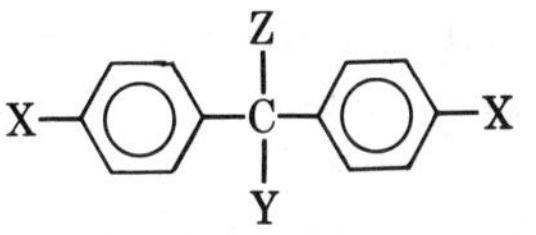

	X	Y	Z
DDE	Cl	=CCl₂	—
DDT	Cl	CCl₃	H
dicofol	Cl	CCl₃	OH
perthane	C₂H₅	CHCl₂	H

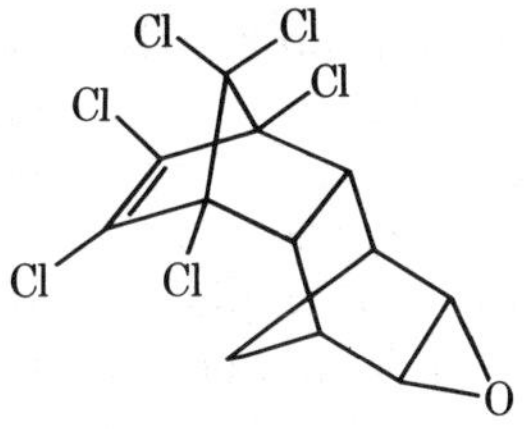

dieldrin

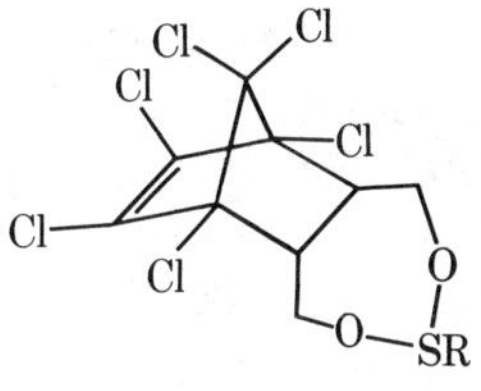

endosulfan sulfate

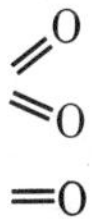

endosulfan I =O

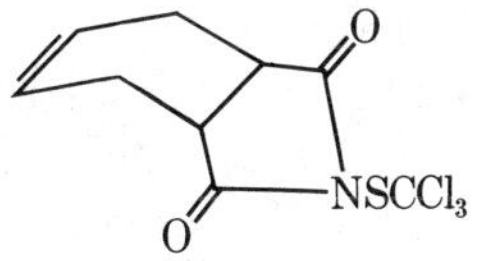

RNHCOR'

captan

chlorpropham

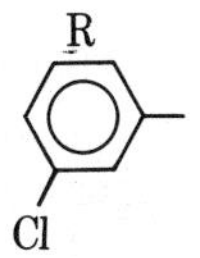

carbaryl CH₃

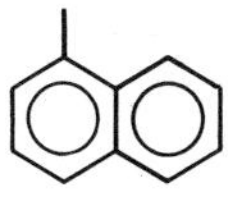

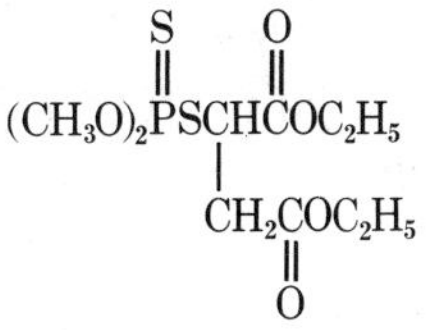

malathion

Several recent monographs deal with trace elements in biochemistry and medicine (107–108). For organic substances, a recent work on pharmaceutical analysis has excellent descriptions of modern methods, including gc, gc–ms, fluorescence and phosphorescence, and ria (9). Recent work has drawn heavily on hplc methods, using electrochemical and fluorimetric detectors. If solute X does not itself fluoresce, a derivative may be made either before or after the hplc separation. In the semiautomated method shown in Figure 11, the fluorogenic reagent is present in the mobile phase, which is then heated post-column to make the fluorescent derivative of X (74).

The field of biomedical trace methods also includes such specialized areas as trace amines in the brain (110).

Some journals of interest in this field include *Clin. Chem.*, *Clin. Chim. Acta*, *J. Chromatogr.*, and *J. Lab. Clin. Med.*

Environmental Studies.   An overview of Federal regulations and criteria for multimedia environmental control is presented in ref. 111. This handbook is an excellent compendium, with the pertinent standards and guidelines (as of mid-1979) for air and air pollutants, water and water pollutants, drinking water, the workplace environment, radiation, toxic substances, pesticides, noise, solid wastes, and marine protection. References to Federal legislation and to relevant CFR sections are given.

In the United States, the EPA has broad responsibilities for controlling and monitoring contaminants in the air, water, and soil, and for preventing the uncontrolled release of hazardous or toxic substances into the environment. Before 1970, when the EPA was established, some of these responsibilities rested with various government departments and agencies, but the legislation of the 1970s increased tremendously the efforts to control and monitor pollutants (111) (see Regulatory agencies).

Air.   Methods of air sampling and analysis have been developed in recent years by an Intersociety Committee representing 13 organizations (including ACGIH and AOAC), under the sponsorship of the APHA which began to publish methods in 1969. The first compilation, in 1972, contained 57 methods; the second edition, in 1977, contained 136 (39). Originally, the Committee considered only ambient air analysis, but now it also considers methods for pollution source monitoring and workplace air. The treatise, "Air Pollution," contains a volume on measuring, monitoring, and surveillance (112).

Most air analysis methods are trace methods. Some are direct methods, but most start with an air sampling or preconcentration step. About half of the procedures in the methods book (39) trap the contaminants of interest in impingers or bubbler tubes containing selected reagents, about one quarter use solid sorbents, and about one quarter use filters (for particulates and aerosols); final measurements are colorimetric in about half the cases, chromatographic (mostly gc) in one quarter, and other methods (including aa, ise, and titration) in the remainder.

For solid particulates, the collection device most used is the "hi-vol" sampler. Alternatively, a cascade-impactor sampler is used to determine particle-size distribution (and collect the size fractions) of aerosols (qv) having mass median diameters of ca 0.5–5 μm. Dichotomous samplers (ie, two size fractions: 0–2.4 and 2.4–20 μm) were used to study the fine-particle aerosol or haze in the Great Smoky Mountains, in September 1978 (113); the haze was found to be dominated by acid sulfates, ap-

Table 10. Selected Trace Determinations on Human Blood Serum

Substance determined	Reference range

A. Some automated colorimetric serum determinations[a]

bilirubin, total	0.2–1.2 mg/dL[b]
calcium	8.5–10.5 mg/dL
cholesterol, enzymatic method (age 40–49)	150–310 mg/dL
glucose, enzymatic method (age >50)	85–127 mg/dL
nitrogen, blood urea (BUN)	10–26 mg/dL
phosphorus, inorganic	2.5–4.5 mg/dL
uric acid	
male	3.9–9.0 mg/dL
female	2.2–7.7 mg/dL
enzymes[c]	
alkaline phosphatase (adults)	30–115 mU/mL[d]
glutamic-oxalacetic transaminase	7.5–40 mU/mL
lactic dehydrogenase	100–225 mU/mL

B. Some radioimmunoassay serum determinations[e]

peptide hormones	
luteinizing hormone (LH), follicular phase	2–30 mIu/mL
follicle-stimulating hormone (FSH), follicular phase	5–20 mIu/mL
thyroid-stimulating hormone (TSH)	up to 8 mIu/mL
prolactin, premenopause female	0–20 ng/mL[f]
human growth hormone (HGH), adult	0–8 ng/mL
insulin, fasting	up to 25 uU/mL
renin, resting	0–2 ng/(mL·h)
steroids	
cortisol, serum (a.m.)	80–270 ng/mL
estriol, unconjugated	
testosterone	
male	300–1000 ng/mL
female	20–76 ng/mL
drug	
digoxin (cardiac glycoside)	0.9–1.7 ng/mL
miscellaneous	
carcinoembryonic antigen (CEA)	<2.5 ng/mL
folate, red blood cells	>200 ng/mL
free thyroxine	0.006–0.018 ng/mL
thyroxine (T_4)	40–110 ng/mL
triiodothyronine (T_3)	0.6–1.9 ng/mL
vitamin B_{12}	0.15–0.55 ng/mL

[a] Ref. 79.
[b] 1 mg/dL =10 mg/L (10 ppm).
[c] Measured in enzyme activity units.
[d] U = activity unit; Iu = international unit.
[e] Ref. 104.
[f] 1 ng/mL = 1 ppb.

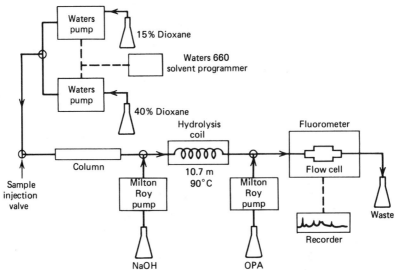

Figure 11. Modular dynamic fluorogenic-labeling liquid chromatograph. OPA = *o*-phthalaldehyde (109). Courtesy of *Analytical Letters*.

proximating NH_4HSO_4, and not by natural organic emissions from the forests as had been earlier suggested. In studying the fallout pattern from a coal-burning power plant, collections of fly ash are differentiated from native dirt particles by looking at concentration ratios, eg, Ni:Al or V:Al (Al being rich in native clays and feldspars). A similar ratio (Mn:V) may be used to trace the origins of acid rain (114).

Knowledge of the composition of particles with <2 μm dia is important to human health, as these particles penetrate deep into the lungs. Methods that determine the speciation of an element or the specific nature of the organic matter are particularly valuable in the search for health–composition correlations. Surface-analysis methods, like secondary-ion mass spectrometry (sims), give some data on the compounds in individual particles (see Figure 12) (115). Also directed toward the inorganic speciation, a recent plasma-emission method heats particulates rapidly from 25 to 1000°C; specific compounds vaporize at characteristic temperatures, giving an emission signal simultaneously from both the metal and the nonmetal of the compound, thus confirming its probable presence in or on the particles (117).

In another study, organic molecules sorbed on workplace air particulates were extracted and analyzed by synchronous fluorescence (both monochromators are scanned, with a $\Delta\lambda$ of 3 nm) (118); eight polynuclear aromatic hydrocarbons were identified from a single scan, and the standard spectra are shown in Figure 13.

EPA-approved methods for pollutants covered by the National Air Quality Standards are given in ref. 119 as follows: SO_2 by the pararosaniline method, after trapping in a tetrachloromercurate solution; suspended particulates, by "hi-vol" sampler; CO by direct, nondispersive ir; ozone by direct chemiluminescence (qv) with added ethylene (ozone replaced the photochemical oxidants standard in 1979); hydrocarbons (excluding methane) by direct flame-ionization detection; NO_2, by direct chemiluminescence with added ozone (see Fig. 14). An earlier method for NO_2, using a trapping solution and colorimetric finish, gave results that differed seriously from supposed true results, delaying the imposition by the EPA of automobile NO_x emission

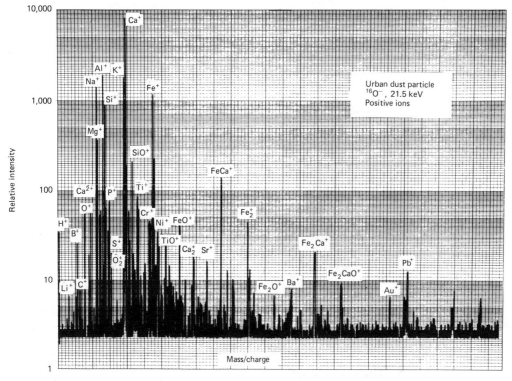

Figure 12. Sims of urban dust particles, positive-ion spectrum from O bombardment (116). Courtesy of Plenum Press.

standards for several years. Permeation tubes are used for the nontrivial task of calibrating gaseous contaminant methods. An apparatus for preparing dilute standards of SO_2 is shown in Figure 15 (see also Air pollution).

In general, methods for gaseous pollutants are designed with chemical specificity in the sorbing, eluting, and determination steps. One exception (122) is the method for organic solvent vapors, which covers at least 50 substances: sorb vapors on charcoal in a tube, elute with CS_2, and determine by gc with flame-ionization detection (CS_2 does not interfere). Personnel monitors used in such a method contain small battery-operated pumps which pull air at a rate of ca 0.2 L/min through pencil-size tubes containing the activated charcoal; after the desired exposure period (eg, 8–10 h), the tubes are broken and the sorbed substances are eluted and determined.

Direct methods are used for determining some air pollutants, especially when continuous monitoring is required. Some examples are particulates determined by scattered light in an integrating nephelometer; C_1–C_5 atmospheric hydrocarbons and peroxyacetyl nitrate by gc; total hydrocarbons by flame ionization detector; CO by long pathlength ir; SO_2 by flame photometric detector (also by Fourier-transform ir, with interferometer, as in monitoring the plume from a distant stack) (39). Methods using lasers and Raman back scattering are being developed.

Water. Standard methods (123) for the analysis of water and wastewaters are available in an Intersociety publication (124) and from the EPA (125). These methods apply to drinking water as well as to surface waters; industrial wastewaters and landfill

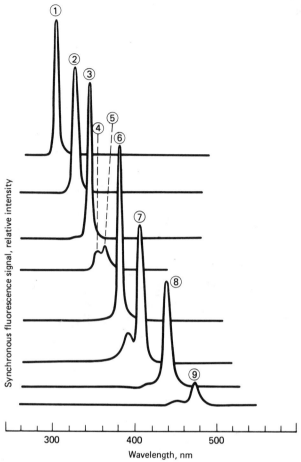

Figure 13. Synchronous fluorescence spectra of polynuclear aromatic hydrocarbons found in workplace air (118). Courtesy of American Chemical Society. ① fluorene; ② acenaphthene/dibenzothiophene; ③ 2,3-benzofluorene; ④ and ⑤ chrysene; ⑥ anthracene; ⑦ benzo[a]pyrene; ⑧ perylene; ⑨ tetracene.

leachates may require special procedures.

As with air analysis, most methods for water parameters are trace methods (concentrations of a few ppm or less). A few parameters can be monitored continuously (pH, pCl, dissolved oxygen, conductance, turbidity), but most concentrations are too low for simple direct monitoring. In unpolluted waters, heavy metals are typically in the ppb range, and toxic organics (pesticides PCBs) are typically <0.1 ppb. Metals can often be determined directly by aa or by the multielement icp–aes method; if preconcentration is necessary, extraction into MIBK using the organic reagent APDC is common (34,55), or a chelating ion exchange resin can be used (44).

Trace organics in water normally must be preconcentrated before determination. Current methods are purge and trap (Fig. 2) or headspace analysis (for volatiles); liquid–liquid extraction, using hexane, methylene chloride, or another suitable solvent; and preenrichment on a short collector column (see Fig. 16), prior to thermal desorption–gc (54) or hplc determination. Reverse osmosis (qv) is also used (32).

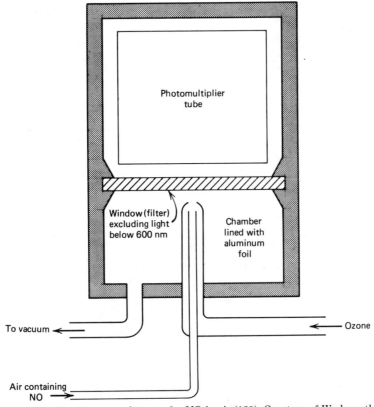

Figure 14.   Chemiluminescence detector for NO in air (120). Courtesy of Wadsworth Publishing Co., Belmont, Calif.

Investigations on the distribution, transport, and fate of organic pollutants, particularly studies on the properties (ie, the chemodynamic properties) that enable predictions to be made about the environmental fate of the compounds, involve much trace analysis. Care and experience are necessary in determining low solubilities and high partition coefficients (K_ps), particularly in analyzing the aqueous phase (67). Pollutants that biodegrade are less persistent in the environment, but such biodegradation may make the determination of fundamental properties more difficult. Biological as well as chemical determinations are necessary in water studies, particularly in regard to using wastewater to recharge aquifers (127) (see also Water, water pollution).

Solid Wastes. Solid materials present many problems today to the analytical chemist. Sometimes trace analyses, eg, of waste chemicals, landfill leachate, or contaminated ground water are vital in assessing the environmental impact of a disposal method. Complex samples usually require elaborate cleanup schemes. Under the 1976 Resources Conservation and Recovery Act (RCRA), standards were established for wells to monitor ground-water contamination near waste-disposal sites; many samples must be taken and analyzed to delineate an underground waste plume.

Sewage sludge has to meet certain EPA and state guidelines to be approved for use on farmlands (128). In this area, there is increasing concern about the possible presence in sludge of toxic organics (especially mutagens) and active viruses, as well as with the toxic metals, particularly Cd which accumulates in many food crops.

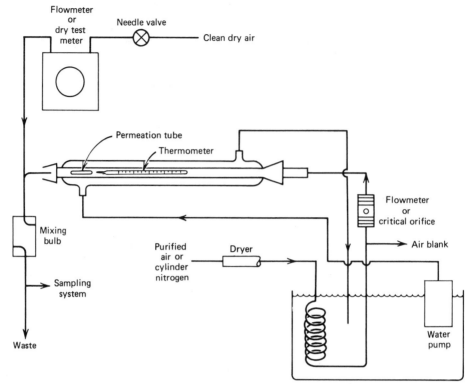

Figure 15.   Permeation tube–gas dilution system for calibrating gaseous pollutant method (121). Courtesy of APHA.

The TSCA in 1976 gave the EPA the authority to control new or existing chemicals that were found to pose an unreasonable risk to the public health or environment. Under TSCA, the EPA banned the manufacture of PCBs and took steps to curtail the production of chlorofluorocarbons. Trace analysis can be crucial in toxicity testing (eg, in classifying chemicals under TSCA); a trace of the extremely toxic substance dioxin in the herbicide 2,4,5-T has made difficult the decision as to the inherent toxicity of the latter compound:

$$Cl\text{—}\underset{Cl}{\overset{Cl}{\bigcirc}}\text{—}OCH_2\overset{\overset{O}{\parallel}}{C}OH$$

2,4,5-T

A number of tests (mostly bioassays, some quantitative as well as qualitative) have been devised to screen compounds for mutagenic and teratogenic potential so that the number of compounds subjected to expensive, long-term animal testing can be kept to a minimum (129). *In vitro* assays, using cell cultures or biochemical materials derived from living cells, may be more versatile and reproducible than other bioassay methods in testing for specific types of hazardous potential in chemicals (130).

There are many specialized journals in this applied area: *Bull. Environ. Contam. Toxicol.; Chem. Geol.; Chemosphere; Environ. Sci. Technol.; Int. J. Environ. Anal. Chem.; J. Air Pollut. Control Assoc.; J. Am. Water Works Assoc.; J. Environ. Sci. Health; Water Pollut. Control Fed. J.; Water, Air, Soil Pollut.;* and *Water Res.*

Industrial Products, Including High Purity Materials.   Trace analysis has been a standby of the metal and alloy industries (ferrous and nonferrous) for decades, as

Preconcentration step

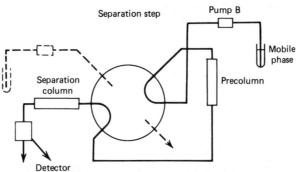

Figure 16. Schematic for precolumn trace enrichment in chromatographic water analysis (126). Courtesy of Academic Press and Gordon and Breach Science Publishers, Inc.

shown by the large number of standard reference materials available for those materials (Table 8). Advances since 1950 in high purity electronic materials also have brought the importance of ultratrace analysis to the force (2,5,131–132). Physical as well as chemical methods often are used to characterize these high purity substances, as in the resistivity fluctuations of germanium shown in Figure 17 (133); nonhomogeneity is a very real problem with materials that have been cast into ingots (see also Fine chemicals; Materials standards and specifications).

There is a relationship between controlled impurity concentrations (as in doped semiconductor materials) and analytical separation methods used in trace analysis. For example, a separation by precipitation or extraction may be "quantitative," with <0.1 wt % of substance A left in the precipitate or extract of B, but if a continuous procedure is used, the trace of A left in B may be essentially constant, reflecting the A:B separation factor for this particular system. By controlling the separation con-

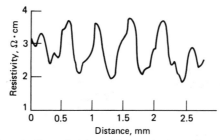

Figure 17. Fluctuations in Ge resistivity with length, from changing impurity concentrations (133).

ditions, one can predetermine the A:B ratio, thus preparing B that contains a controlled trace of A (46,49).

Each year, the ASTM publishes its *Annual Book of Standards* (134), which contains test procedures often cited in Federal and state regulations. This book, in 48 volumes, includes several volumes on analysis, including many trace methods. Noteworthy are Part 12, *Chemical Analysis of Metals and Metal Bearing Ores;* Part 26, with *Atmospheric Analysis;* Part 31, *Water;* Part 42, *Analytical Methods— Spectroscopy, Chromatography, and Computerized Systems.* Other parts, on commercial products, contain some specific trace methods. Many categories of industrial chemicals, eg, plastics, electronic chips, photographic chemicals, have characteristics that can be affected by trace components. With some of these materials, a special approach to analysis is required. As is the case with other professional groups, the Oil Chemists' Society publishes a book of official methods (135).

Only a few of the many aspects of trace analysis for metals, high purity materials, and industrial chemicals can be mentioned. In the metals industries, automated spectrographs and x-ray fluorescence analyzers carry the burden of fast, routine, quality-control analyses (136). In the areas of ferromagnetics and luminophors, electrical and optical methods are prominent (2,132).

In the field of organic chemicals and solvents, trace analysis using preconcentration followed by gc–ms determination is common. Solvents sold commercially for extractions, in procedures involving evaporative preconcentration, must be of ultra high purity (and so certified by analysis).

In the United States, OSHA is authorized to set regulations on toxic and hazardous substances encountered in the workplace. Recommendations are made to OSHA by NIOSH and by other recognized standards-setting bodies, including the ACGIH and ANSI. Using the ACGIH and ANSI lists, OSHA in 1974 set standards for about 400 workplace air contaminants, mostly organic compounds (137). The regulations prescribe permissible levels: 8-h TWA concentrations and, in some cases, ceiling concentrations. Thus, monitoring requirements are different: a composite sampler, like an absorption tube, will suffice for the 8-h average, but a continuous analyzer with an alarm (like a smoke detector) would be needed to ensure compliance with a ceiling standard; special regulations, with proscribed trace-analytical methods, have been established by OSHA for especially hazardous substances: asbestos, cotton dust, coal-tar volatiles and coke-oven emissions, lead, inorganic arsenic, and 16 carcinogenic organics. (A proposed regulation on benzene was revoked by the courts.) Many additional substances are being scrutinized as potential occupational carcinogens (138) (see also Regulatory agencies).

Nearly every field of industrial chemistry has its own professional journal (eg, iron and steel, metallurgy, ceramics, plastics, gas and fuel) that should be consulted for current trace problems within that field.

BIBLIOGRAPHY

1. G. H. Morrison, ed., *Trace Analysis: Physical Methods*, Interscience Publishers, a division of John Wiley & Sons, Inc., New York, 1965.
2. W. W. Meinke and B. F. Scribner, eds., *Trace Characterization, Physical and Chemical*, NBS Monograph No. 100, National Bureau of Standards, U.S. Department of Commerce, Washington, D.C., 1967.

3. I. M. Kolthoff and P. J. Elving, *Treatise on Analytical Chemistry: Part I. Theory and Practice; Part II. Analytical Chemistry of Inorganic and Organic Compounds; Part III. Analytical Chemistry in Industry*, 1st ed., Interscience Publishers, a division of John Wiley & Sons, Inc., New York, 1959–1980.

4. P. W. West and F. K. West, *Anal. Chem.* **40**, 138R (1968).

5. N. B. Hannay in ref. 1, pp. 25–66.

6. W. H. Allaway in ref. 1, pp. 67–102.

7. V. P. Guinn and H. R. Lukens, Jr., in ref. 1, pp. 325–376; D. F. S. Natusch in S. B. Parker, ed., *Encyclopedia of Environmental Science*, McGraw-Hill, New York, 1980.

8. A. A. Smales in ref. 2, pp. 307–336.

9. S. G. Schulman and R. J. Sturgeon in J. W. Munson, ed., *Pharmaceutical Analysis: Modern Methods*, Marcel Dekker, New York, 1981, Part A, pp. 229–323.

10. A. J. Bard and L. R. Faulkner, *Electrochemical Methods. Fundamentals and Applications*, John Wiley & Sons, Inc., New York, 1980, Chapt. 10.

11. H. A. Laitinen in ref. 2, pp. 69–109.

12. A. K. Covington, *Ion-Selective Electrode Methodology*, Vols. I–II, CRC Press, Boca Raton, Fla., 1979.

13. Ref. 1, pp. 136–142.

14. B. J. Kline and W. H. Soine in ref. 9, p. 36.

15. L. R. Snyder and J. J. Kirkland, *Introduction to Modern Liquid Chromatography*, 2nd ed., Interscience Publishers, a division of John Wiley & Sons, Inc., New York, 1979.

16. J. F. Lawrence, *Organic Trace Analysis by Liquid Chromatography*, Academic Press, New York, 1981.

17. O. Samuelson, *Ion Exchange Separations in Analytical Chemistry*, John Wiley & Sons, Inc., New York, 1963.

18. E. Sawicki, J. D. Mulik, and E. Wittgenstein, eds., *Ion Chromatographic Analysis of Environmental Pollutants*, Ann Arbor Science, Ann Arbor, Mich., 1978.

19. A. J. Ahearn, ed., *Trace Analysis by Mass Spectrometry*, Academic Press, New York, 1972.

20. E. J. Cone in ref. 9, pp. 143–227.

21. B. L. Karger, D. P. Kirby, P. Vouros, R. L. Foltz, and B. Hidy, *Anal. Chem.* **51**, 2324 (1979).

22. E. A. Jenne, ed., *Chemical Modeling in Aqueous Systems, Speciation, Sorption, Solubility, and Kinetics*, ACS Symposium Series 93, American Chemical Society, Washington, D.C., 1979.

23. H. P. Guy, *Techniques of Water-Resources Investigation, Book 5, Chapt. C1, Laboratory Theory and Methods for Sediment Analysis*, U.S. Geological Survey, U.S. Department of the Interior, Washington, D.C., 1969.

24. R. M. Perhac, *J. Hydrol.* **15**, 177 (1972).

25. W. P. Bonner, T. Tamura, C. W. Francis, and J. W. Amburgey, Jr., *Environ. Sci. Technol.* **4**, 824 (1970).

26. J. Minczewski, J. Chwastowska, and R. Dybczynski, *Separation and Preconcentration Methods in Inorganic Trace Analysis*, Halsted Press, a division of John Wiley & Sons, Inc., New York, 1982.

27. J. F. Thompson, ed., *Manual of Analytical Methods for the Analysis of Pesticide Residues in Human and Environmental Samples*, U.S. Environmental Protection Agency, Environmental Toxicology Division, Research Triangle Park, N.C., June 1977 revision.

28. R. B. Grieves in ref. 3, 2nd ed., Part I, Vol. 5, 1982, Chapt. 9.

29. *Fed. Regist.* **44**, 69464 (1979): methods 601–603 and 624 for purgeable organics (proposed amendments to 40 CFR Part 136).

30. 40 CFR Part 141.30.

31. S.-T. Hwang and K. Kammermeyer in ref. 3, 2nd ed., Part I, 1982, Chapt. 5.

32. D. C. K. Lin and co-workers, *Prepr. Pap. Nat. Meet. Div. Environ. Chem. Am. Chem. Soc.* **20**(2), 397 (Aug. 1980).

33. G. H. Morrison and H. Freiser, *Solvent Extraction in Analytical Chemistry*, John Wiley & Sons, Inc., New York, 1957.

34. H. Freiser in B. L. Karger and co-workers, eds., *An Introduction to Separation Science*, Interscience Publishers, a division of John Wiley & Sons, Inc., New York, 1973.

35. H. M. N. H. Irving in ref. 3, 2nd ed., Part I, Vol. 5, 1982, Chapt. 11.

36. Ref. 29, Methods 604–613, 625.

37. J. Stary, *The Solvent Extraction of Metal Chelates*, Pergamon Presss, New York, 1964.

38. W. Horwitz, ed., *Official Methods of Analysis*, 13th ed., Association of Official Analytical Chemists, Washington, D.C., 1980.

39. M. Katz, ed., *Methods of Air Sampling and Analysis*, American Public Health Association, Washington, D.C., 1977.

40. R. S. Braman and D. L. Johnson, *Environ. Sci. Technol.* **8**, 996 (1974).

41. R. G. Webb, *Isolating Organic Water Pollutants: XAD Resins, Urethane Foams, Solvent Extraction*, EPA-660/4-75-003, U.S. Environmental Protection Agency, Environmental Research Laboratory, Athens, Ga., June 1975; F. W. Karasek, R. E. Clement, and J. A. Sweetman, *Anal. Chem.* **53**, 1050A (1981).

42. D. G. Goerlitz and L. M. Law, *J. Assoc. Off. Anal. Chem.* **57**, 176 (1974).

43. *Pesticide Analytical Manual. Vol. I, Methods which Detect Multiple Residues; Vol. II, Methods for Individual Pesticide Residues*, Food and Drug Administration, Bureau of Foods, Washington, D.C., 1968, revised periodically.

44. T. M. Florence and G. E. Batley, *Talanta* **23**, 179 (1976); P. Figura and B. McDuffie, *Anal. Chem.* **49**, 1950 (1977).

45. M. De Lafayette, W. E. Bernier, L. Karm, and G. E. Janauer, *Anal. Chim. Acta* **124**, 365 (1981).

46. L. Gordon, M. L. Salutsky, and H. H. Willard, *Precipitation from Homogeneous Solution*, John Wiley & Sons, Inc., New York, 1959.

47. H. Stephen and T. Stephen, eds., *Solubilities of Inorganic and Organic Compounds*, Pergamon Press, Oxford, UK, 1963.

48. W. R. Wilcox and J. Estrin in ref. 3, 2nd ed., Part I, Vol. 5, 1982, Chapt. 8.

49. W. G. Pfann, *Zone Melting*, 2nd ed., John Wiley & Sons, Inc., New York, 1976; L. N. Jones and B. McDuffie, *Anal. Chem.* **41**, 65 (1969).

50. Ref. 30, Appendix C.

51. Ref. 35, p. 533.

52. J. Josephson, *Environ. Sci. Technol.* **16**, 548A (1982).

53. Ref. 39, pp. 168–184.

54. J. P. Ryan and J. S. Fritz in ref. 32, p. 280.

55. J. D. Winefordner, ed., *Trace Analysis: Spectroscopic Methods for the Elements*, Wiley-Interscience, New York, 1976.

56. D. E. Leyden and W. Wegscheider, *Anal. Chem.* **53**, 1059A (1981).

57. E. W. Wolff and co-workers, *Anal. Chem.* **53**, 1566 (1981).

58. Ref. 40, p. 999.

59. 40 CFR Part 50 Appendix A.

60. M. Zief and J. W. Mitchell, *Contamination Control in Trace Element Analysis*, Wiley-Interscience, New York, 1976.

61. ACS Committee on Analytical Reagents, *Reagent Chemicals: American Chemical Society Specifications*, 6th ed., American Chemical Society, Washington, D.C., 1981.

62. Ref. 60, p. 104.

63. H. Irving and J. J. Cox, *Analyst* **83**, 526 (1958).

64. "1982–83 LabGuide," *Anal. Chem.* **54**(10), 242 (1982); section on laboratory chemicals lists 40 vendors of high purity solvents.

65. R. S. Bingham, Jr., *Development and Utilization of a Quality Control Program*, in ref. 3, Part III, Vol. 3, 1976, pp. 127–253.

66. H. F. Beeghley and co-workers, *Standard Materials for Analysis and Testing*, and S. S. Kurtz, Jr., *Standard Methods of Chemical Analysis and Standard-Setting Organizations*, in ref. 3, Part III, Vol. 3, 1976.

67. S. W. Karickhoff and D. S. Brown, *Determination of Octanol/Water Distribution Coefficients, Water Solubilities, and Sediment/Water Partition Coefficients for Hydrophobic Organic Pollutants*, EPA-600/4-79-032, U.S. Environmental Protection Agency, Environmental Research Laboratory, Athens, Ga., April 1979.

68. M. L. Jackson, *Soil Chemical Analysis*, Prentice-Hall, Englewood Cliffs, N.J., 1958.

69. R. F. Korcak and D. S. Fanning, *J. Environ. Qual.* **7**, 506 (1978).

70. B. A. Malo, *Environ. Sci. Technol.* **11**, 277 (1977).

71. T. T. Gorsuch, *Analyst* **84**, 135 (1959).

72. D. C. Bogen, Chapt. 1, and E. C. Dunlop and C. R. Ginnard, Chapt. 2, in ref. 3, 2nd ed., Part I, Vol. 5, 1982.

73. R. W. Moshier and R. E. Sievers, *Gas Chromatography of Metal Chelates*, Pergamon Press, New York, 1965.
74. R. W. Frei and J. F. Lawrence, *Chemical Derivatization in Analytical Chemistry, Vol. 1: Chromatography*, Plenum Press, New York, 1981.
75. D. D. Perrin, *Masking and Demasking of Chemical Reactions*, Interscience Publishers, a division of John Wiley & Sons, Inc., New York, 1970.
76. A. Mizuike in ref. 1, pp. 103–159.
77. J. C. Giddings in ref. 3, 2nd ed., Part I, Vol. 5, 1982, Chapt. 3.
78. E. B. Sandell and H. Onishi, *Photometric Determination of Traces of Metals*, 4th ed., John Wiley & Sons, Inc., New York, 1978.
79. N. W. Tietz, ed., *Fundamentals of Clinical Chemistry*, Saunders, Philadelphia, Pa., 1976.
80. H. H. Willard, L. L. Merritt, Jr., J. A. Dean, and F. A. Settle, Jr., *Instrumental Methods of Analysis*, 6th ed., Van Nostrand, New York, 1981, p. 314.
81. Ref. 80, Chapts. 9–10.
82. S. E. Manahan, *Environmental Chemistry*, 3rd ed., Willard Grant Press, a division of Wadsworth Publishing Co., Belmont, Calif., 1979, p. 227.
83. A. M. Bond, *Modern Polarographic Methods in Analytical Chemistry*, Marcel Dekker, New York, 1980; S. A. Borman, *Anal. Chem.* **54,** 698A (1982).
84. J. B. Flato, *Anal. Chem.* **44**(11), 75A (1972).
85. Ref. 80, pp. 464–478.
86. K. Grob, *Chromatographia* **8,** 426 (1975).
87. M. Novotny, *Anal. Chem.* **50,** 16A (1978).
88. Ref. 29; Methods 613, 624–625 use gc–ms to measure priority pollutants (dioxin, purgeables, and base-neutrals + acids + pesticides, respectively) in aqueous environmental samples after preliminary separation and cleanup procedures.
89. Ref. 20, p. 159.
90. G. H. Morrison, *Anal. Chem.* **43**(7), 22A (1971).
91. *NBS Standard Reference Materials Catalog*, NBS Special Publication 260, 1979–1980 ed., National Bureau of Standards, U.S. Department of Commerce, Government Printing Office, Washington, D.C., 1979; updated to 1982.
92. R. R. Watts, ed., *Analytical Reference Standards and Supplemental Data for Pesticides and Other Organic Compounds*, EPA-600/2-81-011, U.S. Environmental Protection Agency, Health Effects Research Laboratory, Research Triangle Park, N.C., Jan. 1981.
93. *Directory of Certified Reference Materials: Sources of Supply and Suggested Uses*, International Organization for Standardization, Geneva, 1982 (available from ANSI, New York); IAEA, the International Atomic Energy Agency, sponsors collaborative studies and prepares standard nuclear materials, stable isotopes, and environmental and clinical samples.
94. Ref. 64, pp. 242–244, lists many vendors of standards; ref. 38, lists vendors of standard pesticides (p. 72) and mycotoxins (p. 415).
95. Ref. 38, p. xii.
96. W. Horwitz, *Anal. Chem.* **54,** 67A (1982).
97. 21 CFR Chapter 1.
98. 40 CFR Part 180C.
99. R. D. Johnson and co-workers, *Pestic. Monit. J.* **15,** 54 (1981).
100. "U.S. Bans Flame Retardant Used in Children's Sleepwear," *New York Times*, 14 (April 8, 1977).
101. B. Hileman, *Environ. Sci. Technol.* **16,** 543A (1982).
102. C. A. Black, ed., *Methods of Soil Analysis*, Agronomy 9, American Society of Agronomy, Madison, Wisc., 1965; see also ref. 68.
103. W. R. Faulkner and co-workers, eds., *Selected Methods for the Small Clinical Chemistry Laboratory*, Vol. 9, American Association of Clinical Chemists, 1982; also, G. R. Cooper, ed., *Selected Methods of Clinical Chemistry*, Vol. 8, Academic Press, New York, 1977; Vols. 1–7 published as *Standard Methods of Clinical Chemistry*, Academic Press, New York, 1953–1972.
104. W. D. Odell and W. H. Daughaday, *Principles of Competitive Protein-Binding Assays*, Lippincott, New York, 1971; G. E. Abraham, *Handbook of Radioimmunoassay*, Marcel Dekker, New York, 1977.
105. J. A. Lott and J. M. Stang, *Clin. Chem.* **26,** 1241 (1980).
106. G. R. Warnick, J. Benderson, and J. J. Albers, *Clin. Chem.* **28,** 1379 (1982).

107. P. Schramel, ed., *Trace Elements: Analytical Chemistry in Medicine & Biology*, De Gruyter, Hawthorne, N.Y., 1980.
108. Tanaka and DuPont, *Trace Elements in Clinical Biochemistry*, De Gruyter, Hawthorne, N.Y., 1979.
109. H. A. Moye and S. J. Scherer, *Anal. Lett.* **10**, 1052 (1977).
110. E. Uschin and M. Sandler, *Trace Amines in the Brain*, Marcel Dekker, New York, 1976.
111. D. R. Greenwood and co-workers, *A Handbook of Key Federal Regulations and Criteria for Multimedia Environmental Control*, EPA-600/7-79-175, U.S. Environmental Protection Agency, Industrial Environmental Research Laboratory, Research Triangle Park, N.C., Aug. 1979, 288 pp.
112. A. C. Stern, ed., *Air Pollution. Vol. III, Measuring, Monitoring, and Surveillance of Air Pollution*, 3rd ed., Academic Press, New York, 1976.
113. R. K. Stevens and co-workers, *Environ. Sci. Technol.* **14**, 1491 (1980).
114. K. A. Rahn, *Atmos. Environ.* **15**, 1457 (1981).
115. D. E. Newberry in T. Y. Toribara and co-workers, eds., *Environmental Pollutants., Detection, and Measurement*, Plenum Press, New York, 1978, pp. 317–348.
116. *Ibid.*, p. 319.
117. C. F. Bauer and D. F. S. Natusch, *Anal. Chem.* **53**, 2020 (1981).
118. T. Vo-Dinh and co-workers, *Anal. Chem.* **53**, 253 (1981).
119. 40 CFR Part 50.
120. Ref. 82, p. 395.
121. Ref. 39, p. 701.
122. Ref. 39, Method 834.
123. 40 CFR Part 136 (contains a list of methods in EPA regulations).
124. *Standard Methods for the Examination of Water and Wastewater*, 14th ed., American Public Health Association, Washington, D.C., 1975.
125. *Methods for Chemical Analysis of Water and Wastes*, EPA-600/4-79-020, U.S. Environmental Protection Agency, Environmental Monitoring and Support Laboratory, Cincinnati, Ohio, revised ed. March 1979.
126. Ref. 16, p. 225; reset from R. W. Frei, *Int. J. Environ. Anal. Chem.* **5**, 150 (1978).
127. G. E. Janauer in W. J. Cooper, ed., *Chemistry in Water Reuse*, Vol. 1, Ann Arbor Science, Ann Arbor, Mich. 1981, pp. 501–522.
128. *Fed. Regist.* **44**, 53438 (Sept. 13, 1979).
129. *Environmental Assessment. Short-Term Tests for Carcinogens, Mutagens and Other Genotoxic Agents*, EPA-625/9-79-003, U.S. Environmental Protection Agency, Environmental Research Information Center, Cincinnati, Ohio, July 1979.
130. G. M. Williams, *Prepr. Pap. Nat. Meet. Div. Environ. Chem. Am. Chem. Soc.* **20**(2), 377 (Aug. 1980).
131. J. H. Richardson and R. V. Peterson, eds., *Systematic Materials Analysis*, Vols. I–IV, Academic Press, New York, 1974–1978.
132. J. P. Cali, ed., *Trace Analysis of Semi-Conductor Materials*, Pergamon Press, New York, 1964.
133. L. R. Weisberg in ref. 2, pp. 39–74.
134. *1981 Annual Book of Standards*, in 48 vols., American Society Testing Materials, Washington, D.C., 1981.
135. *Official and Tentative Methods*, American Oil Chemists' Society, Champaign, Ill., 1973, and revision.
136. T. S. Harrison, *Handbook of Analytical Control of Iron and Steel*, John Wiley & Sons, Inc., New York, 1979.
137. 29 CFR Part 1910.1000–1045.
138. *Chem. Eng. News*, 23 (Aug. 25, 1980).

General References

References 1, 2, 3, and 80 are also general references.

Methods of trace and residue analysis are included in most treatises on analytical chemistry or in texts on instrumental methods of analysis (see Analytical methods in Volume 2 and Supplement):

F. J. Welcher, ed., *Standard Methods of Analysis*, 6th ed., 5 vols., D. Van Nostrand, Princeton, N.J., 1962–1975.

F. D. Snell and L. S. Ettre, *Encyclopedia of Industrial Chemical Analysis*, 20 vols., Interscience Publishers, a division of John Wiley & Sons, Inc., New York, 1966–1974.

A number of monographs specifically devoted to the subject of trace analysis have appeared:

J. H. Yoe and H. J. Koch, eds., *Trace Analysis*, John Wiley & Sons, Inc., New York, 1957.

V. Valkovic, *Trace Element Analysis*, Halsted Press, a division of John Wiley & Sons, Inc., New York, 1975.

R. D. Reeves and R. R. Brooks, *Trace Element Analysis of Geological Materials*, Wiley-Interscience, New York, 1978.

J. F. Lawrence, *Trace Analysis*, Vols. 1 (Dec. 1981) and 2 (Dec. 1982), Academic Press, New York.

Articles or details on analytical methodology are given in:

D. Purves, *Trace Element Contamination of the Environment*, Elsevier, Amsterdam, 1977.

F. A. Gunther, ed., *Residue Reviews*, 80 vols., Springer-Verlag, Berlin, 1962–1980. One complete volume (Vol. 5, 1964) is on analysis.

Papers on analysis are contained in:

D. D. Hemphill, ed., *Trace Substances in Environmental Health*, Annual Symposia, University of Missouri, Columbia, Mo., Symp. XV, 1981.

Current research is found in the following fundamental journals as well as in journals devoted to types of methods or specific applied fields:

Anal. Chem.

Analyst

Anal. Chim. Acta

Anal. Lett.

Talanta

Rich sources of references covering most trace-analytical methods are *Fundamental Reviews* and *Applications Reviews*, the biennial reviews in *Anal. Chem.*

BRUCE MCDUFFIE
State University of New York at Binghamton

TRACERS. See Radioactive tracers.

TRADEMARKS AND COPYRIGHTS

Trademarks,   348
Copyrights,   360

TRADEMARKS

A trademark, as defined in the Federal statutes, includes any word, name, symbol, or device, or any combination thereof, adopted and used by a manufacturer or merchant to identify his or her goods and distinguish them from those manufactured or sold by others. Related to trademarks are service marks, certification marks, and collective marks. Service marks are names or symbols used to identify the services of one party and distinguish them from the services of others. Certification·marks are used in connection with the products or services of one or more parties other than the owner of the mark to certify the product's origin, material, mode of manufacture, quality, accuracy, or other characteristics, or that the work or labor on the product was performed by members of a union or other organization. Collective marks are trade or service marks used by members of a cooperative, association, or other collective group. They include marks indicating membership in a union, association, or other organization. Principles applicable to service, certification, and collective marks are generally similar to those applicable to trademarks.

Ownership of a trademark confers the exclusive right to use it or authorize its use in connection with the goods of its owner, or goods made by others to the owner's quality standards. Conversely, it also confers the right to prevent others from using the mark (or another mark so similar as to be confused for the original) on similar or related goods. Trademark protection is a type of restraint afforded by the courts against unfair competition.

In the United States, and generally in other common-law countries, the right to protection of a trademark arises independently of statutory provisions. Both the public and the owner have an interest in such protection. The owner's interest lies in protecting the value of the mark as a device for attracting and retaining customers for the goods sold under the mark, based on publicity gained through use of the mark in advertising and the reputation for quality of the goods on which it is used. The public interest lies in avoiding the confusion, deception, and fraud that would be likely as a result of unauthorized use of the mark by others.

The term trade name is often used colloquially as a synonym for trademark. Technically, a trade name includes individual names and surnames, firm names, and other names used by manufacturers, industrialists, merchants, agriculturists, and others to identify their businesses, vocations, or occupations. Thus, "General Electric"—meaning the corporation—is a trade name, but when applied to merchandise, eg, "General Electric refrigerators," it is a trademark.

Trademarks Distinguished from Patents and Copyrights. A trademark is acquired by use on or in connection with the goods of its owner. It is accorded legal protection at common law independent of statutory provisions or registration. Registration, available under the laws of the states as well as under Federal statute, is a procedure whereby protection of the trademark right is enhanced or facilitated, but acquisition of the trademark right by use in connection with the goods is a prerequisite to registration (see Patents, practice and management).

Patents, which afford an exclusive right to use an invention, are creatures of statute. They are acquired by disclosing an invention in an application duly filed and prosecuted in accord with the patent laws. Copyrights, which afford in most cases an exclusive right to control reproduction, adaptation, public performance, or public display of literary or artistic works, arise automatically upon creation of a work, but are dependent on Federal statute (see below).

Authority for protection of patents and copyrights is expressly set out in the Constitution and is the exclusive province of the Federal government. Federal legislation to protect trademarks is based on the authority of Congress under the commerce clause of the Constitution to regulate interstate and foreign commerce of the United States; protection afforded by the individual states is based on their power to regulate intrastate commerce.

History

Trademarks were known in ancient civilizations. Wine jars found in Pompeii bore the mark "Vesuvinum," identifying their contents as a wine produced in the region of Vesuvius. The craftsmen's guilds of Europe adopted marks to identify merchandise originating with their members, and such marks were protected against falsification by penal laws.

An early trademark statute was adopted in 1590. In 1883, the UK statute provided for registration of "fancy" words as trademarks. In the United States, the first Federal statute, adopted in 1870, was declared unconstitutional in 1879. Subsequent Federal statutes were adopted in 1881, 1905, and 1920, and the present comprehensive statute—known as the Lanham Act—was passed in 1946.

Selection of Trademarks

Since it is the function of trademarks to identify the merchandise of the owner of the mark as distinguished from the merchandise of others, it is essential in selecting a mark that it be sufficiently distinct from other trademarks for similar merchandise to avoid confusing or deceiving customers as to the source of origin of the goods. Moreover, any designation that would commonly be used by the public in describing the merchandise or its functions or properties cannot be preempted by one manufacturer to the exclusion of the public. Thus, a mark for which exclusive protection is sought cannot be purely descriptive of the merchandise. Ideally, therefore, a mark should be preferably fanciful or arbitrary and adequately distinguished from marks of others for similar merchandise to avoid likelihood of confusion as to source of origin in commerce (1).

In order to ascertain whether a prospective mark is distinct enough from marks of others, it is customary, before adopting a mark, to make a search in the fields of commerce in which its use is contemplated. Since most marks of importance are registered under the Federal statutes in the U.S. Patent and Trademark Office, this register provides a principal field of search. The search can be made in the Patent and Trademark Office under a classification of the goods for which marks are registered. A complete search requires, in addition, a search of marks registered under state laws and of marks unregistered but in use as indicated in various trade directories, telephone books, and the like. There exist a number of organizations competent to provide a search including these additional fields, and use of their services is highly advisable as a preliminary to adoption of a new mark (see also Patents, literature).

Although a complete search is ideal, is there any way short of that to determine whether or not a mark under consideration for adoption, or any mark resembling it, has already been adopted for goods of a similar character? A useful source of information for this purpose is the annual publication of *The Trademark Register of the United States* by the Trademark Register, Washington, D.C. The editions list the marks on the *Federal Register* (marks registered in the U.S. Patent and Trademark Office) in alphabetical order under the various classes of merchandise and services together with the registration number and date. Full details concerning the goods, application serial number, date of application, and alleged date of first use can be obtained by ordering a copy of the registration from the U.S. Patent and Trademark Office or from the *Official Gazette* of the U.S. Patent and Trademark Office. This weekly publication, available in many libraries, lists all registrations numerically for the week in which they are issued. This procedure can be used as preliminary for screening proposed marks to avoid conflicts. Other reference works that may be helpful in a preliminary survey are the *Handbook of Material Trade Names* (O. T. Zimmerman and I. Lavine, 1953 ed., and Supplements), *Merck Index*, *Thomas Register*, *Physicians' Desk Reference*, and the *Condensed Chemical Dictionary.*

The standards used to determine whether a prospective mark conflicts with an existing third party mark are described below. Principal considerations are twofold: first, similarity of the marks themselves and second, and equally important, the relationship among the kinds of merchandise for which they are used or are intended to be used (2). Without the latter, even identical marks could be unlikely to cause confusion. Similarity between marks may exist in appearance, phonetics, connotation, or any combination of these qualities. Similarity of an accented or initial syllable, or otherwise dominant portion of a mark may be an important factor in such determinations. However, in the final analysis, confusing similarity is determined with respect to the marks as a whole (3). Thus, in a 1968 decision, the Court of Customs and Patent Appeals reversed the Patent and Trademark Office Trademark Trial and Appeal Board in a conflict involving the existing mark Thermopane and a mark Therm-O-Proof sought to be registered, in each case for similar window glass. The court stated that the question was likelihood of confusion not between Pane and Proof but between the names Thermopane and Therm-O-Proof as entireties "taking into account the [purchasing public's] normal fallibility of memory over what may be a considerable period of time." The *Trademark Reporter*, a publication of the U.S. Trademark Association, carries a revealing tabulation in its monthly issues and in its yearly index of marks involved in decisions of the Patent Office and the courts, concerning confusing similarity of the marks as applied to the goods for which they are intended.

It is clear that the nature of the goods for which the mark is used or intended to be used is an essential element in determining the likelihood of confusion between otherwise similar marks. If the goods are of such nature that a purchaser or prospective customer would be likely to assume that the goods have a common source of origin, then a basis for confusion exists.

For example, Iothane, a plastic protective finish for wood surfaces, was held confusingly similar to Diothane, a synthetic resin protective for concrete; RX for plastic thermosetting molding compounds was held confusingly similar to the same mark for latex compounds; Diacon was held similar to Dynacon, both marks for electrical instruments; and Norseman for ice makers was held similar to Norge and Norse for refrigerators.

Pairs of goods held to be such as to give rise to the likelihood of confusing similarity included soft drinks and containers; beauty preparations and soaps, cleaners and lotions; doors and chairs; lighting fixtures and furniture; and guns and steel products. The degree of fame enjoyed by a mark can add to its scope, bringing a broader range of goods and services under its protection.

In addition to the question of possible conflict with other trademarks, it is also necessary to consider whether a proposed new mark has such descriptive character with respect to the goods for which it is intended as to render the mark incapable of distinguishing the goods of the owner from such goods in general. A generic term cannot be preempted as against the right of the public to use it. However, if a term, though descriptive, is used in an unusual sense and long enough and extensively enough that the public understands it as a trademark, then indeed it may function and be protected as a trademark. It is said that the term has acquired secondary meaning. Descriptive marks held registrable include, for example, Lens Bright for eyeglass-cleaning fluid; Twist Off for bottle closures; Kitty Litter for ground clay for small-animal litter; Curv' for permanent-wave solution; Skinvisible for transparent tape. Marks held incapable of trademark function are, for example, jet ignition for stokers; ink sac for a plastic ink dispenser; two 2-year for light bulbs guaranteed to burn for two years; no damage for freight-car service; B-100 for vitamin B complex.

Names of persons—especially surnames—are in the same class as descriptive terms that can nevertheless function as trademarks. When used exclusively for merchandise of a given class to such an extent that the public understands them as trademarks, they acquire secondary meaning and are accorded protection. Thus, Ford and DuPont have long since been recognized as trademarks, despite the fact that they are surnames. Other names or parts of names that have become trademarks are Calvin Klein, Wrigley's and McDonald's.

Marks that are not descriptive, but rather *suggest* the character, quality, or function of the goods are much favored as they are especially useful in advertising and encourage easy recall on the part of the public. Examples include Talon for slide fasteners and Ivory for soap. Such marks are susceptible of protection in the first instance without acquisition of secondary meaning.

The final class of marks is those that are wholly arbitrary with reference to the goods for which they are to be used. They may be words of the language having a meaning that bears no relationship to the goods for which they are used (eg, Camel for cigarettes and Venus for pencils), or they may be synthetic terms having no significance as such (such as Kodak for cameras, Yuban for coffee, or Qiana for fabric). These marks have the advantage of being least likely to lose their trademark status by becoming the generic name of the product, thereby entering the public domain.

In addition to letter combinations or numbers (such as 4711 on cologne and IBM for business machines), a trademark may be a logotype, or a design or symbol such as the flying red horse of the Mobil Oil Company, the configuration or contour of a container such as the Coca-Cola bottle, or a series of notes such as the NBC signature.

The Role of the Trademark as a Basis for Legal Protection

The functions of trademarks may be characterized as an identification of origin, a guarantee of quality, and an advertising device. In identifying origin, a trademark affords a convenience to the owner and customers alike by minimizing the likelihood of confusion of the goods of the trademark owner with those of others. In guaranteeing quality, it assures the purchaser that goods bearing the mark will have the same quality as those previously purchased under the same mark. In both instances, the public interest is involved as a basis for protection by the courts to avoid deception or confusion of the public. In advertising, the trademark offers a convenient shorthand device upon which to focus the attention and, the advertiser hopes, the good will of the public upon the product as a reward for the considerable expense involved in advertising. Protection of the trademark represents protection by the courts of a property right of the trademark owner.

Trademark Registration

As indicated above, the right to trademark protection arises through exclusive use of the mark in connection with the goods of the owner, and this right is protected by common law independent of registration. Registration of trademarks is a statutory creation, which affords a means of publicizing a claim to a trademark right and facilitating its protection. The Federal statute (Trademark Act of 1946) provides a Principal Register for marks fulfilling all requirements for full registration and a Supplemental Register for those lacking certain of the requirements for registration on the Principal Register. Registration on the Principal Register constitutes constructive notice of the claim of ownership of the mark. Federal registration, in most cases, extends protection to all parts of the United States, its territories, and possessions; a common-law trademark right, on the other hand, is effective only in the region where the mark has become known through use. Federal registration on either Register confers jurisdiction of trademark actions on the Federal courts. Thus, substantial advantages arise from Federal trademark registration.

A trademark is registrable on the Principal Register if it is in use in commerce subject to regulation by Congress—ie, interstate and foreign commerce of the United States—provided it does not (4):

1. comprise immoral, deceptive, or scandalous matter, or matter disparaging persons, institutions, beliefs, or national symbols; or

2. comprise the flag or coat of arms or other insignia of the United States or any state, municipality, or foreign nation; or

3. comprise the name, signature, or portrait of an individual without his or her written consent, or the name, signature, or portrait of a deceased President of the United States during the lifetime of his widow without her written consent; or

4. comprise a mark that so resembles an already registered mark or one previously used in the United States and not abandoned, as to be likely, when applied to the goods of the applicant, to cause confusion or mistake or to deceive (In some cases, concurrent registration, generally limited as to territory, may be permitted.); or

5. consist of a mark that, when applied to the goods of the applicant, is mere-

ly descriptive or deceptively misdescriptive, or primarily geographically descriptive or deceptively misdescriptive, or is primarily merely a surname. (Here exceptions are permitted if the mark has become distinctive of the applicant's goods in commerce by use over a period of time.)

The statute expressly provides that the mark may consist of any trademark, symbol, label, package, configuration of goods, name, word, slogan, phrase, surname, geographical name, numeral, device, or any combination of any of the foregoing, provided it is capable of distinguishing the applicant's goods or services. A mark failing to qualify for registration on the Principal Register, but accorded registration on the Supplemental Register, may be registered later on the Principal Register if it can be shown that it has become distinctive of the applicant's goods. Registration on the Supplemental Register, as distinguished from the Principal Register, does not constitute constructive notice of the claim of ownership by the registrant. It may serve, however, like registrations on the Principal Register, as a basis for obtaining registration in some countries that require home registration as a prerequisite, and it provides a priority date for purposes of soliciting registration in foreign countries under international treaties (5).

Concerning descriptive or suggestive character of proposed trademarks, a 1968 decision of the Court of Customs and Patent Appeals is illustrative. Sugar and Spice was held registrable for bakery goods. The court stated, "A mark that merely denotes ingredients, quality, or composition of an article is not capable of being adopted exclusively and used as a trademark, since for policy reasons, descriptive words must be left free for public use. On the other hand, a legal distinction—albeit often obscure—has been drawn between terms that are [merely descriptive and those that are] only suggestive of the goods. The terms sugar and spice, used individually, are well known and well understood by the purchasing public. However, when combined and used on bakery goods, we think they may function as an indication of more than a mere description of the goods" (6).

Registration Procedure. Registration procedure under the Trademark Act of 1946 involves filing an application stating the kind of commerce (interstate or foreign) in which the mark is in use, the date of first use and date of first use in commerce subject to regulation by Congress, the Register (Principal or Supplemental) on which registration is sought, the goods on which the mark is used, and the manner in which it is applied to the goods. The application must include a drawing that shows the mark in the form in which it is used, and five specimens (labels or other reasonable facsimiles such as photographs) that show the mark in the form in which it is applied to the goods. When specimens such as labels are not available, or the manner in which the mark is used does not lend itself to inclusion of specimens thereof in the application, photographs or other facsimiles may be used instead. In the case of a "word" mark, or letter or numerals, for which the design or form of the letters or numerals is not critical, the drawing may be typed, showing the mark in block letters. The application must be executed by the owner. The application is accompanied by the filing fee (7) and is transmitted to the Patent and Trademark Office. If an attorney is to prosecute the application (eg, respond to Patent and Trademark Office objections, file and argue appeals, defend against opposition and cancellation proceedings instituted by others, etc), a power of attorney is included.

After examination, any objections including rejection in view of prior registrations are communicated to the applicant in an office action, requiring a response within a limited period. The response may involve amendment of the application, an argument to overcome the grounds for rejection, or a showing of further facts. Final rejections of applications may be appealed to the Trademark Trial and Appeal Board, and so on, if necessary or desired, as described below in the case of adversary proceedings.

When the application is found to be allowable, it is published in the *Official Gazette* of the Patent Office, and within 30 days—or within such extension thereof as may be granted to a potential opposer—any person may file an opposition alleging the belief that he or she will be damaged by the registration, and the grounds of that belief. The application is then transferred to the jurisdiction of the Trademark Trial and Appeal Board, which fixes terms for taking testimony by the opposer and the applicant. The testimony is taken before an official (court reporter) authorized to do so, and representatives of the adverse party may cross-examine. The depositions and evidence are submitted by the reporter to the Patent and Trademark Office, and a final hearing is held before the Trademark Trial and Appeal Board, which renders a decision based on the testimony presented by the parties. This decision may be appealed to the Court of Appeals for the Federal Circuit or the applicant may have a remedy by civil action in the Federal district courts. If the parties are able to negotiate a settlement while the matter is pending, the opposition can be withdrawn.

If no opposition is filed, or if a decision favorable to the applicant is made in an opposition, the application is allowed, and the registration is issued in due course.

Should a party fail to file an opposition against a published application within the time allowed, the objection can be filed after issuance of the registration. This involves a proceeding for cancellation on grounds similar to those available for opposition. The procedure is similar but, in theory at least, the burden of proof, which devolves upon the party seeking cancellation, is more strictly construed than in an opposition.

Marks registered under the presently effective Trademark Act of 1946 are subject to cancellation within five years after registration. The same time period applies after publication of a notice in the *Official Gazette*, converting a registration under a prior trademark statute to one entitled to protection under the 1946 Trademark Act. Such conversions are allowed except in those instances in which the mark has become the common descriptive name of the article or substance, or has been abandoned, or is disqualified by one of the five criteria listed above. Cancellation after five years is available only in limited cases.

Marks registered only under prior trademark acts can be cancelled at any time, at the instance of someone who claims damage or potential damage by the registration of the mark. Certification marks can be similarly cancelled if the owner fails to exercise control over their use or misuses them (8).

The Trademark Act requires, for maintenance of a registration, that an affidavit or declaration under penalty of perjury be filed within the sixth year of registration showing that the mark is still in use in commerce (interstate or foreign commerce of the United States), subject to regulation by Congress, or that its nonuse is due to circumstances excusing nonuse and not to an intention to abandon the mark. If the document includes an allegation that a mark has been in substantially continuous use for five consecutive years following the date of registration, the right of the registrant to use of the mark for the goods or services for which it has been in such use becomes incontestable, unless it infringes a valid trademark right based upon continuous use antedating the publication of the mark in question, or becomes the generic name of the goods or abandoned, or was improperly registered.

The requirement for the filing of a declaration within the sixth year of registration and the availability of incontestability are limited to marks registered under the Trademark Act of 1946, and those marks registered under earlier acts that have been converted to registrations under the 1946 Act by appropriate application and publication. Marks registered only under earlier acts do not enjoy incontestable status and do not require the filing of an affidavit of use within six years after registration or renewal. All marks, however, require renewal at 20-year intervals after registration. Such renewal is effected by filing an application within six months before expiration or within three months after expiration, identifying the goods or services for which the mark is still in use, or showing that nonuse is due to circumstances excusing such nonuse and not indicating an intent to abandon the mark. The nonuse of marks for alcoholic beverages during the period of prohibition was held to be excused in this sense, and the marks were upheld when prohibition was repealed. Nonuse has also been excused during wartime exigencies and, more recently, during the course of litigation over the right to use the mark.

Effect of Registration.   The registrant of a mark under the Federal statutes is afforded a remedy by civil action in the Federal courts against unauthorized use of a reproduction, counterfeit, copy, or colorable imitation of the mark in connection with any goods or services for which such use is likely to cause confusion or mistake, or to deceive. A certificate of registration on the Principal Register of the 1946 Trademark Act, or republished from the Act of 1881 or 1905, is *prima facie* evidence of the validity of the registration, the distinctiveness of the mark, the registrant's ownership of the mark, and the exclusive right of the registrant to use of the mark in commerce.

If the mark has become incontestable as discussed above, the registration is conclusive evidence of such exclusive right as to the goods and services specified in the declaration conferring incontestability, except under certain well-defined circumstances, including among others, fraudulent registration, abandonment, and generic development of the mark.

The registrant of a trademark can enforce his or her right by civil action in Federal court against one who makes unauthorized use of the mark that is likely to cause confusion or mistake, or to deceive. However, the mere application of the infringing mark to labels, etc, does not afford a right to recover profits or damages, unless the violation was committed with knowledge of the intent to cause confusion or mistake, or to deceive. Printers and publishers are thus also liable for producing infringing labels, etc, but the remedy, if it is established that the infringement was an innocent one, is limited to an injunction against future printing or advertising.

A registrant prevailing in a Federal court action against an infringer is granted an injunction which is enforceable in any District Court of the United States. In addition, the plaintiff may recover the defendant's profits, damages sustained by the plaintiff, and costs. The court may, in its discretion, assess up to triple damages. At the same time, the court may order the destruction of all labels, signs, prints, packaging, wrappers, receptacles, and advertising in the possession of the defendant bearing the infringing mark and all plates, molds, matrices, or the like for producing them.

Whereas registration of a mark on the Principal (but not the Supplemental) Register of the 1946 Trademark Act, as well as registration under the Trademark Act of 1881 or 1905, constitutes constructive notice of the owner's claim to the mark, damages and profits are available only from the time of actual notice to an infringer, unless the owner displays with the mark as used, as notice of registration in the form specified in the Statute, ie, the words "Registered in U.S. Patent and Trademark Office," or the abbreviation "Reg. U.S. Pat. & Tm. Off.," or the letter "R" enclosed in a circle. In such cases, damages and profits may be recovered for infringing acts occurring within three years prior to institution of a civil action for infringement, ie, for those not precluded by the Statute of Limitations.

Another valuable right deriving from registration of a mark under the 1946 Trademark Act is the prohibition against importation of merchandise that copies the registered mark. Under the regulations of the Department of the Treasury, the registration may be recorded with the Customs authorities, and merchandise bearing the infringing mark is subject to seizure and exclusion. The recording procedure includes supplying the Bureau of Customs with a specified number of copies of the registration. These are distributed to the various ports of entry and border stations for reference and notice to the Customs inspectors (9).

Unlike trademarks, trade names are not registrable. The use of a trade name in a business in which such use is likely to cause confusion with a preexisting trademark is subject to remedies similar to those available against a confusingly similar trademark. Conversely, a preexisting trade name may serve as the basis for attack (eg, by opposition, cancellation or suit for unfair competition) upon a confusingly similar trademark.

Under the International Convention for the Protection of Industrial Property, to which the United States is a party, an applicant for registration of a trademark in the United States is entitled to a right of priority in other member countries from the date of the U.S. application if corresponding application is filed in such other country or countries within six months after the date of the U.S. application. Reciprocal rights are granted by the United States to nationals of other member countries. Nationals of such countries applying for U.S. registration based on a prior foreign registration need not allege use in commerce of the United States as a domestic applicant must but, in such case, the registrant may not recover for infringing acts prior to the date of the U.S. registration. Third party rights antedating the foreign priority date are not affected by the registration.

Transfer to Trademark Rights

A trademark and its registration may be assigned, but only together with the good will of the business in which the mark is used, or that part of the good will of the business connected with the use of and symbolized by the mark. The assignment must be in writing and duly executed. A verification sworn to by the party signing the assignment before a notary or other official authorized to administer oaths constitutes *prima facie* evidence of execution. The U.S. Patent Office maintains a record of assignments submitted to it for recording. Unless recorded within three months of its date, an assignment of a registered mark is void as against a good faith purchaser for value if the purchase precedes a subsequent recording in the Patent and Trademark Office.

A trademark right does not exist apart from the good will of the business, or part thereof, for which it is used. Accordingly, an attempt to transfer the right without the associated good will is ineffective. The owner, having attempted such transfer, has divested himself or herself of the right and thus effected an abandonment thereof. The intended transferee acquires no right based on prior ownership of the transferor.

At best, one may establish a new right to the mark based on one's own exclusive use thereof, subject, however, to prior rights of third parties.

A trademark may be licensed, but special rules apply to such licensing by reason of the nature and function of the trademark right. Thus, the public has an interest in the mark as a symbol of origin of the goods or services in the owner, and should be able to rely on the mark as indicating character and quality of the merchandise or services that the owner controls. Essential, therefore, to a valid license is provision for control by the licensor–owner of the quality of the goods or services with which the mark is to be used by the licensee. Not only is it necessary to provide that the licensor has the right to exercise such control, but it must be effectively exercised to validate the license. Without such an arrangement, the license is invalid and constitutes an abandonment of the trademark by the licensor, since the use by the purported licensee is inconsistent with the otherwise exclusive right of the owner to use the mark. On the other hand, when there is a license of which the validity is supported by provision for suitable quality control and by the exercise of such control, use by the licensee inures to the benefit of the licensor (10).

Proper Use of Trademarks

Proper use of a trademark by the owner or by his or her licensee under a valid license is essential to preservation of the exclusive trademark right. The 1946 statute provides, for example, that nonuse of two years constitutes *prima facie* abandonment of the trademark right. However, a trademark may also be lost through other circumstances. Thus, if it comes to be understood by the public, or the class of persons constituting the usual customers for merchandise sold under the mark, as a generic term for the product for which it is used, then the mark enters the public domain, and the exclusive right of the owners is lost. The danger increases as a mark becomes more popular. Examples of marks that have been lost or severely endangered in this manner are Aspirin for acetylsalicylates, Escalator for moving stairways, and Cellophane for regenerated cellulose film. Sometimes, a mark that has become generic may be reestablished by an assiduous advertising campaign, but the road is not an easy one. Thus, for an earlier generation, Singer became largely synonymous with sewing machine, but through vigorous publicity it is now reestablished as a trademark.

Various rules concerning use of trademarks have been devised to reduce the danger of degeneration of the mark into a generic term (11). These rules may be particularly significant with respect to chemical products because their actual generic names are frequently cumbersome or not referred to in common parlance (12). It is essential that a readily usable generic name be developed for new chemical products, and that the rules of proper use be followed in referring to the product and its trademark. The principle of these rules is generally to limit the use of the mark on goods or in advertising to one that is clearly consistent with trademark significance, ie, one that signifies origin of the goods with the trademark owner and that is inconsistent with use of the term to signify the goods themselves. Thus, the mark is appropriately used like an adjective followed by a generic term for the goods. It should be set off from its context by capitalization, distinctive lettering, quotation marks, or the like. It should not be used as a noun, and most important, not as a plural noun. It should not be used as a verb, or as a term characterizing a process or any other than the goods for which it is used. It should be clearly identified as a trademark in each context in which it is used. Thus, if it is registered under the Federal statute, its trademark status may be indicated by a circled "R," or by the legend (or its abbreviation) showing registration in the U.S. Patent and Trademark Office. If the mark is not registered, its trademark status may be indicated by an adjacent "TM" or by the term "Trademark" related, if appropriate, by asterisk to the mark itself. Preferably, the name of the trademark owner should also appear. Emphasis on the trademark status of the mark wherever it is used in commerce is the best means of preventing degeneration of the mark into a generic term for the goods (13).

Questions may arise as to the propriety of use of a mark by parties other than the owner in connection with goods originating with the owner and originally sold under the mark. Dealers are free to use the mark in advertising the goods for sale to customers, but they are not free to use the mark as a designation for a business establishment except to the extent this may be permitted by license from the trademark owner. Use of the mark in the resale of used or reconditioned goods bearing the mark of the manufacturer is not objectionable if the reconditioned or used status is clearly indicated to purchasers. Use of an ingredient mark on a composition containing the ingredient but compounded by a party other than the ingredient mark owner has been held permissible, provided no overemphasized display is given the mark and the goods are marked to show the proportion of the ingredient contained and to indicate that the composition has been manufactured by its producer having no connection with the owner of the ingredient mark.

State Registration

Registration of a trademark under state law is appropriate when the mark is to be used primarily in intrastate commerce. Procedure for registration generally parallels Federal procedure with considerable simplification. The office of the Secretary of State is usually the agency charged with granting trademark registration. State registration may be convenient in some cases—for example, when an owner desires to license or franchise one party in each state of a given territory without extending rights under the mark beyond the state boundary.

A number of states afford remedies against encroachments that may exceed those afforded under the Federal statutes, including especially a remedy against dilution of the mark, eg, by use or adoption of a mark approaching but not confusingly similar to the mark in question or by use of a similar mark on unrelated products (14).

Foreign Trademark Registration

The trademark laws of foreign countries differ in many fundamental respects from U.S. laws. Countries deriving their system of laws from the UK generally recognize use of a mark as the basis for the trademark right. Registration is permitted in some of these countries based on an allegation of intent to use the mark, and continuance of the registration is dependent upon a showing within a limited term that the intended use has been undertaken. Countries that derive their legal systems from continental Europe generally permit registration without use, although nonuse may be a ground for cancellation after a period of years.

Under the International Convention for Protection of Industrial Property, a trademark for which application for registration has been made in one member country can be registered in another member country with priority dating from the application in the first country if the later application is filed within six months and Convention priority is claimed. Twenty countries of the International Convention are also parties to a treaty for international trademark registration, known as the Madrid Arrangement. Under this treaty, a mark registered in the home country of the trademark owner can be internationally registered by filing with the International Trademark Bureau in Berne an application showing the home country registration. The Bureau communicates the resulting international registration to the registration authorities of the other member countries, which may accept or reject it in whole or in part. Subject to such limitations as are thereby imposed, the mark becomes effective as a registered mark in each member country under the provisions of local law governing registered trademarks. The term of an international registration is 20 years. The United States, Canada, the UK, and the Scandinavian countries are not members of the Madrid Arrangement, and registrants whose home registration is in these countries are not entitled to corresponding international registration under the Madrid Arrangement; the members of the latter are, in general, continental European countries and their overseas possessions or former possessions.

BIBLIOGRAPHY

"Trade-Marks" in *ECT* 1st ed., Vol. 14, pp. 281–301, by Robert Calvert, Patent Attorney; "Trademarks" in *ECT* 2nd ed., Vol. 20, pp. 578–596, by Walter G. Hensel, Patent Attorney.

1. United States Trademark Association, *Trademark Management*, Clark Boardman, Ltd., New York, 1981, pp. 1–54.
2. A. D. Brufsky, *Idea* **11,** 7 (1967).
3. R. F. Shryock, *Trade Market Reporter* **57,** 377 (1967).
4. Section 2, Trademark Act of 1946, 15 U.S. Code 1052.
5. *International Convention for the Protection of Industrial Property of 1883*, revised at Washington, D.C., in 1911, at the Hague in 1925, at London in 1934, and at Lisbon in 1958.
6. In Re Colonial Stores, Inc., 157 U.S.P.Q. 382, 394 F. 2d 549 (C.C.P.A. 1968).
7. *Fed. Regist.* **47,** 28063 (June 28, 1982) (37 CRF Parts 2 and 4).
8. Section 14, Trademark Act of 1946, 15 U.S. Code 1064.
9. *Trade Mark Reporter* **59,** 301 (1969).
10. Ref. 1, pp. 109–115.
11. Various companies publish information bulletins and management guides for the use of their trademarks, eg, *Trademarks, The Who What Where and How of Black & Decker Trademarks*, The Black and Decker Manufacturing Company, Towson, Md.; see generally, D. Fey, *The Practical Lawyer* **24**(8), 75 (1968), revised and reprinted by U.S. Trademark Association, New York, 1978.
12. *Chem. Week*, 28 (Jan. 13, 1982).
13. In the drug field, there is a need for simple and useful nonproprietary names for drugs. The USAN (United States Adopted Names) Council chooses each coined generic name for drugs used in the United States. See *USAN and the USP Dictionary of Drug Names*, U.S. Pharmacopeial Convention, Inc., Rockville, Md., 1979.
14. *The State Trademark Statutes*, a looseleaf book (with supplements) containing the trademark laws of the 50 states is available from the United States Trademark Association, New York.

General References

Trademark Rules of Practice of the Patent and Trademark Office with Forms and Statutes, U.S. Department of Commerce, Patent and Trademark Office, on sale by the Superintendent of Documents, U.S. Government Printing Office, Washington, D.C.
Official Gazette, published weekly by U.S. Department of Commerce, Patent and Trademark Office,

Washington, D.C. The trademark section is available separately. It lists marks accepted, subject to opposition, with details of the application and all marks actually registered.

U.S. Patents Quarterly, published weekly by the Bureau of National Affairs, Washington, D.C. It contains decisions of the courts and Patent and Trademark Office in patent, trademark, and copyright cases.

The Trademark Reporter, published monthly by the U.S. Trademark Association.

Journal of the Patent Office Society, published monthly, Patent Office Society, Washington, D.C.

G. Hawley, *Condensed Chemical Dictionary*, Reinhold Company, New York, 1981.

M. Windholz, ed., *Merck Index*, 9th ed., Merck & Co., Inc., Rahway, N.J., 1976.

Major Provisions of Trademark Legislation in Selected Countries, International Bureau of the World Intellectual Property Organization, Geneva, 1977.

Thomas Register of American Manufacturers, Thomas Publishing Company, New York, 1983.

Physicians' Desk Reference for Nonprescription Drugs, Medical Economics Company, Van Nostrand Reinhold, New York, annual publication.

Physicians' Desk Reference for Prescription Drugs, Medical Economics Company, Van Nostrand Reinhold, New York, annual publication.

Publications of the U.S. Trademark Association, New York: *Trademark Management—A Guide for Executives*, 1981; *Trademark Selection—The Management Team Method*, 1960; *Trademark Licensing—Domestic and Foreign*, 1962; *Trademarks in the Marketplace—Selection and Adoption; Proper Use and Protection*, 1964; *Trademarks in Advertising and Selling*, 1966; *1981–82 Trademark Law Handbook* (2 vols.), 1981, 1982; *1980–81 Trademark Handbook*, 1980; *The Protection of Corporate Names: A Country by Country Survey*, 1982; *Trademarks and Brand Management: Selected Annotations*, 1976.

Original (1969) by:
WALTER G. HENSEL
Patent Attorney

Current revisions (1982) by:
JON A. BAUMGARTEN
CHARLES H. LIEB
Paskus, Gordon & Hyman

COPYRIGHTS

Copyright is the right of the creator of a work, or of his or her successors in interest, to control certain uses of that work by others. The works that are the subject of copyright encompass a wide variety of products of intellectual endeavor. The right to control accorded by copyright is principally in the nature of an exclusive right, ie, the right to prohibit or permit a particular use of the subject work. After a work has acquired copyright, certain uses of that work by others are forbidden in the absence of the copyright owner's permission. In a few cases, however, the traditional nature of copyright as an exclusive right has been replaced by the principle of "compulsory license." Under this principle, certain uses of a work are permitted upon compliance with conditions that include royalties or fees established in the law itself, and the copyright owner's consent is not required. Originating, as the name implies, with the right to reproduce or copy, the rights embraced by copyright now extend to a variety of other uses such as performances, broadcasts, and other electronic transmissions.

Since January 1, 1978, copyright in the United States is wholly governed by a single Federal copyright statute, the Copyright Act of 1976. Enactment of this statute brought about major changes in the copyright law of the United States. The earlier Federal copyright statute had, for the most part, remained unchanged since 1909. The program for general revision of U.S. copyright law that culminated in the 1976 Copyright Act consumed more than 20 years of development and debate. It was particularly marked by Congress' attempt to deal with the revolutionary changes in the technologies of creating, reproducing, and disseminating intellectual works that had occurred since the first decade of the twentieth century (1). A symposium held by the American Chemical Society dealt in some detail with the effects of the new law on science and education (see General References).

There are two government agencies in the United States that exercise administrative authority with respect to copyright. The first is the U.S. Copyright Office, a department of the Library of Congress. The Copyright Office is principally responsible for registering claims to copyright and for recording transfers and licenses of rights under copyright. In short, it is an office of public record. The second agency is the Copyright Royalty Tribunal, an independent entity located in the legislative branch of government which draws administrative support from the Library of Congress. The Copyright Royalty Tribunal periodically adjusts the statutory royalties payable under the compulsory licenses established in the Copyright Act. In some cases, it also determines the proper allocation and distribution of those royalties among copyright owners. It is important to recognize that neither the Copyright Office nor the Copyright Royalty Tribunal grants copyrights. As will be described more fully below, copyright attaches to a work automatically upon its creation. The copyright system is thus fundamentally different from that governing patents, which actually are granted to inventors by a government agency, the U.S. Patent and Trademark Office, a division of the Department of Commerce.

The Purpose of Copyright

Congress is empowered to enact copyright legislation by Article I, Section 8, Clause 8 of the Constitution in order to "promote the Progress of Science and useful Arts by securing for limited Times to Authors and Inventors the exclusive Right to their respective Writings and Discoveries." The Constitutional references to science in the broad sense of knowledge (from the Latin *scire*, to know) and to authors form the predicate of copyright law in the United States. Behind the Constitutional authorization is the thought that creativity, and hence societal progress, are fostered by providing economic incentive to authors and publishers (2):

"The economic philosophy behind the clause empowering Congress to grant patents and copyrights is the conviction that encouragement of individual effort by personal gain is the best way to advance public welfare through the talents of authors and inventors in 'Science and useful Arts.' Sacrificial days devoted to such creative activities deserve rewards, commensurate with the services rendered."

Copyrights Differ From Patents. Although both copyright and patent law derive from the same Constitutional provision, they are quite different systems. As noted earlier, patents are granted by the Government; copyrights are not. As is discussed below, copyrightable subject matter need only be *original*, ie, the product of independent creation; patentable subject matter must represent *novelty* and improvement over the prior art. Additionally, copyright law protects only against copying and does not prohibit a second comer from independently creating a similar work; thus, Judge Learned Hand noted that if "by some magic a man who had never known it were to compose anew Keats' *Ode On a Grecian Urn*, he would be an 'author,' and, if he copyrighted it, others might not copy that poem, though they might of course copy Keats" (3). Patents exclude even independent development of the patented subject matter.

Subject Matter of Copyright

The categories of subject matter protected by copyright law in the United States include literary works, ie, works expressed in words, numbers, or other verbal or numerical symbols (including compilations of data in tabular or similar form, computer programs, and data bases); dramatic, musical and dramatico–musical works; pantomimes and choreographic works; pictorial, graphic, and sculptural works; motion pictures and other audiovisual works; and sound recordings. (Sound recordings are the aggregate of sounds resulting from the rendition or performance of a particular literary, dramatic, or musical work as embodied in a phonograph record or audio tape.)

Within these categorical descriptions lies an extraordinarily broad area of subject matter (4):

"Of course, works of literary, musical, fine, cinematographic, and allied 'arts' are included, such as poems, novels, stories, articles, texts and reference books; musical and dramatic compositions; paintings, drawings, sculpture, and photographs; motion pictures, stage presentations, and television programs; and cartographic, choreographic, and pantomimic works. So are many works of 'applied art' and artistic craftsmanship, such as jewelry, toys, fabric designs, and even the embellishments of furniture, architecture, appliances, and a host of other industrial or functional products. But the importance of copyright cannot be limited to the practices of so-called publishing, entertainment, art or design lawyers. At bottom, virtually every concrete, fixed expression of intellect, whether of a personal nature such as private letters and diaries, or business significance—such as directories, manuals, instructions, and specifications; memoranda, reports, studies, and documented, filmed, or taped presentations; schematics, layouts, and models; lists and tabulated or compiled facts, figures, or data; promotional materials and correspondence; and computer programs and automated data base, is subject to copyright protection against certain forms of unauthorized use."

There are two fundamental requirements of copyright protection: First, the work must be fixed in a concrete mode of expression. Works that exist only for an ephemeral or transitory period, such as spontaneous or extemporaneous, and unrecorded, performances or lectures, are not within the ambit of the copyright law (but may be protected under state common law). Second, the work must be original. The concept of originality does not include any criterion of novelty, uniqueness, or improvement over the prior art. It means simply that the work is the product of its creator and has not been copied from a prior source. Neither literary, aesthetic, nor other qualitative merit is required for copyright. The courts recognize, for example, that "it would be a dangerous undertaking for persons trained only to the law to constitute themselves final judges of the worth of pictorial illustrations" (5).

An important principle is that copyright protection pertains only to the *expression* of a work and not to its underlying idea, or to any concept, facts, principle, procedure, or the like embodied in the work. Thus, "Einstein's report of his conclusions on relativity was eligible for copyright protection, but not the resulting famous equation itself" (6). At the same time, however, copyright protection is not limited to verbatim reproductions of the protected work; it includes the right to exclude unauthorized duplication in substantially similar form. The line between a second work that has duplicated the expression of a prior work and that therefore may be an infringement, and one that has only reproduced its underlying ideas or facts and that therefore is lawful, is never easily drawn and is not capable of precise definition. An oft-quoted statement is that of Judge Learned Hand (7):

"Upon any work and especially upon a play, a great number of patterns of increasing generality will fit equally well, as more and more of the incident is left out. The last may perhaps be no more than the most general statement of what the play is about and at times consist of only its title, but there is a point in this series of abstractions where they are no longer protected since otherwise the playwright could prevent the use of his ideas to which apart from their expression his property is never extended."

In a recent decision holding that one organic chemistry textbook did not infringe copyright in another, although both contained similar scientific nomenclature, formulas, chemical reactions, and the like, the court noted (8):

"The objective of the constitutional provision and of the statute is to benefit the public by making available to it the writings of authors and the benefits of invention. The means of accomplishment is the protection of an author's or inventor's exclusive right for a limited time. Inevitably, the question arises as to what constitutes the writings of an author. Writings consist of at least three elements: facts, ideas and expression. It is accepted that ideas are not copyrightable nor is a fact by itself. The special difficulty arises, however, in determining the extent to which the factual element of a writing is protectable under the statute when it constitutes the fruits of research. Competing considerations are at play, both affecting the public interest: if the protectable scope of an author's writings is too narrowly defined he will be discouraged from further writing. If, however, the protectable scope of the original author's work is too broadly defined, creative work by other authors will be discouraged. A balance must be struck."

Attention on this so-called idea (or fact)/expression dichotomy has been the subject of considerable concern with respect to the value of the copyright system for computer-related materials. Although both computer programs and electronic data bases are subjects of copyright, substantial questions exist as to the practical scope of copyright protection for such works. It is unclear, for example, where the line will be drawn between a programmer's expression and an idea (9), and whether the courts will condone unauthorized uses of the discrete facts that comprise the unique value of many data bases (10).

Notwithstanding the broad coverage of copyright law in the United States, there are certain works that are not subject of copyright protection. These include works of *de minimis* authorship, that is, works considered to embody too little intellectual effort to fall within the protection of the copyright statute. However, the *de minimis* principle does not always accord with commercial reality. Thus, advertising slogans, the titles of works, and other short phrases do not fall within the copyright law (they may be protected, under certain circumstances, by trademark law or by the common law of unfair competition). Yet it is well known that the commercial value of a particular catchy advertising slogan or song or book title can often be significant.

Another category that is not protected by copyright comprises works of the U.S. government. This category encompasses all works produced by officers and employees of the Federal government in the course of their duties. The prohibition on copyright in U.S. government works does not extend to works created by government employees outside of the scope of their official duties, even if it involves knowledge or experience gleaned from their employment. It also does not apply to works produced by private parties under Federal government contracts or grants (11). Works generated or resulting from such contracts or grants, including manuscripts, studies, research reports, and the like, are copyrightable unless the contract, grant, or pertinent agency regulation provides otherwise. (It is not uncommon for Federal grants to confirm copyright in subject materials in the grantee, subject to the government's reservation of a nonexclusive right to reproduce the material, royalty-free, for "government purposes." The scope of the quoted reservation is not entirely clear, and is interpreted quite broadly in some agencies.) Apart from statutes, administrative pronouncements, court opinions and similar official edicts, works of foreign, state, and local governments are also copyrightable.

Particularly difficult determinations of copyrightable subject matter arise with respect to works of applied art and industrial design. Under the copyright systems of some countries, the law follows the principle of "unity of art" and makes little meaningful distinction, for copyright purposes, between works of the fine arts and those of applied art, industrial design, or like commercial application. However, U.S. law, and that of other countries, does draw a distinction. Under U.S. law, copyright does not extend to works of artistic craftsmanship unless the artistic element can be physically or conceptually separated from the utilitarian aspects of the useful article. As may be expected, this distinction has not been an easy one for the courts to apply (12).

Acquisition of Copyright

When the new copyright law became effective on January 1, 1978, the United States adopted a principle of automatic copyright. This principle means that works created on and after that date are copyrighted under the Federal copyright statute automatically upon their creation. For these purposes, a work is created as soon as it is fixed in some tangible form of more than transitory stability—for example, written, typed, or drawn on paper or canvas; sculpted or manufactured; or recorded on film or tape. In effect, this means that if a work created after January 1, 1978 is copyrightable, it is copyrighted as soon as it is created. Neither registration with the U.S. Copyright Office nor the use of a copyright marking or notice is a condition to securing copyright under the new law. However, registration and notice remain significant.

Perfection of Copyright—Copyright Formalities

Although copyright is secured automatically upon the creation of a work on or after January 1, 1978, certain statutory conditions, commonly termed formalities, remain very important under the new law. The principal formalities are copyright registration and copyright notice.

Copyright Registration. Registration with the U.S. Copyright Office is not a condition of copyright protection. However, the copyright statute does provide a number of advantages or inducements to prompt registration of claims to copyright. The most important of these pertain to the ability to file suit against infringement and to the evidentiary effect of a certificate of copyright registration.

In most cases, registration must be made for a work before an action for infringement of copyright in that work can be instituted. Since copyright infringement suits may be brought after registration for infringements committed earlier, it is possible—and not uncommon—to defer registration until it becomes necessary to do so to prosecute an infringement action. In the case of an infringement that commenced before registration, however, there are limitations on the remedies that a court may award to a successful copyright owner: no award of statutory damages or attorney's fees generally may be made under such circumstances. Statutory damages are a monetary award that courts may make in lieu of the copyright owner's actual damages from the infringement and the profits of the infringer. Attorney's fees are an amount a court may award to compensate the copyright owner for some of the legal fees incurred in maintaining the infringement action (they generally do not cover the legal fees actually incurred in an action). Because actual monetary loss is frequently difficult to prove in a copyright infringement case, statutory damages can be an important aspect of copyright litigation. Assuring the possibility of an award of statutory damages thus becomes an important inducement to prompt registration.

Although it is sometimes said that copyright registration is a condition to statutory damages and attorney's fees, that description is inaccurate. Courts are not deprived of the authority to award these remedies in the case of: infringements commenced after publication and before registration if registration is made within a grace period of three months after first publication; or infringements commenced after registration, regardless of when registration is made.

A certificate of registration made before or within five years after first publication of a work is *prima facie* evidence of the validity of the copyright and of the facts (eg, copyright ownership) stated in the certificate issued by the Copyright Office. A certificate of registration issued after five years from first publication is entitled to only such evidentiary weight as a court may accord in its discretion. Since the *prima facie* evidentiary effect of registration is frequently important to copyright owners in litigation, this too becomes a substantial inducement to prompt registration.

Copyright registration is not a complicated matter. It is accomplished by submitting a properly completed application for registration, on an official form, to the U.S. Copyright Office, together with an appropriate fee (in 1982, generally $10 per work) and a deposit of the work itself, usually consisting of one or two copies of the work. (The deposit generally becomes available for public inspection. Thus, in the case of highly secure or confidential materials—such as some nonmarketed computer programs—there may be drawbacks to registration unless special arrangements can be made.) The application form calls for the title of the work, the name of its author and copyright owner, the nature of authorship, and related information. The Copyright Office examines the deposit to ensure that it contains copyrightable material; but it does not examine for any novelty or uniqueness in the work nor, generally, compare it with other works. The Office also examines the application to ensure that it is properly filled out and consistent with the information (eg, the name of the copyright owner and author) given on the deposit copies.

Although, as indicated above, copyright registration is usually permissive rather than mandatory, this is not entirely true in the case of works copyrighted before January 1, 1978. As is discussed below, the duration of copyright in such works is divided into two terms. A renewal registration for works copyrighted before January 1, 1978 is required in order to secure the second or renewal term of copyright.

Copyright Notice. The second principal copyright formality is use of a copyright notice. Whenever a work is published—ie, made generally available to the public—a particular marking should be placed on the copies. This marking, or copyright notice, consists of the word Copyright, or the abbreviation Copr., or the symbol ©, together with the name of the copyright owner and year of first publication of the work. (In the case of "phonorecords"—phonograph records and audio tapes—the symbol P enclosed within a circle should be used to protect the particular sounds embodied in the recording.) Under the new statute, the notice may be placed anywhere on a work that is reasonable. The Copyright Office has published regulations giving examples of reasonable placement of notice. These include, for example, the title page, or reverse side of the title page, of a textual work, and the first page on which a particular contribution to a printed collective work (for example, an article in a scientific or technical journal or encyclopedia) appears. (However, the new statute also clears up some uncertainty under the prior law and expressly permits the copyright notice in a particular volume or issue of a collective work to protect the separate copyrights in each contribution to that volume or issue.)

The prior copyright law also included a notice requirement. If a work was published before January 1, 1978, without an appropriate copyright notice, it usually entered the public domain in the United States. This consequence no longer necessarily follows for notice omissions after January 1, 1978. The law now provides that a notice omission on a relatively small number of copies, or in breach of an express agreement, does not invalidate U.S. copyright in the work. Additionally, other notice omissions or defects can be cured by taking two steps: registering the work with the U.S. Copyright Office before or within five years after the publication without notice has occurred; and taking reasonable steps to add the notice to all copies distributed in the United States after the omission has been discovered.

Another aspect of U.S. copyright law that has sometimes been considered a formality is the requirement that certain steps in the manufacturing process of works consisting predominantly of nondramatic, English-language textual material by U.S. authors take place in the United States or Canada in order to assure full protection. These steps include typesetting or plate-making, printing, and binding. The continued validity of this so-called manufacturing clause is open to question.

Copyright Ownership

Copyright belongs in the first instance and upon creation of the work to its author. Usually, this is the particular individual who created the work, and anyone claiming rights in the work must obtain them from that individual. In the case of a work made for hire, however, copyright belongs to the employer (or other person) for whom the work was written, rather than to its natural author, and the natural author has no rights in the work except as may be agreed to by the employer.

Works made for hire include all works created by employees in the course of their duties and certain works produced on special order or commission. Work-for-hire treatment of works made on commission applies only to certain categories specifically mentioned in the Copyright Act, and then only if the parties expressly agree in writing that the work should be considered to be made for hire. These special categories include works commissioned for use as contributions to collective works (eg, sections of encyclopedias or articles in magazines and other serials), as part of a motion picture (eg, background music), as a translation, as a compilation of preexisting data or other works, as a test or as answer material for a test, as an atlas, as a supplementary work (ie, a foreword, afterword, editorial notes, index, or other secondary adjunct to works prepared by others), or as a text prepared for use in the course of systematic instruction. In the case of a work made for hire, the employer or commissioning party is not merely the owner of copyright. It is also considered in law to be the author of the work, and this may have significance under certain provisions in the law that turn on the author's nationality. Certain other provisions of the Copyright Act that permit authors to terminate earlier transfers and licenses of copyright after a period of years do not apply to works made for hire. Thus, in the case of commissioned works that are expected to retain value over several decades and from which the author expects publishing or other royalties, an agreement to consider the work as one made for hire may be particularly disadvantageous.

Copyright owned by one party may be transferred or licensed to others. A copyright license is generally permission given by the copyright owner to another to reproduce, revise, perform, or display the work. The license may be exclusive—in which case, the particular rights will not be granted to any other person—or nonexclusive. Transfers and exclusive licenses must be made in writing.

Copyright is divisible, that is, each particular right or subdivision of a right included in the copyright may be transferred or licensed to different persons. For example, the author of a manuscript may transfer hard-cover publication rights to one party, soft-cover publication rights to another, motion-picture rights to a third, and so on. The rights may even be subdivided further. Thus, an author may assign first serial rights to a magazine, and then transfer various other forms of magazine and book publication rights to others. The particular division of copyright is dependent upon agreements entered into by the author (including, in the case of works made for hire, the employer or other person for whom the work was made).

Agreements and other documents granting transfers or licenses of rights to copyrighted works may be recorded in the Copyright Office. Recording of a transfer or license document refers only to placing the facts in the document itself on the public records maintained by the Copyright Office. It is entirely distinct from registering the claim to copyright in the work itself. Like registration, recording is permissive rather than mandatory, but is subject to several statutory inducements or advantages. From the point of view of the transferee or licensee, these include protection against later transfers of the same rights in the work being granted to other parties. In many cases, recording of a document transferring or licensing copyright must also be made before the transferee or licensee can bring an action for infringement. Copyright in a work is separate from ownership of the particular physical copy in which the work is embodied. Thus, the transfer or delivery of a manuscript does not itself convey rights in the work.

Rights

Copyright is frequently described as a bundle of rights, comprising a number of separate and distinct exclusive prerogatives accorded to the copyright owner. The principal rights embraced under a copyright are

1. The right to reproduce the copyrighted work in any stable form (eg, printed copies, punched cards, audio cassettes and records, disks, magnetic tapes, microforms) from which it can be perceived or communicated.

2. The right to prepare derivative works (eg, revisions, translations and abridgements of articles, new arrangements of musical works, and revised editions of textbooks) based upon the copyrighted work.

3. The right to distribute copies of the copyrighted work to the public. (However, this right is exhausted with respect to each copy after it has first been sold or its ownership otherwise transferred by the copyright owner. Thus, a library, having purchased a copy of a book, is free to lend that copy to others; but an exhibitor that merely leases a motion-picture film is not free to sell, lease, or otherwise transfer that film to others.)

4. The right to perform the copyrighted work publicly (performances include recitations, broadcasts, motion-picture exhibitions, performances from recordings, and live renditions).

5. The right to display the copyrighted work publicly. (This does not prevent the owner of a copy from displaying it to viewers present at the place where the copy is located.)

After setting forth these broad categories of exclusive rights, the Copyright Act then establishes a number of specific exemptions and limitations. These exemptions and limitations are of two types: complete exemption from rights and compulsory licenses.

Complete Exemptions from the Copyright Owner's Exclusive Rights. The statute specifies a number of instances in which the use of a work is totally excluded from the copyright owner's rights. These include, for example, certain performances (but not reproductions) of works in the course of face-to-face teaching activities, instructional broadcasting, religious services, and certain other nonprofit purposes. An exemption is also given to archives and libraries whose collections are open to the public or to nonaffiliated researchers, and which permit their occasional photocopying of textual materials, including journal articles and the like, under certain circumstances for purposes of private scholarship, study, or research. These circumstances do not include direct or indirect commercial purpose, the concerted or related making of multiple copies, systematic photocopying activities, or arrangements whose purpose or effect is to substitute for purchase or subscription (13).

A general exception to the exclusive rights of copyright owners is the doctrine known as fair use. Fair use is not capable of any specific definition. It is intended to encompass limited, reasonable, and customary uses of prior works in connection with such activities as research, scholarship, criticism, comment, teaching, and news reporting. Whether a particular use is or is not fair and within the privilege of this section of the law depends upon a variety of factors in each case, including the purpose and character of the use, the nature of the copyrighted work, the amount and substantiality of the amount used, and, most importantly, the effect of the use upon the potential market for or value of the work. The legislative history of the 1976 Copyright Act includes a set of guidelines that describe the application of fair use to reproductions made for purposes of classroom instruction (14).

Compulsory Licenses. A compulsory license is a privilege given by the Copyright Act that permits the use of a work without the consent of the copyright owner, but only upon compliance with certain conditions established in the statute or implementing regulations, including payment of a royalty fee that has been legislatively or administratively determined. The compulsory license provisions now in the law permit the making of phonograph records and audio tapes of musical works; the retransmission by cable television of copyrighted works embodied in broadcast television programs; the performance of musical works on jukeboxes; and the performance of musical works and exhibition of artistic works by public broadcasting stations.

Effect of Technology

It is within this area of the rights accorded by copyright that the tensions between copyright and developing technology have been keenly felt. As early as 1945, a leading scholar commented upon the relationship between copyright and technological innovation as follows (15):

"Copyright is the Cinderella of the law. Her rich older sisters, Franchises and Patents, long crowded her into the chimney-corner. Suddenly the Fairy Godmother, Invention, endowed her with mechanical and electrical devices as magical as the pumpkin coach and the mice footmen. Now she whirls through the mad mazes of a glamorous ball."

The magical devices discussed by this scholar were motion pictures and radio, but, more recently, copyright has come face-to-face with over-the-air, cable and subscription television; satellite, microwave and laser delivery systems; photocopying and microform reproduction and storage; long-distance facsimile transmission; computer storage and retrieval; and vastly improved modes of audio and video recording. The result for copyright protection has not been one of glamour. These developments have affected copyright in a variety of ways; principally by eroding the copyright owners' rights to control or secure compensation for the use of their works, thus diluting the incentive and economic basis for continued creative effort. More specifically, they have, first, contributed to an enormous public appetite for immediate access to and use of copyrighted works. Second, as typified by the development of photocopying and private audio and video recording, they have made duplication of copyrighted works a simple and inexpensive task and have generated unauthorized reproduction on a massive scale. (A recent, well-publicized aspect of this is in the question of copyright liability for off-air video recording by individuals in private homes (16). A Federal Court of Appeals held that such activity is a copyright infringement. The case has been accepted for review by the Supreme Court.) Third, these developments have caused widespread reproduction to occur in private contexts (such as the use of audio- and video-recording devices, photocopying machines, and like equipment in private homes and business establishments) where detection and enforcement are made difficult. (Again, this is characteristic of the current debate over off-air recording (16). It is also pertinent to the photocopying of scientific and technical journal articles by employees of companies engaged in research and development. A lawsuit claiming copyright infringement by such companies has recently been brought.) Finally, these developments have distorted the traditional roles played by publishers and consumers of copyrighted works. With the aid of an audio- or video-recording device or photocopying machine, the consumer is now capable of serving as the publisher, or creator of copies, as and when needed, on demand. Although Congress has made a number of attempts to accommodate copyright and technology in the new copyright statute, that task is far from completed. Litigation is one way in which copyright owners are seeking to preserve their rights in the face of the onslaught of technology; but there are others. For example, the development of collective mechanisms for the licensing or granting of permission for use of broad classes of works, and the development of techniques that permit instantaneous use, subject to compensation established by the individual copyright owner, rather than by statute or administrative regulation, is another. The latter device has most recently been typified by organization of the Copyright Clearance Center, Inc. (CCC). Publishers of scientific and technical journals entered in the CCC thereby authorize the photocopying of articles from their serials without any need of the user to seek permission in advance, subject to the payment of a fee set by the publisher. (The CCC located in Salem, Massachusetts, has issued a number of explanatory materials pertaining to its operations. Among the founders of the CCC, was the late Michael Harris, former Vice President of John Wiley & Sons, Inc.)

Moral Rights

The rights of copyright owners set forth in the Copyright Act are sometimes denominated as pecuniary rights, as opposed to the moral rights of authors with respect to works they have created. The so-called moral rights include the rights of authors to object to unauthorized distortions of their work, and to claim authorship credit when their works are reproduced. These moral rights are commonly recognized in European copyright systems as a fundamental and inherent aspect of copyright protection. In the United States, apart from the right to control the making of derivative works, they are not recognized by the copyright law as such. However, under certain other legal doctrines, including the laws pertaining to unfair competition and defamation, the moral rights of authors can sometimes be protected in the United States (17).

Duration of Copyright

The rules governing the duration of copyright in the United States have been complicated by the need to coordinate the provisions of the old and new copyright laws. These rules are summarized as follows:

In the case of works created on or after January 1, 1978, copyright endures for a period measured by the life of the author of the work and 50 years after that author's death. In the case of works jointly created by multiple authors, the post-mortem period of 50 years is measured from the death of the last surviving author. Since the duration of copyright in such works is measured by the life of natural persons, this principle requires adjustment in cases where the author of the work is not readily identifiable from the copies (ie, anonymous and pseudonymous works), and where the author might be a corporate or other business entity (ie, works made for hire) rather than a natural person. Accordingly, the new law provides that, in the case of anonymous and pseudonymous works and works made for hire that are created on and after January 1, 1978, the term of copyright is a period of 75 years from the year of first publication of the work, or a term of 100 years from the year of creation of the work, whichever term first expires. However, if before the end of such term, the true identity of one or more of the authors of an anonymous or pseudonymous work is revealed in Copyright Office records, the term of copyright is then converted to a period based upon the life of the author or authors whose identity has been so revealed and a period of 50 years from that author's death.

In the case of works created before January 1, 1978, but not published or registered before that date, the term of copyright is governed by the rules set forth immediately above with respect to works created after January 1, 1978. However, since many of these works would have already been existing for a long period of time, the new law provides that copyright in works created but not published or registered before January 1, 1978 cannot in any event expire before December 31, 2002. Additionally, if such a work is published on or before December 31, 2002, the term of copyright will not expire before December 31, 2027.

In the case of works copyrighted before January 1, 1978, copyright endures for a first term of 28 years from the date it was originally secured. At the expiration of such 28 year term, an additional or renewal term of 47 additional years can be secured if timely application for renewal is made to the Copyright Office before the expiration of the twenty-eighth year of the first term. The copyright law specifies in whose name the renewal application is to be made, and who is entitled to the renewal term. Generally, the renewal is to be made and will be initially owned by the author of the work, or if the author is dead, certain successors designated in the statute (the author's widow, or widower, and children; if there are none, the executor; or if the author died intestate, the author's next of kin). In some special cases, however, the renewal is to be made by the owner of the copyright at the time of renewal. These special cases include works made for hire.

In the case of works copyrighted before January 1, 1978, and already renewed under the prior law, the renewal period, previously a second term of 28 years, is extended by the new law to 47 years.

International Copyright

The foregoing discussion has been directed toward copyright protection for works in the United States. Works originating in the United States are also subject to copyright protection in a great many foreign countries. The phrase international copyright is, however, something of a misnomer, for there is no single law or code of international copyright governing the protection in one country of works originating in others. There are a number of bilateral relations and multilateral conventions pursuant to which works originating in one country are entitled to protection in others. Two principal multilateral conventions are the Universal Copyright Convention, to which the United States is a party, and the Berne Convention, to which the United States does not belong. United States adherence to the Berne Convention is generally understood to be precluded, at the current time, because of the U.S. law's continued emphasis on the "formalities" discussed above in contravention of a fundamental Berne principle that copyright protection shall be free of any formalities (18). (However, works originating in the United States can gain protection in Berne countries by simultaneous publication in the United States and a country that is a member of the Berne Convention.)

Although treaties and conventions do impose certain minimal requirements on member states with respect to the type of protection to be accorded foreign works, in each case the actual nature of protection accorded by a foreign country is ultimately determined by its own law. Members of the Universal Copyright Convention cannot impose formalities as a condition to copyright protection of works originating in the United States that are unpublished or, if published, that contain the notice of copyright described above. (For this purpose, however, the symbol © must be used in the notice.)

The correlation among international copyright protection, political and cultural exchange between nations, and transcontinental communications technology has been widely noted (19). Yet on the international scene, no less than on the domestic, the tensions posed by new technology for copyright are increasingly being felt, and the problems of photocopying, computer usage, satellite and cable-television transmissions, and private audio and video recording are far from settled.

Another element of concern in the international copyright sphere has been the pressure brought by developing countries for severe limitations on the rights accorded to copyright owners. Although ostensibly based upon the need for simplified access to foreign works for purposes of local cultural development, among the more developed countries it is felt that these attitudes are frequently shortsighted. It is pointed out, for example, that if a particular country substantially decreases the amount of protection available to foreign works and thereby encourages use of such works without adequate compensation or the need to seek permission, the end result is likely to be the flooding of that country with foreign works to the detriment of that country's indigenous intellectual and publishing communities. There is a very clear historical example of such a development: the United States. Before 1891, the United States did not accord protection to foreign works and found its markets saturated with unauthorized, inexpensive reprints of works from the UK. The development of indigenous U.S. authorship and publishing quite clearly suffered in this situation. In the words of one contemporary observer (20):

"Why should the Americans write books, when a six-weeks passage brings them in our own tongue, our sense, science, and genius, in bales and hogsheads?"

BIBLIOGRAPHY

1. *Omnibus Copyright Revision—Comparative Analysis of the Issues*, Cambridge Research Institute, American Society for Information Science, Washington, D.C., 1973.
2. Mr. Justice Reed in *Mazer v. Stein*, 347 U.S. 201, 219 (1954).
3. *Sheldon v. Metro-Goldwyn Pictures Corp.*, 81 F. 2d 49, 54 (2d Cir. 1936).
4. J. A. Baumgarten, *District Lawyer*, 21 (Nov.–Dec. 1981).
5. Mr. Justice Holmes in *Bleistein v. Donaldson Lithographing Co.*, 188 U.S. 239, 251 (1903), upholding the copyrightability of a circus poster.
6. C. H. Lieb, *Communications and the Law* **2**, 55, 57 (1980).
7. *Nichols v. Universal Pictures Co.*, 45 F. 2d 119, 121 (2d Cir. 1930).
8. Judge Lasker in *Morrison v. Solomons*, 494 F. Supp. 218, 222 (S.D.N.Y. 1980).

9. *Report No. 94-1476*, House of Representatives, 94th Congress, 2d Session 57, 1976.
10. R. Denicola, *Columbia L. Rev.* **81,** 516 (1981).
11. This was recently confirmed in the case of *Schnapper v. Foley*, 667 F. 2d 101 (D.C. Cir. 1981).
12. Compare, eg, *Esquire, Inc. v. Ringer*, 591 F. 2d 796 (D.C. Cir. 1978) (lighting fixtures); *Kieselstein-Cord v. Accessories by Pearl, Inc.* 632 F. 2d 989 (2 Cir. 1980) (belt buckle).
13. *Photocopying by Academic, Public and Nonprofit Research Libraries*, Association of American Publishers, Inc., and Author's League of America, Inc., New York, May 1978.
14. *Report No. 94-1476*, House of Representatives, 94th Congress, 2d Session 68–71, 1976.
15. Z. Chaffee, *Columbia L. Rev.* **45,** 504 (1945).
16. *Universal City Studios, Inc. v. Sony Corporation of America, Inc.*, 659 F. 2d 963 (9th Cir. 1982).
17. W. Strauss, *U.S. Copr. L. Rev. Study No. 4*, U.S. Government Printing Office, Washington, D.C., 1959.
18. *N.Y.U.J. Int. Law Pol.* **9,** 455 (1977).
19. Eg, A. Ciampi in *Symposium on Practical Aspects of Copyright*, United International Bureau for the Protection of Intellectual Property, Geneva, 1968, p. 81
20. Remarks of Sydney Smith, quoted in R. Nye, *The Cultural Life of the New Nation, 1776–1830*, Harper & Row, New York, 1960, p. 250.

General References

J. Chem. Information and Chem. Sci. **22** (May 1982).
N. Boorstyn, *Copyright Law*, Lawyers Co-Operative Publishing Co., New York, 1981.
D. Johnston, *Copyright Handbook*, R. R. Bowker Co., New York, 1978.
B. Kaplan, *An Unhurried View of Copyright*, Columbia University Press, New York, 1967.
A. Latman, *The Copyright Law*, Bureau of International Affairs, Inc., New York, 1979.
M. B. Nimmer, *Nimmer on Copyright*, Matthew Bender, New York, 1981 (4-vol treatise).
General Guide to the Copyright Act of 1976, U.S. Copyright Office, Library of Congress, Washington, D.C., Sept. 1977; various information circulars, announcements, and application forms available on request.
Bulletin of the Copyright Society of the U.S.A., published by the Copyright Society of the United States of America, New York University Law Center, New York, six times a year.

Copyright Law Revision

Cambridge Research Institute, *Omnibus Copyright Revision—Comparative Analysis of the Issues*, American Society for Information Science, Washington, D.C., 1973.
Copyright Law Revision Studies Nos. 1–34, U.S. Government Printing Office, Washington, D.C., 1960.

History of Copyright

A. Birrell, *Seven Lectures on the Law and History of Copyright in Books*, Cassel and Company, Ltd., London, 1899, Rothman Reprints, Inc., N.J., 1971.
L. R. Patterson, *Copyright in Historical Perspective*, Vanderbilt University Press, Nashville, Tenn., 1968.
H. Ransom, *The First Copyright Statute*, University of Texas Press, Austin, Texas, 1956.
B. Ringer in *200 Years of English and American Patent, Trademark and Copyright Law*, American Bar Association, Chicago, Ill., 1976.

Copyright and New Technology

L. H. Hattery and G. P. Bush, eds., *Reprography and Copyright Law*, American Institute of Biological Sciences, Washington, D.C., 1964.
E. J. Ploman and L. C. Hamilton, *Copyright—Intellectual Property in the Information Age*, Routledge & Kegan Paul, London, 1980.
J. S. Lawrence and B. Timberg, eds., *Fair Use and Free Inquiry*, Ablex Publishing Corp., Norwood, N.J., 1980.
L. E. Seltzer, *Exemptions and Fair Use in Copyright*, Harvard University Press, Cambridge, Mass., 1978.

Final Report of the National Commission on New Technological Uses of Copyrighted Works, Library of Congress, Washington, D.C., 1979.
Copyright in Computer-Readable Works: Policy Impacts of Technological Change, National Bureau of Standards, U.S. Department of Commerce, Oct. 1977.
N. Henry, ed., *Copyright, Congress and Technology: The Public Record*, Oryx Press, Phoenix, Ariz., 1979. Five-vol. compilation of public documents.

International Copyright

J. A. Baumgarten, *U.S.–U.S.S.R. Copyright Relations Under the Universal Copyright Convention*, Practising Law Institute, New York, 1973.
J. A. Baumgarten, *Bull. Copr. Soc. of the U.S.A.* **27**, 419 (1980).
A. Bogsch, *The Law of Copyright Under the Universal Convention*, R. R. Bowker Co., New York, 1972.
M. M. Boguslavsky (N. Poulet, translator), *Copyright In International Relations: International Protection of Literary and Scientific Works*, Australian Copyright Council, Ltd., Sydney, Australia, 1979.
T. R. Kupferman and M. Foner, eds., *Universal Copyright Convention Analyzed*, Federal Legal Publications, Inc., New York, 1955.
S. Nowell-Smith, *International Copyright and the Publisher in the Reign of Queen Victoria*, Clarendon Press, Oxford, 1968.

JON A. BAUMGARTEN
CHARLES H. LIEB
Paskus, Gordon & Hyman

TRAGACANTH.   See Gums.

TRANQUILIZERS.   See Psychopharmacological agents.

TRANSISTORS.   See Semiconductors.

TRANSPORTATION

The transportation of chemicals and related products is unusual in that substantial quantities are moved in packages as well as in bulk. Other materials, such as coal, grain, and ore, are transported in bulk but seldom in packaged form. Moreover, most other bulk commodities, including petroleum and its products, are limited in the diversity of their chemical and physical characteristics and, therefore, do not require as wide a variety of packagings and bulk conveyances as is necessary for the movement of chemicals. Virtually all railroad tank cars are supplied by chemicals producers rather than the railroad companies, which furnish at least a portion of most other types of equipment used in rail transportation. The multiplicity of chemical and physical characteristics, as well as resulting variations in product value, density, volume of movement, and other factors including the type and supply of packaging and conveyances, tend to complicate transport pricing and relations between chemicals shippers and the many transportation carriers that they employ.

Since the late nineteenth century, the Federal government and almost all states have regulated both the supply and pricing of transportation service, and such regulation has had a profound effect on the chemical industry. Although economic regulation was substantially relaxed by legislation enacted in the 1970s and early 1980s, such relaxation was accompanied by more intense regulation of hazardous materials transportation. Because most of the 227×10^6 metric tons annually produced by the chemical industry is classified as hazardous (1), transportation safety has become increasingly important to shippers and carriers of chemicals seeking to comply with a growing body of Federal, state, and even local regulations in an effort to avoid civil and criminal penalties. In an era of increasing litigation and sustained public interest in environmental safety, an additional incentive to such shippers and carriers is the avoidance of civil liability and more burdensome regulation.

This article briefly reviews some of the more significant operational aspects of chemicals transportation in the context of the changing climate of both economic and safety regulation. The technical nature of the latter, especially in connection with hazardous materials and wastes, has necessitated frequent consultation between industrial distribution and technical personnel in the chemical industry.

The chemical industry is one of the largest users of commercial transportation services provided by rail, motor, water, air, and pipeline carriers, including combinations of such carriers, as well as proprietary transportation. Shipments of chemicals are made in a wide variety of containers, such as tank cars and tank trucks, barges, self-propelled vessels, drums, barrels, cylinders, bags, and even small bottles for samples and laboratory specimens.

The cost of transportation has an important effect on the marketability of chemicals. For that reason, transportation, along with numerous other factors, is often a significant consideration in determining the location of chemical-production facilities. In addition, convenient and economical access to water and rail transportation and the interstate highway system, as well as proximity to raw materials and markets, may influence the choice of warehouse and terminal sites for the storage and redistribution of chemical products (see Plant location).

During the past few decades, the concept of transportation management in the chemical industry has been broadened to include such functions as packaging, order processing, sales service, warehousing, and scheduling of inbound raw materials. The expanded concept, commonly referred to as distribution (or physical distribution, as distinguished from market distribution), is in part a consequence of computer technology which has made it possible to relate total distribution costs to individual products, customers, and movements. Transportation, however, continues to be the central concern of most distribution managers in the chemical industry. Table 1 indicates the relative shares of U.S. intercity tonnage (including chemicals) carried by various modes of transportation.

Transportation Modes

Railroads. In 1979, the largest U.S. railroads (Class I) originated over 102×10^6 metric tons of chemicals and allied products (3). Railroads are almost exclusively common carriers which offer their services to the public as transporters of virtually all commodities between all points on their lines, which are privately owned, maintained, and operated.

Rail service may be single-line or joint-line; the former refers to movements that originate and terminate on a single railroad, without intermediate transportation by another rail carrier. Joint-line service occurs when more than one railroad participates in the transportation from origin to destination, generally under agreements among the railroads involved for interchange at specified locations with the necessary facilities. Even where a single railroad serves both the origin and destination cities, joint-line service may be necessary or desirable in some situations, as, for example, where the shipper is served by one carrier and the consignee by another. In most cases, where joint-line service is offered and appropriate routing restrictions are observed by the shipper, joint-line and single-line movements of the same goods between the same points are charged identically, and the carriers who participate in the joint-line service share in divisions of the total revenue. In recent years, the trend toward mergers of large railroads serving broad geographical areas has resulted in fewer joint-line movements.

Traditionally, railroads have furnished boxcars of varying sizes and capacities for general-purpose rail movement of packaged freight, including chemicals and related products in drums, barrels, bags, and other containers. In recent decades, the ability of boxcars to carry greatly increased loads without excessive damage to the lading has

Table 1.   U.S. Intercity Tonnage by Transportation Mode[a]

Year	Rail[b] 10^9 t	%	Truck 10^9 t	%	Oil pipeline 10^6 t	%	Water 10^6 t	%	Air 10^6 t	%
1975	1.33	29.6	1.59	35.1	797	17.7	785	17.5	2.9	0.1
1976	1.34	28.0	1.79	37.4	847	17.7	809	16.8	3.1	0.1
1977	1.33	26.7	1.94	39.1	894	18.0	804	16.1	3.3	0.1
1978	1.34	25.9	2.04	39.6	889	17.2	892	17.2	3.5	0.1
1979	1.45	27.6	2.03	38.6	888	16.8	892	16.9	3.4	0.1
1980	1.44	28.9	1.82	36.4	833	16.7	892	17.9	3.6	0.1

[a] Ref. 2.
[b] Class I and II.

been enhanced by the introduction of energy-absorbing underframes and other technical innovations.

Tank cars and other special-purpose rail cars, such as covered hopper cars for the movement of bulk plastic materials, are generally not supplied by railroads and must be furnished by shippers or receivers, who usually purchase or lease such equipment. Rail-car leases are offered in various forms by car manufacturers and intermediaries, ranging from short-term, full-maintenance rentals to long-term leases requiring outside financing (4). Many chemical shippers have substantial investments or lease commitments in tank cars and similar rail equipment, including cars constructed of or lined with special materials for particular products. Other cars may be thermally insulated to prevent excessive heat build-up in transit or for protection against fire.

The so-called trip lease of private (shipper-furnished) tank cars, used primarily to avoid the assessment of railroad demurrage, is a unique feature of rail transport. Demurrage is a charge made by carriers for the detention of a transport conveyance beyond a period considered sufficient for loading or unloading. Railroad tariffs, however, not only provide for the payment of demurrage on railroad-furnished box cars, flat cars, gondolas, and other freight cars, but also provide that demurrage must be paid even on private cars held on railroad property. Moreover, even where a private car is held on private tracks of other than railroad ownership, demurrage is assessed, unless the ownership of both the private car and the private tracks on which it is held are the same. A car owned or leased by a shipper and consigned to a customer's plant is, therefore, subject to demurrage charges because the car and the customer's plant trackage are owned by different persons. Railroad rules also provide, however, that the lease of a private car or private tracks is, for such purposes, the equivalent of ownership. Thus, to avoid the payment of demurrage by their customers, tank-car shippers usually trip-lease such cars to the consignee for the duration of each trip (5), thereby complying with the rule that ownership of the private car and the private track must be identical.

At many chemical plants, as well as other manufacturing or receiving facilities dependent on rail transportation, railroad tracks are constructed within the plant to permit the shipment or receipt of rail cars. Such tracks, usually called industry or private tracks or sidetracks, connect directly with the tracks of the railroad or railroads serving the plant. Since the sidetrack must be compatible with the railroad track to permit railroad switch engines and crews to enter the plant, the industry and the railroad enter into a written sidetrack agreement (6) which defines their respective rights and obligations with respect to track construction, maintenance, and operation. Included in such agreements are provisions relative to required lateral and overhead track clearances and to the maintenance of hoppers, pits, or other loading or unloading devices.

For the reasons noted above, the storage of private cars on private tracks is generally not subject to the payment of demurrage charges. To avoid such charges when private tracks have insufficient capacity for the number of freight cars required to be stored, shippers or receivers of freight may lease from a railroad additional trackage in the vicinity of the plant (7). Since leased tracks are considered to be private tracks during the term of the lease, demurrage is not payable on private cars of the same ownership held on such tracks, although a reasonable rental for the track lease must be paid. Frequently, tracks located at strategic places remote from a plant facility may be leased for storage of loaded cars in order to have them available for prompt delivery to customers or distributors in the vicinity of the track location. When leased tracks are owned by a railroad, cars stored on such tracks may also be subject to hazardous storage charges if they contain materials classified as hazardous. To avoid such charges, the lessee must have exclusive use of the leased tracks as well as the underlying land (8).

Motor Carriage. During the past 50 years, motor carriage has become an essential part of the U.S. transportation system. Initially confined largely to movements of small shipments or over short distances, motor carriers took advantage of improved public highways, including the interstate system, to develop a network of transportation competitive with railroads in both rates and service. Less capital-intensive and, therefore, more numerous than railroads, and unconfined by the rigidity of tracks, motor carriers demonstrated a flexibility that broke historic patterns of industrial concentration in transportation centers, thereby contributing to the dispersion of manufacturing and other commercial enterprises to suburban and rural areas.

For both economic and legal reasons, individual motor carriers have tended to specialize in the type of services they offer, either in terms of the commodities carried, the areas or locations served, or the type of equipment provided. Although many motor carriers offer to transport general commodities (with specific exceptions) for the public, such service is frequently limited territorially. Some truckers restrict their services to particular categories of materials, such as "Chemicals, in bulk" or "Acids, in packages" or even to single commodities, such as "Acetylene, in cylinders," as well as to specified cities, towns, counties, or states. Thus, the services of common carriage are limited, in the case of trucking companies, to those they are willing or permitted by their governmental franchises to perform. In recent years, as a result of a relaxation of Federal regulations, many motor carriers have expanded the scope of their service although few, if any, are willing or able to provide transportation without some limitations.

In addition to common carriers, a second category widely used in the chemical industry consists of motor carriers who confine their services to one or more particular shippers rather than serving the public. Such carriers may dedicate vehicles and other equipment to the needs of their customers and adapt their services to special requirements, such as pick-ups or deliveries at unusual hours, unusual loading patterns, or similar accommodations. Since such carriers have historically operated under continuing contracts (9) for transportation of a series of shipments over a period of time, they are known as contract carriers or house carriers. In practice, the actual distinctions between common and contract motor carriers have been largely obscured, and the concept of contract carriage retains its significance primarily for certain legal and regulatory purposes.

A third type of motor carriage, of considerable importance to many industries including the chemical industry, is proprietary or private carriage. Such transportation is conducted in furtherance of a primary business other than transportation. Thus, manufacturers transporting goods which they have manufactured or processed or which they will use in such manufacture or process, or for purposes of bona fide sale or purchase, are engaged in private carriage. Contrary to a common misconception, it is usually immaterial whether a private carrier does or does not have legal title to the transported goods and, indeed, when title to goods is acquired solely in an effort to create an appearance of private carriage, the transportation of such goods is considered to be for-hire carriage and, therefore, subject to governmental regulation (10). Furthermore, it is generally not required that a company use only vehicles that it owns, rather than leases, or that it directly employs the drivers of such vehicles, provided that such company actually controls the transport operation and bears its characteristic burdens and financial risks (11).

Recent legislation has reversed a historic rule which held that a parent corporation that performed transportation for compensation on behalf of a subsidiary, or vice versa, was engaged in for-hire transportation and, therefore, was subject to regulation (12). A corollary rule had also prevented, in such cases, the issuance of a required franchise to the transporting corporation even when that corporation was willing to be subjected to such regulation (13). Under the revised legislation (14), corporate members of a single group of corporations may lawfully perform such transportation for a parent or subsidiary, or for a sister subsidiary, provided that the one corporation wholly owns the other or that both are wholly owned by a common parent. Although compliance with certain minimal ICC regulations is required (15), such transportation does not require a franchise and is not subject to other burdensome regulatory requirements. As a result, it has become possible for many corporations to combine in a single vehicle their freight with that of other members of the same group of corporations, thereby improving equipment and labor utilization in consolidated private trucking operations.

Private carriers, by definition, and contract carriers, for legal and other reasons, do not engage in joint-line motor service. Nevertheless, such carriers sometimes deliver to or pick up from common carriers, resulting in a combination of services which, unlike joint-line service, does not depend on prearrangements among the carriers themselves. Common motor carriers, however, frequently offer joint-line service similar to that offered by railroads, although the need for such service is not as apparent as it is in the case of railroads operating on fixed lines. In most cases, shippers prefer single-line motor carriage when it is available.

A wide variety of highway vehicles, including trucks, tractors, trailers, tank vehicles, hopper vehicles, low-boys, vans, and others, are used by common, contract, and private motor carriers. Unlike railroads, commercial motor carriers of bulk liquids or solids in tank trucks or hopper trucks usually offer shippers both power equipment (tractor) and freight-carrying trailers, although shippers frequently supply such trailers under special arrangements. Highway tractors used for long, continuous journeys are usually equipped with sleeper-cabs to allow one driver to rest while a second driver operates the tractor-trailer.

The development of the interstate highway system and more permissive Federal and state legislation have allowed the use of vehicular equipment of increased length and other dimensions, as well as higher weight-carrying capacity, thereby contributing to more economical motor transportation. Such legislation, however, has in turn given rise to disputes concerning highway tolls, fuel taxes, registration fees, and similar assessments, against both for-hire and proprietary truck operators, to permit adequate maintenance of highways. Since highways are used by both passenger and freight-carrying vehicles, such disputes are not readily resolved.

Waterborne Transport. Despite natural limitations, the transportation of chemicals by water has enjoyed substantial growth, especially since the end of World War II. Assisted by governmental development of the inland waterways system, including locks and other navigational aids, water carriers currently transport large quantities of bulk chemicals in barges between inland ports or between such ports and coastal ports. In addition, bulk chemicals are transported by self-propelled tank vessels between U.S. coastal points and between U.S. ports and overseas destinations. In 1979, more than 54×10^6 metric tons of chemicals and related products was transported an average of ca 1500 km in U.S. water carriage (16).

Although water carriers are sometimes classified as common or contract carriers, such distinctions are frequently insignificant, since water carriage of bulk chemicals in the U.S. is essentially unregulated (17). In conformity with long-standing practice in the maritime field, such transportation is often provided under various forms of agreement, such as bareboat charters, time charters, or voyage charters. In a bareboat charter, the owner of a vessel charters (leases) the vessel without crew; in a time or voyage charter, the vessel is leased with crew for a specified period or for a particular voyage. On U.S. inland waterways, chemical shippers sometimes engage towboat operators to tow barges that such shippers either own or charter (lease) from others. In the United States, very little remains of a once flourishing liner trade in the transportation of packaged freight, although such liners are still engaged in such transportation to and from foreign ports.

Barges, like other transportation vehicles, are available in a wide variety of types, sizes, and capacities. On the inland waterways, barges are usually unmanned and without power independent of the towboats which push several barges in a group, or tow. In deepwater, or ocean, transportation, barges sometimes carry a crew and are capable of self-propulsion. Deepwater barges, whether self-propelled or pulled by a hawser (cable) between the barge and towboat, are generally larger than river barges. Deepwater tows rarely consist of more than one or two barges.

As in the case of highways, considerable contention results from public maintenance of the inland waterways for recreation, flood-control, and other purposes, as well as for the transportation of barges and other freight-carrying vessels. Since barge transportation of chemicals is considered essential to their economical distribution, the governmental tolls assessed for such maintenance are of critical interest to the chemical industry. Indeed, until recently, freight transportation by water was entirely free of such tolls.

Most oceangoing vessels, particularly those used between North America and other continents, are self-propelled. For the movement of packaged freight, ocean transportation in recent years has been dominated by container ships designed to load and carry large, trailer-sized containers. Because such ships can be loaded and unloaded more quickly than traditional freight-carrying vessels, the amount of time spent in port is greatly reduced, thereby increasing the number of voyages possible in a given period and reducing operating costs. Other types of ocean vessels include tankers and dry-bulk ships for the transportation of a wide variety of liquid hydrocarbons and chemicals and materials such as coal, coke, and ores, in large quantities. Chemical tankers tend to be smaller in size than petroleum tankers and usually have several compartments, each designed to carry one or more products.

Pipelines. The feasibility of pipeline transportation depends on the availability of very large quantities of compatible materials between locations with sufficient storage facilities. Thus, pipeline transportation is predominantly, but not exclusively, limited to the movement of hydrocarbons, many of which are raw materials in the production of petrochemicals. Although proprietary pipelines, generally of short distances, are not unusual, commercial petroleum pipelines are considered to be common carriers which are available to serve all customers who can tender sufficient quantities of acceptable liquids for transportation between terminals. The transportation of chemicals by pipeline is not economically significant, but the possibility of slurry pipelines may prove to be of value to the chemical industry in the future (see also Pipelines).

Air Transport. Relatively small quantities of chemicals are transported by air, although the availability of such service for the movement of samples, emergency shipments, and radioactive chemicals with a short half-life is important. Both economic and safety considerations impede the development of air carriage as a significant means of transporting a substantial volume of chemicals.

Other Services. Domestic freight forwarders, although sometimes treated as common carriers, do not actually engage in intercity transportation. Instead, they arrange to consolidate multiple small shipments which they then forward in carload or truckload lots to a central location for subsequent distribution to individual destinations. Shipper cooperatives (sometimes called shippers' associations) perform similar consolidations and distribute the resulting savings in freight charges to their members. In the export and import trade, similar services are furnished by commercial operators known as nonvessel operating common carriers (NVOCC).

The diversity and flexibility of a highly developed transportation structure is demonstrated by intermodal transportation, ie, the combination of two or more transportation modes. Traditional combinations, such as rail and water or truck and water, are essentially end-to-end arrangements. More recent combinations of truck and rail facilities include the piggyback transportation of trucks or truck bodies on railroad flat cars, or similar loadings of trucks or containers on ships or barges, or even the movement of loaded barges aboard ships in LASH (lighter-aboard-ship) service. The introduction of such methods, although sometimes plagued with legal and regulatory problems, has led to substantial economies for both carriers and shippers.

Warehouses and Terminals. Warehousing constitutes an integral part of the distribution system of the United States. Although employed primarily to store inventory, warehouses are also used to assure timely deliveries to customers remote from a production facility. Additionally, warehousing may facilitate the aggregation of large shipments, thus reducing the cost of transportation. Warehouses may be owned and operated by individual companies for their own purposes or they may be available to the public for storage of goods. The chemical industry makes extensive use of bulk terminals for storage of liquid- and dry-bulk materials in a wide variety of sizes and types of tanks, silos, bins, and other facilities.

Warehouse and terminal operators, who offer their facilities and services for compensation, are liable for goods in their custody if they are negligent. Many operators limit the amount of their liability by provisions in the warehouse receipt, which is the customary document issued as evidence of goods held in storage. Warehouse charges are generally determined by the amount of space occupied by the stored goods and the period of storage, as well as the ease of handling, hazardous characteristics, and similar considerations.

Shipping

Shipping Terms. Although frequently referred to as shipping terms, terms such as fob, fas, and cif are actually terms of sale because they pertain to the relationship between vendor and vendee, rather than between shipper and carrier. The term fob, for example, means free on board and usually indicates that delivery of the goods to the vendee will occur when the goods, packaged in accordance with the terms of the sales agreement, are delivered aboard a vehicle of the type agreed upon at the fob point named. The risk of loss in transit is usually transferred at the point of delivery. Fob origin means that the vendee or consignee assumes such risk, whereas fob destination means that the vendor or consignor assumes it. In the absence of a contrary agreement between vendor and vendee, freight charges are payable by and are for the account of the party who bears the risk of transit loss.

The selection of shipping terms has a material effect on the sales contract. The party with the risk of loss must decide whether or not to insure against such risk and must prepare and file a claim against the transportation carrier when goods are lost or damaged in transit. Unless otherwise agreed, that party must also pay transportation charges and file any claims for freight overcharges. In export or import transactions, shipping terms such as fas (free alongside ship) or cif (cost, insurance, freight) may also determine the party responsible for preparing required documents, obtaining customs clearances, and similar matters (18).

Shipping Documents.   The document most commonly used in both domestic and international transportation is the bill of lading, which serves as a receipt for goods delivered to a carrier as well as a contract of carriage. Bills of lading may be negotiable documents and, as such, constitute evidence of title or the right to possession of the goods described in the document. The face of the bill of lading identifies the shipper, origin, consignee, destination, vehicle or car number, routing, commodity, containers, quantity shipped, and other information required for the carrier to properly transport and invoice the freight. The contract terms, usually on the reverse side of the bill of lading, specify the rights and obligations of shipper, consignee, and carrier, including, most importantly, limitations on carrier liability, the methods and time limits for submitting damage claims, payment of freight charges, and disposition of the goods in case of nondelivery. The short form bill of lading does not reproduce all contract terms on the reverse side, but refers to such terms as published elsewhere.

In international trade, the ocean bill of lading serves essentially the same purposes although it may differ in form and content and is frequently negotiated in such a manner that payment by the foreign consignee is required before delivery of goods by the carrier. Where a shipper's freight occupies the whole or a substantial portion of a particular vessel, the document used may be a voyage charter, which provides for use of the vessel for a single voyage. For shipments of bulk chemicals by tanker, a shipper may use a specified tank on a particular vessel under an arrangement referred to as a parcel charter. Another document commonly used in ocean shipping is a dock receipt, which is evidence that the goods have been delivered to a dock pending arrival of the vessel on which they will be loaded for transportation overseas.

A freight bill is an invoice issued by a carrier requesting payment for transportation services. Generally, the freight bill contains the information shown on the face of the bill of lading, together with the freight rate and charges and the carrier's invoice, or pro, number. Carriers usually require submission of the original paid freight bill as part of a shipper's claim for freight loss, damage or delay, or for overcharge. A paid freight bill may also be required to prove that the vendor or vendee has paid freight charges, in cases where freight is added to or deducted from the merchandise invoice or is equalized with freight charges from competing shipping points.

A delivery receipt is a document, frequently a copy of the freight bill, which has been signed by the consignee as evidence of delivery of the goods by the carrier. Where no exceptions have been noted on the delivery receipt, it constitutes *prima facie* proof of delivery in full, and in apparent good order and condition.

Interstate and Intrastate Commerce

The applicability of Federal and state transportation laws and regulations depends on whether transportation constitutes interstate or intrastate commerce. Freight rates for interstate and intrastate carriage frequently vary even though movements of identical products and locations are involved.

Except in rare instances, it can be assumed that transportation that requires physical movement across state boundaries is interstate commerce. On the other hand, transportation that takes place wholly within the confines of a single state is not necessarily in intrastate commerce, since such transportation may be a portion of a continuous movement in interstate commerce. As a general rule, where there is an original and persisting intent that transportation be provided from a point in one state to a point in another state or in another country without coming to rest at an intermediate location, all portions of such transportation are considered interstate, even though one or more portions may be performed wholly within a single state (19). It is immaterial that different carriers or even different modes of for-hire transportation may be employed for each portion, or that new bills of lading are issued or separate freight bills rendered.

Thus, for example, where freight is transported by motor carrier from Springfield, Illinois, to a railroad piggyback ramp in Chicago for movement in railroad service to a place in New York, the truck service is in interstate commerce although neither the vehicle nor its driver physically leaves the state of Illinois in the course of such transportation. Consequently, the motor carrier would be subject to Federal franchise requirements and would also be required to publish and assess freight rates applicable to interstate shipments. Similarly, a shipment from Albany, New York, to a New York City pier for export to Europe, is considered interstate (or foreign) transportation and, therefore, subject to Federal regulation, despite the issuance of a new, export bill of lading at the pier.

However, a shipment transported by motor carrier within New York City to a New York City pier for export, although likewise in interstate commerce, would generally not be subject to such regulation because a provision of the Interstate Commerce Act exempts interstate motor carriage within a single municipality or commercial zone (20). An additional variation of the general rule occurs when freight is transported by a private carrier from one state to a second state, where it is given to a for-hire carrier for final delivery within the second state. In such cases, the ICC and the courts have concluded (21) that the for-hire portion of such movement is not subject to Federal regulation even though the freight actually crossed a state line in the course of the through transportation.

Economic Regulation

In the United States, transportation has long been subjected to regulation by both Federal and state governments. Generally, such regulation is directed at operational safety or toward economic concerns such as discrimination in rates and services or excessive competition. At the Federal level, the ICC, the Federal Maritime Commission (FMC), the Civil Aeronautics Board (CAB), and the Federal Energy Regulatory Commission (FERC) are all concerned with economic regulation of various modes of transportation. The ICC regulates interstate railroads, common and contract motor and water carriers, freight forwarders and pipelines (other than oil). The FMC regulates foreign and so-called domestic offshore water carriage (Puerto Rico, Hawaii, Alaska, Virgin Islands). The CAB regulates airlines, and the FERC regulates pipeline transportation of oil and other energy resources.

Generally, economic regulatory statutes provide for the control of entry into the transportation business, the regulation of freight rates and charges, and various finance, accounting, and insurance requirements, although there are numerous exceptions. Among the exceptions of particular importance to the chemical industry is that afforded to water carriage of liquefied and dry-bulk commodities.

In recent years, relaxation of some economic regulatory controls has resulted in substantial change in the transportation industry. Thus, entry into the motor-carrier business has been greatly liberalized, with the result that more carriers are available, with consequent increased competition and reduction in freight rates. Similarly, railroads are now permitted to enter into contracts with individual shippers, providing for guaranteed volumes of movement at reduced rates, improved services or car supply, discounts for routing via specified railroads, and other flexible arrangements previously considered unlawful. Antitrust immunity, which most carriers formerly enjoyed in the collective establishment of rates, has been removed to a substantial extent and, in general, more reliance has been placed on market competition to achieve the objectives of economic regulation.

Regulation, however, has by no means been entirely abandoned and, in many respects, at least the form of railroad and motor carrier regulation remains essentially intact. Thus, for example, both common and contract motor carriers are prohibited from engaging in transportation unless authorized by the terms of operating authority (franchises) issued by the ICC. Although such franchises are more liberally granted than before, their issuance is not a mere formality, and shippers are frequently requested to furnish evidence that the authority sought by the carrier serves a public need. Similarly, railroads must obtain ICC authority in order to extend their lines (although few railroads have undertaken to do so in recent years) or to abandon existing service, though such abandonments are more freely allowed. In most cases, both railroad and motor-carrier consolidations or mergers require ICC approval.

With respect to freight rates, historic rules requiring that rates be reasonable and prohibiting discrimination or preference as between particular shippers or geographic areas remain in effect, although their impact has been largely dissipated by provisions placing increased reliance on competitive forces. In connection with railroad rates, for example, the ICC has lost virtually all of its powers to prescribe maximum reasonable rates, except in cases where railroads exercise market dominance with regard to particular movements (22). In most cases, however, the ICC has tended to make it extremely difficult to prove the existence of such market dominance, to the consternation of transportation managers in the chemical industry who contend that a substantial volume of chemicals railroad traffic is captive to that form of transport.

A significant result of regulatory relaxation is an increase in the authority of the ICC to grant administrative exemptions from railroad regulation (23). The ICC has exercised that authority in a variety of ways, including the virtually complete deregulation of most piggyback transportation (24). Thus, railroads are not treated as regulated carriers when performing piggyback service and need not file with the ICC their rates and charges for such service.

Regulated carriers are required to publish and file with the ICC tariffs or schedules of their rates and charges. Such tariffs must be strictly adhered to, regardless of errors, conflicting promises or agreements, contrary intent, or other circumstances. Any deviation from the terms or provisions of a tariff publication constitutes an illegal rebate, for which the law provides severe criminal penalties. Nevertheless, the rule requiring strict adherence to tariff rates should not be understood to mean that the ICC prescribes such rates and other tariff provisions. Indeed, the ICC only infrequently prescribes freight rates (and is likely to do so even less frequently in the future), and it is common practice for shippers and carriers to negotiate revised rates, although such rates may become effective only after publication upon the requisite period of notice, generally 30 days. The multiplicity and complexity of tariff rules frequently generate errors in their application, and shipper claims of overcharges, as well as carrier claims of undercharges, are not uncommon. In either case, a period of three years is provided for the filing of an appropriate claim or lawsuit (25), after which such claims are irretrievably barred.

Recent Federal legislation permits contracts between shippers and railroads (26). Such contracts must be filed with the ICC but are treated as confidential and withheld from public scrutiny, although tariffs containing the essential terms of such contracts must be filed and kept open for public inspection. Upon approval of such a contract by the ICC, however, transportation performed in accordance with its terms is deemed unregulated and the parties must resolve any contractual disputes among themselves or in the courts, without the aid of the ICC.

Among other aspects of ICC regulation that have survived reform are requirements pertaining to time limitations for the collection of freight charges by common carriers (27) and insurance requirements applicable to motor carriers (28). Thus, ICC credit regulations require railroads to collect all freight charges within five days after issuance of a freight bill, and motor common carriers must collect charges within seven days after billing. Motor carriers must also maintain certain minimal liability insurance coverage for the protection of the public, as well as insurance covering carrier liability for loss of or damage to freight. The latter type of insurance is frequently beneficial to shippers who have difficulty in collecting damage claims because a carrier has become bankrupt.

With some exceptions, state governments have regulated economic activity in intrastate transportation in a manner similar to Federal regulation, but it does not appear that many states have been quick to follow the Federal example of extensive regulatory reform of trucking, although a few, such as Florida and Arizona, have removed all economic restraints on such transportation. With respect to railroads, however, Federal legislation has effectively compelled the states to adopt Federal regulatory requirements or to abandon railroad regulation entirely (29).

Freight Rates and Allowances.   The establishment of freight rates, ie, a transportation price structure, embraces virtually all articles of commerce, in a multitude of packages and quantities, via numerous routes, and between innumerable locations. A variety of intermediate services are included, such as storage or reconsignment in transit and stop-offs to partially load or unload. The classification of freight into various categories or classes is an effort to systemize the various factors considered in fixing a particular rate. Classification is based on freight density, susceptibility to damage or theft, the value of the goods and other factors.

Freight classification also establishes a more-or-less standard nomenclature to identify the numerous products shipped in commerce. Such standardization facilitates the preparation of shipping documents, the determination of freight rates, and the free interchange of freight between connecting or competing carriers. Chemical and freight nomenclatures, however, are frequently different. Thus, for example, many chemicals may be grouped under the single freight description, Chemicals, NOIBN, referring to chemicals which are not otherwise indexed (in the freight classification) by name. On the other hand, a particular product as, eg, acetone, may be specifically listed and, therefore, would not qualify for inclusion in the NOIBN category. Misdescription is a frequent source of freight overcharges and undercharges. Because regulated carriers are required to adhere strictly to the rates and charges published in their tariffs, improper freight descriptions often result in deviations from published charges. Such deviations may sometimes be construed as illegal rebates which are subject to severe criminal penalties. An example of a few listings in the Uniform Freight Classification for railroads is given in Table 2 (30).

Because of the wide diversity of transportation carriers and services, many thousands of tariffs are filed with the regulatory agencies and numerous changes and revisions become effective daily. Large shippers usually find it necessary to maintain libraries of such tariffs and to employ specialists in their use and application. The variety of possible transportation arrangements is virtually without limit, but it may be useful to describe generally a few of the most common types of freight rates and charges.

Less-than-carload or less-than-truckload rates are applicable to quantities of particular commodities less than a specified volume considered to constitute a carload or truckload quantity of such commodities. In most cases, small shipments are also subject to a minimum charge per shipment. However, almost all railroads have abandoned the transportation of less-than-carload freight.

Table 2.   Uniform Freight Classification 6000-A

Item	Article	Less carload rating	Carload minimum, metric tons[a]	Carload rating
23260	carbon tetrachloride			
	in carboys	100	14	45
	in containers in barrels or boxes			
	or in metal cans completely jacketed,			
	or in bulk in steel barrels; also,			
	carload in tank cars	55	16	30
23315	chemicals, phosphoric, in boxes	70	18	35

[a] Classification lists carload minimum in pounds.

Carload or truckload rates are applicable to quantities of a commodity sufficient to constitute a specified minimum carload or truckload volume. Such rates, of course, are substantially lower than less-than-carload or less-than-truckload rates. Freight rates are usually stated in dollars per 100 pounds, although rates on some materials such as coal or gravel may be stated per short ton or similar unit of weight.

Multiple car rates are applicable only when a specified number of carloads is tendered to a railroad for transportation in a single shipment. For commodities such as coal, which move in large volumes, trainload rates may be provided.

Annual (or periodic) volume rates are applicable to individual shipments which are part of an aggregate tonnage of a particular commodity or commodities that a shipper has agreed to ship between specified points in a specified period.

Accessorial charges are charges for services that are ancillary to line-haul transportation, such as switching, demurrage, storage or stopping in transit, reconsignment, and similar services.

All terms and provisions of published freight tariffs must be strictly observed. Thus, any payment by the carrier to the shipper, even as compensation for services performed by the shipper in lieu of the carrier, would be construed as a reduction of the tariff charges and, therefore, an illegal rebate. To avoid such rebates, carriers may publish in their tariffs allowances which they are willing to pay to the shipper, or deduct from the freight bill, for such services. For example, a carrier who includes loading as part of transportation service may publish an allowance to shippers for performing such loading. Similarly, motor carriers may publish allowances for shippers who deliver freight to the carrier's terminal rather than requesting pick-up by the carrier.

Of greatest importance for the chemical industry are the mileage (distance) allowances published by railroads for use of tank cars, hopper cars, and other railroad equipment furnished by shippers for transportation of their products (31). Empty return movements are usually made without charge provided that aggregate loaded and empty distances on each railroad are maintained in equilibrium. Such allowances, paid for loaded miles of car movement, represent large revenues for the industry and are an important consideration in calculating the actual, or net, cost of transporting a given shipment. In recent years, mileage allowances have been substantially increased, especially for newer and larger tank cars, as a result of escalating costs of car leasing and ownership.

Freight Loss and Damage. Under the common law of the U.S., common carriers by land were liable for loss of or damage to goods in their custody, except loss or damage resulting from an act of God, the act of a public enemy (revolution or hostility between governments), an act of governmental authority (such as quarantine), inherent vice or defect of the goods, or the fault or negligence of the shipper. The Interstate Commerce Act codifies such liability for railroads, motor common carriers and domestic freight forwarders and provides that such carriers are liable for the full, actual loss or damage to the goods (32). The act also provides that claims for loss or damage of goods transported via joint through routes may be filed against either the originating or destination carrier or against an intermediate carrier on whose lines the goods are known to have been damaged. Additionally, the Act provides a minimum time limit for filing a claim (9 mo) and commencing suit (2 yr). Despite the 9-mo minimum for filing claims, shippers are well advised, in the case of damage that is not apparent at the time of delivery, to promptly notify the delivering carrier and afford an opportunity for inspection of the damaged goods.

As noted above, common carriers are liable only for full actual loss. Thus, for example, a vendor who has prepaid freight charges that are not separately invoiced to the vendee may not recover both the invoice value of the goods and the freight paid, since such recovery would put the vendor in a more advantageous position than if the goods had been delivered undamaged and the vendee had paid the invoice as rendered.

When goods consigned to a shipper's warehouse or terminal are damaged, disputes frequently arise as to their value, the carrier contending that shippers should not earn profit on sales not made and the shipper contending that it should not be required to produce goods merely to recover its costs. Such disputes are sometimes resolved by payment of the sales price less costs not incurred, such as the cost of delivery from warehouse to consignee.

Common carriers may not limit their liability for loss or damage except for a consideration in the form of a reduced freight rate, the shipper retaining the right to select either full or limited liability. Prior to recent regulatory reform legislation, such limitations generally required the approval of the ICC, which was sparingly given. Since the enactment of that legislation, such approval is not required and shippers and carriers may agree upon liability limitations, although consideration is still required. Freight rates applicable to shipments subject to limited carrier liability are referred to as released rates.

Contract carriers are not held to the same standard of liability as common carriers since they are considered ordinary for-hire bailees and, therefore, are liable only for their failure to exercise a reasonable degree of care for goods in their custody or possession, although such liability may be varied by the contract.

The liability of water carriers is established under principles of traditional admiralty law which, generally, reflect the fundamental concept of liability for negligence, modified to accommodate risks peculiar to the long and dangerous voyages in ancient times. Thus, for example, shipowners may limit their liability to the value of the vessel and cargo after an accident, and cargo jettisoned at sea to save the venture may be compensated by general average charges against owners of the remaining cargo. Many other variations of water-carrier liability for cargo damage can be found upon examination of the multiplicity of charters, bills of lading, tariffs, and contracts employed in connection with such transportation.

The liability of common carriers by land for loss of or damage to freight is sometimes referred to as that of an insurer. This characterization is technically incorrect, however, because carriers are not liable for the fault or negligence of the shipper as, for example, in using faulty or defective packaging or in improper loading. Nevertheless, most transit loss or damage is recoverable from the carrier and, as a result, many shippers find it unnecessary to insure freight transported by land carriers, unless carrier liability has been limited in accordance with the legal principles discussed above. On the other hand, it is common practice for shippers or receivers to insure cargo transported by water, unless the carrier has contractually agreed to purchase such insurance. In some cases, as in transportation by air or in highway carriage of household goods, shippers may be afforded an opportunity to purchase insurance directly from the carrier.

Safety Regulation

Before the creation of the U.S. Department of Transportation (DOT) in 1967, the ICC was authorized to prescribe rules and regulations for rail, truck, and pipeline safety. The Federal Aviation Administration (FAA) was responsible for air safety, and the U.S. Coast Guard for safety on the inland and coastal waterways. Upon the establishment of DOT in 1967, the FAA and the Coast Guard were transferred to that Department and the safety functions formerly administered by ICC were assumed by DOT.

In general, DOT safety regulations fall into two categories. The first pertains to the qualifications and hours of service of carrier employees and the safety of transport operations and equipment. The second, of special concern to the chemical industry, pertains to the transportation of hazardous materials and related commodities.

In connection with motor-carrier safety, DOT has assigned such responsibility to the Federal Highway Administration, which has prescribed extensive regulations regarding drivers' qualifications, the maintenance of drivers' logs, required vehicle equipment and inspections, and accident records and reports (33). Such regulations are applicable to all interstate carriers by highway, and violations are subject to criminal and civil penalties. The National Highway Traffic Safety Administration (NHTSA), a part of DOT, issues motor vehicle safety standards including, eg, standards for tires, brakes and brake fluids, bumper protection, and passenger restraint systems.

Similar regulations for railroads, including requirements pertaining to the quality of tracks, train speeds, and freight-car construction, are prescribed by the DOT's Federal Railroad Administration. The FAA and the Coast Guard continue to regulate safety by air and water, respectively, including aircraft and ship construction, maintenance, and operation. A separate, independent Federal agency, the National Transportation Safety Board, is responsible for investigating serious accidents by all modes of transportation, including pipeline and passenger transportation. This agency makes recommendations to Congress and the regulatory agencies with respect to safe transportation practices and regulatory requirements.

Since the end of World War II, there has been a substantial increase in the volume of hazardous materials transported domestically (34) and internationally by all modes of carriage (estimated at ca 1160×10^4 t·km per year in the United States). Increased concern with environmental protection and public safety has generated widespread interest in the safe transportation of explosives, toxic and radioactive materials, and other products with dangerous potential.

Federal legislation pertaining to the movement of hazardous materials, however, precedes such current concerns. Indeed, as early as 1866, Congress restricted the transportation of such products as nitroglycerin and blasting oil. In 1908 and 1909, the ICC was authorized to issue regulations governing railroad transportation of explosives, and in 1921 such authority was extended to the regulation of flammable liquids and solids, oxidizing materials, corrosive liquids, compressed gases, and poisonous substances (35). Subsequent amendments to the 1921 legislation further extended the ICC's jurisdiction to transportation by contract and private carriers by motor vehicle and certain liquid pipelines, and added etiologic agents and radioactive substances to the categories of hazardous materials subject to such jurisdiction (see also Explosives and propellants).

Following the transfer of safety functions from the ICC, FAA, and Coast Guard to DOT, it became evident that more comprehensive legislation and better coordinated regulation of all modes was required. Accordingly, in 1974, Congress enacted the Hazardous Materials Transportation Act (36), which consolidated the authority of DOT with respect to safety regulation of the various modes, extended its jurisdiction to include manufacturers of containers used for transportation of hazardous materials, greatly increased penalties for violation and provided other enforcement mechanisms, and authorized the regulation of any substance or material that could create an unreasonable risk to health, safety, or property. Under the Hazardous Materials Transportation Act, authority has been assigned to the DOT Materials Transportation Bureau, except for bulk water movements which remain subject to the authority of the Coast Guard.

Generally, the purposes of hazardous materials regulation are to assure adequate containment of such goods in transit, and to inform carrier personnel, cargo handlers, police, fire, and other emergency personnel, and bystanders, of the immediate possible hazards in the event of containment failure. Thus, the DOT hazardous materials regulations, which are published at 49 CFR Parts 171 to 179, prescribe in considerable detail specifications pertaining to the design of shipping containers, tank cars, and tank trucks. To assure the proper transmission of hazard information, the regulations also require that specified information warning of potential dangers be shown on various prescribed labels, placards, and shipping documents. The diamond-shaped red label signifying flammability, for example, is almost universally recognized as indicating the presence of a possible fire hazard (see Fig. 1).

Many private industrial or professional organizations, eg, ASME, the Compressed Gas Association, the CMA, and the Bureau of Explosives, publish standards for containers, materials of construction, tests, and similar matters and such standards are frequently incorporated by reference in the regulations. Contrary to widespread belief, the Bureau of Explosives is not a government authority, but is part of the Association of American Railroads, a private organization of the railroad industry. Because the Transportation of Explosives Act of 1921 specifically authorized the ICC to "utilize the services" of the Bureau, it became a powerful influence in the regulation of hazardous materials. Although some of its authority has recently been withdrawn by DOT,

Figure 1.   Flammable liquid label (37).

the Bureau continues to provide a number of important services, including publication of a tariff which reproduces the DOT hazardous materials regulations and contains carrier restrictions on the acceptance and transportation of hazardous materials.

Although the hazardous materials regulations are frequently characterized as highly complex, their application is greatly facilitated through the use of the Hazardous Materials Table at 49 CFR, §172.101, which lists, alphabetically, hundreds of hazardous materials descriptions by their "proper shipping names" as prescribed by DOT. Some materials are grouped into so-called NOS (not otherwise specified) categories incorporating a variety of products. In addition, the Table contains the hazard class and a standard four-digit identification number for each listed material, the type or types of label prescribed for application to packages, the required packaging and exceptions thereto, and certain restrictions applicable to shipments via cargo or passenger aircraft and water vessels. An Optional Hazardous Materials Table, for use with international water shipments, is provided at 49 CFR, §172.102. International shipments of hazardous materials by water and air are governed by the regulations of the International Maritime Organization (38) and the International Air Transport Association (39–40), respectively. The latter regulations are scheduled to be replaced, beginning in 1984, by those of the International Civil Aviation Organization (40).

The proper selection of chemical shipping descriptions, and the determination of the hazard class, require chemical expertise and familiarity with DOT definitions of such classes, which are provided in Part 173 of the hazardous materials regulations. In some cases, such definitions are sufficiently precise to permit objective determination of the classification of a material under consideration. Other definitions, however, sometimes lead to disagreements among qualified experts as to the appropriate classification of particular products. The proper identification and classification of hazardous materials are critical prerequisites to compliance with the packaging and communications requirements of DOT regulations.

Relatively recent additions to the materials subject to DOT regulations include hazardous wastes as defined by the EPA, and hazardous substances which, if released in certain quantities, must be reported to the U.S. Coast Guard National Response Center.

In addition to the package labels specified in the Hazardous Materials Table and in Subpart E of Part 172 of the regulations, packages and vehicles containing hazardous materials must be marked in accordance with Subpart D of Part 172 of the regulations and vehicles must be placarded in accordance with Subpart F of Part 172. Subpart D requires that certain packages be marked with the proper shipping name prescribed by DOT, the four-digit identification number, the name and address of the consignee or consignor, and package orientation markings in the case of certain liquid materials. Other markings are prescribed for specified materials or vehicles.

Although similar to labels in appearance, placards (see Fig. 2), larger and more durable than labels, are affixed to the exterior of rail cars and other transport vehicles carrying hazardous materials. In accordance with recent changes in the regulations, the identification number (accompanied by the United Nations hazard class number) may be substituted for the word or words (eg, Flammable or Poison) on the placard (Fig. 3), except placards for poison gas, explosives, or radioactive materials. In the latter cases, the number must be displayed on an orange panel of prescribed specifications.

Similar information is provided in a shipping paper prepared by the shipper,

Figure 2.   Flammable gas placard (41).

Figure 3.   The four-digit identification number and the United Nations hazard class number (42).

which must also contain a certification that the shipment has been "properly classified, described, packaged, marked and labeled" in accordance with DOT regulations. Most shippers of hazardous materials have incorporated DOT documentary requirements into their standard form of bill of lading. For hazardous waste, a hazardous-waste manifest must be prepared by the waste generator (shipper) in accordance with EPA regulations (43).

The packaging requirements of the hazardous materials regulations, at 49 CFR Part 173, are extremely detailed and include detailed specifications for shipping containers (49 CFR, Part 178) and materials of construction, tests, test reports, and manufacturer's marks. Part 178 also contains specifications for tank trucks and certain other motor vehicles; tank-car specifications are published in Part 179. The shipper is responsible under the regulations for selecting a type of container authorized for shipment of the hazardous materials to be transported. Generally, in using individual containers of the type selected, shippers may accept the package manufacturer's certification of compliance or identification of the package specification as evidence that such containers conform to DOT requirements for that specification. For containers, such as tank trucks, supplied by a carrier, shippers may rely on the manufacturer's identification plate, specification marking, or certification by the carrier.

Parts 174, 175, 176, and 177 contain loading, handling, and operating requirements for carriage of hazardous materials by rail, air, water, and highway, respectively.

Parts 106 and 107 of 49 CFR contain the procedural rules of the Materials Transportation Bureau (MTB), including provisions pertaining to enforcement and exemptions. Enforcement may result in compliance orders, injunctions, or the assessment of civil or criminal penalties as high as $25,000 and imprisonment for five years for each violation of the hazardous-materials regulations.

Exemptions (formerly referred to as special permits), issued for two years but renewable, authorize specified deviations from prescribed packaging, tests, or other requirements, provided that equivalent safety can be demonstrated (see also Packaging materials). Where the regulations do not specify any packaging for the materials to be transported, the exemption system affords the only mechanism for recognizing an acceptable packaging standard. The highway transportation of certain bulk cryogenic liquids such as hydrogen or natural gas, for example, depends on exemptions because the regulations contain no packaging specification for such transportation (see also Cryogenics).

The recent adoption by MTB of a numbering system for more specific identification of hazardous materials reflects a worldwide effort to improve response to transportation emergencies. Although accidental release of hazardous materials in transit is relatively rare (44), the potential for significant harm is of constant concern to the public and industry and is magnified by the fact that many public emergency-response agencies have had little, if any, training or experience in dealing with chemical emergencies. In an effort to provide immediate and reliable information to carriers and public officials at the scene of an emergency, the Chemical Manufacturers' Association established the Chemical Transportation Emergency Center (CHEMTREC) in Washington, D.C. Since its formation in 1971, CHEMTREC has responded to thousands of emergency calls, providing information from its files containing data on more than 16,000 chemical products. Similarly, the Chlorine Institute has organized a mutual aid program, called CHLOREP, which offers assistance at the scene of emergencies involving chlorine. Industrial response teams are usually available for assistance in connection with clean-up of spills which may be hazardous to the public or the environment (see also Alkali and chlorine products).

Despite continued efforts by industry and the Federal government to improve emergency response, a number of municipalities and communities in the United States have perceived a need to regulate the movement of hazardous materials by enacting routing, prenotification, or permit requirements, curfews, or similar restrictions. The proliferation of such local regulations has tended to burden the free flow of hazardous commodities in commerce with diverse requirements which are often inconsistent with Federal rules. The Federal Hazardous Materials Transportation Act preempts inconsistent state or local regulations, but offers little guidance as to the appropriate relationship between Federal and state or local authorities and provides no convenient mechanism for determining whether particular regulations are inconsistent. As a result, an increasing number of lawsuits (45) challenge the validity of individual state or local enactments, and it is possible that such disputes may eventually be resolved in one or more cases before the U.S. Supreme Court or by additional Federal legislation.

Outlook

Transportation and distribution costs constitute a substantial portion of the total cost of the chemical industry. Most chemical producers, therefore, can be expected to pay continuing attention to the control of such costs and to the maintenance and development of more sophisticated distribution methods. Significant recent changes in the nature and extent of economic regulation in the transportation field promise new challenges to the industrial distribution manager, especially in the area of railroad transportation. Rail carriers, armed with new freedom to price their services, to abandon unprofitable lines, and to merge with other railroads, have already demonstrated a tendency to increase rates on captive chemicals traffic. At the same time, however, the removal of regulatory restraints on contracts between shippers and railroads may, in the foreseeable future, generate a revolution in transport pricing, with benefits of a magnitude as yet unrecognized.

In the motor-carrier field, increased competition resulting from the relaxation of economic regulatory controls seems likely, in the immediate future, to generate considerable instability in the availability and reliability of carrier rates and services, thereby requiring greater vigilance and improved use of computer technology by industrial transportation managers. In time, motor carriers that successfully survive the initial competitiveness of the new transport marketplace may demonstrate a degree of efficiency capable of challenging proprietary transportation in both cost and service. The availability of energy resources and the adequacy of the highway infrastructure may impose substantial constraints on the continued growth of motor transportation.

Without a breakthrough in energy usage, the technology of transportation is not expected to change dramatically in the foreseeable future, and improvement is more likely to be concentrated on the transport infrastructure than on vehicle capabilities. Transportation safety will continue to be of concern to government, especially at state and local levels, but such concern may be directed less at new regulations and restraints and more on the application of existing computer and communications technology to more effective emergency response.

Abbreviations

CAB	= Civil Aeronautics Board
CFR	= Code of Federal Regulations
CGA	= Compressed Gas Association
CHEMTREC	= Chemical Transportation Emergency Center
cif	= cost insurance freight
CMA	= Chemical Manufacturers' Association
DOT	= Department of Transportation
FAA	= Federal Aviation Administration
fas	= free alongside ship
FERC	= Federal Energy Regulatory Commission
FMC	= Federal Maritime Commission
fob	= free on board
ICC	= Interstate Commerce Commission
LASH	= lighter-aboard-ship
MTB	= Materials Transportation Bureau
NOIBN	= not otherwise indexed by name

NOS = not otherwise specified
NVOCC = nonvessel operating common carrier
USC = United States code

BIBLIOGRAPHY

"Transportation" in *ECT* 2nd ed., Vol. 20, pp. 596–610, by W. Blanding, Marketing Publications Incorporated.

1. Chemical Manufacturers' Association, cited in *Toward a Federal/State/Local Partnership in Hazardous Materials Transportation Safety*, U.S. Department of Transportation, Washington, D.C., Sept. 1982, p. 1.
2. *Transportation Facts and Trends*, 17th ed., Transportation Association of America, Washington, D.C., Dec. 1981, p. 4.
3. *Statistics of Railroads, Class I*, Association of American Railroads, Washington, D.C., 1980.
4. Several forms of car leases are shown in S. Hoffman, *Model Legal Forms for Shippers*, Transport Law Research, Inc., Mamaroneck, N.Y., 1970.
5. For a form of trip-lease, see ref. 4, Form 135.
6. Ref. 4, Forms 3–13.
7. Ref. 4, Form 16.
8. See, *ICG R. R. v. Golden Triangle Wholesale*, 423 F. Supp. 679 (U.S. Dist. Ct., N.D., Miss., E.D., 1976); aff'd, 586F. 2d 588 (U.S. Ct. of Appeals, 5th Cct.; 1978).
9. See ref. 4 for various forms of contracts with motor contract carriers.
10. *Wilson—Investigation of Operations*, 82MCC651, 14 Fed. Carrier Cases, Paragraph 34,886 (1960); *Utley Lumber*, 94 MCC (motor-carrier cases) 458, 16 Fed. Carrier Cases, Paragraph 35,723 (1964).
11. *U.S. v. Drum*, 368 U.S. 370 (U.S. Supreme Ct., 1962).
12. *Schenley Distillers v. U.S.*, 326 U.S. 432 (U.S. Supreme Ct., 1946).
13. *Geraci Contract Carrier Application*, 7 MCC 369 (1938); see also, *Toto Purchasing*, 128 MCC 873 (1978).
14. 49 USC §10524(b).
15. 49 CFR, Part 1136.
16. *Waterborne Commerce of the United States, 1979*, Part V, U.S. Corps of Engineers, Washington, D.C., Table 3.
17. 49 USC §10542.
18. For a comprehensive listing of shipping terms and explanations, see "Incoterms," *International Rules for the Interpretation of Trade Terms*, ICC Services S.A.R.L., Paris, 1980.
19. *B&O v. Settle*, 260 U.S. 166 (1922); *Southern Pacific Transportation Co. v. ICC*, 565 F 2d 615 (1977).
20. 49 USC §10526.
21. *Pennsylvania Railroad v. Ohio P.U.C.*, 298 U.S. 170 (1936); *Motor Tsptn.*, 94 MCC 541 (1964), aff'd., 382 U.S. 373 (1966).
22. 49 USC §10701a.
23. 49 USC §10505.
24. Ex Parte No. 230 (Sub.-5), *Fed. Regist.* **46,** 14348 (Feb. 27, 1981).
25. 49 USC §11706.
26. 49 USC §10713.
27. 49 CFR, Parts 1322, 1323, and 1324.
28. 49 CFR, Part 1043.
29. 49 USC §11501.
30. *Uniform Freight Classification 6000-A*, ICC UFC 6000-A, issued Jan. 15, 1981, available from Tariff Publishing Officer, 222 S. Riverside Plaza, Chicago, Ill. 60606.
31. *Mileage Tariff PHJ 6007-H*, ICC PHJ 6007-H, issued March 22, 1982, available from H. J. Positano, Agent, 1250 Broadway, New York, N.Y. 10001.
32. 49 USC §11707.
33. 49 CFR, Parts 390–397.
34. *Regulatory Review and Development Plan and Schedule of Rulemaking Actions (January 1979–January 1980)*, Materials Transportation Bureau, Washington, D.C., p. 4.

35. Act of March 4, 1921, c. 172, 41 Stat. 1444, 1445, later codified to Title 18, U.S. Code, Sections 831–835.
36. 49 USC §1801 *et seq.*
37. 49 CFR §172.419.
38. *International Maritime Dangerous Goods Code*, in 4 vols., incorporating all amendments to and including No. 13, 1976; additional amendments have been issued to and including No. 19, 1982, International Maritime Organization (formerly the Inter-Governmental Maritime Consultative Organization), London, 1976.
39. *Restricted Articles Regulations*, 23rd ed., International Air Transport Association, Montreal, Quebec, Canada, effective Dec. 1, 1980; see also *Dangerous Goods Regulations*, 24th ed., International Air Transport Association, Montreal, Quebec, Canada, effective Dec. 31, 1982, especially the "Important Notice" on the title page and the Preface, page 1, regarding the effective dates of and the relationship between IATA regulations and ICAO regulations.
40. *Technical Instructions for the Safe Transport of Dangerous Goods by Air*, 1982 ed. (Doc. 9284-AN/905), International Civil Aviation Organization (ICAO), Montreal, Quebec, Canada, available in English, French and Spanish editions from INTEREG, International Regulations Publications and Distributing Organization, Chicago, Ill.
41. 49 CFR §172.532.
42. 49 CFR §172.334.
43. 40 CFR, Part 262.
44. *CMA News*, 5 (Sept. 1, 1982).
45. Eg, *City of N.Y. v. Ritter*, 515 F. Supp 663 (U.S. Dist. Ct., S.D., N.Y., 1981).

General References

United States Code Annotated, West Publishing Co., St. Paul, Minn. A compilation of U.S. laws of a general and permanent nature consisting of 50 Titles. Although many provisions of various Titles affect transportation, Titles 49 (Transportation) and 46 (Shipping) are of particular interest. Among other important statutes included in Title 49 are the Interstate Commerce Act, the Revised Interstate Commerce Act (§10101 *et seq.*), the Department of Transportation Act (§1651 *et seq.*), and the Hazardous Materials Transportation Act (§1801 *et seq.*). Title 46 collects various statutes pertaining primarily to water transportation including, among many others, the Shipping Act of 1916 (§801 *et seq.*) and the Jones Act (§688).

Code of Federal Regulations, Office of the Federal Register, National Archives and Records Service, General Services Administration, Washington, D.C. A codification of the rules and regulations published in the *Fed. Regist.* by the departments and agencies of the Federal government. The Code is divided into 50 titles divided into chapters, subchapters, parts, and subparts. Title 49 contains the rules and regulations of the Interstate Commerce Commission, the Department of Transportation, and other Federal agencies concerned with transportation.

Fed. Regist., published daily, Monday through Friday (except official holidays) by the Office of the Federal Register, National Archives and Records Service, General Services Administration, Washington, D.C. Provides a uniform system for making available to the public Federal regulations and proposed regulations, and other information.

Interstate Commerce Commission Reports, Superintendent of Documents, U.S. Government Printing Office, Washington, D.C., cited (*Volume*) ICC (*Page*). A continuing series of decisions of the Interstate Commerce Commission. Volume 1 was published in 1887.

Motor Carrier Cases, Superintendent of Documents, U.S. Government Printing Office, Washington, D.C., cited (*Volume*) MCC (*Page*). A continuing series of ICC decisions pertaining to motor carrier regulation.

Federal Carriers Reporter, Commerce Clearing House, Inc., Chicago, Ill. A loose-leaf service in 4 vols. containing statutes, regulations, forms, and current court and administrative decisions pertaining to motor and water carriers, and domestic freight forwarders.

Federal Carrier Cases, Commerce Clearing House, Inc., Chicago, Ill. A continuing series of volumes selectively reporting decisions of the ICC and Federal and state courts pertaining to motor carrier, water carrier, and domestic freight forwarder regulation.

State Motor Carrier Guide, Commerce Clearing House, Inc., Chicago, Ill. A loose-leaf service in 2 vols. comprising an operating guide to state regulation of motor carriers, with state-by-state digests of laws and regulations.

Interstate Commerce Acts Annotated, Interstate Commerce Commission, Washington, D.C.; 22 vols. plus Advance Bulletins (last published Aug. 1981). A compilation of the Interstate Commerce Act and related Federal laws, with digests of decisions of the Federal courts and the ICC relating to each section of such Acts.

Hawkins Index-Digest-Analysis of Decisions Under the Interstate Commerce Act, Hawkins Publishing Co., Washington, D.C. A loose-leaf service in 10 vols. collecting and digesting decisions under the Interstate Commerce Act pertaining primarily to railroads.

Hawkins Index-Digest-Analysis of Decisions Under Part II and Part IV of the Interstate Commerce Act, Hawkins Publishing Co., Washington, D.C. A loose-leaf service in 4 vols. collecting and digesting decisions under the Interstate Commerce Act pertaining to motor carriers and freight forwarders.

Hawkins Index-Digest-Analysis of Federal Maritime Commission Reports, Hawkins Publishing Co., Washington, D.C.

American Maritime Cases, American Maritime Cases, Inc., Baltimore, Md. Published monthly, except August, under the auspices of the Maritime Law Association of the United States and the Association of Average Adjusters of the United States.

Chemical Regulation Reporter—Hazardous Materials Transportation, The Bureau of National Affairs, Inc., Washington, D.C. A loose-leaf service in 2 vols., containing DOT hazardous materials regulations and news and information on regulatory activity in hazardous materials transport.

J. Guandolo, *Transportation Law*, 3rd ed., Wm. C. Brown Co., Dubuque, Iowa, 1979. A comprehensive review of transportation law and practice, with emphasis on railroads. Regulatory and legislative developments since 1979 suggest caution in the use of this volume.

M. L. Fair and J. Guandolo, *Transportation Regulation*, 7th ed., Wm. C. Brown Co., Dubuque, Iowa, 1972. A comprehensive discussion of transport regulation prior to recent legislative reforms.

G. L. Shinn, *Freight Rate Application*, Simmons-Boardman Publishing Corp., New York, 1948.

J. C. Colquitt, *The Art and Development of Freight Classification*, National Motor Freight Traffic Association, Inc., Washington, D.C., 1956.

J. M. Miller, *Law of Freight Loss and Damage Claims*, 3rd ed. by R. R. Sigmon, Wm. C. Brown Co., Dubuque, Iowa, 1967.

W. J. Augello, *Freight Claims in Plain English*, Shippers National Freight Claim Council, Inc., Huntington, N.Y., 1979.

S. Sorkin, *How to Recover for Loss or Damage to Goods in Transit*, Matthew Bender, New York. A loose-leaf service in 2 vols.

G. Gilmore and C. L. Black, Jr., *The Law of Admiralty*, 2nd ed., The Foundation Press, Inc., Mineola, N.Y., 1975.

W. Poor, *American Law of Charter Parties and Ocean Bills of Lading*, 5th ed., Matthew Bender, New York, 1968.

ICC Practitioners' Journal, published bimonthly by the Association of Interstate Commerce Commission Practitioners, Washington, D.C.

Journal of Maritime Law and Commerce, published quarterly by Jefferson Law Book Co., Division of Anderson Publishing Co., Cincinnati, Ohio.

Traffic World, published weekly by The Traffic Service Corp., Washington, D.C. A widely read news magazine for the transportation industry.

I. L. Sharfman, *The Interstate Commerce Commission*, The Commonwealth Fund, New York, 5 vols., 1931–1937. A classic and scholarly study, largely of historical interest.

T. G. Bugan, *When Does Title Pass*, Wm. C. Brown Co., Dubuque, Iowa, 1951. A valuable and useful discussion of shipping terms, although somewhat outdated.

R. C. Colton and E. S. Ward, *Practical Handbook of Industrial Traffic Management*, The Traffic Service Corp., Washington, D.C., 5th ed. revised by C. H. Wager, 1973.

deregulation of the Transportation Industry, Practising Law Institute, 1981. A course handbook for a program presented March 30–31, 1981, Washington, D.C., K. R. Feinberg, Chairman. A compilation of interesting papers on transport deregulation and its effects.

Hazardous Materials 1980 Emergency Response Guidebook, DOTP 5800.2, U.S. Department of Transportation, Washington, D.C.

L. W. Bierlein, *Red Book on Transportation of Hazardous Materials*, Cahners Books International, Inc., Boston, Mass., 1977.

L. M. Trosten and D. F. Brown, II, *Transportation of Energy Resources*, Callaghan & Co., Wilmette, Ill., 1979. An interesting and useful monograph.

Transportation Agenda for the 1980s: Issues and Policy Directions, U.S. Department of Transportation, Washington, D.C., 1980. Also contains a list of some of the leading transportation organizations.
Courier, a periodic newsletter of the Hazardous Materials Advisory Council, Washington, D.C.
Hazardous Materials Transportation, a newsletter published monthly by *Traffic Management* magazine, Cahners Publishing Co., Boston, Mass.
The Official Railway Guide, North American Freight Service Edition, published bimonthly by National Railway Publication Co., New York. Contains maps of railroads, railroad officers and executives, lists of railroad stations, interchange points and other useful information pertaining to railroads.
Transport of Dangerous Goods, United Nations, 1976. Recommendations of the U.N. Committee of Experts on the Transport of Dangerous Goods.
Handling & Shipping Management, published monthly by Penton-IPC, Cleveland, Ohio.
Transport Topics, weekly newspaper of the American Trucking Associations, Inc., Washington, D.C.
Federal Register Extract Service, Hazardous Materials Advisory Council, Washington, D.C. Extracts of materials originally published in the *Fed. Regist.* pertaining to hazardous materials, hazardous substances, and hazardous wastes.
D. P. Locklin, *Economics of Transportation*, 7th ed., Richard D. Irwin, Inc., Homewood, Ill., 1972. A classic textbook on transportation economics, with strong emphasis on regulation and regulatory history.
C. L. Dearing and W. Owen, *National Transportation Policy*, The Brookings Institution, Washington, D.C., 1949. One of the early postwar studies of national transportation policy calling for reform of regulation, which ultimately led to the major revisions of recent years, chief among which are the *Motor Carrier Act of 1980*, Public Law 96-296 (94 Stat. 793) and the *Staggers Rail Act of 1980*, Public Law 96-448 (94 Stat. 1895).
Hazardous Materials: A guide for State and Local Officials, U.S. Department of Transportation, Washington, D.C., Feb. 1982. A comprehensive guide to DOT hazardous materials regulations, including brief discussions of legislative history and the enforcement and regulatory processes.

Stanley Hoffman
Union Carbide Corporation

TRIAZINETRIOL.   See Cyanuric acid and isocyanuric acids.

TRICRESYL PHOSPHATE.   See Plasticizers; Phosphorus compounds.

TRIETHANOLAMINE.   See Alkanolamines.

TRIETHYLENE GLYCOL DINITRATE.   See Explosives and propellants.

TRIMENE BASE.   See Rubber chemicals.

TRIMETHYLOLETHANE.   See Alcohols, polyhydric.

TRIMETHYLOLETHANE TRINITRATE.   See Explosives and propellants.

TRIMETHYLOLPROPANE.   See Alcohols, polyhydric.

TRIOXANE.   See Formaldehyde.

TRIPENTAERYTHRITOL.   See Alcohols, polyhydric.

TRIPHENYLMETHANE AND RELATED DYES

The triarylmethane dyes are of brilliant hue, exhibit high tinctorial strength, are relatively inexpensive, and may be applied to a wide range of substrates. However, they are seriously deficient in fastness properties, especially fastness to light and washing. Because of these deficiencies, the use of triarylmethane dyes on textiles has decreased as dyes from other chemical classes with superior properties have become available (see also Dyes and dye intermediates). Interest in this class of dyes was revived with the introduction of polyacrylonitrile fibers (see also Acrylonitrile polymers; Fibers, chemical). Triphenylmethane dyes are readily adsorbed on this fiber and show surprisingly high lightfastness and washfastness properties when compared with the same dyes on natural fibers. The durability of acrylic fibers, however, created an even greater demand for fastness properties. Modifications of the classical triarylmethane dyes in order to improve these properties met with only limited success since they were generally accompanied by a reduction in tinctorial strength. Research led to the development of novel dye types from other chemical classes such as the pendant cationic dyes, in which the localized positive charge is isolated from the chromophoric system and which were designed especially to give high lightfastness. Similarly, the diazahemicyanine dyes, which offer both brightness and fastness, gradually replaced triarylmethane dyes for acrylic fibers (see Cyanine dyes; Polymethine dyes). However, the most important commercial black dye for acid-modified fibers is still a mixture of the classical triarylmethane dyes, malachite green [569-64-2] and fuchsine [632-99-5], because of high tinctorial strength and low cost.

The triarylmethane dyes are classified broadly into the triphenylmethanes (CI 42000–43875), diphenylnaphthylmethanes (CI 44000–44100), and miscellaneous triarylmethane derivatives (CI 44500–44535). The triphenylmethanes are classified further on the basis of substitution in the aromatic nuclei as follows: diamino derivatives of triphenylmethane, ie, dyes of the malachite green series, CI 42000–42175; triamino derivatives of triphenylmethane, ie, dyes of the fuchsine, rosaniline, or magenta series, CI 42500–42800; aminohydroxy derivatives of triphenylmethane, CI 43500–43570; and hydroxy derivatives of triphenylmethane, ie, dyes of the rosolic acid series, CI 43800–43875. Monoaminotriarylmethanes are known, but they are not included in the classification since they have little value as dyes.

Chemically, the triarylmethane dyes are derivatives of the colorless compounds triphenylmethane and diphenylnaphthylmethane.

triphenylmethane   diphenylnaphthylmethane

One or more primary, secondary, or tertiary amino groups or hydroxyl groups in the para positions to the methane carbon atom are required to give the spectral absorption characteristics of the dyes. Additional substituents such as carboxyl, sulfonic acid, or halogen may be present on the aromatic rings. The number, nature, and position of these substituents determine both the hue or color of the dye and the application class to which the dye belongs. The hues include vivid reds, violets, blues, and greens. The application classes include pigments (qv) and basic (cationic), acidic (anionic), solvent, and mordant dyes. If no acidic groups are present, the dye is cationic or basic. If sulfonic acid groups are present, the dye is anionic or acidic. Carboxylic groups adjacent to hydroxyl groups in the dye confer mordant-dyeing properties. Pigments are made by combining a cationic triarylmethane dye with an ion of opposite charge derived from the heteropoly acids of phosphorus, molybdenum, and tungsten. In pigments from anionic dyes, the cation is usually barium or calcium.

Structure

The first triarylmethane dyes were synthesized on a strictly empirical basis in the late 1850s. Their structural relationship to triphenylmethane was established by Otto and Emil Fischer (1) with the identification of pararosaniline [569-61-9] as 4,4',4''-triaminotriphenylmethane. Several different structures have been assigned to the triarylmethane dyes (2–4), but none accounts precisely for the observed spectral characteristics. The triarylmethane dyes are therefore generally considered to be resonance hybrids. However, for convenience, usually only one hybrid is indicated, as shown for crystal violet [548-62-9], CI Basic Violet 3.

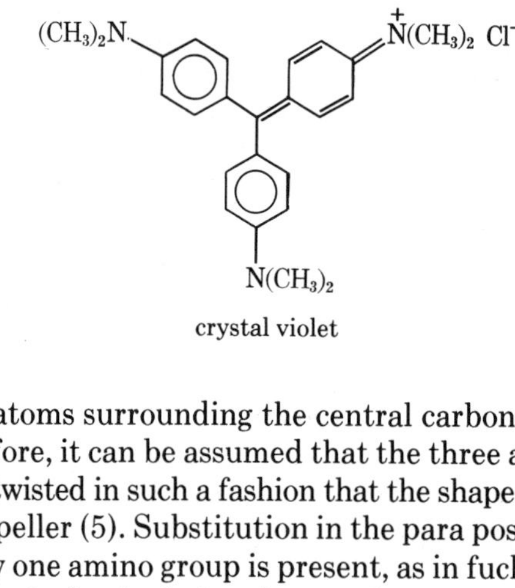

crystal violet

The ortho hydrogen atoms surrounding the central carbon atom show considerable steric overlap. Therefore, it can be assumed that the three aryl groups in the dye are not coplanar, but are twisted in such a fashion that the shape of the dye resembles that of a three-bladed propeller (5). Substitution in the para position determines the hue of the dye. When only one amino group is present, as in fuchsonimine hydrochloride [84215-84-9] (see below), the shade is only a weak orange-yellow.

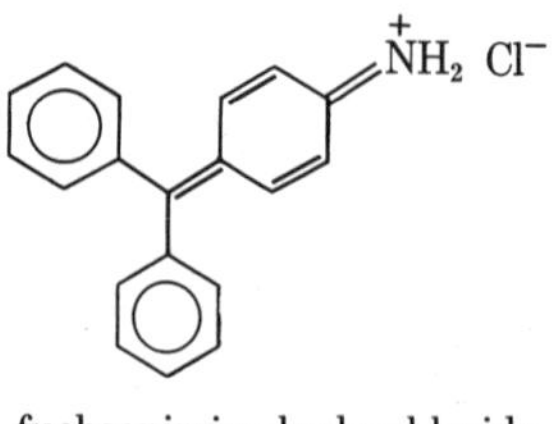

fuchsonimine hydrochloride

However, when two amino groups are present, the resonance possibility is greatly increased, resulting in a much greater intensity of absorption and in a strong bathochromic shift to longer wavelengths. The amino derivatives of commercial value contain two or three amino groups.

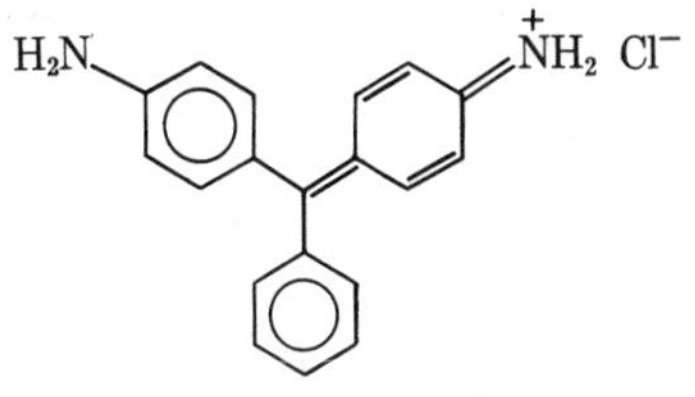

Doebner's violet [3442-83-9]

A further strong bathochromic shift is observed as the basicity of the primary amines is increased by N-alkylation, eg, malachite green.

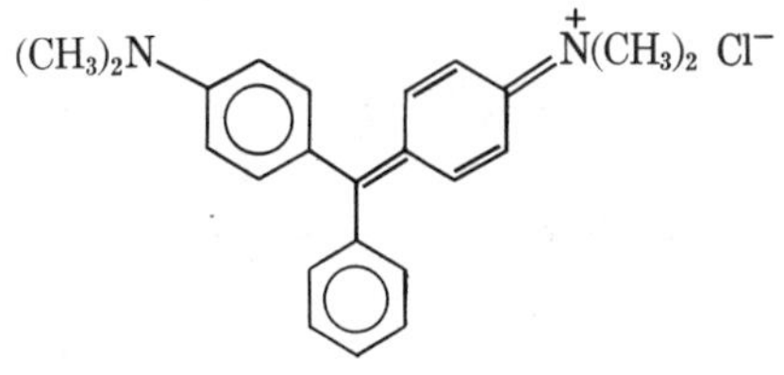

malachite green

Phenylation of the primary amino groups also produces an increased bathochromic shift in the wavelength of absorption with increasing degree of phenylation. Only monophenylation of each amino group is possible, eg,

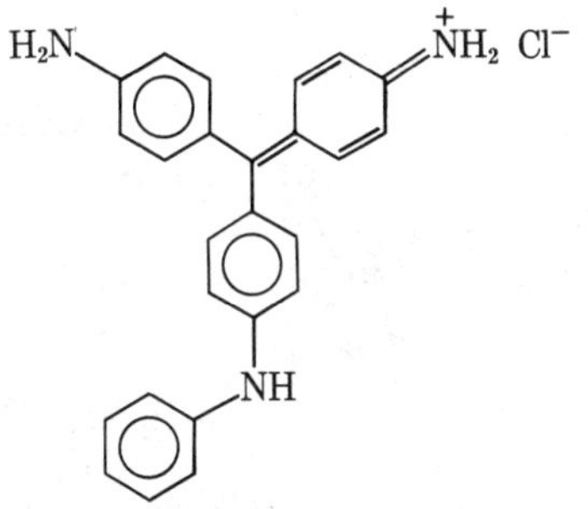

4-[(4-aminophenyl)(4-imino-2,5-
cyclohexadien-1-ylidene)
methyl]-N-phenylaniline,
monohydrochloride
[68966-31-4]

The steric effects of substituents on the color and constitution of triarylmethane dyes has been studied extensively (6–11). Replacement of the hydrogen atoms ortho to the central carbon atom in crystal violet (λ max 589 nm) by methyl groups results in a uniform bathochromic shift (ca 8 nm per methyl group) to the 2,2′,2″-trimethyl derivative [84282-50-8] (λ max 614 nm) and reduced absorptivity values (12). These

phenomena suggest that the axial rotational adjustment needed to accommodate the
o-methyl groups is shared uniformly by the three phenyl rings. The 2,6-dimethyl derivative [85294-29-7], however, shows a much larger bathochromic shift per methyl
group, and it has been suggested that the dimethylaminoxylyl ring undergoes most
of the rotational twist in such a way that the charge is localized on the other two di-
methylaminophenyl rings, thereby causing an increase in the bathochromic shift. The
steric hindrance at the central carbon atom of the 2,6-dimethyl derivative of crystal
violet is evident from the fact that the fuchsone derivative (see below) and not the
corresponding carbinol is formed by the action of a base on the dye.

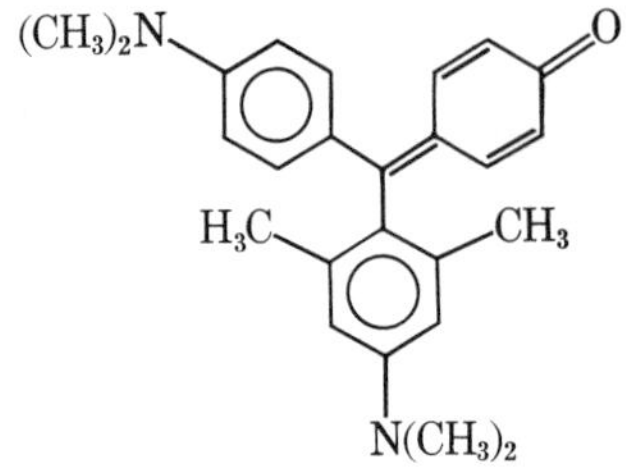

2,6-dimethyl crystal
violet

Steric hindrance facilitates the nucleophilic replacement of the terminal dimethylamino group by the hydroxyl group of the base (13).

Chemical Properties

Dyes in general and triarylmethane dyes in particular are rarely subjected to
chemical processing once they have been formed. The introduction of substituents
is usually carried out during the manufacture of the intermediates where the position
and number of groups introduced may be more precisely controlled. Dyes are sometimes exposed to oxidizing and reducing conditions during application and after-
ward.

Oxidation. Although many triarylmethane dyes are prepared by the oxidation
of leuco bases, they are usually destroyed by strong oxidizing agents. Careful choice
of both the oxidant and the reaction conditions is required to prevent loss of product
during this stage of manufacture. Overoxidation of malachite green gives a quinone
imine identical to that obtained by oxidizing tetramethylbenzidine or Michler's hydrol.

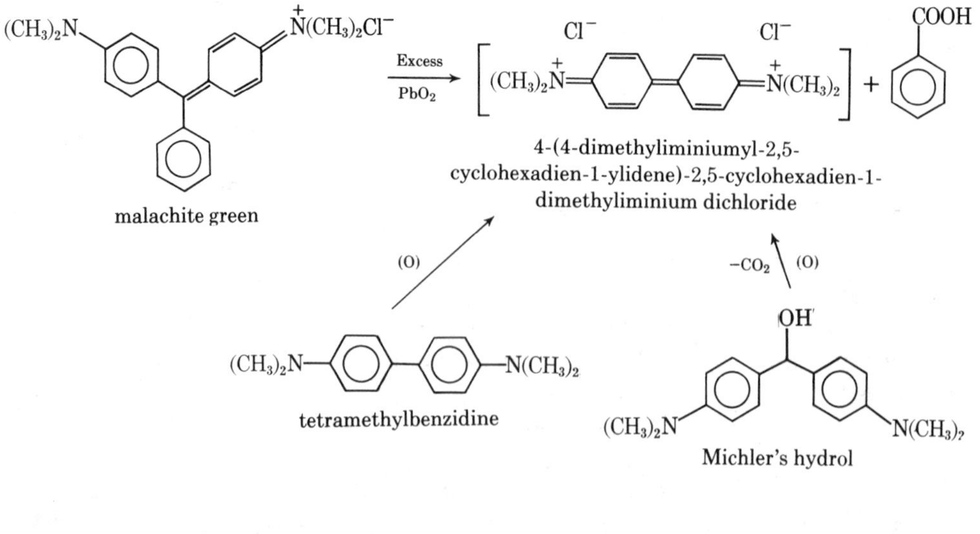

4-(4-dimethyliminiumyl-2,5-
cyclohexadien-1-ylidene)-2,5-cyclohexadien-1-
dimethyliminium dichloride

malachite green

tetramethylbenzidine

Michler's hydrol

Overoxidation may also result in the oxidative cleavage of alkyl groups from the amino substituents. Thus, triarylmethane dyes are destroyed by sodium hypochlorite, which further limits their use as textile dyes.

The triarylmethane dyes are extremely sensitive to photochemical oxidation, which accounts for their poor lightfastness on natural fibers (14–23). The photodecomposition products of malachite green in aqueous solution and on carboxymethylated and sulfatoethylated cellulose have been identified as benzophenone and 4-dimethylaminobenzophenone (14). In addition, p-dimethylaminophenol was found in solution studies. A mechanism is proposed whereby decomposition occurs upon absorption of ultraviolet radiation by the carbinol form. The excited carbinol form either undergoes radical fragmentation followed by reaction with oxygen and water or reacts directly with oxygen and water to produce the products mentioned above.

Similar degradation products have been identified from photooxidation studies on crystal violet using singlet-oxygen sensitizers (17). A mechanism for the formation of these photodecomposition products is proposed whereby the attack of singlet oxygen proceeds through an unstable dioxetane intermediate. Several studies have revealed that N-dealkylation is a contributing factor. Introduction of substituents into the phenyl ring of malachite green produced no marked improvement in the lightfastness because of the presence of the N-alkyl groups in the molecule (23). Replacement with N-aryl groups in the analogues of the indolyldiphenylmethane dye, Wool Fast Blue FBL [6661-40-1], however, raised the lightfastness grade by 1–2 points on the 1–8 Gray scale for evaluating color change. It would appear, therefore, that N-dealkylation is a general phenomenon in dye photochemistry, since it has been observed with thiazine dyes (24), rhodamine dyes (25), and N-methylaminoanthraquinones (26).

Reduction. Triarylmethane dyes are reduced readily to the leuco bases with a variety of reagents including sodium hydrosulfite, zinc and acid (hydrochloric, acetic), zinc dust and ammonia, and titanous chloride in concentrated hydrochloric acid. The reduction with titanium trichloride (Knecht method) is used for rapidly assaying triarylmethane dyes:

$$Ar_3COH + 2\ TiCl_3 + 2\ HCl \rightarrow Ar_3CH + 2\ TiCl_4 + H_2O$$

The $TiCl_3$ titration is carried out to a colorless end point which is usually very sharp (see Titanium compounds, inorganic).

Sulfonation. The direct sulfonation of alkylaminotriphenylmethane dyes gives mixtures of substituted products. Although dyes containing anilino or benzylamino groups give more selective substitution, a sulfonated intermediate such as 3[(N-ethyl-N-phenylamino)methyl]benzenesulfonic acid (ethylbenzylanilinesulfonic acid) is preferred as the starting material. However, Patent Blue V [3536-49-0], CI Acid Blue 3, was made from m-hydroxybenzaldehyde and 2 mol diethylaniline, followed by sulfonation of the leuco base and oxidation to the dye. FD&C Green 2 [5141-20-8], CI Acid Green 5, is still made by trisulfonation of the leuco base using ethylbenzylaniline and benzaldehyde as starting materials.

N-Alkylation and N-Arylation. Dyes containing highly alkylated amino groups are prepared from highly alkylated intermediates and not by direct alkylation of dyes carrying primary amino groups. 4,4′,4″-Triaminotriphenylmethane (pararosaniline) may, however, be N-phenylated with excess aniline and benzoic acid to $N,N′,N″$-

triphenylaminotriphenylmethane hydrochloride [*2152-64-9*], or Spirit Blue, CI Solvent Blue 23.

CI Solvent Blue 23

Spirit blue is one of the few dyes sulfonated as the leuco base. The degree of sulfonation depends upon the conditions. Monosulfonated derivatives, commonly referred to as alkali blues, eg, CI Acid Blue 119 [*1324-76-1*], are used as barium or calcium salts in printing inks. Disulfonated compounds, eg, CI Acid Blue 48 [*1324-77-2*], are employed as sodium or ammonium salts for blueing paper, whereas the trisulfonic derivatives or ink blues, eg, CI Acid Blue 93 [*28983-56-4*], are used in writing inks (qv).

Pigment Formation. Triarylmethane dyes are converted into insoluble compounds, which are used as pigments (qv). Water-soluble cationic dyes are combined with phosphomolybdic acid, phosphotungstomolybdic acid, copper ferrocyanide, and occasionally silicomolybdic acid to form insoluble complexes, so-called pigment lakes, which are used in printing inks. The main use of the barium lakes of alkali blues is in inks based on carbon black where an inexpensive blue component is needed to correct the natural brown tone of the base pigment.

Manufacture

The preparation of triarylmethane dyes proceeds through several stages: formation of the colorless leuco base, conversion to the colorless carbinol base, and formation of the dye by treatment with acid.

Aldehyde Method. The central carbon atom is derived from an aromatic aldehyde or a substance capable of generating on aldehyde during the course of the condensation.

Malachite green, CI Basic Green 4, is prepared by heating benzaldehyde under reflux with a slight excess of dimethylaniline in aqueous acid. The reaction mass is made alkaline, and the excess dimethylaniline is removed by steam distillation. The resulting leuco base is oxidized with freshly prepared lead dioxide to the carbinol base, and the lead is removed by precipitation as the sulfate. Subsequent treatment of the carbinol base with acid produces the dye, which can be isolated as the chloride, the oxalate [*2437-29-8*], or the zinc chloride double salt [*79118-82-4*].

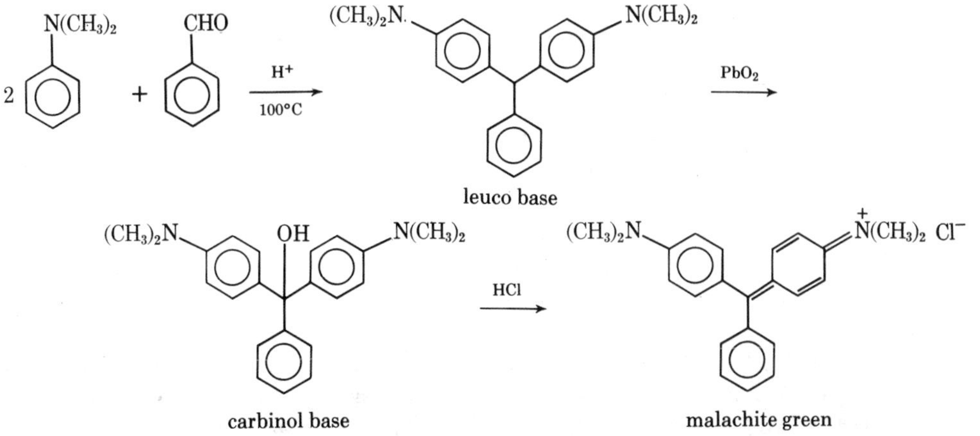

leuco base

carbinol base   malachite green

The starting materials may be sulfonated. For example, CI Acid Blue 9 [2650-18-2] is manufactured by condensing α-(N-ethylanilino)-m-toluenesulfonic acid with o-sulfobenzaldehyde. The leuco base is oxidized with sodium dichromate to the dye, which is usually isolated as the ammonium salt. In this case, the removal of the excess amine is unnecessary.

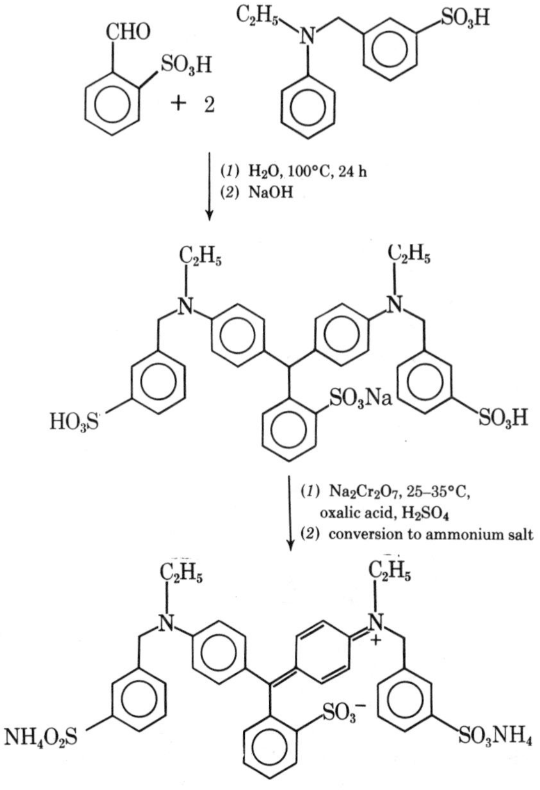

CI Acid Blue 9

This method is generally used for diaminotriphenylmethane dyes or hydroxy-triarylmethane dyes such as CI Mordant Blue 1 [1796-92-5], which is manufactured by condensing o-cresotic acid with 2,6-dichlorobenzaldehyde to give the leuco base, followed by nitrous acid oxidation.

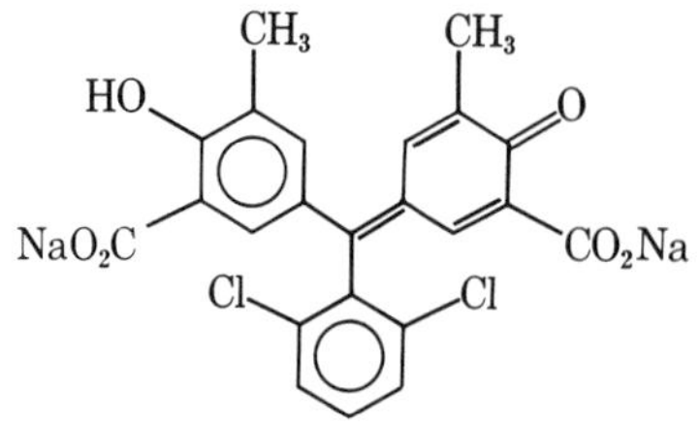

CI Mordant Blue 1

Ketone Method. In the ketone method, the central carbon atom is derived from phosgene (qv). A diarylketone is prepared from phosgene and a tertiary arylamine. It is condensed with another mole of a tertiary arylamine (same or different) in the presence of phosphorus oxychloride or zinc chloride. The dye is produced directly without an oxidation step. Thus, ethyl violet [2390-59-2], CI Basic Violet 4, is prepared from 4,4'-bis(diethylamino)benzophenone with diethylaniline in the presence of phosphorus oxychloride.

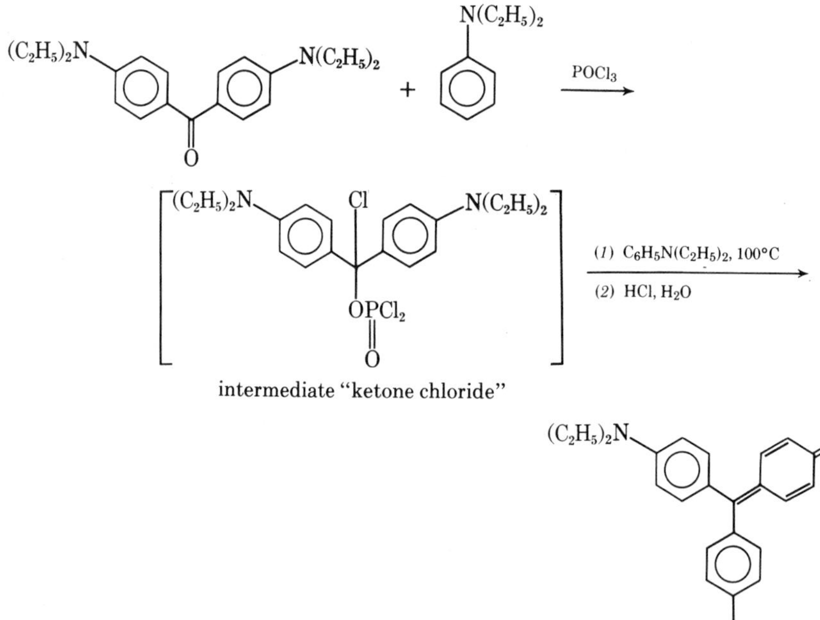

intermediate "ketone chloride"

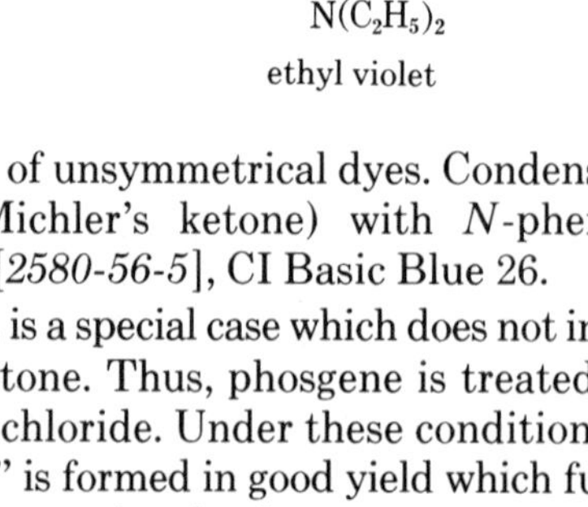

ethyl violet

This reaction is very useful for the preparation of unsymmetrical dyes. Condensation of 4,4'-bis(dimethylamino)benzophenone (Michler's ketone) with N-phenyl-1-naphthylamine gives the dye Victoria Blue B [2580-56-5], CI Basic Blue 26.

The manufacture of crystal violet, however, is a special case which does not involve the isolation of the intermediate Michler's ketone. Thus, phosgene is treated with excess dimethylaniline in the presence of zinc chloride. Under these conditions, the highly reactive intermediate "ketone dichloride" is formed in good yield which further condenses with another mole of dimethylaniline to give the dye.

Diphenylmethane Base Method. In this method, the central carbon atom is derived from formaldehyde which is condensed with two moles of an arylamine to give a sub-

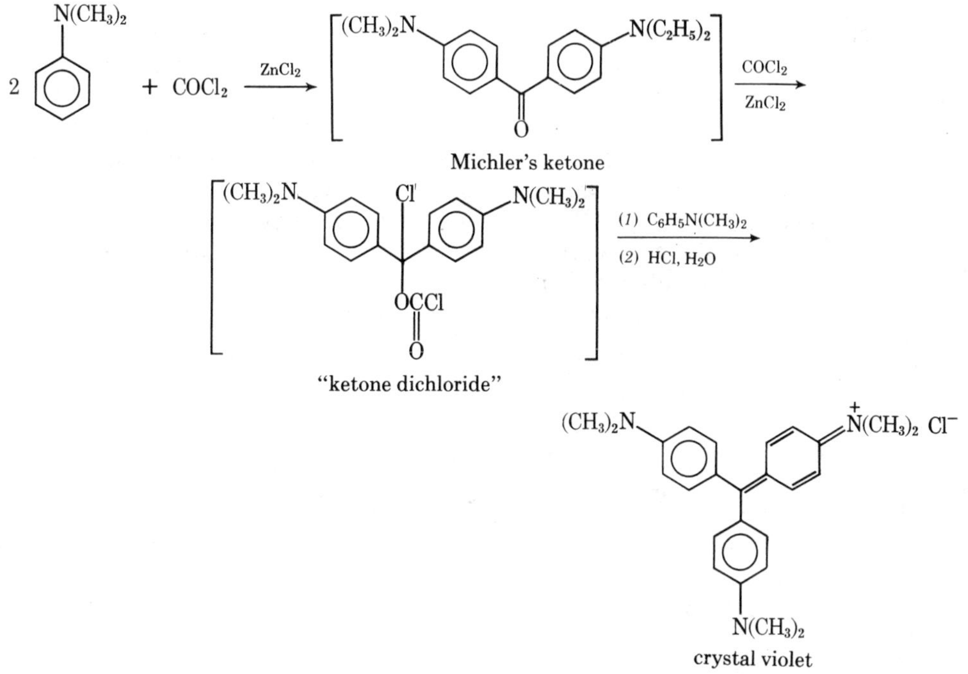

Michler's ketone

"ketone dichloride"

crystal violet

stituted diphenylmethane derivative. The methane base is oxidized with lead dioxide or manganese dioxide to the benzhydrol derivative. The reactive hydrols condense fairly easily with arylamines and comparatively deactivated aromatic derivatives, eg, sulfonated arylamines and sulfonated naphthalenes. The resulting leuco dye is oxidized in the presence of acid.

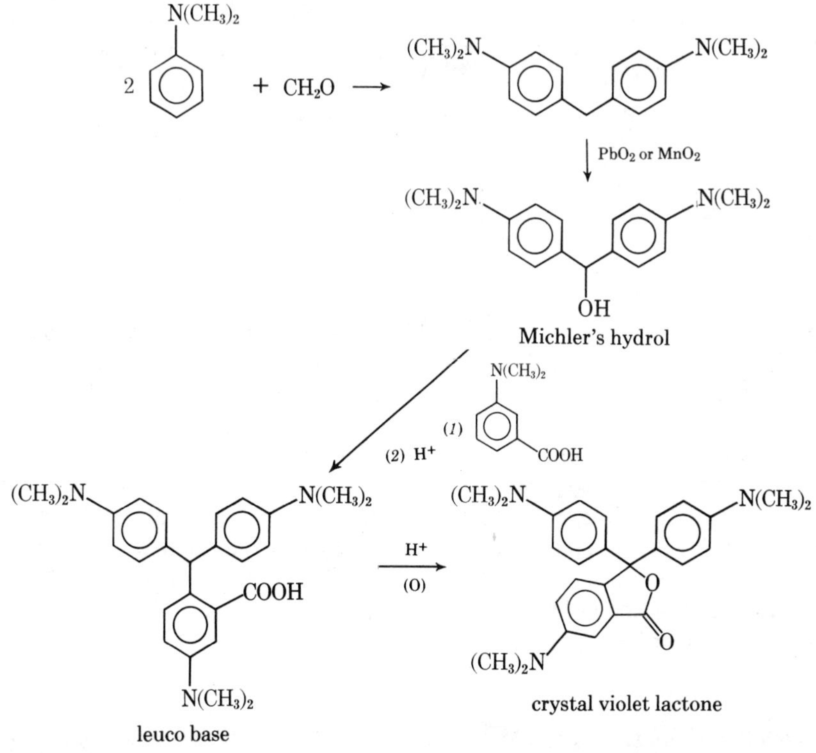

Michler's hydrol

leuco base

crystal violet lactone

Crystal violet lactone [*1552-42-7*], a color former currently used in the manufacture of carbonless copying papers (see Microencapsulation), is prepared from Michler's hydrol and *m*-dimethylaminobenzoic acid.

In a variation of this method, separation of the benzhydrol derivative is not required. The methane base undergoes oxidative condensation in the presence of acid with the same or a different arylamine directly to the dye.

New fuchsine [*3248-91-7*], CI Basic Violet 2, is prepared by condensing 2 mol *o*-toluidine with formaldehyde in nitrobenzene in the presence of iron salts to yield the corresponding substituted diphenylmethane base. This base is not isolated, but undergoes an oxidative condensation with another mole of *o*-toluidine to produce the dye.

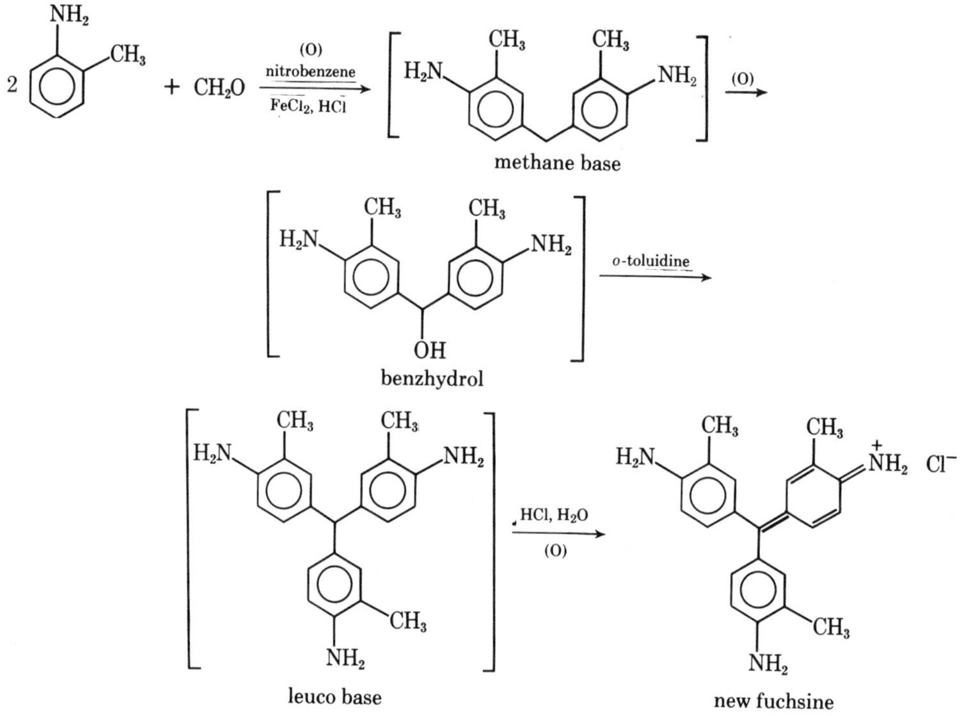

methane base

benzhydrol

leuco base

new fuchsine

Methyl violet [*8004-87-3*], CI Basic Violet 1, is made by the air oxidation of dimethylaniline in the presence of salt, phenol, and a copper sulfate catalyst. It reacts with the dimethylaniline present to produce *N,N,N',N''*-tetramethyldiaminodiphenylmethane, which is oxidized to Michler's hydrol. The hydrol is condensed with *N*-methylaniline formed in addition to the formaldehyde during the initial air oxidation of the dimethylaniline to give the leuco base of methyl violet. Treatment with aqueous acid produces the dye. Since Michler's hydrol may also react with dimethylaniline instead of the *N*-methylaniline to give crystal violet, commercial-grade methyl violet is usually a mixture.

Benzotrichloride Method.   The central carbon atom of the dye is supplied by the trichloromethyl group from *p*-chlorobenzotrichloride. Both symmetrical and unsymmetrical triarylmethane dyes suitable for acrylic fibers are prepared by this method.

4-Chlorobenzotrichloride is condensed with excess chlorobenzene in the presence

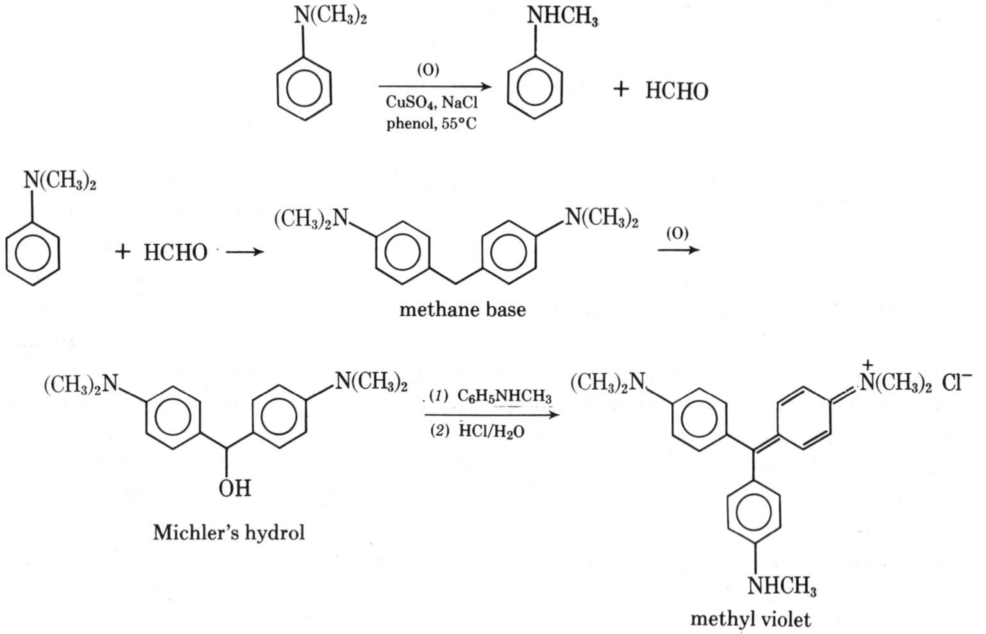

$N(CH_3)_2$ → (O) / $CuSO_4$, NaCl / phenol, 55°C → $NHCH_3$ + HCHO

$N(CH_3)_2$ + HCHO → $(CH_3)_2N$ — — $N(CH_3)_2$ (O) →

methane base

$(CH_3)_2N$ — — $N(CH_3)_2$ / OH

Michler's hydrol

(1) $C_6H_5NHCH_3$ (2) HCl/H_2O →

$(CH_3)_2N$ — — $\overset{+}{N}(CH_3)_2$ Cl^- / NHCH_3

methyl violet

of a Lewis acid such as aluminum chloride to produce the intermediate aluminum chloride complex of 4,4',4''-trichlorotriphenylmethyl chloride. Stepwise nucleophilic substitution of the chlorine atoms of this intermediate is accomplished by successive reactions with different arylamines.

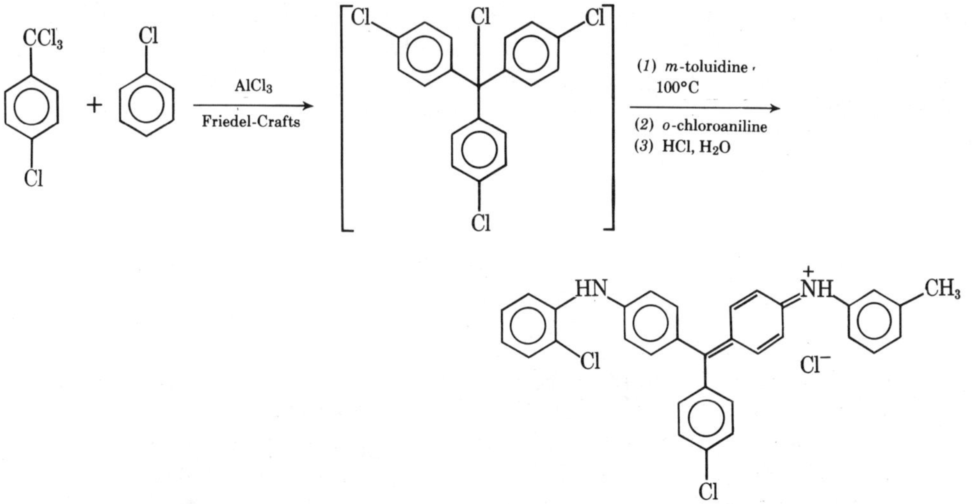

CCl_3 / Cl + Cl → AlCl_3 / Friedel-Crafts → [Cl Cl Cl / Cl]

(1) m-toluidine, 100°C
(2) o-chloroaniline
(3) HCl, H_2O
→

HN — — $\overset{+}{N}H$ — CH_3 / Cl / Cl^- / Cl

[85356-86-1]

Economic Aspects

Since 1973, the U.S. International Trade Commission has reported the manufacture and sales of dyes by application class only. In 1972, the last year for which statistics are available by chemical class, 3800 metric tons of triarylmethane dyes was manufactured, which represents ca 4% of total dyestuff production in the United States. During the last decade, annual dye production, including triarylmethane dyes, has changed very little. Methyl violet, with an annual production of 725 t, is the only triarylmethane dye for which production statistics are available. Some triarylmethane dyes are imported, eg, malachite green (163 t in 1981), followed by methyl violet (40 t), and CI Basic Violet 2 (30 t); the other dye imports total <15 t.

Uses

The present-day usage of triarylmethane dyes is confined mainly to nontextile purposes. Substantial quantities are used for the preparation of organic pigments for printing inks and for the paper-printing trade, where cost and brilliance are more important than lightfastness. Triarylmethane dyes and their colorless precursors, eg, carbinols and lactones, are used extensively in high speed photoduplicating and photoimaging systems. They are also used for speciality applications such as tinting automobile antifreeze solutions and toilet sanitary preparations, in the manufacture of carbon paper, in ink for typewriter ribbons, and jet printing for high speed computer applications.

In addition to the dyeing and printing of natural and acrylic fibers, triarylmethane dyes are suitable for the coloration of other substrates such as leather, fur, anodized aluminum, glass, waxes, polishes, soaps, plastics, drugs, and cosmetics. Several triphenylmethane dyes are used as food colorants and are manufactured under stringent processing controls (see Colorants for foods, drugs, and cosmetics). Triphenylmethane dyes are also used extensively as microbiological stains. Some colorless triarylmethane derivatives are very effective mothproofing agents for wool. Their use as antihalation dyes for photographic materials, infrared absorbers, and indicators is mentioned in the literature.

Related Dyes

Diphenylmethane Dyes. The diphenylmethane dyes are usually classed with the triarylmethane dyes, but in fact have very little in common with them. The dyes of this subclass are ketoimine derivatives, and only two such dyes are of any commercial significance. They are Auramine O [2465-27-2], CI Basic Yellow 2, and CI Basic Yellow 37 [6358-36-7].

R = CH$_3$, Auramine O
R = C$_2$H$_5$, CI Basic Yellow 37

Because of their low cost and the brilliance of their shades, these dyes are still used extensively for the coloration of paper and in the preparation of pigment lakes.

Auramine O is manufactured by heating 4,4′-bis(dimethylaminodiphenyl)-methane with a mixture of urea, sulfamic acid, and sulfur in ammonia at 175°C. The auramine sulfate [52497-46-8] formed in the reaction may be used directly in the dyeing process or can be readily converted into auramine base [492-80-8]. Highly concentrated solutions for use in the paper industry can be prepared by dissolving auramine base in formamide containing sodium bisulfate. The nitrate and nitrite salts exhibit excellent solubility in alcohols which facilitates their use in lacquers and plexographic printing colors. Alkyl and halogen derivatives of N-phenyl(leuco-auramine) are colorless, stable, crystalline compounds that turn dark blue whn in contact with acidic inorganic compounds such as aluminum sulfate, zinc sulfate, bentonite, or kaolin. They are useful in the production of colorless transfer sheets which, on contact with an acidic copying sheet, yield blue prints.

Phthaleins. Dyes of this class are usually considered to be triphenylmethane derivatives, since in their preparation, triphenylmethane derivatives are formed as precursors.

Phenolphthalein [77-09-8] and phenol red [143-74-8] are used extensively as indicators in colorimetric and titrimetric determinations (see Hydrogen-ion activity).

R = CO, phenolphthalein
R = SO$_2$, phenol red

These compounds are prepared by the condensation of phenol with phthalic anhydride or o-sulfobenzoic anhydride, respectively, in the presence of a dehydrating agent.

BIBLIOGRAPHY

"Triphenylmethane and Diphenylnaphthylmethane Dyes" in *ECT* 1st ed., Vol. 14, pp. 302–329, by A. J. Cofrancesco, General Aniline & Film Corp.; "Triphenylmethane and Related Dyes" in *ECT* 2nd ed., Vol. 20, pp. 672–737, by Vincent G. Witterholt, E. I. du Pont de Nemours & Co., Inc.

1. O. Fischer and E. Fischer, *Ber.* **11**, 1079 (1878).
2. R. Wizinger, *Ber.* **60**, 1377 (1927).
3. C. R. Bury, *J. Am. Chem. Soc.* **57**, 2115 (1935).
4. L. Pauling, *Proc. Nat. Acad. Sci. U.S.A.* **25**, 577 (1939).
5. W. Klyne and P. B. D. de la Mare, *Progress in Stereochemistry*, Vol. 2, Academic Press, Inc., New York, 1958, p. 42.
6. C. C. Barker in G. W. Gray, ed., *Steric Effects in Conjugated Systems*, Butterworths, London, 1958, p. 34.
7. C. C. Barker, G. Hallas, and A. Stamp, *J. Chem. Soc.* **82**, 3790 (1960).
8. G. Hallas, *J. Soc. Dyers Colour.* **83**, 368 (1967); **84**, 510 (1968).
9. D. E. Grocock, G. Hallas, and J. D. Hepworth, *J. Soc. Dyers Colour.* **86**, 200 (1970).
10. A. S. Ferguson and G. Hallas, *J. Soc. Dyers Colour.* **87**, 187 (1971); **89**, 22 (1973).
11. S. Ghandi, G. Hallas, and J. Thomasson, *J. Soc. Dyers Colour.* **93**, 451 (1977).
12. C. C. Barker, M. H. Bride, and A. Stamp, *J. Chem. Soc.* **81**, 3957 (1959).
13. C. C. Barker and G. Hallas, *J. Chem. Soc. Part III*, 2642 (1961).
14. J. T. Porter and S. B. Spears, *Text. Chem. Color.* **2**, 191 (1970).
15. N. S. Allen, J. F. McKellar, and B. Mohajerani, *Dyes Pigm.* **1**(1), 49 (1980).
16. I. H. Leaver, *Photochem. Photobiol.* **16**, 189 (1972).
17. N. Kuramoto and T. Kitao, *Dyes Pigm.* **3**(1), 59 (1982).
18. K. Iwamoto, *Bull. Chem. Soc. Jpn.* **10**, 420 (1935).
19. D. Bitzer and H. J. Brielmaier, *Melliand Textilber.* **41**, 62 (1960).
20. J. Wegmann, *Melliand Textilber.* **39**, 408 (1958).
21. E. D. Owen and R. T. Allen, *J. Appl. Chem. Biotechnol.* **22**, 799 (1972).
22. C. H. Giles, C. D. Shah, W. E. Watts, and R. S. Sinclair, *J. Soc. Dyers Colour.* **88**, 433 (1972).
23. N. A. Evans and I. W. Stapleton, *J. Soc. Dyers Colour.* **89**, 208 (1973).
24. H. Obata, *Bull. Chem. Soc. Jpn.* **34**, 1049 (1961).
25. N. A. Evans, *J. Soc. Dyers Colour.* **89**, 332 (1973).
26. C. H. Giles and R. S. Sinclair, *J. Soc. Dyers Colour.* **88**, 109 (1972).

General References

E. N. Abrahart, *Dyes and Their Intermediates*, 2nd ed., Chemical Publishing, New York, 1977.

R. L. M. Allen, *Color Chemistry*, Appleton-Century-Crofts, New York, 1971.

K. Venkataraman, ed., *The Chemistry of Synthetic Dyes*, Vol. 2, Academic Press, Inc., New York, 1952; N. R. Ayyanger and D. B. Tilak, and D. R. Baer in Vol. 4, 1971; J. Lenoir in Vol. 5, 1971; E. Gurr and co-workers in Vol. 7, 1974; N. A. Evans and I. W. Stapleton in Vol. 8, 1978.

P. Bentley and co-workers, *Review in Progress of Color and Related Topics*, Vol. 5, The Society of Dyers and Colourists, Bradford, UK, 1974.

"Manufacture of Triphenylmethane Dyestuffs and Intermediates at Ludwigshafen and Hoechst," *BIOS Final Report 959;* "Manufacture of Triphenylmethane Dyestuffs at Hoechst, Ludwigshafen, and Leverkusen," *BIOS Final Report 1433*, British Intelligence Objectives Subcommittee; "German Dyestuffs and Dyestuff Intermediates," *FIAT Final Report 1313.*

Colour Index, 3rd ed., The Society of Dyers and Colourists, Bradford, UK, and the American Association of Textile Chemists and Colorists, Vols. 1–6, 1971.

P. Rys and H. Zollinger, *Fundamentals of the Chemistry and Applications of Dyes*, Wiley-Interscience, New York, 1972.

H. A. Lubs, ed., *The Chemistry of Synthetic Dyes and Pigments*, American Chemical Society Monograph Series, Reinhold Publishing Corp., New York, 1955.

K. Venkataraman, ed., *The Analytical Chemistry of Synthetic Dyes*, Wiley-Interscience, New York, 1977.

DEREK BANNISTER
JOHN ELLIOTT
CIBA-GEIGY Corporation

TRYPSIN. See Enzymes.

TRYPTOPHAN. See Amino acids.

TUADS. See Rubber chemicals.

TUNG OIL. See Fats and fatty oils; Drying oils.

TUNGSTEN AND TUNGSTEN ALLOYS

Tungsten (wolfram) [7440-33-7] is a silver-gray metallic element with atomic number 74 and appearing in Group VIB of the periodic table below chromium and molybdenum. Its atomic weight is 183.85. The five stable isotopes are given in Table 1. Tungsten has the highest melting point of any metal, 3695 K, a very low vapor pressure, and the highest tensile strength of any metal above 1650°C.

The name tungsten, meaning in Swedish heavy (tung) stone (sten), was first applied to a tungsten-containing mineral in 1755. The mineral was subsequently identified by Scheele in 1781 as containing lime and a theretofore unknown acid which he called tungstic acid. The mineral was then named scheelite. Metallic tungsten was first produced by the carbon reduction of tungstic acid in 1783 in Spain and termed wolfram. This designation became common usage in German. In 1957, the IUPAC chose the English name tungsten and the French name tungstene with wolfram as an alternative. However, W is still used as the chemical symbol.

During the nineteenth century, tungsten remained a laboratory material. The latter half of the century saw the development of high speed tool steels containing tungsten which became the prime use for the metal in the first half of the twentieth century. The pure metal itself was first used as a filament for electric lamps at the beginning of the twentieth century. After some limited success with paste-extruded tungsten powder, the Coolidge process was developed in 1908, by which a pressed and sintered tungsten ingot could be worked at high temperatures by swaging and drawing to form a fine-wire filament. This was a landmark in the development of the incandescent-lamp industry and, later, in the use of tungsten as a welding electrode. In the 1920s, the search for an alternative to expensive diamond dies required for the drawing of the tungsten wire led to the manufacture of cemented carbides now accounting for over half of the tungsten consumption in the world.

Tungsten is the eighteenth most abundant metal having an estimated concentration in the earth's crust of 1–1.3 ppm. Of the more than 20 tungsten-bearing minerals, only four are of commercial importance: ferberite (iron tungstate), huebnerite (manganese tungstate), wolframite (iron–manganese tungstate containing ca 20–80% of each of the pure components), and scheelite (calcium tungstate).

The WO_3 content of wolframite minerals varies from 76.3% in $FeWO_4$ to 76.6% in $MnWO_4$. They are commonly called black ores, as their colors range from black to brown. They occur as well-defined crystals to irregular masses of bladed crystals. They have Mohs hardness 5.0–5.5, a specific gravity 7.0–7.5 and tend to be very brittle. They

Table 1. Isotopes of Tungsten

Isotope	CAS Registry No.	Abundance, %
^{180}W	[14265-79-3]	0.14
^{182}W	[14265-80-6]	26.41
^{183}W	[14265-81-7]	14.40
^{184}W	[14265-82-8]	30.64
^{186}W	[14265-83-9]	28.41

are also weakly magnetic. Scheelite contains 80.6% WO_3. It is white to brown and strongly fluorescent in short-wave ultraviolet radiation. Scheelite occurs as massive crystals and small grains. It has Mohs hardness 4.5–5.0, a specific gravity 5.6–6.1, and is very brittle.

Tungsten deposits occur in association with metamorphic rocks and granitic igneous rocks throughout the world (see Table 2). Deposits in the People's Republic of China constitute >50% of the world reserves and over five times the reserves of the second largest source, Canada.

Table 2. World Tungsten Resources, 10^3 Metric Tons[a]

Country	Reserves	Other[b]	Total
North America			
United States	125	325	450
Canada	270	320	590
Mexico	20	5	25
other	1	2	3
Total[c]	*420*	*650*	*1070*
South America			
Bolivia	39	86	125
Brazil	18	40	58
other	2	2	4
Total[c]	*60*	*130*	*190*
Europe			
Austria	18	55	73
France	16	2	18
Portugal	24	30	54
USSR	210	320	530
UK	0.5	65	65
other	30	9	39
Total[c]	*300*	*480*	*780*
Africa			
Zimbabwe	5	5	10
other	5	14	19
Total[c]	*10*	*18*	*28*
Asia			
Burma	30	75	105
People's Republic of China	1400	2300	3700
Democratic People's Republic of Korea	110	140	250
Republic of Korea	80	80	160
Malaysia	15	30	45
Thailand	20	20	40
Turkey	75	14	89
other	5	5	10
Total[c]	*1700*	*2620*	*4320*
Oceania			
Australia	110	260	370
other	0.5	2	3
Total	*110*	*260*	*370*
World Total[c]	*2600*	*4200*	*6800*

[a] Ref. 1.
[b] Derived in collaboration with the U.S. Geological Survey.
[c] Data may not add to totals shown because of independent rounding.

Physical Properties

Some of the physical properties of tungsten are given in Table 3. Although in general these data are reliable, the original references should be consulted if precise values are required. For further property data, see refs. 12–14. For thermodynamic values, refs. 4, 15–16 should be consulted. Two values are given for the melting point. The value of 3660 K was selected as a secondary reference for the 1968 International Practical Temperature Scale. However, since 1961, the four values that have been reported ranged from 3680 to 3695 and averaged 3688 K.

Chemical Properties

The oxidation states of tungsten range from +2 to +6, and some compounds with zero oxidation state also exist. Above 400°C, tungsten is very susceptible to oxidation. At 800°C, sublimation of the oxide becomes significant and oxidation is destructive. Very fine powders are pyrophoric. Above 600°C, the metal reacts vigorously with water to form oxides. In lamps, in the presence of water vapor, a phenomenon called the water cycle occurs in which the tungsten is oxidized in the hottest part of the filament and then reduced and deposits on the cooler portion of the filament. Tungsten is stable in nitrogen to over 2300°C. In ammonia, nitrides form at 700°C. Carbon monoxide and hydrocarbons react with tungsten to give tungsten carbide at 900°C. Carbon dioxide oxidizes tungsten at 1200°C. Fluorine is the most reactive halogen gas and attacks tungsten at room temperature. Chlorine reacts at 250°C, whereas bromine and iodine require higher temperatures.

Tungsten is resistant to many chemicals. At room temperature, it is only rapidly attacked by a mixture of hydrofluoric and nitric acids. Attack of aqua regia is slow. Hot sulfuric, nitric, and phosphoric acids also react slowly. Sodium, potassium, and ammonium hydroxide solutions slowly attack tungsten at room temperature in the presence of an oxidizing agent such as potassium ferricyanide or hydrogen peroxide. Molten sodium and potassium hydroxide attack tungsten only moderately. The attack is accelerated by addition of an oxidizer. Tungsten resists attack by many molten metals. The maximum temperature of stability for various metals is given below:

Metal	Mg	Hg	Al	Zn	Na	Bi	Li
temperature, °C	600	600	680	750	900	980	1620

In contact with various refractories (qv), tungsten is stable in vacuum as shown below:

Refractory	Al_2O_3	BeO	MgO	ThO_2	ZnO_2
temperature, °C	1900	1500	2000	2200	1600

In reducing atmospheres, these temperatures are lower.

Manufacture

Mining and Beneficiation. Tungsten mines are generally small, producing less than 2000 metric tons of raw ore per day. Worldwide, there are only about 20 mines producing over 300 t/d. Many small mines are inactive at times, depending on the price of tungsten. They are primarily limited by the nature of the ore body and mining is almost exclusively by underground methods. Where open-pit mining has been em-

Table 3. Physical Properties of Tungsten

Property	Value	Ref.
crystal structure, bcc		
lattice constant at 298 K, nm	0.316524	
shortest interatomic distance at 298 K, nm	0.2741	
density[a] at 298 K, g/cm^3	19.254	
melting point, K		
	3660	2
	3695 ± 15	3
boiling point	5936	4
linear expansion per K		5
293–1395 K	$4.266 \times 10^{-6}\,(T-293) + 8.479 \times 10^{-10}\,(T-293)^2$ $- 1.974 \times 10^{-13}\,(T-293)^3$	
1395–2495 K	$0.00548 + 5.416 \times 10^{-6}\,(T-1395) + 1.952 \times 10^{-10}$ $(T-1395)^2 + 4.422 \times 10^{-13}\,(T-1395)^3$	
2495–3600 K	$0.01226 + 7.451 \times 10^{-6}\,(T-2495) + 1.654 \times 15^{-9}$ $(T-2495)^2 + 7.568 \times 10^{-14}\,(T-2495)^3$	
specific heat, C_p, 273–3300 K, J/(mol·K)[b]	$24.94\left(1 - \dfrac{4805}{T^2}\right) + 1.674 \times 10^{-3}\,T + 4.25 \times 10^{-10}\,T^3$	6
enthalpy, $H_T - H_{298}$, J/mol[b]	$24.94\left(T + \dfrac{4805}{T}\right) + 8.372 \times 10^{-4}\,T^2 + 1.062 \times$ $10^{-10}\,T^4 - 7917.8$	6
entropy at 298 K, J/mol[b]	32.66	4
heat of fusion, kJ/mol[b]	46.0	7
heat of sublimation, 298.13 K, kJ/mol[b]	859.8	8
vapor pressure, 2600–3100 K, Pa[c]	$\log P_{Pa} = \dfrac{-45395}{T} + 12.8767$[d]	
thermal conductivity at K, W/(cm·K)		9
0	0	
10	97.1	
50	4.28	
100	2.08	
500	1.46	
1000	1.18	
2000	1.00	
3400	0.90	
electrical resistivity, ρ, 4–3000 K, nΩ·m	$\dfrac{0.04535\,T^{1.2472} - 2.90 \times 10^{-9}\,T^3}{1 + \dfrac{3.442 \times 10^5}{T^{2.98}}} + \rho_0$	10
total emissivity, ϵ_H, 1600–2800 K	$-2.685790 \times 10^{-2} + 1.819696 \times 10^{-4}\,T^4$ $-2.194616 \times 10^{-8}\,T^2$	11

[a] Determined by x-ray.
[b] To convert J to cal, divide by 4.184.
[c] To convert Pa to mm Hg, multiply by 0.0075.
[d] To convert $\log P_{Pa}$ to $\log P_{mm\ Hg}$, subtract 2.1225.

ployed, underground methods are used as the deposit diminishes. Ore deposits usually range from 0.3 to 1.5% WO_3, with exceptional cases as high as 4% WO_3. Because of the low tungsten content of the deposits, all mines have beneficiation facilities which produce a concentrate containing 60–75% WO_3.

Since scheelite and wolframite are both friable, care must be taken to avoid over-grinding which can lead to sliming problems. The ores are crushed and ground in stages, and the fines are removed after each stage. Jaw crushers are employed for the first stage because the tonnages are low. Either jaw- or cone-type crushers are used for the second stage. In some cases, sizes are further reduced by rod milling. After each stage, the fines are removed and the coarse fraction is recirculated. Screening is the preferred method for fairly large particle sizes. Mechanical and hydraulic classifiers can also be used.

Since tungsten minerals have a high specific gravity, they can be beneficiated by gravity separation, usually by tabling. Flotation is used for many scheelites with a fine liberation size but not for wolframite ores. Magnetic separators can be used for concentrating wolframite ores or cleaning scheelites.

Extractive Metallurgy. In extractive metallurgy, a relatively impure ore concentrate is converted into a high purity tungsten compound that can subsequently be reduced to metal powder. This is a particularly important step since high purity is required for all uses of tungsten except as a steel-alloying additive. The two most common intermediate tungsten compounds are tungstic acid, H_2WO_4, and ammonium paratungstate (APT), $(NH_4)_{10}W_{12}O_{41}.5H_2O$ (see Fig. 1). Most commercial processes today use APT. Depending on their source, the impurities in ore concentrates vary considerably, but those of most concern are sulfur, phosphorus, arsenic, silicon, tin, lead, boron, and molybdenum compounds.

The concentrate may be first pretreated by leaching or roasting. In scheelite concentrates, hydrochloric acid leaching reduces phosphorus, arsenic, and sulfur contents. Roasting of either scheelite or wolframites eliminates sulfur, arsenic, and organic residues left from the flotation process.

Next, the concentrate is digested to extract the tungsten. For lower grade scheelites, the high pressure soda process is commonly employed. The concentrate is first ground to <100 μm (−150 mesh) size, and then digested in an autoclave with sodium carbonate at ca 200°C at a pressure of >1.2 MPa (ca 11.9 atm):

$$NaCO_3 + CaWO_4 \rightarrow Na_2WO_4 + CaCO_3$$

The sodium tungstate solution is filtered from the resulting slurry. Similarly, in the alkali roasting process, the concentrate, either scheelite or wolframite, is heated with sodium carbonate in a rotary kiln at 800°C and then leached with hot water to remove the sodium tungstate. Wolframite ores are also decomposed by reaction with a sodium hydroxide solution at 100°C:

$$(Fe,Mn)WO_4 + 2\,NaOH \rightarrow (Fe,Mn)(OH)_2 + Na_2WO_4$$

The insoluble hydroxides are removed by filtration.

In another process, scheelite is leached with hydrochloric acid:

$$CaWO_4 + 2\,HCl \rightarrow CaCl_2 + H_2WO_4$$

In this case, the tungstic acid is insoluble and is removed by filtration and washed. For purification, it is digested in aqueous ammonia to give an ammonium tungstate

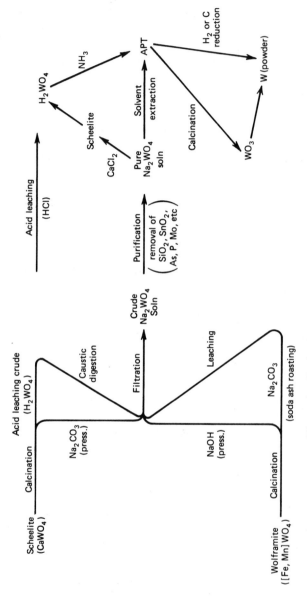

Figure 1. Production of tungsten materials from ores (17).

418

solution. Magnesium oxide is added to precipitate magnesium ammonium phosphates and arsenates. Addition of activated carbon removes collodial hydroxides and silica which are filtered. Evaporation of the ammonium tungstate solution gives APT. The evaporation is not carried to completion so that most of the impurities still present remain in the mother liquor and are removed.

The sodium tungstate from the soda and caustic processes is purified by first adding aluminum and magnesium sulfates to remove silicon, phosphorus, and arsenic, whereas sodium hydrogen sulfide removes molybdenum and other heavy metals. The pH is controlled and the impurities are removed by filtration. Then, the sodium tungstate is converted to ammonium tungstate by means of a liquid ion-exchange process. Since this involves the exchange of anions, most of the impurities which are present as cations are left behind. The ammonium tungstate solution is then evaporated to a fixed specific gravity under constant agitation to give APT.

Reduction to Metal Powder. The metal powder is obtained from APT by stepwise reduction with carbon or hydrogen. The intermediate products are the yellow oxide, WO_3; blue oxide, W_4O_{11} (which is actually a mixture of two oxides, $W_{18}O_{49}$ and $W_{20}O_{58}$); and brown oxide, WO_2. Because carbon introduces impurities, hydrogen is preferred. The reduction is carried out either in tube furnaces or rotary furnaces, heated by gas or electricity and having three separately controlled zones. A tube furnace consists of many tubes, 7–9 m long with a diameter of 80–150 mm. The boats containing the oxide have rectangular or semicircular cross sections and are 375–450 mm long. Both tubes and boats are made of Inconel or a similar heat-resistant alloy (see High temperature alloys). The boats are either manually or automatically stoked through the furnaces. Hydrogen is fed through each tube countercurrent to material flow. The hydrogen is recirculated and is scrubbed, purified, and dried, and new hydrogen added with each cycle. A rotary furnace consists of a large tube 3–>10 m long. It is partitioned into three sections to restrict powder movement down the tube and has longitudinal vanes to carry the powder. The tube is tilted at a small angle and rotated to provide continuous flow of powder through the furnace.

Ammonium paratungstate is decomposed to the yellow oxide by heating in air at 800–900°C. The blue oxide is obtained in a hydrogen atmosphere. Metal powder can be made directly from APT but particle size is better controlled with a two-stage reduction by controlling temperature, bed depth, and hydrogen flow. Since rotary furnaces have effectively a shallow bed, they tend to produce fine powders, and tube furnaces are preferred for the final reduction. Temperatures of 600–900°C are used to produce particle sizes of 1–8 μm.

For the production of lamp-filament wire, aluminum, potassium, and silicon dopants are added to the blue oxide. Some dopants are trapped in the tungsten particles upon reduction. Excess dopants are then removed by washing the metal powder in hydrofluoric acid. For welding electrodes and some other applications, thorium nitrate is added to the blue oxide. After reduction, the thorium is present as a finely dispersed thorium oxide.

Consolidation. Because of its high melting point, tungsten is usually processed by powder metallurgy techniques (see Powder metallurgy). Small quantities of rod are produced by arc or electron-beam melting.

For rod and wire production, ingots ranging in size 12–25 mm square by 600–900 mm are mechanically pressed at ca 200 MPa (20 atm). The bars are very fragile and are presintered at 1200°C in hydrogen to increase their strength. Sintering is done by electric resistance heating. The ingot is mounted between two water-cooled contacts inside a water-jacketed vessel containing a hydrogen atmosphere. A current is passed through the ingot heating it to about 2900°C. This not only sinters the bar to a density of 17.2 to 18.1 g/cm³ but also results in considerable purification by volatilization of the impurities. Larger billets for forging or rolling are isostatically pressed. The powder is placed in a plastisol bag, sealed tightly, and placed in a fluid in a high pressure chamber at 200–300 MPa (20–30 atm). Sintering takes place in an electric-resistance or induction-heated furnace at 2200°C in a hydrogen atmosphere. Densities after sintering are 17.8–18.1 g/cm³. Small parts are made by mechanically pressing powder to which a lubricant has been added, followed by sintering at 1800–2100°C. For some applications, the sintering temperature can be lowered to 1500°C by the addition of small amounts of nickel or palladium which, however, embrittle the tungsten.

Metalworking. Tungsten is unusual as its ductility increases with working. As sintered or after a full recrystallization annealing, it is as brittle as glass at room temperature. For this reason, it is initially worked at very high temperatures and large reductions are required to achieve ductility. Furthermore, the low specific heat of tungsten causes it to cool very rapidly and any working operation requires rapid transfer from furnace to working equipment and frequent reheating during the working operation.

Swaging is historically the oldest process used for the metalworking of tungsten and is the method used first for the manufacture of lamp wire (Coolidge process). Swaging temperatures start at 1500–1600°C and decrease to ca 1200°C as the bar is worked. These temperatures are just below the recrystallization temperature and the working is, therefore, technically cold working. Reductions per pass start as low as 5% but then increase to as high as 40% as the size decreases. Total reductions of ca 60–80% are typical between annealings. For rod and wire production, rod rolling or, more recently, Kocks rolling is also applied, at least in the initial breakdown stages. For rod rolling, oval-to-square sequences are used with reductions of 15–25% per pass. A Kocks mill consists of 8–12 roll stands in sequence, with each stand consisting of three rolls at 120° to each other producing a hexagonal cross section. The material is rolled through the stands at very high speeds to avoid cooling problems. Wire drawing starts at ca 4 mm and at temperatures as high as 1000°C, decreasing to 500°C for fine wire. The graphite lubricant required must be replaced after each draw pass. Reduction per pass is 35–40% at the start of drawing and drops to 7–10% in fine wire. For drawing dies, tungsten carbide is used to a diameter of 0.25 mm; for smaller diameters, diamonds are used.

For larger-diameter rod or plate, rolling is also employed with temperatures starting at 1600°C. However, when rolling large cross sections, most equipment is not powerful enough to handle large reductions and, as a result, large center-to-edge variations develop. These variations lead to center bursting or nonuniform structures after annealing. Large total size reductions overcome this effect. As the material is worked to larger reductions, the temperature is gradually lowered to avoid recrystallization during reheating. Once reductions of >95% are achieved, temperatures can be as low as 300°C.

Forging is also used on tungsten. Hammer forging is generally preferred to press forging because the temperature is better maintained with the higher rate of deformation. Temperature control is also critical here and even more critical with extrusion. Conventional extrusion of glass-coated billets is employed. A rather delicate balance between heat loss in the billet and heat generation in the die is required for good results.

Economic Aspects

Because of fluctuation in the stockpiling policies of the General Services Administration (GSA), tungsten prices in the past 25 years varied greatly, as shown in Table 4. The GSA has redefined its stockpile objective several times and it now stands at 26,810 metric tons, which is about a 3-yr supply for the United States.

World production is given in Table 5. The relatively stable market and the diminished influence of the GSA has improved the atmosphere for the development of new mines and expansion of existing facilities (see Table 6). This activity has led to an increase in the tungsten production capacity in North America from 4900 t in 1976 to a projected 12,065 t in 1985, ca 20% of the world's capacity. The forms of tungsten produced, and the distribution by industry, are given in Tables 7 and 8, respectively. Tungsten carbide products accounted for >65% of tungsten usage.

Primary U.S. tungsten requirements in the year 2000 are estimated to be 23,130 t, which represents an annual growth rate of 4.5%. For the rest of the world, 61,240 t

Table 4. Tungsten Prices, $/kg [a]

Year	WO$_3$	W
late 1950s	2.42	3.05
1962	0.88	1.10
1977	17.60	22.19
1982	10.63	13.40

[a] To convert $/kg to STU (short ton unit), multiply by 9.07. The STU contains 9.07 kg WO$_3$ or 7.19 kg W.

Table 5. World Tungsten Mine Production, 1980 [a]

Country	Production, metric tons
United States (mine shipments)	3,200
Australia	3,400
Austria	1,000
Bolivia	2,800
Brazil	1,400
Burma	500
Canada	3,100
Republic of Korea	2,600
Mexico	200
Portugal	1,400
Thailand	2,000
Turkey	300
other market-economy countries	3,600
People's Republic of China	13,200
USSR	8,900
other Communist countries	2,200
Total	*49,800*

[a] Ref. 18.

Table 6. Tungsten Mine Expansions and Openings, 1975–1985[a]

Location	Mine	Year Opened	Year Expanded	Annual production, 10^6 kg 1975	1980	1985
Canada	Cantung		1976–1979	1.2	3.3	
Australia	King Island		1975–1977	1.2	2.0	
Australia	Mount Carbine		1977	0.1	0.7	
Thailand			1978	0.7	1.8	
Bolivia			1975	0.1	0.7	
Austria	Mittersil	1976			1.2	
Australia	Glen Innes	1979			0.6	
United States, Nevada	Emerson	1978			0.5	0.9
Brazil	Boca DeLaga	1977			0.5	
United States, California	Strawberry		1978		0.4	
Canada	Mount Pleasant	1982				1.5
United States, Nevada	Springer	1982				0.7
Canada	Mactung (estd)	1984				3.2
UK	Hemerdon (estd)	1985				1.8

[a] Ref. 19.
[b] Initial capacity.

Table 7. U.S. Production, Metric Tons Contained Tungsten[a]

Property	1978	1979
metal powder	2454	2738
WC powder	4826	5347
crushed and crystalline WC	759	1009
chemicals	392	363
others	200	149
Total	*8631*	*9606*

[a] Ref. 20.

Table 8. U.S. Tungsten Distribution, Metric Tons[a]

Property	1978	1979
machinery		
metalworking	5,785	6,092
construction and mining	2,161	2,277
transportation	984	1,036
lighting	626	658
electrical	375	395
chemical	169	177
other	111	116
Total	*10,211*	*10,751*

[a] Ref. 1.

are predicted for a 3.2% annual growth. The main area of growth is expected to be in cutting, wear-resistant, and hard-facing applications.

Specifications; Analytical Methods

The 1980 ASTM specifications are given in Table 9.

Tungsten is usually identified by atomic spectroscopy. With optical emission spectroscopy, tungsten in ores can be detected at concentrations of 0.05–0.1%, whereas x-ray spectroscopy detects 0.5–1.0%. Scheelite in rock formations can be identified by its luminescence under ultraviolet excitation. In a wet method, the ore is fired with sodium carbonate and then treated with hydrochloric acid; addition of zinc, aluminum, or tin produces a beautiful blue color.

In the classical method for the quantitative analysis of tungsten in ore concentrates, the ore is digested with acid, the tungsten is complexed with cinchonine, purified, ignited, and weighed. More commonly, x-ray spectrometry is used and its accuracy is enhanced with tantalum as an internal standard. Plasma spectroscopy determines concentrations as low as 0.1 ppm in solutions and 10 ppm in solids. Atomic absorption is a very rapid method but not accurate enough for assay-grade analyses. The thiocyanate–tungsten color complex is specific for tungsten and used for colorimetric analyses for concentrations in the 0.1–1% range. It is not accurate enough for assaying concentrates.

Health and Safety Factors

There are no documented cases of tungsten poisoning in humans. However, numerous cases of pneumoconiosis have been reported in the cemented-carbide industry, but its cause, ie, WC or cobalt, has not been definitely determined. No cases have been reported among workers exposed to tungsten or its compounds. In fact, any hazards encountered seem to be caused by substances associated with the production and use of tungsten, eg, As, Sb, Pb, and other impurities in tungsten ores, Co fumes and dust in the carbide industry, and thoria used in welding electrodes. Exposure limits of 5 mg/m^3 insoluble tungsten and 1 mg/m^3 soluble tungsten have been established.

Uses

Tungsten is used in four forms: tungsten carbide, as an alloy additive, essentially pure tungsten, and tungsten chemicals. Tungsten carbide, because of its high hardness

Table 9. 1980 ASTM Tungsten Specifications

ASTM number	Use
B 410-68	unalloyed sintered billets, bars, rods, and preforms for forging
B 430-76	particle-size distribution of refractory metal powders by turbidimetry
B 459-67	tungsten-base, high density metals
B 482-68	preparation of tungsten and tungsten alloys for electroplating
E 159-68	hydrogen loss of copper, tungsten, and iron powders
E 397-73	chemical analysis
F 73-76	wire for electronic devices and lamps
F 204-76	surface flaws in seal rod and wire
F 269-60	sag of wire
F 288-76	wire for electronic devices and lamps
F 290-68	round wire for winding electron-tube grid laterals

at high temperatures, is used for cutting tools, abrasion-resistant surfaces, and forming tools. This application accounts for ca 65% of tungsten usage, mostly in the form of cemented carbides. Tungsten carbide is produced by the reaction of tungsten powder with carbon black at 1500°C. The tungsten carbide is then milled and blended with 3–25% cobalt and pressed and sintered at ca 1400°C. Addition of tantalum and titanium carbides improves hardness and wear resistance. Cemented carbides are used for cutting tools, mining and drilling tools, forming and drawing dies, bearings, and numerous other wear-resistant applications.

About 16% of tungsten usage is as an alloy additive. Tungsten added to steels forms a dispersed tungsten carbide phase which imparts a finer grain structure and increases the high temperature hardness. The finer grain size improves toughness and produces a more durable cutting edge. For this type of application, up to 3% tungsten is used, usually with 1–4% chromium. For hot-work tool steels, up to 18% tungsten is added. Such steel, when quenched from a very high temperature and tempered, retains its hardness up to a red heat. Tungsten is also used as an additive to nickel- and cobalt-base superalloys. Here again, tungsten imparts high temperature strength and wear resistance (see High temperature alloys).

Metallic tungsten accounts for 16% of tungsten consumption. Frequently, tungsten is used because of its high melting point and low vapor pressure. The best known use is the manufacture of lamp filaments, where potassium, silicon, and aluminum dopants are added to the oxide. After sintering, about 60 ppm potassium is retained in small voids. During working, these voids are stretched into long stringers. Upon heating, they form rows of tiny bubbles which control grain boundary movement resulting in grains that are much longer than the wire diameter and whose boundaries are at a small angle to the wire axis. This structure is very creep resistant and allows the coil to maintain its shape throughout life. Tungsten is widely employed as an electron emitter because it can be used at very high temperatures. Thoria is added to reduce the work function and improve emission. It also improves arc stability and gives longer life to welding electrodes. Tungsten is used as the target in high intensity X-ray tubes. Other high temperature applications include furnace elements, heat shields, vacuum metallizing coils and boats, glass-melting equipment and arc-lamp electrodes. Because of its high elastic modulus and wear resistance, tungsten is used as the target in high modulus and wear resistance, tungsten is used in high speed impact printers. Its low thermal expansion makes it ideal for glass-to-metal seals and as base for silicon semiconductors. There are few tungsten-base alloys. W–Re alloys are used for thermocouple wire and for shock-resistant lamp filaments. Alloys of tungsten with various combinations of iron, nickel, and copper are called heavy alloys. They have the high density of tungsten but in a more machinable form. These are used as counterweights, armor-piercing penetrator cores, x-ray shielding, gyroscope rotors, dart bodies, and other high density applications. A similar material is made by infiltrating porous tungsten with copper or silver. Such alloys are used as electrical contact materials and rocket nozzles. Composite materials with barium and strontium compounds are used in electron-emitting devices.

Nonmetallurgical uses include brilliant organic tungsten dyes and pigments which can be used in a variety of materials. Tungstates are used as phosphors in fluorescent lights, cathode-ray tubes, and x-ray screens. Tungsten compounds are used as catalysts in petroleum refining.

BIBLIOGRAPHY

"Tungsten and Tungsten Alloys" in *ECT* 1st ed., Vol. 14, pp. 353–362, by B. Kopelman, Sylvania Electric Products, Inc.; "Wolfram" in *ECT* 2nd ed., Vol. 22, pp. 334–346, by B. Kopelman and J. S. Smith, Sylvania Electric Products, Inc.

1. P. T. Stafford, *Tungsten*, U.S. Bureau of Mines Preprint 671, U.S. Bureau of Mines, Washington, D.C., 1980.
2. Comité International des Poids et Mesures, *Metrologia* **5**(2), 35 (1969).
3. A. Cezairliyan, *High Temp. Sci.* **4**, 248 (1972).
4. D. R. Stull and H. Prophet, *JANAF Thermochemical Tables*, 2nd ed., NSRDS-NBS37, NBS, Washington, D.C., June 1971.
5. *Thermophysical Properties of Matter, TRPC Data Series*, Vol. 4, *Thermal Expansion*, Plenum Press, New York, 1970.
6. M. Hoch, *High Temp. High Pressures* **1**, 531 (1969).
7. J. W. Shaner, G. R. Gathers, and C. Minichino, *High Temp. High Pressures* **8**, 425 (1976).
8. E. R. Plante and A. B. Sessions, *J. Res. Nat. Bur. Stand.* **77A**, 237 (1973).
9. C. Y. Ho, R. W. Powell, and P. E. Liley, *J. Phys. Chem. Ref. Data* **3**, Suppl. 1 (1974).
10. J. G. Hust, *High Temp. High Pressures* **8**, 377 (1976).
11. R. E. Taylor, *High Temp. High Pressures* **4**, 59 (1972).
12. G. D. Rieck, *Tungsten and Its Compounds*, Pergamon Press, London, 1967.
13. C. J. Smithells, *Tungsten*, Chemical Publishing Co., New York, 1953.
14. S. W. H. Yih and C. T. Wang, *Tungsten*, Plenum Press, New York, 1979.
15. I. Barin and O. Knacke, *Thermochemical Properties of Inorganic Substances*, Springer-Verlag, Berlin, 1973.
16. O. Kubaschewski and C. B. Alcock, *Metallurgical Thermochemistry*, 5th ed., Pergamon Press, New York, 1979.
17. M. J. Hudson, *Chem. Br.* **18**, 438 (June 1982).
18. *Mineral Commodity Summaries 1981*, U.S. Bureau of Mines, Washington, D.C., 1981.
19. C. C. Clark and J. B. Sutliff, *American Metal Market*, Jan. 23, 1981.
20. *Mineral Industry Surveys, Dec. 1979 and 1980*, U.S. Bureau of Mines, Washington, D.C.

JAMES A. MULLENDORE
GTE Product Corp.

TUNGSTEN COMPOUNDS

Tungsten is a Group-VIB transition element with atomic number 74. It has the valence states of 0, +2, +3, +4, +5, or +6 in compounds. However, tungsten alone has not been observed as a cation. Its most stable, and therefore most common, valence state is +6. Tungsten complexes vary widely in stereochemistry and oxidation states, and complex formation is exemplified by the large number of polytungstates. Simple tungsten compounds, such as the halides, are also known.

The chemical uses of tungsten have increased substantially in recent years. Catalyses of photochemical reactions and new types of soluble organometallic complexes for industrially important organic reactions are among the areas of new applications (see also Catalysis).

Tungsten Hexacarbonyl. Tungsten hexacarbonyl [14040-11-0], $W(CO)_6$, may be prepared in yields >90% by the aluminum reduction of tungsten hexachloride [13283-01-7] in anhydrous ether under a pressure of 10 MPa (ca 1 atm) of carbon monoxide at 70°C. It is purified by sublimation or steam distillation. A colorless to white solid, tungsten hexacarbonyl decomposes without melting at ca 150°C, although it sublimes *in vacuo*. It is a zero-valent monomeric compound with a relatively low vapor pressure of 13.3 Pa (0.1 mm Hg) at 20°C and 160 Pa (1.20 mm Hg) at 67°C. It is fairly stable in air, water, or acid, but is decomposed by strong bases and attacked by halogens. Tungsten carbonyl is slightly soluble in organic solvents but insoluble in water (see Carbonyls).

Various applications such as lubricant additives, dyes, pigments, and catalysts are under investigation. Tungsten can be deposited from tungsten hexacarbonyl, but carbide formation and gas-phase nucleation present serious problems (1–2). As a result, tungsten halides are the preferred starting material.

Tungsten Halides and Oxyhalides. Tungsten forms binary halides for all oxidation states between +2 and +6; oxyhalides are only known for oxidation states +5 and +6. In general, tungsten halogen compounds are reactive toward water and oxygen in the air and must therefore be handled in an inert atmosphere. They are all solid-colored compounds at room temperature, except the fluorides, and many decompose on heating before melting. The hexachloride and hexafluoride [7783-82-6] are commercially available and are particularly suitable starting materials for the chemical vapor deposition of tungsten, which is an important process technique for coatings and free-standing parts such as thin-walled tubing (see Film deposition techniques). The resulting structure is generally columnar, but recently a method has been described for obtaining a fine-grained, noncolumnar tungsten structure (3).

Fluorides. *Tungsten hexafluoride* [7783-82-6], WF_6, is a colorless gas at room temperature, sp gr 12.9 with respect to air. At 17.5°C, it condenses into a pale yellow liquid, and at 2.5°C, a white solid is formed. It may be prepared by treating hydrogen fluoride, arsenic trifluoride, or antimony pentafluoride with tungsten hexachloride or by direct fluorination of tungsten:

$$WCl_6 + 6\,HF \rightarrow WF_6 + 6\,HCl$$

$$WCl_6 + 2\,AsF_3 \rightarrow WF_6 + 2\,AsCl_3$$

$$WCl_6 + 3\,SbF_5 \rightarrow WF_6 + 3\,SbF_3Cl_2$$

$$W + 3\,F_2 \rightarrow WF_6$$

Direct fluorination of pure tungsten in a flow system at atmospheric pressure at 350–400°C is the most convenient procedure (4). Tungsten hexafluoride is extremely unstable in the presence of moisture and hydrolyzes completely to tungstic acid [7783-03-1]:

$$WF_6 + 4\,H_2O \rightarrow H_2WO_4 + 6\,HF$$

Tungsten hexafluoride dissolves in benzene or cyclohexane to give a bright red color, in dioxane a pale red, and in ether a violet-brown.

Tungsten pentafluoride [19357-83-6], WF_5, is prepared by the reduction of the hexafluoride on a hot tungsten filament in almost quantitative yield (5).

Tungsten tetrafluoride [13766-47-7], WF_4, is a nonvolatile, hygroscopic, reddish-brown solid. It has been prepared in low yields by the reduction of the hexafluoride with phosphorus trifluoride in the presence of liquid anhydrous hydrogen fluoride at room temperature (6).

Tungsten oxytetrafluoride [13520-79-1], WOF_4, forms colorless plates, mp 110°C, bp 187.5°C. It is prepared by the action of an oxygen–fluorine mixture on the metal at elevated temperatures (7). The compound is extremely hygroscopic and decomposes to tungstic acid in the presence of water.

Tungsten oxydifluoride [14118-73-1], WO_2F_2, is a white solid prepared by the careful hydrolysis of WOF_4 (8). Its chemistry has not been investigated.

Chlorides. *Tungsten hexachloride*, WCl_6, is a blue-black crystalline solid, mp 275°C, bp 346.7°C. It is prepared by the direct chlorination of pure tungsten in a flow system at atmospheric pressure at 600°C. Solidification usually occurs without incident, but further cooling may result in a violent, explosionlike expansion of the solid mass at 168–170°C. This phenomenon may be associated with an $\alpha_2 \rightarrow \alpha_1$ transition. However, tungsten hexachloride may be safely cooled if it occupies not more than one half of the containing vessel. In the presence of moisture or oxygen, some $WOCl_4$ is formed as an impurity. Tungsten hexachloride is very soluble in carbon disulfide but decomposes in water to form tungstic acid. The hexachloride is easily reduced by hydrogen to the lower halides and finally to the metal itself (2).

Tungsten pentachloride [13470-13-8], WCl_5, is a black crystalline deliquescent solid, mp 243°C, bp 275.6°C. It is very slightly soluble in carbon disulfide and decomposes in water to the blue oxide, $W_{20}O_{58}$. Magnetic properties suggest that tungsten pentachloride may contain trinuclear clusters in the solid state, but this structure has not been defined. Tungsten pentachloride may be prepared by the reduction of the hexachloride with red phosphorus (9).

Tungsten tetrachloride [13470-14-9], WCl_4, is obtained as a coarse crystalline deliquescent solid that decomposes upon heating. It is diamagnetic and may be prepared by the thermal-gradient reduction of WCl_6 with aluminum (10).

Tungsten dichloride [13470-12-7], WCl_2, is an amorphous powder. It is a cluster compound and may be prepared by the reduction of the hexachloride with aluminum in a sodium tetrachloroaluminate melt (11).

Tungsten oxytetrachloride [13520-78-0], $WOCl_4$, is a red crystalline solid, mp 211°C, bp 327°C. It is soluble in carbon disulfide and benzene and is decomposed to tungstic acid by water. It may be prepared by refluxing sulfurous oxychloride, $SOCl_2$, on tungsten trioxide (12) and purified after evaporation by sublimation.

Tungsten oxydichloride [13520-76-8], WO_2Cl_2, a pale yellow crystalline solid, mp 266°C, is soluble in cold water and in alkaline solution, although partly decomposed by hot water. It is prepared by the action of carbon tetrachloride on tungsten dioxide at 250°C in a bomb (13).

Tungsten oxytrichloride [14249-98-0], $WOCl_3$, a green solid, is prepared by the aluminum reduction of $WOCl_4$ in a sealed tube at 100–140°C (14).

Bromides. *Tungsten hexabromide* [13701-86-5], WBr_6, bluish-black crystals, mp 232°C, is formed by metathetical exchange reaction of BBr_3 with tungsten hexachloride (15).

Tungsten pentabromide [13470-11-6], WBr_5, violet-brown crystals, mp 276°C, bp 333°C, is extremely sensitive to moisture. It is prepared by the action of bromine vapor on tungsten at 450–500°C (16).

Tungsten tetrabromide [12045-94-2], WBr_4, black orthorhombic crystals, is formed by the thermal-gradient reduction of WBr_5, with aluminum, similar to the reduction of WCl_4 (10).

Tungsten tribromide [15163-24-3], WBr_3, is prepared by the action of bromine on WBr_2 in a sealed tube at 50°C (17). It is a thermally unstable black powder that is insoluble in water.

Tungsten dibromide [13470-10-5], WBr_2, formed by the partial reduction of the pentabromide with hydrogen, is a black powder that decomposes at 400°C.

Tungsten oxytetrabromide [13520-77-9], $WOBr_4$, black deliquescent needles, mp 277°C, bp 327°C, is formed by the action of carbon tetrabromide on tungsten dioxide at 250°C (13).

Tungsten oxydibromide [13520-75-7], WO_2Br_2, light red crystals, is formed by passing a mixture of oxygen and bromine over tungsten at 300°C.

Iodides. *Tungsten tetraiodide* [14055-84-6], WI_4, is a black powder that is decomposed by air. It is prepared by the action of concentrated hydriodic acid on tungsten hexachloride at 100°C.

Tungsten triiodide [15513-69-6], WI_3, is prepared by the action of iodine on tungsten hexacarbonyl in a sealed tube at 120°C (18).

Tungsten diiodide [13470-17-2], WI_2, is a brown powder, sp gr 6.79. It is reported to be prepared by the action of anhydrous hydrogen iodide on tungsten hexachloride at 400–500°C (19). Research by other investigators, however, failed to show the existence of the diiodide, but a stable tungsten oxydiiodide, WO_2I_2, is described (20).

Tungsten oxydiiodide [14447-89-3], is prepared by heating a mixture of tungsten and tungsten trioxide with excess iodine in a 500–700°C temperature gradient for 36 h (21).

Oxides, Acids, and Salts. *Oxides.* Tungsten oxides form a series of well-defined ordered phases to which precise stoichiometric formulas can be assigned (22–29) (see Table 1).

The composition of the tungsten oxides may vary over a fixed range without change in crystalline structure. Thus, the homogeneity ranges are represented by $WO_{2.95-3.0}$, $WO_{2.88-2.92}$ [12165-57-0], $WO_{2.664-2.766}$, and $WO_{1.99-2.02}$. Each tungsten atom is octahedrally surrounded by six oxygen atoms. In WO_3, these WO_6 units are joined through sharing of corner oxygen atoms only; but as the oxygen to tungsten ratio decreases, the WO_6 units become more intricately joined in combinations of corners, edges, and faces to form chains and slabs. The loss of each oxygen atom from the oxide lattice means that two electrons are added to the conduction band of the lattice, and

Table 1. Tungsten Oxides

Oxide	CAS Reg. No.	Phase	O:W, av	Theoretical density, g/cm^3	Color
WO_3	[1314-35-8]	α	3.00	7.29	yellow
$W_{20}O_{58}$	[12037-58-0]	β	2.90	7.16	blue-violet
$W_{18}O_{49}$	[12037-57-9]	γ	2.72	7.78	reddish-violet
WO_2	[12036-22-5]	δ	2.00	10.82	brown
W_3O	[39368-90-6]	(β-W)	0.33	14.4	gray

it is meaningless to speak of pentavalent and tetravalent tungsten atoms in such a lattice.

Tungsten trioxide is a yellow powder. However, the smallest diminution of oxygen brings about a change in color. Tungsten trioxide is pseudorhombic at room temperature, but tetragonal above 700°C. It is usually prepared from tungstic acid or tungstates. It is the most important tungsten oxide and is the starting material for the production of tungsten powder. Tungsten trioxide is reduced to the metal by carbon above 1050°C and by hydrogen as low as 650°C. At lower temperatures, intermediate oxides are formed. Tungsten trioxide is insoluble in water and in acid solutions (except hydrofluoric) but gives tungstate with strong alkali.

$$2\,NaOH + WO_3 \rightarrow Na_2WO_4 + H_2O$$

When heated in a hydrogen chloride atmosphere, WO_3 is completely volatilized at ca 500°C, forming the oxydichloride, WO_2Cl_2.

Tungsten dioxide is a brown powder formed by the reduction of WO_3 with hydrogen at 575°C–600°C. Generally, this oxide is obtained as an intermediate in the hydrogen reduction of the trioxide to the metal. On reduction, first a blue oxide is formed, and then the brown oxide (WO_2) forms. The composition of the blue oxide was in doubt for a long time. However, it now has been resolved that $W_{20}O_{58}$ and $W_{18}O_{49}$ are formed as intermediates. They may also be prepared by the reaction of tungsten with WO_3.

The oxide WO_3 is regarded both as an oxide and a metal phase. It is gray and has a density of 14.4 g/cm^3. It is prepared by the electrolysis of fused mixtures of WO_3 and alkali-metal phosphates. At ca 700°C, it decomposes into W and WO_2; β-tungsten is W_3O.

Tungsten Bronzes. Tungsten bronzes (30–31) constitute a series of well-defined nonstoichiometric compounds of the general formula $M_{1-x}WO_3$, where x is a variable between 0 and 1, and M is some other metal. Generally, M is an alkali metal, although many other metals can be substituted.

The systems most extensively investigated are the sodium tungsten bronzes. These compounds are intensely colored, ranging from golden-yellow to bluish-black, depending on the value of x, and, in crystalline form, exhibit a metallic sheen. They have a positive temperature coefficient of resistance for Na:WO_3 ratios >0.3 and a negative temperature coefficient of resistance at lower ratios. Sodium tungsten bronzes are inert to chemical attack by most acids but may be dissolved by basic reagents. Sodium tungsten bronzes serve as promoters for the catalytic oxidation of carbon monoxide and reformer gas in fuel cells (32) (see Batteries, secondary). In general, these bronzes form cubic or tetragonal crystals, the lattice constants increasing with sodium concentration. They are prepared by electrolytic reduction, vapor-phase deposition, fusion, or solid-state reaction (33–34). The latter method is the most versatile, in which the reagents are finely ground and heated at 500–850°C in vacuum for prolonged periods of time.

Tungsten blue. The mild reduction, for example, by Sn(II), of acidified solutions of tungstates, tungsten trioxide, or tungstic acid in solutions gives intense blue products, which are referred to by the general name of tungsten blues. Thus, they resemble molybdenum blues in many respects. Tungsten trioxide acquires a bluish tint merely on exposure to underwater ultraviolet radiation. If hydrogen is produced in a tungstate solution by means of zinc and hydrochloric acid, blue precipitates form that are stable in air. They are believed to be hydrogen analogues of the tungsten bronzes. These blue hydrogen tungsten bronzes, $H_{1-x}WO_3$, are prepared by the wet reduction of tungstic acid and are structurally related to the alkali tungsten bronzes (35–37). Tungsten blues have a strong tendency to form colloids.

Tungstic Acid and Tungstates. Tungstic acid, H_2WO_4 or $WO_3 \cdot H_2O$, is an amorphous yellow powder that is practically insoluble in water or acid solution, but dissolves readily in a strongly alkaline medium. It may be precipitated from hot tungstate solutions with strong acids. However, if the tungstate solution is acidified in the cold, a white voluminous precipitate of hydrated tungstic acid forms. It has the formula $WO_3 \cdot x H_2O$, where x is ca 2. It is converted to the yellow form by boiling in an acid medium. Both the yellow and white forms tend to become colloidal on washing. Tungstic acid forms a series of stable salts of the types $M(I)_2WO_4$, $M(II)WO_4$, and $M(III)_2(WO_4)_3$, of which some also exist in the hydrated form. Except for tungstates of the alkali metals and magnesium, these salts are generally sparingly soluble in water. They are decomposed by hot mineral acids (except phosphoric) to tungstic acid. The insoluble tungstates are prepared by adding a sodium tungstate solution to a solution of the appropriate salt. Some properties of these tungstates are given in Table 2.

Ammonium tungstate [11140-77-5], $(NH_4)_2WO_4$, cannot be obtained from an aqueous solution since it decomposes when such a solution is concentrated. It is prepared by the addition of hydrated tungstic acid to liquid ammonia.

Anhydrous sodium tungstate, Na_2WO_4, is prepared by fusing tungsten trioxide in the proper proportion with sodium hydroxide or sodium carbonate:

$$WO_3 + 2\,NaOH \rightarrow Na_2WO_4 + H_2O$$
$$WO_3 + Na_2CO_3 \rightarrow Na_2WO_4 + CO_2$$

Table 2. Properties of Normal Tungstates

Compound	CAS Reg. No.	Properties	Sp gr
$BaWO_4$	[7787-42-0]	colorless, tetragonal, $a = 0.564$ nm, $c = 1.270$ nm	5.04
$CdWO_4$	[7790-85-4]	yellow rhombic	
$CaWO_4$	[7790-75-2]	white, tetragonal, $a = 0.524$ nm, $c = 1.138$ nm, n_D^{20} 1.9263	6.06
$Ce_2(WO_4)_3$	[52345-28-5]	yellow monoclinic, $a = 1.151$ nm, $b = 1.172$ nm, $c = 0.782$ nm, $\beta = 109°\ 48'$, mp 1089°C	6.77
$PbWO_4$	[7759-01-5]	colorless, monoclinic, mp 1123°C	8.46
Ag_2WO_4	[13465-93-5]	pale yellow	
Na_2WO_4	[13472-45-2]	white rhombic, mp 698°C	4.179
$Na_2WO_4 \cdot 2H_2O$	[10213-10-2]	white rhombic, loses 2 H_2O at 100°C	3.245
$SrWO_4$	[13451-05-3]	white, tetragonal, $a = 0.540$ nm, $c = 1.190$ nm	6.187

On crystallization from aqueous solution, the dihydrate is generally obtained.

The tungstates are of particular interest in electronic and optical applications. They are also used for ceramics, catalysts, pigments, corrosion, and fire inhibitors, etc.

Polytungstates. An important and characteristic feature of the tungstate ion is its ability to form condensed complex ions of isopolytungstates in acid solution (38). As the acidity increases, the mol wt of the isopolyanions increases until tungstic acid precipitates. The extensive investigations on these systems have been hampered by the lack of well-defined solid derivatives.

The chemistry of tungsten in solution has recently been studied by chromatography and spectroscopy (39–40). Much of the reported work concerns the existence of tungstate species in acid solutions with particular reference to the molar ratio of soluble tungstate species.

If polytungstates are considered as formed by the addition of acid to WO_4^{2-}, then a series of isopolytungstates appears in which the degree of aggregation in solution increases with decreasing pH. The relationships of the species, in order of increasing ratio of $H_3O^+:WO_4^{2-}$, are shown in Table 3 (41).

Metatungstates of the alkali, alkaline-earth, rare-earth, and transition metals have been reported. However, classical synthesis rarely gives high yields of the pure compounds. The rare-earth tungstates, eg, $Ln_2(H_2W_{12}O_{40}).x\,H_2O$, may be prepared by the action of lanthanide carbonates on metatungstic acid, $H_6(H_2W_{12}O_{40})$ [*12299-86-4*]. Other salts are prepared by the action of carbonates or sulfates of the corresponding metal on metatungstic acid or metatungstates. Generally, these compounds are heat sensitive and should be recovered by freeze-drying. Alkali metal and ammonium metatungstates, $M_6(H_2W_{12}O_{40}).x\,H_2O$, may be prepared by the digestion of hydrated tungsten trioxide with the corresponding base (42–47). These salts are generally known for their high solubility in water; the most important is ammonium metatungstate [*12028-48-7*] $(NH_4)_6(H_2W_{12}O_{40})$.

The paratungstates generally are crystallized from slightly basic solutions. By far the most important salt is ammonium paratungstate [*1311-93-9*], $(NH_4)_{10}$-$W_{12}O_{41}.5H_2O$, which is usually known as the heavy form of commercial ammonium paratungstate. It is usually formed by crystallization from a boiling solution. However, if crystallization is allowed to take place slowly at room temperature, an undecahydrate, $(NH_4)_{10}W_{12}O_{41}.11H_2O$ [*12383-34-5*], is formed. This hydrate is known as the light

Table 3. Polytungstates in Order of Increasing Ratio of $H_3O^+:WO_4^{2-}$ [a]

$H_3O^+:WO_4^{2-}$	Polytungstate	CAS Reg. No.	Common name
0.333	$W_{12}O_{46}^{20-}$		para Z
0.667	$W_3O_{11}^{4-}$	[*39898-14-1*]	tritungstate
	$H_4W_3O_{13}^{4-}$		
1.167	$H_{10}W_{12}O_{46}^{10-}$	[*12401-49-9*]	para B
	$HW_6O_{21}^{5-}$	[*11080-77-6*]	para A
1.33	$W_{12}O_{40}^{8-}$		
1.50	$H_2W_{12}O_{40}^{6-}$	[*12207-61-3*]	meta
	$H_3W_6O_{21}^{3-}$	[*12273-48-2*]	pseudo meta
2.00	$WO_3.H_2O\ (H_2WO_4)$	[*7783-03-1*]	tungstic acid

[a] Ref. 41.

form of ammonium paratungstate. Both forms are insoluble in water and decomposed in acid or alkali. They are reduced to the metal by heating in a hydrogen atmosphere. Ammonium paratungstate is widely used as a catalyst. Peroxytungstic acid [41486-83-3], $H_2WO_2(O_2)_2$, may be obtained by treating tungsten trioxide with a solution of hydrogen peroxide. Peroxytungstates are known but tend to be unstable; the instability increases with increasing ratio of oxygen to tungsten (48).

Heteropolyanions are closely related to the isopolyanions and over thirty elements are known to function as the heteroatom with many stoichiometric ratios between the heteroatom and the anion. Both the acids and salts are known and are usually hydrated when crystallized from aqueous solutions. As a class, heteropoly compounds are characterized by a number of properties independent of the heteroatom and the metallic component. Typically, heteropoly tungsten compounds show the following characteristics: high molecular weight, usually >3000; a high degree of hydration; unusually high solubility in water and some organic solvents; strong oxidizing action in aqueous solution; strong acidity in free acid form; decomposition in strongly basic aqueous solutions to give normal tungstate solutions; and highly colored anions or colored reaction products.

Heteropoly anions may be classified according to the ratio of the number of central atoms to tungsten, as shown in Table 4.

Structures of heteropolytungstate and isopolytungstate compounds have been determined by x ray. The anion structures are represented by polyhedra that share corners and edges with one another. Each W is at the center of an octahedron, and an O atom is located in each vertex of the octahedron. The central atom is similarly located at the center of an XO_4 tetrahedron or XO_6 octahedron. Each such polyhedron containing the central atom is generally surrounded by WO_6 octahedra which share corners or edges (or both) with it and with one another; thus, the correct total number of oxygen atoms is utilized. Each WO_6 octahedron is directly attached to a central atom through a shared oxygen atom. In the actual structures, the octahedra are frequently distorted. The oxygens are relatively large spheres, and practically all space within the anion structure is taken up by the bulky oxygens which are close-packed or nearly so.

When the large heteropolytungstate anions are packed together as units in a

Table 4. Principal Species of Heteropolytungstates

Ratio of hetero atoms to W atoms	Principal central atoms, X	Typical formulas	Structure by x ray
1:12	P^{5+}, As^{5+}, Si^{4+}, Ge^{4+}, Ti^{4+}, Co^{3+}, Fe^{3+}, Al^{3+}, Cr^{3+}, Ga^{3+}, Te^{4+}, B^{3+}	$[X^{n+}(W_{12}O_{40})]^{(8-n)-}$	known
1:10	Si^{4+}, Pt^{4+}	$[X^{n+}(W_{10}O_x)]^{(2x-60-n)-}$	unknown
1:9	Be^{2+}	$[X^{2+}(W_9O_{31})]^{6-}$	unknown
1:6	series A:Te^{6+}, I^{7+}	$[X^{n+}(W_6O_{24})]^{(12-n)-}$	isomorphous with 6-molybdates
	series B:Ni^{2+}, Ga^{3+}	$[X^{n+}(W_6O_{24}H_6)]^{(6-n)-}$	known
2:18	P^{5+}, As^{5+}	$[X_2^{n+}(W_{18}O_{62})]^{(12-n)-}$	known
2:17	P^{5+}, As^{5+}	$[X_2^{n+}(W_{17}O_x)]^{(2x-102-2n)-}$	unknown
$1m:6m^a$	As^{3+}, P^{3+}	$[X^{n+}(W_6O_x)]_m^{m(2x-36-n)-}$	unknown

^{a} m unknown.

crystal, the interstices between the anions are very large compared to water molecules or most simple cations. In most compounds, there is apparently no direct linkage between the individual heteropoly anions, eg, in the structure of $K_6CoW_{12}O_{40}.20H_2O$ [37346-54-6] and $K_6P_2W_{18}O_{62}$ [60748-58-5]. Instead, the complexes are joined by hydrogen bonding through some molecules of water of hydrations. These principles are illustrated in the crystal structure of $H_3PW_{12}O_{40}.xH_2O$ [12501-23-4], as determined by x-ray diffraction (49–50).

Heteropoly salts of large cations, eg, cesium, frequently crystallize as acid salts no matter what the ratio of cations to anions is in the mother liquor. Furthermore, salts of these cations are frequently less highly hydrated than salts of smaller cations. Apparently, the larger cations take up so much of the space between the heteropoly anions that there is less room for water. There is often not enough room for the large cations required to form a normal salt. Instead, solvated hydrogen ions fill in to balance the negative charge of the anions, and a crystalline acid salt results.

Commercially, heteropolytungstates, particularly the heteropolytungstates, are produced in large quantities as precipitants for basic dyes, with which they form colored lakes or toners (see also Dyes and dye intermediates). They are also used in catalysis, passivation of steel, etc.

Sulfides. *Tungsten disulfide* [12138-09-9], WS_2, although found in nature, is usually prepared by heating tungsten powder with sulfur at 900°C. It is a soft, grayish-black powder, relatively inert and unreactive, with sp gr 7.5. It is insoluble in water, hydrochloric acid, alkali, and organic solvents or oils, and decomposes in hot, strong oxidizing agents, eg, aqua regia, concentrated sulfuric acid, and nitric acid. Heating in air or in the presence of oxygen yields WO_3. However, its thermal stability in air is ca 90°C higher than that of MoS_2.

Tungsten disulfide forms adherent, soft, continuous films on a variety of surfaces and exhibits good lubricating properties similar to molybdenum disulfide and graphite (51) (see also Lubrication and lubricants). It is also reported to be a semiconductor (qv).

Tungsten trisulfide [12125-19-8], WS_3, is a chocolate-brown powder, slightly soluble in cold water, but readily forming a colloidal solution in hot water. It is prepared by treating an alkali-metal thiotungstate with HCl (52). Tungsten trisulfide is soluble in alkali carbonates and hydroxides.

Tungsten forms thiotungstates corresponding to the tungstates, but one, two, three, or all of the oxygen atoms are replaced by sulfur. These compounds form with solutions of the alkali- or alkaline-earth tungstates saturated with hydrogen sulfide. They vary in color from pale yellow to yellowish-brown and, in general, crystallize well. Acidifying a solution of these salts precipitates tungsten trisulfide.

Potassium tetrathiotungstate [14293-75-5], K_2WS_4, forms yellow rhombic crystals soluble in water. Ammonium tetrathiotungstate [13862-78-7], $(NH_4)_2WS_4$, forms bright orange crystals exhibiting a metallic iridescence. These crystals are stable in dry air and soluble in water. Ammonium tetrathiotungstate is generally prepared by treating a solution of tungstic acid with excess ammonia and saturating with hydrogen sulfide. It is readily decomposed in a nonoxidizing atmosphere to WS_2, for which it is a convenient source.

Interstitial Compounds. Tungsten forms hard, refractory, and chemically stable interstitial compounds with nonmetals, particularly C, N, B, and Si. These compounds are used in cutting tools, structural elements of kilns, gas turbines, jet engines, sandblast nozzles, protective coatings, etc (see also Refractories; Refractory coatings).

Carbides. Tungsten and carbon form two binary compounds, tungsten carbide [12070-12-1], WC, sp gr 15.63, and ditungsten carbide [12070-13-2], W_2C, sp gr 17.15; both are prepared by heating tungsten and carbon at high temperatures. The presence of hydrogen or a hydrocarbon gas promotes the reaction. The relative quantities of the reactants and the temperature determine the phase formed. Tungsten carbide may also be prepared from oxygen-containing compounds of tungsten, but because of the tendency to form oxycarbides, a final heating in vacuum above 1500°C is necessary. Both carbides melt at ca 2800°C and have a hardness approaching that of diamond. Tungsten carbides are insoluble in water, but they are readily attacked by HNO_3—HF.

The most important commercial application is in hard metals. Tungsten carbides are brittle, but combination with, for example, cobalt decreases the brittleness. Approximately 67% of tungsten production is for the manufacture of WC (see Carbides).

Nitrides. The nitrides of tungsten are quite similar to the carbides. Although nitrogen does not react directly with tungsten, the nitrides can be prepared by heating tungsten in ammonia. The two phases, ditungsten nitride [12033-72-6], W_2N, and tungsten nitride [12058-38-7], WN, have been extensively studied (53–54) (see Nitrides).

Borides. Ditungsten boride [12007-09-9], W_2B, and tungsten boride [12007-09-9], WB, are prepared by hot pressing tungsten and boron; ditungsten pentaboride [12007-98-6], W_2B_5, is prepared by heating tungsten trioxide, graphite, and boron carbide *in vacuo*. Tungsten borides are extremely hard and exhibit almost metallic electrical conductivity. Recently, the formation of tungsten boride phases in the manufacture of boron filaments for structural composites for in-space vehicles and aircraft has been reported (55) (see Boron compounds, refractory).

Silicides. Tungsten silicides form a protective oxide layer over tungsten to prevent destructive oxidation at elevated temperatures. The layer fails to protect at lower temperatures, a behavior referred to as disilicide pest. This failure can be explained by the silicon being initially oxidized to SiO_2 at the surface and depleting the surface of Si, forming pentatungsten trisilicide [12039-95-1], W_5Si_3. At high temperatures, a uniform layer of W_5Si_3 is formed, but at lower (pest) temperatures, the attack is not uniform and seems to follow grain boundaries or subgrain boundaries in the disilicide. The next stage is the rapid growth and penetration of the complex oxide into the disilicide layer. This process ultimately consumes the disilicide, causing oxidation of the tungsten substrate. The existence of tritungsten disilicide [12509-47-6], W_3Si_2, and ditungsten silicide [56730-24-6], W_2Si, has been reported (56).

Ditungsten trisilicide [12138-30-6], W_2Si_3, gray, sp gr 10.9, is insoluble in water, acid, or alkaline solutions. It is readily attacked by HNO_3—HF and fused alkali-metal carbonates and hydroxides.

Tungsten disilicide [12039-88-2], WSi_2, forms bluish-gray tetragonal crystals (a = 0.3212 nm, c = 0.7880 nm). It is insoluble in water and melts at 2160°C. The compound is attacked by fluorine, chlorine, fused alkalies, and HNO_3—HF. It may be used for high temperature thermocouples in combination with $MoSi_2$ in an oxidizing atmosphere.

Anionic Complexes. Compounds of tungsten with acid anions, other than halides and oxyhalides, are relatively few in number, and are known only in the form of complex salts. A number of salts containing hexavalent tungsten are known. Potassium octafluorotungstate [57300-87-5], K_2WF_8, can be prepared by the action of KI on $W(CO)_6$ in an IF_5 medium. The addition of tungstates to aqueous hydrofluoric acid gives salts which are mostly of the type $M(I)_2(W_2F_4)$. Similarly, double salts of tungsten oxydichloride are known.

Salts containing pentavalent tungsten may be obtained by the reduction of alkali tungstate in concentrated hydrochloric acid. Salts of the type $M(I)_2(WOCl_5)$ (green), $M(I)(WOCl_4)$ (brown-yellow), and $M(I)(WOCl_4.H_2O)$ (blue) have been isolated. Thiocyanato and bromo salts are also known.

Salts containing tetravalent tungsten have been prepared by various methods. The most important are the octacyanides, $M(I)_4(W(CN)_8)$. They form yellow crystals and are very stable. They are isolated as salts or free acids and can be oxidized by $KMnO_4$ in H_2SO_4 to compounds containing pentavalent tungsten, $M(I)_3(W(CN)_8)$ (yellow).

The only known trivalent tungsten complex is of the type $M(I)_3(W_2Cl_9)$. It is prepared by the reduction of strong hydrochloric acid solutions of K_2WO_4 with tin. If the reduction is not sufficient, a compound containing tetravalent tungsten, $K_2(WCl_5(OH))$ [84238-10-0], is formed (57).

Toxicity

A considerable difference in the toxicity of soluble and insoluble compounds of tungsten has been reported (58). For soluble sodium tungstate, $Na_2WO_4.2H_2O$, injected subcutaneously in adult rats, LD_{50} is 140–160 mg W/kg. Death is due to generalized cellular asphyxiation. Guinea pigs treated orally or intravenously with $Na_2WO_4.2H_2O$ suffered anorexia, colic, incoordination of movement, trembling, and dyspnea.

Orally in rats, the toxicity of sodium tungstate was highest, tungsten trioxide was intermediate, and ammonium tungstate [15855-70-6] least (59–60). In view of the degree of systemic toxicity of soluble compounds of tungsten, a threshold limit of 1 mg of tungsten per m^3 of air is recommended. A threshold limit of 5 mg of tungsten per m^3 of air is recommended for insoluble compounds (61).

Uses

Tungsten compounds, especially the oxides, sulfides, and heteropoly complexes, form stable catalysts for a variety of commercial chemical processes, eg, petroleum processing (62). The tungsten compounds may function as principal catalysts or promoters of other catalysts. The blue oxide, $W_{20}O_{58}$, is an important catalyst in industrial chemical synthesis involving hydration, dehydration, hydroxylation, and epoxidation (63). It is expected that the application of tungsten catalysts will increase greatly since many tungsten compounds are commercially available.

Tungsten hexachloride is used for preparing tungsten metathesis catalysts, which are very interesting because they form double and triple bonds with carbon. It is claimed that these catalysts permit the systematic control of a class of compounds used in the production of petroleum, plastics, synthetic fibers, and detergents. The improved control is expected to cut costs by providing more efficient use of raw materials (64). Films of tungsten deposited on various substrates improve the electrical conductivity of transparent tin oxide coatings on aircraft windows and windshields (see Film deposition techniques). Other uses include fire-retardant catalysts and as a fluxing agent in welding.

Tungsten disulfide forms adherent, soft, continuous films on a variety of substrates and exhibits good lubrication under extreme conditions of temperature, load, and vacuum. Applied as a dry powder, suspension, bonded film, or aerosol, it can be an effective lubricant in wire drawing, metal forming, valves, gears, bearings, packing materials, etc. Oil-soluble tungsten compounds, such as the ammonium salts of tungstate or tetrathiotungstate, are reported to be effective lubricating-oil additives.

Sodium tungstate is used in the manufacture of heteropolyacid color lakes which are used in printing inks, paints, waxes, glasses, and textiles. It is also used as a fuel-cell electrode material and in cigarette filters. Other uses include the manufacture of tungsten-based catalysts, the fireproofing of textiles, and as an analytical reagent for the determination of uric acid.

Calcium tungstate is fluorescent when exposed to ultraviolet radiation and is therefore widely used in the manufacture of phosphors. It is used in lasers, fluorescent lamps, high voltage sign tubes, and oscilloscopes for high speed photographic processes. Small crystals have been used for injection into malignant tumors, thus affording by transillumination a means of x-ray treatment. Other uses include screens for x-ray observations and photographs, luminous paints, and scintillation counters.

Ammonium paratungstate is of commercial significance since it is the precursor of high purity tungsten oxides, tungsten, and tungsten carbide powders. It is slightly soluble in water but reacts with hydrogen peroxide to produce soluble peroxytungsten compounds.

Ammonium metatungstate is of commercial significance because of its high solubility in water. This property as well as its acid characteristics make it a very desirable starting material for catalysts, and the impregnation of catalyst carriers with alkali-free solutions of tungsten. Other uses include nuclear shielding, corrosion inhibitors, and the preparation of other tungsten chemicals.

Tungsten trioxide is the principal source of tungsten metal and tungsten carbide powders. Because of its bright yellow color, it is used as a pigment in oil and water colors (see Pigments). It is used in a wide variety of catalysts with the most recent application in the control of air pollution and industrial hygiene.

Tungsten carbides are widely used in the manufacture of hard carbides for high speed machining tools, wire-drawing dies, wear surfacing, drills, etc.

Heteropoly tungstic acids, particularly the heteropolys, are useful in analytical chemistry and biochemistry as reagents; in atomic-energy work as precipitants and inorganic ion exchangers; in photographic processes as fixing agents and oxidizing agents; in plating processes as additives; in plastics, adhesives, and cements for imparting water resistance; and in plastics and plastic films as curing or drying agents. An important use for 12-tungstophosphoric acid [12067-99-1] and its sodium salts is the manufacture of organic pigments. These compounds are also extensively used for the surface treatment of furs. In the textile industry, the salts are useful as antistatic agents. The acids are used in diverse applications, eg, printing inks, paper coloring, nontoxic paints, and wax pigmentation.

The tungstates and molybdates are good corrosion inhibitors and have been used for some time in antifreeze solutions. In addition, they are used as laser-host materials, phosphors, and for the flameproofing of textiles.

BIBLIOGRAPHY

"Tungsten Compounds" in *ECT* 1st ed., Vol. 14, pp. 363–372, by B. Kopelman, Sylvania Electric Products, Inc.; "Wolfram Compounds" in *ECT* 2nd ed., Vol. 22, pp. 346–358, by M. B. MacInnis, Sylvania Electric Products, Inc.

1. J. J. Lander and L. H. Germer, *Am. Inst. Mining Met. Eng., Inst. Met. Div., Met. Technol.* 14(6), Tech. Publ. 2259 (1947).
2. C. F. Powell, J. H. Oxley, and J. M. Blocher, Jr., *Vapor Deposition*, John Wiley & Sons, Inc., New York, 1966.
3. R. L. Landingham and J. H. Austin, *J. Less-Common Met.* 18(3), 229 (1969).
4. E. J. Barber and G. H. Cady, *J. Phys. Chem.* 60, 505 (1956).
5. R. D. Peacock, *J. Inorg. Nucl. Chem.* 35(3), 751 (1973).
6. T. A. O'Donnell and D. F. Stewart, *Inorg. Chem.* 5, 1434 (1966).
7. G. H. Cady and G. B. Hargreaves, *J. Chem. Soc.*, 1568 (1961).
8. O. Ruff, F. Eisner, and W. Heller, *Z. Anorg. Allgem. Chem.* 52, 256 (1907).
9. G. I. Novikov, N. Y. Andreeva, and O. G. Polyachenok, *Russ. J. Inorg. Chem.* 6, 1019 (1961).
10. R. E. McCarley and T. M. Brown, *Inorg. Chem.* 3, 1232 (1964).
11. W. C. Dorman, *IS-T-510*, National Technical Information Service, Dept. of Commerce, Washington, D.C., 1972, 42 pp.
12. R. Colton and I. B. Tomkins, *Aust. J. Chem.* 18, 447 (1965).
13. E. R. Epperson and H. Frye, *Inorg. Nucl. Chem. Lett.* 2, 223 (1966).
14. G. W. A. Fowles and J. L. Frost, *Chem. Commun.*, 252 (1966).
15. P. M. Druce and M. F. Lappert, *J. Chem. Soc. A* 22, 3595 (1971).
16. R. Colton and I. B. Tomkins, *Aust. J. Chem.* 19, 759 (1966).
17. R. E. McCarley and T. M. Brown, *J. Am. Chem. Soc.* 84, 3216 (1962).
18. C. Djordjevic, R. S. Nyholm, C. S. Pande, and M. H. B. Stiddard, *J. Chem. Soc., A*, 16 (1966).
19. H. E. Roscoe, *Liebigs Ann. Chem.* 162, 366(1872).
20. J. Tillack, P. Eckerlin, and J. H. Dettingmeijer, *Angew. Chem.* 78, 451 (1966).
21. A. Bartecki, M. Cieslak, and S. Weglowski, *J. Less-Common Met.* 26(3), 411 (1972).
22. E. Gebert and R. J. Ackermann, *Inorg. Chem.* 5(1), 136 (Jan. 1966).
23. J. Neugebauer, T. Miller, and L. Imre Tungsram, *Techn. Mitteil.* (2), (Mar. 1961).
24. G. Hagg and A. Magneli, *Rev. Pure Appl. Chem.* 4, 235 (1954).
25. O. Glemser and H. Sauer, *Z. Anorg. Chem.* 252, 144 (1943).
26. L. L. Y. Chang and B. Phillips, *J. Am. Cer. Soc.* 52(10), 527 (1969).
27. G. Hagg and N. Schonberg, *Acta Cryst.* 7, 351 (1954).
28. A. Magneli, *Ark. Kemi* 1, 513 (1950).
29. A. Magneli, *J. Inorg. Nucl. Chem.* 2, 330 (1956).
30. P. G. Dickens and M. S. Whittingham, *Q. Rev. Chem. Soc.* 22(1), 30 (1968).
31. M. J. Sienko, *Adv. Chem. Ser.* 39, 224 (1963).
32. L. W. Niedrach and H. I. Zeliger, *J. Electrochem. Soc.* 116(1), 152 (1969).
33. J. P. Randin, *J. Electrochem. Soc.* 120(3), 378 (1973).
34. V. I. Spitsyn and T. I. Drobasheva, *Zh. Inorg. Khim.* 21(7), 1787 (1976).
35. O. Glemser and C. Naumann, *Z. Anorg. Chem.* 265, 288 (1951).
36. P. G. Dickens and R. J. Hurditch, *Nature* 215, 1266 (1967).
37. E. Schwarzmann and R. Birkenberg, *Z. Naturforsch. B* 26(10), 1069 (1971).
38. D. L. Kepert, *Progr. Inorg. Chem.* 4, 199 (1962).
39. P. Tekula-Buxbaum, *Acta Tech. Acad. Sci. Hung.* 78(3–4), 325 (1974).
40. H. M. Ortner, *Anal. Chem.* 47(1), 162 (1975).
41. T. K. Kim, R. W. Mooney, and V. Chiola, *Sep. Sci.* 3(5), 467 (1968).
42. U.S. Pat. 3,175,881 (Mar. 30, 1965), V. Chiola, J. M. Lafferty, Jr., and C. D. Vanderpool (to Sylvania Electric Products, Inc.).
43. U.S. Pat. 3,591,331 (July 6, 1971), V. Chiola, P. R. Dodds, F. W. Liedtke, and C. D. Vanderpool (to Sylvania Electric Products, Inc.).
44. U.S. Pat. 3,857,928 (Dec. 31, 1974), T. K. Kim, J. M. Lafferty, Jr., M. B. MacInnis, J. C. Patton, and L. R. Quatrini (to GTE Sylvania Incorporated).
45. U.S. Pat. 3,857,929 (Dec. 31, 1974), L. R. Quatrini, T. K. Kim, J. C. Patton, and M. B. MacInnis (to GTE Sylvania Incorporated).

46. U.S. Pat. 3,936,362 (Feb. 3, 1976), C. D. Vanderpool, M. B. MacInnis, and J. C. Patton, Jr. (to GTE Sylvania Incorporated).
47 U.S. Pat. 3,956,474 (May 11, 1976), J. E. Ritsko (to GTE Sylvania Incorporated).
48. A. Chretien and J. Helgorsky, *C. R. Acad. Sci. Paris* **252**, 742 (1961).
49. A. J. Bradley and J. W. Illingworth, *Proc. Roy. Soc. London Ser. A* **49**(157), 113 (1936).
50. R. Signer and H. Gross, *Helv. Chim. Acta* **17**, 1076 (1934).
51. V. R. Johnson, M. T. Lavik, and E. E. Vaughn, *J. App. Phys.* **28**, 821 (1957).
52. O. Glemser, H. Saver, and P. Konig, *Z. Inorg. Chem.* **257**, 241 (1948).
53. A. G. Mattock, R. H. Platt, A. F. Williams, and R. Gancedo, *J. Chem. Soc. Dalton Trans.* **12**, 1314 (1974).
54. L. A. Cherezova and B. P. Kryzhanovskii, *Opt. Spektrosk* **34**(2), 414 (1973).
55. A. L. Buryking, Yu. V. Dzyrdykevich, and V. V. Gorskii, *Poroshk Metall.* **2**, 74 (1973).
56. N. N. Matynshenko, L. N. Efimenko, and D. N. Solonikin, *Fiz. Met. Metalloved.* **8**, 878 (1959).
57. E. Konig, *Inorg. Chem.* **2**, 1238 (1963).
58. *U.S. DHEW (NIOSH) Publication No. 77-127*, DHEW, 1977.
59. F. W. Kinard and J. Van de Erve, *Am. J. Med. Sci.* **199**, 668 (1940).
60. V. G. Nadeenko, *Hyg. Sanit.* **31**, 197 (1966).
61. *Documentation of TLV*, American Conference of Industrial Hygienists, Cincinnati, Ohio, 1966, Appendix C.
62. C. H. Kline and V. Kollonitsch, *Ind. Eng. Chem.* **57**(7), 53 (1965).
63. C. H. Kline and V. Kollonitsch, *Ind Eng. Chem.* **57**(9), 53 (1965).
64. D. N. Clark and R. R. Schrock, *J. Am. Chem. Soc.* **100**, 6774 (1978).

General References

S. W. H. Yih and C. T. Wang, *Tungsten*, Plenum Press, New York, 1979.
G. D. Rieck, *Tungsten and Its Compounds*, Pergamon Press, London, 1967.
K. C. Li and C. Y. Wang, *Tungsten*, Reinhold Publishing Corporation, New York, 1955.
C. J. Smithells, *Tungsten*, Chapman and Hall Ltd., London, 1952.
J. H. Canterford and R. Colton, *Halides of the Transition Elements*, John Wiley & Sons, Inc., New York, 1978.

M. B. MacInnis
T. K. Kim
GTE Products Corporation

TURBIDITY AND NEPHELOMETRY. See Analytical methods.

TURKEY RED OIL. See Castor oil.

TURPENTINE. See Terpenoids.

TYPE METAL. See Lead alloys.

TYROCIDINE. See Antibiotics, peptides.

TYROTHRICIN. See Antibiotics, peptides.

U

ULTRAFILTRATION

Ultrafiltration is a pressure-driven filtration separation occurring on a molecular scale (see also Dialysis; Filtration; Hollow-fiber membranes; Membrane technology; Reverse osmosis). Typically, a liquid including small dissolved molecules is forced through a porous membrane. Large dissolved molecules, colloids, and suspended solids that cannot pass through the pores are retained.

Ultrafiltration separations range from ca 2 to 20 nm. Above ca 20 nm, the process is known as microfiltration. Transport through ultrafiltration and microfiltration membranes is described by pore-flow models. Below ca 2 nm, interactions between the membrane material and the solute and solvent become significant. That process, called reverse osmosis or hyperfiltration, is best described by solution–diffusion mechanisms.

Membrane-retained components are collectively called concentrate or retentate. Materials permeating the membrane are called filtrate, ultrafiltrate, or permeate. It is the objective of ultrafiltration to recover or concentrate particular species in the retentate (eg, latex concentration, pigment recovery, protein recovery from cheese and casein wheys, etc) or to produce a purified permeate (eg, sewage treatment, production of sterile water or antibiotics, etc). Diafiltration is a specific ultrafiltration process in which the retentate is further purified or the permeable solids are extracted further by the addition of water to the retentate. It is analogous to the conventional washing of filter cake.

Membrane filtration has been used in the laboratory for over a century. The earliest membranes were homogeneous structures of purified collagen or zein. The first synthetic membranes were nitrocellulose (collodion) cast from ether in the 1850s.

By the early 1900s, standard graded nitrocellulose membranes were commercially available (1). Their utility was limited to laboratory research because of low transport rates and susceptibility to internal plugging. They did, however, serve a useful role in the separation and purification of colloids, proteins, blood sera, enzymes, toxins, bacteria, and viruses (2).

In the late 1950s and 1960s, a technique was developed that produced highly anisotropic or asymmetric structures, ie, membranes constructed of a very thin, tight surface skin with a porous substructure. The substructure provided the necessary mechanical support for the skin without the hydraulic resistance of previous isotropic structures. Flux rates improved by orders of magnitude, and inherent resistance to plugging increased. A molecule entering a pore through the skin traverses a channel of increasing diameter. Both high flux and plugging resistance are important for achieving an economical membrane performance in industrial applications.

The subsequent improvement of the physical and chemical characteristics of these membranes, their incorporation into machines, and the development of procedures to prevent or clean surface-fouling films were the principal areas of significant advancement in the last two decades. By 1980, the industrial ultrafiltration market had grown to an estimated $(30-50) \times 10^6$.

Media

Most ultrafiltration membranes are porous, asymmetric, polymeric structures produced by phase inversion, ie, the gelation or precipitation of a species from a soluble phase.

Typically, a polymer is first dissolved in a mixture of miscible solvents and nonsolvents. Frequently, this mixture is a better polymer solvent than any of the components (3–4). A film is cast from the deaerated solution. The surface of the film is then placed in contact with a nonsolvent diluent miscible with the solvent. This precipitates or gels the surface almost instantaneously, forming a membrane skin.

Macroscopically, the solvent and precipitant are no longer discontinuous at the polymer surface, but diffuse through it. The polymer film is a continuum with a surface rich in precipitant and poor in solvent. Microscopically, as the precipitant concentration increases, the polymer solution separates into two interspersed liquid phases: one rich in polymer and the other poor. The polymer concentration must be high enough to allow a continuous polymer-rich phase but not so high as to preclude a continuous polymer-poor phase.

The skin is highly stressed because of the polymer consolidation. The surface tears at polymer-poor sites, forming cracks or pores that expose a more fluid internal polymer layer to the precipitant–solvent mixture (5). The pores propagate into so-called fingers by drawing the precipitating polymer from the bottom to the side of the pore (see Fig. 1). Because this process proceeds along a moving boundary into the polymer film, additional pores do not form on the walls. The polymer solution behind these precipitated walls gels into an open-sponge structure (see Fig. 2). The capillary stresses (surface activity) must be low enough to avoid collapsing the structure. Polymers with high elastic moduli and solvents that do not plasticize the polymer are preferred.

Membrane structure is a function of the materials used (polymer composition, molecular weight distribution, solvent system, etc) and the mode of preparation (so-

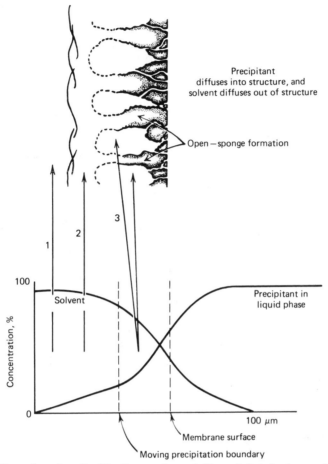

Figure 1. Formation of an ultrafiltration membrane. 1, Unprecipitated polymer solution; 2, polymer solution separating into two phases; 3, pore fingers with precipitant–solvent mixture.

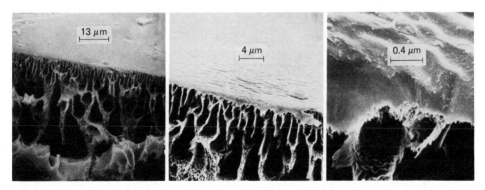

Figure 2. A series of progressively closer SEM (scanning electron microscope) photographs of the same membrane cross section, clearly showing skin and substructure.

lution viscosity, evaporation time, humidity, etc). Commonly used polymers include cellulose acetates, polyamides, polysulfones, dynels (vinyl chloride–acrylonitrile copolymers) and poly(vinylidene fluoride).

Modification of the membranes affects the properties. Cross-linking improves mechanical properties and chemical resistivity. Fixed-charge membranes are formed by incorporating polyelectrolytes into polymer solution and cross-linking after the membrane is precipitated (6), or by substituting ionic species onto the polymer chain (eg, sulfonation). Polymer grafting alters surface properties (7). Enzymes are added to react with permeable species (8–11) and reduce fouling (12–13).

Polyelectrolyte complex membranes are phase-inversion membranes where polymeric anions and cations react during the gelation. The reaction is suppressed before gelation by incorporating low molecular weight electrolytes or counterions in the solvent system. Both neutral and charged membranes are formed in this manner (14–15). These membranes have not been exploited commercially because of their lack of resistance to chemicals (see Polyelectrolytes).

Inorganic ultrafiltration membranes are formed by depositing particles on a porous substrate (16–17). In one form, inorganic particles (alumina, Zr_2SiO_2, etc) of two discrete sizes are deposited. The smaller size can pass through the porous support while the larger size cannot. The mixture forms a controlled porosity film at the entrance to the support's pores. These membranes can be removed and regenerated *in situ*. Alternatively, inorganic or organic binders can be added as stabilizers. Inorganic membranes exhibit good thermal and chemical stability.

Dynamic membranes are concentration–polarization layers formed *in situ* from the ultrafiltration of colloidal material analogous to a precoat in conventional filter operations. Hydrous zirconia has been thoroughly investigated; other materials include bentonite, poly(acrylic acid), and films deposited from the materials to be separated (18).

Track-etched membranes are made by exposing thin films (mica, polycarbonate, etc) to fission fragments from a radiation source. The high energy particles chemically alter material in their path. The material is then dissolved by suitable reagents, leaving nearly cylindrical holes (19) (see Particle-track etching).

Process

Pore-flow models most accurately describe ultrafiltration processes. Other membrane transport mechanisms, which may occur simultaneously although generally at a much lower rate, include dialysis (diffusion), osmosis (solvent by osmotic gradient), anomalous osmosis (osmosis with a charged membrane), reverse osmosis (solvent by pressure gradient larger and opposite to osmotic gradient), electrodialysis (solute ions by electric field), piezodialysis (solute by pressure gradient), electroosmosis (solvent in electric field), Donnan effects, Knudsen flow, thermal effects, chemical reactions (including facilitated diffusion), and active transport.

When pure water is forced through a porous ultrafiltration membrane, Darcy's law states that the flow rate is directly proportional to the pressure gradient:

$$J = \frac{V}{A \cdot t} = \frac{K_m \Delta P}{\mu} \tag{1}$$

where J is permeate flux in units of volume V per membrane area A, at time t, K_m is the membrane hydraulic permeability, μ is the fluid viscosity, and ΔP is the membrane pressure drop between the retentate and permeate.

The membrane hydraulic permeability K_m is a function of the pore size, tortuosity, and length, and any resistance in the substructure. Since ultrafiltration membranes are plastic and can yield (compact) or creep under pressure, K_m is also a function of the pressure history. Dynamic pressure drops from flow through a membrane and static pressure drops from a force applied on a membrane surface (eg, across a fouling film) can both cause compaction. Initial compaction occurs rapidly during startup, whereas long-term compaction occurs slowly over the operating life of the membrane. Swelling agents can sometimes (partially) reverse compaction.

Initial membrane compaction is illustrated by Figure 3. Equation 1 predicts a straight-line response of J to ΔP, or J_3 at P_1. Owing to the compaction, a lower flux J_2 is observed. Once a membrane has been subjected to some pressure (P_1), equation 1 is valid for predicting flux up to that pressure (Fig. 3, curve b). If the membrane is subsequently subjected to higher pressure (P_2), the hydraulic permeability constant is changed (Fig. 3, curve d).

The addition of small membrane-permeable solutes to the water affects permeate transport in the following ways:

1. Solute–solvent interactions change the permeating fluid viscosity.

2. Solute adsorption reduces the apparent membrane-pore diameter (20). Because of high interfacial tension between water and certain materials, the water phase in the pores can be replaced. Dynel and polysulfone membranes, for example, preferentially extract partially soluble alcohols from water. Surfactants suppress hydrophobic adsorption. Adsorption of permeate species is characterized by a lag in permeate concentration as a function of time.

3. The interfacial charge between the membrane-pore wall and the liquid affects permeate transport when the Debye screening length approaches (ca 10%) the mem-

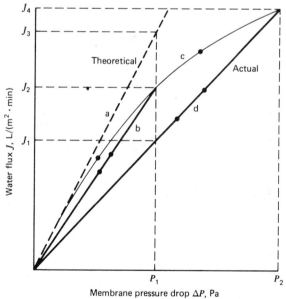

Figure 3. Water flux versus pressure. Equation 1 predicts flux J_3 at P_1. Actual initial water flux follows curve c to flux J_2 at P_1. Subsequent operation at pressure drops less than P_1 follows curve b (eq. 1). If pressure is increased above P_1, flux follows curve c (additional compaction) to P_2. A new value of K_m is used in equation 1. Operation at pressure drops less than P_2 follows curve d. Flux at P_1 is lowered to J_1.

brane-pore size. Flux declines, rejection increases, and electrolyte is retained. Other electrokinetic phenomena become pronounced and may influence fouling (21).

4. High surface tension on hydrophobic membranes forces water molecules to form large clusters in the pores. Water-structuring ions (eg, Na^+, Mg^{2+}, and OH^-) tend to decrease permeability and increase rejection; destructuring ions (eg, Cl^-, NO_3^-, and ClO_4^-) have the opposite effect (22).

5. Solvents, swelling agents, and plasticizers that diffuse into the polymer structure can change the apparent pore size (K_m in eq. 1), or increase the rate of long-term compaction.

If the solute size is approximately the (apparent) membrane-pore size, it interferes with the pore dimensions. The solute concentration in the permeate first increases, then decreases with time. The rejection δ of a solute is defined as:

$$\delta = 1 - \frac{C_{pi}}{C_{Bi}} \qquad (2)$$

where C_{pi} is the permeate concentration of species i and C_{Bi} is the concentration of that species in the retentate. The point of maximum interference is further characterized as a minimum flux. Figure 4 is a plot of retention and flux versus molecular weight. It shows the minimum flux at ca 60–90% retention.

If the solute size is greater than the pore dimensions, the solute is retained by mechanical sieving.

Membrane pores are not of uniform size (23). They are not cylindrical, but rather resemble fissures (24) or cracks (5). Similarly, molecules are not spherical. A long chain of 100,000 mol wt (eg, dextran) may readily pass through a pore which retains a globular protein of 20,000 mol wt. Branching chains may block or plug pores. Frequently, macromolecules change shape as a function of solution pH or ionic strength. The transition between solute passage and rejection is therefore gradual and involves conformational considerations (25–26). The slope of the retention curve of Figure 4

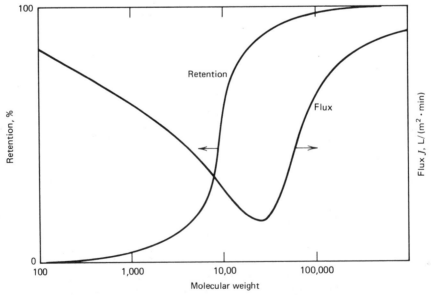

Figure 4. Retention and flux versus molecular weight.

is a measure of the interaction between the pore-size and the solute-size distributions.

Retained species are transported to the membrane surface at the rate:

$$J_i = JC_{Bi} \tag{3}$$

where J is the permeate flux and C_{Bi} is the bulk concentration of the retained species i. They accumulate in a boundary layer at the membrane surface (see Fig. 5). This deposit is composed of suspended particles similar to conventional filter cakes, and more importantly, a slime that forms as retained solutes exceed their solubility. The gel concentration C_g is a function of the feed composition and the membrane-pore size. The gel usually has a much lower hydraulic permeability and smaller apparent pore size than the underlying membrane (27). The gel layer and the concentration gradient between the gel layer and the bulk concentration are called the gel-polarization layer.

Feed–constituent interactions further affect retention (28–29). Dispersing agents and emulsifiers are partially retained because they attach to the dispersed phase. Small molecules may similarly adsorb onto larger particles.

The gel-layer thickness is limited by mass transport back into the solution bulk at the rate:

$$J_i = K \frac{dC_i}{dX} \tag{4}$$

where the mass-transfer coefficient K is multiplied by the concentration gradient.

At steady state,

$$JC_B = K \frac{dC}{dX} \tag{5}$$

where C_B is the bulk concentration of all retained species. Integration gives

$$J = K \cdot \ln \frac{C_g}{C_B} \tag{6}$$

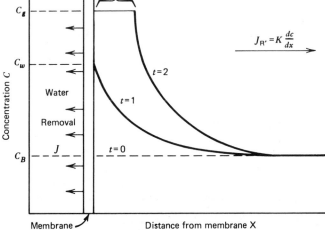

Figure 5. Concentration polarization. C_w = concentration at membrane wall, C_B = bulk concentration, C_{Bi} = bulk concentration of species i, J = flux, and C_g = gel concentration.

In a static system, the gel-layer thickness rapidly increases and flux drops to uneconomically low values. In equation 6, however, K is a function of the system hydrodynamics. Typically, high flux is sustained by moving the solution bulk tangentially to the membrane surface. This action decreases the gel thickness and increases the overall hydraulic permeability. For any given channel dimension, there is an optimum velocity which maximizes productivity (flux per energy input).

A number of analytical solutions have been derived for K as a function of channel dimensions and fluid velocity (30). In practice, the fit between theory and data for K is poor except in idealized cases. Most processes exhibit either higher fluxes, presumably caused by physical disruption of the gel layer from the nonideal hydrodynamic conditions, or lower fluxes caused by fouling (31). In addition, K is a function of the fluid composition.

Ultrafiltration equipment suppliers derive K empirically for their equipment on specific process fluids. Flux J is plotted versus log C_B for a set of operating conditions in Figure 6; K is the slope, and C_g is found by extrapolating to zero flux. Operating at different hydrodynamic conditions yields differently sloped curves through C_g.

The gel-polarization layer has an hydraulic permeability of K_g. Equation 6 states that flux is independent of pressure, and K_g must therefore decrease with increasing pressure. Equation 1 becomes

$$J = \frac{\Delta P}{\mu \left(\dfrac{1}{K_m} + \dfrac{1}{K_g} \right)} = \frac{\Delta P}{\mu (R_m + R_g)} \tag{7}$$

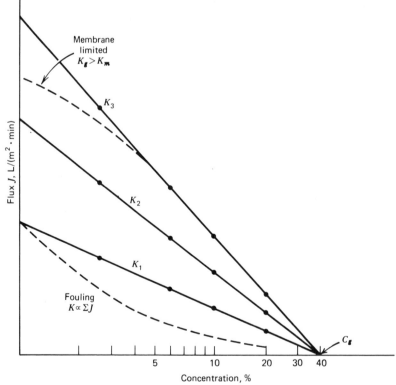

Figure 6.   Flux versus concentration, illustrating the effect of operating conditions on K and deviations from equation 6.

where R_m and R_g are the hydraulic resistances of the membrane gel.

Flux is independent of pressure when the process flux is much less than the water flux ($K_g \ll K_m$). If $K_g > K_m$, the process is limited by the membrane water flux and flux would flatten out at low concentrations of solids (see Fig. 6).

For very small ΔP, flux is linear with pressure. Figure 7 shows a graph of flux versus pressure. Curve a is the pure water flux from equation 1, curve b is the theoretical permeate flux (TPF) for a typical process. As the gel layer forms, the flux deviates from the TPF following equation 7 and curve d. Changing the hydrodynamic conditions changes K_g and results in a different operating curve c.

Fouling.   If the gel-polarization layer is not in hydrodynamic equilibrium with the fluid bulk, the membrane is fouled. Fouling is caused by either adsorption of species on the membrane surface or consolidation (fusing, polymerization, etc) of an existing gel layer. Fouled systems are characterized as follows: flux is a function of total permeate production when hydrodynamic conditions are constant (see Fig 6); if hydrodynamic conditions are changed, hydraulic permeability response of the gel layer is not reversible; and theoretical permeate flux (TPF) changes with time. A sensitive test for predicting fouling or process instability is to measure change in TPF after subjecting the system to process extremes (eg, high pressure with no flow).

Fouling is controlled by selection of proper membrane materials, pretreatment of feed and membrane, and operating conditions. Control and removal of fouling films is essential for industrial ultrafiltration processes.

Suspensions of oil in water (32), such as lanolin in wool scouring effluents, are stabilized with emulsifiers to prevent the oil phase from adsorbing onto the membrane. Polymer latices and electrophoretic paint dispersions are stabilized with surface-active agents to reduce particle agglomeration in the gel-polarization layer.

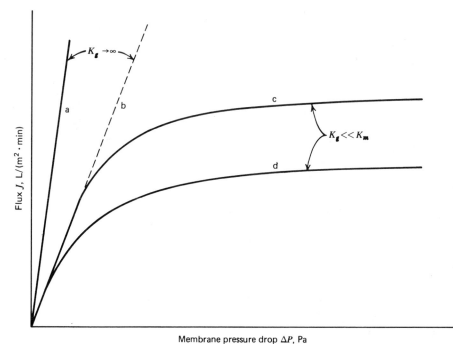

Figure 7.   Flux versus membrane ΔP.

Dairy wheys containing complex mixtures of proteins, salts, and microorganisms rapidly foul membranes. Heat treatment and pH adjustment accelerate the aggregation of β-lactoglobulin with other whey components (33–34). Otherwise, they would interact within the polarization layer (35–36), forming sheet-like fouling gels. These methods also reduce microbial fouling and the formation of apatite gels. Other whey pretreatment methods include demineralization, clarification, and centrifugation (37–38).

Pretreatment of membranes with dynamically formed polarization layers and enzyme precoats have been effective (12–13,39). Pretreatment with synthetic permeates prevents start-up instability with some feed dispersions.

When fouling is present or possible, ultrafiltration is usually operated at high liquid shear rates and low pressure to minimize the thickness of the gel polarization layer.

Cleaning.   Fouling films are removed from the membrane surface by chemical and mechanical methods. Chemicals and procedures vary with the process, membrane type, system configuration, and materials of construction. The equipment manufacturer recommends cleaning methods for specific applications. A system is considered clean when it has returned to its original water flux.

In order to develop an effective cleaning method, it is essential to know the fouling constituents and whether the cleaning agents solubilize or disperse the foulants. Detergents emulsify oils, fats, and grease (40), whereas protein films are dispersed by proteolytic enzymes and alkaline detergents (38). Acids or alkalies solubilize inorganic salts. If the feed contains a mixture of different components, several cleaners may be needed. Depending on the process, cleaning agents may be used in combination or sequentially, separated by rinses.

Dissolved fouling material may pass into the membrane pores. Reprecipitation upon rinsing must be avoided. Membrane-swelling agents, such as hypochlorites, flush out material which may be lodged in the pores.

Cleaning is frequently aided mechanically. Foam balls scour the center of tubes, and hollow-filter systems can be backflushed. Hollow fibers and membranes attached to rigid supports can be back-pressured, thereby eliminating the pressure drop that holds redispersed films on the membrane surface.

Unless redispersed foulants are completely flushed away before using membrane-swelling agents (for sanitizing), they may become entrapped in the membrane structure. Water flux does not recover and the subsequent process fouls faster than usual. This phenomenon is discontinuous and differs from a steady reduction in water flux over many cleaning cycles, which indicates a gradual buildup of a fouling component not attacked by the cleaning composition.

Certain applications require that the equipment meet FDA and USDA sanitary requirements. These requirements ensure that the products are not contaminated by extractables or microorganisms from the equipment. Special considerations are given to the design of such equipment (41–44) (see Sterilization techniques).

Practical Aspects

Since the theoretical models cannot predict flux rates from physical data, plant-design parameters are obtained from laboratory testing, pilot-plant data, or in the case of established applications, performance of operating plants.

Flux response to concentration, cross flow or shear rate, pressure, and temperature should be determined for the allowable plant excursions. Fouling must be quantified and cleaning procedures proven. The final design flux should reflect long-range variables such as feed-composition changes, reduction of membrane performance, long-term compaction, new foulants, and viscosity shifts.

Flux is maximized when the upstream concentration is minimized. For any specific task, therefore, the most efficient (minimum membrane area) configuration is an open-loop system where retentate is returned to the feed tank (see Fig. 8). When the objective is concentration (eg, enzyme), a batch system is employed. If the object is to produce a constant stream of uniform-quality permeate, the system may be operated continuously (eg, electrocoating).

The upstream concentration C_B starts at C_o and ends at C_f, as described by the relationship:

$$C_B = C_o \frac{(V_o)}{V_B} \tag{8}$$

$$V_B = V_o - AJt \tag{9}$$

where V_o is the original volume, A is the membrane area, and t is time. Since J is a function of C_B (eq. 6), the solution can only be approximated.

Open-loop systems have inherently long residence times which may be detrimental if the retentate is susceptible to degradation by shear or microbiological contamination.

A feed–bleed or closed-loop configuration is a one-stage continuous membrane system. At steady state, the upstream concentration is constant at C_f (see Fig. 9). For concentration, a single-stage continuous system is the least efficient (maximum membrane area).

The single-pass system and the staged cascade (see Figs. 10 and 11) have high flux with low residence time. Both trade the concentration dependence of the batch system on time for concentration dependence on position in the system. Thus, a uniform flux is maintained (assuming no fouling) allowing continuous process integration. In practice, the single-pass system is difficult to implement, and therefore most commercial systems are multistaged cascade. The more stages used, the closer the average flux approaches the batch flux. Table 1 compares the flux for batch and staged systems operating on cheese whey.

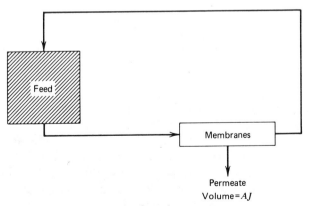

Figure 8.   Open-loop system.

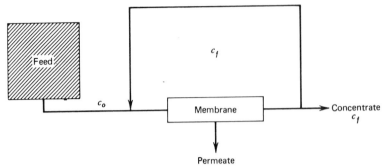

Figure 9.   Closed-loop system (feed bleed).

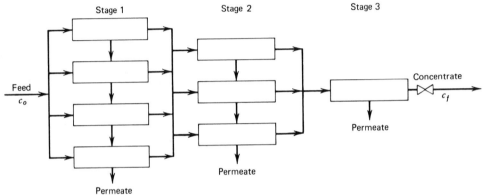

Figure 10.   Single-pass system.

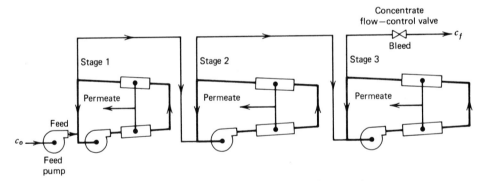

Figure 11.   Continuous multistage (cascade system).

Electroultrafiltration

Electroultrafiltration (EUF) combines forced-flow electrophoresis with ultra-filtration to control or eliminate the gel-polarization layer (45–47). Suspended colloidal particles have electrophoretic mobilities measured by zeta potential [$cm^2/(mV \cdot s)$] (see Flotation). Most naturally occurring suspensoids (eg, clay, PVC latex, and biological systems), emulsions, and protein solutes are negatively charged. Placing an electric field across an ultrafiltration membrane facilitates transport of retained species away from the membrane surface. Thus, the retention of partially rejected solutes can be dramatically improved (see also Electrodialysis).

Table 1. Flux Comparisons Between Batch and Staged Systems Operating on Cheese Whey

Configuration	Relative flux, %
batch open loop	100
single-stage feed bleed	67
two-stage cascade	82
three-stage cascade	87
four-stage cascade	89

Electroultrafiltration has been demonstrated on clay suspensions, electrophoretic paints, protein solutions, oil–water emulsions, and a variety of other materials. Flux improvement is proportional to the applied electric field E up to some field strength E_c, where particle movement away from the membrane is equal to the liquid flow toward the membrane. There is no gel-polarization layer and (in theory) flux equals the theoretical permeate flux. It follows, therefore, that E_c is proportional to ΔP.

At electric-field strengths greater than E_c, flux is proportional to ΔP up to the critical pressure P_c where E becomes E_c (see Fig. 12).

Anodic deposition is controlled by either fluid shear (cross-flow filtration) (48), similar to gel-polarization control, or by continual anode replacement (electrodeposited paints) (46). High fluid shear rates can cause deviations from theory when $E > E_c$ (49). The EUF efficiency drops rapidly with increased fluid conductivity.

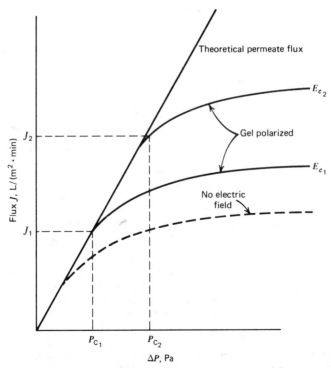

Figure 12. Electroultrafiltration, flux versus ΔP.

Diafiltration

Diafiltration is an ultrafiltration process where water is added to the concentrate and permeate is removed (50). The two steps may be sequential or simultaneous. Diafiltration improves the degree of separation between retained and permeable species.

Constant-volume batch diafiltration is the most efficient process mode. For species that freely permeate the membrane,

$$\ln\left(\frac{C_o}{C_t}\right) = \frac{V_p}{V_o} = N \equiv \text{turnover ratio} \tag{10}$$

where C_o is the permeate concentration at the start of diafiltration, C_t is the instantaneous permeate concentration at time t, V_o is the constant retentate volume, and V_p is the total permeate volume at t, which also equals the added water volume.

The fractional recovery of permeable solids in the retentate is

$$Y_r = \left(\frac{C_t}{C_o}\right) = \exp\left(-N\right) = 1 - Y_p \tag{11}$$

where Y_p is the fractional recovery in the permeate.

For partially retained solutes, equation 10 becomes

$$\ln\left(\frac{C_o}{C_t}\right) = 1 - \delta\left(\frac{V_p}{V_o}\right) \tag{12}$$

Area–time requirements for a specific diafiltration mission is defined as

$$A \cdot t = \frac{V_p}{J} = \frac{NV_o}{J} \tag{13}$$

When flux is independent of C_p:

$$A \cdot t = \frac{K}{C_B \ln\left(\dfrac{C_g}{C_B}\right)} \tag{14}$$

The optimum concentration for any diafiltration (minimum area time) is the minimum of the plot:

$$\frac{1}{J \cdot C_B} \text{ versus } C_B \tag{15}$$

When fouling is absent, the optimum concentration is 0.37 C_g. If the permeate solids are of primary value, it is usually preferable to diafilter at the minimum retentate volume to minimize permeate dilution.

Sequential batch diafiltration is a series of dilution–concentration steps. The concentration of membrane-permeable species is

$$\frac{C_o}{C_t} = \left(1 + \frac{V_p}{V_o}\right)^{n(1-\delta)} \tag{16}$$

where V_p is the permeate volume produced in each of n equal operations, and δ is the rejection of solids. As $n \rightarrow \infty$, equation 16 approaches equation 10.

Continuous diafiltration practiced in one or more stages of a cascade system has the same volume turnover relationship for overall recoveries as sequential batch diafiltration. The residence time however is dramatically reduced. If recovery of permeable solids is of primary importance, the permeate from the last stage may be used as diafiltration fluid for the previous stage. This countercurrent diafiltration arrangement results in higher permeate solids at the expense of increased membrane area.

Membrane Equipment

Commercial industrial ultrafiltration equipment first became available in the late 1960s. Since that time, the industry has focused on five different configurations.

Parallel-Leaf Cartridge. A parallel-leaf cartridge consists of several flat plates, each with membrane sealed to both sides (Fig. 13). The plates have raised (2–3-mm) rails along the sides in such a way that, when they are stacked, the feed can flow between them. They are clamped between two stainless-steel plates with a central tie rod. Permeate from each leaf drains into an annular channel surrounding the tie rod (33).

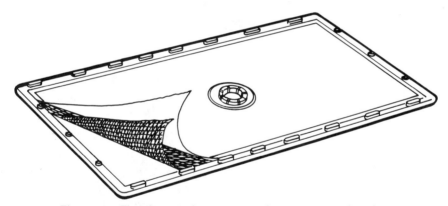

Figure 13. Flat-plate membrane element. Courtesy of Dorr-Oliver Inc.

Another type has several flat plates manifolded into a plastic header. The surface of the laminate is suitable for dip-casting membranes, whereas the interior is several orders of magnitude more porous. Permeate collects in the center of the laminate and drains into the header.

Cartridges are inserted in series into plastic or stainless-steel tubular pressure housings with square cross section (Fig. 14). Feed flows parallel to the leaf surface. A permeate fitting secures each cartridge to the housing wall, which allows permeate egress and facilitates sealing between concentrate, atmosphere, and permeate channels.

Plate and Frame. Plate-and-frame systems consist of plates (Fig. 15) each with a membrane on both sides. The plates have a frame around their perimeter which forms flow channels ca 1-mm wide between the plates when they are stacked. The stack is clamped between two end plates, sealing the frames together (Fig. 16).

Figure 14.   Flat plate cartridge and housing. Courtesy of Dorr-Oliver Inc.

At least one hole near the perimeter of each plate connects the flow channels from one side of the plate to the other. The membrane is sealed around the hole to isolate the permeate from the concentrate. Permeate collects in a drain grid behind the membrane and exits from a withdrawal port on the frame perimeter.

Spiral Wound.   A spiral-wound cartridge has two flat membrane sheets (skin side out) separated by a flexible, porous permeate drainage material. The membrane sandwich is adhesively sealed on three sides. The fourth side of one or more sandwiches is separately sealed to a porous or perforated permeate withdrawal tube. An open-mesh spacer is placed on top of the membrane, and both the mesh and the membrane are wrapped spirally around the tube (see Fig. 17).

Spiral-wound cartridges are inserted in series into cylindrical pressure vessels. Feed flows parallel to the membrane surfaces in the channel defined by the mesh spacer which acts as a turbulence promoter. Permeate flows into the center permeate-withdrawal tube which is sealed through the housing end caps.

Supported Tube.   There are three types of supported tubular membranes: cast in place (integral with the support tube), cast externally and inserted into the tube (disposable linings), and dynamically formed membranes.

The most common supported tubes are those with membranes cast in place (Fig. 18). These porous tubes are made of resin-impregnated fiber glass, sintered polyolefins,

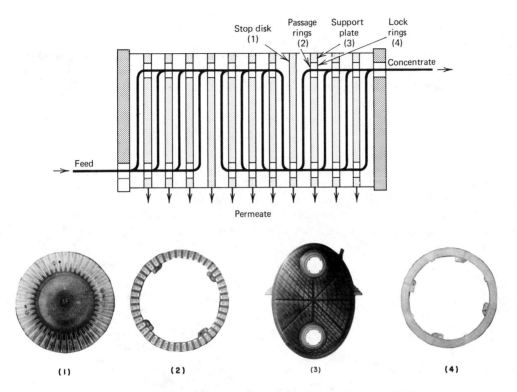

Figure 15. Plate-and-frame ultrafiltration module. Courtesy of DDS.

Figure 16. A 27-m^2 (953 ft^2) ultrafiltration system for concentrating enzymes. Courtesy of DDS.

and similar materials. Typical inside diameters are ca 25 mm. The tubes are most often shrouded to aid in permeate collection and reduce airborne contamination.

Externally cast membranes are first formed on the inside of paper, polyester, or polyolefin tubes. These are then inserted into reusable porous stainless-steel support tubes; inside diameters are ca 12 mm. The tubes are generally shrouded in bundles to aid in permeate collection.

Tubes for dynamic membranes are usually smaller (ca 6-mm ID). Typically, the tubes are porous carbon or stainless steel with inorganic membranes (silica, zirconium oxide, etc) formed in place.

Self-Supporting Tubes. Depending on the membrane material and operating pressure, self-supporting tubes are less than 2-mm ID; inside diameters as small as 0.04 mm are commercially available. Hollow fibers with the skin on the inside are extruded from a set of concentric nozzles. Membrane casting solution is forced through the outer annulus while diluent nonsolvent is pumped through the center (52).

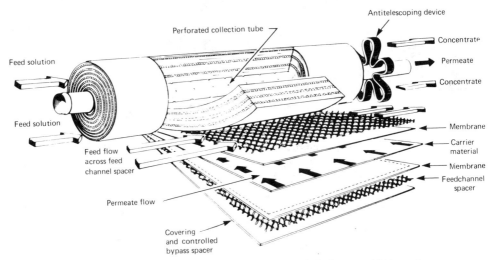

Figure 17.   Spiral-wound membrane configuration (51). Courtesy of Abcor, Inc.

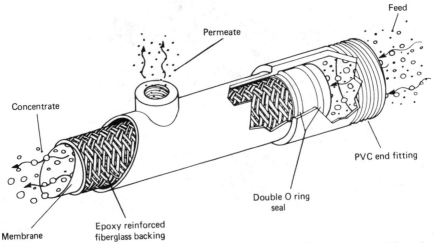

Figure 18.   Tubular membrane element with membrane cast in place. Courtesy of Abcor, Inc.

A large number of fibers are cut to length, and potted in epoxy resin at each end (see Embedding). The fiber bundle is shrouded in a cylinder which aids in permeate collection, reduces airborne contamination, and allows back pressing of the membrane.

Systemization

Each of the aforementioned devices is assembled by connecting the modules into combinations of series, parallel- flow paths, or both. These assemblies are connected to pumps, valves, tanks, heat exchangers, instrumentation, and controls to provide complete systems.

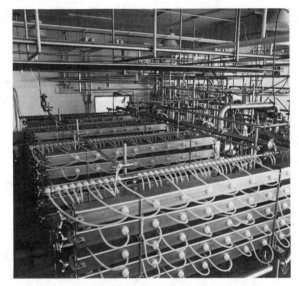

Figure 19.   Partial view of 535-m^2 (5760-ft^2) cartridge clean in place (CIP) ultrafiltration system, processing 545 metric tons (27.2 t/h) of sweet whey in a 20-h process day. Courtesy of Dorr-Oliver Inc.

Figure 20.   A CIP hollow-fiber ultrafiltration module with 334-m^2 (3600-ft^2) membrane area. The module is 6-m long and 0.6-m wide. It can run alone or as a stage in a larger multistage system. Courtesy of Rohmicon Inc.

Because of the broad differences between ultrafiltration equipment, the performance of one device cannot be used to predict the performance of another. Comparisons can only be made on an economic basis and only when the performance of each is known. Figures 19 and 20 are examples of typical ultrafiltration plants.

Uses

Applications of ultrafiltration are given in Table 2.

Table 2.   Ultrafiltration Applications

Application	Process	Kirk-Othmer article	Refs.
electrophoretic paint	control of properties, recovery of solids from rinse systems	Electrodecantation; Paint	52–56
dairy wheys	protein recovery, concentration, purification, diafiltration	Milk and milk products; Synthetic dairy products	33–38, 57–59
milk	cheese and yogurt mfg, 15–20% yield improvement, standardization	Milk and milk products	38, 60–68
oil–water emulsions	concentration	Emulsions	32, 40, 69–70
effluents of wool and yarn scouring	lanolin recovery, pollution abatement	Textiles; Wool	71–72
enzymes	concentration, purification	Enzymes	9–12, 73–78
biological reactors	antibiotic mfg, alcohol fermentation, sewage treatment	Antibiotics; Water, sewage	74, 79–82
vegetable proteins		Fermentation; Foods, nonconventional; Proteins	75, 83–85
latex concentration		Elastomers, synthetic; Latex technology	67
production of pure[a] water		Water, water treatment	51, 86
pulp and paper	lignosulfonate separation from spent liquor	Paper; Pulp; Wood	
blood	fractionation, purification	Blood fractionation	

[a] Virus-free.

Nomenclature

A	= membrane area
C	= concentration
C_B	= bulk concentration of all retained species
C_{Bi}	= concentration of species i in retentate
C_{BR}	= concentration of the bulk of the retentate
C_f	= final concentration
C_g	= gel concentration
C_o	= initial concentration
C_{pi}	= concentration of species i in permeate
C_t	= concentration at time t
C_w	= concentration at membrane wall
E	= electric field
E_c	= critical field strength

i	=	species
J	=	permeate flux on membrane-filtration rate
J_R	=	flux of retentate towards membrane surface
K	=	mass-transfer coefficient
K_g	=	gel hydraulic permeability coefficient
K_m	=	membrane hydraulic-permeability coefficient
P_c	=	critical pressure
R_m, R_g	=	hydraulic resistances of membrane gel
t	=	time
V	=	volume
V_o	=	constant retentate volume
V_p	=	total permeate volume
X	=	distance from membrane
Y_p	=	fractional recovery of permeable solids in the permeate
Y_r	=	fractional recovery of permeable solids in the retentate
μ	=	fluid viscosity
ΔP	=	membrane pressure drop
δ	=	rejection of solute

BIBLIOGRAPHY

1. M. C. Porter, *AIChE Symp. Ser.* **73**, 83 (1977).
2. R. E. Kesting, *Synthetic Polymeric Membranes*, McGraw-Hill, New York, 1971, pp. 8–9.
3. C. M. Hansen, *I&EC Product Research and Development* **8**(1), (March 1969).
4. J. D. Crowley, G. S. Teague, and J. W. Lowe, *J. Paint Technol.* **38**, 270 (May 1966).
5. H. Strathman, K. Kock, P. Amar, and R. W. Baker, *Desalination* **16**, 179 (1975).
6. H. P. Gregor in E. Selegny, ed., *Charged Gels and Membranes*, D. Reidel Publishing Co., Holland, 1976.
7. R. E. Kesting in ref. 2, p. 140.
8. Brit. Pat. 147,4594 (May 25, 1977), P. Klinkowski and E. Ondera (to Dorr-Oliver Inc.).
9. H. P. Gregor and P. W. Rauf, *Enzyme Coupled Ultrafiltration Membranes*, work supported by National Science Foundation Grant GI-32497.
10. R. A. Korus and A. C. Olson, *J. Food Sci.* **42**, 258 (1977).
11. U.S. Pat. 4,033,822 (July 5, 1977), H. P. Gregor.
12. O. Velicangil and J. A. Howell, *Biotechnology and Bioengineering*, Vol. 23, John Wiley & Sons, Inc., New York, 1981, pp. 843–854.
13. S. S. Wang, B. Davidson, C. Gillespie, L. R. Harris, and D. S. Lent, *J. Food Sci.* **45**, 700 (1980).
14. A. Michaels, H. Bixler, R. Hausslein, and S. Flemming, *OSW Research Development*, Report No. 149, U.S. Government Printing Office, Washington, D.C., Dec. 1965; *Ind. Eng. Chem.* **57**, 32 (1945) (brief but more readable).
15. A. Michaels, *Ind. Eng. Chem.* **57**(10), 32 (1965).
16. U.S. Pat. 3,977,967 (Aug. 31, 1976), O. C. Trulson and L. M. Litz (to Union Carbide Corp.).
17. U.S. Pat. 4,060,488 (Nov. 29, 1977), F. W. Hoover and R. K. Iler (to E. I. du Pont de Nemours & Co., Inc.).
18. H. Z. Friedlander and L. M. Litz in M. Bier, ed., *Membrane Processes in Industry and Biomedicine*, Plenum Press, New York, 1971.
19. R. Fleischer, P. B. Price, and E. Symes, *Science* **143**, 249 (1964).
20. C. R. Bennet, R. S. King, and P. J. Petersen, *Pressure Effects on Macromolecular–Water Interactions with Synthetic Membranes*, ACS 181st National Meeting, Atlanta, Ga., March 1981.
21. G. B. Westermann-Clark, Ph.D. thesis, Carnegie-Mellon University, Pittsburgh, Pa., 1981.
22. S. Sourirajan, *Ind. Eng. Chem. Fundam.* **2**(1), 51 (1963).
23. S. Jacobs, *Filtr. Sep.* 525 (Sept.–Oct. 1972).
24. M. P. Freeman, *Colloid Interface Sci.* **5**, 133 (1976).
25. L. Zeman and M. Wales in A. F. Turbak, ed., *Synthetic Membranes*, Vol. 2, ACS Symposium Series 154, American Chemical Society, 1980.
26. A. S. Michaels, *Sep. Sci. Technol.* **15**, 1305 (1980).
27. P. Dejmek, "Permeability of the Concentration Polarization Layer in Ultrafiltration of Macro Mole-

cules," *Proceedings of the International Symposium, Separation Processes by Membranes*, Paris, March 13–14, 1975.

28. Q. T. Nguyen, P. Aptel, and J. Neel, *J. Membr. Sci.* **6**(1), 71 (1980).
29. P. S. Leung in A. R. Cooper, ed., *Ultrafiltration Membranes and Applications*, Plenum Press, New York, 1980, pp. 415–421.
30. W. F. Blatt, A. Dravid, A. S. Michaels, and L. Nelson in J. E. Finn, ed., *Membrane Science and Technology*, Plenum Press, New York, 1970, pp. 47–97.
31. R. F. Madsen and W. K. Nielsen in ref. 29, pp. 423–438.
32. Brit. Pat. 1,456,304 (Nov. 24, 1976) (to Abcor, Inc.).
33. A. C. Epstein and S. R. Korchin, *paper presented at 91st National Meeting*, AIChE, Detroit, Mich., Aug. 1981.
34. M. W. Hickey, R. D. Hill, and B. R. Smith, *N.Z. J. Dairy Sci. Technol.* **15**(2), 109 (1980).
35. D. N. Lee and R. L. Merson, *J. Dairy Sci.* **58**, 1423 (1975).
36. D. N. Lee and R. L. Merson, *J. Food Sci.* **41**, 403 (1976); **41**, 778 (1976).
37. L. L. Muller and W. J. Harper, *J. Agric. Food Chem.* **27**, 662 (1979).
38. W. J. Harper in ref. 29, pp. 321–347.
39. J. A. Howell and O. Velicangil in ref. 29, pp. 217–229.
40. P. A. Bailey, *Filtr. Sep.* **14**(1), 53 (1977).
41. F. E. McDonough and R. E. Hargrove, *J. Milk Food Technol.* **35**(2), 102 (1972).
42. R. G. Semerad, "Sanitary Considerations Involved with Membrane Equipment," *Proceedings of the Whey Product Conference*, Atlantic City, N.J., 1976.
43. N. C. Beaton, *J. Food Protection*, **42**, 584 (July 1979).
44. B. S. Horton, *N.Z. J. Dairy Sci. Technol.* **14**(2), 93 (1979).
45. M. Bier in ref. 18.
46. U.S. Pat. 3,945,900 (March 23, 1976), P. R. Klinkowski (to Dorr-Oliver, Inc.).
47. J. M. Radovich and R. E. Sparks in ref. 29, pp. 249–268.
48. J. D. Henry, Jr., L. Lawler, and C. H. A. Kuo, *AIChE J.* **23**, 851 (1977).
49. J. D. Henry, Jr., and C. H. A. Kuo, *paper presented at the Symposium on Recent Developments in Colloidal Phenomena*, AIChE National Meeting, New Orleans, La., Nov. 1981.
50. P. R. Klinkowski and N. C. Beaton, *J. Sep. Proc. Tech.*, in press; preprint available from Dorr-Oliver, Inc.
51. *Abcor Sanitary Spiral-Wound Ultrafiltration Modules*, Product Bulletin, Abcor, Inc., Wilmington, Mass., 1981.
52. B. R. Breslau, A. J. Testa, B. A. Milnes, G. Medjanis in ref. 29, pp. 109–127.
53. R. A. Scaddan, *Ind. Finish. Surf. Coat.* **27**, 326 (1975).
54. G. E. F. Brewer, *paper presented at Electrocoat 72*, sponsored by the Electrocoating Committee and the National Paint and Coatings Association, Chicago, Ill., Oct. 2–4, 1972.
55. B. J. Weissman in ref. 54.
56. U.S. Pat. 3,663,397 (May 16, 1972), L. R. LeBras and J. Ostrowski (to PPG Industries); U.S. Pats. 3,663,398 and 3,663,402–403 (May 16, 1972), R. Christenson and R. R. Twack (to PPG Industries); U.S. Pats. 3,663,399 and 3,663,404 (May 16, 1972), F. M. Loop (to PPG Industries); U.S. Pats. 3,663,400–401 and 3,663,405 (May 16, 1972), R. Christenson and L. R. LeBras (to PPG Industries); U.S. Pat. 3,633,406 (May 16, 1972), L. R. LeBras and R. R. Twack (to PPG Industries).
57. F. E. McDonough, W. A. Mattingly, and J. H. Vestal, *J. Dairy Sci.* **54**, 1406 (1971).
58. B. S. Horton, R. L. Goldsmith, and R. R. Zall, *Food Technol.* **26**(2) (1972).
59. M. E. Matthews, *N.Z. J. Dairy Sci. Tech.* **14**(2), 86 (1979).
60. F. Lang and A. Lang, *Milk Ind.* **78**(9), 16 (1976).
61. O. J. Olsen in ref. 6, Vol. 2.
62. J. L. Maubois, "Application of Ultrafiltration to Milk Treatment for Cheese Making," in ref. 27.
63. H. R. Covacevich and F. V. Kosikowski, *J. Food Sci.* **42**, 362 (1977).
64. J. E. Ford, F. A. Glover, and K. J. Scott, *Int. Dairy Congress,* 1068 (1978).
65. F. A. Glover, P. J. Skudder, P. H. Stuthart, and E. W. Evans, *J. Dairy Res.* **45**, 291 (1978).
66. S. Jepsen, *J. Cultured Dairy Products* **14**(1), 5 (1979).
67. J. L. Maubois in ref. 29, pp. 305–318.
68. P. R. Poulsen, *J. Dairy Sci.* **61**(6), 807 (1978).
69. R. L. Goldsmith, D. A. Roberts, and D. L. Burre, *J. Water Pollut. Control Fed.* **46**, 2183 (1974).
70. I. K. Bansal, *Ind. Water Eng.* **13**, 6 (Oct.–Nov. 1976).
71. N. C. Beaton, *Textile Institute and Industry* **13**, 361 (Nov. 1975).

72. J. A. C. Pearson, C. A. Anderson, and G. F. Wood, *J. Water Pollut. Control Fed.* **48,** 945 (1976).
73. D. I. C. Wang, A. J. Sinskey, and T. A. Butterworth in ref. 30.
74. N. C. Beaton in ref. 29, pp. 373–404.
75. H. S. Olsen and J. Adler-Nissen in ref. 25, Vol. 2.
76. U.S. Pat. 3,720,583 (March 13, 1973), E. E. Fisher (to A. E. Staley).
77. J. Hong, G. T. Tsao, P. C. Wankat in ref. 12, pp. 1501–1516.
78. W. D. Deeslie and M. Cheryan, *J. Food Sci.* **46,** 1035 (1981).
79. J. Gawel and F. V. Kosikowski, *J. Food Sci.* **43,** 1717 (1978).
80. U.S. Pat. 3,472,765 (Oct. 14, 1969), W. E. Budd and R. W. Okey (to Dorr-Oliver, Inc.).
81. A. G. Fane, C. J. D. Fell, and M. T. Nor in ref. 29, pp. 631–658.
82. A. S. Michaels, C. R. Robertson, and S. N. Cohen, *paper presented at the 180th National Meeting ACS*, San Francisco, Calif., Aug. 1980.
83. G. Jackson, M. M. Stawiarski, E. T. Wilhelm, R. L. Goldsmith, and W. Eykamp, *AIChE Symp. Ser.* **70,** 514 (1974).
84. M. Cheryan in ref. 29, pp. 343–325.
85. J. T. Lawhon, L. J. Manak, and E. W. Lusas in ref. 29.
86. G. Belfort, T. F. Baltutis, and W. F. Blatt in ref. 29, pp. 439–474.

General References

Reference 30 is also a general reference.
N. C. Beaton, "Advances in Enzyme and Membrane Technology," *Institute of Chemical Engineers Symposium Series No. 51*, Institute of Chemical Engineers, London, 1977, pp. 59–70.
N. C. Beaton and H. Steadly in N. Li, ed., *Recent Developments in Separation Science*, Vol. 7, CRC Press, Boca Raton, Fla., 1982, pp. 2–25.
A. R. Cooper, ed., *Ultrafiltration Membranes and Applications*, Proceedings of 178th National ACS Meeting, Washington, D.C., 1979, Plenum Press, New York, 1980.
R. P. deFillipi and R. L. Goldsmith in J. E. Flinn, ed., *Membrane Science and Technology*, Plenum Press, New York, 1970, pp. 33–46.
R.*J. Gross and J. F. Osterle, *J. Chem. Phys.* **49,** 228 (1968).
S. Hwang and K. Kammermeyer, *Membranes in Separations*, John Wiley & Sons, Inc., New York, 1975; good study of Membrane Transport Phenomenon.
R. E. Kesting, *Synthetic Polymeric Membranes*, McGraw-Hill, New York, 1971; good bibliographies.
A. S. Michaels, L. Nelson, and M. C. Porter in M. Bier, ed., *Membrane Processes in Industry and Biomedicine*, Plenum Press, New York, 1971, pp. 197–232.
U.S. Pat. 3,615,024 (Oct. 26, 1971), A. S. Michaels (to Amicon Corporation).
A. S. Michaels, *paper presented at Clemson University Membrane Technology Conference*, March 1979; OWRT Contract No. 14-34-0001-8548.
A. S. Michaels, *Chem. Technol.* **2**(1), 37 (Jan. 1981).
A. S. Michaels, *Desalination* **35,** 329 (1980).
L. Mir, W. Eykamp, and R. L. Goldsmith, *Ind. Water Eng.* **14,** (May–June 1977).
M. C. Porter and L. Nelson in N. N. Li, ed., *Recent Developments in Separation Science*, Vol. 2, CRC Press, Cleveland, Ohio, 1972, pp. 227–267.
M. C. Porter, *Ind. Eng. Chem. Prod. Res. Dev.* **2,** 234 (1972).
M. C. Porter in P. A. Schweitzer, ed., *Handbook of Separation Techniques for Chemical Engineers*, McGraw-Hill, New York, 1979.
R. N. Rickles, *Membranes, Technology and Economics*, Noyes Development Co., Park Ridge, N.J., 1967.
H. Strathman, K. Kock, P. Amar, and R. W. Baker, *Desalination* **16,** 179 (1975).
A. F. Turbak, ed., *Synthetic Membranes*, Vols. 1 and 2, ACS Symposium Series 153 and 154, ACS, Washington, D.C., 1981.
Equipment Available for Membrane Processes, Bulletin No. 115, International Dairy Federation, Brussels, p. 1979 (available through Library of Congress or Harold Wainess and Associates, Northfield, Ill.).

P. R. KLINKOWSKI
Dorr-Oliver, Inc.

ULTRAMARINE.   See Pigments, inorganic.

ULTRASONICS

High power,   462
Low power,   479

HIGH POWER

The object of high power ultrasonic (macrosonic) application is to bring about some permanent physical change in the material treated. This process requires a flow of vibratory power per unit of area or volume. Depending on the application, the power density may range from less than a watt to thousands of watts per square centimeter. Although the original ultrasonic power devices operated at radio frequencies, today most operate at 20–60 kHz.

The piezoelectric sandwich-type transducer driven by an electronic power supply has emerged as the most common source of ultrasonic power; the overall efficiency of such equipment (net acoustic power per electric-line power) is typically >70%. The maximum power from a conventional transducer is inversely proportional to the square of the frequency and is ca 4 kW at 20 kHz. Some applications, such as cleaning, may have many transducers working into a common load.

Ultrasonic-power transducers produce sinusoidal motion and can be coupled to the load directly or through intermediate resonant members. If d = transducer displacement, v = velocity, a = acceleration, and ω = frequency (in rad/s), then $v = \omega \cdot d$ and $a = \omega^2 d$, and the output face of a 20-kHz transducer operating at a peak displacement of 50 μm has a peak velocity of 6.28 m/s and a peak acceleration of 8×10^4 g. Thus, high power ultrasound is characterized by high frequencies, small displacements, moderate point velocities, and very high accelerations.

In most power applications, ultrasonic motion is perpendicular to the plane of the transducer–load interface, and compressional (longitudinal) waves are imparted to the load. The propagation and absorption of these waves depend on the elastic and dissipative characteristics of the medium (1). Whereas only compressional waves are possible in liquids and gases, in solids, because of their ability to transmit shear stresses, other vibrational modes may take place, eg, shear, torsion, or flexure. The ability of ultrasound to propagate in elastic media allows far-field processing where work is performed far from the power source, as in ultrasonic cleaning, remote plastic welding, or acoustic drying.

Most effects of high power ultrasonic work depend on the following vibration-induced phenomena occurring in matter: cavitation and microstreaming in liquids; heating and fatiguing in solids; and surface-instability phenomena occurring at liquid–liquid and liquid–gas interfaces.

The principal applications are in industrial cleaning and plastic welding. Applications such as metal welding, soldering, abrasive machining, inhalation therapy, and cell disruption are of less importance. Emulsification, dispersion of solids, crystal growth, cavitation erosion and fatigue testing, degassing, drying, defoaming, and metal cutting and forming are employed in some industrial establishments. Metal grain refinement, degassing of metals, particle agglomeration, filtering, depolymerization, and many others are still in experimental stage. Although some applications depend entirely on ultrasonic action, others use ultrasound as an aid to conventional processes to reduce time, temperature, and usage of chemicals, or to improve product yield or quality.

Applications in Liquids

In a strong ultrasonic field, most of the energy supplied by the source to the liquid is dissipated in cavitation. Practically all high power uses of ultrasound in liquids depend on cavitation and its secondary effects.

The mechanism of acoustic cavitation is complex and has been probably the most researched topic in power ultrasonics (2–3). Briefly, it pertains to formation in a liquid of gaseous and vapor bubbles which expand and contract in response to high frequency alternating pressure of the sound field. Cavitational bubbles can be either stable and oscillate about their average size over many cycles, or transient when they grow to a certain size and then violently collapse, thereby causing very high momentary local pressures up to several GPa (several tens of thousands of atm). These minute implosions, culminating in local shock waves, are capable of eroding hard metals, puncturing walls of biological cells, fracturing crystals, degrading macromolecules, and performing other unusual feats. Oscillation of stable cavitational bubbles gives rise to acoustic microstreaming (4–5) or formation of miniature eddies that enhance the mass and heat transfer in the liquid. Microstreaming also causes velocity gradients that result in shear forces. Typically, both transient and stable bubbles are present in a cavitating liquid.

The work performed by cavitation depends on the energy level of imploding bubbles as well as on the number of bubbles per unit volume. To increase the cavitation density, it is sufficient to raise the level of ultrasonic excitation. Although raising of hydrostatic pressure suppresses the number of cavitational events, the shock intensity of individual bubbles is substantially increased. Cavitational shock intensity is also a function of bubble size, which increases with the period of excitation frequency. Cavitation is more intense at lower frequencies and ceases altogether in the high MHz range.

Marked differences exist in cavitational intensity of different liquids at equal acoustic pressure (6). Cavitation is also affected by temperature and, typically, there is an optimum temperature for a particular liquid. Cavitation is influenced by vapor pressure, the amount of dissolved gases, and the presence of gaseous and other nuclei that determine its onset.

Cleaning. Industrial ultrasonic cleaning is one of the two largest applications of power ultrasound and is utilized in a large variety of products and industries (7–9). Typical applications include cleaning of car engine parts, ball bearings, castings, medical instruments and glassware, filters, optical lenses, and semiconductors, printed circuits, and other electronic components. Ultrasonic cleaning works best on hard materials such as metal, glass, and plastic which reflect rather than absorb vibration. It is particularly useful when applied to parts of complex geometry with inaccessible or hard-to-reach areas. In addition to saving labor, time, and solvents, ultrasonic cleaning often gives better results than other methods.

The mechanism of ultrasonic cleaning varies somewhat with different detergents and contaminants, which basically are either soluble or insoluble. Depending on cohesion, insoluble contaminants may be dispersed or eroded, or a contaminant coat may be stripped by the combined action of transient cavitation, which cracks the coat, and the pulsating bubbles forming between the coat and the surface, which gradually peel the coat away. Stable cavitation and microstreaming dissolve contaminants by bringing fresh solvent to the surfaces cleaned and contribute to dispersion of grease in aqueous solutions.

Where possible, the ultrasonic cleaning agents should combine the ability to cavitate with chemical action. Aqueous solutions are most common and include both alkali and acid. Freon and other organic solvents are used for removal of greasy contaminants. Ultrasonic cleaning is often combined with other cleaning operations such as degreasing and vapor rinsing.

An ultrasonic cleaning tank (Fig. 1) is energized by transducers attached to its bottom; a resonant pattern forms in the liquid. Parts can be dipped in the tank directly or in perforated baskets. Industrial tanks usually are 5–150-L in size. In recent years, small, inexpensive cleaners have become available that incorporate the tank and the electronic power supply in one unit for use in laboratories and small shops. Most ultrasonic cleaning employs 20–60 kHz with a gradual retreat from the low frequency end because of noise and application considerations.

The arrangement of Figure 1 is effective for coupling vibration to a large volume of liquid and works well for normal cleaning where ultrasonic power density rarely exceeds 5 W/cm^2 of transducer area. An attempt to overdrive the tank results in decoupling, characterized by severe cavitation at the transducer–liquid interface and a diminished activity in the volume of liquid.

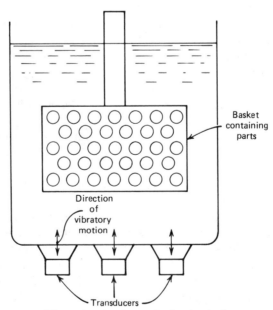

Figure 1. An ultrasonic cleaning tank.

A cylindrical focusing transducer that allows high cavitational intensity for a low power density at the transducer driving surface is shown in Figure 2. It is effective if the processing can be limited to a relatively small region along the transducer axis, such as in continuous wire cleaning.

Erosion. Cavitational implosions at or near the surface of a material can cause erosion of the hardest substance. However, the resistance of materials to cavitational erosion varies greatly and is related to fatigue strength and hardness.

Soldering. Ultrasonic soldering allows tinning without flux and has been of particular interest for aluminum since its oxides cannot be removed with ordinary acid or alkali fluxes. The special fluxes used on aluminum are highly reactive with aluminum itself and are often unacceptable where flux after soldering cannot be completely cleaned. Other uses of ultrasonic soldering are mainly in nonferrous metals to eliminate flux for economic or corrosion considerations, or to achieve faster or more thorough wetting. Applications include tinning of cable ends, conductor termination, and soldering of aluminum window frames and heat exchangers (see also Solders and brazing alloys).

Equipment for high intensity tinning and soldering of small parts is shown in Figure 3. Parts are immersed between two ultrasonic horns, and cavitation in molten solder strips surface oxides, thereby exposing bare metal to solder. For large parts, ultrasonic tanks with externally mounted transducers are used, similar in principle to the cleaner design of Figure 1 (10).

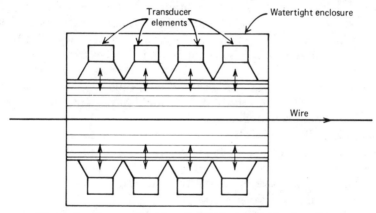

Figure 2.   Cross section of a cylindrical wire-cleaning transducer.

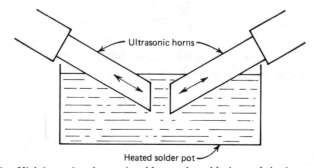

Figure 3.   High intensity ultrasonic solder pot for soldering and tinning small parts.

Deburring. Ultrasonic deburring is not very common and has been mainly of value when applied to delicate precision parts that are damageable by tumbling or other finishing methods (11). Although conventional ultrasonic cleaners can be used, the process works better in pressurized liquids and can be further enhanced by using an abrasive-particle suspension (equal parts of water, glycerol, and abrasive); the optimum size of the abrasive depends on frequency (see also Abrasives).

Erosion Testing. The resistance of materials to cavitational erosion can be tested conveniently and controllably with ultrasonics (12–13). The measureable ultrasonic parameters are the frequency and the output-displacement amplitude of the transducer. Horns are made from the material to be tested directly, or material samples are bonded to the horn. Alternatively, the samples may be stationary and placed near the vibrating surface of the horn. Ultrasonic power can also be monitored if desired. Most of the work to date has been at 20 kHz.

Extraction. ***Cell Disruption.*** Ultrasonic disintegration of biological cells to extract the cell content has been practiced for over two decades in the study of enzymes, lipids, and viruses, and for general preparation of antigenically active extracts (14). A high amplitude ultrasonic horn is used to induce intense cavitation which ruptures the cell walls and releases the cell content into the surrounding liquid.

Ultrasonic disintegration is relatively safe for the substances that are extracted, with typical extraction times ranging from a few seconds to tens of minutes. In the laboratory, solutions are usually treated with a horn in a test tube. The horn amplitude is adjustable to produce a power density up to 100 W/cm². Continuous-flow equipment with optional cooling is available for the production of larger volumes.

Extraction from Plants. Alkaloids, glucosites, aromatic extracts, and other components have been extracted from plant tissues with ultrasonics (15–19). Other uses include extraction of essential oils from hops and of juices from fruit (20) (see Food processing; Fruit juices; Oils, essential). Ultrasonic extraction improves the yield, reduces processing time, lowers temperatures and thus reduces damage to extracts, and prevents loss of volatile components in boiling. At present, however, this process is not significant commercially.

Emulsification. The main advantage of ultrasonic emulsification is the ability to produce emulsions without surfactants or with a reduced surfactant content (21–22). Numerous such uses were reported in the early 1960s in textile, soft-drink, pharmaceutical, cosmetics, and other industries for emulsification of mineral and essential oils, antibiotic dispersions, lotions, etc, using mechanical liquid whistles (23). Although ultrasonic equipment has been greatly improved since that time, it is not used for these applications on a wide scale. Possibly, the benefits of such treatment may have been overestimated at first (see also Emulsions).

Coal-Oil Mixtures. A revival is possible in an energy-related area for preparation of coal-oil mixtures (24–25). Coal powder mixed with oil in concentrations up to 40% can be burned in existing oil-burning facilities after some modifications, but continuous agitation of the mixture is required to prevent settling of coal particles. The other alternatives for maintaining coal in suspension are expensive ultrafine coal grinding or use of surfactants which add to cost and pollution. With an addition of a small amount of water and upon ultrasonic emulsification, a stable coal-oil suspension can be produced that is suitable for storage and transportation. Sample quantities of this fuel are produced in a pilot operation for further evaluation. High power continuous-flow 10-kHz equipment (Fig. 4) is used at flow rates of 80–160 L/min. The success of this application will depend on the future role of coal oil in industry.

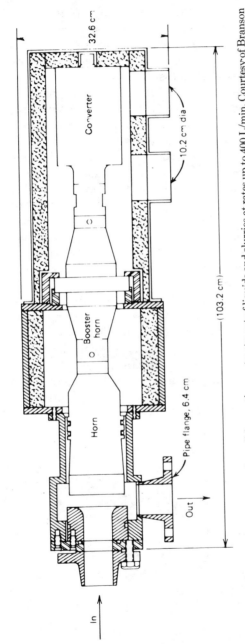

Figure 4. High intensity processing cell for continuous treatment of liquids and slurries at rates up to 400 L/min. Courtesy of Branson Sonic Power Company.

Dispersion of Solids.   Clusters of solid particles can be effectively dispersed in liquids with ultrasound, and uses have been reported in the preparation of dyes, insecticides, antibiotic suspensions, and ointments, and for the dispersion of china clay, oxides, mica, and other materials in rubber and paper production (23,26). Fine, uniform dispersions that are free of flocculation result. Like emulsification uses, however, most of these have not lived up to the early expectations, possibly because of the detrimental effect of abrasive materials on the life of ultrasonic whistles. These problems can be overcome with modern ultrasonic equipment, and dispersion applications appear to hold promise for certain processes in industry.

Magnetic Oxide Dispersion.   Ultrasonic dispersion of magnetic oxides used in the manufacture of audio, video, and computer tapes and disks results in coatings of better quality and may add other advantages to the production process, according to initial investigations (see Magnetic tape; Recording disks).

Treatment of Mineral Slurries.   Ultrasonic treatment of mineral slurries reduces soaking time and improves yields in flotation (qv) and separation processes (27–28). Good results were obtained in delamination of talc, mica, and asbestos, dissolving mineral components from a liquid slurry, precipitation of metal dust, separation of ash and sulfur from coal, and similar applications. In general, ultrasound can be effective in dislodging mechanically interlocked particles, material delamination, cleavage, or defibrillation; the effectiveness usually increases as the processed particles get smaller. Wetting of ultrafine particles and scrubbing of particles to remove a coating are other possibilities.

Sterilization in Liquids.   Significant improvements in the sterilization rates in liquids and in ozone sewage treatment have been observed in the presence of ultrasound (29–31). Such improvement is not owing to direct destruction of bacteria by cavitation, but to dispersion of clumps of bacteria, allowing the conventional reaction to penetrate faster (see also Sterilization techniques).

Flow Enhancement.   Diffusion of liquids through porous media can be enhanced by ultrasound and has been studied in relation to filtering and impregnation.

Filtration.   A common problem with filters is gradual clogging resulting in lowering

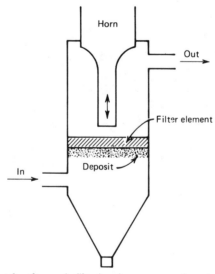

Figure 5.   Arrangement for ultrasonic filtering; in some experiments, the horn is connected to the filter.

of filtration rates. Experiments show that ultrasonic agitation increases filtration rates, reduces the need for filter cleaning, and produces a drier deposit (32–35). Figure 5 shows an arrangement for ultrasonic filtering where the horn may be connected directly to the filter. Sintered brass, stainless steel, sandstone, and other filters were investigated in pore sizes ranging from a few to over 100 μm. Impressive increases in filtering rates were obtained with materials such as contaminated motor oil and coal slurries. No known installations of ultrasonic filters exist, at least in the United States, and it could be questioned whether the filtering rates used in some experiments are representative of those used in the industry. Nevertheless, the subject is interesting because relatively low amounts of ultrasonic power can be effective (see also Filtration).

Atomization. Ultrasonic atomization produces droplets of predictable, relatively uniform size (36–37). It can be applied to viscous liquids that cannot be handled by conventional spraying nozzles. For a given liquid, the droplet size is determined by the frequency; droplet size decreases with increasing frequency.

Inhalation Therapy, Humidification. Ultrasonic technique was first used for medical nebulizers for inhalation therapy. Frequencies of 1–3 MHz were used to produce small (1–5 μm dia) uniform droplets in the absence of gas, which is important for anesthetic systems. Currently, several companies in Japan are marketing ultrasonic humidifiers for home use.

Fuel Atomization. Much work has been done on ultrasonic atomization of home-heating fuel, and uses in automotive carburetion have been considered, but so far neither has proven to be economically attractive.

Drying of Textiles. Recent work indicates that ultrasonic atomization may be of value in the drying of fabric in textile manufacturing. Continuously moving fabric is tensioned over the face of a high amplitude horn (Fig. 6), and the moisture is atomized by vibration. The amount of moisture that must be evaporated by conventional drying is thus reduced, resulting in substantial energy savings. The process works best on thin, nonabsorbing materials. Pilot installations exist in textile plants (see also Drying).

Crystal and Powder Production. Fine crystals of uniform size can be produced

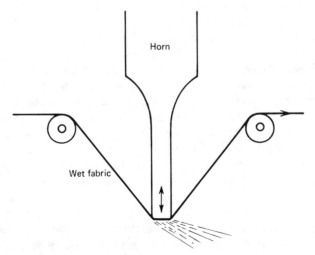

Figure 6. Continuous drying of fabric by atomization. The same arrangement can be used for continuous high intensity cleaning; the horn and the fabric would be submerged in a cleaning solution.

by atomizing a supersaturated solution and freezing the droplets. Experiments have also included atomization of molten metal, glass, and other materials (38–39).

Crystallization. The influence of ultrasound on crystallization has been of interest with regard to the effect on crystal growth in supersaturated solutions and the effect on grain structure in the solidification of metals (40–41).

Crystal Growth. In a strong ultrasonic field, nucleation in organic and inorganic supersaturated solutions can be accelerated and the growth of large crystals suppressed to give finer-grained, more uniform crystals. This result is attributed to cavitation which breaks up and scatters clumps, dendrites, and individual crystals to form a large number of new nuclei. Pharmaceuticals and other materials were investigated, including preparations of progesterone and hydrocortisone (42–45). Several large ultrasonic installations exist in the pharmaceutical industry.

Metal-Grain Refinement. Much research has been done on ultrasonic grain refinement of metals (46). Generally, exposure to ultrasound during solidification results in a finer grain and improved physical characteristics, eg, specific elongation, impact strength, and deformability. However, because of high temperatures involved and the difficulty in treating large volumes, practical implementations have not followed (see also Metal treatments).

Chemical Effects. *Depolymerization.* Macromolecules can be degraded by ultrasound (47–48); this effect is more pronounced for high molecular weight polymers. Solutions of cellulose, gelatin, rubber, proteins, and many others were degraded. High intensity 20-kHz equipment was mostly used, and the effect was usually explained by cavitation.

Degassing. *Beverages.* In an ultrasonic field, gases dissolved in a liquid form cavitational bubbles which grow and coalesce into larger bubbles and rise to the surface (49). Relatively low ultrasonic power is needed, and the process works best in low viscosity liquids. Industrial applications include fobbing of beer and carbonated beverages (qv) where bottles come in contact with an ultrasonic transducer just before capping (50). Photographic solutions have also been degassed by ultrasonic treatment.

Metals. Ultrasonic degassing of molten metals, glass, and other materials to reduce porosity upon solidification has been described in the literature (49).

Applications in Solids

Welding. *Plastic Welding.* Ultrasonic welding of thermoplastics is one of the two largest uses of power ultrasound technology. It includes applications such as welding automotive lenses, dashboards, heater ducts, radio cases, tape cassettes, film cartridges, synthetic textiles, plastic-coated containers, toys, and many others (51–52). The process is fast (most welds are made in less than one second), produces excellent weld strength, does not require skilled operators, and can be easily automated (Table 1).

The principle of ultrasonic plastic welding is shown in Figure 7. The two parts to be welded are clamped together under pressure between an ultrasonic horn and a fixture. Vibration of the horn is imparted to the upper part in contact with the horn. To maximize ultrasonic stress and therefore material heating in the weld zone, the area of contact between the two parts is reduced. This results in selective melting of a small volume of plastic, designated as energy director, calculated to fill the interface. Because of low heat input, ultrasonic welding causes minimal part distortion and material degradation (see also Plastics processing).

Table 1. Weldability of Thermoplastics by Ultrasound[a]

Material	Ultrasonic welding					
	Ease of welding[b]		Swaging and staking	Inserting	Spot welding	Vibration welding
	Near field[c]	Far field[d]				
Amorphous resins						
ABS	E	G	E	E	E	E
ABS–polycarbonate alloy (Cycoloy 800)	E–G	G	G	E–G	G	E
acrylic[e]	G	G–F	F	G	G	E
acrylic multipolymer (XT-polymer)	G	F	G	G	G	E
cellulosics: CA, CAB, CAP	F–P	P	G	E	F–P	E
phenylene oxide-based resins (Noryl)	G	G	G–E	E	G	E
poly(amide–imide)	G	F				G
polycarbonate[f]	G	G	G–F	G	G	E
polystyrene, GP[g]	E	E	F	G–E	F	E
rubber modified	G	G–F	E	E	E	E
polysulfone[f]	G	F	G–F	G	F	E
PVC, rigid	F–P	P	G	E	G–F	G
SAN–NAS–ASA[h]	E	E	F	G	G–F	E
Crystalline resins[i]						
acetal	G	F	G–F	G	F	E
fluoropolymers	P					G–F
nylon[f]	G	F	G–F	G	F	E
polyester, thermoplastic	G	F	F	G	F	E
polyethylene	F–P	P	G–F	G	G	G–F
polymethylpentane (TPX)	F	F–P	G–F	E	G	E
poly(phenylene sulfide)	G	F	P	G	F	G
polypropylene	F	P	E	G	E	E

[a] Courtesy of Branson Sonic Power Company; E = excellent; G = good; F = fair; and P = poor.
[b] Ease of welding is a function of joint design, energy requirements, amplitude, and construction.
[c] Near-field welding refers to joints ≤6.35 mm from area of horn contact.
[d] Far-field welding refers to joints >6.35 mm from contact area.
[e] Cast grades are more difficult to weld because of high molecular weight.
[f] Moisture inhibits welds.
[g] GP = general purpose.
[h] SAN = styrene–acrylonitrile; NAS = styrene–methyl methacrylate copolymer; ASA = acrylonitrile-styrene–acrylic rubber.
[i] Crystalline resins in general require higher amplitudes and higher energy levels because of higher melt temperatures and heat of fusion.

Power density in ultrasonic plastic welding in high, ie, ca 500 W/cm^2 of weld area. High amplitude horns of aluminum and titanium are used, slotted when necessary to ensure uniformity of motion for larger horn areas; power range is 300–3000 W. Most welders operate at 20 kHz, but some work is done at 30–40 kHz because of noise and application considerations.

Plastic films and synthetic fabrics can be welded continuously in an arrangement shown in Figure 8. Large installations exist in the textile industry for quilting and similar operations based on this principle.

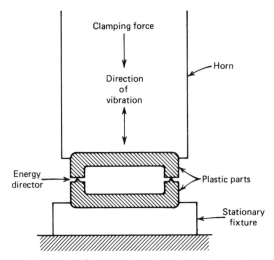

Figure 7.   Cross section of an ultrasonic joint.

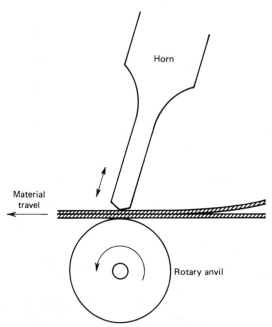

Figure 8.   Continuous welding of fabric and plastic film.

Metal Welding.   The principle of ultrasonic metal welding is illustrated in Figure 9. The horn vibrates parallel to the plane of a stationary anvil and causes the upper part to slip with respect to the lower part, while it applies clamping pressure to keep the parts together. Ultrasonic shear motion breaks up and disperses the oxides and other contaminants at the interface, and the exposed plasticized-metal surfaces form a bond under pressure. The process is relatively cold, and parts usually can be welded below their melting temperatures, which reduces embrittlement due to recrystallization (53–54).

Ultrasonic metal welding is essentially limited to lap welds and is used when other

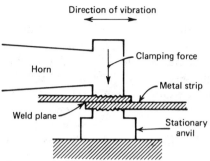

Figure 9.   Ultrasonic shear metal welding.

processes present problems, such as welding of high conductivity and dissimilar metals, or parts of dissimilar electric resistivity or grossly unbalanced heat capacity. Applications are mostly in welding of aluminum and copper conductors and other electrical hardware and of miniature leads to semiconductors (microbonding). Power densities at the weld area can be as high as 5 kW/cm^2; equipment ratings are <100 W to several kW, operating at 10–60 kHz (see Welding).

Forming.   *Plastic Forming.*   Many applications involving ultrasonic plastic welders are actually forming or reforming operations where metals and thermosets are mechanically interlocked with or enclosed in thermoplastics, eg, staking, swaging, and metal-in-plastic insertion. Tubing can be formed into a variety of shapes, and shaping of sheet and texturizing are also possible. The method is limited to forming a relatively low volume of plastic adjacent to the ultrasonic horn (51–52).

Metal Forming.   Despite much research in ultrasonic metal forming over the years, little commercial progress has been made (55–59). Some wire is drawn with ultrasonics, and a few tube-drawing installations may exist. Reduction in drawing forces, faster drawing rates, drawing in fewer passes, and better finish have been demonstrated, but the advantages appear insufficient to justify commercialization.

Machining.   Brittle materials, difficult to machine by conventional methods, can be machined with ultrasound (60). These include ceramics, glass, ferrites, and gemstones.

Rotary Abrasive Machining.   In this method, axial ultrasonic vibration is superimposed on the conventional rotary motion of the tool. Diamond-impregnated or diamond-plated core drills are used, usually cooled by a liquid through the center. Machining rates are faster, tool wear is diminished, and dimensional control is better because of lower forces on the tool compared to conventional operation. Coolant cavitation and tool acceleration prevent loading by abraded particles and result in more efficient cutting. Most applications involve drilling, but other machining operations are possible. In some applications, the vibrating tool is stationary and the workpiece rotates. Such equipment is used for dressing wire-drawing dies and lapidary machining.

Impact Grinding.   Impact grinding is performed by ultrasonic vibration alone. The method allows machining of multiple-hole patterns and irregular and three-dimensional shapes. An abrasive slurry is fed between the tool and the workpiece, and the negative of the horn face is imparted to the work. The method works by accelerating abrasive particles and is relatively slow; 20-kHz equipment is used.

Metal Cutting. Many attempts have been made to assist the conventional metal-cutting operations with ultrasound, and some commercial equipment has been built; however, only marginal benefits have resulted (61).

Fatigue Testing. Ultrasonic fatigue testing allows a drastic reduction in testing time (62–64). For example, testing at 20 kHz is 400 times faster than at 50 Hz, and 1 $\times 10^8$ cycles can be accumulated in less than 2 h compared to 23 d required at the lower frequency. Research has focused on the correlation between low and high frequency testing where significant differences exist for some metals.

Figure 10 illustrates the principle. Although necked-down samples may be used, this is not really necessary because the stress in a uniform section is sinusoidal, and the sample predictably breaks at one quarter of the wavelength from the end. Ultrasonic welding equipment with interchangeable velocity transformers and constant amplitude control is ideally suited for this purpose (see also Nondestructive testing).

An arrangement for fatigue-testing adhesives is shown in Figure 11. A small mass is bonded with the adhesive under test to the end of the horn. The ultrasonic peak force acting on the bond is very nearly given by

$$p = 4\pi^2 f^2 d \cdot w$$

where p = peak force, N (1 N = 0.225 lbf); f = frequency, Hz; d = transducer-peak displacement, m; and w = mass, kg.

Other Solids Applications. *Curing.* Endothermic reactions can be accelerated by the heat induced in the volume of a solid under ultrasonic stress (65). This procedure works well only for relatively thin layers of material, not over ca 1 cm at 20 kHz.

Friction Reduction. High acceleration present at ultrasonic frequencies reduces apparent friction and produces a slippery effect, as if caused by lubrication. The reduction in drawing force in ultrasonically assisted wire drawing mentioned earlier is at least partly because of this effect. An ultrasonically vibrated knife cuts with less effort through spongy, fluffy, or sticky materials, and some success was obtained with cutting angel cake or dough, and in medical surgery. Ultrasonic chutes and sieves have been built to promote flow of sticky powders.

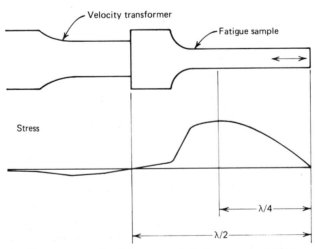

Figure 10. A uniform cross-section fatigue sample. Failure occurs at λ/4 from the free end.

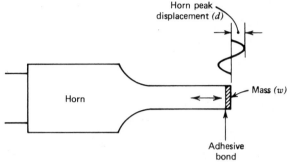

Figure 11.   Adhesive strength test.

Airborne Applications

Airborne high power ultrasonic applications require very high sound intensities, usually >150 dB. The resultant power densities at the area of application, however, are low, typically a fraction of a watt per square centimeter.

Defoaming.   Foams can be broken with high intensity ultrasound, but the effect is fairly local. Some such installations are employed in beverage packaging. Compared to conventional methods, ultrasonic defoaming does not require any contact with the product, which is preferable for sanitary reasons (see Defoamers).

Particle Agglomeration.   Air scrubbing by ultrasound has been investigated (66–67) but cannot compete with other methods. For effective ultrasonic agglomeration, the moisture content of the air should be relatively high (>1 g/m^3), and the agglomerated particles should stick together. The explanation of the agglomeration phenomenon is that ultrasonic motion increases the collision rate between particles. A possible application might be precipitation of conductive materials unsuitable for electric precipitators (see Air pollution control methods).

Drying.   Drying (qv) can be accomplished at lower temperatures in the presence of high intensity airborne sound (68–72). The effect is explained by microstreams reducing the thermal boundary layer. Applications seemed promising in drying powdered materials that are easily damaged by heat, eg, food and pharmaceuticals, and commercial ultrasonic dryers using pneumatic whistles were introduced. However, the market for these devices is limited, and acoustical drying is not much used at present. However, in view of the low efficiency of pneumatic whistles and the progress made in flexural disks and horns, ultrasound has some prospects in drying applications.

Medical Uses

Many attempts have been made at applying power ultrasound to medical uses, eg, drilling of teeth, kidney-stone disintegration, bone welding, cancer treatment, cataract removal, surgery, and many others. The most successful application has been dental-descaling equipment as an alternative to traditional manual scrubbing. A hand-held probe terminated in a small hooked tip is applied under light pressure to the tooth. The scale is removed by lateral movement of the tip vibrating at around 25 kHz, aided by cavitation of the water spray pumped through the center of the tip.

This procedure causes less discomfort to the patient and is more convenient to the operator. The equipment is commonly used by dentists and dental hygienists; low power is required, <50 W.

Economic Aspects

The 1980 U.S. sales of ultrasonic-power equipment reached 8.2×10^7, of which cleaning, welding, drilling, and soldering represent 7.4×10^7 (73). The U.S. suppliers of ultrasound-power equipment are given in Table 2.

The price of ultrasonic equipment is roughly related to power but depends on the application (see Table 3). A 400-W industrial cleaner sells for ca $1500, whereas a welder of comparable power incorporating a mechanical press and controls sells for ca $5000.

Table 2. U.S. Suppliers of Ultrasound-Power Equipment[a]

Company	Location	Equipment[b]
Bendix Corporation	Davenport, Iowa	C
Blackstone Corporation	Jamestown, N.Y.	S
Branson Cleaning Equipment Co.	Shelton, Conn.	C, other
Branson Sonic Power Co.	Danbury, Conn.	W, M, S, B, other
Crest Ultrasonics Corporation	Trenton, N.J.	C
Delta Sonics, Inc.	Paramount, Calif.	C, W
Dukane Corporation	St. Charles, Ill.	W
Heat Systems Ultrasonics, Inc.	Plainview, N.Y.	B
L & R Manufacturing	Kearny, N.J.	C
Lewis Corporation	Oxford, Conn.	C, S, other
Mastersonic, Inc.	Granger, Ind.	W
Sonicor Instrument Corporation	Copiague, N.Y.	C
Sonics & Materials, Inc.	Danbury, Conn.	W
Sonobond Corporation	West Chester, Pa.	W, M, S, other
Ultra Sonic Seal Division	Broomall, Pa.	W
Wave Energy Systems, Inc.	Newtown, Pa.	C, B
Westinghouse Electric Corporation	Sykesville, Md.	C, S

[a] Ref. 73.
[b] C = cleaning; W = welding; M = machining; S = soldering; and B = biological.

Table 3. Power and Prices of Ultrasonic Equipment

Equipment	Power range, W	Price range, $
cleaners		
industrial	400–1,200	1,500–4,000
small laboratory	15–150	70–1,500
welders		
for plastic	300–3,000	3,000–14,000
for metal	1,500–4,000	18,000–25,000
grinders		
rotary	350	7,000–28,000
impact	150–900	5,000–7,500
biological cell disrupters	150–350	1,600–2,500
industrial liquid processors	600–4,000	5,000–36,000

Statistics on breakdown of ultrasonic uses by various industries are not readily available. However, it is known that automotive, electronic, electric, and photo-optical applications represent a substantial portion of the market. The investment in equipment is usually recovered within 20 mo.

Health and Safety Factors

Because ultrasound propagates through the air, its effect on personnel working with industrial ultrasonic equipment must be investigated (75). Absorption of ultrasonic power at levels normally encountered in the industry is negligible, and the only area warranting any serious consideration is that concerning human hearing. According to studies, there is no evidence of danger to hearing at ultrasonic exposures up to at least 110 dB, a level sufficiently high to be of no concern to most industrial applications (see also Noise pollution).

However, some ultrasonic applications generate noise in the audible range of sufficient intensity to cause discomfort or pose a possible hazard to hearing. In the United States, the allowable noise limits for industrial environment are set by OSHA; other industrial countries have similar regulations. In addition, some industries have tighter standards than those imposed by governments. Where noise is a problem, soundproof enclosures are used, or ear-protecting devices are worn by personnel attending the equipment (see Insulation, acoustic).

BIBLIOGRAPHY

"Ultrasonics" in *ECT* 1st ed., Vol. 14, pp. 407–422, by Earnest Yeager and Frank Hovorka, Western Reserve University; "Ultrasonics" in *ECT* 2nd ed., Vol. 20, pp. 773–791, by Alfred Weissler, American University.

1. L. D. Rosenberg, ed., *High-Intensity Ultrasonic Fields*, Plenum Press, New York, 1971.
2. E. A. Neppiras, *Phys. Rep.* **61**(3), 251 (1980).
3. V. A. Akulichev in ref. 1, pp. 203–419.
4. W. L. Nyborg in W. P. Mason, ed., *Physical Acoustics*, Vol. 2, Pt. B, Academic Press, Inc., New York, 1965.
5. W. L. Nyborg, *Proceedings of the First International Symposium on High Power Ultrasonics*, IPC Science and Technology Press Ltd., Guildford, Surrey, UK, 1970, pp. 124–135.
6. B. Niemczewski, *Ultrasonics* **18**(3), 107 (1980).
7. B. A. Agranat, V. I. Bashkirov, and Yu. I. Kitaigorodskii in L. D. Rosenberg, ed., *Physical Principles of Ultrasonic Technology*, Vol. 1, Plenum Press, New York, 1973, pp. 247–376.
8. B. Brown and J. E. Goodman, *High-Intensity Ultrasonics*, Iliffe Books Ltd., London, 1965, pp. 116–135.
9. R. Pohlman, B. Werden, and R. Marziniak, *Ultrasonics* **10**(4), 156 (1972).
10. R. L. Hunicke, *Ultrasonics International 1975 Conference Proceedings*, IPC Science and Technology Press Ltd., Guildford, Surrey, UK, pp. 32–38.
11. Ref. 7, pp. 319–330.
12. *Erosion by Cavitation*, ASTM special technical publication no. 408, American Society for Testing and Materials, Philadelphia, Pa., 1967, pp. 159–219, 239–283.
13. I. Hansson, K. H. Morch, and C. M. Dreece, *Ultrasonics International 1977 Conference Proceedings*, IPC Science and Technology Press Ltd., Guildford, Surrey, UK, pp. 267–274.
14. A. F. McIntosh and R. F. Munro, *Process Biochem.* **6**(3), 22 (1971).
15. M. E. Ovadia and D. M. Skauen, *J. Pharm. Sci.* **54**, 1013 (1965).
16. I. C. Patel and D. M. Skauen, *J. Pharm. Sci.* **58**, 1135 (1969).
17. V. Velichkov and E. Georgiev, *Nauchn. Tr. Vissh. Inst. Khranit. Vkusova Prom. Plovdiv.* **16**(2), 177 (1969); *Chem. Abstr.* **78**, 20103 (1973).

18. C. Srinivasulu, S. C. Srivastava, and S. N. Mahapatra, *Ultrasonics International 1973 Conference Proceedings*, IPC Science and Technology Press Ltd., Guildford, Surrey, UK, pp. 25–27.
19. V. M. Solonko and co-workers, *Farm. Zh.* **30**(1), 84 (1975).
20. Ref. 8, pp. 222–223.
21. V. A. Nosov, *Ultrasonics in the Chemical Industry*, Vol. 2, Consultants Bureau, New York, 1965, pp. 40–43.
22. D. M. Higgins and D. M. Skauen, *J. Pharm. Sci.* **61**(10), 1567 (1972).
23. Ref. 8, pp. 141–159.
24. J. K. O'Neill, D. J. Kenneberg, and M. A. Viola, *Proceedings of the Third International Symposium on Coal-Oil Mixture Combustion*, Vol. 1, Pittsburgh Energy Technology Center, Pittsburgh, Pa., 1981.
25. G. T. Hawkins in ref. 24, Vol. 2.
26. Ref. 21, pp. 43–47.
27. B. A. Agranat and co-workers, *Ultrazvuk v Gidrometallurgii*, Metallurgia, Moskov, 1969, pp. 56–159.
28. W. Kowalski and E. Kowalska, *Ultrasonics* **16**(2), 84 (1978).
29. U.S. Pat. 3,697,222 (Oct. 10, 1972), G. Sierra (to Ontario Research Foundation).
30. T. S. Uko and co-workers, *Yakazaigaku* **33**(2), 101 (1973).
31. D. Gold, *Ultrasonic Sterilization of Pharmaceutical Preparations*, thesis, University of Connecticut, Storrs, Conn., 1962.
32. Ref. 21, pp. 47–48.
33. H. V. Fairbanks in ref. 18, pp. 11–15.
34. A. Semmelink in ref. 18, pp. 7–10.
35. L. Bjorno, S. Gram, and P. R. Steenstrup, *Ultrasonics* **16**(3), 103 (1978).
36. O. K. Eknadiosyants in ref. 7, Vol. 2, pp. 3–88.
37. M. N. Topp and P. Eisendlam, *Ultrasonics* **10**(3), 127 (1972).
38. R. G. Pohlman in ref. 18, pp. 52–55.
39. R. Pohlman, K. Heisler, and M. Cichos, *Ultrasonics* **12**(1), 11 (1974).
40. Ref. 7, Vol. 2, pp. 348–378.
41. Ref. 21, pp. 32–34.
42. D. M. Skauen, *J. Pharm. Sci.* **56**(11), 1373 (1967).
43. S. L. Hem, *Ultrasonics* **5**, 202 (1967).
44. A. V. Kortnev and N. V. Martynovskaya, *Sb. Mosk. Inst. Stali Splavov* **77**, 98 (1974).
45. U.S. Pat. 3,510,266 (May 5, 1970), M. Middler (to Merck and Co., Inc.).
46. O. V. Abrams and I. I. Teumin in ref. 7, Vol. 2, pp. 145–273.
47. A. Basedon and K. H. Ebert, *Angew. Chem. Int. Ed. Engl.* **13**(6), 413 (1974).
48. T. Sato and D. E. Nalepa, *J. Coat. Technol.* **49**, 45 (1977).
49. O. A. Kapustin in ref. 7, pp. 379–509.
50. Ref. 8, pp. 210–213.
51. A. Shoh, *Ultrasonics* **14**(5), 209 (1976).
52. *Ultrasonic Plastics Assembly*, Branson Sonic Power Co., Danbury, Conn., 1979, p. 111.
53. A. M. Mitskevich in ref. 7, pp. 101–243.
54. A. P. Hulst, *Ultrasonics* **10**(6), 252 (1972).
55. B. Langenecker and O. Vodep in ref. 10, pp. 202–205.
56. G. R. Dawson in ref. 10, pp. 206–209.
57. J. Herbertz in ref. 13, pp. 323–328.
58. N. E. Anatasiu, *Ultrasonics* **18**(6), 255 (1980).
59. I. Hansson and A. Thoelen, *Ultrasonics* **16**(2), 57 (1978).
60. V. F. Kazantsev in ref. 7, pp. 3–98.
61. H. V. Fairbanks in ref. 10, pp. 107–109.
62. E. A. Neppiras, *Proc. ASTM* **59**, 691 (1959).
63. P. Bajons and co-workers in ref. 10, pp. 95–101.
64. V. A. Kuzmenko, *Ultrasonics* **13**(1), 21 (1975).
65. T. K. Saksena and N. K. Babbar, *Ultrasonics* **17**(3), 122 (1979).
66. N. L. Shirokova in ref. 7, Vol. 2, pp. 477–539.
67. Ref. 21, pp. 51–56.
68. Yu. Ya Borisov and N. M. Gyukina in ref. 7, Vol. 2, pp. 381–474.
69. Ref. 21, pp. 57–61.

70. K. Seya in ref. 5, pp. 136–139.
71. H. V. Fairbands in ref. 10, pp. 43–45.
72. T. Otsuku, H. Purdum, and H. Fairbanks in ref. 13, pp. 91–93.
73. *1980 Current Industrial Report*, MA-36N, Dept. of Commerce, U.S. Bureau of the Census, Washington, D.C., p. 7.
74. *1982 US Industrial Directory*, Cahners Publishing Company, Stamford, Conn.
75. W. I. Acton, *Proceedings of the International Congress on Noise as a Public Health Problem, Dubrovnik, Yugoslavia, May 13–18, 1973*, Report 550/9-73-008, U.S. Environmental Protection Agency, Washington, D.C.

ANDREW SHOH
Branson Sonic Power Company

LOW POWER

Low power ultrasound is used in many fields of science, engineering, and medicine as a nondestructive, noninvasive testing and diagnostic technique. The term low power ultrasound describes sound waves with frequencies greater than 20 kHz at power levels of milliwatts and below. Typically, the frequencies used are 0.5–20 MHz, although some testing is done as high as 1 GHz. This technique is used industrially for measuring the thickness of machined parts, monitoring corrosion inside pipes and vessels, in finding flaws in welds and castings, and in measuring material properties. In medicine, it is used widely as a diagnostic imaging technique for scanning the abdomen, heart, and other areas to detect disease or abnormalities. Additional information is available from manufacturers of ultrasonic test equipment as well as from the American Society for Nondestructive Testing and the American Institute for Ultrasound in Medicine. Journals are published by these two U.S. societies as well as by a number of international professional ultrasound groups.

Theoretical Considerations

Ultrasonic waves can be generated inside a test material by placing an ultrasonic transducer on the test object as illustrated in Figure 1. Commonly employed ultrasonic transducers use a piezoelectric material as the transducing element. The piezoelectric element converts electrical energy applied to the device into mechanical energy which produces a sound wave in the test material. Conversely, a sound wave that impinges on the transducer from the test material is converted into electrical energy. The basic ultrasonic test principle is illustrated in Figure 1. In this pulse–echo test procedure, an electrical pulse is applied to the transducer which starts the ultrasonic wavefront propagating through the test object. The voltage applied to the transducer can be monitored on an oscilloscope display. The transducer then responds to any return echoes as shown in Figure 1a. If the sound pulse encounters any discontinuity in the

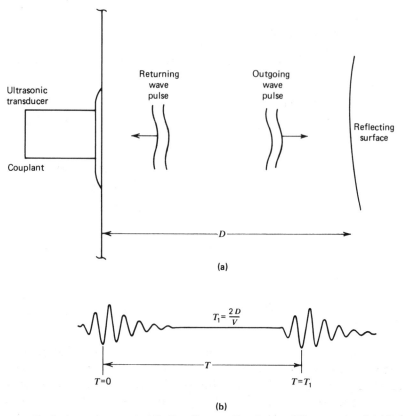

Figure 1.   Basic ultrasonic test principle, D = distance, T = time, and V = wave speed. (**a**) Pulse–echo setup. (**b**) Oscilloscope display.

material, a portion of the sound energy is reflected back toward the transducer. When this echo reaches the transducer, the sound energy is converted back into an electrical signal and can be displayed on the oscilloscope. The distance between the reflector and the transducer can be calculated by knowing the speed of sound in the material being tested and measuring the elapsed time between the transmitted sound pulse and the return echo.

The sound-wave speed of a material is proportional to the square root of the ratio of the material's stiffness to its density. As an example, the wave speed in steel is ca 6 km/s, whereas in soft materials such as biological tissue or water the wave speed is about 1.5 km/s. Typical wave speeds for several common materials are presented in Table 1. The wave speed V is related to the sound frequency f and wavelength λ by $V = f \cdot \lambda$ (1). In some materials, wave speeds correlate with material properties (see Insulation, acoustic).

Typical ultrasonic waveforms are illustrated in Figure 2. Continuous wave excitation is sometimes used in resonance testing and in Doppler analysis. The pulse type waveform, however, is employed most often in pulse–echo ultrasonic analysis, in both nondestructive testing and medical ultrasound.

When an ultrasonic wave encounters an interface between two materials, a portion of the sound energy is transmitted across the interface and a portion is reflected. The relationship between reflection and transmission is illustrated in Figure 3. P_I represents

Table 1. Typical Acoustic Properties for a Variety of Materials

Material	Density, ρ, g/cm^3	Longitudinal wave speed, V, m/s	Shear wave speed, m/s	Acoustic impedance, Z, kg/(cm^2·s)
air	0.001	330		0.03
aluminum	2.7	6,320	3,130	1,710
aluminum oxide	3.7	10,100	5,900	3,740
beryllium	1.82	12,900	8,900	2,300
brass	8.5	4,430	2,120	3,770
bronze	8.86	3,530	2,230	3,130
cast iron	7.7	4,500	2,400	3,470
copper	8.9	4,660	2,260	4,150
cork	0.25	500		10
glass	2.3	5,700	3,400	1,310
glycerol	1.26	1,920		2,420
gold	19.3	3,240	1,200	6,250
ice	0.92	3,980	1,990	370
lead	11.4	2,160	700	2,460
magnesium	1.74	5,790	3,100	1,010
nickel	8.9	6,040	3,000	5,370
motor oil	0.9	1,360		120
poly(methyl methacrylate)	1.18	2,670	1,120	320
polyethylene	0.9	1,950	540	180
polystyrene	1.06	2,350	1,120	250
silver	10.5	3,600	1,600	3,780
steel	7.8	5,800	3,200	4,560
tin	7.3	3,320	1,670	2,420
titanium	4.54	6,070	3,120	2,760
tungsten	19.3	5,220	2,890	10,010
water	1.00	1,480		1,480
zinc	7.13	4,170	2,410	2,970
zirconium	6.53	4,700	2,300	3,070

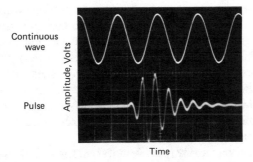

Figure 2. Ultrasonic waveforms.

the incident ultrasonic sound pressure amplitude and P_R and P_T the corresponding reflected and transmitted amplitudes. The reflection factor (R_{12}), and the transmission factor (T_{12}), depend on the acoustic impedance difference between the two materials. The acoustic impedance Z of a material is the product of the material wave speed (V) and density (ρ). It can be seen that if materials 1 and 2 are identical, then all of the

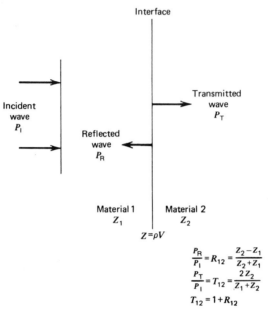

Figure 3.   Reflection (R) and transmission (T) factors at an interface, P = pressure, V = wave speed, Z = acoustic impedance, I = incident, and ρ = material density.

energy is transmitted and there is no reflected echo. If the difference in the impedances of the materials is small, there is a weak echo. The amplitude of these echoes can be used to gauge the impedance difference at the reflecting interface (1–2).

Wave refraction at an interface also affects the direction in which sound waves travel. As an ultrasonic wave impinges on an interface at some oblique angle, a portion of the beam is reflected at an equal angle. The transmitted beam is refracted at a different angle determined by the ratio of the wave speeds on either side of the interface. An incident longitudinal wave can produce two refracted waves, one longitudinal and one shear, a process called mode conversion. Conversely, an incident oblique shear wave can be converted into refracted shear and longitudinal beams. In a longitudinal wave, the actual particle motion in the material is in the same direction as the wave propagation. For shear waves, the particle motion is normal to the propagation direction. In most solid materials, the shear wave speed is approximately 60% of the longitudinal wave speed. Fluids typically do not support shear-wave propagation. The refraction process is illustrated in Figure 4. The refraction formula is defined by Snell's law $V_2 \sin \theta_1 = V_1 \sin \theta_2$, where V is the appropriate wave speed in material 1 or 2, θ_1 is the incident wave angle in material 1, and θ_2 is the refracted beam angle in material 2 for the longitudinal or shear wave of interest.

As a sound wave propagates through a material, the sound amplitude is gradually reduced by attenuation. The sound energy is attenuated either by scattering from many small reflectors, such as material grains, or by viscoelastic energy absorption. In general, attenuation increases as the signal frequency increases. As shown in Figure 5, lower frequency signals can penetrate a greater thickness of material than can high frequency signals before the amplitude decreases to an unusable level (2).

Ultrasonic beams are not confined to straight lines inside a structure. Beam spreading occurs because of interference between portions of the wave as they prop-

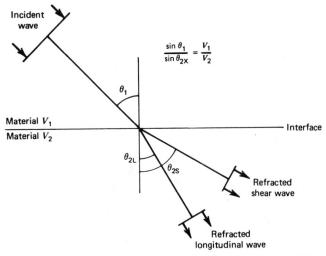

Figure 4. Refraction at an interface, L = longitudinal, S = shear, and X = L or S.

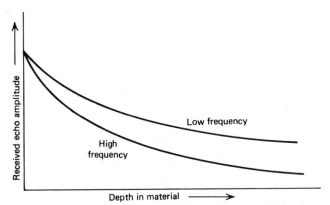

Figure 5. Amplitude decay with increasing material depth.

agate through a structure from various points on the face of the transducer. For each transducer, there is a natural focus or near-field point whose position is determined by the dimensions of the transducer and its frequency. For a circular transducer, the near-field point occurs at a distance of $D^2/4\,\lambda$ where D is the diameter of the transducer element. A series of pressure profiles at various distances from the transducer face are illustrated in Figure 6. Large variations in pressure can be seen within the near field. A smooth pressure profile first occurs at the near-field point and the profile in the far field is also smooth but the beam is broadened. Transducers can be made that focus the sound to a small, well-defined spot. The focus can be produced by varying the phase of the wave across the face of the transducer. This can be accomplished with a curved lens in front of the transducer which produces a focus fixed by the design of the lens. A transducer of a given aperture can only be focused within its natural near field. An alternative method of focusing is to replace the single element and lens with multiple small elements and electronic time-delay circuitry with which the phase of the signal on each element can be adjusted to give a continuously variable focusing effect.

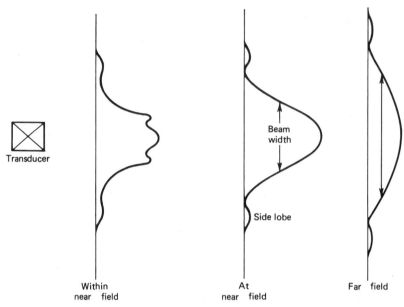

Figure 6. Transverse pressure profiles.

Display Techniques

The time or amplitude information produced by the ultrasonic transducer must be displayed in some fashion to be of use to the test technician or medical ultrasonographer. The A-mode, or amplitude mode, display (Fig. 1**b**) provides a display of echo amplitude versus time. The transducer is held at a single location to acquire data in the pulse–echo test mode. If the reflectors are moving, the echoes on the display move left and right as the target moves toward and away from the transducer. A B-mode display is generated by converting the amplitudes of the echo signals into a dot on the B-mode display. The brightness of the dot is proportional to the amplitude of the echo. The vertical position of each dot in the figure represents distance from the transducer and the horizontal position is determined by the lateral position of the transducer at the time the echo was received. An example of a B-mode display is presented in Figure 7. The B-mode line-of-sight concept can provide a B-scan display, giving a cross-sectional image of the structure under examination. The location of the transducer is determined by a position encoder so that the hundreds of B-mode lines can be superimposed to give the B-scan cross-section image.

In an M-mode presentation, sometimes referred to as TM-mode, echo position as a function of time is recorded. This is done by taking a single line of the B-mode presentation and recording the changing positions of the bright echo dots on a chart recorder. As the recording medium is moved past the oscillating B-mode dots, a time-motion record or M-mode display is produced.

There are several factors that limit the quality of the information displayed. First, as indicated in Figure 6, side-lobe pressure components are produced in the ultrasonic field by the transducer. Side-lobes degrade the image due to artifacts being generated in the image by the transducer sensing reflectors that are not along the transducer axis. It is therefore essential that the side lobes be very small in amplitude compared to the principal lobe so that off-axis echoes will not appear in the image. Side-lobe

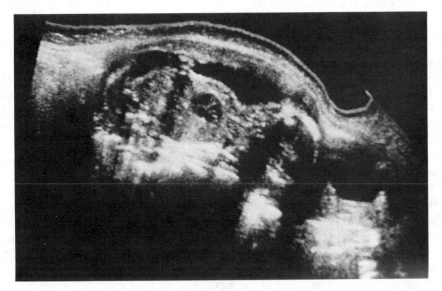

Figure 7.   Gray-scale B-scan image of fetus *in utero*.

amplitude and position can be affected by transducer design and construction and can be reduced by aperture shading or apodization. In addition to side-lobe artifact formation, the actual axial and lateral resolution of the transducer must be considered. Axial resolution is defined as the capability of separating two point reflectors along the axial direction from the transducer. The best possible axial resolution is produced by using as short duration a sound pulse as possible. This is illustrated in Figure 8.

Lateral resolution is the ability to resolve two point reflectors spaced along a lateral direction at some specific distance inside the structure. It is clear that the narrowest possible ultrasonic beam produces the best lateral resolution (2). The narrow beam can be achieved by using a higher frequency, or by using a larger-diameter focused transducer. Beam width and thus lateral resolution varies with depth because of beam divergence. The angle of divergence of the beam from the near-field point is proportional to the sound wavelength divided by the transducer diameter. The best

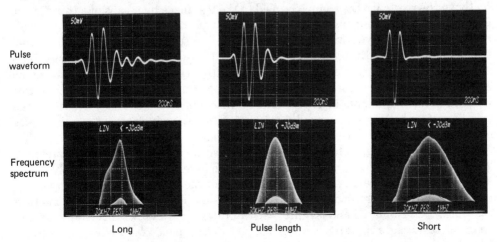

Figure 8.   Pulse length and axial resolution.

lateral resolution occurs at an axial distance equal to either the near-field point or the design focal point of a focused transducer.

Nondestructive Testing

Ultrasonic techniques are used in nondestructive testing to locate and evaluate flaw characteristics in many different industrial materials (see Nondestructive testing). In many industries, it is imperative to determine the integrity of a structure when it is manufactured so that good performance in a field environment can be expected. Ultrasonic testing is also important for flaw location and classification in forgings, castings, welds, plates, and tubes.

Thickness Measurement. One of the most popular uses of ultrasonic nondestructive testing is in thickness measurement; wall-thickness determination of tubing, pressure vessels, and many other test parts is critical in both quality-control testing and for in-service inspection. Thickness variations due to manufacturing variations or in-service degradation can be measured (see Metallic coatings, explosively clad). Thickness measurement applications are important in all sorts of materials including steels, aluminum, plastics, composite materials, etc. A variety of digital readout thickness-gauge instruments is available commercially that simplify thickness measurement. The ultrasonic thickness technique has the distinct advantage of being able to measure a wall thickness even when only one side of the part is accessible (1,3).

Flow Detection. There are several combinations of basic testing techniques which can be used depending on the type of flaws that are being tested for and upon the geometry of the parts being inspected. There are two basic test setups, pulse–echo and through transmission. Pulse–echo testing, which is described above, gives information on flaw size and, by time measurement, gives distance information. Through-transmission testing is done with two transducers on opposite sides of the part, one transmitting and the other receiving. Flaws are detected by a decrease in the received signal amplitude. This method does not give information on the distance of the flaw from the transducer but has the advantage of being able to detect some flaws not detectable with pulse–echo testing.

Contact Scanning. The ultrasound signal in either of these two test methods may be applied to the part being tested in two basic ways. The transducer may be put directly on the part with the sound coupled to the test piece by means of a thin film of liquid couplant. This contact mode has the advantage of being inexpensive, easy to use, portable, and very versatile. Usually, the transducer is moved by hand while the operator observes the screen of the ultrasonic flaw-detector instrument. As echoes are observed on the screen, the operator can position the transducer to maximize the echo and characterize the nature of the reflector. Hand-contact scanning such as this is, however, slow and labor intensive. The liquid couplant is necessary to exclude the thin film of air between the transducer and the test part which would greatly reduce or eliminate the amplitude of the ultrasound signal in the test part.

Immersion Testing. The other sound-application method is immersion testing. The part to be tested and the transducer are both immersed in liquid, usually water. The transducer can be scanned mechanically over the part and the echo signals can be observed by an operator or automatically recorded. This method has the advantage that it can be automated and is quite fast, but it requires a tank for the liquid, a mechanical scanning apparatus, and is limited in the part geometries that can be tested.

The other basic-testing variable is the angle at which the sound enters the part. Most commonly, the sound enters normal to the surface, thus the name normal beam testing. There are many situations, because of part shape or surface obstructions such as the crown of a weld, where the expected flaws cannot be seen with a normal beam (Fig. 9). It is possible to have the sound enter the part at some other angle by means of refraction. In contact testing, the refraction is produced by an angled wedge, usually plastic, between the transducer and the part. In immersion testing, the refraction is caused by angling the transducer relative to the test-part surface during the scanning. In both cases, the path of the sound beam can be determined using Snell's law to determine the location of echo reflectors.

Material Characterization. Material characterization can also be performed with ultrasound. Changes in sound propagation parameters can be used to measure changes in other material properties. For example, the speed of sound in cast iron is dependent on the degree of nodularity of the iron. Thus, the speed can be monitored to check on the processing of the material. Also, the speed of sound in some liquid mixtures and solutions is dependent on the composition of the liquid and may be used as a process control parameter (3) (see also Materials standards and specifications).

Acoustic Emission. Acoustic emission is another method of ultrasonic nondestructive testing. It is passive in that propagating cracks emit the noise and ultrasonic transducers are used to detect it. Triangulation can be used to locate the source. Various integrating and event-counting methods are used to measure the amount of emission. Acoustic emission is particularly useful in proof-testing structures.

Process Monitoring. Ultrasonic signals can be used as the sensing mechanism in a number of process control and monitoring techniques. Pulse–echo equipment can be used to sense liquid level in tanks either by measuring the length of the liquid column or the distance from the tank top to the top surface of the liquid. The temperature dependence of sound speed can be used to measure the temperature of a material. Doppler-shift measurement or differential time-of-flight can be used to measure flow rate in liquids being pumped through pipes (3).

Ultrasonic velocity is a useful predictor of material properties in certain classes of materials in which velocity is correlated with or functionally related to the physical property. Low power ultrasound is useful in process monitoring because the physics of a particular situation shows a variation in an ultrasound propagation parameter as a function of the variable that must be monitored, eg, ultrasonic velocity in a wire is as valid as emf in a thermocouple for thermometry. Wire sonics has been used in nuclear-reactor fuel testing. Ultrasonic attenuation in a slurry can monitor pumping of sludge in a sewage-treatment plant. Ultrasonic velocity can define the edges of

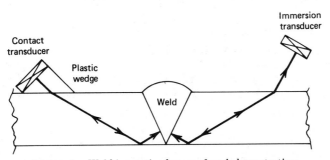

Figure 9.   Weld inspection by use of angle beam testing.

batches of liquid hydrocarbons pumped back-to-back in pipelines. Ultrasonic Doppler effects can measure the pumped mass flow rate in the pipe. Ultrasonic pulse–echo time-of-flight can be turned into a depth gauge in a vat or silo as well as in a navigable body of water (see Liquid-level measurement). Maple syrup can be distinguished from corn syrup by ultrasonic velocity. The possibilities are endless, limited only by physics and cost-effectiveness (4–5).

Signal Processing.   Low power ultrasound is used in the electronic technology field for signal processing. Here, two general types of devices are built using ultrasound: resonators and traveling-wave devices. Various composite and multistage resonators as well as some simple resonators are used as filters, and other simple resonators serve as oscillators. Traveling-wave devices, particularly surface-wave devices, include convolvers, correlators, spectrum analyzers, acousto-optic devices, matched filters, and pulse-compression filters. These devices are used in radar and other electronics, and the resonators are used in these and in telecommunications, TV, CB radios, digital watches, and various frequency-control applications (6–9) (see also Instrumentation and control; Microwave technology).

Future directions in ultrasonic nondestructive testing are towards computer-controlled scanning, data recording, and analysis. Digitally controlled flaw detectors are now available that can utilize many new developments in signal processing and imaging for making objective decisions and evaluations of the parts being tested.

Medical Ultrasound

Low power pulse–echo ultrasound has achieved a role in imaging in diagnostic medicine in most hospitals today and in private practice (10–12). Medical imaging at low power levels can be contrasted to the use of high power ultrasound in thermal therapy. The power densities used in diagnostic ultrasound are on the order of 20 mW/cm^2. One of the early diagnostic applications was the use of A-mode to determine a shift in the midline of the human brain.

Two-dimensional section imaging using B-scan techniques is used to image anatomy, particularly the abdomen where organs like the liver, kidney, and pancreas are studied. Two-dimensional imaging has also become a valuable aid to cardiologists in viewing the functioning of heart valves and chambers. This technique now compliments the older one-dimensional M-mode technique used by cardiologists. Recent improvements in electronics and instruments have improved the gray-scale image quality of B-scans, allowing better differentiation of tissue texture and improved lesion detectability. These improvements have earned ultrasound a complementary role in diagnostic imaging along with x-ray, computed tomography, and nuclear medicine techniques (see Radiopaques; X-ray technology). The nonionizing nature of the energy, and the lack of known biohazards, have led to a key role in obstetrics, gynecology, and in pregnancy monitoring to determine fetal growth and well-being. A cross-sectional B-scan through a fetus and uterus is shown in Figure 7 in which the chambers of the fetal heart can be seen. Advances in the design of transducers have allowed the use of higher frequencies, up to 10 MHz, with adequate penetration for high resolution imaging of small superficial organs including the thyroid gland, the carotid arteries, and the testes.

B-scanners with their mechanically articulated scan arm give a static or snapshot view of the anatomy. A major advance in instrumentation since 1976 has been the use of real-time scanners to observe the movement of anatomical areas much like a movie camera. Using real-time instrumentation, the functioning of the heart chambers and valves can be observed as the heart beats. Three basic schemes have been used to generate scans in real time. The first is the mechanical-sector scanner, where a single-element transducer sweeps a sector scan or several transducers on the periphery of a wheel are rotated. The second is the sector-phased array in which a fixed set of small elements, typically 32–64, are driven with a variable-phase relationship which sweeps the ultrasonic beam from side to side. These techniques produce a pie-shaped field of view and a framing rate of as much as 30 images per second, fast enough to see motion in the body. In the multielement linear array, many, typically 64–128, individual crystal elements are used and the scanning is done electronically by pulsing several elements together and indexing this grouping along the array. This produces a rectangular field of view and can be driven at framing rates high enough to view motion.

M-mode instrumentation, being less expensive than two-dimensional imaging equipment, has become a ubiquitous tool for the cardiologist. Although the probe gathers information along one line of sight, it may be tilted to interrogate the various chambers of the heart. The time history (echocardiography) of the returning echoes plotted on a strip-chart recorder has become a mainstay in the cardiologist's work-up of patients with suspected heart-valve deficiencies (13).

Doppler measurements may be used to gather information about blood flow and areas of vascular stenosis or restriction. The Doppler technique is based on the principle that a moving reflector will shift the frequency of returning echoes in proportion to its velocity. Doppler probes typically use an operating frequency from 2 to 10 MHz. The Doppler signal or frequency shift is on the order of 2 kHz. Modern instruments use digital Fourier-transform techniques to measure this small frequency shift. Continuous-wave (cw) Doppler requires the use of a probe with two elements, a transmitter and a receiver. Cw Doppler detects motion anywhere along the line of sight of the probe, without distinguishing the depth of the moving target in the body. In pulsed Doppler, a short burst of ultrasound is used and electronic range gates are used to analyze signals from specific depths. The probes for pulsed Doppler are similar to other pulse–echo transducers. Conceptually, a pulsed Doppler probe could be scanned and the velocity information recorded and displayed in such a manner as to present a two-dimensional map of motion or blood flow in the body. Doppler techniques have been useful in screening stenotic carotid arteries and are in widespread use for fetal monitoring.

Current medical ultrasonic imaging makes primary use of the amplitude of the returned echoes. There is much information encoded in the returning echoes which is not utilized by present techniques, such as frequency and phase information. Future progress in medical ultrasound will be toward more complex signal processing which can make use of this information with the goal of tissue characterization to distinguish diseased tissue from healthy tissue noninvasively.

Health and Safety Aspects

Because of the increasingly widespread use of ultrasonic scanning for medical diagnosis, much attention has been given to the question of whether there may be long-term or latent undesirable side effects from this exposure. Sound intensities used in diagnostic ultrasound and in industrial ultrasonic testing are typically in the range of 20–100 mW/cm^2 (average) and are well below the levels needed to cause cavitation effects. The main effect biologically is a thermal or heating effect. To date, adverse effects in patients have not been demonstrated (14–15).

BIBLIOGRAPHY

"Ultrasonics" in *ECT* 1st ed., Vol. 14, pp. 407–422, by E. Yeager and F. Hovorka, Western Reserve University; "Ultrasonics" in *ECT* 2nd ed., Vol. 20, pp. 773–791, by A. Weissler, American University.

1. J. Krautkramer and H. Krautkramer, *Ultrasonic Testing of Materials*, 2nd English ed., Springer-Verlag, New York, 1977.
2. J. L. Rose and B. B. Goldberg, *Basic Physics in Diagnostic Ultrasound*, John Wiley & Sons, Inc., New York, 1979.
3. J. R. Frederick, *Ultrasonic Engineering*, John Wiley & Sons, Inc., New York, 1965.
4. L. C. Lynnworth in J. F. Schooley, ed., *Temperature: Its Measurement and Control in Science and Industry*, American Institute of Physics, New York, 1982, pp. 1181–1190.
5. R. L. Parker in F. C. Schwerer, ed., *Physics in the Steel Industry*, AIP Conference Proceedings #84, American Institute of Physics, New York, 1982, pp. 254–271.
6. *Proc. IEEE* **64,** 579 (May 1976).
7. *IEEE Trans. Microwave Theory Tech.* **29,** 393 (May 1981).
8. *IEEE Trans. Sonics Ultrason.* **23,** 115 (May 1981).
9. *Proceedings of the Frequency Control Symposium*, U.S. Army, Fort Monmouth, N.J., published annually by Electronic Industries Association.
10. B. B. Goldberg, ed., *Abdominal Gray Scale Ultrasonography*, John Wiley & Sons, Inc., 1977.
11. A. C. Fleischer and A. E. James, *Introduction to Diagnostic Sonography*, John Wiley & Sons, Inc., New York, 1980.
12. K. R. Erikson, F. J. Fry, and J. P. Jones, *IEEE Trans. Sonics Ultrason.* **21,** (July 1974).
13. A. D. Savakus, K. K. Shung, and N. B. Miller, *J. Clin. Ultrasound* **10,** 413 (1982).
14. R. A. Meyer, *Appl. Radiology/Ultrasound*, 71 (Sept.–Oct. 1982).
15. P. P. Lele, *Ultrasound Med. Biol.* **5,** 307 (1979).

General References

W. N. McDicken, *Diagnostic Ultrasonics*, 2nd ed., John Wiley & Sons, New York, 1981.
P. N. T. Wells, *Biomedical Ultrasound*, Academic Press, London, 1977.
W. P. Mason and R. N. Thurston, eds, *Physical Acoustics: Principles and Methods*, Academic Press, New York, 1964–present.
Ctein, "Auto-focus Looks Sharp," *High Technology*, 53 (Nov./Dec. 1982).
P. D. Edmonds, ed., *Methods of Experimental Physics*, Vol. 19, Academic Press, New York, 1981.
Prenatal Diagnosis, John Wiley & Sons, Inc., New York, published quarterly.

CRAIG S. MILLER
RON E. MCKEIGHEN
Krautkramer-Branson, Inc.

JOSEPH ROSE
Drexel University

ULTRAVIOLET ABSORBERS.   See Uv stabilizers.

UMBER.   See Pigments, inorganic.

UNADS.   See Rubber chemicals.

UNITS AND CONVERSION FACTORS

The barleycorn, inch, foot, yard, rod, furlong, mile, league, ell, fathom, and chain are units of length that the North American colonies inherited from the British. The inch was originally the length of three barleycorns; the yard was the distance from the tip of the nose to the tip of the middle finger on the outstretched arm of a British king (Henry I); the acre was the amount of land plowed by a yoke of oxen in a day. This system of units is very old and may be traced back to ancient Egypt.

A jumble of units existed throughout the world even until the late Eighteenth Century. In 1790, the French National Assembly requested the French Academy of Sciences to work out a system of units suitable for adoption by the whole world. This system was based on the meter as a unit of length and the gram as a unit of mass. Industry, commerce, and especially the scientific community benefited greatly. In 1893, the United States actually adopted the meter and the kilogram as the fundamental standards of length and mass. Although the spellings metre and litre are preferred by the author and ASTM, meter and liter are used in the *Encyclopedia*.

The foundation to international standardization of units was laid with an international treaty, the Meter Convention, which was signed by 17 countries, including the United States, in 1875. This treaty established a permanent International Bureau of Weights and Measures and defined the meter and the kilogram from which evolved a set of units for the measurement of length, area, volume, capacity, and mass. Also established was the General Conference on Weights and Measures (CGPM), which was to meet at regular intervals to consider any needed improvements in the standards. The National Bureau of Standards (NBS) represents the United States in these activities.

From these early beginnings, several metric systems evolved. With the addition of the second as a unit of time, the centimeter–gram–second (cgs) system was adopted in 1881. In the early 1900s, practical measurements in metric units began to be based on the meter–kilogram–second (mks) system. In 1935, the International Electrochemical Commission adopted a proposal to link the mks system of mechanics with the electromagnetic system of units by adding the ampere as a base unit and forming the mksA (meter–kilogram–second–ampere) system.

In 1954, the 10th CGPM added the degree Kelvin as the unit of temperature and the candela as the unit of luminous intensity. At the time of the 11th CGPM in 1960, this new system with six base units was formalized with the title International System of Units. Its abbreviation in all languages is SI, from the French *Le Système International d'Unités*.

Since 1960, various refinements to the system have been made, including redefinition of the second based on the atomic frequency of cesium; change of the name of the unit of temperature from degree Kelvin to the kelvin (symbol K); redefinition of the candela (all in 1967); addition of a seventh base unit, the mole (mol), as the unit of amount of substance; the pascal (Pa) as a special name for the SI unit of pressure or stress, equal to a newton per square meter; the siemens (S) as a special name for the unit of electric conductance, equal to the ampere per volt (all in 1971); addition of two SI units for ionizing radiation, the becquerel (Bq) as the unit of activity, equal to one reciprocal second, and the gray (Gy) as the unit of absorbed dose, equal to one joule per kilogram; prefixes for 10^{18}, exa (E), and 10^{15}, peta (P) (all in 1975); addition of the sievert (Sv) as the unit of dose equivalent, equal to one joule per kilogram; further redefinition of the candela; recognition of both l and L as symbols for liter (all in 1979); and interpretation of the radian and the steradian as dimensionless derived units for which the CGPM allows the freedom of using or not using in expressions for SI-derived units (1980).

Advantages of SI

SI is a decimal system. Fractions have been eliminated, and multiples and sub-multiples are formed by a system of prefixes ranging from exa, for 10^{18}, to atto, for 10^{-18}. Calculations, therefore, are greatly simplified.

Each physical quantity is expressed in one and only one unit, eg, the meter for length, the kilogram for mass, and the second for time. Derived units are defined by simple equations relating two or more base units. Some are given special names, such as newton for force and joule for work and energy.

In an energy-conscious world, SI provides a direct relationship between mechanical, electric, chemical, thermodynamic, molecular, and solar forms of energy. All power ratings are given in watts.

The system is coherent. There is no duplication of units for a quantity, and all derived units are obtained by a direct one-to-one relation of base units or derived units; eg, one newton is the force required to accelerate one kilogram at the rate of one meter per second squared; one joule is the energy involved when a force of one newton is displaced one meter in the direction of the force; and one watt is the power that in one second gives rise to the production of energy of one joule.

The same simplified system of units can be used by the research scientist, the technician, the practicing engineer, and the layman.

The International System of Units

SI rests on seven base units, two supplementary units, and a number of derived units, some of which have special names. A list of these units is given in the introduction to this volume.

Base Units. *Meter.* The meter is the length equal to 1 650 763.73 wavelengths in vacuum of the radiation corresponding to the transition between the levels $2p_{10}$ and $5d_5$ of the krypton-86 atom.

This definition was adopted in 1960 by the 11th CGPM, superseding the old international prototype of a bar of platinum–iridium still kept at the International Bureau of Weights and Measures.

Kilogram. The kilogram is the unit of mass; it is equal to the mass of the international protype of the kilogram.

This international prototype, adopted by the 1st and 3rd CGPM in 1889 and 1901, is a particular cylinder of platinum–iridium kept at the International Bureau of Weights and Measures near Paris. It is the only base unit still defined by an artifact.

Second. The second is the duration of 9 192 631 770 periods of the radiation corresponding to the transition between the two hyperfine levels of the ground state of the cesium-133 atom.

This definition was adopted by the 13th CGPM in 1967 to replace previous definitions based on the mean solar day and, later, the tropical year.

Ampere. The ampere is that constant current which, if maintained in two straight, parallel conductors of infinite length, of negligible circular cross section, and placed 1 meter apart in vacuum, would produce between these conductors a force equal to 2×10^{-7} newton per meter of length.

This definition was adopted by the 9th CGPM in 1948. The electric units for current and resistance had been first introduced by the International Electric Congress in 1893. These international units were replaced officially by so-called absolute units by the 9th CGPM.

Kelvin. The kelvin unit of thermodynamic temperature, is the fraction $^1/_{273.16}$ of the thermodynamic temperature of the triple point of water [which is 0.01°C].

Before the 13th CGPM in 1967, when this definition was adopted, the unit was called the degree Kelvin (symbol °K, now K).

Mole. The mole is the amount of substance of a system that contains as many elementary entities as there are atoms in 0.012 kilogram of carbon-12.

When the mole is used, the elementary entities must be specified and may be atoms, molecules, ions, electrons, other particles, or specified groups of such particles.

This definition was adopted by the 14th CGPM in 1971. Previously, physicists and chemists had based the amount of substance, then called gram–atom or gram–molecule, on the atomic weight of oxygen (by general agreement taken as 16), but with slight differences depending on the isotope used. The 1971 agreement assigned the value of 12 to the isotope 12 of carbon to give a unified scale. At its 1980 meeting, the International Committee for Weights and Measures (CIPM), under the authority of the CGPM, specified that in this definition "it is understood that unbound atoms of carbon-12, at rest and in their ground state, are referred to."

Candela. The candela is the luminous intensity, in a given direction, of a source that emits monochromatic radiation of frequency 540×10^{12} hertz and that has a radiant intensity in that direction of $^1/_{683}$ watt per steradian.

This unit, most recently defined by the 16th CGPM in 1979, replaced the candle and, later, the new candle and a definition of the candela based on the luminous intensity of a specified projected area of a blackbody emitter at the temperature of freezing platinum.

Supplementary Units. *Radian.* The radian is the plane angle between two radii of a circle that cut off on the circumference an arc equal in length to the radius.

Steradian. The steradian is the solid angle which, having its vertex in the center of a sphere, cuts off an area of the surface of the sphere equal to that of a square with sides of length equal to the radius of the sphere.

When these two units were first introduced as part of SI by the 11th CGPM, there was a question as to whether they should be called base units or derived units, and they were assigned to a class called supplementary units. In 1980, the CIPM specified that in SI the quantities plane angle and solid angle should be considered as dimensionless derived quantities. Therefore, the supplementary units radian and steradian are to be regarded as dimensionless derived units which may be used or omitted in the expressions for derived units.

Derived Units. The largest class of SI units, the derived units, consists of a combination of base, supplementary, and other derived units according to the algebraic relations linking the corresponding quantities. When two or more units expressed in base or supplementary units are multiplied or divided to obtain derived quantities, the result is a unit value. The fact that no numerical constant is introduced maintains this coherent system. Special names have been given to 19 derived units. For example, the joule is the name given to the product of a newton and a meter; the siemens is the name given to the quotient of an ampere divided by a volt. A list of derived SI units is given in the introduction to this volume (pp. xiv–xvi). The SI units with special names and their definitions are given in Table 1.

Prefixes. In SI, 16 prefixes are used and are directly attached to form decimal multiples and submultiples of the units (see the introduction to this volume). Prefixes indicate the order of magnitude, thus eliminating nonsignificant digits and providing an alternative to powers of ten; eg, 45 300 kPa becomes 45.3 MPa and 0.0043 m becomes 4.3 mm.

Preferably, the prefix should be selected in such a way that the resulting value lies between 0.1 and 1000. To minimize variety, it is recommended that prefixes representing 1000 raised to an integral power be used. For example, lengths can be expressed in micrometers, millimeters, meters, or kilometers and still meet the 0.1-to-1000 limits. There are three exceptions to these rules:

In expressing area and volume, the intermediate prefixes may be required; eg, hm^2, dL, and cm^3.

In tables of values, for comparison purposes, it is generally preferable to use the same multiple throughout, and one particular multiple is also used in some applications. For example, millimeter is used for linear dimensions in mechanical engineering drawings even when the values are far outside the range 0.1 to 1000 mm.

The centimeter is often used for body-related measurements, eg, clothing.

Compound Units. It is recommended that only one prefix be used in forming a multiple of a compound unit, and that normally it should be attached to the numerator. An exception is the base unit kilogram when it appears in the denominator. Multiples of kilogram are formed by attaching the prefix to the word gram (g). Compound prefixes are not used; eg, 1 pF is correct, not 1 $\mu\mu$F.

Units Used with SI. A number of non-SI units are used in SI (see Table 2).

Time. Although the SI unit of time is the second, the minute, hour, day, and other calendar units may be necessary where time relates to calendar cycles. Automobile velocity is, for example, expressed in kilometers per hour.

Plane Angle. The radian, although the preferred SI unit, is not always convenient, and the use of the degree is permissible. Minute and second should be reserved for special fields such as cartography.

Area. The hectare (ha) is a special name for the square hectometer (10 000 m^2 or hm^2) and is used for large land or water areas.

Volume. The special name liter (L) has been approved for the cubic decimeter, but its use is restricted to volumetric capacity, dry measure, and measure of fluids (both gases and liquids).

Mass. The metric ton (symbol t), equal to 1000 kg, is used widely in commerce, although the megagram (Mg) is the appropriate SI unit.

Units Used Temporarily with SI. Additional non-SI units are used with SI units until the CIPM considers their use no longer necessary (see Table 3).

Energy. The kilowatthour (kWh) is widely used as a measure of electric energy, but it should eventually be replaced by the megajoule (MJ) (1 kWh = 3.6 MJ).

Pressure. Although both bar and torr are widely used for pressure, this use is strongly discouraged in favor of the pascal and its multiples. The millibar is widely used in meteorology (1 mbar = 100 Pa).

Radiation Units. Units in use for activity of a radionuclide, ie, the curie (exposure to x and gamma rays), the roentgen, and the rad (absorbed dose), should eventually be replaced by the becquerel (Bq), coulomb per kilogram (C/kg), and gray (Gy), respectively.

Units to Be Abandoned. Except for the non-SI units referred to in the two preceding sections, a great many other metric units should be avoided in order to maintain the advantages of using one common coherent system of units, eg, units of the cgs system with special names such as the erg, dyne, poise, stokes, gauss, oersted, maxwell, stilb, phot, and angstrom. Other unit names to be deprecated are the kilogram-force, calorie, torr, millimeter of mercury, and the mho (see also pp. xxv–xxvi).

Mass, Force, and Weight. In SI, the basic unit for mass is the kilogram, and the basic unit for force is the newton. The mass of an object is constant and does not change with the gravitational field or acceleration of gravity. Force of gravity, however, depends on the gravitational field or the acceleration of gravity in accordance with Newton's law $F = mg$. Acting under normal Earth's gravitational pull, a mass of 1 kg at rest exerts a force of approximately 9.8 N.

There is considerable confusion in the use of the term weight as a quantity to mean either force or mass. In commercial and everyday use, the term weight nearly always means mass. In science and technology, however, the term weight of a body has usually meant the force that, if applied to the body, would give it an acceleration equal to the local acceleration of free fall (symbol g). For this reason, it is best to avoid the term weight. First, it should be determined whether mass or force is intended, and then kilogram is used for mass or newton for force.

Temperature. The kelvin is the SI unit of thermodynamic temperature, and is generally used in scientific calculations. Wide use is made of the degree Celsius (°C) for both temperature and temperature interval. The temperature interval 1°C equals 1 K exactly. Celsius temperature (t) is related to thermodynamic temperature (T) by the equation: $t = T - 273.15$. The name degree centigrade was dropped in 1948 in favor of the degree Celsius because in some countries the grade has been used as a unit of angular measure.

Pressure and Vacuum. Pressure is usually designated as gauge pressure, absolute pressure, or, if below ambient, vacuum. Pressures are expressed in pascals with appropriate prefixes. When the term vacuum is used, it should be made clear whether negative gauge pressure or absolute pressure is meant. The correct way to express

Table 1. SI Derived Units with Special Names

Quantity	Name	Symbol	Formula	Definition
absorbed dose	gray[a]	Gy	J/kg	the absorbed dose when the energy per unit mass imparted to matter by ionizing radiation is one joule per kilogram
activity	becquerel	Bq	1/s	the activity of a radionuclide decaying at the rate of one spontaneous nuclear transition per second
Celsius temperature	degree Celsius	°C		equal to the kelvin and used in place of the kelvin for expressing Celsius temperature, t, defined by the equation $t = T - T_0$, where T is the thermodynamic temperature and $T_0 = 273.15$ K by definition
dose equivalent	sievert	Sv	J/kg	the dose equivalent when the absorbed dose of ionizing radiation multiplied by the dimensionless factors Q (quality factor) and N (product of any other multiplying factors) stipulated by the International Commission on Radiological Protection is one joule per kilogram
electric capacitance	farad	F	C/V	the capacitance of a capacitor between the plates of which there appears a difference of potential of one volt when it is charged by a quantity of electricity equal to one coulomb
electric conductance	siemens	S	A/V	the electric conductance of a conductor in which a current of one ampere is produced by an electric potential difference of one volt
electric inductance	henry	H	Wb/A	the inductance of a closed circuit in which an electromotive force of one volt is produced when the electric current in the circuit varies uniformly at a rate of one ampere per second

Quantity	Unit	Symbol	Expression in terms of other units	Definition
electric potential, potential difference, electromotive force	volt	V	W/A	the difference of electric potential between two points of a conductor carrying a constant current of one ampere, when the power dissipated between these points is equal to one watt
electric resistance	ohm	Ω	V/A	the electric resistance between two points of a conductor when a constant difference of potential of one volt, applied between these two points, produces in this conductor a current of one ampere, this conductor not being the source of any electromotive force
energy, work, quantity of heat	joule	J	N·m	the work done when the point of application of a force of one newton is displaced a distance of one meter in the direction of the force
force	newton	N	$kg{\cdot}m/s^2$	that force which, when applied to a body having a mass of one kilogram, gives it an acceleration of one meter per second squared
frequency	hertz	Hz	1/s	the frequency of a periodic phenomenon of which the period is one second
illuminance	lux	lx	lm/m^2	the illuminance produced by a luminous flux of one lumen uniformly distributed over a surface of one square meter
luminous flux	lumen	lm	cd·sr	the luminous flux emitted in a solid angle of one steradian by a point source having a uniform intensity of one candela
magnetic flux	weber	Wb	V·s	the magnetic flux which, linking a circuit of one turn, produces in it an electromotive force of one volt as it is reduced to zero at a uniform rate in one second
magnetic flux density	tesla	T	Wb/m^2	the magnetic flux density given by a magnetic flux of one weber per square meter
power, radiant flux	watt	W	J/s	the power which gives rise to the production of energy at the rate of one joule per second
pressure or stress	pascal	Pa	N/m^2	the pressure or stress of one newton per square meter
quantity of electricity, electric charge	coulomb	C	A·s	the quantity of electricity transported in one second by a current of one ampere

497

[a] The gray is also used for the ionizing radiation quantities specific energy imparted, kerma, and absorbed dose index, which have the SI unit joule per kilogram.

Table 2. Units in Use with SI

Unit	Symbol	Value in SI unit
minute	min	1 min = 60 s
hour	h	1 h = 60 min = 3600 s
day	d	1 d = 24 h = 86 400 s
degree	°	$1° = (\pi/180)$ rad
minute	′	$1' = (1/60)°$
		$= (\pi/10\ 800)$ rad
second	″	$1'' = (1/60)'$
		$= (\pi/648\ 000)$ rad
hectare	ha	$1\ \text{ha} = 1\ \text{hm}^2 = 10^4\ \text{m}^2$
liter	L	$1\ \text{L} = 1\ \text{dm}^3 = 10^{-3}\ \text{m}^3$
metric ton	t	$1\ \text{t} = 10^3\ \text{kg}$

Table 3. Units Temporarily in Use with SI

Unit	Symbol	Value in SI unit
kilowatthour	kWh	1 kWh = 3.6 MJ
barn	b	$1\ \text{b} = 10^{-28}\ \text{m}^2$
bar	bar	$1\ \text{bar} = 10^5\ \text{Pa}$
curie	Ci	$1\ \text{Ci} = 3.7 \times 10^{10}\ \text{Bq}$
roentgen	R	$1\ \text{R} = 2.58 \times 10^{-4}\ \text{C/kg}$
rad	rd	1 rd = 0.01 Gy

pressure readings is "at a gauge pressure of 13 kPa" or "at an absolute pressure of 13 kPa."

Energy and Torque. The derived unit for energy (or work) is the joule (J), which is the special name given to 1 N·m of energy; 1 N·m is the work done when the point of application of 1 N of force is displaced through a distance of 1 m in the direction of the force. The unit for torque or bending moment is also the product of force and length, but, when considered vectorially, the concepts are quite different because of the different orientations of force and length. It is important to recognize the difference in using torque and energy, and the joule should never be used to express torque.

Nominal Dimensions. Some dimensions do not have an SI equivalent because their values are nominal, that is, a value is assigned for the purpose of convenient designation. For example, a 1-in. pipe has no dimension that is 25.4 mm. Another common example is the 2-by-4 piece of lumber, which is considerably smaller than 50.8 by 101.6 mm in its finished form.

Dimensionless Quantities. Certain quantities, eg, refractive index and relative density (formerly specific gravity), are expressed by pure numbers. In these cases, the corresponding SI unit is the ratio of the same two SI units, which cancel each other, leaving a dimensionless unit. Units for dimensionless quantities such as percent and parts per million (ppm) may also be used with SI; in the latter case, it is important to indicate whether the parts per million are by volume or by mass.

Density and Relative Density. Density is mass per unit volume and in SI is normally expressed as kilograms per cubic meter (density of water = 1000 kg/m^3 or 1 g/cm^3). The term specific gravity was formerly the accepted dimensionless value describing the ratio of the density of solids and liquids to the density of water at 4°C or for gases to the density of air at standard conditions. The term specific gravity is not an accepted unit in SI, and in some circles it is being replaced by relative density, a more descriptive term.

Style and Usage. If the advantages of SI are to be realized, everyone must use the system in the same manner. Listed below are a number of editorial rules that must be followed:

1. SI symbols are always in roman type, not italics.

2. A space is required between the number and the unit, eg, 150 mm, not 150mm. However, no space is needed between the number and °C, eg, 25°C.

3. A period is not required after a symbol unless the symbol is at the end of a sentence.

4. The plural form of a symbol is the same as the singular. Plurals of unit names are formed by adding an "s," except in henries; hertz, lux, and siemens are not changed.

5. E, P, T, G, and M, the prefixes for 10^6 and above, are capitalized, as are the symbols whose unit names have been derived from proper names, eg, N for newton (Sir Isaac Newton) and Pa for pascal (Blaise Pascal); an exception is the use of L for liter.

6. The product of two or more symbols is indicated by a centered dot and the product of unit names preferably by just a space, eg, N·m for newton meter.

7. A solidus indicates the quotient of two unit symbols and per the division of two unit names: m/s for meter per second. The horizontal line or negative powers are also permissible, eg,

$$\frac{m}{s}$$

or $m \cdot s^{-1}$. The solidus or per is not repeated in the same expression, eg, acceleration as m/s^2 for meter per second squared and thermal conductivity as $W/(m \cdot K)$ for watt per meter kelvin.

8. A prefix is not used in the denominator of a compound unit (except for kg, which is a base unit), eg, V/m, not mV/mm, and MJ/kg, not kJ/g.

9. An exponent attached to a symbol containing a prefix indicates that the multiple of the unit is raised to the power expressed by the exponent, eg, $1 \text{ cm}^3 = (10^{-2} \text{ m})^3 = 10^{-6} \text{ m}^3$.

10. Compound prefixes are not used, eg, pF, not $\mu\mu$F.

11. Since the comma is used as a decimal marker in many countries, a comma should not be used to separate groups of digits. The digits can be separated into groups of three to the left and right of the decimal point, and a space separates the groups, eg, 1 234 567 or 0.123 456. If there are only four digits, the space can be deleted; eg, 1.1234.

12. Because of the difference in the meaning of the word billion in the United States and most other countries, this term must be avoided; the prefix giga is unambiguous.

13. When using powers with a unit name, the modifier squared or cubed is used after the unit name, except for areas and volumes, eg, second squared, gram cubed, but square millimeter, cubic meter.

Pronunciation. Pronunciation of most unit names follows the normal rules, but a number have been pronounced in various ways. The slang expression kilo for kilogram is not acceptable.

Conversion and Rounding. Conversion of quantities should be handled with careful regard to the implied correspondence between the accuracy of the data and the number of digits. In all soft conversions, the number of significant digits retained should be such that accuracy is neither sacrificed nor exaggerated. Following are some examples:

A length is reported as 75 ft. The exact metric conversion is 22.86 m. If the reported length is a value rounded to the nearest 1 ft, it would be more appropriate to round the metric value to the nearest 0.1 m, ie, 22.9 m. If the 75-ft length, however, was rounded to the nearest 5 ft, then the appropriate rounding would be to the nearest 1 m, or 23 m.

Atmospheric pressure at sea level is nominally 14.7 lbf/in.^{2} (psi). The standard atmosphere has been defined by the CGPM as exactly 101.325 kPa. Since the 14.7-lbf/in.^{2} value is a nominal value, 101 kPa might seem to be a good number. However, because of the relatively limited range of standard atmospheric pressure, 101.3 kPa would probably be a better choice for the nominal value.

Significant Digits. Any digit that is necessary to define the specific value or quantity is said to be significant. A problem arises, however, when a value of, eg, 4 in. is given. This may be intended to represent 4, 4.0, 4.00, 4.000 or even more accuracy with a corresponding increase in significant digits (equivalent to 102, 101.6, 101.60, and 101.600 mm, respectively).

Tolerances. *Linear units.* The following procedure is used for converting linear units to the proper number of significant places: The maximum and minimum limits in inches are calculated. The corresponding two values are converted exactly into millimeters by multiplying each by the conversion factor 1 in. = 25.4 mm. The results are rounded in accordance with Table 4.

Temperature. General guidance for converting tolerances from degrees Fahrenheit to kelvins or degree Celsius is given in Table 5.

Pressure or Stress. Values with an uncertainty of more than 2% may be converted without rounding by using the approximate factor 1 lbf/in.^{2} = 7 kPa.

Conversion Factors. Excellent tables of conversion factors are given by ASTM Standard E 380-82 for Metric Practice, where the conversion factors are listed both alphabetically and classified by physical quantity.

The conversion factors are presented for ready adaptation to computer readout and electronic data transmission. The factors are written as a number equal to or greater than one and less than ten with six or less decimal places. The number is fol-

Table 4.   Rounding of Linear Units

Original tolerance, inches		Fineness of rounding, mm
at least	less than	
0.000 04	0.0004	0.0001
0.000 4	0.004	0.001
0.004	0.04	0.01
0.04	0.4	0.1
0.4		1

Table 5.   Temperature Conversion Tolerances

°F	K or °C
2 ± 1	1 ± 0.5
4 ± 2	2 ± 1
10 ± 5	6 ± 3
20 ± 10	11 ± 5.5
30 ± 15	17 ± 8.5
40 ± 20	22 ± 11
50 ± 25	28 ± 14

lowed by E (for exponent), a plus or minus symbol, and two digits which indicate the power of ten by which the number must be multiplied to obtain the correct value. For example:

$$3.523\ 907\ \text{E-02} = 3.523\ 907 \times 10^{-2} = 0.035\ 239\ 07$$

An asterisk (*) after the sixth decimal place indicates that the conversion factor is exact and that all subsequent digits are zero. Where less than six decimal places are shown, more precision is not warranted.

The conversion factors for other compound units not listed can easily be generated from numbers given in the alphabetical list by the substitution of the converted units; eg, to find the conversion factor from lb·ft/s to kg·m/s:

$$1\ \text{lb} = 0.453\ 592\ 4\ \text{kg}$$

$$1\ \text{ft} = 0.3048\ \text{m (exactly)}$$

Substituting,

$$(0.453\ 592\ 4\ \text{kg}) \times (0.3048\ \text{m})/\text{s} = 0.138\ 255\ 0\ \text{kg·m/s}$$

Thus, the factor is 1.382 550 E-01.

To find the conversion factor from oz·in.^{2} to kg·m^2,

$$1\ \text{oz} = 0.028\ 349\ 52\ \text{kg}$$

$$1\ \text{in.}^2 = 0.000\ 645\ 16\ \text{m}^2\ \text{(exactly)}$$

Substituting,

$$(0.028\ 349\ 52\ \text{kg}) \times (0.000\ 645\ 16\ \text{m}^2) = 0.000\ 018\ 289\ 98\ \text{kg·m}^2$$

Thus, the factor is 1.828 998 E-05.

BIBLIOGRAPHY

"Units" in *ECT* 2nd ed., Supplement Volume, pp. 984–1007, by M. L. McGlashan, The University, Exeter, UK.

General References

Standard for Metric Practice E 380-82, ASTM, Philadelphia, Pa., 1982.
The International System of Units (SI), NBS Special Publication 330, Superintendent of Documents, U.S. Government Printing Office, Washington, D.C., 1981.

Canadian Metric Practice Guide, CAN3-Z234.1-79, Canadian Standards Association, Rexdale, Ontario, Canada, 1979.

SI Units and Recommendations for the Use of Their Multiples and of Certain Other Units, ISO/1000-81, available through the American National Standards Institute, New York, 1981.

Metric Editorial Guide, 3rd ed., American National Metric Council, Bethesda, Md., 1981.

L. D. Pedde and co-workers, *Metric Manual*, U.S. Department of the Interior, Bureau of Reclamation, Superintendent of Documents, U.S. Government Printing Office, Washington, D.C., 1978.

ROBERT P. LUKENS
American Society for Testing and Materials

UNSATURATED POLYESTERS.   See Polyesters, unsaturated.

URANIUM AND URANIUM COMPOUNDS

Uranium [*7440-61-1*], U, at no. 92, at wt 238.03, is a member of the actinide series of transition elements (see Actinides and transactinides).

Klaproth, the first professor of chemistry at the University of Berlin, discovered the element in 1789 in a heavy black mineral, *Pechblende* (pitchblende), which is found in Bohemia at St. Joachimsthal (1). He named the element, which he believed he had isolated, Uranit after the planet Uranus, which had just been discovered.

In 1841, the French chemist Péligot, professor of analytical chemistry and glassmaking and director of assays at the Paris mint, questioned the elemental composition of Klaproth's uranit (2–3) and obtained the true elemental uranium as a black powder. Mendeleev corrected Péligot's estimate of the atomic weight and thus gave the element its proper classification.

In 1896, Becquerel discovered that uranium was radioactive (4). A few years later, the Curies isolated the elements polonium and radium from uranium ore.

Although radium turned out to be of great importance in medicine, uranium, which was obtained as a by-product in radium processing, found little practical application; it was used in limited quantities as a coloring agent for glass and ceramics.

The work of Becquerel (4) led rapidly to a basic understanding of the phenomenon of radioactivity, which is the emission of both particles and energy from the nucleus of the atom. In 1934, Fermi produced trace amounts of new radioactive elements by bombarding uranium nuclei with neutrons (5). In Dec. 1938, Hahn and Strassmann discovered nuclear fission (6). In studying the products of the neutron irradiation of uranium, they found by purely chemical techniques that one of the irradiation products, believed to be radium, could be separated from genuine radium but not from barium. Hahn and Strassmann concluded that this irradiation product must indeed be barium with a mass much lower than that of uranium, and since it could have been formed only by splitting the uranium nucleus, it must thus be a fission product. After the uranium nucleus absorbs the neutron, the nucleus splits into two unequal fragments with the emission of more neutrons, which are then available to split additional uranium nuclei. Thus, the elements of a chain reaction are available. Work on an actual chain reaction began at Columbia University in New York and was moved to the University of Chicago in 1942. On Dec. 2, 1942, this work under Fermi created the first chain reactor (7). In the following year, the government started the construction of reactors for the neutron bombardment of ^{238}U to produce ^{239}Np, which in turn decays by β-emission to form ^{239}Pu. Many initial difficulties were encountered, but by 1945, production was large and dependable.

In 1939, ^{235}U, the fissionable natural isotope, was discovered (8). This discovery prompted efforts to separate this isotope from the more abundant nonfissionable ^{238}U, and several separation methods have been developed to the point of industrial plant operations, eg, the gaseous-diffusion process, the centrifuge process, and the electromagnetic-separation process. Other processes were developed to the pilot-plant scale (9) (see Diffusion separation methods).

The uranium isotope separation technology as well as the nuclear reactor technology requires feed materials of highest purity. Purification methods had to be developed for this type of application which presented challenges of a type then unknown to the mining and mineral processing industries (10–11). The importance of uranium as a feed material for the generation of nuclear energy caused a search for new supplies during the late 1940s and 1950s that was unparalleled in mining and metallurgical history (12–14) (see also Nuclear reactors).

By the late 1950s, it became evident that the U.S. government requirements for uranium as well as world requirements could be met by existing mining and processing facilities. Many domestic contracts for the purchase of uranium concentrates by the Atomic Energy Commission (AEC, now the NRC) were scheduled to terminate in 1966, and it appeared that the consequent slack could not be promptly absorbed by private contracts for uranium to be used for nuclear power generation. There was a danger in shutting down a substantial portion of the nation's uranium mines where flooding might cause irreparable damage, and it was felt that it would not be in the best interest of the national defense to permit these mines to become idle for an indefinite period. Accordingly, a stretch-out provision was offered to all producers of uranium concentrates who held AEC contracts. This action provided relief to the uranium-mining industry during the period of transition from the completely government-dominated market to a market served by the privately owned nuclear power industry (15).

There can be little doubt that the demand for uranium as a primary source of power will continue to increase. Figures 1 and 2 and Table 1 give predictions of civilian nuclear-power growth up to 2025 expressed in the uranium requirements for different fuel-cycle strategies (16). However, adverse public reaction to the accident at Three Mile Island and growing concern over nuclear-accident prevention and excessive reactor costs have greatly slowed orders for new nuclear power plants in the United States.

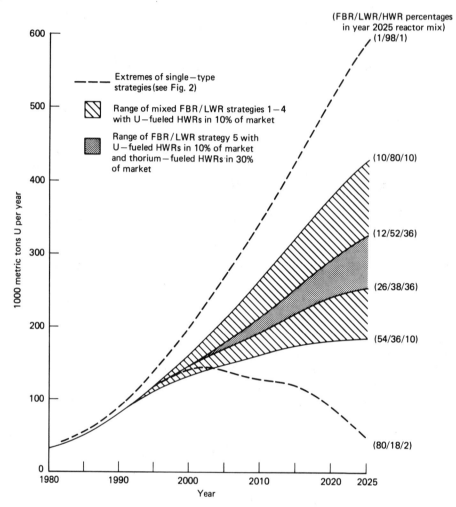

Figure 1. High growth projection of annual natural-uranium requirements in noncommunist world for illustrative mixed strategies (16). The numerals listed at the right of the graph indicate reactor-mix percentages; for example, (12/52/36) = 12% fast-breeder reactor (FBR), 52% light-water reactor (LWR), and 36% heavy-water reactor (HWR) in total strategy.

Occurrence

Uranium is present in the earth's crust at ca 2 ppm. It is thus more abundant than, for instance, Cd, Ag, Hg, or Bi. Significant concentrations have been found in rocks, ocean water, and extraterrestrial materials such as lunar rocks or meteorites. Typical estimates of the concentrations of uranium in various types of geological formations or other matrix materials are given in Table 2.

Acidic rocks with a high silicate content, such as granite, have a uranium content above average, whereas the contents of basic rocks such as basalts are lower than the average. Also below average are sedimentary rocks, with some important exceptions. However, 90% of the world's known uranium resources are contained in conglomerates . and in sandstone. Such sources as marine black shales, phosphate rocks, lignites, coal, and seawater contain significant quantities of uranium, but not enough for economic

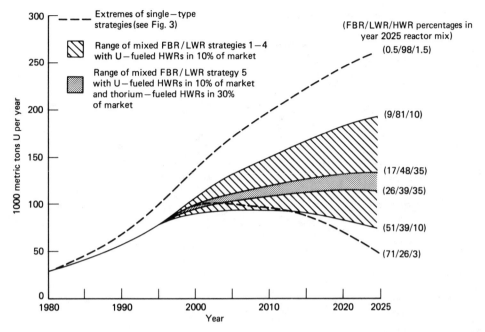

Figure 2. Low growth projection of annual natural-uranium requirements in noncommunist world for illustrative mixed strategies (16). For explanations of numerals at right of graph, see Figure 1.

Table 1. Annual Natural Uranium Requirements for Fuel-Cycle Strategies for Noncommunist World[a], Thousand Metric Tons

Reactor strategy[b]	Nuclear power growth	Year 1980	Year 2000	Year 2025
once-through LWR[c]	high	32	175–200	430–590
	low	28	120–135	190–260
once-through HWR[d]	high	32	170	360–480
	low	28	115	160–220
large-scale FBR[e]	high	32	145	50–240
	low	28	100	50–70
LWR with recycle	high	32	125–140	320–420
	low	28	85–95	140–180
HWR with recycle	high	32	160–175	290
	low	28	110–115	130

[a] Ref. 16.
[b] Tails assay: 0.2% ^{235}U. For additional mixed strategies, see ref. 16.
[c] Light-water reactor.
[d] Heavy-water reactor.
[e] Fast-breeder reactor.

recovery. Recently, attempts have been made to recover uranium from seawater. The ashes of lignite or coal contain uranium.

Uranium is an important constituent of ca 155 minerals; in another 60 minerals, it is a minor constituent or an impurity.

Primary uranium minerals have crystallized from low-melting rocks, which in turn were formed during the last stage of magma solidification. Rocks of this type are

Table 2.   Uranium Occurrence[a]

Occurrence	Ppm	Ref.
igneous rocks		
basalts	0.6	18
granites, normal	4.8	18
ultrabasic rocks	0.03	19
sandstones, shales, limestones	1.2–1.3	20
earth's crust	2.1	18
oceanic	0.64	18
continental	2.8	18
earth's mantle	ca 0.01	18
seawater	0.002–0.003	20
meteorites	0.05	18
chondrites	0.011	18
uraniferous materials		
high grade veins	$(3–8.5) \times 10^5$	
vein ores	$(2–10) \times 10^3$	20
sandstone ores	$(0.5–4) \times 10^3$	20
gold ores[b]	150–600	20
uraniferous phosphates	50–300	20
Chattanooga shales[c]	60	20
uraniferous granites	15–100	20

[a] Refs. 17–20.
[b] South Africa.
[c] United States.

rich in silicates (feldspar, quartz). The primary uranium minerals occur in hydrothermal veins and pegmatites and are usually associated with minerals of other rarer elements, such as Th, Nb, Ta, W, Be, Ti, Zr, and lanthanides. These elements behave in a manner similar to uranium, because their ionic radii prevent them from coprecipitating in the crystal lattice of the more basic, higher-melting plutonic rocks.

Secondary uranium minerals are produced by hydration, metathesis, or oxidation of the primary minerals. Liquid-phase transport of a primary or secondary mineral from its place of origin to a different location yields a secondary mineral. The secondary uranium minerals usually contain hexavalent uranium and are frequently hydrated.

In general, primary uranium minerals are black and contain uranium in a valence <6. Secondary minerals are yellow, greenish-yellow, bright green, or orange, and contain uranium in the valence +6. Silicates, phosphates, arsenates, sulfates, selenites, carbonates, and vanadates are the compounds most frequently encountered. Selected representatives of uranium minerals are given in Table 3.

Anhydrous Oxides. Uraninite and pitchblende differ in physical form; their chemical composition ranges from $UO_{2.0}$ to $UO_{2.67}$. Uraninite (also called ulrichite) is close to $UO_{2.0}$ in its composition, whereas pitchblende usually has a higher oxygen content (25). Uraninite is well crystallized; pitchblende either is amorphous or consists of very fine crystals. An amorphous powdery form, called uranium black, is primarily a black chalky powder found on the weathered surfaces of pitchblende. Depending on the ratio of U:O and on the presence of impurities or alteration products, the uranium content of these oxide minerals ranges from 60 to 92%. Related oxide minerals are cleveite [12197-63-9] and brannerite, which contain, in addition to uranium oxide,

Table 3.   Selected Uranium Minerals[a]

Mineral	CAS Registry No.	Chemical composition[b]
uraninite	[1317-99-3]	$UO_2-UO_{2.67}$
pitchblende[c]	[1317-75-5]	$UO_2-UO_{2.67}$ (amorphous)
euxenite	[1317-53-9]	$(Y,U,Ca,Th,Ce)(Nb,Ta,Ti)_2O_6$
polycrase	[12174-27-9]	
fergusonite	[12173-29-4]	$(Y,Er,Ce,U,Fe)(Nb,Ta,Ti)O_4$
samarskite	[1317-81-3]	$(Y,U,Ca,Th,Fe,Ce,Pb)(Nb,Ta,Ti,Sn)_2O_6$
pyrochlore	[12174-36-6]	$(Na,Ca,U)_2(Ta,Nb,Ti)_2O_6(O,OH,F)$
microlite	[12173-96-5]	
davidite	[12173-20-5]	$(Fe,Ce,U,Y,Ca)_6(Ti,Fe,V,Cr)_{15}(O,OH)_{36}$
brannerite	[12197-37-4]	$(U,Ca,Fe,Th,Y)(Ti,Fe)_2O_6$
carnotite	[60182-49-2]	$K_2(UO_2)_2(VO_4)_2.3H_2O$
autunite	[16390-74-2]	$Ca(UO_2)_2(PO_4)_2.nH_2O, n = 8-12$
tyuyamunite	[12196-95-1]	$Ca(UO_2)_2(VO_4)_2.nH_2O, n = 4-10$
uranophane	[12195-76-5]	$Ca(UO_2)_2So_2O_7.6H_2O$
torbernite	[26283-21-6]	$Cu(UO_2)_2(PO_4)_2.nH_2O, n = 8-12$
coffinite	[14485-40-6]	$U(SiO_4)_{1-x}(OH)_{4x}$
thucholite	[12178-52-8]	uranium oxide and hydrocarbons

[a] Refs. 21–24.

[b] In some double oxides, further substitution with other elements is possible.

[c] A variety of uraninites.

various amounts of thorium and rare-earth oxides. Pitchblende deposits are found all over the world.

Hydrous Oxides.   Gummite [12326-21-5], $UO_3.nH_2O$, and becquerelite [12326-20-4], $2UO_3.3H_2O$, are typical hydrated oxides. They are fairly common, but not important commercially.

Carnotite, $K_2O.2UO_3.V_2O_5.3H_2O$, and tyuyamunite, $CaO.2UO_2.V_2O_5.8H_2O$, are of commercial importance. These minerals are soft with a Mohs hardness of 2–2.5; the specific gravity is ca 3–3.5. They are usually found in the weathering zone of sedimentary rocks enriched with organic residues. Other uranium minerals associated with hydrated oxides are the hydrated calcium uranium silicates, uranophane, $CaO.2UO_3.2SiO_2.6H_2O$, and coffinite, a hydrated silicate, $U(SiO_4)_{1-x}(OH)_{4x}$.

The hydrated oxide uranium minerals are important constituents of the ores in the Colorado Plateau, Wyoming, Utah, and Arizona, and in Katanga, Zaire. At Radium Hill in South Australia, the predominant mineral is davidite, which is a complex hydrated oxide of the type $AB_3(O,OH)_7$, where A = Fe, lanthanides, U, Ca, Zr, Th, and B = Ti, Fe, V, Cr.

Phosphates and Arsenates.   The most important uranium phosphates are autunite, $Ca(UO_2PO_4)_2.8H_2O$, and torbernite, $Cu(UO_2PO_4)_2.12H_2O$. Autunite forms greenish-yellow platelets, and torbernite forms bright green tetragonal platelets. The minerals are soft with a Mohs hardness of 2–2.5; the specific gravity ranges from 3 to 3.6. These minerals are usually associated with the oxidation zones of pegmatites and certain hydrothermal deposits (26–27). They are of some commercial significance in Cornwall, Madagascar, and France. Parsonsite [12137-57-4], $2PbO.UO_3.P_2O_5.H_2O$, is another lead uranium phosphate mineral; uranocircite [12196-92-8], $Ba(UO_2PO_4)_2.8H_2O$, is found in Rosmaneira, Spain. Monazite, $CePO_4$, and xenotime, YPO_4, are rare-earth phosphate minerals containing up to 10% Th and U; they are found in Brazil.

Uranium arsenate minerals found in European deposits are abernathyite [*12005-93-5*], $KUO_2AsO_4.(2-6)H_2O$, novacekite [*12255-29-1*], $Mg(UO_2AsO_4)_2.(0-12)H_2O$, uranospinite [*12255-17-9*], $Ca(UO_2AsO_4)_2.(0-12)H_2O$, heinrichite [*12255-15-1*], $Ba(UO_2AsO_4)_2.(0-12)H_2O$, kahlerite [*12255-23-1*], $Fe(UO_2AsO_4)_2.(8-12)H_2O$, and zeunerite [*12255-21-9*], $Cu(UO_2AsO_4)_2.8H_2O$ (27).

Organic Complexes. In addition to the above-mentioned well-defined minerals, numerous organic uranium complexes are found in sedimentary deposits, eg, thucholite and carburan. They are distinguished from each other by their U:Th ratio. These minerals are found in small, irregularly rounded, asphaltic nodules embedded in feldspar, quartz, or mica, often associated with uraninite and cryolite in pegmatitic veins.

Resources. Uranium resources are classified on the basis of geologic knowledge and the cost of uranium recovery (16) (see Tables 4 and 5). They are further separated into two groups based on the cost of their exploitation. The cost ranges suggested by the International Nuclear Fuel Cycle Evaluation Conference (INFCE) are <$80/kg U and $80–130/kg U. Cost bases include the direct costs of mining and processing as well as the capital costs for providing and maintaining the production unit. Past exploration costs are generally not included, nor is there any return on risk capital, ie, profit.

Reasonable assumed resources (*RAR*) contain deposits of size, grade, and configuration such that recovery is within the given production cost ranges with current mining and processing technology.

This type of resource, recoverable at ca $80/kg U, is estimated at 1.8×10^6 t. More than 84% is found in Australia, Canada, Namibia, Niger, South Africa, and the United States.

Estimated additional resources (*EAR*) is a term that applies to resources that occur as extensions of well-known deposits, little-explored deposits, or undiscovered deposits believed to exist along a well-defined geological trend with known deposits. In the countries mentioned above, the EAR are estimated at 1.5×10^6 t. The RAR exploitable at $130/kg U is believed to amount to 2.6×10^6 t, the EAR at this cost to 2.4×10^6 t.

High grade resources. Sandstone-type resources are found in the United States, Gabon, and Niger, usually in continental sediments near the boundary between oxidized and unoxidized zones. The uranium content is 0.04–0.2% uranium (0.05–0.25% U_3O_8). Higher-grade material may be present. Deposits range from a few hundred to $>6 \times 10^5$ t.

Precambrium quartz-pebble conglomerates host the large uranium deposits of the Elliott Lake area in Canada and the gold–uranium deposits in Witwatersrand, South Africa. Uranium grades and thicknesses are usually uniform and lend themselves to large and systematic mining operations. In Canada, the deposits yield only uranium, whereas in South Africa, uranium is a by-product of gold mining. Uranium values are 0.04–0.13% U at Elliott Lake and ≤0.03% in South Africa. Deposits may be as large as 75,000 t U.

In the Lake Athabasca region of Canada and the Alligator River region in Australia, deposits up to 1.5×10^5 t of uranium that can be mined by open-pit techniques have been found (see also Tar sands). Average grades are high, up to a few percent uranium.

Table 4. Uranium Resources by Continent[a,b], Thousand Metric Tons

Country	Reasonably assured		Estimated additional	
	$80/kg U	$130/kg U[c]	$80/kg U	$130/kg U[c]
North America	752	976	1145	1904
United States	531	708	773	1158
Canada	215	235	370	728
Mexico	6	6	2.4	2.4
Greenland	0	27	0	16
Africa	609	776	139	263
South Africa	247	391	54	139
Niger	160	160	53	53
Namibia	117	133	30	53
Algeria	28	28	0	5.5
Gabon	37	37	0	0
Republic of Central Africa	18	18	0	0
Zaire	1.8	1.8	1.7	1.7
Somalia	0	6.6	0	3.4
Egypt	0	0	0	5
Madagascar	0	0	0	2
Botswana	0	0.4	0	0
Australia	290	299	47	53
Europe	66	391	49	98
France	39.6	55.3	26.2	46.2
Spain	9.8	9.8	8.5	8.5
Portugal	6.7	8.2	2.5	2.5
Yugoslavia	4.5	6.5	5	20.5
UK	0	0	0	7.4
FRG	4	4.5	7	7.5
Italy	0	1.2	0	2
Austria	1.8	1.8	0	0
Sweden	0	301	0	3
Finland	0	2.7	0	0.5
Asia	40	46	1	24
India	29.8	29.8	0.9	23.7
Japan	7.7	7.7	0	0
Turkey	2.4	3.9	0	0
Republic of Korea	0	4.4	0	0
Philippines	0.3	0.3	0	0
South America	97	102	99	105
Brazil	74.2	74.2	90.1	90.1
Argentina	23	28.1	3.8	9.1
Chile	0	0	5.1	5.1
Bolivia	0	0	0	0.5
Total[d]	1850	2590	1480	2450

[a] Ref. 16.

[b] Noncommunist world.

[c] Includes resources at $80/kg U level.

[d] Rounded.

Table 5. World Uranium Resources by Deposit Type[a], Thousand Metric Tons

Deposit	Reasonably assured		Estimated additional	
	$80/kg U	$130/kg U[b]	$80/kg U	$130/kg U[b]
sandstone	812	1047	850	1250
quartz-pebble conglomerate	332	438	257	528
unconformity related	281	293	112	189
igneous and metamorphic	168	206	107	149
vein	180	202	149	310
black shale	0	304	0	0
other	81	100	5	20
Total[c]	*1854*	*2590*	*1480*	*2446*

[a] Ref. 16.

[b] Includes resources at $80/kg U.

[c] Rounded.

Another type of deposit is associated with igneous and metamorphic rocks such as granite pegmatites, carbonatites, alkali syenites, and mica schists, eg, the Rossing deposit in Namibia. Other deposits of this type are known in Greenland, Alaska, and Brazil. They range from 750 to 750,000 t and average 0.025–0.13% uranium.

Low grade resources. Before the sharp rise in both uranium demand and prices in 1973–1974, little attention had been given in the United States and other countries to low grade uranium resources. Further information on such ores, which have been studied thoroughly in the last few years, is available from the INFCE report (16).

Physical Properties

Uranium is a dense, lustrous metal resembling iron; it is ductile and malleable. In air, it tarnishes rapidly, and in a very short time, even a polished surface becomes coated with a dark-colored layer of oxide.

Uranium metal, in the solid state, exists in three allotropic modifications, which are distinguished by their crystallographic properties and their densities. The transformation temperatures and the enthalpies of transformation are given in Table 6. Densities and crystal-structure data are given in Table 7.

The thermodynamic properties of uranium metal have been determined with great accuracy and have been assessed in a recent publication of the International Atomic Energy Agency (IAEA) (28). Thermodynamic properties and free energy functions are compiled in Table 8.

Table 6. Phase-Transformation Temperatures and Enthalpies for Uranium Metal at 101.3 kPa (1 atm)

Transformation	Transformation temperature, °C	Enthalpy of transformation, J/(g·mol)[a]
$\alpha \rightarrow \beta$	667.8 ± 1.3	2791
$\beta \rightarrow \gamma$	774.9 ± 1.6	4757
$\gamma \rightarrow$ liquid	1132.4 ± 0.8	9142

[a] To convert J to cal, divide by 4.184.

Table 7. Crystallographic Properties and Densities of Uranium Modifications[a]

Modification	Reference temperature, °C	Crystal lattice	Unit cell dimensions, nm	Atoms per unit cell	Density, calculated, g/cm³
α	298	orthorhombic	$a = 0.28537$ $b = 0.58695$ $c = 0.49548$	4	19.07
β	720	tetragonal	$a = 1.0763 \pm 0.0005$ $c = 0.5652 \pm 0.0005$	30	18.11
γ	805	bcc	$a = 0.3524 \pm 0.0002$	2	18.06

[a] Ref. 31.

Table 8. Thermodynamic Properties of Uranium Metal[a]

Function or parameter	Value or equation
entropy of α-U, J/K[b]	50.21 ± 0.12
entropy of U(g), J/K[b]	199.6
heat capacity, J/(mol·K)[b]	
$\quad\alpha$-U	$26.92 - 2.5\,T + 2.656 \times 10^{-5}\,T^2 - 7.7 \times 10^4\,T^{-2}$
$\quad\beta$-U	42.42
$\quad\gamma$-U	38.28
$\quad$U(l) (liquid)	48.66
enthalpy of fusion, kJ/mol[b]	8.326 ± 0.54
enthalpy of sublimation, kJ/mol[b]	1062.73
enthalpy, J/mol[b]	$H(T) - H(298) = 48.66\,T - 10137.8$
free energy of vaporization, kJ/mol[b]	
$\quad$solid to gas	$G = 525.3 - 0.137$
$\quad$liquid to gas	$G = 487.6 \pm 0.11$
normal (extrapolated) boiling point, K	3818
vapor pressure of liquid[c]	$\log p\ (\text{kPa}) = -\dfrac{(25{,}230 \pm 370)}{T} + (7.72 \pm 0.17)$

Free energy functions, $-(G_T - H_{298})$

Temperature, K	Ideal gas, J/mol[b]	Condensed phase, J/mol[b]
298	199.66	50.21
600	204.35	55.995
1000	211.71	66.702
1600	220.41	83.566
2000	225.27	93.14
2500	230.79	102.995
3000	235.89	111.189

[a] Ref. 28.
[b] To convert J to cal, divide by 4.184.
[c] To convert kPa to mm Hg, multiply by 7.5.

Linear thermal-expansion coefficients for uranium have been measured in all crystallographic axes (see Table 9). Thermal-conductivity values are given below (10).

temperature, K	309	373	473	573	673
conductivity, W/(cm·K)	0.251	0.263	0.297	0.314	0.326

The resistivity of uranium has been found to be 29 $\mu\Omega$·cm (29). The directional

Table 9. Linear Thermal Expansion Coefficients along the Crystallographic Axes [a]

Temperature interval, °C	Thermal expansion coefficients $\times 10^6$ per °C		
	a_o axis	b_o axis	c_o axis
25–125	+21.7	−1.5	+23.2
25–325	+26.5	−2.4	+23.9
25–650	+36.7	−9.3	+34.2

[a] Ref. 10.

components of the resistivity tensor in α-uranium single crystals have been measured at 273 K (30).

Uranium is weakly paramagnetic (31–35). At 20°C, the magnetic susceptibility has been reported as $\chi = 1.740 \times 10^{-5}$ A/g and at 350°C, $\chi = 1.804 \times 10^{-5}$ A/g (A = 10 emu) (34).

The thermoelectric power of uranium increases monotonically from 2 μV/K at 50 K to 18 μV/K at the $\alpha \rightarrow \beta$ transformation temperature. Discontinuities are observed in the vicinity of the $\alpha \rightarrow \beta$ and the $\beta \rightarrow \gamma$ transformation temperatures (36).

The spectroscopic properties of uranium have been studied in great detail (37–39). In the latest analysis of the uranium spectra, more than 30,000 lines of the arc-and-spark emission spectrum have been measured and catalogued (40); a part could be assigned to transitions belonging to the U(I) and U(II) spectrum. In addition, x-ray spectra (41–43) and spectra of higher ionization states, up to U(VI), have been measured.

The low temperature fluorescence and absorption spectra were studied at the time of the Manhattan Project (38); the latest development of absorption spectroscopy has been summarized in the proceedings of the ACS symposium held in Washington, D.C., Sept. 1979 (39).

Numerous other physical properties of uranium, eg, elastic modulus, tensile strength, hardness, impact strength, creep, and fatigue properties, have been studied in detail (10,31–32,37,44).

Chemical Properties

Of the four oxidation states, +3, +4, +5, and +6, only the +4 and +6 states are stable enough to be of practical importance. Aqueous solutions of uranium(III) may be prepared, but they are readily oxidized to the +4 state with evolution of hydrogen. The +5 state disproportionates into the +4 and +6 states in the presence of water or hydrolytic compounds. The alternation possible between +4 and +6 states has economic significance. The highly stable and disseminated grains of uraninite in igneous rock formations are in the +4 state, but when altered to the +6 state, they are soluble enough to dissolve in circulating groundwater. The solubility of uranium in the +6 state accounts for its wide distribution in seawater, fresh water, and hydrothermal deposits. In aqueous media, the +6 state predominates.

The alternation between the +4 and +6 states is of importance in the extraction of uranium from ores and in purification. The ease of changing oxidation states and the marked amphoteric properties of uranium are the most important factors in the hydrometallurgy of uranium. The ability of uranium to form sulfato-anion complexes, eg, $UO_2(SO_4)_2^{2-}$, is of importance in solvent-extraction recovery processes from sulfate leach liquors.

In pyrochemical processes, such as the metallothermic reduction of uranium tetrafluoride to metal, in molten-salt or other high-temperature systems, the +3 and +4 states are predominant, and U(IV) compounds, eg, UO_2 and UF_4, play an important role.

Chemical and physicochemical properties of uranium are reviewed in refs. 45–48.

The Isotopes

There are fifteen known isotopes of uranium, not counting the isomeric states; three, ^{234}U, ^{235}U, and ^{238}U exist in nature (see Radioisotopes). All isotopes of uranium are instable and, as they decay, emit α or β particles. The most stable and most abundant isotope is ^{238}U. The composition of natural uranium has been measured with great precision and is within the ranges of natural abundance given below (49).

Mass number	Abundance, at %	
	Range	Best value
234	0.0059–0.0050	0.005 ± 0.001
235	0.7202–0.7198	0.720 ± 0.001
238	99.2752–99.2739	99.275 ± 0.002

From these abundances and the mass-spectroscopic isotopic masses, the atomic weight of the natural isotopic mixture is calculated to be 238.0289 ± 0.0001.

The natural isotope composition usually varies within a certain range. However, in the deposit of Oklo, Gabon, an abnormally low ^{235}U concentration has been measured (50). This phenomenon has been subject to much discussion, and it is now believed that, because of favorable conditions, the ore body at Oklo became critical to form a natural nuclear reactor, which burned up part of the primordial ^{235}U (50).

The nuclear properties of uranium isotopes are given in Table 10. Additional data can be found in ref. 51.

Besides ^{235}U, an artificial isotope, ^{233}U, was found to be fissionable. The first processes for its production and isolation were developed at the time of the Manhattan Project and are described in refs. 52–53. Today, uranium-233 is an industrial product and is produced in kilograms by means of the thermal breeding reaction

$$^{232}Th(n,\gamma)^{233}Th \xrightarrow{\beta^-} {}^{233}Pa \xrightarrow{\beta^-} {}^{233}U$$

The ^{233}U is isolated by a radiochemical hot-cell process (see below).

Since most of the commercially available thorium contains ^{230}Th, the isotope ^{232}U is produced as an unwanted by-product by this reaction

$$^{230}Th(n,\gamma)^{231}Th \xrightarrow{\beta^-} {}^{231}Pa(n,\gamma)^{232}Pa \xrightarrow{\beta^-} {}^{232}U$$

The presence of ^{232}U in ^{233}U impedes the handling of the isolated ^{233}U. Special techniques are employed to handle large quantities of ^{233}U and ^{235}U in order to avoid a critical excursion (54). Vessels with special geometry made of neutron-absorbing materials, eg, hafnium, are used with strict controls.

Table 10.   Isotopes of Uranium[a]

Isotope	CAS Registry No.	Half life	Mode of disintegration	Source
^{226}U	[36840-44-5]	0.5 s	α (7.43)[b]	^{232}Th$(\alpha,10n)$
^{227}U	[25724-70-3]	1.1 min	α (6.87)[b]	^{232}Th$(\alpha,9n)$
				^{231}Pa$(p,5n)$
^{228}U	[35788-49-9]	9.1 min	α ($\geq$95%) (6.685; 6.60)[b]	^{232}Th$(\alpha,8n)$
			EC[c] ($\leq$5%)	
^{229}U	[36840-46-7]	58 min	EC[c] (80%)	^{232}Th$(\alpha,7n)$
			α (20%) (6.360; 6.332; 6.297; 6.260;	
			6.223; 6.185)[b]	
^{230}U	[15743-51-8]	20.8 d	α (5.8885; 5.8177; 5.6672; 5.6626;	daughter ^{231}Pa
			5.586; 5.544; 5.534)[b]	
^{231}U	[15700-08-0]	4.2 d	EC[c] ($\geq$99%); α (0.0055%)	^{230}Th$(\alpha,3n)$
^{232}U	[14158-29-3]	71.7 yr	α (5.32030; 5.26354)[b]	daughter ^{232}Pa
^{233}U	[13968-55-3]	1.5911×10^5 yr	α (4.8242; 4.783 + 29 additional α's)[b]	daughter ^{233}Pa
^{234}U (U II)	[13966-29-5]	2.446×10^5 yr	α (4.7758; 4.7237)[b]	natural
^{235}U (AcU)	[15117-96-1]	7.038×10^8 yr	α (4.598; 4.558; 4.503; 4.446; 4.417;	natural
			4.398; 4.367; 4.345; 4.326; 4.267;	
			4.219; 4.158)[b]	
^{235m}U[d]		24.68 min	IT[e]	daughter ^{239}Pu
^{235f}U[f]		20 ns	F[g]	^{234}U(n,γ)
^{236}U	[13982-70-2]	2.3415×10^7 yr	α (4.494; 4.445; 4.331)[a]	daughter ^{240}Pu
				^{235}U(n,γ)
^{236f}U		116 ns	IT[e] (88%); SF[h] (12%)	^{235}U(d,p)
^{237}U	[14269-75-1]	6.752 d	β^- (0.245; 0.09)[b]	^{236}U(n,γ)
				^{238}U$(n,2n)$
^{238}U(U I)	[24678-82-8]	4.4683×10^9 yr	α (4.197; 4.148)[b]	natural
^{238f}U[f]		195 ns	IT[e]; SF[h]	^{238}U(d,pn)
^{239}U	[13982-01-9]	23.54 min	β^- (1.29; 1.21)[b]	^{238}U(n,γ)
^{240}U	[15687-53-3]	14.1 h	β^- (0.326)[a]	^{238}U(n,γ)

[a] Ref. 51 (see also Radioisotopes).
[b] Energy of radiation in MeV.
[c] EC = electron capture.
[d] m = metastable isomer.
[e] IT = internal transition.
[f] f = fission isomer.
[g] F = fission.
[h] SF = spontaneous fission.

Extraction from Ore

The methods used to extract uranium values from ores vary widely, and composition is only one of several factors affecting the choice (10–12,20). The ore may vary from hard, igneous rock to soft, weakly cemented sedimentary rock. The principal gangue mineral may be quartz, which is chemically inactive, or an acid-consuming mineral, such as calcite. Some ores are highly refractory and require intensive processing, whereas others break down between mine and mill.

Preconcentration. In general, conventional ore-dressing techniques have not been successful in the preconcentration of uranium minerals. However, it is usually possible to obtain acceptable concentration ratios. Tailings generally still contain recoverable values.

Where the uranium values occur as masses in pegmatitic rock with large areas of unmineralized pegmatite separating the ore minerals, they are preconcentrated by electronic sorting devices. The ore is fed to a relatively small, slow-moving conveyor or picking belt. A Geiger counter mounted above actuates a cylinder-operated pusher to remove either barren or radioactive pieces from the conveyor.

Successful gravity separations are sometimes possible because uranium minerals are denser than most gangue components. However, uranium minerals tend to concentrate with the fines in the grinding and crushing process of some ores (see also Gravity concentration).

Electrostatic methods generally give low recoveries at low concentrations. Magnetic gangue minerals, eg, magnetite, ilmenite, and garnet, may be separated by magnetic methods, which do not affect the nonmagnetic uranium component (see Magnetic separation).

Flotation (qv) has been thoroughly investigated on a wide variety of ores. A reasonably satisfactory ratio of concentration is usually attainable, but the tailings are not low enough in uranium to discard.

Preconcentration enriches low grade ores to the point where they can be processed economically.

Crushing and Grinding. The sand grains in the sandstone ores are essentially barren. Since they are the hardest and usually the coarsest component of the rock, there is no need to grind the ore any finer than the average grain size of the sand.

Jaw crushers are employed for coarse crushing; smaller jaw crushers, gyratories, or hammer mills are used for secondary crushing. Rod mills, ball mills, and hammer mills are used for grinding. Uranium values are concentrated in the cementing material and in the coating of the sand grains that are separated from the barren sand during the grinding action. In many cases, the ore is so poorly consolidated that there is no need to close the grinding circuit with screens or classifiers.

Roasting and Calcining. A high temperature roasting or calcining operation prior to leaching is frequently desirable and may be useful for several purposes. Carbonaceous material is best removed by an oxidizing roast, which at the same time converts the uranium to a soluble form. An oxidizing roast also converts sulfides or other sulfur compounds to sulfates, which do not interfere in subsequent ion-exchange steps. Other reductants, which also interfere in the leaching step, are removed. Conversion of uranium to the reduced state may be accomplished by a reducing roast, which serves to prevent the dissolution of uranium in a by-product recovery.

The characteristics of many ores are improved by roasting. Clays of the montmorillonite type, for instance, cause thixotropic slurries and thus interfere with leaching, settling, and filtering.

Vanadium-containing ores are roasted with sodium chloride. This treatment converts the vanadium into a soluble form, most likely sodium vanadate, which in turn is believed to form soluble uranyl vanadates (20). Silver, which otherwise might interfere, is converted to silver chloride, and is thus rendered insoluble for easier separation.

Leaching. Treatment with suitable solvents (acids or alkalies) converts uranium contained in the ore to water-soluble species. The uranium and other values are separated by chemical processing, including at least one digestion step with acid or alkaline solution.

Most mills use acid leaching, which completely extracts uranium. Because of its low cost, sulfuric acid is preferred; hydrochloric acid is used where it is a by-product of salt roasting. It is, however, more corrosive than sulfuric acid. Uranium(VI) compounds dissolve readily in H_2SO_4, but minerals such as uranite, pitchblende, or others in which the uranium has a lower valence, do not. They have to be dissolved under oxidizing conditions, provided by addition of suitable oxidants, such as manganese dioxide or sodium chlorate. Typical leach reactions are given below:

$$2\,H_2SO_4 + MnO_2 + UO_2 \rightarrow UO_2SO_4 + MnSO_4 + 2\,H_2O$$

$$3\,H_2SO_4 + NaClO_3 + 3\,UO_2 \rightarrow 3\,UO_2SO_4 + NaCl + 3\,H_2O$$

In practice, the oxidation potential of the solution is determined by measuring the Fe^{3+} to Fe^{2+} ratio. The Fe^{3+} ion is important in the oxidation of U(IV), since it oxidizes the UO_2 component.

$$2\,Fe^{2+} + MnO_2 + 4\,H^+ \rightarrow 2\,Fe^{3+} + Mn^{2+} + 2\,H_2O$$

$$UO_2 + 2\,Fe^{3+} \rightarrow UO_2^{2+} + 2\,Fe^{2+}$$

The effect of the Fe^{3+} concentration on the dissolution rate of UO_2 in the absence of Fe^{2+} is shown in Figure 3. In most ores, sufficient Fe^{3+} is present for this reaction sequence. For some ores, it is necessary to add metallic iron. Most ores of the Colorado Plateau can be leached in ≤ 8 h with adequate agitation slightly above room temperature. Very high leaching efficiencies with H_2SO_4 are common, eg, 95–98% dissolution yield of uranium. The effect of acid concentration on the rate of extraction from pitchblende ore for three different leaches on a single ore at 50% pulp density is shown in Figure 4, and the effect of acid concentration on extraction from two different pitchblende ores is shown in Figure 5.

If acid consumption exceeds 68 kg/t of ore treated, alkaline leaching is preferred. The comparative costs of acid, sodium hydroxide, and sodium carbonate differ widely in different areas and are the determining factor in the long run.

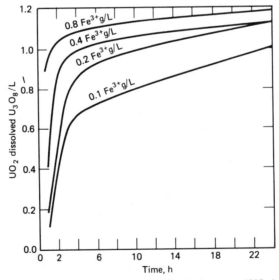

Figure 3.   Effect of ferric iron concentration on the dissolution rate of UO_2 in the absence of ferrous iron. Conditions: 1.0 g UO_2 in 1.0 L H_2O; acidified to pH 1.0 with H_2SO_4; Fe^{2+} added as $Fe_2(SO_4)_3$ and maintained in Fe^{3+} with excess MnO_2 (55).

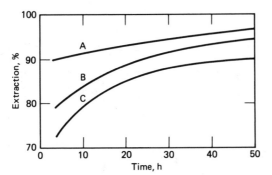

Figure 4.   Effect of acid concentration on rate of extraction of uranium from pitchblende ore. Shown are three leaches on a single ore at 50% pulp density; A, 400 kg H_2SO_4/ton ore; B, 250 kg H_2SO_4/ton ore; and C, 200 kg H_2SO_4/ton ore (56).

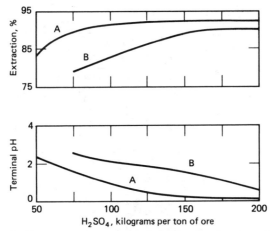

Figure 5.   Effect of acid concentration on extraction of uranium from two-pitchblende ores. Leaching time 16 h (57).

The Carbonate Process.   Alkaline or carbonate leaching is used in the treatment of uranium ores when the lime content results in excessive acid consumption. However, oxidants are needed. Uranium and aqueous carbonate solution give readily soluble carbonate complexes such as $Na_4[UO_2(CO_3)_3]$ [*60897-40-7*]. Under proper oxidizing conditions, extractions with the carbonate yield 90–95% of the uranium values. For this process, fine ores are required.

Carbonate leaching at 101.3 kPa (1 atm) is slow with poor recoveries. Typically, the ore is leached in an autoclave (Fig. 6) or a Pachuca tank (Fig. 7) with air providing most of the needed oxygen. Potassium permanganate is frequently an additional oxidant. The leach liquor is separated from the solid in a countercurrent-decantation system using a series of thickeners. The liquor is clarified in pressure filters, and the solids are washed on drum filters before being discarded as tailings. The uranium is precipitated from the clarified sodium carbonate solution with sodium hydroxide. The excess sodium hydroxide normally present is converted to the carbonate in packed towers through which carbon dioxide is circulated.

If vanadium is present in the ore, the clarified leach liquor is acidified to pH 6, precipitating sodium uranyl vanadate [*54667-49-1*], $NaUO_2VO_4$. The precipitate is

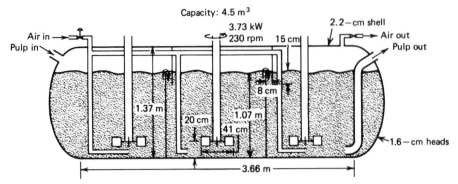

Figure 6.   Diagram of autoclave used in U.S. DOE pilot-plant studies at Grand Junction, Colo. (57). To convert kW to hp, multiply by 1.341.

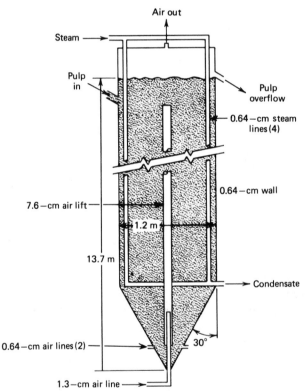

Figure 7.   Diagram of Pachuca tank used in U.S. DOE pilot-plant studies at Grand Junction, Colo. (57).

fused with soda ash and a convenient form of carbon. The uranium is reduced to insoluble UO_2, and the water-soluble sodium vanadate is leached away.

Separation of Uranium from Leach Liquors.   The crude uranium isolated in the leach liquors requires additional purification steps to meet the specifications of both the DOE and private producers of nuclear fuel. All extraction processes have been designed with the primary objectives of high recovery yield and of a purity that meets DOE specifications.

Direct Precipitation. Direct or selective precipitation has not been commercially successful in acid leach systems. The highly complex composition of the liquor makes it virtually impossible to attain specification products by this method. However, precipitation steps may be used to enrich the uranium in side streams for further purification. Precipitation from carbon leach liquors is possible, but the precipitates require additional purification.

Ion Exchange. Anionic sulfato or carbonato complexes of uranium may be absorbed from leach liquors on anion-exchange resins. From the resin, which is rinsed after loading, the uranium is eluted by means of a salt or acid solution. From this solution, it is precipitated and recovered as a fairly pure uranium concentrate. In the ion-exchange process, either stationary columns or continuous ion-exchange contactors may be used (58) (see also Ion exchange). In another technique, the resin is placed into agitated baskets and moved through the leach liquor. If the latter has not been completely clarified, the process is referred to as the resin-in-pulp (RIP) process (see below).

In the treatment of liquors from sulfuric acid leaching, the ion-exchange process rests on the rather unusual capacity of uranium to form anions in sulfate media. These sulfate anions exchange readily with the anion present in the preconditioned resin. The uranium is eluted with sodium chloride or ammonium nitrate.

The equilibrium in a sulfuric acid medium can be illustrated as follows (59):

$$UO_2^{2+} + n\, SO_4^{2-} \rightarrow [UO_2(SO_4)_n]^{2-2n}$$

where $n = 1, 2,$ or 3. The adsorption of uranium from the acid medium takes place by the following two reactions:

$$4\, RCl + [UO_2(SO_4)_3]^{4-} \rightleftharpoons R_4[UO_2(SO_4)_3] + 4\, Cl^-$$

$$2\, RCl + [UO_2(SO_4)_2]^{2-} \rightleftharpoons R_2[UO_2(SO_4)_2] + 2\, Cl^-$$

In carbonato media, the primary reaction is

$$UO_2^{2+} + 3\, CO_3^{2-} \rightarrow [UO_2(CO_3)_3]^{4-}$$

and the reaction with the resin is

$$4\, RCl + [UO_2(CO_3)_3]^{4-} \rightarrow R_4[UO_2(CO_3)_3] + 4\, Cl^-$$

The mechanism in nitrate media is similar. The above reactions indicate that from an acid medium, uranium can be adsorbed either as the disulfato complex or as the trisulfato complex. The high specificity of the anion-exchange process for uranium makes it the most desirable process. In the case of carbonato media, the tricarbonato complex is the only species adsorbed.

The rate of adsorption on such an anion-exchange resin is rapid, and near-equilibrium conditions are obtained in ca 7 min. The precipitate from the eluate of such a resin normally contains ca 80% U_3O_8. Ion-exchange resins, which have been used extensively in uranium recovery, and their capacities are shown in Table 11.

Resin-in-Pulp Ion Exchange. Some uranium ores exhibit extremely poor filtering and settling characteristics after leaching. To avoid large liquid–solid separation equipment, the ion-exchange process has been modified to extract uranium directly from the leach-pulp. This modification is called the resin-in-pulp (RIP) process. The slurry, after treatment with acid, flows through a series of perforated stainless-steel baskets filled with strongly basic ion-exchange resin beads of carefully selected size.

Table 11. Typical Anion-Exchange Resins Used in Uranium Recovery [a]

Operation	Resin	Manufacturer
column	Amberlite IRA-400	Rohm & Haas Co., Philadelphia, Pa.
column	Amberlite IRA-400	Charles Lennig and Co. (G. B.) Ltd., London
column	Deacidite FF	The Permutit Co., London
column	Dowex-1	The Dow Company, Midland, Mich.
column	Permutit (Ionac) SK	Permutit Co., New York
RIP [b]	Amberlite XE-123	Rohm & Haas Co., Philadelphia, Pa.
RIP [b]	Dowex-11	The Dow Company, Midland, Mich.
RIP [b]	Permutit SKB	Permutit Co., New York

[a] Ref. 60.
[b] Resin-in-pulp process.

The baskets (Fig. 8) are alternately raised and lowered into the slurry to bring it into intimate contact of the slurry with the resin. Equilibrium conditions are usually reached in ca 40 min. The elution procedure is similar to that used in fixed-bed ion-exchange systems. Nitrate solutions are preferred because they are more effective and less corrosive than chloride solutions. The eluant consists of 1 M NH_4NO_3 acidified with 0.10–0.20 N H^+.

The RIP process is expensive because of its mechanical features, especially the regeneration of the resin (11).

Solvent Extraction. At the Kerr-McGee mill in Shiprock, N.M., solvent extraction was used for the first time for uranium ore extraction. It has widespread application for uranium recovery from ores.

Contrary to ion exchange, which is a batch process, solvent extraction can be operated in a continuous countercurrent-flow manner. However, the phases cannot be completely separated because of emulsion formation and solubility. These effects

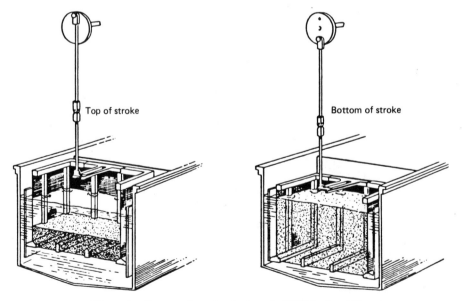

Figure 8.   Cross section of resin-in-pulp (RIP) basket (61).

as well as solvent losses result in financial losses and a potential pollution problem inherent in the disposal of spent leach liquor. For leach solutions with a higher concentration than 1 g U/L, solvent extraction is preferred. For low grade solutions with <1 g U/L and carbonate leach solutions, ion exchange is preferred.

For extraction of uranium from sulfate leach liquors, alkyl phosphoric acids, alkyl phosphates, and secondary and tertiary alkyl amines are used in inert solvents, eg, kerosene. The formation of a third phase is suppressed by addition of long-chain alcohols or neutral phosphate esters. Such compounds also increase the solubility of the amine salt in the diluent and improve phase separation.

The mechanism involved in the amine extraction of sulfate leach liquors is similar to that observed in the anion-exchange separation of uranium from uranyl sulfate solution. Uranyl sulfato complexes are extracted by the alkyl ammonium sulfate cations at pH 1–2; the predominant extracted species in commercial processing is the $[UO_2(SO_4)_3]^{4-}$ complex. At low pH, other uranyl species increase. The amine structure affects selectivity and affinity. Typical commercially available tertiary amines have extraction coefficients of 100–140, whereas for N-benzylheptylamine, a value of 8000 was reported (20).

Other anions, eg, nitrate or chloride, may interfere with the uranium extraction. Nitrate interferes with secondary amines and chloride with tertiary amines. The choice of suitable stripping agents depends on such factors, as does the recycling of solutions. Molybdenum is extracted more readily than uranium concentrates and acts as a poison. At sufficiently high concentration, it may cause serious problems by precipitation at the organic–aqueous interface. The problem may be solved by including one or more specific molybdenum stripping steps in the process (20). Vanadium is extracted to some extent. Various other ions function as salting-out agents for uranium, which is stripped from the organic solvent in their presence. The affinity of nitrate to the amine is so high that the latter has to be scrubbed by means of a hydroxide or carbonate wash before it can be recycled to another extraction run. Chloride does not give this complication, except with secondary amines, which have a high chloride affinity.

Monoalkyl phosphate extractants exhibit good efficiency in the presence of dilute nitrate, sulfate, or chloride, and cause fewer phase-separation problems. However, they are less selective, and other cations present in the leach liquor are coextracted with the uranium from which they must be separated. Even though the term liquid cation exchangers has sometimes been used for monoalkyl phosphates as extractants, the resemblance to cation exchangers is only superficial. The uranium extraction coefficient of monoalkyl phosphates is enhanced by addition of neutral phosphoric acids (synergistic effect). The uranium is usually back-extracted from monoalkyl phosphates by means of carbonate solutions; 10 M HCl is required for stripping from monodecyl phosphate.

The most widely used extractants are di(2-ethylhexyl) phosphate (D2EHPA) and dodecylphosphate (DDPA). The selectivity for uranium is about equal. Fe(III) interferes and has to be quantitatively reduced to Fe(II) in the feed liquor prior to extraction. DPPA has substantially higher solubility losses than D2EHPA, and strong acids are required for back-extraction from DPPA. For D2EHPA, sodium carbonate solution is used as the stripping agent; for DPPA, hydrochloric or hydrofluoric acids are used. The D2EHPA solvent extraction process is generally referred to the DAPEX process (dialkyl phosphate extraction).

The alkyl amines were first used as uranium extractants in Oak Ridge, Tenn., and the process is generally known as the AMEX process.

A flow sheet of a typical present-day uranium-concentration process is shown in Figure 9. It employs a sulfuric acid leach followed by sand-slime separation, clarification, and subsequent extraction and back extraction.

Precipitation of Uranium Concentrates. The product of the extraction processes, whether acid or carbonate leach, is a purified uranium solution that may or may not have been upgraded by ion-exchange or solvent extraction. The uranium in such a solution is concentrated by precipitation and must be dewatered before shipment.

Solutions resulting from carbonate leaching are usually precipitated directly from clarified leach liquor with caustic soda without concentration.

$$2\,[UO_2(CO_3)_3]^{4-} + 2\,Na^+ + 6\,OH^- \rightarrow Na_2U_2O_7 + 6\,CO_3^{2-} + 3\,H_2O$$

Losses are held to a minimum by carbonation of the mother liquor with CO_2 and recycle of the carbonated product to the leach system.

From acid solutions, uranium is usually precipitated by neutralization with ammonia or magnesia. The product obtained with lime precipitation does not meet DOE specifications.

Ammonia gives an acceptable precipitate, for which compositions such as $(NH_4)_2(UO_2)_2SO_4(OH)_4.nH_2O$ were calculated. The ammonium salt is preferred if the product is to be used in the manufacture of fuel-element material.

So-called yellow cake with a higher uranium content can be obtained by precipitation with magnesia. The magnesium sulfate formed is water-soluble, and the uranium compound can be separated by filtration. Yellow cake consists of either ammonium diuranate [7783-22-4] or magnesium diuranate [13568-61-1]. Ammonium di-

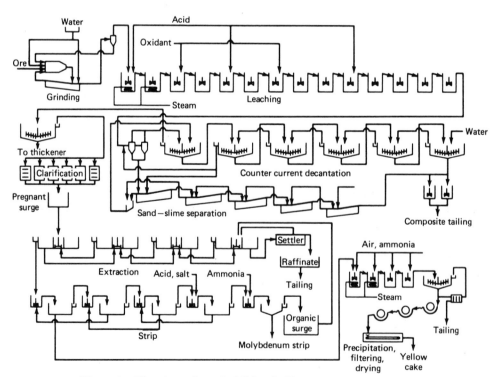

Figure 9. Flow sheet of a typical Colorado Plateau ore-processing plant.

uranate in yellow cake is not a stoichiometric compound, but a mixture of compounds with the formal composition $(NH_4)_2O.x\,UO_3$ ranging from $(NH_4)_2UO_4$ [13597-77-8] to $(NH_4)_2U_8O_{25}$, and having the approximate composition $(NH_4)_2U_2O_7$ (62–63). Sodium uranate [13721-31-4] may be the product from carbonate leach plants. The concentrates vary from 70 to 85% in U_3O_8 content. Moisture in the products from the more modern plants is usually <1%.

Conversion and Purification Processes

The crude product from the refineries is purified to a degree that is usable in nuclear applications. The purified material is converted to uranium dioxide, which may be used directly for reactor ceramics or may, in turn, be converted to uranium tetrafluoride. The latter may either be reduced to uranium metal or be fluorinated to uranium hexafluoride, the basic compound for isotope separation. Detailed descriptions of the conversion chemistry of uranium are given in refs. 10, 44, 47, and 64.

The yellow-cake concentrate is received by the plants in steel drums. Lots are carefully sampled and thoroughly blended. The blended yellow cake is dissolved in nitric acid and purified by extraction with tributyl phosphate (TBP) in hexane diluent. The uranium is back-extracted either with recycled acid concentrate containing <1% HNO_3 or with deionized water. The aqueous product is concentrated to 100 g U/L and to an acid concentration of <0.01 N HNO_3.

The conversion to the next intermediate, uranium trioxide [1344-58-7], UO_3, involves two steps:

1. The $UO_2(NO_3)_2$ [10102-06-4] solution is evaporated to the approximate composition $UO_2(NO_3)_2.6H_2O$, uranyl nitrate hexahydrate [13520-83-7], UNH, which melts at 118°C.

2. The uranyl nitrate hexahydrate is dehydrated and denitrated, yielding UO_3 (65):

$$UO_2(NO_3)_2.x\,H_2O \xrightarrow{\Delta} UO_3 + NO + NO_2 + O_2 + x\,H_2O$$

The pyrolysis may be carried out either in a batch-type reactor or continuously in a fluidized-bed denitrator (Figs. 10 and 11).

Uranium trioxide is reduced to the dioxide [1344-57-6] with hydrogen in a fluidized bed (Fig. 12) (see also Fluidization).

$$UO_3(s) + H_2(g) \rightarrow UO_2(s) + H_2O(g) \qquad \Delta H° = -46.6 \text{ kJ/mol } (-11.14 \text{ kcal/mol})$$

Dissociated ammonia (ie, a hydrogen–nitrogen mixture) has also been used. The reduction is carried out at ca 816°C. A stirred-bed reactor, a vibrating-tray reactor, and a moving-bed reactor have also been used.

The uranium dioxide obtained in this process, when of exact chemical stoichiometry, is a cinnamon-colored powder that may be used directly in the manufacture of pellets for fuel elements.

The next intermediate is uranium tetrafluoride [10049-11-6], UF_4, also called green salt. It is produced in two grades of purity, depending on further use.

Metal-grade UF_4 must contain at least 96% UF_4 to be suitable for calciothermic metal production. The UF_4 content of cascade-grade UF_4, on the other hand, is not fixed, but depends upon the costs of hydrofluorination and the elemental fluorine required to convert the material to UF_6.

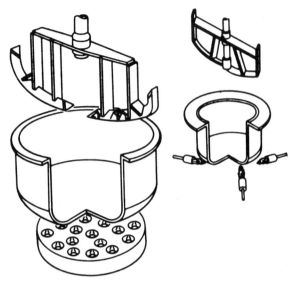

Figure 10. Gas-fired denitration pots for UNH denitration. Large pot: ID 1.68 m, height 0.81 m, heated by three concentric rings of small radiant gas burners. Small pot: ID 76 cm, height 46 cm, heated by four gas burners inside a ceramic furnace (66).

The flow sheet of the process used to convert UO_2 to UF_4 is shown in Figure 13. It is based on the following reaction:

$$UO_2(s) + 4 HF(g) \rightarrow UF_4(s) + 2 H_2O(g) \qquad \Delta H° = -78.1 \text{ kJ/mol } (-18.64 \text{ kcal/mol})$$

A typical continuous stirred-bed reactor, which is used for the reaction, is shown in Figure 14.

The uranium tetrafluoride (green salt) is converted to either uranium hexafluoride [7783-81-5] or the metal (see below).

Uranium hexafluoride is prepared by direct fluorination of UF_4 with elemental fluorine in a fluorination tower (Fig. 15).

$$UF_4(s) + F_2(g) \rightarrow UF_6(g) \qquad \Delta H° = -110.4 \text{ kJ/mol } (-26.4 \text{ kcal/mol})$$

Uranium hexafluoride is a volatile, hygroscopic solid, and all operations must be carried out under strict exclusion of moisture.

Solid UF_4 is fed through suitable locks into the top of the fluorination tower. Filtered and preheated fluorine is introduced into the side of the tower. Unreacted UF_4 is collected in a hopper at the bottom. This material is periodically removed and recycled. Gaseous UF_6 leaves the reactor through a cooler near the bottom of the fluorination tower. The gas then passes through a cyclone for removal of fine solids and, after filtration, is passed through a cold trap or condenser for final removal of solids. The gas then flows to the cylinder-loading station and is shipped in containers such as that shown in Figure 16.

Reconversion of Enriched UF_6. Enriched UF_6 from an isotope-separation plant has to be converted to compounds or the metal for reactor applications (72). The UF_6 is vaporized and hydrolyzed with water vapor:

$$UF_6 + 2 H_2O \rightarrow UO_2F_2 + 4 HF$$
$$[13536-84-0]$$

The resulting UO_2F_2 solution is precipitated with ammonia:

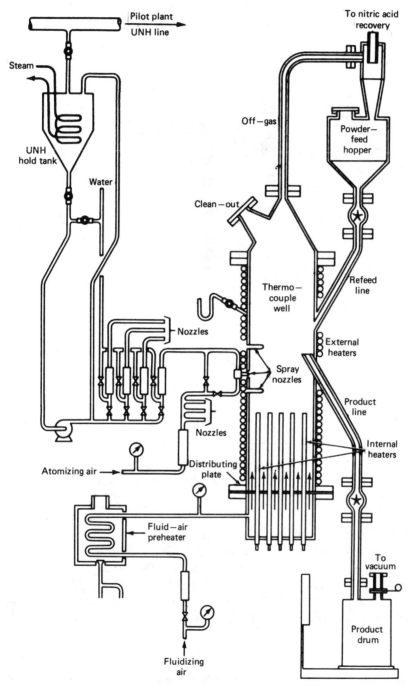

Figure 11. A 25.4-cm pilot-plant fluidized-bed denitrator (67). UNH = uranyl nitrate hexahydrate.

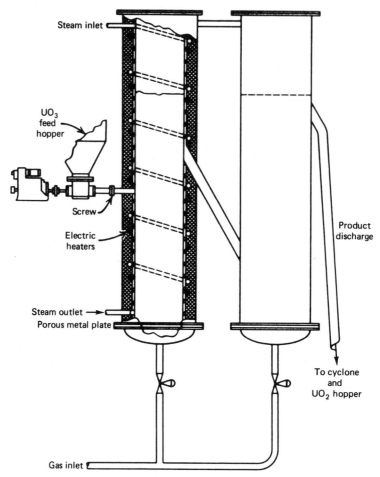

Figure 12.   Fluidized-bed UO_2 reactor (68).

$$2\,UO_2F_2 + 8\,HF + 14\,NH_3 + 3\,H_2O \rightarrow (NH_4)_2U_2O_7 + 12\,NH_4F$$

The ammonium diuranate (ADU), which is of changing stoichiometry, is purified by extractions or precipitation and dried at 200°C. It is converted to U_3O_8 by reduction in an H_2O–H_2 mixture at 500°C and then reduced to UO_2 in H_2 at 500–800°C.

In the AUC process (ammonium uranyl carbonate [18077-77-5]), the UF_6 is fed into a pipe system to which NH_3 and CO_2 gases are added; the mixture is fed into a precipitation vessel where the following reaction takes place:

$$UF_6 + 5\,H_2O + 10\,NH_3 + 3\,CO_2 \rightarrow (NH_4)_4[UO_2(CO_3)_3] + 6\,NH_4F$$

In the IDR process (integrated dry route) which was developed by British Nuclear Fuels Ltd., the vaporized UF_6 is fed into a reactor where it is converted to UO_2F_2 by superheated steam. The resulting UO_2F_2 is reduced to UO_2 with hydrogen.

The UO_2 obtained in the processes described above is reduced to UF_4 with calcium (see following page).

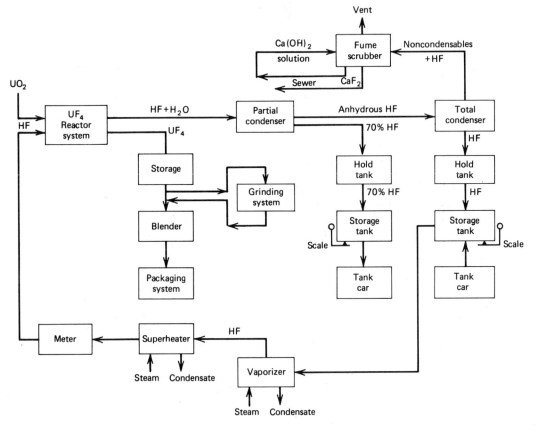

Figure 13.   Flow sheet for UF_4 production (69).

Metal Manufacture

Uranium metal is produced by reduction of uranium tetrafluoride by the Ames process (10,44,47).

$$UF_4(s) + \begin{cases} 2\,Ca \\ 2\,Mg \end{cases} \rightarrow U + \begin{cases} 2\,CaF_2 & \Delta H° = -560.5 \text{ kJ/mol } (-134 \text{ kcal/mol}) \\ 2\,MgF_2 & \Delta H° = -349.4 \text{ kJ/mol } (-83.51 \text{ kcal/mol}) \end{cases}$$

A general flow sheet is shown in Figure 17. The reduction process is carried out in a bomb (Fig. 18).

A mandrel is placed into the proper position inside the reduction bomb, and MgF_2 or CaF_2 powder is vibrated into the space between bomb and mandrel until a solid liner of compacted MgF_2 or CaF_2 powder is formed. The mandrel is then withdrawn, and a charge consisting of anhydrous UF_4 powder and Mg chips (202 kg UF_4 and 32.1 kg Mg in a typical charge) that has been thoroughly mixed is placed into the bomb. It is covered with MgF_2 powder, and the bomb is closed with a screwed-on flange cover.

The charge is ignited spontaneously by heating, and the reduction of the UF_4 proceeds. The control temperature is ca 700°C. The average firing time for a bomb containing a charge of the size given above is ca 4 h ± 30 min, depending on the UF_4 grade. If 98% UF_4 is used, the yield may be 97% pure metal.

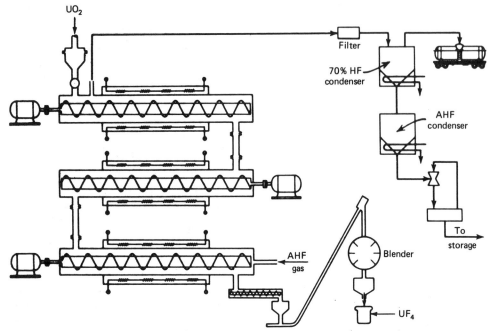

Figure 14.   Continuous stirred-bed hydrofluorination (70). AHF = anhydrous hydrogen fluoride.

The bomb is transferred to a cooling pit and is retained there until the outer temperature has dropped to 480–540°C. After air cooling, the bomb is further cooled in a water tank for 4–6 h until it has cooled to ambient temperature. The water level in the cooling tank is kept at least 7.6 cm below the flange top.

After cooling, the bomb is opened, the slag is loosened, and the bomb shell is inverted over a hopper resting on top of the breakout jolter. The metal ingot, so-called derby, is not suitable for direct machining or extrusion, but must be recast to a suitable bar.

The direct ingot method has been applied to charges of uranium up to 1540 kg of metal (10,74). The dingot bomb is cylindrical in its upper part (see Fig. 19). This cylinder is bolted to a funnel-shaped middle part, which is in turn bolted to the bottom part, which is also cylindrical, but with a smaller diameter than the upper part. The liquid uranium generated in the reduction collects as a pool in the bottom of the bomb and solidifies to an ingot with a diameter of ca 25 cm and a height of 25 cm. An ingot of such dimensions is called a direct ingot or dingot because it can be fed directly to a milling machine or to an extrusion press without intermediate recasting. Magnesium is used as the reductant at ca 1900°C. Dingot metal is of high quality and is directly suitable for fuel-element fabrication.

Uranium-235 Enrichment.   Most nuclear reactors built for the generation of electric power are based on uranium fuel enriched in [235]U (9). Normally for such reactors, the 0.72% natural abundance of [235]U is enriched to 2–3%; in a few special cases, eg, for materials-testing reactors or high flux isotope reactors, enriched [235]U of 96–97% purity has been used.

The natural-uranium reactors, such as the Hanford production reactors or some power reactors in the UK, do not require enriched uranium. For these reactors, the purified natural uranium may be fabricated into metal rods (slugs) or into oxide or carbide shapes for direct use.

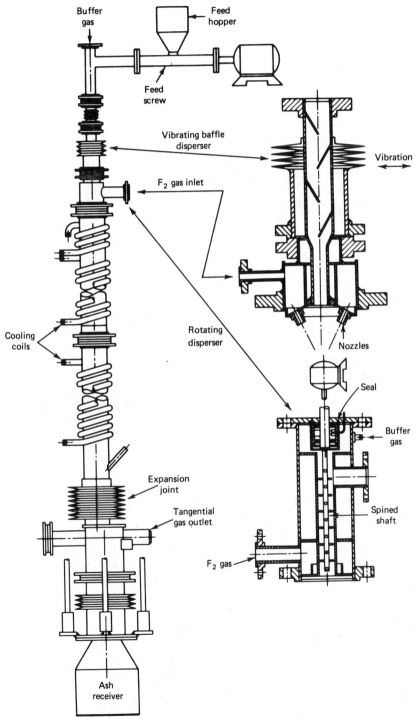

Figure 15. Fluorination tower reactor (71).

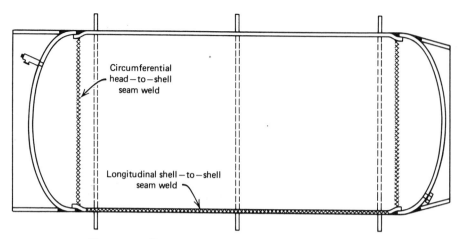

Figure 16.   Shipping container for uranium hexafluoride (72).

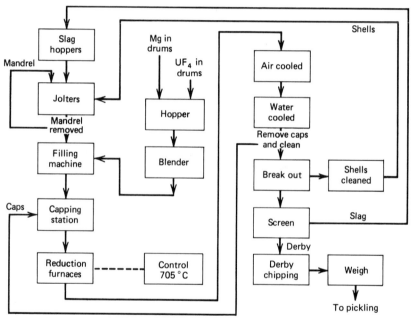

Figure 17.   Flow sheet for the production of uranium metal by reduction of UF_4 with magnesium (73).

Gaseous-Diffusion Process.   This process is the only one used for the separation of ^{235}U and ^{238}U on an industrial scale (75–77). Highly purified gaseous uranium hexafluoride is pumped through a series of diffusion aggregates (diffusers, converters) that are arrayed in cells in a cascade pattern (Fig. 20). The ^{238}U and ^{235}U diffuse through the barrier tubes inside the converters at slightly different rates, and many stages are required to obtain high concentrations of ^{235}U. The diffusion barriers have a pore diameter of ca 10 nm (78). Barriers produced by deposition of polytetrafluoroethylene on a metallic grid are generally satisfactory, but they have also been constructed from fritted alumina, arc-sputtered aluminum, sintered nickel powder, and other metallic or ceramic powders.

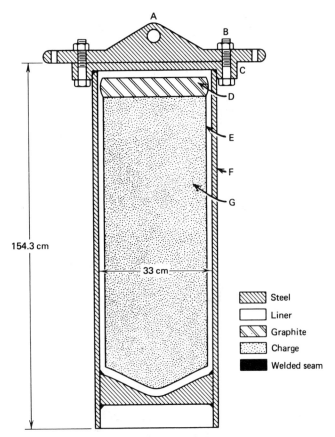

Figure 18.   Reduction bomb for the reduction of UF_4 with Mg by the Ames process. Capacity 144.2 kg uranium metal. A, Steel cover flange with lifting eye; B, bolt and nut; C, top flange of bomb; D, graphite cover; E, liner out of fused dolomitic oxide; F, steel bomb; and G, charge.

Enormous amounts of electric power are required to operate gaseous-diffusion plants (77). The three U.S. gaseous-diffusion plants in Oak Ridge, Tenn., Paducah, Ky., and Portsmouth, Ohio, have a total annual power consumption of 6.06 GW. This is more electric power than the entire annual power consumption of France (see Diffusion separation methods).

Centrifugal Isotope Separation.   The high capital cost and large power requirements of gaseous-diffusion plants have led to an extensive investigation into the possibility of centrifugal separation of ^{235}U and ^{238}U (79). The advantage of centrifugal separation over gas diffusion is that the separation factor is not proportional to the square root of the ratio of the masses of ^{235}U and ^{238}U, but instead to the difference of the masses of these isotopes. Therefore, centrifugal separation is more efficient than gas diffusion. However, the atmosphere is highly corrosive, and careful maintenance is required.

In the United States, a gas-centrifuge enrichment plant (GCEP) is under construction at the site of the Portsmouth gas-diffusion plant (80). This plant is part of the advanced gas-centrifuge program. A pilot plant for the GCEP, the centrifuge plant demonstration facility (CPDF), is located in Oak Ridge, Tenn. In the FRG, the Uranit-Gesellschaft in Jülich operates a centrifuge pilot plant. The centrifuges used in this plant are manufactured by the MAN Neue Technologie in Munich-Karlsfeld, FRG.

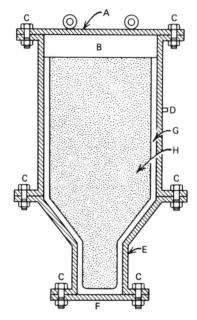

Figure 19.   Section through dingot bomb. A, Cover with lifting eyes; B, cover plate; C, bolts and nuts; D, cylindrical upper part; E, funnel-shaped lower part; F, bottom plate; and G, liner (MgF_2 or dolomitic oxide).

Evaporative current and countercurrent centrifuges with long rotors and ultrahigh speed capability have been extensively studied (81–82). In these centrifuges, a hollow rotor is partially filled with the isotope mixture in liquid form. High velocity is attained under temperature and pressure conditions that favor evaporation. The lighter isotope may be drawn axially from the rotor, whereas the depleted heavier isotope remains in the rotor.

A concurrent type of centrifuge, which does not employ the evaporative principle has been used experimentally. It is claimed that this device is more efficient than the evaporative centrifuge. Increases in the abundance ratio of ^{235}U to ^{238}U of as much as 5.6% have been reported with a single centrifugation (79).

Electromagnetic Separation.   The electromagnetic separation method, one of the most powerful methods, was developed at the University of California Radiation Laboratory and was employed on an industrial scale in the electromagnetic separation plant at Oak Ridge. Because the prototype machines were tested in the Berkeley cyclotron magnets, the Oak Ridge separators were called calutrons (California University Cyclotron). A calutron separator is essentially a 180° mass spectrograph especially designed for a large throughput of ions. The uranium to be separated is converted to the tetrachloride and is charged into the charge bottle of the calutron or is generated in the charge bottle from UO_2 and CCl_4 vapor. The ions generated from the UCl_4 vapor in the ion source are accelerated into the magnetic field and deflected through an angle of 180°. The radius of curvature of the heavy beam (^{238}U) is larger than that of the light beam (^{235}U). Thus, the two beams are impinging on two different locations on the receiver. The receiver, built of graphite, is equipped with pockets in which the beams of the two isotopes are collected. After the run, the receiver is dismantled, and the individual isotopes are worked up separately.

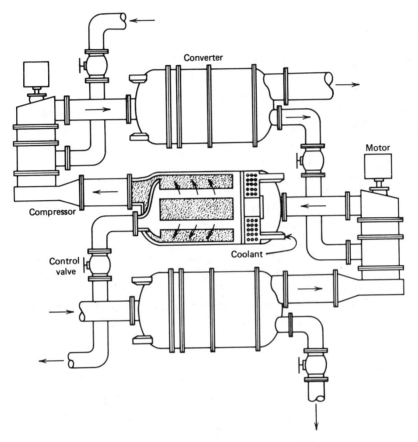

Figure 20.   Gaseous-diffusion converter (75).

At the electromagnetic-separation plant in Oak Ridge, α and a β calutrons were used. The larger α calutrons were charged with natural uranium, in which the natural abundance was enhanced by the calutron separation by a factor of 20. The enriched α product was next charged into the ionic source of the smaller β calutrons, in which it was enhanced by a factor of 40; thus, overall enhancement was $20 \times 40 = 800$. Final β products up to 85% ^{235}U and 15% ^{238}U were obtained in some runs. A comprehensive description of the calutron process has been given in the Division I volumes of the National Nuclear Energy Series (NNES) (83–95).

The calutron technique has been used to separate pure samples of ^{234}U (96), ^{236}U (97–98), and stable isotopes of many other elements. The Y-12 calutron plant at Oak Ridge was shut down in 1980.

Other Methods for Isotope Separation. A number of other methods for separating uranium isotopes have been developed, but none of these has been advanced beyond the pilot-plant stage.

The liquid thermal-diffusion process was installed in a pilot plant in Oak Ridge at the time of the Manhattan Project (99). Uranium hexafluoride, kept liquid at temperatures above its triple point, is subjected to diffusion in a thermal-diffusion column, the center of which is kept at 188–286°C, whereas the wall is maintained at 65°C. By thermal diffusion, the ^{235}U is enriched at the top, while ^{238}U migrates to the bottom of the column. The S-50 plant provided slightly enriched feed material for the calutron plant at Y-12. The S-50 operations were terminated in Sept. 1945, when the K-25 gaseous-diffusion plant went into operation.

Laser excitation appears to be a very promising isotope-separation method. In the atomic vapor laser isotope-separation (AVLIS) process, uranium vapor is ionized by means of a tunable rhodamine-B laser in such a manner that ^{235}U atoms are selectively excited to form positive ions, which are collected on a negative electrode (100). The DOE has announced the construction of a demonstration enrichment plant based on a laser separation process. In the separation-nozzle method (82), uranium hexafluoride vapor is effused out of a nozzle. During the effusion, the light isotope is enriched in certain parts of the gas jet and may be enriched by stripping those parts away from the other parts of the jet. The gas is an H_2–UF_6 mixture containing 5% UF_6. The process has been demonstrated on the pilot-plant scale at the Nuclear Engineering Institute of the Karlsruhe Nuclear Research Center in the FRG.

Isolation of Specific Isotopes. The uranium isotopes ^{232}U, ^{233}U, and ^{234}U are the daughters of the actinide isotopes ^{232}Pa, ^{233}Pa, and ^{238}Pu, respectively, which may be obtained in pure state. After they have decayed for a suitable time, the individual uranium species may be separated by chemical procedures.

Uranium-232. For the production of ^{232}U, a large quantity of protactinium-231 (ca 50 g) is bombarded with neutrons:

$$^{231}Pa(n,\gamma)^{232}Pa \xrightarrow{\beta^-} {}^{232}U$$

The ^{232}U can be separated from the unreacted protactinium by an ion-exchange process that involves elution of Pa with 8 M HCl–0.6 M HF and stripping of ^{232}U with 0.5 M HCl from Dowex 1 × 4. The uranium is further purified by extraction with 20% tributyl phosphate (TBP) with diethylbenzene as diluent, and back extraction with water (101–102).

Uranium-233. Uranium-233 is obtained by bombardment of thorium-232 with slow neutrons.

$$^{232}Th(n,\gamma)^{233}Th \xrightarrow{\beta^-} {}^{233}Pa \xrightarrow{\beta^-} {}^{233}U$$

Milligram quantities were first isolated in 1944 (53). Several extraction processes have been developed for the large-scale isolation of uranium-233, including the hexone-uranium-233 process (103), the Interim 23 process (103), and the Thorex process (103–104). In general, the Thorex process is used today.

The irradiated thorium breeder slugs are dissolved in HNO_3—HF, and the resulting solution is adjusted to the required feed specifications (1.5 M Th, 0.5 M HNO_3, 0.3 g ^{232}U/L, 0.05 N F$^-$, 0.005 M Hg) and extracted in the first extractor with 42.4% TBP in Amsco (commercial petroleum hydrocarbon mixture with 40–90% aromatic hydrocarbons); ^{233}Pa and fission products remain in the aqueous phase. The organic phase, containing ^{233}U and Th, is scrubbed with $Al(NO_3)_3$ solution containing small amounts of Fe^{2+} and PO_4^{3-}. From the combined aqueous streams, ^{233}Pa, or after its decay, additional ^{233}U may be recovered. The organic stream is fed into a second extractor, and Th is stripped into dilute HNO_3. Finally, ^{233}U is stripped into water. From the resulting aqueous solution, the uranium-233 is concentrated on an ion-exchange column for final recovery (103–104).

Uranium-234. Uranium-234 is the daughter product of plutonium-238, which, in turn, is produced by neutron bombardment of neptunium-237:

$$^{237}Np(n,\gamma)^{238}Np \xrightarrow{\beta} {}^{238}Pu \xrightarrow{\alpha} {}^{234}U$$

100 g ^{238}Pu contains 776 mg ^{234}U after decay of 1 yr, and ca 2 g after 3 yr. For the separation, the aged ^{238}Pu is dissolved in concentrated HNO_3 and loaded on a Dowex 1 × 4 column; the ^{234}U is eluted with 7.2 M HNO_3. After evaporation of the HNO_3 solution, the ^{234}U may be further purified by hexane or TBP extraction (105).

Uranium Compounds

Numerous compounds of uranium have been prepared over the years (106–108). Parts C and E of the supplement to *Gmelin* (108) give a comprehensive review and detailed discussion.

Hydrides. Uranium metal, heated to 150–200°C in a hydrogen atmosphere, gives uranium trihydride [*13598-56-6*], UH_3, a black powder. Upon heating in vacuum to 450°C, UH_3 reverts to uranium metal powder, which is prepared by this route. Compact uranium metal is first treated with hydrogen. The hydride is easily powdered in the hydrogen atmosphere and then decomposed in vacuum. The resulting extremely fine metal powder is pyrophoric, highly reactive, and particularly suited for certain reactions.

Rare isotopes can be stored in the form of hydrides. Deuterium or tritium is absorbed on uranium turnings heated to 200°C and may thus safely be stored as a solid. Upon heating to 500°C in vacuum, the deuterium or tritium is released and available as a very pure gas.

Fluorides. Uranium forms seven binary fluorides, ie, UF_3 [*13775-06-9*], UF_4, U_4F_{17} [*12134-52-0*], U_2F_9 [*12134-48-4*], α-UF_5, β-UF_5 [*13775-07-0*], and UF_6. In addition, a large number of ternary fluorine complexes have been prepared.

Uranium trifluoride, UF_3, is prepared from stoichiometric amounts of UF_4 and uranium metal in an evacuated, sealed molybdenum capsule,

$$3\ UF_4 + U \rightarrow 4\ UF_3$$

or by reduction of UF_4 with uranium hydride (109):

$$3\ UF_4 + UH_3 \rightarrow 4\ UF_3 + \tfrac{3}{2}\ H_2$$

The hydride is prepared *in situ* from uranium metal turnings and H_2. Uranium trifluoride is a black, cokelike mass, consisting of small, deep-purple crystals. It is insoluble in water, and inert in weak acids, but dissolves in strong nitric acid with the evolution of nitrogen oxides. It is also soluble in hot sulfuric and perchloric acids. Uranium trifluoride has been used as a component in molten-salt systems.

Uranium tetrafluoride, UF_4 (green salt), is one of the key compounds in uranium conversion. It is used in both uranium metal and uranium hexafluoride production. Uranium tetrafluoride, mp 969°C, density 6.70 g/cm^3, is an emerald-green solid that is insoluble in water. In general, UF_4 is prepared by hydrofluorination of UO_2 with excess gaseous HF at ca 550°C (110):

$$UO_2 + 4\ HF \rightarrow UF_4 + 2\ H_2O$$

It may also be prepared from an aqueous solution of uranyl fluoride in 48% HF by electrolytic reduction, followed by precipitation of hydrated UF_4 (Excer process). At room temperature, $UF_4 \cdot 2.5H_2O$ [10049-15-7] precipitates (111); at 90°C, $UF_4 \cdot 0.75H_2O$ [22981-36-8] forms (112). The hydrates yield anhydrous UF_4 of 97% purity when heated in vacuum to 150–200°C or in a stream of dry nitrogen at 100–250°C (111–112).

Uranium(IV,V) fluorides, U_4F_{17} and U_2F_9, are obtained under specific conditions from UF_4 and UF_6:

$$3\,UF_4 + UF_6 \rightarrow 2\,U_2F_9$$

at 200°C and $p = 2.36$ kPa (17.7 mm Hg). Like U_2F_9, it forms black, cubic crystals, but with a different crystal structure; U_2F_9 and U_4F_{17} are observed in gas centrifuges as decomposition products.

Uranium pentafluoride, UF_5, exists in two polymorphic modifications, α-UF_5 and β-UF_5. The former is obtained by reduction of UF_6 with HBr:

$$2\,UF_6 + 2\,HBr \rightarrow 2\,UF_5 + Br_2 + 2\,HF$$

or from UF_4 and UF_6 at 80–100°C; at 150–200°C, β-UF_5 is obtained

$$UF_4 + UF_6 \rightarrow 2\,UF_5$$

α-UF_5 is grayish-white, β-UF_5 is yellowish-white.

Both compounds are hygroscopic. However, in anhydrous HF containing traces of water, they form a blue solution of U(V).

Uranium hexafluoride, UF_6, is prepared by direct fluorination of UF_4 with fluorine (see Fig. 17). It forms colorless volatile crystals with a high refractive index that sublime at room temperature and atmospheric pressure and melt under elevated pressure. The physical and chemical properties of this compound have been determined in great detail and with great accuracy (71,113–114). Selected properties are given in Table 12.

Uranium hexafluoride is produced on a large scale as feed for gaseous-diffusion, separation-nozzle, and gas-centrifuge processes.

Uranium Fluoride Complexes.   A large number of ternary uranium fluorides are composed of a binary fluoride and an alkali or alkaline earth fluoride.

Uranium hexafluoride forms ternary fluoro complexes with alkali fluorides of the general type MUF_7, M_2UF_8, and M_2UF_9, where M is an alkali metal.

Table 12.   Properties of Uranium Hexafluoride [a]

Property	Value
triple point at 151 kPa[b], °C	64.052
sublimation point, °C	56.4
density	
solid, g/cm³	5.09
liquid, g/mL	6.63
heat of formation, solid at 25°C, kJ/mol[c]	−2158.9
heat of vaporization at 64.01°C, kJ/mol[c]	28.899
heat of fusion at 64.01°C, kJ/mol[c]	19.196
heat of sublimation at 64.01°C, kJ/mol[c]	48.095

[a] Ref. 10.

[b] To convert kPa to mm Hg, multiply by 7.5.

[c] To convert J to cal, divide by 4.184.

Uranium pentafluoride forms three series of fluorine complexes: MUF_5, where M = Li, Na, K, Rb, Cs, NH_4, and Ag; M_2UF_7, where M = K, Rb, Cs, and NH_4; and M_3UF_8, where M = Na, K, Rb, Cs, Ag, and Tl. These compounds can be prepared from alkali fluoride or another metal fluoride with the stoichiometric amount of UF_5 in liquid anhydrous HF. From the blue solutions of UF_5 in slightly moist HF, the acids $HUF_6.2.5H_2O$ [84370-89-8] and $HUF_6.1.25H_2O$ crystallize as blue crystals.

Uranium tetrafluoride forms a large number of fluorine complexes with alkali, alkaline earth, and other metal fluorides (see Table 13). Many of these compounds are utilized in molten-salt reactors and for nonaqueous processing. Their phase diagrams are reported in refs. 115–116.

Uranium trifluoride forms compounds of the types MUF_4 and M_3UF_6, but only a few of these have been prepared and characterized.

Chlorides. In addition to binary chlorides, a number of mixed halides containing up to three different halogens, eg, UClBrI [84370-90-1], have been prepared.

Uranium trichloride [10025-93-1], UCl_3, an olive-green solid, is prepared from uranium hydride and HCl at 250–300°C:

$$UH_3 + 3\,HCl \rightarrow UCl_3 + 3\,H_2$$

Uranium trichloride has some application in the molten-salt electrolytic refinement of uranium metal.

Uranium tetrachloride [10026-10-5], UCl_4, a dark green solid, is best prepared from UO_3 and hexachloropropene:

$$UO_3 + 3\,CCl_3CCl{=}CCl_2 \rightarrow UCl_4 + 3\,CCl_2{=}CClCOCl + Cl_2$$

or by heating UO_2 in a stream of CCl_4 or $SOCl_2$:

$$UO_2 + 2\,CCl_4 \rightarrow UCl_4 + 2\,COCl_2$$

$$UO_2 + SOCl_2 \rightarrow UCl_4 + 2\,SO_2$$

Uranium tetrachloride has application in the calutron process (94). Purification by vacuum sublimation (119) gives large, dark green crystals with a metallic luster.

Uranium pentachloride [13470-21-8], UCl_5, is prepared from UO_3 and CCl_4 or from UCl_4 with chlorine:

$$2\,UO_3 + 5\,CCl_4 \rightarrow 2\,UCl_5 + 5\,COCl_2 + \tfrac{1}{2}\,O_2$$

$$2\,UCl_4 + Cl_2 \rightarrow 2\,UCl_5$$

Table 13. Uranium–Fluorine Complexes [a]

Formula	Metal
M_4UF_8	NH_4
M_3UF_7	Li, Na, K, Rb, Cs
M_2UF_6	Na, K, Rb, Cs, NH_4
$M_7U_6F_{31}$	Na, K, Rb, NH_4
MUF_5	Li, Rb, Cs
MU_2F_9	Na, K
MU_3F_{13}	K, Rb
MU_4F_{17}	Li
MU_6F_{25}	K, Rb, Cs
$M_2U_3F_{14}$	Cs
MUF_6	Ca, Sr, Ba, Pb

[a] Refs. 117–118.

It forms reddish-brown crystals with metallic luster. It is soluble in liquid chlorine from which it may be recrystallized.

Uranium hexachloride [13763-23-0], UCl_6, is a dark green solid obtained by disproportionation of UCl_5 in vacuum:

$$2\,UCl_5 \rightarrow UCl_6 + UCl_4$$

It sublimes at 75–100°C and 0.01 Pa (10^{-4} mm Hg). Like UCl_5, it is only of scientific interest.

Uranium Bromides and Iodides. With bromine and iodine, uranium forms dihalides. Higher bromides and iodides are not stable and decompose upon heating.

Uranium tribromide [13470-19-4], UBr_3, is a reddish-brown crystalline material that is prepared from UH_3 and HBr or directly from the elements:

$$UH_3 + 3\,HBr \rightarrow UBr_3 + 3\,H_2$$

$$2\,U + 3\,Br_2 \rightarrow 2\,UBr_3$$

Purification by gas-phase transport gives black crystals.

Uranium tetrabromide [13470-20-7], UBr_4, forms dark brown, highly hygroscopic crystals. It is prepared by heating clean uranium turnings in a stream of nitrogen saturated with bromine vapor:

$$U + 2\,Br_2 \rightarrow UBr_4$$

The crude compound is purified by vacuum distillation or sublimation in a stream of nitrogen saturated with bromine vapor. Uranium tetrabromide melts at 519 ± 2°C and boils at 765°C and 98.6 kPa (740 mm Hg). It is soluble in numerous organic solvents; a few molten-salt systems have been studied. Highly pure uranium tetrabromide was prepared in the course of the atomic-weight determination of uranium (117).

Uranium pentabromide [13775-16-1], UBr_5, a dark brown hygroscopic solid, is unstable; it is formed by extraction of uranium tetrabromide with liquid bromine at 55°C, followed by recrystallization from liquid bromine or by bromination of uranium turnings in the presence of acetonitrile:

$$2\,UBr_4 + Br_2 \rightarrow 2\,UBr_5$$

$$2\,U + 5\,Br_2 \rightarrow 2\,UBr_5$$

Uranium triiodide [13775-18-8], UI_3, and uranium tetraiodide [13470-22-9], UI_4, are prepared from the elements:

$$U + 2\,I_2 \rightarrow UI_4$$

$$2\,U + 3\,I_2 \rightarrow 2\,UI_3$$

Because UI_4 decomposes to UI_3,

$$2\,UI_4 \rightarrow 2\,UI_3 + I_2$$

the product depends on the temperature and partial pressure of the iodine.

Both iodides form black lustrous crystals. They have been used in attempts to grow uranium metal by the hot-wire method, but these experiments met with little success because the melting point of uranium is too low for crystal-bar growth.

Oxides. Economic and military pressures have contributed to extensive studies of the uranium–oxygen system (46,49,118–119). The oxidation-reduction potential of uranium permits an almost continuous oxygen variation in a given uranium oxide crystal, and the uranium–oxygen system is of an extremely complicated nature. The presence of nonstoichiometric oxygen in the crystal lattice of certain of the oxide phases has been observed.

Alternations of the oxidation state of uranium are significant in recovering the values from ores and also in refining ore concentrates to standards of nuclear-grade purity. Solubility and ion complexing depend largely on the oxidation states of uranium.

Uranium dioxide, UO_2, melts at ca 2800°C. Its crystal structure is fcc, and its density is ca 10.97 g/cm^3. Uranium dioxide is very stable and occurs in nature as uraninite and pitchblende. It is of particular importance in the uranium-refining operation and is the intermediate oxide produced in the manufacture of the metal and the tetrafluoride. Uranium dioxide in both natural and isotopically enriched forms is used widely in the manufacture of fuel pellets for power reactors. Its value in this application is due to its desirable physical properties (118). Average linear coefficient of expansion between 20 and 946°C is 10.8×10^{-6}, which is satisfactory for reactor materials. Volume-expansion coefficients are also satisfactory in reactor temperature ranges. Because of the low heat conductivity of uranium dioxide, it is difficult to remove heat generated within the fuel elements of reactors. Pressed and sintered uranium dioxide is stable when exposed to air or water below 300°C. It is also inert to superheated steam (118).

The reaction of uranium dioxide with hydrogen fluoride to produce uranium tetrafluoride is of great importance as an intermediate step in the production of uranium hexafluoride. With elemental fluorine at elevated temperatures, uranium dioxide forms the hexafluoride. Nonoxidizing chlorinating agents may be used to convert the dioxide to the tetrachloride; with nitric acid, UO_2 gives a solution of uranyl nitrate.

There is no complete agreement as to the maximum oxygen content of UO_2. A value of $UO_{2.25}$ (lower limit for U_4O_9) appears to be the most likely value. Its homogeneity ranges from $O:U = 2.235$ to 2.245 at room temperature; therefore, the compound should be given the formula U_4O_{9-x}.

Spheres may be prepared either by the sol-gel process, or by plasma spheroidization. Single crystals have been grown by sublimation and by electrolysis of molten UO_2Cl_2, uranyl chloride [7791-26-6].

Tetrauranium enneaoxide [12037-15-9], U_4O_9, forms black crystals and is derived from the UO_2 structure at the oxygen-rich side of the UO_2 phase region. It is a nonstoichiometric compound and occurs in three different modifications. Above 1123°C, U_4O_9 decomposes into UO_{2+x} and $UO_{ca\ 2.6}$.

Triuranium heptoxide [12037-04-6]. Four oxides close to the stoichiometry U_3O_7 are known to exist, three tetragonal and one rhombic. Their exact composition depends on the reaction temperature and on the oxygen partial pressure. The β-, γ-, and δ-modification have a more complicated composition, such as $U_{16}O_{37}$ [65982-88-9] (β), $U_{16}O_{38}$ [37217-11-1] (γ), and U_8O_{19} [32217-11-1] (δ). All oxides of the U_3O_7 group are black.

Triuranium octoxide [1344-59-8], U_3O_8, like U_3O_7, is a component of a complicated phase system. The lower and upper phase limits, depending on the oxygen partial pressure, have values from $UO_{2.62}$ to $UO_{2.665}$. Triuranium octoxide is found in pitchblende. The DOE's estimate of U.S. production of uranium is based on U_3O_8 (see Table 14).

Table 14.   U.S. Production and Demand of Uranium[a]

Year	Production, 1000 t U_3O_8	Demand, 1000 t U_3O_8
1982	20.0	14.3
1983	23.0	16.6
1984	24.1	18.7
1985	27.1	19.0
1986	29.1	19.6
1987	27.7	20.5
1988	27.1	20.9
1989	26.3	20.6
1990	25.7	22.1

[a] Ref. 120.

Uranium trioxide, UO_3, exists in no less than six polymorphic modifications that are distinct from each other by their crystallographic properties and their color.

α-UO_3: brown, hexagonal	δ-UO_3: red, cubic
β-UO_3: orange, monoclinic	ϵ-UO_3: brick red, triclinic
γ-UO_3: bright yellow, rhombic	η-UO_3: rhombic

In addition, there exists a substoichiometric phase, $UO_{2.9}$, which has a rhombic structure and is olive green.

Uranium trioxide is readily obtained from the thermal decomposition of various uranyl compounds, such as peroxides, carbonates, oxalates, ammonium uranates, and uranyl nitrate. Most uranyl salts may be converted to UO_3 in an oxygen stream without formation of U_3O_8 under carefully controlled temperature conditions.

Uranium monoxide [12035-97-1], UO, is of no practical importance. At present, it is believed to exist as a thin, brown-to-black film on the surface of highly pure uranium metal, and its existence has been verified only by x-ray diffraction, but it seems that the apparent stoichiometry "UO" is stabilized by N or C (54). Uranium monoxide forms very slowly above 2000°C in mixtures of UO_2 and uranium. This gaseous species is well characterized by mass spectroscopy, thermodynamics, ir spectroscopy, etc.

A pressure–temperature phase diagram of the uranium–oxygen system in the range UO_2–UO_3 is shown in Figure 21 (121).

Uranium peroxide [19525-15-6], $UO_4 \cdot x\,H_2O$, is not known to exist in the anhydrous state, but the hydrated peroxide is important in purification technology. It may be prepared by precipitation from uranyl nitrate solution with hydrogen peroxide. Dehydration gives UO_3.

Other Oxide Phases.   Several hydrated oxides have been extensively studied, including $UO_3 \cdot 2H_2O$ [20593-39-9], $UO_3 \cdot H_2O$ [12060-10-5], and $UO_3 \cdot 0.5H_2O$ [12326-20-4]. Most of the recent work on these compounds has been done at Argonne National Laboratory (121) and in the Netherlands (45).

Nitrates.   Uranyl nitrate hexahydrate, UNH, $UO_2(NO_3)_2 \cdot 6H_2O$, is orthorhombic; it may be prepared by evaporating a neutral solution until it has attained a temperature of 118°C and letting it cool to room temperature. Since 118°C is the melting point of UNH in its water of crystallization upon cooling, $UO_2(NO_3)_2 \cdot 6H_2O$ crystallizes. The heat of formation is −3187.8 kJ/mol (−761.9 kcal/mol). It is soluble in organic solvents, eg, diethyl ether, a property that led to the first known purification of uranium by

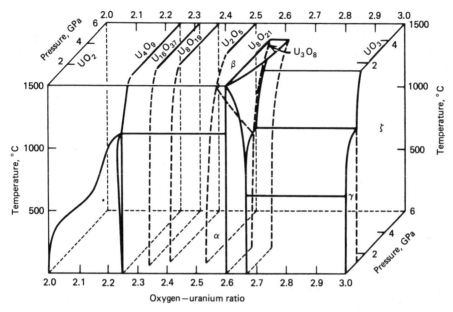

Figure 21. Pressure–temperature phase relationships in the U–O system; UO_2–UO_3 range (123). To convert GPa to kbar, multiply by 10.

Péligot in 1841. The ability of water-immiscible organic solvents to remove uranium from aqueous nitrate solutions is the basis for most current refining processes.

Uranyl nitrate is also of importance in the recovery of uranium from process waste and in the reprocessing of irradiated fuels.

Phosphates and Arsenates. When phosphoric acid, H_3PO_4, or arsenic acid, H_3AsO_4, are added to uranyl salt solutions, $HUO_2PO_4.4H_2O$ [1310-86-7], and $HUO_2AsO_4.4H_2O$ [59305-92-9], respectively, precipitate. The hydrogen can be exchanged for alkali or alkaline-earth or bivalent metal ions:

$$HUO_2XO_4 + M^+ \rightarrow MUO_2XO_4 + H^+$$

$$2\,HUO_2XO_4 + M'^{++} \rightarrow M'(UO_2XO_4)_2 + 2\,H^+$$

where X = P and As; M = Li, Na, K, Rb, Cs, and NH_4; and M' = Mg, Ca, Sr, Ba, and Cu. The resulting salts are identical with a number of natural minerals, such as autunite [12333-86-7, 16390-74-2], $Ca(UO_2PO_4)_2.8$–$12H_2O$, torbernite [26283-21-6], $Cu(UO_2PO_4)_2).12H_2O$, or uranospinite [12255-17-3], $Ca(UO_2AsO_4)_2.10H_2O$, as well as numerous others (26–27,122).

Sulfates. Uranyl sulfate monohydrate [19415-82-8] and trihydrate [12384-63-3] are stable. In ores that are difficult to digest, the high heat stability of uranyl sulfate is used in a rather unique process. The ore is sulfatized at high temperature converting the iron and aluminum, as well as the uranium, to sulfates. Upon heating to 600°C, the sulfates of iron and aluminum are converted to oxides, which are insoluble in water. The soluble uranyl sulfate may be leached away, leaving Fe and Al behind.

Uranyl sulfate trihydrate, $UO_2SO_4.3H_2O$, forms lemon-yellow crystals with green fluorescence. A potassium uranyl sulfate dihydrate [18866-79-0], $K_2UO_2(SO_4)_2.2H_2O$, was the compound used by Becquerel in the discovery of radioactivity.

A solution of uranyl sulfate enriched in ^{235}U has been used in homogenous reactions.

Nitrides. Uranium forms three nitrides, UN [25658-43-9], $UN_{1.5}$ [12033-85-1], and $UN_{1.75}$ [12266-20-5]. The existence of a dinitride, UN_2, is questionable. The highest U:N ratio observed so far is 1:1.9. The nitrides are obtained by synthesis from the elements under varying conditions. Uranium mononitride is the only stable nitride above 1300°C. It exists above 1700°C. The other nitrides exist at lower temperatures.

Uranium mononitride has been considered as a ceramic reactor fuel material (123). It decomposes at 101.3 kPa (1 atm) at 2650 ± 100°C, but may be arc-melted at 293.7 kPa (2.9 atm) in nitrogen.

Carbides. In the system uranium–carbon, the compounds UC [12070-09-6], U_2C_3 [12076-62-9], and UC_2 [12071-33-9] have been identified. These compounds have been considered as ceramic reactor fuel materials (123). Uranium monocarbide, UC (δ-phase), melts at 2350–2400°C. It may be prepared by a number of reactions, such as arc-melting of uranium-carbon compacts, sintering of UO_2–C mixtures, and other techniques. Uranium sesquicarbide, U_2C_3 (ϵ-phase), mp 2417°C, and uranium dicarbide, UC_2 (ζ-phase), mp 2475°C, are obtained from the elements by sintering or arc-melting. The black carbides have been used in high temperature reactors and in gas-cooled reactors.

Uranium dicarbide reacts with water to form a grayish-black solid (probably a hydrated oxide) and a mixture of gaseous and liquid hydrocarbons.

Carbonates. Rutherfordine [12202-79-8], UO_2CO_3, the simple uranyl carbonate, is a naturally occurring mineral. Uranyl carbonates are of importance in high-lime ores, which may be leached by the carbonate process. Tetrasodium uranyl tricarbonate [60897-40-7], $Na_4UO_2(CO_3)_3$, is obtained when ores are pressure-leached at elevated temperatures with soda-ash solutions. The high solubility of alkali and ammonium uranyl tricarbonates in aqueous solutions facilitates removal of impurities, including Fe, Al, Ni, or Cr, that may be precipitated as oxycarbonates or hydroxides. Hexavalent uranium can be extracted from sodium carbonate solutions with basic anion exchange resins. Tetrasodium uranyl tricarbonate decomposes at 400°C to sodium uranate [13510-99-1] and sodium carbonate.

Tetraammonium uranyl tricarbonate, $(NH_4)_4UO_2(CO_3)_3$ forms a series of solid solutions with the hexavalent plutonium carbonate, $(NH_4)_4PuO_2(CO_3)_3$ [72573-65-0]. These solid solutions, $(NH_4)_4(U_xPU_{1-x}O_2)(CO_3)_3$, may be decomposed in hydrogen-donor mixture to form homogenous $(U,Pu)O_2$. They are therefore of great technical importance in fuel element production.

Hexavalent uranium can be reduced in carbonate solutions to the tetravalent state with zinc, hydrosulfate, and ammonia.

Alloys. The alloys of uranium are of importance in reactor technology and have been studied in great detail. Numerous phase diagrams have been established (124).

Numerous other compounds of uranium have been prepared, but most of these are of scientific interest only and have no use in technical applications (see General references).

Health and Safety Factors

Chemical Toxicity. Uranium is not only toxic because of its radiation, but it is also chemically toxic to the same degree as, for example, arsenic (125). LD_{50} values of 40–297 mg/g body weight for male rats have been reported (126). However, the toxicity of uranium compounds varies. Uranium compounds may be ingested, inhaled, or absorbed through the skin. In acute uranium poisoning, kidney lesions, internal hemorrhage, and liver-cell changes were observed. In general, the following degrees of toxicity have been found:

> highly toxic: $UO_2(NO_3)_2.6H_2O$, UO_2F_2, UCl_4, UCl_5
> moderately toxic: UO_3, $Na_2U_2O_7$, $(NH_4)_2U_2O_7$,
> nontoxic: UF_4, UO_2, UO_4, U_3O_8.

Standard laboratory protective measures against chemical poisoning by uranium are mandatory, eg, no pipetting by the mouth; no smoking or eating in the laboratory; protective clothing; surgical gloves; and in operations involving dust formation, face masks, constant ventilation of working areas, and glove boxes.

Radiation Toxicity. The toxicity of uranium caused by its radiation depends on the isotopes present. Such isotopes as ^{232}U, which emits a fairly strong γ-radiation, should be handled in a hot cell, and ^{234}U, ^{235}U, ^{236}U, and ^{233}U should be manipulated in a glove box; ^{235}U and ^{238}U, because of their soft radiation, can be handled on an open laboratory bench. The laboratory should be equipped as an α-laboratory. In ore dressing, the presence of radium, radon, or protactinium, and of the members of the natural decay series should be taken into account.

Pyrophoricity. Finely divided uranium metal, some alloys, and uranium hydride are pyrophoric, ie, they ignite spontaneously in air or oxygen. For such materials, protective atmospheres, for instance in an inert-gas glovebox, should be provided.

Criticality. Large quantities of ^{233}U or ^{235}U that exceed the minimum critical mass may be a hazard because they may be the source of an unexpected critical excursion unless special preventive measures are taken, including uncritical configuration, neutron poisons, and administrative control (54,127–128). A critical excursion is an extremely treacherous phenomenon, and all possible precautions should be taken to prevent it from occurring.

BIBLIOGRAPHY

"Uranium and Uranium Compounds" in *ECT* 1st ed., Vol. 14, pp. 432–458, by J. J. Katz, Argonne National Laboratory; "Uranium and Uranium Compounds" in *ECT* 2nd ed., Vol. 21, pp. 1–36, by V. L. Mattson, Kerr-McGee Corporation.

1. M. H. Klaproth, *Chem. Ann.* **2,** 387 (1789).
2. E. Péligot, *Compt. Rend.* **13,** 417 (1842).
3. E. Péligot, *Ann. Chim. Phys.* **5,** 5 (1842).
4. H. Becquerel, *Compt. Rend.* **122,** 420 (1896).
5. E. Fermi, E. Amaldi, O. D'Agostino, F. Rasetti, and E. Segré, *Proc. R. Soc. London* **A146,** 483 (1934).
6. O. Hahn and F. Strassmann, *Naturwissenschaften* **27,** 11 (1939).
7. C. Allardice and E. R. Trapnell, *The First Pile*, TID-292, U.S. DOE Technical Information Center, Oak Ridge, Tenn., Mar. 1955.
8. A. O. Nier, *Phys. Rev.* **55,** 150 (1939).
9. S. Villani, *Uranium Enrichment*, Springer-Verlag, New York, 1979.
10. C. D. Harrington and A. E. Ruehle, *Uranium Production Technology*, D. Van Nostrand Co., New York, 1959.
11. J. W. Clegg and D. D. Foley, *Uranium Ore Processing*, Addison-Wesley Publishing Co., Inc., Reading, Mass., 1958.
12. *Processing of Low-Grade Uranium Ores, Proceedings of a Panel Convened by the IAEA in Vienna, June 27–July 1, 1966*, IAEA, Vienna, Austria, STI/PUB/146, 1967.

13. R. L. Morgan and G. A. Young, *Nuclear Technology, A Bibliography of Selected Classified Report Literature*, TID-3080, U.S. DOE Technical Information Center, Oak Ridge, Tenn., June 1955.
14. G. A. Young, *Feed Materials, A Bibliography of Classified Report Literature*, Pt. 1, TID-3081, U.S. DOE Technical Information Center, Oak Ridge, Tenn., Nov. 1955.
15. J. J. Katz, L. Morss, and G. T. Seaborg, eds., *The Chemistry of the Actinide Elements*, 2nd ed., Chapman & Hall, London, 1983.
16. *Fuel and Heavy Water Availability*, International Nuclear Fuel Cycle Evaluation (INFCE), IAEA, Vienna, Austria, STI/PUB/534, 1980.
17. J. J. Katz and G. T. Seaborg, *Chemistry of the Actinide Elements*, Methuen, New York, 1957, Chapt. 5.
18. K. S. Heier and J. J. W. Rogers, *Geochim. Cosmochim. Acta* **27**, 137 (1963).
19. S. H. U. Bowie, *Uranium Exploration Geology*, IAEA, Vienna, Austria, STI/PUB/277, 1967, pp. 285–300.
20. R. C. Merritt, *The Extractive Metallurgy of Uranium*, Colorado School of Mines and Research Institute and USAEC, Golden, Colo.
21. J. W. Frondel, M. Fleischer, and R. S. Jones, *Geol. Surv. Bull.* **1250**, 69 pp. (1967).
22. C. Frondel, *Geol. Surv. Bull.* **1064**, 400 (1958).
23. R. D. Nininger, *Minerals for Atomic Energy*, 2nd ed., D. Van Nostrand Co., New York, 1956.
24. A. Maucher, *Die Lagerstätten des Urans*, Friedr. Vieweg & Sohn, Braunschweig, FRG, 1966.
25. *Gmelin Handbuch der Anorganischen Chemie*, 8 Aufl. System-Nr. 55, Verlag Chemie GmbH, Berlin, FRG, pp. 12–17.
26. K. Walenta, *Chem. Erde* **24**, 254 (1965).
27. K. Walenta, *Tschermaks Mineral. Petrogr. Mitt.* **9**, 252 (1965); **9**, 111 (1964).
28. F. L. Oetting, M. H. Rand, and R. J. Ackermann, *The Chemical Thermodynamics of Actinide Elements and Compounds*, Pt. 1, IAEA, Vienna, Austria, STI/PUB/424/1, 1976.
29. S. Arajs and R. V. Coevin, *J. Less-Common Met.* **7**, 54 (1964).
30. M. B. Brodsky, N. J. Griffin, and M. J. Odie, *J. Appl. Phys.* **8**, 895 (1969).
31. A. J. Freeman and J. B. Darby, Jr., *The Actinides: Electronic Structure and Related Properties*, 2 vols., Academic Press, Inc., New York, 1974.
32. A. N. Holden, *Physical Metallurgy of Uranium*, Addison-Wesley Publishing Co., Inc., Reading, Mass., 1958.
33. J. W. Ross and D. J. Lam, *Phys. Rev.* **165**, 617 (1968).
34. F. Bates and D. Hughes, *Proc. Phys. Soc. (London)* **B67**, 28 (1954).
35. P. Gordon, *The Magnetic Susceptibility of Uranium and Some Uranium Alloys with Manganese, Iron, Cobalt, and Nickel*, AECU-1833, thesis, Massachusetts Institute of Technology, Cambridge, Mass., 1952.
36. Ref. 31, Vol. 2, p. 244.
37. *Gmelin Handbuch der Anorganischen Chemie, Ergänzungswerk zur 8 Auflage, System-Nr. 55*, Verlag-Chemie, Weinheim, FRG, Vol. A5.
38. G. Dieke and A. B. F. Duncan, *Spectroscopic Properties of Uranium Compounds*, National Nuclear Energy Series Division III, Vol. 2, McGraw-Hill Book Co., Inc., New York, 1949.
39. N. M. Edelstein, ed., *Lanthanide and Actinide Chemistry and Spectroscopy*, ACS Symposium Series *131*, ACS, Washington, D.C., 1980.
40. J. W. Steinhaus, M. V. Phillips, J. B. Moody, L. J. Radziewsky, Jr., K. J. Fisher, and D. K. Hohn, *The Emission Spectrum of Uranium between 19080 cm^{-1} and 30261 cm^{-1}*, Los Alamos Report LA-4944, Los Alamos, N.M., Aug. 1972.
41. J. A. Bearden, *Rev. Mod. Phys.* **39**, 78 (1967).
42. Y. Cauchois and C. Senemaud, *International Tables of Selected Constants*, Vol. 18, Pergamon Press, Oxford, England, 1978.
43. M. Siegbahn, *Spektroskopie der Röntgenstrahlen*, 2nd ed., Springer Verlag, Berlin, FRG, 1931.
44. W. D. Wilkinson, *Uranium Metallurgy*, Vols. 1 and 2, Wiley-Interscience, New York, 1962.
45. E. Cordfunke, *The Chemistry of Uranium*, Elsevier, Amsterdam, The Netherlands, 1969.
46. J. J. Chernyaev, *Complex Compounds of Uranium*, Oldbourne Press, London, 1966.
47. J. E. Vance and J. C. Warner, *Uranium Technology General Survey*, National Nuclear Energy Series Div. VII, Vol. 2A, USAEC Technical Information Service, Oak Ridge, Tenn., 1953.
48. J. J. Katz and E. Rabinowitch, *The Chemistry of Uranium*, National Nuclear Energy Series Div. VIII, Vol. 5, McGraw-Hill Book Co., New York, 1952.
49. N. E. Holden, *Pure Appl. Chem.* **52**, 2371 (1979).

50. *The Oklo Phenomenon, Proceedings of a Symposium, Libreville, 1975,* IAEA, Vienna, Austria, STI/PUB/405, 1975, 648 pp.

51. C. M. Lederer and V. S. Shirley, *Table of Isotopes,* 7th ed., John Wiley & Sons, Inc., New York, 1977.

52. G. T. Seaborg and L. I. Katzin, *Production and Separation of Uranium-233: Survey, National Nuclear Energy Series Div. IV,* Vol. 17A, USAEC Technical Information Service, Oak Ridge, Tenn., 1951; declassified with deletions as TID-5223.

53. L. I. Katzin, ed., *Production and Separation of Uranium-233: Collected Papers, National Nuclear Energy Series Div. IV,* Vol. 17B, Books 1 and 2, USAEC Technical Information Service, Oak Ridge, Tenn., 1952; declassified with deletions as TID-5223.

54. F. S. Patton, J. M. Googin, and W. L. Griffith, *Int. Ser. Monogr. Nucl. Energy Div. 9* **2,** (1963).

55. Ref. 11, p. 120.

56. Ref. 11, p. 118.

57. Ref. 11, p. 167.

58. I. R. Higgins, *Mechanical Features of the Higgins Continuous Ion Exchange Column,* Oak Ridge National Laboratory Report ORNL-1907, Oak Ridge, Tenn., Oct. 25, 1955.

59. J. H. Gittus, *Uranium, Metallurgy of the Rarer Metals,* Butterworth & Co., Inc., Washington, D.C., 1963.

60. Ref. 11, p. 193.

61. Ref. 11, p. 216.

62. E. H. P. Cordfunke, *J. Inorg. Nucl. Chem.* **24,** 303 (1962).

63. *Ibid.,* **32,** 3129 (1970).

64. J. J. Katz and E. Rabinowitch, *The Chemistry of Uranium: Collected Papers,* TID-5290, Books 1 and 2, Technical Information Service Extension, USAEC, Oak Ridge, Tenn., 1958.

65. Ref. 10, p. 43.

66. Ref. 10, p. 183.

67. Ref. 10, p. 189.

68. Ref. 10, p. 205.

69. Ref. 10, p. 215.

70. Ref. 10, p. 219.

71. Ref. 10, p. 462.

72. W. Bacher and E. Jacob, *Chemikerzeitung* **106,** 117 (1982).

73. Ref. 10. p. 246.

74. W. M. Leaden, *Proceedings of the Metallurgy Information Meeting Held at Oak Ridge, Apr. 11–13, 1955,* TID-7501, Pt. 1, Technical Information Center, Oak Ridge, Tenn., June 1960, pp. 85–108.

75. *AEC Gaseous Diffusion Plant Operations,* ORO-658, Feb. 1968, p. 5.

76. *AEC Gaseous Diffusion Plant Operations,* ORO-684, Jan. 1972.

77. *Data on New Gaseous Diffusion Plants,* ORO-685, U.S. DOE Oak Ridge Operations Office, Oak Ridge, Tenn., Apr. 1972.

78. J. Charpin, P. Plurien, and S. Mommejas, *Proc. 2nd Int. Conf. Peaceful Uses Atomic Energy, Geneva* **4,** 380 (1958).

79. J. W. Beams, A. C. Hagg, and E. V. Murphree, *Developments in the Centrifuge Separation Project, National Nuclear Energy Series Div. X,* Vol. 1, TID-5230, USAEC Technical Information Service, Oak Ridge, Tenn., 1951.

80. *Uranium Enrichment Annual Report 1980,* ORO-822, U.S. DOE Oak Ridge Operations Office, Oak Ridge, Tenn., 1981.

81. J. W. Beams, *Proc. 2nd Int. Conf. Peaceful Uses Atomic Energy, Geneva* **4,** 428 (1958).

82. S. Villani, *Isotope Separation,* American Nuclear Society, Lagrange Park, Ill., 1979, Chapt. 6.

83. A. Guthrie and R. K. Wakerling, *Vacuum Equipment and Techniques, National Nuclear Energy Series Div. I,* Vol. 1, McGraw-Hill Book Co., Inc., New York, 1949.

84. R. K. Wakerling and A. Guthrie, *Magnets and Magnetic Measuring Techniques, National Nuclear Energy Series Div. I,* Vol. 2, TID-5215, USAEC Technical Information Service, Oak Ridge, Tenn., 1951.

85. R. K. Wakerling and A. Guthrie, *Electrical Circuits for Calutrons, National Nuclear Energy Series Div. I,* Vol. 3, TID-5217, USAEC Technical Information Service, Oak Ridge, Tenn., 1952.

86. R. K. Wakerling and A. Guthrie, *Electromagnetic Separation of Isotopes in Commercial Quantities, National Nuclear Energy Series Div. I,* Vol. 4, TID-5217, USAEC Technical Information Service, Oak Ridge, Tenn., 1951.

87. A. Guthrie and R. K. Wakerling, *Characteristics of Electrical Discharges in Magnetic Fields*, National Nuclear Energy Series Div. *I*, Vol. 5, McGraw-Hill Book Co., Inc., New York, 1949.

88. R. K. Wakerling and A. Guthrie, *Sources and Collectors for Uses in Calutrons*, National Nuclear Energy Series Div. *I*, Vol. 6, TID-5218, USAEC Technical Information Service, Oak Ridge, Tenn., 1951.

89. H. W. Savage, *Separation of Isotopes in Calutron Units*, National Nuclear Energy Series Div. *I*, Vol. 7, TID-5233, USAEC Technical Information Service, Oak Ridge, Tenn., 1951.

90. A. H. Barnes, S. M. McNeille, C. Starr, and H. W. Savage, *Problems of Physics in the Ion Source*, National Nuclear Energy Series, Div. *I*, Vol. 8, TID-5219, USAEC Technical Information Service, Oak Ridge, Tenn., 1951.

91. J. D. Trimmer, H. Pearlman, and H. W. Savage, *High Voltage Problems*, National Nuclear Energy Series Div. *I*, Vol. 9, TID-5211, USAEC Technical Information Service, Oak Ridge, Tenn., 1951.

92. C. R. Baldock, E. D. Hudson, and H. W. Savage, *Electrical Equipment for Tanks and Magnets*, National Nuclear Energy Series Div. *I*, Vol. 10, TID-5214, USAEC Technical Information Service, Oak Ridge, Tenn., 1952.

93. C. E. Normand, F. A. Knox, G. W. Monk, A. J. Samuel, and W. R. Perret, *Vacuum Problems and Techniques*, National Nuclear Energy Series Div. *I*, Vol. 11, TID-5210, USAEC Technical Information Service, Oak Ridge, Tenn., 1950.

94. G. A. Akin, H. P. Kackenmaster, R. J. Schrader, J. W. Strohecker, and R. E. Tate, *Chemical Processing Equipment: Electromagnetic Separation Process*, National Nuclear Energy Series Div. *I*, Vol. 12, TID-5232, USAEC Technical Information Service, Oak Ridge, Tenn., 1951.

95. A. E. Cameron, *Determination of the Isotopic Composition of Uranium*, National Nuclear Energy Series Div. *I*, Vol. 13, TID-5213, USAEC Technical Information Service, Oak Ridge, Tenn., 1950.

96. B. Harmatz, *Enrichment of Gram Quantities of U-234*, TID-355, USAEC Technical Information Service, Oak Ridge, Tenn., Oct. 1949.

97. B. Harmatz, H. C. McCurdy, F. N. Case, and R. S. Livingston, *Enrichment of Uranium-236*, ORNL-1169, Oak Ridge National Laboratory, Oak Ridge, Tenn., Dec. 1951.

98. H. W. Savage and P. E. Wilkinson, *The Separation and Collection of U-236 by the Electromagnetic Process*, Y-697, Y-12 Plant Union Carbide Nuclear Division, Oak Ridge, Tenn., Nov. 30, 1950.

99. P. A. Abelson, N. Rosen, and J. I. Hoover, *Liquid Thermal Diffusion*, National Nuclear Energy Series Div. *IX*, Vol. 1, TID-5229, USAEC Technical Information Service, Oak Ridge, Tenn., 1951.

100. R. J. Jensen, O. P. Judd, and J. A. Sullivan, *Los Alamos Sci.* 3(1), 2 (1982).

101. R. Leuze, S. D. Clinton, J. M. Chilton, and V. C. A. Vaughen in D. E. Ferguson, compiler, *Transuranium Quarterly Progress Report for Period Ending Feb. 28, 1962*, ORNL-3290, Feb. 28, 1962, p. 79.

102. R. E. Leuze, S. D. Clinton, J. M. Chilton, and V. C. A. Vaughen in D. E. Ferguson, compiler, *Transuranium Quarterly Progress Report for Period Ending Aug. 31, 1962*, ORNL-3375, Jan. 18, 1963, p. 51.

103. S. M. Stoller and R. B. Richards, *Reactor Handbook*, Vol. 2, Interscience Publishers, a division of John Wiley & Sons, Inc., New York, 1962.

104. A. T. Gresky, M. R. Bennett, S. S. Brandt, W. T. McDuffee, and J. A. Savolainen, *Progress Report: Laboratory Development of the Thorex Process*, ORNL-1367 (rev.), Dec. 17, 1962.

105. P. E. Figgins and R. J. Bernardinelli, *J. Inorg. Nucl. Chem.* **28**, 2193 (1966).

106. F. E. Croxton, *Uranium and Its Compounds—A Bibliography of Unclassified Literature*, K-295, Pt. II, Union Carbide Nuclear Division, Oak Ridge, Tenn., Mar. 1, 1955.

107. K. E. Allen, *Uranium and Its Compounds—A Bibliography of Unclassified Literature*, TID-3041, U.S. DOE Technical Information Center, Oak Ridge, Tenn., June 30, 1953.

108. Ref. 37, Pts. C and E.

109. H. A. Friedman, C. F. Weaver, and W. R. Grimes, *J. Inorg. Nucl. Chem.* **32**, 3131 (1970).

110. *Current Commission Methods For Producing UO_3, UF_4, and UF_6*, TID-5295, Technical Information Service, Oak Ridge, Tenn., Jan. 1956.

111. A. L. Allen, R. W. Anderson, R. M. McGill, and E. W. Powell, *Electrochemical Preparation of Uranium Tetrafluoride*, Pt. 1, K-680, Union Carbide Corporation, Oak Ridge, Tenn., Nov. 10, 1950.

112. A. L. Allen, R. W. Anderson, and E. W. Powell, *Electrochemical Preparation of Uranium Tetrafluoride*, Pt. 2, K-681, Union Carbide Corporation, Oak Ridge, Tenn., Dec. 22, 1950.

113. J. W. Arendt, E. W. Powell, and H. W. Sayler, *A Brief Guide to UF_6 Handling*, K-1323, Union Carbide Corporation, Oak Ridge, Tenn., Feb. 18, 1957.

114. *Procedures for Handling and Analysis of Uranium Hexafluoride*, Vols. 1 and 2, ORO-671-1 and ORO 671-2, USAEC Oak Ridge Operations Office, Oak Ridge, Tenn., Apr. 1972.

115. R. E. Thoma, ed., *Phase Diagrams of Nuclear Reactor Materials*, ORNL-2548, Oak Ridge National Laboratory, Oak Ridge, Tenn., Nov. 6, 1959.

116. R. E. Thoma and W. R. Grimes, *Phase Equilibrium Diagrams for Fused Salt Systems*, ORNL-2295, Battelle Memorial Institute, Columbus, Ohio, June 2, 1958.

117. O. Hönigschmid and W. E. Schilz, *Z. Anorg. Allg. Chem.* **170**, 145 (1928).

118. J. Belle, *Uranium Dioxide: Properties and Nuclear Applications*, Superintendent of Documents, U.S. Government Printing Office, Washington, D.C., 1961.

119. Ref. 37, Vol. C2.

120. *Chem. Week*, 13 (March 2, 1983).

121. H. R. Hoekstra, S. Siegel, and F. X. Gallagher, *J. Inorg. Nucl. Chem.* **32**, 3237 (1970).

122. F. Weigel and G. Hoffmann, *J. Less-Common Met.* **44**, 99 (1976).

123. J. T. Waber, P. Chiotti, and W. N. Miner: *Compounds of Interest in Nuclear Reactor Technology*, Vol. 10 of *Nuclear Metallurgy*, Metallurgical Society of AIME, Ann Arbor, Mich., 1964, pp. 225, 387, 525.

124. A. Rough and A. A. Bauer, *Constitution of Uranium and Thorium Alloys*, BMI-1300, Battelle Memorial Institute, Columbus, Ohio, June 2, 1958.

125. S. Wexler, *Uranium Poisoning*, MDDC-1054, July 9, 1942.

126. C. Voegtlin and H. Hodge, *Pharmacology and Toxicology of Uranium Compounds*, National Nuclear Energy Series Div. *VI*, Vol. 1, Books 1–4, McGraw-Hill Book Co., Inc., New York, 1949.

127. H. C. Paxton and J. T. Thomas, *Critical Dimensions of Systems Containing, U^{235}, Pu^{239}, and U^{233}*, TID-7028, U.S. DOE Technical Information Service, Oak Ridge, Tenn., June 1964.

128. R. D. Carter, G. R. Kiel, and K. R. Ridgway, *Criticality Handbook*, Vols. 1 and 2, ARH-600, June 30, 1968, and ARH-600, May 23, 1969.

General References

Ref. 25 is also a general reference.

Gmelin Handbuch der Anorganischen Chemie, Ergänzungswerk zur 8 Auflage System-Nr. 55, Verlag Chemie, Weinheim, FRG and Springer-Verlag, Heidelberg, FRG; Ergänzungsband: Al, *Uranlagerstätten*, Springer, 1979; A2, *Isotope*, 1980; A3, *Technologie, Verwendung*, 1981; A4, *Irradiated Fuel Reprocessing*, 1981; A5, *Spectra*, 1982; A7, *Analysis, Biology;* C1, *Verbindungen mit Edelgasen und Wasserstoff, System U-Sauerstoff*, 1981; C_2, U_3O_8, UO_3, *Hydroxides, Oxide Hydrates, Peroxides*, 1978; C3, *Ternary and Polynary Oxides*, 1973; C7, *Compounds with Nitrogen*, 1981; C8, *Compounds with Fluorine*, 1980; C9, *Compounds with Chlorine, Bromine, and Iodine*, 1981; C11, *Compounds with Selenium, Tellurium and Boron*, 1981; C14, *Compounds with P, As, Sb, Bi, and Ge*, 1981; D2, *Solvent Extraction*, 1981; D3, *Anion Exchange*, 1982; D4, *Cation Exchange Chromatography*, 1983; E1, *Coordination Compounds*, 1979; E2, *Coordination Compounds*, 1980.

FRITZ WEIGEL
University of Munich

UREA

Urea [57-13-6] was discovered in urine by Rouelle in 1773 and first synthesized from ammonia and cyanic acid by Woehler in 1828. This was the first synthesis of an organic compound from an inorganic compound, and it dealt a deathblow to the vital-force theory. In 1870, urea was produced by heating ammonium carbamate in a sealed tube.

Properties

Urea can be considered the amide of carbamic acid, NH_2COOH, or the diamide of carbonic acid, $CO(OH)_2$. At room temperature, urea is colorless, odorless, and tasteless. Properties are shown in Tables 1–4. Dissolved in water, it hydrolyzes very slowly to ammonium carbamate (1) and eventually decomposes to ammonia and carbon dioxide. This reaction is the basis for the use of urea as fertilizer (qv).

Commercially, urea is produced by the direct dehydration of ammonium carbamate, NH_2COONH_4, at elevated temperature and pressure. Ammonium carbamate is obtained by direct reaction of ammonia (qv) and carbon dioxide (qv). The two reactions are usually carried out simultaneously in a high pressure reactor. Recently, urea has been used commercially as a cattle-feed supplement (see Pet and livestock feeds). Other important applications are the manufacture of resins (see Amino resins and plastics), glues, solvents, and some medicinals. Urea is classified as a nontoxic compound.

At atmospheric pressure and at its melting point, urea decomposes to ammonia, biuret (1), cyanuric acid (qv) (2), ammelide (3), and triuret (4). Biuret is the main and least desirable by-product present in commercial urea. An excessive amount (>2 wt %) of biuret in fertilizer-grade urea is detrimental to plant growth.

(1)	(2)	(3)	(4)
biuret	cyanuric acid	ammelide	triuret

Urea acts as a monobasic substance and forms salts with acids (4). With nitric acid, it forms urea nitrate, $CO(NH_2)_2 \cdot HNO_3$, which decomposes explosively when heated. Solid urea is stable at room temperature and atmospheric pressure. Heated under vacuum at its melting point, it sublimes without change. At 180–190°C under vacuum, urea sublimes and is converted to ammonium cyanate, NH_4OCN (5). When solid urea is rapidly heated in a stream of gaseous ammonia at elevated temperature and at a pressure of several hundred kPa (several atm), it sublimes completely and decomposes partially to cyanic acid, HNCO, and ammonium cyanate. Solid urea dissolves in liquid ammonia and forms the unstable compound urea–ammonia, $CO(NH_2)_2NH_3$, which decomposes above 45°C (2). Urea–ammonia forms salts with

Table 1. Properties of Urea

Property	Value
melting point, °C	135
index of refraction, n_D^{20}	1.484, 1.602
density, d_4^{20}, g/cm^3	1.3230
crystalline form and habit	tetragonal, needles or prisms
free energy of formation, at 25°C, J/mol[a]	−197.150
heat of fusion, J/g[a]	251[b]
heat of solution in water, J/g[a]	243[b]
heat of crystallization, 70% aqueous urea solution, J/g[a]	460[c]
bulk density, g/cm^3	0.74
specific heat, J/(kg·K)[a]	
at 0°C	1.439
50	1.661
100	1.887
150	2.109

[a] To convert J to cal, divide by 4.184.
[b] Endothermic.
[c] Exothermic.

Table 2. Properties of Saturated Aqueous Solutions of Urea

Temperature, °C	Solubility in water, g/100 g solution	Density, g/cm^3	Viscosity mPa·s (= cP)	H$_2$O vapor pressure, kPa[a]
0	41.0	1.120	2.63	0.53
20	51.6	1.147	1.96	1.73
40	62.2	1.167	1.72	5.33
60	72.2	1.184	1.72	12.00
80	80.6	1.198	1.93	21.33
100	88.3	1.210	2.35	29.33
120	95.5	1.221	2.93	18.00
130	99.2	1.226	3.25	0.93

[a] To convert kPa to mm Hg, multiply by 7.5.

alkali metals, eg, NH_2CONHM or $CO(NHM)_2$. The conversion of urea to biuret is promoted by low pressure, high temperature, and prolonged heating. At 10–20 MPa (100–200 atm), biuret gives urea when heated with ammonia (6–7).

Urea reacts with silver nitrate, $AgNO_3$, in the presence of sodium hydroxide, NaOH, and forms a diargentic derivative (5) of a pale-yellow color. Sodium hydroxide promotes the change of urea into the imidol form (6):

which then reacts with silver nitrate. Oxidizing agents in the presence of sodium hy-

Table 3. Properties of Saturated Solutions of Urea in Ammonia[a]

Temperature, °C	Urea in solution, wt %	Vapor pressure of solution, kPa[b]
0	36	405
20	49	709
40	68	952
60	79	1094
80	84	1348
100	90	1267
120	96	507

[a] Ref. 2.
[b] To convert kPa to atm, divide by 101.3.

Table 4. Properties of Saturated Solutions of Urea in Methanol and Ethanol[a]

	Methanol		Ethanol	
Temperature, °C	Urea, wt %	Density, g/cm^3	Urea, wt %	Density, g/cm^3
20	22	0.869	5.4	0.804
40	35	0.890	9.3	0.804
60	63	0.930	15.0	0.805

[a] Ref. 3.

droxide convert urea to nitrogen and carbon dioxide. The latter reacts with sodium hydroxide to form sodium carbonate (8):

$$NH_2CONH_2 + 2\,NaOH + 3\,NaOBr \rightarrow N_2 + 3\,NaBr + Na_2CO_3 + 3\,H_2O$$

The reaction of urea with alcohols yields carbamic acid esters, commonly called urethanes (see Urethane polymers):

$$NH_2\overset{\overset{O}{\|}}{C}NH_2 + ROH \longrightarrow NH_2\overset{\overset{O}{\|}}{C}OR + NH_3$$

Urea reacts with formaldehyde and forms compounds such as monomethylolurea, $NH_2CONHCH_2OH$, dimethylolurea, $HOCH_2NHCONHCH_2OH$, and others, depending upon the mol ratio of formaldehyde, to urea and upon the pH of the solution. Hydrogen peroxide and urea give a white crystalline powder, urea peroxide, $CO(NH_2)_2 \cdot H_2O_2$, known under the trade name of Hypersol, an oxidizing agent.

Urea and malonic acid give barbituric acid (**7**), a key compound in medicinal chemistry (see also Hypnotics, sedatives, and anticonvulsants):

(**7**)

malonyl urea or
barbituric acid

Manufacture

Urea is produced from liquid NH_3 and gaseous CO_2 at high pressure and temperature; both reactants are obtained from an ammonia-synthesis plant. The latter is a by-product stream, vented from the CO_2 removal section of the ammonia-synthesis plant. The two feed components are delivered to the high pressure urea reactor, usually at a mol ratio >2.5:1. Depending upon the feed mol ratio, more or less carbamate is converted to urea and water per pass through the reactor.

The formation of ammonium carbamate and the dehydration to urea take place simultaneously, for all practical purposes:

$$2\ NH_3\ +\ CO_2\ \rightleftarrows\ NH_2\overset{\overset{O}{\|}}{C}ONH_4 \tag{1}$$
ammonium
carbamate

$$NH_2\overset{\overset{O}{\|}}{C}ONH_4\ \rightleftarrows\ NH_2\overset{\overset{O}{\|}}{C}NH_2\ +\ H_2O \tag{2}$$
urea

Reaction 1 is highly exothermic. The heat of reaction at 25°C and 101.3 kPa (1 atm) is in the range of 159 kJ/mol (38 kcal/mol) of solid carbamate (9). The excess heat must be removed from the reaction. The rate and the equilibrium of reaction 1 depend greatly upon pressure and temperature, because large volume changes take place. This reaction may only occur at a pressure that is below the pressure of ammonium carbamate at which dissociation begins or, conversely, the operating pressure of the reactor must be maintained above the vapor pressure of ammonium carbamate. Reaction 2 is endothermic by ca 31.4 kJ/mol (7.5 kcal/mol) of urea formed. It takes place mainly in the liquid phase; the rate in the solid phase is much slower with minor variations in volume.

The dissociation pressure of pure carbamate has been investigated extensively (10–12) and the average values are shown in Table 5.

Ammonium Carbamate. Ammonium carbamate is a white crystalline solid which is soluble in water (2). It forms at room temperature by passing ammonia gas over dry ice. In an aqueous solution at room temperature, it is slowly converted to ammonium carbonate, $(NH_4)_2CO_3$, by the addition of one mol of water. Above 60°C, the ammonium carbonate solution reverts to carbamate solution, and at 100°C, only carbamate is present in the solution. Above 150°C, ammonium carbamate loses a mol of water and forms urea. The specific heat of solid ammonium carbamate is given in Table 6. Ammonium carbamate melts at ca 150°C, and has a heat of fusion of ca 16.74 kJ/mol (4.0 kcal/mol). The conversion of carbamate to urea begins at ≤100°C. To obtain an appreciable amount of urea at 100°C requires 20–30 h. The rate of conversion increases with increasing temperature (13–15); at 185°C, ca 50% of the ammonium carbamate is converted to urea in ca 30 min.

Conversion at Equilibrium. The maximum urea conversion at equilibrium attainable at 185°C is ca 53% at infinite heating time. The conversion at equilibrium can be increased either by raising the reactor temperature or by dehydrating ammonium carbamate in the presence of excess ammonia. Excess ammonia shifts the reaction

Table 5.   Vapor Pressure of Pure Ammonium Carbamate at which Dissociation Begins

Temperature, °C	kPa[a]
40	31
60	106
80	314
100	861
120	2,130
140	4,660
160	9,930
180	15,200; 19,300[b]
200	20,300; 36,500[b]

[a] To convert kPa to atm, divide by 101.3.

[b] The value has been extrapolated because, at temperatures above 170°C, the rate of reaction 2 rapidly increases and it is difficult to determine the carbamate vapor pressure owing to the formation of water and urea and the consequent lowering of the partial pressure of ammonium carbamate.

Table 6.   Specific Heat of Solid Ammonium Carbamate

Temperature, °C	J/(g·K)[a]
20	1.67
60	1.92
100	2.18
140	2.43
180	2.59

[a] To convert J to cal, divide by 4.184.

to the right side of the overall equation:

$$2\,NH_3 + CO_2 \rightleftarrows NH_2CONH_2 + H_2O$$

Water, however, has the opposite effect. Actual equilibrium constants at various temperatures are given in Table 7. A detailed study of the effect of pressure on urea conversion is given in ref. 17.

Once-Through Urea Process.   The main problem in urea manufacture is the separation of the unconverted ammonium carbamate and excess ammonia from the urea solution for recycling.

The unconverted carbamate is decomposed to NH_3 and CO_2 gas by heating the effluent mixture at low pressure. The NH_3 and CO_2 gases escape from the urea solution and are utilized for the production of ammonium salts by absorbing the NH_3 in sulfuric,

Table 7.   Reaction Equilibrium Constant[a]

Temperature, °C	Reaction equilibrium constant, K
140	0.695
150	0.850
160	1.075
170	1.375
180	1.800
190	2.380
200	3.180

[a] Ref. 16.

nitric, or phosphoric acid. Such a plant has a relatively low capital cost but has the drawback of a relatively large amount of NH_3 off-gas.

As the demand for pure solid fertilizer-grade urea increased with the years, the once-through plants became less attractive because of the overproduction of ammonium salts with diminishing returns.

Solution-Recycle Process. The NH_3 and CO_2 gases recovered from the reactor effluent mixture in several pressure-staged decomposition sections are absorbed in water and recycled to the reactor as an ammoniacal aqueous solution of ammonium carbamate. The plant is totally independent of the neutralizer section. Almost half of world urea production is based on this type of process.

Mitsui-Toatsu Total Recycle C-Improved Process. The reactor is operated at ca 25 MPa (ca 246 atm) and ca 195°C at an NH_3-to-CO_2 overall mol ratio (fresh feed plus recycle) of ca 4:1. A relatively high conversion of carbamate to urea per pass of ca 67–70% is reported (18).

The unconverted carbamate and the excess ammonia are recovered from the reactor effluent first in the steam-heated high pressure decomposer at ca 17 MPa (ca 168 atm) and ca 155°C, and then in the steam-heated low pressure decomposer at ca 300 kPa (ca 3 atm) and ca 130°C (see Fig. 1).

The low pressure gas is condensed in the low pressure absorber and the liquid is pumped into the high pressure absorber for absorption of the high pressure decomposer gas. Unabsorbed excess ammonia from the high pressure absorber is condensed in the ammonia condenser and recycled to the reactor, as is the concentrated carbamate solution recovered in the high pressure absorber.

The urea is crystallized under vacuum from the 72–74 wt % solution obtained from the low pressure absorber. The product is centrifuged, washed, dried, and elevated to the top of the prilling tower for remelting. The molten urea product is finely divided into droplets by a spray system, solidified to spherical prills by the countercurrent flow of air, and discharged from the bottom of the prilling tower. The intermediate crystallization step allows the production of urea with <0.5 wt % biuret suitable for foliar application.

Heat is recovered by circulating the hot condensate, used to condense the gas from the high pressure decomposer in the high pressure absorber, to the endothermic vacuum crystallizer.

The reactor is lined with titanium as protection against corrosion. Other pieces of the equipment are made of 316L, 316, 304L, and 304 stainless steel, depending upon operating pressure and temperature. Higher process temperatures and carbamate concentrations require 316L and 316SS equipment, whereas at lower temperatures and carbamate concentrations 304L and 304SS are used. Passivating air is introduced into the high pressure decomposer to prevent corrosion of the stainless-steel equipment.

Through the years, numerous process improvements (19–24) have been made, and a steam usage of 0.9 tons per ton of urea produced as a 72–74 wt % urea solution is reported. A large number of urea plants with capacities of up to 1800 t/d are using this process.

Montedison Urea Process. The reactor is operated at ca 20–22 MPa (ca 197–217 atm) and an overall NH_3-to-CO_2 mol ratio of ca 3.5:1 (fresh feed plus recycle). A 62–63% conversion of carbamate to urea per pass is reported (11). The pressure of the reactor

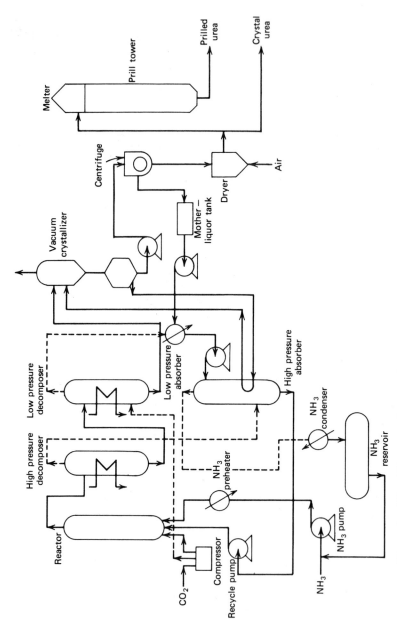

Figure 1. Mitsui Toatsu total recycle C-improved process (18). Courtesy of the British Sulphur Corp.

effluent (see Fig. 2) is reduced to ca 7.5 MPa (ca 74 atm) and steam is heated to recover unconverted NH_3 and CO_2 from the urea solution product. The residual ammonia and CO_2 are recovered in two subsequent pressure-staged decomposers, operating at ca 1.2 MPa (ca 12 atm) and ca 200 kPa (ca 2 atm), respectively.

The 75 wt % aqueous urea solution from the third carbamate decomposer is concentrated to ca 99.5 wt % urea melt in a two-staged high vacuum evaporator system, operating at ca 29 kPa (ca 0.29 atm) and ca 3.4 kPa (ca 0.034 atm). The gas from the third carbamate decomposer is condensed in the water-cooled third absorber and pumped into the second absorber for absorption of the gas from the second carbamate decomposer.

The weak carbamate solution from the second absorber is pumped into the first absorber for absorption of the first carbamate decomposer. The exothermic heat of formation of carbamate is utilized to produce low pressure steam in the first absorber at ca 300 kPa (ca 3 atm) for export from the plant. The reactor is lined with 316L steel; air is injected for passivation.

An improved process, based on the new Isobaric Double-Recycle (IDR) technology, was recently announced (25). The reactor effluent is stripped first with NH_3 gas and then with CO_2 gas, both operating at the synthesis reactor pressure of ca 18–21 MPa (ca 180–210 atm). Considerable reduction in process-steam consumption is reported.

UTI Heat-Recycle Process. In the mid-1970s, a new liquid carbamate recycle gained acceptance by the industry (26–27). It featured fundamentally new concepts:

1. An isothermal reactor is provided with an internal open-ended coil for countercurrent heat transfer from the strongly exothermic process of carbamate formation to the endothermic formation of urea; an increase in reactor conversion per pass is reported.

2. Of the fresh makeup CO_2 feed, stoichiometrically required to produce urea, ca 40–50% is injected into the medium-pressure carbamate decomposition section at ca 2.4 MPa (ca 24 atm), for heat recovery and recycle. A considerable saving in CO_2 gas compression power is attained.

3. More than 70% of the exothermic heat of carbamate formation in the medium-pressure absorption system is exchanged with relatively colder streams within the process for internal recovery of heat and reduction in steam consumption (28). Steam is only used in the first decomposer (see Fig. 3).

The reactor operates at ca 20 MPa (ca 200 atm) and an NH_3-to-CO_2 mol ratio of ca 4.1:1. A 72–74% conversion of carbamate to urea per pass is reported. The excess ammonia is separated from the urea reactor effluent solution at a reduced pressure of ca 2.2 MPa (ca 22 atm). The degassed product is preheated in the first predecomposer before steam heating in the first decomposer; ca 30% of the heat required to decompose the unconverted carbamate is indirectly supplied in the first predecomposer by condensation of the mixture of the first decomposer gas.

The pressure of the urea solution from the first decomposer is reduced to ca 200 kPa (ca 2 atm) and heated in the second decomposer for recovery of residual NH_3 and CO_2 from the urea product solution. The heat required in the second decomposer and in the urea concentrator is recovered from the first decomposer-gas mixture. The urea concentrator operates under mild vacuum to produce an 86–88 wt % product. The gas from the first and second decomposers is condensed and recycled.

The heat recycle process utilizes a unique continuous carbamate analyzer for the automatic water-balance control in the first absorber, thus eliminating the problem

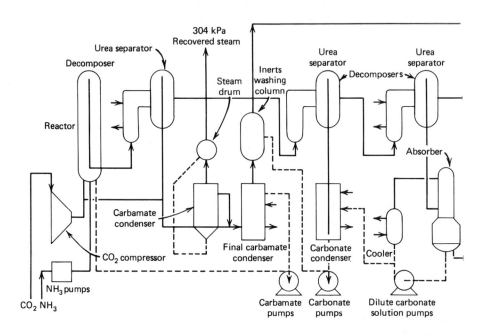

		Synthesis section		
Reaction zone 19.6 MPa	First carbamate decomposition stage, 7.35 MPa		Second carbamate decomposition stage, 1.18 MPa	Third carbamate decomposition stage, 196 kPa

Figure 2. Montedison urea process (18). To convert kPa to psi, multiply by

of carbamate solidification on the water-cooled first condenser tubes.

The reactor is lined with 316L stainless steel. Passivating air is injected into the reactor to prevent corrosion (29). Corrosion is minimized because of the relatively high excess of ammonia employed in the process.

The medium-pressure decomposition and absorption system is made of 316 stainless steel, the low pressure decomposition system of 304L and 304SS. Other equipment is made of conventional material of construction.

Several plants ranging from 300 to 1200 t/d are currently operating or under construction in the United States and around the world. A process steam consumption of 0.5 metric ton per metric ton of urea produced as an 86–88 wt % solution is reported.

High Pressure Gas Stripping. The first such process was developed and commercially applied in the mid-1960s. It is based on high pressure CO_2 gas stripping at reactor pressure and relatively high temperature. The unconverted carbamate is decomposed to NH_3 and CO_2 by the stream of gaseous CO_2 passed through the reactor effluent

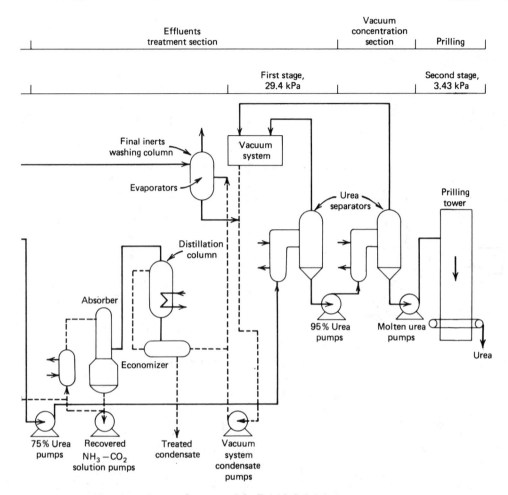

0.145; MPa to psi, multiply by 145. Courtesy of the British Sulphur Corp.

solution at reactor pressure and condensed and recycled to the reactor by gravity. The residual NH_3 and CO_2 in the product are recovered at low pressure by the conventional method of heating carbamate at low pressure. The recovered gas is condensed and recycled.

In a later development, the high pressure ammonia reactor was feed-utilized as the stripping agent for carbamate at reactor pressure.

In contrast to the solution-recycle processes, the stripping process originally required a tall structure to accommodate the reactor, stripper, and condenser. This equipment must be positioned within the structure high enough to ensure the flow of carbamate recycle to the reactor by gravity. However, the more recent development of an eductor for carbamate recycle has greatly reduced the need for a tall structure.

Because of its energy efficiency, the stripping process accounts for almost half of the world's urea production.

Stamicarbon CO_2 Stripping Process. Reactor, carbamate decomposer (stripper), and carbamate condenser each operate at ca 14 MPa (ca 140 atm), at an NH_3-to-CO_2 mol ratio of ca 2.8:1.

The reactor pressure is dictated by the stripper, in which low mol ratio and low

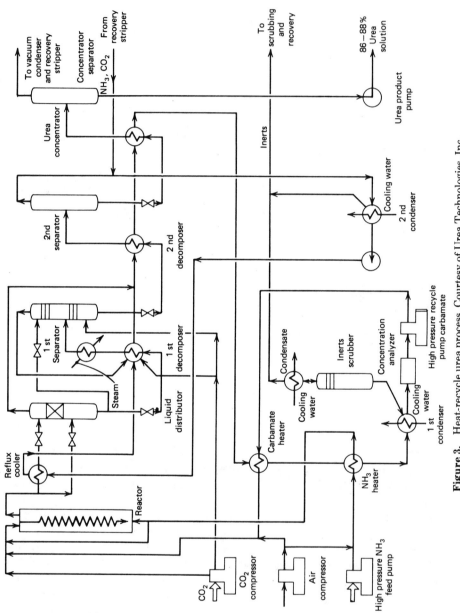

Figure 3. Heat-recycle urea process. Courtesy of Urea Technologies, Inc.

pressure are required in order to facilitate the carbamate decomposition. The bulk of unconverted carbamate is decomposed and recycled to the reactor isobarically at reactor pressure (see Fig. 4); thus, the size of the recycle pump is reduced.

The reactor is provided with perforated trays for better mixing of the liquid NH_3–liquid carbamate recycle from the low pressure absorber, and the gas–liquid mixture from the high pressure condenser. The top of the reactor has a gas pocket for separation of the noncondensable components from the urea product solution.

The noncondensable components, mainly passivating air, are scrubbed by the carbamate solution from the low pressure absorber and vented through the overhead inert-components vent. The liquid from this vent is fed to the carbamate condenser through an ejector operated by the high pressure ammonia reactor feed. The urea product solution overflows into an internal downpipe and is fed to the top of the high pressure stripper.

High pressure steam provides the heat for the carbamate decomposition and maintains a temperature of ca 190°C. High pressure CO_2 is passed through the stripper tubes, countercurrently to the downflowing urea product.

In the presence of excess gaseous CO_2, carbamate is decomposed to NH_3 and CO_2 gas and stripped from the solution. Pressure is reduced in the degassed urea product solution containing some unconverted carbamate and ammonia in order to recover NH_3 and CO_2; subsequently, the product is concentrated to 99.7 wt % urea melt under high vacuum.

After addition of ammonia, the overhead gas from the high pressure stripper is partially condensed to produce low pressure steam for export. The partially condensed mixture flows by gravity from the high pressure condenser back to the reactor. Heat is removed from the high pressure condenser in such a way that a certain amount of uncondensed CO_2 and ammonia gas in the stream is recycled to the reactor in order to maintain the reactor heat balance by further condensation of gas.

The reactor is lined with 316L stainless steel. The high pressure stripper tubes are made of a special urea-grade stainless steel, containing ca 26% chrome.

Process steam is used at the rate of 0.9 t/t urea produced as solid prills. A large number of urea plants with capacities of ca 1200 t/d are using this process successfully (30).

Snamprogetti NH₃ Stripping Process. This process was developed in the late 1960s. A synthesis loop is operated at ca 15 MPa (ca 150 atm) and at an NH_3-to-CO_2 overall mol ratio of 3.8:1. Carbamate conversion to urea per pass is reported to be ca 65–67%.

The reactor effluent is fed to the high pressure stripper for decomposition of unconverted carbamate at reactor pressure (see Fig. 5). The overhead gas from the high pressure stripper is condensed and recycled to the reactor by means of a liquid eductor, operated on high pressure liquid-ammonia reactor feed. Low pressure steam for export is produced in the high pressure condenser.

The residual ammonia in the urea product solution from the high pressure stripper is relatively high and requires two stages of decomposition and recycling, downstream of the synthesis loop.

Passivating air is introduced into the reactor and vented from the high pressure condenser to the medium-pressure absorber for recovery of residual NH_3 and CO_2 at ca 1.5–1.8 MPa (ca 15–18 atm). Uncondensed excess ammonia from the medium-pressure absorber is condensed, mixed with the stream of fresh liquid-NH_3 makeup

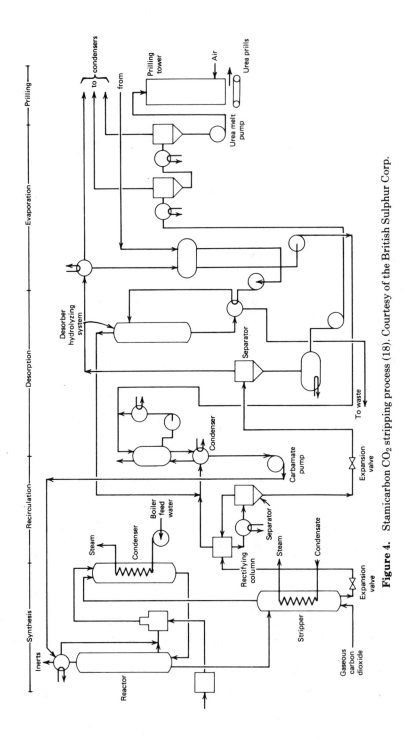

Figure 4. Stamicarbon CO_2 stripping process (18). Courtesy of the British Sulphur Corp.

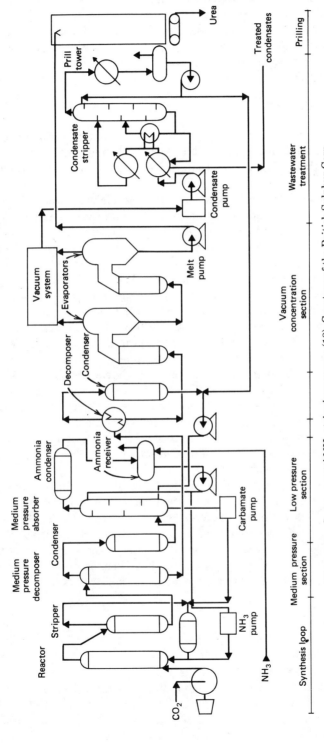

Figure 5. Snamprogetti NH$_3$ stripping process (18). Courtesy of the British Sulphur Corp.

stream, and delivered to the reactor via the carbamate ejector by the high pressure liquid-ammonia pump.

The urea product solution is concentrated under high vacuum to ca 99.7 wt % urea melt and further processed to solid urea by prilling or granulation. A process steam usage of 0.9 t/t urea is reported. Numerous urea plants with capacities up to 1800 t/d are currently using this process worldwide (30).

Wastewater Treatment

Under the pressure of progressively more stringent government regulations with regard to permissible levels of residual NH_3 and urea content in wastewaters, the fertilizer industry made an effort to improve wastewater treatment (see also Water, sewage).

For each mol of urea produced in a total-recycle urea process, one mol of water is formed. It is usually discharged from the urea concentration and evaporation section of the plant. For example, a 1200 t/d plant discharges a minimum of 360 t/d of wastewater. With a barometric condenser in the vacuum section of the evaporation unit, the amount of wastewater is even higher. Small amounts of urea are usually found in wastewaters because of entrainment carry-over.

The problem in reducing the NH_3 and urea content in the wastewaters to below 100 ppm is because it is difficult to remove one in the presence of the other. The wastewater can be treated with caustic soda to volatilize NH_3. However, in a more efficient method, the urea is hydrolyzed to ammonium carbamate, which is decomposed to NH_3 and CO_2; the gases are then stripped from the wastewater.

The Stamicarbon hydrolyzer–desorber dual system and the single-step hydrolyzer stripper by UTI and Vistron are both highly effective in wastewater treatment. These systems hydrolyze urea to ammonium carbamate and strip CO_2 and ammonia from the wastewater. The Stamicarbon system (see Fig. 6) consists of a hydrolyzer for urea and a separate dual-desorption system to strip the ammonia (18). The system is reported to reduce the residual content of ammonia and urea in wastewater to about 70 and 80 ppm, respectively.

The UTI hydrolyzer stripper (see Fig. 7) consists of a single tower, in which urea is hydrolyzed and NH_3 stripped simultaneously by means of steam and CO_2 gas stripping (31). Passivating air prevents corrosion of the 316 stainless-steel tower.

The UTI system is reported to reduce the residual NH_3 and urea in the wastewater to 5–10 and <2 ppm, respectively. A good number of these units, ranging in capacity from 100 to 1200 t/d of treated wastewater, have been in successful operation in the United States and other countries for the past several years.

Finishing Processes

Urea processes provide an aqueous solution containing 70–87% urea. This solution can be used directly for nitrogen-fertilizer suspensions or solutions such as urea–ammonium nitrate solution, which has grown in popularity recently (32). Urea solution can be concentrated by evaporation or crystallization for the preparation of granular compound fertilizers and other products. Concentrated urea is solidified in essentially pure form as prills, granules, flakes, or crystals. Solid urea can be shipped, stored, distributed, and used more economically than in solution. Furthermore, in the solid form, urea is more stable and biuret formation less likely.

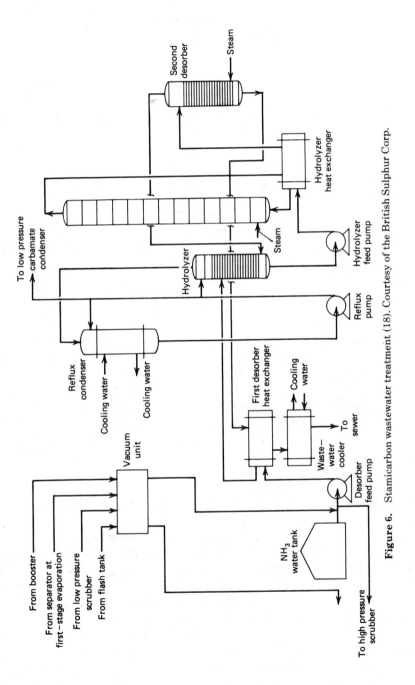

Figure 6. Stamicarbon wastewater treatment (18). Courtesy of the British Sulphur Corp.

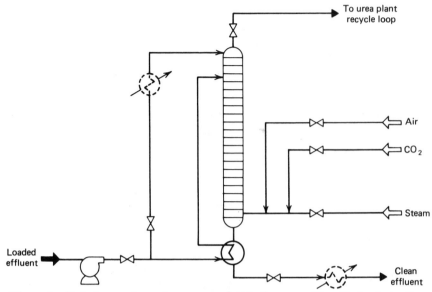

Figure 7. Hydrolyzer-stripper cleanup process (18). Courtesy of Urea Technologies, Inc.

Evaporation. Water is evaporated from steam-heated urea solution either under reduced pressure with or without the addition of hot air as a drying agent (33–35) or by a process of atmospheric air-sweep evaporation (36). Because biuret formation is accelerated by low pressure and high temperature, evaporation (qv) is generally employed only for agricultural-grade urea. The average biuret increase across the evaporation unit is ca 0.4%, depending on the degree of concentration. In granulation processes utilizing a 95–96% solution, biuret formation is minor because of the lower urea concentration in the outlet stream of the evaporator. For most fertilizer uses, biuret content up to 2 wt % is of no consequence; it decomposes in the soil and the nitrogen becomes available to plants (37). Urea foliar spray can have a toxic effect when in contact with seed, citrus plants, or some other crops. In general, biuret concentrations of <0.25 wt % are preferred (38).

Crystallization with Remelting. The presence of biuret is detrimental in the technical-grade urea used in the plastic-manufacturing industry. Solid urea obtained by crystallization is relatively pure and well suited for such applications (see Fig. 8). The urea product solution is fed to a vacuum crystallizer, operated at ca 8.0 kPa (60 mm Hg) and ca 60°C. The water vapor, evaporated from the solution, is condensed in a water-cooled vacuum condenser provided with a vacuum jet. The crystallizer slurry (ca 30 wt % urea) is sent to a continuous pusher-type centrifuge. The crystals are separated from the mother liquor, washed with water, dried, elevated to the top of the prilling tower, and remelted in a steam-heated crystal melter. The urea melt thus obtained usually contains ca 0.3 wt % biuret and ca 0.2 wt % moisture. The mother liquor separated in the centrifuge is steam-heated in the crystallizer heater in order to supply the equivalent amount of heat required to evaporate the water from the solution, and returned to the crystallizer to maintain the proper magma concentration. A small portion of the mother liquor, rich in biuret, is purged from the system in order to prevent biuret buildup.

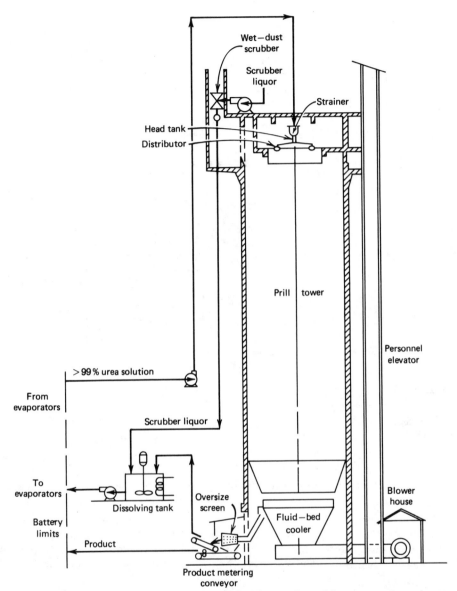

Figure 8. Crystallization with crystal remelting.

Prilling. Until recently, most solid urea was produced by the prilling process (see Fig. 9) (39). Molten urea obtained either by evaporation or by crystal melting is sprayed as droplets from the top of a tall cylindrical tower ca 50–60 m high and allowed to fall countercurrently through a stream of air emanating from a fluid-bed cooler at the base of the tower (see Fluidization). In some plants, the product is screened and off-size material is either remelted or dissolved, reconcentrated, and recycled. The sprays are coarse, usually formed in a rotating conical-shaped bucket with properly sized apertures or some other device which produces a shower of small droplets distributed over the cross-section of the tower. Until recently, the urea droplets usually were small, resulting in a product with a median volume diameter of ca 1.7 mm. Today, taller

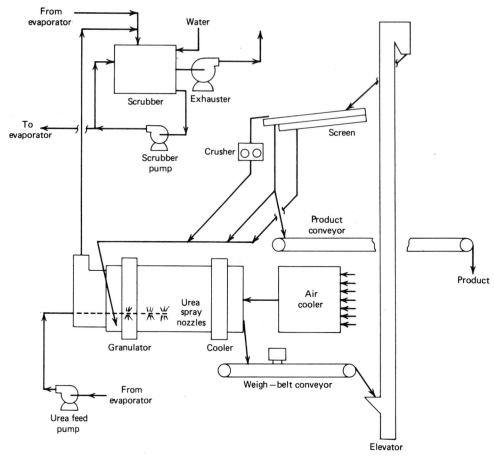

Figure 9.   Flow diagram of prilling unit (39).

towers and other breakthroughs in technology result in urea prills with a median volume diameter of ca 2.1 mm, thus giving material that blends well in bulk with most other solid fertilizers. Prilled urea is generally weaker than granular, both in crushing strength and abrasion resistance even when formaldehyde is incorporated into melt before prilling (40). Stamicarbon has recently made improvements in this area, and prilled urea can be produced with better crushing strength and impact resistance (41) (see Fertilizers).

Prill towers require large quantities of air, and pollution abatement is very expensive. In addition, granulation processes offer far more flexibility in particle size and, in general, produce granules with better physical characteristics.

Granulation.   The U.S. industry is shifting to granulation processes.

C & I/ Girdler–Cominco Granulation Process.   This process is the most widely used in the urea industry. It is proven and dependable, and should continue to be of importance to the industry for many years. Developed by Cominco (Canada), the process (42–43) is a variation of the spray-accretion or spherodizing processes developed by C & I/Girdler in the mid-1950s (see Fig. 10) (44). Urea is concentrated to ca 99.5% and sprayed through nozzles onto a combination of falling granules and a cascading bed of recycled fines in a rotating cylindrical granulation drum, ca 4.3 m in diameter, 15.2 m long, revolving at ca 9.5 rpm. The drum contains flights and retaining dams. Cooling

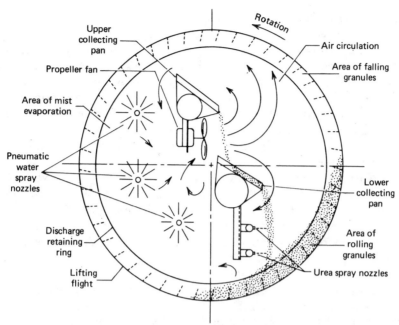

Figure 10. C & I/Girdler—Cominco urea granulation process (44). Courtesy of *Chemical Economy and Engineering Review.*

air, which is refrigerated in some plants, is drawn through the drum countercurrent to the flow of granules. The molten urea solidifies as a coating on the granules, building them up layer upon layer. The cascading and rolling action in the bed helps to maintain the granules in a spherical configuration. About 0.3% formaldehyde, which acts as a conditioner and increases product hardness, is added to the melt before spraying. The velocity of the cooling air in the drum is high and serves to entrain dust and very small seed particles, which are collected in a wet scrubber. The resulting liquor is recycled to the concentration section of the urea unit. Dust loadings are usually ≥10% of the production rate. Spraying occurs primarily in the first third of the granulator; the remaining portion is used mainly as a rotary cooler. The cooled granules discharge from the drum and are screened into oversize, undersize, and product size. The oversize is crushed to form seed and returned to the front of the granulator as recycle along with the undersize. As the recycled fines pass through the granulator, sprayed melt impinges on them and the granules are grown. The ratio of recycle to product is usually ca 2:1 and may be adjusted by directing some of the product to recycle, if necessary. The recycle also acts as a heat sink, which absorbs the heat of crystallization of the sprayed melt until the granules are cooled through contact with the air. Maximum design capacity per granulator is about 381 t/d; some companies have reached production rates of 454 t/d in winter. Utility requirements for granular urea made by this process are reported to be ca 275 kg of steam and 38.6 kW·h of electric energy per metric ton when starting with 99.5 wt % melt (44). These values do not include refrigeration of the cooling air before entering the granulator.

Physical characteristics of the granules are good with usually <0.15% H_2O. Product sizes typically range between 6.7–4 mm or 3.4–1.7 mm, and the granules are hard and spherical. Biuret content varies between plants, but is usually 1–2%.

Pan Granulators. In pan granulation, a concentrated urea solution is sprayed onto a cascading bed of recycle fines in an inclined rotating pan (45–46). Large granules discharge over the rim to be cooled and screened. Oversize granules are crushed and recycled with undersize particles and product size material if necessary. In the TVA (Tennessee Valley Authority) pan granulation process, the temperature and quantity of recycle material are regulated to provide a temperature substantially below the crystallization temperature (132°C) of the molten feed. Urea melt (98.5–99.8 wt %) is sprayed onto the bed; optimum bed temperature is 104.4–107.2°C. A recycle ratio of 2.0:3.0 is required to control granulation temperatures when the temperature of the recycle material is 48.9–54.4°C.

In the Norsk Hydro process, recycle ratios of solids to melt are lower (ca 1:1), and an anhydrous urea melt (99.7 wt %) is sprayed onto the deep part of the pan bed. The temperature is maintained 5–25°C below the crystallization temperature of urea. The granules are discharged to a polishing drum before being cooled and screened. The higher temperature process is reported to give granules with higher crushing strength than the low temperature process, but, according to TVA, process upsets are less likely to occur at the lower temperature. The TVA pan granulation process is used in several plants, whereas two plants of 300 t/d capacity each are starting production in 1982 using the Norsk Hydro process.

Nederlandse Stikstof Maatschappij N.V. (NSM) has an operating 800 t/d granulation unit. TVA and Mitsui-Toatsu/Toyo Engineering should have plants of 300 and 470 t/d capacity, respectively, operational in 1983. These processes are directed at the granulation of urea in a more energy-efficient manner, with lower capital investment, less air-pollution problems, and improved product properties including extreme hardness (crushing strengths of 20–39 N (2–4 kgf) for 3-mm granules), low moisture (<0.1%), low biuret (0.3% max owing to granulation), and product size ranging from small prills (1.5 mm = ca 14 mesh) to jumbo granules (>30 mm).

TVA Falling-Curtain Granulation Process. The falling-curtain evaporative cooling process (47–48) is based on TVA sulfur-coated urea technology (49) in combination with some novel approaches to granulation. A rotary drum is equipped with specially designed internal features (see Fig. 11). Granules are elevated by shallow lifting flights from which they discharge before reaching the apex of the drum. A large number are directed by inclined collecting pans to form a curtain of rapidly falling material. Throughout the length of the drum, atomized sprays of molten urea are directed onto this curtain of falling granules. The fine droplets solidify immediately on contact with the granules, increasing them in size without agglomeration. The heat given off by crystallization is quickly transferred to the air which is being pulled through the drum. Water is atomized directly into the air and, as it evaporates, the air is cooled. Air is recirculated by propeller-type fans through the area of dense falling granules. The drum has an extremely high overall volumetric heat-transfer coefficient, which allows high rates of granulation in a relatively small and lightly loaded drum. The temperature of the granules in the drum is not critical and can range from 60 to 107°C. As shown in Figure 12, granules discharge from the drum to a fluid-bed cooler and continue onward to a screen where oversize and undersize materials are removed from the product stream. Before storage, the product is cooled to below 49°C by passing it through a second side of the fluid-bed heat exchanger. The undersize material is collected in a surge hopper and recycled to the granulator at a metered rate. The recycle rate is typically one half of the production rate.

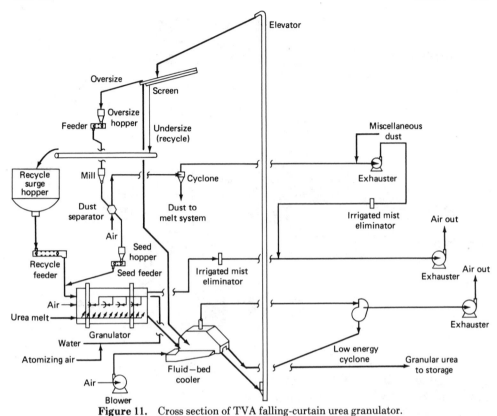

Figure 11.   Cross section of TVA falling-curtain urea granulator.

The air leaving the granulator is washed with recycle scrubber solution, and the resulting spray particles are collected in an irrigated mist eliminator. Dust formation is only ca 2% of the granulation rate, mostly eminating from the crushing of oversize particles to form seed. Particles emitted from the fluid-bed cooler are relatively large and are collected in a low pressure-drop horizontal cyclone.

The process is extremely energy efficient. Based on experimental work in a 1.8 t/h pilot plant, a full-scale plant of ca 608 t/d is expected to use ca 22.1 kW·h/t of product. Steam usage, including concentration of scrubbing solution, is expected to be about 75 kg/t.

No formaldehyde or substitute is added to increase hardness or reduce caking. A product of 3 mm dia has a crushing strength of 29 N (3 kgf) and exhibits good storage properties. Granule diameter can be easily increased from an average of 1.5 to >30 mm.

Nederlandse Stikstof Maatschappij N.V. (NSM) Process.   Granulation is accomplished in a fluid bed divided into several chambers (50–51). Normal bed depth is ca 1 m above a perforated plate through which the fluidizing air is distributed to the granulator. Pneumatic atomizing nozzles are mounted just above the air-distribution plate and oriented to spray a 95–96% urea solution at 130–135°C upward into the active bed. The air atomizing the urea solution is supplied by a blower at 203–253 kPa (2–2.5 atm) and 145°C; particles with a mean drop diameter of 30–60 μm (ca 240–500 mesh) are preferred. The nozzles are designed in conjunction with the bed depth to allow good penetration of the bed by spray particles, yet prevent complete passage through

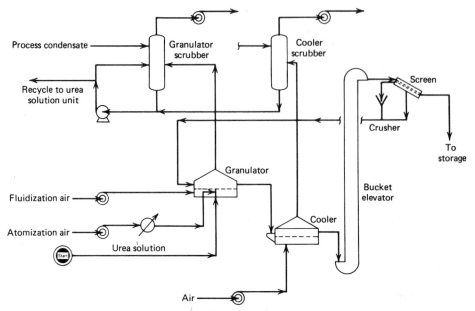

Figure 12. TVA falling-curtain granulation process.

the bed. Seed particles obtained from the controlled crushing of oversize particles are introduced into the first chamber of the granulator. The particles in the granulator grow by accretion when atomized droplets of urea solution strike them and the water evaporates as the concentrated urea crystallizes. Heat released from the crystallizing urea is absorbed by the evaporating water, allowing concentration and granulation in one operation. The particles grow in size and underflow progressively through the chambers in the fluid bed. The product-size granules underflow out of the granulator to a cooler before being screened into oversize, product, and undersize, as shown in Figure 13. The rate of undersize plus seed fed to the granulator is ca one half of the production rate. Air exhausted from granulator and cooler is passed through a wet scrubber to remove the dust and mist. Formaldehyde or a substitute is added to the concentrated urea solution before it is sprayed in order to improve granulation and reduce caking in storage. Urea need not be concentrated above 95%; in this process, biuret content increases only by 0.03%.

Mitsui-Toatsu and Toyo Engineering. Concentrated urea solution is sprayed onto the surface of particles in a fluidizing granulation process; the heat released by solidification of the urea feed is removed by a combination of cooled recycle, fluidizing air, and some water evaporation (52–53). The particles are carried upward by a strong air stream and, on disengaging, fall back into the bed before being elevated again in a cyclic motion. Urea is deposited in thin layers as the solid particles pass through the urea mist sprayed into the bed. The temperature of the bed is not critical and can be 60–110°C. As shown in Figure 14, after leaving the granulator, the granules are cooled in another fluid-bed unit before being screened. Oversize is crushed to seed and mixed with undersize material and then recycled to the granulator at 1.0–1.5 times the production rate to make a product suitable for bulk blending. For a larger product, the screen sizes are changed and recycle and seed rates reduced. The larger dust particles exiting granulator and cooler are caught in cyclones and fed back to the process. A

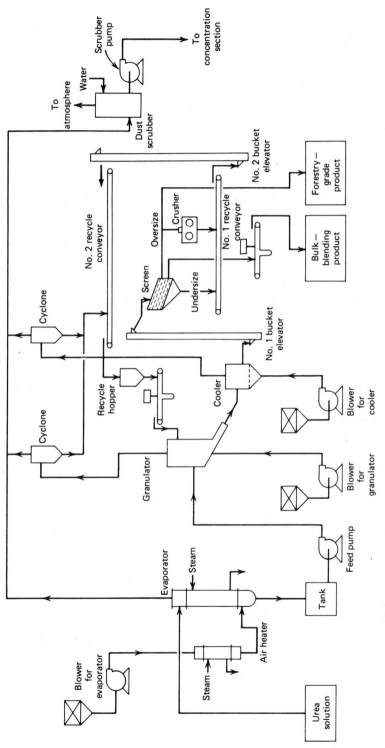

Figure 13. NSM fluidized-bed urea granulation process (51). Courtesy of *Hydrocarbon Processing.*

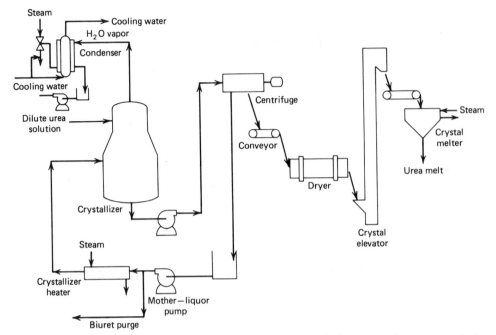

Figure 14. Mitsui Toatsu-Toyo Engineering fluidizing granulation process for urea granulation (53).

wet-scrubbing system removes the fine dust from the air. Hardness may be increased with additives. Granules 3 mm of dia without additives are reported to have a crushing hardness of ca 20 N (2 kgf). A production capacity of 100 t/d per granulator is estimated to be the best size. Multiple granulators are recommended for large plants.

Economic Aspects

The U.S. urea production from 1976 to 1980 (32,54) is given in Table 8. During this period, urea became the most important solid nitrogen fertilizer. Urea solution production almost doubled from 1977 to 1978, proving its importance as a liquid fertilizer.

Table 8.   U.S. Urea Production, 10^6 t[a]

| Year | Fertilizer | | Other | Total |
	Solid	Solution		
1970	1.090	1.278	0.580	2.948
1974	1.302	1.221	0.916	3.439
1976[b]	2.207	1.094	0.779	4.080
1977	2.435	1.337	0.829	4.601
1978	2.868	2.376	0.446	5.690
1979	3.888	1.964	0.498	6.350
1980[b]	3.921	2.583	0.599	7.103

[a] Refs. 32, 54.
[b] Annual growth rate >11.5%.

In 1980, urea consumption in other areas had declined, but not enough to detract from the overall bright future for the material, which had a total U.S. production of 7.103×10^6 metric tons in 1980 in plants with a nameplate capacity of 7.1 t. Estimated urea production in the United States for 1981 is lower than the 1980 production by ca 3%, because of a decline in exports from 1.8 t in 1980 to ca 1.45 t in 1981 (55). Nameplate capacity of urea plants in the United States and Canada is projected to be ca 9.075×10^6 t/yr by 1984 (48). World capacity should be ca 95.150×10^6 t/yr (48). World capacity increased from 27.200×10^6 t/yr in 1970 to ca 77.300×10^6 t/yr in 1980 (56).

Urea prices on the spot market in the spring of 1982 have been reported between \$154/t and 204/t fob U.S. ports or manufacturers' plants. The price depends on the cost of energy and natural gas, which is the chief raw material for ammonia. Typical raw-material and utility consumption for a granular urea plant are presented in Table 9. Utility consumption varies between processes; significant differences in energy consumption occur in finishing processes. The average 1982 investment costs of a commercial total-recycle granular urea plant are given below.

Capacity t/d	Investment, 10^6
300	17.0
600	24.5
1200	38.7

Uses

Solid urea containing 0.8–2.0 wt % biuret is primarily used for direct application to the soil as a nitrogen-release fertilizer. Weak aqueous solutions of low biuret urea (0.3 wt % biuret max) are used as plant food applied as foliage spray.

Mixed with additives, urea is used in solid fertilizers of various formulations, eg, urea–ammonium phosphate (UAP), urea–ammonium sulfate (UAS), and urea–phosphate (urea + phosphoric acid). Concentrated solutions of urea and ammonium nitrate (UAN) solutions (80–85 wt %) have a high nitrogen content but low crystallization point, suitable for easy transportation, pipeline distribution, and direct spray application.

Urea is also used as feed supplement for ruminants, where it assists in the utilization of protein. Urea is one of the raw materials for urea–formaldehyde resins. Urea, (with ammonia) pyrolyzes at high temperature and pressure to form melamine plastics (see also Cyanamides). Urea is used in the preparation of lysine, an amino acid widely used in poultry feed (see Amino acids; Pet and livestock feeds). It also is used in some pesticides.

Partially polymerized resins of urea are used by the textile industry to impart permanent-press properties to fabrics (see also Textiles, finishing).

Table 9.   Raw-Material and Utility Consumption

Material	Amount per metric ton of product
NH_3, t	0.58
CO_2, t	0.75
steam, t	1.4
cooling water, m^3	70
electric power, kWh	148

The consumption of urea for urea–formaldehyde resins has decreased in recent years because of the new findings about the toxicity of formaldehyde slowly released by the resin.

Reagent-grade urea is used in some pharmaceutical preparations. In these applications, urea must meet the purity specifications issued by the ACS.

Clathrates

Urea has the remarkable property of forming crystalline complexes or adducts with straight-chain organic compounds (see Clathration). These crystalline complexes consist of a hollow channel, formed by the crystallized urea molecules, in which the hydrocarbon is completely occluded. Such compounds are known as clathrates. The type of hydrocarbon occluded, on the basis of its chain length, is determined by the temperature at which the clathrate is formed. This property of urea clathrates is widely used in the petroleum-refining industry for the production of jet aviation fuels (see Aviation and other gas-turbine fuels) and for dewaxing of lubricant oils (see also Petroleum, refinery processes). The clathrates are broken down by simply dissolving urea in water or in alcohol.

BIBLIOGRAPHY

"Urea" in *ECT* 1st ed., Vol. 14, pp. 458–466, by F. A. Wolff and D. J. O'Flynn, E. I. du Pont de Nemours & Co., Inc.; "Urea" in *ECT* 2nd ed., Vol. 21, pp. 37–56, by I. Mavrovic, Consulting Engineer.

1. E. Blasiak, *Technology of Nitrogen Compounds*, Vol. 2, State Technical Publisher, Warsaw, 1956, pp. 596–642.
2. E. Janecke, *Z. Elektrochem.* **36,** 645 (1930).
3. A. Seidell, *Solubility of Inorganic Compounds*, 2nd ed., D. Van Nostrand Co., New York.
4. D. F. Du Toit, *Proc. Koninkl. Akad. Wetenschap, Amsterdam* **16,** 555 (1913).
5. R. Escales and H. Kopke, *Chem. Zig.* **33,** 595 (1911).
6. U.S. Pat. 3,232,984 (Feb. 1, 1966), J. A. Finneran (to Pullman, Inc.).
7. U.S. Pat. 3,255,246 (June 7, 1966), I. M. Singer, Jr. (to E. I. du Pont de Nemours & Co., Inc.).
8. Knop and Hufner in A. E. Werner, ed., *Chemistry of Urea*, Longmans, Green and Co., London, 1923, p. 161.
9. C. Matignon and M. Frèjacques, *Bull. Soc. Chim. France* **29,** 21 (1921); **31,** 394 (1922).
10. T. R. Briggs and V. Migrdichian, *J. Phys. Chem.* **28,** 1121 (1924).
11. E. Briner, *J. Chim. Phys.* **4,** 267 (1906).
12. N. F. I. Isambert, *Compt. Rend.* **92,** 919 (1881).
13. F. Fichter and B. Becker, *Ber.* **44,** 3470 (1911).
14. C. Matignon and M. Frèjacques, *Bull. Soc. Chim.* **31,** 394 (1922).
15. K. K. Clark, V. L. Gaddy, and C. E. Rist, *Ind. Eng. Chem.* **25,** 1092 (1933).
16. *Eur. Chem. News* (Jan. 17, 1969).
17. S. Kawasumi, *Bull. Chem. Soc. Jpn.* **24,** 148; **25,** 227; **26,** 218; **27.**
18. U. Zardi, *Nitrogen* **135,** 26 (Jan.–Feb. 1982).
19. U.S. Pat. 3,506,710 (April 14, 1970), S. Inoue and co-workers (to Toyo Koatsu Industries).
20. U.S. Pat. 3,944,605 (March 16, 1976), S. Inoue and T. Kimura (to Mitsui Toatsu Chemicals, Inc.).
21. U.S. Pat. 4,081,469 (March 28, 1978), H. Ono and S. Inoue (to Mitsui Toatsu Chemicals, Inc.).
22. U.S. Pat. 3,573,173 (March 30, 1971), E. Otsuka and co-workers (to Mitsui Toatsu Chemicals, Inc.).
23. U.S. Pat. 3,725,210 (April 3, 1973), E. Otsuka and co-workers (to Mitsui Toatsu Chemicals, Inc.).
24. U.S. Pat. 3,514,483 (May 26, 1970), E. Otsuka and co-workers (to Mitsui Toatsu Chemicals, Inc.).
25. *Eur. Chem. News*, **15** (Aug. 9, 1982).
26. U.S. Pat. 3,952,055 (April 20, 1976).
27. U.S. Pat. 3,759,992 (Sept. 18, 1973).
28. U.S. Pat. 3,886,210 (May 27, 1975), I. Mavrovic.

29. U.S. Pat. 3,574,738 (April 13, 1971), I. Mavrovic.
30. *Hydrocarbon Process.*, 249 (Nov. 1979).
31. U.S. Pat. 3,826,815 (July 30, 1974), I. Mavrovic.
32. J. D. Bridges, *TVA Bulletin Y-150*, 1980.
33. U.S. Pat. 3,171,770 (Nov. 22, 1960), H. J. B. Biekart, M. Bongard, and P. J. C. Kaasenbrood (to Stamicarbon).
34. U.S. Pat. 2,961,464 (Nov. 22, 1960), P. J. C. Kaasenbrood (to Stamicarbon).
35. U.S. Pat. 3,147,174 (Sept. 1, 1964), L. M. Cook (to Chemical Construction Corp.).
36. *Whitlock Mark 99 Concentrator*, Ametek Bulletin 133B, 1977.
37. V. J. Kilmer and O. P. Engelstad, *TVA Bulletin Y-57*, 1973.
38. R. C. Gray, *Proc. Fert. Soc.*, (164) (1977).
39. G. M. Blouin, *TVA Bulletin Y-92*, 1975.
40. G. Hoffmeister, *TVA Bulletin Y-147*, 1979.
41. M. H. Willems, *Fifth Stamicarbon Urea Symposium*, Maastricht (1978).
42. U.S. Pat. 3,232,703 (Feb. 1, 1966), J. B. Thompson and G. C. Hildred (to Cominco, Ltd.).
43. U.S. Pat. 3,398,191 (Aug. 20, 1968), J. B. Thompson and G. C. Hildred (to Cominco, Ltd.).
44. R. M. Reed and J. C. Reynolds, *Chem. Eng. Prog.* **69,** 62 (1973).
45. *Nitrogen* **131,** 39 (1981).
46. I. W. McCamy and M. M. Norton, *Farm Chem.* **140**(2), 61, 64, 68 (1977).
47. U.S. Pat. 4,213,924 (July 1980), A. R. Shirley, Jr. (to Tennessee Valley Authority).
48. A. R. Shirley, Jr., L. M. Nunnelly, and F. T. Carney, Jr., *paper presented at 182nd National Meeting of American Chemical Society, TVA Circular Z-125*, 1981.
49. A. R. Shirley, Jr. and R. S. Meline in J. R. West, ed., *New Uses of Sulfur*, American Chemical Society, Washington, D.C., 1975, pp. 33–54.
50. U.S. Pat. 4,219,589 (Aug. 1980), A. Niks, W. H. P. Van Hijfte, and R. A. J. Goethals (to Compagnie Neerlandaise de l'Azote).
51. J. P. Bruynseels, *Hydrocarbon Process.* **60**(9), 203 (1981).
52. *Nitrogen* **128,** 38 (1980).
53. T. Jojima, T. Kimura, T. Nagahama, M. Nobue, and A. Fukui, *Chem. Econ. Eng. Rev.* **12**(9), 26 (1980).
54. *Inorganic Chemicals Series M28A*, Bureau of the Census, U.S. Department of Commerce, Washington, D.C., annual reports, 1978–1980.
55. *Chem. Eng. News* **60,** 22 (1982).
56. *World Fertilizer Market Information Services*, TVA, National Fertilizer Development Center, Muscle Shoals, Ala.
57. *Farm Chemicals Handbook*, Plant Food Dictionary Section, 1981 ed.

Ivo Mavrovic
Consultant

A. Ray Shirley, Jr.
Applied Chemical Technology

UREA–FORMALDEHYDE RESINS.   See Amino resins and plastics.

URETHANE POLYMERS

The addition polymerization of diisocyanates with macroglycols to produce urethane polymers was pioneered by O. Bayer and his co-workers at the laboratory of I.G. Farbenindustrie in Leverkusen in 1937 (1). The rapid formation of high mol wt urethane polymers from liquid monomers, which occurs even at ambient temperature, is a unique feature of the polyaddition process, and products ranging from cross-linked network polymers to linear fibers and elastomers were produced in Germany in the early 1940s. The enormous versatility of the polyaddition process prompted the manufacture of a myriad of products for a wide variety of applications.

Polyurethanes contain carbamate groups —NHCOO—, also referred to as urethane groups, in their backbone structure. They are obtained by the reaction of a diisocyanate with a macroglycol, a so-called polyol, or with a combination of a macroglycol and a short-chain glycol extender. In the latter case, segmented block copolymers are produced. The macroglycols are based on polyethers (qv), polyesters, or a combination of both. A linear polyurethane polymer has the structure (1), whereas a linear segmented copolymer obtained from a diisocyanate, a macroglycol, and ethylene glycol has structure (2).

$$\left[ROC\overset{\overset{\textstyle O}{\|}}{}NHR'NH\overset{\overset{\textstyle O}{\|}}{C}O\right]_{n}$$

(1)

$$\left[ROC\overset{\overset{\textstyle O}{\|}}{}NHR'NH\overset{\overset{\textstyle O}{\|}}{C}OCH_2CH_2O\overset{\overset{\textstyle O}{\|}}{C}NHR'NH\overset{\overset{\textstyle O}{\|}}{C}O\right]_{n}$$

(2)

In addition to the linear thermoplastic polyurethanes obtained from difunctional monomers, branched or cross-linked thermoset polymers are made with higher functional monomers. Linear polymers have good impact strength, good physical properties, and excellent processability, but owing to their thermoplasticity, thermal stability is limited. Thermoset polymers, on the other hand, have higher thermal stability but lower impact strength. The higher functionality is obtained with higher functional isocyanates, so-called polymeric isocyanates, or with higher functional polyols (see Isocyanates; Glycols; Polyesters). Cross-linking is also achieved by secondary reactions. For example, urea groups are generated in the formation of water-blown flexible foams. An isocyanate group and water give an amino group which immediately reacts with excess isocyanate to form a urea linkage. This reaction is accompanied by the evolution of carbon dioxide, which acts as a blowing agent. Further reaction of the urea group with the isocyanate leads to cross-linking via a biuret group. Water-blown flexible foams contain urethane, urea, and some biuret groups in their network structure (see Foamed plastics). The overall reactions are shown below.

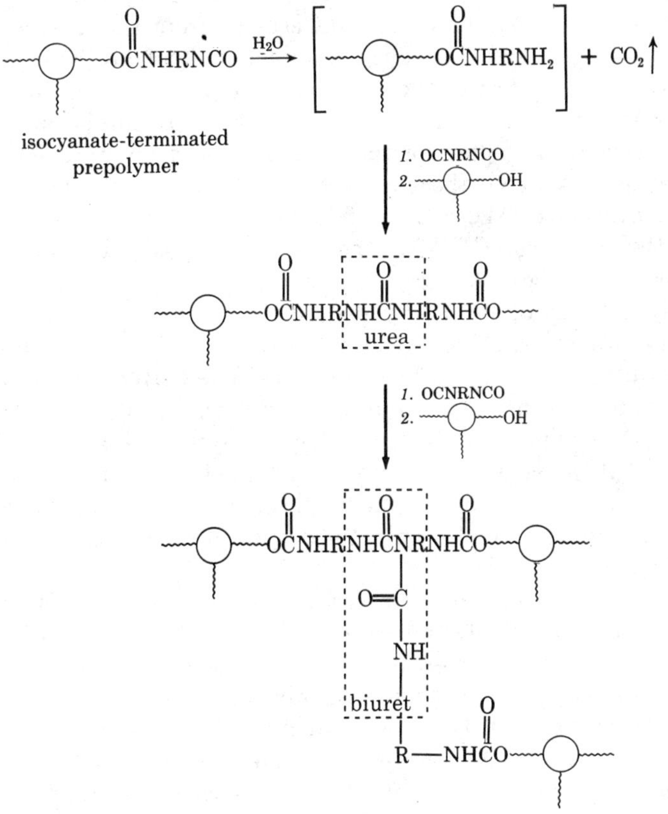

isocyanate-terminated
prepolymer

Water-blown rigid foams are being considered for pour-in-place retrofit insulation applications. Urea-modified segmented polyurethane elastomers are manufactured from diisocyanates, macroglycols, and diamine extenders.

Urethane network polymers are also formed by trimerization of part of the isocyanate groups. This approach is used in the formation of rigid urethane-modified isocyanurate foams with structure (3).

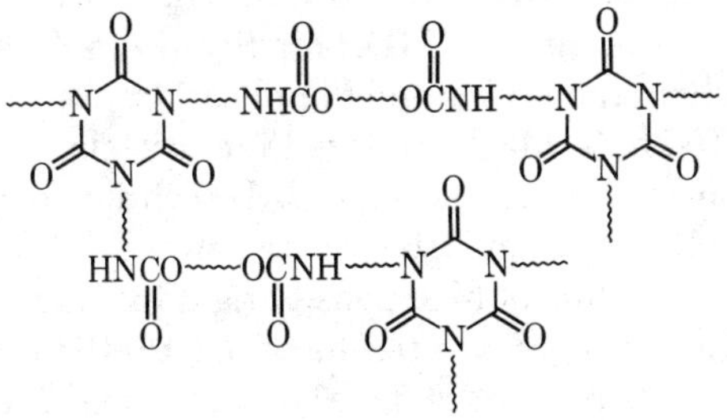

(3)

The early polyurethane products were based on toluene diisocyanate (TDI) and polyester polyols. Rigid foams, flexible foams, millable gums, and polyurethane coatings were prepared from these basic raw materials. An exception was Perlon U, a linear fiber, which was manufactured in Germany from 1,6-hexamethylene diisocyanate (HDI) and 1,4-butanediol. The development of polyurethane technology was delayed by World War II. After the war, urethane polymers became known in the United States, where commercial production of flexible foam began in 1953 (2). In Germany, a toluene diisocyanate consisting of an isomeric mixture of 65% 2,4-isomer and 35% 2,6-isomer was used in the manufacture of flexible foam, whereas in the United States, the less expensive 80:20 isomer mixture was used. In 1956, DuPont introduced poly(tetramethylene glycol) (PTMG), the first commercial polyether polyol; the less expensive polyalkylene glycols appeared in 1957. The availability of these low cost polyether polyols based on both ethylene and propylene oxides provided the foam manufacturers with a broad choice of suitable raw materials which afforded flexible foams with a wide range of desirable properties. The better hydrolytic stability of polyether-based polyurethanes was another advantage. The development of new and superior catalysts, such as Dabco (triethylenediamine) and organotin compounds, led to the so-called one-shot process in 1958, which eliminated an intermediate prepolymer step. Previous to this development, part of the polyol had reacted with excess isocyanate to give an isocyanate-terminated prepolymer. Further reaction with water produced a flexible foam.

In 1958, liquid casting of urethane elastomers was introduced (3), and segmented thermoplastic polyurethane elastomers were developed in the early 1960s. Another development in 1958 was elastomeric spandex fibers (see Elastomers, synthetic; Fibers, elastomeric). Of the original ten products, only two are still produced in the United States by DuPont (4) and the Globe Manufacturing Company. DuPont's synthetic-leather urethane product, Corfam, first marketed in 1963, was a failure, but urethane-type leather products are still produced in Japan and Europe (see Leatherlike materials).

The late 1950s also witnessed the emergence of a new polymeric isocyanate based on the condensation of aniline with formaldehyde. This product was introduced by the Carwin Company (now Upjohn) under the trade name Papi. Similar products were introduced by Bayer and ICI in the early 1960s. The superior heat resistance of rigid foams derived from polymethylene polyphenyl isocyanate (PMDI) was recognized early (5), and today all rigid insulation foam is manufactured from PMDI. The recent emphasis on energy conservation provided the impetus for a spectacular worldwide growth of PMDI products. The large-scale production of PMDI made the coproduct 4,4'-methylenebis(phenyl isocyanate) (MDI) readily available, and today MDI is used almost exclusively in polyurethane elastomer applications. In 1965, Upjohn introduced a liquid MDI product (Isonate 143-L) tailored for cast-elastomer applications. The introduction of reaction-injection molding (RIM) technology (6) in 1974 provided a new market for liquid MDI products (see Plastics processing). The RIM technology is used extensively in the molding of high density rigid foams and automotive bumpers and fascia parts. Work is in progress to reinforce polyurethane elastomers with glass, graphite, boron or aramid fibers, or mica flakes to increase stiffness and reduce thermal expansion (see Laminated and reinforced plastics). The higher modulus thermoset elastomers produced by reinforced reaction-injection molding (RRIM) are used in automotive fender, side panel, hood, and trunk-lid molding.

The availability of PMDI led to the development of polyurethane-modified isocyanurate (PIR) foams in 1967 (7). The PIR foams have superior thermal stability and combustibility characteristics, which extend the use temperature of insulation foam well above 150°C. The PIR foams are used in pipe, vessel, and solar-panel insulation; for example, the insulation of LNG tankers with foil-faced polyisocyanurate laminates. In this application, temperature stability from −180°C to 150°C is essential. Glass-fiber-reinforced, polyisocyanurate roofing panels with superior dimensional stability were developed by the Celotex Division of the Jim Walter Corporation (8).

Recently, inexpensive polyester polyols based on residues obtained in the production of dimethyl terephthalate (DMT) have been used in the formulation of polyurethane and polyisocyanurate rigid-foam products. Although the polyalkylene glycols are still used extensively in combination with TDI in the manufacture of flexible-foam products, a return to polyester polyols in combination with PMDI is the current trend in rigid-foam production.

Properties

The physical properties of polyurethanes are derived from their molecular structure and are determined by the choice of building blocks as well as the supramolecular structures caused by atomic interaction between chains. The ability to crystallize, the flexibility of the chains, and spacing of polar groups are of considerable importance, especially in linear thermoplastic materials. In rigid cross-linked systems, eg, polyurethane foams, other factors such as density determine the final properties.

Thermoplastic Polyurethanes. The unique properties of polyurethanes are attributed to their long chain structure. Melt viscosity depends on the weight average molecular weight $\overline{M}_w$ and is influenced by chain length and branching. Thermoplastic polyurethanes are viscoelastic materials, which behave like a glassy, brittle solid, an elastic rubber, or a viscous liquid, depending on the temperature and time scale of measurement. A typical modulus vs temperature curve is shown in Figure 1. In the relatively high modulus region, the polymers are hard and stiff (glassy). Below the glass transition temperature T_g, the molecular motion is frozen and the material is only able to undergo small-scale elastic deformations. With increasing temperature, the material becomes rubbery because of the onset of molecular motion. At higher temperatures, a free-flowing liquid forms.

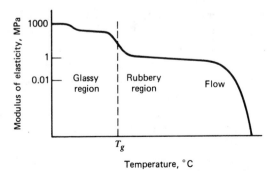

Figure 1. Modulus–temperature curve of an amorphous thermoplastic polyurethane (9). To convert MPa to psi, multiply by 145.

The melt temperature of a polyurethane T_m is important for processability. For linear amorphous polyurethane elastomers, the T_g is ca -50 to $-60°C$. Melting should occur well below the decomposition temperature. The strength in polyurethane elastomers is provided by domain crystallinity caused by polar-group interaction of the hard segments derived from MDI and short-chain diol extenders.

The choice of the macrodiol influences the low temperature performance, whereas the modulus (hardness, stiffness, and load-bearing properties) increases with increasing content of hard segment.

The pseudocross-links generated by the hard-segment interaction are reversed by heating or dissolution. Without the domain crystallinity, thermoplastic polyurethanes would lack elastic character and be more gumlike in nature. In view of the outlined morphology, it is not surprising that many products develop their ultimate properties only on curing at elevated temperature, which allows the soft and hard phase segments to separate.

Thermoset Polyurethanes. The physical properties of rigid urethane foams are usually a function of foam density. A change in strength properties requires a change in density. Rigid polyurethane foams with densities <0.064 g/cm^3, used primarily for thermal insulation, are expanded with a fluorocarbon blowing agent (see Insulation, thermal). High density foams are made by expansion with either a fluorocarbon blowing agent or carbon dioxide. The amount of carbon dioxide is determined by the amount of water used. The amount of water needed in the reaction to produce the desired foam density is equal to the amount of water present in the polyol plus added water. Between 0.6 and 1.8% water is required to produce rigid polyurethane foams in the preferred density range of 0.032–0.064 g/cm^3. Approximately 8–16% fluorocarbon blowing agent is required to produce the same densities.

In addition to density, the strength of a rigid polyurethane foam is influenced by the catalyst, surfactant, polyol, isocyanate, and type of mixing. By changes in the various ingredients, foams can be made that have high modulus, low elongation, and some brittleness (friability), or relative flexibility and low modulus.

Rigid polyurethane foams generally have an elastic region in which stress is nearly proportional to strain. If a foam is compressed beyond its yield point, the cell structure is crushed. Compressive strength values of 10–280 kPa (1–41 psi) can be obtained with rigid polyurethane foams of 0.032 g/cm^3 density. In addition, the elastic-modulus shear strength, flexural strength, and tensile strength increase with density.

The most important property for rigid polyurethane foams is the thermal conductivity (K-factor). It is greatly influenced by the blowing agent, cell size, cell content, and foam density. Most thermal-insulation foams have a density of 0.032–0.048 g/cm^3 and are blown with Fluorocarbon-11 (FC-11, CCl_3F). With good processing technique, they can be made with a fine cell structure and are essentially closed-cell foams. These foams show a change in K-factor on aging, depending on the foam configuration. If the foam is totally enclosed, as for example in an aluminum foil-faced laminate panel, little change occurs in the K-factor.

Temperature variations influence the dimensional stability of polyurethane foams. In closed-cell foams, containing an inert gas as a blowing agent, pressure causes the cells to increase in size slightly as the polymer softens, which results in distortion of the foam. Since foam cells are usually elongated in the direction of foam rise, an increase in pressure tends to make them more spherical, causing uneven growth or distortion. Furthermore, loss of blowing agent with inward migration of air can further aggravate this situation. Low density foams, particularly at densities <0.032 g/cm³, shrink as the polymer structure collapses, since internal pressure no longer maintains the cell structure. Most low density rigid polyurethane foams have a closed-cell content ≥90%. Above 0.032 g/cm³, closed-cell content increases rapidly and is generally >99% above 0.192 g/cm³. Bun foam produced under controlled conditions has a very fine cell structure with cell sizes of 150–200 μm. At foam densities <0.024 g/cm³, it is difficult to maintain a fine-cell structure.

Similar to the rigid thermoset materials, the properties of flexible polyurethane foams are related to density. However, load-bearing properties are also important, especially for flexible foam used in upholstery applications. Under normal service temperatures, flexible foams exhibit rubberlike elasticity to deformations of short duration, but creep under long-term stress. Maximum tensile strength is obtained at densities of ca 0.024–0.030 g/cm³. The densities are controlled by the amount of water in the formulation and may range from 0.045 to 0.020 g/cm³ by raising the amount of water from 2 to 5%. Auxiliary blowing agents, eg, Fluorocarbon 11 or methylene chloride, reduce density and control hardness of flexible foams. The size and uniformity of the cells are controlled by the efficiency of mixing and the nucleation of the foam mix. Flexible foams are anisotropic and the load-bearing properties are best when measured in the direction of foam rise.

Manufacture

Isocyanates. The commodity isocyanates TDI (80:20 of 2,4 and 2,6 isomer, respectively) and PMDI (a polymer isocyanate obtained by the phosgenation of aniline-formaldehyde-derived polyamines) are most widely used in the manufacture of urethane polymers (see also Isocyanates). A coproduct in the manufacture of polymeric MDI or PMDI is 4,4′-methylenebis(phenyl isocyanate) (MDI).

A 65:35 mixture of 2,4- and 2,6-TDI is available, as well as 2,4 isomer. Some TDI producers market TDI distillation residues for use in rigid polyurethane foam applications.

The manufacture of TDI involves the dinitration of toluene, catalytic hydrogenation to the diamines, and phosgenation. Separation of the undesired 2,3-isomer is necessary because its presence interferes with polymerization (10).

Polymeric isocyanates or PMDI are crude products that vary in their exact composition. The main constituents are 40–60% 4,4′-MDI; the remainder is the other isomers of MDI, trimeric species, and higher molecular weight oligomers. Important product variables are functionality and product acidity. Rigid polyurethane foam products are mainly manufactured from PMDI. So-called pure MDI is a low melting solid that is used for high performance polyurethane elastomers and spandex fibers. Liquid MDI products are used in RIM polyurethane elastomers.

The basic raw material for the manufacture of PMDI and its coproduct MDI is benzene. Nitration and reduction afford aniline (see Amines, aromatic). Reaction of aniline with formaldehyde in the presence of hydrochloric acid gives rise to the formation of a mixture of oligomeric amines, which are phosgenated to yield PMDI. The coproduct MDI is obtained by vacuum distillation of PMDI. In contrast to TDI, the yield in the manufacture of PMDI and MDI is quantitative.

Liquid MDI (Isonate 143-L) has found wide application in RIM polyurethane elastomers (11). It is produced by converting some of the isocyanate groups into carbodiimide groups, which react with the excess isocyanate present to give a small amount of the trifunctional four-membered cyclic adduct. This product liquefies the low melting MDI.

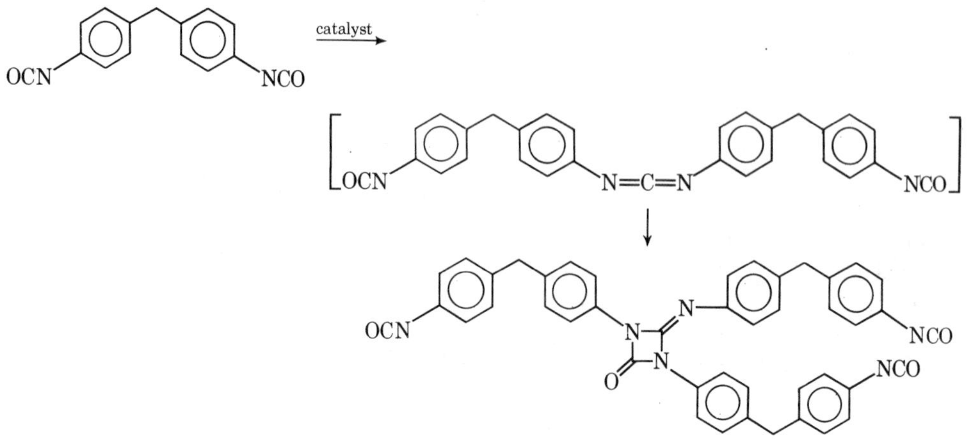

liquid MDI

Liquid MDI products are also made by reaction of the diisocyanate with small amounts of glycols. These products are called prepolymers.

In addition to the commodity isocyanates TDI and MDI, several higher priced specialty aromatic diisocyanates are available in limited quantities; for example, 1,5-naphthalene diisocyanate (NDI) and bitolylene diisocyanate (TODI) (Isonate 136-T) are used in high quality cast elastomers. The corresponding diamine precursors are dye intermediates. These symmetrical high melting diisocyanates give high melting hard segments in polyurethane elastomers.

Urethane polymers derived from aromatic diisocyanates undergo slow oxidation in the presence of air and light, causing a discoloration which is unacceptable in some applications. Polyurethane products derived from aliphatic isocyanates, however, are color stable. The cheapest aliphatic diisocyanate is HDI, which is obtained by phosgenating the nylon intermediate hexamethylenediamine. In view of its low boiling point, HDI is no longer used. However, an HDI-biuret, obtained by treatment of HDI with water, is widely used in rigid-coating applications where its trifunctionality is advantageous. Hydrogenated MDI (methylenebis(cyclohexyl isocyanate)) (H_{12}MDI) and isophorone diisocyanate (IPDI) are also used in coating applications (see Coatings).

Isophorone-based isomeric trimethylhexamethylene diisocyanates (TMDI) are available from Veba Chemie. Trimerized IPDI is marketed as a trifunctional aliphatic isocyanate for hard industrial coatings.

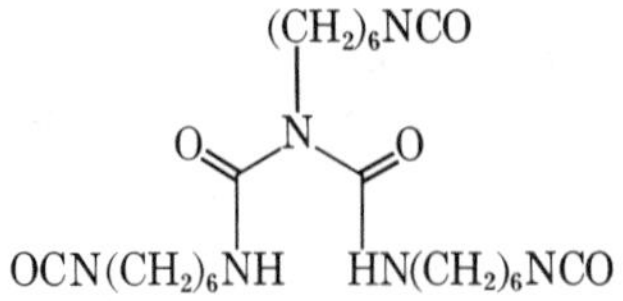

HDI-biuret

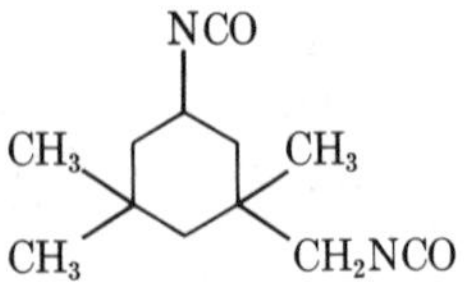

IPDI

Hydrogenated MDI (H$_{12}$MDI) is a coproduct of DuPont's Qiana fiber. The stereoisomeric diamine mixture enriched in the trans-trans isomer is used for fiber production, whereas the cis-cis and the cis-trans diamines are phosgenated to give a liquid diisocyanate mixture.

For nonyellowing coating applications, Takeda in the past has marketed *m*-xylylene diisocyanate (XDI). However, the benzylic isocyanate groups are prone to undergo oxidative discoloration in the presence of light. Reduction of the aromatic ring eliminates this problem, and 1,3-bis(isocyanatomethyl)cyclohexane (H$_6$XDI) is sold by Takeda today.

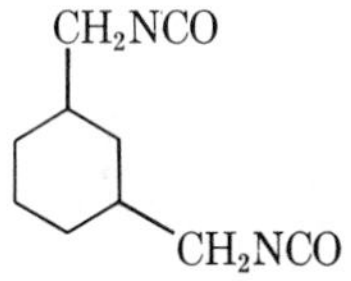

H$_6$XDI

The corresponding fully hydrogenated *p*-xylylene diisocyanate, 1,4-bis(isocyanatomethyl)cyclohexane, is available in development quantities from Eastman and Sun Oil Co. Eastman treats 1,4-cyclohexanedimethanol, an intermediate of Kodel fiber, with ammonia and subsequently phosgenates the diamine to the diisocyanate, whereas Sun Oil Co. uses the same route as Takeda, but starts with *p*-xylene.

Recent innovations in aliphatic isocyanate technology include the development of polymeric aliphatic isocyanates and of bis(cyclic ureas) as ideal masked or blocked aliphatic diisocyanates. Polymeric aliphatic isocyanates are produced by copolymerizing methacrylic acid derivatives, eg, methacrylamide (12) or 2-isocyanatoethyl methacrylate (13), with styrene. Terpolymers of methacrylamide or of 2-isocyanatoethyl methacrylate with styrene and other suitable acrylic acid esters were also prepared.

Difficulties encountered in handling and shipping volatile aliphatic diisocyanates have prompted the development of so-called blocked isocyanates, in which all isocyanate groups are blocked with a suitable blocking agent containing an active hydrogen. Commercial blocking agents include caprolactam, phenol, and acetone oxime. However, as the blocking agent is generated in the thermal unblocking process, ventilation and collection of the volatile chemicals released are required. The overall reaction is shown below with commercial caprolactam-blocked IPDI.

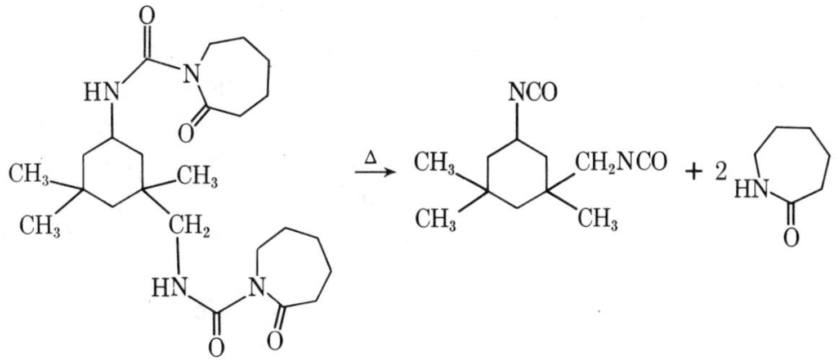

In contrast, bis(cyclic ureas) are ideally blocked aliphatic diisocyanates because no by-products are formed upon thermal release of the reactive isocyanate groups. Cross-linking and hardening of acrylic and epoxy resins containing hydroxyl groups occur readily. This approach is being explored in baked enamels, and wire-coating and powder-coating applications (14). The unblocking of the isocyanate groups involves a ring-opening reaction as shown below:

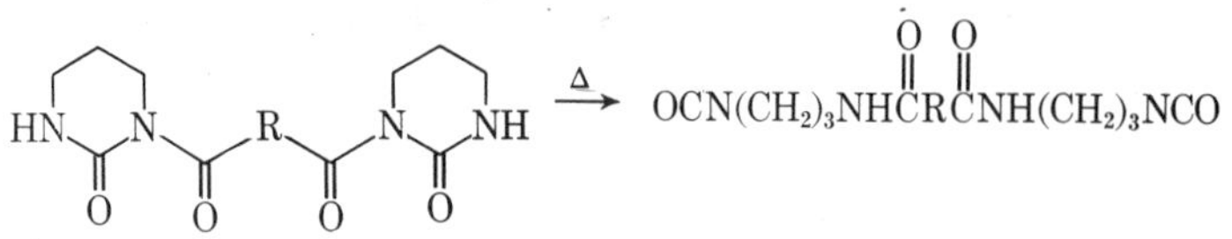

$$\text{HN} \diagdown \text{N} \diagdown \text{R} \diagdown \text{N} \diagdown \text{NH} \quad \xrightarrow{\Delta} \quad \text{OCN(CH}_2)_3\text{NHCRCNH(CH}_2)_3\text{NCO}$$

The properties and manufacturers of some commercial diisocyanates are given in Table 1. Several TDI producers (Allied, DuPont, and Union Carbide) have stopped the production of TDI; DuPont recently sold its H_{12}MDI business to Mobay.

Polyether Polyols. Polyether polyols are addition products derived from cyclic ethers (see Table 2). The alkylene oxide polymerization is usually initiated by alkali hydroxides, especially potassium hydroxide (see Polyesters). For homopolymerization of tetrahydrofuran to poly(tetramethylene glycol) (PTMG), Lewis-acid catalysts, eg, boron trifluoride, are used. Polyether polyols are high molecular weight polyols that range from viscous liquids to waxy solids, depending on structure and molecular weight. Most commercial polyether polyols are based on the less expensive propylene or ethylene oxides or are a combination of the two. Block copolymers are manufactured first by the reaction of propylene glycol with propylene oxide to form a homopolymer. This polymer subsequently reacts with ethylene oxide to give a block copolymer. Since primary hydroxyl groups resulting from ethylene oxide are more reactive than secondary hydroxyl groups, the polyols produced in this manner are more reactive. Random copolymers are obtained by polymerizing mixtures of propylene oxide and ethylene oxide.

With amine initiators so-called self-catalyzed polyols are obtained which are used in systems for rigid-spray foam applications. The rigidity or stiffness of a foam is increased by aromatic initiators such as Mannich bases derived from phenol, phenolic resins, or tolylenediamine.

In polyether polyols used for flexible foam, the viscosity increases with chain length of the polyether branches. In the polyether polyols for rigid foams, the type and functionality of the initiator largely determine the viscosity. The viscosity usually decreases with increasing length of the alkylene oxide chain.

In the manufacture of highly resilient flexible-foam and thermoset RIM elastomers, so-called graft or polymer polyols are used. Graft polyols are dispersions of free-radical polymerized mixtures of acrylonitrile and styrene partially grafted to a polyol. They are available from Union Carbide and BASF. *In situ* polyaddition reaction of isocyanates with amines in a polyol substrate produces so-called PHD (polyharnstoff (urea) dispersion) polyols (15). Interreactive dispersions are used mostly in high resilience-molded flexible foams. The polymer or PHD-type polyols increase the load-bearing properties and stiffness of flexible foams. Interreactive dispersion polyols are also investigated for RIM applications where elastomers of high flexural modulus, low thermal coefficient of expansion, and improved paintability are needed.

Polyester Polyols. Initially, polyester polyols were the preferred raw material for polyurethanes. Today, the less expensive polyether polyols dominate the polyurethane market. Recently, inexpensive aromatic polyester polyols have been introduced in rigid-foam applications. They are obtained from residues of terephthalic acid production or by transesterification of dimethyl terephthalate (DMT) or poly(ethylene terephthalate) (PET) scrap with glycols.

The polyester polyols most frequently used in the production of specialty polyurethane flexible foam and thermoplastic polyurethane elastomers and fibers are based on adipic acid. The 1979 consumption of adipic acid (qv) was 28,000 metric tons, of which ca 65% was used for specialty flexible polyurethane foams, such as textile laminates, garment liners, packaging foam, and reticulated foams for filters and fuel-tank liners. A possible new application would be in automotive roof liners flame-bonded to the metal. The production of thermoplastic polyurethane elastomers uses ca 20% of the total adipate consumption.

The polyester polyols of choice are macroglycols with a low acid number and low water content. For elastomers, linear polyesters of ca 2000 mol wt are preferred. Branched polyesters are used for foam and coatings applications. Phthalates and terephthalates are also used. The glycols include ethylene glycol, propylene glycol, diethylene glycol, dipropylene glycol, 1,4-butanediol, and 1,6-hexanediol. For the manufacture of branched polyesters, triols such as glycerol (qv), trimethylolpropane, and 1,1,1-trimethylolethane are used (see Alcohols, polyhydric; Glycols).

Polyester polyols are also made by the reaction of caprolactone with suitable glycols. Polycaprolactone is used in the manufacture of thermoplastic polyurethane elastomers. The reduction of ester groups in the macromolecule improves the hydrolytic stability of the products (16).

polycaprolactone

Polyurethane Formation. The key to the manufacture of polyurethanes is the unique reactivity of the heterocumulene groups in diisocyanates toward nucleophilic additions. The polarization of the isocyanate group enhances the addition across the carbon–nitrogen double bond, which allows rapid formation of addition polymers from diisocyanates and macroglycols.

$$R—\bar{N}—\overset{+}{C}=\ddot{O} \leftrightarrow RNCO \leftrightarrow RN=\overset{+}{C}—\ddot{\bar{O}}:$$

The monomers, liquid at room temperature, are suitable for rapid bulk-polymerization processes, such as production of continuous block foam, laminate board, and thermoplastic elastomers. Rapid molding of cellular and noncellular polyurethane products from liquid monomers is common practice in the industry.

The polyaddition reaction is influenced by the structure and functionality of the monomers, including the location of substituents in proximity to the reactive isocyanate group (steric hindrance) and the nature of the hydroxyl group (primary or sec-

Table 1. Properties of Commercial Diisocyanates

Name	Structure	CAS Reg. No.	$Bp_{kPa}{}^a$, °C	Mp, °C	Producer
PPDI[b]	NCO — C₆H₄ — NCO	[104-49-4]	110–112$_{1.6}$	94–96	Akzo[c]
TDI	NCO, CH₃ substituted, OCN — C₆H₃	[1321-38-6]		14[d]	BASF, Dow, Mobay, Olin
MDI	OCN—C₆H₄—CH₂—C₆H₄—NCO	[101-68-8]	171$_{0.13}$	37	BASF, Mobay, Rubicon, Upjohn
PMDI	NCO[—C₆H₃—CH₂—]ₙ NCO	[9016-87-9]			BASF, Mobay, Rubicon, Upjohn
NDI	naphthalene with NCO, NCO	[3173-72-6]		130–132	Bayer
TODI	CH₃, CH₃, OCN—biphenyl—NCO	[91-97-4]	160–170$_{0.066}$	71–72	Upjohn
XDI	CH₂NCO — C₆H₄ — CH₂NCO	[3634-83-1]	159–162$_{1.6}$		Takeda
HDI	OCN(CH₂)₆NCO	[822-06-0]	130$_{1.73}$		Bayer
TMDI[e]	OCN—CH₂C(CH₃)₂CH₂CH(CH₃)CH₂CH₂—NCO (CH₃ CH₃, CH₃ CH₃)	[83748-30-5]	149$_{1.33}$		Veba[f]
	OCN—CH₂CH(CH₃)CH₂C(CH₃)₂CH₂CH₂—NCO	[15646-96-5]			
CHDI[g]	NCO — cyclohexane — NCO	[2556-36-7]	122–124$_{1.6}$		Akzo[c]
BDI[h]	CH₂NCO — cyclohexane — CH₂NCO	[10347-54-3]	154–156$_{1.46}{}^f$		Eastman[c], Sun Oil[c]
H₆XDI	CH₂NCO, cyclohexane, CH₂NCO	[38661-72-2]	98$_{0.053}{}^f$		Takeda

586

Table 1 (*continued*)

Name	Structure	CAS Reg. No.	$Bp_{kPa}{}^a$, °C	Mp, °C	Producer
IPDI		[4098-71-9]	$153_{1.33}$		Veba
H$_{12}$MDI		[5124-30-1]	$179_{0.12}{}^f$		Mobay

^{a} To convert kPa to mm Hg, multiply by 7.5.

^{b} 1,4-Diisocyanatobenzene.

^{c} Development product.

^{d} Mixture of 80% 2,4-isomer [584-84-9] and 20% 2,6-isomer [91-08-7].

^{e} 1,6-Diisocyanato-2,2,4-tetramethylhexane and 1,6-diisocyanato-2,4,4-trimethylhexane.

^{f} Mixture of stereoisomers.

^{g} 1,4-Cyclohexanyl diisocyanate.

^{h} 1,4-Cyclohexandebis(methylene isocyanates).

Table 2. Commercial Polyether Polyols

Product	Function-ality	Initiator	Cyclic ether
poly(ethylene glycol) (PEG)	2	water or ethylene glycol	ethylene oxide
poly(propylene glycol) (PPG)	2	water or propylene glycol	propylene oxide
PPG/PEG^a	2	water or propylene glycol	propylene oxide and ethylene oxide
poly(tetramethylene glycol) (PTMG)	2	water	tetrahydrofuran
glycerol adduct	3	glycerol	propylene oxide
trimethylolpropane adduct	3	trimethylolpropane	propylene oxide
pentaerythritol adduct	4	pentaerythritol	propylene oxide
ethylenediamine adduct	4	ethylenediamine	propylene oxide
phenolic resin adduct	4	phenolic resin	propylene oxide
diethylenetriamine adduct	5	diethylenetriamine	propylene oxide
sorbitol adducts	6	sorbitol	propylene oxide or ethylene oxide
sucrose adducts	8	sucrose	propylene oxide

^{a} Random or block copolymers.

ondary). Impurities also influence the reactivity of the system; for example, acid impurities in crude polymeric isocyanates require partial neutralization or large amounts of basic catalysts. Acidity can be neutralized by heat or epoxy treatment of polymeric isocyanates. On the other hand, small amounts of carboxylic acid chlorides lower the reactivity of the crude isocyanates.

The steric effects in isocyanates are best demonstrated by the formation of flexible foams from TDI. In the 2,4-isomer (**4**), the initial reaction occurs at the nonhindered isocyanate group in the 4 position. The unsymmetrically substituted ureas formed in the subsequent reaction with water are more soluble in the developing polymer matrix. Until recently, flexible foams could not be produced from MDI or PMDI. Enrichment of PMDI with the 2,4′-isomer of MDI (**5**) affords a steric environment similar to the one in TDI, which allows the production of flexible foam with good physical properties.

(4)   (5)

The uncatalyzed reaction of diisocyanates with macroglycols is of no significance in the formation of polyurethanes. Nucleophilic reactions of isocyanates are catalyzed by acids and organic bases (17). Triethylamine is a better catalyst than pyridine; strong acids and Lewis acids are also good catalysts. Acid catalysis is of no interest in the formation of polyurethanes, and today only tertiary amines and organometallic compounds, mainly tin derivatives, are used.

The mechanism of the base-catalyzed reaction of isocyanates with alcohols is still under investigation. The two mechanisms under discussion are the nucleophilic catalysis activating the isocyanate (eq. 1) and a general base catalysis activating the alcohol (eq. 2).

$$RNCO + :B \rightleftharpoons RN\!\!\cdots\!\!\overset{O}{\underset{B}{C}} \xrightarrow{R'OH} RNH\overset{O}{\overset{\|}{C}}OR' + :B \qquad (1)$$

$$R'OH + :B \rightleftharpoons R'OH\cdots B \xrightarrow{RNCO} RNH\overset{O}{\overset{\|}{C}}OR' + :B \qquad (2)$$

The reaction of phenyl isocyanate with methanol catalyzed by pyridine follows the general base mechanism (18). However, the mechanism of nucleophilic catalysis (eq. 1) is still widely accepted for the reaction of isocyanates with alcohols.

Tailoring of performance characteristics to improve processing and properties of polyurethane products requires the selection of efficient catalysts. Generally, an increase in base strength in tertiary amines increases the catalytic strength. An exception is Dabco (6) (1,4-diazabicyclo[2.2.2]octane), also called triethylenediamine, in which the nucleophilicity is enhanced by the steric configuration. 1-Azabicyclo[2.2.2]octane (7) and 1,8-diazabicyclo[5.3.0]undec-7-ene (DBU) (8) are also powerful catalysts. The phenol salt of DBU is a heat-activated catalyst.

(6)   (7)   (8)

The catalytic activity of tertiary amines is the result of the free electron pair of the nitrogen; its availability for complexation is more important than its relative base strength, as demonstrated by the superior catalytic activity of Dabco. If crowding or steric hindrance, caused by branched or bulky substituents, exists about the amine nitrogen, the availability of the free electron pair is reduced. Electron-donating substituents enhance catalytic activity. Catalysts containing hydroxyl groups, eg, dimethylaminoethanol, are generally less active than the alkyl-substituted derivatives because of chemical bonding of the catalyst to a growing polymer chain.

Amine catalysts are sold by Air Products and Chemicals (Dabco), Union Carbide Corporation (Niax), Abbott (Polycat), and Jefferson Chemical Company (Thancat).

Many metal compounds catalyze the reaction of isocyanates with macroglycols; for example, di-n-butyltin diacetate was found to be 2400 times more reactive than triethylamine, and synergistic effects were found with combinations of tertiary amines and tin catalysts as shown below for the synergistic catalysis of the phenyl isocyanate–butanol reaction in dioxane at 70°C (19).

Catalyst	Mol %	$K \times 10^4$, L/(mol·s)
triethylamine	0.88	2.4
di-n-butyltin diacetate	0.00105	20
triethylamine/di-n-butyltin diacetate	0.99/0.00098	88

For the reaction of TDI with a polyether triol, bismuth and lead compounds are more reactive catalysts than tin compounds (20). A series of metal acetylacetonates showed the following order of catalytic activity (21):

$$Mn > V > Fe > Cu > Co > Cr$$

Tin catalysts, however, are preferred, mainly because of their slight odor and the low amounts required to achieve high reaction rates. Carboxylic acid salts of calcium, cobalt, lead, manganese, zinc, and zirconium are employed as cocatalysts with tertiary amines, tin compounds, and tin–amine combinations. Carboxylic acid salts reduce cure time of rigid-foam products. Organomercury compounds are used in cast elastomers and in RIM systems to extend cream time, ie, the time between mixing of all ingredients and the onset of a creamy appearance.

The catalysis of the isocyanate reaction with water is especially important in the formation of water-blown cellular products. Based on the rate of carbon dioxide evolution, the rate of the reaction generally increases with increasing base strength of the tertiary amine (see Fig. 2) (22).

Organolead compounds catalyze both the alcohol and the water reaction effectively (23); for example, phenyllead triacetate compared well with stannous dioctoate in the gel test, but was considerably superior in the generation of carbon dioxide from the water reaction (24).

In standard or hot-cure flexible urethane foam production, the tin and amine catalysts must be present in exactly the proper proportions in order for the gelation reaction, catalyzed by the tin, to be in phase with the gas evolution from the water reaction catalyzed by the amine. If the tin concentration is too high, a tight foam is obtained because a large number of closed cells are produced by fast gelation. If the amine concentration is higher than the tin concentration, the foam may collapse from a lack of gel strength. The amine catalyst should be selected to provide the widest range of tin concentration to compensate for minor equipment and metering differences during production. Errors may have significant consequences because flat-topping flexible-foam machines have throughputs as high as 227 kg/min.

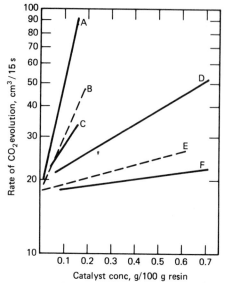

Figure 2.   Effect of catalyst concentration on the isocyanate–water reaction. A, triethylenediamine; B, tetramethyl-1,3-butanediamine; C, triethylamine; D, di-n-butyltin dilaurate; E, N-ethylmorpholine; F, stannous isooctoate.

Highly resilient or cold-cure flexible foams are usually amine catalyzed. However, dibutyltin dilaurate is often added in concentrations <0.05 phr (parts per hundred parts resin) to improve the compression set. Other tin catalysts used in this application include dibutylbis(laurylthio)stannane, dibutyltin bis(isooctylmercaptoacetate), and dibutyltin bis(isooctyl maleate).

Most traditional catalysts promote the release of carbon dioxide at the same time as added fluorocarbon blowing agents vaporize. The recent use of methylene chloride as auxiliary blowing agent, prompted by EPA action on fluorocarbon blowing agents, led to the development of special catalysts, eg, Polycat 12, 72, and 91, and Dabco WT and TL. New mixed-metal organic catalysts contain zinc, antimony, and potassium, eg, CNF-752 and CNF-712.

Delayed-action or heat-activated catalysts are of particular interest in molded-foam applications where mold filling is essential. These catalysts show low activity at room temperature but become active when the exotherm builds up. Most delayed-action catalysts are salts, eg, the phenol salt of DBU.

The rate studies were conducted on model compounds, and test methods were developed that simulate use conditions. For example, gel time in foam systems and the rise rate of the foam are measured.

The recent emergence of urethane-modified isocyanurate foams focused attention on the development of efficient trimerization catalysts. Numerous catalysts, such as lithium oxide; sodium and potassium alkoxides; sodium formate, carbonate, benzoate, and borohydride; potassium and calcium acetates; alkali soaps; lead hydrides; lead salts; metal naphthenates; tertiary amines; N,N-dimethylformamide; and Friedel Crafts-type catalysts have been reported (25). Potassium salts of carboxylic acids, ammonium salts of carboxylic acids, and certain special tertiary amines, such as 2,4,6-tris(N,N-dimethylaminomethyl)phenol (DMT 30), 1,3,5-tris(3-dimethylaminopropyl)hexahydro-s-triazine (HHT), and ammonium salts (**9**) (Dabco TMR), are the catalysts of choice.

Flexible Foam.   Flexible slab or bun foam is typically poured by multicomponent machines at rates ≥45 kg/min. So-called one-shot pouring from traversing mixing

$$\begin{array}{c} \underset{\text{RCO}}{\overset{\text{O}}{\|}}{}^{-} \qquad CH_3-\underset{\underset{CH_3}{|}}{\overset{\overset{CH_3}{|}}{N}}{}^{+}\underset{CH_3}{-}\text{\textbackslash}OH \end{array}$$

(9)

heads is generally used. A typical formulation for furniture-grade foam with a density of 0.024 g/cm^3 includes a polyether triol, TDI, water, catalysts, surfactant, and blowing agent (trichlorofluoromethane, FC-11) (see Table 3). Soft and flexible open-cell foam is formed by the reaction of the diisocyanate with water which generates carbon dioxide. Addition of more FC-11 lowers the density and gives a softer hand.

Table 3.   Typical Furniture-Grade Flexible-Foam Formulation[a]

Ingredient	Parts
polyether triol[b]	100
TDI[c]	50
water	4
catalyst[d]	1
surfactant[e]	1.5
FC-11	3

[a] Ref. 26.
[b] Mol wt 3000, hydroxyl number 56.
[c] 2,4- and 2,6-isomer in an 80:20 ratio.
[d] Stannous octoate, sometimes in combination with tertiary amines.
[e] Silicone copolymers.

Higher density (0.045 g/cm^3) slab or bun foam, also called high resiliency (HR) foam, is produced similarly using polyether triols with a molecular weight of 6000 and a hydroxyl number of ca 28. Special polyether polyols, so-called polymer polyols, improve the load-bearing properties.

Flame retardants are incorporated into the formulations in amounts necessary to satisfy existing requirements (see Flame retardants). Reactive-type polyols are preferred, eg, diethyl N,N-bis(2-hydroxyethyl)aminomethylphosphonate (Fyrol 6). Nonreactive phosphates (Fyrol CEF, Fyrol PCF) are also used, as are mineral fillers such as alumina trihydrate. The mineral fillers are often incorporated into high density flexible-foam products.

Slab flexible foam is produced on continuous bun lines. The bun forms while the material moves down a long conveyor. In so-called flat-top bun lines, the liquid chemicals are dispensed from a stationary mixing head to a manifold at the bottom of a trough. The reactants cream in the trough, expand to a foamed phase, and overflow onto a bottom substrate. Full foam height should occur where the fall plate meets the horizontal conveyor or slightly beyond (Fig. 3).

A high rate of block-foam production (150–220 kg/min) is required in order to obtain large slabs to minimize cutting waste. Bun widths range from ca 1.43 to 2.2 m, and typical bun heights are 0.77–1.25 m. In a flexible-foam plant, scrap can amount to as much as 20%. Most of it is used as carpet underlay and in pillows and packaging (see Packaging materials). The finished foam blocks are stored in a cooling area for at least 12 h and then passed to a storage area or to slitters where the blocks are cut into sheets. In the production unit, the fire risk must be minimized. The danger of spontaneous combustion arises when foam is incorrectly formulated such that an excess of unreacted isocyanate groups is present after manufacture; further reaction can generate more heat.

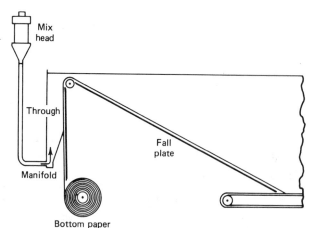

Figure 3.   Flat-top bun line for flexible-foam production.

Most flexible foams produced are based on polyether; ca 5% of the total production is based on polyester. Flexible polyether foams have excellent cushioning properties, are flexible over a wide range of temperatures, and resist fatigue, aging, chemicals, and mold growth. Polyester-based foams are superior in resistance to dry cleaning and can be flame-bonded to textiles.

In recent years, molded flexible-foam products are becoming more popular than contours cut from slab stock. The bulk of the molded flexible urethane foam is employed in the transportation industry, where it is highly suitable for parts such as seat cushions, back cushions, and bucket-seat padding. In earlier years, TDI prepolymers were used in flexible-foam molding in conjunction with polyether polyols. The introduction of organotin catalysts and efficient silicone surfactants facilitates one-shot foam molding which today is the most economical production method. In recent years, the need for heat curing was eliminated by the development of cold-molded or high resiliency foams. These foams are produced from highly reactive polyols and are completely cured under ambient conditions. The polyether triols are 4500–6500 mol wt and are high in ethylene oxide (usually >50% primary hydroxyl content). Reactivity is further enhanced by triethanolamine, liquid aromatic diamines, and aromatic diols. Generally, PMDI, TDI, or blends of PMDI–TDI are used. Load-bearing characteristics are improved by polymer polyols produced by *in situ* polymerization of vinyl monomers in the presence of conventional highly reactive polyols. The products contain up to 20% of dispersed styrene–acrylonitrile copolymers. High resiliency foams exhibit relatively high SAC (support) factors, ie, load ratio, excellent resiliency (ball rebound >60%), and improved flammability properties by conventional test methods.

Proportioned, premixed, machine-ready, flexible polyurethane systems for foam molding are sold by systems producers. In a typical system, the polyols are preblended with surfactant, catalysts, and sometimes flame retardants. The other component is the isocyanate.

Rigid Foam.   Rigid polyurethane foam is mainly used for insulation (qv). The configuration of the product determines the method of production. Rigid polyurethane foam is produced in slab or bun form on continuous lines similar to flexible foam (see Fig. 4) or is continuously laminated between asphalt or tar paper, aluminum, steel fiberboard, or gypsum facings (see Fig. 5). It can be poured or frothed into suitable cavities (pour-in-place application) or sprayed upon suitable surfaces. Applicators can buy formulated chemical systems consisting of isocyanate and polyol components

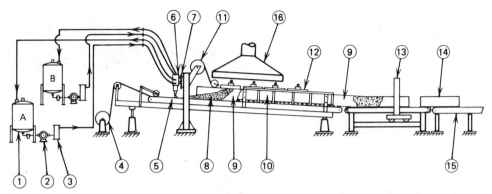

Figure 4. Rigid-bun foam line: 1, material tank with agitators; 2, metering pump; 3, heat exchanger; 4, bottom paper roll; 5, conveyor; 6, mixing head; 7, traverse assembly; 8, rising foam; 9, side paper; 10, side panels (adjustable); 11, top paper roll; 12, top panels (adjustable height); 13, cut-off saw (traversing); 14, cut-foam bun; 15, roller conveyor; 16, exhaust hood.

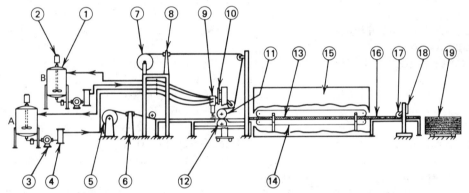

Figure 5. Rigid-foam laminating line: 1, material tank; 2, agitator; 3, metering pump; 4, heat exchanger; 5, bottom facer roll; 6, bottom-facer alignment device; 7, top facer roll; 8, top-facer alignment device; 9, mixing head; 10, traverse assembly; 11, top nip roll; 12, bottom nip roll; 13, take-up-conveyor top belt (adjustable height); 14, take-up-conveyor bottom belt; 15, curing oven; 16, laminate; 17, side-trim saws; 18, cut-off saw (traversing); 19, laminated-panels stack and packaging.

designed to produce foam of the desired density and properties in commercial spraying, pouring, or frothing equipment.

Almost all rigid polyurethane foam is produced from PMDI. Polyols of choice include propylene oxide adducts of polyfunctional hydroxy compounds (penta-erythritol, sorbitol, α-methylglucoside, sucrose, phenol–formaldehyde resins, and others) and propoxylated polyfunctional amino alcohols, diamines, and Mannich bases. Members of the latter group have autocatalytic activity because of their tertiary amino group and are preferred for spray-foam formulations where high reaction rates are required. Crude aromatic polyester diols are used in combination with the multi-functional polyols. The high functionality of the ether polyols combined with the higher functionality of PMDI contributes to the rapid network formation required for rigid polyurethane foams. Flame retardants are sometimes added to meet the existing building-code requirements. The flame retardants contain phosphorus or halogen and may be inert or contain reactive hydroxyl groups. The reactive products are permanently incorporated into the polymer matrix. The most commonly used reactive flame retardants include diethyl *N,N*-bis(2-hydroxyethyl)aminomethylphosphonate (Fyrol 6) and tetrabromophthalate esters (PHT 4-Diol). The reactive flame retardants are combined with the polyol component; storage stability is important. Nonreactive flame retardants include halogenated phosphate esters and highly halogenated aromatic compounds.

Insulation foams are halocarbon blown. In addition to the commonly used trichlorofluoromethane (FC-11, bp, 23.8°C) formulated into the polyol or added to both polyol and isocyanate before polymerization to reduce viscosities, frothing machines meter a separate stream of dichlorodifluoromethane (FC-12, bp, −30°C) to the mixing head and whip the foam chemicals into a froth before dispensing them into a cavity. Frothed foam fills cavities quickly and completely, imparts uniform cells, and exerts lower pressure on cavity walls. Most of the FC-11 is incorporated into the closed foam cells. At a density of 0.032 g/cm³, rigid polyurethane foam is ca 97 vol % gas (FC-11), which accounts for the observed low K-factor; FC-11 has the lowest thermal conductivity of all known gases suitable for foam expansion and suffers little outward diffusion from the foam cells. Small amounts of water are also added to aid in the overall reaction. The heat of the water–isocyanate reaction volatilizes FC-11, and the generated carbon dioxide bubbles act as nucleating centers for the further growth of the FC-11 bubbles.

From the onset of the creaming to the end of the rise during the expansion process, the gas must be retained completely in the form of bubbles which ultimately result in the closed-cell structure. Addition of surfactant facilitates the production of very small uniform bubbles necessary for a fine cell structure. The most effective surfactants are polyoxyalkylene–polysiloxane copolymers which have been specifically designed for rigid foam (see also Surfactants and detersive systems).

The catalysts are tertiary amines and certain organotin compounds, eg, dibutyltin dilaurate. A typical rigid-foam formulation is given in Table 4.

During the molding of high density rigid-foam parts, the dispensed chemicals have to flow a considerable distance to fill the cavities of the corresponding mold. In the filling period, the viscosity of the reacting mixture increases markedly from the initial low value of the liquid mixture to the high value of the final polymerized foam. If the viscosity increases rapidly, incomplete filling results. Chemical factors that influence flow properties are differential reactivity in the polyol components and the addition of water to the formulation. Venting holes allow the escape of air displaced by the rising foam and a moderate degree of overpacking is often advantageous. New

Table 4.   Typical Rigid Polyurethane Formulation[a]

Ingredient	Parts
PMDI	140
polyol[b]	100
fire retardant	15
catalyst[c]	2
surfactant	2
FC-11	11

[a] Ref. 26.

[b] Sucrose- or sorbitol-based polyether polyol with a hydroxyl number of 450.

[c] Mixture of dibutyltin dilaurate and tertiary amines.

high pressure RIM machines have simplified the procedure, and filling intricate molds is no longer a problem.

Premixed machine-ready components for spray-foam applications for roof and tank insulation are sold by suppliers of polyurethane-foam systems. The polyols are preblended with surfactant and catalysts; flame retardants are sometimes included. The other component is the isocyanate. To adjust viscosity, component proportions, and reactivity of the system, a small portion of the polyol component may be worked into the isocyanate to form a partial prepolymer. The blowing agent (FC-11) can be added to either component. Spray-foam systems usually contain reactive polyether polyols based on a nitrogen-containing initiator.

The choice of an aromatic (Mannich base) rather than an aliphatic nucleus on which to initiate the polyether contributes to the stiffness of the urethane polymer. The density of spray insulation foam is usually ca 0.032–0.048 g/cm³.

Recently developed urethane-modified rigid polyisocyanurate foams exhibit superior thermal stability and combustibility characteristics. The former trait was achieved by cyclotrimerizing the isocyanate groups into heterocyclic triisocyanurate groups, which do not revert to the starting materials but rather decompose at higher temperatures with char formation which protects the foam underneath the char. Earlier polyisocyanurate foams showed high friability and poor processability. Modification by incorporating amide, imide, oxazolidinone, carbodiimide, urea, or urethane linkages was attempted to reduce the inherent brittleness. Only the urethane-modified polyisocyanurate foams are used today because the reaction of isocyanates with alcohols is an established technology and the incorporation of hydroxyl-terminated monomers proved to be economically attractive; regular polyurethane-processing equipment could be used. Many different polyol types have been used in the preparation of urethane-modified polyisocyanurate foam, but polyols containing primary hydroxyl groups with low functionality and good compatibility with PMDI are preferred. Although the selection of the polyol is important, the selection of the catalyst is critical because processability and physical properties of the foams are greatly influenced by the catalyst. Control of component temperature in combination with polyol and catalyst selection has led to the development of polyisocyanurate bun, laminate, pour-in-place, and spray foams, which are in all respects comparable to polyurethane foams but are stable up to 150°C; they do not require added flame retardants to meet most regulatory requirements (27).

Minimum polyol concentrations required are ca 0.2 equivalent of polyol per equivalent of isocyanate. A typical polyisocyanurate-foam formulation is shown in Table 5.

In the discontinuous manufacturing of metal-faced insulation panels, polyiso-

Table 5. Typical Rigid Polyisocyanurate Formulation

Ingredient	Parts
PMDI	134
polyol	20
catalyst[a]	5
surfactant	2
FC-11	30

[a] 3:2 mixture of Curithane 52 and Curithane 51.

cyanurate RIM foam provides the required Class I flammability rating, excellent processability, and rapid demolding time.

Polyurethane Elastomers. Initial development of polyurethane elastomers concentrated on products similar in processability to natural rubber (see Elastomers, synthetic). These so-called millable polyurethanes are compounded on a rubber mill or by heavy-duty mixers and are generally cured by vulcanization. The next step led to liquid thermoset cast systems that are polymerized in inexpensive molds. Cast elastomers of this type are characterized by high tensile and tear strength. Development of linear segmented thermoplastic elastomers for injection molding and extrusion added a new dimension of processability to polyurethane elastomers, and more recently, availability of high pressure impingement mixing machines revolutionized the production of thermoset liquid-cast polyurethane elastomers. This process is generally referred to as reaction-injection molding (RIM) or liquid-injection molding (LIM). Addition of fillers increases stiffness and modulus; the process is then called reinforced reaction-injection molding (RRIM). These processes play an important role in the development of lightweight automobiles containing many plastic parts.

Millable polyurethane elastomers are produced by chain extension of linear or lightly branched polyester or polyether polyols with aromatic diisocyanates, eg, TDI and MDI, ca 50,000 mol wt. These products are compounded by either open-mill or heavy-duty mixing equipment and processed further by calendering, extrusion, compression molding, and transfer molding; Shore hardness ranges from 50A to 95A (see Hardness). The hardness or compound modulus is controlled with fillers such as carbon black, silica, or clay. Millable elastomers are generally cured by vulcanization using either sulfur- or peroxide-initiated cross-linking. Physical properties, particularly toughness and tear resistance, are lower than those of comparable cast polyurethanes. However, overall properties compare well to those of natural rubber.

Cast elastomers are usually slightly cross-linked thermoset polymers based on TDI and MDI. The former, eg, DuPont's Adiprene L, is best cured with diamines like methylenebis(o-chloroaniline) (MOCA). Concern over worker exposure to diamines has led to the increased use of MDI-based cast elastomers. Recently, DuPont introduced Adiprene M cast systems based on MDI, PTMG, and butanediol. The polyols can be polyethers or polyesters. Polyethers give lower densities and better hydrolytic stability, fungus resistance, and electrical properties. Polyesters result in higher densities, better tear and abrasion resistance, and less oil and solvent absorption. High hardness and high modulus formulations are of lower molecular weight or contain multifunctional polyols for greater rigidity. Other approaches include branching of polyols by grafting with styrene and acrylonitrile and the incorporation of higher functional isocyanates or trifunctional isocyanurate groups. Increased urethane content resulting from an increase of the hard glycol–urethane segments raises modulus and hardness without significant effect on tensile strength.

Cast polyurethane elastomers are prepared by the one-shot method from MDI prepolymers or partial prepolymers. In the prepolymer process, the isocyanate reacts completely with the polyol, and the chain extender, usually a glycol such as 1,4-butanediol or ethylene glycol, is added. The ratio of reactants can be as high as 12:1. With the MDI partial-prepolymer technique, the reagents can be used in a 1:1 ratio. Only part of the polyol reacts with MDI, and the remainder is mixed with the chain extender. In the one-shot method, isocyanate and polyol plus extender are mixed before reacting. This method is used in the high speed RIM casting of automotive polyurethane elastomers (see Table 6). Differences between the low pressure casting process and the high pressure RIM process are in the speed and efficiency of mixing.

The isocyanates used in the formulation of RIM systems are liquid MDI, MDI prepolymers, and PMDI for rigid RIM applications. Light-stable products are obtainable from liquid aliphatic isocyanates. Polyol development has centered on end-capping polypropylene glycol with ethylene oxide to give primary hydroxyl-terminated polyols with good reactivity.

Recent developments include graft polyols and the use of aromatic glycol and amine extenders to give higher melting hard segments. In view of the high reactivity of aromatic diamines with isocyanates, it was necessary to develop special slower-reacting diamines (see compounds (10)–(12)). Amine extenders improve the green strength of the molded elastomer.

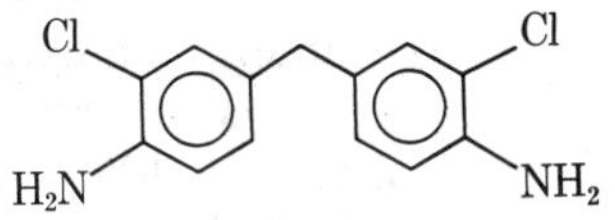

Cl … Cl
H_2N … NH_2

(10) Mbca (MOCA)

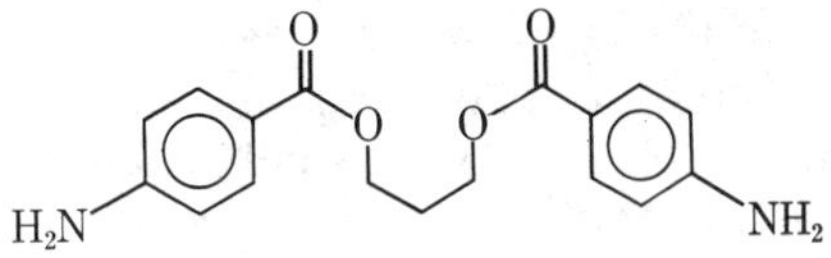

H_2N … O … O … NH_2

(11) Polacure 740M

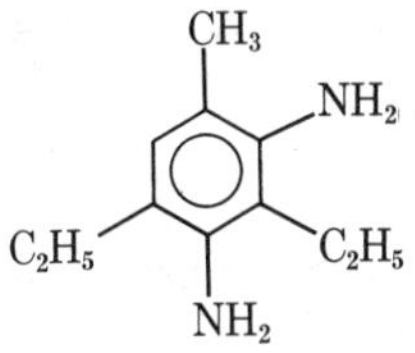

CH_3
NH_2
C_2H_5 … C_2H_5
NH_2

(12) Detda

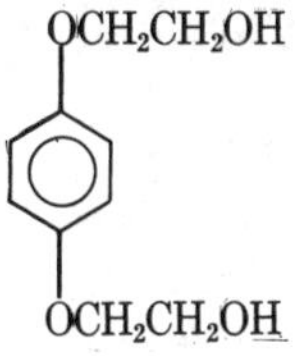

OCH_2CH_2OH
OCH_2CH_2OH

(13) Hqee

OCH_2CH_2OH
OCH_2CH_2OH

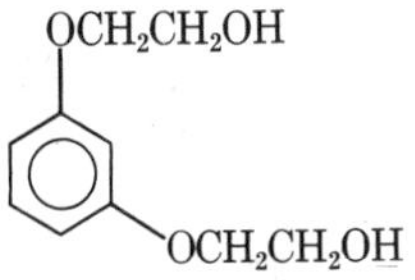

(14) Her

Thermoplastic polyurethane elastomers are segmented linear polymers based on MDI, polyester, or polyether polyols, and glycol extenders. The polyol constitutes the soft segment, whereas the hard segments are formed from the reaction of MDI with the extender glycol. Commonly used soft segments include polyesterdiols, eg, poly-(butylene adipate) and polycaprolactone, and polyether diols, eg, poly(tetramethylene

Table 6. Properties of RIM Systems

Properties	Flexural modulus, MPa[a]		
	0.137–0.517	0.517–1.03	1.37–2.75
tensile elongation (% break)	100–300	50–200	<50
Izod impact (J/m)[b]	534–801	267–801	<267
impact strength	high	medium–high	low
material description	elastomer	pseudo plastic	plastic
automotive application	fascia	fender	hood or deck lid

[a] To convert MPa to psi, multiply by 145.

[b] To convert J/m to ft·lbf/in., divide by 53.38.

glycol), poly(propylene glycol), and mixed polypropylene-polyethylene glycols of ca 1000–2000 mol wt. The hard segments are formed by the reaction of the MDI with the short-chain glycol extender, eg, 1,4-butanediol, 1,6-hexanediol, ethylene glycol, and diethylene glycol.

The physical properties and processing characteristics of the polymers are obtained by adjusting the amounts and ratios of the components. Flexibility is imparted by the soft segments, whereas the degree of hardness is governed by the amount of hard segments present. Thus, polymers with Shore hardness ranging from 70A to 80D are readily formulated. Polyether-based thermoplastic polyurethane elastomers have superior hydrolytic stability, whereas polyester-based products have better oil resistance. The polymers are formulated with isocyanate or hydroxyl end groups. For extrusion-grade products, lower molecular weight resins terminated by hydroxyl are preferred because of their lower melting range.

Within the last few years new and diverse products have been developed with the help of so-called alloy systems. When two compatible thermoplastic polymers are blended, the blend or alloy usually exhibits the properties of each constituent. Polymers compatible with thermoplastic polyurethane elastomers include ABS, PVC, SAN, polyacetal, and cellulose propionate. The blends or alloys are readily produced by dryblending and reextrusion (see Polyblends).

Commercially available blends of ABS and polyurethane elastomers are less expensive and offer increased stiffness or flexural modulus, higher heat-distortion temperature, lower density, and better processability. However, abrasion resistance, tensile strength, and elongation are reduced.

More recently, segmented elastomeric polyurethane fibers (spandex fibers) based on MDI were developed in the United States (see Fibers, elastomeric). DuPont introduced their spandex fiber Lycra in 1962. The generic name spandex fibers designates elastomeric fibers in which the fiber-forming substance is a long-chain polymer consisting of $\geq 85\%$ segmented polyurethane. DuPont is the principal spandex producer, and U.S. consumption in 1979 was 5000 metric tons. Production in Western Europe was 5500 t and in Japan 3600 t.

The elastic behavior of spandex fibers is the result of the alternate arrangements of soft segments, consisting of a macroglycol, such as a polyester polyol (polyadipate), a polyether polyol (polytetramethylene glycol), or a polycaprolactone, and hard segments or blocks containing an aromatic semicarbazid, urea, or urethane groups. The latter are derived from the reaction of an isocyanate-terminated prepolymer with hydrazine, ethylenediamine, or a short-chain glycol. The overall reactions are shown below, using ethylene adipate as the soft block:

$$HOCH_2CH_2O\left[\overset{O}{\overset{\|}{C}}CH_2CH_2CH_2CH_2\overset{O}{\overset{\|}{C}}OCH_2CH_2O\right]_n H \; + \; 2 \quad OCN\text{---MDI---}NCO \longrightarrow$$

poly(ethylene adipate)   MDI

$$OCN\text{---}NH\overset{O}{\overset{\|}{C}}OCH_2CH_2O\left[\overset{O}{\overset{\|}{C}}(CH_2)_4\overset{O}{\overset{\|}{C}}OCH_2CH_2O\right]_n\overset{O}{\overset{\|}{C}}NH\text{---}NCO$$

MDI-terminated polyadipate prepolymer

$$OCNRNCO \; + \; H_2NNH_2 \longrightarrow \left[\overset{O}{\overset{\|}{C}}NHRNH\overset{O}{\overset{\|}{C}}NHNH\right]_n$$

prepolymer   spandex polymer

DuPont uses a dry-spinning process in which the polymer solution in N,N-dimethylformamide is extruded through a spinnerette into a column of circulating hot air. After the solvent has evaporated, the hot spandex filaments are collected, forming a yarn. A variation of wet spinning used by Globe Manufacturing Company is known as reaction spinning. In this case, the isocyanate-terminated prepolymer is extruded into a nonaqueous diamine bath. Although some monofilaments are produced, the individual fibers are usually brought together to form a coalesced multifilament yarn.

Recycling. The amount of waste generated in the manufacture of polyurethane products is estimated to be >90,000 t worldwide in 1980 (28). Refuse material is not included because collection problems prevent efficient utilization of this kind of waste. Flexible polyurethane foam generated in production or recovered after use as packaging material can be recycled by shredding and mixing with a polyurethane binder to produce carpet underlay. Annually, ca 45,000 t of carpet underlay produced in the United States is made from recycled flexible polyurethane foam.

Isocyanate-derived foams can be recycled after hydrolysis with steam, pyrolysis, or glycolysis (see Recycling).

Hydrolysis

$$\text{foam} + H_2O \rightarrow \text{polyol} + \text{polyamine}$$

Pyrolysis

$$\text{foam} \xrightarrow{\Delta} \text{mixture of gaseous and liquid fragments (fuel)}$$

Glycolysis

$$\text{foam} + \text{HOROH} \rightarrow \text{polyol}$$

Since separation of reaction mixtures is difficult, only the glycolysis process developed by Upjohn (28) is used commercially. Cut-up or pulverized rigid foam is fed continuously or in batches into a heated reactor containing glycol at 185–210°C. A clear solution of a urethane polyol in the glycol is obtained. This polyol, in admixture with virgin polyol, can be used for the manufacture of rigid-foam products. In a similar manner, low grade RIM elastomer parts afford a reusable polyol (29).

The chemistry of the glycolysis process involves transesterification of the carbamate groups in the polyurethane foam with the glycol solvent. Thus, three-dimensional network polymers are converted into soluble linear fragments.

Economic Aspects

In 1980, ca 3×10^6 t of urethane polymers was consumed worldwide (see Table 7).

The principal growth of polyurethanes will be in applications related to energy conservation, ie, insulation foams and lightweight, nonrusting automobiles. The anticipated worldwide volume of RIM and RRIM parts in the U.S. automotive industry is ca 70,000 t by 1985. The nonautomotive use for RIM-molded polyurethanes may well exceed automotive uses by the end of the decade. Long glass-fiber reinforcements are fabricated by in-mold, bulk molding (BMC), or sheet-molding compounds (SMC) techniques. Urethane polymers serve basic human needs in many diverse applications ranging from plastics in automobiles to artificial hearts (see Prosthetic and biomedical devices).

Table 7.   Consumption of Polyurethanes in 1980, 1000 t[a]

Region	Flexible foam	Rigid foam	Others[b]
United States	532.0	243.0	158.0
Canada	52.6	20.4	9.1
Latin America[c]	147.8	21.4	26.0
Western Europe	718.2	280.6	175.9
Asia and Pacific Region	220.1	97.4	75.5
Middle East and Africa	130.7	19.9	25.7
Total	*1801.4*	*682.7*	*470.2*

[a] Ref. 30.

[b] Includes elastomers, coatings, fibers, adhesives, and sealants.

[c] Includes Mexico, Central America, and Caribbean.

In 1981, the recession affected U.S. demand for flexible foam (see Table 8). Flexible polyurethane foam is used as cushioning material for upholstered furniture and in foam-core mattresses. Reduction of automobile production in 1981 also affected flexible-foam consumption.

In rigid polyurethane foam consumption, the building and construction market is the main outlet. Principal markets are residential and nonresidential construction, commercial refrigeration, and industrial insulation (see Table 9).

Table 8.   1981 U.S. Flexible-Foam Consumption[a]

Applications	1000 t
furniture	200
transportation	130
bedding	72
carpet underlay	52[b]
textile laminates	16
packaging	11
miscellaneous	34
Total	*515*

[a] Ref. 31.

[b] In addition, >40,000 t of carpet underlay was produced from foam scrap.

Table 9.   1981 U.S. Rigid-Foam Consumption[a]

Application	1000 t
building and construction	140
refrigeration	50
tank and pipe insulation	23
transportation	22
packaging	14
furniture	10
other	11
Total	*270*

[a] Ref. 31.

Other important markets include the transportation and footwear industries. The total 1981 U.S. consumption of polyurethane elastomers in transportation was 30,000 t. The demand for polyurethane shoe soles reached ca 100,000 t in 1979 with the bulk of the production in Italy, the Mideast, Eastern Europe, and Mexico. In addition to the RIM elastomers, ca 25,000 t of thermoplastic polyurethane elastomers, 6,400 t of millable gums, and 5,000 t of elastomeric spandex fibers were consumed in 1981.

In 1981, 40,000 t of polyurethanes was consumed in the United States in a wide variety of surface-coatings applications, either as clear finishes or as pigmented enamels. Emission-free systems, eg, two-component water-based coatings, and powder coatings will grow rapidly, and high solids (>70% solid content) coatings will dominate solvent-based applications. The total 1981 U.S. consumption of polyurethanes in adhesives and sealants (qv) was 59,000 t (30).

The polyurethane industry is dominated by the isocyanate producers who supply this important building block to many polyurethane polymer manufacturers. Several of the principal isocyanate producers (BASF, Bayer, Dow, ICI, Olin, Upjohn) also manufacture polyols to be used with the isocyanates in the production of polyurethanes. Annual production capacities are given in Table 10.

The estimated annual worldwide capacity of polyether polyols for flexible foam is 7.6×10^5 t and for rigid foams it is ca 1.6×10^5 t. Principal U.S. producers are BASF (Pluracol), Dow (Voranol), DuPont (Teracol), Mobay (Multranol), Olin (Poly-G), Quaker Oats (Polymeg), Texaco (Thanol), Union Carbide (Niax), Upjohn (Isonol), and Witco (Fomrez).

The higher demand for polyols for flexible-foam production is caused by the fact that 69 wt % of the flexible foam is polyether polyol. In contrast, only ca 35% of polyether polyol is used in the manufacture of rigid polyurethane foams. For urethane-modified triisocyanurate rigid foams, even less polyol is required.

Health and Safety Factors; Environmental Aspects

Fully cured polyurethanes present no health hazard; they are chemically inert and insoluble in water and most organic solvents. However, dust can be generated in fabrication, and inhalation should be avoided. Because of their inertness, polyurethanes are the polymers of choice in biomedical applications. The artificial heart currently under investigation in several countries and used in one patient in the U.S. is constructed from thermoplastic polyurethane elastomers.

However, some of the chemicals used in the production of polyurethane products, especially the highly reactive isocyanates and the tertiary amine catalysts, must be handled with caution. The polyols and surfactants are relatively inert materials with low toxicity.

Isocyanates. The highly reactive isocyanates must be handled with caution. Exposure to atmospheric moisture must be avoided, and the formation of insoluble ureas is usually the sign of improper handling. Pressure buildup, caused by the evolution of carbon dioxide gas generated by the reaction with water, must be monitored, especially in a closed system. In storage, isocyanates tend to undergo cyclodimerization. Although TDI dimerizes very slowly at ambient conditions, MDI dimerizes quickly and storage under refrigeration is recommended. Aliphatic diisocyanates do not dimerize without catalyst. Specific storage recommendations for individual isocyanates are available from the manufacturers.

Table 10.   1982 Worldwide Isocyanate Capacities, 1000 t

	MDI	TDI
Western Hemisphere		
BASF	45	56
Dow		45
Mobay (Bayer)	135	110
Olin		90
Rubicon (ICI)	56	18
Upjohn	135	
Bayer de Brasil		20
Industrias Cydsa Bayer (Mexico)		12
Total	*371*	*351*
Western Europe		
BASF, Belgium	40	
Bayer, FRG	127	145
Bayer—Shell	24	30
Bayer—Spain	10	14
ICI	70	
Isopor[a]	50	
Montedison	30	40
P.C. Ugine Kuhlmann		35
Progil-Bayer Ugine[b]		45
Quimigal		5
Tolochimie		7
Total	*351*	*321*
Comecon		
Sodaso Hemijski, Yugoslavia		18
State Yugoslavia		20
State East Germany (DDR)	25	8
Total	*25*	*46*
Japan and other Far East		
Mitsui-Nisso	15	30
Nippoly	31	13.5
Sumitomo Bayer	24	13
Takeda		20
Chin Yang (Korea)		10
Total	*70*	*86.5*
World total	*817*	*804.5*

[a] Joint venture Quimigal-Upjohn.

[b] Joint venture Bayer-Rhone-Poulenc.

Thermal degradation of isocyanates occurs on heating above 100–120°C. This reaction, exothermic and when >175°C a runaway reaction, can occur with formation of carbon dioxide gas and carbonization of the residue. The autoignition temperature of TDI is 277°C. In view of the heat sensitivity of isocyanates, it is necessary to melt MDI with caution, and suppliers' recommendations should be followed.

Disposal of empty containers, isocyanate waste materials, and decontamination of spilled isocyanates utilizes the high reactivity of isocyanates. Formulations principally designed for emergency situations include aqueous ammonia (90–95% water, 3–8% conc ammonia solution, and 0.2–5% liquid detergent) and aqueous sodium carbonate (90–95% water, 5–10% sodium carbonate, and 0.2–0.5% liquid detergent). In view of the low miscibility of isocyanates with water, alcoholic solutions may be preferred (50% ethanol, isopropyl alcohol or butanol, 45% water, and 5% conc ammonia solution).

The toxicities of TDI and MDI have been extensively investigated. These isocyanates are well characterized, and their high reactivity with water precludes significant environmental effects. Both the environmental and occupational exposures were determined, and appropriate exposure limits have been set by OSHA at a 20-ppb limit, 8-h TWA. Local exhaust ventilation and the use of respirators for accidental spills protect workers within the OSHA guidelines. A lifetime inhalation study of TDI at concentrations of 0.05 and 0.15 ppm with rats and mice was conducted, and no evidence of carcinogenicity has been found. Teratogenic activity has not been reported.

The total U.S. airborne emissions of the volatile TDI are estimated by the International Isocyanate Institute (III) to be <23 t or 0.008% of the annual U.S. production. Published data show TDI has a $1/3$ life of 8 s in air at 25°C, 50% rh, and a 0.5 s to 3 d half-life in water, depending upon pH and agitation. Without agitation, isocyanates sink to the bottom of the water and react slowly at the interface. Because of this reactivity, there is no chance of bioaccumulation. Tests on TDI gave a 96-h LC_{50} of 164 mg/L for fathead minnows and no significant mortality in grass shrimp exposed to 508 mg/L.

The III has estimated that in the United States, 20,000–30,000 people may be exposed to TDI. A few may develop a bronchial asthmatic hypersensitivity. Excessive exposure can cause respiratory irritation.

Isocyanates are relatively nontoxic (see Table 11). The LC_{50} values show a clear relationship between the acute toxicity and volatility. The type of damage found postmortem following inhalation is consistent with the acute response to corrosive irritants.

In general, diisocyanates cause irritation when applied directly to the skin of rabbits or instilled into their eyes. Repeated skin application leads to an allergic reaction. Subacute toxicity studies on TDI in animals indicate that repeated inhalation causes tracheobronchitis, bronchitis, emphysema, and bronchopneumonia, according to the exposure concentration and frequency of exposure. None of the animal studies showed evidence of sensitization or the asthmalike reaction reported to occur in humans. Inhalation of aliphatic diisocyanates (HDI, IPDI) retarded growth of rats and mice.

Table 11. Acute Toxicity of Diisocyanates in Rats[a]

Isocyanate	LC_{50}, mg/kg	1 h LC_{50}, mg/m^3	Std vapor pressure[b] conc, ppm
HDI	710	310[c]	6.8
IPDI	>2,500	260	0.34
TDI	5,800	58–66	19.6
MDI	>31,600		0.1
NDI	>10,000		0.02[d]

[a] Ref. 32.

[b] Except where otherwise stated.

[c] 4 h.

[d] Vapor pressure at 50°C.

Tertiary Amine Catalysts. The liquid tertiary aliphatic amines used as catalysts in the manufacture of polyurethanes can cause contact dermatitis and severe damage to the eye. Inhalation can produce moderate to severe irritation of the upper respiratory tracts and the lungs, and ventilation, protective clothing, and safety glasses are mandatory when handling these chemicals.

Polyurethanes. The various safety aspects of fully cured polyurethanes have been investigated, and in general, these materials can be considered as safe for human use. However, exposure to dust generated in finishing operations should be avoided. Polyurethane dust irritates eyes and the linings of the nose and throat, and may also affect the lungs. Ventilation, dust masks, and eye protection are recommended in foam-fabrication operations. Extensive studies have been conducted to ascertain the particle-size distribution and concentration of dust to which operators were exposed. Results of midget-impinger, membrane-filter, and elutriator sampling confirm that most of the airborne polyurethane cannot be inhaled. Experiments performed by NIOSH indicate that polyurethane dust introduced by intratracheal injection into rats did not produce cancer; however, emphysema was evident postmortem (33).

Polyurethane or polyisocyanurate dust may present an explosion risk under certain conditions. Experimentally produced dust explosions using 74-μm (200-mesh) polyurethane dust indicate that minimum airborne concentrations of 25–30 g/m^3 of air are required before an explosion occurs.

Since polyurethanes are combustible, they have to be applied in a safe and responsible manner. At no time should exposed foam be used in building construction. An approved fire-resistive thermal barrier must be applied over foam insulation on interior walls and ceilings. Model U.S. building codes specify that foam plastic used on interior walls and ceilings must have a flame-spread rating (determined by ASTM E-84) of ≤ 75 and have smoke generation of ≤ 450. It must be covered with a fire-resistive thermal barrier having a finish rating of not less than 15 min or equivalent to 12.7-mm gypsum board, or having a flame-spread rating of ≤ 25, smoke generation of ≤ 450 (if covered with approved metal facing), and protection by automatic sprinklers. Under no circumstances should direct flame or excessive heat be allowed to contact polyurethane or polyisocyanurate foam. The ASTM numerical flame-spread rating is not intended to reflect hazards presented under actual fire conditions.

Inhalation of thermal decomposition products of polyurethanes should be avoided because carbon monoxide and hydrogen cyanide are among the many products present.

Uses

Flexible Foam. The largest market for flexible foam in the United States is in the furniture and bedding industry. Most furniture cushioning is made of polyurethane foam, predominantly cut from slabs or buns with a density of 0.0192–0.0288 g/cm^3, and polyurethane foam-core mattresses are used increasingly in bedding. Highly resilient flexible foam with a density of 0.040 g/cm^3 is used for seat cushions in higher priced furniture. Widespread use of highly resilient foam, however, has not materialized because of its premium price. The second largest market for flexible foam in the United States is transportation, ie, seats for passenger cars, other motor vehicles, and airplanes. Minor foam uses include carpet underlay and interior padding. Most polyurethane foam molded for the transportation sector is the higher density, highly resilient grade for seating units that consist of frames with prefabricated vinyl-covered molded foam cushions.

Flexible and semi-rigid polyurethane foam products are also used in engineering packaging either in the form of special slab material or by direct foaming; polyester-based flexible foams are used as textile laminates. Specialty applications include reticulated foams for filtration and foams for a variety of consumer products such as sponges, scrubbers, squeegees, and paint applicators. Flexible and semi-rigid polyurethane foam is also used for various gaskets and linings, eg, weather stripping. Horticultural and biomedical applications have also been reported.

Rigid Foam. The bulk of the rigid polyurethane and polyisocyanurate foam production is used in insulation (qv). This application has increased greatly in the last several years as a result of increased energy costs. About half of the rigid foam consumed in 1979 was in the form of board or laminate, and the other half as liquid systems for *in situ* applications (pour-in-place and spray foam). Laminates are used for residential sheathing and board for flat-deck commercial roofing. Most commercial buildings are underinsulated with lightweight decks that are covered with polyurethane spray foam. Spray-foam insulation of residential construction is a recently developed technique. Rigid foam is also used in the insulation of solar-energy systems. Commercial and household refrigeration is another important market for discontinuously poured rigid foam. For commercial refrigeration, PMDI is the isocyanate of choice; for household refrigerators, rigid polyurethane foam based on TDI is preferred. Smaller compressors require more insulation. The low K-factor of fluorocarbon-blown rigid polyurethane foam permits thinner wall sections of equivalent insulation effectiveness, and thus more space is available. Pour-in-place foam is typically integrated in large-scale assembly operations. Almost all rigid polyurethane foam applied *in situ* is consumed in the transportation industry for the insulation of truck trailers, truck bodies, railroad freight cars, and cargo containers. It is generally spray applied, but a good part of the pipe insulation is fabricated from bun stock.

Ships transporting LNG are usually insulated with rigid polyisocyanurate foam laminates providing temperature stability from −180°C to 150°C. The main fuel tank of the Columbia space shuttle has been insulated with polyisocyanurate foam. Rigid polyurethane foam is used in engineered foamed-in-place packaging of industrial or scientific equipment and in the molding of furniture, simulated-wood ceiling beams, and a variety of decorative and structural furniture components. Rigid polyurethane foam is also used for the repair of river barges.

Polyurethane Elastomers. Polyurethane elastomers are used in applications where toughness, flexibility, strength, abrasion resistance, and shock-absorbing qualities are required. They rank high in resistance to ozone, uv radiation, fuels, oils, and many chemicals. Thermoplastic polyurethane elastomers are molded or extruded to produce elastomeric products used in automotive parts, shoe soles, sport boots, roller-skate and skateboard wheels, pond liners, blood bags and tubing, cable jackets, gears, and mechanical goods. Cast and RIM elastomers are used in auto fascia, bumper and fender extensions, printing and industrial rolls, die pads and flexible tooling molds, industrial tires, and industrial and agricultural parts, such as oil-well plugs and grain buckets. Millable gums are used in gaskets, seals, and conveyor belts. Elastomeric spandex fibers are used in hosiery and sock tops (40%), girdles, bras, and support hose (25%), and lightweight knitted swimwear. The use of spandex fibers in outerwear clothing, especially slacks and jeans, is increasing because of its ability to stretch and to return to the original shape (see Fibers, chemical).

Polyurethane Coatings. Polyurethane surface coatings are used wherever applications require abrasion resistance, skin flexibility, fast curing, good adhesion, and chemical resistance (see Coatings). The polyaddition process allows formulation of solventless liquid two-component systems, and ca 50% of commercial polyurethane coatings are of this type. The polyols used in these coatings are either polyesters or polyethers. The isocyanates and isocyanate prepolymers are based mainly on TDI because of the difference in reactivity of the two isocyanate groups. However, MDI-based prepolymers offer the advantage of lower vapor pressure. Aliphatic isocyanates, eg, methylenebis(cyclohexyl isocyanate) (H_{12}MDI), the biuret of hexamethylene diisocyanate, and isophorone diisocyanate (IPDI), offer color stability.

Polyurethane-modified alkyds are made from isocyanates and partially solvolyzed oils, eg, linseed oil (see Alkyd resins). Added driers, such as cobalt naphthenate, catalyze air drying by cross-linking the available double bonds in the polyunsaturated fatty acids (see Driers and metallic soaps). Polyurethane alkyds are used as coatings and foundry-core binders to hold sand cores in shape for metal castings. In recent years, they were replaced by faster amine-cured foundry-core binders based on PMDI and phenolic polyols.

Moisture-cured polyurethane coatings are isocyanate-terminated prepolymers which, after application, are cured by reaction of the residual isocyanate groups with moisture. The amino groups initially formed react with more isocyanate to form urea linkages. Such coatings are applied as architectural finishes.

Polyurethanes are also used in the formulation of water-based coatings. Some suppliers provide aqueous polyurethane dispersions intended for coatings on plastics, metal, wood, concrete, rubber, paper, and textiles. Self-cross-linking aqueous polyurethane dispersions have been developed recently. These coatings are intended for aircraft, appliances, automotive components, and industrial and farm machinery.

Linear polyurethane elastomers dissolved in organic solvents are used in surface coatings of fabric and leather. A small but fast-growing market is coatings based on blocked or masked isocyanates. These products are used in thermoplastic and thermosetting powder-coating applications.

For external applications, aliphatic isocyanates are required because coatings based on aromatic isocyanates discolor under uv. Currently, the biuret of hexamethylene diisocyanate (trade name Desmodur N) and H_{12}MDI (trade name Desmodur W) dominate this market segment. The trifunctional biuret is used in cross-linked hard industrial coatings, whereas the difunctional H_{12}MDI is used in the formulation of soft elastomeric coatings and clear finishes for resilient vinyl flooring and elastomeric automotive coatings.

For baking enamels, wire and powder coatings so-called blocked isocyanates are used. Blocking agents include phenol, caprolactam, and acetone oxime. Heat expels the blocking agent, and the liberated isocyanate reacts with the polyol. Recently, novel masked aliphatic diisocyanates that do not release a blocking agent on heating were introduced.

Unlike the other polyurethane coatings, blocked urethane coatings do not cure below a certain threshold baking temperature. Catalysts, such as tertiary amines and organometallic tin compounds, are often used to lower the curing temperature. Blocked urethane coatings are usually formulated as one-component systems with good storage stability. Because of the dual functionality of its isocyanate groups, IPDI can be used to formulate blocked aliphatic triisocyanates.

A special polyurethane-coating application is in synthetic leather products. These poromeric materials are produced from textile-length fiber mats impregnated and finished with polymeric compositions. The fibers most often used are PET and nylon; polyurethane compositions are incorporated as binders and surface coatings. The initial three-layer structures have been gradually replaced by two and one layer sheets. Permeability to moisture vapor is the key property needed in synthetic leather. The water-vapor permeability can be achieved by a variety of methods ranging from mechanical perforation or inclusion of hydrophilic materials to coagulation in water. The porometic materials are produced in Japan by Kurraray, Kanebo and Teijin, Ltd., and in Europe by Polimex, Enka, and Povair. In addition to shoe applications, poromerics are also used for handbags, luggage, and apparel (see Leatherlike materials).

BIBLIOGRAPHY

"Urethans" in *ECT* 1st ed., Vol. 14, pp. 473–480, by J. A. Garman, Fairfield Chemical Division, Food Machinery and Chemical Corp.; "Urethane Polymers" in *ECT* 1st ed., First Supplement Volume, pp. 888–908, by J. H. Saunders and E. E. Hardy, Mobay Chemical Co.; "Urethan Polymers" in *ECT* 2nd ed., Vol. 21, pp. 56–106, by K. A. Pigott, Mobay Chemical Company.

1. O. Bayer, *Angew. Chem.* **A59,** 257 (1947).
2. A. Hochtlen, *Kunststoffe* **42,** 303 (1952).
3. E. Müller, *Rubber Plast. Age* **39,** 195 (1958).
4. T. B. Marshall, *Text. Ind.* **125,** 75 (1961).
5. V. V. D'Ancicco, *SPE J.* **14,** 34 (1958).
6. H. Ulrich, *Modern Plastics Encyclopedia,* 1978–1979, p. 352.
7. A. A. R. Sayigh, *Polym. News* **2,** 3 (1974).
8. D. G. Gluck, J. R. Hagan, and D. E. Hipchen, *J. Cell. Plast.,* 159 (1980).
9. H. Ulrich, *Introduction to Industrial Polymers,* Carl Hanser, Verlag, München, FRG, 1982.
10. W. J. Schnabel and E. Kober, *J. Org. Chem.* **34,** 1162 (1969).
11. *Isonate 143-L,* Upjohn Polymer Chemicals, LaPorte, Texas.
12. H. J. Wright and K. E. Harwell, *J. Appl. Polym. Sci.* **20,** 3305 (1976).
13. M. R. Thomas, *Preprints of papers presented by the Division of Organic Coatings and Plastic Chemistry,* 183rd National ACS Meeting, Las Vegas, 1982, p. 506.
14. H. Ulrich, *ACS Symp. Ser.* **172,** 519 (1981); P. W. Sherwood, *J. Coat. Technol.* **54**(689), 61 (1982).
15. K. G. Spitler and J. J. Lindsey, *J. Cell. Plast.,* 43 (1981).
16. H. W. Bonk, A. A. Sardanopoli, H. Ulrich, and A. A. R. Sayigh, *J. Elastoplast.* **3,** 157 (1971).
17. D. S. Tarbell, E. C. Mallatt, and J. W. Wilson, *J. Am. Chem. Soc.* **64,** 2229 (1942).
18. R. B. Moodie and P. J. Sansom, *JCS Perkin II,* 664 (1981).
19. E. F. Cox and F. Hostettler, *American Chemical Society Meeting,* Boston, Mass., Apr. 1959.
20. J. W. Britain and P. G. Gemeinhardt, *J. Appl. Polym. Sci.* **4,** 207 (1960).
21. L. B. Weisfeld, *J. Appl. Polym. Sci.* **5,** 424 (1961).
22. H. W. Wolf, Jr., *Catalyst Activity in One-Shot Urethan Foam,* technical bulletin, DuPont, Wilmington, Del., Sept. 1956.
23. G. J. M. Van der Kerk, *Ind. Eng. Chem.* **58,** 29 (1966).
24. H. G. J. Overmars and G. M. van der Want, *Chimia* **19,** 126 (1965).
25. R. Richter and H. Ulrich in *The Chemistry of Cyanates and Their Thio Derivatives,* Pt. 2, John Wiley & Sons, Inc., New York, 1977, p. 619.
26. H. E. Frey and R. L. Maffly in *Economics Handbook,* Stanford Research Institute, Menlo Park, Calif., 1980, pp. 580, 1561A.
27. H. E. Reymore, R. J. Lockwood, and H. Ulrich, *J. Cell. Plast.,* 95 (1978).
28. H. Ulrich, A. Odinak, B. Tucker, and A. A. R. Sayigh, *Polym. Eng. Sci.* **18,** 844 (1978).
29. H. Ulrich, B. Tucker, A. Odinak, and A. R. Gamache, *Elastom. Plast.* **11,** 208 (1979).
30. P. J. Manno, *Polyurethane 81,* The Upjohn Company, Upjohn Polymer Chemicals, LaPorte, Texas.
31. *Mod. Plast.* (Jan. 1982).

32. I. Carney, *Toxicology of Isocyanates*, International Isocyanate Institute, Inc., New Canaan, Conn.
33. K. L. Stemmer, E. Bingham, and W. Barkley, *Environ. Health Perspect.* **11**, 109 (1975).

General References

Ref. 25 is also a general reference.

H. Saunders and K. C. Frisch in *High Polymers*, Vol. XVI, Pts. I and II, Wiley-Interscience, New York, 1962 and 1964.

D. J. David and H. B. Staley in *High Polymers*, Vol. XVI, Pt. III, Wiley-Interscience, New York, 1969.

A. A. R. Sayigh, H. Ulrich, and W. J. Farrissey, Jr. in *Condensation Monomers*, John Wiley & Sons, Inc., New York, 1972.

Precautions for the Proper Usage of Polyurethans, Polyisocyanurates and Related Materials, technical bulletin 107, The Upjohn Company, Chemical Division, Kalamazoo, Mich., 1981.

G. Woods, *Flexible Polyurethane Foams, Chemistry and Technology*, Applied Science Publishers, Ltd., London, 1982.

HENRI ULRICH
The Upjohn Company

URIC ACID

Uric acid [*69-93-2*] (7,9-dihydro-1*H*-purine-2,6,8(3*H*)-trione; 8-hydroxyxanthine) is a member of the purine and xanthine family of compounds and was the first purine to be isolated from a natural source.

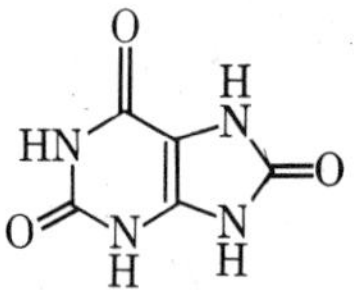

(1)

uric acid

Uric acid (**1**) was discovered independently by Scheele and Bergman in 1776 in human urine and urinary calculi. Fourcroy repeated the work of Scheele in 1793 and named the substance *acide ourique*. He also established the relationship of uric acid with urea by degradation with chlorine water.

Many workers contributed to the work on the properties and synthesis of uric acid; for example, Fischer confirmed a structure proposed in 1875 by synthesis from pseudouric acid (**1**).

Uric acid is widely distributed in nature. It is found in seeds and other plant parts, forms the principal product of nitrogenous metabolism in birds and reptiles, and occurs in the blood and urine of all carnivores.

Bird excrement contains 2–3% ammonium urate; some excreta, eg, the guano of the South Seas, contains nearly 25% ammonium urate.

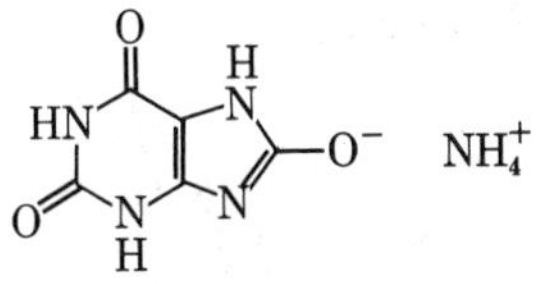

ammonium urate

The uric acid content of boa-constrictor excrement is 90%.

In the human system, uric acid is present in concentrations of 2–6 mg per 100 cm^3 blood (2), and ca 50 mg uric acid is excreted in 100 mL urine daily. Changes in these amounts may indicate diseased states and therefore serve as diagnostic aids. For example, increased uric acid excretion is associated with gout (3).

Uric acid is sold by many chemical suppliers at various prices. In the United States, the Schuykill Chemical Co. produced a few metric tons per year until 1982 at a price of ca $165–198/kg. In Europe, it is produced by Serva Feinbiochemica GmbH, FRG.

Uric acid may be considered a minor pollutant in domestic sewage, where its concentration is typically in the range of 0.2–1.0 mg/L. It is biodegradable by oxidation and hydrolysis (4).

Properties

Uric acid exists as white, colorless, tasteless crystals in the tautomeric forms (2a) and (2b) (5); the enol form (2a) accounts for the acidic properties, and the so-called keto form (2b) accounts for chemical inertness and low solubility (1).

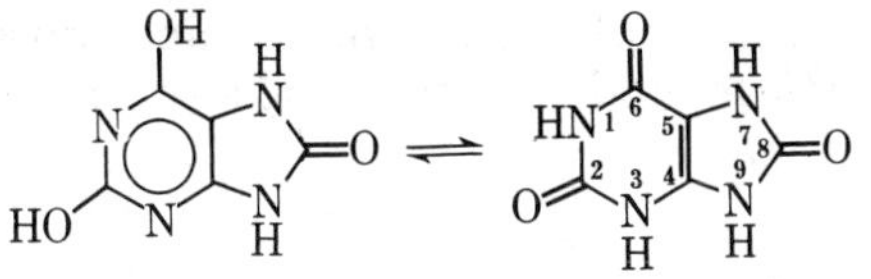

(2a)   (2b)

As a dibasic acid (pK_a 5.75 and 10.3), uric acid forms primary or monobasic salts, $MHC_5H_2N_4O_3$, and secondary or dibasic salts, $M_2C_5H_2N_4O_3$. The primary salts tend to be less soluble in water than the secondary salts. Uric acid decomposes without melting upon heating with evolution of hydrogen cyanide; 1 g dissolves in ca 15,000 parts of cold water and 2,000 parts of boiling water. It is soluble in glycerol, solutions of alkali hydroxides and their carbonates, sulfuric acid (without decomposition), sodium acetate, and sodium phosphate. It is insoluble in alcohol, ether, and chloroform.

Uric acid is best purified by addition of dilute mineral acid to an alkaline solution; the crystalline acid precipitates. After two crystallizations from perchloric acid, ca 99.9% purity may be obtained (6).

Uric acid is stable in both acid and base at moderate conditions but decomposes to ammonia, carbon dioxide, and glycine by vigorous treatment with hydrochloric acid.

The stability under these conditions decreases with increasing nitrogen substitution.

In the uv region, the absorption maxima depend on pH.

Uric acid is stable under reducing conditions, but is degraded to 6-thiouramil (**3**) by heating potassium urate with ammonium sulfate at 160°C. Electrolytic reduction in sulfuric acid gives a variety of products.

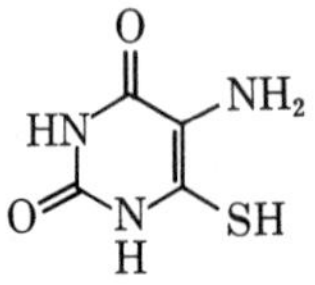

(**3**)

Uric acid is easily oxidized with potassium permanganate to yield allantoin (**4**) (**7**), which upon treatment with HI gives hydantoin (**5**).

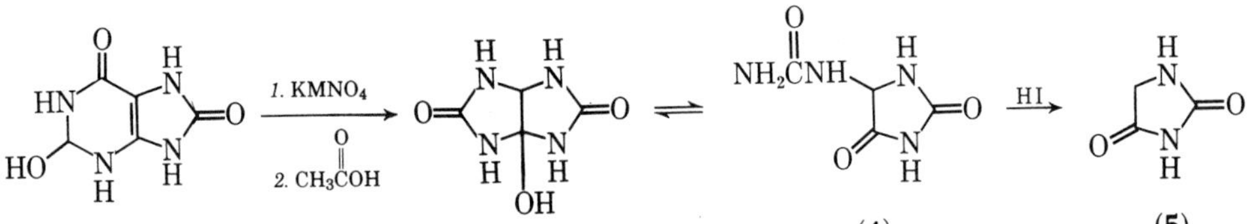

(**1**)

uric acid

(**4**)

allantoin

(**5**)

hydantoin

It is oxidized by 30% hydrogen peroxide under acidic conditions to alloxan (**6**) and further to parabanic acid (**7**) (**7**). Treatment with cold concentrated nitric acid or chlorine gives urea, alloxan (**6**), or alloxan hydrate (**8–9**); hot concentrated nitric acid gives dialuric acid (**8**) and alloxan (**10**). Further heating of this mixture gives alloxantin (**9**). Treatment of alloxantin (**9**) with ammonia gives murexide (**10**) (**11**) (see Fig. 1).

The oxygen at C-8 is easily replaced by opening and subsequent reclosing of the imidazole ring. Thus with formamide, xanthine (**11**) is formed, and with acetic anhydride, 8-methylxanthine (**12**).

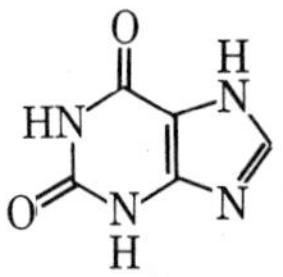

(**11**)

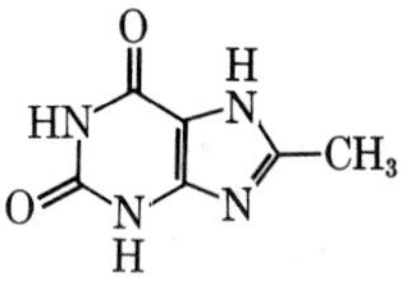

(**12**)

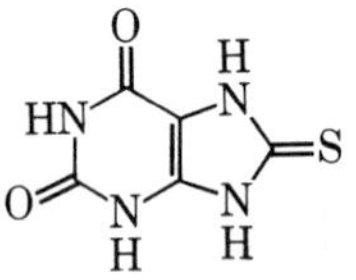

(**13**)

Vigorous treatment with carbon disulfide gives 8-thiouric acid (**13**). Treatment with refluxing phosphorus oxychloride gives 2,6,8-trichloropurine (**14**) and 2,6-dichloro-8-hydroxypurine (**15**). Treatment with hexamethyldisilazane at 200°C for 14 h gives the tetrasilated purine (**16**).

Preparation and Manufacture

Synthesis. Four synthetic methods for uric acid involve interaction of urea with a carbonyl moiety. Three early methods involve the initial formation of a pyrimidine

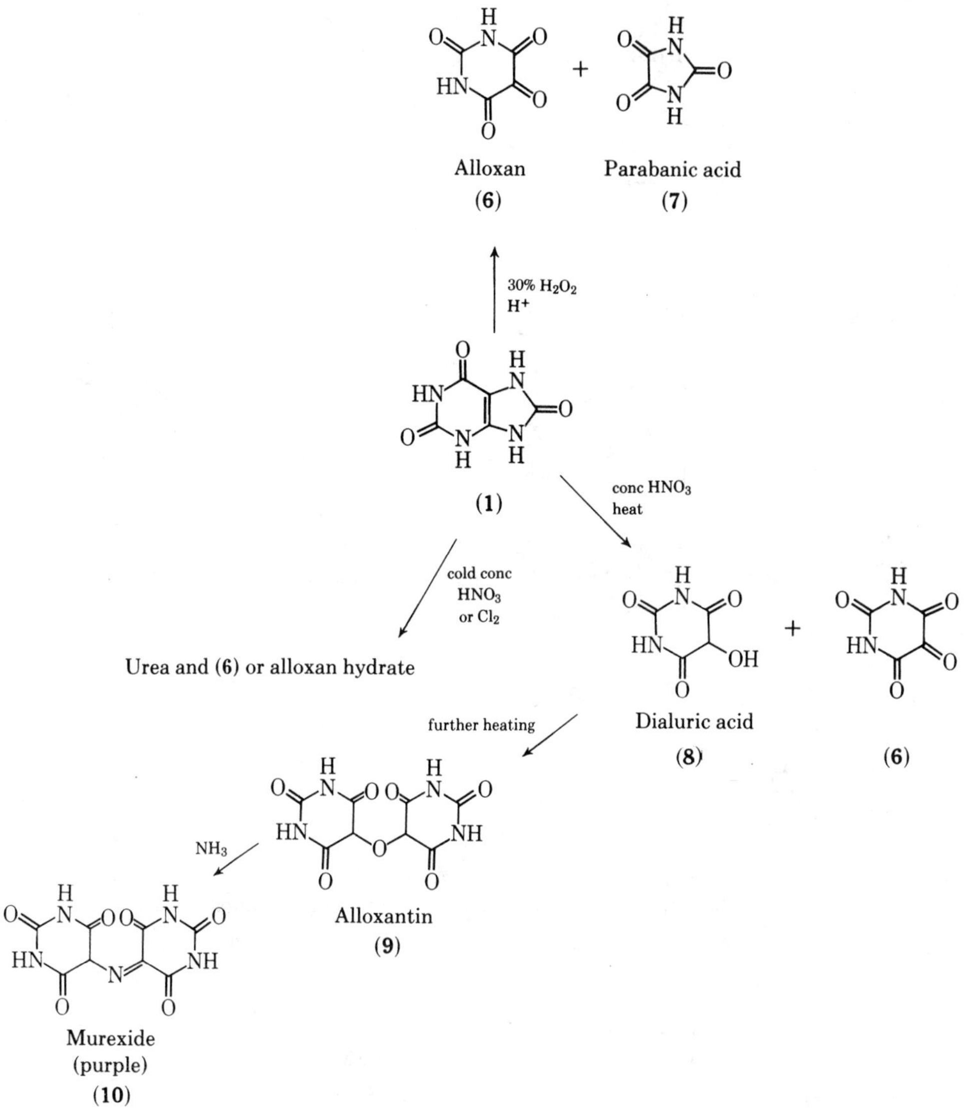

Alloxan
(6)

Parabanic acid
(7)

30% H₂O₂
H⁺

(1)

conc HNO₃
heat

cold conc
HNO₃
or Cl₂

Urea and (6) or alloxan hydrate

Dialuric acid
(8)

(6)

further heating

NH₃

Alloxantin
(9)

Murexide
(purple)
(10)

Figure 1.   Reactions of uric acid.

followed by condensation of the fused imidazole. A more recent synthetic method involves initial formation of the imidazole. The four synthetic procedures are shown below.

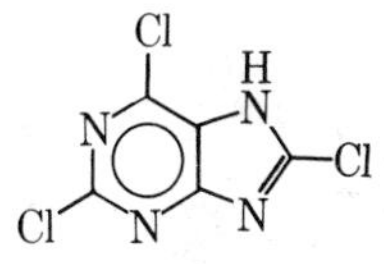

(14)

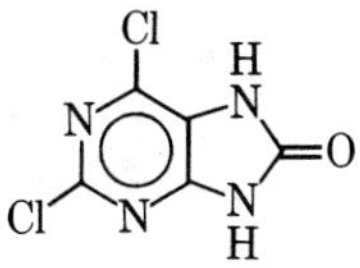

(15)

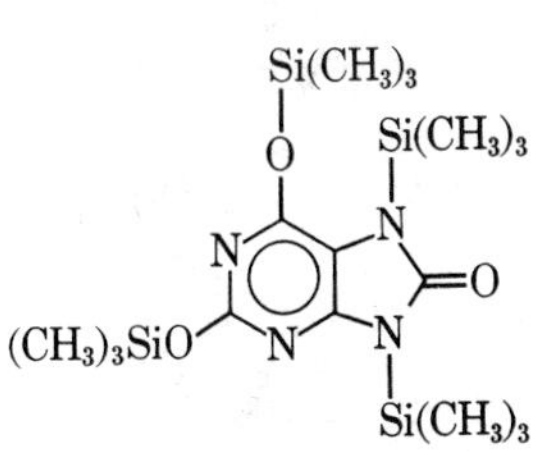

(16)

Urea is treated with acetoacetic acid ester. Further reaction with nitric acid gives isodialuric acid, which with more urea gives uric acid (12).

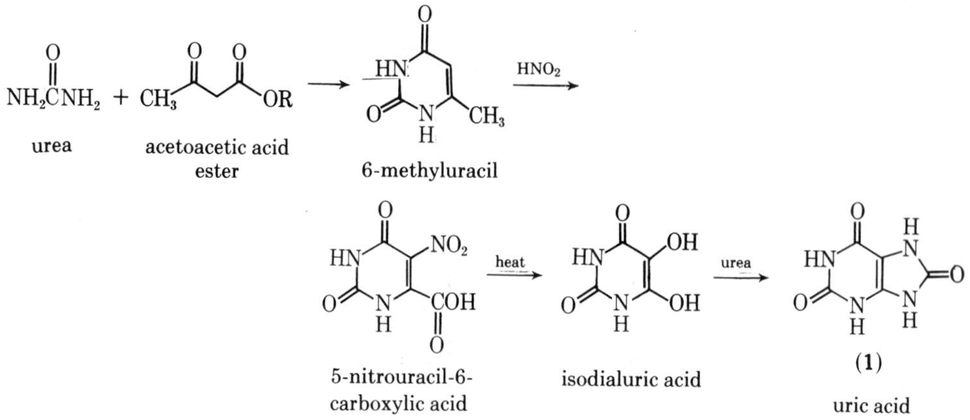
urea
acetoacetic acid ester
HNO_2
6-methyluracil
heat
urea
5-nitrouracil-6-carboxylic acid
isodialuric acid
(1)
uric acid

Uric acid has been synthesized from urea and malonic acid (13).

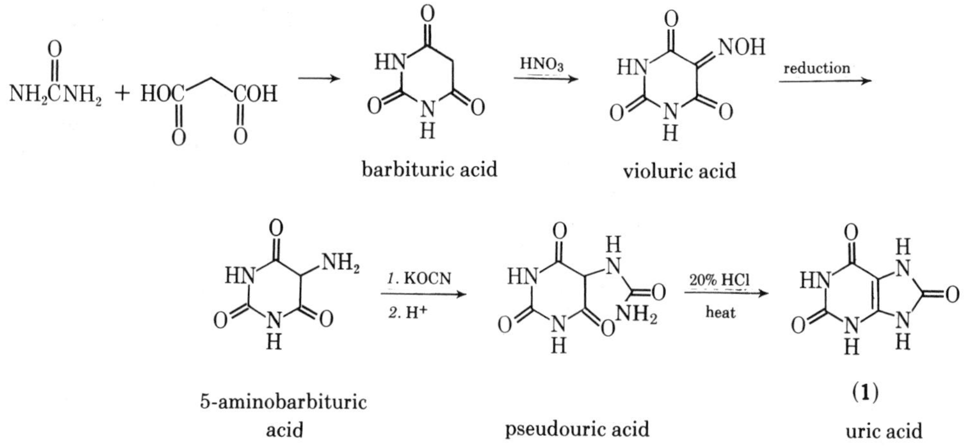
HNO_3
reduction
barbituric acid
violuric acid
1. KOCN
2. H+
20% HCl
heat
5-aminobarbituric acid
pseudouric acid
(1)
uric acid

This synthesis was modified, as shown below (14):

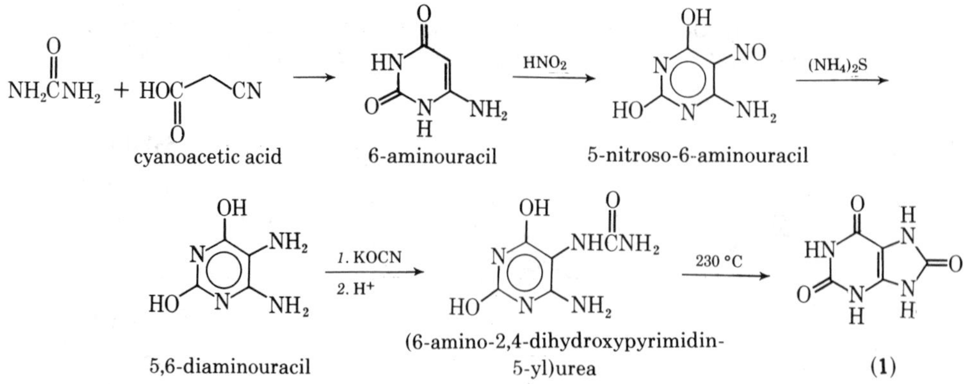
HNO_2
$(NH_4)_2S$
cyanoacetic acid
6-aminouracil
5-nitroso-6-aminouracil
1. KOCN
2. H+
230 °C
5,6-diaminouracil
(6-amino-2,4-dihydroxypyrimidin-5-yl)urea
(1)

The condensation of urea and ethyl α-chloro-β,β-diethoxyacrylate gives an imidazolone which undergoes further ring closure by treatment with base (15).

| ethyl α-chloro-β,β-diethoxyacrylate | (5-ethoxycarbonyl-2-imidazolon-4-yl)urea | (1) uric acid |

Manufacture. Most uric acid is obtained from natural sources by extraction from either reptilian or bird excreta (guano) with sodium hydroxide, precipitation with ammonium chloride, and treatment of the precipitate with mineral acid. Alternatively, the guano may be dissolved by calcium carbonate in alkaline solution and acidified with hydrochloric acid.

In recent procedures, the manures are promptly collected and flash-dried at high temperatures to destroy microorganisms that would degrade uric acid (16). The dried manure is pulverized and sieved, and the material passing through sieves finer than ca 120 μm (120 mesh) is collected. It is then extracted with sodium hydroxide and the lignins precipitated with calcium chloride. Acidification to pH 1–6 precipitates uric acid, which may be further purified (17). Most commercial uric acid is obtained from natural sources.

Analytical Methods

In the murexide test, the sample is treated with concentrated nitric acid and evaporated to dryness on a steam bath. If the sample contains uric acid (or any other purine), a red color develops upon addition of ammonium hydroxide.

Chromatography. Biological fluid is filtered through a cellulose membrane, treated with tungstic and mineral acids for removal of protein, and chromatographed on ion-exchange resin. The fraction containing uric acid is detected at 220–315 nm (18).

Reduction of ferric o-phenanthroline by uric acid in solution may be automated and does not require protein removal. The solutions are buffered, and absorbance is ca 450–700 nm (19).

Blood is treated with alkaline ferricyanide solution, and ferric ions and 5-(2-pyridyl)-2H-1,4-benzodiazepine are added. A purple color develops and is measured at 580 nm against known standards. Removal of protein is necessary (20).

Uses

Uric acid has a wide variety of uses, eg, as a starting material for the production of allantoin, alloxan, alloxantin, parabanic acid, murexide, and others. Because of its stability and inertness in solution and its ability to absorb uv radiation, it is used in cosmetic preparations (see Cosmetics). It has been used as a corrosion inhibitor (21) and a leveling agent in textile dying (22).

Derivatives

Allantoin, 5-ureidohydantoin, mol wt 158.12, is a white powder. It is soluble in ca 190 parts cold water and 500 parts alcohol. It is a product of purine metabolism, where it is formed by the action of uricase. It is synthesized by oxidation of uric acid with alkaline potassium permanganate (7) or by heating urea with dichloroacetic acid (23).

Allantoin has been used against skin ulcers and in hand lotions, lipsticks, and suntan preparations.

Alloxan, 2,4,5,6-(1*H*,3*H*)-pyrimidinetetrone, mol wt 142.07, occurs as white to pink crystals that are acid to litmus, soluble in water and alcohol, slightly soluble in chloroform and petroleum ether, and insoluble in ether. It is prepared by oxidation of uric acid with nitric acid (9) or chlorine (24), oxidation of barbituric acid with chromic acid (25), or oxidation of alloxantin with nitric acid (26). It is used in the production of diabetes in experimental animals, nutritional experiments, and organic synthesis.

Alloxantin, 5,5′-dihydroxy-5,5′-bibarbituric acid, mol wt 206.16, is a white crystalline powder. On exposure to air, it turns red because of ammonia take-up. It is sparingly soluble in cold water, alcohol, and ether; aqueous solutions are acidic. It is prepared by oxidation of alloxan monohydrate (27) or of uric acid with potassium chlorate followed by reduction (28). It has been used in riboflavin (see Vitamins) synthesis and is diabetogenic.

Parabanic acid, imidazolidinetrione, mol wt 114.06, is a white crystalline powder that melts at ca 230°C dec and sublimes at 100°C. It is soluble in ca 20 parts cold water. It is prepared by treatment of uric acid with 30% hydrogen peroxide (29), by oxidation of alloxan, or by condensation of urea with diethyl oxalate (30).

BIBLIOGRAPHY

"Uric Acid" in *ECT* 1st ed., Vol. 14, pp. 480–487, by S. B. Mecca, Schuykill Chemical Company; "Uric Acid" in *ECT* 2nd ed., Vol. 21, pp. 107–114, by S. B. Mecca, Schuykill Chemical Co.

1. *Thorpes Dictionary of Applied Chemistry*, 4th ed., Vol. 9, Longman, Green and Co., London, 1954, p. 802.
2. I. Danishefsky, *Biochemistry for Medical Sciences*, Little, Brown & Co., Boston, Mass., 1980, p. 466.
3. *Ibid.*, p. 524.
4. K. Verschueren, *Handbook of Environmental Data on Organic Chemicals*, Van Nostrand Reinhold Co., New York, 1977, p. 629.
5. A. P. Mathews, *Physiological Chemistry*, 5th ed., William Wood & Co., New York, 1930, p. 765.
6. U.S. Pat. 4,007,186 (Feb. 8, 1977), C. F. Emanuel.
7. W. W. Hartman, E. W. Moffett, and J. B. Dickey in *Organic Syntheses*, Col. Vol. 2, John Wiley & Sons, Inc., New York, 1944, p. 21.
8. S. M. McElvain, *J. Am. Chem. Soc.* **57**, 1303 (1935).
9. Rom. Pat. 65,723 (Oct. 30, 1978), I. Mincu and N. Mihalache.
10. Ref. 6, p. 767.
11. P. B. Hawk, B. L. Oser, and W. Summerson, *Practical Physiological Chemistry*, 12th ed., The Blackiston Co., Inc., New York, 1947, p. 730.
12. R. Behrend and O. Roosen, *Ann.* **251**, 235 (1889).
13. E. Fischer, *Ber.* **28**, 2473 (1895); E. G. Fischer, W. P. Neuman, and J. Roch, *Ber.* **85**, 752 (1952).
14. W. Traube, *Ber.* **33**, 3035 (1900); W. Traube and co-workers, *Ann.* **432**, 266 (1923).
15. C. W. Bills, S. E. Gebura, J. S. Meek, and O. J. Sweeting, *J. Org. Chem.* **27**, 4633 (1962).

16. U.S. Pat. 3,850,930 (Nov. 26, 1974), E. J. Douros and I. T. Warder (to The Gates Rubber Co.).

17. U.S. Pat. 4,196,290 (Apr. 1, 1980), E. J. Douros, I. T. Warder (to The Gates Rubber Co.).

18. Jpn. Pat. 79 150,195 (Nov. 26, 1979), S. Takai and co-workers.

19. U.S. Pat. 3,822,115 (July 2, 1974), L. Morin and J. Prox (to Medico Electronics, Inc.).

20. U.S. Pat. 3,733,177 (May 15, 1973), B. Klein (to Hoffmann-LaRoche).

21. Jpn. Pat. 74 84,934 (Aug. 15, 1974), T. Tamaka and co-workers.

22. Jpn. Pat. 73 32,630 (Oct. 8, 1973), M. Umzawa and co-workers.

23. U.S. Pat. 2,158,098 (May 16, 1939), C. N. Zellner and J. R. Stevens (to Merck & Co.).

24. H. Biltz and M. Heyn, *Ann.* **413,** 60 (1916).

25. A. V. Holmgren and W. Wenner in *Organic Syntheses*, Col. Vol. 45, John Wiley & Sons, Inc., New York, 1963, p. 23.

26. W. W. Hartman and O. E. Sheppard in *Organic Syntheses*, Col. Vol. 3, John Wiley & Sons, Inc., New York, 1955, p. 37.

27. R. S. Tipson in ref. 25, p. 25.

28. D. Nightingale, *Organic Syntheses*, Vol. 23, John Wiley & Sons, Inc., New York, 1943, p. 6.

29. H. Blitz and G. Schiemann, *Chem. Ber.* **59,** 721 (1926).

30. J. I. Murray in ref. 25, p. 744.

L. G. SYLVESTER
CIBA-GEIGY Corporation

UV STABILIZERS

Ultraviolet stabilizers are colorless or nearly colorless organic substances that protect polymeric and other light-sensitive materials from degradation by sunlight and artificial sources of uv (see also Vitamins, Vitamin E).

Ultraviolet radiation may cause polymeric materials to discolor and become brittle; protective coatings to crack, chalk, and delaminate; and dyes and pigments to fade. Several different classes of compounds are used commercially to retard light-induced polymer degradation, including uv absorbers, hindered amines, nickel chelates, hindered phenols, and aryl esters.

In 1981, ca 3000 metric tons of uv stabilizers was sold in the United States, primarily for use in plastics and coatings.

Light Sensitivity of Polymers

The sun emits electromagnetic radiation with wavelengths that range from the x-ray region to the far infrared. The earth's atmosphere, particularly the ozone in the stratosphere, is an excellent absorber for the highly energetic short-wavelength radiation and only light with wavelengths >290 nm reaches the surface of the earth. Although only a relatively small amount of the sunlight energy is in the uv region of 290–400 nm (see Fig. 1), it is responsible for most of the photochemically induced damage (see also Photochemical technology).

The amount of energy in each quantum of solar uv is sufficient to exceed the dissociation of energy of covalent bonds found in polymeric materials, as shown in

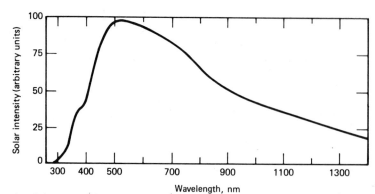

Figure 1. Solar-energy distribution for a midsummer noonday sun in Washington, D.C. (1).

Table 1, and thereby initiate degradation (see Plastics, environmentally degradable in the Supplement Volume).

Theoretically, many polymers, such as polyethylene, polypropylene, and polyacetals, should be very stable to sunlight since the idealized polymers have no chromophores that absorb light with wavelengths longer than 290 nm. However, these polymers degrade quite rapidly when exposed to sunlight. Commercially produced polymers contain uv-absorbing chromophores including catalyst residues, polymer oxidation products, processing aids, colorants, etc.

Photooxidation.   The photooxidation of polymeric materials can be represented by a simplified reaction sequence.

Initiation:

$$ROOH \xrightarrow{h\nu} RO\cdot + \cdot OH \tag{1}$$

$$\underset{\text{photoexcited}}{\underset{\text{chromophore}}{R\overset{O}{\overset{\|}{C}}R}} \xrightarrow{h\nu} R\overset{O^*}{\overset{\|}{C}}R \tag{2}$$

$$R\overset{O}{\overset{\|}{C}}R \longrightarrow R\overset{O}{\overset{\|}{C}}\cdot + \cdot R \tag{3}$$

$$R\overset{O}{\overset{\|}{C}}\cdot \longrightarrow R\cdot + CO \tag{4}$$

Propagation:

$$R\cdot + O_2 \rightarrow ROO\cdot \tag{5}$$

$$ROO\cdot + RH \rightarrow ROOH + R\cdot \tag{6}$$

Chain branching:

$$ROOH \xrightarrow[h\nu]{\Delta \text{ or}} RO\cdot + \cdot OH \tag{7}$$

$$2\,ROOH \rightarrow RO\cdot + ROO\cdot + H_2O \tag{8}$$

In the presence of excess of oxygen, reactions 5 and 6 can be repeated hundreds of times, increasing the concentration of hydroperoxides which cleave homolytically by absorption of thermal or actinic energy to yield additional radicals (eq. 7).

Table 1.   Energy Content of Light in Solar Ultraviolet Compared to Bond Strengths in Organic Compounds[a]

Wavelength, nm	Energy, kJ/ein[b,c]	Bond type	Bond energy, kJ/mol[c]
200	418	C–H	356–418
300	398	C–C	314–335
		C–O	314–335
350	339	C–Cl	293–335
400	297	C–N	251–272

[a] Ref. 2.
[b] 1 Einstein (ein) = Avogadro's number of photons.
[c] To convert J to cal, divide by 4.184.

Stabilization Mechanisms

Initiation of photooxidation can be retarded by additives that function by photophysical mechanisms such as uv absorption and quenching of photoexcited chromophores.

In many polymers, light-initiated oxidation is responsible for most of the light-induced damage. Oxidation can be retarded by antioxidants that scavenge free radicals or destroy hydroperoxides to yield nonradical species.

Many commercially available stabilizers function by more than one mechanism. Thus, 2-hydroxybenzophenones and 2-(2'-hydroxyphenyl)benzotriazoles contribute to stability by absorbing uv, trapping free radicals, and quenching photoexcited chromophores.

Uv Absorbers

Although polymers are degraded most rapidly by radiation with wavelengths <325 nm, degradation takes place also at higher wavelengths. For optimum protection, it is desirable to absorb radiation up to 400 nm. An ideal uv absorber should absorb the radiation between 290 and 400 nm while transmitting all visible light. Absorption directly above 400 nm causes yellowing of the substrate. No uv absorber has this square-wave type of absorption, although substituted 2-(2'-hydroxyphenyl)benzotriazoles approach this ideal most closely. These compounds absorb very strongly throughout most of the uv region and show a rapid decrease in absorbance approaching 400 nm (see Fig. 2).

Compounds with little or no absorbance beyond 400 nm are preferred for most polymer applications.

Strong absorption is only one of a number of important requirements for a uv absorber acceptable for use in polymers and coatings. The absorber itself must be light stable and therefore not destroyed during long-term exposure. The mechanism by which 2-(2'-hydroxyphenyl)benzotriazoles dissipate the energy of absorbed radiant energy involves tautomeric structures (**1**) and (**2**):

                (**1**)   (**2**)

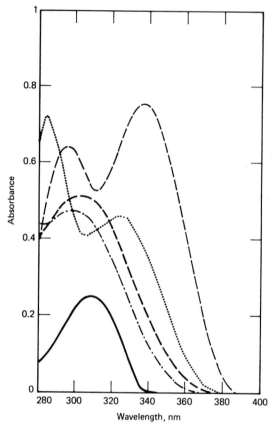

Figure 2. Ultraviolet absorption spectra of some commercial stabilizers dissolved in chloroform (1.0 mg/100 mL, 1-cm path length). − −, 2-(2′-hydroxy-5-methylphenyl)benzotriazole; ·····, 2-hydroxy-4-methoxybenzophenone; - - -, ethyl 2-cyano-3,3-diphenylacrylate; -·-, 2-ethyl-2′-ethoxyoxalanilide; ——, phenyl salicylate.

In the ground state, the phenolic structure (1) is preferred since the electron density on the oxygen atom is much greater than that on the triazole nitrogen. Absorption of uv causes a shift of the electron density away from the oxygen atom toward the triazole ring. The resulting increase in the basicity of the nitrogen atom causes the proton to jump from oxygen to nitrogen. Tautomer (2) is unstable and rapidly loses its energy as heat, reverting back to the ground-state structure (1). This process is highly efficient and accounts for the excellent light stability of 2-(2′-hydroxyphenyl)-benzotriazoles as well as their inactivity as photosensitizers (3).

2-Hydroxybenzophenones dissipate absorbed energy in an analogous manner involving keto-enol tautomerism [eg, (3) ⇌ (4)]:

Aryl esters do not absorb uv strongly but, on prolonged exposure to light, undergo photo-Fries rearrangements to yield strongly absorbing 2-hydroxybenzophenones (4):

The mechanisms by which other types of uv absorbers dispose of actinic energy have not been reported in the literature.

For optimum screening activity, the absorber should be molecularly dispersed in the substrate. Incomplete solubility results in a lower absorbance than the theoretical value calculated from the Beer-Lambert law (5). Poor solubility also results in exudation of the additive to the polymer surface. Generally, modification of the chromophore with appropriate side chains increases solubility. Thus, in the case of 4-alkoxy-2-hydroxybenzophenones, alkyl groups in the C_{10} to C_{16} range are claimed to have optimum compatibility with low density polyethylene (6). Similar polymer compatibility was observed with a homologous series of 2-(2′-hydroxyphenyl)benzotriazoles (5).

Ultraviolet absorbers with extremely low volatility are required for polymers that are extruded at high temperatures. Polycarbonate and poly(ethylene terephthalate) are processed at temperatures of ca 300°C. A considerable amount of the additive is lost when the hot polymers are exposed to the atmosphere unless the additive has a very low vapor pressure. Low volatility is also required for applications such as automotive paints where the stabilizer must suffer only minimal loss during oven drying and outdoor exposure (see Paint). Volatility can be minimized in a number of ways, such as the introduction of appropriate substituents as in the case of 2-[2′-hydroxy-3′,5′-di(α,α-dimethylbenzyl)]phenylbenzotriazole (7). Other approaches are described in the patent and technical literature. Chromophores that absorb uv have been grafted onto the backbones of preformed polymers, built into the polymer chain by condensation, or addition polymerization.

Ultraviolet absorbers should be chemically inert to other additives in the substrate. Those with phenolic groups such as 2-hydroxybenzophenones and 2-(2′-hydroxyphenyl)benzotriazoles may form colored complexes with metal ions such as Co^{2+}, Zn^{2+}, and Cd^{2+}. Such chelate formation can be surpressed to a great extent by bulky substituents that hinder the phenolic group. Thus, 2-(2′-hydroxy-3′,5′-di-*tert*-amylphenyl)benzotriazole is less susceptible to discoloration in the presence of metallic ions than 2-(2′-hydroxy-5′-methylphenyl)benzotriazole. Nonphenolic uv absorbers, such as substituted diphenylacrylates and oxalanilides, do not form colored complexes with metal ions.

According to the Beer-Lambert law, an additive that functions only by absorbing light should be relatively ineffective for protecting substrates with thin cross sections such as fibers, thin film, and protective coatings. According to the following relationship:

$$T = 10^{-abc}$$

[T = transmittance, a = absorbtivity for a particular material at a specific wavelength (L/(g·cm)), b = path length through the sample (cm), and c = concentration (g/L)], the amount of light transmitted approaches 100% as the path length approaches zero.

Ultraviolet absorbers, however, provide levels of stabilization that are greater than can be anticipated from the light absorption alone; for example, 2-(2′-hydroxy-3′,5′-di-t-butylphenyl)-5-chlorobenzotriazole provides better stability than calculated for polypropylene film (8). 2-Hydroxy-4-n-dodecyloxybenzophenone stabilizes polypropylene far more effectively if incorporated into the polymer than when used as an optical filter to shield the specimen from uv (9). The unexpected activity of these stabilizers can be explained either by the quenching of photoexcited chromophores or by an inhibiting oxidative process initiated by radiation. In the case of polypropylene, antioxidant activity is most likely responsible for the enhanced activity of these phenolic uv absorbers.

The effectiveness of 2-(2′-hydroxy-5′-methylphenyl)benzotriazole as a light stabilizer for poly(m-phenyleneisophthalimide) is due to light absorption and quenching of activated chromophores (3). Ultraviolet absorbers derived from 2-hydroxybenzophenone and 2-(2-hydroxyphenyl)benzotriazole also quench photoexcited states of polystyrene (11).

2-Hydroxybenzophenones are consumed during photooxidation of polypropylene. 2-(2′-Hydroxyphenyl)benzotriazoles are also oxidized under these conditions but at a lower rate (10).

Model experiments, in solution, have demonstrated that 2-(2′-hydroxyphenyl)-benzotriazoles, like hindered-phenol antioxidants, are capable of reacting with peroxy radicals (12).

Stabilizers

Hindered-Amine Light Stabilizers. The development of hindered-amine light stabilizers (HALS) represents the most important advance in light-stabilizer technology since the introduction of uv absorbers. The first compound to be commercialized in the United States was bis(2,2,6,6-tetramethyl-4-piperidinyl) sebacate [52829-07-9], a stabilizer highly active in a wide variety of substrates subject to photooxidation, such as polypropylene, polyethylene, styrenic polymers, polyurethane, and acrylic coatings (see Coatings).

The outstanding activity of HALS stabilizers has stimulated a great deal of research efforts directed mainly toward the synthesis of compounds that have superior compatibility and resistance to loss by exudation and volatility. The high molecular weight oligomeric stabilizers Chimassorb 944 and Tinuvin 622 are particularly useful for polyolefin fiber and film.

Although the mechanism by which hindered amines protect polymers against photodegradation has been the subject of many studies, it is not fully explained (13–14). It seems that these compounds function primarily as light-stable antioxidants in which the hindered amine and its oxidation product function as radical scavengers.

In an oxidizing substrate, the hindered amine participates in the following reactions:

$$\underset{\diagdown}{\diagup}\text{NH} \xrightarrow[\overline{O_2}]{\text{radicals}} \underset{\diagdown}{\diagup}\text{NO·} \qquad\qquad (9)$$

$$\underset{\diagdown}{\diagup}\text{NO·} + \text{R·} \longrightarrow \underset{\diagdown}{\diagup}\text{NOR} \qquad\qquad (10)$$

$$\underset{\diagdown}{\diagup}\text{NOR} + \text{ROO·} \longrightarrow \underset{\diagdown}{\diagup}\text{NO·} + \text{ROOR} \qquad\qquad (11)$$

According to this mechanism, the parent $>$NH is consumed by free-radical-induced oxidation, but in the process is converted to species that trap the alkyl radicals (R·) and peroxy radicals (ROO·) that are responsible for propagating thermal oxidation (reactions 5 and 6).

Hindered amines and nitroxyl radicals associate with hydroperoxides, increasing their effectiveness in trapping free radicals produced by photolytic decomposition of the hydroperoxides (15–16):

$$\underset{\diagdown}{\diagup}\text{NO·} + \text{ROOH} \rightleftharpoons \left[\underset{\diagdown}{\diagup}\text{NO----HOOR}\right]$$

$$\underset{\diagdown}{\diagup}\text{NH} + \text{ROOH} \rightleftharpoons \left[\underset{\diagdown}{\diagup}\text{N}\overset{\text{H}}{\underset{\text{---HOOR}}{\diagdown}}\right]$$

It has also been shown that secondary hindered amines can decompose hydroperoxides in polypropylene in the absence of light (17).

Nickel Chelates. Nickel chelates, depending on their structures, contribute to the stabilization of polymeric substances by decomposing hydroperoxides (18), scavenging free radicals (19), absorbing uv radiation (20), and quenching photoexcited chromophores (19). These products, particularly Cyasorb UV 1084, Irgastab 2002, and nickel dialkyldithiocarbamates are highly effective for the stabilization of polyolefins with small cross sections, such as fiber and film. However, the color of nickel chelates limits their utility for white or colorless polymers.

Table 2. Uv Stabilizers

Compound	CAS Registry No.	Trade name	Supplier	Suggested substrates[a]
2-hydroxybenzophenones				
2,4-dihydroxybenzophenone	[131-56-6]	Syntase 100	Neville-Synthèse Organics, Inc.	AC, CE, EP, PES, SC, ST
		Uvinul 400	BASF Wyandotte Corp.	
2-hydroxy-4-methoxybenzophenone	[131-57-7]	Cyasorb UV 9	American Cyanamid Co.	AC, CE, PES, PU, PVC, SC, ST
		Syntase 62	Neville-Synthèse Organics, Inc.	
		UV Absorber 325	Mobay Chemical Co.	
		Acetorb A	Aceto Chemical Co., Inc.	
		Uvinul M-40	BASF Wyandotte Corp.	
2-hydroxy-4-*n*-octyloxybenzophenone[b]	[1843-05-6]	Carstab 700	Carstab Corp.	AC, CE, PES, PO, PU, PVC, SC, ST
		Carstab 705	Carstab Corp.	
		Cyasorb UV 531	American Cyanamid Co.	
		Mark 1413	Argus Chemical Corp.	
		Syntase 800, 805	Neville-Synthèse Organics, Inc.	
		UV Check AM 300	Ferro Corp., Chemical Div.	
		Uvinul 408	BASF Wyandotte Corp.	
2-hydroxy-4-isooctyloxybenzophenone[b]	[33059-05-1]	Carstab 701, 702	Carstab Corp.	CE, PES, PO, PVC, SC, ST
2-hydroxy-4-dodecyloxybenzophenone	[2985-59-3]	Eastman Inhibitor DOBP	Eastman Chemical Products, Inc.	CE, PES, PO, PVC, SC, ST
		UV Check AM 320	Ferro Corp., Chemical Div.	
		Syntase 1200	Neville-Synthèse Organics, Inc.	
2,2′-dihydroxy-4-methoxybenzophenone	[131-53-3]	Cyasorb UV 24	American Cyanamid Co.	AC, CE, PU, PVC, SC, ST
2,2′,4,4′-tetrahydroxybenzophenone	[131-55-5]	Uvinul D 50	BASF Wyandotte Corp.	PES, PU, SC
2,2-dihydroxy-4,4′-dimethoxybenzophenone	[131-54-4]	Uvinul D 49	BASF Wyandotte Corp.	AC, CE, EP, PU, PVC, SC
2-hydroxy-4-methoxy-5-sulfobenzophenone	[4065-45-6]	Cyasorb UV 284	American Cyanamid Co.	SC
		Uvinul MS-40	BASF Wyandotte Corp.	
sodium 2,2′-dihydroxy-4,4′-dimethoxy-5-sulfobenzophenone	[3121-60-6]	Uvinul DS-49	BASF Wyandotte Corp.	SC
polymer of 2-hydroxy-4-(acryloxyethoxy)-benzophenone	[29963-76-6]	Cyasorb UV 2126	American Cyanamid Co.	SC
2-(2′-hydroxyphenyl)benzotriazoles				
2-(2′-hydroxy-5-methylphenyl)benzotriazole[b]	[2440-22-4]	Tinuvin P	CIBA-GEIGY Corp.	AC, EP, PES, PVC, ST
2-(2′-hydroxy-5′-*t*-octylphenyl)benzo-	[3147-75-9]	Cyasorb UV 5411	American Cyanamid Co.	AC, PES, PVC, SC, ST

622

triazole[b]

	CAS No.	Trade name	Manufacturer	Applications
2-[2'-hydroxy-3',5'-(di-t-butyl)phenyl]-benzotriazole[b]	[3846-71-7]	Tinuvin 320	CIBA-GEIGY Corp.	AC, CE, PES, PU, PVC, SC, ST
2-[2'-hydroxy-3',5'-(di-t-amyl)phenyl]-benzotriazole	[25973-55-1]	Tinuvin 328	CIBA-GEIGY Corp.	AC, EP, PES, PO, PU, PVC, SC, ST
2-[2'-hydroxy-3',5'-di-(α,α-dimethyl-benzyl)phenyl]benzotriazole	[70321-86-7]	Tinuvin 900	CIBA-GEIGY Corp.	SC
2-(3'-t-butyl-2'-hydroxy-5'-methyl-phenyl)-5-chlorobenzotriazole[b]	[3896-11-5]	Tinuvin 326	CIBA-GEIGY Corp.	PES, PO, PVC, SC
2-(2'-hydroxy-3',5'-di-t-butyl)-5-chlorobenzotriazole	[3864-99-1]	Tinuvin 327	CIBA-GEIGY Corp.	PES, PO, SC
2-cyano-3,3-diphenylacrylates				
ethyl 2-cyano-3,3-diphenylacrylate	[5232-99-5]	Uvinul N-35	BASF Wyandotte Corp.	AC, CE, EP, PU, PVC, SC
2-ethylhexyl 2-cyano-3,3-diphenylacrylate	[6197-30-4]	Uvinul N-539	BASF Wyandotte Corp.	AC, CE, EP, PES, PVC, SC
oxalanilides				
2-ethyl-2'-ethoxyoxalanilide	[23949-66-8]	Sanduvor VSU	Sandoz Color and Chemicals	AC, CE, PES, PU, PVC, SC, ST
2-ethyl-2'-ethoxy-5'-t-butyloxalanilide	[35001-52-6]	Sanduvor EPU	Sandoz Color and Chemicals	PO, PU, SC
aryl esters				
phenyl salicylate	[118-55-8]	Salol	Dow Chemical Co.	CE, PES, PO, PVC, SC
p(t-butyl)phenyl salicylate	[87-18-3]	Anti UVK	Aceto Chemical Co.	CE, PES, PO, ST
resorcinol monobenzoate	[136-36-7]	Eastman RMB	Eastman Chemical Products	CE, PVC, ST
3,5-di-t-butyl-4-hydroxybenzoates				
2,4-di-t-butylphenyl 3,5-di-t-butyl-4-hydroxybenzoate	[4221-80-1]	UV Check AM 340	Ferro Corp., Chemical Div.	PO
n-hexadecyl 3,5-di-t-butyl-4-hydroxybenzoate	[67845-93-6]	Cyasorb UV 2908	American Cyanamid Co.	PES, PO, ST
nickel chelates				
2,2'-thiobis(4-t-octylphenolato)-n-butylamine nickel	[14516-71-3]	Cyasorb UV 1084	American Cyanamid Co.	PO
nickel bis[O-ethyl(3,5-di-t-butyl-4-hydroxybenzyl)]phosphonate	[30947-30-9]	Irgastab 2002	CIBA-GEIGY Corp.	PO
nickel dibutyldithiocarbamate	[13927-77-0]	Rylex NBC UV Check AM 104 Vanox NBC	E. I. du Pont de Nemours & Co., Inc. Ferro Corp., Chemical Div. R. T. Vanderbilt Co.	PO
hindered-amine light stabilizers structure (5)	[52829-07-9]	Tinuvin 770	CIBA-GEIGY Corp.	PO, PU, ST

623

Table 2 (*continued*)

Compound	CAS Registry No.	Trade name	Supplier	Suggested substrates[a]
structure (6)	[70624-18-9]	Chimassorb 944	CIBA-GEIGY Corp.	PO
structure (7)	[76633-20-0]	Tinuvin 622[b]	CIBA-GEIGY Corp.	PO
structure (8)	[41556-26-7]	Tinuvin 292	CIBA-GEIGY Corp.	SC
structure (9)	[76539-04-3]	Good-rite UV 3032	BFGoodrich	PO
structure (10)	[63843-89-0]	Tinuvin 144	CIBA-GEIGY Corp.	PO, PU, SC, ST

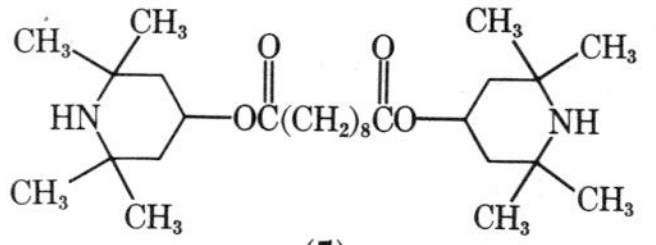

(5)

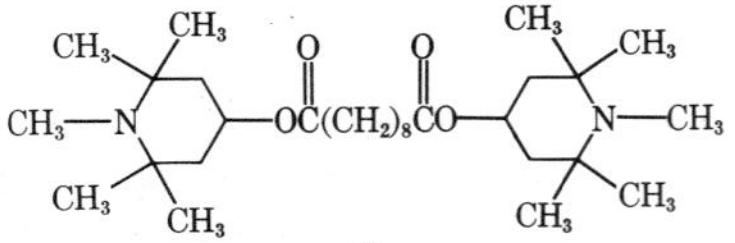

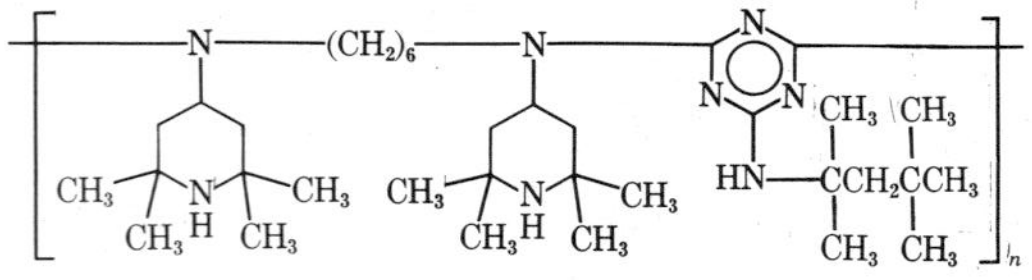

(6)

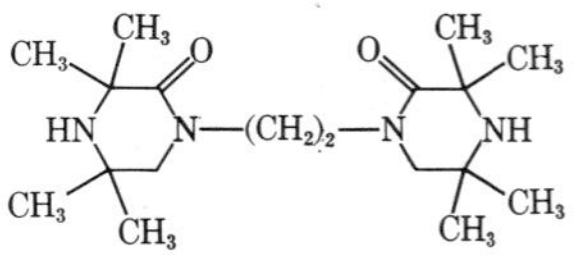

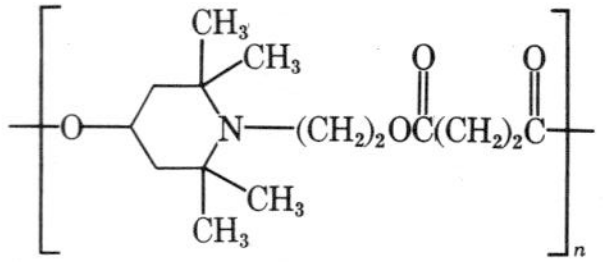

(7)

(8)

(9)

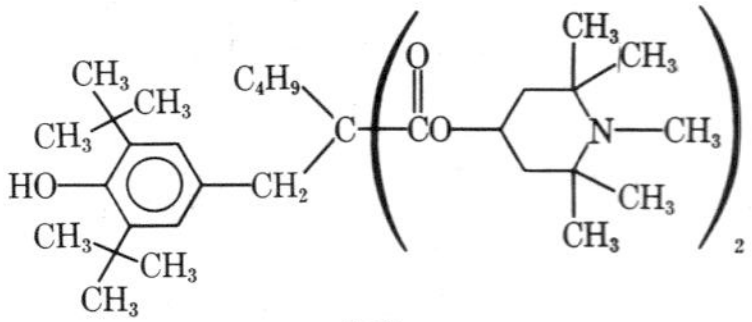

(10)

[a] AC = acrylic polymers; CE = cellulosic polymers; EP = epoxy polymers; PES = polyester; PO = polyolefins; PU = polyurethane; PVC = poly(vinyl chloride); SC = surface coatings; ST = styrenics.

[b] Approved by FDA as food additives as of April 1981. Title 21 of the Code of Federal Regulations gives details.

Esters of 4-Hydroxybenzoic Acid. Esters derived from 3,5-di-t-butyl-4-hydroxybenzoic acid are effective light stabilizers for polyolefins.

These compounds absorb weakly in the near uv and undoubtedly function as antioxidants by trapping free radicals generated during photooxidation. They are considerably more resistant to degradation by light than conventional phenolic antioxidants used to retard thermal oxidation.

Antioxidants. Although thermal antioxidants based on hindered phenols and esters of trivalent phosphorus are not classified as light stabilizers, some markedly improve the light stability of photosensitive polymers. They can increase the light stability of polypropylene by an order of magnitude (21–22); like phosphites (23), they are generally used in combination with uv absorbers.

Commercial uv stabilizers are given in Table 2, including suppliers and suggested substrates.

Testing for Light Stability

Outdoor weathering is the most reliable method for determining the light stability of polymers and other substrates. Exposure sites are usually selected at southern locations, where the strong sunlight and high temperatures increase the rate of photodegradation. In the continental United States, commercial exposure sites are located in southern Florida and Arizona. The same degree of uv exposure causes greater degradation in Florida than in Arizona. This phenomenon is attributed to higher humidity.

Even though outdoor aging provides the most acceptable data on weathering, it presents serious problems. The various weathering factors, such as the amount of uv radiation, temperature, humidity, and rainfall, vary from site to site and season to season. To make meaningful comparisons, specimens are exposed at the same site at the same time. A slow rate of degradation is another disadvantage of outdoor exposure, particularly when evaluating relatively light-stable materials. Years of exposure may be required before significant degradation takes place.

Artificial exposure devices are used to decrease the time required for the determination of light stability and to standardize exposure conditions, such as light intensity, spectral distribution, temperature, humidity, water spray, etc.

Exposure time can be shortened by using light sources with wavelengths below 300 nm or temperatures higher than in actual use. However, the results obtained may not correlate well with outdoor exposures (24).

The filtered xenon arc emits light with a spectral distribution that approximates sunlight. Exposure results obtained with this light source correlate quite well with sunlight. On the other hand, fluorescent black light emits radiation below 300 nm. Degradation is relatively rapid but correlation with outdoor exposure is generally poorer.

Accelerated-weathering devices are used primarily as screening tools. Promising stabilizers are nearly always exposed to outdoor conditions for confirmation of laboratory data.

Health and Safety Factors

Handling safety is assessed by subjecting the light stabilizer to a series of animal toxicity tests, eg, oral, inhalation, eye, and skin tests.

A number of light stabilizers have been regulated by the FDA for use in polymers that come in contact with food (see Food additives). Acceptance is determined by the product's subchronic or chronic toxicity in more than one species and in the concentration expected in the daily diet based on extraction studies.

Only materials that present an insignificant risk to the consumer, based on these tests, are allowed by the FDA.

Several commercial light stabilizers cleared by the FDA for use as indirect food additives are identified in Table 2. The specific FDA regulation must be consulted for details.

Uses

Selection of the type and concentration of uv stabilizers depends upon many factors, such as the composition of the substrate, thickness of the sample, color requirements, processing conditions, and expected service life. Ultraviolet absorbers, HALS, and phosphites are generally used at concentrations of −1.0%. In some applications, where clear film is desired to protect a light-sensitive substrate, uv absorbers may be used at concentrations as high as 3%. The concentration of thermal antioxidants generally is 0.2–0.5 wt % of the polymer.

BIBLIOGRAPHY

"Ultraviolet Absorbers" in *ECT* 1st ed., Suppl. 2, pp. 882–902, by G. M. Gantz and S. M. Roberts, General Aniline & Film Corporation; "UV Absorbers" in *ECT* 2nd ed., Vol. 21, pp. 115–122, by D. A. Gordon, Geigy Industrial Chemicals Division of Geigy Chemical Corporation.

1. L. R. Koller, *Ultraviolet Radiation*, John Wiley & Sons, Inc., New York, 1965.
2. G. R. Lappin in N. M. Bikales, ed., *Encyclopedia of Polymer Science and Technology*, Vol. 14, Interscience Publishers, a division of John Wiley & Sons, Inc., New York, 1971, p. 127.
3. T. Werner, G. Woessner, and H. E. A. Kramer, *Photodegradation and Photostabilization of Coatings*, ACS Symposium Series 15, American Chemical Society, Washington, D.C., 1981.
4. G. C. Newland and J. W. Tamblyn, *J. Appl. Polym. Sci.* **8,** 1949 (1964).
5. H. J. Heller and H. R. Blattman, *Pure Appl. Chem.* **36,** 148 (1976).
6. U.S. Pat. 2,861,053 (Nov. 18, 1958), G. R. Lappin and J. W. Tamblyn (to Eastman Kodak Co.).
7. U.S. Pat. 4,278,589 (July 14, 1981), M. Dexter and R. E. A. Winter (to CIBA-GEIGY Corp.).
8. H. J. Heller and H. R. Blattman, *Pure Appl. Chem.* **30,** 152 (1972).
9. D. J. Carlsson, T. Suprunchuk, and D. M. Wiles, *J. Appl. Polym. Sci.* **16,** 615 (1972).
10. P. Vink in G. Scott, ed., *Developments in Polymer Stabilization*, Applied Science Publishers, Ltd., London, 1980, pp. 119–125.
11. A. P. Pivovarov and A. F. Lukovnikov, *Khim. Vys. Energ.* **2,** 220 (1968).
12. D. K. C. Hodgeman, *J. Polym. Sci. Polym. Lett. Ed.* **16,** 161 (1978).
13. D. W. Grattan, D. J. Carlsson, and D. M. Wiles, *Polymer Degradation and Stability* **1,** 69 (1979).
14. N. S. Allen in N. S. Allen, ed., *Developments in Polymer Photochemistry*, Vol. 2, Applied Science Publishers, Ltd., London, 1981, pp. 239–273.
15. D. W. Grattan, A. H. Reddoch, D. J. Carlsson, and D. M. Wiles, *J. Polym. Sci. Polym. Lett. Ed.* **16,** 143 (1978).
16. J. B. Shilov, J. Petraj, J. Pac, and M. Navratil, *Polymer* **21,** 5 (1980).
17. D. J. Carlsson, K. H. Cohn, J. Durmis, and D. M. Wiles, *J. Polym. Sci. Polym. Chem. Ed.* **20,** 575 (1982).

18. R. P. R. Ranawera and G. Scott, *Eur. Polym. J.* **12,** 825 (1976).
19. D. J. Carlsson and D. M. Wiles, *Makromolecules* **7,** 259 (1974).
20. R. P. R. Ranawera and G. Scott, *Eur. Polym. J.* **12,** 591 (1976).
21. F. Gugumus in G. Scott, ed., *Developments in Polymer Stabilization*, Vol. 2, Applied Science Publishers, Ltd., London, 1979, pp. 267–271.
22. H. J. Heller, *Eur. Polym. J. Suppl.*, 110 (1969).
23. S. L. Fitton, R. N. Haward, and G. R. Williamson, *Br. Polym. J.* **2,** 223 (1970).
24. C. E. Hoey and H. A. Hipwood, *J. Oil Colour Chem. Assoc.* **57,** 151 (1974).

General References

B. Rånby and J. F. Rabek, *Photodegradation, Photooxidation and Photostabilization of Polymers*, John Wiley & Sons, Inc., New York, 1975.
H. Linder in R. Gächter and H. H. Müller, eds., *Taschenbuch der Kunstuff-Additive*, Carl Hanser Verlag, Munchen, Wien, 1979, pp. 122–136.
J. F. McKellar and N. S. Allen, *Photochemistry of Man Made Polymers*, Applied Science Publishers, Ltd., London, 1979.

MARTIN DEXTER
CIBA-GEIGY Corp.

V

VACCINE TECHNOLOGY

A vaccine is a preparation that is used to prevent a specific disease by inducing immunity to that disease in the recipient of the vaccine. The ability to confer resistance to disease by vaccination was described as early as 1798, when Edward Jenner noted that individuals inoculated with cowpox or "vaccinia" were immune to subsequent challenge with smallpox (1). How resistance to disease occurs in nature and how it is induced by vaccination are described in refs. 2–3. An individual who contracts a disease can become resistant to a second episode of this disease by building up immunity as a result of the first illness. This immunity is highly complex and involves the development of humoral antibodies and cellular immunity to the causative agent of the disease. A vaccine is designed to mimic this process by causing the recipient to build up immunity to a specific disease without experiencing the disease itself. This is accomplished by giving the vaccinee a preparation related to the agent of disease that has been rendered nontoxic but still acts as an antigen. Thus, the preparation can still induce the immunity that protects against the disease in question. This process of stimulating the buildup of immunity by vaccination is referred to as active immunization. Passive immunization, in contrast, occurs when an individual is given a preparation, usually serum, that already contains antibodies to the disease in question. This treatment may be used after exposure to a disease, when there is not enough time to build up immunity through active immunization or when the recipient cannot respond to vaccination.

Vaccines can be grouped into two general types: live or killed. With live vaccines, the recipient is given organisms that are alive and able to replicate in the vaccinee. These vaccines have been rendered harmless by attenuation, a treatment that essentially eliminates the virulence or disease-causing ability of the organisms. An example is live poliovirus vaccine. Alternatively, a live vaccine can be a natural agent, closely related to the disease-causing organism, and can induce appropriate immunity without causing disease; an example is the smallpox vaccine. Killed or inactivated vaccines are those preparations that cannot replicate in the recipient. These preparations are generally derived from the virulent agent of disease and are rendered inactive through chemical treatment or heat. An inactivated vaccine can consist of whole organisms or cells, organisms that have been broken apart or split by chemical treatment, or purified subunits from the whole organisms. In addition, extracellular products, eg, toxins, can be used as vaccines after chemical inactivation to yield toxoids.

Licensed Vaccines

Vaccines licensed in the United States as of 1982 are of two types: the first is used in the general population and the second in special populations. All are regulated by the Office of Biologics of the FDA. Recommendations on vaccine usage are made by both the Immunization Practices Advisory Committee (ACIP) of the United States Public Health Service and the Committee on Infectious Diseases of the American Academy of Pediatrics.

Vaccines for the General Population. In the first category are those vaccines recommended for routine immunization of infants and children or adults when appropriate. These vaccines provide protection against poliomyelitis, diphtheria, tetanus, pertussis (whooping cough), measles (rubeola), mumps, and rubella (German measles). The schedule and vaccines used for immunization are shown in Table 1 (4).

Poliomyelitis. Two vaccines for the prevention of poliomyelitis are currently licensed for use in the United States. The live, attenuated oral poliovirus vaccine (OPV) is recommended by the ACIP for primary immunization of children. The killed or inactivated poliomyelitis vaccine (IPV) is recommended for immunization of unvac-

Table 1. Schedule for Immunization of Normal Infants and Children[a]

Age	Immunization
2 mo	DTP[b], TOPV[c]
4 mo	DTP, TOPV
6 mo	DTP[d]
15 mo	measles, mumps, rubella[e]
18 mo	DTP, TOPV
4–6 yr	DTP, TOPV
14–16 yr	Td[f], repeat every 10 yr

[a] Ref. 4.

[b] DTP, diphtheria and tetanus toxoids combined with pertussis vaccine.

[c] TOPV, trivalent oral poliovirus vaccine.

[d] TOPV is optional but may be given in areas where poliomyelitis is endemic.

[e] May be given as measles–rubella or measles–mumps–rubella combined vaccines.

[f] Td, combined tetanus and diphtheria toxoids (adult type), which contains a smaller amount of diphtheria antigen.

cinated adults at increased risk of exposure to poliomyelitis and of immunodeficient patients and their household contacts (5). Both OPV and IPV are effective in preventing poliomyelitis and each has its advantages and disadvantages (6).

Indications for use. Poliovirus vaccine is indicated for use in the prevention of poliomyelitis caused by poliovirus serotypes 1, 2, and 3. Except as noted above, infants starting at the age of 6–12 wk, all children, and adolescents through the age of 18 yr may receive either OPV or IPV (4,7).

Composition and methods of manufacture. The live vaccine is a mixture of three types of attenuated polioviruses that have been propagated in cultures of monkey kidney cells or human diploid cell lines. The cells are grown in the presence of a nutrient medium consisting of inorganic salts, amino acids, vitamins, dextrose, phenol red (pH indicator), and sodium bicarbonate and are supplemented with antibiotics and calf serum (8). The final vaccine is prepared by dilution of the three virus types to the desired concentration with a modified cell-culture medium and stabilizer, eg, sorbitol (7). Inactivated poliomyelitis vaccine is a mixture of three types of polioviruses grown in monkey kidney tissue cultures and chemically inactivated with formaldehyde. The final vaccine may contain a preservative.

Standardization and testing. Requirements that the vaccine must meet are described in refs. 9–10. Testing procedures have been established to ensure the safety of the vaccine for human use. For OPV and IPV, both *in vivo* (small animals, monkeys) and *in vitro* (cell culture) tests are performed at several steps during the manufacture of the vaccine to control for safety, potency, and the presence of adventitious agents, eg, other viruses or microorganisms.

Diphtheria, Tetanus, and Pertussis (DTP). *Indications for use.* Diphtheria and tetanus toxoids combined with pertussis vaccine are routinely used for active immunization of infants and young children against diphtheria, tetanus, and pertussis simultaneously. Immunizations should begin at two-to-three months of age and be completed no later than the age of six years. Booster doses for tetanus and diphtheria should be with Td, which contains a smaller amount of diphtheria toxoid compared to the pediatric vaccine. Children over the age of six years are not normally injected with pertussis vaccine (4).

Composition and methods of manufacture. The diseases of diphtheria and tetanus are caused by toxins synthesized by the organisms *Corynebacterium diphtheriae* and *Clostridium tetani*, respectively. Diphtheria and tetanus vaccines contain purified antigens or toxoids, which are prepared by inactivation of these toxins with formaldehyde.

Tetanus toxin can be obtained by growing *Cl. tetani* in a complex medium specifically formulated for production of high yields of toxin (11). The medium contains beef heart infusion, an enzymatic digest of casein, dextrose, sodium chloride, and other essential nutrients. The medium for the production of *C. diphtheriae* toxin is also a complex liquid medium. After growth of both organisms, the culture fluid is harvested, and intact cells and cell debris are removed by filtration. Detoxification of the toxins is accomplished with formaldehyde under controlled conditions of time, temperature, pH, and formaldehyde concentration. Because many impurities are contained in the crude toxoid preparation, the antigens are further isolated and purified. The alcohol fractionation method described in ref. 12 is widely used for both. This involves two sequential precipitations of the toxoid with methanol at acid pH. The methanol is then removed by freeze-drying and the resulting powder is dissolved in buffer containing glycine and is sterilized by filtration.

The pertussis vaccine component of DTP is killed whole cells of *Bordetella pertussis* bacteria. The nutrient medium for growth of the organisms may contain an acid hydrolysate of casein, minerals, and other growth factors. Concentrated bacteria are killed and detoxified either by heating, by the addition of a chemical agent, eg, sodium ethylmercurithiosalicylate (thimerosal) and appropriate aging, or by a combination of these methods (13).

The final vaccine contains the pertussis cells and diphtheria and tetanus toxoids, plus buffer. It may also contain an adjuvant, ie, a substance that increases the response to an antigen when combined with the antigen, such as aluminum hydroxide or aluminum potassium sulfate.

Standardization and testing.   Requirements for DTP are given in refs. 9 and 13. Standardization of potency for diphtheria and tetanus toxoids relies on antigenic and flocculation tests. In principle, the antigenic tests are conducted to measure the ability of the vaccine to induce specific antibodies in guinea pigs. The flocculation test provides a quantitative estimate of the amount of toxoid in the vaccine. The toxoids react qualitatively and quantitatively with specific antisera to produce a visible precipitate in a test tube.

The U.S. Standard Pertussis Vaccine is used to determine the potency of pertussis vaccine. The number of protective units in the vaccine is estimated for each lot from the results of simultaneous intracerebral mouse-protection tests of the vaccine being studied and the U.S. Standard (13).

Measles, Mumps, and Rubella.   *Indications for use.*   Live, attenuated vaccines are used for simultaneous or separate immunization against measles, mumps, and rubella in children from 15 mo of age to puberty. Because of the presence of maternal antibodies in the circulation of infants younger than 6 mo, and because of low measles seroconversion rates in infants under 15 mo, combined vaccine should not be given at earlier ages (4).

Composition and methods of manufacture.   The combined vaccine for simultaneous immunization is a mixture of the three live, attenuated viruses: measles, Moraten strain; mumps, Jeryl Lynn strain; and rubella, RA 27/3 strain (14–16). The measles and mumps virus strains are grown in cultures of primary chick-embryo cells, whereas rubella is propagated in the WI-38 strain of human diploid cells. In either case, the cell cultures are usually grown in a nutrient medium similar in composition to that described for cultures of monkey kidney cells. After virus propagation, the cell-culture fluids are clarified and the viruses are mixed to the desired concentration. A stabilizing solution, eg, a mixture of partially hydrolyzed gelatin, sorbitol, cell-culture medium, and buffer, may be added (17).

Standardization and testing.   General requirements for measles, mumps, and rubella vaccines are given in refs. 9–10. The concentration of the virus in the vaccine constitutes the measure of potency. All titrations are performed in tissue culture and are run in parallel with a U.S. Reference virus as a titration control.

As with other live, attenuated, viral vaccines, eg, OPV, *in vivo* and *in vitro* tests are performed to ensure freedom from adventitious agents and safety for humans.

Vaccines for Special Populations.   Vaccines used for special populations include influenza, pneumococcal polysaccharide, hepatitis, and others (see Table 2).

Influenza.   *Indications for use.*   The ACIP recommends annual influenza vaccination for all persons who are at increased risk from infections of the lower respiratory tract and for all older persons, particularly those over the age of 65 (27). Immunization of normal infants, adolescents, and young adults is not indicated.

Table 2.   Selectively Used Vaccines

Type	Composition	Use	Refs.
viral			
rabies	inactivated rabies virus grown in cultures of human diploid cells	post-exposure immunization for treatment of animal bites; preexposure immunization for those at high risk	18
yellow fever	live, attenuated yellow fever virus grown in embryonated chicken eggs	required for international travel to certain areas	19–20
adenovirus	live adenovirus, types 4 and 7	prevention of acute respiratory disease caused by types 4 and 7	21
smallpox	live vaccinia virus grown in embryonated chicken eggs	indicated only for laboratory workers directly involved with smallpox or closely related pox viruses	22
bacterial			
meningococcal polysaccharide; group A or group C or groups A and C	purified capsular polysaccharides of *Neisseria meningitidis*, serogroups A and C	to control outbreaks of disease caused by serogroup A or C; serogroup C vaccine given routinely to military recruits	23
cholera	inactivated *Vibrio cholera*, strains Inaba and Ogawa	required for international travel to certain areas	20, 24
typhoid	inactivated *Salmonella typhi* organisms	for travel to areas where there is a recognized risk of exposure to typhoid because of poor food and water sanitation	25
plague	inactivated *Yersinia pestis* organisms	for individuals at high risk of exposure, eg, laboratory personnel working with *Y. pestis*	26

Composition and methods of manufacture.   Two types of influenza viruses (A and B) are responsible for causing periodic outbreaks of febrile respiratory disease. The manufacture of an effective vaccine is complicated by antigenic variation or drift, which can occur from year to year within the two virus types, so that immunization to one strain may not induce immunity to a distantly related strain of the same type. Each year, antigenic characterization of current strains is important in selecting the virus strains to be included in the vaccine.

Two types of inactivated-influenza vaccine are prepared from viruses grown in embryonated chicken eggs: whole virion (whole virus) and subvirion (split virus). Whole virion vaccines consist of influenza viruses purified from allantoic fluid harvests and inactivated by formaldehyde or β-propiolactone. Subvirion vaccines are prepared by further chemical treatment with detergents or organic solvents to disrupt or split the virus particle.

Chemicals useful for splitting viruses include cetyltrimethylammonium bromide, sodium dodecyl sulfate and Triton X-100, Triton N101, butyl acetate and ethyl acetate, and Tween 80 (poly(oxyethylene sorbitan monooleate)) and tri-*n*-butyl phosphate (28–32).

Standardization and testing. General requirements for influenza vaccine are given in refs. 9–10. The two glycoproteins, hemagglutinin (H) and neuraminidase (N), are the main immunogenic components of influenza virus (33). Because the hemagglutinin appears to be the more important of the two for inducing resistance to infection, standardization of influenza-vaccine potency relies on the quantitation of hemagglutinin. The single radial immunodiffusion (SRID) technique is used to standardize the antigenic mass of hemagglutinin in influenza vaccines (34–35). In SRID assays, detergent-disrupted (eg, with sodium sarcosyl sulfate, Emulphogene BC-720, Triton X-100, or Nonidet P-40) reference and test antigens or vaccines are placed in circular wells of agarose slab gels containing the appropriate antiserum. After suitable incubation, the gels are dried and then stained, eg, with Coomassie R-250. Circular zones of antigen–antibody complexes are measured and vaccine potency is computed based on reference and test-vaccine dose response curves.

Pneumococcal Polysaccharide. *Indications for use.* Pneumococcal polysaccharide vaccine may be used for immunization of persons two years of age or older who are at an increased risk of pneumococcal disease (36).

Composition and methods of manufacture. The vaccine consists of a mixture of purified capsular polysaccharides from 14 pneumococcal types that are responsible for at least 80% of serious pneumococcal disease in the world (37). Each of the pneumococcal polysaccharide types is produced separately and treated to remove impurities. The latter is commonly achieved by alcohol fractionation, eg, with methanol, ethanol, or isopropyl alcohol; centrifugation; treatment with cationic detergents (eg, cetyltrimethylammonium bromide), proteolytic enzymes (eg, trypsin), nucleases, or activated charcoal; diafiltration; and lyophilization (38–39). The 14 purified polysaccharides are combined in amounts to give 50 micrograms of each type per dose. The vaccine also contains a preservative, eg, phenol or thimerosal.

Standardization and testing. Because immunogenicity of polysaccharides is directly proportional to the molecular weight of the polysaccharide, the Office of Biologics has established standards for the apparent size of all types of pneumococcal polysaccharides used in vaccines (40–41). Additional tests rely on chemical evaluation of the polysaccharides to be sure that they are free of nucleic acid and proteins and that they contain the correct amounts of sugar components. In addition, *in vitro* immunological tests are run to confirm the identity of the polysaccharides.

Hepatitis B. *Indications for use.* Hepatitis B vaccine may be used for persons at high risk of contracting hepatitis B, eg, certain groups of health-care workers or hemodialysis patients. Three doses of vaccine are recommended for complete immunization (42).

Composition and methods of manufacture. The vaccine is prepared by isolation of the viral antigen from sera of infected patients. This process involves isopycnic banding, ammonium sulfate fractionation, and density-gradient centrifugation. Digestion with pepsin and urea treatment removes extraneous liver and plasma components. Gel filtration and then inactivation with formaldehyde follows. Aluminum hydroxide is used as an adjuvant (43).

Standardization and testing. Vaccines are standardized to contain a specific amount of hepatitis B surface antigen. Safety testing for the presence of live virus is done in various animals, including marmosets and chimpanzees (42).

Vaccines Being Developed

Despite the tremendous advances since the 1960s in the biomedical fields, including the total eradication of smallpox and reduction of mortalities resulting from various diseases, there remains a large number of diseases that are endemic in many parts of the world. The third world or developing countries bear the brunt of several of these, eg, malaria, trypanosomiasis, and schistosomiasis. In developed countries, such diseases as gonorrhea are becoming increasingly prevalent. Vaccines for many of the etiological agents that still cause disease have not been prepared for several reasons. These include a lack of understanding of how immunity can be artificially induced and an inability to grow sufficient quantities of these agents to produce vaccines. Vaccine developments are described in refs. 44–46.

Gonorrhea. Gonorrhea, which is caused by *Neisseria gonorrheae*, is the most commonly reported communicable disease in the United States. Approximately 10^6 cases were reported to the Center for Disease Control (CDC) in 1979, but actual cases could be two to three times higher (47). In addition, an increasing number of strains are becoming resistant to penicillin, the antibiotic that is usually used against this disease.

Development of a vaccine is problematic since natural infection does not necessarily provide immunity. Whether this results from a poor immunological response or to strain differences is not certain. At any rate, studies are being conducted on various structural components of the gonococcal bacterium, including pili, outer membrane proteins, lipopolysaccharide, and the outer capsule, in an effort to develop a vaccine. One of the more promising approaches involves a vaccine made with pili. These structures are responsible for attachment of the gonococci to mucosal surfaces, the first step necessary for infection to occur. Antisera against pili may prevent disease by preventing this attachment. One method for obtaining pili involves growth of the gonococci in liquid culture followed by mechanical shearing of the pili from the surface of the bacterium (48). Pili are further purified by differential centrifugation and ammonium sulfate precipitations. This type of preparation was shown to yield a protein pili vaccine that is immunogenic in human volunteers (49). Additional human studies indicate that a pili vaccine stimulates antibody formation that is 50–100 times the prevaccination level and is effective in preventing disease after challenge (50).

Haemophilus Influenzae. The *Haemophilus influenzae* b serotype is the most common cause of bacterial meningitis in the United States. It has a fatality rate of nearly 10% and is responsible for ca (1500–2000) deaths per year. In ca 30% of the cases that recover, some neurological damage is sustained. The peak incidence is at ca six months of age (3).

Because of the importance of the capsule of this organism in pathogenesis, vaccine studies have mainly been oriented toward raising an antibody to this bacterial component. This polysaccharide capsule contains ribose, ribitol, and phosphate and is designated PRP (polyribitol phosphate). Purification is accomplished by a series of chemical treatments involving hexadecyltrimethylammonium bromide (cetavlon), ethanol, and hydroxylapatite (51). Vaccines made from PRP have been used and their safety is well established (52). However, PRP is not immunogenic in children less than two years old; hence, it is not suitable for the target population of infants. Recent studies indicate that this problem may be overcome by combining PRP with pertussis or DTP as an adjuvant (53).

Hepatitis.   There are three distinct types of virus that cause hepatitis: type A (infectious hepatitis), type B (serum hepatitis), and type C (non-A, non-B hepatitis). The type B is of special interest since chronic carriage of this agent is possible, and it has been linked to liver carcinoma. In the United States, ca 16×10^6 persons are at high risk of contracting hepatitis B. Approximately $(8-10) \times 10^4$ cases are reported each year; ca 8×10^5 are chronic carriers. Worldwide, there are ca 200×10^6 chronic carriers (54).

Prevention of hepatitis A through vaccination is hampered by the difficulty in growing large quantities of this virus. However, methods of isolation from infected animals or human sera do exist. One process for isolation of hepatitis A virus involves purification after growth in marmoset liver (55). The livers are perfused with buffered saline and ground to release the virus. Purification is then accomplished through density-gradient centrifugation with cesium chloride, sodium bromide, sodium tartrate, or sucrose. Inactivation is accomplished by incubation with formaldehyde. Another process involves isolation of hepatitis A from the stools of infected patients (56). Sonication and ether extraction in addition to density-gradient centrifugation are used in the purification process. This preparation is also treated with formaldehyde.

A comprehensive review of the biology of the hepatitis B virus is given in ref. 57. Investigations leading to the recognition of the Australia antigen (hepatitis B surface antigen) and isolation of the virus particle (Dane particle) are given in refs. 58–61. The vaccine in use was licensed in 1981 and is derived from the plasma of chronic carriers. Although it has been shown to be safe and efficacious (62–64), it is difficult and expensive to prepare. Consequently, a second-generation vaccine is being developed. The gene for the surface antigen has been cloned in *E. coli* by means of recombinant-DNA techniques and has been produced in yeast (46,65–69) (see Genetic engineering).

Pregnancy.   A review of immunological methods examined in relation to pregnancy prevention is given in ref. 70. A candidate vaccine has been developed that contains the *beta* subunit of human chorionic gonadotrophin hormone (HCG) linked to tetanus toxoid. This preparation is immunogenic in humans, and produces antibodies to HCG six-to-eight weeks after vaccination. After six months, however, detectable antibodies to HCG are no longer present (71). Because the antibody raised by the *beta* subunit of HCG also reacts with human luteinizing hormone (LH), other subunits or chemical modifications are being studied with some success in eliminating this cross-reactivity (see Hormones, anterior pituitarylike).

Malaria.   Malaria infection occurs in over 30% of the world's population and almost exclusively in developing countries. Approximately 150×10^6 cases occur each year, with 10^6 deaths occurring in African children (72). The majority of the disease in man is caused by four different species of the malarial parasite. Vaccine development is problematic for several reasons. First, the parasites have a complex life cycle. They are spread by insect vectors and go through many different stages and forms (intracellular and extracellular; sexual and asexual) as they grow in the blood and tissues (primarily liver) of their human hosts. In addition, malaria is difficult to grow in large quantities outside the natural host (44). Despite these difficulties, vaccine development has been pursued for many years. An overview of the state of the art is given in refs. 44 and 73. The first vaccines were directed against sporozoites, the form of the parasite that is first injected into the host by a mosquito. The sporozoites are isolated from mosquitos' salivary glands and are inactivated by irradiation, formaldehyde, or freeze-drying. These vaccines have been used in mice and monkeys and have been shown to induce immunity, but only to the sporozoite stage.

Preparations involving merozoites (a form that infects the red blood cells) plus an adjuvant confer protection in monkeys against several malarial species. The main difficulty here is the inability to obtain large quantities of merozoites. Finally, vaccines containing the reproductive stages of the parasite are somewhat effective in animal systems, but large quantities cannot be readily obtained. The latest approaches to the problem of a malarial vaccine involves the use of monoclonal antibodies. Studies indicate that this tool facilitates identification of the protective antigens, and several potential vaccine antigens have been identified this way (44). Once appropriate vaccine antigens are identified, it may be possible to produce large quantities by recombinant-DNA technology or chemical synthesis.

Herpes Simplex. There are two types of Herpes simplex (HSV) that infect man. Type I causes orofacial lesions and ca 30% of the U.S. population suffers from recurrent episodes. Type II is responsible for genital disease and anywhere from (3×10^4)–(3×10^7) cases per year (including recurrent infections) occur. The primary source of neonatal herpes infections, which are severe and often fatal, is the mother infected with type II. In addition, there is evidence to suggest that cervical carcinoma may be associated with HSV-II infection (74–75).

Vaccine development is hampered by the fact that recurrent disease is common. Thus, natural infection does not provide immunity and the best method to induce immunity artificially is not clear. The genome of these viruses is also able to cause transformation of normal cells, thus conferring on them one of the properties attributed to cancerous cells. Vaccines made from herpes viruses must, therefore, be carefully purified and screened to eliminate the possibility of including any active genetic material.

Vaccine preparations that are efficacious against skin infections in mice have been made (76). Prevention of lethal infection in mice has also been accomplished with a detergent-soluble extract of virus-infected cells (77). Some human studies with inactivated herpes viruses have been reported but the efficacy of the preparations is not clear (78).

Future Technology

Vaccines for many diseases are unavailable because of an inability to determine the appropriate method for vaccination or difficulty in obtaining large quantities. Genetic engineering and monoclonal antibody production offer tremendous potential in solving these problems.

Genetic Engineering. Genetic engineering (recombinant-DNA technology) is reviewed in depth in refs. 79–84. It involves preparation of DNA fragments (the passengers) coding for the substance of interest, inserting the DNA fragments into vectors (the cloning vehicles), and introducing the recombinant vectors into living host cells where the passenger DNA fragments replicate and are expressed, ie, transcribed and translated, to yield the desired substance. The passenger DNA fragments can be obtained from natural DNA molecules by treatment with restriction endonucleases (enzymes that cut DNA at specific sites) or by mechanical shearing (85–86). They can also be synthesized either from messenger RNA (mRNA) through the actions of reverse transcriptase and DNA polymerase or by pure chemical methods (87–93). The vectors are autonomously replicating DNA molecules (replicons), eg, plasmids, bacteriophages, and animal viruses. Small plasmids and bacteriophages are the most suitable vectors, because their maintenance does not require integration into the host genome and their DNA can be isolated readily in an intact form (84). Many plasmid and bacteriophage vectors of improved qualities have been constructed by addition to and deletion of some of their genetic elements. Insertion of passenger DNA fragments into cloning vehicles can be carried out by one of three methods: ligation of cohesive ends produced by restriction endonuclease, homopolymer tailing, and blunt-end ligation. *Escherichia coli* has been exclusively used as the host cells for cloning. Recently, however, other microorganisms, eg, *Bacillus subtilis* and *Saccharomyces cerevisiae*, have also been used successfully (94–95). Introduction of the recombinant vectors into host cells (transformation) can be accomplished by different methods, depending on the vector–host cell system used. A calcium–heat-shock treatment has been used exclusively in the plasmid–*E. coli* system (96). Successfully transformed host cells can be selected from the whole population using the drug resistance and nutritional markers carried by plasmid vectors, the plaque-forming abilities of phage vectors, immunochemical methods by means of antibodies directed at the substance of interest, or nucleic acid hybridization methods (97–98).

Genetic engineering (qv) offers new and, in some instances, safer and more effective methods for development and production of vaccines of consistently high quality for prophylaxis against viral, bacterial, mycotic, and parasitic infection (see also Genetic engineering, Supplement Volume). It facilitates the study of the pathogenicity and immunology of viruses, eg, Epstein-Barr virus, which grows very slowly and whose yield is very low (99). It has provided a large quantity of interferon, an antivirial and potential anticancer reagent, for clinical trial (100–101). It also has been applied to the development of vaccines against hepatitis B virus, which is difficult to grow *in vitro* (65,102–106), and against HSV (107). Genetic engineering may eventually replace the conventional methods of vaccine production because of the potential economics to be gained or the safer production methods resulting from not having to grow large quantities of pathogenic agents. An example of this is the recent development by recombinant techniques of a vaccine against foot-and-mouth disease (108–110). The possibility of using this technique in constructing a virulent poliovirus and HSV as vaccines has also been reported (111–113).

Monoclonal Antibodies. Monoclonal antibodies are antibodies produced by a culture derived from a single cell. These antibodies recognize a single determinant or structure of a given antigen. Methods for producing monoclonal antibodies and practical applications have been reviewed (114–117). Monoclonal-antibody technology was first introduced in 1975 (118). The general procedure is as follows: A mouse is immunized with a purified antigen or a mixture of antigens. Spleen cells, which produce antibodies but cannot be cultured *in vitro*, from the immunized mouse are fused in the presence of poly(ethylene glycol) with mouse myeloma cells deficient in the enzyme hypoxanthine-guanine phosphoribosyltransferase (HPGRT). These myeloma cells can be cultured *in vitro* but die in the presence of a medium containing hypoxanthine-aminopterin-thymidine (HAT). After the fusion, the cells are distributed into individual cell cultures containing the selective HAT medium and are incubated. Only the cells that have been successfully fused grow. Two-to-four weeks later, culture medium from the growing cells (the hybrids) is examined for the presence of the specific antibody of interest by using a very sensitive method, eg, solid-phase radioimmunoassay or enzyme-linked immunosorbent assay (ELISA). Cultures positive for the desired antibody, which are designated hybridomas, are cloned or purified by the limiting dilution method or the soft agar plate method. The clones are stored frozen, grown in large tissue cultures to produce the antibody, or injected into animals to produce myelomas that secrete the antibody.

Monoclonal antibody technology provides a practically unlimited supply of uniform, highly specific antibodies. Monoclonal antibodies could potentially be of use in the therapy of certain infectious diseases for which there are no suitable prophylatic or therapeutic measures or in instances when it is too late to use these measures, eg, rabies or tetanus. They may also be used in place of current polyclonal antibodies to provide better quantitation of the antigens in vaccine products. Their discriminating power makes it possible to analyze the antigenic variation in infectious agents, eg, influenza and rabies viruses, as well as to select protective antigens in, eg, the malaria parasite (73,118–122).

Monoclonal antibodies can be coupled to resins and used to purify antigens, eg, interferon, by immunoaffinity chromatography (see Chromatography, affinity). This technique can also be used to purify and concentrate an important antigen present only in low quantities from a biological mixture. Immunoaffinity chromatography and recombinant-DNA technology will greatly enhance future vaccine development.

Economic Aspects

The U.S. vaccine market in 1978 included distribution of ca 100×10^6 doses, with sales of ca $\$100 \times 10^6$ (see Table 3). The current figures are somewhat higher because of inflation and a growing increase in the immunization rates. The latest estimates by the Center for Disease Control indicate that ca 95% of the population is being vaccinated against polio in accordance with recommendations for general immuni-

Table 3.   Vaccine Doses Distributed in 1978[a]

Vaccine[b]	Number of doses, × 1000
DTP	17,992
DT (pediatric)	823
Td (adult)	9,191
T	10,971
polio (live)	23,211[c]
measles	8,931
mumps	4,649
rubella	7,553
smallpox	4,649
influenza	20,410
rabies	181[d]

[a] Ref. 123.
[b] DTP = diphtheria and tetanus toxoids combined with pertussis vaccine, DT = diphtheria and tetanus toxoids, Td = combined tetanus and diphtheria toxoids, and T = tetanus toxoid.
[c] 1977.
[d] 1976.

zations (125). This figure was 67%, at best, in 1974 (126). The majority of vaccines are made by five manufacturers: Connaught, Lederle, Merck, Parke-Davis, and Wyeth (124). Because most of the vaccines distributed are used in children, the market is not likely to change dramatically, unless the birth rate changes or new vaccines are introduced.

Costs of vaccine development include the following: research to determine what type of vaccine is to be produced; development of a manufacturing process and scaleup; *in vitro* testing for potency, safety, stability, and other relevant biochemical properties; animal studies; and human studies to assess safety and efficacy. Complete, precise costs for the development of vaccines currently on the market would be difficult to determine, but some estimates have been made. The Salk vaccine (inactivated) for polio was introduced in 1955 and was estimated to cost 41×10^6 in research and field trials (127). The cost of research and clinical studies for the development of one of the pneumococcal polysaccharide vaccines on the market was ca 13×10^6; approximately half was government funds (128).

Costs of vaccine manufacture vary according to the type of vaccine produced and how it is supplied. Live vaccines generally are less expensive, since the quantitative mass to be given to the recipient is less than that in an inactivated vaccine. Acellular components, subunits, or purified fractions used as vaccines are more costly, since more processing is involved in their manufacture. The number of strains of organisms or antigens also affect the price. Packaging the vaccine is a very expensive part of the process, and the fewer the number of doses supplied in each container, the higher is their cost. Based on the sales and number of vaccine doses, vaccines are relatively inexpensive. Immunizations also include the cost of administration of the vaccine, and this varies considerably, depending upon the clinical setting for vaccination. A review of immunization costs for some vaccines is given in Table 4 (129). The cost of immunization against hepatitis B has been estimated at $90–120 for the three-dose series (130). In general, the newer vaccines, ie, hepatitis and pneumococcal polysaccharide, are more expensive.

Another important economic aspect of vaccine technology is the cost–benefit relationship between preventive vaccination and disease treatment. Generally, the cost savings are high. For the period 1955–1961, the net savings in medical case costs as a result of polio immunization were ca 327×10^6. If loss of income were added, the cost savings of vaccination rises to ca 1×10^9/yr for that period. Measles vaccination was estimated to have saved 100×10^6/yr in medical costs, and lost work and lost school days during the period 1963–1967 (127). Other cost-effectiveness and cost-

Table 4. Vaccination Costs[a]

Vaccine	Base period	Cost per person[b], $
pertussis	1977	0.30
poliomyelitis	1957	0.81
measles	1963–1968	3.00
rubella	1975	3.00
pneumococcal pneumonia	1977–1979	8.61–11.37
influenza	1960	3.00

[a] Ref. 129.
[b] Includes cost of administration.

benefit analyses confirm the substantial savings to be gained from these two vaccines and others in the childhood-immunization series (129). Studies of vaccines for special populations, eg, influenza and pneumococcal polysaccharide, are cost-effective in improving health at low cost and in avoiding certain medical-care expenses but do not necessarily result in net medical-care savings (129–130).

Economic aspects of vaccine technology also include liability. In any immunization campaign, there are inevitably a small number of vaccine recipients that have an associated injury. This can be as mild as local redness or as severe as paralysis. Expenses related to such cases are given in Table 5 (131). In the United States, compensation is not generally available from public funds. Exceptions to this are the state of California, where a compensation program for injuries related to mandatory vaccinations does exist, and the Swine Flu Program in 1976 in which the government agreed to compensate victims of adverse reactions. Although not directly involved in vaccine administration, manufacturers can be held liable for failure to warn of potentially associated adverse reactions. In one of the lawsuits that set precedent for this policy (Reyes vs Wyeth, 1974), the court said that the manufacturers should assume the loss and make it a part of the cost of doing business. This general precedent plus the unpredictability and high settlements in some lawsuits have been stated as partly responsible for the declining number of vaccine manufacturers over the last few years (nine U.S. manufacturers in 1972, five in 1982). This issue is discussed in refs. 123 and 132, as are recommendations for a national compensation program for vaccine-related injuries.

Table 5.   Ranges of Total Costs for Some Vaccine-Associated Injuries for Persons Less Than One Year of Age (Assuming a Discount Rate of 2.5%)[a,b]

Vaccine	Cost, $
DTP	95–891,963
poliomyelitis	422–903,160
measles	1,313–927,533
rubella	34–869,229
mumps	575–462,315
influenza	11,405–922,461

[a] Ref. 131.
[b] Costs include present and future expenses associated with the most mild (low value) or most severe (high value) reactions recorded as being associated with a particular vaccine.

BIBLIOGRAPHY

1. *Chem. Week*, 42 (Nov. 24, 1982).
2. W. K. Joklik, H. P. Willett, and D. B. Amos, *Zinsser Microbiology*, 17th ed., Appleton-Century-Crofts, New York, 1980.
3. B. Davis, R. Dulbecco, H. Eisen, and H. Ginsberg, *Microbiology*, 3rd ed., Harper and Row, New York, 1980.
4. *Report of the Committee on Infectious Diseases*, 18th ed., American Academy of Pediatrics, Evanston, Ill., 1977.
5. *Morbidity and Mortality Weekly Report* **31**(3), 22 (1982).
6. J. L. Melnick, *Bull. W.H.O.* **56**(1), 21 (1978).
7. *ORIMUNE® Package Insert*, Lederle Laboratories, Pearl River, N.Y., revised Sept. 1980.
8. H. Eagle, *Science* **122**, 501 (1955).

9. CFR, Title 21, Part 610.
10. CFR, Title 21, Part 630.
11. J. H. Mueller and P. A. Miller, *J. Immunol.* **56,** 143 (1947).
12. L. Pillemer, D. B. Grossberg, and R. G. Wittler, *J. Immunol.* **54,** 213 (1946).
13. CFR, Title 21, Part 620.
14. M. R. Hilleman, E. B. Buynak, R. E. Weibel, J. Stokes, Jr., J. E. Whitman, Jr., and M. B. Leagus, *J. Am. Med. Assoc.* **206,** 587 (1968).
15. E. B. Buynak and M: R. Hilleman, *Proc. Soc. Exp. Biol. Med.* **123,** 768 (1966).
16. S. A. Plotkin, J. D. Farquhar, M. Katz, and F. Buser, *Journal of Diseases of Children* **118,** 178 (1969).
17. U.S. Pat. 4,147,772 (April 3, 1979), W. J. McAleer and H. Z. Markus (to Merck & Co., Inc.).
18. *Morbidity and Mortality Weekly Report* **29,** 265 (1980).
19. *Morbidity and Mortality Weekly Report* **27,** 268 (1978).
20. *HSS Publication No. (CDC) 81-8280*, U.S. Government Printing Office, Washington, D.C., 1981.
21. F. H. Top, Jr., *Yale J. Biol. Med.* **48,** 185 (1975).
22. *Morbidity and Mortality Weekly Report* **29,** 417 (1980).
23. *Morbidity and Mortality Weekly Report* **27,** 327 (1978).
24. *Ibid.*, 173 (1978).
25. *Ibid.*, 231 (1978).
26. *Ibid.*, 255 (1978).
27. *Morbidity and Mortality Weekly Report* **30,** 279 (1981).
28. U.S. Pats. 4,064,232 (Dec. 20, 1977) and 4,140,762 (Feb. 20, 1979), H. Bachmayer and G. Schmidt (to Sandoz, Ltd.).
29. U.S. Pat. 4,029,763 (June 14, 1977), E. D. Kilbourne (to Mt. Sinai School of Medicine of the City University of New York).
30. U.S. Pat. 4,158,054 (June 12, 1979), I. G. S. Furminger and M. I. Brady (to Duncan Flockart & Co., Ltd.).
31. U.S. Pat. 4,000,257 (Dec. 28, 1976), F. R. Cano (to American Cyanamid Co.).
32. U.S. Pat. 3,962,421 (June 8, 1976), A. R. Neurath (to American Home Products Corp.).
33. E. D. Kilbourne, ed., *The Influenza Viruses and Influenza*, Academic Press, New York, 1975.
34. G. C. Schild, M. Aymard, and H. C. Pereira, *J. Gen. Virol.* **16,** 231 (1972).
35. J. M. Wood, G. C. Schild, R. W. Newman, and V. A. Seagroatt, *J. Biol. Stand.* **5,** 237 (1977).
36. *Morbidity and Mortality Weekly Report* **27,** 25 (1978).
37. *PNU IMUNE® Package Insert*, Lederle Laboratories, Pearl River, N.Y., revised Jan. 1981.
38. U.S. Pat. 4,242,501 (Dec. 30, 1980), F. R. Cano and J. S. C. Kuo (to American Cyanamid Co.).
39. Eur. Pat. 0002404A1 (June 13, 1979), D. J. Carlo, K. H. Nolstadt, T. H. Stoudt, R. B. Walton, and J. Y. Zeitner (to Merck & Co., Inc.).
40. J. G. Howard, H. Zola, G. H. Christie, and B. M. Courtenany, *Immunology* **21,** 535 (1971).
41. F. A. Kabat and A. E. Bezer, *Arch. Biochem. Biophys.* **78,** 206 (1958).
42. S. Krugman, *J. Am. Med. Assoc.* **247,** 2012 (1982).
43. U.S. Pat. 4,129,646 (Dec. 12, 1978), W. J. McAleer and E. H. Wasmuth (to Merck & Co., Inc.).
44. A. Mizrah, I. Hertman, M. A. Klingberg, and A. Kohn, eds., *Progress in Clinical and Biological Research*, Vol. 47, Alan R. Liss, Inc , New York, 1980.
45. A. Sabin, *J. Am. Med. Assoc.* **246,** 236 (1981).
46. B. P. Marmion, *Philos. Trans. R. Soc. London Ser. B* **290,** 395 (1980).
47. J. I. Ito, Jr., *Ariz. Med.* **38,** 626 (1981).
48. B. T. M. Buchanan, Jr., *Experimental Medicine* **141,** 1470 (1975).
49. J. D. Nelson and C. Grasi, eds., *Current Chemotherapy and Infectious Disease*, ASM, Washington, D.C., 1980, p. 1239.
50. *New Drug Commentary* **8,** 17 (1981).
51. U.S. Pat. 4,196,192 (April 1, 1980), J. S. C. Kuo (to American Cyanamid Co.).
52. E. R. Moxon, P. Anderson, D. H. Smith, B. Adrian, G. G. Graham, and R. S. Baker, *Bull. W.H.O.* **52,** 87 (1975).
53. S. D. King, H. Wynter, A. Ramlal, K. Moodie, D. Castle, J. S. C. Kuo, L. Barnes, and C. L. Williams, *Lancet* **2,** 705 (Oct. 3, 1981).
54. C. K. Jatta, *W.V. Med. J.* **77,** 166 (1981).
55. U.S. Pat. 4,031,203 (June 21, 1977) and 4,029,764 (June 14, 1977), P. J. Provost, O. L. Ittensohn, and M. R. Hilleman (to Merck & Co., Inc.).

56. U.S. Pat. 4,017,601 (April 12, 1977), M. R. Hilleman, W. J. Miller, and P. J. Provost (to Merck & Co., Inc.).
57. P. Tiollais, P. Charmay, and C. N. Vyas, *Science* **213,** 406 (1981).
58. B. S. Blumberg, H. J. Alter, and S. Visnich, *J. Am. Med. Assoc.* **191,** 541 (1965).
59. D. S. Dane, C. H. Cameron, and M. Briggs, *Lancet* **1,** 695 (1970).
60. W. S. Robinson, *American Journal of Medical Sciences* **270,** 151 (1975).
61. A. J. Czaja, *Mayo Clin. Proc.* **54,** 721 (1979).
62. W. Szmuness, C. E. Stevens, E. J. Harley, E. A. Zang, W. R. Oleszko, D. C. Williams, R. Sadovsky, J. M. Morrison, and A. Kellner, *N. Engl. J. Med.* **303,** 833 (1980).
63. A. J. Zuckerman, *Nature (London)* **287,** 483 (1980).
64. A. J. Zuckerman, *J. Infect. Dis.* **143,** 301 (1981).
65. P. Charmay, M. Gervair, A. Louise, F. Galibert, and P. Tiollar, *Nature (London)* **286,** 893 (1980).
66. S. Z. Hirschman, P. Prince, E. Garfinkel, J. Christman, and G. Als, *Proc. Natl. Acad. Sci. U.S.A.* **77,** 550 (1980).
67. J. C. Edman, R. A. Hallewell, P. Valenzucla, H. M. Goodman, and W. J. Rultz, *Nature (London)* **291** (1981).
68. *Sci. News* **120**(6), 84 (1981).
69. *Genetic Eng. News* **2**(1), 24 (1982).
70. J. Segal, *Contraception* **13,** 125 (1976).
71. G. P. Talwar, S. K. Dubey, M. Salahuddin, and C. Das, *Contraception* **13,** 237 (1975).
72. S. Cohen, *Proc. R. Soc. London Ser. B* **203,** 323 (1979).
73. F. E. G. Cox, *Nature (London)* **994,** 612 (1981).
74. A. L. Notkins, R. A. Bankowski, and S. Baron, *J. Infect. Dis.* **127,** 117 (1973).
75. F. Rapp, *Conn. Med.* **44**(3), 131 (1980).
76. R. J. Klein, E. Buimovici-Klein, H. Moses, R. Moucher, and J. Hilfenhaus, *Arch. Virol.* **68,** 73 (1981).
77. Y. Ohashi, Y. Sakaue, S. Kato, T. Wada, and K. Sato, *Biken Journal* **23,** 199 (1980).
78. T. Nasemann and S. W. Wasselew, *Br. J. Vener. Dis.* **55**(2), (1979).
79. S. N. Cohen, *Sci. Am.* **233**(1), 24 (July 1975).
80. R. L. Sinsheimer, *Annu. Rev. Biochem.* **46,** 415 (1977).
81. R. P. Novick, *Sci. Am.* **243**(6), 102 (1980).
82. W. Gilbert and L. Villa-Komaroff, *Sci. Am.* **243**(4), (1980).
83. P. Chambon, *Sci. Am.* **244**(5), 60 (1981).
84. N. G. Carr and co-workers, eds., *Principles of Gene Manipulation: An introduction to genetic engineering*, University of California Press, Berkeley, Calif., 1980.
85. P. C. Wensink and co-workers, *Cell* **3,** 315 (1974).
86. L. Clarke and J. Carbon, *Cell* **9,** 91 (1976).
87. E. Y. Friedman and M. Rosbash, *Nucleic Acids Res.* **4,** 3455 (1977).
88. G. N. Buell and co-workers, *J. Biol. Chem.* **253,** 471 (1978).
89. M. P. Wickens and co-workers, *J. Biol. Chem.* **253,** 2483 (1978).
90. K. Itakura and co-workers, *Science* **198,** 1056 (1977).
91. D. V. Goeddel and co-workers, *Proc. Natl. Acad. Sci. U.S.A.* **76,** 106 (1979).
92. H. G. Khorana, *Science* **203,** 614 (1979).
93. K. Itakura and A. D. Riggs, *Science* **209,** 1041 (1981).
94. P. S. Lovett and K. M. Keggins, *Methods Enzymol.* **68,** 342 (1979).
95. A. Hinnen and co-workers, *Proc. Natl. Acad. Sci. U.S.A.* **75,** 1929 (1978).
96. M. Mandel and A. Higa, *J. Mol. Biol.* **63,** 159 (1970).
97. S. Broome and W. Gilbert, *Proc. Natl. Acad. Sci. U.S.A.* **75,** 2246 (1978).
98. M. Grunstein and D. S. Hogness, *Proc. Natl. Acad. Sci. U.S.A.* **72,** 3961 (1975).
99. S. Yano and co-workers, *Gene* **13,** 203 (1981).
100. J. K. Dunnick and co-workers, *J. Infect. Dis.* **143,** 297 (1981).
101. S. Maeda and co-workers, *Proc. Natl. Acad. Sci. U.S.A.* **77,** 7010 (1980).
102. P. Mackay and co-workers, *Proc. Natl. Acad. Sci. U.S.A.* **78,** 4510 (1981).
103. *Prog. Med. Virol.* **27,** 88 (1981).
104. R. Sutherland and co-workers, *Proc. Natl. Acad. Sci. U.S.A.* **78,** 4510 (1981).
105. *Sci. News* **120,** 84 (1981).
106. A. Kilejian, *Am. J. Trop. Med. Hyg.* **29,** 1125 (1980).
107. R. J. Watson and co-workers, *Science* **218,** 381 (1982).

108. *Sci. News* **119,** 405 (1981).

109. J. B. Brooksby, *Nature (London)* **289,** 535 (1981).

110. *Chem. Week*, 29 (July 7, 1982).

111. M. Karanikas, *Genetic Eng. News* **2,** 8 (Jan.–Feb. 1982).

112. E. S. Mocarski and co-workers, *Cell* **22,** 243 (1980).

113. S. Kit, *Infect. Dis.* **12**(10), 4 (1982).

114. D. E. Yelton and M. D. Scharff, *Annu. Rev. Biochem.* **50,** 657 (1981).

115. C. Milstein, *Sci. Am.* **243**(4), 66 (1980).

116. R. H. Kennett and co-workers, *Monoclonal Antibodies*, Plenum Press, New York, 1980.

117. J. W. Yowdell and W. Gerhard, *Annu. Rev. Microbiol.* **35,** 185 (1981).

118. G. Kohler and C. Milstein, *Nature (London)* **256,** 495 (1975).

119. W. Gerhard and R. G. Webster, *J. Exp. Med.* **148,** 383 (1978).

120. W. G. Laver and co-workers, *Virology* **98,** 226 (1979).

121. T. J. Wiktor and K. H. Koprowski, *J. Exp. Med.* **152,** 99 (1980).

122. T. J. Wiktor and K. H. Koprowski, *Proc. Natl. Acad. Sci. U.S.A.* **75,** 3938 (1978).

123. *Compensation for Vaccine Related Injuries*, Technical Memoranda, Office of Technological Assessment, Washington, D.C., Nov. 1980.

124. *Bus. Week*, 2529 (April 10, 1978).

125. *Morbidity and Mortality Weekly Report* **31,** 22 (1982).

126. *Evaluation of Poliomyelitis Vaccines*, Institute of Medicine, National Academy of Sciences, Washington, D.C., 1977.

127. H. H. Fudenberg, *J. Lab. Clin. Med.* **79,** 353 (1972).

128. M. A. Riddiough and J. S. Willems, *J. S. Science* **209,** 563 (1980).

129. J. S. Willems and C. R. Sanders, *J. Infect. Dis.* **144,** 486 (1981).

130. Food-Drugs-Cosmetics, F-D-C Reports, Inc., Chevy Chase, Maryland, Feb. 8, 1982.

131. *U.S. Department of Health and Human Services, Washington, D.C., Publication No. (PHS) 81-3272*, April 1981.

132. M. Sun, *Science* **211,** 906 (1981).

V. A. Jegede
K. J. Kowal
W. Lin
M. B. Ritchey
Lederle Laboratories
American Cyanamid Co.

VACUUM TECHNOLOGY

Vacuum technology concerns the means to predict, effect, and control subatmospheric environments (vacuum) (1). Increasingly, each vacuum environment must be safe and also cost-, energy-, and materials-effective.

Vacuum production was essential in the early steam-actuated pumps used for pumping water. In these engines, steam pressure raised a piston against atmospheric pressure. The work stroke in the engine was created by condensing the steam, thus allowing atmospheric pressure to perform the output of the engine. These vacuum-force-type engines were replaced by the steam engine of James Watt. Philosophical notions of void and vacuum developed over centuries turned out to be more realistic than the idea that subatmospheric pressure in a gas characterizes a vacuum.

The vacuum environment offers a great range and diversity of uses (see Table 1). Construction and control represent a challenge to the engineer.

Vacuum systems can be grouped into crude (CR), rough (R), controlled (C), highly controlled (HC), and ultracontrolled (UC) categories (2). These, in some instances, correspond to the traditional categories where vacuum is referred to as low, medium, high, ultrahigh, and beyond ultrahigh. The traditional categorization focuses on the magnitude of pressure rather than on the parameters and their magnitudes that are essential to a given use. The degree of control of a vacuum environment demonstrated by the base pressure of the system is regarded as a balance between the rate at which molecules enter the gas phase and the rate at which molecules are pumped away. There is no explicit recognition of the types of molecules present, and the base pressure of a system at room temperature is not sufficient information to predict the behavior of the system when a dynamic process or experiment is attempted in the chamber. Regarding vacuum as a molecular environment from the beginning can help to keep in view the importance of any possible consequence between the process and vacuum system as a whole.

Within vessels, vacuum environments comprise gaseous molecular phase(s) in contact but not necessarily in equilibrium with condensed molecular phases (2). The condensed phases consist of desirable and undesirable aggregations of molecules and structures such as grain boundaries, dislocations, vacancies, and defects. Typically, these are found in configurations that include undifferentiated bulk at interfacial, thin-film, and surface locations. The gaseous phase, especially under extreme dynamic conditions, may include neutral, excited, metastable, electrically charged cluster, and particulate species.

Nonmolecular species, including radiant quanta, electrons, holes, and phonons may interact with the molecular environment. In some cases, the electronic environment (3), in a film for example, may be improved by doping with impurities (4). Contamination by undesirable species must at the same time be limited. In general, depending primarily upon temperature, molecular transport occurs in and between phases (5), but it is unlikely that the concentration ratios of molecular species is uniform from one phase to another or that, within one phase, all partial concentrations or their ratios are uniform. Molecular concentrations and species that are anathema in one application may be tolerable or even desirable in another.

Toxic and other types of dangerous gases are handled or generated in vacuum systems. References 6–7 discuss safety.

Through its committees, divisions, and chapters, the American Vacuum Society has produced a nearly complete bibliography (to 1982) (8); a dictionary of terms (9); a monograph series; and a number of other useful publications, eg, ref. 10. Another source of information is the Association of Vacuum Equipment Manufacturers. A history of vacuum ideas and technology development from the Middle Ages to Newton is given in ref. 11.

Vacuum Dynamics

Units and Concentration. In the gaseous as well as the condensed phases, molecular concentration by molecular species is of prime importance. By convention, total pressure in a Maxwellian gas is used as though it indicates the quality of the vacuum and as though Maxwellian gases were the rule rather than the exception (12). In general in dynamic systems, gas pressure is neither isotropic nor an adequate indicator of molecular significance.

For a Maxwellian gas in steady state, one standard atmosphere is defined as being equal to 101,323.2 N/m^2; 1 N/m^2 = 1 pascal (Pa) $\rightarrow$ 2.6 $\times$ 10^{20} molecules per cubic meter; 1 torr = 133.32 Pa $\cong$ 1 mm Hg = 1000 micrometers; 1 millibar = 100 Pa; 1 in. Hg = 3386.33 Pa; and 1 $lb/in.^2$ = 6895.3 Pa.

Condensed-Phase vs Surface-Phase Concentration. There are ca 4 $\times$ 10^{18} molecules/m^2 in a monolayer on the surface of a plane, depending upon substrate and adsorbed species. At bulk impurities of 10^{-6} and a bulk thickness of 1 mm, ca 7 $\times$ 10^{19} molecules/m^2 diffuse to the surface as a function of temperature and time (see below). Thus, in a vessel of 1 m^3 at 100 μPa (7.5 $\times$ 10^{-7} mm Hg) and 300 K, the reservoir of molecules in the gas phase is 2.6 $\times$ 10^{16}, whereas surface and bulk are apt to hold ca 2 $\times$ 10^{19} and >4 $\times$ 10^{20} impurity molecules, respectively.

Interaction between Gaseous and Condensed Phases. In a closed vessel of volume V containing a nonionized, unexcited molecular gas with total number of molecules N, the change in the pressure P in the gas can be often predicted if the steady-state absolute temperature T is changed to another steady, constant level.

$$PV = NkT \tag{1}$$

where k = the Boltzmann constant, relating the steady-state absolute temperature T and the equilibrium pressure P in the gas.

However, it is not practical to set the gas temperature in steady state without equally setting the temperature of the surface and bulk phases bounding the gas. Consideration of the response of the system as a vacuum environment can then provide a sufficiently precise prediction of the pressure P and the surface coverage θ at temperature T for molecules of a known species in a known state on a known surface. For example, an isotherm is established between the surface of the condensed and the gaseous phases, depending, eg, on the heat of desorption Q. For submonolayer coverage on a known surface, the pressure P is likely to be an exponential function of T. Among several isotherms, the Temkin isotherm (13) may be used to predict a specific pressure:

$$P = C \cdot \exp Q_\theta/RT \, (\exp Q/RT)^{-1} \tag{2}$$

where C is a constant. Changing the temperature can follow or significantly depart from equation 1, the kinetic formula predicting the dependency of P on T.

Table 1. Vacuum Applications

Category[a]	Gas pressure, ~Pa[b]	Pump[c]	Use
C, HC	<10^3	A, B, D	aneroid barometers
all	all	all	annealing
HC, UC	<10^{-3}	D	arc circuit breakers and switches
CR, R, C	<1	A, B, D	arc furnaces
HC	<10^{-4}	D	betatrons
C	3	B	blood-plasma dehydration
CR, R, C	10^3	A, D	capacitors
CR, R	all	A, D	casting
R, C	60	A	citrus-juice dehydration
CR	5×10^4	A	cleaning
C		A	coffee packing
CR	5×10^4	A	concrete casting
all	all	all	cooling
R, C	10^{-5}	cryo	cryogenic wind tunnels
R, C	3	B	dehydration of antibiotics
R, C	3	B	distillation of plasticizers
C, HC	<10^{-2}	D	electron- and ion-beam lithography
C	<10^{-2}	D	electron-beam furnaces
C	<10^5	A, B, D	electron-beam welding
HC, UC	<10^{-3}	D	electron-diffraction cameras
HC, UC	<10^{-3}	D	electron and ion linear accelerators
C, HC, UC	<10^{-2}	D	electron microscopes
C	50	A	essential-oil distillation
CR, R, C	10^3	A	filtration
C	<2×10^{-2}	D	fluorescent lights
R, C	2	A, B, D	freeze-drying foods and pharmaceuticals
C, HC, UC	<10^{-2}	A, B, D	fusing analysis
HC, UC	<10^{-1}	D	fusion-power research
HC	<5×10^{-3}	D	geiger-counter tubes
C, HC, UC	10^{-2}	D	helium-leak detectors
CR, R	20	A	impregnation of cables
CR, R	20	A	impregnation of capacitors
CR, R	50	A	impregnation of castings
CR, R	10	A	impregnation of wood
C	<1	B, D	incandescent lamps
all	<10^{-1}	B, D	induction melting
R, C	<10^{-1}	A, D	infrared spectrometers
C, HC, UC	<10^{-2}	D	mass spectrometers
HC, UC	2	A, B, D	mercury switches
C, HC	2	A, B	mercury thermometers
CR, R	10^{-1}	A, B	metalizing capacitor paper
all	<10^{-2}	D	metal evaporation
C, HC	<10^{-1}	B, D	metal sputtering (triode)
R, C, HC, UC	10^{-1}	A, B, D	molecular distillation
C, HC, UC	<10^{-2}	A, D	molecular, ion, and electron beams
CR	5×10^4	A	milking machines
R, C	10^{-2}	D	neon signs
C, HC, UC	<10^{-4}	D	isotope separators
R, C	10^3	A	oil deodorizers
C	<10^{-2}	D	optics coating
CR	10^4	A	paper-mill equipment

646

Table 1 (*continued*)

Category[a]	Gas pressure, ~Pa[b]	Pump[c]	Use
CR, R	10^4	A	petroleum distillation
C, HC	10^{-4}	D	photoelectric cells
C, HC, UC	$<10^{-5}$	D	photomultiplier tubes
C, HC	2	B, D	radio-receiving tubes
C, HC, UC	$<10^{-3}$	D	ratio-transmitting tubes
C, HC	3	B, D	radiofrequency diode sputtering
R, C	3	A, B	refrigeration units
C, HC	1	B, D	relays
C, HC, UC	1	B, D	sealed resistors
C	5	B	serum ampules
all	$<10^{-1}$	D	sintering
R	10	A	solvent recovery
CR, R	5	B	smelting
all	all	all	space simulation
R, C	$<10^{-1}$	all	spectrophotometers
CR, R	2×10^3	A	steam-turbine exhaust
CR, R, C	50	A	steel degassing
HC, UC	$<10^{-8}$	D	storage-ring and colliding-beam machines
CR, R	10^4	A	sugar-evaporating pans
R, C	<50	A, cryo	supersonic wind tunnels
C, HC, UC	$<10^{-3}$	D	synchrotrons: electron, ion
R, C	2×10^{-3}	D	thermocouples
C	2×10^{-2}	D	thermos bottles
C	5×10^{-5}	D	television tubes
HC, UC	$<10^{-5}$	D	thin-film circuits
CR	50	A	transformer-oil drying
CR, R	$<10^{-2}$	D	ultracentrifuge
C, HC	$<10^{-2}$	D	ultraviolet instruments
C	2	B, D	vaccines
C, HC	<1	B, D	vapor lamps
CR	5×10^4	A	vehicular-transportation, brakes, engines, etc
CR, R, C	$<10^4$	A, B, D	wind tunnels
HC, UC	$<10^{-4}$	D	x-ray tubes

[a] CR, crude; R, rough; C, controlled, HC, highly controlled, UC, ultracontrolled.

[b] To convert Pa to mm Hg, multiply by 0.0075.

[c] A, mechanical pump or steam ejector; B, booster pump; D, cryo, turbomolecular, sorption, ion, or trapped diffusion pumps.

An example would be a cubical chamber (1000 cm^3) at 298 K constructed of metal, ceramic, or glass that is evacuated (pumped down) and then filled with hydrogen to a pressure of ca 13 mPa (9.75×10^{-5} mm Hg). On cooling to 77 K, the H_2 pressure drops to ca $77/298 \times 1.3 \times 10^{-2} = 3.4$ mPa (25.5×10^{-4} mm Hg) in agreement with the kinetic theory. If a clean tungsten surface is deposited by evaporating tungsten, and then enough H_2 gas is admitted intermittently to give a nonvarying gas-phase pressure at 298 K of ca 13 mPa (9.75×10^{-5} mm Hg), a concentration of $N_{H_2} = 3 \times 10^{15}$ is obtained. Then the tungsten surface (projected area only) has absorbed ca 5.6×10^{18} hydrogen molecules/m^2, or $>3.4 \times 10^{17}$ molecules (100 times the total in the gas phase). If the chamber is again cooled to 77 K, the hydrogen partial pressure in the gas phase falls to <10 fPa ($<75 \times 10^{-18}$ mm Hg), as predicted by the Temkin isotherm (13).

Kinetics Modified by Dynamic Interaction.   The kinetic theory of gases is a valuable tool for vacuum technology. The unmodified kinetic theory must not be applied when the gas interacts significantly with itself or the molecular phases that bound it. When interaction occurs, as it does for many molecular species in the systems considered here, the kinetic predictions must be modified by dynamic considerations. The condensed phase dominates the behavior of the gaseous phase in almost every respect under free molecular conditions. In general, measuring vacuum is not equivalent to measuring any single parameter (1).

Partial Concentration.   The sum of the partial concentrations (pressures) in a free molecular gas is equal to the total concentration (pressure). However, all gaseous components, at the same partial pressure or absolute pressure or ratios thereof, are not likely to have the same significance to any or all vacuum applications. The significance of the condensed-phase concentrations must be determined.

Essential Parameters.   Traditionally, all vacuum environments are characterized in terms of one parameter, ie, pressure in the gaseous phase. However, when costs, energy, safety, hazardous wastes, and other requirements are taken into account, each system must be characterized by a host of parameters. Their magnitudes must be determined in order to judge system performance.

The role of a component as a function of position, use, history, and time may change. For example, a gas-pumping system is always a source of contamination, which may be negligible when the system is in good condition. The significance of a given component in a system at a given time must be determined.

Electrical Breakdown.   The electrical breakdown between parallel planar vacuum electrodes is a function of gas species and pressure (14–15). Markedly higher a-c and d-c voltages can be held off at gas pressures of 100 μPa to 1 Pa (7.5×10^{-7} to 7.5×10^{-3} mm Hg) (Fig. 1). The composition of the surface molecular phase is a key factor.

Zinc Coating of Capacitors.   In the zinc coating of paper strip for capacitors, the paper strip is fed from air through locks into a vacuum environment. There, it is coated by thermally evaporated zinc. The rate of evaporation is so high that contamination of the zinc vapor is excluded. The paper is fed at the maximum rate permitted by its own strength.

Electron Phenomena.   Electron (r-f) phenomena are erased in electron linacs by an improved vacuum environment. Before the development of the 3000-m-long Stanford Linear Accelerator (SLAC), electron linacs were plagued by so-called multipactoring because secondary electrons were trapped in the r-f cavities of the accelerator when voltage was first applied. No beam could be accelerated through the linac until these cavities were conditioned. The vacuum achieved was thought not to contribute to multipactoring. The vacuum thought to be satisfactory for the accelerator was based upon the prevention of beam interaction with the residual gas. This mean-free-path consideration is species dependent, but by custom the vacuum was specified in terms of total pressure in the gas phase only. When the SLAC was planned, each klystron r-f power tube was to be open to the accelerator vacuum because reliable r-f windows transmitting the power required had not yet been developed. Thus, SLAC was built to have a vacuum good enough for a klystron tube. However, r-f windows were developed but the accelerator vacuum environment was kept clean as a klystron. When a beam was first atempted at SLAC, no multipactoring occurred, and the beam passed through from electron source to target the first time r-f voltage was applied.

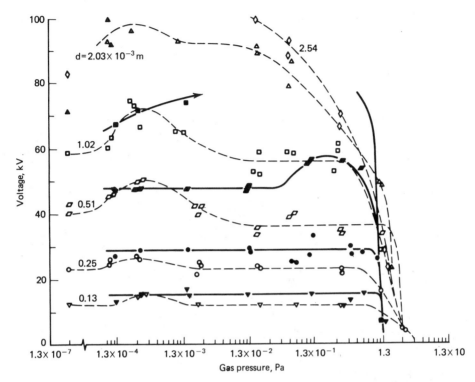

Figure 1. Currents are d-c (broken lines and empty squares) and peak a-c (solid lines and solid squares) breakdown voltage vs gas pressure in nickel for various gaps (16). To convert Pa to mm Hg, multiply by 0.0075.

Film Contamination from the Bulk Phase. The contamination of an epitaxial film of GaAs from an oven charge can be corrected by doping (4).

Oil Contamination of Helium Gas. For more than 20 years, helium gas has been used in a variety of nuclear experiments to collect, carry, and concentrate fission-recoil fragments and other nuclear reaction products. Reaction products, often isotropically distributed, come to rest in helium at atmospheric concentration by collisional energy exchange. The helium is then allowed to flow through a capillary and then through a pinhole into a much lower vacuum. The helium thus collects, carries, and concentrates products that are much heavier than itself, electrically charged or neutral, onto a detector that may be a photographic emulsion. If the helium is contaminated with pump oil, the efficiency of delivery to the detector is markedly increased. Oil contamination, anathema in some systems, is desirable here.

Field Emission of Electrons. Nonthermionic emission from Spindt-type arrays of cold cathode tips is affected by gas-phase concentration and species (see Fig. 2) (17). The time response of emission at constant voltage to changes, and reverse, from negligible to significant concentrations of water, hydrogen, and oxygen indicates that this emission may depend upon diffusion rates into and out of the bulk phase in addition to cathode surface adsorption–desorption. Emission from Spindt-type tips was also increased by other active gases tested, eg, NH_3, CH_4, and H_2S (not O_2), but was unchanged by corresponding concentrations of He, Ne, or Ar.

The bulk and surface phases of the anode receiving the electron emission must be degassed at ca 1200 K in order to avoid electrical breakdown.

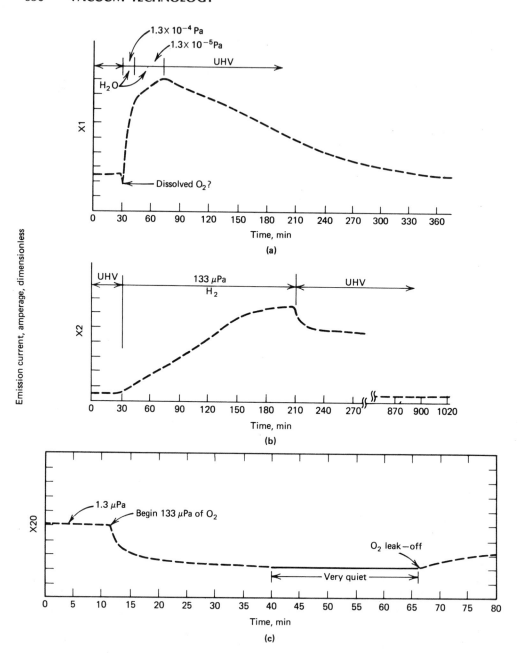

Figure 2. Behavior of electron-field emission at room temperature from Spindt-type arrays of 5000 tips per mm^2, beginning and ending with UHV (ultrahigh vacuum), ie, UC (ultracontrol). Dashed line indicates noise. (**a**) Water; (**b**) hydrogen; and (**c**) oxygen. To convert Pa to mm Hg, multiply by 0.0075.

Action of Vacuum on Spacecraft Materials. For service beyond the atmosphere, the vacuum environment allows materials to evaporate or decompose under the action of various forces encountered (1,18–19). These forces include the photons from the sun, charged particles from solar wind, and dust. The action of space environment on materials and spacecraft can be simulated by a source–sink relationship in a vacuum environment. Thus, for example, the lifetime of a solar panel in space operation may be tested (see Photovoltaic cells).

A vacuum system can be constructed that includes a solar panel, ie, a leak-tight, instrumented vessel with a hole through which a gas vacuum pump operates. An approximate steady-state base pressure is established without test parts. It is assumed that the vessel with the test parts can be pumped down to the base pressure. The chamber is said to have an altitude potential corresponding to the height from the surface of the earth where the gas concentration is estimated to have the same approximate value as the base pressure of the clean, dry, and empty vacuum vessel.

In general, the test object cannot be heated above its operating temperature in space. As free molecular conditions are obtained around the object, it outgases and, if solar-spectrum photons impinge on the object, release of gas increases. Since the object is in a vessel and the area of the hole leading to the gas pump is small compared with the projected interior area of the vessel, molecules originating from the test object can return to the test object provided that they do not interact in some manner with the vessel walls and the other components of the molecular environment. The object inside the vessel establishes an entirely different system than the clean, dry, and empty vacuum vessel. The new system no longer has the capability to reach the "clean, dry, and empty" base pressure within a reasonable time.

This simulation can be achieved in terms of a source–sink relationship. Rather than use the gas concentration around the test object as a target parameter, the test object can be surrounded by a sink of ca 2π solid angle. The solar panel is then maintained at its maximum operating temperature and irradiated by appropriate fluxes, such as those of photons. Molecules leaving the solar panel strike the sink and are not likely to come back to the panel. If some molecules return to the panel, proper instrumentation can determine this return as well as their departure rates from the panel as a function of location. The system may be considered in terms of sets of probabilities associated with rates of change on surfaces and in bulk materials.

Electronic Vacuum Tube. In special electronic vacuum-diode tubes, with spacing between the cathode and anode of 10 μm, high gas concentrations of some types are beneficial to the operation of the tube under proper control.

Pump-Down

Many problems encountered in producing a highly controlled vacuum are due to the system's design, history, contents, use, and maintenance.

Initially, the vessel is filled with ambient air. Any given macrosample of air may contain at least 1600 substances that have been identified (20). Among these are the gases and vapors listed in Table 2. Most important is usually water; others include viable and nonviable particulates (21), aerosols (21), and cluster species of molecules that can be in states of excitation or ionization. Aerosols originate from both man-made and natural sources; in urban air, the former far outweigh the latter. Typical contaminants include cigarette smoke, lead, and asbestos. Gasoline and diesel fuel contribute significantly to contamination. Effluents and exudates are present, including skin particles, hair, and innumerable other substances from human sources (see also Air pollution).

Table 2. Typical Average Diurnal Concentrations of Molecular Species in Nonpolluted Ambient Air[a]

Species	Concentration, molecules/m³
More chemically reactive and surface adsorbing	
O_2	5×10^{24}
H_2O	ca 2.5×10^{23}
O_3	ca 1×10^{18}
NO	ca 1.5×10^{17}
HO_2	ca 6.5×10^{14}
HO	ca 4×10^{11}
Cl	ca 2×10^{10}
N_2O	ca 5×10^{19}
H_2	ca 5×10^{19}
Less chemically reactive and surface adsorbing	
N_2	2.1×10^{25}
CO_2	8.9×10^{21}
CH_4	ca 5×10^{19}
Chemically inert, but slightly adsorbing	
Ar	2.5×10^{23}
Ne	4.9×10^{20}
He	1.4×10^{20}
Kr	3.0×10^{19}
Xe	2.3×10^{18}
Ra	2×10^{4}

[a] Ref. 20.

Microstructure on Surfaces. Gross cracks and voids are usually lined with microstructure, as indicated by Figure 3. As the depth–width ratio of a crack is held constant but the dimensions approach molecular dimensions, the crack becomes more retentive. At room temperature, gaseous molecules can enter such a crack directly and by two-dimensional diffusion processes. The amount of work necessary to completely remove the water from the pores of an artificial zeolite can be as high as 400 kJ/mol (95.6 kcal/mol). The reason is that the water molecule can make up to six H-bond attachments to the walls of a pore when the pore size is only slightly larger. In comparison, the heat of vaporization of bulk water is 42 kJ/mol (10 kcal/mol), and the heat of desorption of submonolayer water molecules on a plane, solid substrate is up to 59 kJ/mol (14.1 kcal/mol). The heat of desorption appears as an exponential in the equation correlating desorption rate and temperature.

Turbulent Gas Flow (Rough Pumping). An oil-sealed mechanical pump in good condition, with vented or trapped exhaust, is gas purged by running for several hours. A liquid-nitrogen (LN) trap is between pump and vessel. It can consist of a U-shaped

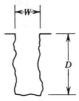

Figure 3.   Crack of width W and depth D; ratio D/W = constant (22).

tube of thin-walled stainless steel clad externally with heavy-wall copper along its vertical legs (23). In each riser leg of the U-trap, a twisted piece of copper of width d (d = ID of tube) is inserted to ensure that under contaminate-flow conditions, oil cannot pass without encountering a liquid-nitrogen-cooled surface. An in-line hot (450 K) all-metal valve on the system side of the trap is connected in the line above the copper-clad leg by a 0.02-m-long stainless-steel neck (2.5×10^{-4}-m wall thickness). Thus, the rate of boil-off of the liquid nitrogen from the trap is kept reasonable. The trap can be filled automatically from a local reservoir, from a built-in LN supply line, or by hand. By keeping the all-metal valve always at 450 K, it is possible to close it, allow the trap to warm up, when needed, refill the trap and re-open the valve. This arrangement provides satisfactory control of contamination from an oil-sealed mechanical pump (23).

Using absorbent material is time consuming and more expensive, and can contribute minute, solid pieces of the sorbent into the system.

With valves open, the air in the vessel is exhausted through the U-trap by the oil-sealed pump. As the pressure falls, the composition of the gas in the vessel begins to change. At a pressure of ca 13 Pa (0.097 mm Hg), water is the dominant species in the gas phase. The surface phases then change appreciably, although initially, water was the dominant species on the surfaces. The bulk phase is unlikely to contain any water molecules as such, except in voids and gross defects. Water is desorbed from glass as a result of OH radicals changing to H_2O at the surface.

Diffusion-Pump System. After the pump line and trap have been shut off, a large valve is opened slowly enough that the mass flow of gas from the chamber through the valve into the oil-diffusion pump system does not disrupt the top jet of the diffusion pump (DP) (see Fig. 4). When the liquid nitrogen is replenished after the trap has been operated for some time, release of previously trapped gas must be avoided. So-called ionization-gauge response pips at the start of the liquid-nitrogen replenishment are an indication of trap ineffectiveness.

The trap fill line is separated from the high vacuum region by an overhanging copper skirt, a so-called creep barrier, that also serves to keep the interior surface exposed to the working environment at a temperature independent of the liquid level in the trap. Oil creepage in two dimensions along surfaces is effectively inhibited at <200 K. Contribution to creepage by the liquid-nitrogen trap is usually small compared with the contamination delivered during the filling of the trap and reduction of LN level.

Speed Factor. Pumping-speed efficiency depends on trap, valve, and system design. For gases with velocities close to the molecular velocity of the DP top jet, system-area utilization factors of 0.24 are the maximum that can be anticipated: eg, less than one quarter of the molecules entering the system can be pumped away where the entrance area is the same as the cross-sectional area above the top jet (Fig. 4). The system speed factor can be quoted together with the rate of contamination from the pump set. Utilization factors <0.1 for N_2 are common.

The rate of contamination from the pump set is <10^9 molecules/($m^2 \cdot s$) for molecular weights >44 (23). This is the maximum contamination rate for routine service for a well-designed system that is used constantly, with automatic liquid-nitrogen filling and routine maintenance.

A fraction of gas, depending on species, is pumped by the diffusion pump or trapped on the cold surfaces of the LN trap; for example, 0.05 for H_2 and 0.9 for H_2O,

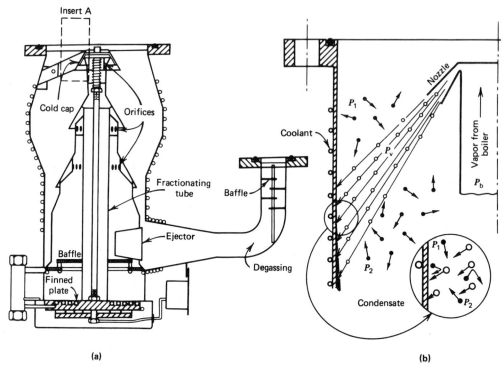

Figure 4. (a) Multistage diffusion pump. (b) Insert A, jet spray. Closed circles indicate gas molecules P_1; open circles indicate vapor-jet molecules, P_2; and P_b, boiler pressure (23).

respectively, have been measured. With enough flow, multilayers of gas can build up on the liquid-nitrogen-cooled surface, thereby evaporating at the bulk vapor pressure of the gas at LN temperature. To cope with such build-ups, a well-designed valve is required immediately between the system and the LN-cooled surface. Designs incorporating this valve within the LN trap itself have been marketed for 20 years. The valve can be closed and the trap warmed sufficiently to purge accumulated gases such as carbon dioxide. Furthermore, the trap can be allowed to warm to room temperature without delivering contamination into crannies and surfaces of the valve exposed to the chamber when the valve is open.

Leaks

A vacuum system can be stalled by gas leaks (4,6,25). Traditionally, leaks are categorized as real or virtual. A real leak refers to permeation processes or cracks or holes that allow external gas (air) to seep into the vacuum environment. Atmospheric gases such as helium and hydrogen permeate glass equipment, especially at elevated temperature. The noble gases do not permeate metals, but hydrogen does. Virtual leaks refer to gases that originate from within, eg, from trapped volumes, the gauges, pumps and the bulk and surface-phase species. For example, carbon in bulk stainless steel may precipitate along grain boundaries and then combine with surface oxygen to give CO which is then desorbed into the gas phase (26). Proper instruments readily distinguish real leaks from virtual leaks.

In practice, it is often necessary to take readings from hot-filament ionization gauges or other devices. Figure 5 gives pump-down curves for six different types of pumping equipment on the same vacuum chamber (23). The shape of curve 1 indicates that a real leak could be responsible for the zero slope demonstrated by the Bayard-Alpert gauge (BAG). The shape of the other curves could be due to combinations of real and virtual leaks.

In fact, the leveling-off slope of curve 1 was entirely owing to gas issuing from the pumping equipment itself, and there were no other sources of leakage. Curves 2–6 all resulted from combinations of virtual and real leakage. Most of the leakage in curves 2–4 originated in the pumping equipment. This was also true for the early part of curve 5. The phenomenology of the early stage of curve 6 is due to chamber-wall outgassing. The latter stages of curve 6 show a combination of wall outgassing and a small leakage from the pump itself.

Figure 6 is instructive regarding the magnitude of leakage from pump 5 shown in Figure 5 after ca 5 h of pumping. Pump 5 was a commercial Orbitron made by the National Research Corporation. After ca 4 h from the start of the pump-down, the electric power to the pump was turned off, whereas the power to the BAG remained on. The total pressure in the system rose very rapidly, and this response is recorded in Figure 6. Within a short time, however, this rise reached a maximum and assumed

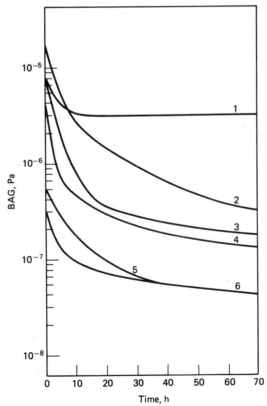

Figure 5.   Plots of pump-down performance for pumps operating on 0.1 m dia × 0.43-m-long stainless-steel tubing. Curves 1–4 are sputter-ion pumps of different makes; curve 5, Orbitron type; and curve 6, LN-trapped oil DP. Pressure measured with Bayard-Alpert gauge (BAG) (27). To convert Pa to mm Hg, multiply by 0.0075.

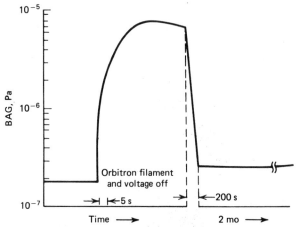

Figure 6.   Bayard-Alpert gauge response vs time upon cutting off power to Orbitron pump (28). To convert Pa to mm Hg, multiply by 0.0075.

a small slope downward. After ca 200 s, the reading on the BAG was close to the reading achieved before the Orbitron power had been turned off. It remained essentially constant for two months after the initial pump-down.

Figure 6 can be interpreted as follows: The Orbitron pump is delivering a virtual leak back into the vacuum environment. As with many other pumps, this quantity of gas is significant. The operational speed of the Orbitron just before the electric power was turned off was zero. Thus, an equilibrium was established between the source–sink properties of the vacuum environment. When the Orbitron power is turned off, a slug of electron-bombarded, evaporating titanium remains hot for some time and continues to outgas. As the slug cools down sufficiently, however, the BAG intrinsic speed is sufficient to balance the decreasing rate of gas evolved. The intrinsic steady speed for the BAG is known to be ca 100 cm^3/s. Thus, after ca 4 min from the time the electric power was switched off to the Orbitron, the operational speed of the system is provided by the ionization gauge and the sorption properties of the system. The ionization gauge was thus able to provide almost the same equilibrium base pressure as did pump 5 which had a rated speed nearly 4000 times greater than that of the BAG. The significance of equilibrium base pressure is use dependent. The category virtual leak must include evolution from the pumping system and the condensed phases, as well as from atmospheric air and gas trapped in cracks and voids. The question of virtual leak vs real leak may be important in every system.

Molecular Transport

Molecular transport concerns the mass motion of molecules in condensed and gaseous phases. The mass motions are driven primarily by temperature. As time progresses, the initial mass motion results in concentration gradients. In the condensed phase, flow along concentration gradients is described by Fick's law.

Standard texts may be consulted on the topic of diffusion in solids (6,12–13). Some generalizations, however, are possible. No noble gas permeates a metal. Metals are, however, permeated readily by hydrogen. Stainless steel, for example, can be permeated by hydrogen from concentrations likely in air. The least permeable material for hydrogen is carbon. Glasses are permeable, especially by the light noble gases at elevated temperatures.

After a bake-out of 600–700 K, the bulk phase is likely to far exceed the surface phase as a source of atomic (molecular) impurities that desorb into the gas phase (30). Bake-out at 1300 K greatly reduces bulk-phase impurities.

Gas Transport. Initially, in a vessel containing air at atmospheric pressure, mass motion takes place when temperature differences exist and especially when a valve is opened to a gas pump. Initial flow in practical systems is discussed in ref. 31; Monte Carlo methods to treat shockwave, turbulent, and viscous flow phenomena under transient and steady-state conditions in ref. 5.

Viscous Transport. Low velocity viscous laminar flow in gas pipes is commonplace. Practical gas flow can be based upon pressure drops of <50% for low velocity laminar flow in pipes whose length-to-diameter ratio may be as high as several thousand. Under laminar flow, bends and fittings add to the frictional loss as do abrupt transitions.

Free Molecular Transport. The free molecular gas regime is illustrated by Figure 7. A duct of maximum transverse dimension D and length L connects two chambers, each of minimum interior dimension $\gg D$. Free molecular transport (Knudsen flow) is often sufficiently approximate when $\lambda \geq D$. For a right circular cylinder of length $L = D$, the diameter, the internal pressure drop in free molecular flow is ca 50% (see Figs. 8–10). In free molecular flow at steady state (Fig. 7), the temperature of the gas entering a duct determines the rate of passage through the duct, not the temperature (other than zero) of the duct itself. Volumetrically, Knudsen flow is proportional to gas entering velocity only; it is constant, independent of gas concentration or gradient. Free molecular flow can be described in terms of a statistically valid number of molecules interacting with the surfaces of a duct provided that the entering spatial and ongoing reflection distributions from the walls by the molecules are known.

If a Maxwellian gas at steady state is entering end 1 in Figure 8, then the duct conductance

$$C = \frac{1}{4} \, \bar{v} A_1 W_{1 \to 2} \tag{3}$$

where $\bar{v}$ is the average velocity of the Knudsen Maxwellian gas entering end 1 (before wall encounter); A_1 is the area of end 1; and the transmission probability of a Maxwellian gas (typically reflecting from the walls by a cosine distribution) is $W_{1 \to 2}$ = number in end 1 per number out through end 2.

Under isothermal conditions where energy is not added or removed from the system, the second law of thermodynamics obtains, and

$$A_1 W_{1 \to 2} = A_2 W_{2 \to 1} \tag{4}$$

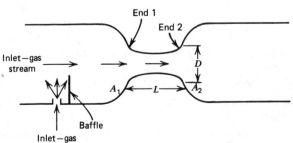

Figure 7. Duct connecting two volumes of dimension $\gg D$.

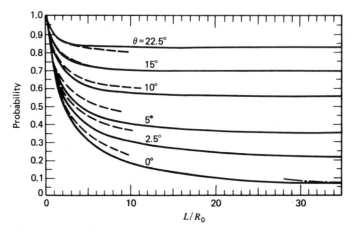

Figure 8.   Conical ducts with length L, small end of radius $= R_o$, and half angle θ (32). Dashed line, Monte Carlo method (31,34); broken line, long-tube asymptote. The divergence of solid and dashed lines is because of an error in the Monte Carlo Method (32,34).

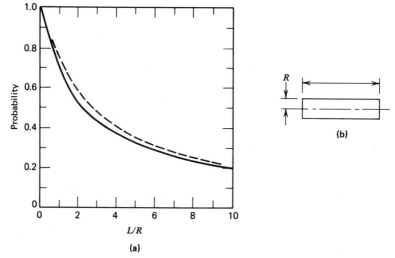

(a)

Figure 9.   Molecular-transmission probability (**a**) for a cylindrical tube (**b**) as a function of the ratio of length to radius, L/R. The conductance is $C = \frac{1}{4} \pi R^2 \bar{v} W_{1\rightarrow2}$. The solid curve is from Clausing's calculation; the dashed curve corresponds to the approximation $W = 1/(1 + (3L/8R))$ (35).

Thus, the probability of transmission must be directional for $A_1 \neq A_2$, but the conductance C cannot be directional (see Thermodynamics). If C is directional, energy must be supplied from external sources and C is a pumping speed. Thus, work is performed by pumping but not by conductance. Under free molecular flow, the volumetric rates of transport in the gas phase are independent of the pumps being on or off; bends in a duct little alter the probability of passage over a straight duct with the same axial length. The intuitive idea that the walls of transition sections must be "fared in" (smooth flow) does not apply in free molecular flow. For example, conical transitions (see Fig. 9) (32) and the area above the top jet in Figure 4 always have smaller transmission probabilities than right circular cylinders with one restricted end where

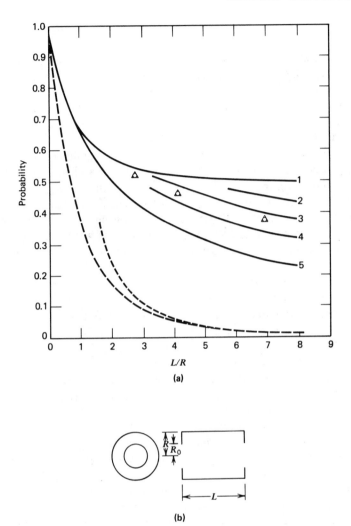

Figure 10. Molecular-transmission probability (**a**) for circular cylinder (**b**) with two restricted ends. Solid line, fraction transmitted without wall encounter; broken line, prediction of this fraction by the formula R_c^2/L^2 vs L/R_o. $(R/R_o)^2$: 1, ∞; 2, 3; 3, 2; 4, 1.5; 5, 1.0. (○, Argon, $(R/R_o)^2 = 2$; and △, nitrogen $(R/R_o)^2 = 2$ (36).

diameters and length are the same as in a given conic frustum. A right circular cylinder without a restricted end has a free molecular conductance that is greater than that of the conic frustum. However, there can be other reasons than conductance for using conical transitions in a given instance.

Wall Geometries. Rougher-than-rough wall geometries can reduce transmission probabilities in Knudsen flow by as much as 25% compared to so-called rough walls (33,37). For this and other reasons, conductance calculations that claim accuracy beyond a few percent may not be realistic.

In free molecular flow, if gaseous conductance were not independent of the flow direction, a perpetual-motion machine could be constructed by connecting two large volumes by a pair of identical ducts with a turbine in front of one of the ducts. A duct with asymmetrically shaped grooves on its wall surface could alter the probability of molecular passage in such a way that for a tube of equal entrance and exit areas, the probability of passage would be made directional.

On purely kinetic grounds, however, the use of the term random must be used carefully in describing a Maxwellian gas. The probability of a Maxwellian gas entering a duct is not a random function. This probability is proportional to the cosine of the angle between the molecular trajectory and the normal to the entrance plane of the duct. The latter assumption is consistent with the second law of thermodynamics, whereas assuming a random distribution entry is not.

The probability of passage is independent of the entrance velocity of free molecules and the subsequent velocity ($v \neq 0$) of these molecules within the tube. It depends upon the entering angular and wall-reflection distributions of the molecules. It is difficult to predict the distribution functions of molecules reflecting from single-crystal surfaces (33–34,37–38) and thus also from engineering surfaces. For engineering surfaces and gases at room temperature, reasonable results within ±10 percent are obtained by assuming that a statistical number of molecules impinging on a surface exhibits a cosine distribution upon reflection from the surface. Some engineering surfaces may be classed as rougher than rough, however, and the distribution of reflected molecules from an element of surface shows a maximum near the normal to the wall of the tube (33) rather than the spherical shape predicted by the cosine law. Thus, cooling or heating a structure such as a trap does not alter the probability of molecular passage through the structure. This statement assumes that the spatial distribution of molecules reflecting from the walls of the trap is not a function of temperature and that a Maxwellian gas in steady state is entering the trap. In actuality, arranging constant temperature differences in steady-state flow can work on a free molecular gas and cause a pumping effect. This pumping effect is distinguished from gaseous conductance because net work is performed on the free molecular gas. In other words, in special cases, series arrays of ducts can be caused to pump when temperature differences are maintained in the arrays. These constant temperature differences are maintained because energy is supplied to the system from external sources to maintain the temperature differences. This energy from the external sources keeps the system operating at steady state with an end-to-end pressure (concentration) difference. In free molecular flow, all tubes of a geometrically similar shape have the same free molecular probability of passage for the same entering gas distribution. When we are dealing with Maxwellian gases, therefore, we can plot the probability of passage through a duct and use this probability to calculate the conductance for all pipes of that shape (see Figs. 8–12). Under free molecular flow, bends in the tube of equal entrance and exit areas little alter the probability of molecular passage compared to a straight tube with the same axial length. In a few cases, the probability of passage through a duct can be obtained by inspection; for example, through a thin orifice, it is 1; through a large box that has a pinhole entrance and a pinhole exit, it is 0.5. If a small plate is placed in the box obscuring the line of sight between the pinholes, the probability of passage is still 0.5.

Combining Conductances. Combining conductances may be difficult because, if a free molecular gas that is Maxwellian in steady state enters conductance 1 (length $\neq 0$), the gaseous distribution is no longer Maxwellian at exit 1. This corresponds to the so-called beaming effect. The overall conductance can be estimated if the probabilities of passage of the individual components are known and if the juxtaposed components do not vary more than about a factor of two in cross-sectional areas (6,36).

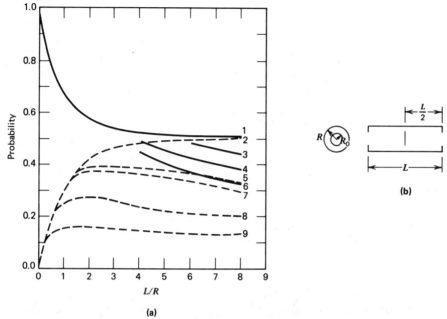

(a)

Figure 11. Molecular-transmission probability (**a**) of a round pipe (**b**) with (solid line) and without (broken line) blocking plate and restricted entrance and exit apertures. $(R/R_o)^2$: 1, ∞; 2, ∞; 3, 3; 4, 2; 5, 2.25; 6, 1.5; 7, 2.0; 8, 1.5; and 9, 1.25 (39).

In general, exact methods of calculating conductance depend upon knowing the spatial distribution of molecules encountering and reflecting from real surfaces.

Pumping Speed. If the standard formulas for gas flow in vacuum are applied (eg, pumping speed $S = Q/P$, where Q = mass of gas transported, P = pressure), it is assumed that a Maxwellian free molecular gas is entering the pump. The pump must be used on a chamber with a cross-sectional area much larger than that of the cross-sectional entrance of the pump. The gas concentration is then sufficiently uniform in the chamber and across the entrance to the pump. As the chamber is made smaller, finally coinciding with the pumps' entrace area, the gaseous distribution entering the pump is increasingly non-Maxwellian. This does not present a problem, however, provided that this fact is taken into account when assessing the system. For example, the behavior of a free molecular gas from point to point can be predicted mathematically, whether it is Maxwellian or not (5). The calculations can be augmented by experimental measurement. When using only experimental measurements, the data are treated like those obtained with Maxwellian gases. Calculating the probability of passage and pumping speed for more complex shapes may require high speed computers, but once a code is obtained, only a limited amount of machine time may be needed (5). Similar codes have been worked out for neutron and radiation problems.

Instead of calculations, practical work can be done with scale models (36). In any case, calculations should be checked wherever possible by experimental methods. Using a Monte Carlo method, for example, on a shape that was not measured experimentally, the sample size in the computation was allowed to degrade in such a way that the results of the computation were inaccurate (see Fig. 8) (32,34). Reversing the computation or augmenting the sample size as the calculation proceeds can reveal or eliminate this source of error.

Bulged elbow	L/D	Probability, experimental
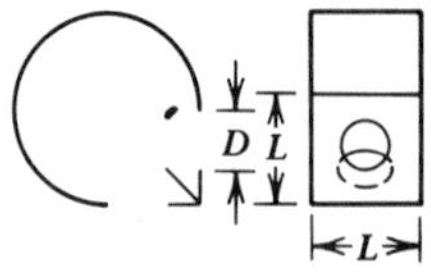 Plain	2.00	0.44
	1.33	0.39
	1.00	0.32
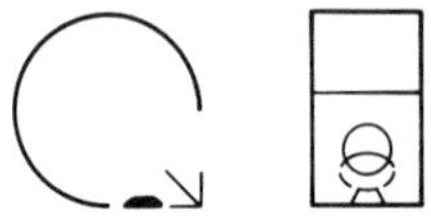 With jet cap	2.00	0.33
	1.33	0.30
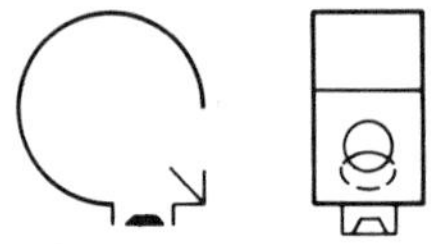 On diffusion pump	2.00	0.32
	1.33	0.27
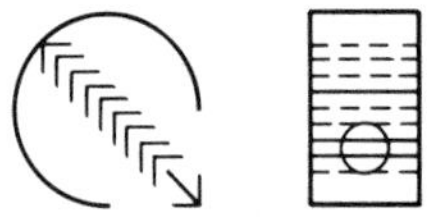 With Chevron (L/D = 5)	2.00	0.38
	1.66	0.35
	1.33	0.31

Figure 12. Probability for bulged-elbow geometries (34).

System Pumping Speed. The operational speed of the pump is a systems effect. All current pumps perform more than one function in the system. Pumps are sinks and at the same time sources for molecules. Pumping speed relates to work on the gas phase. Each type of pumping system occupies a projected area on the vacuum vessel wall. This area may be a hole in the wall, a surface that getters gas, or a cold surface that condenses gas. In the case of cryogenic or gettering surfaces, there can be some advantage in making the actual area of the surface larger than its projected area. If the primary interest is to increase the pumping speed of the system as distinct from the molecular handling capacity, the speed can be increased for molecular species with sticking coefficients <0.3.

Increasing the specific area of a soid surface that retains the molecules that strike it impedes a free molecular gas in reaching more deeply into the extended surface area. This conductance effect is geometrical and thus nearly independent of the dimensions of the increased area. However, depending upon temperatures and species, when molecular-size pores such as encountered in sorbent material are used, molecules that arrive at the pores initially can plug the pores and prevent subsequent molecules from entering. Depending upon the temperature of the surface and the molecular species, however, some two-dimensional adjustment can take place where molecules can penetrate into higher specific-area regions by two-dimensional flow along surfaces. The system pumping speed for holes cut in the side of a vacuum system can be evaluated in terms of efficiency and cost for this pumping. For Maxwellian gases in free molecular flow, the chance of molecules being pumped through a hole is independent of the concentration. This pumping probability has been incorrectly called the Ho coefficient for the system (6). The probability of a molecule exiting from a system through a hole and not returning is equivalent to the Clausing factor for Maxwellian free molecular gases. Currently, pumping systems, whether based on turbomolecular, ionic, getter, cryogenic, or diffusion-pump principles, are not able to fully utilize holes and also be of reasonable size and cost. If the area of a hole in the vessel is equal to the cross-sectional entrance area of a pump operating on that hole for the light gases hydrogen and helium, the system pumping probability is ca 1:10. Thus, theoretically, another pump could be placed on the same hole and deliver ten times the pumping speed available. This inability to utilize holes and still keep the cost of a pump within reasonable bounds stems from several difficulties.

Diffusion pumps, whether they are oil or mercury, require a trap to prevent the working fluid of the pump from contaminating the vacuum environment. Liquid-nitrogen traps are usually employed. Their temperature, however, is not low enough to condense any hydrogen or helium gas even on sorbents of large specific area. In addition to the trap, a valve is often essential in the system (see Vacuum Brazing Furnace below). Analytical estimates of the maximum system probability realizable for a diffusion pump showed it to be less than 3:10 for molecular species with velocities near to or less than the molecular velocity of the top jet of the diffusion pump (Fig. 4). So far, it appears impractical to increase the molecular velocity of the top jet. The pumping of light gases such as hydrogen and helium is fundamentally limited by the velocity of working fluid molecules in the top jet. From a user's point of view, there should be a sizeable market for pumping equipment that would utilize the potential of wall area better without increasing the system cost.

Turbomolecular pumps operating on holes are faced with the same fundamental problem as the diffusion pump. The tip speed of the turbine rotor in commercial pumps, because of the strength-to-density ratio of the turbine blade material, is limited to $<6 \times 10^2$ m/s. The average velocity for helium at room temperature is ca 1×10^2 m/s and for hydrogen, 2×10^2 m/s.

Patents have been filed related to a novel vacuum pump (40). Magnetically suspended self-balancing rotating arrays of fibers permitting tip speeds of ca 3×10^3 m/s are self-cutting and of low mass (light weight).

Instrumental Measurement

Especially important under dynamic conditions, the role of a system and each component can be disclosed by appropriate measurements. Thus, it can be established when the system environment is ready for a dynamic use and if the pump is likely to perform as a molecular sink, a source, or some combination of these.

Gauges. At present, there is no way to measure a true molecular vacuum environment except in terms of its use. Readings related to gas-phase concentration are provided by diaphragm, McCleod, thermocouple, Pirani gauges, and hot and cold cathode ionization gauges (manometers).

Ionization gauges (IG) do not give pressure in the gaseous phase but are set to provide readings proportional to the concentration of molecules in the gaseous and, to a lesser extent, the condensed phases. These concentrations are translated to units of gaseous pressure. The hot and cold cathode ionization gauges, such as shown in Figures 13 and 14, provide information about vacuum environment as a host of parameters. Perhaps largely by trial and error, selection of a gauge readout indicates when

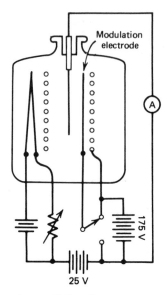

Figure 13.   The introduction of a so-called modulator electrode in a Bayard-Alpert gauge permits the pressure range to be lowered by a factor of 10^{-1} to 133 pPa (10^{-12} mm Hg) (41).

to begin the process or experiment. Turning the gauge off and on provides the so-called flash-filament gauge response. When the filament is turned off, it cools, and molecules from the gaseous environment impinge and stick to the surface of the cooling filament. When the filament is turned on again and its temperature rises to incandescence, these molecules are desorbed. Sufficient electrons are provided by the filament to indicate the gas-phase concentration increase from this desorbed material. An abruptly rising pip with a longer decaying tail can be recorded from the output of the gauge. The area under this desorption peak is a measure of the integrated pumping effect of the cold bare filament. The filament is bombarded by the free molecular gas over a ca $2\,\pi$ solid angle. The flux of particles striking the filament is given by $n \cdot \bar{v}$, where n is the molecular concentration in the gas phase and $\bar{v}$ is the average velocity of this Maxwellian gas. By changing the power supplied to the filament in its heating phase and varying the time interval when the filament is left cold, approximate but useful information can be obtained (13).

   Residual Gas Analyzers.   A gaseous molecular phase is analyzed with a mass spectrometer (43) (see Analytical methods). If heat is delivered to the condensed phase or if electrons are caused to strike its surface, molecular desorption provides a signal for the mass spectrometer analysis. Direct electron desorption does not involve heating. For contaminated surfaces, the probability of a molecule being desorbed per incident electron can be as high as 10^{-3}. For submonolayer coverage of some species, this probability may decrease to $<10^{-7}$. These probabilities are large enough to allow a significant signal from the mass spectrometer. The electron current delivered to the surface can be raised to deliver intense heat, providing information about the bulk phase near the surface. Photons desorb molecules directly from surfaces (except for water vapor).

   Ultrasound frequencies can be introduced into the walls of the vacuum system. If a source of ultrasound is placed on the wall of an ultrahigh vacuum system, a large hydrogen peak is observed (see Ultrasonics). Related phenomena, presumably from

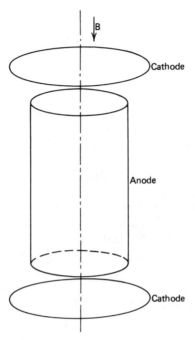

Figure 14. Geometry of a single cell of a Penning gauge. Space charge of the trapped, circulating electrons equalizes the axis potential with that of the cathode. Thus, the electric field is radial. Electron density is at a maximum a short distance from the anode. Electrons progress radially toward the anode only as they lose kinetic energy, mainly through inelastic (ionizing) collisions with molecules (42). B = magnetic field.

frictional effects, are observed if the side of a vacuum system is tapped with a hammer; a desorption peak can be seen. Mechanical scraping of one part on another also produces desorption.

Vacuum Systems and Equipment

Wall Materials. Glass is often the material of choice for small laboratory systems and sealed systems in commercial practice. Glass has a wide range of useful properties, including high compressive strength but relatively low tensile strength, requiring careful selection of the glass and design. Evacuated glass tubes, such as photomultiplier tubes, are exposed to temperatures as high as 720 K. In some applications, glass has been displaced by high alumina ceramic. Borosilicate glasses, in the form of industrial glass pipe, have been used in large experimental systems which frequently require demountable joints.

Industrial glass pipe and other wall materials are connected to other materials with demountable seals made of elastomers or soft metal wire, eg, lead, indium, tin and their alloys. The soft-wire seals, however, have led to difficulties. High temperature bake-out must be avoided, and glow-discharge cleaning must not reach the gasket. Furthermore, the gasket material can creep under load and admit a leak, except in the case of indium which creeps but does not usually leak (see also Packing materials.)

The choice of metals for vacuum walls is largely based upon the ease of fabrication of the metal, machining, cleaning (26), welding, etc; aluminum alloys are the material of choice for out-gassing at room temperature.

Demountable joints are commercially available in great variety in stainless steel, but not in aluminum alloy or related materials. Experimental joints have been made with aluminum flanges employing aluminum foil as gasket material.

Vacuum Brazing Furnace. The cross sections of a vacuum brazing furnace of the bell-jar type are shown in Figures 15 and 16. Contamination is very low even when rapidly cycled from one work load to the next. The bell jar can serve on one hearth, while another hearth is being loaded for the next operation.

Rapid cooldown by helium heat transfer is made possible at an interior of ca 100 Pa (0.1 atm); a convective fan transfers heat efficiently from the interior hot surfaces of the furnace to water-cooled base and wall parts.

During start-up, the base-plate heat shields are lowered, thereby allowing pump-down through a large gap. The heat shields can then be raised to produce a narrow gap when maximum temperature is required.

Offset design gives ready access along the axis of the hot zone. This design permits routine operation and cycling of the furnace without sacrificing control of contamination, access, and speed for condensables or noncondensables.

An optimized relationship is obtained between the bell jar, 60° swing-leaf valve,

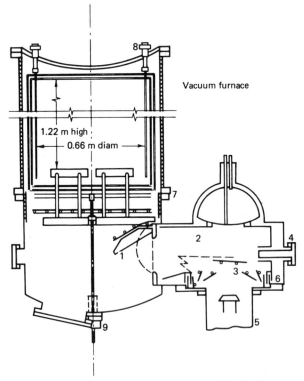

Figure 15. A low-contamination vacuum brazing furnace for brazing large metal-to-ceramic joints. 1, Valve, open and shut (- - -); 2, LN trap; 3, water-cooled baffle; and 4, valve-actuator feedthroughs for lowering the bottom heat shields; 5, diffusion pump oil, 1.25 cm dia; 6, creep barrier; 7, flange separation for opening the furnace; 8, electrical feedthroughs for lowering the bottom heat shields; 9, seal for raising and lowering of the furnace (44).

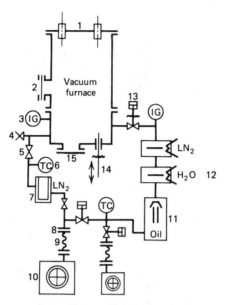

Figure 16.   Schematic representation of the vacuum furnace shown in Figure 15. The graphic symbols used are among those contained in the American Vacuum Society Standard 7.1. 1, Electrical feedthrough; 2, viewport; 3, vacuum-gauge hot-filament ionization; 4, air valve; 5, valve; 6, thermocouple vacuum gauge; 7, thimble trap; 8, demountable coupling; 9, flexible line; 10, two-stage liquid-sealed mechanical pump; 11, DP; 12, water-cooled baffle; 13, pneumatic valve with sealed bellows; 14, linear motion feedthrough with sealed bellows; and 15, blind-flange port (45).

LN trap, baffle for the oil, and the plane of action for the diffusion pump (DP) top jet. The valve open area equals 0.38 of the cross-sectional area of the inside diameter of the furnace. The volumetric speed factor for water vapor is thus $0.38 \times 0.9 \simeq 0.34$ (where 0.9 = Clausing factor).

No gas pips occur during the filling of the LN trap, and the temperature of the trap surfaces does not vary more than 0.01 K with the liquid-nitrogen level variation in the reservoir. The high conductance water-cooled baffle provides a minimum restriction of flow to the DP yet returns back-streamed pump oil along the walls of the pump attached at its room temperature edge only. A thin stainless-steel anticreep barrier, cooled by radiation to the LN-chilled annulus, prevents surface migration of pump oil to the furnace. The bottom heat shields may be lowered by a rod during the warming of the furnace to provide high conductance from the hot zone to the trap inlet. A hot-zone temperature of >1500 K has been achieved.

Liquid-Nitrogen Traps.   The principal reason that cold traps are frequently ineffective in preventing the passage of oil or mercury is the warming of the trap and its internal filling lines when LN is added and/or as LN depletes; however, some designs eliminate this problem (Fig. 15). A liquid-nitrogen trap need not have an active chemical surface, but in some cases, it is advantageous to cool a chemically active surface to liquid-nitrogen temperature in order to obtain effective trapping of methane and other low molecular weight species. This method, however, has a predictable finite life determined by the formation of monolayers on the chemically active surface. A plain metal surface cooled to liquid-nitrogen temperature in a trap has two primary functions, namely, to act as a cryopump for water vapor and to prevent contamination from DP working fluid from reaching a given vacuum environment. A well-designed LN trap can provide a pumping speed of at least 10^2 m^3/s per m^2 pf system entrance area for water vapor (at room temperature, under free molecular conditions) and confine oil contamination to negligible levels. Measurements have shown that oil molecules and cracked-oil products with molecular weights >60 pass through a particular design at rate <10^9 molecules/(m^2·s) (23). These rates correspond to a vapor pressure at room temperature of less than ca 0.1 pPa (7.5 × 10^{-16} mm Hg).

Typically, a baffle condenses a vapor flow to a liquid in such a way that liquid can drain off for recycle. A well-designed and useful trap, on the other hand, catches and retains condensables such as water and the higher vapor-pressure fraction of the working fluid of a pump. Although the terms baffle and trap seem to describe the same type of function, namely, to slow down but not prevent the rate at which some species of molecules pass through, traps can completely stop some molecular species (23).

Molecules arrive at the surfaces of traps and baffles by volume flow and surface creep. Molecules are trapped in vacuum systems by binding with energies much greater than kT of the surface, or by lowering the temperature of the surface in such a way that kT is less than the heat of physisorption of a molecular species on a surface (where k is Boltzmann's constant and T the absolute temperature).

If a trap is placed directly over a DP top jet (Fig. 4) without baffling action, all of the working fluid is caught by the trap in the time predicted by the rate at which the fringe of the top jets feed the working fluid to the trap. This rate may be many orders of magnitude greater than the rate predicted from the equilibrium vapor pressure of the working fluid at DP wall temperature.

Valve, trap, and baffle can be combined in such a manner that elastomers can be used in the valve and contamination is controlled from the valve-actuator mechanisms and from the gasket of the valve-plate seal (Fig. 15) (44).

The role, design, and maintenance of creep-proof barriers in traps, especially those in oil DPs, remain to be fully explored. In general, uncracked oil from a DP is completely inhibited from creeping by a surface temperature <223 K. On the other hand, a cold trap, to perform effectively in an ordinary vacuum system, must be <173 K because of the vapor pressure of water, and ≤78 K because of the vapor pressure of CO_2. For ultracontrolled vacuum environments, LN temperature or lower is required; CO_2 accumulation on the trap surface must be less than one monolayer.

The effectiveness of a LN trap can be observed by the absence of pressure pips on an ionization gauge when LN is replenished in the reservoir.

The author is in general agreement with the view that DP systems can be shut down when they are not in use to conserve energy. If a liquid-nitrogen trap is incorporated, the manner in which this trap is warmed up and the DP is cooled down should be determined by the presence or absence of a valve between the chamber and the liquid-nitrogen trap. In critical systems, this head valve can be included in order to permit rapid shutdown and rapid return to operation. The assertion that dry nitrogen gas can be used to sweep contamination from traps and pumps in such a manner that oil contamination is prevented from running counter to the nitrogen-sweeping flow direction is questionable. Proper placement of valves can eliminate the need of a sweep gas.

Process Equipment

An up-to-date survey of process equipment widely used in chemical engineering today is given in ref. 46.

The pressure in the gaseous phase characterizes the likely vacuum process environment. Rough vacuum ranges from 101 kPa to ca 100 Pa (760 to 0.75 mm Hg). Medium vacuum ranges from 100 Pa to ca 0.1 Pa (0.75 to 0.00075 mm Hg). The chemical engineer is likely to work in the rough vacuum region in which distillation, evaporation, drying, and filtration are normally conducted. The medium vacuum range is employed in molten-metal degassing, molecular distillation, and freeze drying.

Vacuum equipment requires strength to withstand the pressure of the surrounding atmosphere; the full load is ca 101.3 kPa when the internal gas pressure in the system is sufficiently reduced.

Steam Ejectors. Ejectors are simple vacuum pumps. They have no moving parts but accomplish compression through fluid-momentum transfer. A high pressure motive fluid enters the ejector and expands through the converging and diverging section of the nozzle. The initial pressure energy is converted to velocity. This increase in velocity entrains the load to be handled, which enters through the suction inlet. The motive and suction fluids are then mixed and recompressed through the defuser which discharges this mixture into a higher intermediate pressure. By convention, an ejector represents a single-point design and is most efficient at a single set of conditions. Ejectors are either single-stage or multistage, depending on the suction pressures required. A practical compression ratio for single-stage ejectors is ca 6:1, when discharging to atmospheric pressure. This compression ratio can be as high as 10^6:1 for six-stage ejectors. Multistage steam ejectors are usually equipped with direct-contact or surface-type condensors between stages in order to prevent contamination of cooling fluid. When designing a multistage steam ejector, condensables and noncondensables in the intake must be distinguished. Booster stages followed by a condensor are used to handle large quantities of condensables. Intercondensers are sized for both the motive steam from the booster and the process vapors. Thus, the load to the ejector stages downstream is significantly reduced. Such an ejector system uses considerably less steam, and purchase and installation costs are lower than those of a system designed to handle process vapors as condensables.

Liquid-Ring Pumps. In a liquid-ring pump, the rotor is the only moving part. The liquid ring performs all the functions normally done by mechanical pistons or vanes. There are a number of variations of this design, but all operate on the principle that before start-up the pump casing is partially filled with a sealant liquid. When the rotating impeller is turned on, the liquid is caused to centrifugally contact the periphery of the casing. Thus, a liquid ring is formed that seals off the cylindrical pump body. Since the rotor axis is offset from the body axis, a piston action is established as the liquid fills and then almost empties each of the chambers between the rotor blades. Compression ratios as high as 10:1 can be achieved when discharging to atmospheric pressure for a single-stage pump. It is apparently difficult to rate a liquid-ring pump on the basis of swept volume, and it is classed as an isothermal machine rather than a true positive displacement compressor. The liquid ring acts as a heat sink to maintain constant-temperature operation. Assuming a normal sealing-liquid flow, a temperature rise in compressing air from ca 26.6 to 101.3 kPa (ca 0.25 to 1 atm) is ca 3–6 K. For comparison, an uncooled adiabatic compressor would cause a temperature rise exceeding ca 90 K. In sizing the pump, both evaporative and condensing effects must be considered. Evaporative cooling takes place whenever dry gases are introduced at temperatures higher than those of the sealed liquid. Condensation occurs when pumping gas that is saturated with vapor, ie, the pump, behaves like a direct-contact condensor.

Rotary-Piston Pumps. Positive-displacement, oil-sealed machines that isolate a specific volume of gas with each revolution and compress and exhaust it to the atmosphere, are the rotary-piston pumps. The oil-sealed piston revolves and traps the aspirated gas ahead of it by closing the inlet port. The gas is compressed, the discharge valve opens, and the gas is exhausted to the atmosphere. Compression ratios can be as high as more than 10^6:1 for a single-stage pump. These pumps operate in an internal oil bath which lubricates the pump and seals against backstreaming from the exhaust into the intake. The rate of flow and the distribution of oil through the pump are the important features of the design. The piston must be sufficiently lubricated or it fails. Failures are often caused by breakdowns of the oil-distribution systems.

Rotary-piston pumps are frequently stalled by condensation of process vapors in the lubricating fluid. In addition, condensate can accumulate in the pump oil, resulting in mechanical failure and permanent damage. Water vapor is almost always present in vacuum-processing operations and is a persistent source of oil contamination. Higher alcohols and other solvents normally encountered also have a tendency to condense in the lubricating oil during compression. A number of techniques have been developed to prevent condensation of process vapors in oil-sealed pumps. Gas ballast, the most common technique, involves drilling a hole in the head of the pump to admit air or other gas into the cylinder during the latter portion of the compression stroke. Ballasting takes effect while the gas being compressed is sealed off from the intake by the piston. The ballasting method reduces the partial pressure of the condensible vapor and thus reduces or eliminates condensation within the oil. The introduction of gas-ballast air into the pump increases the pressure differential across the seals between intake and exhaust. Ballasting results in increased leakage past the seals, which can significantly reduce the capacity of rotary-piston pumps operating below ca 100 Pa (ca 0.1 atm). However, the effect on pump capacity is negligible in most processing applications. Below the processing range, however, ballasting when handling the water in ordinary atmospheric air is not needed.

Rotary-Vane Pumps. Rotary-vane pumps are positive-displacement machines with spring-loaded vanes that contact the inside of the pump casing. Gas entering the pump is trapped between adjacent blades, compressed, and forced out to the atmosphere through the discharge point. Maintenance requirements have severely limited the use of these pumps in process applications. Rotary-vane pumps are still found in laboratory applications; they can achieve compression ratios well above 10^6:1 for discharge directly into the atmosphere. Oil contamination by process fluids causes a deterioration in the pump performance and can necessitate frequent cleaning and reassembly of the pump.

Rotary-Blower Pumps. This type of pump employs two interlocking rotors to trap and compress gases. The rotors are prevented from touching one another, and there is no sealing liquid in the pump. The gears and rotor bearings are lubricated with oil, but they are external to the rotors. Clearance between the rotors is generally 25–100 μm. Typically, these pumps operate at high speeds of 3000–4000 rpm. Because there is no positive seal between the rotors, the rotary blower is limited to small compression ratios but can be designed for higher throughput than any other mechanical pump. In process and most other applications, the rotary blower is limited to operation in conjunction with other mechanical pumps. Inherently, rotary blowers are potentially subject to overheating, because of the lack of a discharge valve separating the heated gas. Because of overheating, the compression ratio of single-stage blowers is limited to ca 2.3:1. However, if the rotary blower discharges into a rotary-piston oil-sealed pump, the combination can exceed 10^6:1. When operating below 133 Pa (1 mm Hg), overheating need not be a consideration because the work done in compressing the process load is small.

Economic Aspects

Up-to-date worldwide dollar volume sales and profits are not available. In the United States, total sales of high vacuum components, supplies, instruments, and systems amounted to ca 505×10^6 in 1980 and 504×10^6 in 1981. Systems, largely for the semiconductor industry, are the main source of sales (see Semiconductors). The sales of all vacuum equipment, pumps, valves, sensors, etc, in the United States, including applications not in vacuum systems, are much higher.

A reasonably comprehensive list of high vacuum manufacturers is supplied by the American Vacuum Society's Exhibitor's list. In Europe, a special issue of the journal *Vacuum* serves similarly.

Capital investment, capital costs, operating costs, return on investment, and energy conservation are discussed in ref. 6. In the economic analysis, the speed of each type of pump considered is normalized to 1 m^3/s as a common basis.

With regard to fixed amortized investment (47), utilization of the wall area of the chamber to be evacuated with a given pumping method must be considered. Depending upon the process, it may or may not be possible to expose gettering or cryogenic pumping surfaces directly. Many uses would contribute dynamically to a gettering or cryogenic surface, making it uneconomic to handle the energy flux, sputtering processes, etc, which direct exposure might entail (see also Cryogenics).

Energy costs are not directly related to the energy efficiency of the process (6,44). Even if the thermal efficiency of a steam ejectory, for example, is less than that of mechanical equipment run by an electric motor, the overall cost of the energy to run the steam ejector may be less.

BIBLIOGRAPHY

"Vacuum Technique" in *ECT* 1st ed., Vol. 14, pp. 503–536, by B. B. Dayton, Consolidated Vacuum Corporation; "Vacuum Technology" in *ECT* 2nd ed., Vol. 21, pp. 123–157, by B. B. Dayton, The Bendix Corporation.

1. N. Milleron in F. J. Clauss, ed., *Surface Effects on Space Craft Materials, 1st Symposium 1959*, John Wiley & Sons, Inc., New York, 1960, pp. 260, 303, 325–342.
2. N. Milleron in J. A. Dillon, Jr., and V. J. Harwood, eds., *Experimental Vacuum Science and Technology*, Marcel Dekker, Inc., New York, 1973.
3. J. D. Dow, R. E. Allen, O. F. Sankey, J. P. Buisson, and H. P. Hjalmarson, *J. Vac. Sci. Technol.* **19,** 502 (1981).
4. P. D. Kirchner, J. M. Woodall, J. L. Freeouf, D. J. Wolford, and G. D. Pettit, *J. Vac. Sci. Technol.* **19,** 604 (1981).
5. G. A. Bird, *Molecular Gas Dynamics*, Oxford University Press, Inc., Clarendon, Oxford, UK, 1976.
6. J. F. O'Hanlon, *A User's Guide to Vacuum Technology*, Wiley-Interscience, New York, 1980.

7. L. C. Beavis, V. J. Harwood, and M. T. Thomas, *Vacuum Hazards Manual*, American Vacuum Society, New York, 1975.

8. P. Holloway, *Vacuum Book Bibliography*, American Vacuum Society, New York, 1982.

9. M. S. Kaminsky and J. J. Lafferty, *Diction of Terms for Vacuum Science and Technology, Surface Science, Thin Film Technology, Vacuum Metallurgy and Electronic Materials*, American Vacuum Society, New York, 1980.

10. J. L. Vossen, *Bibliography on Metallization Materials and Techniques for Silicon Devices*, Vols. 6 (1980), 7 (1981), and 8 (1982), American Vacuum Society, New York.

11. E. Grant, *Much Ado About Nothing: Theories of Space and Vacuum from the Middle Ages to the Scientific Revolution*, Cambridge University Press, New York, 1981.

12. G. L. Weissler and R. W. Carlson, eds., *Vacuum Physics and Technology*, Vol. 14 of Marton, ed., *Methods of Experimental Physics*, Academic Press Inc., New York, 1979, p. 4.

13. P. A. Redhead, J. P. Hobson, and E. V. Kornelsen, *The Physical Basis of Ultra High Vacuum*, Chapman and Hall, London, 1968.

14. R. V. Latham, *High Voltage Insulation*, Academic Press, Inc., New York, 1981.

15. J. M. Lafferty, ed., *Vacuum Arcs Theory and Application*, John Wiley & Sons, Inc., New York, 1980.

16. R. Hackam and L. Altcheh, *J. Appl. Phys.* **46**(2), 631 (1975).

17. C. A. Spindt, *Development Program on a Cold Cathode Electron Gun*, NASA CR 159570, Stanford Research Institute, Menlo Park, Calif., 1979, pp. 37–38.

18. I. J. Scialdone, *NASA TN D-7250*, Goddard Space Flight Center, Greenbelt, Md., 1972.

19. N. Milleron in R. L. Chuan, *Res. Dev.* **11a**, 44 (Jan. 1964).

20. T. E. Graedel, *Chemical Compounds in the Atmosphere*, Academic Press, Inc., New York, 1978.

21. A. D. Zimon, *Adhesion of Dust and Powder*, Consultants Bureau, New York and London, 1982.

22. Ref. 12, p. 425.

23. N. Milleron, *IEEE Trans. Nucl. Sci.* **14**(3), 794 (1967).

24. Ref. 12, p. 143.

25. N. G. Wilson and L. C. Beavis, *Handbook of Vacuum Leak Detection*, American Vacuum Society, New York, 1976.

26. D. J. Mattox, *Surface Cleaning in Thin Film Technology*, American Vacuum Society, New York, 1975.

27. Ref. 23, p. 800.

28. Ref. 23, p. 801.

29. Ref. 12, Vol. 14.

30. L. C. Beavis, *J. Vac. Sci. Technol.* **20**, 972 (1982).

31. D. S. Miller, *Internal Flow*, Cranfield British Hydro. Mechanical Research Association, 1971.

32. E. M. Sparrow and V. K. Jansson, *AIAA J.* **1**(5), 1081 (1963).

33. D. Davis, L. L. Levenson, and N. Milleron, *J. Appl. Phys.* **35**, 529 (1964).

34. D. H. Davis, L. L. Levenson, and N. Milleron in L. Talbot, ed., *Rarified Gas Dynamics*, Academic Press, Inc., New York, 1961, p. 99.

35. Ref. 12, p. 19.

36. L. L. Levenson, N. Milleron, and D. H. Davis, *Transactions of the American Vacuum Society*, Pergamon Press, Inc., Elmsford, N.Y., 1961.

37. F. O. Goodman and H. Y. Wachman, *Dynamics of Gas Surface Scattering*, Academic Press, Inc., New York, 1976.

38. R. H. Edwards, *Low Density Flows through Tubes and Nozzles*, Vol. 51, Pt. 1 of J. K. Potter, ed., *Progress in Astronautics and Aeronautics*, American Institute of Aeronautics and Astronomy, New York, 1977.

39. D. H. Davis, *J. Appl. Phys.* **31**, 1169 (1960).

40. N. Milleron and D. N. Frank, *J. Vac. Sci. Technol.* **20**, 1052 (Apr. 1982).

41. Ref. 12, p. 71.

42. Ref. 12, p. 219.

43. M. J. Drinkwine and D. Lichtman, *Partial Pressure Analysis*, American Vacuum Society, New York, 1977.

44. Ref. 12, p. 277.

45. Ref. 12, p. 278.

46. J. L. Ryans and S. Croll, *Chem. Eng.*, 73 (Dec. 14, 1981).

47. E. P. De Garmo, J. R. Canada, and W. G. Sullivan, *Engineering Economy*, 6th ed., Macmillan-Collier, New York, 1979.

NORMAN MILLERON
EMR Photoelectric

VANADIUM AND VANADIUM ALLOYS

Vanadium

Vanadium [*7440-62-2*] (V, at no. 23, at wt 50.942) is a member of both group VB of the periodic system and the first transition series. It is a gray, body-centered-cubic metal; in its high purity form, it is very soft and ductile. Because of its high melting point, it is referred to as a refractory metal, like niobium, tantalum, chromium, molybdenum, and tungsten. The principal use of vanadium is as an alloying addition to iron (qv) and steel (qv), particularly in high strength steels and, to a lesser extent, in tool steels and castings. It is also an important beta stabilizer for titanium alloys. Interest in the intermetallic compound V_3Ga [*12024-15-6*] for superconductor applications could lead to expanded use in the future (see Superconducting materials). During the 1970s, vanadium alloys were considered for use as cladding material for the fuel in liquid-metal-cooled fast reactors. However, most development programs involving vanadium for this purpose have been reduced because of insufficient funding.

Vanadium was first discovered in 1801 by del Rio while he was examining a lead ore obtained from Zimapan, Mexico. The ore contained a new element and, because of the red color imparted to its salts on heating, it was named erythronium (redness). The identification of the element vanadium did not occur until 1830 when it was isolated from cast iron processed from an ore from mines near Taberg, Sweden. It was given the name vanadium after Vanadis, the Norse goddess of beauty. Shortly after this discovery, vanadium was shown to be identical to the erythronium that del Rio had found several years earlier.

Occurrence. Vanadium is widely distributed throughout the earth but in low abundance, ranking 22 among the elements of the earth's crust. The lithosphere contains ca 0.07 wt % vanadium and few deposits contain more than 1–2 wt %. Vanadium occurs in uranium-bearing minerals of Colorado, in the copper, lead, and zinc vanadates of Africa, and with certain phosphatic shales and phosphate rocks in the western United States. It is a constituent of titaniferous magnetites, which are widely distributed with large deposits in the USSR, South Africa, Finland, the People's Republic of China, eastern and western United States, and Australia. At one time, the largest and most important vanadium deposits were the sulfide and vanadate ores from the Peruvian Andes, but these are depleted. Most of the vanadium reserves are in deposits in which the vanadium would be a by-product or coproduct with other minerals, including iron, titanium, phosphate, and petroleum.

Trace amounts of vanadium have been found in meteorites and seawater, and it has been identified in the spectrum of many stars including the earth's sun. The occurrence of vanadium in oak and beech trees and some forms of aquatic sea life indicates its biological importance.

There are over 65 known vanadium-bearing minerals, some of the more important of which are listed in Table 1. Patronite, bravoite, sulvanite, davidite, and roscoelite are classified as primary minerals, whereas all of the others are secondary products which form in the oxidizing zone of the upper lithosphere. The carnotite and roscoelite ores in the sandstones of the Colorado Plateau have been important sources of vanadium as well as of uranium.

The metallic vanadates of lead, copper, and zinc, which occur in Namibia (Southwest Africa) and Zambia, are also a large resource of vanadium-bearing ores, as are the phosphatic shales and rocks of the phosphoria formation in Idaho and Wyoming. Vanadium salts are obtained as by-products of the phosphoric acid and fertilizer industries (see Fertilizers). Large reserves of vanadium in Arkansas and Canada and the titaniferous magnetite ores will probably become increasingly significant as sources of production. Certain petroleum crude oils, especially those from South America, contain varying amounts of vanadium compounds. These accrue as fly ash or boiler residues upon combustion of the crude oils and they can be reclaimed (see Air pollution control methods).

Table 1. Important Minerals of Vanadium

Mineral	CAS Registry No.	Color	Formula	Location
patronite	[12188-60-2]	greenish black	$V_2S + nS$	Peru
bravoite	[12172-92-8]	brass	$(Fe,Ni,V)S_2$	Peru
sulvanite	[15117-74-5]	bronze-yellow	$3Cu_2S.V_2S_6$	Australia, United States (Utah)
davidite	[12173-20-5]	black	titanate of Fe, U, V, Cr, and rare earths	Australia
roscoelite	[12271-44-2]	brown	$2K_2O.2Al_2O_3(Mg,Fe)O.-3V_2O_5.10SiO_2.4H_2O$	United States (Colorado and Utah)
carnotite	[1318-26-9]	yellow	$K_2O.2U_2O_3, V_2O_5.3H_2O$	southwest United States
vanadinite	[1307-08-0]	reddish brown	$Pb_5(VO_4)_3Cl$	Mexico, United States, Argentina
descloizite	[19004-61-6]	cherry-red	$4(Cu,Pb,Zn)O.V_2O_5.H_2O$	Namibia, Mexico, United States
cuprodescloizite	[12325-36-9]	greenish brown	$5(Cu,Pb)O.(V,As)_2O_5.2H_2O$	Namibia
vanadiferous phosphate rock			$Ca_5(PO_4)_3(F,Cl,OH)$; VO_4 ions replace some of the PO_4 ions	United States (Montana)
titaniferous magnetite			$FeO.TiO_2–FeO.(Fe,V)O_2$	USSR, People's Republic of China, Finland, Union of South Africa

Physical Properties. Vanadium is a soft, ductile metal in pure form, but it is hardened and embrittled by oxygen, nitrogen, carbon, and hydrogen (1–2). Selected metal additions lead to higher strength alloys which maintain a reasonable level of ductility (3). Its thermal conductivity is significantly lower than that of copper. Important physical properties of vanadium are listed in Table 2. Some of these properties depend upon the purity of the material used for the determinations. Although a purity level of 99.99% has been achieved experimentally, such high purity material has not been used for all of the determinations listed (7).

Chemical Properties. Vanadium has oxidation states of +2, +3, +4, and +5. When heated in air at different temperatures, it oxidizes to a brownish black trioxide, a blue-black tetroxide, or a reddish orange pentoxide. It reacts readily with chlorine at fairly low temperatures (180°C) forming VCl_4 and with carbon and nitrogen at high

Table 2. Physical Properties of Vanadium Metal[a]

Property	Value
melting point, °C	1890 ± 10
boiling point, °C	3380
vapor pressure (at 1393–1609°C), kPa[b]	$R \ln P = \dfrac{121{,}950}{T} - 5.123 \times 10^{-4}\,T + 38.3$
crystal structure	bcc
lattice constant, nm	0.3026
density, g/cm³	6.11
specific heat (at 20–100°C), J/g[c]	0.50
latent heat of fusion, kJ/mol[c]	16.02
latent heat of vaporization, kJ/mol[c]	458.6
enthalpy (at 25°C), kJ/mol[c]	5.27
entropy (at 25°C), kJ/(mol·°C)[c]	29.5
thermal conductivity (at 100°C), W/(cm·K)	0.31
electrical resistance (at 20°C), $\mu\Omega$·cm	24.8–26.0
temperature coefficient of resistance (at 0–100°C), ($\mu\Omega$·cm)/°C	0.0034
magnetic susceptibility, m³/mol[d]	0.11
superconductivity transition, K	5.13
coefficient of linear thermal expansion, °C⁻¹	
at 20–720°C (x ray)	$(9.7 \pm 0.3) \times 10^{-6}$
at 200–1000°C (dilatometer)	8.95×10^{-6}
thermal expansion (at 23–100°C), $\times 10^{-6}$/°C	8.3
recrystallization temperature, °C	800–1000
modulus of elasticity, MPa[e]	$(1.2–1.3) \times 10^5$
shear modulus, MPa[e]	4.64×10^4
Poisson ratio	0.36
thermal neutron absorption, m²/at.[f]	$(4.7 \pm 0.02) \times 10^{-28}$
capture cross section for fast (1 MeV) neutrons, m²/at.[f]	3×10^{-31}

[a] Refs. 4–6.
[b] To convert kPa to atm, divide by 101.3. In Antoine equation, R = gas constant, T = K, and P = pressure (kPa).
[c] To convert J to cal, divide by 4.184.
[d] To convert m³/mol to cgs units, multiply by $4\,\pi \times 10^{-6}$.
[e] To convert MPa to psi, multiply by 145.
[f] To convert m² to barns, multiply by 1×10^{28}.

temperatures forming VC and VN, respectively. The pure metal in massive form is relatively inert toward oxygen, nitrogen, and hydrogen at room temperature.

Vanadium is resistant to attack by hydrochloric or dilute sulfuric acid and to alkali solutions. It is also quite resistant to corrosion by seawater but is reactive toward nitric, hydrofluoric, or concentrated sulfuric acids. Galvanic corrosion tests run in simulated seawater indicate that vanadium is anodic with respect to stainless steel and copper but cathodic to aluminum and magnesium. Vanadium exhibits corrosion resistance to liquid metals, eg, bismuth and low oxygen sodium.

Manufacture

Ore Processing.   Vanadium is recovered domestically as a principal mine product, as a coproduct or by-product from uranium–vanadium ores, and from ferrophosphorus as a by-product in the production of elemental phosphorus. In Canada, it is recovered from crude-oil residues and in the Republic of South Africa as a by-product of titaniferous magnetite. Whatever the source, however, the first stage in ore processing is the production of an oxide concentrate.

The principal vanadium-bearing ores are generally crushed, ground, screened, and mixed with a sodium salt, eg, NaCl or Na_2CO_3. This mixture is roasted at ca 850°C and the oxides are converted to water-soluble sodium metavanadate, $NaVO_3$. The vanadium is extracted by leaching with water and precipitates at pH 2–3 as sodium hexavanadate, $Na_4V_6O_{17}$, a red cake, by the addition of sulfuric acid. This is then fused at 700°C to yield a dense black product which is sold as technical-grade vanadium pentoxide. This product contains a minimum of 86 wt % V_2O_5 and a maximum of 6–10 wt % Na_2O.

The red cake can be further purified by dissolving it in an aqueous solution of Na_2CO_3. The iron, aluminum, and silicon impurities precipitate from the solution upon pH adjustment. Ammonium metavanadate then precipitates upon the addition of NH_4Cl and is calcined to give vanadium pentoxide of greater than 99.8% purity.

Vanadium and uranium are extracted from carnotite by direct leaching of the raw ore with sulfuric acid. An alternative method is roasting the ore followed by successive leaching with H_2O and dilute HCl or H_2SO_4. In some cases, the first leach is with a Na_2CO_3 solution. The uranium and vanadium are then separated from the pregnant liquor by liquid–liquid extraction techniques involving careful control of the oxidation states and pH during extraction and stripping.

In the Republic of South Africa, the recovery of high vanadium slags from titaniferous magnetites has been achieved on a large scale (8). The ore, containing about 1.75 wt % V_2O_5, is partially reduced with coal in large rotary kilns. The hot ore is then fed to an enclosed, submerged-arc electric smelting furnace which produces a slag containing substantial amounts of titania and pig iron containing most of the vanadium that was in the ore. After tapping from the furnace and separation of the waste slag, the molten pig iron is blown with oxygen to form a slag containing up to 25 wt % V_2O_5. The slag is separated from the metal and may then be used as a high grade raw material in the usual roast–leach process.

Solvent extraction following roasting and leaching is a promising processing method for dolomitic shale from Nevada (9).

Ferrovanadium. The steel industry accounts for the majority of the world's consumption of vanadium as an additive to steel. It is added in the steelmaking process as a ferrovanadium alloy [12604-58-9], which is produced commercially by the reduction of vanadium ore, slag, or technical-grade oxide with carbon, ferrosilicon, or aluminum. The product grades, which may contain 35–80 wt % vanadium, are classified according to their vanadium content. The consumer use and grade desired dictate the choice of reductant.

Carbon Reduction. The production of ferrovanadium by reduction of vanadium concentrates with carbon has been supplanted by other methods in recent years. An important development has been the use of vanadium carbide as a replacement for ferrovanadium as the vanadium additive in steelmaking. A product containing ca 85 wt % vanadium, 12 wt % carbon, and 2 wt % iron is produced by the solid-state reduction of vanadium oxide with carbon in a vacuum furnace.

Silicon Reduction. The preparation of ferrovanadium by the reduction of vanadium concentrates with ferrosilicon has been used but not extensively. It involves a two-stage process in which technical-grade vanadium pentoxide, ferrosilicon, lime, and fluorspar are heated in an electric furnace to reduce the oxide; an iron alloy containing ca 30 wt % vanadium but undesirable amounts of silicon is produced. The silicon content of the alloy is then decreased by the addition of more V_2O_5 and lime to effect the extraction of most of the silicon into the slag phase. An alternative process involves the formation of a vanadium–silicon alloy by the reaction of V_2O_5, silica, and coke in the presence of a flux in an arc furnace. The primary metal then reacts with V_2O_5 yielding ferrovanadium.

A silicon process has been developed by the Foote Mineral Company and has been used commercially to produce tonnage quantities of ferrovanadium (10). A vanadium silicide alloy containing less than 20 wt % silicon is produced in a submerged-arc electric furnace by reaction of vanadium-bearing slags with silica, flux, and a carbonaceous reducer followed by refinement with vanadium oxide. This then reacts with a molten vanadiferous slag in the presence of lime yielding a ferrovanadium alloy (Solvan) containing ca 28 wt % vanadium, 3.5 wt % silicon, 3.8 wt % manganese, 2.8 wt % chromium, 1.25 wt % nickel, 0.1 wt % carbon, and the remainder iron. A unique feature of this process is its applicability to the pyrometallurgical process of vanadium-bearing slags of the type described in the preceding section.

Aluminum Reduction. The aluminothermic process for preparing a ferrovanadium alloy differs from the carbon and silicon reduction processes in that the reaction is highly exothermic. A mixture of technical-grade vanadium oxide, aluminum, iron scrap, and a flux are charged into an electric furnace and the reaction between aluminum and vanadium pentoxide is initiated by the arc. The temperature of the reaction is controlled by adjusting the size of the particles and the feed rate of the charge by using partially reduced material or by replacing some of the aluminum with a milder reductant, eg, calcium carbide, silicon, or carbon. Ferrovanadium containing as much as 80 wt % vanadium is produced in this way.

Ferrovanadium can also be prepared by the thermite reaction, in which vanadium and iron oxides are co-reduced by aluminum granules in a magnesite-lined steel vessel or in a water-cooled copper crucible (11) (see Aluminum and aluminum alloys). The reaction is initiated by a barium peroxide–aluminum ignition charge. This method is also used to prepare vanadium–aluminum master alloys for the titanium industry.

Pure Vanadium.   Vanadium, like its sister group VB elements, dissolves significant quantities of oxygen, nitrogen, hydrogen, and carbon interstitially into its lattice. In so doing, a severe loss of ductility results. Formation of a ductile pure metal or alloy requires that contamination by these elements is carefully controlled. Generally, small quantities of some or all of these elements, particularly oxygen, are tolerated in a compromise between increased strength and loss of ductility. The production method for pure vanadium and vanadium-based alloys must be tailored with this need for high purity and for economic and engineering considerations in mind. Most reduction processes suffer from inabilities to reduce the amount of all impurities simultaneously to the desired level. Consequently, one or more purification methods are used to overcome the limitations of the original reduction.

Vanadium metal can be prepared either by the reduction of vanadium chloride with hydrogen or magnesium or by the reduction of vanadium oxide with calcium, aluminum, or carbon. The oldest and most commonly used method for producing vanadium metal on a commercial scale is the reduction of V_2O_5 with calcium. Recently, a two-step process involving the aluminothermic reduction of vanadium oxide combined with electron-beam melting has been developed. This method makes possible the production of a purer grade of vanadium metal, ie, of the quality required for nuclear reactors (qv).

Calcium Reduction.   High purity vanadium pentoxide is reduced with calcium to produce vanadium metal of ca 99.5% purity. The exothermic reaction is carried out adiabatically in a sealed vessel or bomb. In the original process, calcium chloride was added as a flux for the CaO slag (12). The vanadium metal was recovered in the form of droplets or beads. A massive ingot or regulus has been obtained by replacing the calcium chloride flux with iodine (13). This latter reaction became the basis of the first large-scale commercial process for producing vanadium. The reaction is initiated either by preheating the charged bomb or by internal heating with a fuse wire embedded in the charge. Calcium iodide formed by the reaction of calcium with iodine serves both as a flux and as a thermal booster. Thus, sufficient heat is generated by the combined reactions to yield liquid metal and slag products. The resulting metal contains ca 0.2 wt % carbon, 0.02–0.8 wt % oxygen, 0.01–0.05 wt % nitrogen, and 0.002–0.01 wt % hydrogen. Two factors that contribute to the relative inefficiency of this process are the rather low metal yields (75–80%) and the required amount of calcium reductant (50–60% excess of stoichiometric quantity).

Vanadium powder can be prepared by substituting V_2O_3 for the V_2O_5 as the vanadium source. The heat generated during the reduction of the trioxide is considerably less than for the pentoxide, so that only solid products are obtained. The powder is recovered from the product by leaching the slag with dilute acid.

Aluminothermic Process.   In the development of the liquid-metal fast-breeder reactor, vanadium has been considered for use as a fuel-element cladding material (see Nuclear reactors, fast-breeder reactors). Difficulty was encountered in the fabrication of alloys prepared from the calcium-reduced metal, a factor attributable to the high interstitial impurity content. An aluminothermic process was developed by the AEC (now the NRC) in order to meet the more stringent purity requirements for this application (14). In this process, vanadium pentoxide reacts with high purity aluminum in a bomb to form a massive vanadium–aluminum alloy. Use of proprietary additions to either increase the reaction temperature, decrease the melting point of the slag or metal, or increase the fluidity of the two phases leads to formation of a solid metallic regulus that is relatively free of slag. The alloyed aluminum and dissolved oxygen are subsequently removed in a high temperature, high vacuum processing step to yield metal of greater than 99.9% purity.

Purified V_2O_5 powder and high purity aluminum granules are charged into an alumina-lined steel crucible. The vessel is flushed with an inert gas to minimize atmospheric contamination and then is sealed. The reaction is initiated by a vanadium fuse wire. Sufficient heat is generated by the chemical reaction to produce a molten alloy of vanadium containing ca 15 wt % aluminum; a fused aluminum oxide slag also forms. The liquid alloy separates from the alumina slag and settles to the bottom of the crucible as a massive product. The feasibility of carrying this reaction out in a water-cooled copper crucible, thus eliminating the alumina liner which is a source of some contamination, has been demonstrated.

Examination of the metallic product (regulus) of such aluminothermically produced vanadium metal reveals the presence of oxide phases in the metal matrix. This suggests that there is a decreasing solubility for aluminum and oxygen below the melting point. To date, no purification processes have been developed that take advantage of the purification potential of this phenomenon.

The vanadium alloy is purified and consolidated by one of two procedures, as shown in the flow diagram of the entire aluminothermic reduction process presented in Figure 1. In one procedure, the brittle alloy is crushed and heated in a vacuum at 1790°C to sublime most of the aluminum, oxygen, and other impurities. The aluminum

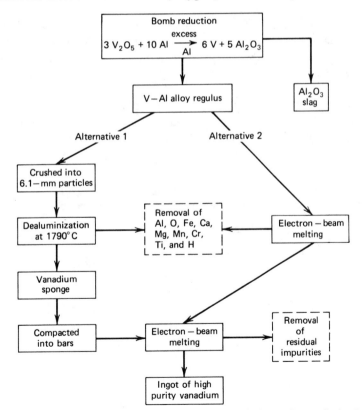

Figure 1. Flow diagram for aluminothermic process showing alternative methods of aluminum removal from alloy regulus.

facilitates removal of the oxygen, which is the feature that makes this process superior to the calcium process. Further purification and consolidation of the metal is accomplished by electron-beam melting of pressed compacts of the vanadium sponge.

The alternative procedure involves direct electron-beam melting of the vanadium–aluminum alloy regulus. Two or more melting steps are required to achieve the desired levels of aluminum and oxygen in the final ingot. Chemical analyses of two ingots of vanadium metal prepared from the identical vanadium–aluminum alloy and processed by the two methods described above are presented in Table 3. Comparable purities are obtained by these procedures. Ingots weighing up to 454 kg have been prepared by this process involving direct electron-beam melting of the alloy.

Refining of Vanadium. In addition to the purification methods described above, vanadium can be purified by any of three methods: iodide refining (van Arkel-deBoer process), electrolytic refining in a fused salt, and electrotransport.

Metal of greater than 99.95% purity has been prepared by the iodide-refining method (15). In this process, an impure grade of vanadium metal reacts with iodine at 800–900°C forming vanadium diiodide and the volatilized iodide is thermally decomposed and deposited on a hot filament at ca 1300°C. The refining step is carried out in an evacuated and sealed tube. The main impurities removed in the process are the gaseous elements and those metals that form stable or nonvolatile iodides. Vanadium metal containing 5 ppm nitrogen, 150 ppm carbon, and 50 ppm oxygen has been prepared in this way.

An electrolytic process for purifying crude vanadium has been developed at the U.S. Bureau of Mines (16). It involves the cathodic deposition of vanadium from an electrolyte consisting of a solution of VCl_2 in a fused KCl–LiCl eutectic. The vanadium content of the mixture is 2–5 wt % and the operating temperature of the cell is 650–675°C. Metal crystals or flakes of up to 99.995% purity have been obtained by this method.

The highest purity vanadium reported to date has been purified by an electrotransport technique (17). A high density current is passed through a small rod of electrolytically refined metal, heating it to 1700–1850°C. Under these conditions,

Table 3. Vanadium Prepared by Dealuminization with Vacuum Heating Compared to that Prepared by Direct Electron-Beam Melting of Vanadium Alloy Containing 15 wt % Aluminum

	By dealuminization, ppm		By direct melting, ppm	
Impurity	Sponge	Electron-beam-melted ingot	First melt	Second melt
C	100	100	100	100
O	50	60	200	150
N	40	45	40	50
H	<1	<1	<1	<1
Al	>1000	300	ca 1000	100
Ca	<10	<10	<10	<10
Cu	<20	<20	<20	<20
Fe	70	70	60	60
Mg	<20	<20	<20	<20
Mn	<20	<20	<20	<20
Ni	60	60	60	60
Si	300	300	300	300

interstitial solute atoms, eg, carbon, oxygen, and nitrogen, migrate to the negative end of the bar, which results in a high degree of purification along the remainder of the rod. Small amounts of vanadium containing less than 10 ppm of carbon, oxygen, and nitrogen and having a resistance ratio, $R_{300\ K}:R_{4.2\ K}$, of greater than 1100 have been prepared by this technique.

Consolidation, Fabrication, and Properties

Since no process has been developed for selectively removing impurities in vanadium and vanadium alloys in the metallic state, it is essential that all starting materials, in aggregate, be pure enough to meet final product purity requirements. In addition, the consolidation method must be one that prevents contamination through reaction with air or with the mold or container material.

Consolidation. Consolidation by the consumable-electrode electric-arc melting technique is ideally suited for vanadium and is used extensively for preparation of ingots of most of the reactive and refractory metals (18–19). An electrode consisting of carefully weighed portions of each alloy constituent is prepared by combining various sizes and shapes of starting materials. For example, an alloy composed of vanadium plus 15 wt % titanium and 7.5 wt % chromium might consist of an electron-beam purified vanadium ingot to which has been welded high purity titanium plat and electrolytically produced chromium granules. Welding is performed either in a vacuum or under an inert gas, eg, argon, helium, or a mix of the two.

The electrode is attached mechanically and electrically to a system that feeds it downward as well as in all horizontal directions so as to center the electrode in the crucible. The electrode is then placed in a water-cooled copper crucible whose internal diameter is 2.5–7.6 cm greater than the circle circumscribing the electrode. Chips or blocks of vanadium or of the alloy to be melted are placed in the bottom of the crucible, forming a starter pad for the arc melting. The furnace assembly is sealed and the chamber evacuated and then backfilled with a pressure of helium and argon, which varies from between a few millipascals (10^{-5} mm Hg) to nearly 101.3 kPa (= 1 atm). An arc is struck between the electrode and the starter pad; voltage and amperage are varied to control the energy input and, thus, the rate at which the electrode is melted and consumed to form the alloy ingot.

Multiple-arc melting for a minimum of two melts is conventionally used to ensure a homogeneous ingot. Although conventional arc-melt practice involves a negative electrode, improved alloying is achieved with a positive electrode for at least one of the several melts and usually the first melt.

The as-cast ingot generally exhibits a rough side wall because of entrapment of porosity as the molten metal contacts the water-cooled copper crucible. Also, a shrinkage cavity usually forms near the top of the ingot as a result of volumetric changes during solidification. Both of these artifacts are normally removed prior to fabrication, since most vanadium alloys exhibit only limited ductility in the as-cast condition, and the porosity, if not removed, would lead to cracking. Machining is the conventional surface-conditioning method.

Fabrication. Primary or initial fabrication is generally performed by either forging or extrusion at 1000–1200°C (19). Since the alloy oxidizes quite rapidly at elevated temperatures and since the pentoxide melts at 690°C, the machined ingot is clad and sealed in a mild-steel container. Following the initial hot-working sequence, subsequent working can be performed at RT to 500°C, depending upon the alloy and the stage of processing. Intermediate and final recrystallization annealing are performed at 650–1000°C. Annealing is done in a vacuum or inert gas to reduce metal loss and contamination of the metal which would occur if heating were done in air.

The fabrication of most vanadium alloys is difficult because of increased strength and decreased ductility, especially at low temperatures. Generally, higher temperatures are used for each step of fabrication. Also, processes, eg, extrusion in which the forces are largely compressive, are used for the initial ingot breakdown.

Properties. Most of the alloys developed to date were intended for service as fuel cladding and other structural components in liquid-metal-cooled fast-breeder reactors. Alloy selection was based primarily upon the following criteria: corrosion resistance in liquid metals, including lithium, sodium, and NaK, and a mixture of sodium and potassium; strength; ductility, including fabricability; and neutron considerations, including low absorption of fast neutrons as well as irradiation embrittlement and dimensional-variation effects. Alloys of greatest interest include V 80, Cr 15, Ti 5 [*39308-80-0*]; V 80, Ti 20 [*12611-15-3*]; V 77.5, Ti 15, Cr 7.5 [*51880-37-6*]; and V 86.65, Cr 9, Fe 3, Zr 1.3, C 0.05 [*84215-85-0*].

Based upon considerations of economics, fabricability, and performance, reasonable allowable levels of interstitial elements are as follows:

Element	Allowable level, ppm wt
carbon	200
nitrogen	100
oxygen	300
hydrogen	100

Because of the effects of impurity content and processing history, the mechanical properties of vanadium and vanadium alloys vary widely. The typical RT properties for pure vanadium and some of its alloys are listed in Table 4. The effects of alloy additions on the mechanical properties of vanadium have been studied and some alloys that exhibit room-temperature tensile strengths of 1.2 GPa (175,000 psi) have strengths of up to ca 1000 MPa (145,000 psi) at 600°C. Beyond this temperature, most alloys lose tensile strength rapidly.

As in the case of many metal–alloy systems, weld ductility is not as good as that of the base metal. Satisfactory welds can be made in vanadium alloys provided the

Table 4.   Typical Room Temperature Properties of Vanadium and Vanadium Alloys[a]

Metal	Tensile strength, MPa[b]	Yield strength, MPa[b]	Elongation, %
pure vanadium			
annealed or hot-worked	380–550	410–480	20–27
cold-worked	910	760	2–7
alloys			
V 87.5, Ti 15, Cr 7.5 annealed	730	620	30
V 80, Cr 15, Ti 5 annealed	600	500	28
various high strength	1210		2–5
alloys, warm rolled			

[a] Refs. 3, 20.

[b] To convert MPa to psi, multiply by 145.

fusion zone and the heat-affected zone (HAZ) are protected from contamination during welding. Satisfactory welds can be made by a variety of weld methods, including electron-beam and tungsten-inert-gas (TIG) methods. It is also likely that satisfactory welds can be made by advanced methods, eg, laser and plasma techniques (see Lasers; Plasma technology, Supplement Volume).

Economic Aspects

The United States dominated world vanadium production for all uses until the late 1960s when several countries, notably the USSR, expanded production significantly. At about the same time, the United States shifted from being a net exporter to a net importer; this situation continues. In 1978, the United States supplied 15% of the total world production but consumed 23%. World production values for 1978 and anticipated capacities are shown in Table 5 (21); U.S. production and demand for the period 1958–1979, as well as forecasts, are shown in Table 6 (21).

Usually vanadium is produced as a coproduct or by-product of other materials, including uranium, phosphorus, iron, crude oil, or tars. As such, pricing and availability depends on the aggregate supply–demand relationships of several other commodities. In 1979, the price for vanadium oxide was ca $13.90/kg of contained vanadium. The current price for high purity vanadium metal ingot is ca $661/kg, but this could decrease markedly if consumption were to increase. U.S. import duties vary from 3% for unwrought, ie, cast or unworked, alloys to 45% for wrought metals, depending upon the most-favored-nation status.

Because of the strategic nature of many of the uses, vanadium is one of the materials designated in the National Defense Stockpile Inventory. The goals for 1980 for vanadium-containing materials was 907 metric tons of contained vanadium in

Table 5. World Vanadium Production (1978) and Capacity (1978, 1979, and 1985), Metric Tons

Country	Production 1978	Capacity 1978	Capacity 1979	Capacity 1985
North America				
United States	4,721[a]	7,711	7,983	10,886
South America				
Chile	ca 689	1,087	1,087	1,087
Europe				
Finland	2,805	3,131	3,131	3,131
Norway	ca 463	1,179	1,179	1,179
USSR	ca 9,526	14,518	14,518	17,282
Total	*12,794*	*18,828*	*18,828*	*21,592*
Africa				
Republic of South Africa	ca 11,249	12,955	14,225	17,237
Nambia (Southwest Africa)	440	726	726	726
Total	*11,689*	*13,681*	*14,951*	*17,963*
Asia				
People's Republic of China	ca 1,996	2,631	8,709	8,709
Oceania				
Australia			635	3,084
World Total	*31,889*	*43,938*	*52,193*	*63,321*

[a] Recovered vanadium.

Table 6.   Comparison of U.S. Vanadium Production and Demand: 1958–1979, 1990, and 2000, t

Year	U.S. primary demand	U.S. primary production
1958	1,270	2,532
1961	2,315	5,277
1964	4,280	4,580
1967	5,523	5,372
1970	6,410	5,075
1973	7,756	4,413
1976	8,872	5,622
1979	7,755	5,224
1990	11,521 estd	8,437 estd[a]
2000	16,692 estd	8,800 estd[a]

[a] 21-year trend.

ferrovanadium, and 6985 t of contained vanadium in vanadium pentoxide. As of March 1981, the inventory consisted of 491 t of contained vanadium in vanadium pentoxide; there was no ferrovanadium in the inventory (22).

Health and Safety Factors, Toxicology

In the consolidated form, vanadium metal and its alloys pose no particular health or safety hazard. However, they do react violently with certain materials, including BrF_3, chlorine, lithium, and some strong acids (23). As is true with many metals, there is a moderate fire hazard in the form of dust or fine powder or when the metal is exposed to heat or flame. Since vanadium reacts with oxygen and nitrogen in air, control of such fires normally involves smothering the burning material with a salt.

Vanadium compounds, including those which may be involved in the production, processing, and use of vanadium and vanadium alloys, are irritants chiefly to the conjuctivae and respiratory tract. Prolonged exposure may lead to pulmonary complications. However, responses are acute, never chronic. Toxic effects vary with the vanadium compound involved. For example, LD_{50} (oral) of vanadium pentoxide dust in rats is 23 mg/kg of body weight (24).

The toxicity of vanadium alloys may depend upon other components in the alloy. For example, the V_3Ga alloy requires precautions related to both vanadium and gallium, and gallium is highly toxic. Similarly, alloys with chromium may require precautions associated with that metal.

The adopted values for TWAs for airborne vanadium, including oxide and metal dusts of vanadium, is 0.5 mg/m^3; the values for fumes of vanadium compounds is 0.05 mg/m^3. These limits are for normal 8-h workday and 40-h work-week exposures. The short-term exposure limit (STEL) is 1.5 mg/m^3 for dusts (25). A description of health hazards, including symptoms, first aid, and organ involvement, personal protection, and respirator use has been published (26).

The ammonium salts of vanadic acid and vanadium pentoxide have been listed as toxic constituents in solid wastes under the Resource Conservation and Recovery Act (27).

Uses

The most important use of vanadium is as an alloying element in the steel industry where it is added to produce grain refinement and hardenability in steels. Vanadium is a strong carbide former, which causes carbide particles to form in the steel, thus restricting the movement of grain boundaries during heat treatment. This produces a fine-grained steel which exhibits greater toughness and impact resistance than a coarse-grained steel and which is more resistant to cracking during quenching. In addition, the carbide dispersion confers wear resistance, weldability, and good high temperature strength. Vanadium steels are used in dies or taps because of their deep-hardening characteristics and for cutting tools because of their wear resistance. They are also used as constructional steel in light and heavy sections; for heavy iron and steel castings; forged parts, eg, shafts and turbine motors; automobile parts, eg, gears and axles; and springs and ball bearings. Vanadium is an important component of ferrous alloys used in jet-aircraft engines and turbine blades where high temperature creep resistance is a basic requirement (see High temperature alloys).

The principal application of vanadium in nonferrous alloys is the titanium 6–4 alloy (6 wt % Al–4 wt % V), which is becoming increasingly important in supersonic aircraft where strength-to-weight ratio is a primary consideration. Vanadium and aluminum impart high temperature strength to titanium, a property that is essential in jet engines, high speed air frames, and rocket-motor cases. Vanadium foil can be used as a bonding material in the cladding of titanium to steel. Vanadium is added to copper-based alloys to control gas content and microstructure. Small amounts of vanadium are added to aluminum alloys to be used in pistons of internal combustion engines to enhance the alloys' strength and reduce their thermal expansion coefficients. Because of its low capture cross section for fast neutrons as well as its resistance to corrosion by liquid sodium and its good high temperature creep strength, vanadium alloys are receiving considerable attention as a fuel-element cladding for fast-breeder reactors. Vanadium is a component in several permanent-magnet alloys containing cobalt, iron, sometimes nickel, and vanadium. The vanadium content in the most common of these alloys is 2–13 wt %. Vanadium and several vanadium compounds are also used as catalysts in certain chemical and petrochemical reactions.

Liquid-Metal Systems. The liquid-metal fast-breeder reactor (LMFBR) program in the United States and corresponding fast-reactor programs in other countries have considered the use of vanadium as a fuel cladding since its neutron economy, high temperature strength, and corrosion resistance in liquid metals promises higher operating temperatures than stainless steel, which is the reference cladding material.

Interstitial mass transfer is the dominant mechanism of corrosion in vanadium-based alloy–liquid alkali metal systems, eg, lithium or sodium, at 500–650°C (28). Vanadium loses oxygen but absorbs carbon and nitrogen when in contact with lithium at these temperatures; in sodium, vanadium absorbs all three elements. Since varying absorption and loss mechanisms exist for the liquid metals in contact with other structural metals in a system, these effects can be quite severe. Since all three elements contribute to vanadium's strength and ductility, large changes in properties can occur. Alloy additions to vanadium do not change the direction of interstitial transfer but can alter the rate of transfer as well as the morphology of the resultant structure. Formation of surface layers or intermetallic compounds can lead to spalling.

Specific restraints of use in the LMFBR program precluded consideration of all alloy systems. Accordingly, it is reasonable to expect that a much broader range of property improvements is possible through a broader use of alloying additions. No significant programs aimed at achieving these properties have been reported, largely since no commercially attractive results are anticipated.

Superconductivity. One potential future use of vanadium is in the field of superconductivity. The compound V_3Ga exhibits a critical current at 20 T (20 × 10^4 G), which is one of the highest of any known material. Although niobium–zirconium and Nb_3Sn have received more attention, especially in the United States, the vanadium compound is being studied for possible future application in this field since V_3Ga exhibits a critical temperature of 15.4 K as opposed to 18.3 K for Nb_3Sn (see Superconducting materials).

Like Nb_3Sn, V_3Ga has a Type A15 structure, which exhibits low ductility. Consequently, efforts at producing the compound in fine-grained fibrous structures have involved unique processing techniques (29). One method consists of forming a billet consisting of vanadium rods surrounded by a Cu–Ga bronze alloy. The billet is fabricated to wire by conventional extrusion, rod-rolling, or swagging and wire-drawing techniques. Following production of a wire, a heat treatment is performed which leads to diffusion of Ga to the fine V filaments, thereby forming the V_3Ga intermetallic compound.

Another technique, developed originally for Cu–Nb alloys, begins by rapidly chilling a liquid Cu–V alloy. During solidification, a fine dispersion of primary vanadium grains form in the matrix of the Cu–V alloy. The cast billet is fabricated by the methods described above. Following formation of the wire, gallium is plated on it and thermally diffuses, forming the V_3Ga which is in long, thin platelets ca 0.01–0.1 μm dia.

Fusion Reactors. The development of fusion reactors requires a material exhibiting high temperature mechanical strength, resistance to radiation-induced swelling and embrittlement, and compatibility with hydrogen, lithium and various coolants. One alloy system that shows promise in this application, as well as for steam-turbine blades and other applications in nonoxidizing atmospheres, is based upon the composition $(Fe,Co,Ni)_3V$ (30). Through control of an ordered–disorder transformation, the yield strength of these alloys increases at elevated temperatures above the room-temperature yield strength. One composition, for example, exhibits a yield strength of 480 MPa (70,000 psi) at ca 750°C compared to its room temperature value of 345 MPa (50,000 psi). The alloys also show good resistance to radiation-induced swelling (see Fusion energy).

These alloys can be fabricated, eg, by rolling, at 1000–1100°C and are then heat-treated at 800–1100°C.

BIBLIOGRAPHY

"Vanadium and Vanadium Alloys" in *ECT* 1st ed., Vol. 14, p. 583, by Jerome Strauss, Vanadium Corporation of America; "Vanadium and Vanadium Alloys" in *ECT* 2nd ed., Vol. 21, pp. 157–167, by O. N. Carlson and E. R. Stevens, Ames Laboratory of the U.S. Atomic Energy Commission.

1. S. A. Bradford and O. N. Carlson, *ASM Trans. Q.* **55**, 493 (1962).
2. R. W. Thompson and O. N. Carlson, *J. Less-Common Met.* **9**, 354 (1965).
3. D. L. Harrod and R. E. Gold, *International Metals Reviews*, 163 (1980).
4. C. A. Hampel, ed., *The Encyclopedia of the Chemical Elements*, Reinhold Publishing Corp., New York, 1968, p. 790.

5. C. A. Hampel, ed., *Rare Metals Handbook*, 2nd ed., Reinhold Publishing Corp., New York, 1961, p. 634.
6. *Metals Handbook*, 9th ed., Vol. 2, American Society for Metals, Metals Park, Ohio, 1979, p. 822.
7. O. N. Carlson in H. Y. Sohn, O. N. Carlson, and J. T. Smith, eds., *Extractive Metallurgy of Refractory Metals*, The Metallurgical Society of AIME, Warrendale, Pa., 1980, p. 191.
8. T. J. McLeer, Foote Mineral Co., Exton, Pa., personal communication, July 1969.
9. P. T. Brooks and G. M. Potter, *Recovering Vanadium from Dolomitic Nevada Shale*, Bureau of Mines RI 7932, U.S. Bureau of Mines, Washington, D.C., 1974, 20 pp.
10. U.S. Pat. 3,420,659 (Oct. 11, 1967), H. W. Rathmann and R. T. C. Rasmussen (to Foote Mineral Co.).
11. F. H. Perfect, *Trans. Metall. Soc. AIME* **239**, 1282 (1967).
12. J. W. Marden and M. N. Rich, *Ind. Eng. Chem.* **19**, 786 (1927).
13. R. K. McKechnie and A. U. Seybolt, *J. Electrochem. Soc.* **97**, 311 (1950).
14. O. N. Carlson, F. A. Schmidt, and W. E. Krupp, *J. Met.* **18**, 320 (1966).
15. O. N. Carlson and C. V. Owen, *J. Electrochem. Soc.* **108**, 88 (1961).
16. T. A. Sullivan, *J. Met.* **17**, 45 (1965).
17. F. A. Schmidt and J. C. Warner, *J. Less-Common Met.* **13**, 493 (1967).
18. R. W. Huber and I. R. Lane, Jr., *Consumable-Electrode Arc Melting of Titanium and Its Alloys*, Bureau of Mines R.I. 5311, U.S. Bureau of Mines, Washington, D.C., 1957, 36 pp.
19. R. W. Buchman, Jr., *International Metals Reviews*, 158 (1980).
20. Ref. 6, pp. 822–823.
21. G. A. Morgan, "Vanadium" in *Mineral Facts and Problems*, Bureau of Mines Bulletin 671, U.S. Bureau of Mines, Washington, D.C., 1980, 10 pp.
22. L. O. Giuffrida, *Stockpile Report to the Congress, October 1980–March 1981*, P&P-1, Federal Emergency Management Agency, Washington, D.C., Nov. 1981, 30 pp.
23. N. I. Sax, *Dangerous Properties of Industrial Materials*, 5th ed., Van Nostrand Reinhold Company, New York, 1979, p. 1082.
24. Ref. 23, p. 1083.
25. *Threshold Limit Values for Chemical Substances in Workroom Air Adopted By ACGIH for 1981*, American Conference of Governmental Industrial Hygienists, Cincinnati, Ohio, 1981.
26. *NIOSH/OSHA Pocket Guide to Chemical Hazards*, American Optical Corporation, Southbridge, Mass., 1978, pp. 108–109.
27. *Fed. Regist.* **45**, 33121, 33133 (May 19, 1980).
28. R. L. Ammion, *International Metals Reviews*, 255 (1980).
29. B. N. Das, J. E. Cox, R. W. Huber, and P. A. Meussner, *Metall. Trans.* **8A**, 541 (1977).
30. C. T. Liu, *J. Nucl. Mat.* **85/86**, 907 (1979).

General References

R. Rostoker, *The Metallurgy of Vanadium*, John Wiley & Sons, Inc., New York, 1965.
T. E. Dietz and J. W. Wilson, *Behavior and Properties of Refractory Metals*, Stanford University Press, Stanford, Calif., 1965.
M. E. Weeks, *Discovery of the Elements*, 6th ed., Mack Printing Company, Easton, Pa., 1956.
C. A. Hampel, ed., *Rare Metals Handbook*, 2nd ed., Reinhold Publishing Corp., New York, 1961, p. 634.
Economic Analysis of the Vanadium Industries, U.S. Department of Commerce Document PB-176 471, U.S. Department of Commerce, Washington, D.C., June 1967.
D. R. Spink, G. L. Rempel, and C. O. Gomez-Bueno in H. Y. Sohn, O. N. Carlson, and J. T. Smith, eds., *Extractive Metallurgy of Refractory Metals*, The Metallurgical Society of AIME, Warrendale, Pa., 1980, p. 147.
G. Gabra and I. Malinsky in H. Y. Sohn, O. N. Carlson, and J. T. Smith, eds., *Extractive Metallurgy of Refractory Metals*, The Metallurgical Society of AIME, Warrendale, Pa., 1980, p. 167.
Economic Analysis of the Vanadium Industries, U.S. Department of Commerce Document PB-176 471, U.S. Department of Commerce, Washington, D.C., June 1967, p. 70.
R. C. Svedberg and R. W. Buchman, Jr., *International Metals Reviews*, 223 (1980).
R. E. Gold and D. L. Harrod, *International Metals Reviews*, 232 (1980).
U.S. Pat. 4,002,504 (Jan. 11, 1977), D. G. Howe.
D. G. Howe, T. L. Francavilla, and D. U. Gubser, *IEEE Trans. Magn.* **13**, 815 (Jan. 1977).

EDMUND F. BAROCH
International Titanium, Inc.

VANADIUM COMPOUNDS

Vanadium is widely dispersed in the earth's crust at an average concentration of ca 150 ppm. Deposits of ore-grade minable vanadium are rare. Vanadium is ordinarily recovered from its raw materials in the form of the pentoxide, but sometimes as sodium and ammonium vanadates. These initial compounds have catalytic and other chemical uses (see Catalysis). For such uses and for conversion to other vanadium chemicals, granular V_2O_5 usually is made by decomposing ammonium metavanadate. For metallurgical uses, which represent ca 90% of vanadium consumption, the oxides are prepared for conversion to master alloys by fusion and flaking to form glassy chips. The preparation and application of alloying materials, eg, ferrovanadium and the proprietary products, Carvan, Nitrovan, and Ferovan, are described elsewhere (see Vanadium and vanadium alloys). Vanadium–aluminum master alloy is made for alloying with titanium metal. A part of such vanadium-rich raw materials as slags, ash, and residues is smelted directly to master alloys or to alloy steels. Figure 1 shows the prevailing process route (1).

Possibly because of price and performance competition from chromium, titanium, and other transition elements, only about a dozen vanadium compounds are commercially significant; of these, vanadium pentoxide is dominant.

Physical Properties

Some properties of selected vanadium compounds are listed in Table 1. Detailed solubility data are given in ref. 3. Physical constants of other vanadium compounds are listed in ref. 4. Included are the lattice energy of several metavanadates and the magnetic susceptibility of vanadium bromides, chlorides, fluorides, oxides, and sulfides (5).

Vanadium, a typical transition element, displays well-characterized valence states of 2–5 in solid compounds and in solutions. Valence states of -1 and 0 may occur in solid compounds, eg, the carbonyl and certain complexes. In oxidation state 5, vanadium is diamagnetic and forms colorless or pale-yellow compounds. In lower oxidation states, the presence of one or more $3d$ electrons, usually unpaired, results in paramagnetic and colored compounds. All compounds of vanadium having unpaired electrons are colored, but because the absorption spectra may be complex, a specific color does not necessarily correspond to a particular oxidation state. As an illustration, vanadium(IV) oxy salts are generally blue, whereas vanadium(IV) chloride is deep red. Differences over the valence range of 2–5 are shown in Table 2. The structure of vanadium compounds is discussed in refs. 6–7.

Chemical Properties

The chemistry of vanadium compounds is related to the oxidation state of the vanadium. Thus, V_2O_5 is acidic and weakly basic, VO_2 is basic and weakly acidic, and V_2O_3 and VO are basic. Vanadium in an aqueous solution of vanadate salt occurs as the anion, eg, $(VO_3)^-$ or $(V_3O_9)^{3-}$, but in strongly acid solution, the cation $(VO_2)^+$ prevails. Vanadium(IV) forms both oxyanions $(V_4O_9)^{2-}$ and oxycations $(VO)^{2+}$.

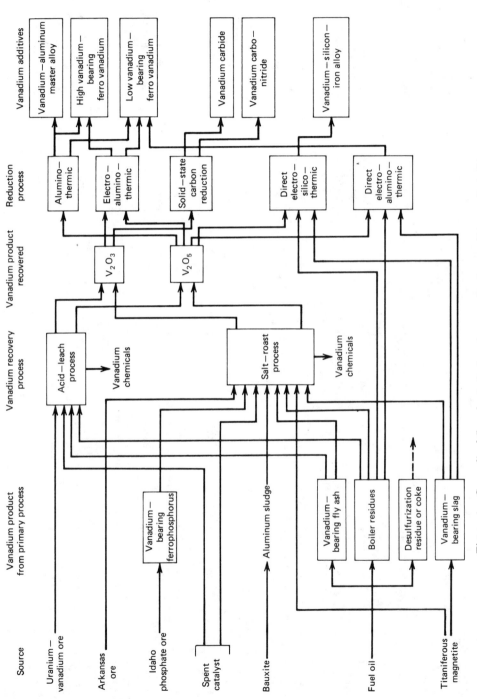

Figure 1. Generalized flow sheet of minerals processing to vanadium products (1).

Table 1. Physical Properties of Some Industrial and Other Selected Vanadium Compounds[a]

Compound	CAS Reg. No.	Formula	Appearance	Mol wt	Density, g/cm³	Mp, °C	Bp, °C	Solubility
vanadic acid, meta	[13470-24-1]	HVO_3	yellow scales	99.95				soluble in acid and alkali
ammonium metavanadate	[7803-55-6]	NH_4VO_3	white–yellowish or colorless crystals	116.98	2.326	200 dec		slightly soluble in H_2O
potassium metavanadate	[13769-43-2]	KVO_3	colorless crystals	134.04				soluble in hot H_2O
sodium metavanadate	[13718-26-8]	$NaVO_3$	colorless, monoclinic prisms	121.93		630		soluble in H_2O
sodium orthovanadate	[13721-39-6]	Na_3VO_4	colorless, hexagonal prisms	183.94		850–856		soluble in H_2O
sodium pyrovanadate	[13517-26-5]	$Na_4V_2O_7$	colorless, hexagonal prisms	305.84		632–654		soluble in H_2O
vanadium carbide	[12070-10-9]	VC	black cubic	62.95	5.77	2810	3900	insoluble in H_2O; soluble in HNO_3 with decomposition
vanadium nitride	[24646-85-3]	VN	black cubic	64.95	6.13	2320 dec		soluble in *aqua regia*
vanadium(III) trichloride	[7718-98-1]	VCl_3	pink crystals, deliquescent	157.301	3.00			soluble (deliquescent) in H_2O; soluble in methanol and ether
vanadium(IV) tetrachloride	[7632-51-1]	VCl_4	red-brown liquid	192.75	1.816	28 ± 2	148.5	soluble (deliquescent) in H_2O; soluble in methanol, ether, and chloroform
vanadium(V) oxytrichloride	[7727-18-6]	$VOCl_3$	yellow liquid	173.30	1.829	−77 ± 2	126.7	soluble (deliquescent) in H_2O; soluble in methanol, ether, acetone, and acid
vanadium(II) oxide	[12035-98-2]	VO	light green crystals	66.95	5.758	ignites		soluble in acid
vanadium(III) oxide	[1314-34-7]	V_2O_3	blue crystals	149.88	4.87	1970		soluble in HNO_3, HF, and alkali in presence of oxide
vanadium(IV) oxide	[12036-21-4]	VO_2	blue crystals	82.94	4.339	1967		soluble in acid and alkali
vanadium(V) oxide	[1314-62-1]	V_2O_5	yellow-red rhombohedra	181.88	3.357	690	1750 dec	slightly soluble in H_2O; soluble in acid and alkali
vanadium(IV) disilicide	[12039-87-1]	VSi_2	metallic prisms	107.11	4.42			soluble in HF
vanadium(III) acetylacetonate	[13476-99-8]	$V(C_5H_7O_2)_3$	brown crystals	348.27	0.9–1.2	178–190		soluble in methanol, acetone, benzene, and chloroform
biscyclopentadienyl-vanadium chloride	[12083-48-6]	$(C_5H_5)_2VCl_2$	pale green crystals	252.04		decomposes at <250		soluble in methanol and chloroform

[a] Ref. 2.

690

Table 2. Magnetism and Color of Vanadium Compounds

Compound	CAS Registry No.	Vanadium valence	No. of d electrons	Magnetic moment, $J/T \times 10^{-23}$ [a]	Color
$VOCl_3$	[7727-18-6]	5	0	0	yellow
$VOSO_4.5H_2O$	[12439-96-2]	4	1	1.60	blue
$(NH_4)V(SO_4)_2.12H_2O$	[29932-01-2]	3	2	2.60	blue
$VSO_4.7H_2O$	[36907-42-3]	2	3	3.47	violet

[a] To convert J/T to μ_β, divide by 9.274×10^{-24}.

Compounds of vanadium(III) and (II) in solution contain the hydrated ions $[V(H_2O)_6]^{3+}$ and $[V(H_2O)_6]^{2+}$, respectively.

Coordination compounds of vanadium are mainly based on six coordination, in which vanadium has a pseudooctahedral structure. Coordination number four is typical of many vanadates. Coordination numbers five and eight also are known for vanadium compounds, but numbers less than four have not been reported. The coordination chemistry of vanadium has been extensively reviewed (8–12) (see Coordination compounds).

Aqueous pentavalent vanadium is readily reduced to the quadrivalent state by iron powder or by SO_2 gas. A stronger reducing agent, eg, zinc amalgam, is needed to yield divalent vanadium. Divalent and trivalent vanadium compounds are reducing agents and require storage under an inert atmosphere to avoid oxidation by air.

Interstitial and Intermetallic Compounds. In common with certain other metals, eg, Hf, Nb, Ti, Zr, Mo, W, and Ta, vanadium is capable of taking atoms of nonmetals into its lattice. Such uptake is accompanied by a change in the packing pattern to a cubic close-packed structure. Carbides, hydrides, and nitrides so formed are called interstitial compounds. Their composition is determined by geometrical packing arrangements rather than by valence bonding. As all possible vacant lattice sites need not be filled, the compositions display a range of nonmetal content up to the theoretical limit. A range for the vanadium nitrogen compound is $VN_{0.71–1.00}$. Compounds corresponding to V_2C [12012-17-8], VC, and $VH_{1.8}$ [12713-06-3] have been formed. Vanadium borides also are known. In VB [12045-27-1], the boron atoms form a zigzag chain; whereas, in VB_2 [12007-37-3], the interstitial boron atoms are linked in a layer of hexagons.

Large-atomed nonmetals, eg, Si, Ge, P, As, Se, and Te, form compounds with vanadium that are intermediate between being interstitial and intermetallic. The interstitial and intermetallic vanadium compounds are very hard crystalline solids (9–10 on Mohs scale), have high melting points (2000–3000°C), and generally are resistant to attack by mineral acids. Except for carbides that are used in impure form, eg, Union Carbide's Carvan, in making alloy steels, this group of compounds is little used. Vanadium disilicide, VSi_2, which is made by the reaction of vanadium pentoxide with silicon metal at ca 1200°C, has limited use as a refractory material (see Refractories). Preparation of vanadium nitride from the reaction of vanadium oxides with ammonia was investigated by the U.S. Bureau of Mines (13). In subsequent studies, the nitrides were to be converted to vanadium metal.

Vanadium Oxides. Vanadium pentoxide (V_2O_5) is intermediate in behavior and stability between the highest oxides of titanium, ie, TiO_2, and of chromium, ie, CrO_3. It is thus less stable to heat than TiO_2 and more heat-stable than CrO_3. Also, V_2O_5

is more acidic and a stronger oxidant than TiO_2, but less so than CrO_3. An excess of aluminum is capable of reducing V_2O_5 to a V–Al alloy, but only calcium can reduce the oxide to vanadium metal. Solubility in water is 0.1–0.8 g/100 g H_2O and is lowest for crystals solidified from the molten state. The pentoxide readily dissolves in both acids and alkalies. In strongly alkaline solutions (pH >13), simple mononuclear vanadate ions are present. In strongly acid solutions, the main species is the dioxyvanadium(V) cation $(VO_2)^+$. More complex ions are present in solutions of intermediate strength alkali or acid; these include $(VO_3)^-$, $(HVO_4)^{2-}$, $(V_3O_9)^{3-}$, $(V_4O_{12})^{4-}$, $(V_{10}O_{28})^{6-}$, and others.

Vanadium(IV) Oxide. Vanadium(IV) oxide (vanadium dioxide, VO_2) is a blue-black solid, having a distorted rutile (TiO_2) structure. It can be prepared from the reaction of V_2O_5 at the melting point with sulfur or carbonaceous reductants such as sugar or oxalic acid. The dioxide slowly oxidizes in air. Vanadium dioxide dissolves in acids to give the stable $(VO)^{2+}$ ions and in hot alkalies to yield vanadate(IV) species, eg, $(HV_2O_5)^-$.

Vanadium(III) Oxide. Vanadium(III) oxide (vanadium sesquioxide, V_2O_3) is a black solid, having the corundum (Al_2O_3) structure. It can be prepared by reduction of the pentoxide by hydrogen or carbon. Air oxidation proceeds slowly, but oxidation by chlorine to give $VOCl_3$ and V_2O_5 is rapid.

Vanadium(II) Oxide. Vanadium(II) oxide is a nonstoichiometric material with a gray-black color, metallic luster, and metallic-type electrical conductivity. Metal–metal bonding increases as the oxygen content decreases, until an essentially metal phase containing dissolved oxygen is obtained (14).

Vanadates. Ammonium metavanadate and, to a lesser extent, potassium and sodium metavanadates, are the main vanadates of commercial interest. The pure compounds are colorless crystals. Vanadates(V) are identified as the meta (MVO_3), ortho (M_3VO_4), and pyro ($M_4V_2O_7$) compounds, and the nomenclature commonly used is that applied to the phosphates (M is univalent). Orthovanadates contain discrete tetrahedral $(VO_4)^{3-}$ anions. The pyrovanadates also contain discrete ions $(V_2O_7)^{4-}$ and have two VO_4 tetrahedra sharing a corner. The metavanadate structure is different for the anhydrous and hydrated salts. In the anhydrous salts KVO_3 and NH_4VO_3, the vanadium atoms are four-coordinate, with VO_4 tetrahedra linking through two oxygen atoms. In the hydrated form, $KVO_3.H_2O$, the vanadium atoms are five-coordinate, with three shared oxygen atoms per vanadium and two terminal oxygen atoms.

Vanadium Halides and Oxyhalides. Known halides and oxyhalides of vanadium, their valences, and their colors are listed in Table 3. Only vanadium(V) oxychloride ($VOCl_3$) and the tetrachloride (VCl_4) have appreciable commercial importance. The trichloride (VCl_3) is of minor commercial interest. The absence of pentavalent vanadium halides, other than the fluoride, is attributed to the relative weakness of the fluoride–fluoride bond compared with analogous bonds in other halides. Even VCl_4 is somewhat unstable, and VBr_4 decomposes at $-23°C$. The halides and oxyhalides are well-characterized in ref. 6.

Vanadium(V) Oxytrichloride. Vanadium(V) oxytrichloride ($VOCl_3$) is readily hydrolyzed and forms coordination compounds with simple donor molecules, eg, ethers, but is reduced by reaction with sulfur-containing ligands and molecules. It is completely miscible with many hydrocarbons and nonpolar metal halides, eg, $TiCl_4$, and it dissolves sulfur.

Table 3.   Vanadium Halides and Oxyhalides

Vanadium valence	Formula	CAS Registry No.	Appearance[a]
Halides			
V	VF_5	[7783-72-4]	white
IV	VF_4	[10049-16-8]	green
	VCl_4	[7632-51-1]	red-brown liquid
	VBr_4[b]	[13595-30-7]	magenta
III	VF_3	[10049-12-4]	green
	VCl_3	[7718-98-1]	red-violet
	VBr_3	[13470-26-3]	gray
	VI_3	[15513-94-7]	dark brown
II	VF_2	[13842-80-3]	blue
	VCl_2	[10580-52-6]	pale green
	VBr_2	[14890-41-6]	orange-brown
	VI_2	[15513-84-5]	red
Oxyhalides			
V	VOF_3	[13709-31-4]	pale yellow
	$VOCl_3$	[7729-18-6]	yellow liquid
	$VOBr_3$	[13520-90-6]	leep red liquid
	VO_2F	[14259-82-6]	brown
	VO_2Cl	[13759-30-3]	orange
IV	VOF_2	[13814-83-0]	yellow
	$VOCl_2$	[10213-09-9]	green
	$VOBr_2$	[13520-89-3]	yellow-brown
III	$VOCl$	[13520-87-1]	brown
	$VOBr$	[13520-88-2]	violet

[a] At room temperature; solid unless identified as liquid.
[b] Decomposes at −23°C.

Vanadium(IV) Chloride. Vanadium(IV) chloride (vanadium tetrachloride, VCl_4) is a red-brown liquid, is readily hydrolyzed, forms addition compounds with donor solvents such as pyridine, and is reduced by such molecules to trivalent vanadium compounds. Vanadium tetrachloride dissociates slowly at room temperature and rapidly at higher temperatures, yielding VCl_3 and Cl_2. Decomposition also is induced catalytically and photochemically. This instability reflects the difficulty in storing and transporting it for industrial use.

Vanadium(III) Chloride. Vanadium(III) chloride (vanadium trichloride, VCl_3) is a pink-violet solid, is readily hydrolyzed, and is insoluble in nonpolar solvents but dissolves in donor solvents, eg, acetonitrile, to form coordination compounds. Chemical behavior of the tribromide (VBr_3) is similar to that of VCl_3.

Vanadium Sulfates. Sulfate solutions derived from sulfuric acid leaching of vanadium ores are industrially important in the recovery of vanadium from its raw materials. Vanadium in quadrivalent form may be solvent-extracted from leach solutions as the oxycation complex $(VO)^{2+}$. Alternatively, the vanadium can be oxidized to the pentavalent form and solvent-extracted as an oxyanion, eg, $(V_3O_9)^{3-}$. Pentavalent vanadium does not form simple sulfate salts.

Vanadium(IV) Oxysulfate. Vanadium(IV) oxysulfate pentahydrate (vanadyl sulfate, $VOSO_4 \cdot 5H_2O$) is an ethereal blue solid and is readily soluble in water. It forms from the reduction of V_2O_5 by SO_2 in sulfuric acid solution. Vanadium(III) sulfate

[*13701-70-7*] ($V_2(SO_4)_3$) is a powerful reducing agent and has been prepared in both hydrated and anhydrous forms. The anhydrous form is insoluble in either water or sulfuric acid. Vanadium(II) sulfate heptahydrate ($VSO_4.7H_2O$) is a light red-violet crystalline powder that can be prepared by electrolytic reduction of $VOSO_4$. The powder is oxidized by air and dissolves in water to give a red-violet solution.

Manufacture

Primary industrial compounds produced directly from vanadium raw materials are principally 98 wt % fused pentoxide, air-dried (technical-grade) pentoxide, and technical-grade ammonium metavanadate (NH_4VO_3). Much of the fused and air-dried pentoxides produced at the millsite is made by thermal decomposition of ammonium vanadates. Prior to 1960, the main vanadium mill products were fused technical-grade pentoxide (black cake) containing 86–92 wt % V_2O_5 and air-dried technical-grade pentoxide (red cake) containing 83–86 wt % V_2O_5, both being high in alkali content. An historical review of the manufacture of vanadium compounds until 1960 is included in ref. 15. Modern milling practices for the production of primary vanadium compounds from ores are reviewed in ref. 16.

Vanadium raw materials are processed to produce vanadium chemicals, eg, the pentoxide and ammonium metavanadate (AMV) primary compounds, by salt roasting or acid leaching. Interlocking circuits, in which unfinished or scavenged material from one process is diverted to the other, are sometimes used. Such interlocking to enhance vanadium recovery and product grade became more feasible in the late 1950s with the advent of solvent extraction.

Salt Roasting. Iron ore concentrate, uranium–vanadium ores, ferrophosphorus from manufacture of elemental phosphorus, vanadiferous shale, and assorted slag, ash, fumes, residues, and depleted catalysts, singly or in combination, are suitable feed for the salt-roast process. Sometimes, substitution of sodium carbonate for part or all of the salt results in improved vanadium recovery. The presence of calcium and magnesium carbonates is deleterious, because these form vanadates that are insoluble in water. Interference from calcium carbonate can be ameliorated by acidulating the ore with sulfuric acid before roasting or by adding pyrite to the charge, which results in formation of calcium sulfate which is innocuous in salt roasting. Calcium vanadates, present in a salt-roast calcine, can be dissolved by acid leaching the calcine. Limestone of low magnesium content must be added to the charge when salt roasting ferro-phosphorus (as at Soda Springs, Idaho) to combine with the phosphate that forms.

The ore is ordinarily ground to pass through a ca 1.2-mm (14-mesh) screen, mixed with 8–10 wt % NaCl and other reactants that may be needed, and roasted under oxidizing conditions in a multiple-hearth furnace or rotary kiln at 800–850°C for 1–2 h. Temperature control is critical because conversion of vanadium to vanadates slows markedly at ca 800°C, and the formation of liquid phases at ca 850°C interferes with access of air to the mineral particles. During roasting, a reaction of sodium chloride with hydrous silicates, which often are present in the ore feed, yields HCl gas. This is scrubbed from the roaster off-gas and neutralized for pollution control, or used in acid-leaching processes at the mill site.

Hot calcine from the kiln is water-quenched or cooled in air before being lightly ground and leached. Air cooling allows back reactions, which adversely affect vanadium extraction for some ores. Leaching and washing of the residue is by percolation in vats or by agitation and filtration. Extraction of vanadium is 65–85%. Vanadium solution from water leaching of the calcine has a pH of 7–8 and a vanadium content of ca 30–50 g V_2O_5/L. If the water-leach residue is leached subsequently with sulfuric acid solution, as for uranium extraction, as much as 10–15 wt % more vanadium may dissolve. Originally, such acid-leached vanadium precipitated from the uranium solvent extraction raffinate as an impure sludge that was recycled to the salt-roast kiln. Currently, the vanadium generally is recovered from the uranium raffinate by solvent extraction (see Uranium and uranium compounds).

Recovery of Vanadium from Salt-Roast Leach Solution.

When recovery of vanadium as sodium red cake was practiced, the leach solution pH was adjusted to 2.7 by adding sulfuric acid. Any vanadium not already in the pentavalent form was oxidized by addition of sodium chlorate, and the solution was boiled for 2–3 h to precipitate substantially all of the vanadium as shotlike particles of sodium red cake. Although melting and flaking converted the highly alkaline red cake to black cake suitable for most metallurgical uses, extensive purification, usually by conversion to ammonium metavanadate, was required to prepare compounds for catalytic and chemical use. The process routes now favored for recovering vanadium from the leach solution are solvent extraction and precipitation of ammonium vanadates or vanadic acid. Sometimes both solvent extraction and direct precipitation are used in integrated circuits.

For solvent extraction of pentavalent vanadium as an oxyvanadium anion, the leach solution is acidified to ca pH 3 by addition of sulfuric acid. Vanadium is extracted in about four countercurrent mixer–settler stages by a 3–5 wt % solution of a tertiary alkyl amine in kerosene. The organic solvent is stripped by soda-ash solution, and addition of ammoniacal salts to the rich vanadium strip liquor yields ammonium metavanadate. A small part of the metavanadate is marketed in that form and some is decomposed at a carefully controlled low temperature to make air-dried or fine granular pentoxide, but most is converted to fused pentoxide by thermal decomposition at ca 450°C, melting at 900°C, then chilling and flaking.

For solvent extraction of a quadrivalent vanadium oxyvanadium cation, the leach solution is acidified to ca pH 1.6–2.0 by addition of sulfuric acid, and the redox potential is adjusted to −250 mV by heating and reaction with iron powder. Vanadium is extracted from the blue solution in ca six countercurrent mixer–settler stages by a kerosene solution of 5–6 wt % di-2-ethylhexyl phosphoric acid (EHPA) and 3 wt % tributyl phosphate (TBP). The organic solvent is stripped by a 15 wt % sulfuric acid solution. The rich strip liquor containing ca 50–65 g V_2O_5/L is oxidized batchwise initially at pH 0.3 by addition of sodium chlorate; then it is heated to 70°C and agitated during the addition of NH_3 to raise the pH to 0.6. Vanadium pentoxide of 98–99% grade precipitates, is removed by filtration, and then is fused and flaked.

For direct precipitation of vanadium from the salt-roast leach liquor, acidulation to ca pH 1 without the addition of ammonia salts yields an impure vanadic acid; when ammonium salts are added, ammonium vanadates precipitate. The impure vanadic acid ordinarily is redissolved in sodium carbonate solution, and ammonium metavanadate precipitates upon addition of ammonium salts. Fusion of the directly precipitated ammonium salts, in one instance, yields pentoxide containing ca 88 wt % V_2O_5, 11 wt % V_2O_4, and ca 1 wt % combined impurities. Amine solvent extraction is sometimes used to recover 1–3 g/L of residual V_2O_5 from the directly precipitated tail liquors.

New Developments. Recovery of vanadium from a dolomitic shale by salt roasting, acid leaching, and amine solvent extraction is reported in ref. 17. A patent was issued for recovery of vanadium-bearing solution from silica containing titaniferous magnetic ore by roasting a mixture of ore, a sodium salt, and cryolite at <1350°C for 30 min to 2 h and then leaching the calcine (18). Roasting of iron ores with limestone to form calcium vanadate [14100-64-2] and leaching of the vanadate with ammonium carbonate or bicarbonate solution have also been patented (19). A patent was granted for heating a mixture of slag and sodium carbonate in a converter at 600–800°C in the presence of oxygen to solubilize the vanadium, leach with water, and recover vanadium from the leach solution (20). Another patent for vanadium recovery from slag calls for an oxidizing roast, followed by leaching of the calcine with phosphoric acid to dissolve the vanadium, and recovery of vanadium from the leach solution (21).

Acid Leaching. Direct acid leaching for vanadium recovery is used mainly for vanadium–uranium ores and less extensively for processing spent catalyst, fly ash, and boiler residues. Although V_2O_5 in spent catalysts dissolves readily in acid solutions, the dissolution of vanadium from ores and other feed materials requires leaching for ca 14–24 h in strong, hot, oxidizing sulfuric acid solutions. Ore is ground to ca 0.60 mm (28 mesh) and is leached at 50–55 wt % solids and 75°C in a series of about four agitated tanks. Enough sulfuric acid and sodium chlorate are added to the first tank in the series to maintain ca 70 g/L of free acid in the second tank and a terminal redox potential of at least −430 mV. Such intensive leaching conditions dissolve ca 75% of the vanadium and 95% of the uranium. Excess acid in the leach liquor, after liquid–solids separation, is neutralized by reaction with fresh ore. A second liquid–solids separation produces a clarified leach liquor of ca pH 1 for solvent extraction. Vanadium is recovered from the uranium solvent-extraction raffinate.

For vanadium solvent extraction at the Atlas Minerals' mill at Moab, Utah, iron powder is added to reduce pentavalent vanadium to quadrivalent and trivalent iron to divalent at a redox potential of −150 mV. The pH is adjusted to 2 by addition of NH_3, and an oxyvanadium cation is extracted in four countercurrent stages of mixer–settlers by a diesel-oil solution of EHPA. Vanadium is stripped from the organic solvent with a 15 wt % sulfuric acid solution in four countercurrent stages. Addition of NH_3, steam, and sodium chlorate to the strip liquor results in the precipitation of vanadium oxides, which are filtered, dried, fused, and flaked (22).

At the White Mesa Mill in Blanding, Utah, vanadium is extracted from oxidized uranium raffinate by solvent extraction with a tertiary amine, and ammonium metavanadate is produced from the soda-ash strip liquor. Fused and flaked pentoxide is made from the ammonium metavanadate (23). Union Carbide, at Uravan, Colo., recovers the vanadium from its uranium ion-exchange tail liquor by amine solvent extraction and sends the loaded soda-ash strip liquor to its Rifle, Colo. plant for precipitation of ammonium metavanadate.

Australian Vanadium–Uranium Ore. A calcareous carnotite ore at Yeelirrie, Australia, is ill-suited for salt roasting and acid leaching. Dissolution of vanadium and uranium by leaching in sodium carbonate solution at elevated temperature and pressure is being tested on a pilot-plant scale; commercial operation is scheduled to begin in 1985 or 1986 (24). Other soda-ash leaching schemes are also being examined.

Halides and Oxyhalides. Vanadium(V) oxytrichloride is prepared by chlorination of V_2O_5 mixed with charcoal at red heat. The tetrachloride (VCl_4) is prepared by chlorinating crude metal at 300°C and freeing the liquid from dissolved chlorine by repeated freezing and evacuation. It now is made by chlorinating V_2O_5 or $VOCl_3$ in the presence of carbon at ca 800°C. Vanadium trichloride (VCl_3) can be prepared by heating VCl_4 in a stream of CO_2 or by reaction of vanadium metal with HCl.

Production

The bulk of world vanadium production is derived as a by-product or coproduct in processing iron, titanium, and uranium ores, and, to a lesser extent, from phosphate, bauxite, and chromium ores and the ash, fume, or coke from burning or refining petroleum. Total world production of vanadium was ca 3.6×10^4 metric tons in 1980. Estimated world production from ores and concentrates, by country, is given in Table 4. Vanadium resources, production, and demand are comprehensively discussed in ref. 1.

Most U.S. production of primary vanadium compounds has been as by-products or coproducts of uranium and of ferrophosphorus derived from smelting Idaho phosphates. However, vanadium as a sole product from Arkansas shale has contributed significantly to domestic supply since ca 1969 (25). Union Carbide's Arkansas plant also treats spent catalysts, slags, and residues. Foote Mineral Company in Cambridge, Ohio, produces vanadium chemicals from oil residues and imported slags. Gulf Chemical and Metallurgical Company was reported to be expanding its plant for recovery of pentoxide from spent catalysts at Texas City, Texas. Murex, Ltd. (a subsidiary of Engelhard) started processing vanadiferous ash from U.S. east coast oil-burning power plants at its mill in Bartlesville, Okla. in 1980. Despite recent sharp curtailments in U.S. uranium production, by-product vanadium output from vanadium–uranium operations has not significantly declined.

Most foreign vanadium is obtained as a coproduct of iron and titanium. South

Table 4. World Production of Vanadium from Ores and Concentrates [a], Metric Tons of Contained Vanadium

Country	1970	1975	1980
Australia			600
Chili (in slag)	600	600	400
People's Republic of China (in slag)			4,500
Finland (in vanadium pentoxide)	1,300	1,400	2,800
Namibia (South-West Africa) (in lead vanadate concentrate)	400	600	
Norway	1,100	500	500
Republic of South Africa			
(in pentoxide, vanadates)	2,400	4,800	4,000
(in slag)	4,000	5,800	8,500
USSR		8,000	10,000
United States	4,800	4,400	4,300
Total	*14,600*	*26,100*	*35,600*

[a] Data rounded from U.S. Bureau of Mines Minerals Yearbook; does not include Japanese production, estimated as 400 t in 1980, nor U.S. production from slags and residues of foreign origin (ca 2500 t).

Africa, the world's largest supplier, has three operating firms. Norway and Finland each has one operating company, but Finland recovers the pentoxide from an iron–titanium operation and a vanadium mine. Chile produces slag from an iron operation. Australia's first vanadium operation started producing fused pentoxide flake from a vanadium mine in 1980. The USSR and the People's Republic of China produce slag and pentoxide from iron–titanium ores.

Economic Aspects and Specifications

Prices are published only for vanadium pentoxide and vanadium oxytrichloride. In Jan. 1982, the following equivalent prices per kilogram of V_2O_5 or $VOCl_3$ were reported (26): vanadium pentoxide (technical-grade granular, 250-kg drums, works), $7.83; vanadium pentoxide (fused or flake, 250-kg drums, works), $7.83; and vanadium oxytrichloride (1361-kg cylinders, works), $11.02–11.25.

Some reported 1980 prices and specifications are as follows (all prices are fob works and per kilogram of V_2O_5 unless otherwise stated) (27): ammonium metavanadate (technical grade, V_2O_5 77.0 wt % min, SiO_2 0.05 wt % max, alkali 0.50 wt % max, 0.5 mm and smaller, 36-kg fiber drums), $9.54; pentoxide (technical-grade flake, V_2O_5 98.5 wt % min, SiO_2 0.3 wt % max, alkali 1.0 wt % max, 2–4 mm thick flake, 272-kg steel drums), $9.06; pentoxide (technical-grade granular, V_2O_5 99.2 wt % min, SiO_2 0.02 wt % max, alkali 0.50 wt % max, Fe 0.1 wt % max, 1.6 mm and smaller, 181-kg steel drums), $9.06; pentoxide (high purity granular, V_2O_5 99.6 wt % min, SiO_2 0.07 wt % max, Na_2O 0.03 wt % max, K_2O 0.02 wt % max, Fe 0.03 wt % max, 1.6 mm and smaller, 227-kg steel drums), $13.36; vanadium oxytrichloride (V 29.0 wt % min, Cl 60.0 wt % min, 1361-kg returnable cylinders), price per kilogram of liquid tetrachloride, $12.35.

The actual prices of V_2O_5 with prices in constant 1980 dollars at five-year intervals from 1955 to 1980 are listed in Table 5.

Product data issued by a second producer of vanadium catalysts include the following specifications. Vanadium(III) acetylacetonate, $V(C_5H_7O_2)_3$ 98.0 wt % min, $V(C_5H_7O_2)_2$ 2 wt % max, must be maintained under inert atmosphere (nitrogen containing <10 ppm O_2 is recommended). Vanadium oxytrichloride, $VOCl_3$ 95.0 wt % min, V^{5+} 28.5 wt % min, Cl 60.5–61.5 wt % Fe 0.06 wt % max, must be maintained under inert atmosphere (N_2 containing <10 ppm O_2); exposure to moisture in air results in formation of HCl and HVO_3. Vanadium tetrachloride, vanadium(IV) 22.5 wt % min, vanadium(V) 3.0 wt % max, free Cl_2 ca 3.0 wt %, chloride 73.0 wt %; store under inert atmosphere and away from heat; reacts vigorously with water.

Energy Use in the Manufacture of Fused Pentoxide. The energy required to produce fused pentoxide from a vanadium–uranium ore containing 1.3 wt % V_2O_5 and 0.20 wt % U_3O_8 is ca 360 MJ/kg (155,000 Btu/lb) (28). Treatment is assumed to be by salt

Table 5. Time–Price Relationships for Vanadium Pentoxide [a], Average Annual U.S. Price, $/kg

	1955	1960	1965	1970	1975	1980
actual price	2.93	3.04	2.32	5.40	6.09	7.81
price based on 1980 dollars	8.57	7.84	5.77	10.49	8.59	7.81

[a] Adapted from U.S. Bureau of Mines.

roasting, water and acid leaching, uranium and vanadium solvent extraction, and production of fused pentoxide. Vanadium recovery from ore is ca 80%. By current standards, the hypothetical carnotite ore selected for the energy estimation is relatively rich in vanadium and uranium. Processing of lower grade feed would require more energy per unit of product.

Imports and Exports. The United States has long been a significant importer of vanadium slags, but imports of pentoxide were negligible until they rose quickly to 850 metric tons in 1974, and 2000 t in 1975 (mostly from the Republic of South Africa). Pentoxide imports then declined to 1400 t in 1980 with Finland being the main and South Africa the minor suppliers. In recent years, U.S. imports of ammonium and potassium vanadates and of other vanadium compounds have been 100–200 t/yr, mainly from the UK, the FRG, and the Republic of South Africa.

Annual U.S. exports of the pentoxide and other compounds were 1300–1400 t in 1978 and declined to 800–900 t in 1979 and 1980. The anhydrous pentoxide accounted for roughly three-fourths of the compounds exported.

Import duties imposed on vanadium compounds by the United States are shown in Table 6. Ore, concentrate, slag, and residues are admitted free.

U.S. Stockpile. A new U.S. Government stockpile goal for vanadium pentoxide of 6985 t contained vanadium was announced on May 1, 1980. This is equivalent to 12,470 t of V_2O_5. At the time of the announcement, the stockpile contained only 491 t of vanadium in the form of the pentoxide (29). Physical requirements are that V_2O_5 be supplied as broken flake, all of a size to pass a 2.54-cm screen and not more than 5 wt % to pass a 4.7-mm screen. Packaging in polyethylene film inside 208-L steel drums and marking of the drums is described in detail in ref. 30. Chemical requirements for the two grades of pentoxide to be stockpiled are listed in Table 7.

Analytical and Test Methods

A delicate qualitative test for the presence of vanadium is the formation of brownish-red pervanadic acid upon addition of hydrogen peroxide to a solution of a vanadate. Although titanium reacts similarly, its color disappears when fluoride or phosphate ions are added (31). Quantitative determinations over a wide range of vanadium content are readily performed by atomic absorption spectroscopy. Acetylene or nitrous oxide flames are ordinarily used. A highly sensitive atomic-absorption technique involving a carbon-filament atom reservoir has a detection limit of 100 pg V in a 1-μL sample. Volumetric, colorimetric, and spectrographic methods for vana-

Table 6. U.S. Import Duties[a]

| Tariff item | Tariff no. | Most favored nations (MFN) | | Non-MFN |
		Jan. 1, 1980	Jan. 1, 1981	Jan. 1, 1980
vanadium carbide	422.58	5.8	4.2	25
vanadium pentoxide (anhydride)	422.60	16.0	16.0	40
vanadium compounds, other	422.62	16.0	16.0	40
vanadium salts	427.22	16.0	16.0	40

[a] %, ad valorem.

Table 7.   Chemical Specifications for Fused, Flake Vanadium Pentoxide for U.S. Stockpile, June 25, 1981[a], Max % Permissible Except Min % V_2O_5

Substance	Grade A	Grade B
V_2O_5	98.0	98.0
P	0.05	0.03
S	0.04	0.05
SiO_2	0.50	0.25
As	0.05	0.03
Fe	0.20	0.15
Na + K	0.75	0.50
B		0.005
Cr		0.10
Cu		0.05
Pb		0.01
Mn		0.05
Mo		0.10
Ni		0.05
N		0.05
Si		0.12
W		0.01

[a] Ref. 30.

dium are well-developed (32). Conversely, gravimetric methods are seldom used. X-ray absorption spectroscopy is a convenient means for identifying traces of vanadium in coal (33).

Health, Safety, and Environmental Considerations

The effect of vanadium compounds in the workplace and in ambient air on human health and safety is extensively reviewed in ref. 34. In humans, toxic effects have been observed from occupational exposure to airborne concentrations of vanadium compounds that were probably several milligrams or more per cubic meter of air. Direct irritation of the bronchial passageways results from such exposure and is accompanied by coughing, spitting, wheezing, and eye, nose, and throat irritation. Some workers exhibit weakness, neurasthenia, and slight anemia, which suggests chronic toxic effect from vanadium absorption. Threshold limits for V_2O_5 in the air of the workplace have been established by OSHA as 0.5 mg V/m^3 for dust and 0.05 mg V/m^3 for fumes.

Oral vanadium toxicity in humans is minimal. Ingestion of 4.5 mg/d of vanadium has been without effect, but higher doses produce gastrointestinal distress and the greentongue associated with excessive inhalation of vanadium (35). The concentration of vanadium in vegetation varies from undetectable to 4 ppm in alfalfa and in animal tissues from 0.25 to slightly over 1 ppm. Drinking water contains up to 220 ppb (parts per billion (10^9)) of vanadium. Although contamination of water supplies by seepage from vanadium processing wastes seems possible, evidence of such contamination has not been found. Postulated to be an essential trace element for human well-being, the function of vanadium and its limiting concentrations have yet to be established (36) (see Mineral nutrients).

In the United States, the largest concentration of atmospheric vanadium occurs over eastern seaboard cities where residual fuels of high vanadium content from

Table 8.   U.S. Consumption of Vanadium in Catalysts and Other Chemicals

Data source and item	1972	1974	1975	1977	1979	1980	Vanadium compound used
U.S. Bureau of Mines Minerals Yearbook							
catalysts	132	230	215	129	73	53	
other[a]	170	81	59	39	52	29	
Total consumption	*302*	*311*	*274*	*168*	*125*	*82*	
National Research Council Report[b]							
catalysts							
sulfuric acid	90	150	170				pentoxide and ammonium metavanadate
adipic acid	73	74	66				ammonium metavanadate
EPDM[c] rubber	60	83	67				vanadium oxytrichloride, some tetrachloride
maleic anhydride	10	11	9				ammonium metavanadate
phthalic anhydride	1	1	1				fluid-bed catalyst containing ca 2 wt % V or imported, fixed-bed catalyst containing 0.3 wt % V
organic polymers	3	3	3				vapor phase: oxides or vanadates combined with other metal salts; liquid phase: vanadium(III) acetylacetonate and VCl_3
organic oxidation	1	1	1				liquid phase: vanadium naphthenate (contains 3 wt % V)
Total catalysts	*238*	*323*	*317*				
ceramic pigments	59	67	50				vanadate–zirconia compositions, blues and yellows
electronic and others	14	19	15				yttrium vanadate–europium (red phosphor in TV tubes)
Total consumption	*311*	*409*	*382*				

[a] Includes ceramic pigments, electronics, magnetic alloys, and unspecified uses.
[b] Ref. 1.
[c] Ethylene propylene diene monomer.

Venezuela are burned in utility boilers. Coal ash in the atmosphere also contains vanadium (37). Ambient air samples from New York and Boston contain as much as 600–1300 ng V/m^3, whereas air samples from Los Angeles and Honolulu contained 1–12 ng V/m^3. Adverse public health effects attributable to vanadium in the ambient air have not been determined. Increased emphasis by industry on controlling all plant emissions may have resulted in more internal reclamation and recycle of vanadium catalysts. An apparent drop in consumption of vanadium chemicals in the United States since 1974 may be attributed, in part, to such reclamation activities.

Uses

Conversion of fused pentoxide to alloy additives is by far the largest use of vanadium compounds. Air-dried pentoxide, ammonium vanadate, and some fused pentoxide, representing ca 10% of primary vanadium production, are used as such, purified, or converted to other forms for catalytic, chemical, ceramic, or specialty applications. The dominant single use of vanadium chemicals is in catalysts (see Catalysis). Much less is consumed in ceramics and electronic gear, which are the other significant uses (see Batteries and electric cells). Many of the numerous uses reported in the literature are speculative, proposed, obsolete, or in such small quantities as to be generally reported under such consolidated headings as miscellaneous or other.

U.S. consumption of vanadium in catalysts and other chemicals from 1972 to 1980 are listed in Table 8. The NRC Committee's estimates of catalyst use for 1972, 1974, and 1975 were 40–80% higher than the BOM's, and the former's estimates for overall use, including ceramics, electronics, and unspecified uses, were 3–39% higher. Data published by the BOM through 1980 show a precipitous decline in total annual use from ca 311 t in 1974 to ca 82 t in 1980. The decline appears to have been caused mainly by unfavorable business conditions and conservation efforts, but part of the apparent decline may stem from incomplete collection of data.

Catalytic uses result in little consumption or loss of vanadium. The need to increase conversion efficiency for pollution control from sulfuric acid plants, which require more catalyst, and expanded fertilizer needs, which require more acid plants, were factors in the growth of vanadium catalyst requirements during the mid-1970s. Use was about evenly divided between initial charges to new plants and replacements or addition to existing plants.

Among minor uses of vanadium chemicals that generally are included in the Other category of Table 8 are preparation of vanadium metal from refined pentoxide or vanadium tetrachloride; liquid-phase organic oxidation reactions, eg, production of aniline black dyes for textile use and printing inks; color modifiers in mercury-vapor lamps; vanadyl fatty acids as driers in paints and varnish; and ammonium or sodium vanadates as corrosion inhibitors in flue-gas scrubbers.

Uses reported in the early literature, but which were insignificant in recent years at least in the United States, include refractories, V_2O_3 for coloring glass green, and V_2O_5 for ultraviolet screening in glass. Developments that may lead to future uses include an expanded role as catalyst in such applications as vapor-phase polymerizations of ethylene and propylene, ammoxidations to form acrylonitrile and terephthalonitrile, and hydrodesulfurization of crude oils; V_3Ga [12024-15-6] as a superconductor; VO_2 as a thermal or light-activated resistor–conductor; vanadate glasses as electrooptical switches; rare-earth vanadites as magnetic materials; and addition of ca 5 wt % V_2O_5 to silicon carbide refractories for increased oxidation resistance at high temperatures; the latter is reportedly being practiced in the FRG (38).

BIBLIOGRAPHY

"Vanadium Compounds" in ECT 1st ed., Vol. 14, pp. 594–602, by H. E. Dunn and C. M. Cosman, Vanadium Corporation of America; "Vanadium Compounds" in ECT 2nd ed., Vol. 21, pp. 167–180, by G. W. A. Fowles, University of Reading, Reading, England.

1. *Vanadium Supply and Demand Outlook*, Report No. NMAB-346, National Research Council, Washington, D.C., 1978, 125 pp.
2. R. C. Weast, ed., *Handbook of Chemistry and Physics*, 62nd ed., CRC Press, Inc., Boca Raton, Fla., 1981–1982.
3. Seidell, *Solubilities—Inorganic and Metal-Organic Compounds*, 4th ed., Vol. 2, American Chemical Society, Washington, D.C., 1965.

4. Ref. 2, pp. B-162 and C-689.
5. Ref. 2, pp. D-84 and E-123.
6. R. J. H. Clark, *The Chemistry of Titanium and Vanadium*, Elsevier Publishing Co., Amsterdam, The Netherlands, 1968.
7. *Transition Metal Compounds*, Vol. 4 of A. F. Trotman-Dickenson, ed., *Comprehensive Inorganic Chemistry*, Pergamon Press, Ltd., Oxford, UK, 1973.
8. D. Nichols, *Chem. Rev.* **1**, 379 (1966).
9. S. F. Ashcroft and C. T. Mortimer, *Thermochemistry of Transition Metal Complexes*, Academic Press, Inc., New York, 1970.
10. A. E. Martel, ed., *Coordination Chemistry*, Vol. 1, ACS Monograph 168, Van Nostrand Reinhold Co., New York, 1971.
11. R. G. Wilkins, *The Study of Kinetics and Mechanisms of Reactions of Transition Metal Complexes*, Allyn and Bacon, Inc., Boston, 1974.
12. A. E. Martel, ed., *Coordination Chemistry*, Vol. 2, ACS Monograph 174, American Chemical Society, Washington, D.C., 1978.
13. R. A. Guidotti, G. B. Atkinson, and D. G. Kesterke, *Nitride Intermediates in the Preparation of Columbium, Vanadium, and Tantalum Metals*, Pt. I, RI-8079, U.S. Bureau of Mines, Washington, D.C., 1975, 25 pp.
14. Ref. 7, p. 352.
15. P. M. Busch, *Vanadium*, IC-8060, U.S. Bureau of Mines, Washington, D.C., 1961, 95 pp.
16. J. B. Rosenbaum, *Vanadium Ore Processing, Meeting of High Temperature Metal Committee*, preprint A71-52, AIME, New York, 1971, 14 pp.
17. P. T. Brooks and G. M. Potter, *Recovering Vanadium from Dolomitic Nevada Shale*, RI-7932, U.S. Bureau of Mines, 1974, 20 pp.
18. U.S. Pat. 3,733,193 (May 15, 1973), J. S. Fox and W. H. Dresher (to Union Carbide Corp.).
19. U.S. Pat. 3,853,982 (Dec. 10, 1974), C. B. Bore and J. W. Pasquali (to Bethlehem Steel Corp.).
20. U.S. Pat. 3,929,460 (Dec. 30, 1975), F. J. W. M. Peters, S. Middelholk, and A. Rijkelboer (to Billiton Research, B.V.).
21. U.S. Pat. 4,039,614 (Aug. 2, 1977), N. P. Slotvinsky-Sidak and N. V. Grinberg.
22. L. White, *Eng. Min. J.*, 87 (Jan. 1976).
23. C. E. Baker and D. K. Sparling, *Min. Eng.*, 382 (Apr. 1981).
24. *Eng. Min. J.*, 105 (Feb. 1979).
25. I. R. Taylor, *Min. Eng.*, 82 (Apr. 1969).
26. *Chem. Market. Rep.*, 39 (Jan. 25, 1982).
27. *Metal Chemical Prices*, Metals Division, Union Carbide Corporation, New York; effective Jan. 2, 1980, replaces Apr. 1, 1979.
28. *Energy Use Patterns in Metallurgical and Non-Metallic Mineral Processing, Phase 6—Low Priority Commodities*, final report to U.S. Bureau of Mines, Battelle Columbus Laboratory, Columbus, Ohio, July 1976.
29. *Met. Week* **51**(18), 5, 10 (May 1980).
30. *National Stockpile Purchase Specifications*, P-58-R-2, U.S. Department of Commerce with approval of the Federal Emergency Management Agency, Washington, D.C., June 25, 1981, 4 pp.
31. H. H. Willard and H. Diehl, *Advance Quantitative Analysis*, D. Van Nostrand Company, New York, 1943, p. 239.
32. W. J. Williams, *Handbook of Anion Determination*, Butterworth & Co., London, 1979, pp. 251–261.
33. D. H. Maylotte, J. Wong, R. L. St. Peters, F. W. Lytle, and R. B. Greegor, *Science* **214**, 554 (Oct. 1981).
34. *Medical and Biological Effects of Environmental Pollutants—Vanadium*, National Academy of Sciences, Washington, D.C., 1974, 117 pp.
35. V. W. Oehme, *Toxicity of Heavy Metals in the Environment*, Pt. 2, Marcel Dekker, Inc., New York, 1979.
36. W. Mertz, *Science* **213**, 1332 (Sept. 18, 1981).
37. R. D. Smith, J. A. Campbell, and W. D. Felix, *Min. Eng.*, 1603 (Nov. 1980).
38. Ref. 1, p. 115.

General References

Vanadium in *Mineral Facts and Problems*, Bull. 671, U.S. Bureau of Mines, Washington, D.C., 1980.

JOE B. ROSENBAUM
Consultant

VANILLIN

Vanillin [*121-33-5*] is the common name for 3-methoxy-4-hydroxybenzaldehyde.

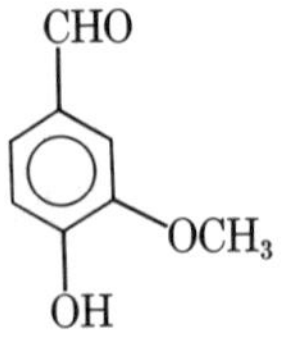

Vanillin occurs in nature as a glucoside, which hydrolyzes to vanillin and sugar. It has been identified in many oils, balsams, resins, and woods. The best-known natural source of vanillin is the vanillin plant (*Vanilla plansfolia*), a member of the orchid family. The vanilla bean was used by the Mexican Indians at the time of the Spanish conquests and was brought to Europe at the beginning of the sixteenth century. Since then, it has been a favorite food flavor. Vanilla beans are now grown in Mexico, Madagascar, Java (Indonesia), Reunion, and Tahiti.

Vanillin is also produced synthetically, but it is derived principally from lignin (qv), the main component in the spent sulfite liquors from sulfite pulp (qv) mills. Besides being a very popular flavor in the food industry, it is also useful in the synthesis of drugs; 40% of the vanillin is consumed in manufacturing drugs such as Aldomet, L-dopa, and Trimethaprim. Vanillin is also used in the perfume and metal-plating industries.

Physical and Chemical Properties

Vanillin crystallizes from water, water–alcohol, and organic solvents in the form of monoclinic prismatic needles (1). For crystallographic data, see reference 2. The physical properties of vanillin are listed in Table 1. Data have been compiled on the absorption spectra of vanillin in the vapor state and in the following solutions: ethyl alcohol, ethyl alcohol and sodium ethoxide, ethyl alcohol and sodium hydroxide, and hexane (3–4).

Vanillin is slightly soluble in water and freely soluble in ethyl alcohol, chloroform, ethyl ether, carbon disulfide, glacial acetic acid, pyridine, and caustic solutions.

Table 1. Physical Properties of Vanillin

Property	Value
melting point, °C	81–83
boiling point, °C	
at 101.3 kPa (1 atm)	284 (dec)
at 133 Pa (1 mm Hg)	127
density, d_4^{20}, g/cm^3	1.056
heat of soln in H$_2$O at infinite dilution, J/mol[a]	21.76
heat of neutralization[b], J/mol[a]	38.74
heat of combustion[c], J/mol[a]	3.83
solubility, wt %	
water, 14°C	1
water, 75°C	5
glycerol, 25°C	4.5
propylene glycol, 25°C	23.0

[a] To convert J to cal, divide by 4.184.
[b] When 1 mol vanillin in 30 L H$_2$O is neutralized with 0.2 N NaOH.
[c] Constant pressure and volume.

Vanillin is not considered a toxic compound. The LD for mice and rats is 4333 mg/kg and 4730 mg/kg, respectively. Inhalation of its vapors from a saturated solution shows no acute toxicity. However, application on the skin and mucous membrane causes local irritation. Vanillin causes changes in the blood indexes, nervous system, cardiovascular system, kidney, and adrenal gland. It has a low cumulative effect (5).

Reactions

Vanillin has both a phenolic and an aldehydic group and is capable of undergoing three different types of reactions: those of the aldehyde group, the phenolic hydroxyl, and the aromatic nucleus. The aldehyde group undergoes certain typical aldehyde condensation reactions that allow various substitutions for the aldehyde group. The aldehyde group can also be partially or completely reduced. However, as a p-hydroxybenzaldehyde, vanillin does not undergo some very common aromatic aldehyde reactions, such as the Cannizzaro reaction, the benzoin condensation, and oxidation with Fehling's solution to the corresponding acid, vanillic acid. If the hydroxyl group in vanillin is protected, oxidation to vanillic acid derivatives readily occurs. As a phenol, vanillin forms esters and ethers, and the nucleus is easily substituted by halogen and nitro groups. In comparison with most other aldehydes, vanillin is notable for its stability.

Condensation. With acetone in the presence of alkali, vanillin forms vanillylideneacetone, which can be reduced to zingerone (mp 41°C), the chief flavoring agent of ginger.

vanillylideneacetone   zingerone

Condensation of vanillin with hydroxylamine yields vanillin oxime, which is reduced with sodium amalgam and acetic acid to vanillylamine.

vanillin oxime   vanillylamine

The amides derived from vanillylamine are pungent. Capsaicin (*trans*-8-methyl-*N*-vanillyl-6-nonenamide), for example, has a potency one thousand times that of zingerone.

capsaicin

With phenylhydrazines, vanillin forms hydrazones. With heptaldehyde in dilute alcoholic potassium hydroxide, vanillin yields α-amyl-3-methoxy-4-hydroxycinnamaldehyde, which has a stronger aroma than jasminal, a fraction of oil of jasmine (see Oils, essential).

α-amyl-3-methoxy-4-hydroxycinnamaldehyde   jasminal

In a Perkin condensation, the *p*-acetate of ferulic acid, ie, 3-methoxy-4-acetyloxycinnamic acid, can be prepared by prolonged boiling of vanillin with sodium acetate and acetic anhydride. Vanillylidene cyanohydrin can be made by the action of potassium cyanide on a sodium bisulfite solution of vanillin. Vanillylidenenitromethane is prepared from nitromethane by using methylamine hydrochloride and sodium carbonate or ethylamine as the condensing agent.

3-methoxy-4-acetoxycinnamic acid   vanillylidene cyanohydrin   vanillylidenenitromethane

Vanillin condenses with acyl derivatives of glycine to form azlactones. This condensation furnishes the most useful method for the synthesis of compounds that may in turn be converted into aromatic amino and ketonic acids such as phenylalanine. Thus, vanillin reacts with acetylglycine in the presence of acetic anhydride and sodium acetate to produce an azlactone, an intermediate needed in one of the several routes to L-dopa (6).

4-[[4-(acetyloxy)-3-methoxyphenyl]-
2-methyl-5(4H)-oxazolone]

Reduction.   Vanillin can be reduced by means of platinum black in the presence of ferric chloride to give vanillyl alcohol (mp 115°C) in excellent yield. Vanillin can also be reduced to vanillyl alcohol with sodium amalgam in water. The yields are poor, and there are a number of by-products. Hydrovanilloin is a product of this reduction. High yields of vanillyl alcohol have been obtained by electrolytic reduction (7).

vanillyl alcohol   hydrovanilloin

Oxidation.   When vanillin is fused with alkali, it undergoes oxidation and/or demethylation, thereby yielding vanillic acid, protocatechuic acid, or both.

vanillic acid   protocatechuic acid

Exposure to air also causes vanillin to oxidize slowly to vanillic acid. When vanillin is exposed to light in an alcoholic solution, a slow dimerization takes place with the formation of dehydrodivanillin. This compound is also formed in other solvents.

Optimum yields of methoxyhydroquinone, which has antioxidant properties, are obtained by the oxidation of vanillin with alkaline hydrogen peroxide (see Antioxidants).

dehydrodivanillin   methoxyhydroquinone

Demethylation of vanillin to protocatechualdehyde has been accomplished using an aluminum chloride suspension in pyridine with a solution of vanillin in methylene dichloride and refluxing for 24 h (8).

Etherification. Methylation with either dimethyl sulfate or methyl chloride and caustic yield veratraldehyde. Ethyl, propyl, isopropyl, and benzyl ethers have also been prepared.

protocatechualdehyde

Esterification. The hydroxyl group of vanillin can be acetylated with acetic anhydride to produce vanillin acetate. Prolonged heating with acetic anhydride with or without a catalyst such as sulfuric acid causes the aldehyde group to react, forming vanillin triacetate.

vanillin acetate   vanillin triacetate

Substitution. When vanillin is halogenated or nitrated, substitution generally occurs in the 5-position. If the hydroxyl group is blocked by esterification or etherification, substitution takes place in the 2- and 6-positions. Bromination of vanillin acetate gives high yields of a 6-bromo derivative; nitration of the acetate gives high yields of the 2-nitro derivative and small amounts of a 6-nitro derivative. A variety of chlorine-substituted products of vanillin has been prepared, including the monochloro derivatives with substitution in the 2-, 5-, and 6-positions (9).

With ethyl Grignard reagent, vanillin gives good yields of isoeugenol. Vanillin reacts with acetaldehyde in the presence of alkali under controlled conditions to give coniferaldehyde (ferulaldehyde). Similarly, vanillin condenses readily with other aliphatic aldehydes.

Vanillin also readily forms the usual carbonyl derivatives, which can be used for identification, and the sodium bisulfite addition compound. This latter compound is useful in analysis and also in separating vanillin from acetovanillone in the manufacture of vanillin by the lignin processes.

isoeugenol   coniferaldehyde   acetovanillone

Although *ortho*- and *para*-hydroxybenzaldehyde do not undergo the Cannizzaro reaction under ordinary conditions, this reaction can be carried out with vanillin in the presence of silver.

Manufacture

Vanillin is obtained principally from natural sources, but 10–20% is made synthetically from guaiacol. The lignin processes are based on an alkaline air oxidation of a fermented spent-waste liquor from a sulfite pulp mill. The source of alkali is either sodium hydroxide or a combination of lime and soda ash which produces sodium hydroxide. The reaction is generally run at 160–175°C and 1.1–1.2 MPa (150–160 psig) for ca 2 h. The reaction product is treated differently by each manufacturer to isolate and purify vanillin. Ontario Paper Company, the largest vanillin producer, filters the calcium carbonate for recycle, acidifies, extracts with toluene, and then back-extracts with a caustic solution. The crude sodium vanillinate [57531-76-7] is then treated with SO_2, filtered to remove acetovanillone and other phenolic impurities, and acidified to precipitate the vanillin, which is filtered, vacuum distilled, and crystallized.

Monsanto and ITT extract the reaction product with either 2-propanol or butanol. When butanol is used, the extract is back extracted with caustic solution, allowed to react with SO_2, filtered to remove acetovanillone, and acidified to precipitate vanillin which is then purified. With 2-propanol, the extract is stripped, acidified, extracted with toluene, stripped of solvent, distilled, and crystallized.

Borregaard (Sarpsburg, Norway), which operates a lignin vanillin plant based on technology purchased from Monsanto Company, has been a leader in the investigation of ultrafiltration (qv) to upgrade the quality of the spent sulfite liquor (10). The main feature of this membrane separation is that it removes most of the undesirable low molecular weight compounds in the fermented liquors that cannot be oxidized to vanillin. Ultrafiltration has the advantages of increasing capacity of a vanillin plant by as much as 25% through a yield improvement and reducing the consumption of caustic and other raw materials. Borregaard reportedly has operated its vanillin plant on ultrafiltered liquors. This is consistent with their announcement that their plant capacity has been increased from 850 to 1000 metric tons per year (11).

Lignin vanillin is also produced in Japan, the People's Republic of China, and Poland (12). Interest in the production of vanillin continues to increase throughout the world; much as been published on this subject in recent years, especially in the Japanese and USSR literature (13–20).

In recent years, a number of patents have been issued for improvements of various aspects of the vanillin process (21–27). Several of these have been issued to Canadian International Paper Company.

Synthetic Routes

In a chemical complex built by Rhone Poulenc during the 1970s, a family of products is made from hydroquinone and catechol, the latter obtained by hydrogen peroxide oxidation of phenol. This family of products includes vanillin and ethyl vanillin. The vanillin is synthesized from catechol by first alkylating it to guaiacol. The guaiacol is then converted to vanillin via a modification of the Riedel process (28). This process involves the condensation of a salt of glyoxylic acid with guaiacol, followed by the air oxidation of the condensate. Then vanillin is released by acidifying the oxidation mass.

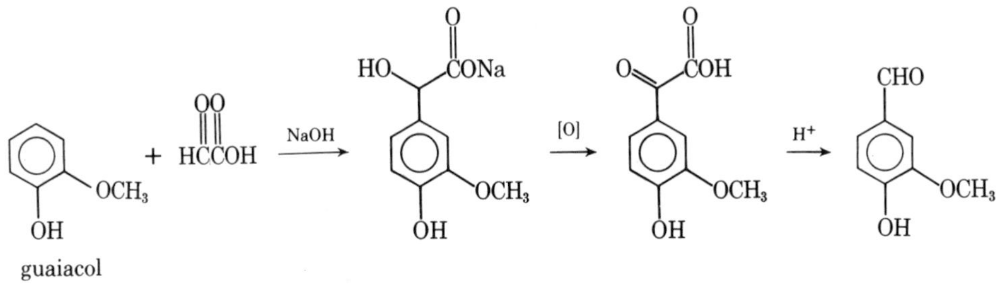

guaiacol

Either sodium or potassium salts of glyoxylic acid may be used. Potassium may be preferable because of its greater solubility.

Of the other methods reported for the preparation of vanillin, the best known are discussed below.

From Eugenol. Oil of cloves contains 85–95% eugenol (see Hydroquinone, resorcinol, and catechol), which provided a source of material of historical and practical significance. When eugenol is treated with strong alkali, it rearranges by migration of the double bond to form isoeugenol. Oxidation produces vanillin.

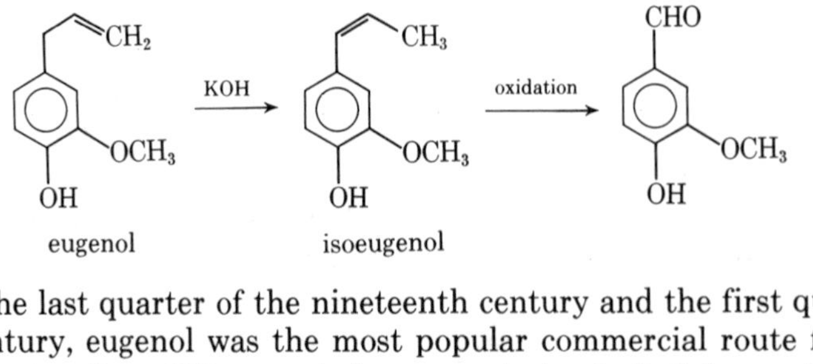

eugenol   isoeugenol

During the last quarter of the nineteenth century and the first quarter of the twentieth century, eugenol was the most popular commercial route for synthetic vanillin production. One of the accepted ways was to treat oil of cloves with potassium hydroxide at 200°C, thereby transforming the eugenol to isoeugenol. The reaction mixture was acidified with sulfuric acid and extracted with benzene. The isoeugenol from the benzene extract was acetylated to protect the phenolic group, and oxidation to vanillin was accomplished with ozone, potassium permanganate, or potassium dichromate.

From Guaiacol. Guaiacol (o-methoxyphenol, sometimes called catechol monomethyl ether) was first obtained from wood tar or coal tar in the destructive distillation of wood and coal. It was also made from o-dichlorobenzene. The ortho compound was treated with alkali to form catechol, which upon methylation gave a good yield of guaiacol. Starting with guaiacol, numerous vanillin syntheses are possible.

The Reimer-Tiemann reaction can be used to introduce an aldehyde group into the benzene ring. Resin formation is one disadvantage of this method.

The Gatterman synthesis involving guaiacol and hydrocyanic acid has also been used. In this case, unwanted vanillin isomers are formed as by-products.

| guaiacol | | isovanillin [621-59-0] | o-vanillin [148-53-8] |

Fries rearrangement of guaiacol acetate is reported to give good yields. The resulting acetovanillone (3-methoxy-4-hydroxyacetophenone) was then oxidized with nitrobenzene to vanilloylformic acid, which in turn was decarboxylated without resinification in a solution of dimethyl-p-toluidine.

guaiacol acetate

acetovanillone   vanilloylformic acid

The Sandmeyer reaction can be applied in a modified form to guaiacol to prepare vanillin. In this case, guaiacol is treated with formaldehyde and *p*-nitrosodimethyla- niline by heating a methanol solution on a water bath for several hours while bubbling in gaseous hydrochloric acid. The products are vanillin and *p*-aminodimethylani- line.

vanillyl alcohol

From Safrole. Safrole is obtained from camphor oil. Treated with alkali, it converts to isosafrole by migration of the double bond. Isosafrole can be oxidized to give pi- peronal (called heliotropine in the perfume industry) which, on treatment with PCl_5, gives protocatechualdehyde (3,4-dihydroxybenzaldehyde). Methylation with dimethyl sulfate in alkali yields vanillin and isovanillin (29).

safrole   isosafrole   piperonal

protocatechualdehyde   vanillin   isovanillin

Economic Aspects

Vanillin capacity in the West increased sharply in the early 1970s and again in the late 1970s. Capacity is estimated at 9,500–10,000 t/yr. The latest increase in ca- pacity occurred when Rhone Poulenc brought their new 2000-t plant on stream in 1978, followed in the next year by an increase in ITT's capacity.

The demand for vanillin increased steadily from 1970 to 1977 at a rate of ca 10%/yr, reaching a peak of 7000 t in 1977. Since then, demand has declined to a level of 5700–5900 t/yr in the period of 1978–1981.

Vanillin is sold in two grades, ie, USP and technical. The selling price of USP vanillin has dropped from its high of $12.75/kg in 1977–1978 to <$8.60/kg, but is now increasing again with a price that approached $11.00/kg in 1982. The selling price of technical vanillin follows the lead of USP and generally is ca $1/kg below USP.

Vanilla beans were in short supply in 1979–1980, meeting only about half the world demand. As a result, spot prices for Bourbon vanilla beans from Madagascar were $120–$140/kg in Mar. 1980, and Java beans were quoted at $100–$120/kg. Because of these high prices, many American processors turned to vanillin substitutes to meet their flavoring needs. It is estimated that a third of the vanilla bean–pure vanilla extract market has been lost to reformulation. In Mar. 1981, lessening of demand caused the bean prices on spot quotes to drop to $68/kg for Bourbon and $61.50/kg for Java beans, ie, about half the price of the 1980 spot quotes (30).

Grades and Specifications

Both USP and FCC (Food Chemical Codex) grades of vanillin must contain not less than 97.0% $C_8H_8O_3$, calculated on a dry basis. The USP grade should lose no more than 1.0%, and the FCC grade no more than 0.5% of its weight on drying for four hours over silica gel. Both specifications require that the residue on ignition should not be more than 0.05%. The limits of impurities for arsenic and heavy metals are 3 ppm max and 10 ppm max, respectively.

The technical grade vanillin as compared to USP contains foreign odors and color or related organic material from lignin that make this grade undesirable for flavoring. In addition, its assay of >98% and melting point of 80.5°C are lower than those for the USP grade.

Analysis

Identification. When a solution of ferric chloride is added to a cold, saturated vanillin solution, a blue color appears and changes to brown upon warming to 20°C for a few minutes. On cooling, a white to off-white precipitate (dehydrodivanillin) of silky needles forms. Vanillin can also be identified by the white to slightly yellow precipitate formed by addition of lead acetate to a cold aqueous solution of vanillin.

Determination. Various techniques are used for the analyses of vanillin, including colorimetric, gravimetric, spectrophotometric, and chromatographic (tlc, gc, and hplc) methods. The latest USP and FCC (*Food Chemical's Codex*) prescribe ultraviolet spectrophotometry for identifying and assaying vanillin. However, more vanillin analyses today are made by either gc or hplc. An outline of other methods formerly used is given in reference 31.

Gas chromatography is a widespread method for analyzing vanillin products and manufacturing process streams, vanilla extract, and compounded flavors. It is ideal for detecting many trace impurities associated with either lignin vanillin or guaiacal vanillin manufacture. It can determine levels of impurities in the vanillin finished product in ppm levels. Many commercial chromatographic column packings are available for analyzing vanillin. Generally, 2-m glass columns are preferred to stainless steel. Since there are so many potential methods for gc analysis, generally, each laboratory has its own gc techniques. However, a number of methods have been published (31–38), including one for separating, identifying, and assaying vanillin and other vanillin-related compounds (39). Information has also been published on tlc (40–41). Liquid chromatography (hplc) is just now becoming widely used in the flavor–fragrance industry to measure vanillin and other phenolic compounds (42–47).

Uses

Since 1970, the use of vanillin as a chemical intermediate in the production of pharmaceutical products has surpassed the quantities used for flavoring purposes (see Flavor and spices). Technical-grade vanillin is used in these pharmaceutical applications. The single largest use for vanillin is as a starting material for the manufacture of Aldomet by Merck. Aldomet is the trademark for an antihypertensive drug that has the chemical name of L-methyldopa or *levo*-3-(3,4-dihydroxyphenyl)-2-methylalanine. Merck has been assigned many patents to cover its process.

L-Dopa and Trimethoprim are two other drugs that are made from vanillin; L-dopa is used for treatment of Parkinson's disease, and Trimethaprim for upper respiratory-tract infections and some strains of venereal disease. Monsanto has patented a process for manufacture of L-dopa based on a unique asymmetric reduction step (5) (see Pharmaceuticals, optically active). Trimethaprim is a trademark of Burroughs-Wellcome.

Papaverine, used to treat heart patients, is another drug that was made originally from vanillin. Now it is made from veratrole, 1,2-dimethoxybenzene (see Cardiovascular agents).

Hydrazones of vanillin have been shown to have a herbicidal action similar to that of 2,4-D (48), and the zinc salts of dithiovanillic acid (made by the reaction of vanillin and ammonium polysulfide in alcoholic hydrochloric acid) which is a vulcanization inhibitor (49). 5-Hydroxymercurivanillin, 5-acetoxymercurivanillin, and 5-chloromercurivanillin have been prepared and found to have disinfectant properties. Vanillin itself has some bacteriostatic properties and has therefore been used in formulations to treat dermatitis (see Disinfectants and antiseptics).

A new potential use for vanillin is as a ripening agent to increase the yield of sucrose in sugarcane by treatment of the cane crop a few weeks before harvest (50).

Other uses for vanillin include prevention of foaming in lubricating oils, as a brightener in zinc coating baths (51), as an activator for electroless plating of zinc (52), as an aid to the oxidation of linseed oil, as an attractant in insecticides (53), as an agent to prevent mouth roughness caused by smoking tobacco (54), in the preparation of syntans for tanning, as a solubilizing agent for riboflavin, and as a catalyst to polymerize methyl methacrylate (55).

Derivatives and Related Compounds

Ethyl Vanillin. Ethyl vanillin, 3-ethoxy-4-hydroxybenzaldehyde, bourbonal, vanillal (mp 77–78°C), forms white crystalline needles completely soluble in ethanol. Ethyl vanillin is a synthetic flavoring and fragrance compound similar to vanillin but three to four times more intense. It is sold under such trade names as Ethavan (Monsanto Co.) and Vanaldol (Fries Bros.). In early 1982, it was selling for $28.90/kg.

Ethyl vanillin is made synthetically by Rhone Poulenc by the Reidel process using ethacol as the starting material.

Vanillic Acid. Vanillic acid, 3-methoxy-4-hydroxybenzoic acid, forms colorless needles from water (mp 207°C) and sublimes on heating. It is soluble to the extent of 0.12 g and 2.5 g in 10 mL water at 14°C and 100°C, respectively. Vanillic acid is very soluble in ethyl alcohol and ethyl ether.

Controlled potassium hydroxide fusion of vanillin gives high yields of vanillic acid, and vanillin is oxidized quantitatively to the acid in caustic soda with silver oxide.

Oxidation of vanillin has also been effected with mercuric and auric oxides (56). A good yield of vanillic acid can also be obtained by treatment of vanillin oxime with acetic anhydride followed by hydrolysis (57). In addition, vanillic acid is a by-product of the air oxidation of lignosulfonates in spent sulfite waste liquor. However, none of the lignin-vanillin manufacturers isolates this product from the oxidized lignin liquors.

Vanillic Acid Esters. Vanillic acid esters are easily prepared in high yield from vanillic acid and the appropriate alcohol. These esters have been used as nontoxic food preservatives, slime-control agents, and disinfectants, in pharmaceuticals, and in sun creams.

Ethyl vanillate forms colorless needles (mp 44°C, bp 293°C). It is insoluble in water, very soluble in ethyl alcohol and ether, and soluble in alkalies. Ethyl vanillate is less toxic than sodium benzoate when administered in oil. During World War II, the FDA considered the use of ethyl vanillate in amounts up to 0.10% as a preservative when such use would permit the delivery of acceptable food products to the U.S. Armed Forces. Ethyl vanillate has remarkable preservation properties in foods such as salt fish and fresh fruit (58).

Acetovanillone. Acetovanillone, 4-hydroxy-3-methoxyacetophenone, apoxynin, crystallizes from water as colorless prisms (mp 115°C, bp 295–300°C); it is very soluble in ether and benzene and insoluble in petroleum ether. Crude vanillin containing a small amount of acetovanillone as an impurity is available commercially and may be substituted for pure vanillin for many chemical purposes.

Veratraldehyde. Veratraldehyde is made by methylating vanillin, using dimethyl sulfate and sodium hydroxide. It is used as a metal brightener in the plating industry and as a perfume.

Isovanillin. Isovanillin, 4-methoxy-3-hydroxybenzaldehyde, crystallizes from water (mp 115–117°C). It is soluble in alcohol and ether and slightly soluble in carbon disulfide. The most attractive process for making isovanillin is based on first producing veratraldehyde and then deblocking the methoxy group in the 3-position (59). Isovanillin has been considered for making a synthetic sweetener (see Sweeteners) with the structure:

BIBLIOGRAPHY

"Vanillin" in *ECT* 1st ed., Vol. 14, pp. 603–611, by D. M. C. Reilly, Food Machinery and Chemical Corp. (Chemical Div.); "Vanillin" in *ECT* 2nd ed., Vol. 21, pp. 180–196, by Donald G. Diddams, Sterling Drug Inc., and Jack K. Krum, The R. T. French Co.

1. M. B. Jacobs, *Am. Perfum. Essent. Oils Rev.* **57,** 45 (July 1952).
2. W. C. McCrone, *Anal. Chem.* **22,** 500 (1950).
3. A. Lioxin, *Combined Phenol and Aldehyde*, Ontario Paper Co., Ltd., Thorold, Can., 1953, p. 16.
4. Slavo Chemical Co., Subsidiary Sterling Drug, Appleton, Wisc. 1950.
5. M. I. Makuruk, *Gig. Sanit.* **6,** 78 (1980); *Chem. Abstr.* **93,** 108350y (1980).
6. U.S. Pat. 4,005,127 (Jan. 25, 1977), W. S. Knowles, W. S. Sabacky, and B. D. Vineyard (to Monsanto Co.).
7. P. L. Sharma and J. N. Gaeuz, *J. Appl. Electrochem.* **1**(2), 173 (1981).
8. R. G. Lang, *J. Org. Chem.* **27,** 2037 (1962).
9. L. C. Raiford and J. G. Lichty, *J. Am. Chem. Soc.* **52,** 4576 (1930).
10. U.S. Pat. 4,151,207 (Apr. 24, 1974), H. Evju (to Borregaard Co.).
11. *Chem. Age*, 16 (July 31, 1981).
12. H. Wawrzyniak, *Przeglad Paper* **22**(9), 292 (1966).
13. Jpn. Kokai 74 25,934 (July 4, 1974), (to Soda Sangyo Co.).
14. H. Kawakami and T. Kanda, *Kami Pa Gikyoshi* **30**(3), 65 (1976).
15. E. I. Kovolenko and co-workers, *Nov. Electrokhim. Org. Soedin* **8,** 24 (1973).
16. Y. G. Mileshkevidc, M. V. Latosh, and V. M. Reznikov, *Khim. Der.* **12,** 21 (1972).
17. USSR Pat. 497,281 (Dec. 30, 1975), N. V. Mikhailov and co-workers.
18. Jpn. Kokai 73 10,035 (Feb. 8, 1973), (to Takasago Perfumery).
19. S. Kagawa and M. Rokugawa, *Kami Pa Gikyoshi* **25**(10), 506 (1971).
20. USSR Pat. 335,231 (Apr. 11, 1972), E. N. Viseleva and co-workers.
21. U.S. Pat. 3,600,442 (Aug. 17, 1971), D. G. Diddums and C. A. Hoffman (to American Can Co.).
22. U.S. Pat. 3,790,637 (Feb. 5, 1974), Chang-Tsing Yang (to American Cyanamid).
23. U.S. Pat. 3,920,750 (Nov. 18, 1975), V. B. Ralpkatzin, T. L. Diebold, and Mistemake (Canadian International Paper).
24. U.S. Pat. 4,021,493 (May 3, 1977), F. W. Major and F. M. A. Nicolle (to Canadian International Paper).
25. U.S. Pat. 4,075,248 (Feb. 21, 1978), H. B. Marshall and D. L. Vincent (to Domtar).
26. U.S. Pat. 4,090,922 (May 23, 1978), K. Bauer and H. W. Brandt, Jr. (to Bayer).
27. U.S. Pat. 4,277,626 (July 7, 1981), K. G. Foss, E. T. Talka, and K. E. Fremer.
28. Brit. Pat. 401,562 (Nov. 16, 1933), J. D. Reidel (to E. de Häen A.G.).
29. C. T. M. Bach and B. M. Taenius, *Rev. Cent. Cienc. Saude* (1–2), 137 (1979).
30. J. Rice, *Food Process.*, 40 (Nov. 1981).
31. J. Mendez and F. Stevenson, *J. Gas. Chromatogr.* **4**(12), 483 (1966).
32. A. C. Shaw and K. T. Waldock, *Pulp Pap. Mag. Can.* **68**(3), T118 (1967).
33. J. Mendez, *Gas Chromatogr.* **6**(3), 168 (1968).
34. D. Lindquest and T. K. Kirk, *Acta Chem. Scand.* **25**(3), 889 (1971).
35. I. A. Pearl and A. Olcay, *Tappi* **54**(10), 1656 (1971).
36. M. G. S. Chua and M. Wayman, *Tappi* **62**(3), 103 (1979).
37. R. H. Dyer and G. E. Martin, *Am. J. Enol. Vitic.* **31**(1), 3709 (1980).
38. N. G. Johansen, *J. Gas Chromatogr.* **3**(6), 202 (1965).
39. J. M. Pepper, M. Manotopoulo, and K. Burton, *Can. J. Chem.* **40,** 1976 (1962).
40. G. M. Barton, *J. Chromatogr.* **26**(1), 320 (1967).
41. H. C. Dass and G. M. Weaver, *J. Chromatogr.* **67**(1), 105 (1972).
42. A. B. McKogue, *J. Chromatogr.* **208**(2), 287 (1981).
43. E. Roggendorf and R. Spatz, *J. Chromatogr.* **204,** 263 (1981).
44. G. B. Shreis, *Sevost'yanov Y. A. Khromatogr. Khim. Drev.*, 61 (1975); *Chem. Abstr.* **84,** 32832a (1976).
45. H. W. Lange and K. Hempel, *Beckman Rep.* (1), 18 (1971).
46. G. E. Martin, G. G. Guinand, and D. M. Figert, *J. Agric. Food Chem.* **21**(4), 544 (1973).
47. J. Klimes and D. Lamparsky, *Int. Flavours Food Addit.* **7**(6), 272 (1976).
48. H. S. Reed and co-workers, *Compt. Rend.* **230,** 2317 (1950).

49. G. Bruni and T. G. Levi, *Atti Acad. Lincei* **32 i,** 5 (1923).
50. U.S. Pat. 3,994,715 (Nov. 30, 1976), L. G. Nickell (to Hawaiian Sugar Planters Assoc.).
51. K. Ohkubo, G. Suyama, and S. Toriba, *Nagamo Ken Seimitsu Kogyo Shikenjo Jigyo Gaiyo*, 86 (1976); *Chem. Abstr.* **88,** 67146n.
52. Fr. Demande 2,278,798 (May 21, 1976), F. Popesau.
53. Ger. Offen. 2,547,874 (Apr. 28, 1977), N. Weigand and co-workers (to Schuelke and Mayer).
54. Ger. Offen. 2,400,512 (July 17, 1975), J. F. Banks (to Brown and Williams' Tobacco Co.).
55. M. Imoto, T. Maeda, and T. Ouchi, *Chem. Lett.*, (2), 153 (1978).
56. U.S. Pat. 2,431,419 (Nov. 25, 1974), I. A. Pearl (to Sulfite Products Co.).
57. U.S. Pat. 2,433,277 (Dec. 23, 1974), H. F. Lewis and I. A. Pearl (to Sulfite Products Co.).
58. I. A. Pearl, *Food Ind.* **17,** 1173 (1945); I. A. Pearl and F. J. McCoy, *Food Ind.* **17,** 1458 (1945).
59. U.S. Pat. 3,367,972 (Feb. 6, 1968), W. B. Gitchel, E. M. Pagalals, and E. W. Schoeffel (to Sterling Drug).

J. H. VAN NESS
Monsanto Company

VAPOR-LIQUID EQUILIBRIA.   See Absorption; Distillation.

VARNISH.   See Insulation, electric (properties); Resins, natural.

VEGETABLE OILS

Vegetable oils are obtained as a tree crop or from the seed of annually grown crops. They include most of the fatty-acid esters of glycerol (qv), commonly called triglycerides, which provide the world with its supplies of edible oils and fats. The designations fat and oil commonly distinguish substances that are, respectively, solid and liquid at ambient temperature (see Fats and fatty oils).

Vegetable oils have a long history of use in human nutrition, but they were also extensively used as illuminants and lubricants and for soapmaking in ancient civilizations (see Soap). The advent of large-scale exploitation of mineral oils in the nineteenth century rapidly reduced the role of vegetable oil in illumination and lubrication, and the introduction of synthetic detergents shortly before World War II had a similar though less severe effect on the quantities of vegetable oils and fats used for soapmaking.

Until the middle of the last century, vegetable oils were recovered by mechanical or hydraulic pressing, a technique believed to have been in use in Asia ca four thousand years ago.

Of the extensive range of plants known to have oil-rich fruit or seeds, fewer than twenty are exploited commercially on a significant scale, and >90% of world vegetable-oil production is accounted for by nine oils (Table 1), of which soybean oil has for many years contributed the largest volume. Oleic (*cis*-9-octadecenoic) acid and stearic

Table 1. World Production and Exports of Vegetable Oils, 1000 Metric Tons

Oil	1969–1970		1974–1975		1979–1980	
	Production	Export	Production	Export	Production	Export
Edible oils						
cottonseed	2,345	341	3,114	410	3,063	440
groundnut	2,933	862	3,037	699	3,042	724
soybean	6,990	2,998	8,208	3,417	14,054	6,854
sunflower	3,430	700	3,802	687	5,705	1,569
rape	1,559	415	2,391	686	3,308	1,091
sesame	632	103	673	89	676	na
olive	1,380	237	1,552	205	1,559	245
corn	342	30	426	50	632	na
safflower	171	47	228	38	332	na
palm	1,433	704	2,648	1,695	4,405	2,752
palm kernel	424	310	474	385	587	426
coconut	2,289	1,038	2,729	1,305	2,944	1,300
miscellaneous	653	138	782	122	932	400
Total	*24,581*	*7,923*	*30,064*	*9,788*	*41,239*	
Nonedible oils						
linseed	1,020	433	691	252	868	
castor	337	244	432	153	339	

(octadecanoic) acid are the most common fatty acids in vegetable oils, but the range of fatty acids present in significant amounts in oils in general use extends from octanoic acid, normally present at a level of 5–10% in coconut oil, to erucic (*cis*-13-docosenoic) acid, which may be present at levels exceeding 50% in certain varieties of rapeseed oil. Unsaturation in the fatty acid occurs principally but not exclusively in those having an 18-carbon atom chain. Oleic, linoleic (*cis*,*cis*-9,12-octadecadienoic) and linolenic (*cis*,*cis*,*cis*-9,12,15-octadecatrienoic) acids are widely distributed in vegetable oils (see Carboxylic acids). Detailed fatty acid analyses of many oils are described in ref. 1, and the nomenclature used in the literature is explained in this Encyclopedia (see Fats and fatty oils). Oils are classified according to their principal fatty acid in ref. 3, and for present purposes it is useful to adapt and extend this classification (Table 2). Various fatty acids of unusual structure are to be found in a number of oils (2), usually at very low levels, but 12-hydroxy-9-octadecenoic (ricinoleic acid), which constitutes 80–90% of the fatty acids present in castor oil, is an exception.

For prediction of the triglyceride composition on the basis of the fatty acids present, the method based on 2-random 1,3-random distribution (3–4) is currently considered the most informative, although it cannot be applied in all cases.

Free (unesterified) fatty acids constitute the most important class of minor components in vegetable oils and normally must be removed in order to make the oil acceptable for edible purposes. The levels of mono- and diglycerides found in oils are related to the quality of the oil since hydrolytic breakdown of the triglycerides owing to improper handling of the seed or oil leads to the formation of these partial glycerides as well as fatty acids (5–6). The phosphatides are important components because of their value in the form of lecithin and its derivatives as well as their effect on oil processing (7). In addition, oils contain a considerable number of unsaponifiable components (8) (Table 3), of which the sterols and tocopherols have potential value as by-products. The tocopherols are responsible for resistance to oxidation of the oils in which they are found (see Vitamins, Vitamin E).

Table 2. Principal Vegetable Oils and Their Sources

Principal fatty acid	Oil	Content of principal fatty acid, %	Source	Principal regions of cultivation	Oil content of oil-bearing material, %
lauric	coconut	44–52	*Cocos nucifera*	Philippines, Indonesia, India, Sri Lanka	65–68[b]
palmitic	palm kernel	46–52	*Elaeis guineensis*	Malaysia, Indonesia, W. Africa	45–50[c]
	palm	32–47	*Elaeis guineensis*	Malaysia, Indonesia, W. Africa	45–50[d]
oleic	olive	65–86	*Olea europaea*	Spain, Italy, Greece, Tunisia	15–40[e]
	groundnut	42–72	*Arachis hypogaea*	India, China, United States	45–55[f]
	rapeseed	48–60	*Brassica campestris, Brassica napus*	Canada, China, India	40–50[g]
linoleic (medium)	sesame	34–45[a]	*Sesamum indicum*	China, India, Burma	44–54[g]
	soybean	52–60	*Glycina max*	United States, Brazil, China	18–20[g]
	cottonseed	40–55	*Gossypium hirsutum*	USSR, China, United States	15–24[h]
	corn	34–62	*Zea mays*	United States	33–39[i]
linoleic (high)	sunflower	58–67	*Helianthus annus*	USSR, United States, Argentina	22–36[g]
	safflower	78	*Carthamus tinctorius*	United States, Mexico	25–44[g]
linolenic	linseed	30–60	*Linum usitatissimum*	United States, Argentina, India	35–44[g]
ricinoleic	castor	80–90	*Ricinus communis*	Brazil, India, USSR	35–55[g]

[a] Linoleic acid content in same range.
[b] On dried meat.
[c] On dried kernel.
[d] In pericarp.
[e] On undried fruit.
[f] On kernels.
[g] On seed.
[h] On whole seed.
[i] Of germ.

Table 3. Minor Components in Crude Vegetable Oils

Oil	Phosphatides, % (P × 30)	Total unsaponifiable matter, %	Sterols, %	Tocopherols, ppm
coconut	trace	0.2–0.6	0.06–0.22	30–80
corn	1–3	0.8–2.9	0.8–1.5	870–2500
cottonseed	0.7–0.9	0.5–0.8	0.26–0.31	1000–1200
groundnut	0.3–0.4	0.2–0.8	0.19–0.47	220–600
olive	ca 0.1	0.5–1.3	0.16–0.6	30–300
palm	0.07–0.12	0.2–0.5	0.03–0.26	500–800
palm kernel	na	0.2–0.8	0.06–0.3	trace
rapeseed	0.1	0.6–1.2	0.35–0.84	550–700
soybean	1.1–3.2	0.5–1.6	0.15–0.42	1000–2800
sunflower	<1.5	0.3–1.6	0.25–0.75	700

The color of oils is primarily due to the presence of carotenoids and chlorophyll and its derivatives (see Dyes, natural; Terpenoids). Carotenoid content of most oils is <50 ppm (9), the outstanding exception being palm oil, which contains 500–700 ppm carotene in the case of Malaysian plantation-produced oils and may contain up to 1500 ppm carotene when produced in primitive conditions (6). The carotenoid pigments can generally be rendered colorless at temperatures >200°C provided oxygen is absent. The chlorophyll content of oils seldom exceeds 10 ppm, although seed exposed to frost during maturation may give an oil containing higher levels (10). The pigmentation of crude cottonseed is due to the presence of high levels of gossypol.

Metals are present in many oils. Copper (generally <1 ppm) and iron (generally <10 ppm) are of greatest significance because of their potentially adverse effect on product quality. Heavy metal contents of oils have been monitored (11). Sulfur is found in rapeseed oil at levels of up to 30 ppm (12) and must be removed in order to avoid subsequent processing difficulties. Pesticides are found at low levels in many oils because of their widespread use in intensive agriculture (13) (see Trace and residue analysis).

Spoilage of fats occurs as a result of hydrolysis or oxidation (14–15). In either case, the fat becomes unsuitable for human consumption. To reduce the danger of spoilage, steps are taken throughout the process, from seed harvesting to product storage, to eliminate the relevant causes. These steps include the destruction or inactivation of harmful microorganisms, the preservation of natural antioxidants in oils, the removal of pro-oxidants, and the exclusion of oxygen during processing and of water during storage. The oxidation of linolenic acid is best avoided by selective hydrogenation (see Hydrogenation below).

Physical Properties

Density, viscosity, and melting behavior are the physical properties of oils that are of most general interest to processers and users. Because of the close similarities between triglyceride molecules found in different oils, the densities and viscosities of most oils lie within a narrow band, with the notable exception of castor oil (qv), which is characterized by a much higher viscosity and higher density. Comprehensive sets of data are given in refs. 1 and 16.

Dilatometric measurements and, more recently, nmr determinations (17) constitute important techniques for measuring the amount of crystallized material in a fat. Dilatometry is also used in the form of statistical dilatation equivalents for fat blending (18) and isodilatation diagrams for assessing the compatibility of different fats (19). Because the refractive index of an oil is a function of molecular weight and unsaturation, refractometry is used in quality and process control (20). A more direct measure of unsaturation is, however, obtained by determination of the iodine value (21), a long-established analytical method that is to some extent being supplanted by various chromatographic techniques (3).

The triglycerides are relatively apolar; consequently, solvents such as hexane are completely miscible with the completely liquid oils. Acetone, although more polar, is also a useful oil solvent and is used in fractional crystallization. The short-chain aliphatic alcohols show some solubility for fatty acids but very much less for triglycerides. They can therefore be used to extract fatty acids from the crude oils (22). Data on the vapor pressure of the fatty acids are given in ref. 23. Studies on potential applications of supercritical gas extraction to oil extraction and processing are beginning to provide data on the solubility of fatty acids and various glycerides in a number of gases at values of temperature and pressure above their critical point (24).

Oil Production

The growth in demand for fats and oils as a result of increasing world population and rising living standards has been met overwhelmingly by a growing output of vegetable oils. A comparison of production statistics (Table 1) for the period 1970–1980 shows an increase of >60% in vegetable oil output, whereas animal and marine oil production was increasing in volume by only ca 9%. Four oils, ie, soybean, palm, rapeseed, and sunflower oils, are largely responsible for this growth in overall production.

Production of soybean oil has grown rapidly in the United States in the 1970s and recently also in Brazil and Argentina (see Soybeans and other seed proteins). The growth in palm oil output is primarily due to the systematic extension of oil palm plantations in Malaysia, but a growing number of other countries have raised their output of the oil as well (25). Rapeseed oil production, which has advanced significantly in Canada and a number of European countries, has benefited from the availability of new varieties of the seed that give an oil low in erucic acid, as now widely required when the oil is to be used as food (26). In the countries of the European Economic Community (EEC), production has responded well to the policy of supporting domestically produced rapeseed (27). Rapeseed oil rich in erucic acid is now produced exclusively for industrial purposes. The production of sunflower oil has risen sharply following upon the availability of commercially useful hybrids of oil-enriched seed, and the United States has become a major exporter (28–29).

Advances in the techniques of plant genetics continue to benefit oil production. The position of palm oil in the vegetable-oil market is likely to be further strengthened by the work on oil-palm cloning using tissue-culture methods (30–31). Tissue-culture studies on rapeseed have also been reported (32). In the case of the soybean plant, genetic studies have focused on the factors responsible for the spoilage tendency of the oil (33) (see Genetic engineering).

The potential for increase in the production of cottonseed oil, an oil that is widely valued for its stability, is uncertain owing to the impact of the synthetic-fibers industry on the natural-fibers market, as well as difficulties in pest and plant-disease control (34). The introduction of glandless cotton cultivars, on the other hand, offers the prospect of an oil free of the undesirable pigment gossypol, provided lint yields are maintained.

A number of other oils are of considerable regional importance. Of these, olive oil is the most important. Its production, which is largely confined to the Mediterranean zone, has remained relatively static in the last decade at ca 1.5×10^6 metric tons per year. On a far smaller scale, ricebran oil is indigenous to India and Japan, and India also produces and consumes other indigenous oils that are largely of forest origin.

Vegetable oil prices on the international market are determined by the interaction of the forces of world supply and demand. Interchangeability between oils leads to a tendency for all prices to move together (35), although the particular supply–demand situation of an individual oil may cause its price to move more or less than that of other oils.

Oil price fluctuations, besides reflecting planting decisions and environmental effects, are subject to complex influences stemming from the interactions between oil, meal, and oilseed prices since soybeans are dominant in both the world vegetable oil and oilcake market. Price trends (Fig. 1) will depend on whether world production of edible oils rises fast enough over the long term to meet the growth in world needs, now ca 4.5% per year.

Recovery of Oil from Seed and Fruit. For the harvested oilseed, storage conditions play an important part in determining the quality of the oil produced, particularly since the environmental conditions prior to and during harvesting may be less than ideal. Moisture content and temperature of the seed must be controlled if deterioration is to be avoided, but a low moisture content can cause dehulling difficulties. Ventilated

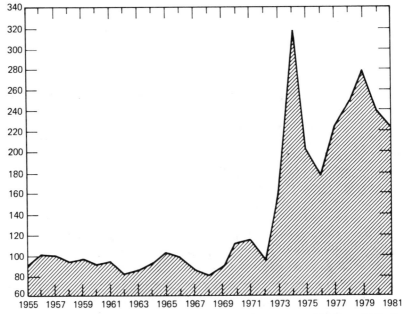

Figure 1. Index of world edible-oil prices, 1954 = 100. Mean 1981 oil price = $600. Courtesy of Economics Dept., Unilever PLC, UK.

silos are the preferred method of storage, but even in good conditions deterioration may occur on long storage (36–38).

The stored seed frequently contains potentially harmful quantities of extraneous matter from harvesting, making seed cleaning essential. Decortication, the removal of hulls, is applied to cottonseed, sunflower seed, and soybeans, but cottonseed must be delinted before dehulling (39) (see Cotton). Soybean dehulling serves to improve the protein content of the meal and may be omitted where this is not required, but the omission of sunflower seed dehulling leads to increased wax content of the oil produced (40). Rapeseed dehulling has proved impractical in view of the difficulties encountered in obtaining a clean separation of hulls (41). Before the cleaned and, where appropriate, decorticated seed passes to the extractor, it is flaked to ensure satisfactory extraction (42). Mechanical pressing as a means of oil recovery is virtually obsolete except in combination with solvent extraction, where prepressing of oilseeds of high oil content facilitates extraction plant operation. A typical arrangement of an extraction plant using prepress of seed is shown in Figure 2.

A number of extractor types are available for oilseed extraction; most are based on percolation rather than on total immersion of the material to be extracted (43) (see Extraction, liquid–solid). A combined percolation and immersion process, with intermediate desolventizing and flaking, is sometimes employed for material containing

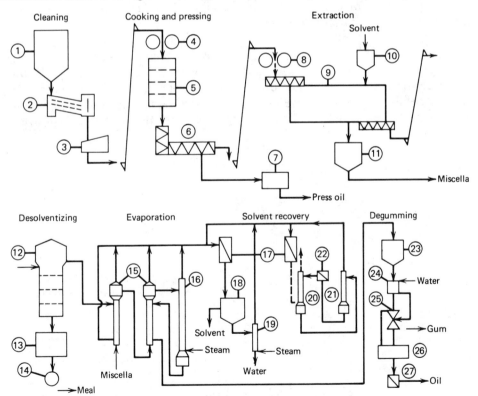

Figure 2. Flow diagram of a pre-press extraction plant. 1, Silo; 2, cleaner; 3, stoner; 4, crushing mill; 5, cooker; 6, press; 7, filter; 8, flaking mill; 9, extractor; 10, solvent tank; 11, miscella tank; 12, desolventizer; 13, meal cooler; 14, mill; 15, long tubular evaporators; 16, stripping column; 17, condensers; 18, separator tank; 19, stripper; 20, absorber; 21, stripping column; 22, heat exchanger; 23, tank; 24, mixer; 25, degumming separator; 26, oil dryer; 27, oil cooler. Courtesy of Elsevier (41).

>40% oil (44). Industrial hexane is used throughout the industry as extraction solvent, largely because of its toxicological acceptability, relative selectivity for glycerides, and ease of recovery (see Solvents, industrial).

The extractor is normally operated to give a final miscella containing 25–30 wt % oil. Desolventizing of oil and meal and the post-extraction toasting of the meal, required on nutritional grounds, make the extraction process energy-intensive; soybean extraction normally requires 300–350 kg steam and 110–150 MJ power per metric ton of seed input. Extended drainage of the defatted meal at the end of the extraction can significantly reduce the amount of solvent to be removed from the meal and thus save energy (45).

The recovery of palm oil of good quality from the bunches of palm fruit as harvested requires rapid deactivation of the lipase present in the fruit. This is achieved by sterilization of the fruit bunch with live steam. A digester is used to prepare the fruit, loosened from the bunch during sterilization, for pressing. The pressed oil is freed from remaining solid matter and free moisture by centrifugal separators, and a final vacuum drying stage reduces the moisture content of the oil to ca 0.1% (46). Apart from the damaging effect of lipolytic activity on the free fatty acid content of the oil, and thus on the yield of refined oil, poor handling also has an adverse effect on the ease of color removal, a most important criterion for an oil rich in carotene (47). In contrast to the energy requirements of oilseed extraction, a palm oil mill can produce enough energy to meet its own needs (46).

Olive oil is recovered from the fruit in hydraulic presses operated at up to ca 39 MPa (5700 psi) (48) and, increasingly, by more recent techniques for effecting separation of oil from the pulp (49). The oil thus recovered (virgin oil) is sold unrefined if its organoleptic and acidity standards are considered adequate. Conventional refining is applied to virgin oil of lesser quality. The press residue is ground further, dried, and solvent-extracted to give a husk oil that is refined and may then be blended with virgin oil.

Oil Processing

Refining. Refining (see Fig. 3) transforms the crude oil, which contains a number of minor components that affect the acceptability of the product, into an oil sufficiently

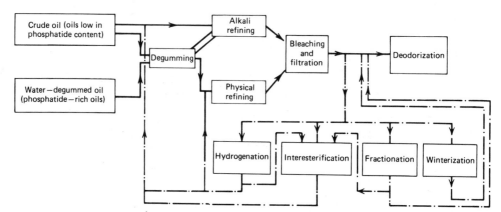

Figure 3. Sequence of oil refining and modification processes.

purified for the application in question. This requires the removal primarily of free fatty acids, phosphatides, pigments, and volatile components. The refining process must also remove artifacts introduced by the process itself, eg, the soap formed in alkaline neutralization of free fatty acids, as well as components such as phosphatides, metals, and sulfur that have an adverse effect on the process.

The refining process is primarily operated for product quality, but the economics of refining, which is strongly affected by neutral oil yield and increasingly by effluent and energy charges, is also a principal factor in shaping the nature of process improvements.

Degumming and Other Pretreatments. Removal of phosphatides from oils rich in these lipids (5) is essential because they affect oil-processing efficiency and product quality (50–52). Some of the phosphatides present in oils (Table 3) are readily hydrated (and therefore precipitated by water) whereas others, mainly the calcium and magnesium salts of phosphatidic or lysophosphatidic acid, are not hydrated by water but can be hydrated by alkali or acid (53). Water degumming, often carried out in the extraction plant in order to reduce the risk of phosphatides hydrating and precipitating during storage, reduces the level to 0.3–0.6%, equivalent to 100–200 ppm phosphorus. Live steam is sometimes used to degum oils if water proves to be an ineffective degumming agent.

Conventionally, degumming is combined with alkali refining. The phosphatides not precipitated by water can be precipitated by reaction with concentrated phosphoric acid or other acids immediately prior to alkaline neutralization of the free fatty acids. Other agents for removing phosphatides from soybean oil have also been proposed (54).

Alternatively, citric acid is used to precipitate the phosphatides, and if separated only after the temperature of the oil is lowered, the residual phosphatide contents are reduced to 20–40 ppm (as phosphorus) (55). Membrane filtration of hexane miscella has also been shown to be effective in removing phosphatides and iron from oils (56).

Virtually total removal of phosphatides is particularly important when steam stripping is used to remove fatty acids, and in this case, the reduction of iron content is equally essential. This twin objective is achieved by degumming with concentrated phosphoric acid, a technique also used in the pretreatment of oils, eg, palm oil, that have low phosphatide content but contain substantial amounts of iron and, in some cases, copper (57).

The need to remove waxes (qv) applies only to a few oils, the outstanding example being sunflower oil. Dewaxing may be carried out on the miscella from a seed extraction plant (58) or in combination with neutralization (59), but the traditional method of dewaxing as applied to the refined oil is similar to fractional crystallization (winterization) (40) (see also Fractional Crystallization). Prior removal of phosphatides leads to significantly higher rates of filtration of the suspension of wax particles (60), apparently because larger crystals form in the absence of the phosphatides (61). Filtration is normally carried out in the presence of a filter aid.

Deacidification. Deacidification with alkali (alkali refining) is widely practiced using batch or continuous operation (62–63); the latter is based largely on centrifugal separators for separating oil and soapstock, whereas gravity separation of the phases is used in batch operation. Its ability to deal effectively with a wide range of oil qualities gives alkali refining a clear advantage over other processes, but oils of poorer quality suffer serious neutral oil losses, and the process generates considerable quantities of effluent. Neutral oil losses and effluent formation can be reduced by miscella refining (64), an option best suited to a refinery closely integrated with a seed extraction plant.

Physical refining, or deacidification by steam stripping, constitutes the principal alternative to alkali refining and is always carried out in a semicontinuous or continuous manner. This process, normally operated at 0.2–1.2 kPa (1.5–9 mm Hg), requires temperatures of 240–270°C for oils containing C_{16} or C_{18} fatty acids, but the lauric oils (C_{12}) may be stripped at 200–220°C. Neutral oil losses in stripping are significantly lower than those quoted for alkaline deacidification (65), although oil losses in pretreatment must be taken into account in making this comparison (66). Effluent formation is also lower in physical refining (67). The plants used are in many cases identical with those used for deodorization, and up to 5 wt % of stripping steam may be used. The stripped fatty acids are recovered by means of a partial condenser or in a scrubber.

At the stripping temperature, carotenoids decompose rapidly, but chlorophyll is resistant to thermal decomposition and must be removed by adsorption. The high temperature also leads to some loss of tocopherol from the oil (51). Despite the process advantages, physical refining is not suitable for all oils, particularly those containing high levels of oxidized triglycerides or pigments likely to discolor at high temperature (68).

Bleaching. Bleaching of edible oils has traditionally been carried out with natural or activated bleaching clay (69–72) using temperatures of 90–110°C and preferably operating under vacuum. The amount of bleaching clay required depends on the history and prior processing of the oil, but it is normally 0.5–2.0 wt %. Where deacidification has been carried out with alkali, the earth removes residual soap as well as pigments and various minor components. Reduction in pigment content as a function of contact time for rapeseed oil was used to characterize the kinetics of adsorptive bleaching (73). Because of its high cost and high oil retention, activated carbon is little used in bleaching, but its ability to adsorb certain hydrocarbons is valuable in some cases (74).

The reduction of bleaching clay levels is an important objective for the refiner, since higher levels mean higher clay costs, greater losses of oil in spent clay, and lower filter-press capacity. However, the reduction of the labor content of the process has also been important, and thus the industry in Europe has increasingly used one or other of the automatic centrifugal-discharge filters, which are available in sizes up to 100 m^2 filtration area (75) (see Filtration).

Deodorization. The volatile components present in the deacidified, bleached oil are removed by deodorization, conventionally carried out at 0.2–1.2 kPa (1.5–9 mm Hg) and 175–270°C. The actual conditions used depend on the oil to be processed and the utilities available. Aldehydes and ketones formed by oxidative and thermal breakdown of unsaturated fatty acids predominate in the volatile fraction, but many other compounds have also been found (76). Removal of pesticides from the oil requires a deodorization temperature of at least 240°C (77), but in some cases, eg, in processing cocoa butter, it is essential to use a low temperature, preferably <180°C. A small amount of concentrated citric acid solution is normally added during deodorization to scavenge any metal present in the oil.

Deodorization is carried out batchwise by sparging steam into the oil through a distributor in the lower part of an upright cylindrical vessel. This distributor has many 2–3 mm openings to give a steam velocity at the point of entry into the oil phase sufficient to form bubbles of 3–5 mm dia. A head of ca 2 m of oil is commonly provided above the distributor, which leads to a rapid change of pressure as the bubble passes through the oil. Bubble expansion and fragmentation produce a continuous thick oil froth at the liquid surface, and transfer of the volatile components to the steam is facilitated by the interfacial area thus created.

The principle of steam dispersion in a continuous oil phase is also applied in the design of semicontinuous deodorizers (Fig. 4). These consist of a series of stages with the oil being held for a fixed time in each. In the first two stages, the oil is heated to the deodorizing temperature, in part by exchanging heat with deodorized oil. The oil then passes to one or more deodorizing trays. The final tray is used to cool the oil to a temperature at which it can be discharged from the deodorizer. The temperature used in semicontinuous operation, 220–270°C, facilitates removal of the volatile fraction and thus makes reduction of the cycle time possible. The conditions employed and constructional considerations lead to the use of shallower oil layers in semicontinuous deodorizers, and in some designs the steam injection device is modified to give improved steam dispersion and more vigorous oil circulation (78). These principles are equally applicable to the design and operation of continuous deodorizers, but in this case, a steam-continuous steam-oil contacting system may also be used, eg, in the form of a packed column (79), thus gaining the advantage of low pressure drop when operating countercurrently.

Deodorization is the most energy-intensive of the processes in the refining sequence, and reduction of the energy requirement has attracted much attention. Energy is used to heat the oil to the operating temperature, to provide the live steam used for stripping, and to provide the motive steam normally used to operate the vacuum

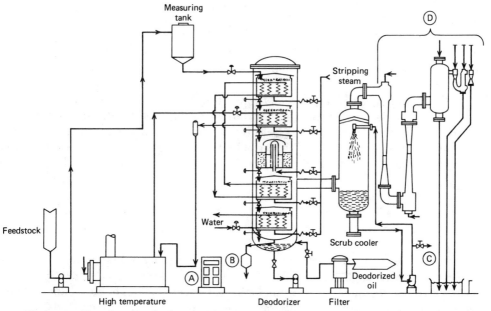

Figure 4. Flow sheet for deodorizing plant using a Votator semicontinuous deodorizer with heat recovery and distillate recovery units. A, Instrument panel; B, shell drain condensate collection tank; C, deodorizer distillate; D, steam-jet ejector system. Courtesy of Chemetron Corporation.

system. Batch deodorizers offer limited scope for energy conservation, as the recoverable energy is of a low grade. In semicontinuous deodorizers, heat recovery of up to 70% has been claimed (80), and similar high levels of recovery of the energy used to heat the oil can be expected in fully continuous operation. Overall energy requirement data cited (81) vary over a wide range, but levels of 200–400 kg steam per metric ton oil including, where necessary, the equivalent of other forms of energy used for oil heating, are most common for semicontinuous and continuous operation.

Oil Modification

Hydrogenation, interesterification, and fractional crystallization provide means of modifying the relationship between solids content and temperature of oils and their blends to give products of specific functional properties. They have consequently created great flexibility for vegetable oils. Hydrogenation provides this flexibility by virtue of the wide range of process conditions applicable, whereas interesterification is used to modify individual oils or blends of oils. Fractional crystallization, on the other hand, is a separation process able to create fractions differing in melting behavior, and this characteristic has a variety of uses.

Hydrogenation. Hydrogenation, the most widely used of the oil-modification processes, reduces the degree of unsaturation in the fatty-acid groups of the glycerides (82–83), but at the same time it causes isomerization, both cis–trans and positional, of the residual unsaturated fatty acids (84). The reduction in unsaturation has two important consequences, ie, higher melting-range temperatures and enhanced resistance to oxidation and flavor deterioration.

It is convenient to consider the process as a series of consecutive pseudo-first-order reactions, represented by

$$\text{Le} \xrightarrow[k_{\text{Le}}]{+\,H_2} \text{L} \xrightarrow[k_{\text{L}}]{+\,H_2} \text{O} \xrightarrow[k_{\text{O}}]{+\,H_2} \text{S}$$

where Le = triene (linolenic acid and isomers); L = diene (linoleic acid and isomers); O = monoene (oleic acid and isomers); S = saturate (stearic acid); and k_{Le}, k_{L}, and k_{O} are reaction constants.

The importance of preferentially converting trienes and dienes has directed attention to the selectivity of the catalyst and operating conditions. Linolenic acid selectivity, sometimes referred to as Selectivity II, is given by the quotient $k_{\text{Le}}/k_{\text{L}}$. Selectivity I is defined as $k_{\text{L}}/k_{\text{O}}$. The rate of hydrogen uptake by the oil may be limited by the rate of mass transfer of the gas into the oil (physically controlled hydrogenation) or by the rate of reaction (chemically controlled process), but in most cases, mass-transfer and reaction rates are comparable in magnitude, except in the latter stages of the process, when the rate of reaction can be expected to be dominant. The selectivity can be computed from analytical data (85) or from the process operating conditions (86).

Nickel, deposited on a support (normally kieselguhr) to maximize its catalytic activity, is the principal catalyst used in oil hydrogenation, but other catalysts have been explored on account of poor linolenic acid selectivity (II) of nickel and because of the coupling of Selectivity I and the isomerization reactions. Copper, mostly in the form of copper chromite, is more selective than nickel (87) but requires far more stringent posthydrogenation refining because of the prooxidant properties of even very low levels of the metal. Attempts to develop catalysts for homogeneous-phase hydrogenation have yielded some promising results (82,88).

Selectivity as defined above is important because of its effect on oil stability and on the solids content–temperature relationship, but the latter characteristic is also strongly affected by cis–trans isomerization of the unsaturated fatty acids. The coupling of Selectivity I and isomerization is well illustrated by data on the hydrogenation of Canola (low-erucic rapeseed) oil (89).

The effect of the important process conditions on hydrogenation rate, selectivity, and isomerization can be seen in Table 4 and more specifically in ref. 83.

Batch hydrogenation is generally preferred to continuous operation of the process because it is more flexible and offers greater control over product quality.

Either a dead-end or gas-recirculation system of hydrogen supply may be used. In the former, gas is dispersed into the oil by means of a sparger in the lower part of the vessel and the agitator serves to redisperse the gas leaving the oil surface. In the recirculation system, gas is withdrawn from the hydrogenation vessel, purified, and recycled to the sparger together with fresh gas. Details of batch plant and the operating conditions normally applied are given in ref. 83. Catalyst reuse in vegetable oil hydrogenation is generally practiced and limits consumption to 0.1–0.2% (as nickel) on the mass of oil.

The high rate of reaction in the early stages of hydrogenation requires adequate cooling surface in the autoclave (4–5 m^2/t oil) to remove the exothermic reaction heat of ca 4.5 MJ/t oil per unit change of iodine value. In continuous hydrogenation, the exotherm can be used more effectively; it is claimed that a continuous-hydrogenation plant requires heating steam only during startup (90). The hydrogen used, which can be produced by electrolysis or steam reforming, must have a purity of at least 99.5%, since contamination with carbon monoxide is a cause of catalyst deactivation, particularly at low hydrogenation temperature (see Catalysis).

The geometrical and positional isomers formed during hydrogenation have been extensively studied to elucidate their dietary effects. The trans isomer of oleic acid, ie, elaidic acid, is the main isomer present in hydrogenated fats, but lesser amounts of isomerized dienes also occur (91–92). Of these, *trans,trans*-linoleic acid has attracted special attention (91).

Table 4. Effect of Process Variables on Overall Hydrogenation Rate

Process variable	Physically controlled process[a]		Chemically controlled process[a]	
increasing pressure	rate	+	rate	+
	selectivity	−	selectivity	−
	isomerization	−	isomerization	−
increasing agitation	rate	+	rate	0
	selectivity	−	selectivity	0
	isomerization	−	isomerization	0
increasing catalyst	rate (per unit of catalyst)	−	rate (per unit catalyst)	0
	selectivity	+	selectivity	0
	isomerization	+	isomerization	0
increasing temperature	rate	±	rate	+
	selectivity	+	selectivity	±
	isomerization	+	isomerization	±

[a] + = positive effect, − = negative effect, ± = small effect, and 0 = no effect.

Studies on nutritional effects have shown high levels of digestibility for partially hydrogenated fats, thus distinguishing them from the saturated fats which are less well absorbed (93). Multigeneration studies on animals have also shown that fats containing high levels of trans fatty acids have no significant effect on growth, longevity, and reproduction (94), and autopsy results (95) cast doubt on the hypothesis that the trans fatty acids may be more atherogenic than other fatty acids (96).

Serum-cholesterol level has been found in some cases to be affected by the presence of elaidic acid to an extent intermediate between that of oleic acid and the shorter-chain saturated fatty acids (97).

Although there is a recognized tendency for elaidic acid to be preferentially incorporated in phospholipids, the availability of polyunsaturated fatty acids in the diet is believed to be the overriding factor affecting membrane structure and functionality (91), although alternative analyses of findings have been put forward (98). Research into various aspects of the presence of trans fatty acids in fats has also been comprehensively reviewed (99).

Acknowledging the divergent findings and interpretations reported, a recent report on special margarines (100) makes specific recommendations on acceptable levels of trans monoenes and trans,trans dienes, as well as calling for *cis,cis*-linoleic acid to be present at a level sufficient to provide the required linoleic acid intake.

Interesterification.   Interesterification is used extensively to rearrange the fatty acid groups in triglyceride mixtures, thus modifying their melting behavior (101–102). In most applications, the goal is randomization of the fatty acid distribution, but by coupling interesterification with a separation process (eg, fractional crystallization), directed or nonrandom interesterification is obtained (103). The sodium alcoholates, sodium metal, and a sodium–potassium alloy have been the catalysts most extensively used for this reaction, but sodium hydroxide and sodium glycerolate have increasing application. The catalyst requirement, 0.05–1.0% of the fat charge, is minimized by processing well-refined and dried fats (101). Batch interesterification can be carried out in refinery vessels, but stirred-tank and tubular reactors have found favor in continuous operation (104–105). The operating temperature is then raised from 90–110°C to 130–160°C. Reaction time is thereby reduced from 2–4 h to a few minutes, leading to much smaller reactor volumes.

Catalyst residues and artifacts such as the alkyl esters of fatty acids, formed when using sodium alcoholate catalysts, are readily removed in postrefining. Lipolytic enzymes can be used in place of conventional catalysts in interesterification (106), but their use requires reaction temperatures <60°C which makes it advisable to carry out the reaction in solvent.

Fractional Crystallization.   Fractional crystallization or fractionation, referred to as winterization when removing high melting components such as waxes, offers a means of modifying the melting properties of fats without resorting to chemical change. It is also used to improve the stability of an oil by hydrogenating the most unsaturated fraction and removing the high-melting triglycerides formed by crystallization.

Crystallization may be from the melt, commonly referred to as dry fractionation, or from solvent. The slurry produced in dry fractionation is separated by filtration or by transfer of the crystal aggregates to a surfactant solution by inducing wetting of the aggregates (107–108) (Fig. 5). A recent development makes use of 2-propanol to separate crystallized material from the oil phase (109). Hydraulic pressing as a method of separating saturated from unsaturated glycerides is now virtually obsolete

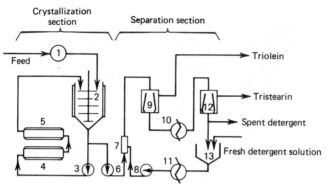

Figure 5. Continuous detergent fractionation. 1, 10, Heat exchangers; 2, crystallization vessel; 3, 6, 8, pumps; 4, 5, scraped-surface heat exchangers; 7, mixer; 9–12, centrifuges; 11, cooler; 13, detergent solution storage. Courtesy of *Revue Française des Corps Gras.*

because of the high labor requirement. In fractional crystallization from solvent (solvent fractionation), several stages of crystallization and separation may be employed if product purity warrants this.

Like other long-chain molecules, the glycerides exhibit polymorphism when crystallizing (110), and use of mild crystallization conditions generally leads to the formation of the most stable polymorph. In crystallization from the melt, rates of cooling that give low degrees of supersaturation are therefore favored, since these conditions also reduce occlusion of noncrystallizing material and improve filtration rates. High rates of cooling are acceptable when detergent wetting of the crystal mass (detergent fractionation) is used to produce crystals that will wet readily.

In dry fractionation, generally carried out batchwise, considerable importance is attached to control of coolant temperature (111) and to low rates of agitation for the crystallizer vessel, which promote crystal growth. Filtration rates are 35–500 kg/(m²·h) for filter presses and filters operating under reduced pressure (112–113). The lower values mainly pertain to winterization, when waxes present retard filtration by blinding the filter cloth.

Continuous crystallization is applied in detergent fractionation because of the higher acceptable rates of cooling. The detergent solution is added immediately after crystallization. High shear agitation is required at this stage to ensure complete wetting of the crystal mass. The volume of detergent solution added to the oil, normally in the ratio 0.5–1.5:1, is related to the solids content of the oil. Centrifugal separation of oil from the aqueous dispersion of crystallized material is followed by reheating of the aqueous phase to above the melting point of the solid fraction. The consequent phase separation permits recovery of the solid fraction by a further centrifugal operation. The detergent solution is recycled a number of times.

Crystallization from solvent is capable of giving sharper separations than processes based on crystallization from the melt, an advantage particularly important in the production of confectionery fat components, for which even low levels of contamination by the more unsaturated components of the oil can have a strongly adverse effect on product quality. Solvent-to-oil ratios of 1–5:1 are used in the crystallizer, normally of the scraped-surface type, and the slurry produced is best filtered on a horizontal belt filter in order to facilitate cake washing.

The choice of fractional crystallization process for a given separation is largely governed by considerations of quality requirement and the yield–cost relationship.

This relationship, in addition to taking direct processing costs into account, must also reflect the value of the secondary product (114). Solvent crystallization is far more energy-intensive (115) and also entails much greater capital expenditure than the solvent-free processes, but its greater separating power permits the production of fractions of higher value. Detergent fractionation offers a better yield of the liquid fraction than the straightforward dry fractionation (Fig. 6), an advantage somewhat offset by the greater capital investment required and higher operating costs in the form of power and chemicals charges (Table 5). Regulations governing the use of solvents or processing aids in food processing (116) must also be taken into consideration when making a process choice.

Shortenings, Margarine and Related Spreads

Shortenings, margarines, and other emulsified spreads (eg, low-fat spreads) have evolved from traditional fatty foods but have been developed further to offer the user improved functional properties. Historically, lard was the source of shortenings for domestic and industrial baking, but factors such as vegetable fat availability and flexibility have enabled vegetable fats to displace lard as the principal source of shortenings. These are available in liquid, fluid, and plastic form (117). Of these, the first contains no solid fat at ambient temperature; the readily pumpable fluid shortening consists of crystallized fat dispersed in a liquid oil; a plastic shortening normally contains 15–30% crystallized fat in the ambient temperature range and is produced

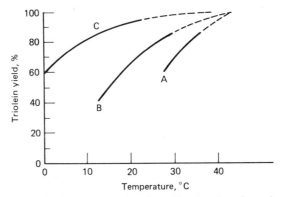

Figure 6. Fractional crystallization of palm oil. A, Dry fractionation using vacuum filtration; B, detergent fractionation; C, solvent fractionation with palm oil:acetone ratio of 1:5. Courtesy of *Revue Française des Corps Gras.*

Table 5. Energy Requirements in Fractional Crystallization

Energy source	Crystallization from the melt		Solvent crystallization (hexane, single-stage)
	Dry fractionation	Detergent fractionation	
steam (300 kPa or 44 psi), kg/t oil input	80	100	300
electricity, kW·h/t oil input	22	27–44	35
water, m³/t	0.1	2	2
reference	111	107	115

in solid form. Shortenings, which are sometimes aerated in order to give the products a lighter texture and improve their color, may contain emulsifiers to enhance their functionality, but they must also be formulated and processed for good oxidative stability.

In margarine production, the dominant trend in recent years has been toward products that offer good spreadability at low (refrigerator) temperature, contain high levels of polyunsaturated oil, and maintain high standards of bacteriological and taste stability. In view of its widespread use in domestic frying, margarine must also display good emulsion stability at high temperature. The development of low-fat spreads, which have approximately half the fat content of margarine, has provided the industry with a further opportunity for diversification and has obliged regulatory authorities to frame appropriate regulations for the control of such products (118). These spreads are normally unsuitable for cooking and baking owing to their lower fat content and more limited emulsion stability. In the United States, where products containing 60% fat have been marketed, any product containing <80% fat must be labelled as an imitation margarine (119).

Composition.   In most countries, the manufacture of margarine is subject to strict control, particularly with regard to total fat content, which is generally required to be at least 80%, and water content, which in most cases must not exceed 16 wt % (120). In addition, the use of various minor additives, eg, emulsifiers, flavor components, vitamins (qv), color, and preservatives, is usually controlled by legislation (see Food additives; Colorants for foods, drugs, and cosmetics).

The functional properties of the product largely determine the specification of the solids content–temperature relationship of the fat to be used (Fig. 7), but these properties are also affected by the polymorphism of the triglycerides, since fats whose stable polymorph is the β' form have smoother texture and better aeration properties than those that are stable in the β form. The latter, however, are more suitable for liquid shortenings.

Emulsifiers are used in shortenings as well as in margarines and related spreads. Their function in shortenings is to facilitate aeration in the product application; in margarine, the emulsifier creates and stabilizes the initial w/o emulsion. The amount of monoglyceride emulsifiers (121) most commonly used for this purpose is 0.2–0.5%, a level that must be increased when the dispersed phase content is increased, as in

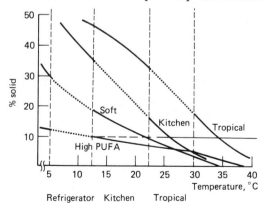

Figure 7.   % solid vs temperature for some margarines. PUFA = polyunsaturated fatty acids. Dotted section of curves indicate solids content range in conditions of use. Courtesy of the *Journal of American Oil Chemists Society.*

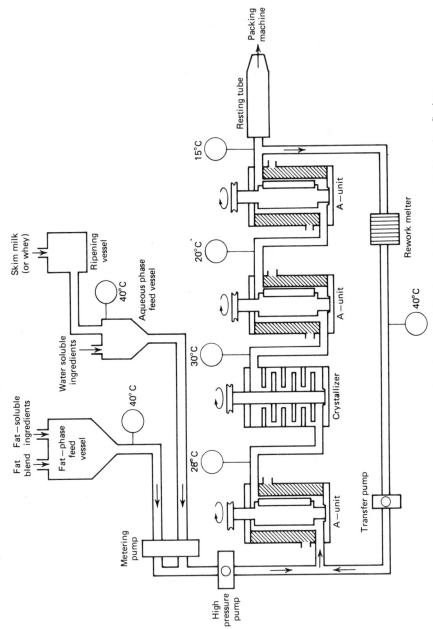

Figure 8. Production of margarine. Courtesy of the *Journal of American Oil Chemists Society.*

low-fat spreads; other stabilizers may then also be used to enhance emulsion stability. This stability may be adversely affected by the presence of protein (122). Lecithin is added to margarines to reduce spattering when the product is used for frying and forms part of the fat phase (see Emulsions).

Milk, reconstituted milk, or whey is used mostly as the aqueous phase and flavor-producing substrate in margarine products (see Milk and milk products). The flavor is generated by souring the milk enzymatically or chemically under controlled conditions or by addition of a commercially available flavor. The salt content of a margarine or spread is normally 0.25–2%, depending on regional taste preference, the higher concentration in the aqueous phase also serving to maintain bacteriological stability.

Processing. Processing of the shortening or margarine blend (Fig. 8) must allow for the rheology as well as the crystallization kinetics of the feedstream, a requirement met by the use of scraped-surface heat exchangers to remove sensible and latent heat and crystallizers, both agitated and static, to provide the time required for crystallization (Fig. 9). The processing conditions also influence the rheology of the product.

The scraped-surface heat exchanger, of which the Votator A-unit is the most common example, must remove energy dissipated by the shaft and scraper blades in the narrow (5–10 mm) annulus and transfer enthalpy from the feedstream to the refrigerant. Studies of heat transfer and power consumption in the Votator (123–124) show that maximum enthalpy reduction is obtained by increasing the number of scraper blades rather than the rotor speed (125). In practice, A-units having 0.5–1.0 m^2 cooling surface are common. These have installed power of 75–100 kJ/kg for a throughput rate of 3–4 t/(m^2·h).

The agitated crystallizers used in the process play an important role in determining the structure of the product, as they enable the fat phase to assume a more stable polymorphic form than would be produced in the heat exchanger under the influence of its high temperature gradients. When the crystallizer residence time is insufficient to permit recrystallization, this process continues during storage and may

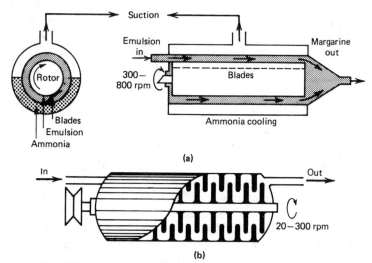

Figure 9. (a) Scraped-surface heat exchanger, Votator A-unit. (b) Crystallizer. Courtesy of the *Journal of American Oil Chemists Society.*

lead to increased product hardness. Agitation during crystallization provides the important benefits of stabilizing the emulsion and of avoiding product brittleness by preventing the formation of rigid crystal networks, but excessive shear at this stage can cause exudation of oil during storage (18). Energy dissipation in the agitated crystallizer is far lower than in the A-unit.

The resting tube placed after the last A-unit provides further holding time for crystallization, but in this case without agitation. Because margarine that is to be wrapped requires a longer residence time than tub margarine, the resting tube for wrapped margarine is larger than that in a tub-filling production line.

The separation of heat transfer and crystallization functions in the margarine and shortening production process, though incomplete, permits the use of different arrangements of the individual units. Thus, it becomes possible to optimize operating conditions for formulations in which the solids content of the product, the rate of crystallization of the fat, and the consistency of the product cover a broad spectrum. The recirculation of crystallized material, normally from immediately before or after the final crystallizer, is one such alternative that has been used.

In the production of margarine and other emulsified spreads, provision must be made for the handling of rework, which may be produced during startup and shutdown as well as at times of packing-machine failure. The material, which must be remelted before it can be reprocessed, becomes bacteriologically vulnerable as its temperature is increased. It is therefore common practice to repasteurize this stream. For obvious reasons, rework from a shortening line poses fewer problems.

Storage at a temperature above a product's packaging temperature can lead to slow recrystallization and crystal growth in the fat phase, which may affect the product adversely. In the case of certain types of margarines, such as those designed for the bakery trade, such changes have the opposite effect of conferring product advantages, particularly with regard to plasticity and creaming properties, and for these products, storage conditions are selected accordingly. The practice is referred to as tempering and is also applied to shortenings.

Quality control of shortenings and spreads is strongly linked to customer assessment and, for margarine-type products, to microbiological control. Oxidative stability of the fats used can be monitored directly (126) or, as in the case of margarine, is assessed as part of the subjective assessment of taste stability. Consistency, which has important functional significance in a variety of products, is extensively used in product quality control in the form of a yield value measured by penetrometry (127).

Applications

The relative simplicity and versatility of the hydrogenation process have made possible the use of vegetable oils in every type of edible oil or fat application. This has led to the domination of the edible oil market by vegetable oils that have been refined and, where required, subsequently modified by hydrogenation, interesterification, and fractional crystallization. Olive oil, by contrast, is substantially used as a salad oil in the unrefined state. For oils that are liquid at ambient temperatures, the end uses are mainly salad and frying oils and dressings such as mayonnaise. Fats or blends of fats and oils are used for margarine and related spreads, shortenings, and speciality fats. A recent development has been the introduction of spreads consisting of a mixture of vegetable oil and butterfat (128). Products required to be rich in linoleic acid may be formulated to contain high levels of safflower, sunflower, or other oils abundant in this acid. Oils rich in oleic acid display greater oxidative stability than those containing the more unsaturated acids and therefore are widely used in frying oil. In the industrialized countries, fat consumption has been shifting toward vegetable oils, particularly those containing high levels of polyunsaturated fatty acids, at the expense of animal and marine fats, including butter (Fig. 10).

The vegetable butters constitute a distinctive class because of their sharp melting characteristics, which give them a special position in confectionery fat formulation. Cocoa butter, the principal member of the group, contains >50% 2-oleopalmitostearin. Illipe fat, sheanut butter, and sal fat, which are less rich in symmetrical triglycerides, may be processed for removal of diunsaturated triglycerides prior to use. The confectionery fat industry also makes considerable use of lauric oils, eg, coconut and palm kernel oils, as well as fractions of hardened soybean and cottonseed oils (129).

The oils generally classified as nonedible, mainly linseed, castor, and tung oils, account for <5% of the total vegetable oil output, but several oils produced mainly for edible use also find industrial applications. Fatty acids recovered from vegetable

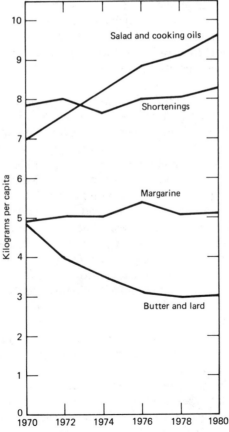

Figure 10. U.S. use of fats and oils in products for civilian consumption, 1970–1980. Courtesy of USDA.

oil refining are another source of feedstock for nonedible products.

Oils for nonedible purposes may be used in the form of crude or refined triglycerides, as the fatty acids, or as one of the many fatty acid derivatives (see Carboxylic acids, manufacture). The surface-coatings industry makes substantial use of various unsaturated oils in the production of alkyd resins (qv) and of paints and varnishes; linseed oil and soybean oil are the oils mainly in use. For their use in paints and varnishes, these oils may be modified by thermal or oxidative treatment (130).

Castor oil (qv), dehydrated to give it the characteristics of a drying oil such as linseed oil, is also used in surface coatings (131). Fatty acids, produced by hydrolysis of oils or of the soapstock that is a refinery by-product, are often preferred to the triglycerides for specific functional properties in this industry.

The soap (qv) industry, now much smaller than that of synthetic detergents, shares with the surface-coatings industry the ability to use fatty acids or the oils from which they have been derived, although in the case of fatty acids, recovery of glycerol is essential on cost grounds. The lauric oils, mainly palm kernel and coconut oils, are of greatest interest to the soap manufacturer because of their contribution to the lathering properties and hardness of the product.

The fatty acids not only form an important market in their own right, but they also provide the raw material for most of the fatty acid derivatives used in various industries. Fatty acids and their uses have been reviewed in detail (132), as has the potential of the oleochemical industry in competition with petrochemical-derived fatty acid derivatives (133). The use of the fatty acids of rapeseed oil, and in particular of erucic acid, is reviewed in ref. 134. Fatty alcohols, derived from fatty acids by high-pressure hydrogenation (135), are important to the detergents industry in the form of the sulfates, but these face strong competition from alkyl sulfates from petrochemical feedstocks.

The friction-reducing properties of vegetable oils, their fatty acids, and derivatives find widespread application in lubrication and lubricants (qv). The demand for high-temperature, high-pressure lubrication has provided opportunities for the very high-melting 12-hydroxystearic acid, produced by careful hydrogenation of ricinoleic acid obtained from castor oil, and in particular, for the alkali metal salts of this acid. Fatty-acid amides meet the demand for friction-reducing additives in the production of various polymers in sheet form. Erucamide has been described as superior to oleamide in this respect. Crambe oil is a preferred source of erucic acid for this type of application since its content of oleic and particularly linoleic acid is appreciably lower than that of rapeseed oil (136).

Tall oil (qv), a by-product of the paper industry in its processing of wood pulp, contains approximately equal proportions of resin and fatty acids when crude and is normally distilled to give a fraction enriched in fatty acids. Tall oil fatty acids are particularly valuable in the formulation of alkyd resins where a fatty acid mixture low in linolenic acid is required. Jojoba oil (137) has in recent years evoked much interest for its potential value relative to more common vegetable oils. In fact, the oil is essentially a liquid wax with a very low glyceride content, and applications are in the industrial rather than the edible field.

Recently, vegetable oils have been discussed as fuels, particularly in admixture with diesel fuels (138–139) (see Fuels from biomass). The industrial significance of such studies depends heavily on the relative prices of vegetable and mineral oils, though a limited application of the concept is known in producer countries, particularly those without indigenous supplies of fossil fuels.

BIBLIOGRAPHY

1. N. O. V. Sonntag in D. Swern, ed., *Bailey's Industrial Oil and Fat Products*, 4th ed., Vol. I, Wiley-Interscience, New York, 1979, Chapt. 1.
2. A. Langstraat, *J. Am. Oil Chem. Soc.* **53**, 241 (1976).
3. C. Litchfield, *Analysis of Triglycerides*, Academic Press, Inc., London, 1972.
4. A. Fincke and F. B. Padley, *Fette Seifen Anstrichm.* **83**, 461 (1981).
5. M. Loncin, *Rev. Ferment. Ind. Aliment. Bruxelles* **8**, 61 (1953).
6. J. A. Cornelius, *Prog. Chem. Fats Other Lipids* **15**, 5 (1977).
7. R. M. Williams and D. Chapman, *Prog. Chem. Fats Other Lipids* **11**, 1 (1970).
8. S. P. Kochar and M. L. Meara, *Scientific and Technical Surveys*, no. 98, Leatherhead Food R.A., UK, 1977.
9. J. A. G. Box and H. A. Boekenoogen, *Fette Seifen Anstrichm.* **69**, 724 (1967).
10. G. R. List and co-workers, *J. Am. Oil Chem. Soc.* **54**, 8 (1977).
11. A. Thomas, *Fette Seifen Anstrichm.* **74**, 141 (1976).
12. J. K. Daun and F. W. Hougen, *J. Am. Oil Chem. Soc.* **53**, 169 (1976).
13. A. M. Parsons, *Prog. Chem. Fats Other Lipids* **11**, 243 (1970).
14. H. Pardun, *Z. Lebensm. Technol. Verfahrenstech.* **32**, 109, 149 (1981).
15. R. Krishnamurthy in D. Swern, ed., *Bailey's Industrial Oil and Fat Products*, Vol. II, 4th ed., Wiley-Interscience, New York, 1982, Chapt. 5.
16. M. L. Meara, *Scientific and Technical Surveys*, no. 110, Leatherhead Food R.A., UK, 1978.
17. J. C. van den Enden and co-workers, *Fette Seifen Anstrichm.* **80**, 180 (1978).
18. A. J. Haighton, *J. Am. Oil Chem. Soc.* **53**, 397 (1976).
19. J. B. Rossel, *Chem. Ind. (London)*, 832 (1973).
20. M. W. Formo in ref. 1, Chapt. 3.
21. W. E. Link, ed., *Official and Tentative Methods of the American Oil Chemists' Society*, American Oil Chemists' Society, Champaign, Ill., 1977 AOCS Official Method CD 1-25.
22. W. Hamm in T. C. Lo, H. I. Baird, and C. Hanson, eds., *Handbook of Solvent Extraction*, Wiley-Interscience, New York, 1983.
23. H. Stage, *Riv. Ital. Sostanze Grasse* **52**, 291 (1975).
24. S. Peter, G. Brunner, and R. Riha, *Ger. Chem. Eng.* **1**, 26 (1978).
25. D. Colon, *Oleagineux* **34**, 163 (1979).
26. R. Ohlson in L. A. Appelqvist and R. Ohlson, *Rapeseed Cultivation, Composition, Processing and Utilisation*, Elsevier, Amsterdam, The Netherlands, 1972, Chapt. 2.
27. A. Lysons, *Long-term Development in the World Oils and Fats Market*, paper presented to Society of Chemical Industry, London, Dec. 3, 1981.
28. J. J. L. van Waalwijk van Doorn, *Fette Seifen Anstrichm.* **80**, 416 (1978).
29. *J. Am. Oil Chem. Soc.* **57**, 264A (1980).
30. C. Lioret and co-workers, *Oleagineux* **36**, 113 (1981).
31. R. H. V. Corley and co-workers in *The Oil Palm in Agriculture in the Eighties*, symposium, Kuala Lumpur, Malaysia, July 1981 (to be published).
32. S. S. Radwan, *Fette Seifen Anstrichm.* **78**, 70 (1975).
33. G. W. Chapman and co-workers, *J. Am. Oil Chem. Soc.* **56**, 54 (1976).
34. L. A. Jones, *J. Am. Oil Chem. Soc.* **57**, 24A (1980).
35. A. Lysons, *Chem. Ind.*, 297 (1979).
36. E. H. Gustafson, *J. Am. Oil Chem. Soc.* **53**, 248 (1976).
37. E. H. Gustafson, *J. Am. Oil Chem. Soc.* **55**, 751 (1978).
38. L. A. Appelqvist and B. Loof in ref. 26, Chapt. 5.
39. J. P. Galloway, *J. Am. Oil Chem. Soc.* **53**, 271 (1976).
40. W. Kehse, *Fette Seifen Anstrichm.* **81**, 463 (1979).
41. K. Anjou in ref. 26, Chapt. 9.
42. R. P. Hutchins, *J. Am. Oil Chem. Soc.* **53**, 279 (1976).
43. E. D. Milligan, *J. Am. Oil Chem. Soc.* **53**, 286 (1976).
44. A. E. Bernardini, *The New Oil and Fat Technology*, 2nd ed., Technologie, Rome, Italy, 1973.
45. H. P. J. Jongeneelen, *J. Am. Oil Chem. Soc.* **53**, 291 (1976).
46. B. de Ramecourt, *J. Am. Oil Chem. Soc.* **53**, 256 (1976).
47. B. Jacobsberg and D. Jacqmain, *Oleagineux* **28**, 25 (1973).
48. E. Fedeli, *Prog. Chem. Fats Other Lipids* **15**, 57 (1977).

49. G. Jacini and C. Carola, *Riv. Ital. Sostanze Grasse* **41**, 290 (1964).
50. T. L. Mounts, *J. Am. Oil Chem. Soc.* **58**, 51A (1981).
51. R. T. Sleeter, *J. Am. Oil Chem. Soc.* **58**, 239 (1981).
52. G. R. List, T. L. Mounts, and A. J. Heakin, *J. Am. Oil Chem. Soc.* **55**, 280 (1978).
53. A. Hvolby, *J. Am. Oil Chem. Soc.* **48**, 503 (1971).
54. O. L. Brekke in D. R. Erickson and co-eds, *Handbook of Soy Oil Processing and Utilization*, American Soybean Association and American Oil Chemists' Society, Champaign, Ill., 1980, Chapt. 6.
55. Brit. Pat. 1,541,017 (Feb. 21, 1979), H. J. Ringers and J. Segers (to Unilever Ltd.).
56. Brit. Pat. 1,564,403 (Apr. 10, 1980), A. K. Sen Gupta (to Unilever Ltd.).
57. A. Athanassiadas in D. A. Earp and W. Newall, eds., *International Developments in Palm Oil, Proceedings Malaysian International Symposium on Palm Oil Processing and Marketing, Kuala Lumpur, Malaysia, June 1976*, Incorporated Society of Planters, 1977, pp. 399–409.
58. W. H. Morrison and J. K. Thomas, *J. Am. Oil Chem. Soc.* **52**, 485 (1976).
59. Fr. Pat. 2,460,996 (July 9, 1980), A. B. Pallmar and K. W. H. Sarebjork (to Alfa Laval AB).
60. B. Ostric-Matijasevic and J. Turkulov, *Ref. Fr. Corps Gras* **20**, 5 (1973).
61. M. Rac in *Proceedings 5th International Sunflower Conference*, Clermont-Ferrand, France, 1972, p. 365.
62. B. Braae, *J. Am. Oil Chem. Soc.* **53**, 353 (1976).
63. R. A. Carr, *J. Am. Oil Chem. Soc.* **53**, 347 (1976).
64. G. C. Cavanagh, *J. Am. Oil Chem. Soc.* **53**, 361 (1976).
65. G. E. Sullivan, *J. Am. Oil Chem. Soc.* **53**, 358 (1976).
66. J. M. Klein, *Rev. Fr. Corps Gras* **28**, 309 (1981).
67. D. C. Tandy and W. J. McPherson in ref. 54, Chapt. 13.
68. J. Lau, *Fette Seifen Anstrichm.* **83**, 532 (1981).
69. H. Eicke, *Seifen Ole Fette Wachse* **97**, 712 (1971).
70. H. B. W. Patterson, *J. Am. Oil Chem. Soc.* **53**, 339 (1976).
71. E. H. Goebel, *J. Am. Oil Chem. Soc.* **58**, 199 (1981).
72. L. L. Richardson, *J. Am. Oil Chem. Soc.* **55**, 777 (1978).
73. U. I. Brimberg, *Fette Seifen Anstrichm.* **83**, 184 (1981).
74. H. H. R. H. Wendt, *Fette Seifen Anstrichm.* **83**, 541 (1981).
75. J. H. Perry and C. H. Chilton, eds., *Chemical Engineers' Handbook*, 5th ed., McGraw-Hill, Inc., New York, 1973, Sect. 19.
76. G. Hoffmann, *Chem. Ind. (London)*, 729 (1970).
77. J. P. Wolff, *Rev. Fr. Corps Gras* **21**, 161 (1974).
78. U.S. Pat. 3,693,322 (Sept. 26, 1972), D. D. Lineberry and F. A. Dudrew (to Chemetron Corporation).
79. U.S. Pat. 3,506,696 (Apr. 14, 1970), J. S. Baker and J. B. Edwards (to Procter and Gamble Co., Inc.).
80. A. Athanassiadis, *Fette Seifen Anstrichm.* **82**, 219 (1980).
81. H. Stage, *Seifen Oele Fette Wachse* **105**, 395 (1979).
82. J. W. E. Coenen, *J. Am. Oil Chem. Soc.* **53**, 382 (1976).
83. R. R. Allen in ref. 15, Chapt. 1.
84. H. J. Dutton in E. A. Emken and H. J. Dutton, eds., *Geometrical and Positional Fatty Acid Isomers*, American Oil Chemists' Society, Champaign, Ill., 1979, pp. 1–16.
85. C. R. Scholfield, R. O. Butterfield, and H. J. Dutton, *J. Am. Oil Chem. Soc.* **56**, 664 (1979).
86. A. H. Chen, D. D. McIntire, and R. R. Allen, *J. Am. Oil Chem. Soc.* **58**, 816 (1981).
87. L. E. Johansson and S. T. Lundin, *J. Am. Oil Chem. Soc.* **56**, 974 (1979).
88. U.S. Pat. 4,038,295 (July 26, 1977), R. Stern (to Institut Français du Pétrole and Institut des Corps Gras (ITERG)).
89. Y. E.-Shattory, L. de Man, and J. M. de Man, *Can. Inst. Food Sci. Technol. J.* **14**, 53 (1981).
90. G. Leuteritz, *Fette Seifen Anstrichm.* **71**, 441 (1969).
91. U. M. T. Houtsmuller, *Fette Seifen Anstrichm.* **80**, 162 (1978).
92. J. M. de Man and Y. El-Shattory, *Chem. Mikrobiol. Technol. Lebensm.* **7**, 33 (1981).
93. T. H. Applewhite, *J. Am. Oil Chem. Soc.* **58**, 260 (1981).
94. R. B. Alfin-Slater and L. Aftergood in ref. 84, pp. 53–74.
95. H. Heckers and co-workers, *Atherosclerosis* **28**, 389 (1977).
96. F. A. Kummerow and co-workers, *Artery* **4**, 360 (1978).
97. A. J. Vergroesen, *Proc. Nutr. Soc.* **31**, 323 (1972).

98. M. G. Enig and co-workers, *Proceedings of the Maryland Nutrition Conference for Food Manufacturers*, Mar. 1979, pp. 9–17.
99. R. T. Holman, *Chem. Ind. (London)* (20), 704 (1981).
100. *Report of the Ad Hoc Committee on the Composition of Special Margarines*, Ottawa, Dec. 5–7, 1979, Minister of Supply and Services, Canada, 1980.
101. *Fette Seifen Anstrichm.* **75,** 467 (1973); *Fette Seifen Anstrichm.* **75,** 587 (1973).
102. N. O. V. Sonntag in ref. 15, Chapt. 2.
103. U.S. Pat. 2,442,531 (June 1, 1948), E. W. Eckey (to Procter and Gamble Co. Inc.).
104. Brit. Pat. 1,236,233 (June 23, 1971), W. H. de Groot and M. H. Hilder (to Unilever Ltd.).
105. L. H. Going, *J. Am. Oil Chem. Soc.* **44,** 414A (1967).
106. Brit. Pat. 1,577,933 (Oct. 29, 1980), M. H. Coleman and A. R. Macrae (to Unilever Ltd.).
107. J. P. Seuge and H. F. Vinconneau, *Oleagineux* **30,** 25 (1975).
108. G. Haraldsson, *Riv. Ital. Sostanze Grasse* **58,** 491 (1981).
109. L. Koslowsky, *Oleagineux* **29,** 421 (1974).
110. E. S. Lutton, *J. Am. Oil Chem. Soc.* **49,** 1 (1972).
111. F. Tirtiaux, *Oleagineux* **31,** 279 (1976).
112. G. M. Neumunz, *J. Am. Oil Chem. Soc.* **55,** 396A (1978).
113. W. Kehse, *Rev. Fr. Corps Gras* **27,** 569 (1980).
114. J. W. E. Coenen, *Rev. Fr. Corps Gras* **21,** 343 (1974).
115. E. Bernardini and M. Bernardini, *Oleagineux* **30,** 121 (1975).
116. *Evaluation of Food Additives, 14th Report of Joint FAO/WHO Expert Committee on Food Additives*, WHO Technical Report Series 462, Geneva, Switzerland, 1971.
117. A. E. Thomas, *J. Am. Oil Chem. Soc.* **55,** 830 (1978).
118. H. Wessels, *Fette Seifen Anstrichm.* **83,** 82 (1981).
119. L. H. Wiedermann, *J. Am. Oil Chem. Soc.* **55,** 823 (1978).
120. *Fed. Regist.* **38,** 10952 (May 3, 1973).
121. J. B. Lauridsen, *J. Am. Oil Chem. Soc.* **53,** 400 (1976).
122. U.S. Pat. 4,071,634 (Jan. 31, 1978), I. E. M. Wilton and co-workers (to Lever Bros. Co.).
123. A. M. Trommelen and S. Boerema, *Trans. Inst. Chem. Eng.* **44,** T329 (1966).
124. A. M. Trommelen, *Trans. Inst. Chem. Eng.* **45,** T176 (1967).
125. O. Moller and A. M. Trommelen, *Fette Seifen Anstrichm.* **72,** 235 (1970).
126. W. E. Link, ed., *Official and Tentative Methods of the American Oil Chemists' Society*, American Oil Chemists' Society, Champaign, Ill., 1977, Official Method Cd 12-57.
127. A. J. Haighton, *J. Am. Oil Chem. Soc.* **36,** 345 (1959).
128. T. Alsafar, *Can. Inst. Food Sci. Technol. J.* **7,** 220 (1974).
129. F. R. Paulicka, *J. Am. Oil Chem. Soc.* **53,** 421 (1976).
130. M. W. Formo in ref. 1, Chapt. 10.
131. F. C. Naughton, *J. Am. Oil Chem. Soc.* **51,** 65 (1974).
132. E. H. Pryde, *J. Am. Oil Chem. Soc.* **56,** 849 (1979).
133. H. Klimmek, *Chem. Ind. (Dusseldorf)* **33,** 136 (1981).
134. R. Ohlson in ref. 26, Chapt. 12.
135. J. A. Monick, *J. Am. Oil Chem. Soc.* **56,** 853A (1979).
136. N. M. Molnar, *J. Am. Oil Chem. Soc.* **51,** 84 (1974).
137. J. Wisniak, *Prog. Chem. Fats Other Lipids* **15,** 167 (1977).
138. G. Martin, *Oleagineux* **36,** 280 (1981).
139. R. P. Morgan and E. B. Shultz, *Chem. Eng. News* **59**(36), 69 (1981).

General References

Refs. 1, 15, and 44 are also general references.
J. Baltes, *Gewinnung und Verarbeitung von Nahrungsfetten*, Vol. 17 of *Grundlagen und Fortschritte der Lebensmitteluntersuchung*, Verlag Paul Parey, Berlin, 1975.
A. J. C. Andersen and P. N. Williams, *Margarine*, 2nd ed., Pergamon Press, Oxford, UK, 1965.

W. HAMM
Unilever Research
Colworth Laboratory, UK

VELVETEX. See Surfactants and detersive systems.

VERMICULITE. See Insulation, thermal.

VERMILION. See Pigments, inorganic.

VETERINARY DRUGS

With the acceleration of technology and research in veterinary medicine over the past 25 years, the use of pharmaceuticals by the practicing veterinarian in the treatment and prevention of animal diseases has greatly expanded. Knowledge collected across the entire spectrum of research, from the cellular level to population dynamics, has opened numerous therapeutic avenues and created both drugs and uses that did not exist a few years ago. Today's veterinarian, whether in companion-animal practice, equine service, feedlot medicine, dairy work, zoo management, or any medical or surgical specialty, has a wide range of products from both the human and veterinary pharmaceutical industries for managing the needs of patients and clients (1–5).

Antimicrobial Agents

The use of drugs to control infection is considerably older than the recognition of the causes of or complex physiological responses to infection. Such unlikely and diverse agents as vinegar (wine), copper salts, and honey were prescribed well over 2000 years ago. With the discoveries of sulfanilamide and penicillin in the 1930s and 1940s, the use of antimicrobial agents based on the knowledge of specific causative organisms and antimicrobial activities has ushered in the golden age of antimicrobial therapy (see Antibacterial agents, synthetic; Antibiotics).

The selection of the most appropriate antimicrobial agent depends on an accurate diagnosis and identification of the offending organism. In vitro culture and sensitivity testing of isolated organisms are routine methods for determining the antimicrobial of choice. The spectrum of activity of most antibiotics is broadly described in terms of activity against gram-positive or gram-negative organisms. This classification is based on staining characteristics with a blue primary stain of crystal violet with iodine and a red counterstain, usually safranin. The biochemical foundation for an organism retaining the blue, gram-positive color or not is related to the physical and chemical characteristics of the cell wall or cytoplasmic membrane. On an empirical basis, the functional activity of many antibacterials correlates to some degree with the gram-staining reaction. The medical trend is toward use of relatively narrow-spectrum therapeutics with strong activity against specific organisms as opposed to use of broad-spectrum agents without the supporting diagnostics. When time is critical, as in life-threatening conditions, therapy with a broad spectrum agent may be started before the culture and sensitivity testing. In addition to organism identification, consideration is given to whether an agent kills or inhibits the organism or its growth. The route of administration, dosage rate, frequency of treatment, and overall duration of treatment must also be considered. Microbes are constantly undergoing genetic change. Some variant strains have the ability to deactivate certain antimicrobials or grow in the presence of antimicrobials to which earlier generations were sensitive. Under antimicrobial therapy, the resistant strains may survive and render the agent less effective. Because of this phenomenon of selection, several agents representative of structurally different antimicrobial families may be used consecutively or, less often, concurrently for an evolving microbial population.

Antimicrobial agents are also used as prophylactics during surgery or at generally lower levels of administration to promote an animal's ability to withstand pathogenic challenge when under stress. In addition to the families discussed below, antimicrobial agents include carbadox [6804-07-5], the cephalosporins, chloramphenicol [56-75-7], nitrofurans, nitroimidazoles, and tylosin [1401-69-0].

Sulfonamides. The sulfonamides (sulfas) are derivatives of *para*-aminobenzenesulfonamide [63-74-1]. These agents are active against a broad spectrum of gram-positive and gram-negative organisms. Their mode of action is by competitive antagonism of *para*-aminobenzoic acid (PABA), a folic acid precursor. As mammalian cells do not synthesize folic acid, as do the bacteria that are sensitive to the sulfas, mammalian toxicity is low. The antibacterial activity of the sulfas can be augmented by concurrent use of trimethoprim [738-70-5] which blocks another step in folic acid synthesis.

Although the antibacterial spectrum is similar for many of the sulfas, chemical modifications of the parent molecule have produced compounds with a variety of absorption, metabolism, tissue distribution, and excretion characteristics. Administration is typically oral or by injection, usually intravenous. When absorbed, they tend to distribute widely in the body, be metabolized by the liver, and excreted in the urine. Toxic reactions or untoward side effects have been characterized as blood dyscrasias, crystal deposition in the kidneys, especially with insufficient urinary output, and allergic sensitization. Selection of organisms resistant to the sulfonamides has been observed, but has not been correlated with cross-resistance to other antibiotic families.

Penicillins. Since the discovery of penicillin in 1928 as an antibacterial elaborated by a mold, *Penicillium notatum*, the global search for better antibiotic-producing organism species, radiation-induced mutation, and culture-media modifications have been used to maximize production of the compound. These efforts have resulted in the discovery of a variety of natural penicillins differing in side chains from the basic molecule, 6-aminopenicillanic acid [551-16-6]. These chemical variations have produced an assortment of drugs with diverse pharmacokinetic and antibacterial characteristics (see Antibiotics, β-lactams).

The mechanism of antibacterial activity is through inhibition of gram-positive bacterial cell-wall synthesis; thus, the penicillins are most effective against actively multiplying organisms. Since mammalian cells do not have a definitive cell-wall structure like bacteria, the mammalian toxicity of the penicillins is low. Allergic phenomena in patients following sensitization may occur.

The penicillins as natural and semisynthetic agents are used primarily against susceptible *Pasteurella sp.*, staphylococci, streptococci, clostridia, and *Corynebac-*

terium sp. Penicillin is widely used for therapeutic purposes against these organisms and in animal feeds as a growth promoter. The latter effect is considered to be a result of subtle and reversible effects on the gastrointestinal microflora.

Aminoglycosides. The aminoglycosides, such as streptomycin [128-46-1], neomycin [119-04-0], kanamycin [59-01-8], and gentamycin [1403-66-3], have a hexose nucleus joined to two or more amino sugars (see Antibiotics, aminoglycosides). They all tend to be poorly absorbed from the gastrointestinal tract but are absorbed well following parenteral administration. They are rapidly bactericidal by inhibiting intracellular protein synthesis. The active transport mechanism which allows intracellular access strongly depends on pH, divalent cations, osmolality, and oxygen tension. The latter renders many anaerobes resistant to the aminoglycosides which typically have very broad activity spectra with greater activity against gram-negative bacteria. For this reason, a penicillin may be complementary to an aminoglycoside and provide, overall, a broader spectrum. Toxicity following exaggerated or prolonged dosage schedules is characterized by renal failure or damage to the eighth cranial nerve (auditory) with auditory- or vestibular-balance dysfunction.

Tetracyclines. The tetracyclines, including chlortetracycline [57-62-5] and oxytetracycline [79-57-2], are produced as fermentation products (see Antibiotics, tetracyclines). They have a broad antibacterial spectrum including gram-positive and gram-negative organisms, rickettsiae, *Chlamydia sp.* and *Mycoplasma sp.* In addition, the tetracyclines are commonly employed at low dosages as growth promoters in the main food-producing species. Most tetracyclines are incompletely absorbed following oral administration and are antagonized through chelation primarily with divalent cations. The mechanism of action is like the aminoglycosides, by intracellular inhibition of protein synthesis. Toxicity is rare. Although seldom of clinical significance, the tetracyclines are incorporated into metabolically active calcified tissues resulting in discoloration. This phenomenon is most prominent in rapidly growing bone, eg, fetal or juvenile skeletal and dental systems.

Growth Promoters. The tetracyclines and penicillin when administered to food-producing species (poultry, swine, and cattle) during active growth improve significantly the rate of weight gain and efficiency of feed utilization (see also Pet and livestock feeds). Other antibacterials are also used for this purpose, including avoparcin [37332-99-3], monensin [17090-79-8], bacitracins, virginiamycin [11006-76-1], lincomycin [154-21-2], tylosin, and flavomycin [11015-37-5]. These effects are not related specifically to prevention or treatment of bacterial diseases, but to subtle shifts in enteric processes (6).

Antifungal Agents. Fungi and related organisms are encountered most commonly in superficial infections of the skin and less commonly as systemic or deep mycoses affecting internal organs (see Chemotherapeutics, antimycotic). Superficial lesions may range from minor hair loss (ringworm) to severe, generalized hair loss and marked pathological changes in the skin with secondary bacterial infections. The systemic diseases are typically refractory to most therapeutics and are frequently fatal. Many of the mycotic infections have, to some highly variable degree, zoonotic potential. Because they tend to be spore-forming organisms, mycoses also tend to be periodically recurrent or persistent in a given environment (see also Fungicides).

Favorable responses of the superficial infections have been observed following exposure to sunlight or administration of vitamin A [68-26-8] (qv). Some infections remit spontaneously as a young animal matures. More often, topical application of an antifungal such as nystatin [34786-70-4] or cuprimyxin [28069-65-0] or systemic griseofulvin [126-07-8] over six to 12 weeks is justified. The systemic mycoses are sensitive to very few therapeutic agents. Some, such as actinomycosis and actinobacillosis of cattle, will respond to sulfa therapy but others (cryptococcosis, blastomycosis, etc) may show response only to amphotericin B [1397-89-3], a relatively toxic antibiotic. Animals with systemic infections are frequently euthanized because of the history of limited therapeutic success and the zoonotic disease potential.

Parasiticides

Parasiticides can be roughly divided according to parasites, host species, or chemical classification (see Chemotherapeutics, anthelmintic and antiprotozoal). By any classification, they are ubiquitous in the management and control of parasites of both companion and food-producing animals (5,7).

Organophosphates and Carbamates. The main pharmacologic action of this group is the inhibition of the cholinesterase enzymes, primarily acetylcholinesterase (AChE) (see Cholinesterase inhibitors; Enzymes, immobilized). Generally, acetylcholine (ACh) is responsible for transmission of neural impulses at voluntary neuromuscular junctions, at the sympathetic ganglia synapses, and throughout the parasympathetic system (see Choline). Under normal conditions, it is rapidly hydrolyzed and inactivated by AChE. In the presence of AChE inhibitors, the enzyme is phosphorylated with the consequent pharmacologic and toxic actions produced by excessive accumulations of ACh. Because ACh is an integral part of insect and helminth physiology, the antiparasitic utility spectrum of the drug family is immense.

Organophosphates and carbamates are typically lipid-soluble and are, as a consequence, rapidly absorbed following inhalation or oral, parenteral, or topical administration. Once absorbed, metabolism is primarily by hepatic hydrolysis or oxidation. Various organophosphates (O-Ps) and carbamates are used against virtually all animal parasites. The use of O-Ps and carbamates is widespread, both as animal antiparasiticides and as agricultural and home-use pesticides (see Insect control technology). In addition, concurrent exposure to more than one agent results in cumulative physiologic effect and therefore the incidence of toxic effect is relatively high. Atropine is an excellent antidote by blocking the action of ACh within the parasympathetic nervous system. It is neither a complete antagonist nor does it modify the rate at which the enzyme is regenerated. Pralidoxime chloride is another antidote frequently used as an adjunct to atropine for specifically O-P toxicity. It acts by regenerating the enzyme throughout the system, but may exacerbate toxicity in cases with reversible carbamate–esterase bonding.

Avermectins. The avermectins [65195-52-0, 65195-58-6] are fermentation products derived from *Streptomyces avermitilis*. They are macrocyclic lactones with a very broad spectrum of insecticidal and anthelmintic activity. First commercially available in 1981, they are highly active at doses of ca 0.2 mg/kg body weight against the internal parasites of cattle and horses as well as against lice, internally migrating fly larvae (warbles), and mites. They are active in dogs against larval stages of heartworm disease and intestinal parasites, with the exception of tapeworms.

Levamisole. The racemic mixture of the *d* and *l* isomers of tetramisole [6649-23-6] was first described in the *Janssen Pharmaceutica* in 1966. It is used as an anthelmintic against a wide variety of nematodes, including lungworms, of ruminants, swine, horses, dogs, and poultry. Anthelmintic activity resides in the *l* isomer, levamisole [*14769-73-4*], the form currently used.

Benzimidazoles. The benzimidazoles include a large family of anthelmintics, eg, thiabendazole [*148-79-8*], albendazole [*54965-21-8*], cambendazole [*26097-80-3*], fenbendazole [*43210-67-9*], mebendazole [*31431-39-7*], oxfendazole [*53716-50-0*], and oxibendazole [*20559-55-1*]. Administration is oral, and the spectrum of activity is broad against nematode parasites of the intestinal tract. The usual dosage of thiabendazole is 50–110 mg/kg body weight. Dosages of other benzimidazoles are 2–30 mg/kg. The activity of the individual compound varies against specific parasite species; none, however, is effective against lungworms. Benzimidazoles have the advantage of a low mammalian toxicity, ca 10–30 times the recommended dosage. Absorption is rapid, parent compound and metabolites are excreted in the urine. There has been some indication of teratogenic effects with albendazole and cambedazole.

Other Parasiticides. The parasiticides described below have a relatively limited usage owing to a narrow spectrum of antiparasitic activity or because of the recent introduction of inherently safer or more effective products.

Sodium thiacetarsamide [*14432-82-0*] is an arsenical for intravenous use only in dogs against the adult stage of heartworm infection.

Carbon disulfide is used, in combination with other orally administered anthelmintics, by stomach tube for bots (*Gastrophilus sp.* larvae) and ascarids (roundworms) of horses.

Coccidiosis. Coccidiosis, caused by protozoans of the genera Eimeria and Isospora, may be present in any domesticated animal species but is ubiquitous in the poultry industry with serious consequences. The life cycle involves both asexual and sexual intracellular parasitic stages characterized by a rapid development and multiplication of infective stages with consequent destruction of, primarily, the intestinal lining of the host. This leads to growth retardation and, when severe, a high mortality. Anticoccidial agents are added routinely as feed components throughout the life of broiler chickens. A partial list of additives includes amprolium [*121-25-5*], ethopabate [*59-06-3*], robenidine [*25875-50-7*], arprinocid [*55779-18-5*], monensin, lasalocid [*25999-31-9*], chlortetracycline [*57-62-5*], and the sulfa compounds. The most widely used additives are representatives of the ionophore antibiotics, ie, monensin and lasalocid. Historically, the appearance of resistance by the coccidia to anticoccidial agents has been rapid and frustrating. This resistance has not yet been observed to any notable degree with the ionophores even after more than a decade of extensive use. A program of rotation, where anticoccidials with differing modes of action are used in succession minimizes the impact of resistance.

Diethylcarbamazine [*98-89-1*] is a piperazine derivative which is given daily as a prophylactic for canine heartworm disease (*Dirofilaria immitis*) or as a therapeutic for roundworms. *D. immitis* is transmitted only by mosquitoes, and therefore the period of administration varies geographically, depending on temperatures and humidity which regulate the mosquito life cycle.

Phenothiazine [*58-37-7*] (thiodiphenylamine) is used orally against intestinal nematodes of ruminants and horses. It is used with occasional gastrointestinal upset, hemolytic processes, and photosensitivity. It is used routinely at low concentrations on horse farms to suppress the egg production of intestinal parasites (strongyles) and thus limit pasture contamination and transmission (8).

Hexachloroethane [*67-72-1*] has, like carbon tetrachloride [*56-23-5*], been used to remove liver flukes from ruminants. These have been replaced by albenzadole, previously mentioned as a benzimidazole, and clioxanide [144327-41-3], oxychozanide [2277-92-1] or rafoxanide [22662-39-1].

Niclosamide [50-65-7] (2',5-dichloro-4'-nitrosalicylanilide) is commonly used against tapeworms in small animals (4). Although tapeworms (cestodes) are frequently refractory to anthelmintics highly active against other intestinal parasites, they are sensitive to niclosamide as well as bunamidine hydrochloride [1055-55-6] and the recently introduced praziquantel [55268-74-1].

Anti-inflammatory Agents

Inflammation is a defense mechanism of the body that plays a key role in fighting disease and initiating wound healing. It is clinically characterized by local redness, swelling, pain, and heat. Because of these symptoms, inflammation can be more detrimental than beneficial to normal body function and at such times the practitioner chooses to slow it down. This group of compounds is used on an individual animal basis and tends to be used more in the companion animal and equine specialities (8) (see also Analgesics, antipyretics, and anti-inflammatory agents).

The classic example of an anti-inflammatory drug is aspirin [50-78-2], acetosalicylic acid, an effective analgesic for many years. It is well tolerated by the dog and the horse, but is relatively toxic to cats. Under the proper clinical circumstances, it can be used for prolonged therapy in chronic inflammatory diseases such as arthritis.

Pyrazolone derivatives, specifically phenylbutazone [50-33-9] and, for limited conditions, dipyrone [5907-38-0], are very popular with the equine practitioner and are particularly useful in managing cases of lameness and controlling inflammation after trauma or surgery (8). Dipyrone is an analgesic for cases of equine colic. Phenylbutazone is an effective anti-inflammatory drug, but is more toxic than the salicylates which limits its long-term use. However, for short-term usage, its ease of administration as injectable or oral preparations makes it a popular product for the equine or small-animal specialist.

The most widely used group of anti-inflammatory drugs are the corticosteroids and their synthetic analogues. This group of compounds has several physiologic actions, including effects on sodium retention and liver glycogen deposition as well as inhibitory effects on wound healing and, more recently recognized, proliferation of cancer cells (see below).

The natural compounds cortisol [50-23-7], cortisone [53-06-5], and corticosterone [50-22-6] vary only slightly in structures and pharmacologic properties (see Steroids). The synthetic analogues in use today, prednisolone [52438-85-4], dexamethasone [50-02-2], and triamcinolone [124-94-7], have greater anti-inflammatory potency and their effects on sodium retention tend to be less severe.

The uses of corticosteroid anti-inflammatory drugs in veterinary medicine are many and varied. In the intact animal, the glucocorticoids and mineralcorticoids are produced in the adrenal glands. Exogenous compounds are, therefore, used for their glucogenic physiologic effect in cases where the animal is unable to produce sufficient quantities of these compounds. When given at pharmacologic dosage, their effects include anti-inflammatory aspects useful in controlling healing and inflammation following trauma or surgery; controlling inflammation in severe dermatologic cases, thereby improving effective treatment of the cause of the problem; and in helping to control allergic reactions. On a cellular level, these compounds exert less well-defined effects in helping to preserve cell-membrane integrity and improve cellular metabolism. These effects, in addition to the effects on the microcirculation, are the basis for corticosteroid use in shock-syndrome therapy, which, however, is controversial.

Hormones

Hormones (qv) as naturally occurring, semisynthetic, or synthetic compounds are used to regulate reproductive cycles, gestation, and parturition, as therapeutics for hormonal imbalances and responsive physical or physiological abnormalities, or as growth promoters in ruminants. The application of other than sex-related hormones is not as complex as in human therapy because of the relatively short life spans of animals and high cost.

Hormones can either delay or induce estrus. The therapeutic manipulations of normal, or abnormal, estrus sequences are based on the intricate biological feedback relationship between the pituitary gland and the gonads. In broad terms, follicle-stimulating hormone [9034-38-2] releases estrogens. The estrogens cause a decrease in FSH and an increase in luteinizing hormone [9002-67-9] (LH) which causes ovulation and formation of an ovarian corpus luteum (CL). The CL releases natural progesterone, the level of which will either help maintain pregnancy or, at some point, reinitiate the cycle. The therapeutic applications are based on adjusting or creating the normal sequence of events. Administration of progesterone [57-83-0] or progestogens (mibolerone [3704-09-4] or megestrol acetate [3562-63-8]) simulates the hormonal action of the corpus luteum and, in so doing, delays the onset of an estrus. Estrus can be terminated with progesterone, LH, or some prostaglandins with a resulting fertile ovulation. An artificial estrus, which does not produce a fertile ovulation can be induced by exogenous estrogens such as diethylstilbestrol [56-53-1]. Pregnant mare serum, a functional gonadotrophin, the chorionic gonadotrophins and FSH create superovulation with an increased number of developed ova. The estrogens are frequently used to prevent zygote implantation and thus pregnancy, when given shortly after an unintended mating. Oxytocin [50-56-6] of pituitary origin, stimulates sensitive uterine muscle at parturition. Oxytocin and prolactin [12585-34-1] facilitate milk production and let-down in lactating animals. Cases of enlarged prostate glands in males are frequently responsive to estrogen therapy. The progestins, notably megestrol acetate, have been used successfully in the management of a variety of dermatitides and behavioral problems in small animals and discrete clinical syndromes are associated with estrogen and testosterone imbalances in both males and females. Medical therapy, in light of the multisystem effects of these compounds, is always conservative and adverse effects related to feminization of males or aggressive masculinization of females is frequent. Therapy is often an adjunct to surgical ovariectomy or castration.

Estrogens, testosterone [58-22-0], or compounds such as zeranol [26538-44-3] or trenbolone [10161-33-8] which can mimic their effects, have shown utility in accelerating the rate of weight gains and decreasing the amount of feed required to produce these gains in food-producing animals (9). The potential for human consumption of these compounds via the food supply has come under severe regulatory scrutiny and most of the drugs used for this purpose are administered as an implant or pellet in a part of the body, usually the ear, which is discarded at the time of slaughter. Extended withdrawal periods between the time of administration and the allowable date of slaughter depend on the release characteristics of the implant and may range from one month to one year. Dosages are relatively low, allowing drug release of 2–5 mg/d.

Tranquilizers and Anesthetics

Tranquilizers find their niche in veterinary medicine in the management of excitement in individual animals (10) (see Psychopharmacological agents). This group of compounds allows the practitioner to examine the frightened or injured patient with less chance of further damage or injury to the animal, the owner, and the veterinarian. They are also useful in the management of stress to avoid injury during the shipping of animals (see also Anesthetics; Hypnotics, sedatives, and anticonvulsants).

Acepromazine [61-00-7], a phenothiazine, is used in most animal species in both oral and injectable forms. It can be used at varying dosages to provide the state of tranquilization desired by the veterinarian. The product has a good margin of safety and has been used successfully by the veterinary profession for many years. Xylazine hydrochloride [23076-35-9] is another product used for both large and small animals. A thiazine compound unrelated to the phenothiazines, it acts primarily as a sedative. This compound is especially useful in examining fractious horses under field conditions (8). Both acepromazine and xylazine can be combined with other anesthetics for varying degrees of anesthesia or tranquilization.

Tranquilizers are employed for restraint in minor surgical procedures. The cardiovascular system, respiratory system, and blood chemistry are all greatly altered by general anesthesia. In larger animals, such as cattle and horses, the weight of the animal's body alone resting on its side can be a physiologically adverse stress on the heart and lower lung field that the veterinarian would prefer to avoid. If the surgery is of a confined or local nature, such as the repair of a superficial laceration, the animal can be tranquilized and analgesics provided to the wound area by use of a specific nerve block or by infiltrating the area around the site with a local anesthetic. The latter are synthetic non-narcotic cocaine substitutes, the first anesthetic ever used. The synthetic substitutes include procaine hydrochloride [51-05-8], tetracaine hydrochloride [136-47-0], lidocaine [137-58-6], and mepivacaine hydrochloride [1722-62-9]. Lidocaine is preferred in veterinary medicine.

These agents are often combined with a vasoconstrictant such as epinephrine [51-43-4]. With such a combination, the local anesthetic is held in the area for a longer period of time and its effect extended; hemorrhage is minimized, blood loss prevented, and a better surgical repair obtained.

A drug combination in popular use in dogs is a mixture of fentanyl [437-38-7], a narcotic analgesic, and droperidol [548-73-2], a butyrophenone tranquilizer. This combination produces a state of neuroleptanalgesia in which sedation and analgesia are achieved. The mixture is sold commercially and can be administered by both subcutaneous and intramuscular injection. Since the combination contains a narcotic, it has the advantage of being rapidly reversible with narcotic antagonists such as naloxone [465-65-6] and nalorphine [62-67-9] once the effects are no longer needed.

Another injectable anesthetic widely used in feline and primate practice is ketamine hydrochloride [1867-66-9]. Ketamine, a derivative of phencyclidine, can be chemically classified as a cyclohexamine and pharmacologically as a dissociative agent. Analgesia is produced along with a state that resembles anesthesia but in man has been associated with hallucinations and confusion. For these reasons, ketamine is often combined with a tranquilizer. The product is safe when used in accordance with label directions, but the recovery period may be as long as 12–24 h.

Another group of anesthetics is comprised of barbituates. By substituting various side chains on the basic structure, anesthetic activity can be greatly altered with regard to onset and duration of action. Short-acting barbituates, such as thiopental [77-27-0] often provide only a few minutes of sedation. These products are useful for induction to other types of general anesthesia, trachea intubation, and minor manipulations, examinations, and procedures. Pentobarbital [57-33-0] is longer acting and can be useful in more extensive or time-consuming surgery or in procedures requiring an extended sedation. Long-acting barbiturates, such as barbital and phenobarbital, have a prolonged effect in the animal, but also have a delayed onset of activity. They have generally been replaced in veterinary medicine by inhalation anesthetics.

In veterinary medicine, the list of inhalation anesthetics generally includes only two agents, halothane [151-67-7] and methoxyflurane [76-38-0]. Although ether (ethyl ether) is still used extensively in experimental work with laboratory animals, the risks associated with its use and the advantages of halothane and methoxyflurane have removed it from general use by the practitioner.

Halothane and methoxyflurane are volatile and are used in a vaporizer and delivered to the animal via an oxygen carrier. Both agents can be delivered with nitrous oxide [14522-82-8], a mild anesthetic that when combined with halothane or methoxyflurane can induce anesthesia faster than halothane or methoxyflurane alone. The recovery is faster because of the low solubility of nitrous oxide in the blood. Nitrous oxide can also be used alone, but must be supplemented with a barbituate or a narcotic.

It must be remembered that all anesthetics and tranquilizers are used by the practitioner following a risk–benefit evaluation. General anesthesia, even being administered by an experienced practitioner, can result in death through cardiac or respiratory depression. The veterinarian is acutely aware of these risks and chooses the drug and method of administration considering the patient's health status, the nature of and need for the procedure, and the likelihood of success.

Cancer Chemotherapy

In the veterinary patient, as in the human patient, neoplasms are often metastatic and widely disseminated throughout the body. Surgery and irradiation are limited in their use to well-defined neoplastic areas and, therefore, chemotherapy is becoming more prevalent in the management of the veterinary cancer victim (see Chemotherapeutics, antimitotic). Because of the expense and time involved, such management must be restricted to individual animals for which a favorable risk–benefit evaluation can be made and treatment seems appropriate to the practitioner and the owner. In general, treatment must be viewed not as curative, but as palliative.

The purpose of cancer chemotherapy, most briefly put, is to kill specific cells. The compounds are most active against rapidly growing and dividing cells, ideally the neoplastic cells, but all dividing cells can be attacked. For this reason, toxic signs such as alopecia, anemias and leukopenias, anorexia, vomiting, and other gastrointestinal signs may be indicative of undesirable effects of therapy. Periods of rest are often built into the treatment regimen to allow the animal's body a chance to recover and reestablish normal function.

Chemotherapeutic agents are grouped by cytotoxic mechanism. The alkylating agents, such as cyclophosphamide [50-18-0] and melphalan [148-82-3] interfere with normal cellular activity by alkylation of DNA (deoxyribonucleic acid). Antimetabolites, interfering with complex metabolic pathways in the cell, include methotrexate [59-05-2], 5-fluorouracil [51-21-8], and cytosine arabinoside hydrochloride [69-74-9]. Antibiotics such as bleomycin [11056-06-7] and doxorubicin [23214-92-8] have been used as have the plant alkaloids vincristine [57-22-7] and vinblastine [865-21-4].

Since the compounds described above vary in their specific mechanism of action and often have different effects on the individual patients, they are generally used in combinations, eg, corticosteroids with an alkylating agent, or an antimetabolite with a plant alkaloid in a rotating schedule.

Immunostimulation

The body's immune mechanism, both humoral and cell-mediated, affords a primary defense against invasion by foreign substances, ie, exogenous entities that the body may encounter including viruses, bacteria, chemicals, drugs, grafts, and transplants. The reaction by the immune system kills, neutralizes, or rejects the entity. The mechanisms involved in this complex system are under intense investigation, and a better understanding of the immune system will, in the future, permit the control of disease by means only speculated about today.

Human and veterinary practitioners have been manipulating the immune system for many years with bacterins and virus vaccines, in order to induce a response in the immune system. The animal forms antibodies which destroy the antigen. When the same or similar antigen is encountered again, as during exposure to the disease organism, the immune system is activated more quickly through an anamnestic response, thereby preventing the disease. Vaccines and bacterins are widely used in veterinary medicine for most domestic and exotic species (11) (see Vaccine technology).

The prophylactic stimulation of the immune system with vaccines and bacterins is time-consuming. Of even greater value would be the ability to activate the system to combat a disease attack already underway, or to be able to increase the response to abnormal cells and neutralize neoplasia in any organ of the body. Several compounds, some unique entities, some already in use for other purposes, have shown potential utility as such nonspecific immune stimulants.

In 1971, levamisole, an anthelmintic compound widely used in cattle and swine, was shown to improve the effects of an experimental *Brucella abortus* vaccine in mice. Since that time, the veterinarians and physicians have explored the effects of levamisole in such diverse areas as arthritis, lupus erythematosis, cancer therapy, respiratory diseases, Newcastle disease, foot-and-mouth disease, mastitis, and vaccine potentiation. Although the exact mechanism of action has as yet not been determined there is substantial evidence that, under defined circumstances, levamisole can augment the animal's natural immune response (12). Definite dosages and treatment regimes have not yet been established.

Although discovered in 1957, only recently has a group of natural substances called interferons been the subject of therapeutic interest. Interferons are glycoproteins synthesized by cells when under attack by a virus. Interferon seems to be an integral part in the body's basic defense mechanism, but more recent work indicates broad therapeutic activity and the possibility of cross-species efficacy. The emergence of recombinant DNA technology might allow sufficient quantities of interferon to be produced at a reasonable cost and thus may make interferon therapy a practical reality (see Genetic engineering).

Governmental Regulations

Compounds developed for use in veterinary medicine are subject to governmental regulations. The following information must be provided:

Acute toxicity in laboratory animals and target species, including LD_{50}, eye and skin irritation and toxicity, and inhalation toxicity.

Subacute toxicity in laboratory animals by 28- and 90-day feedings.

Chronic toxicity in laboratory animals by two-year or lifetime feedings in two species of laboratory animals, multiple-generation teratogenicity, and one-year feeding of dogs.

Specialized in vitro mutagenicity tests.

Overdose or extended treatment studies in the target species.

Efficacy to justify label claims.

Drug stability studies to determine rate of compound degradation and incompatibilities with other compounds or feed ingredients according to the anticipated conditions of use.

Metabolism studies to identify site of metabolism and principal metabolites.

Tissue residues in food-producing animals to document persistence in edible tissues.

Manufacturing methods to assure product consistency and the safety of personnel exposed to the drug or process intermediates.

Environmental effects, including effect on methanogenic and nitrifying bacteria, persistance in the environment, and projections of possible liability to relevant ecosystems.

Comprehensive labeling and directions.

The exact requirements for regulatory approval of a given compound vary from country to country. Generally, product development requires at least eight years and an investment of large sums of money, eg, 10^7. Both prescription and over-the-counter products are subject to regulations, which reflects safety and ease of product administration.

BIBLIOGRAPHY

"Veterinary Drugs" in *ECT* 2nd ed., Vol. 21, pp. 241–254, by A. L. Shor and R. J. Magee, American Cyanamid Co.

1. D. C. Blood and J. A. Henderson, *Veterinary Medicine*, 5th ed., Lea and Febiger, Philadelphia, Pa., 1979.
2. A. G. Gilman, L. S. Goodman, and A. Gilman, *The Pharmacologic Basis of Therapeutics*, 6th ed., Macmillan Publishing Co., Inc., New York, 1980.

3. L. M. Jones, N. H. Booth, and L. E. McDonald, *Veterinary Pharmacology and Therapeutics*, 4th ed., The Iowa State University Press, Ames, Iowa, 1977.
4. R. W. Kirk, *Current Veterinary Therapy, Small Animal Practice*, 7th ed., W. B. Saunders Co., Philadelphia, Pa., 1980.
5. O. H. Siegmund, *The Merck Veterinary Manual*, 5th ed., Merck and Co., Inc., Rahway, N.J., 1979.
6. *Feed Additive Compendium*, Miller Publishing Co., Minneapolis, Minn., 1982, published annually.
7. J. R. Georgi, *Parasitology for Veterinarians*, 3rd ed., W. B. Saunders Co., Philadelphia, Pa., 1980.
8. E. J. Catcott and J. F. Smithcors, *Equine Medicine and Surgery*, 2nd ed., American Veterinary Publications, Inc., Wheaton, Ill., 1972.
9. J. L. Howard, *Current Veterinary Therapy, Food Animal Practice*, W. B. Saunders Co., Philadelphia, Pa., 1981.
10. L. R. Soma, *Textbook of Veterinary Anesthesia*, The Williams and Wilkins Company, Baltimore, Md., 1971.
11. S. Krakowka, *Mod. Vet. Pract.* **62,** 447 (1981).
12. J. Symoens and M. Rosenthal, *Journal of the Reticuloendothelial Society* **21,** 175 (1977).

DAVID M. PETRICK
R. B. DOUGHERTY
American Cyanamid Co.

VETIVER. See Oils, essential.

VINEGAR

Vinegar is the liquid condiment or food flavoring used to give a sharp or sour taste to foods. It is also used as a preservative in pickling. The word vinegar is derived from Latin via old French as vinaigre meaning eager wine. In old English and old French, the word eager (*aigre*) meant sour or sharp. Thus, vinegar is a sharp or sour wine. Vinegar results from the action of the enzymes of bacteria of the genus *Acetobacter* on dilute solutions of ethyl alcohol such as cider, wine, beer, or diluted distilled alcohol (1) (see also Ethanol). Most vinegars for table use, such as the dressing of salads, derive from the acetic acid-bacterial fermentation of wine or cider (see Acetic acid). These, in turn, are produced by alcoholic fermentation (qv) of dilute sugar solutions such as grape juice, cider, or malt. *Saccharomyces cerevisiae* is the yeast usually involved in this enzymatic conversion of fermentable sugars to dilute alcoholic solutions (see Yeasts). Although fruits and honey are the most commonly employed sources of fermentable sugar for vinegar production, barley malt and, in Japan, rice, after hydrolysis of starch, are frequently used (see also Food processing; Fruit juices). Some raw materials are given in Table 1.

In the United States, standards of identity for vinegar date back to the Federal Food and Drug Act of 1906, where six types of vinegar are defined as follows: "vinegar,

Table 1.   Raw Materials and Fermentation Products

Raw material	Products	Ref.
tropical fruits	wines, vinegars	1
mango waste	syrups, wines, vinegars	2
whey	vinegar	3
rejected bananas	vinegar	4
palm sap	vinegar	5
dates	vinegar	6
white soy sauce	rice vinegar	7
enzyme preparation	rice vinegar	8

cider vinegar, apple vinegar, is the product made by the alcoholic and subsequent acetous fermentations of the juice of apples, and contains, in 100 cubic centimeters (20°C), not less than 4 grams of acetic acid." The other five types of vinegar are defined in the same terms except that cider vinegar is replaced by wine vinegar, malt vinegar, sugar vinegar, glucose vinegar, or spirit vinegar. In the case of the malt vinegar, a hydrolyzed starch solution is fermented to ethanol. The dilute beer, without concentration, is immediately oxidized by *Acetobacter* to vinegar (see also Beer). The quantity of 4 g in 100 cm³ of the quoted Federal regulation is equivalent to 40 g/L acetic acid or 40-grain strength in the terms used by vinegar producers (see also Beverage spirits, distilled). The *Federal Register* carries regulatory announcements concerning vinegar production at frequent intervals. Production figures for the United States are given in Table 2.

A number of factors govern the composition of vinegar, including the nature of the raw material, the substances added to promote alcoholic fermentation and the growth and activity of *Acetobacter*, the procedure used for the acetification, and finally the aging, stabilization, and bottling operations. Vinegars are made from natural solutions containing fermentable sugars such as fruit juices and honey; solutions in which the sugars are produced by hydrolysis of starch, such as beers and sake; and solutions of distilled ethyl alcohol. Grape juices and sweet ciders usually contain all the trace nutrients required by *Saccharomyces* for fermentation of sugars to alcohol. Other fruit

Table 2.   U.S. Vinegar Production, 1000 m³ [a]

Year	Number of plants	White distilled, 10%[b]	Cider, 5%[b]	Wine	Malt	Other	Total
1972	46	406.0	99.3	19.0	4.4	5.6	534.3
1973	45	407.0	93.3	18.0	4.7	13.3	536.2
1974	48	423.3	93.4	19.1	4.4	13.8	554.0
1975	34	432.5	92.7	18.9	4.0	13.8	562.0
1976	28	400.6	95.8	21.8	5.1	19.8	543.0
1977	25	400.4	83.2	19.4		24.2	527.2
1978	28	405.7	87.0	20.2		29.9	542.8
1979	27	408.5	84.8	19.5		32.0	544.7
1980	23	372.1	67.3	11.0		20.4	470.7

[a] As reported by the Vinegar Institute. Although subject to some uncertainty, these figures are the only ones available on U.S. vinegar production. To convert m³ to gal, multiply by 264.
[b] Percent acetic acid.

juices and diluted honey, as well as barley malt and rice extract, frequently need additions of nitrogen, phosphate, and potassium compounds together with some autolyzed yeast to facilitate the yeast growth necessary to fermentation. Stimulation of *Acetobacter* growth frequently requires the addition of autolyzed yeast, vitamin B (qv) complex, and phosphates. The character and composition of the vinegar is greatly influenced by the method used for the acetification and the subsequent processing steps (see below).

Vinegar is produced by surface or submerged-culture oxidation. Surface processes are older and are typified by the Orleans process for traditional wine vinegar. Vinegar production is accelerated by increasing the surface of liquid and bacteria exposed to oxygen. This speed-up is usually accomplished by passing the liquid through a column, tank, or vat packed with inert material on which the bacteria are adsorbed. Air is passed up through the vessel countercurrent to the liquid flow, thus providing for a much greater effective surface for a given volume of liquid. In the submerged process, air is introduced as very fine bubbles at the bottom of a tank in which an agitator mixes the contents with the gas. The *Acetobacter* are suspended in the liquid. This is, in effect, a modified surface process since bacteria are in contact both with the liquid and air at the bubble surface as it slowly rises through the bulk of the liquid.

Manufacture

Starch Hydrolysis and Alcoholic Fermentation. In general, yeasts cannot utilize starch as a carbon source for growth, and the starch must be hydrolyzed to sugar before it can be utilized. Malt vinegars, commonly used as table vinegar in the UK, are made from malted barley or a mixture of the malted barley with other starchy grains. The enzymes of the malted barley hydrolyze the starch to glucose and maltose which are readily fermented by *Saccharomyces* yeast. In Japan, where vinegars are made from rice, a mixture of hydrolyzing enzymes produced by the fungus *Aspergillus oryzae* converts rice starches to sugars (see also Enzymes, immobilized). A small amount of cooked rice is cultured with the fungus and then added to a larger quantity of cooled steamed rice. Frequently, the yeast that converts the sugar to alcohol is added at the same time resulting in a dilute alcoholic solution rather than a dilute sugar solution. For the alcoholic fermentation itself, strains of *Saccharomyces cerevisiae* are used in most cases. The malt alcoholic solutions usually contain 5–10 vol % of ethanol; rice-derived alcoholic solutions, ie, sake, may contain 15–20%. These alcoholic solutions are immediately converted to vinegar. If they are to be stored, the pH should be low enough to discourage contamination by undesirable organisms. In particular, lactic bacteria can cause problems at relatively high pH. Addition of 20–30 mg SO_2 per liter of vinegar prevents undesirable bacterial infections.

In alcoholic fermentation, glucose from hydrolyzed starch or the glucose and fructose from fruits or honey are converted by yeast enzymes into ethyl alcohol and carbon dioxide (see also Ethanol; Sugar). According to the Gay-Lussac equation, $C_6H_{12}O_6 = 2 CO_2 + 2 C_2H_5OH$, 1000 g glucose or fructose yields 0.5114 g ethanol. The theoretical yield is never obtained because of side reactions in addition to other factors such as the temperature, agitation, condition of the yeast population, etc. Yields of 88–94% are considered good commercial practice.

The alcoholic fermentation is frequently conducted in two phases, although in a modern vinegar plant it can be conducted in one. The first phase is a vigorous fermentation during which the rapid evolution of carbon dioxide protects the alcoholic solution from air. The second or slower phase is fermentation of the residual sugar at a lower rate during which, again, protection from air is required. In the first phase, 50–100 mg SO_2 is added to 1 L of the sugar-containing mash, as well as 1–3% of an actively fermenting pure-culture starter of *Saccharomyces cerevisiae*. The fermentation process is monitored for disappearance of sugar and increase in temperature. Rates of alcoholic fermentations are highest at ca 25–30°C; higher temperatures damage the enzyme systems. The decrease in sugar content is measured hydrometrically and is usually expressed in degree Brix. Degree Brix refers specifically to weight percentage of sucrose in a sucrose–water mixture. However, the sugar content of fruit juice of corresponding density is only slightly less than the equivalent percentage of glucose and fructose. Some error is introduced by the presence of nonsugar solids. As alcohol is produced during the fermentation, the degree Brix decreases more rapidly than the sugar since ethanol is less dense than water. Temperatures during the fermentation are measured at least twice daily, and the vessel is cooled if necessary. Cooling jackets or internally mounted cooling coils are used, although in older installations the fermenting medium is pumped from the tank through an external heat exchanger and returned to the tank. When the vigorous fermentation has slowed to Brix readings of ca +0.5 to 0°, the partially fermented alcoholic medium is siphoned off and placed in another tank for final fermentation. This tank is equipped with a water seal or a valve permitting the escape of CO_2 and preventing entry of air. The residual sugar is fermented slowly in this vessel. In a modern plant equipped with closed stainless-steel tanks, the fermentation can be conducted as a single-stage process in one tank. When all the sugar has been fermented (negative degree Brix reading and <1 g reducing sugars per liter), the wine or beer may be acetified immediately or stored protected from air. For prolonged storage, 50 mg SO_2/L prevents growth of lactic organisms.

Both the fermentation of hexose sugars to ethanol and carbon dioxide and the oxidation of ethanol to acetic acid are exothermic processes (see Sugar). The first reaction is expressed as follows:

$$180 \text{ g } C_6H_{12}O_6 \rightarrow 92 \text{ g } C_2H_5OH + 88 \text{ g } CO_2 + 234 \text{ kJ (55.6 kcal)}$$

The yeast enzymes capture ca 92 kJ (22 kcal) of this energy for the formation of adenosine triphosphate (ATP), and the actual waste heat per mol wt of sugar fermented is ca 142 kJ (33.9 kcal). Depending upon the size of the fermentor and the rates of fermentation and aeration, waste heat is lost from the fermenter through radiation, conduction, and vaporization of water and ethanol plus carbon dioxide. Although small fermenters may require no cooling, large ones require external cooling devices.

Wines have a low sodium and high potassium content and also contain tartaric, malic, and succinic acids, a wide spectrum of amino acids, phenolic materials, and trace quantities of vitamins and growth factors. These substances are also found in wine vinegar. Vinegar materials from other fruits, honey, and sugar-containing natural materials also contain a wide spectrum of nutrients from the base material. Contamination with pesticide residues must be avoided.

Acetic Acid.   Ethyl alcohol is converted to acetic acid by air oxidation catalyzed by the enzymes within bacteria of the genus *Acetobacter*:

$$46 \text{ g } C_2H_5OH + 32 \text{ g } O_2 \rightarrow 60 \text{ g } CH_3COOH + 18 \text{ g } H_2O + 487.2 \text{ kJ (116.4 kcal)}$$

One gram ethanol should yield 1.304 g acetic acid. Practical yields are 77–85%. The excess heat must be removed during the course of the oxidation. For example, oxidation of 1 m^3 (264 gal) of a solution of ethanol (10% by volume) yields 836.7 MJ (ca 2×10^5 kcal). If this oxidation occurs over a 4-d period, and the heat is liberated at a uniform rate, heat production amounts to 2.42 kW.

In contrast with the well-known Embden-Meyerhof-Parnass glycolysis pathway for the conversion of hexose sugars to alcohol, the conversion of ethanol to acetic acid remains in doubt. Certainly, ethanol is first oxidized to acetaldehyde and water (2). For further oxidation, two alternative routes are proposed: more likely, hydration of the acetaldehyde gives $CH_3CH(OH)_2$, which is oxidized to acetic acid. An alternative is the Cannizzaro-type disproportionation of two molecules of acetaldehyde to one molecule of ethyl alcohol and one molecule of acetic acid. Possibly, *Acetobacter* initiates both reactions (3). The disproportionation reaction takes place at pH 8, whereas the dehydrogenation occurs more readily under acidic conditions. It seems likely that the pH at the enzyme surface within the cell is relatively low and that, therefore, the dehydrogenation reaction is preferred over the disproportionation reaction.

Orleans Process. Early mention of vinegar is found in the Talmud in accounts of wine and beer turning to vinegar in ancient times (4). The Babylonians, about 5000 BC, made and used vinegar as a flavor enhancer and as a pickling agent or preservative. Production of vinegar by the Greeks and Romans is described in numerous writings, but it was not until the fourteenth century in Modena, Italy, and near Orleans, France, that the process was further developed. In the Orleans process, the wine oxidizes slowly in a barrel where it is covered with a film of *Acetobacter*. Holes that are covered with screens as protection against insects are bored in each head to permit access to air oxygen. Wine is added through the bung hole with a long-stemmed funnel below the surface of the bacterial film and without disturbing the film. A spigot is mounted in the barrel head near the bottom. In operation, vinegar is removed through the spigot for consumption and is replaced by an equivalent quantity of wine through the funnel. Wines with 10–12% ethanol give vinegars of 8–10% acetic acid concentration. Orleans vinegars are characterized by a relatively high concentration of ethyl acetate, detected by its strong odor. The Orleans process is no longer much in use, because it is very slow, although the product is highly desirable. Furthermore, the vinegar in the barrels tends to become slimy from the production of exocellular bacterial cellulose generated by *Acetobacter xylinum*. This slimy cellulose, called mother of vinegar, encapsulates the bacterial cells and dramatically slows down the rate of production. Significant amounts of red-wine vinegar are produced in Europe by the Pasteur modification of the Orleans process. A wooden grating is placed at the surface of the liquid in the partially filled barrels or in shallow tanks to support the film of vinegar bacteria.

In 1973, a multistage surface-fermentation process was patented in Japan for the production of acetic acid (5); eight surface fermenters were connected in series and arranged in such a way that the mash passed slowly through the series without disturbing the film of *Acetobacter* on the surface of the medium. This equipment is reported to produce vinegar of 5% acidity and 0.22% alcohol with a mean residency time in the tanks of 22 h.

Generator Processes. References to the quick or generator processes for vinegar production are found as early as the seventeenth century (6). Usually, generators are packed with shavings of beech wood which tend to curl and thus provide packing that does not consolidate but allows open spaces for the free flow of liquid and air. In addition, beech wood does not contribute undesirable flavors or impurities to the vinegar. In the modern generator, a recirculating pump transfers the partially acetified alcoholic mixture from the bottom section of the generator to a distributing system at the top of the packed section. The air is measured into the upper part of the storage section below the packed section of the vat and exhaust air is vented from the top of the packed section to the outside, although in some generators it is recirculated. Cooling coils may be located in the packed section, but more frequently are placed at the bottom of the receiver section or are incorporated in the line for recirculating the mash. Packing materials include beech-wood shavings, coke, grape twigs, rattan bundles, corn cobs, or unglazed ceramic saddles. A pilot-plant Frings generator produces 20.4 μg/L per second (7).

The various species and many strains of *Acetobacter* are used in vinegar production (3,8). Aeration requirements have to be considered and, in general, cider or wine vinegars require no added nutrients. Addition of nutrients is essential for vinegar production from dilute distilled alcohol.

Submerged-Culture Generators. Success during the early 1940s in adapting the surface-film growth procedure for producing antibiotics to an aerated submerged-culture process led vinegar producers to try the same technique (9). A mechanical system keeps the bacteria in suspension in the liquid in the tank in intimate contact with fine bubbles of air. The excess heat must be removed and the foam broken down which accumulates at the top of the tank. The most widely used submerged-culture oxidizer is the Frings acetator (10). It uses a bottom-driven hollow rotor turning in a field of stationary vanes arranged in such a way that the air which is drawn in is intimately mixed with the liquid throughout the whole bottom area of the tank (11–12). In the United States, continuous cavitator units are used widely for cider-vinegar production.

A strain of thermophilic *Acetobacter* was patented in Japan for oxidizing ethanol in a submerged-culture oxidizer at temperatures as high as 37°C with considerable savings in cooling water. Another thermophilic strain of *Acetobacter* maintained full activity at 35°C and 45% of its maximum activity at 38°C.

A Frings acetator consisting of a 48-m^3 (12,700 gal) tank produces 12 m^3 (3,200 gal) of 10% acetic acid vinegar per day. It requires 2.2 L of cooling water per second at 15°C and an energy input of ca 36 MW (8600 kcal/s) (2). Thus, the submerged-culture oxidizer is capable of producing vinegar at ca twice the rate of the best generator. Furthermore, submerged-culture oxidizers are smaller for a given amount of production and, most important, they are more flexible in their operation. Thus, it is possible to change from one vinegar type to another with different feed-alcohol concentrations and nutrient requirements quicker than with a generator.

Submerged-culture oxidizers are usually operated on a semicontinuous basis. In most cases, ca half of the liquid in the tank is removed every 1–2 d when the alcohol concentration has dropped to 0.1–0.2 vol %. The removed vinegar is replaced with a wine or mash of richer ethanol and lower acetic acid concentration giving a mixture in the tank of 5–6 vol % ethanol and 6–8 vol % acetic acid. These are the optimum conditions for *Acetobacter* growth. Wine or concentrated cider does not require the addition of nutrients, but diluted distilled alcohol solutions need ca 10–15 g of inorganic substances such as diammonium acid phosphate, potassium chloride, and traces of other metals per L of alcohol and 30–50 g of organic materials such as glucose, autolyzed yeast, citric acid, and powdered whey per L of alcohol. The pH of the fermenter mixture should be 3.9–5.0, the ideal temperature between 28 and 31°C. Since the new charge of mash or wine to the oxidizers lowers the temperature, cooling may be interrupted until the temperature again reaches 28–31°C. The rate of aeration depends on the surface of contact between air and liquid and is an inverse function of the bubble size. At optimum aeration rate, 60 millimol/h O_2 is introduced into the solution per liter of mash, ie, 1870 cm^3 air/(s·m^3 mash). The maximum value for aeration recommended in ref. 2 is 1100 cm^3/(s·m^3 mash).

Foam production is most troublesome under conditions adverse to bacterial growth and can thus be minimized by keeping nutrient, ethanol, and acetic acid concentrations in the optimum ranges (see Defoamers). Temperature and aeration rate are also critical. Dead or dying cells seem to promote foam formation. Even under optimum conditions, some type of foam breaker mounted in the top of the oxidizer is needed; the foam is usually broken down by centrifugal force. Food-grade silicone antifoaming agents are employed under some circumstances.

Submerged-culture oxidizers can also be operated on a continual basis, with continuous analysis of ethanol and acetic acid concentrations, and control of feed and withdrawal streams. Optimum production, however, is achieved by semicontinuous operation because the composition of vinegar desired in the withdrawal stream is so low in ethanol that it impedes vigorous bacterial growth. Bacterial concentrations up to 100×10^6 cells/cm^3 have been reported for high acetic acid concentrations.

A submerged-culture oxidizer with instrumentation to accurately control the oxygen concentration of the mash and with a heat-transfer system that efficiently controls the temperature is described in ref. 3. In a process patented in Japan, bacterial cells are filtered from the new vinegar as it is withdrawn from the oxidizer and immediately returned (13–14). Glycerol catalyzes the production of vinegar from the alcoholic solution obtained from malt wort (15), and its degradation pathways have been elucidated. Certain strains of *Saccharomyces cerevisiae* produce enough SO_2 to impede the initiation of the oxidation by *Acetobacter* (16). A scrubber has been patented which greatly increases the efficiency of vinegar production by recycling ethanol and acetic acid vapors normally lost with the exhaust air stream (17).

Tower Fermenters. The tower fermenter or oxidizer does not have any packing: the liquid is held on a porous plate by the pressure of air introduced below the plate (18–19). Air bubbles penetrate the plate and keep the *Acetobacter* in suspension and active for the ethanol oxidation in the liquid phase. Addition of a hydrous titanium–cellulose chelate suspension to the liquid phase improves the performance (18,20–21). The increased oxidation rate is the result of a greater cell mass per unit volume.

Concentrated Vinegars. The U.S. regulations require at least 4 g acetic acid/100 cm^3 vinegar. Commercial vinegar and many quality table vinegars are significantly more concentrated. Submerged-culture oxidizers easily give acetic acid concentrations of 10–13 g per 100 cm^3. Production rate is somewhat lower than for lower acetic acid concentrations. Recently, submerged-culture oxidizers that produce vinegars with acetic acid concentrations ranging from 15 to 20% have been patented (22–26). Continuous aeration and careful stepwise addition of ethanol as it is oxidized seem to be the key to successful operation. In order to obtain even higher acetic acid concentrations, the water is removed after the generation step by freezing (27–29). The ice crystals are removed by filtration or centrifugation. Concentrated vinegars are of particular value in the pickling industry where dilution of the vinegary, spiced mixture by the water from the cucumbers is a serious and costly problem (3). A 12% vinegar can be obtained from a more dilute vinegar by freezing and centrifuging the ice (30). The concentrated wine vinegar is stabilized with bentonite, silica gel, or $K_4Fe(CN)_6$. In another process, the water is removed by formation of a hydrate of trichlorofluoromethane. The solid hydrate is separated from the concentrated vinegar and the fluorocarbon is recovered and recycled (see also Clathration).

Vinegar Eels and Mother of Vinegar. The nematode *Anguilla aceti* grows readily in packed-tank vinegar generators. Although it is esthetically undesirable, it is not usually harmful. These nematodes, known as vinegar eels, may actually be of some assistance in consuming dead bacteria from the surface of the packing material in the tank, and thus aid in prolonging the operation of the system. Vinegar eels may also make nutrients more readily available to *Acetobacter* (31). Vinegar eels are removed from the raw vinegar by filtration and pasteurization before it is sold or used further in pickling or other processes.

Mother of vinegar is the term given to the cellulosic slime that coats the bacterial cells and is produced by a strain of *Acetobacter xylinium*. Different strains and different medium compositions result in different consistencies and crystalline forms of the cellulosic slime. Although of no problem in submerged-culture oxidizers, the slime can effectively block the passage ways in packed-tank generators. High concentrations of acetic acid in the vinegar and the generator discourage the production of mother of vinegar slime.

Processing and Preparation for Marketing

Clarification. Raw vinegars as removed from the production unit vary widely in stability, depending upon the raw material and the type of generator or oxidizer employed. Table vinegars from wine, cider, malt, or other natural materials frequently contain unstable phenolic materials, pectins, and traces of proteins which form clouds or deposits. Vinegars from distilled alcohol are more stable, but still might contain traces of unstable materials. Generator vinegars are relatively free of *Acetobacter* cells in contrast to the submerged-culture vinegars which carry a high and cloudy suspension of bacterial cells. Clarification and stabilization of vinegars generally follow the standard practice of the wine industry. Bentonite is used as clarifier and, occasionally, a proprietary formulation of potassium ferrocyanide is used to remove traces of heavy metals. Submerged-culture vinegar is clarified with mixed suspensions of bentonite and alginic acid (32); treatment with bentonite prepared with $NaHCO_3$ has been patented (33). Japanese patents describe vinegar stabilization with alumina or silica gels (34) or polyvinylpyrrolidinone, cellulose, and Dowex A-1 for the same purpose. Activated carbon adsorbs some compounds causing clouding or precipitates in bottled vinegars. Vinegars are usually given a rough filtration on plate-and-frame filters with pads coated with diatomaceous earth or the more recent leaf filters. Immediately before bottling, the vinegar is filtered through more retentive pads or possibly membranes of pore size small enough to exclude all yeasts cells and bacteria. Membrane filtration can be combined with aseptic bottling to provide a vinegar free of all microorganisms, but the process is expensive and not essential to the stability of bottled vinegars. A membrane can be used for continuous microfiltration of cloudy vinegar (35). The surface is kept clean with a vigorous flow of liquid across the membrane parallel to its surface. This method has been successful in both laboratory and pilot plant.

Aging and Blending. Wine, cider, and malt vinegars, unless made by an Orleans-type process, benefit from some aging during which there is a smoothing and complexing of the vinegar character. Particularly if the fresh vinegar contains some residual ethanol, the reaction with acetic acid giving ethyl acetate increases the fruitiness of the aroma. Generator vinegars and submerged-culture vinegars, used primarily for pickling, are normally not aged, although there are subtle changes in the aroma and taste during a few months of aging. Vinegars from different production batches are blended to a constant acetic acid strength and color before marketing. Products used for commercial pickling or as condiments in packaged foods, do not require constant composition and appearance.

Sterilizing and Packing. Vinegars bottled for table use or pickling are pasteurized before shipment. In the lower strength vinegars, the cellulose-producing acetic bacteria and certain strains of lactic bacteria may create problems. The former cause clouding, whereas the latter alter the flavor. Small amounts of sulfur dioxide are frequently added to minimize lactic-organisms growth. Sterile filtration through very tight pads or through membranes followed by aseptic bottling is possible but difficult in the case of bacterial contaminants. For pasteurization, the vinegar is heated in bulk to 65–70°C, filled hot into bottles, sealed, and cooled slowly. The filled and sealed bottles are pasteurized by heating to 65–70°C. Sterilization of many submerged-culture vinegars with high bacterial cell concentrations requires a pasteurization temperature of 77–78°C. The vinegar must be protected from exposure to bacterial or yeast contaminants and iron or copper in all processing steps following stabilization, ie, the filling equipment must be of stainless steel, plastic, or glass (see also Sterilization techniques).

BIBLIOGRAPHY

"Vinegar" in *ECT* 1st ed., Vol. 14, pp. 675–686, by M. A. Joslyn, University of California; "Vinegar" in *ECT* 2nd ed., Vol. 21, pp. 254–269, by M. A. Joslyn, University of California.

1. F. Pandl, *Szeszipar* **26**(1), 3 (1978).
2. G. Keszthelyi, *Mitt. Hoeheren Bundeslehr Versuchsanst. Wein Obstbau Klosterneuburg* **24,** 445 (1974).
3. H. A. Connor and R. J. Allgeier, *Adv. Appl. Microbiol.* **20,** 81 (1976).
4. E. Huber, *Dtsch. Essigind.* **31**(1), 12 (1927); **31**(2), 28 (1927).
5. Brit. Pat. 1,305,868 (Feb. 7, 1973), (to Kewpie Jozo Kabushiki Kaisha).
6. C. A. Mitchell, *Vinegar: Its Manufacture and Examination*, 2nd ed., Griffin, London, 1926.
7. R. J. Allgeier, R. T. Wisthoff, and F. M. Hildebrandt, *Ind. Eng. Chem.* **44,** 669 (1952); **45,** 489 (1953); **46,** 2023 (1954).
8. G. B. Nickol in H. J. Peppler and D. Perlman, eds., *Microbial Technology*, 2nd ed., Vol. 2, Academic Press, New York, 1979, pp. 5–72.
9. G. J. Fowler and V. Subramaniam, *J. Indian Inst. Sci.* **6,** 147 (1923).
10. O. Hromatka and H. Ebner, *Enzymologia* **13,** 369 (1949).
11. U.S. Pat. 2,997,424 (Aug. 22, 1961), E. Mayer (to Hunt Foods and Industries); E. Mayer, *Food Technol.* **17,** 582 (1963).
12. U.S. Pat. 2,913,343 (Nov. 17, 1959), A. C. Richardson (to California Packing Corp.).
13. Jpn. Kokai Tokyo Koho 80 08,150 (March 1, 1980), H. Okumura, I. Ohmori, H. Kunimatsu, and H. Masai.
14. Jpn. Kokai Tokyo Koho 80 54,890 (April 22, 1980), (to Nakano Vinegar Co., Ltd.).
15. Ger. Offen. 2,215,456 (Oct. 4, 1973), R. N. Greenshields and D. D. Jones.
16. C. Zambonelli, M. E. Guerzoni, M. Nanni, and G. Gianstefani, *Riv. Vitic. Enol.* **25,** 214 (1972).
17. Jpn. Kokai Tokyo Koho 80 08,149 (March 1, 1980), H. Masai, H. Yamada, and Nishimura.

18. Brit. Pat. 1,514,425 (June 14, 1978), S. A. Barker, R. N. Greenshields, J. D. Humphreys, and J. F. Kennedy (to Gist-Brocades N.V.).
19. Ger. Offen. 2,205,638 (Aug. 23, 1973), R. N. Greenshields.
20. J. F. Kennedy, *Enzyme Eng.* **4**, 323 (1978).
21. J. F. Kennedy, J. D. Humphreys, S. A. Barker, and R. N. Greenshields, *Enzyme Microb. Technol.* **2**, 209 (1980).
22. Jpn. Kokai Tokyo Koho 81 18,190 (April 27, 1981), (to Nakano Vinegar Co., Ltd.).
23. Ger. Offen. 2,657,330 (June 30, 1977), H. Ebner and A. Enenkel (to Frings G.m.b.H. und Co. K.-G.).
24. Jpn. Kokai Tokyo Koho 80 09,706 (Jan. 23, 1980), A. Mori, H. Suzue, and Y. Kowatani (to Kyu-Pi Co., Ltd., Kyu-Pi Jozo K.K., Toshoku K.K., and Frings G.m.b.H und Co. K.-G.).
25. Fr. Demande 2,374,416 (July 13, 1978) (to Frings G.m.b.H. und Co. K.-G.).
26. Ger. Offen. 3,005,099 (Aug. 21, 1980), Y. Kunimatsu, H. Okumura, H. Masai, K. Yamada, and M. Yamada (to Nakano Vinegar Co., Ltd.).
27. F. K. Lawler, *Food Eng.* **23**, 68, 82 (1961).
28. U.S. Pat. 2,800,001 (July 23, 1957), E. P. Wenzelberger (to Commonwealth Engineering Co. of Ohio).
29. J. R. Dooley and D. D. Lineberry in *Symposium on New Developments in Bioengineering—Minneapolis*, preprint, American Institute of Chemical Engineering, New York, 1965, 12 pp.
30. T. A. Tonchev and G. K. Bambalov, *Lozar. Vinar.* **14**(8), 31 (1965).
31. R. C. Zalkan and F. W. Fabian, *Food Technol.* **7**, 453 (1953).
32. Jpn. Kokai Tokyo Koho 74 108,295 (Oct. 15, 1974), H. Masai and K. Yamada (to Nakano Vinegar Co., Ltd.).
33. Czech. Pat. 151,118 (Nov. 15, 1973), J. Vyslouzil.
34. Jpn. Kokais Tokyo Koho 81 11,430 and 81, 11,431 (March 14, 1981), (to Nisshin Flour Milling Co., Ltd.).
35. H. Ebner, *Chem. Ing. Tech.* **53**(1), 25 (1981).

General References

References 1, 2, 3, and 8 are also general references.
M.-H. Lai, W. T. H. Chang, and B. S. Luh in B. S. Luh, ed., *Rice; Production and Utilization*, AVI, Westport, Conn., 1980, pp. 712–735.
R. N. Greenshields, *Econ. Microbiol.* **2**, 121 (1978).
H. Ebner in E. Bartholome, E. Biekert, and H. Hellmann, eds., *Ullmanns Encykl. Tech.*, *4 Aufl.*, Vol. 11, Verlag Chem., Weinheim, FRG, 1976, pp. 41–55.
H. Ito, *Nippon Jozo Kyokai Zasshi* **73**, 200 and 453 (1978).
H. Masai, *Nippon Shokuhin Kogyo Gakkai-Shi* **25**(2), 104 (1978).
H. Masai, Y. Kawamura, and K. Yamada, *Nippon Nogei Kagaku Kaishi* **52**(8), R103 (1978).
H. Masai, *Nippon Jozo Kyokai Zasshi* **74**, 798 (1979); **75**, 888 (1980).
H. Masai, *Proceedings, Oriental Fermented Foods*, Food Industry Research Development Institute, Hsinchu, Republic of China, 1980.
F. Yanagita and Y. Koizumi, *Nippon Jozo Kyokai Zasshi* **75**, 854 (1980).
E. Levonen and C. Llaguno, *Rev. Agroquim. Technol. Aliment.* **18**, 289 (1978).
M. Perez, M., *Bibliografia de los Vinagres*, Departmento de Informacion Tecnica, Ministerio de Industrias, Havana, Cuba, 1963.
C. Stella, *Vini Ital.* **18**, 269 (1976).
A. D. Webb and W. Galetto, *Am. J. Enol. Vitic.* **16**(2), 79 (1965).
H. Mizumoto, H. Minamisone, S. Mori, and K. Azuma, *Kagoshima-Ken Kogyo Shikenjo Nempo* **22**, 67 (1975); *Chem. Abstr.* **87**, 132439f.
O. R. Garcia, E. A. Carballido, and M. Castana Torres, *An. Bromatol.* **25**(2), 121 (1973).
M. A. Suarez, M. C. Polo, C. Llaguno, and J. L. Andreu, *Rev. Agroquim. Technol. Aliment.* **16**, 531 (1976).
G. Yamaguchi and H. Masai, *Agric. Biol. Chem.* **39**, 1903 (1975).
F. Yanagida, K. Takashima, Y. Yamamoto, H. Nishizima, and K. Suminoe, *Nippon Jozo Kyokai Zasshi* **66**, 1185 (1971).
F. Yanagida, K. Takashimai, Y. Yamamoto, N. Kaneko, and K. Suminoe, *Nippon Jozo Kyokai Zasshi* **68**, 130 (1973).

S. Furukawa, N. Takenaka, and R. Ueda, *Hakko Kogaku Zasshi* **51,** 321 and 327 (1973).
A. Carballido and M. T. Valdehita, *An. Bromatol.* **29**(2), 103 (1975).
J. H. Kahn, G. B. Nickol, and H. A. Conner, *J. Agric. Food Chem.* **20,** 214 (1972).
C. Llaguno, *Sem. Vitivinic.* **27**(1.367–1.368), 4309, 4311, 4313 (1972).
F. Mecca and L. Di Vecchio, *Riv. Soc. Ital. Sci. Aliment.* **6**(3), 177 (1977).
M. Ferrer Gimenez and R. Clotet Ballus, *An. Bromatol.* **31**(2), 109 (1979).
N. Nakamura and A. Mori, *Nippon Jozo Kyokai Zasshi* **74,** 471 and 479 (1979).
M. L. Gil De la Pena, M. D. Garrido, and C. Llaguno, *Rev. Agroquim. Tecnol. Aliment.* **16,** 413 (1976).
S. Fukano, F. Ushio, K. Nishida, M. Doguchi, and T. Kani, *Tokyo Toritsu Eiseni Kenkyusho Kenkyu Nempo* **27,** 203 (1976).
F. Mecca and P. Spaggiari, *Sci. Aliment.* **17,** 235 (1971).
E. R. Schmid, I. Fogy, and E. Kenndler, *Z. Lebensm. Unters. Forsch.* **163,** 121 (1977); **166,** 221 (1978).

A. D. WEBB
University of California, Davis

VINYLBENZENE.   See Styrene.

VINYL CHLORIDE.   See Vinyl polymers.

VINYL ETHER.   See Anesthetics; Vinyl polymers.

VINYLIDENE CHLORIDE AND POLY(VINYLIDENE CHLORIDE)

The most valuable property of poly(vinylidene chloride) [9002-85-1] (PVDC) and its copolymers is low permeability to a wide range of gases and vapors. The most serious deficiency is thermal instability at melt-processing temperatures. The techniques used to overcome the instability, ie, copolymerization and plasticization, were developed at The Dow Chemical Company during the 1930s (1). The commercialization of these polymers under the trade name Saran began in 1939.

In the United States, Saran is a generic term for high VDC-content polymers. (A letter following the name is used by Dow to indicate the type of comonomer, eg, Saran A for the homopolymer, B for copolymers with vinyl chloride (VC), and F for copolymers with acrylonitrile (AN).) In other countries, Saran is still a trademark of The Dow Chemical Company. The synonymous use of poly(vinylidene chloride) and Saran has resulted in many materials being identified in the literature as poly(vinylidene chloride) that were actually copolymers of unknown composition. Although it has valuable properties, the homopolymer has not been used to any extent because of its difficult fabrication. Of the many copolymers that have been prepared, only three types are commercially important: vinylidene chloride–vinyl chloride copolymers [9011-06-7], vinylidene chloride–alkyl acrylate and methacrylate copolymers, and vinylidene chloride–acrylonitrile copolymers [9010-76-8] and methacrylonitrile copolymers [9010-80-4]. The literature on vinylidene chloride polymers through 1972 is reviewed in ref. 2. Older reviews may also be of interest (3–4).

Vinylidene Chloride

Properties. Pure vinylidene chloride [75-35-4] (1,1-dichloroethylene) is a colorless, mobile liquid with a characteristic sweet odor. Its properties are summarized in Table 1. Vinylidene chloride is soluble in most polar and nonpolar organic solvents. Its solubility in water (0.25 wt %) is nearly independent of temperature at 16–90°C.

Manufacture. Vinylidene chloride monomer can be conveniently prepared in the laboratory by the reaction of 1,1,2-trichloroethane with aqueous alkali:

$$2\ CH_2ClCHCl_2 + Ca(OH)_2 \xrightarrow{90°C} 2\ CH_2{=}CCl_2 + CaCl_2 + 2\ H_2O$$

Other methods are based on bromochloroethane, trichloroethyl acetate, tetrachloroethane, and catalytic cracking of trichloroethane (7). Catalytic processes produce by-product HCl rather than less valuable salts, but yields of vinylidene chloride have been too low for commercial use of these processes. However, good results have been reported with metal salt catalysts (8–10).

Vinylidene chloride is prepared commercially by the dehydrochlorination of 1,1,2-trichloroethane with lime or caustic in slight excess (2–10%) (6,11). A continuous liquid-phase reaction at 98–99°C yields ca 90% VDC. Caustic gives better results than lime. Vinylidene chloride is purified by washing with water, drying, and fractional distillation. It forms an azeotrope with 6 wt % methanol (12). Purification can be by distillation of the azeotrope followed by extraction of the methanol with water; an inhibitor is usually added at this point. Commercial grades contain 200 ppm of the monomethyl ether of hydroquinone (MEHQ). Many other inhibitors for the polymerization of vinylidene chloride have been described in patents, but MEHQ is the

Table 1. Properties of Vinylidene Chloride Monomer[a]

Property	Value
molecular weight	96.944
odor	pleasant, sweet
appearance	clear, liquid
color (APHA)	10–15
solubility of monomer in water, at 25°C, wt %	0.25
solubility of water in monomer, at 25°C, wt %	0.035
normal boiling point, °C	31.56
freezing point, °C	−122.56
flash point (tag closed-cup), °C	−28
flash point (tag open-cup), °C	−16
flammable limits in air (ambient conditions), vol %	5.6–16.0
autoignition temperature, °C	513[b]
Q value	0.22
e value	0.36
latent heat of vaporization, ΔH_v°, kJ/mol[c]	
at 25°C	26.48 ± 0.08
at normal boiling point	26.14 ± 0.08
latent heat of fusion (at freezing point), ΔH_m°, J/mol[c]	6514 ± 8
heat of polymerization (at 25°C), ΔH_p°, kJ/mol[c]	−75.3 ± 3.8
heat of combustion, liquid monomer (at 25°C), ΔH_c°, kJ/mol[c]	1095.9
heat of formation	
liquid monomer (at 25°C), ΔH_f°, kJ/mol[c]	−25.1 ± 1.3
gaseous monomer (at 25°C), ΔH_f°, kJ/mol[c]	1.26 ± 1.26
heat capacity	
liquid monomer (at 25°C), C_p°, J/(mol·K)[c]	111.27
gaseous monomer (at 25°C), C_p°, J/(mol·K)[c]	67.03
critical temperature, T_c, °C	220.8
critical pressure, P_c, MPa[d]	5.21
critical volume, V_c, cm³/mol	218
liquid density, g/cm³	
−20°C	1.2852
0°C	1.2499
20°C	1.2137
index of refraction, n_D	
10°C	1.43062
15°C	1.42777
20°C	1.42468
absolute viscosity, mPa·s (= cP)	
−20°C	0.4478
0°C	0.3939
20°C	0.3302
vapor pressure, t, °C (P measured from 6.7–104.7 kPa)[e]	$\log P_{kPa} = 6.1070 - 1104.29/(t + 237.697)$

[a] Refs. 5–6.
[b] Inhibited with methyl ether of hydroquinone.
[c] To convert J to cal, divide by 4.184.
[d] To convert MPa to atm, divide by 0.101.
[e] To convert kPa to mm Hg, multiply by 7.5 (add 0.875 to the constant to convert $\log_{kPa}$ to $\log_{mm\ Hg}$).

one most often used. The inhibitor can be removed by distillation or by washing with 25 wt % aqueous caustic under an inert atmosphere at low temperatures.

For many polymerizations, MEHQ need not be removed; instead, polymerization initiators are added. Vinylidene chloride from which the inhibitor has been removed should be refrigerated at $-10°C$, under a nitrogen atmosphere, in the dark, and in a nickel or baked phenolic-lined storage tank. If not used within one day, it should be reinhibited.

Laboratory Polymerization. Vinylidene chloride polymerizes by both ionic and free-radical reactions. Processes based on the latter are far more common (13). Vinylidene chloride is of average reactivity when compared with other unsaturated monomers. The chlorine substituents stabilize radicals in the intermediate state of an addition reaction. Since they are also strongly electron withdrawing, they polarize the double bond, making it susceptible to anionic attack. For the same reason, a carbonium ion intermediate is not favored.

The 1,1-disubstitution of chlorine atoms causes steric interactions in the polymer as is evident from the heat of polymerization (see Table 1) (14). When corrected for the heat of fusion, it is significantly less than the theoretical value of -83.7 kJ/mol (-20 kcal/mol) for the process of converting a double bond to two single bonds. The steric strain apparently is not important in the addition step, because VDC polymerizes easily. Nor is it sufficient to favor depolymerization; the estimated ceiling temperature for PVDC is ca 400°C.

Homopolymerization. The free-radical polymerization of VDC has been carried out by solution, slurry, suspension, and emulsion methods.

Solution polymerization in a medium that dissolves both monomer and polymer has been investigated (15). The kinetic measurements lead to activation energies and frequency factors in the normal range for free-radical polymerizations of olefinic monomers. The kinetic behavior of VDC is abnormal when the polymerization is heterogeneous (16). Slurry polymerizations are usually used only at the laboratory level. They can be carried out in bulk or in common solvents, eg, benzene. Poly(vinylidene chloride) is insoluble in these media and separates from the liquid phase as a crystalline powder. The heterogeneity of the reaction makes stirring and heat transfer difficult; consequently, these reactions cannot be easily controlled on a large scale. Aqueous emulsion or suspension reactions are preferred for large-scale operations. Slurry reactions are usually initiated by the thermal decomposition of organic peroxides or azo compounds. Purely thermal initiation can occur, but rates are very slow (17).

The spontaneous polymerization of VDC, so often observed when the monomer is stored at room temperature, is caused by peroxides formed from the reaction of VDC with oxygen. Very pure monomer does not polymerize under these conditions. Irradiation by either uv or gamma rays (16,18) also induces polymerization of VDC.

The heterogeneous nature of the bulk polymerization of VDC is apparent from the rapid development of turbidity in the reaction medium following initiation. The turbidity results from the presence of minute PVDC crystals. As the reaction progresses, the crystalline phase grows and the liquid phase diminishes. Eventually, a point is reached where the liquid slurry solidifies into a solid mass. A typical conversion-time curve is shown in Figure 1 for a benzoyl-peroxide-catalyzed mass polymerization (19). The first stage of the reaction is characterized by rapidly increasing rate, which levels off in the second stage to a fairly constant value; this is often called the

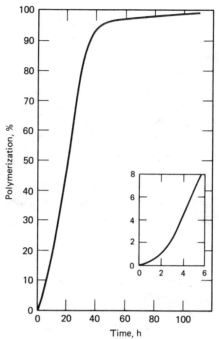

Figure 1. Bulk polymerization of vinylidene chloride at 45°C with 0.5 wt % benzoyl peroxide as initiator (19).

steady-state region. Throughout the first two stages, monomer concentration remains constant because the polymer separates into another phase. In the third stage, there is a gradual decrease in rate to zero as the monomer supply is depleted. Because the mass solidifies while monomer is still present (usually at conversions below 20%), further polymerization generates void space. The final solid, therefore, is opaque and quite porous. A similar pattern of behavior is observed when vinylidene chloride is polymerized in solvents, eg, benzene, that do not dissolve or swell the polymer. In this case, however, the reaction mixture may not solidify if the monomer concentration is low.

Heterogeneous polymerization is characteristic of a number of monomers, including vinyl chloride and acrylonitrile. A completely satisfactory mechanism for these reactions has not been determined. For VDC, two early kinetic studies have been reported: in one, uv initiation of polymerization both in mass and in cyclohexane was studied (16); in the other, the reaction initiated by benzoyl peroxide was investigated (20). Neither investigation was broad enough to elucidate the mechanism. A more recent study was undertaken to explore the effect of solid-phase morphology on the kinetics of heterogeneous, free-radical polymerization of VDC (21). Papers on the kinetics of polymerization of vinyl chloride and on acrylonitrile emphasize the importance of the polymer properties and morphology (22–24). The similarities in morphology of as-polymerized poly(vinyl chloride) (PVC) and polyacrylonitrile (PAN) are suggestive of a common mechanism of polymerization in solution followed by precipitation or solid-stage growth. The morphology of as-polymerized PVDC is quite different (21). Roughly spherical aggregates form in the first-mentioned systems, whereas very anisotropic growth takes place in the latter case. The VDC kinetic data do not support the mechanism of polymerization followed by crystallization and require that the polymerization and crystallization steps are somehow coupled. The proposed model is not sufficiently detailed to permit the kinetics for the polymerization and crystallization processes to be distinguished.

Emulsion and suspension reactions are doubly heterogeneous; the polymer is insoluble in the monomer and both are insoluble in water. Suspension reactions are similar in behavior to slurry reactions. Oil-soluble initiators (qv) are used, so the monomer–polymer droplet is like a small mass reaction. Emulsion polymerizations are more complex. Since the monomer is insoluble in the polymer particle, the simple Smith-Ewart theory should not apply (25).

A kinetic model for the particle growth stage for continuous-addition emulsion polymerization has been proposed (26). Below the monomer saturation point, the steady-state rate of polymerization R_p depends upon the rate of monomer addition R_a according to the reciprocal relationship:

$$\frac{1}{R_p} = \frac{1}{K} + \frac{1}{R_a}$$

where K depends on the number of particles and the propagation rate constant. A later study explores the kinetics of emulsion polymerization of nonswelling and swellable latex particles to define the locus of polymerization (27). There are no significant differences between the behavior of swelling and nonswelling emulsion particles and neither follows Smith-Ewart kinetics. The results indicate strongly that polymerization takes place at the particle–water interface or in a surface layer on the polymer particle.

Redox initiator systems are normally used in the emulsion polymerization of VDC to develop high rates at low temperatures. Reactions must be carried out below ca 80°C to prevent degradation of the polymer. Poly(vinylidene chloride) in emulsion is also attacked by aqueous base. Therefore, reactions should be carried out at low pH.

The instability of PVDC is one of the reasons why ionic initiation of VDC polymerization has not been used extensively. Many of the common catalysts either react with the polymer or catalyze its degradation. For example, butyllithium polymerizes VDC by an anionic mechanism, but the product is a low molecular weight, discolored polymer with a low chlorine content (28). Cationic polymerization of VDC seems unlikely in view of its structure (29). Some available data, however, suggest the possibility. In the low temperature, radiation-induced copolymerization of VDC with isobutylene, reactivity ratios vary markedly with temperature, indicating a change from a free-radical mechanism (30). Coordination-complex catalysts may also polymerize VDC by a nonradical mechanism. Again, this speculation is based on copolymerization studies. Poly(vinylidene chloride) telomers can be prepared with chlorine as the initiator and chain-transfer agent (31). Plasma polymerization of VDC in a radio-frequency glow discharge yields cross-linked polymer, which is partially degraded compared to the conventional polymer (32).

Copolymerization. The importance of VDC as a monomer results from its ability to copolymerize with other vinyl monomers. It most easily copolymerizes with acrylates, but it also reacts, though more slowly, with other monomers, eg, styrene, that form highly resonance-stabilized radicals. Reactivity ratios (r_1 and r_2) with various monomers are listed in Table 2. Many other copolymers have been prepared from monomers for which the reactivity ratios are not known. The commercially important

Table 2.   Reactivity of Vinylidene Chloride with Important Monomers[a]

Monomer	r_1[b]	r_2
styrene	0.14	2.0
vinyl chloride	3.2	0.3
acrylonitrile	0.37	0.91
methyl acrylate	1.0	1.0
methyl methacrylate	0.24	2.53
vinyl acetate	6	0.1

[a] Ref. 33.
[b] VDC.

copolymers include those with vinyl chloride (VC), acrylonitrile (AN), or various al-kylacrylates, but many commercial Saran polymers contain three or more components, VDC being the principal one. Usually one component is introduced to improve the processability or solubility of the polymer; the others are added to modify specific use properties. Most of these compositions have been described in the patent literature and a list of various combinations has been compiled (34). A typical terpolymer might contain 90 wt % VDC with the remainder made up of acrylonitrile and an acrylate or methacrylate monomer.

Bulk copolymerizations yielding high VDC-content copolymers are normally heterogeneous. Two of the most important pairs, VDC–VC and VDC–AN, are heterogeneous over most of the composition range. In both cases and at either composition extreme, the product separates initially in a powdery form; however, for intermediate compositions, the reaction mixture may only gel. Copolymers in this composition range are swollen but not completely dissolved by the monomer mixture at normal polymerization temperatures. Copolymers containing more than 15 mol % acrylate are normally soluble in the monomers. These reactions are therefore homogeneous and, if carried to completion, yield clear, solid castings of the copolymer. Most copolymerizations can be carried out in solution because of the greater solubility of the copolymers in common solvents.

During copolymerization, one monomer may add to the copolymer more rapidly than the other. Except for the unusual case of equal reactivity ratios, batch reactions carried to completion yield polymers of broad composition distribution. More often than not, this is an undesirable result.

Vinylidene chloride copolymerizes randomly with methyl acrylate and nearly so with other acrylates. Very severe composition drift occurs, however, in copolymerizations with vinyl chloride or methacrylates. Several methods have been developed to produce homogeneous copolymers regardless of the reactivity ratio (for examples, see ref. 35). These methods are applicable mainly to emulsion and suspension processes where adequate stirring can be maintained (see Commercial Methods of Polymerization and Processing). Copolymerization rates of VDC with small amounts of a second monomer are normally lower than its rate of homopolymerization. The kinetics of the copolymerization of VDC and VC have been studied (36–39).

Copolymers of VDC can also be prepared by methods other than conventional free-radical polymerization. Copolymers have been formed by irradiation and with various organometallic and coordination-complex catalysts (18,34,40–43). Graft copolymers have also been described (44–48).

Health and Safety Factors. Vinylidene chloride is highly volatile and, when free of decomposition products, has a mild sweet odor. Its warning properties are ordinarily inadequate to prevent excessive exposure. Inhalation of vapor presents a hazard, which is readily controlled by observance of precautions commonly taken in the chemical industry (5). A short exposure to a high concentration of vinylidene chloride vapor, eg, 4000 ppm, rapidly causes intoxication, which may progress to unconsciousness on prolonged exposure. However, prompt and complete recovery from the anesthetic effects occurs when the exposure is of short duration. A single, prolonged exposure and repeated short-term exposures can be dangerous, even when the concentrations of the vapor are too low to cause an anesthetic effect. They may produce organic injury to the kidneys and liver. For repeated exposures (8 h/d, 5 d/wk), the vapor concentration of vinylidene chloride should be below a time-weighted average of 10 ppm. If a person should be affected by or overcome from breathing vinylidene chloride vapors, the victim should be removed to fresh air at once, made to rest, kept warm, and given immediate medical attention. Investigations of vinylidene chloride exposure have been reviewed (49–50).

Vinylidene chloride is hepatotoxic, but it does not appear to be a carcinogen (51–56). Pharmacokinetic studies indicate that the behavior of vinyl chloride and vinylidene chloride in rats and mice is substantially different (57). No unusual health problems were observed in workers exposed to vinylidene chloride monomer over varying periods (58). The air-pollution hazards of vinylidene chloride in the United States have also been assessed (59). Emissions were estimated to be 589 metric tons per year. However, the monomer degrades rapidly in the atmosphere; therefore, these emissions are not likely to be a problem. Worker exposure is the main concern. Sampling techniques for monitoring worker exposure to vinylidene chloride vapor are being developed (60).

Precautions should be taken to prevent skin contact with vinylidene chloride. The liquid is irritating to the skin after only a few minutes of contact. The inhibitor MEHQ may be partly responsible for this irritation. If contact does occur, the affected skin area should be washed with soap and water. Inhibited vinylidene chloride is moderately irritating to the eyes. Contact causes pain and conjunctival irritation and, possibly, some transient corneal injury and iritis; permanent damage is not likely. Chemical safety goggles should be worn by those handling vinylidene chloride.

In the presence of air or oxygen, uninhibited vinylidene chloride forms a violently explosive complex peroxide at temperatures as low as −40°C. Decomposition products of vinylidene chloride peroxides are formaldehyde, phosgene, and hydrochloric acid. A sharp acrid odor indicates oxygen exposure and probable presence of peroxides. This is confirmed by the liberation of iodine from a slightly acidified dilute potassium iodide solution. Formation of insoluble polymer may also indicate peroxide formation. The peroxide adsorbs on the precipitated polymer and separation of the polymer may result in an explosive composition. Any dry composition containing more than ca 15 wt % peroxide detonates from a slight mechanical shock or from heat. Vinylidene chloride containing peroxides may be purified by washing several times, either with 10 wt % sodium hydroxide at 25°C or with a fresh 5 wt % sodium bisulfite solution. Residues in vinylidene chloride containers should be handled with great care, and the peroxides should be destroyed with water at room temperature.

Copper, aluminum, and their alloys should not be used in handling vinylidene chloride. Under the proper conditions, copper can react with acetylenic impurities forming copper acetylides, and aluminum can react with the vinylidene chloride forming aluminum chloralkyls. Both of these compounds are extremely reactive and potentially hazardous.

Poly(Vinylidene Chloride)

Structure and Properties. *Chain Structure.* The chemical composition of poly-(vinylidene chloride) has been confirmed by various techniques including elemental analysis; x-ray diffraction analysis; ir, Raman, and nmr spectroscopy; and degradation studies. The polymer chain is made up of vinylidene chloride units added head-to-tail:

$$-CH_2CCl_2CH_2CCl_2CH_2CCl_2-$$

Since the repeat unit is symmetrical, no possibility exists for stereoisomerism. Variations in structure can come about only by head-to-head addition, branching, or degradation reactions that do not cause chain scission. This includes such reactions as thermal dehydrochlorination, which creates double bonds in the structure to give, eg,

$$-CH_2CCl_2CH=CClCH_2CCl_2-$$

and a variety of ill-defined oxidation and hydrolysis reactions which generate carbonyl groups.

The infrared spectra of PVDC often show traces of unsaturation and carbonyl groups. The slightly yellow tinge of many of these polymers comes from the same source; the pure polymer is colorless. Elemental analyses for chlorine are normally slightly lower than theoretical value of 73.2%.

The high crystallinity of PVDC indicates that no significant amounts of head-to-head addition or branching can be present. This has been confirmed by nmr spectra (61). Studies of well-characterized oligomers with degrees of polymerization (dp) of 2–10 offer further nmr evidence (31). There is only one peak in the spectrum from the methylene hydrogens. Either branching or another mode of addition would product nonequivalent hydrogens and a more complicated spectrum. However, nmr cannot detect small amounts of such structures. The ir and Raman spectra can also be interpreted in terms of the simple head-to-tail structure (62–63).

Molecular weights of PVDC can be determined directly by dilute solution measurements in good solvents (64). Viscosity studies indicate that polymers with degrees of polymerization from 100 to more than 10,000 are easily obtained. Dimers and polymers with DP <100 can be prepared by special procedures (31). Copolymers can be more easily studied because of their solubility in common solvents. Gel-permeation chromatography studies indicate that molecular weight distributions are typical of vinyl copolymers.

Crystal Structure. The crystal structure of PVDC is fairly well established. Several unit cells have been proposed, but probably the best representation is that described in ref. 65. The unit cell contains four monomer units with two monomer units per repeat distance. The calculated density is higher than the experimental values, which are 1.80–1.94 g/cm³ at 25°C, depending on the sample. This is usually the case with crystalline polymers, because samples of 100% crystallinity usually cannot be obtained. A direct calculation of the polymer density from volume changes during polymerization yields a value of 1.97 g/cm^3 (66). If this value is correct, the unit-cell densities may be low.

The repeat distance along the chain axis (0.468 nm) is significantly less than that calculated for a planar zigzag structure. Therefore, the polymer must be in some other conformation (67–69). Based on ir and Raman studies of PVDC single crystals and normal vibration analysis, the best conformation appears to be $TXTX'$ where T, the skeletal angle, is 120° and the torsional angles X, X' of opposite sign are 32.5°. This conformation is in agreement with theoretical predictions (70).

The melting point (T_m) of PVDC is independent of molecular weight above DP = 100. However, as shown in Figure 2, it drops sharply at lower molecular weights. Below the hexamer, the products are noncrystalline liquids.

The properties of PVDC (see Table 3) are usually modified by copolymerization. Copolymers of high VDC content have lower melting points than PVDC. Copolymers containing more than ca 15 mol % acrylate or methacrylate are amorphous. Substantially more acrylonitrile (25%) or vinyl chloride (45%) is required to destroy crystallinity completely.

The effect of different types of comonomers on T_m varies. The VDC–MA co-

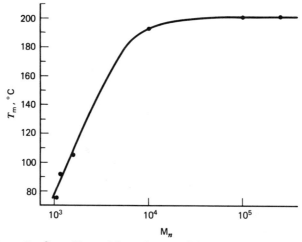

Figure 2. Crystalline melting points of poly(vinylidene chloride) (31).

Table 3. Properties of Poly(Vinylidene Chloride)

Property	Best value	Reported values
mp, °C	202	198–205
T_g, °C	−17	−19 to −11
transition between T_m and T_g, °C	80	
density (at 25°C), g/cm^3		
amorphous	1.775	1.67–1.775
unit cell	1.96	1.949–1.96
crystalline		1.80–1.97
refractive index (crystalline), n_D	1.63	
heat of fusion (ΔH_m), J/mol[a]	6275	4600–7950

[a] To convert J to cal, divide by 4.184.

polymers obey Flory's melting-point-depression theory and give a reasonable value for the heat of fusion; the VDC–VC system and the VDC–AN system do not (65,71–72). This is probably because the comonomer units can enter into the PVDC crystal structure as defects. Consequently, they are less effective in lowering T_m.

The glass-transition temperatures of Saran copolymers have been studied extensively (73–75). The effect of various comonomers on the glass-transition temperature is shown in Figure 3 (75). In every case, T_g increases with the comonomer content at low comonomer levels, even in cases where the T_g of the other homopolymer is lower. The phenomenon has been observed in several other copolymer systems as well (76). In these cases, a maximum T_g is observed at intermediate compositions. In others, where the T_g of the other homopolymer is much higher than the T_g of PVDC, the glass temperatures of the copolymers increase over the entire composition range. The glass-transition temperature increases most rapidly at low acrylonitrile levels but changes the slowest at low vinyl chloride levels. This suggests that polar interactions affect the former; but the increase in T_g in the VDC–VC copolymers may simply result from loss of chain symmetry. Because of these effects, the temperature range in which copolymers can crystallize is drastically narrowed. Crystallization induction times are prolonged and subsequent crystallization takes place at a low rate over a long period of time. Plasticization, which lowers T_g, decreases crystallization induction times significantly. Copolymers with lower glass-transition temperatures also tend to crystallize more rapidly (77).

Crystallization curves have been determined for 10 mol % acrylate copolymers

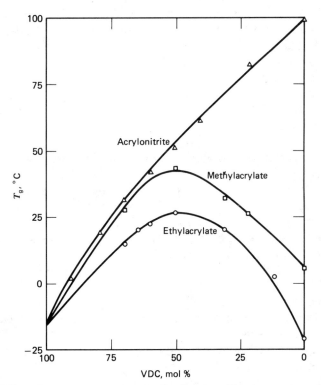

Figure 3.   Effect of comonomer structure on the glass-transition temperature of VDC copolymers (75).

of varying side-chain length. Among the acrylate copolymers, the butyl acrylate co-polymer has a T_g of 8°C; the octyl acrylate, −3°C; and the octadecyl acrylate, −16°C. The rates of crystallization of these copolymers are inversely related to the glass-transition temperatures. Apparently, the long alkyl side chains act as internal plas-ticizers, lowering the melt viscosity of the copolymer even though the acrylate group acts to stiffen the backbone.

The maximum rates of crystallization of the more common crystalline copolymers occur at 80–120°C. In many cases, these copolymers have broad composition distri-butions containing both fractions of high VDC content that crystallize rapidly and other fractions that do not crystallize at all. Poly(vinylidene chloride) probably crystallizes at a maximum rate at 140–150°C, but the process is difficult to follow because of severe polymer degradation. The copolymers may remain amorphous for a considerable period of time if quenched to room temperature. The induction time before the onset of crystallization depends on both the type and amount of comonomer; PVDC crystallizes within minutes at 25°C.

Orientation or mechanical working accelerates crystallization and has a pro-nounced effect on morphology. Crystals of uniaxially oriented filaments are oriented along the fiber axis (65). The long period, as determined by small-angle scattering, is 7.6 nm and decreases with comonomer content. The fiber is 43% crystalline and has a melting point of 195°C with an average crystal thickness of 4.5 nm. The crystal size is not greatly affected by comonomer content, but both crystallinity and melting point decrease.

Copolymerization also affects morphology under other crystallization conditions. Copolymers in the form of cast or molded sheets are much more transparent because of the small spherulite size. In extreme cases, crystallinity cannot be detected optically, but its effect on mechanical properties is pronounced. Before crystallization, films are soft and rubbery with low modulus and high elongation. After crystallization, they are leathery and tough with higher modulus and lower elongation.

Significant amounts of comonomer also reduce the ability of the polymer to form lamellar crystals from solution. In some cases, the polymer merely gels the solution as it precipitates rather than forms distinct crystals. At somewhat higher VDC content, it may precipitate in the form of aggregated, ill-defined particles and clusters.

Morphology and Transitions. The highly crystalline particles of PVDC precipi-tating during polymerization are aggregates of thin lamellar crystals (78). The sub-structures are 5–10 nm thick and 100 or more times larger in other dimensions. In some respects, they resemble the lamellar crystals grown from dilute solution (79–81). The single crystals are better characterized than the as-polymerized particles. They are highly branched crystals with branching angles of 65–70°; the angle appears to be associated with a twin plane in the crystal (82).

Melting points of as-polymerized powders are high, ie, 198–205°C as measured by differential thermal analysis (dta) or hot-stage microscopy (78). Two peaks are usually observed in dta curves: a small lower temperature peak and the main melting peak. The small peak seems to be related to polymer crystallized by precipitation rather than during polymerization.

As-polymerized PVDC does not have a well-defined glass-transition temperature because of its high crystallinity. However, a sample can be melted at 210°C and quenched rapidly to an amorphous state at $< −20$°C. The amorphous polymer has a glass-transition temperature of −17°C as shown by dilatometry (73). Glass-transition temperature values of -19 to $-11°C$, depending both on method of measurement and on sample preparation, have been determined.

Once melted, PVDC does not regain its as-polymerized morphology when subsequently crystallized. The polymer recrystallizes in a spherulitic habit. Spherulites between crossed polaroids show the usual Maltese cross and are positively birefringent. The size and number of spherulites can be controlled. Quenching and low temperature annealing generate many small nuclei which, on heating, grow rapidly into small spherulites. Slow crystallization at higher temperatures produces fewer but much larger spherulites. The melting point T_m and degree of crystallinity of recrystallized PVDC also depend on crystallization conditions. The melting point increases with crystallization temperature, but the as-polymerized value cannot be achieved. There is no reason to believe that even these values indicate the true melting point of PVDC; it may be as high as 220°C. Slow, high temperature recrystallization and annealing experiments are not feasible because of the thermal instability of the polymer (83). Other transitions in PVDC have been observed by dynamic mechanical methods.

Solubility and Solution Properties.   Poly(vinylidene chloride), like many high melting polymers, does not dissolve in most common solvents at ambient temperatures. Copolymers, particularly those of low crystallinity, are much more soluble. However, one of the outstanding characteristics of Saran polymers is resistance to a wide range of solvents and chemical reagents. The insolubility of PVDC results less from its polarity than its high melting point. It dissolves readily in a wide variety of solvents above 130°C (83).

The polarity of the polymer is important only in mixtures with specific polar aprotic solvents; many solvents of this general class solvate PVDC strongly enough to depress the melting point by more than 100°C. Solubility is normally correlated with cohesive energy densities or solubility parameters. For PVDC, a value of 40 ± 1.3 $(J/cm^3)^{1/2}$ (10 ± 0.3 $(cal/cm^3)^{1/2}$) has been estimated from solubility studies in nonpolar solvents; the value calculated from Small's relationship is 42.89 $(J/cm^3)^{1/2}$ (10.25 $(cal/cm^3)^{1/2}$). The use of the solubility parameter scheme for polar crystalline polymers like PVDC is of limited value. A typical nonpolar solvent of matching solubility parameter is tetrahydronaphthalene. The lowest temperature at which PVDC dissolves in this solvent is 140°C. Specific solvents, however, dissolve PVDC at much lower temperatures. A list of good solvents is given in Table 4. The relative solvent activity is characterized by the temperature at which a 1 wt % mixture of polymer in solvent becomes homogeneous when heated rapidly.

Poly(vinylidene chloride) also dissolves readily in certain solvent mixtures (84). One component must be a sulfoxide or N,N-dialkylamide. Effective cosolvents are less polar and have cyclic structures. They include aliphatic and aromatic hydrocarbons, ethers, sulfides, and ketones. Acidic or hydrogen-bonding solvents have an opposite effect, rendering the polar aprotic component less effective. Both hydrocarbons and strong hydrogen-bonding solvents are nonsolvents for PVDC.

As-polymerized PVDC is not in its most stable state; annealing and recrystallization can raise the temperature at which it dissolves (80). Low crystallinity polymers dissolve at a lower temperature forming metastable solutions. However, on standing at the dissolving temperature, they gel or become turbid, indicating precipitation.

Copolymers with a high enough vinylidene chloride content to be quite crystalline behave much like PVDC. They are more soluble, however, because of their lower melting points. The solubility of amorphous copolymers is much higher. The selection

Table 4. Solvents for Poly(Vinylidene Chloride) [a]

Solvents	T_m [b], °C
nonpolar	
1,3-dibromopropane	126
bromobenzene	129
1-chloronaphthalene	134
2-methylnaphthalene	134
o-dichlorobenzene	135
polar aprotic	
hexamethylphosphoramide	−7.2
tetramethylene sulfoxide	28
N-acetylpiperidine	34
N-methylpyrrolidinone	42
N-formylhexamethyleneimine	44
trimethylene sulfide	74
N-n-butylpyrrolidinone	75
diisopropyl sulfoxide	79
N-formylpiperidine	80
N-acetylpyrrolidinone	86
tetrahydrothiophene	87
N,N-dimethylacetamide	87
cyclooctanone	90
cycloheptanone	96
di-n-butyl sulfoxide	98

[a] Ref. 83.
[b] Temperature at which a 1 wt % mixture of polymer in solvent becomes homogeneous.

of solvents in either case varies somewhat with the type of comonomer. Some of the more common types are listed in Table 5. Solvents that dissolve PVDC also dissolve the copolymers at lower temperatures. The identification of solvents that dissolve PVDC at low temperatures makes possible the study of dilute solution properties. Both light-scattering and intrinsic-viscosity studies have been reported (64). Intrinsic viscosity–molecular weight relationships are listed below for the three solvents investigated ([η] in dL/g):

$$[\eta] = 1.31 \times 10^{-4}\, \overline{M}_v^{\,0.69} \qquad N\text{-methylpyrrolidinone (MP)}$$
$$[\eta] = 1.39 \times 10^{-4}\, \overline{M}_v^{\,0.69} \qquad \text{tetramethylene sulfoxide (TMSO)}$$
$$[\eta] = 2.58 \times 10^{-4}\, \overline{M}_v^{\,0.65} \qquad \text{hexamethylphosphoramide (HMPA)}$$

Table 5. Common Solvents for Vinylidene Chloride Copolymers

Solvents	Copolymer type	Temperature, °C
tetrahydrofuran	all	<60
methyl ethyl ketone	low crystallinity	<80
1,4-dioxane	all	50–100
cyclohexanone	all	50–100
cyclopentanone	all	50–100
ethyl acetate	low crystallinity	<80
chlorobenzene	all	100–130
dichlorobenzene	all	100–140
dimethylformamide	high acrylonitrile	<100

The relative solvent power (HMPA > TMSO > MP) agrees with solution temperature measurements. The characteristic ratio C_∞ is ca 8 ± 1, which is slightly larger than that of a similar polymer, polyisobutylene.

The dilute solution properties of copolymers are similar to those of the homopolymer. The intrinsic viscosity–molecular weight relationship for a VDC–AN copolymer (91 wt % AN) is (85):

$$[\eta] = 1.06 \times 10^{-4}\,\overline{M}_w^{\,0.72}$$

The characteristic ratio of $C_\infty = 8.8$ for this copolymer.

An extensive investigation of the dilute solution properties of several acrylate copolymers has been reported (82). The behavior is typical of flexible-backbone, vinyl polymers. The length of the acrylate ester side chain has little effect on properties.

Intrinsic viscosity–molecular weight relationships have been obtained for copolymers in methyl ethyl ketone. The value for a 15 wt % ethyl acrylate (EA) copolymer is $[\eta] = 2.88 \times 10^{-4}\,\overline{M}_w^{\,0.6}$.

In older literature, the molecular weights of PVDC and VDC copolymers were characterized by the absolute viscosity of a 2 wt % solution in o-dichlorobenzene at 140°C. The exact correlation between this viscosity value and molecular weight is not known. Gel-permeation chromatography is the preferred method for characterizing molecular weight. Studies of Saran copolymers have been reported (86–87).

Mechanical Properties. Because of the difficulty of fabricating PVDC into suitable test specimens, very few direct measurements of its mechanical properties have been made. In many cases, however, the properties of copolymers have been studied as functions of composition, and the properties of PVDC can be estimated by extrapolation. Some characteristic properties of high VDC-content unplasticized copolymers are listed in Table 6. The performance of a given specimen is very sensitive to morphology, including the amount and kind of crystallinity, as well as orientation. Tensile strength increases with crystallinity whereas toughness and elongation drop. Orientation, however, improves all three properties. The effect of stretch ratio applied during orientation on properties of VDC–VC (vinyl chloride) monofilaments is shown in Table 7.

The dynamic mechanical properties of VDC–VC copolymers have been studied in detail. The incorporation of VC units in the polymer results in a drop in dynamic

Table 6. Mechanical Properties of Poly(Vinylidene Chloride)

Property	Range
tensile strength, MPa[a]	
unoriented	34.5–69.0
oriented	207–414
elongation, %	
unoriented	10–20
oriented	15–40
softening range (heat distortion), °C	100–150
flow temperature, °C	>185
brittle temperature, °C	−10 to 10
impact strength, J/m[b]	26.7–53.4

[a] To convert MPa to psi, multiply by 145.

[b] To convert J/m to ft·lbf/in., divide by 53.38 (see ASTM D 256).

Table 7. Effect of Stretch Ratio Upon Tensile Strength and Elongation of a VDC–VC Copolymer[a,b]

Stretch ratio	Tensile strength, MPa[c]	Elongation, %
2.50:1	235	23.2
2.75:1	234	21.7
3.00:1	303	26.3
3.25:1	268	33.1
3.50:1	316	19.2
3.75:1	330	21.8
4.00:1	320	19.7
4.19:1	314	16.2

[a] Ref. 88.
[b] Average of five determinations, using the Instron test at 5 cm/min.
[c] To convert MPa to psi, multiply by 145.

modulus because of the reduction in crystallinity. However, the glass-transition temperature is raised; the softening effect observed at room temperature, therefore, is accompanied by increased brittleness at lower temperatures. Saran B copolymers are normally plasticized in order to avoid this. Small amounts of plasticizer (2–10 wt %) depress T_g significantly without loss of strength at room temperature. At higher levels of VC, the T_g of the copolymer is above room temperature and the modulus rises again. A minimum in modulus or maximum in softness is normally observed in copolymers in which T_g may be above room temperature. A thermomechanical analysis of VDC–AN (acrylonitrile) and VDC–MMA (methyl methacrylate) copolymer systems shows a minimum in softening point at 79.4 and 68.1 mol % VDC, respectively (89).

In cases where the copolymers have substantially lower glass-transition temperatures, the modulus decreases with increasing comonomer content. This results from a drop in crystallinity and in glass-transition temperature. The loss in modulus in these systems is, therefore, accompanied by an improvement in low temperature performance. At low acrylate levels (<10 wt %), however, T_g increases with comonomer content. The brittle points in this range may, therefore, be higher than that of PVDC.

The long side chains of the acrylate ester group can apparently act as internal plasticizers. Substitution of a carboxyl group on the polymer chain increases brittleness. A more polar substituent, eg, an N-alkyl amide group, is even less desirable. Copolymers of VDC with N-alkylacrylamides are more brittle than the corresponding acrylates even when the side chains are long (90). Side-chain crystallization may be a contributing factor.

Barrier properties. Vinylidene chloride polymers are more impermeable to a wider variety of gases and liquids than other polymers. This is a consequence of the combination of high density and high crystallinity in the polymer. An increase in either tends to reduce permeability (see Barrier polymers). A more subtle factor may be the symmetry of the polymer structure. It has been shown that both polyisobutylene and PVDC have unusually low permeabilities to water compared to their monosubstituted counterparts polypropylene and PVC (91). The values listed in Table 8 include estimates for the completely amorphous polymers. The estimated value for highly crystalline PVDC was obtained by extrapolating data for copolymers.

The effect of copolymer composition on gas permeability is illustrated by the data in Table 9. The inherent barrier in VDC copolymers can best be exploited by using

Table 8.　Comparison of the Permeabilities of Various Polymers to Water Vapor[a]

Polymer	Density, g/mL		Permeability[b]	
	Amorphous	Crystalline	Amorphous	Crystalline
ethylene	0.85	1.00	200–220	10–40
propylene	0.85	0.94	420	
isobutylene	0.915	0.94	90	
vinyl chloride	1.41	1.52	300	90–115
vinylidene chloride	1.77	1.96	30	4–6

[a] Refs. 33, 91–92.

[b] In g/(h·100 m^2) at 7.1 kPa (53 mm Hg) pressure differential and 39.5°C for a film 25.4 μm (1 mil) thick.

Table 9.　Effect of Composition on the Permeability of Various Gases Through Saran[a]

Polymer	Gas	T, °C	$10^{13} P$, (cm^3·cm)/(cm^2·s·kPa)[b]
PVDC	O_2	25	1.5
	N_2	25	0.75
	CO_2	25	9.0
90/10 VC	He	25	50.0
	H_2	25	57.0
	O_2	25	3.2
	N_2	25	0.75
	CO_2	25	22.0
	CH_4	25	0.19
	H_2S	30	23
85/15	He	34	225
	O_2	25	9.0
	CO_2	20	45.0
70/30	O_2	25	8.1
50/50	O_2	25	27.0
80/20 AN	O_2	25	3.2
	N_2	25	0.4
	CO_2	25	7.95
60/40 AN	O_2	25	16.0
	N_2	25	1.9
	CO_2	25	35.0

[a] Ref. 93.

[b] To convert kPa to cm Hg, multiply by 0.75.

films containing little or no plasticizers and as much VDC as possible. However, the permeability of even completely amorphous copolymers, eg, 60 VDC–40 AN or 50 VDC–50 VC, is low compared to that of other polymers. The primary reason is that diffusion coefficients of molecules in VDC copolymers are very low. This factor together with the low solubility of many gases in VDC copolymers and the high crystallinity results in very low permeability. Permeability is affected by the kind and amounts of comonomer as well as crystallinity. A change from PVDC to 50 wt % VC or 40 wt % AN increases permeability tenfold but has little effect on selectivity.

A more polar comonomer, eg, an AN comonomer, increases the water-vapor transmission more than VC, other factors being constant. For the same reason, AN copolymers are more resistant to penetrants of low cohesive energy density. All VDC copolymers, however, are very impermeable to aliphatic hydrocarbons. Comonomers that lower T_g and increase the free volume in the amorphous phase increase permeability more than the polar comonomers; this seems to be the way in which the higher acrylates act. Plasticizers increase permeability for similar reasons.

The effect of plasticizer on the permeability of VDC–VC copolymers has been carefully investigated (94–95). Oxygen permeability through a VDC–VC copolymer as a function of plasticizer content and temperature is shown in Figure 4. The T_g of this polymer is $-2°C$; at 7.2 wt % plasticizer, it drops to $-9°C$. When all measurements are above T_g, the permeability is doubled by addition of 1.6–1.85 parts of liquid additive.

The effect of plasticizer on the moisture-vapor transmission rate (MVTR) is shown by the data in Table 10. The moisture-vapor transmission rate is somewhat less sensitive to liquid additive levels; 3.5 parts are required to double it compared to 1.7 parts for O_2 permeability. This permeability also is less sensitive to relative humidity than is the permeability to oxygen.

Degradation Chemistry. Poly(vinylidene chloride) is thermally unstable and, when heated above 125°C, evolves HCl. Degradation occurs at lower temperatures when the polymer is exposed to radiation (uv, x ray, γ ray) or is treated with alkaline reagents, active metals, and Lewis acids. The common feature of these processes is that chloride is removed from the polymer either as HCl or in salt form, leaving a carbonaceous residue. The nature of the residue is very dependent on the mode of decom-

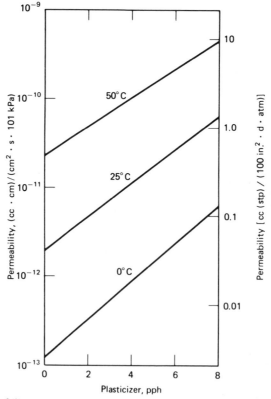

Figure 4. Permeability of oxygen through VDC copolymer with 12 wt % vinyl chloride as a function of liquid additive level. Plasticizer = Citroflex A-4 and Paraplex G-60 (94).

Table 10. Permeability of Vinylidene Chloride Copolymer Films to Water at Selected Temperatures and Relative Humidities [a]

Plasticizer			Permeability $\times 10^{13}$, $(g \cdot cm)/(cm^2 \cdot s \cdot kPa)$ [b]		
Acetyl tributyl citrate, pph	Epoxidized soybean oil, pph	rh, %	30°C	38°C	45°C
0.5	0.0	87	1.11	3.17	4.04
		80	1.39	3.39	4.32
		62	0.82	3.29	4.03
2.7	0.0	87	2.47	4.82	5.96
		80	2.11	4.37	6.69
		62	1.56	5.63	6.46
4.9	0.0	87	3.83	7.19	11.36
		80	6.49	7.58	12.47
		62	5.01	9.43	8.82
7.2	0.0	87	5.93	8.14	16.45
		80	7.13	10.02	15.08
		62	5.39	10.40	11.96
0.5	1.9	87		4.35	5.73
		80	3.35	3.46	5.44
		62	2.83	3.77	4.95
0.5	5.9	87	6.05	6.89	11.42
		80	5.50	8.93	11.83
		62	4.55	8.64	9.69

[a] Ref. 95.
[b] To convert kPa to cm Hg, multiply by 0.75.

position and the temperature. Also, most PVDC degradation reactions are heterogeneous, because the polymer is present in the reaction mixture as a crystalline solid. In such cases, the course of the reaction is influenced by polymer morphology and the interaction between phases. Heterogeneous degradation differs significantly from melt or solution reactions. The literature through 1973 on poly(vinylidene chloride) degradation and carbonization is surveyed in ref. 96.

Thermal decomposition in the solid state (T < 200°C) has been the most widely studied reaction (97–98). The reaction is normally described as a two-step process:

formation of a conjugated polyene:

$$+CH_2CCl_2+_n \xrightarrow[\Delta]{fast} +CH=CCl+_n + n\ HCl$$

and carbonization:

$$+CH=CCl+_n \xrightarrow[\Delta]{slow} 2\ n\ C + n\ HCl$$

Although this simplified description illustrates the main features of the reaction, it should not be interpreted as a mechanism; the process is far more complicated, being affected by physical changes in the solid (annealing effects) and by the method of preparation and purity of the polymer. The presence of oxygen in the polymerization reaction and selection of initiator also affect stability (99).

However, all PVDC polymers, even those prepared with pure monomer in an inert atmosphere, degrade in a characteristic manner. The polymer discolors gradually from yellow to brown and finally to black. In the very early stages (<1 wt % HCl loss), it becomes insoluble, indicating cross-linking. The melting point decreases, and bands characteristic of double bonds are detected in the ir spectrum. The polymer eventually becomes infusible and the PVDC crystal structure as detected by x-ray diffraction disappears even though the gross morphology is retained (100). Free radicals can be detected by esr measurements. If further reaction is brought about by raising the temperature above 200°C, aromatic structures form. Finally, at very high temperatures (>700°C), complete carbonization occurs.

The separate reactions can be followed by use of stepwise heating methods. If a continuously increasing temperature is used, as in a dta experiment, the polymer may melt before significant degradation occurs. At rates greater than 10°C/min, decomposition occurs violently at 225°C with rapid evolution of HCl gas.

Poly(vinylidene chloride) carbons are of great interest in surface science and catalysis because of their unique structure and high purity. They exhibit molecular sievelike properties and unique adsorptive capabilities. New methods of synthesis, structure studies, and properties have been described (101–103).

The stability of VDC copolymers depends on the comonomer. Copolymers with VC and the acrylates degrade slowly. Acrylonitrile copolymers degrade more rapidly and release HCN as well as HCl (104). The degradation of PVDC in solution in non-polar solvents is much slower than the solid-state reaction (105). In both cases, however, a free-radical mechanism is evident. The rate of dehydrochlorination increases markedly with solvent polarity. In a very strong solvent, eg, hexamethylphosphoramide, the rates exceed those of solid-state reactions (106–107). Under these conditions, an ionic mechanism is more likely.

The reaction of PVDC with strong bases is an ionic process, involving both elimination and substitution reactions. In nonaqueous media, carbon-forming reactions usually occur (108–110). Aqueous bases have a limited effect on PVDC, primarily because the polymer is scarcely wet or swollen by water. However, it is decomposed by hot concentrated caustic over time (111). Weak bases, eg, ammonia, amines, or polar aprotic solvents, also accelerate the decomposition of PVDC. The products from these reactions do not appear to have a simple polyene structure. In many cases, substitution reactions leading to bound nitrogen and oxygen occur (112).

Amines can also swell the polymer, leading to very rapid reactions. Pyridine, for example, would be a fairly good solvent for VDC copolymer if it did not attack the polymer chemically. However, when pyridine is part of a solvent mixture that does not dissolve the polymer, pyridine does not penetrate into the polymer phase (79). Studies of single crystals indicate that pyridine removes HCl only from the surface. Kinetic studies and product characterizations suggest that the reaction of two units in each chain fold can easily take place; further reaction is greatly retarded either by the inability of pyridine to diffuse into the crystal or by steric factors.

Transition-metal salts, eg, $AlCl_3$, $ZnCl_2$, and $FeCl_3$, and other Lewis acids, catalyze the thermal decomposition of PVDC (111,113). The catalysis probably occurs by a carbonium-ion mechanism. Chain scission occurs in the process yielding low molecular weight, unsaturated polymers. Traces of metal salts in PVDC can produce instability with slow degradation taking place even at ambient temperatures. Pure PVDC does not appear to degrade at a measurable rate in the dark below 100°C. When exposed to uv light or sunlight, however, it discolors. Hydrochloric acid is eliminated in the process and cross-linking takes place. The photodegradation reaction has been studied (114–115).

Unlike uv light, higher energy irradiation, ie, with γ rays, of PVDC causes chain scission. Copolymers of VDC and VC undergo cross-linking and chain scission (PVC itself cross-links), but the net result depends on polymer morphology as well as on copolymer composition (116–117).

Stabilization. The art of stabilizing VDC-type polymers is highly developed. Although detailed mechanisms are lacking, some general principles have been established. The ideal stabilizer system should achieve the following: absorb or combine with evolved HCl irreversibly under conditions of use, but not strip HCl from the polymer chain; act as a selective uv absorber; contain a reactive dienophilic molecule capable of preventing discoloration by reacting with and disrupting the color-producing conjugated polymer sequences; possess antioxidant activity so as to prevent the formation of carbonyl groups and other chlorine-activating structures; and to be able to chelate metals, eg, iron, in order to prevent the formation of metal chlorides, catalysts for polymer degradation.

Acid acceptors are of three general types: alkaline-earth and heavy-metal oxides or salts of weak acids, eg, barium, cadmium, or lead fatty-acid salts; epoxy compounds, eg, epoxidized soybean oil or glycidyl ethers and esters; and organotin compounds, eg, salts of carboxylic acids and organotin mercaptides. Effective light stabilizers have a chemical configuration that leads to hydrogen bonding and ring chelation, showing exceptional resonance stability and very good uv absorption properties. The principal compounds of commercial interest are derivatives of salicylic acid, resorcylic acid, benzophenone, and benzotriazole. Examples of dienophiles that have been used are maleic anhydride and dibasic lead maleate.

Antioxidants are generally of two types: those that react with a free radical to stop a radical chain and those that reduce hydroperoxides to alcohols (see Antioxidants and antiozonants). Phenolic antioxidants, eg, 2,6-di-*tert*-butyl-4-methylphenol and substituted bisphenols, are of the first type. The second is exemplified by organic sulfur compounds and organic phosphites. The phosphites, ethylenediaminetetraacetic acid (EDTA), citric acid, and citrates can chelate metals (see Chelating agents). The ability of organic phosphites to function as antioxidants and as chelating agents illustrates the dual role of many stabilizer compounds. It is common practice to use a combination of stabilizing compounds to achieve optimum results (118–119). Synergism of stabilizing action frequently occurs. An excellent list of patents on stabilization of vinyl chloride and vinylidene chloride polymers and copolymers has been tabulated (120).

Commercial Methods of Polymerization and Processing

Processes that are essentially modifications of laboratory methods allowing operation on a larger scale are used for commercial preparation of vinylidene chloride polymers. The intended use dictates the polymer characteristics and, to some extent, the method of manufacture. Emulsion polymerization and suspension polymerization are the preferred industrial processes (see Polymerization mechanisms and processes). Either process is carried out in a closed, stirred reactor, which should be glass-lined and jacketed for heating and cooling. The reactor must be purged of oxygen and the water and monomer must be free of metallic impurities to prevent an adverse effect on the thermal stability of the polymer.

Emulsion Polymerization. Emulsion polymerization is used commercially to make vinylidene chloride copolymers; they are utilized in two ways. In some applications, the latex which results is used directly, usually with additional stabilizing ingredients, as a coating vehicle to apply the polymer to various substrates; in others, the polymer is first isolated from the latex before use. In applications where the polymer is not used in latex form, the emulsion process is chosen over alternative methods. The polymer is recovered in dry powder form, usually by coagulating the latex with an electrolyte, followed by washing and drying. The principal advantages of emulsion polymerization as a process for preparing vinylidene chloride polymers are twofold. First, high molecular weight polymers can be produced with reasonable reaction times; this especially pertains to copolymers with vinyl chloride. The initiation and propagation steps can be controlled more independently than in the suspension process. Second, monomer can be added during the polymerization to maintain copolymer composition control. The disadvantages of emulsion polymerization result from the relatively high concentration of additives in the recipe. The water-soluble initiators, activators, and surface-active agents generally cause the polymer to have greater water sensitivity, poorer electrical properties, and poorer heat and light stability. Recovery of the polymer by coagulation, washing, and drying to some extent improves these properties over those of the polymer deposited in latex form.

A typical recipe for batch emulsion polymerization is shown in Table 11 (121). A reaction time of 7–8 h at 30°C is required for 95–98% conversion. A latex is produced with an average particle diameter of 100–150 nm. Other modifying ingredients may be present, eg, other colloidal protective agents such as gelatin or carboxymethyl cellulose; initiator activators, such as redox types; chelates; plasticizers; stabilizers; and chain-transfer agents.

Commercial surfactants are generally anionic emulsifiers, alone or in combination with nonionic types (see Surfactants and detersive systems). Representative anionic emulsifiers are the sodium alkylaryl sulfonates, the alkyl esters of sodium sulfosuccinic acid, and the sodium salts of fatty alcohol sulfates. Nonionic emulsifiers are of the ethoxylated alkylphenol type. Free-radical sources other than peroxysulfates may be used, eg, hydrogen peroxide, organic hydroperoxides, peroxyborates, and peroxycarbonates. Many of these are used in redox pairs, in which an activator promotes the decomposition of the peroxy compound. Examples are peroxysulfate or perchlorate activated with bisulfite, hydrogen peroxide with metallic ions, and organic hydro-

Table 11.   Typical Recipe for Batch Emulsion Polymerization[a]

Ingredient	Parts by wt
vinylidene chloride	78
vinyl chloride	22
water	180
potassium peroxysulfate	0.22
sodium bisulfite	0.11
Aerosol MA[b] (80 wt %)	3.58
nitric acid (69 wt %)	0.07

[a] Ref. 121.

[b] Aerosol MA (trademark of American Cyanamid Co.) = dihexyl sodium sulfosuccinate.

peroxides with sodium formaldehydesulfoxylate. The use of activators causes the decomposition of the initiator to occur at lower reaction temperatures, which allows the preparation of a higher molecular weight polymer within reasonable reaction times. This is an advantage, particularly for copolymers of vinylidene chloride with vinyl chloride. Oil-soluble initiators are usually effective only when activated by water-soluble activators or reducing agents.

To ensure constant composition, the method of emulsion polymerization by continuous addition is employed. One or more components are metered continuously into the reaction. If the system is properly balanced, a steady state is reached in which a copolymer of uniform composition is produced (122). A process of this type can be used for the copolymerization of VDC with a variety of monomers. A flow diagram of the apparatus is shown in Figure 5; a typical recipe is shown in Table 12. The monomers are charged to the weigh tank A, which is kept under a nitrogen blanket. The emulsifiers, initiator, and part of the water are charged to tank B; the reducing agent and some water to tank C. The remaining water is charged to the reactor D and

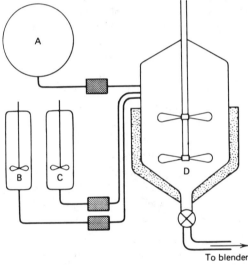

Figure 5. Apparatus for continuous-addition emulsion polymerization of a VDC–acrylate mixture (122). A, weigh tank; B and C, tanks; D, reactor.

Table 12. Recipe for Emulsion Polymerization by Continuous Addition[a]

Ingredient	Parts by wt
vinylidene chloride	468
comonomer	52
emulsifiers	
Tergitol[b] NP 35	12
sodium lauryl sulfate (25 wt %)	12
initiator	
ammonium peroxysulfate	10
sodium metabisulfite ($Na_2S_2O_5$, 5 wt %)	10
water	436

[a] Ref. 122.

[b] Registered trademark of Union Carbide for their nonionic wetting agent.

the system is sealed and purged. The temperature is raised to 40°C and one tenth of the monomer and initiator charges are added; then, one tenth of the activator is pumped in. Once the reaction begins, as indicated by an exotherm and pressure drop, feeds of A, B, and C are started at programmed rates which begin slowly and gradually increase. The emulsion is maintained at a constant temperature during the run by cooling water which is pumped through the jacket. When all components are in the reactor and the exotherm begins to subside, a final addition of initiator and reducing agent completes the reaction.

Suspension Polymerization.   Suspension polymerization of vinylidene chloride is used commercially to make molding and extrusion resins. The principal advantage of the suspension process over the emulsion process is the use of fewer ingredients that might detract from the polymer properties. Stability is improved and water sensitivity is decreased. Extended reaction times and the difficult preparation of higher molecular weight polymers are disadvantages of the suspension process compared to the emulsion process, particularly for copolymers containing vinyl chloride.

A typical recipe for suspension polymerization is shown in Table 13 (123). At a reaction temperature of 60°C, the polymerization proceeds to 85–90% conversion in 30–60 h. Unreacted monomer is removed by vacuum stripping; then it is condensed and reused after processing. The polymer is obtained in the form of small (590–149-μm (30–100 mesh)) beads, which are dewatered by filtration or centrifugation and then dried in a flash dryer or fluid-bed dryer. Suspension polymerization involves monomer-soluble initiators, and polymerization occurs inside suspended monomer droplets which form by the shearing action of the agitator and are prevented from coalescence by the protective colloid. It is important that the initiator be uniformly dissolved in the monomer before droplet formation. Unequal distribution of initiator causes some droplets to polymerize faster than others, leading to monomer diffusion from slow droplets to fast droplets. The fast polymerizing droplets form polymer beads that are dense, hard, glassy, and extremely difficult to fabricate because of their inability to accept stabilizers and plasticizers. Common protective colloids that prevent droplet coalescence and control particle size are poly(vinyl alcohol), gelatin, and methyl cellulose. Organic peroxides, peroxycarbonates, and azo compounds are used as initiators for vinylidene chloride suspension polymerization.

The batch suspension process does not compensate for composition drift. Constant-composition processes have been designed for emulsion or suspension reactions as previously described. It is more difficult to design controlled-composition processes by suspension methods. In one approach (124), the less reactive component is removed continuously from the reaction to keep the unreacted monomer composition constant. This method has been used effectively in VDC–VC copolymerization, where the slower reacting component is a gas and can be released during the reaction to maintain con-

Table 13.   Recipe for Suspension Polymerization

Ingredient	Parts by wt
vinylidene chloride	85
vinyl chloride	15
deionized water	200
400 mPa·s (= cP) methyl hydroxypropyl cellulose	0.05
lauroyl peroxide	0.3

stant pressure. In many other cases, there is no practical way of removing the slower-reacting component.

Economic Aspects

Vinylidene chloride monomer is produced commercially in the United States by The Dow Chemical Company and PPG Industries. The monomer is produced in Europe by Imperial Chemical Industries, Ltd., in the UK; Badische Anilin-und Soda-Fabrik and Chemische Werk-Huels in the FRG; Solvay et Cie and Amaco et Compagnie in France, and The Dow Chemical Company in the Netherlands. The monomer is produced in Japan by the Asahi Chemical Company, Kureha Chemical Industries, and Kanto Denka Kogyo Company. Production in the United States exceeds 90,700 metric tons per year and is estimated at 45,400 t/yr in Europe and 22,700 t/yr in Japan. Approximately 68,000 t of monomer is converted to polymer annually in the United States. The Dow Chemical Company makes over half of the VDC copolymers.

Although Saran is a generic name for VDC copolymers in the United States, Saran is a Dow trademark in most foreign countries. Other trade names include Daran (W. R. Grace), Amsco Res (Union Oil), Serfene (Morton Chemical) in the United States and Haloflex (Imperial Chemical Industries, Ltd.), Diofan (BASF), Ixan (Solvay), and Polyidene (Scott-Bader) in Europe. Production of vinylidene chloride monomer and polymers is expected to increase by 7% annually. The monomer is of particular economic interest because its cost increases slower than the hydrocarbons because of its high content of low cost chlorine.

Uses

Molding Resins. Vinylidene chloride–vinyl chloride copolymers were originally developed for thermoplastic molding applications, and small amounts are still used for this purpose. The resins, when properly formulated with plasticizers and heat stabilizers, can be fabricated by common methods, eg, injection, compression, and transfer molding (see Plastics processing). Conventional or dielectric heating can be used to melt the polymers. Rapid hardening is achieved by forming in heated molds to induce rapid crystallization. Cold molds result in supercooling of the polymer; it remains soft and amorphous, and the part cannot be easily removed from the mold without distortion. Mold temperatures of up to 100°C allow rapid removal of dimensionally stable parts. The range of molding temperatures is rather narrow because of the crystalline nature of the resin and thermal sensitivity. All crystallites must be melted to obtain low polymer-melt viscosity, but prolonged or excessive heating must be avoided to prevent dehydrochlorination.

Thermal degradation is a problem even when the resin is formulated with the very best stabilizers. Molding equipment is designed to alleviate this problem by having all passages through the heating cylinder streamlined to prevent plastic buildup. Any plastic that remains in the heating cylinder for longer than a few minutes decomposes, thereby releasing HCl and forming carbon. The carbon may build up or break off and contaminate molded parts with occlusions. It is especially important that an injection-molding heating cylinder not be shut down when loaded with molten resin. The cylinder must be purged with a more stable resin, eg, polystyrene.

The metal parts of the injection molder, ie, the liner, torpedo, and nozzle, that contact the hot molten resin must be of the noncatalytic type to prevent accelerated decomposition of the polymer. In addition, they must be resistant to corrosion by HCl. Iron, copper, and zinc are catalytic to the decomposition and cannot be used, even as parts of alloys. Magnesium is noncatalytic but is subject to corrosive attack, as is chromium when used as plating. Nickel alloys, eg, Duranickel and Hastelloy B, are recommended as construction materials for injection-molding metal parts. These and pure nickel are noncatalytic and corrosion-resistant; however, pure nickel is rather soft and is not recommended.

The injection mold need not be made of noncatalytic metals; any high grade tool steel may be used, because the plastic cools in the mold and undergoes little decomposition. However, the mold requires good venting to allow the passage of small amounts of acid gas as well as air. Vents tend to become clogged by corrosion and must be cleaned periodically.

Molded parts of vinylidene chloride copolymer plastics are used to satisfy the industrial requirements of chemical resistance and extended service life. They are used in such items as gasoline filters, valves, pipe fittings, containers, and chemical process equipment. Complex articles are constructed from molded parts by welding; hot-air welding at 200–260°C is a suitable method. Molded parts have good physical properties but have lower tensile strength than films or fibers, because crystallization is random in molded parts. Higher strength is developed by orientation in films and fibers. Physical properties of a typical molded vinylidene chloride copolymer plastic are listed in Table 14.

Extrusion Resins. Extrusion of VDC–VC copolymers, which is the main fabrication technique for filaments, films, rods, and tubing or pipe, involves the same concerns for thermal degradation, streamlined flow, and noncatalytic materials of construction as described for injection-molding resins (88,126). The plastic leaves the extrusion die in a completely amorphous condition and is maintained in this state by quenching in a water bath to ca 10°C, thereby inhibiting recrystallization. In this state, the plastic is soft, weak, and pliable. If it is allowed to remain at room temperature, it hardens

Table 14.   Properties of Saran 502 Resin for Injection-Molding Applications[a]

Typical resin properties	Test	Value
ultimate tensile strength, MPa[b]	ASTM D 638	24.1–34.5
yield tensile strength, MPa[b]	ASTM D 638	19.3–26.2
ultimate elongation, %	ASTM D 638	160–240
modulus of elasticity in tension, MPa[b]		345–552
Izod impact strength, J/m[c] of notch	ASTM D 256	21.35–53.38
density, g/cm^3	ASTM D 792	1.65–1.72
hardness, Rockwell M	ASTM D 785	50–65
water absorption, % in 24 h	ASTM D 570	0.1
mold shrinkage, cm/cm (injection-molded)	ASTM D 955	0.005–0.025
limiting oxygen index, %	ASTM D 2863	60.0[d]
UL 94	UL 94 Test	V–O[d]

[a] Ref. 125.

[b] To convert MPa to psi, multiply by 145.

[c] To convert J/m to ft·lbf/in. of notch, divide by 53.38.

[d] The results of small-scale flammability tests are not intended to reflect the hazards of this or any other material under actual fire conditions.

gradually and recrystallizes partially at a slow rate with a random crystal arrangement. Heat treatment can be used to recrystallize at controlled rates.

Crystal orientation is developed in the supercooled extrudate by plastic deformation and heat treatment. In the manufacture of filaments, the stretching produces orientation in a single direction and develops unidirectional properties of high tensile strength, flexibility, long fatigue life, and good elasticity. The filaments are removed from the supercooling tank, wrapped several times around smooth take-off rolls, and then wrapped several times around orienting rolls, which have a linear speed about four times that of the take-off rolls. The difference in roll speeds produces the mechanical stretching and causes orientation of crystallites along the longitudinal axis while the polymer is crystallizing. Heat treatment may be used during or after stretching to affect the degree of crystallization and control the physical properties of the oriented filaments.

A variation of the preceding process is used to produce oriented vinylidene chloride copolymer films. The plastic is extruded into tube form and then is supercooled and subsequently biaxially oriented in a continuous bubble process. The supercooled tube is flattened and passed through two sets of pinch rolls, which are arranged so that the second set of rolls travels faster than the first set. Between the two sets, air is injected into the tube to create a bubble which is entrapped by the pinch rolls. The entrapped air bubble remains stationary while the extruded tube is oriented as it passes around the bubble. Orientation is produced in the transverse and the longitudinal directions, creating excellent tensile strength, elongation, and flexibility in the film. The commercial procedure is described in ref. 126.

Unoriented film can be formed by extruding through a slit die. The temperature must be controlled to promote crystallization before winding on a roll. Extruded monofilaments in diameters of 0.13–0.38 mm have been widely used in the textile field as furniture and automobile upholstery, drapery fabric, outdoor furniture, venetian-blind tape, filter cloths, etc. Chemically resistant tubing and pipe liners are extruded. The pipe liner is inserted into an oversized steel pipe, which is swaged to size, and lengths are connected by flanged joints and vinylidene chloride copolymer gaskets. Pipe fittings are lined with injection-molded liners.

The biaxially oriented extruded films are used in packaging applications where their excellent resistance to water vapor and most gases makes them ideal transparent barriers. Because they are highly oriented, these films exhibit some shrinkage when exposed to higher-than-normal temperatures. Preshrinking or heat-setting can be performed to minimize residual shrink, or the shrinkage may be used to advantage in the heat-shrinking of overwraps on packaged items. Shrink bags made from high shrink Saran film are used in the packaging of cuts of fresh red meat. Films for packaging are used in tube form or as flat film for overwraps or conversion to film bags on modified bag-making machinery. The electronic or dielectric seal is the most satisfactory type for sealing the film to itself, although hot-plate sealers or cement-type seals produce an airtight seal on overwraps without fusing of the material.

Multilayer Film. A significant new application for Saran resins is in the construction of multilayer film and sheet (127–128). This innovation permits the design of a packaging material with a combination of properties not obtainable in any single material. A VDC copolymer layer is incorporated into multilayer film for perishable-food packaging because it provides a barrier to oxygen. A special high barrier resin is supplied specifically for this application. Typically, multilayer packaging films contain outer layers of a tough, low cost polymer such as high density polyethylene (HDPE) with VDC copolymer as the core layer. The film is made in a special coextrusion process. The properties of a 0.05-mm film are listed in Table 15 (see Film and sheeting materials).

Solid-Phase Forming. Solid-phase forming is a new technology for plastics fabrication. It is ideally suited for the fabrication of thermally unstable polymers, eg, VDC copolymers, and for the forming of composite structures because of its essentially scrapless operation. Either preformed sheets or briquettes of compressed polymer powders can be used (129–130). The latter can be used to fabricate pure PVDC and unplasticized high VDC copolymers without degradation. Solid-phase compaction has also been studied (131).

Table 15.   Comparative Physical Properties [a]

Property	Saranex 21, 0.05-mm value	0.921 Density polyethylene, 0.05-mm value	Test method
yield tensile strength, MPa[b]	MD 14 TD 13	MD 12.1 TD 9.7	ASTM D 882-61T
ultimate tensile strength, MPa[b]	MD 24 TD 17	MD 20.0 TD 17	ASTM D 882-61T
tensile modulus, MPa[b]	MD 170 TD 150	MD 180 TD 180	ASTM D 882-61T
elongation, %	MD 400 TD 400	MD 325 TD 550	ASTM D 882-61T
Elmendorf tear strength, g	MD 800 TD 650	MD 200 TD 250	ASTM D 1922
gas transmission at 24°C, mL/(100 cm^2·24 h·101.3 kPa)[c]			ASTM D 1434-63
oxygen	1.9	210	
carbon dioxide	6.0	1250	
nitrogen	0.4	110	
water-vapor transmission at 95% rh and 38°C, g/(100 cm^2·24 h)	0.33	0.65	ASTM E 96-63T

[a] Ref. 127.
[b] To convert MPa to psi, multiply by 145.
[c] To convert kPa to atm, divide by 101.3.

Table 16.   Properties of a Typical Barrier Latex

Properties	Value
total solids, wt %	60–62
viscosity at 25°C, mPa·s (= cP)	25
pH	4.5–5.0
color	white–cream
particle size, nm	ca 250
density, g/cm^3	1.38
mechanical stability	excellent
storage stability	excellent
chemical stability	not stable to di- or trivalent ions

Table 17. Polymer Film Properties

Property	Value
water-vapor transmission (for 0.013 mm at 38°C and 95% rh), g/(24 h·100 cm^2)	0.573[a]
grease resistance	excellent
scorability and fold resistance	moderate
oxygen permeability (at 25°C), (cm^3·cm)/(cm^2·s·kPa)[b]	0.3 × 10^{-13}
heat sealability[c]	excellent
light stability	fair
density, g/cm^3	1.60
color	water–white
clarity	excellent
gloss	excellent
odor	none

[a] Values 0.37 g/(24 h·100 in.^{2}) for 0.5 mils.
[b] To convert kPa to cm Hg, multiply by 0.75.
[c] Face-to-face.

Lacquer Resins. Vinylidene chloride polymers have several properties which are valuable in the coatings industry, ie, excellent resistance to gas and moisture-vapor transmission, good resistance to attack by solvents and by fats and oils, high strength, and the ability to be heat-sealed (132–133). These characteristics result from the highly crystalline nature of the very high vinylidene chloride content of the polymer, which ranges from ca 80 wt % to more than 90 wt %. Minor constituents in these copolymers generally are vinyl chloride, alkyl acrylates, alkyl methacrylates, acrylonitrile, methacrylonitrile, and vinyl acetate. Small concentrations of vinylcarboxylic acids, eg, acrylic acid, methacrylic acid, or itaconic acid, are sometimes included to enhance adhesion of the polymer to the substrate. The ability to crystallize and the extent of crystallization are reduced with increasing concentration of the comonomers; some commercial polymers fail to crystallize at all. The most common lacquer resins are terpolymers of VDC–methyl methacrylate–acrylonitrile (134–135). The VDC level and the methyl methacrylate–acrylonitrile ratio are adjusted for the best balance of solubility and permeability. These polymers exhibit a unique combination of high solubility, low permeability, and rapid crystallization (136).

Acetone, methyl ethyl ketone, methyl isobutyl ketone, dimethylformamide, ethyl acetate, and tetrahydrofuran are solvents for vinylidene chloride polymers used in lacquer coatings; methyl ethyl ketone and tetrahydrofuran are most extensively employed. Toluene is used as a diluent for either. Lacquers prepared at 10–20 wt % polymer solids in a solvent blend of two parts ketone and one part toluene have a viscosity of 20–1000 mPa·s (= cP). Lacquers can be prepared from polymers of very high vinylidene chloride content in tetrahydrofuran–toluene mixtures and stored at room temperature. Methyl ethyl ketone lacquers must be prepared and maintained at 60–70°C or the lacquer forms a solid gel. It is critical in the manufacture of polymers for lacquer application to maintain a fairly narrow compositional distribution in the polymer to achieve good dissolution properties.

The lacquers are applied commercially by roller coating, doctor and dip coating, knife coating, and spraying. Spraying is useful only with lower viscosity lacquers and solvent balance is important to avoid webbing from the spray gun. Solvent removal is difficult from heavy coatings and multiple coatings are recommended where a heavy film is desired; sufficient time is allowed between coats to avoid lifting of the previous coat by the solvent. In the machine coating of flexible substrates, eg, paper and plastic films, the solvent is removed by infrared heating or forced-air drying at 90–140°C. Temperatures of 60–95°C promote the recrystallization of the polymer after the solvent has been removed. Failure to recrystallize the polymer leaves a soft, amorphous coating that blocks or adheres between concentric layers in a rewound roll. A recrystallized coating can be rewound without blocking. Handling properties of the coated film are improved with small additions of wax as a slip agent and of talc or silica as an antiblock agent to the lacquer system. The concentration of additives is kept low to prevent any serious detraction from the vapor-transmission properties of the vinylidene chloride copolymer coating. For this reason, plasticizers are seldom, if ever, used.

A primary use of vinylidene chloride copolymer lacquers is the coating of films made from regenerated cellulose or of board or paper coated with polyamide, polyester, polyethylene, polypropylene, poly(vinyl chloride), and polyethylene to impart resistance to fats, oils, oxygen, and water vapor (137). These are used mainly in the packaging of foodstuffs, where the additional features of inertness, lack of odor or taste, and nontoxicity are required. Vinylidene chloride copolymers have been used extensively as interior coatings for ship tanks, railroad tank cars, and fuel-storage tanks, and for coating of steel piles and structures (138–139). The excellent chemical resistance and good adhesion result in excellent long-term performance of the coating. Brushing and spraying are suitable methods of application (see Coatings processes).

The excellent adhesion to primed films of polyester combined with good dielectric properties and good surface properties make the vinylidene chloride copolymers very suitable as binders for iron-oxide-pigmented coatings for magnetic tapes (140–142). They perform very well in audiotapes, videotapes, and computer tapes.

Vinylidene Chloride Copolymer Latex.   Vinylidene chloride polymers are often made in emulsion, but normally they are isolated, dried, and used as conventional resins. Stable latexes have been prepared and can be used directly for coatings (143–148) (see also Latex technology). The principal applications for these materials are as barrier coatings on paper products and, more recently, on plastic films. The heat-seal characteristics of VDC copolymer coatings are equally valuable in many applications. Vinylidene chloride copolymer latexes have also been used with cement to make high strength mortars and concretes (149–150). They are also used as binders for paints and nonwoven fabrics (151) (see Paint; Nonwoven textiles). The use of special VDC copolymer latexes for barrier laminating adhesives is growing, and the use of vinylidene chloride copolymers in flame-resistant carpet backing is well known (152–155) (see Adhesives; Flame retardants).

Poly(vinylidene chloride) latexes can be easily prepared by the same methods but have few uses because they do not form films. Copolymers of high VDC content are film-forming when freshly prepared but soon crystallize and lose this desirable characteristic. Since crystallinity in the final product is very often desirable, eg, in barrier coatings, a significant developmental problem has been to prevent crystallization in the latex during storage and to induce rapid crystallization of the polymer after coating. This has been accomplished by using the proper combination of comonomers with VDC.

Most vinylidene chloride copolymer latexes are made with varying amounts of acrylates, methacrylates, acrylonitrile, and minor amounts of vinyl carboxylic acids, eg, itaconic and acrylic acids. Low foam latexes with high surface tension are prepared with copolymerizable sulfonate monomers (153,155–156). The total amount of comonomer ranges from ca 8 wt % for barrier latexes to as high as 60 wt % for binder and paint latexes. The properties of a typical barrier latex used for paper coating are listed in Table 16. Barrier latexes are usually formulated with antiblock, slip, and wetting agents. They can be deposited by conventional coating processes, eg, with an air knife (157–158). Coating speeds in excess of 305 m/min can be attained. The latex coating can be dried in forced-air or radiant-heat ovens (159–160). Multiple coats are applied, particularly in paper coating, to reduce pinholing (161). A precoat is often used on porous substances to reduce the quantity of the more expensive VDC copolymer latex needed for covering (162). The properties of a typical coating are listed in Table 17.

Vinylidene Chloride Copolymer Foams. Low density, fine-celled VDC copolymer foams can be made by extrusion of a mixture of vinylidene chloride copolymer and a blowing agent at 120–150°C (163) (see Foamed plastics). The formulation must contain heat stabilizers and the extrusion equipment must be made of noncatalytic metals to prevent accelerated decomposition of the polymer. The low melt viscosity of the VDC copolymer formulation limits the size of the foam sheet that can be extruded.

Expandable VDC copolymer microspheres are prepared by a microsuspension process (164). The expanded microspheres are used in reinforced polyesters, blocking multipair cable, and in composites for furniture, marble and marine applications (165–168). Vinylidene chloride copolymer microspheres are also used in printing inks and paper manufacture (169) (see Printing processes; Papermaking additives).

Vinylidene Chloride Copolymer Flame Retardants. The role of halogen-containing compounds in ignition and flame suppression has been studied for many years (170–175). (The following discussion is not intended to reflect hazards of VDC copolymers or any material under actual fire conditions.) Vinylidene chloride copolymers are an abundant source of organic chlorine, eg, often above 70 wt %. Vinylidene chloride emulsion copolymers are used in a variety of flame-retardant binding applications (176–177). Powders dispersible in nonsolvent organic polymer intermediates, eg, polyols, are used for both reinforcement and flame retardance in polyurethane foams; VDC copolymer powder is also used as a flame-retardant binder for cotton batt (178–180) (see Flame retardants, halogenated).

The halogenated polymers generate significantly more smoke compared to polymers with aliphatic backbones, even though the presence of the halogen does increase the limiting oxygen index. Heavy-metal salts retard smoke generation in halogenated polymers (180–181). A VDC emulsion copolymer with a high acrylonitrile graft can be used to make flame-retardant acrylic fibers (182). A rubber-modified VDC copolymer combines good burning resistance with good low temperature flexibility (183–184). The rubber-modified VDC-copolymer is being evaluated in wire coating where better flame retardance and lower smoke generation are needed.

Materials are also blended with VDC copolymers to improve toughness (184–187). Vinylidene chloride copolymer blended with ethylene–vinyl acetate copolymers improves toughness and lowers heat-seal temperatures (188–189). Adhesion of a VDC copolymer coating to polyester can be achieved by blending the copolymer with a linear polyester resin (190).

BIBLIOGRAPHY

"Vinylidene Polymers (Chloride)" in *ECT* 2nd ed., Vol. 21, pp. 275–303, by R. Wessling and F. G. Edwards, The Dow Chemical Company.

1. R. A. Wessling, *Polyvinylidene Chloride*, Gordon and Breach Science Publishers, New York, 1977, Chapt. 1.
2. Ref. 1, entire book.
3. C. E. Schildknecht, *Vinyl and Related Polymers*, John Wiley & Sons, Inc., New York, 1952, Chapt. 8.
4. R. A. Wessling and F. G. Edwards in N. M. Bikales, ed., *Encyclopedia of Polymer Science and Technology*, Vol. 14, John Wiley & Sons, Inc., New York, 1971, p. 590.
5. *Vinylidene Chloride Monomer Safe Handling Guide*, #102-232-80, The Dow Chemical Co., Midland, Mich., 1980.
6. L. G. Shelton, D. E. Hamilton, and R. H. Fisackerly in E. C. Leonard, ed., *Vinyl and Diene Monomers, High Polymers*, Vol. 24, Interscience Publishers, a division of John Wiley & Sons, Inc., New York, 1971, pp. 1205–1282.
7. U.S. Pat. 2,238,020 (April 8, 1947), A. W. Hanson and W. C. Goggin (to The Dow Chemical Co.).
8. U.S. Pat. 3,760,015 (Sept. 18, 1973), S. Berkowitz (to FMC Corp.).
9. U.S. Pat. 3,870,762 (March 11, 1975), M. H. Stacey and T. D. Tribbeck (to Imperial Chemical Industries, Ltd.).
10. U.S. Pat. 4,225,519 (Sept. 30, 1980), A. E. Reinhardt III (to PPG Industries).
11. P. W. Sherwood, *Ind. Eng. Chem.* **54,** 29 (1962).
12. U.S. Pat. 2,293,317 (Aug. 18, 1942), F. L. Taylor and L. H. Horsley (to The Dow Chemical Co.).
13. G. Talamini and E. Peggion in G. E. Ham, ed., *Vinyl Polymerization*, Vol. 1, Marcel Dekker, New York, 1967, Part 1, Chapt. 5.
14. P. J. Flory, *Principles of Polymer Chemistry*, Cornell University Press, Ithaca, New York, 1953, Chapt. 6.
15. W. H. Stockmayer, K. Matsuo, and G. W. Nelb, *Macromolecules* **10,** 654 (1977).
16. J. D. Burnett and H. W. Melville, *Trans. Faraday Soc.* **46,** 976 (1950).
17. C. E. Bawn, T. P. Hobin, and W. J. McGarry, *J. Chim. Phys.* **56,** 791 (1959).
18. W. J. Burlant and D. H. Green, *J. Polym. Sci.* **31,** 227 (1958).
19. R. C. Reinhardt, *Ind. Eng. Chem.* **35,** 422 (1943).
20. W. I. Bengough and R. G. W. Norrish, *Proc. R. Soc. London Ser. A* **218,** 149 (1953).
21. R. A. Wessling and I. R. Harrison, *J. Polym. Sci. Part A-1* **9,** 3471 (1971).
22. V. V. Mazurek, *Vysokomol. Soedin.* **8,** 1174 (1966).
23. J. D. Cotman, M. F. Gonzalez, and G. C. Claver, *J. Polym. Sci. Part A-1*, **5,** 1137 (1967).
24. O. G. Lewis and R. M. King, Jr., in N. A. J. Platzer, ed., *Addition and Condensation Polymerization Processes*, American Chemical Society, Washington, D.C., 1969, Chapt. 2.
25. J. L. Gardon in C. E. Shildknecht, ed., *Polymerization Processes*, John Wiley & Sons, Inc., New York, 1977, Chapt. 6.
26. R. A. Wessling, *J. Appl. Polym. Sci.* **12,** 309 (1968).
27. R. A. Wessling and D. S. Gibbs, *J. Macromol. Sci. Chem. A7,* 647 (1973).
28. A. Konishi, *Bull. Chem. Soc. Jpn.* **35,** 197 (1962).
29. *Ibid.*, 193 (1962).
30. A. P. Sheinker and co-workers, *Dokl. Akad. Nauk SSSR* **124,** 632 (1959).
31. D. R. Roberts and R. H. Beaver, *J. Polym. Sci. Polym. Lett. Ed.* **17**(3), 155 (1979).
32. A. R. Westwood, *Eur. Polym. J.* **7,** 377 (1971).
33. J. Brandrup and E. H. Immergut, eds., *Polymer Handbook*, 2nd ed., John Wiley & Sons, Inc., New York, 1975.
34. J. F. Gabbett and W. Mayo Smith in G. E. Ham, ed., *Copolymerization*, John Wiley & Sons, Inc., New York, 1964, Chapt. 10.
35. T. C. Chiang, C. H. Graillat, J. Guillot, Q. T. Pham, and A. Guyot, *J. Polym. Sci. Polym. Chem. Ed.* **15,** 2961 (1977).
36. K. Matsuo and W. H. Stockmayer, *Macromolecules* **10,** 658 (1977).
37. W. I. Bengough and R. G. W. Norrish, *Proc. R. Soc. London Ser. A* **218,** 155 (1953).
38. C. Pichot, Q. T. Pham, and J. Guillot, *J. Macromol. Sci. Chem.* **12,** 1211 (1978).
39. C. Pichot and Q. T. Pham, *Eur. Polym. J.* **15,** 833 (1979).

40. N. Yamazaki, K. Sasaki, T. Nisiimura, and S. Kambara, *Polym. Prepr. Am. Chem. Soc. Div. Polym. Chem.* **5,** 667 (1964).
41. A. Konishi, *Bull. Chem. Soc. Jpn.* **35,** 395 (1962).
42. B. L. Erusalimskii, I. G. Krasnosel'skaya, V. V. Mazurek, and V. G. Gasan-Zade, *Dokl. Akad. Nauk SSSR* **169,** 114 (1966).
43. Brit. Pat. 1,119,746 (July 10, 1968), (to Chisso Corp.).
44. Can. Pat. 798,905 (Nov. 12, 1968), R. Buning and W. Pungs (to Dynamit Nobel Corp.).
45. U.S. Pat. 3,366,709 (Jan. 30, 1968), M. Baer (to Monsanto Co.).
46. U.S. Pat. 3,509,236 (April 28, 1970), H. G. Siegler, R. B. Oberlar, and W. Pungs (to Dynamit Nobel Aktiengeselfichaft Co.).
47. U.S. Pat. 3,655,553 (April 11, 1972), R. C. DeWald (to Firestone Tire and Rubber Co.).
48. M. Pegoraro, E. Beati, and J. Bilalov, *Chim. Ind. (Milan)* **54,** 18 (1972).
49. T. J. Haley, *Clinical Toxicol.* **8,** 633 (1975).
50. H. S. Warren and B. E. Ricci, *Oak Ridge National Lab/Tox Information Response Center Report #77/3,* NITS, Washington, D.C., 1978.
51. P. L. Viola and A. Caputo, *Environ. Health Perspect.* **21,** 45 (1977).
52. C. C. Lee and co-workers, *J. Toxicol. Environ. Health* **4,** 15 (1978).
53. V. Ponomarkov and L. Tomatis, *Oncology* **37,** 136 (1980).
54. R. D. Short and co-workers, *EPA Report #PB 281713,* Environmental Protection Agency, Washington, D.C., 1977.
55. T. R. Blackwood, D. R. Tierney, and M. R. Piana, *EPA Report #PB 80-146442,* Environmental Protection Agency, Washington, D.C., 1979.
56. J. M. Norris, private communication, The Dow Chemical Co., 1982.
57. M. J. McKenna, P. G. Watanabe, and P. J. Gehring, *Environ. Health Perspect.* **21,** 99 (1977).
58. M. G. Ott and co-workers, *J. Occup. Med.* **18,** 735 (1976).
59. J. Hushon and M. Kornreich, *EPA Report PB 280624,* Environmental Protection Agency, Washington, D.C., 1978.
60. D. Foerst, *Am. Ind. Hyg. Assoc. J.* **40,** 888 (1979).
61. T. Fisher, J. B. Kinsinger, and C. W. Wilson, *Polym. Lett.* **5,** 285 (1967).
62. M. Meeks and J. L. Koenig, *J. Polymer Sci. A-2* **9,** 717 (1971).
63. S. Krimm, *Fortschr. Hochpolym. Forsch.* **2,** 51 (1960).
64. K. Matsuo and W. H. Stockmayer, *Macromolecules* **8,** 660 (1975).
65. K. Okuda, *J. Polymer Sci. A-2,* 1749 (1964).
66. E. J. Arlman and W. M. Wagner, *Trans. Faraday Soc.* **49,** 832 (1953).
67. M. M. Coleman and co-workers, *J. Macromol. Sci. Phys.* **15,** 463 (1978).
68. M. S. Wu and co-workers, *J. Polym. Sci. Polym. Phys. Ed.* **18,** 95 (1980).
69. *Ibid.,* 111 (1980).
70. R. H. Boyd and L. Kesner, *J. Polym. Sci. Polym. Phys. Ed.* **19,** 393 (1981).
71. B. Dumont, P. Berticat, and J. Guillot, *Makromol. Chem.* **182,** 1207 (1981).
72. A. Douillard, B. Dumont, and J. Guillot, *Makromol. Chem.* **182,** 1283 (1981).
73. R. F. Boyer and R. S. Spencer, *J. Appl. Phys.* **15,** 398 (1944).
74. K-H. Illers, *Kolloid Z.* **190,** 16 (1963).
75. R. A. Wessling, F. L. Dicken, S. R. Kurowsky, and D. S. Gibbs, *Appl. Polym. Symp.* **25,** 83 (1974).
76. N. W. Johnston, *Rev. Macromol. Chem.* **C14,** 215 (1976).
77. G. R. Riser and L. P. Witnauer, *Polym. Prepr. Am. Chem. Soc. Div. Polym. Chem.* **2,** 218 (1961).
78. R. A. Wessling, J. H. Oswald, and I. R. Harrison, *J. Polym. Sci. Polym. Phys. Ed.* **11,** 875 (1973).
79. I. R. Harrison and E. Baer, *J. Colloid Interface Sci.* **31,** 176 (1969).
80. R. A. Wessling, D. R. Carter, and D. L. Ahr, *J. Appl. Polym. Sci.* **17,** 737 (1973).
81. A. F. Burmester and R. A. Wessling, *Bull. Am. Phys. Soc.* **18,** 317 (1973).
82. M. Asahina, M. Sato, and T. Kobayashi, *Bull. Chem. Soc. Jpn.* **35,** 630 (1962).
83. R. A. Wessling, *J. Appl. Polym. Sci.* **14,** 1531 (1970).
84. *Ibid.,* 2263 (1970).
85. M. L. Wallach, *Polym. Prepr. Am. Chem. Soc. Div. Polym. Chem.* **10,** 1248 (1969).
86. A. Revillion, B. Dumont, and A. Guyot, *J. Polymer Sci. Chem. Ed.* **14,** 2263 (1976).
87. A. Revillion, *J. Liq. Chromatogr.* **3,** 1137 (1980).
88. E. D. Serdensky in H. Mark, E. Cernia, and S. M. Atlas, eds., *Man-Made Fibers,* John Wiley & Sons, Inc., New York, 1968, p. 319.
89. G. S. Kolesnikov, L. S. Fedorova, B. L. Tsetlin, and N. V. Klimentova, *Izv. Akad. Nauk SSSR Otd. Khim. Nauk,* 731 (1959).

90. E. F. Jordan, G. R. Riser, B. Artymyshyn, W. E. Parker, J. W. Pensabene, and A. N. Wrigley, *J. Appl. Polym. Sci.* **13,** 1777 (1969).
91. S. W. Lasoski, *J. Appl. Polym. Sci.* **4,** 118 (1960).
92. S. W. Lasoski and W. H. Cobbs, *J. Polym. Sci.* **36,** 21 (1959).
93. H. J. Bixler and O. S. Sweeting in O. J. Sweeting, ed., *The Science and Technology of Polymer Films,* Vol. 2, Interscience Publishers, a division of John Wiley & Sons, Inc., New York, 1971, Chapt. 1.
94. P. T. DeLassus, *J. Vinyl Technol.* **1,** 14 (1979).
95. P. T. DeLassus and D. J. Grieser, *J. Vinyl Technol.* **2,** 195 (1980).
96. Ref. 1, Chapts. 9–10.
97. D. H. Davies, D. H. Everett, and D. J. Taylor, *Trans. Faraday Soc.* **67,** 382 (1971).
98. R. D. Bohme and R. A. Wessling, *J. Appl. Polym. Sci.* **16,** 1761 (1972).
99. D. R. Roberts and A. L. Gatzke, *J. Polym. Sci. Chem. Ed.* **16,** 1761 (1972).
100. A. Bailey and D. H. Everett, *J. Polym. Sci. Part A-2,* **7,** 87 (1969).
101. L. L. Ban, D. Crawford, and A. Marsh, *J. Appl. Crystallogr.* **8,** 415 (1975).
102. G. J. Howard and Sknutton, *J. Appl. Polym. Sci.* **19,** 697 (1975).
103. D. D. Schmidt and R. G. Melcher, *Ext. Abst. 13th Bienn. Conf. Carbon* **13,** 257 (1977).
104. V. Rossbach, M. Bert, J. Guillot, and A. Guyot, *Angew. Makromol. Chem.* **40–41,** 291 (1974).
105. D. E. Agostini and A. L. Gatzke, *J. Polym. Sci. Chem. Ed.* **11,** 649 (1973).
106. D. H. Davies and P. M. Henheffer, *Trans. Faraday Soc.* **66,** 2329 (1970).
107. D. H. Grant, *Polymer* **11,** 581 (1970).
108. E. Tsuchida, C-N. Shih, I. Shinohara, and S. Kambara, *J. Polym. Sci. Part A.* **2,** 3347 (1964).
109. S. S. Barton, G. Boulton, B. H. Harrison, and W. Kemp, *Trans. Faraday Soc.* **67,** 3534 (1971).
110. S. S. Barton, J. R. Dacey, and B. H. Harrison, *Am. Chem. Soc. Div. Org. Coat. Plast. Chem. Pap.* **31,** 768 (1971).
111. R. A. Wessling, *Am. Chem. Soc. Div. Org. Coat. Plast. Chem. Pap.* **34,** 380 (1976).
112. T. Nakagawa, *Kogyo Kagaku Zasshi* **71,** 1272 (1968).
113. U.S. Pat. 3,852,223 (Dec. 3, 1974), R. D. Bohme and R. A. Wessling (to The Dow Chemical Co.).
114. G. Oster, G. K. Oster, and M. Kryszewski, *J. Polym. Sci.* **57,** 937 (1962).
115. M. Kryszewski and M. Mucha, *Bull. Acad. Pol. Sci. Ser. Sci. Chim.* **13,** 53 (1965).
116. D. E. Harmer and J. A. Raab, *J. Polym. Sci.* **55,** 821 (1961).
117. C. Chen, M. Igbal, C. V. Pittinon, Jr., and J. N. Helbert, *Makromol. Chem.* **179,** 2109 (1978).
118. U.S. Pat. 3,882,598 (May 13, 1975), R. T. Marzolf (to The Dow Chemical Co.).
119. U.S. Pat. 3,882,081 (May 6, 1975), A. W. Baker (to The Dow Chemical Co.).
120. F. Chevassus and R. deBroutelles, *The Stabilization of Polyvinyl Chloride,* St. Martin's Press, New York, 1963.
121. U.S. Pat. 3,033,812 (May 8, 1962), P. K. Isacs and A. Trofimow (to W. R. Grace & Co.).
122. D. M. Woodford, *Chem. Ind. (London)* (8), 316 (1966).
123. U.S. Pat. 2,968,651 (Jan. 17, 1961), L. C. Friedrich, Jr., J. W. Peters, and M. R. Rector (to The Dow Chemical Co.).
124. U.S. Pat. 2,482,771 (Sept. 27, 1944), J. Heerema (to The Dow Chemical Co.).
125. *Saran Resins,* #190-289-79, The Dow Chemical Co., Midland, Mich., 1979.
126. Ref. 93, Chapt. 6.
127. D. L. Roodvoets in P. F. Bruin, ed., *Packaging with Plastics,* Gordon and Breach, New York, 1974, p. 85.
128. *SARANEX® Films,* The Dow Chemical Company, Midland, Mich., 1979.
129. U.S. Pat. 3,739,050 (June 12, 1973), R. E. Ayres, K. J. Cleereman, and W. J. Schrenk (to The Dow Chemical Co.).
130. U.S. Pat. 4,161,502 (July 17, 1979), R. A. Wessling and E. F. Gurnee (to The Dow Chemical Co.).
131. R. J. Crawford and D. Paul, *J. Mater. Sci.* **14,** 2693 (1979).
132. S. F. Roth, *Am. Chem. Soc. Chem. Market Econ. Div. Symp. N.Y.,* 29-6 (1976).
133. *Saran F Resin,* Technical Bulletin, The Dow Chemical Co., Horgen, Sw. Europe, 1969.
134. U.S. Pat. 3,817,780 (June 18, 1974), P. E. Hinkamp and D. F. Foye (to The Dow Chemical Company).
135. U.S. Pat. 3,879,359 (April 22, 1975), P. E. Hinkamp and D. F. Foye (to The Dow Chemical Company).
136. U.S. Pat. 4,097,433 (June 27, 1978), W. P. Kane (to E. I. du Pont de Nemours & Co., Inc.).
137. U.S. Pat. 2,462,185 (Feb. 22, 1949), P. M. Hauser (to E. I. du Pont de Nemours & Co., Inc.).
138. W. W. Cranmer, *Corrosion (Houston)* **8**(6), 195 (1952).
139. R. L. Alumbaugh, *Mater. Prot.* **3**(7), 34, 39 (1964).

140. U.S. Pat. 3,144,352 (Aug. 11, 1964), J. P. Talley (to Ampex Corporation).
141. U.S. Pat. 3,865,741 (Feb. 11, 1975), F. J. Sischka (to Memorex Corporation).
142. U.S. Pat. 3,894,306 (July 1, 1975), F. J. Sischka (to Memorex Corporation).
143. L. J. Wood, *Mod. Packag.* **33,** 125 (1960).
144. R. F. Avery, *Tappi* **45,** 356 (1962).
145. A. D. Jordan, *Tappi* **45,** 865 (1962).
146. B. J. Sauntson and G. Brown, *Rep. Prog. Appl. Chem.* **56,** 66 (1972).
147. P. S. Bryant, *European Flexographic Technical Association Barrier Coatings and Laminations Seminar, Manchester, England* **1,** 7 (1977).
148. G. H. Elschnig, A. F. Schmid, K. Goetz, and F. Witt, *Pop. Plast.* **17**(2), 19 (1972).
149. K. Fisher, *Proc. Br. Ceram. Soc.* **24,** 59 (1975).
150. H. W. H. West, R. W. Ford, and J. F. Goodwin, *Proc. Br. Ceram. Soc.* **24,** 49 (1975).
151. U.S. Pat. 3,787,232 (Jan. 22, 1974), B. K. Mikofalvy and D. P. Knechtges (to B. F. Goodrich Co.).
152. R. G. Jahn, *Adhes. Age* **20**(6), 37 (1977).
153. U.S. Pat. 3,946,139 (March 23, 1976), M. Bleyle, W. D. Waltham, R. DelVecchio, and A. Trofimow (to W. R. Grace & Co.).
154. U.S. Pat. 3,850,726, D. R. Smith and H. Peterson (to A. E. Staley Co.).
155. U.S. Pat. 3,617,368 (Nov. 2, 1971), D. S. Gibbs and R. A. Wessling (to The Dow Chemical Co.).
156. Brit. Pat. 1,233,078 (May 26, 1971), H. Gould and J. A. Zaslowsky (to Alcolac Chemical Co.).
157. G. H. Elschnig and A. F. Schmid, *Pop. Plast.* **17**(3), 36 (1972).
158. *Ibid.,* **17** (4), 17 (1972).
159. *Ibid.,* **17**(6), 17 (1972).
160. G. H. Elschnig and A. F. Schmid, *Paintindia* **22**(6), 22 (1972).
161. F. C. Caruso, *Test. Pap. Synth. Conf.,* TAPPI, Atlanta, Ga., 1974, p. 167.
162. E. A. Chirokas, *Tappi* **50,** 59A (1967).
163. U.S. Pat. 3,983,080 (Sept. 28, 1976), K. S. Suh, R. E. Skochdopole, and M. E. Luduc (to The Dow Chemical Co.).
164. U.S. Pat. 3,615,972 (Oct. 26, 1971), D. S. Morehouse and R. J. Tetreault (to The Dow Chemical Co.).
165. D. S. Morehouse and H. A. Walters, *SPE J.* **25,** 45 (1969).
166. T. E. Cravens, *Am. Chem. Soc. Div. Org. Coat. Plast. Chem. Pap.* **33,** 74 (1973).
167. R. C. Mildner, K. F. Nacke, E. W. Veazey, and P. C. Woodland, *Mod. Plast.* **47**(5), 98 (1970).
168. T. F. Anderson, H. A. Walters, and C. W. Glesner, *J. Cell. Plast.* **6**(4), 171 (1970).
169. *Mater. Plast. Elastomeri* **10,** 468 (Oct. 1980).
170. D. L. Chamberlain in W. C. Kuryla and A. J. Papa, eds., *Flame Retardancy of Polymeric Materials,* Vol. 2, Marcel Dekker, New York, 1973, pp. 109–168.
171. D. W. Van Krevelen, *Polymer* **16,** 615 (1975).
172. L. G. Imhoff and K. C. Stueben, *Polym. Eng. Sci.* **13,** 146 (1973).
173. E. R. Larsen, *J. Fire Flammability Fire Retardant Chemistry* **1,** 4 (1974).
174. *Ibid.,* **2,** 5 (1975).
175. R. C. Kidder, *Fire Retardant Chemicals Association Semi-Annual Meeting,* Oct. 30, 1977, pp. 45–51.
176. J. Knightly and J. C. Bax, *Fire Retardation, Proceedings of the European Conference on Flammability Fire Retardance, 1st 1977,* 1979, pp. 75–83.
177. J. R. Goots and D. P. Knechtges, *Polym. Plast. Technol. Eng.* **5,** 131 (1975).
178. C. V. Neywick, R. E. Yoerger and R. F. Peterson, *J. Cell. Plast.* **16,** 171 (1980).
179. U.S. Pat. 4,232,129 (Nov. 4, 1980), D. S. Gibbs, J. H. Benson, and R. T. Fernandez (to The Dow Chemical Co.).
180. U.S. Pat. 4,002,597 (Jan. 11, 1977), E. D. Dickens (to B. F. Goodrich Co.).
181. U.S. Pat. 4,055,538 (Oct. 25, 1977), W. J. Kroenke (to B. F. Goodrich Co.).
182. U.S. Pat. 4,186,156 (Jan. 29, 1980), D. S. Gibbs (to The Dow Chemical Co.).
183. *Plast. Technol.* **26**(1), 13 (1980).
184. U.S. Pat. 4,206,105 (June 3, 1980), O. L. Stafford (to The Dow Chemical Co.).
185. U.S. Pat. 4,239,799 (Dec. 16, 1980), A. S. Weinberg (to W. R. Grace & Co.).
186. U.S. Pat. 3,840,620 (Oct. 8, 1974), R. Gallagher (to Stauffer Chemical Co.).
187. U.S. Pat. 3,513,226 (May 19, 1970), T. Hotta (to Kureha Kagaku Kogyo Kabushiki Kalsha Co.).
188. U.S. Pat. 3,565,975 (Feb. 23, 1971), F. V. Goff, F. Stevenson, and W. H. Wineland (to The Dow Chemical Co.).
189. U.S. Pat. 3,558,542 (Jan. 26, 1971), J. W. McDonald (to E. I. du Pont de Nemours & Co., Inc.).

190. U.S. Pat. 3,896,066 (July 22, 1975), R. O. Ranck (to E. I. du Pont de Nemours & Co., Inc.).

DALE S. GIBBS
R. A. WESSLING
Dow Chemical U.S.A.

VINYLIDENE POLYMERS, POLY(VINYLIDENE FLUORIDE) ELASTOMERS.
See Fluorine compounds, organic.

VINYL POLYMERS

Poly(vinyl acetal)s, 798
Poly(vinyl acetate), 817
Poly(vinyl alcohol), 848
Vinyl chloride and poly(vinyl chloride), 865
Vinyl ether monomers and polymers, 937
N-Vinyl monomers and polymers, 960

POLY(VINYL ACETAL)S

Poly(vinyl acetal)s are resins made from the reaction products of poly(vinyl alcohol) and aldehydes (qv) (see Vinyl polymers, poly(vinyl alcohol)). The first known poly(vinyl acetal) was prepared in Germany in 1924 from benzaldehyde and poly(vinyl alcohol) [9002-89-5] (1). During the 1930s and 1940s, much technical effort was spent in Canada and the United States on developing the process and commercializing the poly(vinyl acetal) resins (2–9). The main driving force for the commercialization of these polymers was the use of plasticized poly(vinyl butyral) [63148-65-2] as the interlayer for laminated safety glass in automobiles and other transportation vehicles, where it replaced the plasticized cellulose acetate in use at that time (see Laminated materials, glass). This application is the largest worldwide use of poly(vinyl acetal) resins.

Two other poly(vinyl acetal)s, made from acetaldehyde and formaldehyde, were commercialized in the late 1930s and early 1940s, respectively. The acetaldehyde derivative, called Alvar, was originally developed by Shawinigan Chemicals, Ltd., Can., as a replacement for shellac (qv). Early uses also included injection-molded articles for a variety of applications and, in solution form, as paper and textile coatings. Production peaked in the early 1950s and then gradually decreased as a result of competition from less expensive resins, eg, poly(vinyl chloride) (see Vinyl polymers, vinyl chloride and poly(vinyl chloride)). Alvar is no longer manufactured. The formaldehyde derivative, when combined with phenolics and other resins and chemicals, was used extensively as a tough, heat-resistant magnet-wire coating; its impressive properties made it the leading magnet-wire coating for over 30 yr (4,10–11). It is still used for that and other applications. Its use in electrical insulation is the second largest worldwide use for any poly(vinyl acetal) (see Insulation, electric—electric wire and cable coverings). General reviews of the background, chemistry, physical properties, manufacture, applications, and other aspects of poly(vinyl acetal)s are given in references 12–14.

Monomeric acetals form from the reaction of one molecule of aldehyde and two molecules of alcohol in the presence of acid.

$$\text{RCHO} + 2\ \text{R}'\text{OH} \xrightarrow{\text{H}^+} \text{RCH(OR}')_2 + \text{H}_2\text{O}$$

$$\text{aldehyde} \quad \text{alcohol} \quad\quad \text{acetal}$$

Similarly, poly(vinyl acetal)s are prepared by the reaction of an aldehyde and two hydroxyl groups on the poly(vinyl alcohol) chain. The poly(vinyl alcohol) is in turn prepared from poly(vinyl acetate) by hydrolysis or alcoholysis. The poly(vinyl alcohol)s vary in molecular weight and structure because of the manner in which the poly(vinyl acetate)s are made and contain varying percentages of hydroxyl and acetate groups, depending on the degree of hydrolysis (see Fig. 1).

When a poly(vinyl alcohol) is acetalized with an aldehyde, the conditions of the acetal reaction are closely controlled to form a poly(vinyl acetal) containing a predetermined proportion of acetate groups, hydroxyl groups, and acetal groups. The final poly(vinyl acetal) may be represented as shown in Figure 2 where the three basic units are randomly distributed along the molecule.

The poly(vinyl acetal) unit is a cyclic acetal because the hydroxyl groups are attached to the same chain. A synthetic polymer containing three basic groups or monomer units was unusual in the 1930s. It was only in the 1950s that polymer molecules containing three or more monomeric units became available commercially. The residual hydroxyl content of poly(vinyl acetal) resins can never be reduced to zero in the acetalization reaction (15).

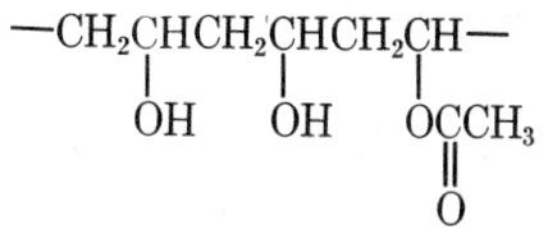

$$-\text{CH}_2\text{CHCH}_2\text{CHCH}_2\text{CH}-$$

Figure 1.   A poly(vinyl alcohol) structure.

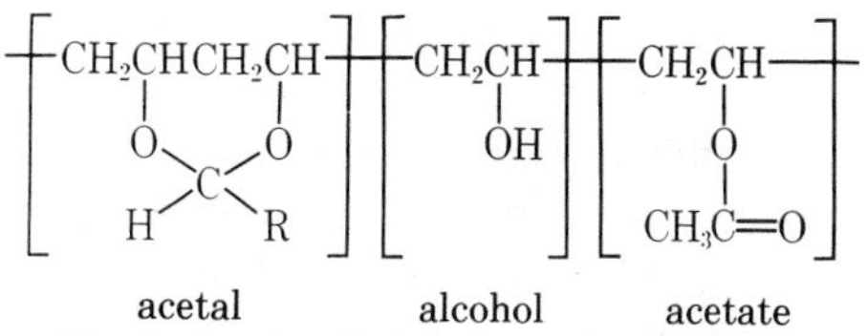

acetal   alcohol   acetate

Figure 2.   A poly(vinyl acetal) structure.

Properties

The principal types of poly(vinyl formal)s available in the United States and their properties are listed in Table 1. The letter E indicates a material most suitable for wire enamels. Many of the previously available types are no longer manufactured commercially. Grades 5/95E, 6/95E, 7/95E, and 15/95E have the same acetate and formal content but differ in molecular weight. They have practically the same properties except viscosity. However, grade 12/85, because of its higher acetate and low formal content, has certain significantly different mechanical and thermal properties. It is also soluble in a far greater number of solvents, as shown in Table 2.

The principal types of poly(vinyl butyral) available in the United States and their properties are listed in Table 3. Grades B-76 and B-79, which differ only in mol wt, have a high butyral content and a lower hydroxyl content than all of the other grades, which also differ only in mol wt. As a consequence of this difference in composition, B-76 and B-79 show lower water absorption, lower mechanical and thermal properties, and significant differences in electrical properties. They are the only grades soluble in toluene.

The solubilities of poly(vinyl formal) [9003-33-2] and poly(vinyl butyral) in a representative list of solvents are shown in Table 2. In general, the choice of pure solvents for poly(vinyl formal) is limited even for the most soluble grade, Formvar 12/85. This has restricted its use in certain applications. However, poly(vinyl butyral) is much more soluble in common solvents with the low hydroxyl grades, ie, B-76 and B-79, showing the broadest range. Both poly(vinyl formal) and poly(vinyl butyral) are soluble in certain blends of polar and nonpolar solvents. A good blend is 40:60 ethyl alcohol (95 wt %)–toluene by weight. The low temperature flexibility of films, coatings, adhesives, etc, made from poly(vinyl acetal) resins can be improved by incorporation of plasticizers (qv); many of the common phthalate, phosphate, and polyester types are used. The poly(vinyl acetal)s are compatible with a variety of resins and natural gums. By suitable compounding, the physical and chemical properties as well as the cost of a formulation can be altered to suit a given application. Tables showing the compatibilities of plasticizers, other resins, and natural gums with the various poly(vinyl acetal) resins are given in reference 16.

Although poly(vinyl acetal) resins are thermoplastic and soluble in a range of solvents, the fact that they all contain secondary hydroxyl groups permits cross-linking reactions with a variety of thermosetting resins, eg, phenolics, ureas, melamines, epoxies, etc, and with chemicals, eg, dialdehydes and diisocyanates. In this manner, incorporation of even small amounts of the poly(vinyl acetal) resin into thermosetting compositions markedly improves the toughness, flexibility, and adhesion of the cured composition. Alternatively, incorporation of smaller quantities of thermosets often gives the poly(vinyl acetal)s a better balance of properties. Figures 3–7 illustrate the probable mechanism of cross-linking poly(vinyl acetal)s with various materials (16).

Any aldehyde can be used in a poly(vinyl acetal) reaction. Increasing the length of the acetal side chain tends to give more flexible chains (17). Poly(vinyl formal) is the toughest of the resins, whereas poly(vinyl butyral), although less tough, is more flexible. The percentage of acetal can be varied from very little to as much as can react chemically (15). Thus, there can be a wide variation in the properties of the acetal resins. For example, in one series of poly(vinyl butyral) resins with increasing percentage of acetalization, it is possible to produce water-soluble products (by partial acetalization or with >50 wt % poly(vinyl alcohol) content) as well as products that are completely soluble in toluene (by high acetalization or with <13 wt % poly(vinyl alcohol) content). In general, the commercial polymers are those with high degrees of acetalization. Coacetals, where two and even three aldehydes react with the same poly(vinyl alcohol) chain, have been made but are not sold commercially. Their manufacture generally involves processing techniques that are prohibitively expensive. Ketones, because they contain the reactive carbonyl group, can be used in place of aldehydes in reactions with poly(vinyl alcohol). The resulting poly(vinyl ketal)s have been made in the laboratory but they have not become commercially successful.

Manufacturing and Processing

Poly(vinyl acetal)s can be prepared by simultaneous, sequential, or semisequential reactions. In the simultaneous or one-stage reaction, the hydrolysis of the poly(vinyl acetate) and acetalization of the resulting poly(vinyl alcohol) are carried out at the same time in the same kettle with an acid catalyst. In the sequential or two-step reaction, the alcoholysis and acetalization reactions occur separately. In the semisequential process, the aldehyde is added after the hydrolysis has progressed to a certain point but without separation of the hydrolysis product. The kinetics of the hydrolysis and acetalization reactions have been studied by manufacturers of the products as well as by other investigators (18–19).

Commercial poly(vinyl formal) is manufactured by a simultaneous process in acetic acid (20). Poly(vinyl acetate) of the proper molecular weight is dissolved in a mixture of acetic acid, water, and formaldehyde. Sulfuric acid catalyst is added, and the mixture is maintained at 75–85°C until the desired reactions are complete. In this case, the hydrolysis of poly(vinyl acetate) is the rate-controlling step. The sulfuric acid catalyst is then neutralized in solution, and the resin precipitates as fine porous granules upon blending with water under rapid agitation. After careful washing to remove salts and organics and further neutralizing as necessary, the resin is centrifuged and dried. Varying the poly(vinyl acetate) molecular weight, the charge composition, and the reaction conditions yields the poly(vinyl formal) resins listed in Table 1. In all cases, the polymers contain the three basic units, ie, hydroxyl, acetate, and acetal.

Commercial poly(vinyl butyral)s are manufactured by sequential processes. Although there are many variations of this type of process, it is believed that the bulk is made in one or the other of the following ways. In one case, the poly(vinyl acetate) is dissolved in alcohol. At reflux temperature and with an acid catalyst and strong agitation, the alcoholysis reaction produces the poly(vinyl alcohol) in powder form (9). The poly(vinyl alcohol) is then suspended in ethyl alcohol and, with the addition of acid catalyst and butyraldehyde, is acetalized at >70°C; a solution forms at the completion of the reaction. Water is then added to the solution while it is agitated, resulting in precipitation of the poly(vinyl butyral) which is carefully washed, neutralized, and dried. In the other case, the poly(vinyl alcohol) usually made by alkaline catalysis is dissolved in water, and the acetalization reaction is carried out in the presence of an acid catalyst and butyraldehyde. The resin precipitates during the course of the reaction; at the end of the reaction, the resin is washed, neutralized, and

Table 1. Properties of Commercial Poly(Vinyl Formal) Formvar[a]

Property	Formvar 12/85	Formvar 5/95E	Formvar 6/96E	Formvar 7/95E	Formvar 15/95E	ASTM
Physical						
form			white, free-flowing powder			
volatiles (max as packed), wt %	1.5	1.5	1.0	1.0	1.0	[b]
M_w	26,000–34,000	10,000–15,000	14,000–17,000	16,000–20,000	24,000–40,000	[c]
solution viscosity (15 wt %), mPa·s (= cP)	500–600	100–200	200–300	300–500	3,000–4,500	[c]
resin viscosity, mPa·s (= cP)	18–22	8–12	12–15	15–20	37–53	[c]
sp gr (±0.002), 23°/23°	1.219	1.227	1.227	1.227	1.227	D 792-50
burning rate, cm/min	1.3	2.0	2.3	2.3	2.5	D 635-56T
refractive index (±0.0005)	1.495	1.502	1.502	1.502	1.502	D 542-50
water absorption (24 h), %	1.0	1.2	1.2	1.2	1.2	D 570-59aT
hydroxyl content expressed as % poly(vinyl alcohol)	5.5–7.0	5.0–6.5	5.0–6.5	5.0–6.5	5.0–6.0	D 1396-58[d]
acetate content expressed as % poly(vinyl acetate)	22–30	9.5–13.0	9.5–13.0	9.5–13.0	9.5–13.0	D 1396-58[d]
formal content expressed as % poly(vinyl formal)	ca 68	ca 82	ca 82	ca 82	ca 82	
Mechanical						
tensile strength, MPa[e]						
yield	57–64	59–66	59–66	59–66	59–66	D 638-58T
break	45–52	52–59	52–59	52–59	52–59	D 638-58T
elongation, %						
yield	5	7	7	7	7	D 638-58T
break	30	50	50	50	50	D 638-58T
modulus of elasticity (apparent), GPa[e]	2.75–2.90	2.75–3.10	2.75–3.10	2.75–3.10	2.75–3.10	D 638-58T
flexural strength (yield), MPa[e]	115–120	117–124	117–124	117–124	117–124	D 790-59T
hardness, Rockwell						
M	155	150	150	150	150	D 785-51
E	75	65	65	65	65	D 785-51
impact strength Izod, notched 1.25 cm × 1.25 cm, J/m[f]	55	70	70	70	70	D 256-56
Thermal						
flow temp (at 6.9 MPa[e]), °C	145–150	140–150	140–150	140–150	160–170	D 569-59
glass temp (apparent), °C	92–100	103–113	103–113	103–113	103–113	D 1043-51[g]
heat-distortion temp, °C	72–74	83–87	85–90	85–90	87–93	D 648-56
heat-sealing temp, °C	90	96	96	99	107	[h]

Electrical						ASTM method
dielectric constant						D 150-59T
50 Hz	3.5	3.2	3.2	3.2	3.4	
1 kHz	3.1	3.3	3.3	3.3	3.0	
1 MHz	2.9	3.1	3.1	3.1	2.8	
10 MHz	2.9	3.0	3.0	3.0	2.8	
dissipation factor						D 150-59T
50 Hz	8.7×10^{-3}	8.1×10^{-3}	8.1×10^{-3}	8.1×10^{-3}	8.7×10^{-3}	
1 kHz	9.0×10^{-3}	10.0×10^{-3}	10.0×10^{-3}	10.0×10^{-3}	10.0×10^{-3}	
1 MHz	18.0×10^{-3}	21.0×10^{-3}	21.0×10^{-3}	21.0×10^{-3}	21.0×10^{-3}	
10 MHz	160.0×10^{-3}	190.0×10^{-3}	190.0×10^{-3}	190.0×10^{-3}	180.0×10^{-3}	
dielectric strength (3.2 mm), V/μm						D 149-59
short time	13	24	13	13	12	
step-by-step	11	12	12	12	13	

^{a} Formvar is a registered trademark of Monsanto Co. (16).

^{b} Molecular weight was determined by fractionating reacetylated samples of the poly(vinyl alcohol)s used for production of the various Butvar and Formvar resins. Distributions were based upon intrinsic viscosities of the various fractions, and weight-average molecular weights were calculated from these distribution curves.

^{c} Solution viscosity was determined in 15 wt % solutions in 60:40 toluene–ethanol at 25°C with a Brookfield viscometer. Resin viscosity (5 g resin made to 100 mL with ethylene dichloride) was measured at 20°C with an Ostwald viscometer.

^{d} The ASTM method noted for hydroxyl content and acetate content refers specifically to poly(vinyl butyral) resins. However, the same method is applicable to poly(vinyl formal) resins.

^{e} To convert MPa to psi, multiply by 145; for GPa, by 145,000.

^{f} To convert J/m to lbf/in., divide by 53.38.

^{g} Apparent glass temperature (T_g) was determined by ASTM D 1043-51 and by differential scanning calorimetry (DSC). Results by DSC are 5–8°C higher than results from the ASTM method.

^{h} Heat-sealing temperature was determined for a 25-μm dried film on paper, cast from a 10% solution in 60:40 toluene–ethanol. A dwell time of 1.5 s at 414-kPa (60-psi) line pressure was used for the heat sealer.

Table 2.   Solvents for Poly(Vinyl Acetal) Resins[a,b]

	Poly(vinyl formal)		Poly(vinyl butyral)	
Solvent	Formvar 5/95E, 6/95E, 7/95E, and 15/95E	Formvar 12/85	Butvar B-76 and B-79	Butvar B-72, B-73, B-74, B-90, and B-98
acetic acid (glacial)	sol	sol	sol	sol
acetone	insol	insol	sol	insol
butanol	insol	insol	sol	sol
butyl acetate	insol	insol	sol	insol
carbon tetrachloride	insol	insol	insol	insol
cresylic acid	sol	sol	sol	sol
cyclohexanone	insol	sol	sol	sol
diacetone alcohol	insol	sol	sol	sol
diisobutyl ketone	insol	insol	sol	insol
dioxane	sol	sol	sol	sol
N,N-dimethylacetamide	sol	sol	sol	sol
N,N-dimethylformamide	sol	sol	sol	sol
ethanol, 95 wt %	insol	insol	sol	sol
ethyl acetate, 99 wt %	insol	insol	sol	insol
ethyl acetate, 85 wt %	insol	insol	sol	sol
ethyl Cellosolve	insol	insol	sol	sol
ethylene chloride	sol	sol	sol	sol
hexane	insol	insol	insol	insol
methyl acetate	insol	insol	sol	sol
methanol	insol	insol	insol	sol
methyl Cellosolve	insol	sol	sol	sol
methyl Cellosolve acetate	insol	sol	sol	insol
2-methyl-3-butyn-2-ol	sol	sol	sol	sol
3-methyl-1-pentyn-3-ol	sol	sol	sol	sol
methyl ethyl ketone	insol	insol	sol	insol
methyl isobutyl ketone	insol	insol	sol	insol
N-methyl-2-pyrrolidinone	sol	sol	sol	sol
nitropropane	insol	sol	insol	insol
2-propanol, 95 wt %	insol	insol	sol	sol
toluene	insol	insol	sol	insol
toluene–ethanol (60:40 by wt)	sol	sol	sol	sol
xylene	insol	insol	insol	insol
xylene–butanol (60:40 by wt)	insol	insol	sol	sol

[a] Ref. 16.
[b] sol = completely soluble; insol = insoluble or not completely soluble.

dried. Using either process and varying the poly(vinyl acetate), the charge, and the reaction conditions yield a series of poly(vinyl butyral) resins with varied properties. The bulk density, grain structure, solvent and plasticizer absorption rates, and even transparency can be controlled by the conditions at precipitation. For all practical purposes, these resins contain only two basic units, ie, hydroxyl and acetal. The acetate unit generally comprises <2.5 wt %.

Aqueous Poly(Vinyl Butyral) Dispersion

In addition to poly(vinyl acetal)s in powder form, aqueous dispersions containing poly(vinyl butyral) have been developed commercially. One version available in the

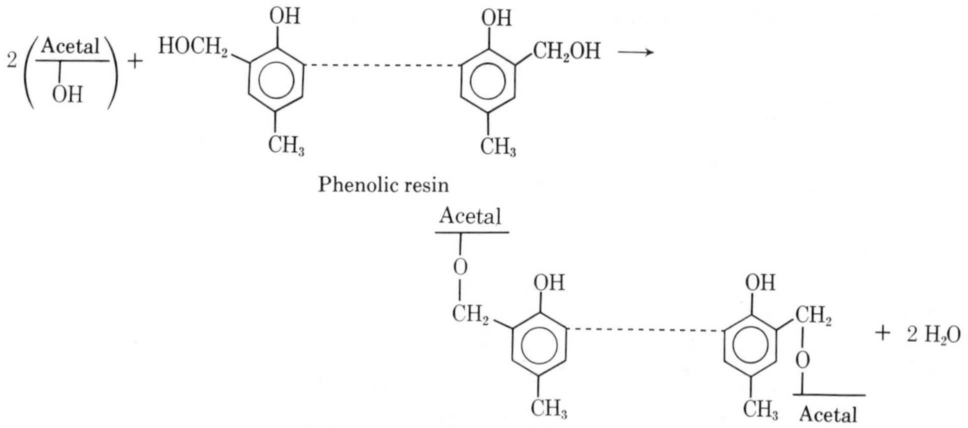

Figure 3.   Reaction with phenolics.

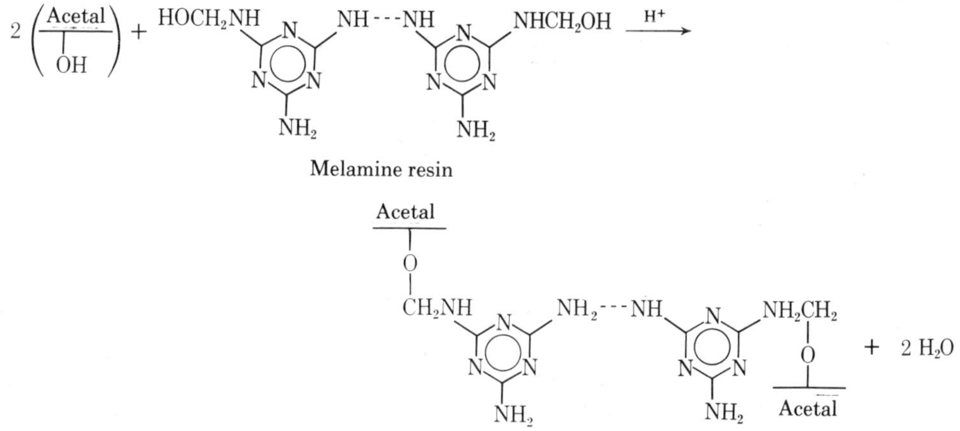

Figure 4.   Reaction with melamines.

United States is Butvar dispersion BR resin, which contains a high molecular weight Butvar, 40 parts of plasticizer per 100 parts of resin, and 50% total solids and has a particle size <1 μm (21). It is anionic and has a pH of ca 9–10. The dispersions are usually produced by a mixer process somewhat similar to that used for making rubber latex from reclaimed stock (22–24). They have been made with plasticizer amounts of 0–50 phr (parts per hundred rubber).

Films of poly(vinyl butyral) dispersion exhibit the toughness, flexibility, and transparency for which the resin is noted. Physical properties vary with plasticizer

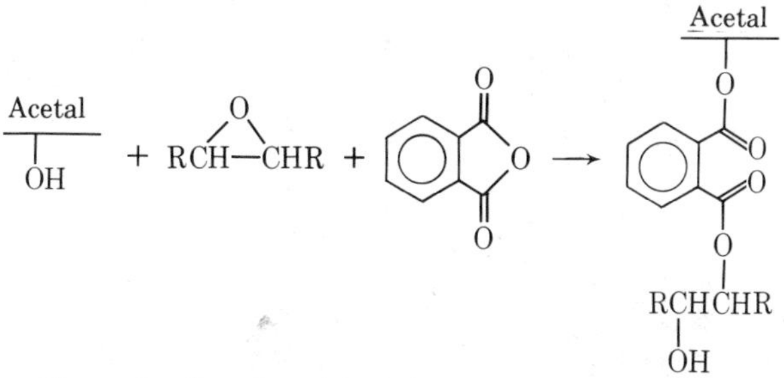

Figure 5.   Reaction with epoxy groups (anhydride cure).

Table 3. Properties of Commercial Poly(Vinyl Butyral) (Butvar)[a]

Property	Butvar B-72	Butvar B-74	Butvar B-73	Butvar B-76	Butvar B-79	Butvar B-90	Butvar B-98	ASTM method
Physical								
form			white, free-flowing powder					
volatiles (max), wt %	3.0	3.0	3.0	5.0	5.0	5.0	5.0	[b]
$\overline{M}_w$	180,000–270,000	100,000–150,000	50,000–80,000	45,000–55,000	34,000–38,000	38,000–45,000	30,000–34,000	[c]
soln viscosity (15 wt %), mPa·s (= cP)	8,000–18,000	4,000–8,000	1,000–4,000	500–1,000	100–200	600–1,200	200–450	
soln viscosity (10 wt %), mPa·s (= cP)	ca 1,570	ca 700	ca 400	ca 175	ca 55	ca 195	ca 75	[c]
sp gr (±0.002), 23°/23°	1.100	1.100	1.100	1.083	1.083	1.100	1.100	D 792-50
burning rate, cm/min	2.5	2.5	2.5	2.5	2.5	2.3	2.0	D 635-56T
refractive index (±0.0005)	1.490	1.490	1.490	1.485	1.485	1.490	1.490	D 542-50
water absorption (24 h), %	0.5	0.5	0.5	0.3	0.3	0.5	0.5	D 570-59aT
hydroxyl content expressed as % poly(vinyl alcohol)	17.5–21.0	17.5–21.0	17.5–21.0	9.0–13.0	9.0–13.0	18.0–20.0	18.0–20.0	D 1396-58[d]
acetate content expressed as % poly(vinyl acetate)	0–2.5	0–2.5	0–2.5	0–2.5	0–2.5	0–1.0	0–2.5	D 1396-58[d]
butyral content expressed as % poly(vinyl butyral)	ca 80	ca 80	ca 80	ca 88	ca 88	ca 80	ca 80	
Mechanical								
tensile strength, MPa[e]								D 638-58T
yield	47–54	47–54	45–52	40–47	40–47	43–50	43–50	
break	48–55	48–55	41–48	32–39	32–39	39–46	39–46	
elongation, %								D 638-58T
yield	8	8	8	8	8	8	8	
break	70	75	80	110	110	100	110	
modulus of elasticity (apparent), GPa[e]	2.28–2.35	2.28–2.35	2.20–2.28	1.93–2.00	1.93–2.00	2.07–2.14	2.14–2.21	D 638-58T
flexural strength (yield), MPa[e]	83–90	83–90	79–86	72–79	72–79	76–83	76–83	D 790-59T
hardness, Rockwell								D 785-51
M	115	115	115	100	100	115	110	
E	20	20	20	5	5	20	20	
impact strength Izod (notched 1.25 cm × 1.25 cm), J/m[f]	59	59	55	43	43	48	37	D 256-56

806

								ASTM method
Thermal								
flow temperature (at 6.9 MPa [e]), °C	145–155	135–145	125–130	110–115	110–115	125–130	105–110	D 569-59
glass temperature (apparent), °C	62–68	62–68	62–68	48–55	48–55	62–68	62–68	D 1043-51 [g]
heat-distortion temperature, °C	56–60	56–60	55–59	50–54	50–54	52–56	45–55	D 648-56
heat-sealing temperature, °C	105	105	99	93	93	96	93	[h]
Electrical								
dielectric constant								D 150-59T
50 Hz	3.2	3.2	3.0	2.7	2.7	3.2	3.3	
1 kHz	3.0	3.0	2.7	2.6	2.6	3.0	3.0	
1 MHz	2.8	2.8	2.6	2.6	2.6	2.8	2.8	
10 MHz	2.7	2.7	2.5	2.5	2.5	2.7	2.8	
dissipation factor								D 150-59T
50 Hz	6.4×10^{-3}	6.4×10^{-3}	5.8×10^{-3}	5.0×10^{-3}	5.0×10^{-3}	6.6×10^{-3}	6.4×10^{-3}	
1 kHz	6.2×10^{-3}	6.2×10^{-3}	5.5×10^{-3}	3.9×10^{-3}	3.9×10^{-3}	5.9×10^{-3}	6.1×10^{-3}	
1 MHz	27×10^{-3}	27×10^{-3}	22×10^{-3}	13×10^{-3}	13×10^{-3}	22×10^{-3}	23×10^{-3}	
10 MHz	31×10^{-3}	31×10^{-3}	22×10^{-3}	15×10^{-3}	15×10^{-3}	23×10^{-3}	24×10^{-3}	
dielectric strength (3.2 mm thickness), V/μm								D 149-59
short time	17	17	19	19	19	18	16	
step-by-step	16	16	16	15	15	15	15	

[a] Butvar is a registered trademark of Monsanto Co. (16).

[b] Molecular weight was determined by fractionating reacetylated samples of the poly(vinyl alcohol)s used for production of the various Butvar and Formvar resins. Distributions were based upon intrinsic viscosities of the various fractions, and weight-average molecular weights were calculated from these distribution curves.

[c] Solution viscosity was determined in 15 wt % solutions in 60:40 toluene–ethanol at 25°C with a Brookfield viscometer. Resin viscosity (5 g resin made to 100 mL with ethylene dichloride) was measured at 20°C with an Ostwald viscometer.

[d] The ASTM method noted for hydroxyl content and acetate content refers specifically to poly(vinyl butyral) resins. However, the same method is applicable to poly(vinyl formal) resins.

[e] To convert MPa to psi, multiply by 145; for GPa, by 145,000.

[f] To convert J/m to lbf/in., divide by 53.38.

[g] Apparent glass temperature (T_g) was determined by ASTM D 1043-51 and by differential scanning calorimetry (DSC). Results by DSC are 5–8°C higher than results from the ASTM method.

[h] Heat-sealing temperature was determined for a 25-μm dried film on paper, cast from a 10% solution in 60:40 toluene–ethanol. A dwell time of 1.5 s at 414-kPa (60-psi) line pressure was used for the heat sealer.

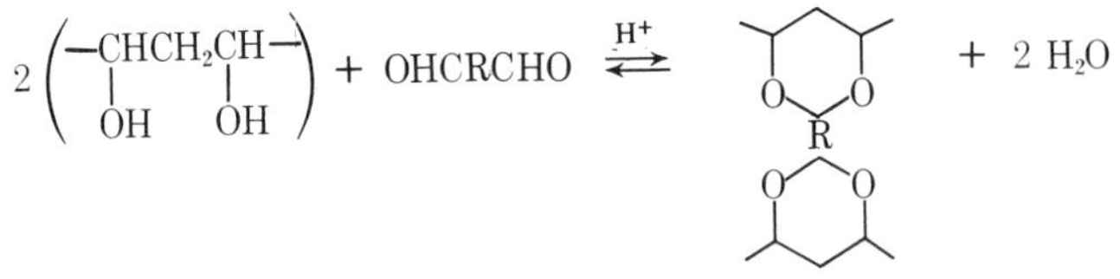

Figure 6.   Reaction with dialdehydes.

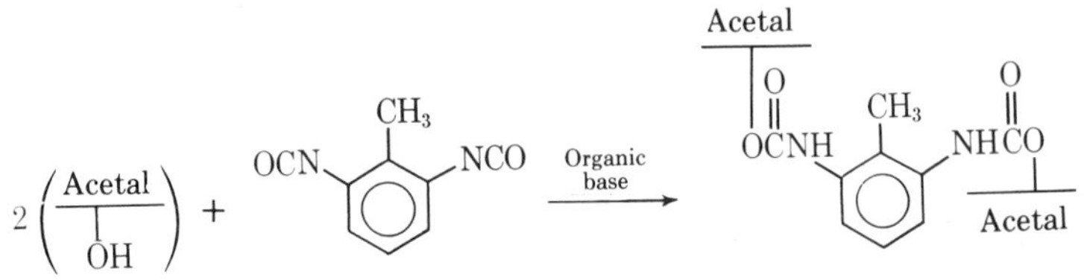

Figure 7.   Reaction with isocyanates.

content. The tensile strengths of unplasticized dispersions are 41–48 MPa (6000–7000 psi), whereas those containing 40–50 phr plasticizer break at ca 14 MPa (2000 psi). Elongations at break show a similar variation. The plasticizer content of poly(vinyl butyral) dispersions can be increased by emulsifying the required amount of plasticizer in water and adding it to the dispersion with gentle agitation. An emulsifying agent that is anionic or nonionic must be used to avoid coagulation of the dispersion. The mixed emulsion should be allowed to stand overnight to achieve uniform plasticization. Reactive thermosetting resins, eg, water-soluble or dispersible phenolics and melamines, can be incorporated to modify properties further. Poly(vinyl butyral) dispersions are widely used in the textile industry to impart increased abrasion resistance, durability, strength, and slippage control, and reduced color crocking, ie, reduced color transfer when rubbed.

Economic Aspects

Annual production of poly(vinyl formal) is ca 3500–4000 metric tons worldwide. Monsanto Company, with its trademark Formvar poly(vinyl formal) resins, is the only U.S. producer. The products are available as white powders at ca $4.50/kg.

Annual production of poly(vinyl butyral) is ca $(4.5–5) \times 10^4$ metric tons worldwide with DuPont and Monsanto the only manufacturers in the United States. DuPont makes poly(vinyl butyral) solely for captive use as a plasticized interlayer for safety-glass laminates. Monsanto also makes the plasticized interlayer and sells Butvar poly(vinyl butyral) resin for use in many other applications. The resin is available as a white, free-flowing powder at ca $5.50/kg.

Health and Safety Factors, Toxicology

In acute toxicity studies with rats and rabbits, Butvar poly(vinyl butyral) resin is practically nontoxic by ingestion in single oral doses ($LD_{50} > 10.0$ g/kg) or by single dermal applications ($LD_{50} > 7.94$ g/kg) (16). Butvar resin is only slightly irritating to the eyes (2.8 on a scale of 0–110) and nonirritating to the skins (0 on a scale of 0–8) of rabbits tested in standard FHSA tests for irritation. It seems to have no acute toxicological properties that would require special handling other than the good hygienic practices employed with any industrial chemicals.

In addition, curable coatings containing either Formvar or Butvar resin can be formulated to meet the extractibility requirement of the FDA. Butvar resins can be used in accordance with CFR regulations 175.105, 175.300, 176.170, and 176.180 as ingredients of can enamels, adhesives, and components of paper and paperboard in contact with aqueous and fatty foods.

Uses

Primers and Surface Coatings. Poly(vinyl formal) and poly(vinyl butyral) are used in primers and surface-coating formulations for metal, wood, plastic, concrete, and leather substrates. The films, mostly in combination with other resins, can be air-dried, baked, or cured at room temperature by proper compounding. The hydroxyl group in the polymer molecule furnishes a reactive site for chemical combination with thermosetting resins.

The best known application in protective coatings is in wash primers, which are also referred to as metal conditioners. Compared with other corrosion-inhibiting materials, wash primers are unique because in a single treatment they offer several means of preventing corrosion (see Corrosion and corrosion inhibitors). The action of wash primers over steel is discussed in reference 16. The U.S. Navy and Air Force have long recognized the need for the use of wash primers as a surface pretreatment for metals prior to subsequent painting; these wash primers are specified by them in MIL-P-15328C and MIL-C-8514B, respectively. Research in this area continues (25).

Poly(vinyl acetal) resins are used in a wide variety of metal-coating applications in combination with other resins, eg, phenolics, melamines, epoxies, isocyanates, etc, to improve coating uniformity, adhesion, toughness, and flexibility, and to minimize cratering. They can be compounded and formed into baked coatings that have good chemical resistance and withstand postforming. Drum and can linings can be formulated to meet the extractibility requirements of the FDA. Weldable corrosion-resistant primers for iron and steel have been developed (26). The electrostatic application of poly(vinyl butyral) powder compositions by spray and fluidized bed on metal has been described in USSR literature (27) (see Powder coatings).

Poly(vinyl butyral) is widely used as a component of wash coats and sealers in wood-finishing operations (16). The poly(vinyl butyral) confers good holdout, intercoat adhesion, moisture resistance, flexibility, toughness, impact resistance, and protection against wood discoloration. Details of uses on other substrates are given in references 28–31.

Safety Laminations. Much of the poly(vinyl butyral) resin produced in the world is plasticized, extruded into sheet, and used as the interlayer in safety-glass laminates (32). For this application, high mol wt polymer with 20 ± 5 wt % poly(vinyl alcohol) units is used. For many years, the plasticizer used almost universally was triethylene glycol bis(2-ethyl butyrate) (3). Recently, this has been supplanted by adipates, tetraethylene glycol derivatives, dibutyl sebacate, ricinoleates, and others (33–39). The laminates are made by bonding the interlayer between glass or plastic layers under heat and pressure; they have excellent properties of transparency, penetration resistance, and permanence over a wide temperature range (40) (see Laminated materials, glass).

The largest use for laminated safety glass is in automobile windshields. For this application, the strength of the adhesive bond between the glass and the interlayer is carefully controlled with additives, usually salts, to achieve the desired balance between high penetration resistance and retention of glass fragments upon impact (41–47). The interlayer is often printed with a color band to control sun glare (48–52).

The fastest growing application area is architectural, where poly(vinyl butyral) is used with pigments, dyes, and other additives to control light, energy, and sound transmission, as well as to provide safety (53–56). Poly(vinyl butyral) is also used increasingly and is often laminated with transparent plastics, eg, poly(methyl methacrylate) and polycarbonate, in locations where security and bullet resistance are important (57). When plastics are laminated, it is often necessary to choose special plasticizers to avoid crazing (38,58). A specially formulated interlayer is used in aircraft and certain military applications. A new use for plasticized poly(vinyl butyral) sheet is in solar cells and collectors (59).

Adhesives, Binders, and Sealants. The principal adhesive uses of poly(vinyl acetal)s are in high performance thermosetting adhesives and in hot-melt formulations (see Adhesives). The poly(vinyl acetal)s are combined with other components and provide toughness, flexibility, and high adhesive strength (16,60). Their use with olefins is described in refs. 61–62.

Combinations of poly(vinyl formal) or poly(vinyl butyral) with phenolic resins were among the first synthetic resin adhesives uses for the structural bonding of metals; they replaced rivets and provided stronger joints and reduced weight (63). Poly(vinyl acetal)s are compatible with many epoxys and can serve both as coreactants and flexibilizers. The formulations, properties, and uses are discussed in references 64–70.

Poly(vinyl butyral) is an excellent base for hot-melt adhesives, particularly for difficult-to-bond surfaces, and is usually formulated with plasticizers, waxes, and resins (16,71–72). Among other adhesive applications are mastics, caulking compounds, and propellant binders (73–75) (see Sealants).

Blends, Composites, Molded Parts, Fibers, and Films. Fibers with good dyeability or fire resistance have been patented in Japan (76–77) (see Flame retardants in textiles). Leather substitutes and leatherlike material are reported in USSR and Japanese literature (78–79) (see Leather; Leatherlike materials). The toughness of molding compounds based on trioxane copolymers, epoxies, and polyesters is improved by poly(vinyl acetal)s (80–82). Other examples of the use of poly(vinyl butyral) in these applications are discussed in references 83–84.

Foams, Filters, Membranes, and Sponges. Foams from the acetalization of poly-(vinyl alcohol) yield tough, resilient, and soft synthetic sponges (85–86) see Foamed plastics). A commercially available model was based on poly(vinyl formal). Applications in addition to household sponges include cigarette filters and powder puffs (86–87). Semipermeable membranes for reverse osmosis (qv) or ultrafiltration (qv) and ion-exchange membranes can also be made from poly(vinyl acetal)s (88–89) (see Ion exchange).

Electrical Insulation. The first and most extensive use for poly(vinyl formal) resins is in electrical insulation for magnet wire. A typical formulation for 105°C performance first described in 1943 consists of 100 parts of Formvar 15/95E and 50 parts of a cresol–formaldehyde resin dissolved in a mixture of cresylic acid and solvent naphtha (10). The wire enamels are coated on copper or aluminum wire and cured in ovens at elevated temperatures to give cross-linked film coatings with excellent electrical, physical, and chemical properties.

Poly(vinyl formal) resins have been formulated into many different wire enamels for specific application requirements. Typical developments are solderable enamels, high cut-through enamels, Freon-resistant enamels for hermetic motor applications, multicoat magnet wires for 155–180°C performance, and varnished or encapsulated magnet wire for higher temperature performance (11,90–91). Other developments include work on improving the overload resistance of magnet wires coated with Formvar and the use of poly(vinyl formal) and poly(vinyl butyral) to overcoat magnet wire so that coils made from them can be adhered with heat or solvents (92–94). Many new patents describe improvements in electrical insulation by enamels containing poly(vinyl formal) (95–99).

Reprographics, Toners, Dielectrics, and Photography. Poly(vinyl acetal)s are used for electrostatic reprographic copying in toners, as binders and coatings for organic and inorganic photoconductors, and as dielectric coatings in the direct electrostatic recording process (see Electrophotography). For example, a toner or developer for xerographic images may be composed of ca 5 wt % pigment dispersed in a tough, low molecular weight poly(vinyl butyral) blended with a phenol–formaldehyde resin. Descriptions of these and other related applications are given in references 100–109.

Printed Circuits. The poly(vinyl acetal) resins are used in printed circuits in a number of ways. They often are contained in the bonding varnishes used in the construction of the board itself. They are practically always used in the adhesives that bond the copper film to the board, and they occasionally are used in the photoresist layer that coats the copper film. Dielectric properties, peel strength, and blister resistance are all very important contributions made by the poly(vinyl acetal)s. Test methods for these type of properties have been standardized by the National Electrical Manufacturers Association and the American Society for Testing Materials. In recent years, the Japanese have published almost all of the research on printed-circuit boards (110–116) (see Integrated circuits).

Inks, Dyes, and Printing Plates. Poly(vinyl butyral) is used in flexographic, letterpress, and gravure inks to provide improved toughness, flexibility, abrasion resistance, and adhesion, and to produce good color yields in dye preparations for transfer printing of textiles (117–118) (see Dyes). Various poly(vinyl acetal)s, including the acetals of 2,4-disulfobenzaldehyde [88-39-1] and butyraldehyde [123-72-8], are used in light-sensitive and offset-printing plates.

Paper and Textiles. Formulations based on poly(vinyl butyral) have long been used to waterproof fabrics and make them stain-resistant without noticeably affecting appearance, feel, drape, or color (119) (see Textiles). The resin is usually compounded with a plasticizer and cross-linking resins, eg, urea–formaldehyde, phenol–formaldehyde, melamines or isocyanates, and (optionally) pigments (60). The coating is applied by calendering or knife coating, and is cured at 120°–175°C (16). Almost any fairly tightly woven fabric, including those based on cotton, wool, silk, nylon, viscose rayon, and other synthetics, can be made waterproof and stain-resistant in this way (16).

Poly(vinyl butyral) has been used in heat-sensitive paper, poly(vinyl formal) in water-resistant paper, poly(vinyl acetal) for restoring and preserving papers, and poly(vinyl butyral) in carbon paper (120–123) (see Papermaking additives). Aqueous dispersions of plasticized poly(vinyl butyral) are described in refs. 23–24. They provide tough, transparent films on textiles when dried at room temperature, and they impart increased abrasion resistance, durability, strength, and reduced color crocking (21).

Ceramics, Mold Additives, Cements, and Refractories. Poly(vinyl butyral) is used to toughen the phenolic binder for casting sand (124). With cement (qv), it is used as a coating to improve the corrosion resistance of concrete and as a sealant on asbestos-cement sheets (125–126). Machinable refractory articles of silicon nitride are made with plasticized poly(vinyl butyral) (127). With ceramics, a recent USSR patent describes the use of poly(vinyl acetal) and finely divided silver for metallization, but the largest application in this area is the use of plasticized poly(vinyl butyral) as the binder in the production of ceramic laminates for printed circuits (128–129) (see Ceramics).

Miscellaneous. Poly(vinyl acetal)s continue to be evaluated in many applications because of their unique combination of properties, eg, adhesion, film toughness and flexibility, pigment binding properties, etc. A few of the many miscellaneous applications that illustrate the poly(vinyl acetal) versatility are combustible cartridges, wound dressings, surgical sutures (qv), polarizing lenses, fluxes for welding and soldering, liquid crystals (qv), and phonograph-record cleaners (130–137).

Outlook

The poly(vinyl acetal) resins have been in use for >40 yr and can be considered commercially mature products. The bulk of published research has shifted from the United States and Western Europe to Japan and Eastern Europe, especially the USSR. Research of the use of poly(vinyl acetal)s in primers and coatings is greater than in any other application area. Basic research is second, followed closely by safety laminations and adhesives applications. The USSR has more publications of basic research than all other countries combined, followed by Japan.

Basic Research. The differences in kinetics and in statistical chain dimensions with the acetals formed from the isotactic and syndiotactic portions of the poly(vinyl alcohol) chains are discussed in references 138–140. Interaction of poly(vinyl butyral) chains in concentrated solutions were studied, and strong chain interactions in ethanol solution were determined even at 1 mg/cm^3 concentration (141). The viscosity of concentrated poly(vinyl butyral) solutions in different solvents depends not only on the polymer characteristics and concentration and temperature, but also on the molecular weight of the solvent; an equation relating these factors has been developed (142). The mechanical relaxation behavior of poly(vinyl acetal)s below their glass-transition temperatures is discussed in reference 143. With poly(vinyl formal), the dependence of creep on stress–strain curves was studied by the uniaxial compression method. As the stress–strain curve rises, the creep curve declines and reaches a limit, which depends on temperature and stress (144).

There are many articles and patents on new processes and techniques for preparing poly(vinyl acetal)s. Among these are three procedures that start with the poly(vinyl acetate) as an aqueous dispersion and two with poly(vinyl acetate) polymerized in acetic acid (145–146). Three processes of aqueous acetalization are described; in two of the processes, emulsifiers are added (147). One process claims improved plasticizer compatibility by acetalizing in a mixture of water and isobutyl alcohol in the presence of the anionic additives, eg, $ZnCl_2$, and another aqueous acetal process is claimed to provide better solubility and particle structure for the poly(vinyl butyral) as a result of the use of a strong acid catalyst in the presence of 0.1–3 wt % NH_4CNS (148–149). In one process, the continuous precipitation of poly(vinyl butyral) from solution is detailed (150).

In a series of articles from Poland, the reaction of poly(vinyl butyral) with aluminum alcoholates via the hydroxyl side chain is described. The kinetics of the reactions as well as thermal properties were studied (151). Several patents cover the use of phenol derivatives and phosphate buffers to improve the thermal stability of poly(vinyl butyral) (152–153). The analysis of poly(vinyl formal) by pyrolysis gas chromatography is described in ref. 154. Studies of the infrared spectra of poly(vinyl formal) and poly(vinyl butyral) were carried out in India and the USSR, respectively (155).

BIBLIOGRAPHY

"Poly(Vinyl Acetals)" in *ECT* 2nd ed., Vol. 21, pp. 304–317, by George O. Morrison, Technical Consultant.

1. Ger. Pat. 480,866 (July 20, 1924) and Ger. Pat. 507,962 (Apr. 30, 1927), W. O. Herrmann and W. Haehnel (to Consortium for Elektrochemische Industrie).
2. U.S. Pat. 2,036,092 (Mar. 31, 1936), G. O. Morrison, F. W. Skirrow, and K. G. Blaikie (to Canadian Electro Products); Reissue 20,430 (June 29, 1937).
3. U.S. Pat. 2,167,678 (June 13, 1939), H. F. Robertson (to Carbide and Carbon Chemical Corp.).
4. U.S. Pat. 2,114,877 (Apr. 19, 1938), R. W. Hall (to General Electric Co.).
5. U.S. Pat. 2,168,827 (Aug. 8, 1939), G. O. Morrison and A. F. Price (to Shawinigan Chemicals Ltd.).
6. U.S. Pat. 2,258,410 (Oct. 7, 1941), J. Dahle (to Monsanto Chemical Co.).
7. U.S. Pat. 2,396,209 (Mar. 5, 1946), W. H. Sharkey (to E. I. du Pont de Nemours & Co., Inc.).
8. U.S. Pat. 2,397,548 (Apr. 2, 1946), W. O. Kenyon and W. F. Fowler, Jr. (to Eastman Kodak Co.).
9. U.S. Pat. 2,496,480 (Feb. 7, 1950), E. Lavin, A. T. Marinaro, and W. R. Richard (to Shawinigan Resins Corp.).
10. U.S. Pat. 2,307,063 (Jan. 5, 1943), E. H. Jackson and R. W. Hall (to General Electric Co.).
11. E. Lavin, A. H. Markhart, and R. W. Ross, *Insulation* 8(4), 25 (1967).
12. N. Platzer, *Mod. Plast.* 28(10), 142 (1951).
13. H. Warson in S. A. Miller, ed., *Ethylene*, E. Benn Ltd., London, 1969.
14. M. K. Lindemann in N. Bikales, ed., *Encyclopedia of Polymer Science and Technology*, Vol. 14, Interscience Publishers, a division of John Wiley & Sons, Inc., New York, 1971, pp. 208–239.
15. P. J. Flory, *J. Am. Chem. Soc.* **61**, 1518 (1939).
16. *Butvar, Poly(Vinyl Butyral) and Formar, Poly(Vinyl Formal)*, Technical Bull. No. 6070D, Monsanto Co., St. Louis, Mo., June 1977.
17. A. F. Fitzhugh and R. N. Crozier, *J. Polym. Sci.* 8(2), 225 (1952).
18. Y. Ogata, M. Oksano, and T. Ganke, *J. Am. Chem. Soc.* **78**, 2962 (1956).
19. G. Smets and B. Petit, *Makromol. Chem.* **33**, 41 (1959).
20. A. F. Fitzhugh, E. Lavin, and G. O. Morrison, *J. Electrochem. Soc.* **100**, 8 (1953).
21. *Butvar Dispersion BR Resin*, Monsanto Data Sheet No. 6019-B, Monsanto Co., May 1977.
22. Brit. Pat. 233,370 (May 7, 1925), W. B. Pratt.
23. U.S. Pats. 2,455,402 (Dec. 7, 1948); 2,532,223 (May 28, 1950); 2,611,755 (Sept. 23, 1952), W. H. Bromley, Jr. (to Shawinigan Resins Corp.).
24. U.S. Pat. 3,234,161 (Feb. 8, 1966), J. A. Snelgrove and W. Whitney (to Monsanto Co.).
25. Jpn. Pat. 78 00,410 (Jan. 9, 1978), N. Hirota (to Mitsubishi Heavy Industries Ltd.).
26. U.S. Pat. 3,325,432 (June 13, 1967), M. D. Kellert and R. V. DeShay (to Monsanto Co.).; Brit. Pat. 1,093,200 (Nov. 20, 1967), W. Borkenhayen and co-workers (to VEB Lada-und Lackkunstharzfabrik.).
27. A. D. Yakovlev and co-workers, *Mater. Tech. Poluch Pokrytii Aerodispersii Polim.*, 10 (1975); *Chem. Abstr.* **85**, 161997a (1976); N. Y. Shakhov and co-workers, *Silnye Elektr. Polya Teknol. Protsessakh* **3**, 151 (1979); E. G. Ilina and L. S. Koretskaya, *Lakokras. Mater. Ikh Primen.*, (1), 36 (1981).

28. Ger. Pat. 2,144,233 (Mar. 8, 1973), H. Hinichs, J. Peter, and W. D. Schuessler (to Reichhold-Albert Chemie A.-G.).
29. U.S. Pat. 3,313,651 (Apr. 11, 1967), R. J. Burns (to Union Carbide Corp.).
30. M. A. Veber and co-workers, *Sb. Tr. Leningr. Inzh. Stroit. Inst.* **86,** 55 (1973).
31. T. P. Shvetsova and T. I. Zhdanova, *Kozh-Obuvn. Promst.* **18**(6), 22 (1976).
32. C. E. Schildknecht, *Vinyl and Related Polymers*, John Wiley & Sons, Inc., New York, 1952, pp. 323–385.
33. U.S. Pat. 3,920,876 (Nov. 18, 1975), R. H. Fariss and J. A. Snelgrove (to Monsanto Co.).
34. U.S. Pat. 4,144,217 (Mar. 13, 1979), J. A. Snelgrove and D. I. Christensen (to Monsanto Co.).
35. U.S. Pat. 3,841,955 (Oct. 15, 1974), A. W. M. Coaker, J. R. Darby, and T. C. Mathis (to Monsanto Co.).
36. U.S. Pat. 4,230,771 (Oct. 28, 1980), T. R. Phillips (to E. I. du Pont de Nemours & Co., Inc.).
37. Jpn. Pat. 71 42,901 (Dec. 18, 1971), K. Takaura, T. Misaka, and S. Ando (to Sekisui Chem. Co. Ltd.).
38. U.S. Pat. 4,128,694 (Dec. 5, 1978), D. A. Fabel, J. A. Snelgrove, and R. H. Fariss (to Monsanto Co.).
39. Eur. Pat. Appl. 11,577 (May 28, 1980), D. Dages (to Saint-Gobain Industries S.A.).
40. Jpn. Pat. 80 51,740 (Apr. 15, 1980), N. Nurishi and K. Azuma (to Nippon Sheet Glass Co. Ltd.).
41. U.S. Pat. 3,262,837 (July 26, 1966), E. Lavin, G. E. Mont, and A. F. Price (to Monsanto Co.).
42. U.S. Pat. 3,249,487 (May 3, 1966), F. T. Buckley and J. S. Nelson (to Monsanto Co.).
43. Jpn. Pat. 75 121,311 (Sept. 23, 1975), I. Karasudani, T. Takashima, and Y. Honda (to Sekisui Chem. Co., Ltd.).
44. Ger. Pat. 2,410,153 (Sept. 4, 1975), R. Beckmann and W. Knackstedt (to Dynamit Nobel A.-G.).
45. U.S. Pat. 3,718,516 (Feb. 27, 1973), F. T. Buckley, R. F. Riek, and D. I. Christensen (to Monsanto Co.).
46. Ger. Pat. 2,904,043 (Aug. 9, 1979), H. K. Inskip (to E. I. du Pont de Nemours & Co., Inc.).
47. Ger. Pat. 2,646,280 (Apr. 20, 1978), H. D. Hermann and J. Ebigt (to Hoechst A.-G.).
48. U.S. Pat. 3,982,984 (Sept. 28, 1976), D. B. Baldridge (to Monsanto Co.).
49. U.S. Pat. 3,973,058 (Aug. 3, 1976), J. L. Grover and W. H. Power (to Monsanto Co.).
50. Ger. Pat. 2,837,768 (Mar. 15, 1979), J. R. Mannheim.
51. U.S. Pat. 3,591,406 (July 6, 1971), R. E. Moynihan (to E. I. du Pont de Nemours & Co., Inc.).
52. Ger. Pat. 2,841,287 (Apr. 3, 1980), D. S. Postupack (to PPG Industries, Inc.).
53. U.S. Pat. 3,523,847 (Aug. 11, 1970), J. W. Edwards (to Monsanto Co.).
54. Ger. Pat. 1,301,022 (Aug. 14, 1969), F. T. Buckley, I. L. Seldin, and R. B. Wojcik (to Monsanto Co.).
55. U.S. Pat. 3,932,690 (Jan. 13, 1976), G. Gliemeroth (to Jenaer Glaswerk Scott und Gen.).
56. Ger. Pat. 2,034,998 (Feb. 3, 1972), R. Quenett (to Deutsche Tafelglas A.-G.).
57. Ger. Pat. 2,903,115 (Mar. 13, 1980), H. Rodemann, H. D. Funk, and G. Breitenbuerger (to BFG Glassgroup).
58. U.S. Pat. 4,251,591 (Feb. 17, 1981), H. K. Chi (to Monsanto Co.).
59. A. M. Lindrose and T. R. Guess, *Natl. SAMPE Symp. Exhib.* **23,** 386 (1978).
60. E. Lavin and J. A. Snelgrove in I. Skeist, ed., *Handbook of Adhesives*, 2nd ed., Van Nostrand Reinhold Co., New York, 1977, Chapt. 31.
61. A. D. Yakovlev and co-workers, *Plast. Massy*, (8), 75 (1975).
62. K. Tenchev and co-workers, *Adhaesion*, (11), 368 (1971).
63. Brit. Pat. 577,823 (June 3, 1946), N. A. de Bruyne.
64. *Technical Note No. 144*, Ciba (A.R.L.) Ltd., Dec. 1954.
65. *The Tropical Durability of Metal Adhesives*, RAF Technical Note No. Chem. 1349, Royal Air Force, England, U.K., Feb. 1959.
66. H. W. Eichner, *Environmental Exposure of Adhesive-Bonded Metal Lapjoints*, WADC Technical Report 59, Pt. I, Wright Air Development Center, Dayton, Ohio, 1960, p. 564.
67. W. Whitney and S. C. Herman, *Adhes. Age* **34**(1), 22 (1960).
68. Jpn. Pat. 76 105,344 (Sept. 17, 1976), K. Ito (to Aica Kogyo Co. Ltd.).
69. Jpn. Pat. 79 96,541 (July 31, 1979), T. Nakamura, K. Iwata, and A. Horike (to Teijin, Ltd.).
70. Jpn. Pat. 79 139,884 (Oct. 30, 1979), T. Miyazaki, T. Tanaka, and S. Sakurada (to Koyo Sangyo Co., Ltd.).
71. Czech. Pat. 155,587 (Dec. 15, 1974), M. Schatz, K. Salz, and J. Volek.
72. Jpn. Pat. 74 13,247 (Feb. 5, 1974), M. Sera.
73. Rom. Pat. 54,052 (Feb. 2, 1972), M. V. Mateescu and co-workers.

74. Jpn. Pat. 77 109,533 (Sept. 13, 1977), T. Miyanaga (to Lion Dentrifice Co. Ltd.).
75. U.S. Pat. 3,960,088 (June 1, 1976), W. L. Greever (to U.S. Dept. of the Navy).
76. Jpn. Pat. 73 91,383 (Nov. 28, 1973), K. Hirakawa and K. Ohno (to Kuraray Co. Ltd.).
77. Jpn. Pat. 74 118,999 (Nov. 13, 1974), M. Sumi and co-workers (to Unitika Co. Ltd.).
78. M. M. Bernshtein and co-workers, *Izv. Vyssh. Uchebn. Zaved. Tekhnol. Legk. Promsti.*, (5), 22 (1974).
79. Jpn. Pat. 72 46,896 (Nov. 27, 1972), N. Michima and co-workers (to Kanebo. Co. Ltd.).
80. Ger. Pat. 2,233,813 (Jan. 31, 1974), G. Sextro and co-workers (to Farbwerke Hoechst A.-G.).
81. Y. K. Esipov and co-workers, *Elektrotekhnika*, (6), 39 (1979).
82. Jpn. Pat. 81 10,515 (Feb. 3, 1981), (to Toshiba Chemical K.K.).
83. USSR Pat. 562,536 (June 25, 1977), Y. F. Sokolov, G. M. Khutortsov, and V. F. Kurdenkov.
84. U.S. Pat. 3,382,298 (May 7, 1968), H. R. Larsen and R. S. Zalkowitz (to Union Carbide Canada Ltd.).
85. Brit. Pat. 973,951 (Nov. 4, 1964), Kalle A.G.
86. Jpn. Pat. 74 64,586 (Oct. 24, 1974), K. Matsuzaki (to Kanebo Co., Ltd.).
87. Jpn. Pat. 75 37,712 (Dec. 4, 1975), N. Fujisawa (to Kanebo Co., Ltd.).
88. Fr. Pat. 2,245,676 (Apr. 25, 1975), (to Babcock and Wilcox, Ltd.).
89. Jpn. Pat. 76 93,791 (Aug. 17, 1976), K. Motani and co-workers (to Tokuyama Soda Co. Ltd.).
90. U.S. Pat. 3,069,379 (Dec. 18, 1962), E. Lavin, A. F. Fitzhugh, and R. N. Crozier (to Shawinigan Resins Corp. and Phelps Dodge Copper Products Corp.).
91. U.S. Pat. 3,104,326 (Sept. 17, 1963), E. Lavin, A. H. Markhart, and R. F. Kass (to Shawinigan Resins Corp.).
92. R. V. Carmer and E. W. Daszewski, *National Conference on the Application of Electrical Insulation*, Chicago, Ill., 1960.
93 C. F. Hunt, A. F. Fitzhugh, and A. H. Markhart, *Electro-Technology* **60**(3), 131 (1962).
94. U.S. Pat. 3,516,858 (June 23, 1970) and U.S. Pat. 3,516,858 (Jan. 11, 1972), A. F. Fitzhugh and J. A. Snelgrove (to Monsanto Co.).
95. Jpn. Pat. 75 96,628 (July 31, 1975), Y. Ueba and M. Kowaguchi (to Somitono Electric Industries Ltd.).
96. Sp. Pat. 412,082 (Jan. 1, 1976), J. C. Oromi.
97. U.S. Pat. 4,129,678 (Dec. 12, 1978), M. Seki and co-workers (to Hitachi Ltd.; Hitachi Cable Ltd.).
98. U.S. Pat. 4,254,007 (Mar. 3, 1981), R. G. Flowers and W. A. Fessler (to General Electric Co.).
99. USSR Pat. 753,878 (Aug. 7, 1980), E. Ya Shvaitsburg and co-workers.
100. U.S. Pat. 2,753,308 (July 3, 1956), R. B. Landrigan (to the Haloid Corp.).
101. Jpn. Pat. 80 159,453 (Dec. 11, 1980) and Jpn. Pat. 81 01,948 (Jan. 10, 1981), (to Hitachi Metals Ltd.).
102. U.S. Pat. 3,290,147 (Dec. 31, 1966), J. A. Mattor, D. B. Henderson, and B. Millard (to S. D. Warren Co.).
103. Jpn. Pat. 73 26,536 (Aug. 11, 1973), K. Maki and T. Watanabe (to K. K. Canon).
104. U.S. Pat. 3,639,640 (Feb. 1, 1972), M. E. Gager (to Plastic Coatings Corp.).
105. U.S. Pat. 3,951,882 (Mar. 8, 1973), A. H. Markhart and D. R. Cahill (to Monsanto Co.).
106. R. H. Windhager and G. D. Sinkovitz, *TAPPI Fall Coat. Graphic Arts Division Week Prepr.*, Philadelphia, Pa., 1975, p. 15.
107. R. B. Rosenfeld, *Res. Discl. (England)* **148**, 27 (1976).
108. U.S. Pat. 3,653,902 (Apr. 4, 1972), N. T. Notley and I. M. Senentz (to Kalvar Corp.).
109. Ger. Pat. 2,315,233 (Oct. 11, 1973), Y. Takegawa and co-workers (to Oriental Photo Industrial Co.).
110. Jpn. Pat. 76 33,180 (Mar. 22, 1976), K. Kuroiwa and K. Kitsugi (to Shin Etsu Polymer Co. Ltd.).
111. Jpn. Pat. 74, 99,636 (Sept. 20, 1974), H. Takahashi and A. Yamanaka (to Hitachi Chemical Co. Ltd.).
112. Jpn. Pat. 73 90,331 (Nov. 26, 1973), H. Takahashi and Y. Nakana (to Hitachi Chemical Co. Ltd.).
113. Jpn. Pat. 79 112,939, 79 112,940, and 79 112,941 (Sept. 4, 1979), N. Uozu and S. Takanezawa (to Hitachi Chemical Co. Ltd.).
114. Jpn. Pat. 69 73,241 (June 9, 1979), Y. Uozu and S. Takanezawa (to Hitachi Chemical Co. Ltd.).
115. Jpn. Pat. 76 39,025 (Apr. 1, 1976), E. Hasegowa and M. Murata (to Fugi Photo Film Co.).
116. Ger. Pat. 2,730,725 (Jan. 12, 1978), J. V. Crivello (to General Electric Co.).
117. USSR Pat. 804,681 (Feb. 15, 1981), N. M. Dzyuba, S. A. Satushev, and N. S. Zadernovskaya.
118. Ger. Pat. 2,547,862 (May 6, 1976), K. Taubert (to Sandoz-Patent-GmbH).

119. P. S. Plumb, *Ind. Eng. Chem.* **36**, 1035 (1944).
120. USSR Pat. 184,609 (July 21, 1966), B. B. Gutman and E. A. Goldfarb.
121. Brit. Pat. 1,283,405 (July 26, 1972), Kuraray Co. Ltd.
122. U.S. Pat. 3,862,916 (Jan. 28, 1975), C. B. Hayworth (to World Patent Dev. Corp.).
123. Ger. Pat. 2,016,725 (Oct. 21, 1971), G. Kaase and H. Lobschat (to Pelikan-Werke).
124. Jpn. Pat. 76 45,616 (Apr. 19, 1976), R. Takahashi and S. Okazaki (to Hitachi Metals Ltd.).
125. M. A. Veber and co-workers, *Sb. Tr. Leningr. Inzh. Stroit. Inst.* **86,** 55 (1973).
126. U.S. Pat. 3,413,140 (Nov. 26, 1968), W. A. Heausler and R. M. Johnson (to National Gypsum Co.).
127. Brit. Pat. 1,274,212 (May 17, 1972), C. O. Silverstone and J. S. O'Neill (to U.K. Atomic Energy Authority).
128. USSR Pat. 744,741 (June 30, 1980), D. T. Kostin and co-workers.
129. Ger. Pat. 2,227,343 (Jan. 18, 1973), L. C. Anderson, R. W. Nufer, and F. G. Pugliese (to I.B.M. Corp.).
130. U.S. Pat. 3,474,702 (Oct. 28, 1969), R. F. Remaly, W. P. Shefcik, and M. B. Nelson (to U.S. Dept. of the Army).
131. Ger. Pat. 1,939,916 (Feb. 4, 1971), H. Mueller (to Beiersdorf A.-G.).
132. Fr. Pat. 1,589,917 (May 15, 1970), (to Henkel and Cie GmbH).
133. S. Fritsch, *Pharmazie* **22**(1), 41 (1967).
134. U.S. Pat. 3,300,436 (Jan. 24, 1967), A. M. Marks and M. M. Marks.
135. USSR Pat. 360,187 (Nov. 28, 1972), V. P. Makinov and V. P. Lezhnikov; USSR Pat. 359,117 (Dec. 3, 1972), I. A. Khuzman and co-workers.
136. U.S. Pat. 4,161,557 (Mar. 28, 1978), F. K. Susuki and T. W. Thomas (Liquid Crystal Products, Inc.).
137. Jpn. Pat. 80 157,698 (Dec. 10, 1980), (to Shin-Etsu Polymer Co., Ltd.).
138. K. Fugii, J. Ukida, and M. Matsumoto, *Macromol. Chem.* **65**, 86 (1963).
139. K. Shibatani and co-workers, *IUPAC Symposium on Macromolecular Chemistry*, Tokyo, Japan, 1966, No. 5-5-03.
140. H. Matsuda and H. Inagaki, *J. Macromol. Sci. Chem.* **A2**(1), 191 (1968).
141. V. I. Irzhak and co-workers, *Dokl. Akad. Nauk SSSR* **198**(3), 626 (1971).
142. V. A. Stolyarova, A. D. Yakovlev, and I. S. Okhrimenko, *Zh. Prikl. Khim. (Leningrad)* **46**(3), 595 (1973).
143. G. A. Olshanik and co-workers, *Mater. Vses. Soveshch. Relaksatsionnym Yareniyam Polim. 2nd 1971* **2**, 144 (1974); *Chem. Abstr.* **85**, 78508w (1976).
144. V. N. Borsenka, A. B. Sinani, and V. A. Stepanov, *Mekh. Polim.* **4–5**, 787 (1968).
145. A. G. Sayadyan and co-workers, *Arm. Khim. Zh.* **22**(6), 535 (1969); R. A. Kulman and co-workers, *Kolloid Zh.* **30**(6), 860 (1968); USSR Pat. 418,487 (Mar. 5, 1974), V. M. Ostrovskaya and co-workers.
146. E. A. Akopyan and co-workers, *Arm. Khim. Zh.* **22**(8), 727 (1969); L. A. Sarkisyan and co-workers, *Prom. Arm.*, (4), 48 (1974).
147. Jpn. Kokai 74 90,792 (Aug. 29, 1974), Y. Kodera and co-workers (to Sekisui Chem. Co. Ltd.); Ger. Pat. 2,838,025 (Mar. 15, 1979), P. Dauveigne (to Saint-Gobain Industries S.A.); Ger. Pat. 2,732,717 (Feb. 8, 1979), H. D. Herman, J. Ehigt, and M. U. Hutten (to Hoechst A.-G.).
148. N. I. Tyazhlo and co-workers, *Zh. Prikl. Khim. (Leningrad)* **47**(10), 2285 (1974).
149. Pol. Pat. 96,247 (May 31, 1978), H. Pietkiewicz, M. Knypl, and A. Madeja.
150. Jpn. Pat. 73 72,244 (Sept. 29, 1973), K. Mizuno, H. Asahara, and J. Nakamura (to Unitika Ltd.).
151. R. Hippe, *Bull. Acad. Pol. Sci. Ser. Sci. Chim.* **21**(7–8), 537 (1973); **23**(1), 77 (1975); **23**(4), 355 (1975).
152. Jpn. Pat. 79 122,398 (Sept. 21, 1979), L. Kurosudani and A. Marahara (to Sekisui Chemical Co. Ltd.); USSR Pat. 759,534 (Aug. 30, 1980), Z. R. Uspenskaya and co-workers.
153. Ger. Pat. 2,636,336 (Feb. 24, 1977), G. E. Mont and J. A. Snelgrove (to Monsanto Co.).
154. I. Takahashi and co-workers, *Kogyo Kagaku Zasshi* **73**(2), 303 (1970).
155. M. R. Padhye and B. B. Iyer, *Curr. Sci.* **41**(14), 528 (1972); V. I. Grachev and co-workers, *Vysokomol. Soedin. Ser.* **14**(6), 462 (1972).

EDWARD LAVIN
JAMES A. SNELGROVE
Monsanto Company

POLY(VINYL ACETATE)

Vinyl Acetate Monomer

$$O$$
$$\|$$

Vinyl acetate (VA), CH_2=$CHOCCH_3$, is a colorless, flammable liquid having an initially pleasant odor which quickly becomes sharp and irritating. Its chief use is as a monomer for making poly(vinyl acetate) [9003-20-7] (PVAc) and vinyl acetate copolymers, which are widely used in water-based paints, adhesives (qv), paper coatings or nonwoven binders, and applications not requiring service at extreme temperatures. Poly(vinyl acetate) is the precursor for poly(vinyl alcohol) and poly(vinyl acetal) resins (PVA) (see Vinyl polymers, poly(vinyl alcohol); Vinyl polymers, poly-(vinyl acetals)). Vinyl acetate is also copolymerized as the minor constituent with vinyl chloride and with ethylene to form commercial polymers and with acrylonitrile (qv) to form acrylic fibers (see Acrylic and modacrylic fibers).

Vinyl acetate was first reported in 1912 in a patent describing the preparation of ethylidene diacetate from acetylene and acetic acid; the vinyl acetate was a minor by-product (1). Industrial interest in vinyl acetate monomer and poly(vinyl acetate) developed by 1925 and processes for their production were perfected (2). World production of vinyl acetate has increased steadily and rapidly since ca 1950. In 1969–1970, ca 4.5×10^5 metric tons of monomer was produced worldwide (3).

Physical Properties. Some physical properties of vinyl acetate are listed in Table 1. The values of vapor pressure, heat of vaporization, vapor heat capacity, liquid heat capacity, liquid density, vapor viscosity, liquid viscosity, surface tension, vapor thermal conductivity, and liquid thermal conductivity over temperature ranges have been calculated and graphed (9).

Vinyl acetate is completely miscible with organic liquids but not with water. At 20°C, a saturated solution of vinyl acetate in water contains 2.0–2.4 wt % vinyl acetate, whereas a saturated solution of water in vinyl acetate contains 0.9–1.0 wt % water (4–7). At 50°C, the solubility of vinyl acetate in water is 0.1 wt % more than at 20°C, but the solubility of water in VA doubles to ca 2 wt %. Vinyl acetate is soluble in a dilute (2.0 wt %) solution of sodium dodecyl sulfate to the extent of 4.0 wt % at 30°C (10). Azeotropes containing vinyl acetate are listed in Table 2.

Polymerization can be initiated by organic and inorganic peroxides; azo compounds; redox systems, which may include organometallic components; light; and high energy radiation (see Initiators). Polymerization is inhibited or strongly retarded by aromatic hydroxyl, nitro, or amine compounds and by oxygen, quinone, crotonaldehyde, copper salts, sulfur, conjugated polyolefins, and enynes. Recently compiled tabulations of quantitative information, eg, polymerization-rate constants, chain-transfer constants, and activation energies for the polymerization reactions of vinyl acetate, are given in refs. 12–13. There often is disagreement among the findings of different investigators.

Vinyl acetate has been polymerized in bulk, suspension, solution, and emulsion. It copolymerizes readily with some monomers but not with others. Some reactivity

Table 1. Physical Properties of Vinyl Acetate

Property	Value	Ref.
boiling point, °C	72.7	4–7
melting point, °C	−100, −93	4–7
specific gravity, 20/20	0.9338	4–7
refractive index, n_D^{20}	1.3952	4–7
viscosity 20°C, mPa·s (= cP)	0.42	4–7
flash point, tag open cup, °C	0.5–0.9	4–7
heat of vaporization at 72°C, J/g[a]	379.1	4–7
heat of combustion, kJ/g[a]	24.06	4–7
heat of polymerization, kJ/mol[a]	89.12	6, 8
critical temperature, °C, estd	252	9
critical pressure, kPa[b]	4200	9
critical density, g/cm^3, estd	0.324	9

[a] To convert J to cal, divide by 4.184.
[b] To convert kPa to psi, multiply by 0.145.

Table 2. Vinyl Acetate Azeotropes[a]

Second component	Azeotropic boiling point, °C	Vinyl acetate, wt %
water	66.0	92.7
methanol	58.9	63.4
2-propanol	70.8	77.6
cyclohexane	67.4	61.3
heptane	72.0	83.5

[a] Ref. 11.

Table 3. Reactivity Ratios in Vinyl Acetate Copolymerizations[a]

Comonomer	R_{VA}	r_M	Temperature, °C
acrylic acid	0.1	2	70
acrylonitrile	0.07	6	70
chloroprene	0.01	50	65
diallyl phthalate	0.72	2.0	
diethyl fumarate	0.01	0.44	60
ethylene	1	1	90–150
ethyl vinyl ether	3	0	60
isobutyl methacrylate	0.025	30	60
isopropenyl acetate	1	1	75
maleic anhydride	0.07	0.01	
styrene	0.01	55	60
vinyl chloride	0.6	1.4	40
N-vinylpyrrolidinone	0.2	3.3	50

[a] Refs. 12–13.
[b] M = monomer.

818

ratios are presented in Table 3 for common.comonomers (see Copolymers). The Q (monomer reactivity factor) and e (electronic factor) values are 0.026 and −0.22, respectively.

Pure vinyl acetate does not absorb uv significantly at wavelengths longer than 250 nm in ethanol solvent or than 253 in hexane solvent (ie, log e = 0 at the points given, and uv absorption rapidly ends at shorter wavelengths) (14–15).

Chemical Properties. The most important chemical reaction of vinyl acetate is free-radical polymerization. The scientific, as distinguished from the technological, aspects of the subject have been reviewed (12). The reaction is shown in equation 1:

$$n\ \mathrm{CH_2{=}CHOCCH_3} \longrightarrow \underset{\substack{|\\ \{CH_2CH\}_n}}{\overset{\substack{O\\ \|\\ CH_3CO}}{}} \tag{1}$$

Other chemical reactions of vinyl acetate are those common to an ester and a compound containing a double bond; that the alcohol portion of the ester is unsaturated does, however, lend the molecule characteristics unlike those in ordinary esters. Thus, vinyl acetate is hydrolyzed by acidic and basic catalysis forming acetic acid; the expected alcohol, ie, vinyl alcohol, is unstable and acetaldehyde instantaneously forms by tautomeric rearrangement:

$$\mathrm{CH_2{=}CHOCCH_3} \xrightarrow[\mathrm{H^+,\,OH^-}]{\mathrm{H_2O}} [\mathrm{CH_2{=}CHOH}] + \mathrm{CH_3COH} \tag{2}$$

$$\downarrow$$

$$\mathrm{CH_3CHO}$$

Thus, the hydrolysis rate of vinyl acetate is 1000 times that of its saturated analogue, ethyl acetate, in alkaline media (16). The rate of hydrolysis is minimal at pH 4.44 (17).

Chemical reactions involving the addition of a reagent to the double bond are known. Chlorine or bromine adds readily giving the 1,2-dihaloethyl acetate, which can be distilled under vacuum without decomposition. Hydrogen chloride or hydrogen bromide adds readily and quantitatively at low temperatures yielding the 1-haloethyl acetate (18). Some 2-haloethyl acetate forms at higher temperatures. The ozonide of VA is explosive when dry.

Vinyl acetate can be used to prepare other vinyl esters by transfer of its vinyl group to another organic acid; a mercuric salt is necessary as catalyst (19–20). Numerous vinyl carboxylates have been prepared by this method, as shown in equation 3:

$$\mathrm{CH_3COCH{=}CH_2} + \mathrm{RCOH} \xrightarrow[\mathrm{H^+}]{\mathrm{Hg^{2+}}} \mathrm{RCOCH{=}CH_2} + \mathrm{CH_3COH} \tag{3}$$

Acetic acid adds to the double bond with acid catalysis giving ethylidene diacetate,

$$CH_3CH(O\overset{O}{\overset{\|}{C}}CH_3)_2$$

Oxidation of vinyl acetate with palladium acetate at 60°C produces 1,4-diacetoxyl-1,3-butadiene and significant amounts of 1,1,4-triacetoxy-2-butene, 1,1,4,4-tetraacetoxybutane, and 1-acetoxy-1,3-butadiene (21). By use of diiodomethane with a zinc–copper couple, a methylene is added to vinyl acetate yielding cyclopropyl acetate (22). Polyhaloalkanes, eg, CCl_4 and CCl_3Br, react with vinyl acetate by a free-radical mechanism to yield 1:1 adducts or telomers having the fragments of the polyhalomethane as end groups (23). Numerous other molecules that can participate in free-radical chains similarly add across the double bond or are active in telomerization with vinyl acetate. Vinyl acetate undergoes the oxo reaction forming acetoxypropionaldehydes (see Oxo process). Thiols add across the double bond with Lewis acid catalysis and by a free-radical-catalyzed reaction to form thioacetates. Similar additions are known for ammonia, amines, silanes, and metal carbonyls (17). Vinyl acetate also serves as a weak dieneophile in Diels-Alder reactions, generally requiring forcing conditions.

Manufacture.   The manufacturing process most widely used for vinyl acetate is the oxidative addition of acetic acid to ethylene in the presence of a palladium catalyst (eq. 4). Liquid-phase and vapor-phase processes are used commercially:

$$CH_3\overset{O}{\overset{\|}{C}}OH + H_2C{=}CH_2 + \tfrac{1}{2}\,O_2 \longrightarrow CH_3\overset{O}{\overset{\|}{C}}OCH{=}CH_2 + H_2O \qquad (4)$$

The oldest process, which is still used, is the addition of acetic acid to acetylene (eq. 5):

$$CH_3\overset{O}{\overset{\|}{C}}OH + HC{\equiv}CH \longrightarrow CH_3\overset{O}{\overset{\|}{C}}OCH{=}CH_2 \qquad (5)$$

The third method of commercial synthesis is a ten-step process, in which acetic anhydride is combined with acetaldehyde forming ethylidene diacetate, which is then pyrolyzed to vinyl acetate and acetic acid (eqs. 6–7):

$$CH_3CHO + (CH_3\overset{O}{\overset{\|}{C}}O)O \longrightarrow CH_3CH(O\overset{O}{\overset{\|}{C}}CH_3)_2 \qquad (6)$$

$$CH_3CH(O\overset{O}{\overset{\|}{C}}CH_3)_2 \longrightarrow CH_2{=}CHO\overset{O}{\overset{\|}{C}}CH_3 + CH_3\overset{O}{\overset{\|}{C}}OH \qquad (7)$$

Acetic Acid Addition to Ethylene.   Addition of acetic acid to ethylene is catalyzed by palladium salts and came into widespread usage in the 1970s (see Acetaldehyde). Producers estimate that the use of ethylene instead of acetylene for vinyl acetate synthesis led to a 20% reduction in raw-material costs. As a result, essentially all post-1980 capacity for monomer will be based on ethylene feed.

As with acetylene, vinyl acetate processes based on both liquid-phase and vapor-phase ethylene feed are known. Liquid-phase processes have been developed by Hoechst (FRG), Imperial Chemicals Industries, Ltd. (UK), and Nippon Gosei (Japan). The vapor-phase processes developed by Bayer and Hoechst cooperatively (FRG) and National Distillers and Chemicals (U.S.I. Chemicals, United States) are being used exclusively on a commercial basis.

In the liquid-phase process (see Fig. 1), a mixture of ethylene and oxygen at ca 3 MPa (30 atm) is fed into the single-stage reactor, which contains acetic acid, water, and catalyst at 100–130°C. The products, ie, vinyl acetate and acetaldehyde, are separated from the exiting gas stream in a series of distillation columns (24). Acetic acid and water are also fed into the reactor; the proportion of water in the catalytic solution determines the ratio of acetaldehyde to vinyl acetate formed. The acetaldehyde can be converted to acetic acid, which is fed back to the vinyl acetate process, thus providing a method for producing monomer from ethylene and oxygen. A product ratio of 1.14 mol of acetaldehyde per mol of vinyl acetate is optimum for such production and is within the limits of variability available by control of the water feed.

The catalyst solution contains palladium salts at a concentration of 30–50 mg Pd^{2+}/L and soluble copper salts at a concentration of 3–6 g Cu^{2+}/L. Chloride ion is also necessary; its presence in the form of HCl used to cause serious corrosion problems downstream from the reactor when the liquid-phase processes were first operated, but use of proper materials of construction, ie, titanium, resin–graphite composites, and ceramics, has overcome this problem (25).

The ethylene–oxygen feed is maintained outside the ignition limits of the mixture, ie, at 5.5% oxygen or less, and this limits the conversion of ethylene per pass to 2–3%. The unreacted gas, however, carries the product out of the bubble-column reactor and is cleaned for recycle without much reduction of pressure. Too little oxygen leads to precipitation of cuprous chloride in the reactor.

Overall yields are 90% based on ethylene and 95% based on acetic acid. Capital costs are ca 50% higher than required by the acetylene vapor-phase process and energy consumption is higher. However, as long as the current large price difference between

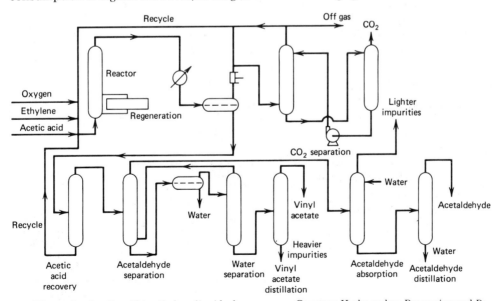

Figure 1.   Acetic acid to ethylene liquid-phase process. Courtesy *Hydrocarbon Processing* and *Petroleum Refiner* (24).

ethylene and acetylene exists, this route may be advantageous economically. The main chemical reactions occurring in this process are shown by equations 8–11:

$$CH_2{=}CH_2 + CH_3\overset{O}{\overset{\|}{C}}OH + PdCl_2 \longrightarrow CH_2{=}CHO\overset{O}{\overset{\|}{C}}CH_3 + Pd + 2\,HCl \qquad (8)$$

$$CH_2{=}CH_2 + H_2O + PdCl_2 \rightarrow CH_3CHO + Pd + 2\,HCl \qquad (9)$$

$$Pd + 2\,CuCl_2 \rightarrow PdCl_2 + 2\,CuCl \qquad (10)$$

$$2\,CuCl + 2\,HCl + \tfrac{1}{2}\,O_2 \rightarrow 2\,CuCl_2 + H_2O \qquad (11)$$

The reaction of equation 8 occurs in a molecular complex of Pd^{2+} with acetate and ethylene, where acetate displaces hydride which is simultaneously transferred to the central palladium atom (26). Equations 9–10 show the *in situ* regeneration of the catalytic system by oxidation of the precipitated palladium metal by cupric ion and oxidation of the cuprous ion by oxygen. The overall change is shown in equation 8.

The vapor-phase process differs from that described above in that very little acetaldehyde forms. The extensive licensing by Japanese interests of the vapor-phase processes follows from the use pattern of poly(vinyl acetate) in Japan: ca 80% is converted to poly(vinyl alcohol). The acetic acid is then available for recycling for monomer synthesis, as shown in Figure 2. A gaseous mixture of acetic acid, ethylene, and oxygen is blown over the catalyst in a tubular reactor and the exit stream, containing vinyl acetate, unreacted starting materials, water, and small amounts of acetaldehyde, carbon dioxide, and other by-products, is separated by a combination of scrubbers and distillation stages. Ethylene at ca 0.6–1.1 MPa (5–10 atm gauge) is saturated with acetic acid at ca 120°C and is preheated before entering the reactor. Oxygen is added

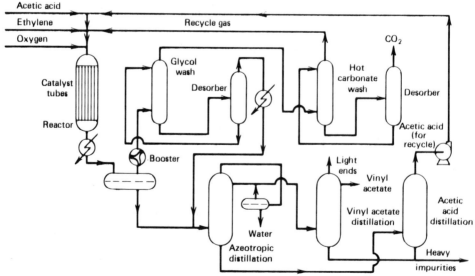

Figure 2.   Acetic acid to ethylene vapor-phase process. Courtesy *Hydrocarbon Processing* and *Petroleum Refiner* (24).

just in front of the reactor; the amount that can be added is limited by the explosive limit. At 785 kPa (8.74 atm gauge), 7 vol % oxygen is the maximum amount, thus limiting attainable conversions of ethylene and acetic acid. The gases react over a catalyst of 0.1–2 wt % palladium metal on an inert support at 175–200°C. The catalyst is held in a multitube reactor, and the heat of reaction is used to generate steam for other process needs. Oxygen reacts to 60–70% per pass, 20% of acetic acid reacts and 10% of the ethylene, with space–time yields of over 200 g vinyl acetate/L catalyst per hour. About 10% of the reacted ethylene is converted to carbon dioxide. When the gases exit from the reactor, they are cooled and the liquid and gaseous phases separated. The gaseous portion is scrubbed with propylene glycol to separate vinyl acetate and with hot carbonate solution to remove CO_2 and then is recycled. The combined liquids, ie, the initial condensate and desorbed vinyl acetate, are distilled and the acetic acid is recycled. In contrast to the liquid-phase, ethylene-based process, corrosion is not a problem and the normal materials required for hot acetic acid are used in construction.

Acetic Acid Addition to Acetylene. A minor portion of the world's vinyl acetate production in 1980 was based on acetic acid addition to acetylene. The catalytic vapor-phase process operates at 180–210°C with an acetylene-to-acetic-acid molar feed ratio of (4–5):1 (27). The catalyst is zinc acetate supported on about three times its weight of activated carbon. A flow diagram is shown in Figure 3.

During operation of the process, acetylene is sparged through acetic acid at a temperature that gives the desired ratio of the reactants in the exiting vapor. Typically, operating conditions of 70–80°C at 129–136 kPa (4–5 psig) are used. The vaporized charge is preheated to 180–210°C before being passed through the reactor in which the catalyst is contained. The preheat temperature depends on the age of the catalyst, with the higher temperatures used toward the end of the catalyst life. The reactor may consist of a parallel array of 5-cm dia, 4-m long tubes contained in a shell, through which a heat-exchange fluid is circulated. The reaction operates at 115–122 kPa (2–3 psig) and at a space velocity of 300–400 h^{-1} giving acetic acid conversions of 80% or more. Gas temperatures increase 5–10°C during passage through the reactor. The overall yields are 92–98% based on acetylene and 95–99% based on acetic acid. Catalyst life is a few months. The vinyl acetate product is purified by distillation in a series of columns which also separate the unreacted acetylene and acetic acid for recycle and

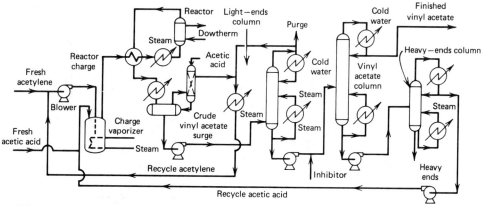

Figure 3. Acetic acid to acetylene vapor-phase vinyl acetate synthetic process. Courtesy *Petroleum Refiner* (27).

the principal by-products, acetaldehyde and ethylidene diacetate. Stainless steel is used extensively because of the corrosive operating conditions; contact of cuprous metals with monomer after distillation is avoided because of the potent inhibiting action of copper on polymerization, but copper still pots are used to prevent polymerization during distillation. Cadmium salts have also been used as catalysts for the vapor-phase, acetic acid–acetylene reaction as have other supports, eg, silica gel or activated alumina.

The liquid-phase process for the acetic acid–acetylene route to vinyl acetate involves a mercuric salt catalyst, which precipitates in the reaction medium upon addition of sulfuric acid, oleum, or phosphoric acid (13). Acetylene gas is blown into the reactor and the vinyl acetate product is entrained by unreacted acetylene and is carried to the separation section. The vinyl acetate should not remain in the reaction vessel for too long as it would react further forming ethylidene diacetate. Ethylidene diacetate forms in ca 10–30% yields. Reaction temperatures of 30–75°C have been used and batch or continuous operation is possible. The process is commercially obsolete.

Acetaldehyde and Acetic Anhydride.   The one remaining commercial acetaldehyde and acetic anhydride process for vinyl acetate is operated by the Celanese Corporation in Mexico (see Fig. 4) (28). Acetaldehyde and acetic anhydride react in the presence of a catalyst at an elevated temperature forming ethylidene diacetate. This product is passed to a cracking tower where pyrolysis occurs and vinyl acetate and acetic acid form. Separation and purification are carried out in a number of distillation steps.

Other Methods.   Vinyl acetate can be made in excellent yields by the reaction of vinyl chloride and sodium acetate in solution at 50–75°C in the presence of catalytic amounts of palladium chloride (29) (eq. 12):

$$CH_2{=}CHCl + CH_3\overset{\overset{\displaystyle O}{\|}}{C}O^- \xrightarrow{\ PdCl_2\ } \cdot CH_2{=}CHO\overset{\overset{\displaystyle O}{\|}}{C}CH_3 + Cl^- \qquad (12)$$

Economic Aspects.   U.S. vinyl acetate production by the acetic-acid-addition-to-ethylene route in 1980 is shown in Table 4. U.S. production of vinyl acetate by vapor-phase reaction of ethylene, acetic acid, and oxygen is expected to total 8.6 ×

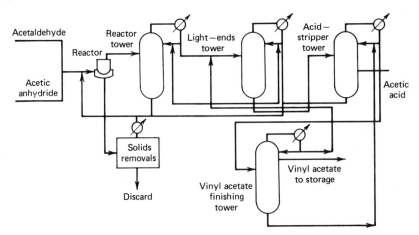

Figure 4.   Process for vinyl acetate via ethylidene diacetate. Courtesy *Hydrocarbon Processing* and *Petroleum Refiner* (24).

Table 4.   U.S. Production of Vinyl Acetate Monomer by Acetic Acid–Ethylene Process[a]

Company	Location	Process	Production, 1000 metric tons
Celanese	Bay City, Texas and Clear Lake, Texas	modified Bayer	385.9
Union Carbide	Texas City, Texas	modified Bayer	204.3
E. I. du Pont de Nemours & Co., Inc.	La Porte, Texas	modified Bayer	181.6
National Distillers	Strange, Texas	National Distillers	272.4
Borden	Geismar, La.	Borden/Blau-Knox	68.1
National Starch	Long Mott, Texas	Wacker	27.2
Total			*1140*

[a] Ref. 30.

10^5 t in 1982 (31). With the shutdown of Borden Chemical's vinyl acetate unit, which is based on acetylene, total U.S. nameplate capacity has declined to a bit more than 1.0×10^6 t/yr. Union Carbide's plant will increase its capacity to ca 4.5×10^4 t by the end of 1982. U.S. exports in 1981 exceeded 2.7×10^5 t.

Capacity in Japan in 1978 was ca 681,000 t, 80% ethylene-based Western European overall capacity, 75% of which is ethylene-based was ca 545,000 t in 1978.

In 1978, 7.7×10^5 t of vinyl acetate was produced in the United States, 4.54×10^5 t in Western Europe and 4.5 t in Japan (Fig. 5). The U.S. list price of the monomer in 1970 was 27¢/kg (railroad tank cars, delivered); monomer was sold under contract at 18¢/kg. As shown in Figure 5, prices continued to fall until 1973, when oil prices in the United States took a dramatic upturn. From 1960 when vinyl was 33¢/kg, it reached a low of 16¢/kg in 1972, then rebounded to 44¢/kg in 1978 and to ca 62¢/kg in 1980 (see Table 5).

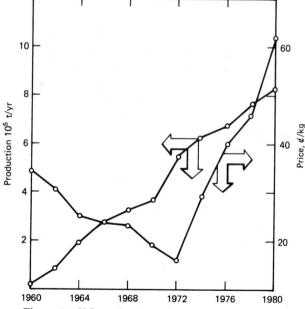

Figure 5.   U.S. production and prices of vinyl acetate.

Table 5.   Economics of U.S. Vinyl Acetate Production[a]

Year	Production, 1000 t	Price, ¢/kg
1960	113.95	34.0
1962	144.37	30.9
1964	199.76	25.3
1966	275.12	23.8
1968	325.97	23.6
1970	364.56	18.9
1972	549.34	15.6
1974	636.96	28.7
1976	671.92	40.2
1978	763.17	44.4
1980	848.98	62.2

[a] Ref. 32.

Specifications and Standards.   Vinyl acetate monomer is supplied in three grades, which differ in the amount of inhibitor they contain but otherwise have identical specifications. A low p-hydroquinone grade, containing 3–7 ppm p-hydroquinone, is preferred if it is expected to be used within two months of delivery. For monomer stored up to four months before polymerization, the grade containing 12–17 ppm p-hydroquinone is used. When indefinite storage is anticipated, a grade containing 200–300 ppm diphenylamine is supplied. The diphenylamine must be removed by distillation before the monomer is used for polymerization; the p-hydroquinone-inhibited grades need not be distilled before use. Typical manufacturers' specifications are given in Table 6.

Vinyl acetate is commonly stored in carbon-steel tanks. Baked-phenolic-coated steel, aluminum, glass-lined tanks, and stainless steel are also suitable construction materials. Copper and its alloys cause discoloration as well as inhibition of polymerization if brought into solution and should not be used unless the monomer is distilled before use. The flammable limits of vinyl acetate vapors in air are 2.6–13.4 vol %. At normal storage temperatures, an explosive vapor–air mixture can exist in the vapor space of a storage tank unless air is excluded. Thus, nitrogen blanketing should be considered. All tanks and lines should be electrically bonded and grounded, and explosion-proof equipment (UL class 1, group D) should be used.

Tests indicate that inhibited vinyl acetate has good storage stability at normal temperatures, ie, below 30°C. For aboveground tanks, water-spray cooling or exterior white paint is sometimes used to minimize storage temperatures during hot summer

Table 6.   Typical Manufacturers' Specifications for Vinyl Acetate

Specification	Value
vinyl acetate, wt %, min	99.8
boiling point, °C	72.3–73.0
acidity as acetic acid, wt %, max	0.007
carbonyls as acetaldehyde, wt %, max	0.013
water, wt %, max	0.04
color, APHA system	0–5
suspended matter	none

weather. The unloading and storage precautions used are those common to the handling of a volatile, flammable liquid having a low flash point and autoignition temperature.

Vinyl acetate requires a red precautionary label for shipment. It should not be stored without an inhibitor or transferred between vessels by means of air pressure. It can be obtained in tank cars, tank trucks, and drums. Additional information concerning the safe handling of vinyl acetate is provided in ref. 33.

Analytical and Test Methods. Gas chromatography is an excellent method for determining vinyl acetate and its volatile impurities simultaneously. If wet chemical techniques are used, vinyl acetate is assayed by adding excess bromine to an aliquot of material, followed by addition of excess potassium iodide and titration with standard sodium thiosulfate. Acidity is determined by direct titration in methanol solution with standard caustic; aldehydes are determined by the addition of excess sodium bisulfite followed by titration with standard iodine solution; and water is determined by the Karl Fischer method. p-Hydroquinone and diphenylamine can be determined by standard titration techniques, or spectrophotometrically in the uv region, after evaporation of vinyl acetate. Several companies describe in their product brochures empirical procedures for determining the polymerization activity of the monomer. In these tests, a given amount of monomer is polymerized under standard conditions and the rate of temperature increase or volume shrinkage and the induction period are measured. Appropriate ASTM specifications and test procedures are D 2190, D 2191, D 2193, and D 2083.

Health and Safety Factors, Toxicology. Vinyl acetate is only moderately toxic if ingested or if absorbed through the skin. It is not a severe irritant on contact with skin or eyes; however, prolonged or repeated contact with the skin should be avoided. The vapor is not especially irritating to the nose and throat, but good industrial hygiene practices should be followed and adequate ventilation provided. Repeated or prolonged exposure should be avoided.

The LD_{50} for oral ingestion in rats is 2.9 g/kg body weight; for absorption through the skin, the LD_{50} in rats is more than 5 mL/kg in 24 h. Breathing a vapor concentration of 1000 ppm in air was not fatal to any animals tested. At 4000 ppm, half of the test animals died within two hours. First-aid procedures to be followed in the event of overexposure to vinyl acetate are

Type of exposure	Treatment
ingestion	induce vomiting
inhalation	provide fresh air; keep victim warm and quite, apply artificial respiration if necessary
skin contact	wash with water
eye contact	flush with water for 15 min

Uses. Polymerization is the only use for vinyl acetate. Products that retain their character as poly(vinyl acetate) make up 55–62% of monomer production. Most of this use is in emulsion polymerization processes. Poly(vinyl alcohol) requires 18–20% of the monomer produced, the poly(vinyl acetal)s ca 8%, and vinyl chloride copolymers ca 8%. Ethylene–vinyl acetate copolymers are a small use but are growing at 5%/yr. Additional quantities are used in polymeric lube-oil additives and in acrylonitrile copolymers for acrylic fibers.

Vinyl Acetate Polymers

Properties. Poly(vinyl acetate)s vary with increasing values of molecular weight from viscous liquids and from low melting solids to tough, horny materials. They are neutral, water-white to straw-colored, tasteless, odorless, and nontoxic. The resins have no sharply defined melting points but become softer with increasing temperature. They are soluble in organic solvents, eg, esters, ketones, aromatics, halogenated hydrocarbons, carboxylic acids, etc, but are insoluble in the lower alcohols (excluding methanol), glycols, water, and very nonpolar liquids such as ether, carbon disulfide, aliphatic hydrocarbons, oils, and fats. Alcohols, eg, ethyl, propyl, and butyl, containing 5–10 wt % water dissolve PVAc; butyl alcohol or xylene, both of which only swell the polymer at normal temperatures, dissolve it when heated. Some physical properties are listed in Table 7. The electrical properties are strongly affected by the ability of poly(vinyl acetate) to absorb water. Whereas dried resin has a dielectric constant ϵ of 3.3 and a loss factor ϵ' of 0.08 ± 0.02 at 35°C and 60 Hz, after exposure to 100% rh the figures are 10 and 2.7, respectively (34–35).

As with many thermoplastic resins, strength properties increase with molecular weight; tensile strengths up to 50.3 MPa (7300 psi) may be obtained. The softening point, as determined by the ring-and-ball method or the Kraemer and Sarnow method, also increases with molecular weight, as shown in Table 8. Poly(vinyl acetate) is commercially available in pure, dry form as beads, granules, or lumps and is graded according to the viscosity at 20°C of a solution containing 86.09 g or one mole of resin dissolved in benzene to make one liter; the viscosity grades correspond to the molecular weights shown in Table 7. In Europe, the Fikentscher K value, also derived from viscosity measurements, is used to characterize commercial resins.

When heated above room temperature, all viscosity grades become very flexible; and at 50°C, they become limp. They can be heated at 125°C for hours without changing, but at 150°C they gradually darken and at over 225°C they liberate acetic acid forming a brown insoluble resin which carbonizes at a much higher temperature. On cooling below room temperature, ie, at 10–15°C, poly(vinyl acetate)s become brittle. The brittle point may be lowered by plasticization or copolymerization. The products of thermal decomposition of PVAc are acetic acid at 150–200°C, and aromatic compounds, ie, benzene, toluene, naphthalene, etc, at 300–350°C (38). Aging qualities of PVAc are excellent because of its resistance to oxidation and its inertness to the effects of uv and vis radiation.

The nmr spectrum of PVAc in carbon tetrachloride solution at 110°C shows absorptions at 4.86 δ (pentad) of the methine proton; 1.78 δ (triad) of the methylene group; and 1.98 δ, 1.96 δ, and 1.94 δ which are the resonances of the acetate methyls in isotactic, heterotactic, and syndiotactic triads, respectively. Poly(vinyl acetate) produced by normal free-radical polymerization is completely atactic and noncrystalline. The nmr spectra of ethylene–vinyl acetate copolymers have also been obtained (39). The ir spectra of the copolymers of vinyl acetate differ from that of the homopolymer depending on the identity of the comonomers and their proportion.

The chemical properties of PVAc are those of an aliphatic ester. Thus, acidic or basic hydrolysis produces PVA and acetic acid or the acetate of the basic cation. Industrially, poly(vinyl alcohol) is produced by a base-catalyzed ester interchange with methanol, where methyl acetate forms in addition to the polymeric product. The chemical properties of PVAc can be modified by copolymerization. When a comonomer

Table 7. Physical Constants of Poly(Vinyl Acetate)

Property	Value
absorption of water at 20°C for 24–144 h, %	3–6
coefficient of thermal expansion, K^{-1}	
cubic	6.7×10^{-4}
linear, below T_g	7×10^{-5}
above T_g	22×10^{-5}
cohesive energy density, $(MJ/m^3)^{1/2}$ [a]	18.6–19.09
compressibility, $cm^3/(g \cdot kPa)$ [b]	17.8×10^{-6}
decomposition temperature, °C	150
density, g/cm^3	
at 20°C	1.191
25°C	1.19
50°C	1.17
120°C	1.11
200°C	1.05
dielectric constant (at 2 MHz)	
at 50°C	3.5
150°C	8.3
dielectric dissipation factor (at 2 MHz), tan δ	
at 50°C	150
120°C	260
dielectric strength, V/L	
at 30°C	0.394
60°C	0.307
dipole moment, $C \cdot m$ [c] per monomer unit	
at 20°C	2.30
150°C	1.77
dynamic mechanical loss peak (at 100 Hz), °C	70
elongation at break (at 20°C and 0% rh), %	10–20
glass-transition temperature, T_g, °C	28–31
pressure dependence, °C/100 MPa [d]	0.22
hardness (at 20°C), Shore units	80–85
heat capacity (at 30°C), J/g [a]	1.465
heat distortion point, °C	50
heat of polymerization, kJ/mol [a]	87.5
refraction index, n_D	
at 20.7°C	1.4669
30.8°C	1.4657
52.1°C	1.4600
80°C	1.4480
142°C	1.4317
interfacial tension, mN/m (= dyn/cm)	
at 20°C with polyethylene	14.5
20°C with polydimethylsiloxane	8.4
20°C with polyisobutylene	9.9
20°C with polystyrene	4.2
internal pressure, MJ/m^3 [a]	
at 0°C	255
28°C	397.8
60°C	418.7
20°C	284.7
40°C	431.3
modulus of elasticity, GPa [d]	1.275–2.256
notched impact strength, J/m [e]	102.4
softening temperature, °C	35–50
specific volume, L/kg	

Table 7 (*continued*)

Property	Value
at t = 100–200°C	$0.823 + (6.4 \times 10^{-4})\,t$
t = 28°C (T_g)	0.84
surface resistance (ohm/cm)	5×10^{11}
surface tension, mN/m (= dyn/cm)	
at 20°C	36.5
140°C	28.6
180°C	25.9
tensile strength, MPa[d]	29.4–49.0
thermal conductivity, mW/(m·K)	159
Young's modulus, MPa[d]	600

[a] To convert J to cal, divide by 4.184.
[b] To convert kPa to atm, divide by 101.3.
[c] To convert C·m to debye, divide by 3.336×10^{-30}.
[d] To convert MPa to psi, multiply by 145; GPa to psi, multiply by 145,000.
[e] To convert J/m to lbf/in., divide by 53.38.

Table 8. Softening Points and Molecular Weights of Commercial Poly(Vinyl Acetate)

Grade viscosity[a], mPa·s (= cP)	Fikentscher[b], K value	Softening point, °C		$\overline{M}_w$
		Kraemer and Sarnow[c]	Ring-and-ball	
1.5	13	65	75	11,000
2.5	19	81	90	18,000
7	32	106	116	45,000
15	42	131	139	90,000
25	48	153	163	140,000
60	58	196		300,000
100	62	230		500,000
800	79			1,500,000

[a] A 1 M soln in benzene at 20°C.
[b] Ref. 36.
[c] Ref. 37.

having a carboxylic acid group or a sulfuric acid group is used, the copolymer becomes soluble in dilute aqueous alkali or ammonia. These copolymers also adhere better to metals than homopolymers or neutral copolymers because of the interaction between the acid groups and the metal surface.

Many properties of PVAc and various copolymer emulsions are determined not by the characteristics of the pure polymer it contains but by the aqueous phase and its contents. The specific gravity at 20°C for all the emulsions is about 1.1. There generally is a slight odor of residual monomer but if this is removed by some means, such as steam stripping, the emulsions are virtually odorless. The traces of acetic acid can be neutralized with a base, eg, ammonia, sodium bicarbonate, or triethanolamine, and when the free monomer has been removed, the pH of the resulting emulsion can be adjusted to remain constant at ca 7. By the judicious selection of monomer concentration, protective colloid, emulsifying and wetting agents, method of polymerization and post-treatment, the properties can be varied to suit the polymer's intended use. Such properties include average particle size and particle size range, polymer molecular weight, pH, emulsion viscosity, particle charge, adhesion, speed of tack, solvent tolerance, film characteristics and water resistance, and stability to storage, freezing, dilution, mechanical action, and compounding. Various resins, plasticizers, thickening agents, solvents, pigments, extenders, and some dyes may be added, but caution is necessary and it is advisable to follow closely the manufacturer's directions. Plasticizers are often added to provide increased flexibility to the dried film, but these usually are not incorporated before the final formulation is made, since their presence may adversely affect stability, particularly in cold weather.

When the emulsion is applied to a surface, water is lost by evaporation and by absorption if the surface is porous. The particles stick together and eventually coalesce to form a tough and somewhat clear continuous coating. The clarity is often improved by the presence of plasticizer, which also enhances the film's resistance to water. Films from most emulsions containing poly(vinyl alcohol) as the protective colloid are likely to re-emulsify on contact with water, unless they contain relatively large quantities of plasticizer or solvent.

Poly(vinyl acetate) emulsion films adhere well to most surfaces and have good binding capacity for pigments and fillers. Plasticized films are strong and flexible at ordinary temperatures. The films are unaffected by light, oxygen, chlorine in moderation, and dilute solutions of acids, alkalies, and salts. The films are also inert to oils, fats, waxes, and greases, unless these are mostly aromatic. Solvents, eg, acetone, alcohol, ethyl acetate, benzene, and toluene dissolve or at least swell the film. An important property of a PVAc film is its permeability to water vapor. This allows the film to be laid down on a damp surface with trapped moisture gradually passing through the film without lifting or blistering it. The permeability to saturated water vapor at 40°C is 2.1 $g/(h \cdot m^2)$ for a film 0.025 mm thick.

Polymerization Processes. Vinyl acetate has been polymerized industrially in bulk, solution, suspension, and emulsion (40). Perhaps 90% of the material identified as poly(vinyl acetate) or copolymers that are predominantly vinyl acetate are made by emulsion techniques. The patent literature abounds in descriptions of this technology, and recipes and procedures are readily available in the monomer brochures of producing companies as well as in the literature (4,7,17,40).

The emulsions (qv) are milk-white liquids containing ca 55 wt % PVAc, the balance being water and small quantities of wetting agents or protective colloids. Their use eliminates the need for expensive, flammable, odorous, or toxic solvents and the need for the recovery of such solvents. They are easy to apply and the equipment is easy to clean with water, if done promptly. Emulsions also offer the advantage of high solids content with fluidity, since the viscosity of the emulsion is independent of the molecular weight of the resin.

An emulsion recipe, in general, contains monomer, water, protective colloid or surfactant, initiator, buffer and, perhaps, a molecular weight regulator. The monomer may consist of 30–60 wt % of the charge, but most commercially available emulsions contain ca 55 wt % solids. Several monomers are copolymerized commercially with vinyl acetate in emulsion polymerization and numerous others have been copolymerized on a laboratory scale. Among the comonomers most commonly used industrially in emulsion copolymerization with vinyl acetate are ethylene; dibutyl maleate; bis(2-ethylhexyl) maleate; ethyl, butyl, and 2-ethylhexyl acrylates; vinyl laurate; and Vinyl Versatate. Vinyl hydrogen maleate and vinyl hydrogen fumarate have also been used as comonomers, as have acrylic acid and sodium ethylenesulfonate, in order to incorporate an acidic or ionic group in the polymer. The neutral comonomers are added primarily to decrease the brittle temperature of the polymer below commonly encountered ambient temperatures, since many uses of PVAc require some degree of flexibility in service. The monomers that contain acidic groups are added primarily to make the copolymer soluble in basic media, eg, aqueous ammonia. Copolymerization, compared to softening which can also be accomplished by addition of plasticizers such as dibutyl phthalate, tricresyl phosphate, etc, to the preformed polymer, gives a polymer which is innately and permanently plasticized. Plasticization, whether internal (by copolymerization) or external (with additives), is also extremely important for proper performance at the time of application. The ease of coalescence and the wetting characteristics of the polymer emulsion particles are related to their softness and the chemical nature of the plasticizer. The most efficient plasticizing comonomer on a weight or price basis is ethylene; copolymerization of vinyl acetate with ethylene is the most significant in the adhesive market (41).

Many different combinations of surfactant and protective colloid are used in emulsion polymerizations of vinyl acetate (see Surfactants and detersive systems). The properties of the emulsion and of the polymeric film which forms from it depend to a large extent on the identity and quantity of the emulsifiers. These properties include average value and distribution of particle size; stability of the emulsion under conditions of mechanical shear, change of temperature, compounding, and time; characteristics of the coalesced resin after application, eg, smoothness, opacity, water resistance, etc, and characteristics of the emulsion as it is being used, eg, flow properties and setting time (42).

Poly(vinyl acetate) emulsions can be made with a surfactant alone or with a protective colloid alone, but the usual practice is to use a combination of the two. Normally, up to 3 wt % emulsifiers may be included in the recipe, but when water sensitivity or wet tack of the film is desired, as in some adhesives, more may be included. The most commonly used surfactants (qv) are the anionic sulfates and sulfonates, but cationic emulsifiers and nonionics are also suitable. Indeed, some emulsion compounding formulas require the use of cationic or nonionic surfactants for stable formulations. The most commonly used protective colloids are poly(vinyl alcohol) and hydroxyethyl cellulose, but again, there are many others, natural and synthetic, which are usable if not preferable for a given application.

In general, the greater the quantity of emulsifiers in a recipe, the smaller the particle size of the emulsion. At very low levels of emulsifier, eg, 0.1 wt %, the polymer does not form a creamy dispersion that stays indefinitely suspended in the aqueous phase but is in the form of small beads that settle and may be easily separated by filtration. This suspension or pearl polymerization process has been used to prepare polymers for adhesive and coating applications and for conversion to PVA. Products in bead form are available from several commercial suppliers of PVAc resins. At the higher emulsifier levels (≥1 wt %), used for emulsion polymerization, the polymer forms tiny particles that do not settle but remain indefinitely suspended. Particle sizes resulting from high surface-active-agent-to-low-protective-colloid ratios may be 0.005–1 μm; such emulsions contain lower solids and have lower viscosities. Commercial PVAc emulsions usually are made with higher ratios of protective colloids in the recipes and usually contain ca 55 wt % solids. The average particle sizes are 0.2–10 μm and the viscosity of the emulsion is 400–5000 mPa·s (= cP). These latter compositions, ie, with higher ratios of protective colloids, are occasionally described as stable dispersions rather than true emulsions, the latter designation is reserved for the former compositions, but this distinction is generally ignored in commercial characterization (4,7,43–47). The term latex is also used to denote these products, particularly those with small, ie, <0.2 μm, particle size.

The initiators or catalysts used in vinyl acetate polymerizations are the familiar free-radical types, eg, hydrogen peroxide, peroxysulfates, benzoyl peroxide, and redox combinations. Emulsion polymerizations are conducted with water-soluble catalysts; benzoyl peroxide has been used in emulsion polymerizations with water-soluble catalysts, especially where monomer has been added continuously during the reaction. Suspension polymerizations, on the other hand, are run with monomer-soluble initiators predominantly.

Buffers are frequently added to emulsion recipes and serve two main purposes. The rates of decomposition of some initiators are affected by pH and the buffer is added to stabilize those rates, since decomposition of catalyst frequently causes changes in pH. The rate of hydrolysis of vinyl acetate and of some other monomers is also pH-sensitive, and it is desirable to minimize hydrolysis of monomer since it produces acetic acid which can affect the catalyst and acetaldehyde which may lower the molecular weight of the polymer undesirably. Emulsion recipes are usually buffered to pH 4–5, eg, with phosphate or acetate, but buffering at neutral pH with bicarbonate also gives excellent results. The pH of most commercially available emulsions is 4–6.

Often a chain-transfer agent is added to vinyl acetate polymerizations, whether emulsion, suspension, solution, or bulk, to control the polymer molecular weight. Aldehydes, thiols, carbon tetrachloride, etc, have been added; when solution polymerization is carried out, the solvent acts as a chain-transfer agent and, depending on its transfer constant, has an effect on the molecular weight of the product. The rate of polymerization is also affected by the solvent but not in the same way as the degree of polymerization: the reactivity of the solvent-derived radical plays an important part here. Chain-transfer constants for solvents in vinyl acetate polymerizations have been tabulated (13). Some emulsion procedures call for the recipe to include a quantity of preformed PVAc emulsion, and sometimes antifoamers must be added.

A polymerization may be carried out simply by charging all ingredients to the reactor, heating to reflux, and stirring until the reaction is over; however, this simple procedure is seldom followed. Typically, only a portion of the monomer and catalyst is initially charged and the remainder is added during the course of the reaction. Better control of the rate of polymerization can be maintained in this fashion, which is particularly important in industrial-scale operations where large quantities of material are involved and heat-dissipation capacity may be limited. Continuous monomer addition in emulsion polymerization also leads to smaller particle size and a more stable dispersion. On the molecular level, delayed monomer feed results in more branches in the polymer chain. Consequently, the rate of monomer addition has some effect on final film properties. Copolymerizations usually must be conducted with a continuous monomer feed to obtain homogeneous polymer compositions. Emulsifiers also may be added in increments. After the polymerization has been run at a given temperature, the temperature is usually raised so as to finish the reaction and ensure the absence of unreacted monomer. Industrially, polymerizations are carried out to over 99% conversion and thus there is no need to strip unreacted monomer. Most poly(vinyl acetate) emulsions contain a maximum of 0.5 wt % unreacted vinyl acetate, which minimizes development of acetic acid and acetaldehyde by monomer hydrolysis on long storage.

In some continuous emulsion homopolymerization processes, materials are added continuously to a first kettle and partially polymerized, then passed into a second reactor where, at a higher temperature and with additional catalyst, the reaction is concluded. Continuous emulsion copolymerizations of vinyl acetate with ethylene have been described (48–50). Large-scale bulk polymerizations are difficult to conduct because of the increased viscosity as polymer forms and the consequent difficulty in removing heat. Low molecular weight polymers have been made in this fashion, however, the continuous processes are known (51).

All of the preceding processes are operated at atmospheric pressure in conventional glass-lined or stainless-steel kettles or reactors. The ethylene–vinyl acetate copolymer (EVA) processes must of necessity be operated under high pressure (41). The low vinyl acetate EVA copolymers, ie, those containing 10–40 wt % vinyl acetate, are made by processes similar to those used to make low density polyethylene for which pressures are usually ≥ 103 MPa (15,000 psi). A medium, ie, 45 wt %, vinyl acetate copolymer with rubberlike properties is made by solution polymerization in t-butyl alcohol at 34.5 MPa (5000 psi). The 70–95 wt % vinyl acetate emulsion copolymers are made in emulsion processes under ethylene pressures of 2.07–5.2 MPa (300–750 psi).

At the molecular scale, vinyl acetate polymerizations generally are understood as free-radical polymerizations, but they are characterized in particular by a relatively large amount of chain transfer (12,23). This high reactivity of the PVAc growing chain radical is attributed to its low degree of resonance stabilization. The high reactivity of the vinyl acetate radical also contributes to the high rate constant for propagation in vinyl acetate polymerization compared to styrene, the acrylates, and the methacrylates. Chain transfer to monomer and to other small molecules leads to lower molecular weight products, but when polymerization occurs in the relative absence of monomer and other transfer agents, such as solvents, chain transfer to polymer becomes more important. As a result, toward the end of batch-suspension or emulsion polymerizations, branched polymer chains tend to form. In suspension and emulsion processes where monomer is fed continuously, the products tend to be more branched than when polymerizations are conducted in the presence of a plentiful supply of monomer. Chain transfer also occurs with the emulsifying agents, leading to their permanent incorporation into the product.

Investigation has shown that chain transfer to polymer occurs predominantly on the acetate methyl group in preference to the chain backbone; one estimate of the magnitude of the predominance is 40-fold (52–53). In the study of chain transfer to polymer and after determination of the molecular weight, the polymer is hydrolyzed to PVA and then is reacetylated to reform PVAc. The branches whose starting points were located on the acetate group are no longer present after this procedure, and the number of branches per molecule of poly(vinyl acetate) polymerized at 60°C is ca 3 at 80% conversion; it rises rapidly thereafter and is ca 15 at 95% conversion and 1–2 $\times 10^4$ number-average degrees of polymerization.

Chain transfer to monomer is an extremely important factor controlling the mol wt (15). Several determinations of the transfer constant to monomer C_m, ie, the ratio of rate constants of the transfer reaction and the propagation step k_{tr}/k_p, show that C_m increases from 1×10^4 to 3×10^4 from 0° to 75°C (12,54–55).

An assessment of the best values of the rate constants for propagation and termination is given below (23):

$$\text{Propagation: } k_\text{p} = 3.2 \times 10^7 \exp\left[-3150 \text{ L/(mol·K·s)}\right]$$

$$\text{Termination: } k_\text{t} = 3.7 \times 10^9 \exp\left[-1600 \text{ L/(mol·K·s)}\right]$$

For example, at 60°C, $k_\text{p} = 2300$ and $k_\text{t} = 2.9 \times 10^7$. An estimate of kinetic chain lifetime, ie, the time from initiation to termination by reaction with another radical, is 1–2 s at 50°C and 4% per hour rate of polymerization (15). If there are five chain-transfer steps in the course of the kinetic chain, then a PVAc molecule forms in 0.2–0.4 s. Faster rates of conversion give shorter kinetic chain lifetimes in inverse proportion, but increased percent of conversion leads to longer chain lifetimes. At 75% conversion and at 60°C, the radical lifetime is ca 10 s.

Vinyl acetate polymerizes chiefly in the usual head-to-tail fashion, but some of the monomers orient themselves head-to-head and tail-to-tail as the chain grows. The fraction of head-to-head addition increases with temperature. A 1.15 mol % head-to-head structure and a 1.86 mol % structure were determined at 15°C and 110°C, respectively (56).

Because of its considerable industrial importance as well as its intrinsic interest, emulsion polymerization of vinyl acetate in the presence of surfactants has been extensively studied (17,57). The Smith-Ewart theory, which describes emulsion polymerization of several other monomers, does not fit the behavior of vinyl acetate. One reason for this is the substantial water solubility of vinyl acetate monomer; as a result, much initiation of polymerization occurs at the dissolved monomer in the water phase as well as in the micelles. Since even low polymers are water-insoluble, phase separation soon occurs when chain growth begins. The rate of polymerization is proportional to $(10^{0.7-1})$ times the initiator concentration; the Smith-Ewart theory predicts $R_\text{p} = K[I]^{0.4}$. The rate of polymerization shows small or virtually no dependence on emulsifier concentration, depending on the study, whereas the Smith-Ewart theory predicts the rate to be proportional to $10^{0.6}$ of the emulsifier concentration. Another very important feature of vinyl acetate emulsion polymerization distinguishing it from the Smith-Ewart theory is chain transfer to monomer, which results in the formation of a small free-radical species that migrates out of the particle and reinitiates polymerization in the aqueous phase.

Poly(vinyl acetate) chains are also stabilized as aqueous-soluble anionic species by complexation with a surfactant. The charge on the water-soluble species prevents their absorption into the particle. These and other recent findings are summarized in ref. 58.

The degree of grafting of PVAc or PVA during emulsion polymerization strongly affects latex properties, eg, viscosity, rheology, and polymer solubility (59). The composition and structure of PVA significantly affects vinyl acetate emulsion polymerization. Thus, the presence of acetaldehyde condensates as a PVA impurity strongly retards polymerization (60).

The chain distribution of residual acetate groups in poly(vinyl alcohol) is very important to successful vinyl acetate emulsion polymerization. A blocklike distribution is preferred over a random distribution (61–64).

Emulsion polymerization of vinyl acetate in the presence of ethylene-oxide- or propylene-oxide-based surfactants and protective colloids also are characterized by the formation of graft copolymers of vinyl acetate on these materials. The degree of grafting influences latex properties (65).

The kinetics of vinyl acetate emulsion polymerization in the presence of nonyl-phenyl–polyethanol surfactants of various chain lengths indicate that part of the emulsion polymerization occurs in the aqueous phase and part in the particles (66). A study of the emulsion polymerization of vinyl acetate in the presence of sodium lauryl sulfate reveals that a water-soluble poly(vinyl acetate)–sodium dodecyl sulfate polyelectrolyte complex forms, and that latex stability, polymer hydrolysis, and mol wt are controlled by this phenomenon (67).

The effects of emulsion polymerization process type or latex and polymer properties have been studied. Thus, emulsion copolymerization of vinyl acetate–butyl acrylate comonomer systems by a delayed monomer addition process yields a core-shell structure particle, in which poly(butyl acrylate)-rich copolymer forms the core and vinyl acetate-rich copolymers the shell (68). The structure of poly(vinyl acetate-*co*-vinyl alcohol) emulsion particles has been determined by electron microscopy. The latex particle is formed of a great many smaller particles packed together by flocculation processes (69–70).

Continuous emulsion copolymerization processes for vinyl acetate and vinyl acetate–ethylene copolymer have been reported. Cyclic variations in the number of particles, conversion, and particle-size distribution have been studied and controlled by on-line analyses and the use of preformed latex seed particles (48,50). A symposium on the emulsion polymerization of vinyl acetate is given in ref. 71. Suspension copolymerization processes for the production of vinyl acetate–ethylene bead products have been described, and the properties of the copolymers determined (72).

New block copolymers of vinyl acetate with methyl methacrylate, acrylic acid, acrylonitrile, and vinyl pyrrolidinone have been prepared by copolymerization in viscous, poor solvents for the vinyl acetate macroradical (73). Similarly, the copolymerization of vinyl acetate with methyl methacrylate is enhanced by the solvents acetonitrile and acetone and is decreased by propanol (74). Copolymers of vinyl acetate containing cyclic functional groups in the polymer chain have been prepared by copolymerization of vinyl acetate with *N*,*N*-diallylcyanamide and *N*,*N*-diallylamine (75–76).

Alternating equimolar copolymers of vinyl acetate and ethylene and alternating copolymers of vinyl acetate and acrylonitrile have been reported (77–78). Vinyl acetate and certain copolymers can be produced directly as films on certain metallic substrates by electroinitiation processes in which the substrate functions as one electrode (79).

New terpolymers of vinyl acetate with ethylene and carbon monoxide have been prepared and their uses as additives to improve the curing and flexibility of coating resins, eg, nitrocellulose, asphalt, phenolics, and polystyrene, have been described (80–82). In vinyl acetate polymerizations, the molecular weights of the products increase with the extent of conversion; the ratio of weight-to-number-average-degree-of-polymerization also changes, becoming larger at higher conversions (83–84). The dilute solution viscosity of poly(vinyl acetate) can be related to the molecular weight by the following equations (34):

$$[\eta] = 0.0102 \, \overline{M}^{0.72} \qquad \text{acetone solvent, 30°C}$$
$$[\eta] = 0.314 \, \overline{M}^{0.60} \qquad \text{methanol solvent, 30°C}$$

These equations apply to linear polymers of $\overline{M}_w/\overline{M}_n = 2.0$.

Economic Aspects.   Prices for PVAc polymers depend on the form of the polymer, ie, whether it is resin or emulsion, homopolymer or copolymer, as well as on the specific product. As of 1980, emulsion prices were $1.00–1.50/kg dry resin. Prices of copolymer emulsions tend to be higher than those of the homopolymer by 4–6¢/kg. Bead resins and other dry forms are priced at $1.60–2.20/kg. Specialty copolymers generally have a premium price; alkali-soluble resins for example, cost ca 22¢/kg more than the homopolymer. These price ranges are for large shipments.

Growth in PVAc consumption is illustrated in Figure 6. The emulsions continue to dominate the adhesives and paint markets (see Paints). The companies listed in Table 9 are among the major suppliers of poly(vinyl acetate)s and vinyl acetate copolymers, but there are numerous other suppliers. Many other companies produce these polymers and consume them internally in the formulation of products.

Specifications and Standards.   Typical specifications of the commercially available emulsions are tabulated in Table 10. However, there are exceptions to the ranges given. For example, most emulsions contain 55–56 wt % solids but some are available at 46–47 wt % with viscosities of 10–15 mPa·s (= cP), and others at 59 wt % solids with viscosities of 200–4500 mPa·s. Specialty copolymer emulsions are available containing 65 wt % solids.

Borax stability is an important property in adhesive, paper, and textile applications where borax is a frequently encountered substance. Other emulsion properties tabulated by manufacturers include tolerance to specific solvents, surface tension, minimum filming temperature, dilution stability, freeze–thaw stability, percent soluble polymer, and molecular weight. Properties of films cast from the emulsions are also sometimes listed; they include clarity, gloss, light stability, water resistance, flexibility, heat-sealing temperature, specific gravity, and bond strength.

Homopolymer resin specifications usually include viscosity grade (1.0 molar solution in benzene at 20°C), 1.5–800 mPa·s (= cP) ± 10%; volatiles, 1–2 wt %; acidity as acetic acid, 0.1–0.3 wt %; and softening point (see Table 7). Data are also given on heat-sealing temperature, tensile strength, elongation and abrasion resistance.

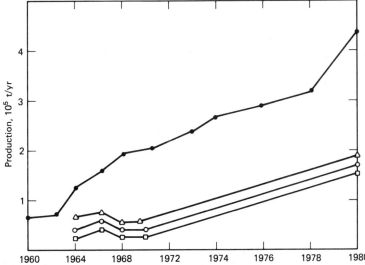

Figure 6.   Poly(vinyl acetate) production. ● = Total production, △ = adhesives, ○ = paints, and □ = paper (73).

Table 9.   Some Large Poly(Vinyl Acetate) Producers

Country	Company	Trade names
United States	Air Products	Vinac, Flexbond, Airflex
	Borden Chemical	Polyco
	E. I. du Pont de Nemours & Co., Inc.	Elvacet, Elvace, Elvax
	W. R. Grace (Dewey and Almy)	Darex, Daratak, Evenflex
	Monsanto	Gelva
	National Starch and Chemical	Resyn
	Reichhold Chemical	Wallpol
	Union Carbide Corp.	Bakelite, Ucar, Co-Mer
	Union Oil of California	
	U.S.I. Chemicals	Ultrathene
Japan	Kobunshi Chem	
	Nippon Carbide	
	Nippon Gosei	
	Sekisui	
	Showa Denko	
	Sumitomo	Sumikaflex
UK	Imperial Chemical Industries, Ltd.	
	Revertex	Emultex
FRG	Farbwerke Hoechst	Mowilith
	Wacker Chemie	Vinnapas
Switzerland	Ebnoether	
	Lonza	Vipolit
Canada	Shawinigan	Gelva
France	Rhone-Poulenc	Rhodopas
Italy	Monte-Edison	Vinavil

Table 10.   Poly(Vinyl Acetate) Emulsion Specifications

Property	Range
solids, wt %	48–55
viscosity, mPa·s (= cP)	200–4500
pH	4–6
residual monomer, % max	0.5
particle size	0.1–3.0
particle charge	neutral or negative
density at 25°C, g/cm^3	0.92
borax stability	stable or unstable
mechanical stability	good or excellent

Poly(vinyl acetate) is also available as spray-dried emulsion solids with average particle sizes of 2–20 μm; the product can be reconstituted to an emulsion by addition of water or it can be added directly to formulations, eg, concrete. Solutions of resin in methyl and ethyl alcohol at 25–50 wt % solids are also available.

There are many ASTM tests used to evaluate the functional class of materials, eg, paints and lacquers. Poly(vinyl acetate) is nontoxic and is approved by the FDA for food-packaging applications.

Emulsions are shipped in 19-L pails, 209-L drums, 11.4–18.9-m^3 (3000–5000 gal) tank trucks, and 38–76-m^3 (10,000–20,000 gal) railroad tank cars. Large shipments are insulated against cold. Precautions should be taken to avoid the formation of foam during unloading. Latexes corrode ordinary steel and must be stored in tanks with stainless-steel, glass, plastic, or coated surfaces. Suitable coatings are baked phenolic, epoxy phenolic, or PVC. Storage temperatures should be about room temperature, since excessive exposure to high or low temperatures may cause phase separation. Containers should be closed to prevent evaporation of water, which leads to skin formation. Diaphragm pumps and screw pumps are preferred to centrifugal pumps. Dry resins should be stored at or below room temperature to prevent caking.

Uses. *Adhesives.* The main areas of poly(vinyl acetate) adhesive use are packaging and wood gluing (86–87). The emulsion form of PVAc is especially suitable for adhesives because of several properties that are peculiar to emulsion systems. The stability of PVAc emulsions allows them to accept many types of modifying additives without being damaged. For example, solvents, plasticizers, tackifying resins, and fillers can be added directly to homopolymer and copolymer emulsions without requiring preemulsification, unlike the elastomeric adhesive latexes. Homopolymer emulsions containing partially hydrolyzed poly(vinyl alcohol) as a protective colloid can accept greater amounts of these modifying additives than any other type of emulsion without coagulating.

The stability of the emulsions further permits them to be compounded in simple liquid-blending vessels by means of agitators, eg, marine-type propellers, paddles, or turbines. The adhesives can be adapted to any type of machine application, ie, from spray guns to rollers to extruder-type devices. Different applicators are fairly specific in their viscosity requirements; so are the various substrates receiving the adhesive.

Poly(vinyl acetate) emulsions can be used in high speed gluing equipment. In contrast to aqueous solutions of natural or synthetic products which lose water slowly, resulting in setting or formation of an adhesive bond, the emulsions quickly lose water and invert and set rapidly.

Homopolymers adhere well to porous or cellulosic surfaces, eg, wood, paper, cloth, leather, and ceramics. Homopolymer films tend to creep less than copolymer or terpolymer films. They are especially suitable in adhesives for high speed packaging operations.

Copolymers wet and adhere well to nonporous surfaces, such as plastics and metals. They form soft, flexible films, in contrast to the tough, horny films formed by homopolymers, and are more water-resistant. As the ratio of comonomer to vinyl acetate increases, the variety of plastics to which the copolymer adheres also increases. Comonomers containing functional groups often adhere to specific surfaces; for example, carboxyl-containing polymers adhere well to metals.

Cross-linked polymer emulsions accept water-miscible solvents better than straight-chain emulsions. In the film state, the former resist water and organic solvents better and tend to have higher heat-sealing temperatures, which signify a greater resistance to blocking and cold flow. Straight-chain polymers can have high molecular weights, which contribute to high heat-sealing temperatures. Varying the conditions of polymerization results in either straight-chain or cross-linked polymer emulsions.

Tack enables an adhesive to form an immediate bond between contacting surfaces when they are brought together. It permits the alignment of an assembly and prevents the adherends from separating before the adhesive sets. Tack is differentiated into two types: wet tack, also called grab or initial tack, is the tack of an adhesive before the liquid carrier, ie, organic solvent or water, has fully evaporated; dry tack, also called residual tack or pressure sensitivity, is the tack remaining after the liquid carrier has evaporated. Wet tack is often necessary in paper-converting operations. Applicators with little or no pressure on the combining section require emulsion adhesives with enough wet tack to bond strongly at the slightest touch. Emulsions containing poly-(vinyl alcohol) as a protective colloid have stronger wet tack than those protected with a cellulose derivative. If still more wet tack is required of the poly(vinyl alcohol)-protected adhesive, both a solvent and a plasticizer can be added and the solids content can be increased. Dry tack is needed where two nonporous surfaces are to be bonded. The nature of the adherends does not permit the water in the adhesive to escape by penetration or evaporation, so the adhesive must dry before the adherends are joined, yet they still retain enough tack to form a permanent bond. Film-to-film and film-to-foil laminations are good examples of applications requiring dry tack.

The setting speed of an adhesive is the time during which the bond formed by the adhesive becomes permanent. Before the setting of an emulsion adhesive can occur at all, inversion of the emulsion must take place; that is, it must change from a dispersion of discrete polymer particles in an aqueous, continuous phase to a continuous polymer film containing discrete particles of water. The point at which the adhesive has actually set is that point at which an assembly, whether joint or lamination, can no longer be disassembled without damaging one or more of the adherends. Rapid setting speeds are necessary in many types of packaging adhesives intended for use with high speed gluing equipment. The rapid inversion possible with PVAc emulsions and their low viscosities allow them to be compounded into adhesives that not only set rapidly but also machine easily at high speeds. This machinability permits their application by practically any means, including glue guns, rollers, and spray guns.

Increases in setting speed can be achieved by increasing the solids content; that is, by increasing the amount of water-insoluble substances contained in the emulsions. This crowds the aqueous phase of the emulsion, hastening inversion and setting. Adding tackifying resins is one way to crowd the emulsion; adding plasticizers and solvents is another, but these additives mainly increase the setting speed by softening the polymer particles and hastening their coalescence. Surface-active agents also increase setting speed by helping the water in an adhesive to penetrate porous surfaces more rapidly; they can also retard it by stabilizing the emulsion. The usual way to prepare a high speed adhesive is to add both a solvent and a plasticizer to the base emulsion.

Poly(vinyl acetate) emulsions are excellent bases for water-resistant paper adhesives destined for use in manufacturing bags, tubes, and cartons. Glue-lap adhesives, which require a moderate-to-high resistance to water, exemplify this type. When routine water resistance is required, a homopolymer vinyl acetate emulsion containing a cellulosic protective colloid is effective for most purposes. Next are emulsions containing fully hydrolyzed PVA as a protective colloid, followed by those containing partially hydrolyzed PVA.

When more than routine water resistance is required, a copolymer vinyl acetate emulsion can be used. The plasticizing comonomer in the polymer particles increases their intrinsic coalescing ability; thus, they can coalesce more readily than homopolymer particles to a film that resists water more stubbornly. This resistance to water does not extend to the organic solvents, however, which are better resisted by homopolymer films. The soft copolymers have lower solubility parameters than homopolymers and are more readily attacked by solvents of low polarity, eg, hydrocarbons.

Despite their high resistance to water, copolymer emulsions are seldom chosen to bond paper to paper, as the less expensive homopolymer emulsions are effective enough. When copolymers are used in paper adhesives, it is done so chiefly to join coated or uncoated papers to films so as to take advantage of the lower critical surface tension required for wetting the copolymers. Since most packaging adhesives must be able to form strong paper-to-paper bonds, homopolymer vinyl acetate emulsions containing cellulosic protective colloids are always the first choice. The water resistance of any emulsion can be increased by improving its coalescence into a film either by adding solvents and plasticizers to it or by adding specific coalescing agents, eg, diethylene glycol monoethyl ether or diethylene glycol monoethyl ether acetate. Emulsions containing completely hydrolyzed PVA can also be incorporated with cross-linking agents, which further insolubilize them. Cross-linking agents, however, can make a formulation unstable, so that it must often be used within a few hours after it has been compounded.

For any adhesive to be effective, it must first thoroughly wet the surface to be bonded; hence, it must be fluid at the time it is applied. Fluidity is not a problem with emulsion adhesives; wetting, however, can be one, especially when slick or coated surfaces are to be joined. An example is the manufacture of high gloss, clay-coated cartons or the adhering of waxed papers.

Copolymer emulsions tend to wet slick surfaces better than homopolymer emulsions because of the extra mobility and softness given to the polymer particles by the plasticizing comonomer. Their dried films also tend to conform better with this type of substrate. Homopolymer emulsions containing poly(vinyl alcohol) as a protective colloid wet paper surfaces well and are the first choice for paper-to-paper packaging adhesives. The setting ability of these emulsions can be increased by incorporating either nonionic wetting agents or partially hydrolyzed PVA, or both. Small amounts of partially hydrolyzed PVA present in an adhesive also helps it to wet lightly waxed papers.

The viscosity of an adhesive directly influences its penetration into a substrate: as the viscosity increases, the penetrating power decreases. It also determines the amount of mileage or spread that can be obtained. An optimum viscosity exists for each substrate and each set of machine conditions and must be achieved in order to manufacture an efficient adhesive. Poly(vinyl acetate) emulsions are frequently too low in viscosity to be metered efficiently or to perform well as adhesives by themselves. They must be bodied to working viscosities, eg, by adding thickeners.

Blocking and cold flow are undesirable properties. Blocking refers to dried adhesive surfaces that become sticky, causing unwanted bonding. Cold flow occurs with a polymer having a low softening point; if the temperature of the adhesive assembly warms to the softening range of the polymer, the bond slips. The temperature at which a dried adhesive film forms an instantaneous bond between two surfaces when heat is applied is its heat-sealing temperature. This property is closely related to the blocking temperature and to the temperature at which cold flow or creep can occur.

The heat-sealing temperature of an emulsion is related to the thermoplasticity of the poly(vinyl acetate) particles dispersed in it. Thermoplastic polymers are softened by heat; those of relatively high molecular weight or those that are cross-linked soften at higher temperatures than those of low molecular weight. In addition, homopolymers soften at higher temperatures than do copolymers of like molecular weight. Another factor which affects the heat-sealing temperature of an emulsion is the amount of poly(vinyl alcohol) in it, if any, since poly(vinyl alcohol) has a high melting point. High heat-sealing temperatures, which are desirable in an emulsion adhesive since they indicate resistance to blocking and cold flow, are usually attainable with emulsions containing large amounts of poly(vinyl alcohol) as a protective colloid.

Fillers (qv) are added to emulsion adhesives to build the total solids content, to reduce penetration into a porous substrate, and to lower costs. Plasticizers are added to emulsion adhesives to modify several properties of both the emulsion and the finished adhesive film. By softening the polymer particles dispersed in the emulsion and increasing their mobility, plasticizers cause them to flow together more easily. This usually increases the viscosity of the emulsion and tends to destabilize it for faster breaking and setting speeds at the time it is applied. In addition, the increased softness and mobility help the emulsion to wet smooth, nonporous surfaces, eg, films, foils, and coated papers, thereby increasing its adhesion to them. Also, the softened polymer particles coalesce more rapidly and at a lower temperature than is possible with the unplasticized emulsion. This improved coalescence increases the water resistance of the adhesive film. Plasticizers are usually high boiling esters, eg, phthalates. Phosphate esters are useful as fire-retardant plasticizers (qv).

Solvents are frequently used to perform several functions in emulsion adhesives: they promote adhesion to solvent-sensitive surfaces; they increase the viscosity of the emulsion and intensify the tack of the wet adhesive; and they improve the coalescing properties of the film. Low boiling solvents impart only wet tack to the adhesive film, whereas high boiling solvents confer both dry and wet tack and lower the heat-sealing temperature. The solvent can impart the necessary speed to the wet adhesive but, because it evaporates, it does not cause the dried bone to creep, which often happens with a plasticized film that ages under stress. Solvents promote adhesion to solvent-sensitive adherends by softening and partially dissolving them, thus allowing the adhesive to wet or penetrate the surface or both.

Tackifiers are used to increase the tackiness and the setting speed of adhesives. They increase tackiness by softening the poly(vinyl acetate) polymer in the wet and the dry adhesive film. Tackifiers are usually rosin or its derivatives or phenolic resins. Other additives frequently needed for specific application and service conditions are antifoams, biocides, wetting agents, and humectants.

Specialized copolymer latexes, which are inherently and permanently tacky, are available as pressure-sensitive emulsions. They are mechanically stable and have excellent machinability. They are compatible with many other PVAc latexes and, therefore, can be easily blended with other resins for modification of surface tack, peel strength, and creep (7).

Poly(vinyl acetate) dry resins and EVA copolymers are used in solvent adhesives, which can be applied by typical industrial techniques, eg, brushing, knife-coating, roller-coating, spraying, or dipping. Proper allowances must be made for evaporation of solvent during or before bonding. Poly(vinyl acetate) resins and EVA copolymers containing 21–40 wt % vinyl acetate are widely used in hot-melt adhesive applications. Homopolymers are compounded with 25–35 wt % plasticizer and 25–30 wt % extender resins. These additives increase fluidity and adhesion as well as reduce cost. Ethylene–vinyl acetate resins are mixed with waxes, rubbers, and resin to make the hot-melt adhesive compound. Hot-melt adhesive application processes may be extremely rapid both in application and setting speed, cause no problems resulting from solvent or water evaporation, and provide bonds of high water resistance. The adhesive materials have indefinite shelf lives. Hot-melts are largely used in packaging, laminating, and bookbinding. For the latter application, alkali-soluble copolymers are frequently used to facilitate the reclamation of scrap paper.

Paints. Poly(vinyl acetate) emulsion paints form flexible, durable films with good adhesion to clean surfaces, including wood, plaster, concrete, stone, brick, cinder blocks, asbestos board, asphalt, tar paper, wallboards, aluminum, and galvanized iron (88) (see Coatings, industrial; Paint). Adherence is also good on painted surfaces if the surfaces are free from dirt, grease, and rust.

Poly(vinyl acetate) latex paints are the first choice for interior use (89). Their ability to protect and decorate is reinforced by several advantages belonging exclusively to latex paints: they do not contain solvents so that physiological harm and fire hazards are eliminated; they are odorless; they are easy to apply with spray gun, roller-coater, or brush; and they dry rapidly. The paint can be thinned with water, and brushes or coaters can be cleaned with soap and tepid water. The paint is usually dry in 20 min–2 h, and two coats may be applied the same day.

Poly(vinyl acetate) latex paints are also widely used as exterior paints. Their durability, particularly their resistance to chalking, far surpasses that of any conventional oleoresinous paints, which chalk soon after application. The good non-chalking properties of PVAc paints result from their resistance to degradation by uv light. Latex paints that are correctly formulated from quality PVAc latexes develop little or no chalk, thus giving maximum tint retention, and they last a long time before repainting becomes necessary. The blister resistance of PVAc paints is another important advantage for their exterior use. Latex-paint films are permeable enough to permit water vapor to penetrate them, which prevents blistering and peeling. Their film formation is not impaired if they are painted on damp surfaces or are applied under very humid conditions.

Several types of plasticizing comonomers can be copolymerized with vinyl acetate to produce latexes suitable for manufacturing paints. The older, less common comonomers are the alkyl maleates and fumarates, eg, dibutyl maleate and dibutyl fumarate. Acrylic esters, eg, 2-ethylhexyl acrylate, as well as ethylene, are the most widely used comonomers today. The chief reason for using a comonomer in vinyl acetate polymers is to increase the deformability of the paint film permanently, thus permitting it to expand and contract as the dimensions of the substrate change with changes in temperature. The plasticizing comonomer also softens the polymer particles. For a film to form from a latex paint, the polymer particles must deform and fuse to form a continuous film. This coalescing ability is directly related to the amount of comonomer present. In both interior and exterior paints, the improvement in coalescence that is obtainable by using high comonomer levels results in a general improvement in all the properties necessary for interior and exterior paints.

High levels of comonomers, such as ethylene or 2-ethylhexyl acrylate, must be added with care. If high comonomer levels are used with low molecular weight polymers, the resulting paint film suffers from excessive dirt pickup at high temperatures because it becomes soft and tacky. It is required, therefore, to prepare high molecular weight polymers, since it is only with these that a sufficient amount of comonomer can be incorporated to give the tint retention without dirt pickup that is required in exterior paints.

The strength of a polymer is directly proportional to its molecular weight; consequently, the toughest paint films are formed from latex polymers having the highest molecular weights. For exterior paints, there is a Federal specification of a minimum intrinsic viscosity of 0.45 dL/g. The minimum intrinsic viscosity of commercial paint polymers usually is 0.60–1.0 dL/g and above.

Special vinyl acetate copolymer paints have been developed with greatly improved resistance to blistering or peeling when immersed in water. This property allows better cleaning and use in very humid environments. These latexes exhibit the water resistance of higher priced acrylic resins (90).

The critical pigment volume concentration (CPVC) of vinyl acetate paints is 48–<60%. Generally, the smaller and softer the polymer particles, the higher the CPVC and the greater the pigment-binding capacity of the latex. The most widely used white pigment is titanium dioxide in the form of rutile or of anatase. The color pigments used in latex paints are of two types: the organic pigments (see Pigments, organic), which are usually hydrophobic, and the inorganic pigments (see Pigments, inorganic), which are usually hydrophilic. The hydrophilic pigments are relatively easy to incorporate into water-base paints. The hydrophobic pigments are more difficult to incorporate, but this can be overcome by choosing the correct blend of surfactants for a particular pigment.

The organic colorants commonly used are toluidine red, Hansa yellow, phthalocyanine blue and green, pigment green B, and carbon black, among others. Examples of acceptable inorganic colorants are iron oxide red, brown, yellow, and black; and chrome oxide green. Colorants that should not be used are those that are reactive, partially water-soluble, or sensitive to pH changes.

In addition to latex and pigment, paint formulations contain dispersants and wetting agents (both surfactants), defoamers, thickeners and protective colloids, freeze–thaw stabilizers, coalescing agents, and biocides.

Paper. Poly(vinyl acetate) emulsions and resins are used as the binder in coatings for paper (qv) and paperboard. The coatings may be clear, colored, or pigmented, and are glossy, odorless, tasteless, greaseproof, nonyellowing, and heat-sealable. Conventional paper-coating equipment is used; formulations normally contain 60–65 wt % solids with a pigment-to-binder ratio of 1:5. Printing quality and ink-pick resistance are excellent. In papermaking, emulsions applied as wet-end additions to the furnish improve the strength and durability of the final product (see Papermaking additives). Plasticized emulsions may be used to give the product toughness and flexibility.

Emulsions used in paper coatings must meet special requirements: the particle size must be small (ca 1 μm) and its distribution rather narrow. These properties provide good pigment binding efficiency and high gloss. They must have exceptionally good mechanical stability and freedom from off-sized latex particles or grit, and they must exhibit no trace of dilatant flow in a pigment slurry. These properties are required for modern high speed paper-coating applications. They should also be compatible with starch and alginate natural resins frequently used as cobinders (91).

Textiles. Poly(vinyl acetate) emulsions are widely used as textile finishes because of their low cost and good adhesion to natural and synthetic fibers (see Textiles). In textile piece goods finishing, dispersions diluted with water to 1–3 wt % resin are most often used to obtain a stiff or crisp hand on woven cotton fabrics. Concentrations of

2–20 wt % emulsion are recommended for bodying, stiffening, and bonding. Principal applications include the stiffening of felts and the binding of nonwoven fabrics (see Nonwoven textile fabrics). Finishes to improve the snag resistance and body or hand of nylon hosiery are based on PVAc emulsions.

Poly(vinyl acetate) emulsions are used to prime-coat fabrics to improve the adhesion of subsequent coatings or to make them adhere better to plastic film. Plasticized emulsions are applied, generally by roller-coating, to the backs of finished rugs and carpets to bind the tufts in place and to impart stiffness and handle. For upholstery fabrics woven from colored yarns, PVAc emulsions may be used to bind the tufts of pile fabrics or to prevent slippage of synthetic yarns.

The emulsion formulations are generally applied to cloth by padding from a bath and squeezing off the excess. Modifying a formulation in the pad box, eg, to increase or decrease firmness, can be easily done by adding an emulsion or softener. The alkali-soluble vinyl acetate copolymers previously mentioned can be used as warp sizes during weaving.

The use of vinyl acetate copolymers as binding agents for nonwoven fabrics has grown rapidly. Vinyl acetate–ethylene copolymer latexes have been particularly successful in part because of their low cost but also because of their excellent adhesion to a wide range of substrates, their water and alkali resistance, and their low flammability (92–93). Self-cross-linking polymers are available as are polymers requiring external curing agents.

Other. Vinyl acetate resins are useful as antishrinking agents for glass-fiber-reinforced polyester molding resins (94–95). Poly(vinyl acetate)s are also used as binders for numerous materials, eg, fibers, leather, asbestos, sawdust, sand, clay, etc, to form compositions that can be shaped with heat and pressure. The compressive and tensile strength of concrete is improved by addition of PVAc emulsions to the water before mixing. A polymer-solids-to-total-solids ratio of ca 10:90 is best. The emulsions also aid adhesion between new and old concrete when patching or resurfacing. Joint cements, taping compounds, caulks, and fillers are other uses.

Emulsions containing added PVA and bichromate are used to make light-sensitive stencil screens for textile printing and ceramic decoration. The resins are used in printing inks, nitrocellulose lacquers, and special high gloss coatings. Inks made with PVAc and metallic pigments look like foil since the formulations have a high leafing power, do not induce tarnish, and contribute no unwanted color or aging.

Vinyl acetate polymers have long been used as chewing-gum bases. They have recently been studied as controlled release agents for programmed administration of drugs and as a base for antifouling marine paints (96–97).

BIBLIOGRAPHY

"Vinyl Acetate" in *ECT* 1st ed., under "Vinyl Compounds," Vol. 14, pp. 686–691, by T. P. G. Shaw; pp. 691–698, by K. G. Blaikie and T. P. G. Shaw; pp. 699–709, by K. B. Blaikie and M. S. W. Small, Shawinigan Chemicals, Ltd., "Poly(Vinyl Acetate)" in *ECT* 2nd ed., under "Vinyl Polymers," Vol. 21, pp. 317–353, by D. Rhum, Air Reduction Co.

1. Ger. Pat. 271,381 (June 22, 1912), F. Klatte (to Chemische Fabriken Grieshiem-Electron).
2. C. A. Schildknecht, *Vinyl and Related Polymers*, John Wiley & Sons, Inc., New York, 1952, p. 323.
3. *BDSA Quarterly Industry Report*, *Chemicals*, U.S. Department of Commerce, Washington, D.C., April 1969, p. 26.
4. *Vinyl Acetate*, Bulletin No. S-56-3, Celanese Chemical Co., New York, 1969.

5. *Vinyl Acetate Monomer F-41519*, Union Carbide Corp., New York, June 1967.
6. *Vinyl Acetate Monomer BC-6*, Borden Chemical Co., New York, 1969.
7. *Vinyl Acetate Monomer*, Air Reduction Co., New York, 1969.
8. F. S. Dainton and K. J. Irwin, *Trans. Faraday Soc.* **46,** 331 (1950).
9. R. W. Gallant, *Hydrocarbon Process.* **47**(10), 115 (1968).
10. S. Okamura and I. Motoyama, *J. Polym. Sci.* **58,** 221 (1962).
11. L. H. Horsley, *Azeotropic Data, II, Advances in Chemistry Series*, No. 35, American Chemical Society, Washington, D.C., 1962.
12. M. K. Lindemann in G. E. Ham, ed., *Vinyl Polymerization*, Vol. 1, Marcel Dekker, Inc., New York, 1967, Part 1, Chapt. 4.
13. J. Brandrup and E. H. Immergut, eds., *Polymer Handbook*, Interscience Publishers, a division of John Wiley & Sons, Inc., New York, 1966.
14. W. P. Paist and co-workers, *J. Org. Chem.* **6,** 280 (1941).
15. M. S. Matheson and co-workers, *J. Am. Chem. Soc.* **71,** 2610 (1949).
16. A. F. Rekusheva, *Russ. Chem. Rev.* **37,** 1009 (1968).
17. M. K. Lindermann in N. M. Bikales, ed., *Encyclopedia of Polymer Science and Technology*, Vol. 15, John Wiley & Sons, Inc., New York, 1971, p. 636.
18. G. O. Morrison and T. P. G. Shaw, *Trans. Electrochem. Soc.* **63,** 425 (1933).
19. D. Swern and E. F. Jordan, Jr., in N. Rabjohn, ed., *Organic Syntheses*, Collective Vol. 4, John Wiley & Sons, Inc., New York, 1963, p. 977.
20. H. Hopff and M. A. Osman, *Tetrahedron* **24,** 2205 (1968).
21. C. F. Kohll and R. van Helden, *Rec. Trav. Chim.* **86,** 193 (1967).
22. H. E. Simmons and R. D. Smith, *J. Am. Chem. Soc.* **81,** 4256 (1959).
23. C. Walling, *Free Radicals in Solution*, John Wiley & Sons, Inc., New York, 1957, Chapt. 6.
24. *Hydrocarbon Process.* **46**(4), 146 (1967).
25. R. Ramirez, *Chem. Eng.* **75**(17), 94 (1968).
26. R. VanHelden and co-workers, *Rev. Trav. Chim.* **87,** 961 (1968).
27. *Petr. Refiner* **38,** 304 (1959).
28. *Hydrocarbon Process.* **44**(11), 287 (1965).
29. H. C. Volger, *Rev. Trav. Chim.* **87,** 501 (1968).
30. *Chemical Purchasing* 61, (Feb. 1981).
31. *Chem. Eng.*, 11 (Aug. 2, 1982).
32. *International Trade Commission Reports*, Washington, D.C., 1960–1980.
33. *Chemical Safety Data Sheet SD-75: Manual Sheet TC-4*, Manufacturing Chemists' Association, Washington, D.C.
34. D. N. Mead and R. M. Fuoss, *J. Am. Chem. Soc.* **63,** 2832 (1941).
35. S. O. Morgan and Y. A. Yager, *Ind. Eng. Chem.* **32,** 1519 (1940).
36. H. Fikentscher, *Cellul. Chem.* **13,** 71 (1932).
37. E. O. Kraemer, *Ind. Eng. Chem.* **30,** 1200 (1938).
38. A. Ballisteri and co-workers, *J. Polym. Sci. Polym. Chem. Ed.* **18,** 1147 (1980).
39. T. Okada, *Polym. J.* **9,** 121 (1977).
40. H. Bartl in E. Muller, ed., *Methods of Organic Chemistry (Houben- Weyl), Macromolecular Materials*, Georg Thieme Verlag, Stuttgart, FRG, 1961, Part 1, pp. 905–918.
41. M. K. Lindemann, *Paint Manuf.* **38**(9), 30 (1968).
42. E. Levine, W. Lindlaw, and J. Vona, *J. Paint Technol.* **41,** 531 (1969).
43. *Product List*, Air Reduction Co., New York, 1968.
44. *Elvacet Poly(Vinyl Acetate) Brochure*, E. I. du Pont de Nemours & Co., Inc., Wilmington, Del., 1968.
45. Thermoplastics Division Product Director, Borden Chemical Co., New York, 1968.
46. *Gelva Poly(Vinyl Acetate) Technical Bulletin*, Publication No. 6103, Monsanto Chemical Co., St. Louis, Mo., 1969.
47. E. Tromsdorff and C. E. Schildknecht in C. E. Schildknecht, ed., *Polymer Processes*, Interscience Publishers, Inc., New York, 1956, pp. 105–109.
48. U.S. Pat. 4,164,489 (Aug. 14, 1979), W. E. Lenney and W. E. Daniels (to Air Products and Chemicals, Inc.).
49. U.S. Pat 4,035,329 (Nov. 19, 1975), H. Wiest and co-workers (to Wacker Chemie, G.m.b.H.).
50. C. Kipparissides, J. F. MacGregor, and A. E. Hamiliec, *Can. J. Chem. Eng.* **58,** 1, 48 (1980).
51. R. D. Dunlop and F. E. Reese, *Ind. Eng. Chem.* **40,** 654 (1948).

52. S. Imoto, J. Ukida, and T. Kominami, *Kobunshi Kagaku* **14**, 101 (1957).
53. D. Stein and G. V. Schultz, *Makromol. Chem.* **52**, 249 (1962).
54. S. P. Pontis and A. M. Deshpande, *Makromol. Chem.* **125**, 48 (1969).
55. W. W. Graessley, W. C. Uy, and A. Gandhi, *Ind. Eng. Chem. Fundam.* **8**, 697 (1969).
56. P. J. Flory and F. S. Leutner, *J. Polym. Sci.* **3**, 880 (1948); **5**, 267 (1950).
57. J. W. Breitenbach, H. Edelhauser, and R. Hochrainer, *Monatsh. Chem.* **99**, 625 (1968).
58. V. T. Stannett, *Proc. R. Aust. Chem. Inst.* **42**, 232 (1975).
59. I. Gavat, V. Dimonie, and D. Donescu, *J. Polym. Sci. Polym. Symp.* **64**, 125 (1978).
60. V. T. Shiriniyan and co-workers, *Plast. Massy* **8**, 15 (1974).
61. K. Noro, *Br. Polym. J.* **2**, 128 (1970).
62. M. Shiraishi, *Br. Polym. J.* **2**, 135 (1970).
63. S. S. Mnatskanov and co-workers, *Vysokmol. Soyed.* **A14**, 4,851 (1972).
64. V. T. Shirininyan and co-workers, *Vysokmol. Soyed.* **A17**, 1,182 (1975).
65. D. Donescu, *Rev. Roum. Chim.* **24**, 9, 1399 (1979).
66. G. F. Lundardon, G. P. Talamini, and V. Grosso, *Eur. Polym. J.* **11**, 437 (1975).
67. P. K. Isaacs and H. A. Edelhauser, *J. Appl. Polym. Sci.* **10**, 171 (1966).
68. S. C. Misra, C. Pichot, M. El-Aasser, and J. W. Vanderhoff, *J. Polym. Sci. Lett. Ed.* **17**, 567 (1979).
69. M. Furuta, *J. Polym. Sci. Polym. Lett. Ed.* **11**, 113 (1973).
70. *Ibid.*, **12**, 459 (1974).
71. M. El-Aasser and J. W. Vanderhoff, eds., *Emulsion Polymerization of Vinyl Acetate*, Applied Science Publishers, Inc., Englewood, N.J., 1981.
72. V. T. Shiriniyan and co-workers, *Zh. Prikl. Khim. (Leningrad)* **44**, 1345 (1977).
73. R. B. Seymour and G. A. Stahl, *J. Macromol. Sci. Chem.* **11**(1), 53 (1977).
74. W. K. Busfield and R. B. Low, *Eur. Polym. J.* **11**, 309 (1975).
75. A. G. Sayadyan and D. A. Simonyan, *Arm. Khim. Zh.* **21**, 1041 (1968).
76. U.S. Pat. 4,260,533 (April 7, 1981), J. G. Iacoviello and W. E. Daniels (to Air Products and Chemicals, Inc.).
77. T. Yatsu, S. Moriuchi, and H. Fuji, *Makromolecules* **10**, 243 (1977).
78. C. H. Chen, *J. Polym. Sci.* **14**, 2109 (1976).
79. B. Tidswell and A. W. Train, *Br. Polym. J.* **7**, 409 (1975).
80. U.S. Pat. 4,137,382 (Jan. 30, 1979), C. J. Vetters (to National Distillers and Chemical Corp.).
81. U.S. Pat. 4,172,939 (Oct. 30, 1979), G. J. Hoh (to E. I. du Pont de Nemours & Co., Inc.).
82. *Research Disclosure #13816*, Industrial Opportunities, Ltd., Hampshire, UK, Oct. 1975.
83. D. Stein, *Makromol. Chem.* **76**, 170 (1964).
84. M. Matsumoto and I. Ohyang, *J. Polym. Sci.* **46**, 441 (1960).
85. *Proceedings of Society of Plastic Industry Conferences*, 1972–1980.
86. *Adhesives Handbook*, Airco Chemical and Plastics Division of Air Reduction Co., Inc., New York, 1969.
87. R. A. Weidener in R. L. Patrick, ed., *Treatise on Adhesion and Adhesives*, Vol. 2, Marcel Dekker, Inc., New York, 1969, Chapt. 10, pp. 432–447, 467–471.
88. *Paint Handbook*, Airco Chemical and Plastic Division of Air Reduction Co., Inc., New York, 1969.
89. *Am. Paint. J.* **53**, 7, 58 (1968).
90. *Am. Paint/Coating J.*, 66 (Nov. 13, 1978).
91. U.S. Pat. 4,228,047 (July 31, 1981), W. E. Daniels and W. H. Pippen (to Air Products and Chemicals, Inc.).
92. J. R. Halker, *Formed Fabric Industry*, 26 (June 1976).
93. P. L. Rosamilia, *Tappi Paper Synthetics Proceedings*, 251 (1979).
94. K. E. Atkins and B. W. Lipinsky, *Avk. Offentliche Jahrestagung Der. 13 Int. Tagung*, Freudenstadt, Oct. 1976, p. 39/1-4.
95. B. Pelzer, G. Kampf, H.-W. Schultz in ref. 94, p. 38/1-7.
96. H. Leeper and H. Benson, *SPE 2nd Ann. Conf.*, Seattle, Wash., Aug. 1976, pp. 141–149.
97. U.S. Pat. 4,143,015 (Jan. 21, 1977), E. Soeterik.

WILEY DANIELS
Air Products and Chemicals, Inc.

POLY(VINYL ALCOHOL)

Poly(vinyl alcohol) [9002-89-5] (PVA) is a polyhydroxy polymer and, consequently, a water-soluble synthetic resin. It is produced by the hydrolysis of poly(vinyl acetate); the theoretical monomer, $CH_2=CHOH$, does not exist (see Vinyl polymers, poly(vinyl acetate)). Discovery of PVA was credited to German scientists W. O. Herrmann and W. Haehnel in 1924, and the polymer was commercially introduced into the United States in 1939 (1).

Poly(vinyl alcohol) is one of the very few high molecular weight commercial polymers that is water-soluble (see Resins, water-soluble). It is a dry solid and is available in granular or powdered form. Grades include both the fully hydrolyzed form of poly(vinyl acetate) and products containing residual, ie, unhydrolyzed, acetate groups. Resin properties vary according to the molecular weight of the parent poly(vinyl acetate) and the degree of hydrolysis. A wide range of grades is offered by PVA manufacturers.

The wide range of chemical and physical properties of PVA resins has led to their broad industrial use. They are excellent adhesives (qv) and highly resistant to solvents, oil, and grease. Poly(vinyl alcohol) forms tough, clear films that have high tensile strength and abrasion resistance. Its oxygen-barrier qualities are superior to those of any known polymer (see Barrier polymers); however, PVA must be protected from moisture, which greatly increases the gas-transmission rates through it. Poly(vinyl alcohol) also contributes to emulsification and stabilization of aqueous colloidal dispersions.

The main uses of PVA in the United States are in textile and paper sizing, in adhesives, and as an emulsion-polymerization aid (see Textiles; Papermaking additives; Emulsions). Significant volumes are also used in such diverse applications as joint cements for building construction, water-soluble film for hospital laundry bags, emulsifiers in cosmetics (qv), temporary protective films to prevent scratching of highly polished surfaces, and soil binding to control erosion. Poly(vinyl alcohol) is an intermediate in the production of poly(vinyl butyral), the adhesive interlayer in laminated safety glass (see Laminated materials, glass; Vinyl polymers, poly(vinyl acetal)s). Outside the United States, PVA is also used for textile fiber, although it must be chemically treated to become water-insoluble (2). Poly(vinyl alcohol) fiber is produced in Japan and the People's Republic of China for captive use.

Physical Properties

The physical properties of PVA are controlled by molecular weight and the degree of hydrolysis. The upper portion of Figure 1 shows the variation in properties with molecular weight at a constant degree of hydrolysis (3). Hydrolysis effects at constant molecular weight are given in the lower portion of the figure. Since hydrolysis and molecular weight can be independently controlled in the manufacturing process, a product matrix has evolved that provides the property balance needed for different applications. The PVA product matrix has four important molecular weight ranges and three key hydrolysis levels, although intermediate products are available (see Specifications). Various physical properties are listed in Table 1.

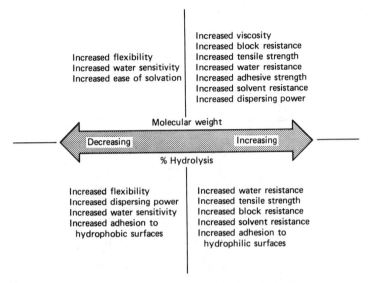

Figure 1. Properties of poly(vinyl alcohol) (3).

Table 1.   **Physical Properties of Poly(Vinyl Alcohol)**

Property	Value
appearance	white-to-cream granular powder
specific gravity	
of solid	1.27–1.31
of 10 wt % sol at 25°C	1.02
thermal stability	gradual discoloration above 100°C; darkens rapidly above 150°C; rapid decomposition above 200°C
refractive index (film) at 20°C	1.55
thermal conductivity, W/(m·K)[a]	0.2
electrical resistivity, ohm·cm	$(3.1–3.8) \times 10^7$
specific heat, J/(g·K)[b]	1.5
melting point (unplasticized), °C	230 for fully hydrolyzed grades; 180–190 for partially hydrolyzed grades
T_g, °C	75–85
storage stability (solid)	indefinite when protected from moisture
flammability	burns similarly to paper
stability in sunlight	excellent

[a] To convert W/(m·K) to (Btu·in.)/(h·ft^2·°F), divide by 0.1441.
[b] To convert J to cal, divide by 4.184.

Solubility.   All commercial PVA grades are soluble in water, the only practical solvent. The ease with which PVA can be dissolved is controlled primarily by the degree of hydrolysis. Figure 2 shows the effect of degree of hydrolysis on solubility with other variables held constant. Fully hydrolyzed products must be heated close to the atmospheric boiling point of water to dissolve completely. Lower temperatures are required as the degree of hydrolysis is decreased until 75–80% hydrolysis is reached, at which point the product is fully cold-water soluble but precipitates upon heating. The

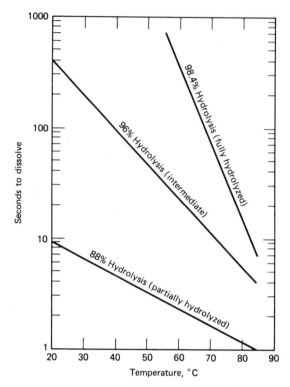

Figure 2.   Solubility of 0.4-mm cast film. Courtesy of Air Products and Chemicals, Inc.

hydrolysis range of 87–89% is considered optimum for both cold- and hot-water solubility. Products with this optimum degree of hydrolysis are commonly referred to as partially hydrolyzed PVA. Regardless of the degree of hydrolysis, all commercial PVA grades remain dissolved upon cooling.

Solubility is also influenced by particle size, ie, surface area, molecular weight, and crystallinity. Decreasing particle size and molecular weight improve the solubility rate of PVA. Crystallinity is induced by heat treatment and retards the solubility rate (4). Because the presence of residual acetate groups reduces the extent of crystallinity, low hydrolysis grades are much less sensitive to heat treatment.

Poly(vinyl alcohol) solutions show a high tolerance toward many electrolytes, as shown in Table 2. Small additions of strong inorganic acids and bases do not precipitate PVA from solution, but the hydrolysis reaction continues to completion at extreme pHs.

Poly(vinyl alcohol) cannot be dissolved by most common organic solvents, eg, gasoline, kerosene, benzene, xylene, trichloroethylene, carbon tetrachloride, methanol, ethylene glycol, acetone, and methyl acetate (5). It has limited solubility in dimethyl sulfoxide; the solubility is in proportion to the residual acetate content. Although there are no good solvents for PVA other than water, up to 50% of lower alcohols can be added to PVA solutions without causing precipitation.

Solution Viscosity.   The viscosity of a PVA solution is controlled by molecular weight, concentration, and to a lesser degree, temperature. Degree of hydrolysis does not strongly affect viscosity, although the viscosity is proportional to degree of hydrolysis at constant molecular weight. Viscosity relationships for low, medium, and

Table 2.   Maximum Salt Concentration in which PVA is Soluble, % in Water[a,b]

Electrolyte	PVA degree of hydrolysis	
	98%	88%
Na_2SO_4	5	4
$(NH_4)_2SO_4$	6	5
$Na_2HPO_4.7H_2O$	8	5
$Na_3PO_4.12H_2O$	8	6
$Na_2HPO_4.H_2O$	9	6
$NaHCO_3$	9	7
$Al_2(SO_4)_3.16H_2O$	10	6
$Na_2S_2O_3.5H_2O$	10	8
$ZnSO_4.7H_2O$	13	10
NaCl; KCl	14	10
$CuSO_4.5H_2O$	15	10
$CH_3COONa.3H_2O$	23	15
$NaNO_3$	24	20

[a] Courtesy of Air Products and Chemicals, Inc.
[b] Determined by adding a 10% solution of PVA dropwise to 50 mL of the salt
solution at increasing concentration until precipitation is observed.

high molecular weight grades are shown in Figure 3. Viscosity, rather than solubility, limits the concentration of PVA solutions. With conventional batch-mixing equipment, the practical concentration limits for low, medium, and high molecular weight resins are ca 30 wt %, 20 wt %, and 15 wt %, respectively.

The viscosities of partially hydrolyzed PVA solutions remain stable if the solutions are stored at high temperatures over a wide range of concentrations. However, viscosities of concentrated solutions of fully hydrolyzed PVA gradually increase over a period of days when stored at room temperature, and gelation occurs in products that contain <1 mol % acetate groups. This viscosity increase or gelation can be reversed by reheating. Lower solution concentrations and lower degrees of hydrolysis eliminate viscosity instability associated with long-term solution storage.

Strength.   The tensile strength of unplasticized, fully hydrolyzed PVA is 55–69 MPa (8,000–10,000 psi) at 50% rh. Tensile strength decreases linearly with increasing humidity and falls to 30–35 MPa (4400–5100 psi) at 80% rh. At constant test conditions, tensile strength varies with degree of hydrolysis and degree of polymerization (Fig. 4). For commercial PVA grades, the tensile response to molecular weight is nonlinear. The greatest variation is between low and medium viscosity resins; tensile differences between medium and high viscosity resins are small (see Specifications). Plasticizers reduce the tensile strength of PVA in proportion to the amount added (see Extrudability).

Tensile elongation of PVA is extremely sensitive to humidity and ranges from ≤10% when completely dry to 300–400% at 80% rh. Plasticizer addition can double these elongation levels. Elongation of PVA films is relatively independent of degree of hydrolysis but is proportional to molecular weight. The tear strength of PVA varies similarly with relative humidity. Addition of small amounts of plasticizer can greatly improve the tear resistance.

Adhesion.   Poly(vinyl alcohol) is well known for its exceptional adhesion to cellulosic surfaces and its binding power in cement formulations. All PVA grades show good adhesion to hydrophilic materials, but the fully hydrolyzed products are con-

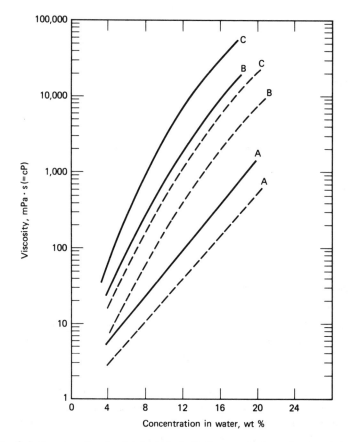

Figure 3. Solution viscosity of poly(vinyl alcohol). Degree of polymerization: A, 800; B, 2000; and C, 2400. Solid line, at 20°C; broken line, at 65°C. Courtesy of Air Products and Chemicals, Inc.

sidered best. Partially hydrolyzed PVA, however, is the only grade that gives good adhesion to hydrophobic substrates, eg, glass, metal, and many plastics. The peel strength of PVA coated on polyester film sharply decreases above 95% hydrolysis (see Fig. 5) (6). Molecular weight does not significantly affect PVA adhesion to a substrate, although it is important for controlling penetration when bonding porous materials.

Chemical modification of PVA with boric acid produces aqueous solutions that exhibit excellent wet tack (7). Adhesives formulated with PVA and boric acid must be acidic to prevent gelation. The solution pH is also critical in developing optimum wet tack. As shown in Figure 6, wet tack for commercially available tackified PVA is optimal at pH 4.7 and negligible at pH 6.0 (3).

Solvent Resistance. Poly(vinyl alcohol) is virtually unaffected by greases, petroleum hydrocarbons, and animal or vegetable oils. Resistance to organic solvents increases with degree of hydrolysis.

Gas Barrier. The oxygen-barrier properties of PVA at low humidity are unsurpassed by any other synthetic resin (see Barrier polymers). Fully hydrolyzed PVA has transmission rates that are 2–4 orders of magnitude lower than poly(vinylidene chloride), the next best barrier polymer. However, as a water-sensitive polymer, PVA exhibits rapid loss of barrier performance above 50% rh (Fig. 7). Other than physically

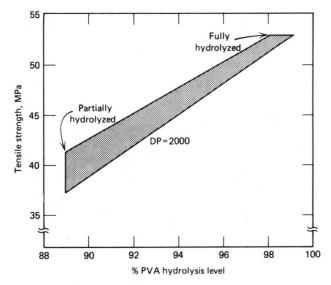

Figure 4. PVA film tensile strength as a function of degree of hydrolysis. Conditions of PVA hydrolysis: 0.4-mm dry-film thickness equilibrated to 22°C at 50% rh, determined by Instron (6). To convert MPa to psi, multiply by 145.

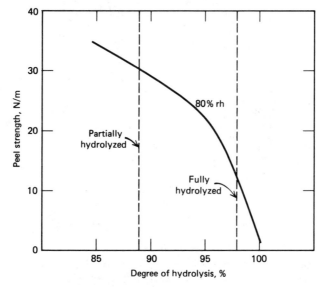

Figure 5. Adhesion expressed as peel strength of poly(vinyl alcohol) film on polyester film (6). To convert N/m to lbf/in., multiply by 5.714×10^{-3}.

protecting the PVA from moisture exposure, there are no known additives or chemical modifiers that effectively reduce this moisture sensitivity.

The gas-barrier performance of PVA is affected by degree of hydrolysis and rapidly diminishes below the 98% hydrolysis level. High ash (as sodium acetate) content also increases the gas-transmission rates and must be <0.1 wt % for optimum performance.

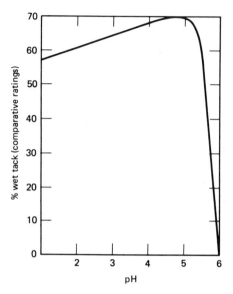

Figure 6. Wet tack of poly(vinyl alcohol) treated with boric acid. Courtesy of Air Products and Chemicals, Inc.

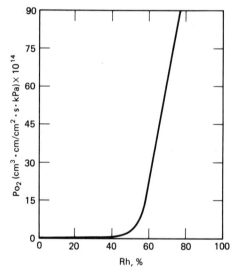

Figure 7. Oxygen permeability of poly(vinyl alcohol) as a function of humidity (99.9% hydrolyzed PVA with 4% glycerol plasticizer). To convert kPa to mm Hg, multiply by 7.5.

Extrudability and Heat Sealing. Unplasticized PVA is not considered a thermoplastic, because the degradation temperature is below the 230°C melting point for fully hydrolyzed grades (see Plastics processing). Although the melting point of PVA decreases with degree of hydrolysis, all grades require plasticizer addition to develop adequate melt flow without severe degradation. High boiling, water-soluble organic compounds containing hydroxyl groups are the most effective plasticizers (qv). Glycerol is one of the most widely used plasticizers, as are poly(ethylene glycol)s. Plasticizers commonly used in the plastics industry are not effective with PVA. Water is the best plasticizer for PVA, but it produces a foamy product as it flashes at the extruder outlet. High boiling methylol compounds, eg, pentaerythritol and 1,2,6-hexanetriol, are preferred for high temperature processing (3). Heat-sealing operations also require plasticized PVA for good performance, although amounts of plasticizer need not be as high as those required for extrusion.

Other. The surface tension of PVA solutions varies linearly with degree of hydrolysis from 88 to 98%, but then increases sharply above 98% (Fig. 8). Partially hydrolyzed (87–89%) grades are commonly used as oil-in-water emulsifiers, especially in the polymerization of poly(vinyl acetate). Poly(vinyl alcohol) film has exceptional clarity and transparency. It also has high transmission values in the near-infrared and ultraviolet ranges. The equilibrium moisture content of PVA is ca 4 wt % at room temperature and 50% rh. Moisture content doubles to ca 8 wt % at 80% rh.

Chemical Properties

Poly(vinyl alcohol) undergoes chemical reactions in a manner similar to other secondary polyhydric alcohols. One of the most important reactions is with aldehydes to form acetals. Poly(vinyl butyral) is formed by the reaction of PVA with butyraldehyde. Another example is the reaction of PVA with formaldehyde to form poly(vinyl formal). Other basic reactions include the formation of esters upon reaction with acids or anhydrides and ether-forming reactions. Examples of the latter include cyanoethylation (qv) of PVA with acrylonitrile and reaction with ethylene oxide to form hydroxyethyl groups. An excellent summary of these PVA reactions is given in ref. 8.

Cross-Linking. Poly(vinyl alcohol) can be readily cross-linked for improved water resistance. The most practical means of cross-linking PVA is with chemical additives, eg, glyoxal, urea–formaldehydes, and melamine–formaldehydes (see Amino resins). Trimethylolmelamine is often preferred if a low temperature is required. An acid catalyst, eg, ammonium sulfate or ammonium chloride, is necessary with the formal-

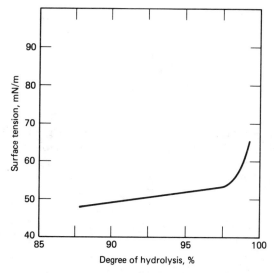

Figure 8. Surface tension of aqueous poly(vinyl alcohol) solution at 20°C (1.5% solution concentration of PVA with a 2000 DP). To convert mN/m (= dyn/cm) to lbf/in., multiply by 5.714 × 10⁻⁶. Courtesy of Air Products and Chemicals, Inc.

dehyde cross-linkers. Metal compounds can also be effective insolubilizers for PVA. These additives include strongly chelating metal salts of copper and nickel, eg, cupric ammonium complexes; chromium complexes; organic titanates; and dichromates. The heat treatment during drying of PVA films or coatings is generally sufficient to complete the cross-linking reaction, although when dichromates are used, the reaction is best catalyzed by ultraviolet light. Cross-linking slowly takes place even at room temperature, and prolonged storage of the treated PVA solution should be avoided. Cross-linking can also be accomplished by simply heating dry PVA to >100°C; this dehydrates the polymer and yields an unsaturated carbon backbone. Intermolecular reactions between unsaturated groups in adjacent polymer chains form permanent cross-links. However, thermal cross-linking is not considered a practical reaction because it is accompanied by polymer decomposition.

Although PVA film can be rendered insoluble by cross-linking, it swells in water and loses strength upon extended exposure. Complete water insensitivity cannot be achieved, although it improves with degree of hydrolysis. Fiber-grade PVA has a degree of hydrolysis of at least 99.9%; it exhibits negligible swelling in water, although it absorbs moisture like many natural fibers.

Gelation. The controlled gelation of PVA solution is important where penetration into a porous substrate is undesirable, eg, paper coatings and adhesives. Boric acid and borax react strongly with PVA and are widely used industrially as gelling agents (see Boron compounds, boric acid and esters). Poly(vinyl alcohol) is extremely sensitive to borax, which causes gelation by forming a bisdiol complex as shown below.

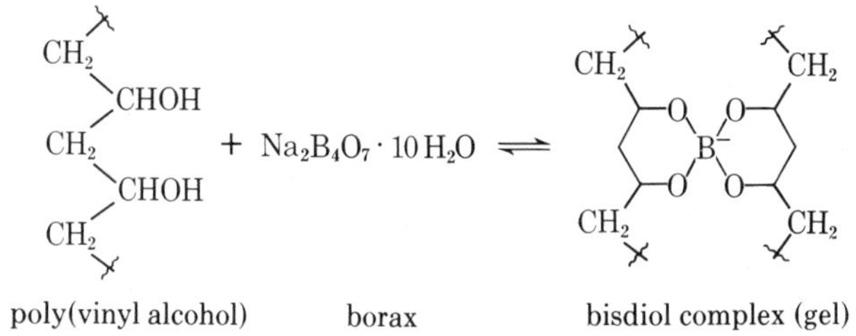

poly(vinyl alcohol)   borax   bisdiol complex (gel)

As little as 0.1% borax, based on solution weight, can cause thermally irreversible gelation. Boric acid forms a weaker monodiol complex and is preferred for controlled, partial gelation of PVA (7). The reaction is very sensitive to pH, and full gelation occurs above pH 6.

Congo red is an excellent gelling agent for PVA in applications where colored films or coatings can be tolerated. Colorless, thermally reversible gels form upon the addition of resorcinol, catechol, gallic acid, and 2,4-dihydroxybenzoic acid to PVA solutions. However, the use of organic gelling agents with PVA is limited industrially.

Graft Copolymers. Poly(vinyl alcohol) can be grafted with other monomers to modify its properties. The graft reaction is usually carried out in aqueous solution with free-radical or ionic catalysts. The industrial importance of PVA graft copolymers has not been in the production of new products; rather, it arises because their formation during emulsion polymerizations, in which PVA is used as a stabilizer, affects the final product performance (9). Apparently, no comprehensive survey of grafting reactions has been published, but a wide range of monomers has been successfully grafted to PVA; these include vinyl acetate, styrene, acrylamide, acrylic acid, 2-hydroxyethyl methacrylate, acrylonitrile, 1,3-butadiene, methyl methacrylate, methacrylic acid, vinylidene chloride, vinyl chloride, and a variety of acrylate esters.

Manufacture

All PVA manufacture involves poly(vinyl acetate) as the starting material. The theoretical monomer, vinyl alcohol (CH_2=CHOH), does not exist except as an insignificantly small component of samples of its keto tautomer, acetaldehyde (CH_3CHO). Although other poly(vinyl ester)s can be used as starting materials, economics favors the use of poly(vinyl acetate). Conversion of poly(vinyl acetate) to PVA is generally accomplished by base-catalyzed methanolysis (10); sodium hydroxide is the usual base.

The alcoholysis reaction can also be catalyzed with a strong acid, eg, sulfuric or hydrochloric. However, this process is not practical industrially because of the relatively slow reaction rate and severe corrosion problems (10).

Polymerization of vinyl acetate and conversion of the polymer to PVA are illustrated below.

$$n \quad \begin{matrix} O \\ \parallel \\ OCCH_3 \\ | \\ CH{=}CH_2 \end{matrix} \quad \xrightarrow[\text{catalyst}]{\text{free-radical}} \quad \left[\begin{matrix} O \\ \parallel \\ OCCH_3 \\ | \\ CHCH_2 \end{matrix} \right]_n$$

vinyl acetate   poly(vinyl acetate)

$$\text{poly(vinyl acetate)} + n\,CH_3OH \xrightarrow{\text{NaOH}} \left[\begin{matrix} OH \\ | \\ CHCH_2 \end{matrix} \right]_n + n\,CH_3OCCH_3 \begin{matrix} O \\ \parallel \\ \\ \end{matrix}$$

poly(vinyl alcohol)   methyl
(PVA)   acetate

Poly(vinyl acetate) polymerization is done by conventional processes, eg, solution, bulk, or emulsion polymerization (see Polymerization mechanisms and processes). Solution polymerization is favored because the subsequent alcoholysis reaction requires solvent addition (10). The polymerization step controls the ultimate molecular weight of the PVA. Catalyst selection, temperature, and solvent control the degree of polymerization; acetaldehyde is an effective chain-transfer agent, as are agents commonly used in polymerization of vinyl monomers.

The degree of hydrolysis of PVA is controlled during the alcoholysis reaction and is independent of molecular-weight control. Fully hydrolyzed PVA is obtained if methanolysis is allowed to go to completion. The reaction can be terminated by neutralizing or removing the sodium hydroxide catalyst. The addition of small amounts of water to the reactants promotes saponification of poly(vinyl acetate), which consumes sodium hydroxide. The extent of hydrolysis is inversely proportional to the amount of water added. A disadvantage of water addition is an increase in by-product sodium acetate, which is present as ash in all commercially available grades of PVA. The alcoholysis reaction can be carried out in a highly agitated slurry process; a fine precipitate forms as the poly(vinyl acetate) converts into PVA. The product is then washed with methanol and is filtered and dried. A moving-belt process allows the PVA to form a gel and subsequently to be cut into granular form (11).

The alcoholysis process yields methyl acetate as a by-product. The methyl acetate can be used as a solvent or it can be processed to recover methanol and acetic acid. One such process involves mixing the methyl acetate with water and passing them through a cation-exchange resin to catalyze the hydrolysis reaction (10):

$$\begin{matrix} O \\ \parallel \\ CH_3OCCH_3 \end{matrix} + H_2O \rightleftharpoons CH_3OH + \begin{matrix} O \\ \parallel \\ HOCCH_3 \end{matrix}$$

Methanol recovered from this process can be totally recycled to the methanolysis step, and the acetic acid is sold as a by-product.

Waste Disposal. Poly(vinyl alcohol) can be effectively biodegraded in acclimatized, activated sludge wastewater systems. If the waste-treatment system has not included PVA, however, it may take several weeks to acclimate. The BOD_{30} (biological oxygen demand, 30-day measurement) of PVA has been measured at 1.2 mg/mg, vs a COD (chemical oxygen demand) value of 1.64 mg/mg (theoretical 1.82 mg/mg) (12). Over 90% of the PVA is typically degraded in a properly functioning wastewater-treatment system. The presence of PVA does not appear to interfere with the treatment of other biodegradable materials normally present in the wastewater (see Water, sewage).

Production

There are three U.S. producers of PVA, as shown in Table 3. Capacity of PVA in 1980 in noncommunist countries was >360,000 metric tons. Over half of this is in Japan, which also has more than twice the total U.S. capacity. Western European capacity is collectively less than that of the United States, and all remaining producers have <10% of total capacity. A large portion of the Japanese production is captively consumed for fiber. Commercial development of PVA fiber outside Japan has been undertaken, but competitive synthetics have been more economically attractive except in the People's Republic of China, where an unknown quantity of PVA fiber is produced.

Economic Aspects

The PVA process is very capital intensive, requiring separate facilities for production of poly(vinyl acetate), alcoholysis of poly(vinyl acetate) to PVA, and recovery of by-product acetic acid from the ester formed during alcoholysis. Capital costs are far in excess of those associated with the traditional polymerization of vinyl resins for which the monomer exists.

Because it is a petroleum-based product, vinyl acetate costs have followed the sharp rise in crude-oil costs during 1970–1980. The 1980 price of vinyl acetate monomer was ca $0.62/kg. Energy is also a large factor in total PVA manufacturing costs. The selling price of ca $0.77/kg PVA in 1970 increased to $2.20/kg in 1980. Future prices should continue to reflect petroleum costs.

Table 3.   Worldwide Poly(Vinyl Alcohol) Producers

Producer	Trade name
United States	
Air Products and Chemicals, Inc.	Vinol
DuPont	Elvanol
Monsanto Company	Gelvatol
Other	
Hoechst, FRG	Mowiol
Kuraray, Japan	Poval
Nippon Gohsei, Japan	Gohsenol

Specifications and Regulations

Three important commercially available types of PVA are distinguished by the mole percent residual acetate groups in the resin, ie, fully hydrolyzed (1–2 mol % acetate), intermediate hydrolyzed (3–7 mol %), and partially hydrolyzed (10–15 mol %) PVA. Poly(vinyl alcohol)s with other degrees of hydrolysis are produced, but collectively, they have a much smaller market share than any of the three principal grades. When no reference is made to the degree of hydrolysis in describing PVA, it is generally assumed to be a fully hydrolyzed grade.

Poly(vinyl alcohol) is produced in four general molecular weight ranges, as shown in Table 4. Several other molecular weight resins are produced, but they have only a minor market share. Industry practice expresses the molecular weight of a particular grade in terms of its 4% aqueous solution viscosity. There is no limit to the viscosity that can be generated by customer blending of the available molecular weights. Products of different degrees of hydrolysis can also be blended to develop a particular performance characteristic that is between the grades, eg, intermediate solubility. Blended products have a broad distribution in molecular weight and degree of hydrolysis, which may be undesirable in some applications.

FDA regulations on the use of PVA as an indirect food additive are listed in Table 5.

Table 4. Molecular Weight of Main Commercial Poly(Vinyl Alcohol) Grades[a]

Viscosity grade	Nominal M_n	4% solution viscosity, mPa·s (= cP)[b]
low	25,000	5–7
intermediate	40,000	13–16
medium	60,000	28–32
high	100,000	55–65

[a] Courtesy of Air Products and Chemicals, Inc.
[b] Measured at 20°C with Brookfield viscometer.

Table 5. FDA Regulations for Poly(Vinyl Alcohol) as an Indirect Food Additive

Regulation	Description
181.30	prior sanctioned substances used in manufacture of paper and paperboard products used in food packaging for fatty foods only
175.105	adhesives, no limitations
176.170	components of paper and paperboard in contact with aqueous and fatty foods, extractive limitations
176.180	components of paper and paperboard in contact with dry food, no limitations
177.1200	cellophane coating, no limitations
177.1670	poly(vinyl alcohol) film
177.2260	filters, resin-bonded where filter fiber is cellulose
177.2600	filters, resin-bonded; extractables must be less than 0.08 mg/cm² (0.5 mg/sq in.)
175.300	resinous and polymeric coatings
175.320	resinous and polymeric coatings for polyolefin films; net extractable less than 0.08 mg/cm² (0.5 mg/sq in.)
177.2800	textiles and textile fibers, for dry foods only
178.3910	surface lubricants in the manufacture of metallic articles

Storage and Handling

Poly(vinyl alcohol) is an innocuous material with unlimited storage life in dry form. It is most commonly supplied in 22.7-kg bags that have moisture barriers to prevent caking from water absorption. It is also available in bulk form. During transport and handling, granular PVA forms an explosive mixture with air with a low severity rating of 0.1 on a scale in which coal dust has a rating of 1.0 (Bureau of Mines rating) (13). However, the explosive hazard is highly dependent upon particle size, and extremely fine dust has a stronger explosive rating of 1.0–2.0. Residual methanol and methyl acetate present in all commercially available grades of PVA can also accumulate in the air space of bulk storage tanks, especially at elevated temperatures. Precautions should be taken to ventilate the air space in large vessels and to eliminate spark-producing equipment in the area. Handling of PVA in bags poses no significant explosion risk.

Solution Preparation and Handling. Poly(vinyl alcohol) must be completely dispersed in water at room temperature or lower prior to heating. Good agitation is important to prevent lump formation during addition of the solid PVA to water. Water should never be added to the dry solid. A large-diameter, low speed agitator is preferable to provide good turnover of the liquid surface without excessive air entrainment. Agitation requirements increase with solution concentration (viscosity) and decreasing degree of hydrolysis, which is associated with greater cold-water solubility and a tendency to lump.

After dispersion of the PVA, the slurry temperature should be increased by direct steam injection, if possible. This eliminates the potential for PVA buildup on steam coils or heated surfaces caused by localized water evaporation. When heat-set, this PVA may be difficult to remove, which greatly reduces the heat-transfer rate. About 15% of the total water charge should be withheld from the slurry step if direct steam injection is used for heating.

Aqueous solutions of PVA are stable upon storage but must be protected from rust contamination and bacterial growth. Stainless-steel or plastic containers are recommended for long-term storage. Formaldehyde is an effective and inexpensive biocide if added at 500 ppm to PVA solutions. Many other biocides are effective for use with PVA, including several FDA-approved compounds (3).

Prolonged heating of a PVA solution has a negligible effect on its properties. However, the addition of a strong acid or base to solutions of partially hydrolyzed PVA can lead to an increase in degree of hydrolysis over a period of several days. Fully hydrolyzed grades are essentially unaffected by pH.

Health and Safety Factors, Toxicology

Poly(vinyl alcohol) is not a hazardous material according to the American Standard for Precautionary Labeling of Hazardous Industrial Chemicals (ANSI Z129.1-1976). Short-term inhalation of PVA dust has no known health significance but can cause discomfort and should be avoided in accordance with industry standards for exposure to nuisance dust. The dust is mildly irritating to the eyes. There are no known dermal effects arising from short-term exposure to either solid PVA or its aqueous solutions; it can easily be removed by washing with water. Poly(vinyl alcohol) has a low oral toxicity rating; the acute oral LD_{50} (rats) is >10,000 mg/kg.

Poly(vinyl alcohol) may be used in many food-contact applications, including food-packaging adhesives, and as a coating for paper and polymeric films (see Table 5). It is listed in *United States Pharmacopeia XX* as a pharmaceutical aid (14). Many cosmetics also are based on PVA; the total cosmetic formulation must meet FDA regulations.

Uses

The largest U.S. application for PVA is in textile sizing. In 1978, this application alone accounted for close to half the total U.S. consumption of PVA, excluding poly-(vinyl butyral) production (15). Poly(vinyl alcohol) is also widely used as a binder in adhesive formulations, as a polymerization aid, and as a paper size or coating, as shown in Table 6 (15).

Miscellaneous uses for PVA are as main components of water-soluble film, non-woven fabric binders, thickeners for latex coatings, binders for phosphorescent pigments in television picture tubes, and principal constituents of cosmetic face masks. Overall U.S. consumption is forecast to grow at ca 4%/yr into the mid-1980s. About two thirds of the PVA consumed in the United States consists of fully hydrolyzed grades.

On a worldwide basis, textiles warp sizing is also the largest application for PVA. Because of Japanese consumption of PVA for fiber, this market is larger than those of adhesives, polymerization, and paper coatings.

Textiles. *Warp Sizing.* Poly(vinyl alcohol) is an excellent textile warp size because of its superior strength, adhesion, flexibility, and film-forming properties. With the proper hydrolysis level, PVA adhesion can be tailored to any natural or synthetic fiber or fiber blend. For example, adhesion to hydrophobic fibers, eg, polyester or glass, is enhanced as the degree of PVA hydrolysis is reduced (Fig. 5). Fully hydrolyzed grades adhere well to cotton and other hydrophilic fibers and impart the highest tensile strength for equivalent amounts added (equivalent add-on) relative to partially hydrolyzed grades; therefore, to achieve equivalent strength levels, less fully hydrolyzed PVA can often be applied. Until ca 1970, fully hydrolyzed PVA was the preferred grade for textile warp sizing and accounted for two thirds of the PVA used. However, there has been a trend away from fully hydrolyzed PVA because of its poor adhesion to many synthetic fibers and the difficulty in desizing heat-set fabric (16). Poly(vinyl alcohol) products with solubility characteristics of the intermediate degree of hydrolysis (see Fig. 2) represent the largest segment of PVA use in this application (17). Further reductions in degree of hydrolysis improve the adhesion and solubility characteristics

Table 6.   U.S. Consumption of Poly(Vinyl Alcohol) in 1978[a,b], 1000 t

textiles	28
adhesives	15
polymerization aid	8
paper coatings	5
miscellaneous	4
Total	*60*

[a] Ref. 15.
[b] Excluding poly(vinyl butyral) production.

at the expense of reduced tensile strength. The market share for partially hydrolyzed grades has grown significantly. Nevertheless, the best hydrolysis for warp sizing is specific to each manufacturing operation and is controlled by factors such as yarn quality, yarn construction, and speed and type of loom.

Machine conditions and fabric construction also determine the molecular weight required for good weaving performance. Generally, the highest molecular weight grades provide the best protection at equivalent add-on. However, they also require the lowest sizing-solution concentrations to maintain a manageable viscosity, and this increases drying costs and limits production rates. The most popular warp-size grades are in the medium molecular weight range; these provide a balance in performance and production rate. Low molecular weight PVA has some application in sizing of filament yarn, where minimum strength enhancement is required.

Poly(vinyl alcohol) can be recovered from desize wash water by means of commercial ultrafiltration equipment. Recovery rates and effluent losses are inversely proportional to PVA solution viscosity and independent of degree of hydrolysis.

Typical warp-sizing formulations contain a lubricant wax and other processing aids at 5–10 wt % of the PVA solids. Size-application levels are highly dependent upon both fabric and machine variables, but usually are 5–15%.

Miscellaneous. Poly(vinyl alcohol) is used in wash-and-wear finishes for knit and woven fabrics (3) (see Textiles, finishes). Superhydrolyzed PVA is used in this application because of its ability to be heat-set into a permanent finish. Poly(vinyl alcohol) can also serve as a binder for nonwoven fabrics (see Nonwoven textiles). The main factor that limits greater use of PVA in this application is the stiffness which it imparts to the fabric. Plasticizers, eg, glycerol, are used where stiffness is a problem. Poly(vinyl alcohol) has greatest applicability in nonwoven products where solvent resistance is important.

Adhesives. Poly(vinyl alcohol) is best known as a superior adhesive for hydrophilic materials. Adhesive formulations based upon fully hydrolyzed PVA are used in the manufacture of paperboard, sealed cases and cartons, book padding, wound tube stock, and laminated solid fiberboard. Where water resistance is important, superhydrolyzed (>99% hydrolysis) PVA is preferred. Higher molecular weight grades also improve water resistance.

Partially hydrolyzed PVA is widely used as an emulsifier and protective colloid for polymerization of emulsion adhesives, especially poly(vinyl acetate) emulsions. It also promotes wetting and penetration of the adherents and improves bond strength. Postpolymerization addition of PVA to an emulsion allows further optimization of its performance. High concentrations of partially hydrolyzed PVA can be used for remoistenable adhesive formulations and for adhesives requiring easy cleanup. High molecular weight grades increase the viscosity without significantly increasing solids content. All grades of PVA promote the acceptance of starch or clay fillers into a formulation, which prevents excessive adhesive penetration into porous surfaces. Poly(vinyl alcohol) also improves solvent resistance and can be used to control the drying rate and speed of set in commercial adhesive formulations.

The controlled addition of boric acid to PVA can further improve the wet tack of unmodified resin, and it can reduce adhesive penetration into porous surfaces (7). Tackified PVA products are commercially available in a wide range of wet-tack levels (3). The tackified resins have higher solution viscosities than unmodified PVA and must be maintained at pH 4.6–4.9 for optimum wet adhesion. Alkaline pH must be avoided to prevent irreversible gelation (see Chemical Properties). Dilute (5–10%) solutions of boric acid-treated PVA slowly increase in viscosity upon prolonged storage (greater than one week) but do not readily gel. Higher concentrations, ie, initial viscosities of ≥ 10 Pa·s (100 P) at room temperature, can gel within a day, but the gel is thermally reversible. Since tackified PVA resins can be maintained at elevated temperatures without chemical change, viscosity drift can be avoided by high temperature storage. However, care must be taken to prevent water evaporation.

Paper Coating. Poly(vinyl alcohol) is used for paper sizing to improve strength, increase grease and solvent resistance, and reduce porosity (18). Strength factors, eg, fold resistance, mullen burst, tensile strength, and tear strength, can be dramatically improved with a surface application of PVA. The excellent film-forming ability of PVA reduces porosity and provides a solvent- and abrasion-resistant surface. Water-resistant coatings can be achieved by addition of PVA insolubilizers to the size formulations, which cure during the drying operation. Boric acid treatment of PVA is used to reduce size penetration on an extremely porous surface. The size coatings can also be mixed with pigments to improve optical properties, eg, opacity, brightness, and gloss. Poly(vinyl alcohol)-coated paper can be calendered to develop exceptional smoothness and surface finish. Finished grade specifications can often be met with low quality paper stock that has been sized with PVA. Increased recycling of paper products leads to property losses which can be at least partially offset by PVA surface sizing (see Papermaking additives; Recycling).

Paper coated with PVA is also used for packaging foods and chemicals where grease, oil, and oxygen-barrier performance is important. It is used as a release coating on paper for casting poly(vinyl chloride) plastisol film; when the film is cured, it can be peeled from the paper, which is then reused. Poly(vinyl alcohol) is also an important component in the manufacture of release paper for pressure-sensitive adhesive labels and tape (see Abherents). The release paper is sized with PVA, which prevents absorption of expensive silicone release agents.

Building Products. Poly(vinyl alcohol) is widely used in the formulation of cement coatings and finishes. Partially hydrolyzed grades are best in this application because of their superior adhesion to cement materials. Addition of 1–5 wt % PVA to a cement mix greatly enhances its strength, improves water retention, and promotes bonding to a variety of substrates. For many dry-mix products, it is important that the product have rapid cold-water solubility, which can be achieved with finely ground PVA. Partially hydrolyzed grades are commercially available in both standard and fine-particle-size forms (3) (see also Cement).

The addition of PVA to dry-wall joint cements is the most common building-products application. It is also used in stucco finishes, thin-bed tile mortars, cement paint and roof coatings, and cement toppings for repairs.

Fiber. Poly(vinyl alcohol) fiber has good strength characteristics and an attractive feel in fabrics. Only superhydrolyzed PVA is used to develop sufficient water insensitivity; chemical cross-linking is also required. Japan is the only noncommunist country that widely uses PVA fiber in textiles, although its use has been considered in the United States. Tire cord (qv) is another potential application for PVA fiber, but economics have favored competitive materials. Performance advantages of PVA fiber over other synthetics have not balanced the greater difficulties in extrusion, which requires a plasticizer, and the necessary chemical treatment to build water resistance. Only the particular economics and raw-material availability in Japan continue to make it an attractive fiber in that country.

Other. Poly(vinyl alcohol) film can be produced by solution casting or extrusion. The latter is difficult because of the close proximity of melting point and decomposition temperature, which requires the use of a plasticizer. Film casting is most common, because unplasticized films of all molecular weights and extents of hydrolysis can be readily produced. Water solubility of the films can be controlled by selection of the proper degree of hydrolysis (see Physical Properties). These films can be used for packaging preweighed bleach, insecticides, and other materials that are to be dissolved in water without removal of the package. Hospital laundry bags produced from PVA eliminate the need for handling contaminated linen when it is placed in the washing machine.

Poly(vinyl alcohol) is useful as a temporary protective coating for metals, plastics, and ceramics. Such a coating reduces damage from mechanical or chemical agents during manufacture, transport, and storage. The protective films can be removed by peeling or washing with water.

The ultraviolet cross-linking of PVA with dichromates is the basis for its use in photoengraving, photogravure, screen printing, printed-circuit manufacture, and color television tube manufacture.

The emulsifying, thickening, and film-forming properties of PVA are utilized in many cosmetic applications. Partially hydrolyzed grades are generally used because of their superior water solubility and, thus, ease of removal. Poly(vinyl alcohol) also is used as a viscosity builder for aqueous solutions or dispersions.

BIBLIOGRAPHY

"Polyvinyl alcohol" in *ECT* 1st ed., Vol. 14, pp. 710–715, by T. P. G. Shaw and L. M. Germain, Shawinigan Chemicals, Limited; "Poly(Vinyl Alcohol)" in *ECT* 2nd ed., Vol. 21, pp. 353–368, by Morton Leeds, Air Reduction Company, Ltd.

1. U.S. Pat. 1,672,156 (June 5, 1928), W. O. Herrmann and W. Haehnel (to Consortium fur Elektrochemische Industrie); Ger. Pat. 450,286 (Oct. 5, 1927), W. O. Herrmann and W. Haehnel (to Consortium fur Elektrochemische Industrie).
2. U.S. Pat. 3,084,989 (Apr. 9, 1963), Hitoshi Ave and Yasuji Ono ($\frac{1}{4}$ to Air Products and Chemicals, Inc., $\frac{3}{4}$ to Kurashiki Rayon Co., Ltd.).
3. *Vinol Product Handbook*, Air Products and Chemicals, Inc., Allentown, Pa., 1980.
4. R. K. Tubbs, H. K. Inskip, and P. M. Subramanian, *Society of Chemical Industry*, *Monograph 30*, London, 1968, pp. 88–103.
5. E. S. Peierls, *Mod. Plast.* **18**(6), 53 (1941).
6. *Vinol WS-53 Product Bulletin*, Air Products and Chemicals, Inc., Allentown, Pa., 1980.
7. U.S. Pat. 3,135,648 (June 2, 1964), R. L. Hawkins (to Air Products and Chemicals, Inc.).
8. C. A. Finch in C. A. Finch, ed., *Polyvinyl Alcohol*, John Wiley & Sons., Inc., New York, 1973, pp. 183–202.
9. E. V. Gulbekian and G. E. Reynolds in ref. 8, pp. 427–460.
10. Y. Chin, *Polyvinyl Acetate and Polyvinyl Alcohol*, private report by Process Economics Program, Stanford Research Institute, Menlo Park, Calif., 1976, Report 57A.
11. R. Demry, *Polyvinyl Acetate and Polyvinyl Alcohol*, private report by Process Economics Program, Stanford Research Institute, Menlo Park, Calif., 1970, Report 57.
12. J. P. Casey and D. G. Manly, *Polyvinyl Alcohol Biodegradation by Oxygen-Activated Sludge*, *Proceedings of the 3rd International Biodegradation Symposium, University of Rhode Island, August 1975*, Applied Science Publishers, Ltd., London, 1975.
13. *Material Safety Data Sheet for Polyvinyl Alcohol*, Air Products and Chemicals, Inc., Allentown, Pa., 1981.
14. *The United States Pharmacopeia XX (USP XX–NF XV)*, The United States Pharmacopeial Convention, Rockville, Md., 1980.

15. H. E. Frey, *Chemical Economics Handbook Marketing Research Report on Vinyl Acetate, Polyvinyl Acetate, and Polyvinyl Alcohol*, Stanford Research Institute, Menlo Park, Calif., Nov. 1979.

16. J. E. Moreland, *Text. Chem. Color.* **12**(4), 21 (Apr. 1980).

17. D. L. Nehrenberg, *AATCC RA73 Research Committee Warp Size Symposium, Atlanta, March 7, 1981*, Atlanta, Ga., published by DuPont Co., Wilmington, Del.

18. G. Davidowich and G. D. Miller, *Pulp Pap.* **50**(8), 118 (July 1976).

General References

Ref. 8 is also a general reference.

J. G. Pritchard, *Poly(Vinyl Alcohol) Basic Properties and Uses*, Gordon and Breach, Science Publishers, Inc., New York, 1970.

R. L. Davidson, ed., *Handbook of Water-Soluble Gums and Resins*, McGraw-Hill Book Company, New York, 1980, Chapt. 20.

M. K. Lindemann in N. M. Bikales, ed., *Encyclopedia of Polymer Science and Technology*, Vol. 14, John Wiley & Sons, Inc., New York, 1971, p. 149.

DAVID L. CINCERA
Air Products and Chemicals, Inc.

VINYL CHLORIDE AND POLY(VINYL CHLORIDE)

Vinyl chloride,   865
Poly(vinyl chloride),   886

VINYL CHLORIDE

Vinyl chloride [*75-01-4*], CH_2=CHCl, by virtue of the wide range of applications for its polymers in both flexible and rigid forms, is one of the largest commodity chemicals in the United States and is an important item of international commerce. Growth in vinyl chloride production is directly related to demand for its polymers and, on an energy-equivalent basis, rigid poly(vinyl chloride) [*9002-86-2*] (PVC) is one of the most energy-efficient construction materials available (See Engineering plastics). Initial development of the vinyl chloride industry stemmed from the discovery that, with plasticizers, PVC can be readily processed and converted into a rubbery product (1). However, it was not until after World War II that vinyl chloride production grew rapidly as a result of the increased volume of PVC products for the consumer market.

Vinyl chloride was first prepared from the reaction of dichloroethane with alcoholic potash (2). On prolonged exposure to sunlight in a sealed tube, white flakes deposited from the vinyl chloride. This white solid was studied in 1872 and was described as *Kaperenchorid*, empirical formula $(C_2H_3Cl)_n$ (3). The reaction between hydrogen chloride and acetylene was studied and, in 1912, a patent was obtained for the use of mercuric chloride as a catalyst for this reaction and established an effective industrial route to vinyl chloride (4). The acetylene-based process has been supplanted by a balanced process from ethylene and chlorine in which vinyl chloride is made by pyrolysis of ethylene dichloride (1,2-dichloroethane) (see also Chlorocarbons and chlorohydrocarbons).

Vinyl chloride, chloroethylene, is a colorless gas at normal temperature and pressure. Industrially, vinyl chloride is handled as the liquid (bp, −13.4°C). However, no human contact with the liquid is allowed. Vinyl chloride is an OSHA-regulated material.

Physical Properties

The physical properties of vinyl chloride are listed in Table 1. Vinyl chloride is slightly soluble in water, 0.11 g/100 g H_2O at 25°C. The solubility of water in vinyl chloride at −15°C is 0.03 g/100 g $CH_2{=}CHCl$. Vinyl chloride is soluble in hydrocarbons, oil, alcohol, chlorinated solvents, and most common organic liquids.

Table 1. Physical Properties of Vinyl Chloride

Property	Value	Refs.
molecular weight	62.499	5
melting point, °C	−153.8	5
boiling point, °C	−13.4	5
specific heat, J/(kg·K)[a]		
vapor at 20°C	858	6
liquid at 20°C	1352	6
critical temperature, °C	156.6	5
critical pressure, MPa[b]	5.60	5
critical volume, cm³/mol	169	5
compressibility factor	0.265	5
Pitzer's acentric factor	0.122	5
dipole moment, C·m[c]	5.0×10^{-30}	5
latent heat of fusion, J/g[a]	75.9	
latent heat of evaporation, J/g[a]	330	5
standard enthalpy of formation, kJ/mol[a]	35.18	5
standard Gibbs energy of formation, kJ/mol[a]	51.5	5
vapor pressure, kPa[b]		
−30°C	50.7	5, 7
−20°C	78.0	
−10°C	115	
0°C	164	
viscosity, mPa·s (= cP)		
−40°C	0.3388	5
−30°C	0.3028	
−20°C	0.2730	
−10°C	0.2481	
explosive limits in air, vol %	4–22	8
self-ignition temperature, °C	472	8
flash point (open-cup), °C	−77.75	8
liquid density (at −14.2°C), g/cm³	0.969	5

[a] To convert J to cal, divide by 4.184.
[b] To convert MPa to psi, multiply by 145.
[c] To convert C·m to D, divide by 3.336×10^{-30}.

Reactions

Polymerization. The most important reaction of vinyl chloride is its polymerization and copolymerization in the presence of a radical-generating catalyst (see below).

Substitution at the Carbon–Chlorine Bond. Vinyl chloride is generally considered to be inert to nucleophilic replacement compared to alkyl halides. However, recent work has shown that the chlorine can be exchanged rapidly under nucleophilic conditions in the presence of palladium and other transition metals (9–15). Vinyl acetates, alcoholates, vinyl esters, and vinyl ethers can be readily produced from these reactions. The mechanism for the reaction generally is thought to proceed through an addition–elimination path with initial formation of a π-complex (9,16).

Reaction of vinyl chloride and carbon monoxide with low valent transition metals yields acryloyl chloride (16). The mechanism for this reaction may proceed by oxidative addition of a low valent metal. Use of alcohol as a solvent for carbonylation with reduced Pd catalysts gives vinyl esters (17–18). Addition of radical scavengers to the reaction solutions does not inhibit the reactions, showing that the products do not form by a free-radical path during the oxidative-addition reaction.

Reaction of vinyl chloride with butyllithium and then with CO_2 in ether, at low temperatures, gives high yields of α,β-unsaturated carboxylic acids (19). When vinyl chloride reacts with borane in THF (tetrahydrofuran) solution, chlorine is replaced by hydrogen to yield ethylene (20).

Vinylmagnesium chloride (Grignard reagent) can be directly prepared from vinyl chloride (21). This compound can be used to add a vinyl anion to numerous organic functional groups in the normal Grignard reaction (qv) sequence. The vinylmagnesium compound can be coupled with cuprous chloride at $-60°C$ to give butadiene (22). It also adds to α,β-unsaturated compounds giving γ,δ-unsaturated compounds (23). Vinyl ketones and alcohols can also be prepared by the addition of vinylmagnesium chloride to organic acids (24–25).

A useful vinyllithium compound can be formed directly from vinyl chloride by means of a lithium dispersion containing 2 wt % sodium at 0–10°C (26). The vinyllithium compound is a reactive intermediate for the formation of vinyl alcohols from aldehydes, vinyl ketones from organic acids, vinyl sulfides from disulfides, and monosubstituted alkenes from organic halides (27–29). It can also be converted to a vinyl copper compound, which can be used to introduce a vinyl group stereoselectively into a variety of α,β-unsaturated systems (30). Vinyllithium reagents also can be converted to secondary alcohols with trialkylboranes (31) (see also Hydroboration).

Ethyl vinyl ether is produced from vinyl chloride and sodium ethoxide (32) (see Vinyl polymers, vinyl ether, monomers and polymers). A wide range of alcoholates can take part in this reaction. The reaction of vinyl chloride with hydrogen fluoride over a Cr_2O_3-on-Al_2O_3 catalyst at 380°C yields vinyl fluoride (33).

Oxidation. The oxidation of vinyl chloride by a chlorine-atom-sensitized reaction yields 74 vol % ClCHO and 25 vol % CO in the gas phase with 30–32% conversion (34). The reaction proceeds by a nonchain path at high oxygen–chlorine ratios. The reaction with triplet oxygen $[O(3p)]$ atoms gives high yields of CO and chloroacetaldehyde, (CH_2ClCHO); the rate of this reaction with $O(3p)$ atoms has been reported (35). Oxidation of vinyl chloride with oxygen in the gas phase above 250°C produces no C_2 carbonyl compounds; the main products are CO, HCl, HCO_2H, and ClCHO. This oxidation reaction proceeds by a nonradical path which is unique to vinyl chloride among the chloroolefin compounds. The ozone reaction with vinyl chloride can be used to remove it from gas streams in a vinyl chloride production plant (36–37). Ozonolysis in liquid or gas phase gives formic acid and formyl chloride. At $-15°$ to $-20°C$, vinyl chloride reacts with oxygen, with uv initiation, to give a peroxide $+OCH_2CHClO+_n$ (38). On heating to $35°C$, this peroxide decomposes to formaldehyde, CO, and HCl. In aqueous solution at pH 10, it is possible to oxidize vinyl chloride to CO_2 with $KMnO_4$. This oxidation can be used for wastewater purification (39–40) (see Water, water reuse). The reaction of hypochlorous acid with vinyl chloride yields chloroacetaldehyde (41).

Addition.   The chlorination of vinyl chloride can proceed by either an ionic or a radical path. In the liquid phase and in the dark, 1,1,2-trichloroethane forms by an ionic path when a metal catalyst ($FeCl_3$) is used. The same product forms in radical reactions at up to $250°C$ (42). The photochemically initiated chlorination also produces 1,1,2-trichloroethane by a radical path (43). Above $250°C$, the chlorination of vinyl chloride gives unsaturated chloroethylenes produced by dehydrochlorination of 1,1,2-trichloroethane. The presence of small amounts of oxygen greatly accelerates the rate of the radical-chain chlorination reaction at above $250°C$ (44). Other halogens can be added to vinyl chloride to form similar 1,2-addition products but have not been thoroughly studied. Sulfuryl chloride, (SO_2Cl_2), in the presence of pyridine reacts with vinyl chloride to give 1,1,2-trichloroethane and 1,2-dichloroethanesulfonyl chloride. The addition reaction of hydrogen chloride and hydrogen iodide to vinyl chloride proceeds by an ionic mechanism (45–46). The addition of hydrogen bromide involves a chain reaction in which a bromine atom is the chain carrier. The product of the addition of hydrogen halides to vinyl chloride by either mechanism is the 1,1-adduct. Zinc chloride on Celite is an effective catalyst for the addition of HCl to vinyl chloride in the gas phase.

Various vinyl chloride adducts can be formed under acid-catalyzed Friedel-Crafts (qv) conditions. Vinyl chloride can be condensed with ethyl chloride to yield 1,1-dichloroethane and 1,1,3-trichlorobutane (47). The reaction of 2-chloropropane with vinyl chloride yields 1,1-dichloro-3-methylbutane (47). At $0–5°C$, vinyl chloride reacts with benzene resulting in a mixture of 1-chloroethylbenzene and 1,1-diphenylethane (48–49). With aromatics, the initially formed product reacts faster with a second aromatic molecule to produce the 1,1-substituted products. The rapid reaction of the initially formed product with toluene leads to a 68% yield of 1,1-ditolylethane, whereas anisole (methoxybenzene) gives a 42% yield of 1,1-di-p-anisylethane (50–51). Phenol also reacts to give p-vinylphenol (52). Condensation of vinyl chloride with formaldehyde and hydrogen chloride (Prins reaction) yields 3,3-dichloro-1-propanol and 2,3-dichloro-1-propanol (53–54). In the presence of iron pentacarbonyl, bromoform ($CHBr_3$) adds to vinyl chloride (55). Vinyl chloride can be hydrogenated over a 0.5% Pt/Al_2O_3 catalyst to ethyl chloride and ethane (56). This reaction is zero order in olefin and first order in hydrogen.

Pyrolysis.   Vinyl chloride is more stable than saturated chloroalkanes to thermal pyrolysis. Because of this, one of the principal reactions for its production is the thermal dehydrochlorination of 1,2-dichloroethane. When vinyl chloride is heated to $450°C$, small amounts of acetylene form (57). Although little conversion of vinyl chloride occurs, even at $525–575°C$, the main products are chloroprene and acetylene. The use of HCl during the pyrolysis of vinyl chloride lowers the amount of chloroprene formed.

The combustion of vinyl chloride in air at 510–795°C produces mainly carbon dioxide and hydrogen chloride along with carbon monoxide. A trace of phosgene also forms (58). When dry and in contact with metals, vinyl chloride does not decompose below 450°C. However, in the presence of water, vinyl chloride can corrode iron, steel, and aluminum because of the presence of trace amounts of hydrogen chloride. This hydrogen chloride may result from the hydrolysis of the peroxide formed between oxygen and vinyl chloride (38).

Manufacture

Vinyl chloride monomer was first produced commercially in the 1930s from the reaction of hydrogen chloride with acetylene (qv) derived from calcium carbide (see Acetylene-derived chemicals). As demand for vinyl chloride increased, more economical feedstocks were sought. After ethylene (qv) became plentiful in the early 1950s, commercial processes were developed to produce vinyl chloride from ethylene and chlorine. These processes included direct chlorination of ethylene to produce 1,2-dichloroethane (ethylene dichloride, EDC) and pyrolysis of EDC to produce vinyl chloride. However, since the EDC cracking process also produced HCl as a coproduct, the industry did not expand immediately except in conjunction with acetylene-based technology. The development of ethylene-oxychlorination technology in the late 1950s encouraged new growth in the vinyl chloride industry. In this process, ethylene reacts with HCl and oxygen to produce ethylene dichloride. Combining the component processes of direct chlorination, EDC pyrolysis, and oxychlorination provided the so-called balanced process for production of vinyl chloride from ethylene and chlorine with no net consumption or production of HCl.

Although a small fraction of the world's vinyl chloride capacity is still based on acetylene or mixed acetylene–ethylene feedstocks, most production is conducted by the balanced process based on ethylene and chlorine (59). The reactions for each of the component processes are shown in equations 1–3 and the overall reaction is given by equation 4:

Direct chlorination $\qquad\qquad$ $CH_2{=}CH_2 + Cl_2 \rightarrow ClCH_2CH_2Cl$ $\qquad\qquad$ (1)

Oxychlorination $\qquad$ $CH_2{=}CH_2 + 2\,HCl + \tfrac{1}{2}\,O_2 \rightarrow ClCH_2CH_2Cl + H_2O$ $\qquad$ (2)

Ethylene dichloride pyrolysis $\quad$ $\underline{2\,ClCH_2CH_2Cl \rightarrow 2\,CH_2{=}CHCl + 2\,HCl}$ $\qquad$ (3)

Overall reaction $\qquad$ $2\,CH_2{=}CH_2 + Cl_2 + \tfrac{1}{2}\,O_2 \rightarrow 2\,CH_2{=}CHCl + H_2O$ $\qquad$ (4)

In a typical balanced plant producing vinyl chloride from ethylene dichloride, all the HCl produced in ethylene dichloride pyrolysis is normally used as the feed for oxychlorination. On this basis, EDC production is about evenly split between direct chlorination and oxychlorination, and there is no net production or consumption of HCl. The three principal operating steps used in the balanced process for ethylene-based vinyl chloride production are shown in the block flow diagram in Figure 1, and a schematic of the overall process for a conventional plant is shown in Figure 2. A typical material balance for this process is given in Table 2.

Direct Chlorination of Ethylene. Direct chlorination of ethylene to ethylene dichloride is conducted by mixing ethylene and chlorine in liquid EDC. The reaction is a homogeneous catalytic reaction in the liquid phase, but under typical process conditions the reaction rate is controlled by mass transfer with absorption of ethylene

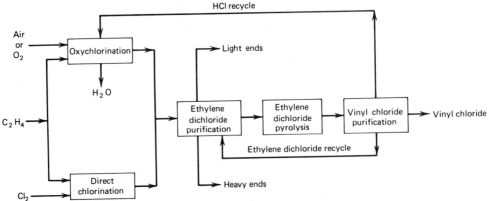

Figure 1.   Principal steps in balanced vinyl chloride process.

as the limiting factor (61). Ferric chloride is a highly selective and efficient catalyst for this reaction and is used in most commercial processes (59). The addition reaction proceeds through a polar mechanism, by which the catalyst may polarize chlorine, as shown in equation 5. The polarized chlorine molecule then acts as an electrophilic

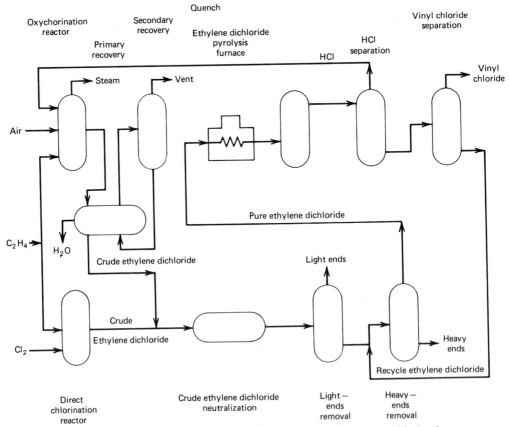

Figure 2.   Typical balanced vinyl chloride process with air-based oxychlorination.

Table 2. Typical Material Balance for Vinyl Chloride Production by the Air-Based Balanced Ethylene Process[a]

Components, kg	Raw materials	Interme-diates	By-products	Aqueous streams	Direct chlori-nation[b]	Oxychlori-nation	Distill-ation columns	Product
C_2H_4	0.4656				0.0025			
Cl_2	0.5871				0.0001		0.0001	
N_2	0.5782					0.5779	0.0003	
O_2	0.1537					0.0214		
CO_2	0.0003					0.0116		
CO						0.0032		
$ClCH_2CH_2Cl$		1.6370[c]	0.0029		0.0016	0.0017	0.0045	
HCl		0.6036						
H_2O	0.0171		0.1438	0.1196		0.0413		
NaOH				0.0008				
NaCl				0.0014				
lights			0.0029		0.0003	0.0025		
heavies			0.0023					
$CH_2{=}CHCl$			0.0008		0.0001	0.0012	0.0024	1.0000
Total, kg/kg vinyl chloride	*1.8020*	*2.2406*	*0.1527*	*0.1218*	*0.0046*	*0.6608*	*0.0073*	*1.0000*

[a] Ref. 60.

[b] Inerts present in chlorine feed are emitted in this vent stream.

[c] Represents EDC necessary for a stoichiometric balance, including that converted to by-products but no recycled EDC.

reagent to attack the double bond of ethylene, thus facilitating chlorine addition (eq. 6):

$$FeCl_3 + Cl_2 \rightleftarrows FeCl_4^- \text{---} Cl^+ \tag{5}$$

$$FeCl_4^- \text{---} Cl^+ + CH_2{=}CH_2 \rightarrow FeCl_3 + ClCH_2CH_2Cl \tag{6}$$

The direct chlorination reaction may be run with a slight excess of ethylene or chlorine, depending on the methods available for handling effluent gases from the reactor. Conversion of the limiting component is essentially 100%, and selectivity to ethylene dichloride is greater than 99% (59). The main by-product is 1,1,2-trichloroethane, which probably forms through radical reactions beginning with homolytic dissociation of a small fraction of the chlorine. However, oxygen, which is frequently present as an impurity in chlorine, tends to increase selectivity to EDC by inhibition of free-radical reactions that produce 1,1,2-trichloroethane. Amides, eg, N,N-dimethylformamide also increase selectivity to EDC (62).

The heat of reaction is removed either through conventional water cooling for reactors that operate at moderate temperatures of 50–65°C, or by operating the reactor at the boiling point of ethylene dichloride and allowing the pure product to vaporize (63–65). For reactors equipped with liquid product removal, the EDC is usually treated to remove ferric chloride. The latter, which would lead to rapid fouling of the EDC cracking reactor, can be removed by washing with water or by adsorption on a solid (66–67). With dry feedstocks and good temperature control, carbon steel can be used in the direct chlorination reactor and auxiliary equipment (68). Compared with direct chlorination, the oxychlorination process is characterized by higher capital investment, higher operating costs, and less-pure EDC product. However, the use of the oxychlorination process is dictated by the need to consume the HCl generated in EDC pyrolysis.

Oxychlorination of Ethylene.   In the oxychlorination process, ethylene reacts with dry hydrogen chloride and either air or pure oxygen to produce EDC and water. Various commercial processes for oxychlorination differ somewhat because they were developed independently by several vinyl chloride producers (59,69). In general, however, the reaction is carried out in the vapor phase in either a fixed- or fluid-bed reactor containing a modified Deacon catalyst. However, oxychlorination of ethylene occurs readily at temperatures well below that required for the oxidation of hydrogen chloride in the Deacon process for chlorine production (70). Oxychlorination catalysts generally contain copper chloride as the main active ingredient, but they may also contain numerous additives impregnated on a porous support, eg, alumina, silica–alumina, diatomaceous earth, etc (71–72). Although the detailed catalytic mechanism is not known, copper(II) chloride is generally recognized as the active chlorinating agent. The copper(I) chloride produced during the ethylene chlorination step is rapidly reconverted to copper(II) chloride under reaction conditions, and the presence of some copper(I) is thought to be advantageous because it readily complexes with ethylene, bringing it into contact with copper(II) chloride for a long enough time for chlorination to occur (59). A very simple representation of this heterogeneous catalytic cycle is given in equations 7–9.

$$CH_2=CH_2 + 2\ CuCl_2 \rightarrow 2\ CuCl + ClCH_2CH_2Cl \tag{7}$$

$$\tfrac{1}{2}\ O_2 + 2\ CuCl \rightarrow CuOCuCl_2 \tag{8}$$

$$2\ HCl + CuOCuCl_2 \rightarrow 2\ CuCl_2 + H_2O \tag{9}$$

Another suggested mechanism involves initial formation of ethylene oxide as the possible rate-limiting step in the reaction (73).

Since commercial oxychlorination processes differ somewhat with either fluid- or fixed-bed reactors and either air or oxygen feed, the operating conditions, feed ratios, conversions, and yields also vary, depending on the particular combination used in production and on the methods employed for secondary recovery of feedstock and product. For any particular combination of reactor type and oxidant, however, good temperature control of this highly exothermic reaction is essential for efficient production of ethylene dichloride. Increasing temperatures in the reactor lead to increased by-product formation, mainly through increased combustion of ethylene to carbon oxides and increased cracking of EDC. Cracking, ie, dehydrochlorination, of EDC results in the formation of vinyl chloride, and subsequent oxychlorination and cracking steps lead progressively to by-products with higher levels of chlorine substitution. High temperatures can also lead to deactivation of the catalyst through increased sublimation of copper(II) chloride.

Fluidized-bed reactors typically are cylindrical vessels equipped with internal cooling coils for heat removal and either external or internal cyclones to minimize catalyst carry-over (see Fluidization). Fluidization of the catalyst assures intimate contact between feed and product vapors, catalyst, and heat-transfer surfaces and results in a uniform temperature within the reactor (74). Reaction heat can be removed by the generation of steam within the cooling coils or by some other heat-transfer medium. An operating temperature of 220–235°C and reactor gauge pressures of 150–500 kPa (22–73 psig) are typical for oxychlorination with a fluidized catalyst. With these operating conditions, fluid-bed reactors are commonly constructed with a carbon-steel shell and a corrosion-resistant alloy for the internal parts (68).

Fixed-bed reactors resemble multitube heat exchangers with the catalyst packed in vertical tubes held in a tubesheet at top and bottom. Uniform packing of catalyst within the tubes is important to ensure uniform pressure drop, flow, and residence time through each tube. Reaction heat can be removed by the generation of steam on the shell side of the reactor or by some other heat-transfer fluid. However, temperature control is more difficult in a fixed-bed reactor than in a fluid-bed reactor because localized hot spots tend to develop in the tubes. The tendency to develop hot spots can be minimized by packing the reactor tubes with active catalyst and inert diluent mixtures in certain proportions so that there is low catalyst activity at the inlet but that the activity steadily increases to a maximum at the outlet (59). Another method is packing the tubes with catalysts having progressively higher loadings of copper chloride so as to provide an activity gradient along the length of the tubes. Multiple reactors are also used in fixed-bed oxychlorination primarily to control heat release by staging oxidant feed, ie, air or oxygen. Each successive reactor may also contain catalyst with progressively higher loadings of copper chloride. These methods of staging oxidant and of grading the catalyst activity tend to flatten the temperature profile and allow improved temperature control. Compared with the fluid-bed oxychlorination process, the fixed-bed process generally operates at higher temperatures (230–300°C) and gauge pressures (150–1400 kPa (22–203 psig)). With these operating conditions, a corrosion-resistant alloy is used for the reactor tubes; tubesheets and reactor heads are clad with nickel on steel; and the reactor shell is constructed of carbon steel (68).

In the air-based oxychlorination process with either fluid- or fixed-bed reactors, ethylene and air are fed in slight excess of stoichiometric requirements to ensure high conversion of HCl and to minimize losses of excess ethylene that remains in the vent gas after product condensation. Under these conditions, typical feedstock conversions are 94–99% for ethylene and 96–99% for HCl with EDC selectivities of 93–96%. Downstream product recovery involves cooling the reactor exit gases by either direct quench or with a heat exchanger and condensation of the ethylene dichloride and water, which are then separated by decantation (Fig. 2). The remaining gases still contain 1–5 vol % EDC, so they are further processed in a secondary recovery system involving either solvent absorption or a refrigerated condenser. In air-based processes operating with high ethylene conversion, the dilute ethylene remaining in the vent is generally incinerated; but in those operating at lower conversion, various schemes are first used to recover unconverted ethylene, usually by direct chlorination to EDC (75–76).

The use of oxygen instead of air in the oxychlorination process with either fixed- or fluid-bed reactors permits operation at lower temperatures and results in improved operating efficiency and product yield (69,76). Unlike the air-based process, ethylene is generally fed into an oxygen-based reactor in somewhat larger excess over stoichiometric requirements. Reactor exit gases are cooled, purified from traces of unconverted HCl, separated from EDC and water by condensation, recompressed to the reactor inlet pressure, reheated, and recycled to the oxychlorination reactor. Recycle of the effluent gases permits lower ethylene conversion per pass through the reactor with minimal loss in overall ethylene yield. A small amount of reactor off-gas, typically 1–5 vol %, is continuously purged from the system to prevent accumulation of impurities, eg, carbon oxides, nitrogen, argon, and unreacted hydrocarbons, which either form in the oxychlorination reactor or enter the process as impurities in the feed streams. An important advantage of oxygen-based oxychlorination technology over air-based operation is the drastic reduction of the vent-gas volume discharged by the oxychlorination process. Since nitrogen is no longer present in the reactor feed streams, only a small amount of purge gas is vented. On a volume comparison, the reduced purge-gas stream would typically amount to only 1–5% of the vent-gas volume for air-based operation. Air-based processes release significant quantities of vent gases to the atmosphere, generally after treatment by incineration and scrubbing. Typically, for every kilogram of EDC produced by oxychlorination, ca 0.7–1.0 kg of vent gases is emitted from the air-based process (see Table 2). Therefore, for an air-based, balanced, vinyl chloride plant with a rated capacity of 450,000 metric tons per year of vinyl chloride, the total vent-gas volume released to the atmosphere would be 7–10 m^3/s (245–353 ft^3/s). However, the vent gas consists mainly of nitrogen, some unconverted oxygen, and small amounts of carbon oxides. Depending on the type of oxychlorination process involved, however, there are differing levels of undesirable impurities, ethylene, and chlorinated hydrocarbons in the oxychlorination vent gas.

Chlorinated by-products of ethylene oxychlorination typically include 1,1,2-trichloroethane; chloral (trichloroacetaldehyde); trichloroethylene; 1,1-dichloroethane; cis- and trans-1,2-dichloroethylenes; ethyl chloride; vinyl chloride; 1,1-dichloroethylene (vinylidene chloride); mono-, di-, tri-, and tetrachloromethanes (methyl chloride, methylene chloride, chloroform, and carbon tetrachloride); and higher boiling compounds. All of these by-products present problems and their production should be minimized in order to lower raw-material costs, lessen the difficulties in ethylene dichloride purification, prevent fouling in the pyrolysis reactor, and minimize by-product handling and disposal. Chloral, in particular, should be removed because it tends to polymerize in the presence of strong acids forming solids which foul and clog operating lines and controls (59). Oxychlorination-reactor feed purity can also contribute to by-product formation. Normally, however, the only problem is with low levels of acetylene present in the HCl from the EDC cracking process. Since acetylene in the feed causes the formation of highly chlorinated by-products and tars, selective hydrogenation of this acetylene to ethylene and ethane is practiced by many companies (77).

Purification of Ethylene Dichloride for Pyrolysis. By-products contained in EDC from the three main processes must be removed; these include by-products in EDC from direct chlorination, and oxychlorination and recovered EDC from the cracking process. 1,2-Dichloroethane used for pyrolysis to vinyl chloride must be of high purity, ie, usually greater than 99.5 wt %, because the ethylene dichloride cracking process is exceedingly susceptible to inhibition and fouling by trace quantities of impurities (59,78). It must also be dry (no separate water phase and less than 10 ppm of total dissolved water) to prevent excessive corrosion downstream of the pyrolysis unit. Inadvertent moisture pickup, however, is always possible; in such cases, the corrosion of steel equipment tends to be the greatest in reboilers, the bottom section of distillation columns, bubble caps, plates, condensers, water separators, valves, pumps, and fittings (68).

Direct chlorination generally produces EDC with a purity greater than 99.5 wt % and, except for removal of the FeCl$_3$, little further purification is necessary. Ferric chloride can be removed by adsorption on a solid, or the EDC can be distilled from the FeCl$_3$ in a boiling reactor. Alternatively, the ferric chloride can be removed by washing with water, usually in conjunction with EDC from the oxychlorination process.

Ethylene dichloride from the oxychlorination process is generally less pure than direct-chlorination EDC and, thus, is usually washed with water and.then with caustic solution to remove chloral and other water-extractable impurities (79). Low boiling impurities and water are taken overhead in a first (light-ends) distillation column, and then pure dry EDC is taken overhead in a second (heavy-ends) column (see Fig. 2).

Ethylene dichloride recovered from the cracking process contains an appreciable number of impurities. Two of these, trichloroethylene and chloroprene (2-chloro-1,3-butadiene), are not readily removable by distillation and necessitate the use of other treatments (59). Chloroprene, if not altered by chemical treatment, concentrates in the light-ends column where it can polymerize to solid or rubbery materials which seriously foul this column (see Chlorocarbons, chloroprene). Trichloroethylene forms an azeotrope with EDC, boiling very close to EDC; if it is allowed to accumulate, it leads to inhibition of the cracking reaction and increased fouling rates (see Chlorocarbons, trichloroethylene). Both impurities can be removed by subjecting the recycle ethylene dichloride stream to chlorination prior to distillation (80–82). Treatments with HCl and hydrogenation have also been patented as methods for removal of chloroprene (83–86).

Ethylene Dichloride Pyrolysis to Vinyl Chloride. Thermal cracking of EDC to vinyl chloride and hydrogen chloride occurs as a homogeneous, first-order free-radical chain reaction. The accepted general mechanism involves the four steps shown in equations 10–13 (87–89):

Initiation $\qquad$ $ClCH_2CH_2Cl \rightarrow ClCH_2\overset{\cdot}{C}H_2 + Cl\cdot$ $\hspace{3cm}$ (10)

Propagation $\qquad$ $Cl\cdot + ClCH_2CH_2Cl \rightarrow ClCH_2\overset{\cdot}{C}HCl + HCl$ $\hspace{2cm}$ (11)

$\qquad\qquad\qquad$ $ClCH_2\overset{\cdot}{C}HCl \rightarrow CH_2{=}CHCl + Cl\cdot$ $\hspace{2.5cm}$ (12)

Termination $\qquad$ $Cl\cdot + ClCH_2\overset{\cdot}{C}H_2 \rightarrow ClCH{=}CH_2 + HCl$ $\hspace{2cm}$ (13)

Reactions 11 and 12 are the chain-propagation steps, because each elementary step consumes one of the two chain carriers and simultaneously produces the other. The net effect of reactions 11 and 12 is continuation of the chain by conversion of EDC to vinyl chloride. Thus, the two chain carriers are chlorine atoms and 1,2-dichloroethyl radicals. In general, anything consuming a chain carrier is an EDC cracking inhibitor and anything producing a chain carrier is a promoter. That is, any molecular or radical species that consumes a chain carrier without simultaneously producing either 1,2-dichloroethyl radicals or chlorine atoms is an EDC cracking inhibitor, eg, propylene. The allylic hydrogen atoms of propylene can be easily abstracted by one of the chain carriers, either 1,2-dichloroethyl radicals or chlorine atoms. The resulting allyl radical can then combine with a chlorine atom forming allyl chloride. Since the same sequence can occur two more times, one molecule of propylene can consume up to six chain carriers. Reaction initiators or accelerators include carbon tetrachloride, chlorine, bromine, iodine, or oxygen (87). More recently, however, exclusion of oxygen is claimed to result in considerably less fouling on the pyrolysis tube walls (90).

The endothermic cracking of ethylene dichloride is relatively clean at atmospheric pressure and temperatures of 425–550°C. Commercial operations, however, generally operate at gauge pressures up to 2.5–3.0 MPa (360–435 psig) and temperatures of 500–550°C in order to provide better heat transfer, reduced equipment size, and easier separation of HCl from vinyl chloride by fractional distillation. Ethylene dichloride conversion levels per pass through the pyrolysis reactor are normally maintained at 50–60% at residence times of 2–30 s, with selectivities to vinyl chloride of 96 to >99% (59). Increasing cracking severity gives somewhat higher EDC conversion but also results in lower selectivities to vinyl chloride. Since some of the by-products generated during cracking act as inhibitors to the free-radical sequence, increasing cracking severity leads to progressively smaller increases in EDC conversion and progressively larger problems with pyrolysis-tube coking and downstream-product purification.

An important processing requirement in EDC cracking is rapid cooling or quenching of the reaction mixture. If cooling is done too slowly, substantial yield losses to heavy ends and tars result (84,91–95). Therefore, the hot effluent gases are normally quenched and partially condensed by direct contact with cold EDC in a quench tower. Although each producer of vinyl chloride has its own minor modifications in the HCl–vinyl chloride recovery section, in general the quench column effluent is distilled to remove first HCl and then vinyl chloride (Fig. 2). The vinyl chloride is generally further treated to produce specification product, recovered HCl is sent to the oxychlorination process, and unconverted EDC is purified for removal of light and heavy ends prior to recycle to the cracking furnace. The light and heavy ends either are further processed or are disposed of by incineration or other methods. By-products from EDC pyrolysis can include acetylene, ethylene, methyl chloride, butadiene, vinylacetylene, benzene, chloroprene, vinylidene chloride, 1,1-dichloroethane, chloroform, carbon tetrachloride, 1,1,1-trichloroethane, and other chlorinated hydrocarbons (59). Most of these impurities remain in the unconverted ethylene dichloride fraction and are subsequently removed in EDC purification as light and heavy ends. Ethylene and acetylene codistill with the HCl and are routed back to the oxychlorination reactor after optional hydrogenation of the acetylene to ethylene. Methyl chloride and butadiene tend to codistill with the vinyl chloride, depending on the efficiency of the vinyl chloride fractional distillation system. Addition of chlorine or carbon tetrachloride to the cracker feed is claimed to suppress methyl chloride formation (96). Removal of butadiene, a contaminant which can interfere with polymerization of vinyl chloride, has been done by treatment with chlorine, anhydrous HCl, or selective hydrogenation (97–99).

By-Product Disposal. Disposal of by-products from vinyl chloride manufacturing processes involves a number of methods because a variety of gaseous, organic liquid, and aqueous streams must be handled. Vent-gas streams from various units may contain small amounts of HCl, vinyl chloride, chlorine, ethylene, methane, and carbon monoxide. These streams can sometimes be treated chemically or by scrubbing, sorption, or other methods to recover some chemicals when economically justified. For objectionable components remaining in the vent-gas streams, however, the common cleaning technique is either incineration or catalytic combustion followed by removal of HCl from the vent gases (see Incinerators; Exhaust control, industrial). Organic liquid streams include the light and heavy ends from ethylene dichloride purification (see Fig. 2 and Table 2). The light ends contain mainly ethyl chloride, *cis*- and *trans*-1,2-dichloroethylene, chloroform, and carbon tetrachloride. The heavy ends contain mostly 1,1,2-trichloroethane, lesser concentrations of tetrachloroethanes, chlorinated butanes, chlorinated aromatics, and many other compounds present in smaller amounts. These streams are sometimes fractionated to recover useful components, and the remaining by-products are incinerated and scrubbed to remove chlorine. Another method involves combining all liquid by-product streams and passing them along with air into a fluidized-bed, catalytic oxidation reactor (100). The resulting combustion product stream, consisting essentially of HCl, H_2O, N_2, O_2, and carbon oxides, is fed directly into an oxychlorination reactor where the HCl content is recovered as ethylene dichloride. Further, the heat of combustion is recovered as high pressure steam in a manner similar to that in fluid-bed oxychlorination processes. Process water streams from vinyl chloride manufacture are typically steam-stripped to remove volatile organics, neutralized, and then treated in an activated sludge system to remove nonvolatile organics remaining in the water (101).

Economic Aspects

Yearly U.S. production volumes and prices of vinyl chloride are listed in Table 3. The cost of vinyl chloride is not as vulnerable to the increasing cost of hydrocarbon, since 56% of the compound is chlorine. The cost of chlorine has not increased as rapidly as the hydrocarbon portion. The lower relative cost of poly(vinyl chloride) has allowed it to compete effectively with metals in the housing and automobile industries.

U.S. production and capacities are listed in Table 4. Worldwide production capacity of each country for vinyl chloride is listed in Table 5. With the global increase in vinyl chloride capacity in the industrialized nations, there will be little opportunity for significant increase in the exporting of vinyl chloride as 1,2-dichloroethane from the United States. Also, many poly(vinyl chloride) producers are manufacturing vinyl chloride.

Environmental Considerations

Vinyl chloride emissions in the balanced process occur from a number of sources, but the main ones are the vinyl chloride purification system vents and the product-loading facility vents. Losses from the purification system are continuous and those from the loading system are intermittent. In 1974, the EPA estimated typical vinyl chloride plant emissions in terms of the source and kilograms of vinyl chloride lost per 100 kg vinyl chloride produced as follows: vinyl chloride finishing column, 0.24

Table 3. U.S. Vinyl Chloride Production and Prices[a]

Year	Production, 1000 t	List price, ¢/kg
1955	240	23.1
1960	470	27.6
1965	907	17.6
1970	1833	10.5
1975	1903	19.8–26.5
1979	3422	33.1
1980	2933	48.5
1981	3005	48.5

[a] Ref. 102.

Table 4.   U.S. Producers of Vinyl Chloride Monomer[a]

Company	Capacity, 1000 t	Remarks
Borden, Inc. (Borden Chemical Division)		
Geismar, La.	277	became sole owner of Monochem capacity, Jan. 1982
Conoco, Inc. (Conoco Chem. Co.)		
Lake Charles, La.	318	Conoco purchased by E. I. du Pont de Nemours & Co., Inc., 1981
Dow Chemical USA		
Freeport, Texas	68	
Oyster Creek, Texas	340	
Plaquemine, La.	567	
Ethyl Corporation (Chemicals Group)		
Baton Rouge, La.	136	
Formosa Plastics U.S.A.		
Baton Rouge, La.	136	purchased Imperial Chemical Industries, Ltd. plant, Dec. 1980
Point Comfort, Texas	240	
Georgia Pacific Corp. (Chem. Division)		
Plaquemine, La.	454	
The BFGoodrich Co. (BFGoodrich Chemical Group)		
Calvert City, Ky.	454	
LaPorte, Texas	454	purchased Diamond-Shamrock LaPorte plant, Jan., 1982
Convent, La.	0	scheduled startup of a 726,000-t/yr vinyl chloride plant at Convent delayed
PPG Industries, Inc. (Chemicals Group, Chemical Division, U.S.)		
Lake Charles, La.	408	
Shell Chemical Co.		
Deer Park, Texas	381	
Norco, La.	318	
Total	*4551*	

[a] Ref. 102.

kg; fugitive, 0.1215 kg; EDC finishing column, 0.05 kg; oxychlorination process, 0.0364 kg; and process water, 0.0007 kg. Because of the toxicity of vinyl chloride, the EPA presented the following regulations in 1975 as the emission standards for vinyl chloride manufacture: emissions from all point sources except oxychlorination must be reduced to 10 ppm vinyl chloride; emissions from oxychlorination must be reduced to 0.02 kg vinyl chloride per 100 kg EDC product from the oxychlorination process; preventable relief-valve discharges are not permitted; and fugitive emissions are to be minimized by enclosure of the emission sources and collection of the emissions (104). These regulations were predicated upon reduction of the 1974 estimated typical vinyl chloride plant emissions by 94% by means of the best available technology. Compliance testing of these facilities started in the last quarter of 1978. Additional EPA and state actions were initiated in mid-1977 also to reduce hydrocarbon emissions from vinyl chloride plants in nonattainment regions. These actions were directed primarily against the ethylene content of oxychlorination vent gas from air-based units. The effect of these various regulations has been to increase substantially the scope of add-on technology in vinyl chloride production plants, such as: installation of primary and redundant incineration facilities for vinyl chloride point-source and collected fugitive emissions, except oxychlorination; installation of HCl scrubbing and neutralization or recovery units in conjunction with the incinerators; installation of closed-process sewers, collection systems and larger or redundant wastewater strippers; replacement of single mechanical seals on pumps and agitators with double mechanical seals; leak-detection systems and portable monitors; enclosed sampling and analytical systems; and vapor-recovery systems for vinyl chloride loading, unloading, and equipment clearing (59).

Technology Trends. Recent developments in commercial vinyl chloride processes based on the balanced ethylene-feedstock route, which is used overwhelmingly worldwide, include boiling-liquid reactors for direct chlorination, a trend toward oxygen-based oxychlorination, and efforts to improve conversion and minimize by-product formation in the ethylene dichloride (EDC) pyrolysis process (59,63–65,69,76,105–106). Direct chlorination in liquid-boiling reactors offer advantages of energy savings and reduced product purification requirements, compared with conventional processes operating at lower temperatures. Energy savings arise by essentially using the reactor as a reboiler for the conventional EDC purification system. With this modification, the heat of reaction is used to provide most of the required column vapor load for fractionation. This results in lower steam and cooling-water usage requirements compared with the conventional process. In addition, the EDC is distilled from dissolved catalyst, thus facilitating product purification. Use of pure oxygen instead of air for the oxychlorination process offers a number of advantages, including improvements in ethylene and hydrogen chloride yields, reduced equipment requirements for secondary recovery of ethylene and EDC, and reduced emissions (69,76). The main advantage is the drastic reduction of the vent-gas volume discharged by the process. Therefore, destruction of the environmentally objectionable compounds in this stream is more manageable, and savings in incineration can be sufficient to pay for the oxygen raw-material requirement. With emission-control standards becoming more severe, it is expected that the ratio of oxygen-based to air-based plants will increase. For existing air-based plants, conversion to oxygen will depend on local emission standards, oxygen availability and cost, and alternative add-on systems available for cleaning the vent gas, eg, catalytic oxidation or sorption techniques. In terms of ethylene dichloride pyrolysis chemistry, it is expected that companies will continue their efforts in developing cracking promoters, by-product inhibitors, and improved feed purification techniques (89,93). Since current cracking technology limits EDC conversion to 50–60%, considerable energy and cost savings could be achieved through increased conversion levels without concurrent losses of EDC to undesirable side reactions and coking problems. Recent development of a laser-induced EDC cracking technique at the Max Planck Institute (Göttingen, FRG), for example, has provoked interest in the vinyl chloride industry (94–95). At temperatures comparable to those used in commercial operation, this laser-induced cracking process is claimed to increase conversion while concurrently decreasing by-product formation.

Vinyl chloride processes based on acetylene or mixed acetylene–ethylene streams are limited by local availability and relative costs of such feedstocks. However, the development of lower cost routes to acetylene will renew interest in this original manufacturing process. For example, development of a relatively new crude-oil

Table 5. Worldwide Vinyl Chloride Capacity as of January 1982, 1000 t/yr[a]

Country	Capacity, 1000 t/yr
North America	
Canada	408
Mexico	70
United States	4,390
Total	*4,868*
Europe	
Western Europe	
Belgium	820
Finland	62
France	1,150
FRG	1,645
Greece	22
Italy	950
Norway	300
Portugal	15
The Netherlands	500
Spain	370
Sweden	100
Switzerland	15
UK	620
Total	*6,569*
Eastern Europe	
Bulgaria	185
Czechoslovakia	215
GDR	200
Hungary	199
Poland	84
Romania	226
USSR	542
Yugoslavia	100
Total	*1,751*
South America	
Argentina	51
Brazil	262
Colombia	9
Peru	8
Venezuela	60
Total	*390*
Asia	
Far East	
India	84
Indonesia	00
Japan	2,064
Pakistan	5
People's Republic of China	80
Philippines	10
Republic of China	600
Republic of Korea	210
Total	*3,053*
Middle East	
Iran	63
Israel	100
Turkey	54
Total	*217*

Table 5 (*continued*)

Country	Capacity, 1000 t/yr
Africa	
Algeria	40
Egypt	0
Libya	62
Morocco	27
Republic of South Africa	180
Senegal	0
Total	*309*
Australia	62
Grand total	*17,219*

[a] Ref. 103.

cracking process, which involves very high temperature steam as a heat-transfer fluid, produces substantial yields of acetylene along with ethylene (107). High temperature naphtha cracking also produces considerable amounts of acetylene and ethylene. Therefore, under certain economic and geographic conditions, these processes may provide acetylene or mixed hydrocarbon feedstocks for vinyl chloride production. Typical conditions for the hydrochlorination of acetylene are temperatures of 150–180°C, total pressures of ca 500–1500 kPa (5–15 atm), and the use of a carbon-supported mercuric chloride catalyst (108–109). With stoichiometric quantities of reactants, essentially complete conversion is obtained with selectivities of 98%. Ethylene does not react under these conditions; thus, mixed acetylene–ethylene streams can be used as feeds. Ethylene is easily recovered from the vinyl chloride product by fractional distillation and is then chlorinated yielding 1,2-dichloroethane. In addition to the three general types of vinyl chloride processes in commercial use, ie, those based on ethylene, acetylene, and mixed-gas feedstocks, various other process schemes based on ethane, ethylene, and methane have been described in the patent literature in recent years (59,110–115).

Specifications

Technical-grade vinyl chloride should not contain more than the amounts of impurities listed in Table 6.

Health and Safety Factors

Vinyl chloride is an OSHA-regulated material (117). Current OSHA regulations require that no employee be exposed to vinyl chloride concentrations greater than 1.0 ppm over any 8-h period, or 5.0 ppm averaged over any period not exceeding 15 min. Monitoring is required at all facilities where vinyl chloride is produced or PVC is processed. The monitoring may be discontinued for any employee only when at least two consecutive determinations, made not less than five working days apart, show exposures at or below the action level of 0.5 ppm. Contact with liquid vinyl chloride is prohibited. Chronic exposure to vinyl chloride at concentrations of 100 ppm or more is reported to have produced Raynaud's syndrome, lysis of the distal bones of the fingers, and a fibrosing dermatitis. However, these effects are probably related to

Table 6.   Impurity Levels in Vinyl Chloride[a]

Impurity	Maximum level, ppm
acetylene	2.0
acidity, as HCl by wt	0.5
acetaldehyde	0.0
alkalinity, as NaOH by wt	0.3
butadiene	6.0
1-butene	3.0
2-butene	0.5
ethylene	4.0
ethylene dichloride (EDC)	10.0
nonvolatiles	150.0
propylene	8.0
water	200.0
iron, by wt	0.25

[a] Ref. 116.

continuous intimate contact with the skin. Chronic exposure is also reported to have produced a rare cancer of the liver (angiosarcoma) in a small number of workers after continued exposure for many years to large amounts of vinyl chloride gas (118). Toxicology data on vinyl chloride, eg, TC_{Lo} (human), TC_{Lo} (rat), LD_{50} (rat), and threshold limit values, are reported in ref. 8. Exposures to vinyl chloride can be readily reduced by feasible engineering controls or work practices to below the OSHA acceptable levels. Use of closed systems or laboratory hoods that have protection factors adequate to prevent worker exposure are recommended. Whenever exposure is above the permissible OSHA limit and cannot be reduced by feasible engineering practices, respirators are required and must be used in accordance with a standard respirator program (118).

Vinyl chloride is flammable when exposed to heat, flame, or oxidizing agents. Large fires of the compound are very difficult to extinguish. Vapors represent a severe explosion hazard. Peroxides can form on standing in air, especially in the presence of iron impurities. Vinyl chloride is generally transported in railroad tank cars and in tank trucks (119). Because of possible peroxide formation, vinyl chloride should be transported or handled under an inert atmosphere. The presence of peroxide from vinyl chloride and air can initiate polymerization of stored vinyl chloride; however, stabilizer can be added to prevent polymerization (48). Because of the worldwide production and transport of vinyl chloride, the higher temperatures of certain climates may require that small amounts of phenolic or other stabilizers be added to the vinyl chloride.

Uses

Vinyl chloride has gained worldwide importance because of its industrial use as the precursor to poly(vinyl chloride). It is also used in a wide variety of copolymers. The inherent flame-retardant properties, wide range of plasticized compounds, and the low cost of the polymers from vinyl chloride have made it a major industrial chemical (see Flame retardants, halogenated fire retardants). The use of vinyl chloride as a starting material for the synthesis of other industrial compounds will be increasingly important because of the large volume of vinyl chloride available at a cost which is only partially dependent on the increasing cost of petroleum.

BIBLIOGRAPHY

"Vinyl Chloride" under "Chlorine Compounds, Organic" in *ECT* 1st ed., Vol. 3, p. 786, by J. Werner, General Aniline & Film Corp., General Aniline Works Division; "Vinyl Chloride" under "Vinyl Compounds" in *ECT* 1st ed., Vol. 14, pp. 723–726, by C. H. Alexander and G. F. Cohan, B.F. Goodrich Chemical Co.; "Vinyl Chloride" under "Chlorocarbons and Chlorohydrocarbons" in *ECT* 2nd ed., Vol. 5, pp. 171–178, by D. W. F. Hardie, Imperial Chemical Industries, Ltd.

1. U.S. Pats. 2,188,396 (Jan. 30, 1940) and 1,929,453 (Oct. 10, 1933), W. L. Semon (to B.F. Goodrich Co.).
2. V. Regnault, *Ann. Chim. Phys.* **58,** 307 (1835).
3. E. Baumann, *Liebigs Ann.* **163,** 308 (1872).
4. Ger. Pat. 278,249 (Oct. 11, 1912), (to Chemische Fabrik Griesheim-Electron).
5. R. C. Reid, J. M. Prausnitz, and T. K. Sherwood, *The Properties of Gases and Liquids*, 3rd ed., McGraw-Hill Publishing Co., New York, 1977.
6. R. W. Gallant, *Physical Properties of Hydrocarbons*, Gulf Publishing Co., Houston, Texas, 1968.
7. For further vapor pressure data refer to J. Timmermans, *Physico-Chemical Constants of Pure Organic Liquids*, Vol. 1, Elsevier Publishing Co., London, 1950, p. 276; Vol. 2, 1965, p. 234; *Vapor Pressures and Critical Points of Liquids VII: Halogenated Ethanes and Ethylenes*, Item 76004 Engineering Sciences Data Unit, London, 1976.
8. N. I. Sax, *Dangerous Properties of Industrial Materials*, 5th ed., Van Nostrand Reinhold Co., New York, 1979.
9. P. M. Henry, *J. Org. Chem.* **37,** 2443 (1972).
10. A. Misona, Y. Uchida, and K. Furuhata, *Bull. Chem. Soc. Jpn.* **43,** 1243 (1970).
11. M. Tamura and T. Yasui, *Koy. Kayaka Zass.* **72,** 572 (1969).
12. C. F. Kohll and R. VanHelden, *Rec. Trav. Chim. Pays-Bas* **87,** 481 (1968).
13. J. Rajaram, R. G. Pearson, and J. A. Ibers, *J. Am. Chem. Soc.* **96,** 2103 (1974).
14. E. W. Stern, M. L. Spector, and H. P. Leftin, *J. Catal.* **6,** 152 (1966).
15. U.S. Pat. 3,784,646 (Jan. 8, 1974), J. M. Holovka (to Marathon Oil Co.).
16. U.S. Pat. 3,627,827 (Dec. 14, 1971), J. A. Scheben (to National Distillers and Chemical Corp.).
17. U.S. Pat. 3,991,101 (Nov. 9, 1976), J. F. Knifton (to Texaco, Inc.).
18. O. N. Temkin and co-workers, *Kinet. Katal.* **11,** 1592 (1970).
19. G. Kobrich, H. Trapp, and I. Hornke, *Tetrahedron Lett.*, 1131 (1964); G. Kobrich and K. Flory, *Tetrahedron Lett.*, 1137 (1964).
20. A. Arase, M. Hoshi, and Y. Masuda, *Bull. Chem. Soc. Jpn.* **54,** 299 (1981).
21. H. Normant, *Bull. Soc. Chim. Fr.*, 728 (1957); *Compt. Rend.* **239,** 1510 (1954); *Adv. Org. Chem.* **2,** 1 (1960).
22. T. Kauffmann and W. Sohm, *Angew. Chem. Int. Ed. Engl.* **6,** 85 (1967).
23. J. Hooz and R. B. Layton, *Can. J. Chem.* **48,** 1626 (1970).
24. S. Watanabe and co-workers, *Can. J. Chem.* **50,** 2786 (1972); S. Watanabe, K. Suga, and Y. Yamaguchi, *J. Appl. Chem. Biotechnol.* **22,** 43 (1972).
25. J. H. Babler and D. O. Olsen, *Tetrahedron Lett.*, 351 (1974).
26. R. West and W. H. Glaze, *J. Org. Chem.* **26,** 2096 (1961).
27. H. Neumann and D. Seebach, *Tetrahedron Lett.*, 4839 (1976).
28. J. C. Floyd, *Tetrahedron Lett.*, 2877 (1974).
29. J. Millon, R. Lorne, and G. Linstrumelle, *Synthesis* **7,** 434 (1975).
30. E. J. Corey and R. L. Carney, *J. Am. Chem. Soc.* **93,** 731 (1971); **94,** 4395 (1973).
31. H. C. Brown, A. B. Levy, and M. M. Midland, *J. Am. Chem. Soc.* **97,** 5017 (1975).
32. Brit. Pat. 332,605 (March 22, 1929), (to I. G. Farbenindustrie, A.G.).
33. A. Akramkhodaev, T. S. Sirlibaev, and K. U. Usmanov, *Uzb. Khim. Zh.* **1,** 429 (1981).
34. E. Sanhueza, I. C. Histaune, and J. Heicklen, *Chem. Rev.* **76,** 801 (1976).
35. R. Atkinson, and J. N. Pitts, Jr., *J. Chem. Phys.* **67,** 2488 (1977).
36. U.S. Pat. 4,045,316 (May 27, 1975), R. W. Legan (to Shintech, Inc.).
37. U.S. Pat. 3,933,980 (Dec. 19, 1974), L. A. Smalheiser (to Stauffer Chemical Co.).

38. M. Lederer, *Angew. Chem.* **71,** 162 (1959).
39. U.S. Pat. 4,062,925 (Jan. 21, 1977), D. E. Witenhafer, C. A. Daniels, and R. F. Koebel (to BFGoodrich Co.).
40. USSR Pat. 734,544 (May 15, 1980), V. A. Alferov and co-workers.
41. U.S. Pat. 2,060,303 (Nov. 10, 1936), H. P. A. Groll and G. Hearne (to Shell Oil Co.).
42. G. V. Sukhanov, A. F. Revzin, and V. Y. Shtern, *Kinet. Katal.* **15,** 551 (1974).
43. F. S. Dainton, D. A. Lomax, and M. Weston, *Trans. Faraday Soc.* **57,** 308 (1961); P. B. Ayscough, A. J. Cocker, F. S. Dainton, and S. Hirst, *Trans. Faraday Soc.* **57,** 318 (1961).
44. A. I. Subbotin, V. S. Etlis, and V. N. Antonov, *Kinet. Katal.* **9,** 490 (1968).
45. R. G. Rinker and W. H. Corcoran, *Ind. Eng. Chem. Fundam.* **6,** 333 (1967).
46. M. S. Kharasch, J. A. Norton, and J. F. R. Mayo, *J. Am. Chem. Soc.* **62,** 81 (1940).
47. L. Schmerling, *J. Am. Chem. Soc.* **68,** 1653 (1946).
48. J. M. Davidson and A. Lowry, *J. Am. Chem. Soc.* **51,** 2979 (1929).
49. J. Boeseken and co-workers, *Rec. Trav. Chim.* **32,** 184 (1911).
50. I. P. Tsukervanik and K. Y. Yuldashev, *Zh. Obshch. Khim.* **31,** 858 (1961).
51. M. S. Malinovskii, *Zh. Obshch. Khim.* **17,** 2235 (1947).
52. U.S. Pat. 2,006,517 (July 2, 1935), G. W. Seymur (to Celanese Corp. of America).
53. E. Arundale and L. A. Mikeska, *Chem. Rev.* **51,** 505 (1965).
54. U.S. Pat. 2,124,851 (July 26, 1938) and Brit. Pat. 465,467 (May 3, 1937), W. Fitzky (to I. G. Farbenindustrie A.G.).
55. T. T. Vasil'eva and co-workers, *Izv. Akad. Nauk SSR Ser. Khim.* **7,** 1584 (1980).
56. A. H. Weiss and K. A. Krieger, *J. Catal.* **6,** 167 (1966).
57. D. H. R. Barton and K. E. Howlett, *J. Chem. Soc.,* 165 (1949).
58. M. M. O'Mara, L. B. Crider, and R. L. Daniel; *J. Am. Ind. Hyg. Assoc.* **32,** 153 (1971).
59. R. W. McPherson, C. M. Starks, and G. J. Fryar, *Hydrocarbon Process.,* 75 (March 1979).
60. M. Sittig, *Vinyl Chloride and PVC Manufacture, Process and Environmental Aspects,* Noyes Data Corp., Park Ridge, N.J., 1978, p. 75.
61. S. N. Balasubramanian and co-workers, *Ind. Eng. Chem. Fundam.* **5,** 184 (1966).
62. U.S. Pat. 3,338,982 (Aug. 29, 1967), H. S. Leach (to Monsanto Chemical Co.).
63. U.S. Pat. 3,911,036 (Oct. 7, 1975), L. DiFiore and B. Calcagno (to Soc. Italiana Resine).
64. U.S. Pat. 3,917,727 (Nov. 4, 1975), U. Tsao (to Lummus Co.).
65. U.S. Pat. 3,941,568 (March 2, 1976), B. D. Kurtz and A. Omelian (to Allied Chemical Co.); Ger. Pat. 3,024,610 (Jan. 1, 1982), E. Birnbaum and co-workers (to BASF A.G.).
66. Ger. Pat. 2,540,292 (March 17, 1977), W. Opitz and H. Hennen (to Hoechst A.G.).
67. U.S. Pat. 4,000,205 (Dec. 28, 1976), R. G. Campbell (to Stauffer Chemical Co.).
68. C. M. Schillmoller, *Hydrocarbon Process.,* 89 (March 1979).
69. W. E. Wimer and R. E. Feathers, *Hydrocarbon Process.,* 81 (March 1976).
70. *Chlorine—Its Manufacture, Properties and Uses,* ACS Monograph Series No. 154, American Chemical Society, Washington, D.C., 1962, pp. 250–260.
71. Eur. Pat. Appl. 41,330 (Sept. 12, 1981), R. A. Kearley (to PPG Industries).
72. Jpn. Pats. 81 158,148 (Dec. 5, 1981) and 82 22,224 (Jan. 7, 1982), (to Tokuyama Soda Co.).
73. R. V. Carruba and J. L. Spencer, *Ind. Eng. Chem. Process Des. Dev.* **9,** 414 (1970).
74. U.S. Pat. 3,488,398 (Jan. 6, 1970), A. E. Van Antwerp, J. W. Harping, R. G. Sterbenz, and T. L. Kang (to B.F. Goodrich Co.).
75. U.S. Pat. 4,046,822 (Sept. 6, 1977), F. T. Severino (to Stauffer Chemical Co.).
76. P. Reich, *Hydrocarbon Process.,* 85 (March 1976).
77. Brit. Pat. 1,189,815 (April 29, 1970), K. Miyauchi (to Mitsui Toatsu Chemical Co.).
78. Ref. 60, pp. 37–39.
79. U.S. Pat. 3,966,300 (Dec. 7, 1976), R. C. Ahlstrom, Jr. (to The Dow Chemical Co.).
80. U.S. Pat. 3,935,286 (Jan. 27, 1976), J. C. Strini and J. R. Costes (to Rhone-Progil).
81. U.S. Pat. 4,060,460 (Nov. 29, 1977), E. W. Smalley, B. E. Kurtz, and B. Bandyopadhyay (to Allied Chemical Corp.).
82. Brit. Pat. 1,266,676 (March 15, 1972), (to Knapsack Co.).
83. Fr. Pat. 1,602,522 (Jan. 29, 1971), (to Solvay et Cie).
84. U.S. Pat. 3,484,493 (Dec. 16, 1969), A. Krekeler (to Knapsack Co.).
85. Brit. Pat. 956,618 (1964), A. G. Jacklin (to Imperial Chemical Industries, Ltd.).; Fr. Pat. 1,343,801 (Nov. 22, 1963), (to Imperial Chemical Industries, Ltd.).
86. Ger. Pat. 2,217,694 (Oct. 18, 1973), W. Froelich (to Hoechst A.G.).

87. D. H. R. Barton, *J. Chem. Soc.*, 148 (1949).
88. K. E. Howlett, *Trans. Faraday Soc.* **48,** 25 (1952).
89. P. G. Ashmore and co-workers, *J. Chem. Soc. Faraday Trans. 1*, 657 (1982).
90. U.S. Pat. 3,896,182 (July 22, 1975), D. P. Young (to B.P. Chemicals, Ltd.).
91. Brit. Pat. 938,824 (Oct. 9, 1963), (to BFGoodrich Co.).
92. Jpn. Pat. 67 22,921 (June 30, 1967), (to Mitsui Chemical Industries).
93. Ger. Pat. 3,024,156 (Jan. 21, 1982), A. Czekay and co-workers (to Hoechst A.G.).
94. Ger. Pat. 2,938,353 (Sept. 21, 1979), J. Wolfrum (to Max Planck Institute (Göttingen, FRG)).
95. Ger. Pat. 3,008,848 (March 7, 1980), J. Wolfrum (to Max Planck Institute (Göttingen, FRG)).
96. Brit. Pat. 1,168,329 (Oct. 22, 1969), (to Monsanto Chemical Co.).
97. U.S. Pat. 3,125,607 (March 17, 1964), H. M. Keating (to Monsanto Chemical Co.).
98. U.S. Pat. 3,142,709 (July 28, 1964), E. H. Gause (to Monsanto Chemical Co.).
99. U.S. Pat. 3,125,608 (March 17, 1964), D. W. McDonald (to Monsanto Chemical Co.).
100. J. S. Benson, *Hydrocarbon Process.*, 107, 108 (Oct. 1979).
101. U.S. Pat. 3,557,229 (Jan. 19, 1971), H. Riegel (to Lummus Co.).
102. *Chemical Economics Handbook, Marketing Research Report on Polyvinyl Chloride Resins*, Stanford Research Institute International, Menlo Park, Calif., April 1982.
103. *1982 Vinyl Chloride Report*, World Petrochemicals Program, Stanford Research Institute International, Menlo Park, Calif., 1982, pp. 113-7, 113-8.
104. *U.S. Environmental Protection Agency Report No. EPA-450/2-75-009*, EPA, Research Triangle Park, N.C., 1975.
105. Ref. 60, pp. 8–15.
106. Ref. 60, pp. 72–89.
107. Ger. Pat. 2,217,694 (Oct. 18, 1973), W. Froelich (to Hoechst A.G.).
108. Brit. Pat. 977,578 (Dec. 9, 1964) and 1,068,793 (May 17, 1967), (to Kureha Chemical Industries); E. Ger. Pat. 150,985 (Sept. 30, 1981), J. Glietsch and co-workers.
109. Ref. 60, pp. 42–46.
110. Ref. 60, pp. 7, 15, 36, 47–68, 333–339.
111. U.S. Pat. 3,670,037 (June 13, 1972), J. J. Dugan (to Esso Research and Engineering Co.).
112. U.S. Pat. 3,799,998 (March 28, 1974), D. G. Mead (to Imperial Chemical Industries, Ltd.).
113. U.S. Pat. 4,115,323 (Sept. 19, 1978), C. G. Vinson, M. F. Lemanski, and F. C. Leitert (to Diamond Shamrock).
114. U.S. Pat. 4,102,935 (July 25, 1978), W. J. Kroenke, R. T. Carroll, and A. J. Magistro (to BFGoodrich Co.).
115. U.S. Pat. 4,100,211 (May 17, 1976), A. J. Magistro (to BFGoodrich); Can. Pat. 1,111,454 (Oct. 10, 1981), T. P. Li (to Monsanto Co.).
116. G. R. Black and D. B. Schrock, personal communications, BFGoodrich Co., Chemical Group; generally accepted, industry-wide specifications.
117. *OSHA Regulations* 1910.1017, June 19, 1980.
118. *Prudent Practices for Handling Hazardous Chemicals in Laboratories*, National Academy Press, Washington, D.C., 1981, pp. 150–152.
119. *Vinyl Chloride Monomer—Handling and Properties*, 3rd ed., PPG Industries Brochure, Pittsburgh, Pa., Sept. 1977.

General References

S. Patai, *Chemistry of the Carbon–Halogen Bond*, Parts 1–2, John Wiley & Sons, New York, 1973.
Vinyl Chloride Process Economics Program, Report Series No. 5, Stanford Research Institute, Menlo Park, Calif., May 1965.

J. A. COWFER
A. J. MAGISTRO
BFGoodrich Co.

POLY(VINYL CHLORIDE)

Poly(vinyl chloride) [9002-86-2] (PVC), one of the few synthetic polymers that has wide application in commerce, has a sales volume between polyethylene and polystyrene. By the year 2000, the Stanford Research Institute predicts that, in the United States, PVC will be the leader with an annual volume of 17×10^9 metric tons. This widespread use arises from a high degree of chemical resistance and a truly unique ability to be mixed with additives to give a large number of reproducible PVC compounds with a wider range of physical, chemical, and biological properties than any other plastic material. Thus, with the help of properly chosen additives, a PVC formulation can be used as wire insulation, rigid pipe, or house siding. It is this unsurpassed versatility that has given PVC its great utility.

Produced by the free-radical polymerization of vinyl chloride, PVC has the following basic structure:

$$\left[\begin{array}{c} CH_2CH \\ | \\ Cl \end{array} \right]_n$$

(1)

where the degree of polymerization n ranges from 300 to 1500. Poly(vinyl chloride) is thermoplastic, ie, it softens and melts at elevated temperatures, but upon cooling regains its original properties (see Elastomers, synthetic, thermoplastic). Thus, by the application of heat and pressure, PVC may be extruded or molded into any desired shape. Processing can be from <150 to >200°C, depending upon the molecular weight of the resin and formulation. By coextruding different PVC formulations or PVC with other thermoplastics, products may be fabricated in which the component parts have widely different properties. For example, house siding may be coextruded with a thin, weather-resistant, highly pigmented so-called skin containing costly stabilizers and pigments, whereas the body may contain a less expensive compound. Similarly, through the coextrusion of hard and soft PVC compounds, a product can be made that is both rubberlike and yet contains rigid elements that impart strength and stiffness. Like all thermoplastics, PVC loses strength at high temperatures and has little application above 100°C. Although special high temperature compounds that can be used between 100 and 150°C are marketed, they may have only a limited service life.

The formation of PVC was first observed in 1835 following the exposure of 1,2-dichloroethane to sunlight (1). It was correctly reported that the white material produced had the elemental formula C_2H_3Cl and a density of 1.406 g/cm^3 (2). The polymeric nature and commercial importance of this material, however, were not recognized until the early 1900s when the application of PVC was first described in a patent for producing fibers, films, and lacquers (3). The technique of copolymerization and the use of peroxides as polymerization initiators were introduced subsequently (4) (see Initiators). Wide areas of nonrigid applications were opened up through the discoveries of plasticization and heat stabilization (5–6) (see Plasticizers; Heat stabilizers). During World War II, natural-rubber supplies were greatly curtailed, and PVC was used in wire and cable insulation, where it is superior in many ways to rubber (see Insulation, electric). The growth of PVC products has continued to the present.

Structure

Morphology. Most general-purpose PVC resin is produced by mass or suspension polymerization. However, emulsion polymerization is of some commercial importance. In emulsion polymerization, the medium is water (see Polymerization mechanisms and processes). When the polymerization is completed, the polymer is in the form of colloidal, spherical particles dispersed in water. The diameters of the spheres range in size from ca 0.05 to 2.0 μm (7–8). The particle size can be controlled by the polymerization conditions. If they are properly chosen, latexes with extremely narrow particle-size distributions are obtained (see Fig. 1).

The PVC latexes shrink in the beam of the transmission electron microscope, and unless precautions are taken, the measured diameters are too low. In a polymerized latex, the individual submicrometer particles are dispersed evenly throughout the water phase and show little evidence of flocculation. If the latex is dried, however, agglomerates form, as, for example, in dispersion resins. Dispersion resins are essentially latex polymer that has been dried; they are sold as a powder. Dispersion resins are characterized by a primary particle size and an agglomerate size. The former is the size of the original latex particle and is generally <1 μm. These primary particles form the agglomerates which may be well over 20 μm dia. Since a dispersion resin is rarely totally agglomerated, the measured size distribution is very broad with a range of diameters from <1 μm to >20 μm. A typical dispersion resin is shown in Figure 2. A cumulative particle-size distribution curve is shown in Figure 3 for a typical dis-

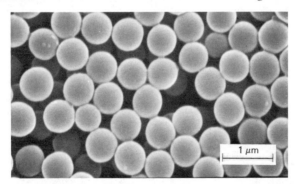

Figure 1. An electron micrograph of narrow particle-size-distribution PVC latex; average particle size ca 0.5 μm.

Figure 2. An electron micrograph of a typical PVC dispersion resin.

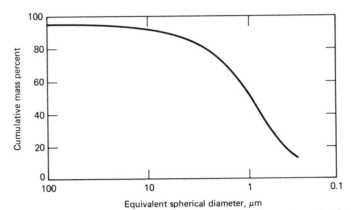

Figure 3. Typical particle-size-distribution curve for a PVC dispersion resin. Data obtained on Micrometrics Sedigraph.

person resin. The exceedingly broad distribution is due to the presence of agglomerates.

General-purpose resins, made by either the mass or suspension process, have a particle size typically of 80–200 μm. A typical suspension-resin grain has a smooth exterior and is irregular in shape. It has a porous interior composed of primary particles of 1–3 μm size (Fig. 4). Of particular interest is the skin or pericellular membrane which covers the exterior (see Fig. 5). It is ca 0.5–1 μm thick and consists mostly of PVC; in addition, it usually contains a small amount of chemically grafted dispersant (9). The

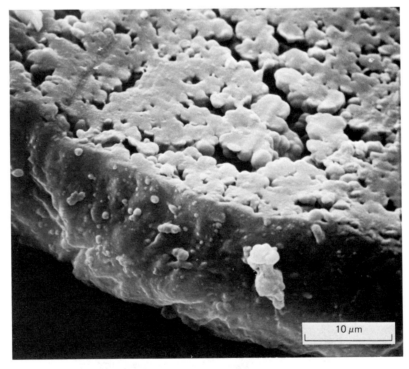

Figure 4. An electron micrograph of a cross section of a commercial PVC suspension resin showing the porous interior and the primary particles.

Figure 5. Electron-micrograph cross section of a commercial PVC suspension resin showing the surface-skin pericellular membrane.

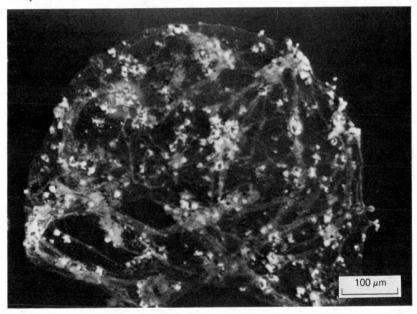

Figure 6. Electron micrograph of the isolated pericellular membrane ghost from a PVC suspension-resin grain.

presence of this graft copolymer can be easily demonstrated by dissolving the PVC with tetrahydrofuran. The pericellular membrane dissolves completely, except for a 20-nm thick so-called ghost which has been demonstrated to be a PVC–dispersant copolymer (see Fig. 6) (9). Such structures were first reported in 1970 (10).

The peculiar and complex morphology of suspension PVC is owing to the insolubility of PVC in its monomer and to the high PVC density (1.40 g/cm³) compared with that of the liquid vinyl chloride (0.910 g/cm³). Furthermore, vinyl chloride is soluble in water to the extent of 1–2%.

In suspension polymerization, 100 parts water and ca 100 parts vinyl chloride are charged to an autoclave with a small amount (less than one part) of dispersant, typically poly(vinyl alcohol) (PVA) or carboxymethyl cellulose. Agitation gives a suspension of vinyl chloride. Polymerization is initiated by addition of an oil- or monomer-soluble free-radical catalyst. Materials such as benzoyl peroxide and lauroyl peroxide are typical oil-soluble catalysts which have been employed to manufacture suspension PVC. The droplets of vinyl chloride fill with tiny particles of PVC insoluble in the monomer. At the same time, polymerization is initiated in the water phase primarily by chain transfer from the monomer.

Submicrometer-sized particles of a graft copolymer of PVC and the dispersant form. The copolymer migrates to the surface of the vinyl chloride droplets and builds up the skin from the outside. Thus, the pericellular membrane is built up from both directions during the early stages of the polymerization (see Figs. 7 and 8). At ca 1–2% conversion, the pericellular membrane forms a complete shell around the polymerizing droplet (Fig. 8(**b**)), while the inside of the monomer droplet is filled with micrometer-sized particles of PVC in Brownian motion. These particles are electrically charged (9). Thus, at 1–2% conversion, a double colloid exists with the 100 μm dia vinyl chloride droplets mechanically stabilized against flocculation by the pericellular membrane. The monomer droplet contains a stable sol of PVC in vinyl chloride that is electrically stabilized.

The fate of this colloid system depends upon agitation conditions within the reactor. In a nonagitated system or quiescent polymerization, the density difference between PVC and monomer causes a gradual collapse of the monomer droplet, and eventually the electrically stabilized grains of PVC inside the monomer droplet are forced into contact. The density difference between PVC and monomer causes the familiar dimpling of the surface of PVC particles (Fig. 9). If agitation conditions within the reactor are mild, the electrically stabilized PVC grains flocculate and an irregular structure is created inside the droplet (see Fig. 8(**d**)). More vigorous agitation coalesces the encased monomer droplets, and the familiar pluracellular structure of PVC forms. Thus, the interior and exterior morphology of a suspension PVC grain is the result of complex agglomeration operating with the polymerization process.

Mass polymerization is based upon the fact that PVC is monomer insoluble and that by suitable control of agitation, a particulate product resembling suspension PVC can be manufactured.

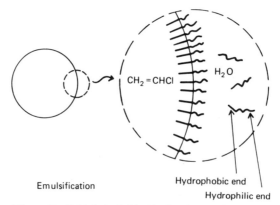

Figure 7.   Initial vinyl chloride droplet emulsification.

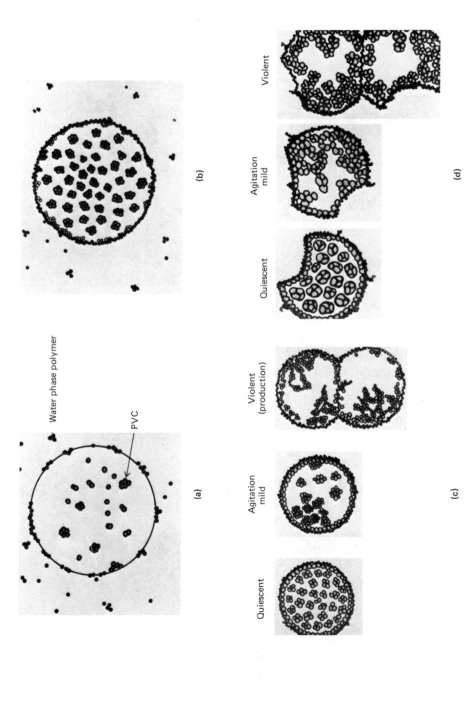

Figure 8. Growth of a PVC suspension-resin grain. (**a**) Below 1–2% conversion; (**b**) early conversion 1–2%; (**c**) early conversion 2–4%; and (**d**) ca 10% conversion.

891

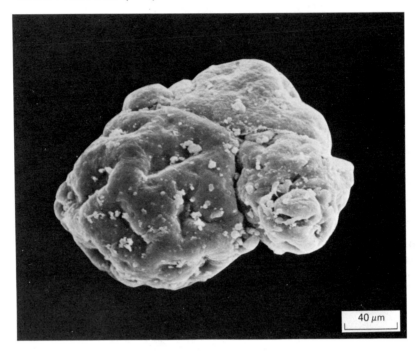

Figure 9.   Electron micrograph of a commercial PVC suspension-resin grain.

Such resins were originally claimed to be skinless (see Fig. 10). The resin in Figure 11, however, clearly shows evidence of a skin around a mass-resin particle. Such a skin is not a pericellular membrane and contains only PVC. The mechanism for the formation of such skin is at present not understood.

Molecular Structure.   Addition of the vinyl chloride monomer units during polymerization can occur either in head-to-tail fashion, resulting in 1,3 positions for the chlorine atoms,

$$\text{+CH}_2\text{CHClCH}_2\text{CHCl+}_n$$
(2)

or head-to-head, tail-to-tail, placing the chlorine atoms in 1,2 positions.

$$\text{+CH}_2\text{CHClCHClCH}_2\text{+}_n$$
(3)

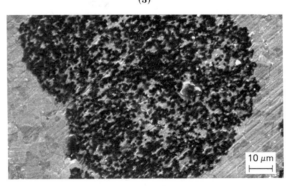

Figure 10.   An electron micrograph of a cross section of a commercial mass-polymerized PVC resin.

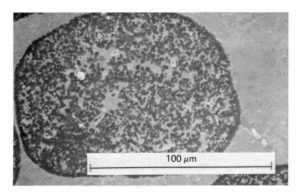

Figure 11.   An electron micrograph of a cross section of a commercial mass PVC resin.

Dehalogenation of structure (**3**) would be expected to be complete, for each chlorine atom can be removed by reaction with a chlorine on an adjacent carbon atom. In structure (**2**), however, if dechlorination starts at random positions along the chain, 13.5% of the chlorine is left (11). Treatment of dilute solutions of PVC with zinc actually removes up to 87% (not 100%) of the chlorine, proving the essentially head-to-tail arrangement of the polymer (12). A product corresponding to structure (**3**) can be prepared by chlorination of *cis*-1,4-polybutadiene to a chlorine concentration of 56.6% (56.8%, theoretical). The ir spectrum of this product differs greatly from that of conventional PVC, especially in the 1333-cm^{-1} region.

There are three possible categories of end groups. Saturated groups are formed by chain transfer to monomer and polymer and by termination through disproportionation:

$$-CH_3; \quad -CH_2Cl; \quad -CHCl_2$$
$$\quad(4)\qquad\;(5)\qquad\;\;(6)$$

Unsaturated chain ends result from termination by disproportionation and chain transfer to monomer:

$$-CH_2CHClCH{=}CHCl; \quad -CH_2CHClCCl{=}CH_2; \quad -CH_2CHClCH{=}CH_2$$
$$\qquad(7)\qquad\qquad\qquad\quad(8)\qquad\qquad\qquad\quad(9)$$

Initiator or solvent (chain-transfer agent) fragments, represented by R, can be incorporated in the terminal group:

$$-CH_2R; \quad -CHClR$$
$$\;(10)\qquad\;\;(11)$$

Because of the high transfer activity of the monomer, ca 60% of the polymer molecules are estimated to have unsaturated end groups (13). For the same reason, the percentage of chain ends containing initiator fragments is low (14); the amount of solvent fragments depends upon transfer activity.

Long-chain branching can be caused by the incorporation of the terminal double bond of a polymer molecule into a growing chain:

$$\sim\!\!\sim CH_2\overset{\cdot}{C}HCl \;+\; \sim\!\!\sim CH{=}CHCl \rightarrow CH_2CHClCH\overset{\cdot}{C}HCl \tag{1}$$

or by intermolecular chain transfer to polymer:

$$\sim\!\!\sim CH_2\overset{\cdot}{C}HCl \;+\; \sim\!\!\sim CH_2CHCl\!\sim\!\!\sim \;\rightarrow\; \sim\!\!\sim CH_2CH_2Cl \;+\; \sim\!\!\sim CH_2\overset{\cdot}{C}Cl \tag{2}$$

Intramolecular chain transfer (backbiting) leads to formation of short side chains:

$$
\begin{array}{c}
\text{(structure)} \longrightarrow \text{(structure)}
\end{array}
\qquad (3)
$$

For the determination of total branching, the polymer is hydrogenated, and the chlorine removed with lithium aluminum hydride. The ratio of methyl to methylene groups is determined by ir spectroscopy, using the bands at 1378 and 1350, or 1370 and 1386 cm^{-1}, respectively. In conventional resins polymerized in bulk or suspension at 50–90°C, there are 0.2 to 2 branches per 100 carbon atoms (14). The distribution between long and short chain branching is not known. With lower temperature, the number of branches decreases, because the enthalpy of activation for propagation is smaller than that for chain transfer.

The tactical placement of the monomer units with respect to their next neighbors affects the properties of the resulting polymer.

The propagation rate coefficients are k_i for isotactic and k_s for syndiotactic placement. The potential energy for syndiotactic conformations is 4.2–8.4 kJ/mol (1–2 kcal/mol) lower than that for isotactic placement (15–16). Consequently, the free energy of activation for syndiotactic placement ΔG_s must be lower than ΔG_i, and from absolute reaction-rate theory it follows that the ratio k_s/k_i must increase with decreasing temperature.

$$k_s/k_i = \exp\left[(\Delta G_i - \Delta G_s)/RT\right]$$

$$k_s/k_i = \exp\left[-(\Delta S_i - \Delta S_s)/R\right] \exp\left[(\Delta H_i - \Delta H_s)/RT\right]$$

where ΔS = entropy of activation and ΔH = enthalpy of activation. Thus, with decreasing polymerization temperature, the degree of syndiotacticity of PVC should increase for purely thermodynamic reasons without necessarily requiring adsorption or complexing by a catalyst. The extent of syndiotactic placement can be determined by nmr (17) or ir (18). Measurements on several series of polymers prepared in bulk or suspension by radical initiation at 120 to −80°C show some discrepancies but clearly demonstrate an increase in syndiotactic placement from ca 50 to 65%. Experimental values for the difference in activation enthalpy ($\Delta H_i - \Delta H_s$) range from 1.3 to 2.5 kJ/mol (310–600 cal/mol), whereas the difference in activation entropy, ($\Delta S_i - \Delta S_s$), is ca 2.5 J/(mol·K) [0.6 cal/(mol·K)] (19).

The syndiotactic sequences in PVC are capable of crystallization. The degree of crystallinity can be determined from x-ray patterns and correlates with the degree of syndiotacticity. The unit cell is orthorhombic with axes of 1.06, 0.54, and 0.51 nm (20); the dimensions of the crystallites are 5.0–10.0 nm (19). It is also possible to prepare single crystals of the homopolymer of vinyl chloride (21). The degree of crystallinity of a sample depends on its history; for conventional resins, it is as high as 5%, and for low temperature resins, it is 25–30%. Higher degrees of crystallinity are found in samples polymerized in the presence of aliphatic aldehydes (22). This is probably owing to the exceptionally low degree of polymerization and branching obtained under these conditions (23).

The degree of crystallinity in percent X_c can also be determined from the specific volume (reciprocal density) of an unknown sample $\bar{v}$ if the specific volumes of a completely crystalline $\bar{v}_c$ and a completely amorphous sample $\bar{v}_a$ are known.

$$X_c = 100 \frac{\bar{v} - \bar{v}_a}{\bar{v}_c - \bar{v}_a}$$

Experimental values reported in the literature are 0.694 for $\bar{v}_c$ (20) and 0.722 for $\bar{v}_a$ (24); the latter is in good agreement with a theoretical value calculated under the assumption of additivity of group increments for molar volumes (25). Other authors obtained 0.671 for $\bar{v}_c$ and 0.709 for $\bar{v}_a$ with a different annealing technique (26). The degree of crystallinity can also be determined from differential scanning calorimetry measurements (dsc). The molecular structure of PVC has been reviewed (27).

Physical Properties

Solution Properties. It is practically impossible to obtain a truly molecular-disperse PVC solution. In the case of suspension resins, the pericellular membrane ghosts are always present; microgels are also observed (26). Measurements of properties are often performed on cyclohexanone or tetrahydrofuran solutions. High boiling solvents have commercial application as plasticizers for PVC (see under Compounding).

The molecular weight of PVC is commonly determined by measurement of a solution property, usually the intrinsic viscosity (IV). The standard ASTM procedure (ASTM D 1243) uses cyclohexanone and a dilute-solution viscometer of the Ubbelohde type (28). The intrinsic viscosity $[\eta]$ (dL/g) is defined as follows:

$$[\eta] = \frac{\ln \eta_r}{c} \lim c \to 0$$

where η_r (relative viscosity) is the viscosity of the solution t divided by that of the pure solvent t_o. The quantity $\ln \eta_r/c$ is known as the inherent viscosity. In practice, η_r is determined at several concentrations. A plot of $\ln \eta_r/c$ vs c (concentration) gives the intrinsic viscosity at the zero concentration intercept.

Commercially, PVC resins are referred to in terms of intrinsic viscosity directly, although the molecular weight (viscosity average) can be calculated from the Mark-Houwink equation:

$$[\eta] = kM^a$$

The constants k and a must be determined from samples that have been characterized by an absolute method of molecular weight determination, such as light scattering or any method that gives a weight-average molecular weight, since this quantity is close to the viscosity average. Different values of these constants are used because of solution anomalies (29–49). The following equation, proposed for cyclohexanone at 25°C, is based on the work of several authors (49):

$$[\eta] = (2.69 \times 10^{-2})(M_w)^{0.72}$$

For the determination of molecular weight distribution, gpc is commonly employed using tetrahydrofuran as a solvent (50). Since these instruments are usually calibrated with standard polystyrene samples, PVC data are of relative significance only.

Thermal Properties. The glass-transition temperature T_g of poly(vinyl chloride) depends upon the polymerization temperature of the resin. Table 1 presents T_g values measured by dilatometry (51), differential-scanning calorimetry (52), and dynamic-mechanical methods (53–54) on samples polymerized at the indicated temperatures. The data show an approximately linear decrease with increasing polymerization temperature t_{pol}, showing the effects of molecular weight and the amount of crystalline material present:

$$T_g = 93.4 - 0.20 \times t_{pol}$$

The glass-transition temperature of pure commercial PVC resins is ca 81°C (53,55). In the case of copolymers, T_g can be approximately determined by a weighted average of the transition temperatures, T_{g1} and T_{g2} of the homopolymer components (56–57):

$$\frac{1}{T_g} = \frac{w_1}{T_{g1}} + \frac{w_2}{T_{g2}}$$

where w_1 and w_2 represent the weight fractions. Examples for measured values are 90°C for a copolymer of 60% vinyl chloride and 40% acrylonitrile (58); 63 and 59°C for copolymers with 5 and 10% vinyl acetate, respectively (59); and 60 and 38°C for copolymers with 10 and 20% ethylene, respectively (60).

Table 1.   Glass-Transition (T_g) and Melting-Point (T_m) Temperatures of Poly(Vinyl Chloride)

Polymerization temperature, °C	T_g, °C	T_m, °C	Refs.
125	68	155	53
90	75		53
60	85		24
40	80	220	53
−10	90	265	53
−15	105	285	24
−20	95		54
−30	100		52
−40	105		52, 54
−50	106		52
−60	110		54
−75		310	24
−80	100	>300	53

The melting point T_m of PVC cannot be measured directly because of the thermal instability of the resin (see below). It is usually determined from the melting temperature T_m^* of solutions of the polymer in plasticizers, according to the relation (61):

$$\frac{1}{T_m} = \frac{1}{T_m^*} - \left(\frac{R}{\Delta H_u}\right)\left(\frac{V_u}{V_1}\right)(v_1 - X_1 v_1^2)$$

where R is the gas constant, ΔH_u the heat of fusion per mole of repeating units of the polymer, V_u and V_1 are the molar volumes of repeating unit and diluent, v_1 is the volume fraction, and x_1 the interaction parameter of the diluent. A value for ΔH_u of 3.28 kJ/mol (785 cal/mol) has been reported (51). Table 1 shows that the melting-point temperatures thus obtained again increase linearly with decreasing polymerization temperature t_{pol} of the resin, according to the approximate relation:

$$T_m = 257.5 - 0.81 \times t_{pol}$$

The linear coefficients of expansion of PVC are $(6-8) \times 10^{-5}/°C$ below T_g and $(20-22) \times 10^{-5}/°C$ above T_g. The specific heat values at constant pressure, c_p, in these two ranges are 1.046–1.255 and 1.757 J/(g·K) (0.25–0.30 and 0.42 cal/(g·K)), respectively. Thermal conductivity at 20°C is 0.1588 W/(m·K) (62). For detailed thermodynamic functions, see reference 63.

At temperatures >100°C, PVC begins to decompose at a noticeable rate (64–72). Upon heating, hydrogen chloride is evolved and the resin becomes discolored, brittle, and finally insoluble. The rate of decomposition depends upon many variables, including the surrounding atmosphere, temperature, and molecular weight of the polymer. In oxygen, the rate is higher than in nitrogen, but the resin shows less discoloration at an equal extent of reaction. In an inert atmosphere, the molecular weight of the polymer increases from the beginning of the reaction, whereas in oxygen, it goes through a minimum. At constant oxygen pressure, the rate is proportional to the square root of the partial pressure of the gas (73). The activation enthalpy in this case is ca 100.8 kJ/mol (24 kcal/mol), whereas in a nitrogen atmosphere it is as high as 138 kJ/mol (33 kcal/mol) (73–74). The effect of temperature upon the evolution of HCl and the limiting-viscosity number is shown in Table 2 (75).

The rate of HCl evolution at a given temperature increases with decreasing number-average molecular weight of the starting resin, according to the following relation (74):

$$\mu\text{mol HCl/(g·h)} = a + b/M_n$$

Table 2. Effect of Temperature upon Decomposition of Poly(Vinyl Chloride)[a,b]

Temperature, °C	Mol HCl per mol resin	$[\eta]$, mL/g
130	0.10	110
150	0.53	110
170	4.77	120
190	27.5	160

[a] Ref. 75.

[b] Starting with a resin of $[\eta] = 110$, after 4 h exposure at the indicated temperature.

In nitrogen, $a = 0$ and $b = 7 \times 10^5$; in air, $a = 6$ and $b = 6 \times 10^5$; and in oxygen, $a = 18$ and $b = 3 \times 10^5$. All values are measured at 182°C.

The mechanism of the thermal degradation of PVC is not fully understood. It decomposes more readily than any of its model compounds. The decomposition temperatures (T_ds) of model compounds for PVC are 3-chloro-2-pentene, 400°C; 2,4-dichloropentane, 360°C; 2-chloropropane, 340°C; 4-chloro-1-hexene, 325°C; 3-chloro-1-pentene, 280°C; 2-methyl-2-chloropropane, 240°C; and 4-chloro-2-hexene, 150°C. T_d is defined as the temperature of initiation of decomposition, as measured by ir spectroscopy (76).

At 150°C, PVC and 4-chloro-2-hexene have the same initial decomposition-rate constant (64), indicating that allylic and tertiary chlorines may be the temperature-sensitive groupings in PVC. They can be formed in the polymer by disproportionation or chain transfer to monomer (structures (7), (8), and (9)) and by branching reactions (eqs. 2 and 3). Once dehydrohalogenation of an allyl chloride end group has started, new allylic groupings are formed and the reaction can proceed in a zipper fashion, leading to polyene structures which are largely responsible for discoloration (75):

$$—CH_2CHClCH_2CHClCH{=}CH—$$

$$\xrightarrow{\text{--HCl}} —CH_2CHClCH{=}CHCH{=}CH—$$

$$\xrightarrow{\text{--HCl}} —CH{=}CHCH{=}CHCH{=}CH—$$

Networks are formed by elimination of HCl between chains. In the presence of oxygen, additional reactions take place, eg, formation of peroxides and β-chloroketones, as well as chain scission and network formation via radical mechanisms. In addition, unimolecular elimination and ionic mechanisms are claimed in the decomposition reactions of PVC. The HCl formed during decomposition has an accelerating effect on further degradation (77). Chlorides of certain metals, such as iron, barium, and zinc, are used as catalysts for the dehydrohalogenation.

The frequency distribution of the polyene sequences can be determined by uv spectroscopy. At low degree of dehydrohalogenation and a reaction temperature of 180°C, there are average sequence lengths of 5–10 conjugated double bonds. The number of longer sequences decreases continually, with maximum lengths of 25–30 conjugated double bonds (74,78). With increasing temperature and time of degradation, the frequency distribution shifts toward lower sequences (79). At high degrees of conversion, aromatic decomposition products, eg, benzene and toluene, are formed (80).

The thermal degradation of PVC can be delayed or slowed down by the addition of heat stabilizers which are essential in PVC compounds.

Chemical Properties. The primary objectives of chemical modification of PVC include: increased solubility in inexpensive solvents, increased heat-distortion temperature, increased resistance to hot melt flow, introduction of ion-exchange capacity, preparation of polymer structures not otherwise available, improved stability to light and heat, and improved melt flow (81).

The most important chemical reaction is chlorination. The process can be carried out in an organic medium, eg, carbon tetrachloride, at moderate temperatures under the influence of uv irradiation. With increasing chlorine content, the product becomes soluble and is later recovered by precipitation with methanol (82). The reaction can also take place in aqueous suspension with the addition of a swelling agent, eg, chloroform or carbon tetrachloride, catalyzed by uv irradiation (83) or an oil-soluble acyl peroxysulfonate (84).

Other patents describe the chlorination of PVC in the dry state (85) or after addition of small amounts of chloroform (86). The chlorine content can be raised to ca 73 wt %, corresponding to the introduction of an additional chlorine atom per monomer unit. Higher halogen contents can only be obtained with difficulty; by carrying out the reaction in thionyl chloride, up to 75.6 wt % chlorine is reported (87). The chlorine content can be calculated from the measured density d, in g/cm^3, of a press-polished sheet containing 2 wt % organotin stabilizer (88), as follows:

$$wt \% \; Cl = (65.12 \times d) - 33.37$$

The T_g of a given resin increases with increasing degree of chlorination; it follows approximately the relation (in °C), as shown (89):

$$T_g = (3.8 \times wt \% \; Cl) - 136$$

The thermal stability of highly chlorinated poly(vinyl chloride) resins, as measured by weight loss upon heating, is considerably greater than that of the original polymer (90). The chlorinated products have higher mechanical strength and lower impact strength than the starting resin (91).

The structure of chlorinated poly(vinyl chloride) can be elucidated by chemical means (92), ir (93), nmr (94), and pyrolysis-gas chromatography (95). Chlorine can be incorporated into the polymer chains in the following ways:

$$\sim\sim CH_2CHCl\sim\sim \; + \; 1/2 \; Cl_2 \left[\begin{array}{l} \longrightarrow \sim\sim CHClCHCl \sim\sim \\ \qquad\qquad (12) \\ \longrightarrow \sim\sim CH_2CCl_2 \sim\sim \\ \qquad\qquad (13) \end{array} \right.$$

leading either to a 1,2-dichloroethylenic (12) or a 1,1-dichloroethylenic (13) unit. The measured ratio of 1,2- to 1,1- units formed is higher than the ratio of 2:1 statistically expected from the availability of hydrogen in the methylene and methyne groups. The experimental ratio is only slightly higher than 2:1 at low degrees of chlorination and reaches 5:1 at high chlorine contents. Smaller amounts of trisubstituted 1,1,2- units exist at >67 wt % chlorine. These ratios are identical for all chlorination techniques. However, some physical properties vary with the method used: resins chlorinated in solution have greater solubility and lower glass-transition temperatures than resins treated in suspension in the presence of swelling agents. The difference is probably because of a different sequence distribution of the halogenated units (96).

Nucleophilic displacement of chlorine from PVC using thiol compounds has been carried out under various conditions to give products with enhanced physical properties (97–98). Investigation of nucleophilic displacement with different sodium thiolates (RS^-Na^+) showed that the nucleophilicity increased if an ether linkage was in a β position to the thiol group, $[RO(CH_2)_2S^-Na^+]$ (99). Up to 33% of the Cl atoms is replaced by the thiolate, and the polymer behaves like an internally plasticized PVC. Substitution of chlorine by acetoxy groups has been studied, using potassium acetate in tetrahydrofuran together with the crown ether (18-crown-6), 1,4,7,10,13,16-hexaoxacyclooctadecane (100–101). Depending upon reaction conditions, up to 4.8% of the chlorine can be replaced by acetoxy groups. This extent of substitution is far greater than previously reported for labile chlorine in PVC (102).

The preparation of both anionic and cationic ion-exchange resins from PVC is included in the review in ref. 81 (see Ion exchange). Anion-exchange resins are prepared by treating the resins with amines, such as *m*-toluidine, aniline, and aliphatic diamines. The sulfonated polymer prepared by reaction with chlorosulfuric acid or sulfuryl chloride produces cation-exchange resins. Reaction of PVC with aromatic diamines improves the tensile properties by cross-linking the polymer chains (103–104). Other cross-linking reagents mentioned in the patent literature for improved heat stability include allyl *tert*-butyl peroxycarbonate (105).

Reductive dechlorination of PVC with lithium aluminum hydride followed by spectroscopic examination has been extensively studied as a method for determining the microstructure of the polymer (106). At 100°C in THF, ca 98–99% PVC is reduced, although side reactions impede structure determination. Tri-*n*-butyltin hydride, however, gives 99.7% reduction, and this method, in conjunction with ^{13}C nmr, has proven useful for determining methyl branch and long branch frequencies (107). Over a wide range of polymerization temperatures, 43–75°C, there is little effect upon branch frequency. The chloromethyl-branch frequency over this temperature range is reported as 2.1–2.8 and the long-branch frequency as 0.4–0.6 for every 1000 carbon atoms in the polymer chain (107). Chloromethyl branches are formed after occasional head-to-head monomer additions followed by 1→2-chlorine migration and propagation. The formation of anomalous structures in PVC (ie, branch structures) and their influence on thermal stability are reviewed in ref. 108.

Polymerization

Mass Polymerization. The polymerization of vinyl chloride by a mass or bulk procedure, where a free-radical initiator is added to the liquid monomer, is normally difficult to control because of the heterogeneous reaction products (109) (see also Polymerization mechanisms and processes). At high conversions, the mixture becomes extremely viscous, impeding heat removal and leading to low molecular weight and broad molecular weight distribution. A two-stage process that overcomes these problems is the basis of the commercial process originally developed by Saint Gobain (France) in the 1940s and now licensed by the successors, Chloé Chimie (110). The first stage of the process is carried out in a prepolymerizer, a vertical reactor equipped with a flat-blade turbine stirrer and baffles, where the monomer is polymerized for ca 1–1.5 h with 7–10% conversion. Below 7% conversion, the grains are not sufficiently cohesive to be transferred without breaking up, whereas above 10% conversion, the mixture is too viscous for the system to remain homogeneous. The grains act as skeleton seeds for growing polymer in the second stage of the reaction. The structure of the seed has great influence upon the final properties of the product. Indications are that the number of grains remains constant throughout the entire reaction (111).

In the second stage, the mixture from the prepolymerizer together with more monomer and initiator is transferred into the autoclave (Figs. 12 and 13). The autoclave may be either a horizontal or vertical reactor [ca 15–47 m^3 (4,000–12,500 gal)], twice the size of the prepolymerizer, equipped with slowly rotating agitator blades. In the horizontal reactor, the blades are cagelike in structure and rotate closely to the reactor wall. In the vertical autoclave, a screw runs from the top to the bottom circulating the

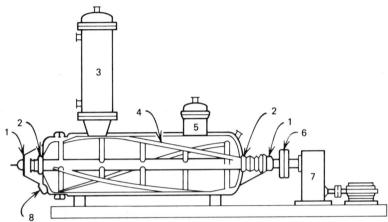

Figure 12. Mass polymerization, horizontal second-stage autoclave assembly. 1, Agitator bearings; 2, agitator packing seals; 3, condenser; 4, agitator; 5, degassing filter; 6, agitator coupling; 7, gear box; and 8, PVC unloading port. Courtesy of *Hydrocarbon Processing* (112).

powder; a blade in the base of the reactor prevents settling and feeds the screw (Fig. 14). The reaction proceeds through the liquid stage and, at ca 25% conversion, becomes a powder. Reaction continues until the pressure in the autoclave starts to fall and free liquid monomer is no longer available for heat removal by the condenser. Considerable heat is evolved, ca 71–111 kJ/mol (16.9–26.5 kcal/mol) (113). It is estimated that 60% of this heat is removed through the condenser, 30% through the jacket, and 10% through the cooled agitator shaft (112). Reaction time (3–9 h) depends upon the product. The initiators are similar to those employed in suspension polymerization. Unreacted monomer is removed by vacuum and recovered by vapor compression and condensation in the recycle condenser. The resin is transferred to the receiver by means of an air eductor.

Mass resins exhibit high porosity with good fusion characteristics and high film clarity. The spherically shaped grains and the narrow size distribution give resins with high bulk density and fast extrusion rates. From a process standpoint, the mass operation has low utility costs with minimum water consumption, no drier operation, and few raw materials. However, suspension polymerization offers more flexibility.

Suspension Polymerization. The monomer is first finely dispersed in water by vigorous agitation. Polymerization is started by means of monomer-soluble initiators. Addition of suspension stabilizers and other suspending agents minimizes coalescence of the growing grains by forming a protective coating. The hydrophobic–hydrophilic properties and, hence, monomer and water solubility of the suspending agents, are key factors in determining resin properties and controlling grain agglomeration (114).

The kinetics of suspension polymerization is identical with that of mass polymerization (115), which shows increasing rate behavior. The molecular weight of the resulting polymer is practically independent of the concentration of the initiator and exhibits only a slight increase with increasing conversion (116). Molecular weight decreases with increasing temperature, which is therefore used to control the molecular weight. For very low molecular weight products, chain-transfer agents, eg, trichloroethylene and methyl-substituted olefins, are useful (see under Copolymerization).

Suspension polymerization, of great technical importance, is used for an estimated

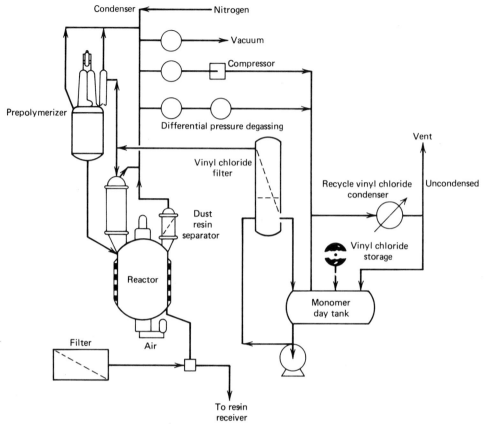

Figure 13. Two-step mass PVC plant with prepolymerizer and vertical autoclave. Courtesy of *Hydrocarbon Processing* (112).

82% of U.S. production. A flow sheet for a typical batch process is shown in Figure 15 (117). Most current reactors are water-jacketed and lined with glass or stainless steel to minimize polymer buildup on the walls. Significant developments in clean-reactor technology have taken place. The prevention of polymer buildup on the reactor walls has increased reactor productivity by reducing cleaning, as well as resulting in improved product quality (118–119). Reactor sizes vary from ca 7.57 to 190 m^3 (2,000–50,000 gal). Ratios of water to vinyl chloride are (1.2–4):1. Low ratios allow higher monomer charges for a given reactor, whereas high water content facilitates temperature control and permits higher conversions. After monomer and water have been metered into the reactor, initiator, suspending agents, and buffer are added. The agitated mixture is heated to reaction temperature, usually 45–75°C, and the heat of polymerization removed by cooling water. The reaction is usually carried out to ca 85% conversion or a given pressure drop. At ca 70% weight conversion, no more free monomer is available. The partial pressure is thus reduced, and the reaction pressure continues to decline with increasing conversion. After polymerization, the mixture is transferred to a dump tank where a large portion of unreacted monomer is recovered. To remove the remaining vinyl chloride monomer from the resin, in the mid-1970s, BFGoodrich developed a stripping column that removes residual vinyl chloride almost completely (120). In this process, resin slurry is passed through a vertical column

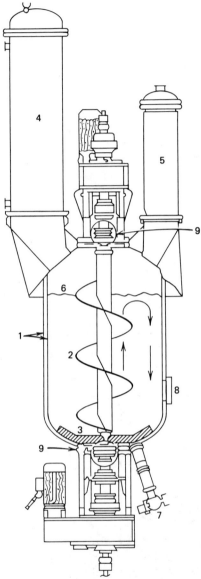

Figure 14. Mass polymerization, 47.3 m³ (12,500 gal) vertical autoclave–stripper assembly. 1, Autoclave shell and jacket; 2, upper screw agitator; 3, lower scraper agitator; 4, reflux condenser; 5, degassing filter; 6, maximum PVC resins level; 7, PVC unloading valve; 8, manhole; and 9, packing seal. Courtesy of *Hydrocarbon Processing* (112).

through which steam is passed. The resin slurry then passes to the centrifuge where the water is separated; the resin is dried in a stream of hot air with a rotary or fluid-bed dryer. The product is separated from the wet air stream in the cyclone separator, from which it is screened and sent to storage. The wet air stream containing the fines is passed through a filter. The process is licensed in the United States and many other countries. Various types of stripping operations are reviewed in ref. 121.

Suspending agents include maleic anhydride–vinyl acetate copolymers, urea–formaldehyde condensation products, precipitated inorganic carbonates, phosphates

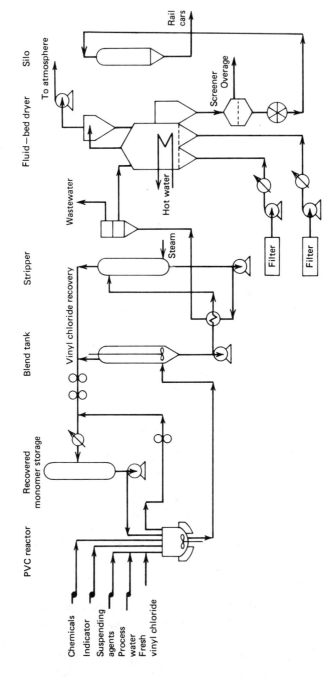

Figure 15. Suspension-polymerization plant. Courtesy of *Hydrocarbon Processing* (117).

and silicates, water-soluble cellulose derivatives, and methacrylic acid copolymers (see Table 3) (122–140). Partially hydrolyzed poly(vinyl acetate) (PVAc), gelatin, and methyl cellulose are preferred, usually in concentrations of 0.05–0.5 parts per hundred parts of monomer (phm). The minimum amount of hydrolyzed PVAc required is determined by the molecular weight; low molecular weight polymer is an effective suspending agent at 0.005 phm, whereas high molecular weight material at this concentration is ineffective (141). The degree of hydrolysis of the PVAc is also important in imparting suspension stability and resin properties. Suspension resins usually have grain sizes of 120–150 μm dia. Microsuspension resins are produced with grain sizes of 30–50 μm dia and are used for blending in plastisols. The grain size is reduced by increasing the concentration of suspending agents.

The net activation energy of the polymerization is mainly determined by the activation energy for the creation of free radicals. The activation energies of the propagation and termination steps are negligible. Because chain transfer should not affect the polymerization rate, the net activation energy is primarily due to the thermal decomposition of the initiator into free radicals. An activation energy of decomposition for a peroxydicarbonate initiator has been reported as 84.1 kJ/mol (20.1 kcal/mol) and 91.2 kJ/mol (21.8 kcal/mol) for benzoyl peroxide (142–143). Generally speaking, manufacturers report activation energies for peroxyesters and peroxydicarbonates from 105 to 126 kJ/mol (25–30 kcal/mol) in n-decane or trichloroethylene. Lower values for redox initiation, 27 kJ/mol (6.5 kcal/mol) (144), and photoinitiation, 14.7 kJ/mol (3.5 kcal/mol) (145), have been reported. Initiators are employed in concentrations of 0.03–0.1% based on weight of monomer. In general, peroxydicarbonates or peroxyesters are employed as initiators, eg, bis(2-ethylhexyl) peroxydicarbonate and bis-(*sec*-butyl) peroxydicarbonate, or t-butyl peroxyneodecanoate and t-butyl peroxypivalate, respectively. The reactivity of peroxyesters

is strongly affected by the type of substitution on the α carbon; different R' substituents create initiators with various half-lives. However, peroxydicarbonate initia-

Table 3. **Suspending Agents**

Suspending agent	Refs.
magnesium silicate	122
maleic acid–vinyl acetate copolymer	123
urea–formaldehyde condensate	124
cellulose derivatives[a]	125
inorganic phosphates	126–128
inorganic carbonates	128
partially hydrolyzed poly(vinyl acetate)	129
gelatin	130
polyvinylpyrrolidinone	131
polyglycerol esters	132

[a] Carboxymethyl, hydroxypropyl, hydroxyethyl.

tors, because of the greater distance between the oxygen–oxygen bond and the α carbon, show very similar half-lives. The selection of initiators is important from the productivity point of view, where combinations of initiators may be employed to achieve more uniform and linear reaction rates. In addition, initiators show appreciable water solubility (10–1000 mg/kg water), which influences reactor buildup and resin structure. Initiator hydrolysis and impurities affect resin color and electrical properties and contamination of recovered vinyl chloride monomer. New formulations of initiator dispersions in water offer greater margins of safety and automatic reactor charging (146). Typical polymerization times are 4–10 h, depending upon the mol wt of the resin being prepared, as well as the heat-removal capacity of the reactor system.

Mass and Suspension Kinetics and Mechanisms. The kinetics and mechanisms of mass and suspension polymerization are generally treated together because the suspension droplet is simply considered to be a mass polymerization on a small scale. In vinyl chloride polymerization, the polymer precipitates out from the monomer when the chain has reached 10–20 monomer repeat units in length. The precipitated polymer is swollen by monomer, and the reduced termination rate in the swollen-gel phase may be considered responsible for the observation that with increasing conversion, the polymerization rate increases. Furthermore, polymerization is accompanied by ca 35% shrinkage. The reactions taking place are described below:

Reactions in the liquid phase

initiation $M \xrightarrow{k_i} M\cdot$

propagation $R\cdot + n\ M \xrightarrow{k_p} R(M)_n\cdot$

chain transfer to a monomer $R\cdot + M \rightarrow P + M\cdot$

termination $R\cdot + R\cdot \xrightarrow{k_t} P$

Reactions in the polymer particles

radical trapped in polymer particle $R\cdot + P \rightarrow (R\cdot) + P$

propagation $(R\cdot) + n\ M \rightarrow (R(M)_n\cdot)$

chain transfer to a monomer $(R\cdot) + M \rightarrow P + M\cdot$

termination $(R\cdot) + (R\cdot) \rightarrow P$

$\qquad\qquad (R\cdot) + M\cdot \rightarrow P$

Reactions of the monomeric radicals in the polymer particles

propagation $M\cdot + M \rightarrow (R\cdot)$

chain transfer to polymer $M\cdot + P \rightarrow (R\cdot)$

escape into the liquid phase $M\cdot \rightarrow R\cdot$

where $R\cdot$ = chain radical in the liquid phase, M = monomer molecule, P = polymer molecule, $M\cdot$ = monomer radical in the liquid phase, and $(R\cdot)$ = chain radical trapped in a polymer particle. The symbols k_i, k_p, and k_t indicate the reaction rate coefficients for initiation, propagation, and termination. Absolute values for k_p and k_t at 60°C are 1.23×10^5 and 2.3×10^{10} L/(mol·s), respectively (145).

The presence of two phases, a monomer-swollen-polymer phase and the dilute-liquid monomer phase, forms the basis for all kinetic model descriptions (147–148). The polymerization rate is slower in the dilute phase than in the swollen-polymer phase because of lower mobility and decreased termination rate of the growing polymer chain. It has been calculated that the volume fraction of polymer is only 0.001 in the dilute-monomer phase and 0.6 in the concentrated monomer-swollen-polymer phase (147). The summation of the rates in the two phases led to the following expression which successfully describes experimental data up to 30–50% conversion (115,149–150).

$$c = \frac{1}{q} \exp\left[(q k_i [I]^{1/2} t) - 1\right] \tag{4}$$

where $q = -(1 + A - AQ)$; A = monomer–polymer ratio in the concentrated phase; Q = the ratio of the polymerization rates in the concentrated and the dilute phase; and I = initiator.

The two-phase kinetic model for vinyl chloride polymerization has been developed (151). It takes into account volume change with conversion and the depletion of initiator with time. The polymerization rate in the dilute monomer-rich phase R_2 is treated by the theory of homogeneous kinetics where:

$$R = k_p (f k_d / k_t I)^{1/2} [M] \tag{5}$$

and related to the rate in the polymer-rich phase, R_1, by $R_2 = P R_1$, where P is a constant. The final expression relating conversion x with time t describes polymerization behavior over a wide range of temperature and conversion and different initiators (152).

$$\frac{dx}{dt} = \frac{1 + Qx}{\sqrt{1 - Bx}} k_1 I_o{}^{1/2} \exp\left(\frac{-k_d t}{2}\right) \tag{6}$$

where $Q = (P(1 - xf) - 1)/xf$ and the change of $V = V_o(1 - Bx)$ where $B = (C_p - C_m/C_p)$, and C_p and C_m are the polymer and monomer densities, respectively; I_o is the original concentration of initiator and k_d is its first-order decomposition constant; xf is the highest conversion that is found to fit the experimental data and ranges from 80% at 30°C to 72% at 70°C. It increases with a decrease in temperature and represents the composition of the polymer-rich phase. This kinetic treatment also describes molecular weight distributions in vinyl chloride polymerizations (151). Chain transfer to monomer is the predominant transfer reaction affecting the molecular weight distribution.

The chain-transfer constant to the monomer, C_m, defined as the ratio of the rate coefficient of transfer to monomer to that of chain propagation is 6.25×10^{-4} at 30°C and 2.38×10^{-3} at 70°C and closely corresponds to $C_m = 5.78 \exp(-2768.1/T)$ (151,153–155). Thus, at 30°C, on the average, one in every 1600 reactions between monomer and chain radical leads to termination of the chain and transfer of the radical activity to the monomer. At 70°C, the chain-transfer reaction is significantly increased and only 420 reactions between monomer and chain radical occur on average before transfer to monomer. A more recent explanation of chain transfer to monomer considers that head-to-head addition of chain radicals with monomer takes place and that splitting off of Cl· or H· radicals is the most likely mechanism (156–157).

Emulsion Polymerization. In this technique, vinyl chloride is emulsified in water by means of surface-active agents. Most of the monomer is thus present as emulsion droplets, although a small amount is dissolved in a fraction of the laminar soap micelles, ie, colloidal aggregates formed above a critical emulsifier concentration. Water-soluble initiators are added, and polymerization begins when a radical enters a monomer-swollen micelle (158–159). Additional monomer is supplied to the growing latex particle by diffusion through the aqueous phase from the monomer emulsion droplets; the stability of the latex is maintained by adsorption of additional detergent from the remaining micelles. Chain termination takes place within each latex particle by the usual radical–radical interaction. This technique offers the possibility of obtaining high molecular weight polymers at very rapid rates. A theory for this true emulsion polymerization is proposed in ref. 160. It postulates that under certain assumptions, the overall reaction rate is proportional to the number N of latex particles per mL of emulsion; this number, in turn, depends upon the 0.6 power of the soap concentration $[S]$ and the 0.4 power of the initiator concentration $[I]$. The average number of radicals per latex particle $\bar{n}$ is given as 0.5. A sharp maximum in the rate of polymerization is expected at the time when the micelles disappear from the system (161).

The kinetics of the emulsion polymerization of vinyl chloride does not follow this scheme. The reaction rate shows the same increasing behavior as in mass polymerization (162–163). The number N depends on $[S]$, but the functionality varies greatly with type and concentration of the emulsifier (164–165). However, the rate of polymerization is practically independent of N (163,166); the order with respect to $[I]$ is found to be between 0.5 and 0.7 for various initiators (165–166). Very low values for $\bar{n}$ (10^{-1} to 5×10^{-4}) (163) are obtained, and termination seems to occur without the participation of a second radical (167). Vinyl chloride at its saturation pressure has a solubility of 0.6 wt % in water at 50°C. This relatively high solubility explains the strong deviations from true emulsion-polymerization kinetics by surmising that a large portion (166) or practically all of the polymerization (168) takes place in the aqueous solution. Arguments against this interpretation (169) derive from the observation that a reduction in the pressure of vinyl chloride leads to an increase in the reaction rate (163) and the fact that monomers with higher solubility in water than vinyl chloride show only minor deviations from the theory postulated in ref. 160. It is also suggested that the insolubility of PVC in its own monomer affects process kinetics.

Recent explanations to account for the observed behavior and low values of $\bar{n}$ are based on rapid desorption and readsorption of radicals formed by chain transfer (170). The rate expression for emulsion polymerization of vinyl chloride may be expressed as:

$$R_p = \frac{-dm}{dt} = \frac{k_p M_m^p}{N_a} \bar{n} \tag{7}$$

where R_p = polymerization rate; k_p = propagation-rate constant; M_m^p = monomer concentration in latex particles; N_a = Avogadro's number; and $\bar{n}$ = average number of radicals per latex particle. In equation 7, $\bar{n}$ may be substituted using the following expression:

$$\bar{n} = 1/N \, Ci^{1/2} \left(\frac{V_p}{2 \, k_{tp}} + \frac{N}{2 \, k_d} \right)^{1/2} \tag{8}$$

where Ci = the rate of radical production in the aqueous phase; V_p = the total volume of the particles; N = the number of particles per unit volume of water; and k_{tp} = the termination constant in the particles. The desorption constant k_d was shown experimentally to be related as follows:

$$k_d = k_{dm} \frac{k_f}{k_{pm}} \tag{9}$$

where k_{dm} = the desorption constant for monomer radicals; k_f = the rate constant for chain transfer to monomer; and k_{pm} = the propagation-rate constant for the monomer radical. Using the substituted form of expression 7, it describes vinyl chloride polymerization over a wide range of initiator and emulsifier concentrations.

Emulsion polymerization has considerable technical importance in the production of PVC, although the expected advantages in rate and molecular weight are not realized. Emulsion latexes up to 0.2 μm in particle size are often sold in liquid form for use in water-based paints, printing inks, and finishes for paper and fabrics. For improved coalescence of the polymer droplets, comonomers are frequently added during polymerization. Modifications produce particle diameters of ca 2–12 μm and give dispersion or plastisol resins. Isolation of dispersion resins or paste resins may be carried out by spray drying or coagulation followed by grinding.

A recent patent for the preparation of plastisols describes the isolation of paste resin by coagulation and filtration without the need for spray drying (171). After steam stripping, the latex containing 30% solids is coagulated under acidic conditions with aluminum sulfate. The slurry is filtered through a filter press and then ground at 55°C.

A wide variety of emulsifier and initiator systems are reported in the literature (172–173). Preferred detergents are alkyl sulfates, alkanesulfonates, and fatty-acid soaps. Sulfonated phthalates and benzoate-emulsifying agents impede polymer buildup in the reactor (174). The type and quantity of emulsifier used in the polymerization process affect the resin quality in later processing.

Residual soaps affect the polymer's clarity, viscosity, electrical resistivity, water absorption, and heat stability. Typical initiators are hydrogen peroxide, organic peroxides, peroxydisulfates, and redox systems. Oxygen is excluded, and the pH of the mixture is maintained at 6–8. Conversion is carried to a relatively high percentage, corresponding to a final solids contents of 50–55% in continuous or batch reactors. The mass, suspension, and emulsion PVC polymerization process have recently been reviewed (175).

Copolymerization. Vinyl chloride can be copolymerized with a large variety of monomers (M_2), such as other unsaturated halogenated hydrocarbons, styrene and its halogenated derivatives, vinyl esters and ethers, olefins, dienes, esters and other derivatives of acrylic and methacrylic acids, olefinic dicarboxylic acids and esters, and various heterocompounds. Review articles present details on the preparation, properties, applications, and tables of copolymerization ratios r, and Q and e values (176–178). The r_1, r_2, and Q and e values for the more important comonomers are given in Table 4.

The most important product is the copolymer of vinyl chloride and vinyl acetate. Commercial resins contain 3–20 wt % vinyl acetate; 75% of U.S. production is manufactured by the suspension process, the remainder by emulsion. The industry has aimed to reduce the vinyl acetate content below ca 13 wt % and to increase the molecular weight for improved physical properties and uniformity. Because of the higher

Table 4. Copolymerization Parameters of Vinyl Chloride

M_2	r_1	r_2	e	Q	Temperature, °C	Refs.
acrylic acid	0.107	6.8	0.77	1.15	60	136
acrylonitrile	0.04	2.7	1.20	0.60	60	137
butadiene	0.035	8.8	−1.05	2.39	50	138
n-butyl acrylate	0.07	4.4	1.06	0.50	45	122
diethyl fumarate	0.12	0.47	1.25	0.61	60	130
dimethyl itaconate	0.053	5.0	1.34	1.03	50	139
ethylene	3.21	0.21	−0.20	0.015	50	123
isobutylene	2.05	0.08	−0.96	0.033	60	124, 179
isoprene			−1.22	3.33		
maleic anhydride	0.296	0.008	2.25	0.23	75	129
methacrylic acid	0.034	23.8	0.65	2.34	60	136
methacrylonitrile	0.11	2.38	0.81	1.12	60	180
methyl acrylate	0.12	4.4	0.60	0.42	50	125
methyl methacrylate	0.1	10	0.40	0.74	68	126
octyl acrylate	0.12	4.8	1.07	0.35	45	122
propylene	2.27	0.3	−0.78	0.002		127, 179
styrene	0.02	17	−0.80	1.0	60	128
vinyl acetate	1.68	0.23	−0.22	0.026	60	140
N-vinylcarbazole	0.17	4.8	−1.40	0.41	50	131
vinyl chloride			0.20	0.044		
vinylidene chloride	0.3	3.2	0.36	0.22	60	132
vinyl isobutyl ether	2.0	0.02			50	141
N-vinylpyrrolidinone	0.53	0.38	−1.14	0.14	50	142

reactivity of vinyl chloride, any given mixture of two monomers leads to copolymers with increasing vinyl acetate content. If a homogeneous chemical composition is preferred, a portion of the vinyl chloride must be added continuously during the polymerization, thus keeping the monomer ratio constant. Molecular weight of the product may be controlled by the addition of chain-transfer agents, such as trichloroethylene. Applications of vinyl chloride–vinyl acetate copolymers [9003-22-9], in order of commercial importance, are flooring-industry products, phonograph records, and coatings.

Copolymers with olefins have been produced commercially. A vinyl chloride–propylene copolymer [25119-90-9] was first introduced for a glass-clear plastic-bottle application (181). It possesses a lower melt-extrusion temperature range than the homopolymer, 145–180°C as compared with 190–210°C, respectively. Furthermore, this copolymer has better heat stability than the homopolymer because the propylene units are randomly distributed in small amounts and break up the zipper dehydrochlorination reaction. Vinyl acetate acts in a similar way; although the heat stability is not as good as that of propylene, it is reported to be better than the homopolymer (182). Olefins act as chain-transfer agents by transfer of allylic hydrogens on the methyl groups. This is a very effective reaction; propylene is reported to have a chain-transfer activity ten times greater than vinyl chloride (183). Olefins have been investigated as agents for preparing low molecular weight suspension resin vinyl chloride copolymers (184–185).

Other commercial copolymers made by emulsion or suspension techniques are acrylonitrile, 2-ethylhexyl acrylate, vinylidene chloride, and vinyl isobutyl ether.

Graft copolymerization is carried out either by reacting various monomers in the presence of poly(vinyl chloride) or by polymerizing vinyl chloride onto other polymeric substrates. In the first group, comonomers, eg, acrylonitrile or ethyl acrylate and acrylonitrile, are used (186–187). In the second group, vinyl chloride is, for example, grafted onto an ethylene–vinyl acetate copolymer producing an impact-strength modifier for compounding with PVC (188). The synthesis, properties, and applications of graft copolymers with a backbone of PVC are reviewed in ref. 189, the synthesis and properties of block and graft copolymers of PVC with vinyl monomers in ref. 190.

Compounding

Poly(vinyl chloride) is never used alone; it is always mixed with other ingredients before being processed (191). A thermal stabilizer is usually required because PVC is sensitive to heat.

Compounding depends on the physical form and type of the PVC. Compound requirements and procedures vary for latex, dispersion, or general-purpose PVC (see Table 5).

Rigid PVC compounds are used in the manufacture of bottles, pipe, and siding. In general, they contain, in addition to PVC (100 parts), a heat stabilization system (ca 2 parts), a lubricant (ca 3 parts), a processing aid (0–3 parts), an impact modifier (0–15 parts), and up to 30 parts of fillers and pigments.

Flexible PVC compounds are used in wire insulation and wall coverings and contain >25% of a plasticizer system in addition to stabilizers, lubricants, and pigments to give the desired properties. Compounds that contain <25% plasticizer are referred to as semirigid compounds.

Heat Stabilizer.　The choice of a thermal stabilizer must go beyond merely protecting the PVC from thermal degradation. Other important factors include the tendency of the stabilizer to plate out on the walls of the processing equipment; the effect on melt rheology; the effect on the fusion rate; possible interactions with other compounding ingredients; and finished product characteristics, such as toxicity, odor, heat stability, heat distortion, clarity, and electrical characteristics.

Stabilizers are based on metallic salts of inorganic acids, organic acids, and phenols, organometallics (eg, tin and antimony), epoxy compounds, and phosphates. These materials are often used in combination. For example, barium–cadmium–zinc and phosphates are used in combination to achieve certain processing and product properties. Organotins, eg, tin mercaptides, are efficient heat stabilizers and provide good initial color, long-term stability, and excellent product clarity. However, the sulfur may impart an objectional odor to the product. Dibutyltin dilaurate and dibutyltin maleate avoid the odor problem. Organotins are more expensive than other stabilizers, but they are generally used in lower concentrations.

Lead stabilization systems are used in wire and cable applications which require good electrical insulating properties and heat stability. They are not suitable for transparent compounds and are not compatible with sulfur-containing materials; possible lead sulfide formation may result in black staining.

Epoxy compounds act as secondary stabilizers when used in conjunction with barium–cadmium or certain organotin stabilizers.

Stabilizers approved by the FDA are required in PVC compounds intended for food- and beverage-packaging applications, eg, flexible compounds for meat wraps

Table 5. Compounding of PVC and PVC Copolymers

Factor	Vinyl latex	Dispersion resin	General-purpose resins
% of total PVC production	1	9	80
mode of manufacture	emulsion polymerization	emulsion polymerization or modification thereof	mass (9%) and suspension polymerization (79%)
particle size	0.05–1 μm	0.05–1 μm	50–200 μm
form sold	colloid dispersion of polymer in water	dry powder[a]	powder
form used	colloidal dispersion in water	colloidal dispersion in plasticizer (plastisol) or in mixture of plasticizers and solvent (organosol)	powder or precompounded fused (6-mm) cubes or pellets
uses	modification of the properties of other materials primarily by coating procedures	modification of properties of other materials by coating; used for 100% vinyl products by casting and molding techniques	for 100% vinyl products by extrusion, injection molding, blow molding, and compression molding; extruded coatings, eg, wire insulation; calendered films; blown films; powder coating
compounding considerations	ingredients must be in colloidal form; colloidal stability of latex must not be destroyed; colloid properties of compound are important	rheology of the plastisol must be matched to the application; fusion and gelation characteristics of plastisol or organosol highly important	heat stability of the compound; compound should be fused at sufficiently high temperature to obtain optimum properties
additives	water; thickeners; surfactants; temporary plasticizers; defoamers; humectants; salts, acids, or bases; plasticizers; heat stabilizers; antiblocking agents; flame retardants; pigments; and fillers	plasticizers; solvents (thinners); heat stabilizers; fillers; pigments; specialty ingredients: viscosity modifiers, viscosity depressants, nonvinyl polymer modifiers, and air-release enhancers; chemical blowing agents; and flame retardants	heat stabilizers; plasticizers; lubricants; processing aids; impact modifiers; pigments; fillers; fungicides; flame retardants; chemical blowing agents; and antistatic agents

[a] Containing agglomerates of primary particles.

and bottle-cap sealants and rigid compounds for meat trays and mouthwash bottles. These stabilizers include certain dioctyltins in systems based on organic salts of calcium, magnesium, and zinc in conjunction with epoxy compounds. Since these systems are generally not as efficient, processing of nontoxic PVC compounds can be more difficult.

Lubricants. Lubricants, either external or internal, form an essential part of PVC formulations that are extruded or injection molded. External lubricants serve to reduce the polymer's tendency to stick to the hot metal surfaces of the processing machinery, whereas internal lubricants increase the flow of the individual PVC resin particles over one another within the melt. The choice depends upon the processing methods and the type and concentration of other compounding ingredients. Typically, lubricants are used at 0.1–4 phr. Internal lubricants are metal stearates, eg, stearic acid and fatty acid esters, whereas external lubricants are low molecular weight polyethylene, paraffin oils, and paraffin waxes. With a correctly chosen and balanced lubricant system, optimum processing rates give products with good physical properties and surface appearance (see also Lubrication and lubricants).

Plasticization. In 1926 at BFGoodrich, it was discovered that solutions of PVC prepared at elevated temperatures with high boiling solvents possess unusual properties when cooled to room temperature (192). Such solutions are, in fact, flexible and elastic and exhibit a high degree of chemical inertness and solvent resistance.

This rather unusual behavior is due to unsolvated crystalline regions in PVC that act as cross-links and allow the PVC to accept large amounts of solvent. Thus, the rigid PVC is transformed to a rubberlike material with stable properties over a wide temperature range (193–196). The high boiling solvents are known as plasticizers (qv), and plasticized PVC products constituted the first commercial application of PVC (192). Plasticizers are added to PVC at 15–20 phr for semirigid compounds and at >100 phr for soft flexible compounds, eg, gasketing material.

A plasticizer must be compatible with the resin in order to resist migration and extraction by liquids, such as water. It should be nonvolatile and boil above 400°C at atmospheric pressure. Other desirable properties include nonflammability, good heat and light stability, good low and high temperature properties, lack of toxicity, compatibility with other compounding ingredients, and of course, low cost.

A few plasticizers impart specific properties needed for a particular application. For example, citrate esters, such as diethyl citrate, are used in food-contact applications, whereas benzoates, such as diethylene glycol dibenzoate, are employed where a high degree of stain resistance is required. Similarly, chlorinated hydrocarbons improve flame resistance and electrical properties.

Primary plasticizers can be used alone and are highly compatible with PVC at concentrations as high as 150 phr. Chemically, plasticizers in this class are commonly esters of alcohols containing 8–10 carbons (see Table 6). The acids include phthalic anhydride, phosphorus oxychloride, sebacic, azelaic, adipic, and fatty acids. Perhaps the most commonly used are the phthalate esters. They offer, at low cost, good compatibility with other compounding ingredients, good low temperature properties, low volatility, and good processing characteristics. Plasticization of PVC and the use of different plasticizers have recently been reviewed (see Plasticizers).

Aliphatic diesters offer excellent low temperature flexibility but are expensive. They are often used in conjunction with less expensive phthalates to improve low temperature properties.

Phosphate esters improve flame resistance, but have poor low temperature properties and can adversely affect the heat stability.

Trimellitates are used when a high resistance to heat aging is required without loss in properties. They are often used in a blend with other plasticizers because they are expensive (see Table 6). Epoxy plasticizers such as soybean and linseed oils have low volatility and excellent heat and light stability (see also Epoxidation).

Polymeric plasticizers have a high degree of compatibility with PVC and do not migrate easily into other materials. They give compounds with excellent high temperature resistance but with relatively poor low temperature properties. Typically,

Table 6.   Primary PVC Plasticizers

Ester	Abbreviation
phthalates[a]	
di(2-ethylhexyl)	DOP
diisooctyl	DIOP
diisodecyl	DIDP
butyl benzyl	BBP
butyl octyl	BOP
ditridecyl	DTDP
diundecyl	DUP
phosphates[b]	
trioctyl	TOP
cresyl diphenyl	CDP
tricresyl	TCP
trimellitates	
tris(2-ethylhexyl)	TOTM
triisooctyl	TIOTM

[a] Normal dialkyl phthalates are also used.
[b] Impart flame resistance.

they are esters of polyhydric alcohols, like propylene glycol, and dibasic acids, such as adipic and sebacic or azelaic.

Secondary plasticizers are often employed in PVC formulations but, because of limited compatibility, cannot be used alone. In general, they are used to confer some special property to the compound or to reduce cost. Examples are chlorinated paraffins, esters of fatty acids, and alkylated benzene and petroleum fractions.

Various attempts have been made to rate plasticizers between good and poor. Boiling point is one simple criterion; a good plasticizer boils above 400°C and is highly miscible with the resin. Miscibility is measured with Flory-Huggins interaction parameter X using an equilibrium-swell technique (197). The results generally agree with observed compatibility. Materials with X values >0.5 exude, whereas those with X values <0.3 are designated good compatible plasticizers. Plasticizer efficiency can also be determined by measuring the quantity of plasticizer required to give a certain value of a certain property. Elongation (197–199), torsional modulus (200), and resilience (201) have all been used as criteria. The depression of the glass-transition temperature for a given concentration of plasticizer is another useful criterion to determine efficiency.

Pigments and Colorants.   Pigment systems are available as dry powders and liquid dispersion or pelletized concentrates. Both organic and inorganic pigments can be used, although the heat resistance of organic pigments generally is lower than that of inorganic pigments. Concentrations range from 0.1 phr of a toner for clarity up to 20 phr of a pigment for weatherability. Dyes are sometimes used at low concentrations, usually in transparent products. Dyes have a tendency to migrate and can be easily extracted from PVC products (see Colorants for plastics).

Fillers.   Fillers are used both to reduce cost and gloss. The most commonly used filler is calcium carbonate. It is available in a variety of forms with an average particle size from 0.07 to well over 50 μm. Some forms are surface treated with stearic acid and other ingredients to aid in processing.

Clay fillers, such as calcined clay, improve electrical properties. Talc, mica, asbestos, diatomaceous earth, and barium sulfate also find some application as fillers for PVC.

A high filler content significantly increases the specific gravity, but affects physical properties and weakens chemical resistance. Furthermore, the abrasive nature of these materials attacks the processing equipment. All of these factors must be carefully considered if fillers are used as a cost-reduction measure.

Biocides. Although PVC itself and most rigid PVC compounds are resistant to attack by microorganisms, flexible PVC products are not. The selection of a biocide should include a consideration of its effect on heat stability, toxicity, compatibility, and weatherability of the PVC compound.

Impact Modifiers. In rigid applications, toughness can be increased with an impact modifier. These are generally materials of low modulus with limited compatibility with PVC, such as chlorinated polyethylenes, poly(acrylonitrile-*co*-butadiene-*co*-styrene) (ABS), poly(methacrylate-*co*-butadiene-*co*-styrene) (MBS), poly(ethylene-*co*-vinyl acetate), and acrylic polymers. Impact modifiers are expensive and are generally used below 15 phr.

Processing Aids. Processing aids increase the melt strength of rigid PVC compounds during processing. Thus, they prevent edge tear in the extrusion of complex profiles. They are typically styrene–acrylonitrile (SAN) copolymers and acrylic polymers and are used at ca 1.0–5.0 phr.

Flame Retardants. Since PVC contains nearly half its weight of chlorine, it is inherently a flame retardant (qv). However, as PVC is compounded with other materials, the overall percentage of chlorine decreases and flame retardancy may be reduced. Rigid and semirigid PVC compounds exhibit better flame retardancy than flexible compounds that contain >30 phr of plasticizer. The latter may require additional flame retardants, such as antimony oxide, phosphate-type plasticizers, and chlorinated or brominated hydrocarbons (81, 202).

Foaming and Blowing Agents. Cellular PVC in both open- and closed-cell forms may be made by a variety of techniques, such as physically whipping air into a plastisol or dispersion of fine-particle PVC in plasticizer; incorporation of a gas under pressure in a vinyl extrusion system near the die while the compound is in the melt state; or use of chemical blowing agents that volatize or release gas upon decomposition while the PVC compound is in the form of a hot melt.

Commonly used organic chemical blowing agents include azo compounds, eg, 1,1′-azobisformamide (ABFA), which like all azo compounds yields nitrogen when heated to decomposition.

Other organic blowing agents include *N*-nitroso, sulfahydrazo, and azedo

$$(-N{\underset{N}{\overset{N}{\diagdown}}}\,)$$

compounds. Inorganic blowing agents, eg, ammonium bicarbonate, find some application. Since the decomposition reactions are usually exothermic, additional heat stabilization may be necessary (see also Foamed plastics).

Liquids boiling below 100°C, such as fluorinated aliphatic hydrocarbons, trichloroethylene, or cyclohexane, find application in plastisol technology. Vaporization is endothermic and hence, when used in conjunction with chemical blowing agents, they lower the heating associated with the decomposition exotherm.

Flexible Compounding.   Flexible compounds generally contain plasticizer, stabilizers, lubricants, fillers, and pigments (see Table 7).

For specific applications, the formulation might include: for wire and cable, 5 parts by weight of a lead stabilizer, such as tribasic lead sulfate, instead of the Ba–Cd; a less volatile plasticizer, such as diisodecyl phthalate (DIDP), for interior automotive trim to decrease window fogging; low migration plasticizers, such as polyesters and biocide, for a refrigerator-door gasket; and FDA-approved ingredients for a transparent beverage container. The last application must be certified by the appropriate Federal agencies.

Rigid Compounding.   A typical rigid PVC formulation includes PVC resin, stabilizers, impact modifiers, processing aids, lubricants, pigments, and fillers (see Table 8).

Economic Aspects

Like other plastics over the last three years, PVC prices have reflected changes in monomer costs, but the value added differential between monomer and polymer has remained fairly constant. Polyethylene and polypropylene have maintained larger

Table 7.   Basic General-Purpose Flexible PVC Formulation

Component	Parts by weight
PVC resin, high mol wt[a]	100.0
plasticizer	
dioctyl phthalate	30.0–80.0
processing aid, epoxidized soybean oil	5.0
stabilizer, barium–cadmium	3.0
filler, calcium carbonate	≤30.0
lubricant, stearic acid	0.5
pigment	≤3.0

[a] Inherent viscosity, >0.95.

Table 8.   Rigid-Pipe Formulation, Powder

Component	Parts by weight
PVC resin, medium-high mol wt[a]	100.0
impact modifier, MBS type	0–10.0
processing aid, acrylic type	0–3.0
stabilizer, a tin mercaptide	0.2–2.5[b]
lubricant	
internal, calcium stearate	0.4–2.0[b]
external	
oxidized polyethylene	0–0.5[b]
paraffin wax[c]	0.4–1.2[b]
pigment, titanium dioxide	0.5–3.0
filler, stearate-coated calcium carbonate	0–5.0

[a] Inherent viscosity 0.90–0.94.
[b] Depending upon the type of extruder employed, twin or single screw.
[c] 74°C.

value added differential prices than PVC, but recently, the gap between the monomer prices has narrowed (see Fig. 16). The sales volume of PVC is reflected by the general economy and the market outlets (see Table 9). The construction and automotive industries have traditionally accounted for large-volume sales but have not been recession-proof as observed in 1974–1975 and 1980–1981.

U.S. PVC production experienced an exponential growth during the period 1955–1974, with a 25% decline in 1975 (see Fig. 17). The growth rate recovered but was never fully restored. In 1935–1938-dollar terms, PVC prices steadily declined until 1970; then they increased slowly, reflecting increased feedstock costs. Actual prices have risen more steeply since 1972; however, PVC is less dependent on oil prices than other plastics. Producers and their capacities are given in Table 10.

The energy consumption of PVC manufacture is lower than that of other synthetic

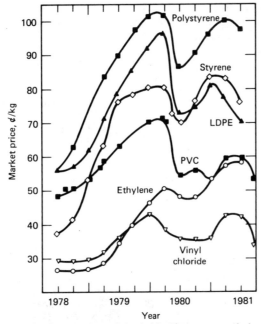

Figure 16. Monomer and polymer prices of vinyl chloride, styrene, ethylene, and low density polyethylene (UDPE) (203).

Table 9.   1981 Breakdown of the Plastic Market, % [a]

Application	PVC	Polyethylene	Polypropylene
appliances	9.1	1.0	10.9
building	39.8	9.0	0.3
electrical	22.5	26.8	0.7
furniture	16.9	2.2	4.0
houseware	4.1	48.3	16.8
packaging	5.6	60.7	7.9
transportation	6.7	1.7	15.9
toys	11.4	36.5	15.4

[a] Ref. 204.

Table 10.   1982–1983 World PVC Producers and Capacities, Thousand Metric Tons [a]

North America			Belgium	325
Canada	286		BASF	100
Diamond Shamrock—Alberta Gas	100		Solvic	225
ESSO Chemical Canada	50		France	1,175
BFGoodrich (2) [b]	136		Société Artesienne de Vinyl	160
Mexico	218		ATO Chimie	125
Industrias Resistol (2)	40		Produits Chimiques Ugine	125
Plasticas Omega	2		Kuhlmann	
Policyd (2)	107		Chloe Chimie (3)	375
Polimeros de Mexico	30		Shell Chimie	140
Promociones Industriales Mexicanas	37		Solvic	250
Aldeva, S.A.	2		FRG	1,486
United States	3,818		BASF	180
Air Products and Chemicals (2)	182		Chemische Werke Huels	410
Borden (2)	236		ICI	115
Certain-Teed	86		Solvay	180
Conoco Chemicals (2)	324		Hoechst (2)	240
Diamond Shamrock (2)	90		Lonza	32
Ethyl	82		Wacker-Chemie (2)	329
Formosa (3)	107		Italy	1,055
General Tire & Rubber (2)	84		Societa Chimica Ravenna	120
Georgia-Pacific	318		Liquichimica Ferrandina	50
BFGoodrich (7)	875		Montedison (3)	550
Goodyear Tire & Rubber	32		Rumianca Sud	115
Great American Chemical	34		Societa Italiana Resine	120
Occidental (Hooker) (4)	413		Solvic	100
Keysor-Century	30		Netherlands	345
Pantasote	64		DSM	170
Shintech	308		Shell	175
Tallyrand	35		UK	525
Tenneco Chemicals (3)	438		Norsk Hydro	150
Union Carbide	80		ICI	375
Total	4,322		*Total*	4,911
Central America			*Other Western European Nations*	
Nicaragua			Austria	
Polycasa	30		Halvic Kunstatoffwerke	60
Total	30		Finland	
South America			Pekema Oy	60
Argentina	57		Greece	107
Electroclor SACI	29		ESSO Pappas Chemical	47
Indupa	22		Northern Greece Chemical Industry	60
Vinisia	6		Norway	
Brazil	314		Norsk Hydro	70
Brasivil	55		Portugal	
C.P.C.—Petroquimica	140		Companhia Industrial de Resinas	60
S.A. Geon do Brasil	40		Sinteticas	
Industrias Quimicas Electro Chloro	79		Spain	345
Chile			Hispavic Industrial (2)	135
Petroquimica—Dow	15		Aiscondel, S.A.	90
Colombia	50		Rio Rodano (2)	120
Colombiana de Carburo y Derivados	8		Sweden	
Petroquimica Colombiana	42		Kemanord (2)	140
Peru			Switzerland	
Sociedad Paramonga	11		Lonza	35
Venezuela			*Total*	877
Plasticas Petroquimica	40		*Eastern Europe*	
Total	487		Bulgaria	
Western Europe, EEC			State Industry (2)	130

Table 10 (*continued*)

Czechoslovakia		Nissan Petrochemicals	24
State Industry (2)	240	Ryo—Nichi	116
GDR		Shin—Etsu Chemical (2)	218
VEB Chemische Werke	555	Sumitomo Chemical (2)	120
Hungary		Sun Arrow Chemical	60
State Industry (2)	193	Toagosei Chemical	47
Poland		Tokuyama Sekisui	49
State Industry (4)	406	Toyo Soda	105
Romania		North Korea	
State Industry (4)	270	State Industry	20
USSR		South Korea	*310*
State Industry (5)	706	KPIC (6)	280
Yugoslavia	270	Lucky	30
Hemijska Ind. (2)	80	Malaysia	*24*
Jugovinil	50	Malayan Electro Chemical	12
Organsko Hemijska Ind.	95	Synthetic Resin	12
Vinyl Plastika	45	Pakistan	*27*
Total	*2,770*	Arokey	5
Middle East		Fauji	22
Iran		Philippines	*48*
Abadan	60	Mabuhay Vinyl	28
Israel		Philippine Vinyl Consortium	20
Electrochemical Industries	100	Singapore	
Turkey		Singapore Polymer	15
Petkim (2)	52	Taiwan	*606*
Total	*212*	Cathay Plastics	66
Far East and Australia		China Gulf Plastics	72
People's Republic of China	239	Formosa Plastics	420
India	*135*	Ocean Plastics	48
Delhi Cloths. Gen. Mills	20	Thailand	*72*
Chemicals and Plastics India	20	Bankok	42
NOCIL	20	Samuthprokaan	30
Plastics Resins and Chemicals	40	Australia	*192*
Shiram Vinyl	20	BFGoodrich	87
Koyali	15	ICI	105
Indonesia	*56*	*Total*	*3,629*
P.T. Eastern Polymer	20	*Africa*	
P.T. Standard Toyo Polymer	36	Algeria	
Japan	*1,885*	Sonatrach	35
Asahi Glass	24	Libya	
Central Chemical	24	National Organization for	60
Chisso (2)	85	Industrialization	
Chisso Petrochemical	51	Morocco	
DENKA (3)	171	SNEP	25
Kanegafuchi Chemical (3)	226	South Africa	*205*
Kawasaki Organic Chemicals	28	Sentrachem	25
Kureha Chemical	120	AE & CI and Sentrachem	180
Mitsubishi Monsanto	108	*Total*	*325*
Mitsui Toatsu	117	*Grand total*	*17,563*
Nippon Zeon	192		

[a] Ref. 117.
[b] Number in parentheses indicates the number of plants.

polymers (see Table 11) (208). Between 1970 and 1980, the energy costs of production increased on the average of 10.8 times, with PVC showing an increase of 12 times over the 1970 costs. With these staggering increases, in 1980, the energy cost as a percentage

and penetration; water-miscible organic liquids added as temporary plasticizers, defoaming agents, or humectants, eg, glycols, glycerol, and Carbitol acetate; water-soluble salts, acids, and bases added to adjust pH, alter flow properties, aid compound dispersion, and stabilize the polymer against heat and light, eg, sodium hydroxide, dicyandiamide, EDTA, sodium carbonate, and tetrasodium pyrophosphate (TSPP); waxes, tackifiers, oil-soluble stabilizer, and antifoaming agents; plasticizers; and pigments and fillers.

These materials are either added as aqueous solution, eg, pigments and fillers, or as an oil in water emulsion. The latex, oil ingredient, water, and surfactant are mixed by stirring and then passed through a suitable colloid mill, homogenizer, or high shear mixer. The particle size of the final emulsion should be <3 μm to ensure uniformity.

Dispersion Resins. Latex technology is based on the fact that a colloidal dispersion of polymer in water can be dried and heated causing the polymer particles to fuse and form a continuous adhesive film. Latex technology thus offers a convenient way to beneficially modify the properties of a material by the introduction of a relatively small amount of polymeric material.

Dispersion-resin technology is also based on the conversion of a PVC in dispersed colloidal form to a solid material.

Dispersion resins, like latex, are usually produced by emulsion polymerization, but are sold in dry powder form. In use, they are mixed with plasticizer to form a colloidal dispersion. Such dispersions are known as plastisols and are easily handled and readily pourable. However, when heated to 148–177°C, the plastisol is transformed to a homogeneous hot melt which, upon cooling to below 50°C, results in a tough flexible PVC product. In some applications, the plastisol is diluted with a volatile solvent that evaporates during the fusion process. Such a diluted plastisol is known as an organosol.

Using coating technology, plastisols and organosols, like latex, offer a convenient way to modify the properties of existing materials through the introduction of small amounts of plasticized vinyl polymer. They can, however, be used to fabricate relatively massive flexible vinyl items through a variety of molding techniques. The fusion of a plastisol results in a ca 1% volume change of the system and, furthermore, no volatiles are normally given off during the fusion process. In addition, a plastisol formulation must contain a stabilizer and may contain pigments, thinners, and fillers.

Blowing Agents. Blowing agents are added to create foamed products. Although cellular vinyl foams can be produced by mechanically incorporating air into the plastisol, blowing agents that decompose under heat to generate gas are often used. Typically, blowing agents (0.1–10 parts by wt) result in foams with 5–95 wt % gas, respectively. These foams have densities of 0.064–1.2 g/cm^3. Both closed- and open-cell foams can be produced.

Cellular vinyl exhibits high tensile and tear properties; good abrasion resistance; resistance to most acids, bases, and chemicals; inertness to water and moisture; good aging characteristics and heat and light stability; excellent dimensional stability (open-cell foams); and easy heat sealing by dielectric heating.

Plastisols. Plastisols are prepared with low speed, high shear mixers equipped with water jackets for temperature control; only enough shear is needed to break up loose agglomerates. The temperature should be kept at 20–40°C to avoid premature gelation.

Fusion. After the plastisol has been applied in its final coated or molded shape, it must be fused into a homogeneous solid; ie, the crystallite structures in the polymer particles is melted, followed by solution of the molten polymer in the plasticizing vehicle. Since no chemical cure takes place, the stock fuses rapidly as it reaches the fusion temperature. When the melt is slightly cooled, crystallinity is reestablished and this network formation gives a tough flexible product.

The fusion temperature of plasticized vinyl depends on molecular weight, amount of comonomer, plasticizer content, and polymer–plasticizer interaction parameter (208). Since dispersions are fused by heat alone without mechanical working, particle size, particle size distribution, and nonvinyl constituents at particle surfaces affect the fusion temperature.

As the PVC crystallites melt over a fairly wide temperature range, some strength is developed at temperatures below the fusion point. This cannot be increased by increasing the heating time.

In practice, the fusion temperature of a dispersion system is determined empirically from a curve of physical properties (usually the ultimate tensile strength and ultimate elongation at break) of cast film heated at various temperatures (Fig. 18). The true fusion temperature is considered to be the temperature that produces maximum properties.

Vinyl-Dispersion Processes. Plastisols and organosols may be applied by spread coating or molding to form the final product. In addition, specialized processes are available, such as strand coating, spray coating, and extrusion.

Spread coating is used to make a wide range of products from roll-goods flooring to apparel fabric and automotive padding. Coated paper products include packaging board, shelf-lining papers, shoe liners, auto door panel finishes, and masking tape. Knife coaters and reverse-roll coaters are probably the most commonly used equipment. Speeds up to 30.48 m/min can be used on knife machines, and 304.8 m/min on reverse-roll coaters.

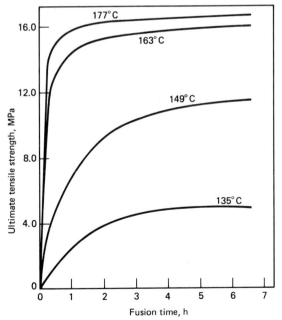

Figure 18. Effect of time on plastisol fusion-temperature–tensile-strength relation. To convert MPa to psi, multiply by 145.

In die coating, the plastisol or organosol is pumped on the surface of a thin die at the supply side. Excess plastisol is removed as the strand passes through the die, resulting in a smooth even coating. After coating, the strand passes into a fusion tunnel where the plastisol or organosol is fused to form a tough vinyl coating. Alternatively, the strand may be passed vertically upward through a bath. By careful adjustment of operating conditions, an even coating without runback into the bath can be achieved.

Both organosols and plastisols can be sprayed as an alternative to painting. Coating thicknesses from 0.05 to >1.5 mm can be obtained.

In addition, plastisols may be extruded as a means of preparing partially fused materials for subsequent molding after producing finished goods. The extrusion of plastisols offers a convenient method for making extremely soft compounds.

General-Purpose Resins. General-purpose resins are made by the mass and suspension processes outlined previously. They comprise ca 80% of all the PVC resin produced and are used chiefly to make so-called 100% vinyl products by a variety of molding and extrusion techniques. Extrusion is also used in conjunction with these materials to coat various other materials, eg, wire insulation (flexible PVC) and vinyl-clad wooden window frames (rigid PVC).

A resin is chosen according to its molecular weight. High molecular weight resins are tough and have greater strength, chemical resistance, and resistance to temperature extremes. However, they are harder to process because of high melt viscosity. A compromise between properties and processability has been made by the vinyl industry, and as a result, 80% of general-purpose PVC resins sold have an intrinsic viscosity >0.90 with a range of ca 0.6–1.15 covering almost the entire spectrum of available resins.

Resins intended for flexible applications should have good uptake of plasticizer in a dry-blending operation. Similarly, homopolymer resins intended for flexible films should be free of nonporous particles, which tend to cause imperfections or fisheyes in the final film.

The resin is mixed with the various compounding ingredients required. The mixture may be a dry free-flowing powder, similar in consistency to the original resin, or melted and fused cubes (0.256 cm^3) or pellets of various dimensions. The powder is produced in a powder-mixing line, the cubes or pellets in the melt-mixing equipment. A melt-mixed compound offers the fabricator the advantage of a precompounded PVC which is dust-free and easy to use, store, and handle. However, in some product lines, such as rigid PVC pipe, powder-mix compounds are preferred.

The powder can be mixed in low shear mixers, such as a paddle or ribbon blender. High shear mixers, such as the Henschel (HPM Corp.) or the Welex, are usually preferred. These devices are similar in concept and design to food blenders and generate considerable shear heat. In high speed mixing, the temperature of the powder can quickly reach 100°C, and upon completion of the mixing cycle, the compound must be transferred to a cooler to avoid thermal degradation. In the cooler, generally a low shear mixer with cooled walls, the powder is cooled to ca 40°C. After cooling, the compound is ready for use.

If the plasticizer is present in amounts that fill the resin's internal porosity, heat must be applied during the mixing in order to cause sufficient diffusion of the plasticizer into the PVC to give a dry free-flowing powder. In high speed powder mixing, there is enough frictional (shear) heat for this diffusion. In low speed mixers, the walls must be heated. A typical powder-mixing line is illustrated in Figure 19. Powder mixing is a batch process with a cycle time of typically 2–4 min for high speed mixers that can handle up to 550 kg.

Melt mixers can be of either continuous or batch type. Cycle times for batch operations are ca 2–5 min, with batches of 70–180 kg. The mixer is a rugged heated steel chamber in which two counter-rotating mixing blades shear and melt the powder into a homogeneous melt. After mixing, the melt is dropped onto a two-roll mill and rolled into a sheet. The sheet is cooled and cut into cubes ready for use.

Typical batch-melt mixers or intensive internal mixers, as they are often called, are the Banbury and the Intermix (see also Mixing and blending).

In continuous mixers, similar to extruders, the powder is compressed, fused, and melted, and then transferred to a cooling line. Typical examples are the twin-rotor Farrel continuous mixer (FCM), the twin-screw Kombiplast, and the Ko-Kneader. The largest of these machines can mix >6.8 t of plasticized compound and 3.6 t of rigid compound per hour.

Compounds can also be made in twin-screw extruders, and occasionally with single-screw extruders. This operation differs only in detail, rather than in concept, from a conventional extrusion with screw designs tailored to provide good mixing at controlled temperatures.

A general melt-mixing line is illustrated in Figure 20. The resin is transferred from the weigh bin to an intensive powder mixer where the compounding ingredients are added. The premixed compound is cooled before being transferred to the melt mixer. The melt is rolled into a sheet on a 2-roll mill. The cooled compound is transferred directly to the extruder.

Most PVC products made from general-purpose resins are extruded for melt preparation. In conventional extrusion, the melt is forced through a die. In sheet,

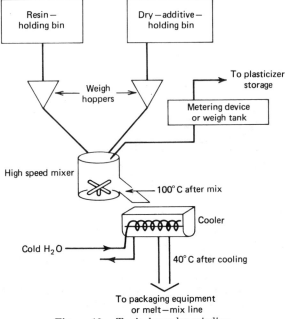

Figure 19.   Typical powder-mix line.

sheet for cladding, house siding, and vacuum-formed articles are a few of the more common uses. Sheet is produced in widths up to 190.5 cm (see Film and sheeting materials).

Rigid-Profile Extrusion.   Rigid profiles are extruded from melt-mixed cubes or pellets. Since profile shapes can be quite complex, they are generally extruded at low temperature since high temperature, low viscosity melts are difficult to handle. The profile is cooled by air with or without special supports to maintain shape; cooling must be uniform to avoid warping. Rigid PVC profiles have a wide variety of applications; house siding and window frames are two large-volume products.

Flexible Wire and Cable Insulation.   In wire coating, a cross-head die is employed in which the extruder is at an angle to the direction of wire travel. Often, 90° dies are used, although other angles (30–40°) are used for efficient plant layout. In a wire die, the wire is surrounded by melt or it leaves the die and carries the melt with it. In a die designed to coat a jacket insulation over a cable, a tubing die is used. In this type of die, a vacuum is applied to pull the extrudate over the substrate. The wire is cooled in a water trough after coating.

After the wire is coated and the coating cooled, continuity of the insulation is checked with a spark tester (see Insulation, electric).

Flexible Film.   For flexible film, cross-head dies are used with appropriately sized die and mandrel. The die is usually oriented such that the extrudate is directed vertically upward. Air is introduced through the rear of the mandrel or one of the spider legs, and a bubble is blown. The bubble is maintained by a collapsing frame operated in conjunction with a set of pull-off pinch rolls which effectively close the end of the bubble, permitting control of the bubble by the amount of air fed into it. The film generally is slit on both sides to make two rolls. Flexible film finds wide applications as food wrap. Its breathability makes it ideal for red meat and poultry packaging (see Film and sheeting material).

Injection Molding.   In injection molding, an extruder is run intermittently and used to fill a cooled mold. Special attention must be paid to the heat stability of the compound and the residence time in the molding machine. Both flexible and rigid PVC can be injection molded. Injection-molded rigid-PVC pipe fittings are perhaps the largest volume product of injection molding.

Calendering.   In calendering, a hot PVC melt is formed into a thin sheet by means of heated rolls. Sheet or films from >0.25 to 0.875 mm thick are produced. Thicker sheet is generally made by extrusion. Calendering is often used to produce a vinyl top coating on a substrate. Typical examples are floor coverings, wall coverings, and upholstery fabrics. Unsupported vinyl sheet finds application in tank or reservoir linings, agricultural mulching materials, raincoats, clothing, and shower curtains. Any number of roll configurations can be used with two, three, or four rolls making up the calender. Generally, a calendering line is fed by a melt-mix line similar to that shown in Figure 20. The hot melt, upon leaving the melt mixer and two-roll mill, directly enters the calendering line.

Despite high initial capital investment, calendering is preferred to plastisol coating in the laminate and flooring industry where very hard films that contain little plasticizer are required. These films cannot be readily made by plastisol-coating techniques.

Powder Coating. In powder coating (qv), a flexible PVC powder compound is fused onto a substrate. Generally, a narrow particle-size distribution vinyl resin is employed with a particle-size range of 65–90 μm. After compounding, the powder coating may be applied to the substrate by a variety of methods, such as electrostatic spraying, spread coating, or dipping in a fluidized bed. In dipping operations, the substrate should be heated above the fusion temperature of the compound. A postheat oven is often used to complete fusion. In electrostatic spraying and spread coating, the powder is applied to a cold substrate and subsequently fused in a hot-air oven.

Postforming. Since PVC is a thermoplastic, it may be partly formed by a variety of processes, eg, vacuum forming. Phonograph records are made by compression molding of PVC–vinyl acetate copolymer sheet (see Recording disks). Various types of laminating equipment are also used in conjunction with film or sheet. Because of its strongly polar structure, PVC is easily sealed with dielectric heat. Adhesive joining may be accomplished by solvent alone or by cement.

Chlorinated Poly(Vinyl Chloride). PVC is readily chlorinated, and PVCs with chlorine contents >70% (CPVC) have been made.

Increasing the chlorine content lowers the degree of crystallinity and significantly increases the heat-distortion temperature. The principal use of CPVC is in pipe and pipe fittings, especially those intended for hot-water service. Such applications require a compound with a heat-distortion temperature of >100°C; PVC compounds usually have a heat-distortion temperature of 65°C. High heat-distortion compounds require special consideration in processing, although in general, CPVC is processed on equipment similar to that used for PVC.

Compounds in the heat-distortion range of 82–87°C have superior processing characteristics and find application in extruded applications in areas requiring slightly greater heat-distortion temperatures than available from standard PVC compounds.

Environmental Aspects

In terms of ecological and environmental considerations, PVC has much in its favor (210). Poly(vinyl chloride) resins are manufactured from petroleum hydrocarbon derivatives that were once considered by-products and also were at one time burned or otherwise disposed of. Consumption of these hydrocarbons is not expected to significantly affect the short-term depletion of such natural resources. Technology exists for control of water and air pollution at the resin-manufacturing level.

Poly(vinyl chloride), during its life, from production through fabrication, use, and eventual disposal, compares favorably with other materials in total energy requirements, and the risk of polluting water or air is minimal. Industrial waste generated during the fabrication can be conveniently disposed of through recycling (qv) into other products or sanitary landfills.

As PVC products enter the disposal stage, they are easily collected and can be disposed of by landfill, incineration, or recycling. In a sanitary landfill, PVC wastes do not contribute to instability and, unlike biodegradable materials, do not give off toxic decomposition products. In incineration, PVC burns much like other organic materials, such as grass, leaves, paper, wood, or food wastes. Although incineration of PVC emits some HCl, it is not considered a problem.

Proper incineration operation and design can control HCl and other acidic

93. H. Germar, *Makromol. Chem.* **86,** 89 (1965).
94. G. Svegliado and F. Zilio Grandi, *J. Appl. Polym. Sci.* **13,** 1113 (1969).
95. S. Tsuge, T. Okumoto, and T. Takeuchi, *Macromolecules* **2,** 277 (1969).
96. W. Trautvetter, *Kunstst. Plast.* **13**(2), 54 (1966).
97. Y. Nakamura, K. Mori, and M. Saito, *Kobunshi Ronbunshu* **36**(8), 523 (1979).
98. W. H. Starnes, I. M. Plitz, D. C. Hische, D. J. Freed, F. C. Schilling, and M. L. Schilling, *Polym. Prep. Am. Chem. Soc. Div. Polym. Chem.* **19**(1), 623 (1978).
99. S. Marian and G. Levin, *J. Appl. Polym. Sci.* **26**(10), 3295 (1981).
100. J. Lewis, M. K. Naqvi, and G. S. Park, *Makromol. Chem. Rapid Commun.* **1**(2), 119 (1980).
101. *Ibid.*, **1**(6), 411 (1980).
102. A. H. Frye and R. W. Horst, *J. Polym. Sci.* **40,** 419 (1959).
103. C. V. Oprea and M. Popa, *Rev. Roum. Chim.* **26**(2), 291 (1981).
104. C. V. Oprea and M. Popa, *Colloid Polym. Sci.* **258**(4), 371 (1980).
105. Jpn. Kokai Tokkyo Koho, 81 50, 945 (May 8, 1981), (to Toa Gosei Chem. Ind.).
106. K. B. Abbas, F. A. Bovey, and F. C. Schilling, *Makromol. Chem. Suppl.* **1,** 227 (1975).
107. W. H. Starnes, F. C. Schilling, K. B. Abbas, I. M. Plitz, R. L. Hartless, and F. A. Bovey, *Macromolecules* **12**(1), 13 (1979).
108. T. Hjertberg, *Anomalous Structures in PVC*, dissertation, Chalmers University of Technology, Goteborg, Sweden, 1982.
109. C. E. Schildknecht in C. E. Schildknecht, ed., *Polymer Processes*, Interscience Publishers, Inc., New York, 1956, Chapt. 2.
110. Fr. Pat. 1,382,072 (Dec. 18, 1964), J. C. Thomas (to Produits Chemiques Péchiney-Saint Gobain).
111. D. N. Bort, Y. Y. Rylov, N. A. Okladnov, and V. A. Kargin, *Polym. Sci. USSR* **9**(2), 334 (1967).
112. N. Fischer and L. Gioran, *Hydrocarbon Process.* **60**(5), 143 (1981).
113. K. J. Ivin in J. Brandrup and E. H. Immergut, eds., *Polymer Handbook*, 2nd ed., John Wiley and Sons, Inc., New York, 1975, Sect. 2, p. 421.
114. M. H. Lewis and G. R. Johnson, *J. Vinyl Technol.* **3**(2), 102 (June 1981).
115. A. Crosato-Arnaldi, P. Gasparini, and G. Talamini, *Macromol. Chem.* **117,** 140 (1968).
116. E. J. Arlman and W. M. Wagner, *J. Polym. Sci.* **9,** 581 (1952).
117. J. B. Cameron, A. J. Lundeen, and J. H. McCulley, *Hydrocarbon Process.* **59**(3), 39 (1980).
118. U.S. Pat. 4,024,330 (May 17, 1977), M. G. Morningstar and H. J. Kehe (to BFGoodrich Company).
119. U.S. Pat. 4,080,173 (Mar. 21, 1978), L. Cohen (to BFGoodrich Company).
120. Ger. Pat. 2,628,700 (Jan. 20, 1977), R. J. Davis, A. R. Berens, G. R. Huddleston, and D. E. Witenhafer (to BFGoodrich Company).
121. R. Lukas, M. Kolinsky, and I. Horacek, *Int. Polym. Sci. Technol.* **8**(3), T84 (1981).
122. U.S. Pat. 2,440,808 (May 4, 1948), H. T. Neher and F. J. Glavis (to Rohm & Haas Company).
123. U.S. Pat. 2,476,474 (July 19, 1949), M. Baer (to Monsanto Chemical Company).
124. U.S. Pat. 2,543,094 (Feb. 27, 1951), C. A. Brighton and J. J. P. Staudinger (to The Distillers Company).
125. Brit. Pat. 1,275,395 (May 24, 1972), (Borden Inc.).
126. Brit. Pat. 1,391,597 (Apr. 23, 1975), G. J. Gammon and P. Lewis (to B. P. Chemicals).
127. Brit. Pat. 1,391,598 (Apr. 23, 1975), G. J. Gammon and P. Lewis (to B. P. Chemicals).
128. Ger. Pat. 2,165,369 (July 27, 1972), J. Desilles (to Aquitaine-Organico).
129. U.S. Pat. 2,546,207 (Mar. 27, 1951), D. Bandel (to Mathieson Chemical Company).
130. U.S. Pat. 2,524,627 (Oct. 3, 1950), W. P. Hohenstein (to Polytechnic Institute of Brooklyn).
131. Brit. Pat. 1,298,636 (Dec. 6, 1972), G. J. Gammon and P. Lewis (to B. P. Chemicals).
132. U.S. Pat. 3,813,373 (May 28, 1974), I. Ito, T. Sekihara, and T. Emura (to Sumitomo Chemical Company).
133. H. Kainer, *Polyvinylchlorid und Vinylchlorid-Mischpolymerisate*, Springer-Verlag, Berlin-Heidelberg, New York, 1965.
134. U.S. Pat. 2,133,257 (Oct. 11, 1938), D. E. Strain (to E. I. du Pont de Nemours & Company).
135. U.S. Pat. 3,108,044 (Feb. 15, 1936), J. W. C. Crawford and J. McGrath (to Imperial Chemical Industries, Ltd.).
136. Brit. Pat. 444,257 (Mar. 17, 1936), A. Renfrew, J. W. Walter, and W. E. F. Gates (to Imperial Chemical Industries, Ltd.).
137. U.S. Pat. 2,300,566 (Nov. 3, 1940), H. J. Hahn and E. Brown (to General Aniline & Film Corporation).

138. U.S. Pat. 2,450,000 (Sept. 28, 1948), B. W. Howk and F. L. Johnston (to E. I. du Pont de Nemours & Company).
139. U.S. Pat. 2,486,855 (Nov. 1, 1949), E. Lavin and C. L. Boyce (to Shawinigan Resins Corporation).
140. Brit. Pat. 1,396,703 (June 4, 1975), G. J. Gammon and P. Lewis (to B. P. Chemicals).
141. F. H. Winslow and W. Matreyek, *Ind. Eng. Chem.* **43,** 1108 (1951).
142. E. Farber and M. Koral, *Polym. Eng. Sci.* 8(1), 11 (1968).
143. W. I. Bengough and R. G. W. Norrish, *Proc. Roy. Soc. (London), Ser. A* **200,** 301 (1950).
144. A. Konishi and K. Nambu, *J. Polym. Sci.* **54,** 209 (1961).
145. G. M. Burnett and W. W. Wright, *Proc. Roy. Soc. (London) Ser. A* **221,** 28 (1954).
146. U.S. Pat. 3,825,509 (July 23, 1974), R. K. Miller (to BFGoodrich Company).
147. C. W. Johnston in ref. 81, Vol. 1, Chapt. 3, p. 33.
148. J. Ugelstad, P. C. Mork, and F. K. Hansen, *Pure Appl. Chem.* **53,** 323 (1981).
149. G. Talamini and G. Vidotto, *Makromol. Chem.* **50,** 129 (1961); **53,** 21 (1962).
150. G. Talamini, *J. Polym. Sci.* **A2**(4), 535 (1966).
151. Ahmed H. Abdel-Alim and A. E. Hamielec, *J. Appl. Polym. Sci.* **16,** 783 (1972).
152. *Ibid.*, **18,** 1603 (1974).
153. J. W. Breitenbach, *Makromol. Chem.* **8,** 147 (1952).
154. F. Danusso, G. Pajaro, and D. Sianesi, *Chim. Ind. (Milan)* **37,** 278 (1955).
155. *Ibid.*, **41,** 1170 (1959).
156. A. Rigo, G. Palma, and G. Talamini, *Makromol. Chem.* **153,** 219 (1972).
157. G. S. Park and M. Saleem, *Polym. Bull.* **1,** 409 (1979).
158. H. Fikentscher and G. Hagen, *Angew. Chem.* **51,** 433 (1938).
159. W. D. Harkins, *J. Am. Chem. Soc.* **69,** 1428 (1947).
160. W. V. Smith and R. H. Ewart, *J. Chem. Phys.* **16,** 592 (1948).
161. J. G. Watterson, A. G. Parts, and D. E. Moore, *Makromol. Chem.* **116,** 1 (1968).
162. G. Talamini and E. Peggion in G. E. Ham, ed., *Vinyl Polymerization,* Vol. 1, Pt. 1, Marcel Dekker, Inc., New York, 1967, Chapt. 5.
163. J. Ugelstad, P. C. Mork, P. Dahl, and P. Rangnes, *J. Polym. Sci.* C **27,** 49 (1969).
164. J. T. Lazor, *J. Appl. Polym. Sci.* **1,** 11 (1959).
165. E. Peggion, F. Testa, and G. Talamini, *Makromol. Chem.* **71,** 173 (1964).
166. K. Giskehaug, *Society of Chemical Industry Monograph No. 20,* London, 1966, p. 235.
167. H. Gerrens, *DECHEMA Monogr.* **49,** 53 (1963).
168. B. Jacobi, *Angew. Chem.* **64,** 539 (1952).
169. H. Cherdron, *Kunststoffe* **50**(10), 568 (1960).
170. J. Ugelstad, P. C. Mork, and F. K. Hansen, *Pure Appl. Chem.* **53,** 323 (1981).
171. U.S. Pat. 4,292,424 (Sept. 29, 1981), G. R. Huddleston, J. W. Turner, and K. D. Konter (to BFGoodrich Chemical Co.).
172. D. E. M. Evans in R. N. Burgess, ed., *Manufacture and Processing of PVC,* Macmillan Publishing Co., Inc., New York, 1982, Chapt. 3.
173. F. A. Bovey, I. M. Kolthoff, A. I. Medalia, and E. J. Meehan, *Emulsion Polymerization,* Interscience Publishers, a division of John Wiley & Sons, Inc., New York, 1965.
174. U.S. Pat. 4,273,904 (June 16, 1981), C. N. Bush, C. A. Daniels, and R. F. Koebel (to BFGoodrich Company).
175. M. Clark in ref. 27, Chapt. 1.
176. G. E. Ham, ed., *Copolymerization,* Interscience Publishers, a division of John Wiley & Sons, Inc., New York, 1964.
177. M. W. Kline and E. N. Skiest in ref. 81, Vol. 1, Chapt. 4, p. 109.
178. J. V. Koleski and L. H. Wartman, *Poly(Vinyl Chloride),* Gordon & Breach Science Publishers, Inc., New York, 1969.
179. A. R. Cain, Polym. Preprints 11(1), 312 (1970).
180. J. Brandrup and E. M. Immergut, *Polymer Handbook,* 2nd ed., John Wiley & Sons, Inc., New York, 1975.
181. *Chem. Eng. News,* 31 (Mar. 7, 1966).
182. L. Weintraub, J. Zufall, and C. A. Heiberger, *Polym. Eng. Sci.* **8,** 64 (1968).
183. C. A. Heiberger and R. Phillips, *Rubber Plast. Age* **48,** 636 (1967).
184. M. Langsam, *J. Appl. Polym. Sci.* **21,** 1057 (1977).
185. *Ibid.*, **23,** 867 (1979).

from iodine-initiated polymerizations in polar solvents; the molecular weight distributions narrow with decreasing solvent polarity for polymerizations initiated at 0°C (8). The iodine-initiated copolymerization of butyl vinyl ether with 1,4-butanediol divinyl ether [3891-33-6] (B1DDVE) or with diethylene glycol divinyl ether [764-99-8] (DEGDVE) produces soluble copolymers (9). This is somewhat surprising in that stannic chloride-initiated and free-radical-initiated copolymerization of these divinyloxy monomers with α-methylstyrene or vinyl chloride yields cross-linked gels (10–11). Homopolymerizations of B1DDVE and DEGDVE at 0°C with iodine yield soluble saturated polymers, whereas iodine-initiated polymerizations at 20°C produce insoluble gels (9).

Homopolymers. *Properties.* Physical properties of the poly(alkyl vinyl ether)s depend on molecular weight, the nature of the alkyl group, the nature of the initiator, stereospecificity, and crystallinity. Physical characteristics of the homopolymers range from viscous liquids, through sticky liquids and rubbery solids, to brittle solids. Polyethers with long alkyl side chains are waxy. The characteristics of some amorphous homopolymers that have been commercially available at one time or another are given in Table 1.

As shown in Table 2, the glass-transition temperatures of the amorphous straight-chain alkyl vinyl ether homopolymers decrease with increasing length of the side chain. Also, the melting points of the semicrystalline poly(alkyl vinyl ether)s increase with increasing side-chain branching.

The degradative effects of γ radiation on diethyl ether solutions of poly(vinyl ether)s have been studied under a variety of conditions (13). All alkyl vinyl ether polymers, except t-butyl vinyl ether, exhibit similar degradative behavior showing

Table 1.   Some Commercial Vinyl Ether Homopolymers

Vinyl ether polymer[a]	CAS Registry No.	Polymer specific viscosity (η_{sp})	Trademark (manufacturer)	Uses
methyl	[34465-52-6]			
viscous liquid (balsamlike)		0.68	Lutonal M (BASF)	plasticizer for coatings; aqueous tackifier
viscous liquid (balsamlike)		0.3–0.5	Gantrez M (GAF)	
ethyl	[25104-37-4]			
viscous liquid		1.0	Lutonal A (BASF)	plasticizer for cellulose nitrate and natural-resin lacquers
elastomeric solid high polymer		solid and solutions supplied	PVEE (Union Carbide)	pressure-sensitive adhesive base
isobutyl	[55605-93-1]			
viscous liquid		1.0	Lutonal I (BASF)	tackifier for adhesives
viscous liquid		0.1–0.5	Gantrez B (GAF)	tackifier for adhesives
elastomeric solid		2–6	Oppanol C (BASF)	pressure-sensitive adhesive base
octadecyl	[9003-96-7]			
waxy solid		[b]	V-Wax (BASF)	polishing and waxing agents

[a] Some viscous liquid polymers are also supplied as high solids solutions, eg, 70% in toluene.
[b] Low degree of polymerization, mp 50°C.

Table 2. Glass-Transition Temperatures of Amorphous Poly(Vinyl Ether)s and Melting Points of Crystalline Poly(Vinyl Ether)s[a]

Poly(vinyl ether)	CAS Registry No.	T_g, °C	Mp, °C
methyl	[34465-52-6]	−34	144
ethyl	[25104-37-4]	−42	
isopropyl	[25585-49-3]	−3	191
n-butyl	[25232-87-5]	−55	
isobutyl	[9003-44-5]	−19	170
2-ethylhexyl	[29160-05-2]	−66	
n-pentyl		−66	
n-hexyl	[25232-88-6]	−77	
n-octyl	[25232-89-7]	−80	
t-butyl	[25655-00-9]		238

[a] Ref. 12.

free energy of scission (G_{sc}) values of 0.3–0.9 scissions per 100 eV at 0°C. Degradation is much more pronounced for poly(t-butyl vinyl ether), which has a G_{sc} of 3.6 at 0°C.

Poly(methyl vinyl ether) (PVM) is the only remaining commercially available vinyl ether homopolymer from a U.S. manufacturer, namely GAF Corporation. Properties of the three different molecular weight, commercial, amorphous methyl vinyl ether homopolymers in water and toluene solutions are summarized in Table 3.

Homopolymerizations. *Monovinyl ethers.* Cationic initiators, eg, Lewis acids, are by far the preferred catalysts for preparing alkyl vinyl ether homopolymers. Only low molecular weight homopolymers (viscous oils) result when free-radical initiators, uv radiation, or heat alone is used for initiating these polymerizations. Large yields of high molecular weight alkyl vinyl ether polymers result from polymerizations initiated by high energy (β and γ) radiation (15) (see Initiators).

Atactic homopolymers. Atactic alkyl vinyl ether homopolymers are readily prepared by initiation with Friedel-Crafts type catalysts, eg, boron fluoride, aluminum trichloride, stannic chloride, etc, either in bulk or in inert dry solvents. Impurities in either monomer or solvent tend to reduce both the rate of propagation and polymer

Table 3. Properties of Commercial Poly(Methyl Vinyl Ether)s[a]

Product	Gantrez M-154	Gantrez M-574	Gantrez M-555	Gantrez M-550
solvent	water	toluene	toluene	toluene
solids, %	50	70	50	50
specific viscosity (1 g/100 mL benzene)	0.47	0.47	0.77	
viscosity, mPa·s (= cP)	ca 40,000	ca 30,000	ca 15,000	
bulk density, g/cm³	1.03	0.96	0.94	0.94
flash point (Tagliabue open cup), °C	none	22	22	22

[a] Ref. 14.

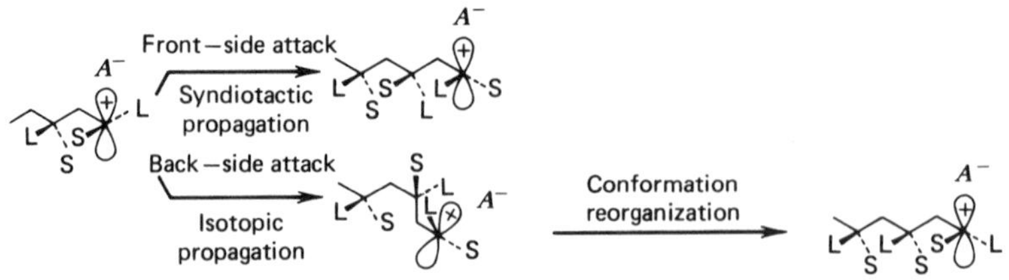

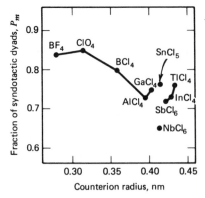

Figure 1.   Effect of ion pairing on stereochemistry of propagation for alkyl vinyl ethers. L = larger substituent; S = smaller substituent (39).

approach by side-chain substituent placement. Thus, front-side attack leading to syndiotactic polymers is favored in highly polar solvents where more separated ion pairs are stabilized. However, nonpolar solvents, eg, toluene, favor tight association of the propagating cation and its counterion leading to back-side approach and an isotactic polymer. When the growing ion-pair end is crowded by a bulky substituent, such as t-butyl, interference between the bulky side chain and the counterion may result, thereby lowering coulombic attraction between ion-pair components which would favor front-side attack, ie, syndiotactic placements. Figures 2 and 3, respectively, show how the fraction of syndiotactic dyads, P_m in poly(t-butyl vinyl ether) decreases with increasing counterion size at 0.25–0.40 nm and increasing polymerization solvent dielectric strength (polarity) (32).

The effects of alkyl substitution on either the α or β carbons of alkyl vinyl ether monomers have also been examined. Polymerization of the α-methyl vinyl alkyl ethers tends to yield principally syndiotactic products even in nonpolar solvents; thus, it was concluded that the previously described mechanism was consistent with this result (29).

The presence of a methyl substituent on the β carbon of a vinyl ether monomer enhances its reactivity (16). cis-Propenyl ethers are several times more reactive than $trans$-propenyl ethers in nonpolar solvents, whereas in polar solvents, the cis and trans isomers show comparable reactivity. When a bulky alkoxy group is present, the trans isomer is the more reactive in polar solvents.

When used alone, phosphoryl chloride, thionyl chloride, and chromyl chloride only char vinyl ether monomers; however, in the presence of triethylaluminum, their behavior is significantly modified and they can initiate stereoregular polymerization

Figure 2.   The relation between the steric structure and the counterion radius. Polymerization condition: $-76°C$, 7:3 (vol:vol) CH_2Cl_2–toluene (32).

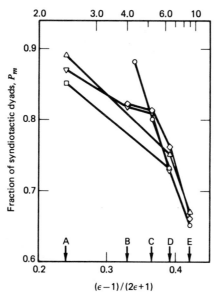

Figure 3. The relation between the structure (P_m = fraction of isotactic dyads) and the polymerization condition. The abscissa is the dielectric constant ϵ and the measure of polarity by Kirkwood ($\epsilon -$ 1)/(2 ϵ + 1). Solvents: A, toluene; B, 7:3 (vol:vol) toluene–CH_2Cl_2; C, 1:1 (vol:vol) toluene–CH_2Cl_2; D, 3:7 (vol:vol) toluene–CH_2Cl_2; E, CH_2Cl_2. Counteranion: ○, BCl_4^-; □, $AlCl_4^-$; △, $GaCl_4^-$; ▲, $InCl_4^-$; ▽, $TlCl_4^-$ (32).

(30). Aging IBVE and triethylaluminum together prior to addition of the oxychloride increases stereoregularity at the expense of conversion, whereas the reverse effect is obtained when oxychloride and triethylaluminum are aged together first.

Heterogeneous initiator systems frequently provide alkyl vinyl ether polymers having greater stereoregularity than homogeneous initiators, especially near room temperature. Pioneering research on complex catalysts prepared from sulfuric acid and an aluminum salt showed how MVE polymers (PVM) with crystallinities up to 48% and even higher could be prepared (40–43). Recently, a Fe_2O_3 initiator was prepared by exposing $Fe(OH)_3$ from ammonia-hydrolyzed $Fe(NO_3)_3$ or $FeCl_3$ to 1 N H_2SO_4, followed by calcining at 500°C for 3 h. High yields of semicrystalline MVE and ethyl vinyl ether homopolymers were obtained in <90 min at −20–0°C.

Cyclohomopolymerizations of divinyl ethers. Divinyl ether [*109-93-3*] and the divinyl ethers of ethylene glycol [*764-78-3*], diethylene glycol (DEGDVE), and 1,4-butanediol (B1DDVE) can be polymerized in dilute solution to low conversions with either free-radical or cationic initiators to give soluble polymers with highly cyclized, low mol wt structures (44–48). Gelation occurs when polymerizations are carried out at high concentrations, at high temperatures, and to high conversions, ie, >30–35%. Based on ^{13}C-nmr spectra, the divinyl ether homopolymer contains a tetrahydrofuran unit with a pendant vinyloxy group and a bicyclic unit (a dioxabicyclo[3.3.0]octane system) in a 1:1 ratio (44). The free-radical cyclopolymerization process for divinyl ethers is summarized in Figure 4.

homopolymer [*9003-19-4*] of divinyl ether

Economic Aspects.   The 1981 bulk price of a 50% solution of poly(methyl vinyl ether) in toluene was $15.50/kg.

Health and Safety Factors, Toxicology.   Dry PVM polymer or aqueous solutions on 200 human subjects showed PVM as neither a primary irritant nor a sensitizer. In acute oral toxicity tests on white rats and guinea pigs, doses of up to 90–100 cm^3/kg of a 25% solids aqueous solution of PVM were tolerated with no visible effects (60).

Uses.   Applications for the poly(alkyl vinyl ether)s include adhesives (qv), surface coatings, lubricants, greases, elastomers, molding compounds, fibers, films, and chemical processing (61) (see Coatings; Lubrication and lubricants; Elastomers, synthetic; Fibers). Applications reported since 1970 are summarized in Table 8.

Poly(methyl vinyl ether) is soluble in water in all proportions at room temperature because of hydrogen bonding of water to the polymer ether linkage. Heating aqueous PVM solutions destroys the ether linkage hydration and decreases water solubility. Thus, PVM polymer comes out of water solution at a specific temperature (33°C) known as the cloud point. Addition of water-miscible solvents or surfactants, eg, the sodium salt of the sulfated adduct of nonylphenol, can raise or even eliminate the cloud point. Strong bases and inorganic salts, however, either raise or lower the cloud point depending on their effectiveness in competing with the polymer ether oxygen for water.

Table 8.   Recent Vinyl Ether Homopolymer Applications

Vinyl ether polymer	Application	Remarks	Ref.
C_{20} straight-chain poly(alkyl vinyl ether)	fuel oil pour-point depressant	mp, 47–50°C; $\overline{M}_n$ 2000–3000 0.1 wt % C_{20} vinyl ether polymer plus 5–10% asphalt residuum depresses the pour point of Arabian Gas Vacuum oil by 16–21°C	62
poly(ethyl vinyl ether)	moisture-permeable, nonirritating surgical cast	a mixture of 80 wt % poly(ϵ-caprolactone) and 20 wt % PVE is calendered into bands; bands are applied to broken member, are heated to 60°C, and are allowed to cool	63
lower alkyl vinyl ether homopolymers	imparts dry-film flexibility and builds viscosity of a photochemical resist-coating composition	poly(methyl vinyl ether) with a molecular weight of K-40 and K-60 is compatible with remaining resist components and does not interfere with resist development processes	64
poly(methyl vinyl ether) (PVM)	pressure-sensitive adhesive composition that adheres to wet surfaces and is removable with water;	PVM 97 wt %; 9:1 acrylic acid–butyl acrylate copolymer 0.5 wt %; poly(vinyl alcohol) 2 wt %;	65
	extends desalination efficiency of reverse-osmosis membranes	service life of reverse-osmosis membranes is extended 70% by treatment of the membranes first with 100 ppm aqueous PVM and then with 10 ppm PVM	66
poly(2-chloroethyl vinyl ether) [29160-80-5]	x-ray-beam resists based on poly(2-chloroethyl vinyl ether) are three times more sensitive and give twice the resolution of poly(ethyl vinyl ether)-based resists	chlorine substituents on poly(vinyl ether) alkoxy side chains enhance sensitivity and resolution of poly(vinyl ether)-based x-ray-beam resists	67

Cloud-point lowering results from reducing the hydration of the ether oxygens (68). A practical application for this negative PVM solubility coefficient with temperature is sensitization of polymer latices so that latex coagulation and gelation occurs when the latex is heated to a specific critical temperature.

Poly(methyl vinyl ether) functions as a nonmigrating plasticizer, tackifier, or both and thus improves adhesion, wet-out, and flow. Because it is soluble in organic solvents as well as water, the polymer exhibits high adhesion to both high and low free-surface-energy substrates. It promotes the adhesion of nonadhering materials to glass, metal, and plastics. Monomolecular layers of moisture on highly polar substrates do not interfere with its adhesion-promoting properties. Poly(methyl vinyl ether) has remarkably low hygroscopicity for a water-soluble polymer. Whereas many water-soluble polymers pick up as much as 25 wt % water when exposed to 90% rh, PVM picks up only 5%.

The three important commercial applications for PVM are for viscosity control and dry-film flexibility in uv photoresist coating solutions; as a tackifier and adhesion promoter for acrylic pressure-sensitive adhesives used in tapes, labels, and decals; and as a semipermeable membrane for reverse osmosis (qv) (66). The polymer is compatible with a wide variety of other polymers and copolymers. Blends of polystyrene (PS) and PVM, in particular, have been widely studied and exhibit interesting phase-transformation behavior. For example, this pair of polymers is miscible in all proportions in hydrocarbon solvents, eg, toluene, but not in trichloroethylene (69). A film of PS–PVM cast from trichloroethylene appears cloudy and shows two T_gs. The same polymer blend cast from toluene is transparent and exhibits only a single T_g. When the heterogeneous PS–PVM film cast from trichloroethylene is heated to 85°C for 12 h, it becomes an optically clear, single-T_g, homogeneous mixture. On heating to 170°C, these homogeneous, transparent films once again become cloudy (69). The cloud point temperatures are dependent on molecular weight and molecular weight distribution. One application of these properties involves writing with chloroform on a homogeneous film of 1:1 PS–PVM to give imaged areas. After etching the imaged areas with water, the unimaged areas are unaffected, and the imaged and etched areas clearly display the original writing (71).

Copolymers. *Properties and Uses.* Trade names, suppliers, and principal applications for the most important commercial vinyl ether copolymers are presented in Table 9. Other potential applications reported for vinyl ether copolymers are presented in Table 10.

Interpolymers with maleic anhydride are the most significant commercial class of vinyl ether copolymers and, within this class, the methyl vinyl ether copolymer is by far the most important. The copolymer is available in four grades: Gantrez AN-119 (η_{sp} 0.1–0.5, low mol wt grade); Gantrez AN-139 (η_{sp} 1.0–1.4, medium mol wt grades); Gantrez AN-149 (η_{sp} 1.5–2.0, medium mol wt grades); and Gantrez AN-169 (η_{sp} 2.6–3.5, high mol wt grade) (GAF) (82).

Poly(methyl vinyl ether-*co*-maleic anhydride) (PVM–MA) is supplied as a white, fluffy powder and is soluble in ketones, esters, pyridine, lactams, and aldehydes. It is insoluble in aliphatic, aromatic, or halogenated hydrocarbons, ethyl ether, and nitroparaffins. When the copolymer dissolves in water or alcohols, the anhydride group is cleaved, forming the polymeric free acid or the half esters of the alcohol, respectively. The viscosities of aqueous PVM–MA solutions decrease with time, particularly when exposed to light. Water-soluble uv absorbers (see Uv stabilizers), eg, Uvinul MS-40

Table 10. Vinyl Ether Copolymer Applications

Copolymer	CAS Registry No.	Application	Remarks	Ref.
poly(methyl vinyl ether-*co*-maleic anhydride) (PVM–MA)	[9011-16-9]	nonsilver-based photographic film prepared from reaction of PVM–MA with *p*-N₃SO₂C₆H₄NHCO₂CH₂CH₂OH followed by neutralization of the product with trimethylamine or NaOH	exposure of polyethylene films coated with the salt through a pattern to uv radiation for 2 min gives green images after development in acetone–DMF and immersion in a malachite green bath	72
poly(isobutyl vinyl ether-*co*-mono-ethyl maleate)	[36572-31-3]	aerosol hairspray resin	requires neutralization with 5 mol % aminomethylpro-panediol	73
poly(octadecyl vinyl ether-*co*-maleic anhydride) (POVE-MA)	[28214-64-4]	release coat for pressure-sensitive adhesive-tape backings	PVOE-MA in toluene reacts with octadecylamine for 8 h at 70–75°C; the product is diluted with toluene to 1% solids and is applied to the back side of a cellophane adhesive tape	74
poly(isobutyl vinyl ether-*co*-vinyl chloride)	[25154-85-2]	electrostatic image-developing powders, ie, toner; binding resin;	100 parts of the polymer and 10 parts carbon black are kneaded, cooled, and pulverized to give a toner with improved electrostatic characteristics;	75
		adhesive tape for joining and sealing PVC single-ply roofing membranes	adhesive composition consists of 40 parts of the copolymer (Caroflex MP-45, BASF), 60 parts of 70% nitrile rubber–30% PVC homopolymer, 45 parts dioctyl phthalate, 3 parts epoxy resin, 2 parts barium cadmium stabilizer, and 10 parts CaCO₃	76
poly(cetyl vinyl ether-*co*-octadecyl vinyl ether-*co*-N-vinylpyrrolidi-none)	[29351-75-5]	protective colloid in suspension polymerization of vinyl chloride	PVC prepared in the presence of 0.5 parts per 100 parts of monomer of the copolymer has a particle size 87% ≤74 mm (200 mesh); particle size of PVC prepared in the presence of the same amount of PVP K-90 had only 7.4% ≤74 mm (200 mesh)	77
poly(methyl vinyl ether-*co*-mono-	[50935-57-4]	copolymer coating on polyester film provides improved	copolymer is applied to cellulose acetate–	78

Table 10 (*continued*)

Copolymer	CAS Registry No.	Application	Remarks	Ref.
ethyl maleate		printability	butyrate-coated polyester film to give a mat surface with improved ink uptake	
poly(hexadecyl vinyl ether-*co*-octa-decyl vinyl ether-*co*-styrene-*co*-maleic anhydride)	[40472-29-5]	aqueous-solution thickening agent for heavy-duty cleaners and ammonia-based fertilizers	inclusion of long-chain vinyl ethers in the copolymer provides greater thickening effectiveness	79
poly(2-chloroethyl vinyl ether-*co*-*N*-vinylpyrrolidinone)	[54803-58-6]	photochemical cross-linking of polymers to immobilize enzymes	copolymer reacts with stilbazole, producing a photosensitive polymer; irradiation of a film of the water-soluble polymer yields an insoluble material, in which enzymes are entrapped	80
poly(methyl vinyl ether-*co*-mono-butyl maleate)	[54578-91-5]	thermographic copying material	a gel-like complex can be prepared by mixing an ethanolic solution of the copolymer with an equal weight of PVP K-90; a DMF solution of the complex is applied to polyester film, passed with an original through a Thermofax copier, and immersed in a 0.5 wt % crystal violet solution to yield a violet copy	81

provides a stable electron-beam resist resin that is also a good film former and is insensitive to visible and uv light of wavelengths >200 nm (97).

A lightly cross-linked version of PVM–MA copolymer is available for use in textile print-paste systems. This copolymer (Gaftex PT) belongs to the class of water-soluble polymers known as microgels. Although the polymer is not soluble in many organic solvents because of crosslinking, when it is neutralized in water, the polymer chains become fully extended behaving as though they were dissolved. This produces very high viscosities at low concentrations of polymer.

Vinyl ethers also form useful products when copolymerized with monomers other than maleic anhydride. For example, impact-resistant plastics with high flexural strength, high softening points, and good gas and vapor resistance result from the copolymerization of butyl vinyl ether with acrylonitrile and indene in the presence of acrylonitrile–butadiene latex (98). A plastic that resists yellowing when injection-molded is obtained by graft polymerizing a 95:5 methyl vinyl ether–butadiene copolymer [29697-46-9] with 70 wt % of a 25:75 acrylonitrile–styrene mixture (99). Co-

Table 11.   Typical Properties of Alcohol Solutions of the Half Esters of PVM–MA [a]

Property	Gantrez ES-225	Gantrez ES-335-I	Gantrez ES-425	Grantrez ES-435
alkyl group of monoester	ethyl	isopropyl	butyl	butyl
physical form	clear, viscous liquid	clear, viscous liquid	clear, viscous liquid	clear, viscous liquid
activity, % solids	50 ± 2	50 ± 2	50 ± 2	50 ± 2
solvent	ethanol	2-propanol	ethanol	2-propanol
acid number (100% solids)	275–300	255–285	245–275	245–275
density, g/cm^3	0.983	0.957	0.977	0.962

[a] Ref. 60.

polymers of dimethylaminoethyl vinyl ether [3622-76-2] and acrylamide and their salts are useful as pigment-retention aids and flocculating aids (100) (see Pigments, organic; Flocculating agents). The most interesting and widely studied vinyl ether copolymer in the biochemical and medical sciences fields is the maleic anhydride–divinyl ether copolymer [27100-68-1] (DIVEMA) frequently referred to as Pyran copolymer. It has been demonstrated that DIVEMA modulates a variety of biologic responses related to bacteria, fungi, and viruses; enhances immune responsiveness; inhibits adjuvant arthritis; and alters the phagocytic activity of the recticuloendothelial system (101–104). Probably the most important properties of DIVEMA are its antiviral and antitumor effects (102,105).

Copolymerizations.   Monovinyl ethers.   Reports of copolymerizations of alkyl vinyl ethers with one or two other monomers that have appeared in the literature since 1970–1971 are listed in Table 12. The copolymerization behavior of vinyl ether monomers is not well understood. Cationic initiation of the polymerization of a mixture of different vinyl ether monomers or a mixture of a vinyl ether monomer with even another electron-donating monomer does not always produce true copolymers. Frequently, a homopolymer forms first, with the other monomer entering the polymer only when a higher temperature is reached or more initiator is added. However, alternating copolymers of alkyl vinyl ethers and acrylates can be prepared by cationic initiation if care is taken in the selection of initiator, solvent, comonomer, and temperature. For example, an alternating copolymer of methyl methacrylate and butyl vinyl ether (BVE) has been prepared in good yield in toluene at 0°C in 3 h with trichlorotriethyldialuminum (124–125). Block copolymers of IBVE and substituted styrenes can be prepared, for example, by the cationic polymerization of *p*-methoxystyrene with iodine in CCl$_4$ at -15°C, which results in a living polycation that readily initiates polymerization of IBVE monomer added later (121). The proportion of added IBVE incorporated in the block copolymer averages 30–40% at 32% conversion of the added IBVE. The living polycation does not form when the polymerization is carried out in a more polar solvent, eg, CH$_2$Cl$_2$, unless a common ion salt such as tetrabutylammonium iodide is added to suppress dissociation of the propagating ion pair. Polymerization by MX$_n$ (M = metal, X = halide) initiators fails to give block copolymers in either CCl$_4$ or CH$_2$Cl$_2$.

Copolymers of vinyl ethers produced by ionic initiation have never become important industrially. It is the free-radical-initiated copolymerization of the vinyl ethers with monomers bearing strong electron-withdrawing substituents leading to uniform, high molecular weight, high conversion copolymers that is of greatest commercial interest. Here, cationic initiation of polymerization is far less desirable than free-radical initiation because of the tendency of the vinyl ethers to homopolymerize in the presence of cationic species. Such electron-withdrawing monomers include, for example, maleic, itaconic, and citraconic anhydrides, and their partial and full esters and fumaronitrile, alkyl vinyl sulfones, maleimides, divinyl sulfone, vinylene carbonate, vinyl chloroacetate, and nitroethylene. There is abundant experimental evidence that these free-radical vinyl ether copolymerizations proceed via a charge-transfer complex when the electron-withdrawing characteristics of the substituents on the comonomer are sufficiently strong (117–118). The equilibrium constant for charge-transfer-complex formation decreases with increasing solvent polarity, since in polar solvents free monomers are better stabilized than the complex of the monomers. The decreasing rate of polymerization and decreasing conversions observed in vinyl ether–maleic anhydride copolymerizations with increasing solvent polarity are consistent with the lower concentration of the complex in the more polar solvents.

Free-radical-initiated terpolymerizations of butyl vinyl ether, maleic anhydride, and acrylonitrile in $CHCl_3$ have been studied (108). The charge-transfer complex of the vinyl ether with maleic anhydride participated in the copolymerization as if it were a single monomeric species (113). Reactivity ratios for the complex ($r_{complex} = 83$) with acrylonitrile ($r_{acrylonitrile} = 0.009$) were used to estimate Alfrey-Price Q-e values (126) for the complex of Q = 35 (a measure of resonance stabilization of monomer) and e = 0.66 (a measure of polarity of monomer) (127).

When the electron-withdrawing effect of substituents on an olefin comonomer is extremely strong, eg, in the case of vinylidene cyanide, simultaneous, spontaneous, and explosive anionic initiation of the homopolymerization of vinylidene cyanide and cationic homopolymerization of IBVE occurs when the vinyl ether monomer is added (128). The copolymerization of the somewhat less reactive nitroethylene with IBVE has been studied for a better understanding of the nature of the processes occurring in this explosive vinylidene cyanide–IBVE system (122). Equimolar quantities of nitroethylene and IBVE spontaneously react at 0°C to yield an isolatable cycloadduct, 2-nitrocyclobutyl isobutyl ether [54850-00-9]. This cycloadduct is so reactive that at room temperature it not only undergoes spontaneous terminationless cationic homopolymerization of itself to give an alternating copolymer of IBVE and nitroethylene, but the adduct also initiates the anionic homopolymerization of nitroethylene at 0°C. The copolymerization behavior of tetracyanoquinodimethane (TCNQ) with vinyl ethers is anomalous. This strong electron-withdrawing monomer initiates the cationic homopolymerization of BVE and IBVE, but 2-chloroethyl vinyl ether [110-75-8] and phenyl vinyl ether [766-94-9] undergo alternating, spontaneous copolymerization with TCNQ by a free-radical process (111).

Reactivity-ratio measurements for the free-radical copolymerization of common vinyl monomers with methyl, ethyl, octyl, dodecyl [765-14-0], and octadecyl [930-02-9], vinyl ethers give r_2 values near zero (see Table 13).

Divinyl ethers. Radical-initiated copolymerizations of divinyl ether (DVE) and substituted divinyl ethers with maleic anhydride are described in refs. 119–122. Completely cyclized, alternating copolymers containing a divinyl ether–maleic anhydride ratio of 1:2 have been obtained. The originally proposed structure for these copolymers has been confirmed, and the proposed copolymerization mechanism is presented in Figure 5 (129–130).

Table 12. Vinyl Ether Copolymers

Copolymer, poly(M$_1$-co-M$_2$)	Mole ratio	CAS Reg. No.	Initiator	Temperature, °C	Solvent	Conversion, %	Reactivity ratios r_1	r_2	$\bar{M}_n$	[η], dL/g	Polymer properties	Refs.
poly(isobutyl vinyl ether-co-trichloroethylene)	50:50	[51033-82-0]	γ rays	5	bulk	32	0.045	0	2,500		hard, pale-yellow rubber; softens at 40°C	106
poly(l-menthyl vinyl ether-co-vinylene carbonate)	50:50	[72265-26-6]	AIBN	50	C$_6$H$_6$	20	0.185	0.16	very low		mp, 65–110°C	107
poly(l-menthyl vinyl ether-co-indene)	71:29	[69860-91-9]	AIBN	50	C$_6$H$_6$	9	0.4	9.0			mp, 158–169°C	107
poly(vinyl isobutyl ether-co-cyclopentadiene)	94:6	[39921-95-4]	BF$_3$·(C$_2$H$_5$)$_2$O	−78	CH$_2$Cl$_2$	60				0.70		108
poly(isobutyl vinyl ether-co-α-methylstyrene)	70:30	[35744-20-8]	C$_2$H$_5$AlCl$_2$·H$_2$O	−50	CH$_2$Cl$_2$	20	11.6	0.17	15,600			109–110
poly(phenyl vinyl ether-co-tetracyanoquinodimethane)	52:48	[84642-74-0]	AIBN	60	CH$_3$CN	29						111
poly(chloroethyl vinyl ether-co-tetracyanoquinodimethane)	52:48	[78747-51-0]	none	50	CH$_3$CN	4	0.0	0.0		0.45		111
poly(phenyl vinyl ether-co-maleic anhydride)	90:10	[27495-64-3]	AIBN	70	bulk	75			60,000		mp, 275°C	112
poly(butyl vinyl ether maleic anhydride) (charge-transfer complex-co-acrylonitrile)	50:50	[26298-63-5]	AIBN	60	CHCl$_3$	5	83	0.009				113
poly(ethyl vinyl ether-co-carbon dioxide)		[55993-87-8]	tributoxyaluminum	80	bulk	14						114–115
poly(isobutyl vinyl ether-co-vinylene carbonate)	50:50	[30919-92-7]	γ rays	40–80	bulk		0.148	0.118				116

Copolymer	Ratio	CAS No.	Initiator	Temp (°C)	Solvent	% conv	r_1	r_2		MW	Properties	Ref
poly(isobutyl vinyl ether-co-phenyl vinyl sulfide)	10:90	[84642-75-1]	$BF_3 \cdot (C_2H_5)_2O$	0	nitroethane		7.2	0.15				117
poly(2,3-dihydropyran-co-maleic anhydride)	50:50	[26950-71-0]	benzoyl peroxide	60	$CHCl_3$	12					amorphous resins	118
poly(methyl vinyl ether-co-carbon dioxide)		[55993-88-9]	spontaneous initiation	80	bulk	27				600	transparent amorphous resin; soluble in water, C_6H_6, $CHCl_3$, and $(C_2H_5)_2O$	119
poly(ethyl vinyl ether-co-divinyl sulfone)	51:49	[84642-76-2]	benzoyl peroxide	60	C_6H_6	11	0.012	0.158				120
poly(2,3-dihydropyran-co-divinyl sulfone)	54:56	[84642-77-3]	AIBN	70	C_6H_6	13	0.011	0.533				120
isobutyl vinyl ether-p-methoxystyrene block copolymer		[69644-35-5]	iodine	-15	$CHCl_3$	70				16,800		121
poly(isobutyl vinyl ether-co-nitroethylene)	50:50	[84642-78-4]	spontaneous initiation at 25°C	25–35	$C_6H_5CH_3$	25			1.5			122
poly(isobutyl vinyl ether-co-vinyl chloride)	38:72	[25154-85-2]	$K_2S_2O_8$	30	water-emulsion polymerization	96					tough, amorphous films; tensile strength 22.9 mPa (3.32×10^4 psi)	123
poly(2-chloroethyl vinyl ether-co-isobutylene-co-butyl acrylate)	6:43:51	[31049-61-3]	$C_2H_5BI_2$	-20					3.55		white amorphous solid	124

a Many of the copolymerizations were deliberately stopped at low conversions so good reactivity ratio data could be obtained.

Table 13. Reactivity Ratios of the Free-Radical Copolymerization of Alkyl Vinyl Ethers (M_2) with Other Vinyl Monomers [a,b]

Comonomer (M_1)	r_1	r_2
acrylonitrile	0.8–1	0
butyl maleate	0–0.1	0
maleic anhydride	0.0	0
methylacrylate	2.7–3.0	0
methyl methacrylate	10	0
styrene	>50	0
vinyl acetate	3.4–3.7	0
vinyl chloride	1.7–2.2	0
vinylidene chloride	1.3–1.5	0

[a] Ref. 124.
[b] Bulk copolymerization with methyl, octyl, dodecyl, and octadecyl vinyl ethers with benzoyl peroxide as initiator at 40–100°C.

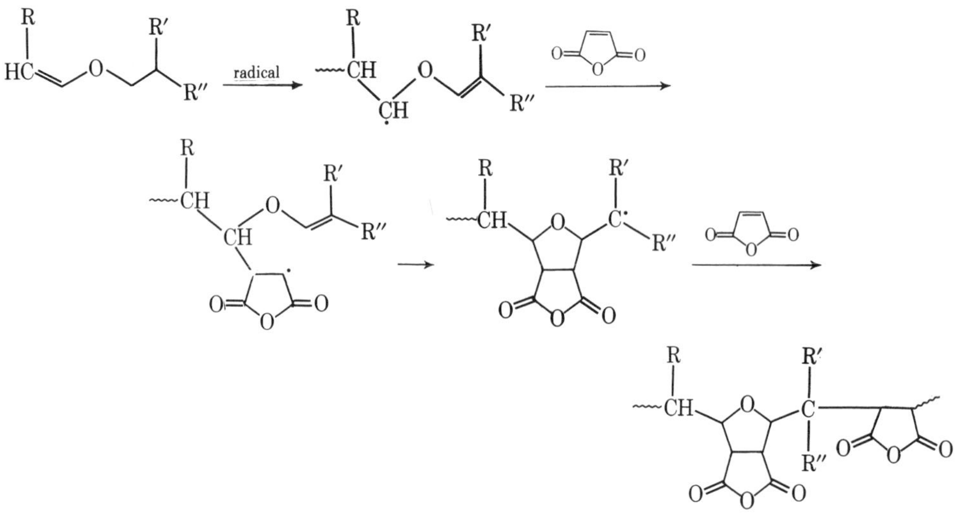

R = R′ = R″, divinyl ether (DVE)
R = R′ = CH$_3$ R″ = H, dipropenyl ether (DPE)
R = R′ = H R″ = CH$_3$, propenyl vinyl ether (PVE)
R = H R′ = R″ = CH$_3$, 2-methyl propenyl vinyl ether (CH$_3$-PVE)

Figure 5. Mechanism for the copolymerization of substituted divinyl ether and maleic anhydride (129).

Radical copolymerization of 1,4-butanediyl divinyl ether with maleic anhydride proceeds very rapidly and produces insoluble gels (131). Heating 1,4-butanediyl divinyl ether with maleic anhydride in the absence of solvent and initiator can result in an explosive polymerization (132). This copolymer [*33520-84-2*] is useful as a carrier for enzymes (131).

It has been suggested that the radical copolymerization of monovinyl ethers with multiple-double-bond monomers bearing strong electron-withdrawing substituents, eg, divinyl sulfone, may proceed by a process analogous to that presented in Figure 5 (120). Ethyl vinyl ether copolymerizes with divinyl sulfone to give soluble alternating copolymers (120). The reaction occurs at 60°C in C_6H_5 and gives a 1:1 ratio of vinyl ether–divinyl sulfone in the polymer. Evidence for formation of a charge-transfer complex can be detected by uv spectroscopy; λ_{max} for the complex is ca 394 nm.

Economic Aspects. Bulk prices in 1981 for poly(methyl vinyl ether-co-maleic anhydride), a 50 wt % ethanol solution of poly(methyl vinyl ether-co-monoethyl maleate), and a 50 wt % ethanol solution of poly(methyl vinyl ether-co-monobutyl maleate) were, respectively, $5.15/kg, $3.96/kg, and $3.80/kg.

Health and Safety Factors, Toxicology. Poly(methyl vinyl ether-co-maleic anhydride) and the monoalkyl ester of poly(methyl vinyl ether-co-maleic anhydride) have been shown (on rabbits) to be neither primary irritants nor primary sensitizers. The acute oral toxicities (on white rats) of the two copolymers are, respectively, 29 g/kg body weight and 25 g/kg body weight.

BIBLIOGRAPHY

"Reppe Chemistry" in *ECT* 1st ed., Vol. 11, p. 651, by J. M. Wilkinson, Jr., Jesse Werner, H. B. Haas, and Hans Beller, GAF; "Vinyl Ether Monomers and Polymers" in *ECT* 2nd ed., Vol. 21, pp. 412–426, by C. E. Schildknecht, Gettysburg College.

1. I. W. J. Still, J. N. Reed, and K. Turnbull, *Tetrahedron Lett.* **17,** 1481 (1979); U.S. Pat. 3,991,125 (Nov. 9, 1976), J. N. Labovitz and C. A. Henrick (to Zoecon); S. S. Sabirov and M. D. Babakhanova, *Khim. Tadzh.*, 48 (1973); *Chem. Abstr.* **84,** 58538 (1976).
2. R. W. Aben and H. W. Scheeren, *J. Chem. Soc. Perkin Trans.* **1,** 3132 (1979).
3. Y. Yoshida and S. Inoue, *Chem. Lett.* **11,** 1375 (1977).
4. U.S. Pat. 4,057,575 (Nov. 8, 1977), D. L. Klass (to Union Oil of California); U.S. Pat. 4,161,610 (July 17, 1979), D. L. Klass (to Union Oil of California).
5. Jpn. Kokai Tokkyo Koho 80 02,416 (Jan. 19, 1980), K. Tagaki and C. Motobashi (to Sumitomo).
6. J. Wislicenus, *Ann.* **192,** 106 (1878).
7. A. F. Johnson and R. N. Young, *J. Polym. Sci. Polym. Symp.* **56,** 211 (1976).
8. T. Ohtori, Y. Hirokawa, and T. Higashimura, *Polym. J.* **11,** 471 (1979).
9. S. L. N. Seung and R. N. Young, *J. Polym. Sci. Polym. Lett. Ed.* **16,** 367 (1978).
10. I. Itoh, S. Itoh, I. Matsumura, and H. Maruyama, *J. Polym. Sci. Part C* **33,** 135 (1971).
11. W. Heitz, F. Kraffczyk, K. Pfitzner, and D. Randau in E. Kovats, ed., *Column Chromatography, 5th International Symposium on Separation Methods, 1969*, Sauerlaender A.G., Aarau, Switzerland, 1970, pp. 126–127.
12. N. D. Field and D. H. Lorenz in E. C. Leonard, ed., *Vinyl and Diene Monomers*, Pt. 1, Wiley-Interscience, New York, 1970, pp. 365–411.
13. Y. Suzuki, J. M. Rooney, and V. Stannett, *J. Macromol. Sci. Chem.* **A12,** 1055 (1978).
14. *Gantrez M*, technical bulletin 8740-001, GAF Corporation, 1970.
15. N. M. Bikales, "Vinyl Ether Monomers," in N. M. Bikales, ed., *Encyclopedia of Polymer Science and Technology*, Vol. 14, Wiley-Interscience, New York, 1971, pp. 511–521.
16. T. Higashimura and K. Yamamoto, *Makromol. Chem.* **175,** 1139 (1974).
17. M. Biswas and N. C. Maity, *J. Macromol. Sci. Chem.* **A15,** 1153 (1981).
18. T. Fueno, T. Okuyama, and J. Fuyukawa, *J. Polym. Sci. Part A-1* **7,** 3219 (1969).
19. C. E. Schildknecht, *Vinyl and Related Polymers*, John Wiley & Sons, Inc., New York, 1952, pp. 622–625.
20. A. Ledwith and H. J. Woods, *J. Chem. Soc. B*, 310 (1970).
21. T. Fueno, T. Okuyama, T. Matsumura, and J. Furukawa, *J. Polym. Sci. Part A-1* **7,** 1447 (1969).
22. M. Biswas, G. M. A. Kabir, and S. S. Bhagawan, *Makromol. Chem.* **179,** 1209 (1978).
23. C. E. Schildknecht, A. O. Zoss, and C. McKinley, *Ind. Eng. Chem.* **39,** 180 (1947).
24. C. E. Schildknecht, S. T. Gross, and A. O. Zoss, *Ind. Eng. Chem.* **41,** 1998 (1949).
25. C. E. Schildknecht, A. O. Zoss, and F. Grosser, *Ind. Eng. Chem.* **41,** 2891 (1949).
26. N. Oguni, S. Sano, M. Fujimura, and H. Tani, *Polym. J.* **4,** 607 (1973).
27. S. Okamura, T. Kodama, and T. Higashimura, *Makromol. Chem.* **53,** 180 (1962); T. Higashimura, K. Suzuoki, and S. Okamura, *Makromol. Chem.* **86,** 259 (1965).

28. D. J. Sikkema and H. Angad-Gaur, *Makromol. Chem.* **181,** 2259 (1980).

29. K. Matsuzaki, S. Okuzono, and T. Kanai, *J. Polym. Sci. Polym. Chem. Ed.* **17,** 3447 (1979).

30. M. Biswas and G. M. A. Kabir, *Polymer* **19,** 357 (1978).

31. N. Tsubokawa, N. Takeda, and K. Kudoh, *Carbon* **18,** 163 (1980).

32. T. Kunitake and K. Takarabe, *Makromol. Chem.* **182,** 817 (1981).

33. M. Hino and K. Arata, *J. Polym. Sci. Polym. Chem. Ed.* **18,** 235 (1980).

34. M. Hino and K. Arata, *Chem. Lett.* **8,** 963 (1980).

35. J. Lal and J. E. McGrath, *J. Polym. Sci. Part A-2,* 3369 (1964).

36. T. Tanaka, *Kagaku Kogyo* **25,** 257 (1974) (in Japanese); *English Translation WT-81-45,* Associated Technical Services, Inc., Glen Ridge, N.J., 1981, pp. 257–268.

37. S. Murahashi, S. Nozakura, and K. Matsumura, *J. Polym. Sci. Part B* **4,** 59 (1965); S. Murahashi, S. Nozakura, M. Sumi, H. Yuki, and K. Hatada, *J. Polym. Sci. Part B* **4,** 65 (1965); S. Murahashi, S. Nozakura, and M. Sumi, *J. Polym. Sci. Part B* **3,** 245 (1965).

38. T. Kunitake and C. Aso, *J. Polym. Sci. Part A-1* **8,** 665 (1970); T. Kunitake and S. Tsugawa, *Macromolecules* **8,** 709 (1975).

39. A. Ledwith, E. Chiellini, and R. Solaro, *Macromolecules* **12,** 240 (1979).

40. U.S. Pat. 2,549,921 (Apr. 24, 1951), S. A. Mosely (to Union Carbide).

41. U.S. Pat. 3,284,426 (Nov. 8, 1966), E. J. Vandenberg (to Hercules).

42. U.S. Pat. 3,157,626 (Nov. 17, 1964), R. F. Heck (to Hercules).

43. U.S. Pat. 3,159,613 (Dec. 1, 1964), E. J. Vandenberg (to Hercules).

44. M. Tsukino and T. Kunitake, *Macromolecules* **12,** 387 (1979).

45. M. Tsukino and T. Kunitake, *Polym. J.* **13,** 657 (1981).

46. T. Nishikubo, T. Iizawa, and A. Yoshinaga, *Makromol. Chem.* **180,** 2793 (1979).

47. S. L. N. Seung and R. N. Young, *J. Polym. Sci. Polym. Lett. Ed.* **16,** 367 (1978).

48. M. Guaita, G. Camino, and L. Trossarelli, *Makromol. Chem.* **149,** 75 (1971).

49. A. M. Goineau, J. Kohler, and V. Stannett, *Makromol. Sci. Chem.* **A11,** 99 (1977).

50. Y. Suzuki, A. Chudgar, J. M. Rooney, and V. Stannett, *J. Macromol. Sci. Chem.* **11,** 115 (1977).

51. W. C. Hsieh, H. Kubota, D. R. Squire, and V. Stannett, *J. Polym. Sci. Polym. Chem. Ed.* **18,** 2773 (1980).

52. Ka. Hayashi, Ko. Hayashi, and S. Okamura, *J. Polym. Sci. Part A-1* **9,** 2305 (1971).

53. Y. J. Chung, J. M. Rooney, D. R. Squire, and V. Stannett, *Polymer* **16,** 527 (1975).

54. F. Subira and P. Sigwalt, private communication, 1981; Ref. 51, p. 2774.

55. C. C. Ma, H. Kubota, J. M. Rooney, D. R. Squire, and V. Stannett, *Polymer* **20,** 317 (1979).

56. A. Ledwith, E. Lockett, and D. C. Sherrington, *Polymer* **16,** 31 (1975).

57. V. R. Desai, Y. Suzuki, and V. Stannett, *J. Macromol. Sci. Chem.* **A11,** 133 (1977).

58. A. Deffieux, W. C. Hsieh, D. R. Squire, and V. Stannett, *Polymer* **22,** 1575 (1981).

59. T. A. Du Plessis, V. Stannett, and A. M. Goineau, *J. Polym. Sci. Polym. Chem. Ed.* **12,** 2457 (1974).

60. *Gantres ES Monoester Resins,* technical bulletins 9642-052 and 9642-068, GAF Corporation, 1977.

61. *Alkyl Vinyl Ethers,* technical bulletin 7543-055, GAF Corporation, 1966.

62. Can. Pat. 1,006,351 (Mar. 8, 1977), J. B. Biasotti, S. Herbstman, and G. S. Saines (to Texaco).

63. Ger. Offen. 2,135,995 (Jan. 27, 1972), B. Phillips, D. F. Pollart, and J. V. Koleske (to Union Carbide).

64. U.S. Pat. 3,634,082 (Jan. 11, 1972), C. W. Christensen (to Shipley).

65. Brit. Pat. 1,245,410 (Sept. 8, 1971), (to Kuramoto Sangyo).

66. Jpn. Kokai 78 28,083 (Mar. 15, 1978), S. Nakamura and Y. Misaka (to Kurita Water Industries).

67. S. Imamura, S. Sugawara, and K. Murase, *J. Electrochem. Soc. Solid State Sci. Technol.* **30,** 1139 (1977).

68. R. A. Horne, J. P. Almeida, A. F. Day, and N. Yu, *J. Colloid Interface Sci.* **35,** 77 (1971).

69. D. D. Davis and T. K. Kwei, *J. Polym. Sci. Polym. Phys. Ed.* **18,** 2337 (1980).

70. T. Nishi and T. K. Kwei, *Polymer* **16,** 285 (1975).

71. U.S. Pat. 3,765,968 (Oct. 16, 1973), J. W. Leggingwell, C. Theis, and D. W. Werkmeister (to NCR).

72. U.S. Pat. 3,784,527 (Jan. 8, 1974), A. A. R. Sayigh, F. A. Stuber, and H. Ulrich (to Upjohn).

73. Ger. Offen. 2,138,736 (Feb. 10, 1972), D. H. Lorenz and K. J. Valan (to GAF).

74. Jpn. Kokai 74 23,237 (Mar. 1, 1974), Y. Nakamura and K. Matsuoka (to Mitsubishi Pencil Co.).

75. Jpn. Kokai Tokkyo Koho 78 106,040 (Sept. 14, 1978), Y. Nomura, S. Nemoto, and M. Aoki (to Ricoh).

76. Eur. Pat. Appl. 79,300,852.5 (May 17, 1979), E. R. Kreutzer (to Braas & Co.).

77. U.S. Pat. 3,839,310 (Oct. 1, 1974), N. D. Field, E. M. Smolin, and E. P. Williams (to GAF).
78. U.S. Pat. 3,870,549 (Mar. 11, 1975), A. P. Ruygrok (to GAF).
79. U.S. Pat. 3,723,375 (Mar. 27, 1973), N. D. Field and E. P. Williams (to GAF).
80. K. Ichimura and S. Watanabe, *J. Polym. Sci. Polym. Chem. Ed.* **18**, 891 (1980).
81. U.S. Pat. 3,912,844 (Oct. 21, 1975), I. Endo and co-workers (to Canon).
82. *Gantrez AN*, technical bulletin 9653-023, GAF Corporation, 1965.
83. U. P. Strauss, B. W. Barbieri, and G. Wong, *J. Phys. Chem.* **83**, 2840 (1979).
84. P. J. Martin and U. P. Strauss, *Biophys. Chem.* **11**, 397 (1980).
85. U. P. Strauss and M. S. Schlesinger, *J. Phys. Chem.* **82**, 571 (1978).
86. U.S. Pat. 3,654,164 (Apr. 4, 1972), R. L. Sperry (to Petroleum Solids Control).
87. U.S. Pat. 4,169,818 (Oct. 2, 1979), R. N. DeMartino (to Celanese).
88. U.S. Pat. 4,288,493 (Sept. 9, 1981), J. E. Kropp (to 3M).
89. E. Chalhoub, H. A. M. El-Shibini, and N. A. Daabis, *Sci. Pharm.* **48**, 24 (1980).
90. L. Goldstein, A. Lifshitz, and M. Sokolovsky, *Int. J. Biochem.* **2**, 448 (1971).
91. E. Chalhoub, H. A. M. El-Shibini, and N. A. Daabis, *Pharm. Ind.* **38**, 844 (1976).
92. J. Patschorke, *J. Signalaufzeichnungsmaterialen* **5**, 317 (1977).
93. Ger. Offen. 2,232,682 (Sept. 27, 1973), J. Patschorke (to Celfa).
94. Ger. Offen. 2,830,958 (Feb. 1, 1979), E. J. Masters (to Mallinckrodt).
95. U.S. Pat. 4,261,969 (Apr. 14, 1981), J. Heller (to World Health Org.).
96. Ger. Offen. 2,250,731 (Apr. 26, 1973), T. S. Mestetsky (to GAF).
97. H. S. Cole, D. W. Skelly, and B. C. Wagner, *IEEE Trans. Electron Devices* **ED22**, 417 (1975).
98. U.S. Pat. 4,082,819 (Apr. 4, 1978), G. S. Li and G. W. Dirks (to Standard Oil of Ohio).
99. Ger. Offen. 2,238,967 (Feb. 28, 1974), F. Haaf and G. Heinz (to BASF).
100. U.S. Pat. 3,899,471 (Aug. 12, 1975), D. H. Lorenz and E. P. Williams (to GAF).
101. W. Regelson and A. E. Munson, *Ann. N.Y. Acad. Sci.* **173**, 831 (1970).
102. P. S. Morahan, W. Regelson, and A. E. Munson, *Antimicrob. Agents Chemother.* **2**, 16 (1972).
103. D. S. Breslow, *Pure Appl. Chem.* **46**, 103 (1976).
104. A. E. Munson, W. Regelson, W. Lawrence, Jr., and W. R. Wooles, *RES J. Reticuloendothel. Soc.* **7**, 375 (1970).
105. P. S. Morahan, W. Regelson, L. G. Baird, and A. M. Kaplan, *Cancer Res.* **34**, 506 (1974).
106. T. A. Du Plessis and A. C. Thomas, *J. Polym. Sci. Polym. Chem. Ed.* **11**, 2681 (1973).
107. M. Kurokawa and Y. Minoura, *J. Polym. Sci. Polym. Chem. Ed.* **17**, 3297 (1979).
108. U.S. Pat. 3,853,829 (Dec. 10, 1974), A. Priola, S. Cesca, and G. Ferraris (to Snam Progetti).
109. P. D. Trivedi, *Polym. Bull.* **1**, 433 (1979).
110. P. D. Trivedi, *J. Macromol. Sci. Chem.* **A14**, 589 (1980).
111. S. Iwatsuki and T. Itoh, *Macromolecules* **12**, 208 (1979).
112. R. Vukovic, V. Kuresevic, and D. Fles, *J. Polym. Sci. Polym. Chem. Ed.* **15**, 2891 (1977).
113. K. Fujimori and N. A. Wickramasinghe, *Aust. J. Chem.* **33**, 189 (1980).
114. K. Soga, S. Hosoda, Y. Tazuke, and S. Ikeda, *J. Polym. Sci. Polym. Lett. Ed.* **13**, 265 (1975).
115. Jpn. Kokai 76 38,386 (Mar. 31, 1976), S. Ikeda (to Tokyo Institute of Technology).
116. N. G. Schnautz, *J. Polym. Sci. Polym. Chem. Ed.* **14**, 1045 (1976).
117. H. Inoue and T. Otsu, *J. Polym. Sci. Polym. Chem. Ed.* **14**, 845 (1976).
118. T. Kokubo, S. Iwatsuki, and Y. Yamashita, *Macromolecules* **1**, 482 (1968).
119. K. Soga, M. Sato, S. Hosoda, and S. Ikeda, *J. Polym. Sci. Polym. Lett. Ed.* **13**, 543 (1975).
120. K. Fujimori, *J. Macromol. Sci. Chem.* **A10**, 999 (1976).
121. T. Higashimura, M. Mitsuhashi, and M. Sawamoto, *Macromolecules* **12**, 178 (1979).
122. N. Kushibiki, M. Irie, and K. Hayashi, *J. Polym. Sci. Polym. Chem. Ed.* **13**, 77 (1975).
123. U.S. Pat. 3,741,946 (June 26, 1973), W. E. Daniels (to GAF).
124. U.S. Pat. 3,752,788 (Aug. 14, 1973), M. Hirooka and K. Mashita (to Sumitomo).
125. Ger. Offen. 2,065,345 (May 3, 1973), M. Hirooka, K. Takeya, Y. Uno, A. Yamane, and K. Maruyama (to Sumitomo).
126. T. Alfrey, Jr., and C. C. Price, *J. Polym. Sci.* **2**, 101 (1947).
127. G. E. Ham in N. M. Bikales, ed., *Encyclopedia of Polymer Science and Technology*, Vol. 4, Wiley-Interscience, New York, 1966, pp. 165–244.
128. H. Gilbert and co-workers, *J. Am. Chem. Soc.* **78**, 1669 (1956).
129. M. Tsukino and T. Kunitake, *Polym. J.* **13**, 671 (1981).
130. R. J. Samuels, *Polymer* **18**, 452 (1977).
131. Ger. Offen. 2,008,990 (Sept. 9, 1971), K. Pfitzner, W. Bruemmer, and M. Klockow (to Merck).

132. B. H. Waxman, private communication, GAF Corporation, Wayne, N.J., 1981.

EUGENE V. HORT
R. C. GASMAN
GAF Corporation

N-VINYL MONOMERS AND POLYMERS

MONOMERS

N-Vinylamines

In 1888, Gabriel reported the synthesis of vinylamine by dehydrobromination of 2-bromoethylamine with silver oxide (1). Subsequently, however, it was shown that the product obtained was not vinylamine, but its isomer, ethyleneimine (2). Many other attempts to isolate vinylamine [593-67-9] have been unsuccessful, although its presence as a transient intermediate has been detected by spectroscopic methods (3). Like its oxygen analogue, poly(vinyl alcohol), polyvinylamine [26336-38-9] has been prepared by modifying another polymer. Typical of such preparations is hydrolysis of polyvinylacetamide (4):

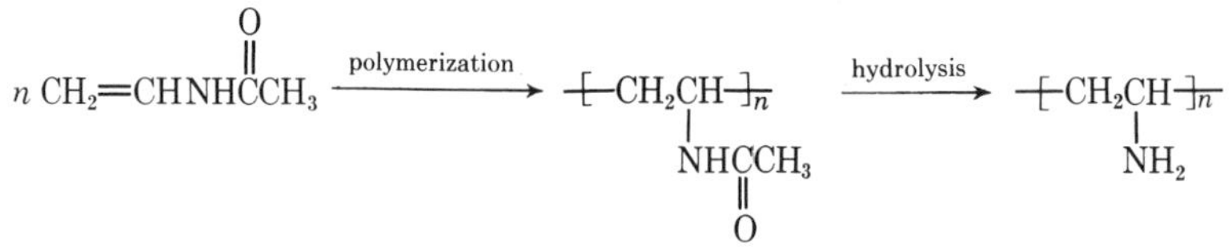

$$n\ CH_2{=}CHNHCCH_3 \xrightarrow{\text{polymerization}} {\left[CH_2CH \right]}_n \xrightarrow{\text{hydrolysis}} {\left[CH_2CH \right]}_n$$

The monoalkylvinylamines have a similar history. In 1956, a series of monoalkylvinylamines was supposedly prepared by the vinylation of alkylamines with acetylene (5). Both alkylvinylamines and alkyldivinylamines were described as products. A few years later, these products were shown not to be vinylamines, but the isomeric vinylideneamines, which can also be obtained by condensing aldehydes with amines (6):

$$CH_3CHO + C_2H_5NH_2 \rightarrow CH_3CH{=}NHC_2H_5 + H_2O$$

Dialkylvinylamines are moderately stable at low temperature in the absence of oxygen. They are prepared by treatment of a secondary amine with acetaldehyde (7) or with acetylene (8) (see Acetylene-derived chemicals; Amines, lower aliphatic):

$$CH_3CHO + (C_2H_5)_2NH \rightarrow (C_2H_5)_2NCH{=}CH_2 + H_2O$$

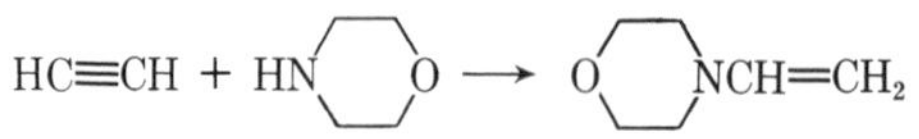

$$HC{\equiv}CH + HN\underset{}{\bigcirc}O \rightarrow O\underset{}{\bigcirc}NCH{=}CH_2$$

Trialkylamines react with acetylene to give quaternary vinyltrialkyl ammonium compounds; for example, acetylene and trimethylamine give the naturally occurring base, neurine (9). Excess amine serves as catalyst:

$$(CH_3)_3N + HC\equiv CH + H_2O \rightarrow (CH_3)_3\overset{+}{N}CH=CH_2\ OH^-$$

$$(CH_3)_3\overset{+}{N}H\ Cl^- + HC\equiv CH \rightarrow (CH_3)_3\overset{+}{N}CH=CH_2\ Cl^-$$

Weakly basic aromatic secondary amines, such as pyrrole, indole, and carbazole, give relatively stable vinylamines (see Amines, aromatic). Such compounds are prepared by direct vinylation (10–11), dehydration of the hydroxyethyl derivatives (12), or by exchange with another vinyl compound such as a vinyl ether (13):

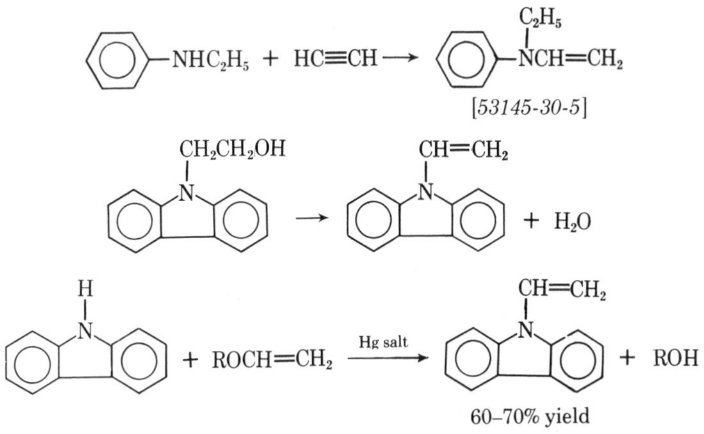

[53145-30-5]

60–70% yield

The only commercially available vinylamines are *N*-vinylcarbazole and 1-vinylimidazole. They are manufactured on a low volume basis by BASF in the FRG by vinylation of the heterocyclic bases with acetylene.

Boiling and melting points of some vinylamines are given in Table 1.

Table 1. Physical Properties of Selected Vinyl Amines

Compound	CAS Registry No.	Bp, $°C_{kPa}{}^a$	Mp, °C
N,N-dimethylvinylamine	[5763-87-1]	37–38	
N,N-diethylvinylamine	[6053-97-0]	65–71	
N-methyl-*N*-phenylvinylamine	[7025-99-2]	$98–99_{2.1}$	
N,N-diphenylvinylamine	[4091-13-8]	$105–110_{0.2}$	52–54
1-vinylpyrrole	[13401-81-5]	122	
1-vinylindole	[1557-08-0]	$71–72_{0.13}$	29–30
N-vinylcarbazole	[1484-13-5]	$175–178_{2.0}$	64
1-vinylimidazole	[1072-63-5]	$78–79_{1.7}$	

^{a} To convert kPa to mm Hg, multiply by 7.5. Pressure = 101.3 kPa if not shown.

N-Vinyl Isocyanate

A particularly interesting N-vinyl monomer is vinyl isocyanate, bp 32°C, d_4^{20} 0.9388, n_D^{20} 1.4167. It is prepared by heating acryloyl chloride with sodium azide (14), by combining hydrocyanic acid and acetylene (15), or by cracking carbamoyl chlorides (16) (see Isocyanates, organic):

$$CH_2{=}CHCOCl + NaN_3 \rightarrow CH_2{=}CHNCO + N_2 + NaCl$$
$$HNCO + HC{\equiv}CH \rightarrow CH_2{=}CHNCO$$

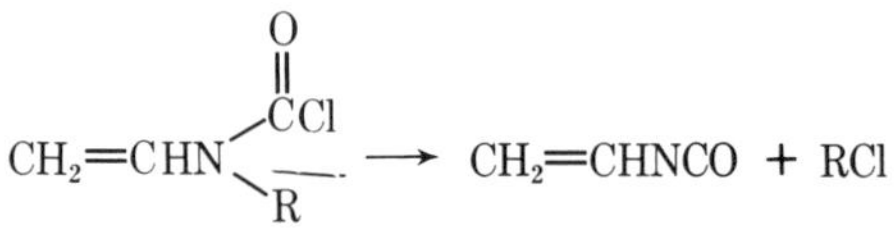

Vinyl isocyanate is polymerized either through the vinyl groups to give a polymer with pendant isocyanate groups or through the isocyanate to give a polymer with pendant vinyl groups (14) (see under Polymers for monomer properties).

N-Vinylamides and N-Vinylimides

Amides and imides are readily vinylated with acetylene (17) by dehydration of hydroxyethyl substituents (18), by pyrolysis of ethylidenebisamides (4), or by vinyl exchange (19), among other methods; the monomers are stable:

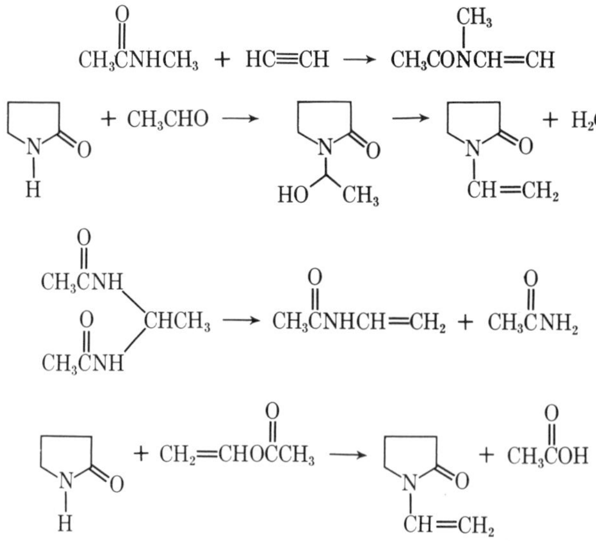

Only N-vinyl-2-pyrrolidinone is of commercial importance. Vinylcaprolactam is available on a small scale as a developmental chemical.

Boiling and melting points of some vinylamides and vinylimides are given in Table 2.

Table 2.   Physical Properties of Selected Vinylamides and Vinylimides

Compound	CAS Registry No.	Bp, °C$_{kPa}$^{a}	Mp, °C
N-vinylacetamide	[5202-78-8]	107–109	
N,*N*-methylvinylacetamide	[3195-78-6]	70$_{3.3}$	
N-vinylacetanilide	[4091-14-9]	102–105$_{0.13}$	52
N-vinyl-2-piperidinone	[4370-23-4]	125–126$_{3.3}$	45
N-vinylcaprolactam	[2235-00-9]	129–130$_{2.7}$	34.5
N-vinylphthalimide	[3485-84-5]	128–130$_{3.3}$	86.5
N-vinyl-2-oxazolidinone	[4271-26-5]	77–78$_{0.067}$	
N-vinyl-5-methyl-2-oxazolidinone	[3395-98-0]	105–108$_{0.33}$	

^{a} To convert kPa to mm Hg, multiply by 7.5. Pressure = 101.3 kPa if not shown.

N-Vinyl-2-Pyrrolidinone

N-Vinyl-2-pyrrolidinone, commonly called vinylpyrrolidone or VP, was developed in Germany at the beginning of World War II (20). The present method of manufacture is the same as the original synthesis, vinylation of 2-pyrrolidinone. It is mainly used to prepare the homopolymer, poly(*N*-vinyl-2-pyrrolidinone) [9003-39-8] (PVP), but considerable amounts are also used as a comonomer.

Properties.   Physical properties are given in Table 3.

The chemistry of *N*-vinyl-2-pyrrolidinone is characterized by reactions of the vinyl double bond.

Hydrogenation.   Catalytic hydrogenation gives essentially quantitative yields of *N*-ethyl-2-pyrrolidinone (21).

Addition of Hydroxyl Compounds.   Hydroxyl compounds add to the double bond in the expected Markovnikov direction. Alcohols require acid catalysis, but phenols, because of their acidity, react exothermically without a catalyst (22–23):

Table 3.   Properties of *N*-Vinyl-2-Pyrrolidinone

Property	Value
freezing point, °C	13.5
boiling point, °C$_{kPa}$^{a}	46$_{0.1333}$
	88$_{1.33}$
	147$_{13.3}$
	219$_{101.3}$
density, g/cm^3 at 25°C	1.04
refractive index, n_D^{25}	1.511
viscosity at 25°C, mPa·s (= cP)	2.07
flash point, open cup, °C	98.4
fire point, °C	100.5

^{a} To convert kPa to mm Hg, multiply by 7.5.

In aqueous acid, a hemiacetal type of adduct presumably forms and readily decomposes to 2-pyrrolidinone and acetaldehyde (24):

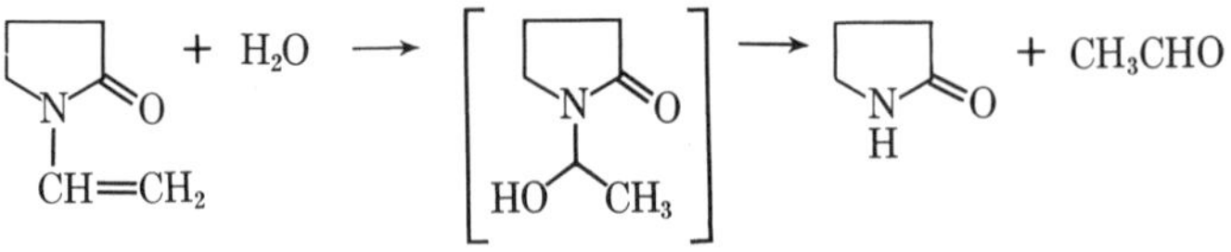

Addition of Amides.   Amides add across the vinyl double bond in the same fashion as hydroxyl groups. Of particular interest is the addition of 2-pyrrolidinone. Treatment of *N*-vinyl-2-pyrrolidinone with aqueous mineral acids also gives this adduct because of partial hydrolysis of the starting material which provides 2-pyrrolidinone (25):

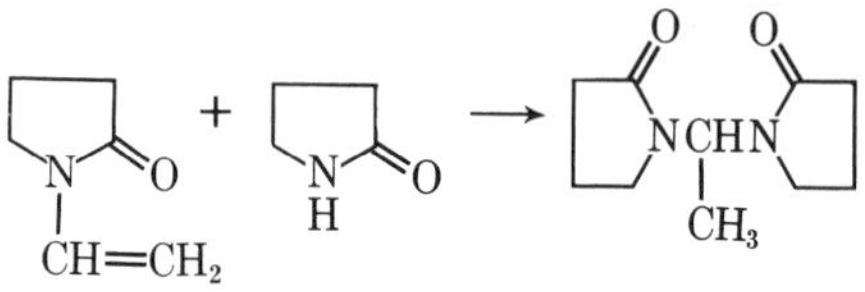

Dimerization.   Fluoroacetic acid catalysis at 85–90°C gives a >90% yield of the dimer, 1,1′-(3-methyl-1-propene-1,3-diyl)bis-2-pyrrolidinone (26):

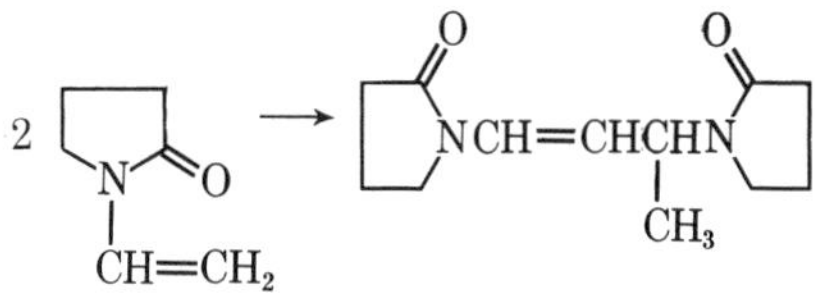

Manufacture.   The principal manufacturers of *N*-vinyl-2-pyrrolidinone are GAF and BASF. Both consume most of their production captively as a monomer for the manufacture of PVP and copolymers. The vinylation of 2-pyrrolidinone is carried out under alkaline catalysis analogous to the vinylation of alcohols. 2-Pyrrolidinone is treated with ca 5% potassium hydroxide, then water and some pyrrolidinone is distilled at reduced pressure. A ca 1:1 mixture (by vol) of acetylene and nitrogen is heated at 150–160°C and ca 2 MPa (22 atm). Fresh 2-pyrrolidinone and catalyst are added continuously while product is withdrawn. Conversion is limited to ca 60% to avoid excessive formation of by-products. The *N*-vinyl-2-pyrrolidinone is distilled at 70–85°C at 670 Pa (5 mm Hg) and the yield is 70–80% (27).

Shipment and Storage; Specifications.   *N*-Vinyl-2-pyrrolidinone is available in tank cars and tank trailers and in drums of various sizes. Shipping containers are normally steel or stainless steel. Tank cars are provided with heating coils to facilitate unloading in cold weather. Rubber, epoxy, and epoxy–phenolic coatings are attacked and must be avoided. Carbon steel has been successfully used for storage tanks, but stainless steel preserves product-quality better. Aluminum and certain phenolic coatings are also satisfactory.

Commercial *N*-vinyl-2-pyrrolidinone is stabilized either with sodium hydroxide flake (from which it may readily be decanted or filtered) or with an amine stabilizer, eg, *N,N'*-di-*sec*-butyl-*p*-phenylenediamine.

Purity, as determined by gas chromatography, is specified as 98.5% minimum; and water content, as determined by Karl-Fischer analysis, is specified as 0.2% maximum. As sold, the freezing point is 13.0°C minimum and APHA color 60 maximum. *N*-Vinyl-2-pyrrolidinone tends to darken on standing.

Health and Safety Factors. Contamination of *N*-vinyl-2-pyrrolidinone with strong acids must be avoided to prevent exothermic polymerization reactions.

Tests indicate that *N*-vinyl-2-pyrrolidinone is neither a skin sensitizer nor a primary irritant. The small amount of flake caustic used as stabilizer is corrosive, however, and.must not be allowed to come in contact with skin or eyes. Because of its low vapor pressure, *N*-vinyl-2-pyrrolidinone does not ordinarily present a vapor hazard. Acute oral toxicity for white rats has been determined as follows:

Dosage	mL/kg
LD_0	1.0
LD_{50}	1.5
LD_{100}	2.5

Uses. The principal use of *N*-vinyl-2-pyrrolidinone is as a monomer for the preparation of PVP homopolymer and various copolymers. In copolymers, *N*-vinyl-2-pyrrolidinone frequently imparts hydrophilic properties and is useful for cosmetic applications of polymers, viscosity index improvers, and soft contact lenses (qv). Its use as a component in radiation-curable monomer operations is growing rapidly. It serves as a reactive diluent which reduces viscosity and increases curing rates.

Because of monomer instability, poly(*N*-vinylamine) and its monoalkyl analogues can not be derived from their respective monomers. Poly(*N*-vinylamine) has been prepared by hydrolysis of poly(*N*-vinylphthalimide) [26809-43-8] (28) or poly(*N*-vinyl carbamate)s (29). It is a strong base, it is oxidized in air, and readily reacts with acid chlorides, isocyanates, and acid anhydrides (29).

Disubstituted *N*-vinylamines are more stable than the monosubstituted ones; aryl substitution imparts more stability than alkyl groups. Among aromatic, disubstituted *N*-vinylamines, the best characterized is *N*-vinylcarbazole (NVC). It is easily polymerized by radical or cationic catalysis; high polymers are obtained when initiation is by a 1% solution of boron fluoride-dibutyl ether complex in trichloroethylene or by benzoyl peroxide, sodium peroxide, sodium perborate, or sodium chromate in alkaline emulsion (30).

Poly(*N*-vinylcarbazole) [25064-59-8] can be cast into films from chloroform, toluene, methylene chloride, and ethylene chloride. It may also be extruded at 260–280°C into fibrous powders used for molding articles of improved strength (30). The polymers of *N*-vinylcarbazole possess excellent electrical properties, particularly in regard to specific resistance and dielectric constant. The physical properties of poly(*N*-vinylcarbazole) are compared with those of its styrene copolymers and polystyrene in Table 4. Poly(*N*-vinylcarbazole) is resistant to boiling water, dilute acids and bases, alcohols, mineral oils, turpentine, and carbon tetrachloride. It is, however, readily attacked by strong oxidative acids (see Styrene plastics).

Table 4. Physical Properties of Poly(*N*-Vinylcarbazole) and Its Styrene Copolymers

Property	Poly(*N*-vinyl-carbazole)	Polystyrene	NVC/S (85:15)	NVC/S (70:30)
density, g/cm^3	1.2	1.1	1.2	1.2
coefficient of thermal expansion, $10^{-6} \times °C^{-1}$	40	60–80	40	40
dielectric constant (10^3–10^6 Hz)	3	2.5	3	3
specific resistance by volume, 50% rh, 25°C, (Ω-cm)	10^{15}–10^{16}			

POLYMERS

N-Vinyl monomers are enamines (or derivatives thereof) and possess high electron density at the β-carbon. It is not surprising, therefore, that e values cited for most *N*-vinyl monomers are negative, as indicated in Table 5 (see Copolymers).

The *N*-vinylimides and *N*-vinylamides, especially the cyclic analogues, where one or two electron-withdrawing moieties are attached to the nitrogen atom, are relatively stable and readily polymerized. *N*-Vinylimides, despite two carbonyl groups attached to the nitrogen atom, still possess negative e values. In general, they are readily polymerized by radical catalysts but their ability to polymerize under cationic catalysis is poor. *N*-Vinylamides, most notably *N*-vinylpyrrolidinone, are readily polymerized by both cationic and free-radical catalysis and yield stable polymers with a variety of molecular weights. They also readily copolymerize under free-radical conditions with monomers of negative e value, eg, vinyl acetate, $e = -0.22$, and with positive e value, eg, methyl methacrylate, $e = 0.40$.

N-Vinylcaprolactam polymerizes readily in the presence of Hg^{2+}, Sb^{3+}, and Sb^{5+} chlorides or with free-radical initiators such as cumene hydroperoxide, t-butyl peroxide, t-butyl perbenzoate, or azobisisobutyronitrile (33) (see Initiators). It has been copolymerized with vinyl formate, vinyl acetate, methyl acrylate, methyl methacrylate, acrylamide, acrylonitrile, methyl vinyl ether, and isopropenyl methyl ether (34). Because it possesses many of the substantive and solubility parameters of its lower molecular weight analogue, poly(*N*-vinyl-2-pyrrolidinone), *N*-vinylcaprolactam is useful in the vat dye, photographic, cosmetic, and paper industries. Poly(*N*-vinylcaprolactam)

Table 5. Properties of *N*-Vinyl Monomers

Compound	CAS Registry No.	Reactivity, Q	Electron acceptance, e	Ref.
N-vinylcarbazole	[1484-13-5]	0.41	−1.40	31
N-vinyl-2-pyrrolidinone	[88-12-0]	0.14	−1.14	31
vinyl isocyanate	[3555-94-0]	0.16	−0.70	31
N-vinylacetanilide	[4091-14-9]	0.11	−2.08	32
N-vinylsuccinimide	[2372-96-5]	0.13	−0.34	31
N-vinylphthalimide	[3485-84-5]	0.36	−1.52	32
N-vinylcaprolactam	[2235-00-9]	0.081	−1.55	33
N,*N*-divinylaniline	[7178-41-8]	0.19	−1.54	31
N-vinylpyridinium fluoroborate	[658-21-9]	0.12	−2.12	31

[*25189-83-7*], unlike PVP, precipitates from aqueous solution above 30–35°C. It is water soluble, but it is significantly less hygroscopic than PVP at high relative humidity, and is effective as a hairspray resin (34) (see Hair preparations).

Poly(N-Vinyl-2-Pyrrolidinone)

Poly(*N*-vinyl-2-pyrrolidinone) (PVP) is undoubtedly the best characterized and most widely studied *N*-vinyl polymer. It derives its commercial success from its biological compatibility, low toxicity, film-forming and adhesive characteristics, unusual complexing ability, and relatively inert behavior toward salts, acids, and thermal degradation in solution. These properties have suggested many medicinal applications.

In the United States, PVP is sold in both pharmaceutical and technical grades. The pharmaceutical grades, known generically as povidone, are marketed under the trade names Plasdone (GAF Corporation) and Kollidon (BASF). The technical grades are manufactured under a variety of names, such as PVP, Peregal ST, Albigen A, and Luviskol. Special cross-linked grades of PVP known as Crospovidone, PVPP, and polyvinylpyrrolidone are sold to the pharmaceutical industry under the trade names Polyplasdone XL (GAF) and Kollidon CL (BASF), and to the beer (qv) and wine (qv) industries under the trade name Polyclar.

First developed in Germany at I. G. Farben during the 1930s, PVP was subsequently widely used in Germany as a blood-plasma substitute and extender during World War II. In the United States, it has been manufactured since 1956 by GAF, the sole domestic supplier of PVP. GAF also makes the starting chemicals for its production at Calvert City, Kentucky, and Texas City, Texas; the latter plant became operational in 1969.

Properties. Poly(*N*-vinyl-2-pyrrolidinone) is described in the *United States Pharmacopia* (35) as consisting of linear *N*-vinyl-2-pyrrolidinone groups of varying degrees of polymerization. The molecular weights of PVP samples are determined by osmometry, ultra-centrifugation, light-scattering photometry, and solution viscosity techniques (36). The most frequently employed method of determining and reporting the molecular weight of PVP samples utilizes solution viscosity and, therefore, the viscosity–average molecular weight $\overline{M}_v$ is obtained.

A frequently used and commonly recognized method of distinguishing between different molecular weight grades of PVP is the K value. Its nomenclature is accepted by the USP, FDA, and other authoritative bodies worldwide. The K value is usually determined at 1% wt/vol of a given PVP sample in aqueous solution. The relative viscosity is obtained with an Ostwald-Fenske or Cannon-Fenske capillary viscometer, and the K value is derived from Fikentscher's equation (37):

$$\log \frac{\eta_{rel}}{c} = \frac{75\,K_o^2}{1 + 1.5\,K_o c} + K_o$$

where $K = 1000\,K_o$, rel = relative, and c = concentration of the solution in g/100 mL. Solving directly for K, the Fikentscher equation is converted to:

$$K = [\sqrt{300\,c \log Z + (c + 1.5\,c \log Z)^2} + 1.5\,c \log Z - c]/(0.15\,c + 0.003\,c^2)$$

where $Z = \eta_{rel}$.

The intrinsic viscosity $[\eta]$ is defined as:

$$[\eta] = \lim_{c \to o} \left(\frac{\ln \eta_{rel}}{c} \right)$$

and is usually determined by measuring the relative viscosity at a number of concentrations and extrapolating $\ln \eta_{rel}/c$ to zero concentration. It may, however, be approximated from the Fikentscher equation by:

$$[\eta] = 2.303 \, (0.001 \, K + 0.0000 \, 75 \, K^2)$$

where $[\eta]$ = intrinsic viscosity and K = K value of sample.

Utilizing the Mark-Houwink equation (38–39):

$$[\eta] = k\overline{M}_v^a$$

where k = the Mark-Houwink constant, it is possible to relate the viscosity–average molecular weight $(\overline{M}_v)$ to the K value.

In the past, the two principal worldwide suppliers of PVP, GAF and BASF, have reported different $\overline{M}_v$ for products with equivalent K values. The difference does not arise from differences in the actual material, but rather from the use of different values for the constants k and a in the Mark-Houwink equation. Previously, GAF employed values for k and a of 1.6×10^{-5} and 0.9, respectively, whereas BASF used the values $k = 1.4 \times 10^{-4}$ and $a = 0.7$. Recently, all manufacturers have elected to adopt the constants $k = 1.4 \times 10^{-4}$ and $a = 0.7$ (38–39). A comparison of the viscosity–average molecular weight for various K-value grades of PVP is given in Table 6. The old values reflect the earlier GAF determinations and the new values reflect the use of recently proposed constants (38–39); the latter are currently used by both GAF and BASF.

Because many earlier disclosures utilized the older GAF determinations of $\overline{M}_v$, it is suggested that the reader employ the K value when attempting to repeat or verify earlier experiments or formulations. Table 7 lists PVP grades with their K values and calculated $\overline{M}_v$ values.

Table 6. PVP K Values and Viscosity–Average Molecular Weight ($\overline{M}_v$)

K Value	$[\eta]^a$	Old GAF $\overline{M}_v^{\,b}$	New GAF $\overline{M}_v^{\,c}$
10	0.0403	6,010	3,260
12	0.0525	8,070	4,760
15	0.0734	11,700	7,680
17	0.0891	14,500	10,100
20	0.1151	19,300	14,600
25	0.1655	28,900	24,500
28	0.1999	35,600	32,100
29	0.2120	38,000	34,900
30	0.2245	40,500	37,900
31	0.2373	43,100	41,000
32	0.2505	45,800	44,300
50	0.5469	109,000	135,000
60	0.7599	157,000	216,000
80	1.2894	283,000	460,000
85	1.4434	321,000	541,000
90	1.6061	360,900	630,000
100	1.9572	450,000	836,000

^{a} $[\eta] = 2.303 \, (0.001 \, K + 0.000075 \, K^2)$.

^{b} $\overline{M}_v = \left(\dfrac{[\eta]}{1.6 \times 10^{-5}}\right)^{1/0.9}$

^{c} $\overline{M}_v = \left(\dfrac{[\eta]}{1.4 \times 10^{-4}}\right)^{1/0.7}$

Table 7.   Viscosity–Average Molecular Weights of Commercial Grades of PVP

Grade	Manufacturer	Calculated $\overline{M}_v$
Technical		
PVP K-15	GAF	7,680
K-30	GAF	37,900
K-60	GAF	216,000
K-90	GAF	630,000
Luviskol 90	BASF	630,000
Peregal ST, K-30	GAF	37,900
Albigen A, K-30	BASF	37,900
Pharmaceutical		
Plasdone C-15[a]	GAF	7,680
C-30[a]	GAF	37,900
K-25	GAF	24,500
K-26/28	GAF	29,400
K-29/32	GAF	40,000
K-90	GAF	630,000
Kollidon K-12 PF	BASF	4,760
K-17 PF	BASF	10,000
K-25	BASF	24,500
K-30	BASF	37,900
K-90	BASF	630,000

[a] Pyrogen-free.

Poly(*N*-vinyl-2-pyrrolidinone) is soluble in a variety of organic solvents. The solubility behavior of PVP toward organic solvents is given in Table 8; 5% solutions in heptane, Stoddard's solvent, kerosene, and toluene may be prepared from a 25% solution of PVP in butanol. Solutions in the chlorofluoroalkane propellants can be made by using a 20–30% PVP in ethanol.

Solubility in water is limited only by the viscosity of the resulting solution. The heat of solution is -4.81 kJ/mol (-1.15 kcal/mol) (40); aqueous solutions are slightly acidic (pH 4–5).

Table 8.   Solubility of PVP in Organic Solvents

Soluble[a]	Insoluble
alcohols	hydrocarbons
acids	ethers
esters	esters
ethyl lactate	ethyl acetate
ketones	*sec*-butylacetate
methylcyclohexanone	ketones
chlorinated hydrocarbons	2-butanone
methylene dichloride	acetone
chloroform	cyclohexanone
ethylene dichloride	chlorinated hydrocarbons
amines	chlorobenzene
glycols	
lactams	
nitroparaffins	

[a] Minimum of 10 wt % PVP dissolves at room temperature.

Figure 1 illustrates the kinematic viscosity of three grades of PVP in aqueous solution. The kinematic viscosity of PVP K-30 in various organic solvents is given in Table 9. Poly(*N*-vinyl-2-pyrrolidinone) is a gelling agent for polar solvents such as halogenated alkanes, halogenated alkenes, tertiary amines, ketones, and esters when mixed with C_1–C_5 alcohols (41).

Under ordinary conditions, PVP is stable as a solid and in solution (42). The solid

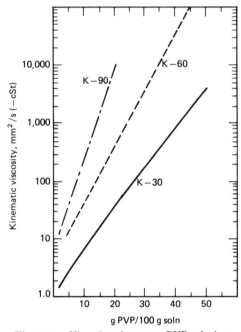

Figure 1.   Viscosity of aqueous PVP solutions.

Table 9.   Viscosity of PVP K-30 in Organic Solvents

Solvent	2% PVP, mm²/s (= cSt)	10% PVP, mm²/s (= cSt)
acetic acid, glacial	2	12
1,4-butanediol	101	425
butyrolactone	2	8
cyclohexanol	80	376
diacetone alcohol	5	22
diethylene glycol	39	165
ethanol, absolute	2	6
ethyl lactate	4	18
ethylene glycol	24	95
ethyl ether	3	12
glycerol	480	2046
2-propanol	4	12
methylcyclohexanone	3	10
N-methyl-2-pyrrolidinone	2	8
methylene dichloride	1	3
nonylphenol	3300	
propylene glycol	66	261
triethanolamine	156	666

tolerates heating in air for 16 h at 100°C, but darkening and loss in solubility occurs at 150°C. At pH 12, the polymer gels irreversibly within four hours at 100°C. In strong acid solution, PVP is unusually stable with no change in appearance or viscosity for two months at 24°C in 15% HCl. Studies of various thickening agents for acid gelling showed only PVP to be stable in 15% HCl at 107°C (43). However, viscosity increases in concentrated hydrochloric acid, and in concentrated nitric acid PVP forms a stable gel (44).

The glass-transition temperature of PVP is 175°C (45). The melt viscosity is too high for typical thermoplastic forming operations.

Films of PVP are clear, transparent, glossy, and hard. They can be cast from water, methyl alcohol, chloroform, or ethylene dichloride. Dried film from PVP K-30 has a specific gravity of $d_4^{25} = 1.25$ and a refractive index of $n_D^{25} = 1.53$ (46). It is comparatively hygroscopic and is between carboxymethyl cellulose (CMC-70) and poly(vinyl alcohol) (PVA) in absorption of water at 30–90% rh, with CMC-70 > PVP > PVA. At 70% rh, PVP films become tacky, and at 50% rh, they contain 18% moisture.

A number of synthetic and natural resins can be combined with PVP to yield clear solutions and films. Among these compatible resins are ethyl cellulose, polyethylene, poly(vinyl chloride), poly(vinyl alcohol), poly(vinyl methyl ether), shellac, corn dextrin, and polyacrylonitrile (1:3) (see also Resins, water-soluble). Incorporation of high molecular weight PVP increases the transparency of polyamides (47) and improves dye receptivity of cellulose derivatives (48). Combinations of cellulose and PVP are also used as hemodialysis membranes (49).

The single most attractive property of PVP is its binding capability. This property has permitted utilization in numerous commercial applications. Small quantities of PVP stabilize aqueous emulsions (qv) and suspensions, apparently by its absorption as a thin layer on the surface of individual colloidal particles. Thus, PVP K-90 is effective in controlling the particle size in the suspension polymerization of styrene (50) and vinyl chloride (51), and it is used as a suspending agent in granulated feed (52). Its suspending ability is a chief reason for its wide use in pill tableting and capsule granulation (53).

Because of its strong complexing ability, PVP improves dye receptivity (48). It stabilizes analytical reagents against oxidation (54). Recent investigations into the nature of the complexation of PVP are contradictory regarding the significance of hydrophobic interactions in the binding of monomolecular species to the polymer (55–57). A loss in complexing ability of PVP occurs at molecular weights of 1000–5000 (57).

A study of the interaction of PVP with albumin and salicylic acid indicates that the PVP reduces the binding of albumin to salicylic acid (58). Poly(*N*-vinyl-2-pyrrolidinone) stabilizes hydrogen peroxide (59). Complexation with PVP controls the release of CO_2 (60). Heavy-metal salt complexes with PVP (61) have been used in computed tomographic (CT) diagnosis of tumors (62) (see Chelating agents; Medical diagnostic reagents; X-Ray technology). The single most widely studied and best characterized PVP complex is that of PVP–iodine [*25655-41-8*]. Hydrogen triiodide forms a complex with PVP that is so stable that there is no appreciable vapor pressure. It is superior to tincture of iodine as a germicide (63–65) (see Disinfectants and antiseptics).

Polymerization and Manufacture. *N*-Vinyl-2-pyrrolidinone (VP) is readily polymerized with cationic, eg, boron trifluoride (21), or anionic initiators, eg, potassium amide (66) and alkali metals and their oxides (67). Anionic initiators yield a crosslinked polymer which is insoluble in water, strong mineral salts, caustic, and organic solvent. It is sold as Polyclar to the beverage trade where it is used as a clarifier, and to the pharmaceutical trade as Polyplasdone, a tablet disintegrant.

Poly(*N*-vinyl-2-pyrrolidinone) is manufactured by bulk, solution, or suspension polymerization under free-radical catalysis. Bulk polymerization was first employed in the FRG with hydrogen peroxide as catalyst. The reaction is highly exothermic, the temperature rises rapidly from 110 to 190°C and leads to discoloration (46). Therefore, this process was subsequently refined using a 30% aqueous solution of *N*-vinyl-2-pyrrolidinone catalyzed by 0.2% hydrogen peroxide and 0.1% ammonia (46). The rate expression (68) for this polymerization is

$$\text{rate} = k \,[\text{HOOH}]^{1/2}[\text{NH}_3]^{1/4}[\text{VP}]^{3/2}$$

The rate of polymerization is proportional to the VP concentration up to 50%; beyond 60%, the rate falls off rapidly. The polymerization is relatively insensitive to pH in the range 7–12. Above pH 12, the rate slows and above 13 it is inhibited. The molecular weight of the polymer increases with the VP concentration up to 30%.

In polymerization between 40 and 70°C catalyzed with azobisisobutyronitrile, the rate of propagation (K_p) is expressed as $K_p = 1.87 \times 10^3 \times C^{-7100/RT}$ and the rate of termination (K_t) is expressed as $K_t = 1.00 \times 10^9 \times C^{-1600/RT}$ (69). Studies with various solvents indicate that the energies of termination and propagation differ significantly and are related to the complex formation between the VP and the solvent (see Table 10) (70).

A kinetic study of VP polymerization in aqueous medium showed an interaction energy of ca 8 kJ/mol (1.9 kcal/mol), which indicates a weak hydrogen bond. Hydrogen bonding interactions between water and VP apparently increases the reactivity of VP. This influence is exerted up to a ratio of 75:25 VP:H$_2$O by volume or a 1:2 molar ratio of VP:H$_2$O. More water acts as diluent, retards the propagation rate, and has no influence on $\overline{M}_v$ (71).

N-Vinyl-2-pyrrolidinone has been copolymerized with a variety of comonomers in both solution and emulsion systems (see Table 11). The reactivity ratios and propensity to copolymerize may be significantly affected by the solvent (72).

A number of *N*-vinyl-2-pyrrolidinone–vinyl acetate (VA) copolymers [25086-89-9] are marketed as hairspray resins, tablet excipients, and adhesives. The water solubilities of these resins increase with the VP–VA ratio (see Table 12).

Poly(*N*-vinyl-2-pyrrolidinone) has been used to form graft copolymers by poly-

Table 10. Termination and Propagation Activation Energies for 50% VP Solutions

Solvent	Energy of termination, kJ/mol[a]	Energy of propagation, kJ/mol[a]
water	19	68
isopropyl alcohol	13	40 ± 2
methanol	9	32 ± 3
ethyl acetate	8	23

[a] To convert J to cal, divide by 4.184.

Table 11.   Reactivity Ratios (*r*) for Free-Radical Copolymerization of *N*-Vinyl-2-pyrrolidinone (M_1)[a]

Comonomer (M_2)	r_1	r_2
acrylonitrile	0.06 ± 0.07	0.18 ± 0.07
allyl alcohol	1.0	0.0
allyl acetate	1.6	0.17
allylidene diacetate	0.92	0.94
crotonic acid[b]	0.85	0.02
maleic anhydride	0.16 ± 0.03	0.08 ± 0.03
methyl methacrylate	0.005 ± 0.05	4.7 ± 0.05
trichloroethylene	0.54 ± 0.04	<0.01
tris(trimethylsiloxy)vinylsilane	4.0	0.1
vinyl chloride	0.38	0.53
vinyl cyclohexyl ether	3.84	0.0
vinyl phenyl ether	4.43	0.22
vinylene carbonate	0.4	0.7

[a] From ref. 72, except where otherwise indicated.
[b] From ref. 73.

Table 12.   VP–Vinyl Acetate Copolymers

Designation BASF	Designation GAF	VP, %	VA, %	Solvent
	PVP/VA E-335	30	70	ethanol
	PVP/VA E-535	50	50	ethanol
	PVP/VA E-635	60	40	ethanol
	PVP/VA E-735	70	30	ethanol
Luviskol VA-37	PVP/VA I-335	30	70	2-propanol
	PVP/VA I-535	50	50	2-propanol
	PVP/VA I-735	70	30	2-propanol
Luviskol VA-64[a]	PVP/VA S-630	60	40	

[a] Powder.

merization techniques (74–76). The graft copolymers are available as latices from GAF under the trade name Polectron; polymers [*25085-37-4*] of PVP with ethyl acrylate are Polectron 130 and, with vinyl acetate, Polectron 8252 and 9452. These copolymers are compatible with synthetic and natural resins and impart oil and grease resistance, dye receptivity, improved machining qualities, and increased adhesion to a number of substrates. They are typically employed in the adhesives, cleaning agents, textile, paper, particle-binder, leather, metal, and cosmetic industries. Alternatively, alkylated homopolymers may be obtained by grafting an alkyl group onto PVP (77) and functional polymers by the simultaneous polymerization and aminoalkylation of *N*-vinyl-2-pyrrolidinone (78). These alkylated *N*-vinyl-2-pyrrolidinones, exemplified by Ganex V-516 [*53240-90-7*], are suspending aids in the paint, radiation-curing, and polymerization industries, in both aqueous and solvent medium. Gafquat 734 and 755N, copolymers [*30581-59-0*] of *N*-vinyl-2-pyrrolidinone and dimethylaminoethyl methacrylate, are widely used as additives in hair-care products.

Specifications and Analysis. The specifications for technical and pharmaceutical grades of PVP are given in Tables 13 and 14.

In powders, moisture content is determined by Karl Fischer reagent; the Cenco moisture balance is used with aqueous solutions. Residual N-vinylpyrrolidinone is determined by iodometric titration; ash by ignition; heavy metals by spectrographic emission; arsenic by standard USP method; nitrogen by Kjeldahl or Dumas methods; and acetaldehyde by hydroxylamine method. The K value is measured as specified by the second supplement of the USP (35).

Health and Safety Factors. The acute oral lethal dose (LD_0) of PVP K-30 is reported to be >100 g/kg. It is not a skin or eye irritant, or a skin sensitizer. Toleration of PVP K-30 is good by intraperitonal, intramuscular, and intravenous routes. Its use as a plasma volume expander is predicated on this tolerance. Cancer occurrence in humans has not been demonstrated for assimilation of any molecular weight of PVP by any route.

Apparently, PVP is not absorbed from the gastrointestinal tract. Studies of intravenous application indicate that the lower molecular weight material is readily excreted through the kidneys; higher molecular weight material is more slowly eliminated. Poly(N-vinyl-2-pyrrolidinone) with molecular weights over 100,000 is not readily removed and apparently is phagocytized by the cells of the reticuloendothelial system and deposited in storage sites in the liver, spleen, lung, etc. Such storage usually is not associated with pathological changes.

Table 13.   Specifications of Technical PVP Grades

Designation	Form	K range	Water, % max	Ash, % max	Residue monomers, % max
PVP K-15	powder	12–18	5	0.02	1.0
PVP K-30	powder	26–35	5	0.02	1.0
PVP K-60	aqueous solution	50–62	55	0.02	1.0
PVP K-90	aqueous solution	80–100	80	0.02	1.0
PVP K-90	powder	80–100	5	0.02	1.0
Polyclar AT	powder	cross-linked	5		

Table 14.   Specifications of Pharmaceutical PVP Grades

Assay	Value
K value	
10–15	85–115% of stated value
16–90	90–107% of stated value
moisture, % max	5
pH[a]	3.0–7.0
residue on ignition, %, max	0.02
aldehydes, %[b], max	0.02
N-vinyl-2-pyrrolidinone, %, max	0.20
lead, ppm, max	10
arsenic, ppm, max	1
nitrogen, %	11.5–12.8

[a] Of a 5% solution in distilled water.

[b] Calculated as acetaldehyde.

Uses. *Cosmetics and Toiletries.* Poly(N-vinyl-2-pyrrolidinone) and its copolymers are widely used in the hair- and skin-care industries, not merely as additives but as an integral part because of their nontoxic and sorptive behavior as well as emulsifying, thickening, emollient, and dye-solubilizing abilities (see also Cosmetics; Hair preparations).

Photographic Industry. Poly(N-vinyl-2-pyrrolidinone) acts as a protective colloid (79) and silver halide suspending agent (80). More recently, it has been claimed as a processing aid in the development of silver halide film and is used to eliminate the occurrence of dichroic stain (81). As a coating aid, PVP in silver halide emulsions reduces viscosity and increases the covering power of the developed image (82) (see also Photography).

Oil-Recovery Industry. Poly(N-vinyl-2-pyrrolidinone) is useful in various areas of oil recovery (see Petroleum, chemicals for enhanced recovery). It has been employed as an additive to cement formulations to increase viscosity and setting time while decreasing fluid loss (83–84) (see Cement).

It is an exceptionally stable acid-gelling agent for acid fracturing (43,85–91). Because of its relative insensitivity to salt concentration and environmental degradation, PVP has been claimed as a valuable tool in polymer flooding using high salt concentrations in areas containing water-sensitive clays (92–93). Studies of PVP in the area of surfactant flooding indicate that injection of an aqueous solution of PVP into the formation before injection of the surfactant greatly reduces the loss of the surfactant by adsorption on the formation (94).

Textiles. Incorporation of PVP into hydrophobic fibers such as polyacrylonitrile (95–96), polyesters (97–98), nylon (99–100), and cellulosic material (48) greatly increases their dyeability. Poly(N-vinyl-2-pyrrolidinone) has also been utilized as an anti-soil-redeposition agent (101–103), stripping agent (104), and pigment-shock reducer (105). Graft copolymers of PVP with nylon (57) exhibit improved wet-crease recovery and moisture regain (106–107).

Detergents. Poly(N-vinyl-2-pyrrolidinone) is compatible in clear, liquid, heavy-duty detergent formulations (108). It has been formulated with borax in a pretreat washing formulation (109). Owing to its detoxifying behavior, it is utilized as an essential component in formulations containing phenolic sanitizer cleaners.

Beverages. The ability of PVP to complex with certain polyphenolic compounds (tannins and others) has led to its use in the clarification and chillproofing of fruit beverages (110–111). The addition of 0.01–0.02% of soluble PVP to the brew kettle improves taste and reduces chill haze (112–113). Poly(N-vinyl-2-pyrrolidinone) is similarly employed in wines (114), vinegar, etc (115).

Pharmaceuticals. The lack of toxicity and high solubility of poly(N-vinyl-2-pyrrolidinone) have made it ideally suited for a number of pharmaceutical applications. Special grades of pyrogen-free PVP are marketed under the label Plasdone C and are available in K-15 ($\overline{M}_v$ 8000) and K-30 ($\overline{M}_v$ 38,000) molecular weights.

Poly(N-vinyl-2-pyrrolidinone) was first employed as a blood extender during World War II. It is nonantigenic, requires no cross-matching, and avoids the danger of infectious diseases inherent in blood.

As a tablet binder, PVP's solubility in both aqueous and organic solvents enables it to be used in virtually all formulations (116). The combination of high solubility and workable viscosities, enables the reduction of the volume of granulating solution, resulting in decreased drying times and cost. Moisture-sensitive drugs can be successfully granulated by utilizing PVP in anhydrous solvent (117). Water-soluble PVP (eg, Plasdone) can be dry-blended with the powder mix and then wetted with an appropriate solvent during granulation. The use of PVP aids in the production of free-flowing, compressible granulation which produces hard tablets with good dissolution rates. It is also used in sustained-release formulations. Poly(N-vinyl-2-pyrrolidinone) has also found wide acceptance as an ingredient in aqueous and solvent-based tablet-coating solutions. It improves the adhesion of the film to the tablet surface; modifies the disintegration times of films based on hydrophobic materials; acts as a plasticizer, stabilizer, and dispersant; and increases the spreadability of pigment-containing solutions.

An important application is derived from the ability to complex and solubilize poorly soluble drugs and hence increase their bioavailability and efficiency. Poly(N-vinyl-2-pyrrolidinone) is coprecipitated with a hydrophobic drug by dissolving both in a common solvent and then evaporating to dryness. The resulting complex is powdered and formulated in a dosage form. A wide number of drugs have been formulated in this manner (118–125). Alternatively, drugs with poor water solubility may be prepared as aqueous solutions by dissolving or suspending the drug in aqueous solutions of PVP (116). In tablet formulations, PVP increases the stability of certain drugs. For example, it decreases the volatility in nitroglycerin and the hydrolysis of aspirin (88–90).

Highly cross-linked poly(N-vinyl-2-pyrrolidinone), generically termed Crospovidone, N.F., and sold in the United States as Polyplasdone XL (GAF), is a tablet excipient. It is a water-insoluble but still highly hydrophilic form of the polymer, which is utilized in both wet and direct compression tableting. When exposed to water, this PVP swells and causes high stress on the surrounding tablet components and rapid disintegration (126).

A recent advance in medical technology is the development of transdermal application, which utilizes patches containing drugs capable of being absorbed through the skin. One such system employs nitroglycerin for the treatment of angina pectoris (127). Poly(N-vinyl-2-pyrrolidinone) is used as a binder to create a gel-like matrix for the drug, facilitating diffusion (see Pharmaceuticals, controlled release).

BIBLIOGRAPHY

"Polyvinylpyrrolidone" in *ECT* 1st ed., Vol. 10, pp. 759–764; "Polyvinylpyrrolidone" in *ECT* 2nd ed. under "Vinyl Polymers," Vol. 21, pp. 427–440, by A. S. Wood, GAF Corporation.

1. S. Gabriel, *Ber.* **21,** 1049 (1888).
2. C. C. Howard and W. Marckwald, *Ber.* **32,** 2036 (1899).
3. F. J. Lovas and F. O. Clark, *J. Chem. Phys.* **62,** 1925 (1975).
4. D. J. Dawson, R. D. Gless, and R. E. Wingard, Jr., *J. Am. Chem. Soc.* **98,** 5996 (1976).
5. W. Reppe and co-workers, *Ann.* **601,** 128 (1956).
6. C. W. Kruse and R. F. Kleinschmidt, *J. Am. Chem. Soc.* **83,** 213 (1961).
7. G. Laban and R. Mayer, *Z. Chem.* **7**(1), 12 (1967).
8. U.S. Pat. 3,179,661 (April 20, 1965), N. Blumenkopf and O. F. Hecht (to GAF Corporation).
9. W. Reppe and co-workers, *Ann.* **601,** 136 (1956).
10. V. P. Pivnenko, O. I. Domrin, and V. V. Dudka, *Zh. Obshch. Khim.* **44,** 1385 (1974).
11. W. Reppe and co-workers, *Ann.* **601,** 132 (1956).
12. A. Lattes and M. Riviere, *C. R. Acad. Sci. Ser. C* **262,** 1797 (1966).
13. V. D. Filimonov, E. E. Sirotkina, I. L. Gaibel, and V. I. Kulachenko, *Zh. Org. Khim.* **10,** 1790 (1974).

14. C. G. Overberger and C. J. Podsiadly, *Macromol. Synth.* **4**, 87 (1972).
15. U.S. Pat. 3,898,258 (Aug. 5, 1975), R. Van Helden and A. J. Mulder (to Shell Oil Co.).
16. U.S. Pat. 3,862,201 (Jan. 21, 1975), K.-H. Koenig and H. Kiefer.
17. W. Reppe and co-workers, *Ann.* **601**, 134 (1956).
18. J. Falbe and H. J. Schulze-Steinem, *Brennst. Chem.* **48**, 136 (1967).
19. Jpn. Pat. 74 20,583 (May 25, 1974), H. Ito and K. Kimura (to Toa Gosei Chemical Industries Co.).
20. Fr. Pat. 865,354 (May 3, 1940), H. Weese, G. Hecht, and W. Reppe (to I. G. Farbenind. A.G.).
21. C. E. Schildnecht, A. O. Zoss, and F. Grosser, *Ind. Eng. Chem.* **41**, 2891 (1949).
22. W. Reppe and co-workers, *Ann.* **601**, 81 (1956).
23. M. F. Shostakovskii and co-workers, *Izv. Akad. Nauk SSSR Otd. Khim. Nauk*, 482 (1961); *Chem. Abstr.* **55**, 27267 (1961).
24. M. F. Shostakovskii, F. P. Sidel'kovskaya, and M. G. Zelenskaya, *Izv. Akad. Nauk SSSR, Otd. Khim. Nauk*, 689 (1954); *Chem. Abstr.* **49**, 10853 (1955).
25. J. W. Breitenbach and co-workers, *Monatsh. Chem.* **87**, 580 (1956).
26. Ger. Pat. 2,239,918 (March 1, 1973), E. V. Hort (to GAF Corporation).
27. S. A. Miller, *Acetylene, Its Properties, Manufacture, and Uses*, Vol. 2, Academic Press, New York, 1965, pp. 338–339.
28. D. D. Reynolds, W. O. Kanyon, *J. Am. Chem. Soc.* **69**, 911 (1947).
29. R. Hart, *Makromol. Chem.* **32**, 51 (1959).
30. C. E. Schildknecht, *Vinyl and Related Polymers*, John Wiley & Sons, Inc., New York, 1952, pp. 658–659.
31. J. Brandrup and E. H. Immergut, *Polymer Handbook*, Interscience Publishers, a division of John Wiley & Sons, Inc., New York, 1967.
32. S. Musakaski and A. Umehara, *PVC + Polymers* **7**(1), 19 (Jan. 1967).
33. N. Cobianu and co-workers, *Mater. Plast. (Bucharest)* **10**(2), 75 (1973).
34. U.S. Pat. 3,145,147 (Aug. 18, 1964), S. A. Glickman (to GAF Corporation).
35. *USP XX Second Supplement*, The United States Pharmacopeial Convention, Inc., Rockville, Md., p. 122, released Jan. 15, 1981.
36. *PVP: An Annodated Bibliography 1951–1966*, Vol. 1, GAF Corporation, New York, 1967.
37. H. Fikentscher and K. Herrle, *Mod. Plast.* **23**, 157, 212, 214, 216, 218 (1945).
38. W. Scholtan, *Makromol. Chem.* **7**, 209 (1951).
39. J. Hengstenberg and E. Schuch, *Makromol. Chem.* **7**, 236 (1951).
40. G. Silva, *An. Fac. Quim. Farm. Univ. Chile* **18**, 126 (1966).
41. U.S. Pat. 3,566,969 (March 2, 1971), A. R. Henrickson and B. L. Atkins (to The Dow Chemical Co.).
42. *Poly Vinyl Pyrrolidone*, Technical Bulletin 2303-112, GAF Corporation, Wayne, N.J., 1982.
43. C. W. Crowe, R. C. Martin, and A. M. Michaelis, *paper presented at 55th Annual Fall Technical Conference and Exhibition of the SPE of AIME*, Dallas, Texas, Sept. 21–24, 1980.
44. Belg. Pat. 668,325 (Dec. 1, 1965), K. Buchholz (to Badische Aniline-und Soda-Fabrik).
45. Y. Y. Tan and G. Challa, *Polymer* **17**, 739 (1976).
46. C. E. Schildknecht, *Vinyl and Related Polymers*, John Wiley & Sons, Inc., New York, p. 674.
47. U.S. Pat. 3,564,075 (Feb. 16, 1971), K. H. Hermann and K. Schneider (to Farbenfabriken Bayer A.G.).
48. U.S. Pat. 3,377,412 (April 9, 1968), N. E. Franks (to American Enka Corporation).
49. U.S. Pat. 3,847,822 (Nov. 12, 1974), H. F. Shuey (to U.S. Secretary, Department of Health, Education, and Welfare).
50. Belg. Pat. 668,325 (Dec. 1, 1967), K. Buchholz (to Badische Aniline-und Soda-Fabrik).
51. U.S. Pat. 2,840,549 (June 24, 1958), P. G. McNulty and R. I. Leininger (to Diamond Alkalai Co.).
52. U.S. Pat. 3,553,313 (Jan. 5, 1971), L. P. Tort.
53. U.S. Pat. 3,558,543 (Jan. 26, 1971), G. Tohy (to Borden, Inc.).
54. U.S. Pat. 3,546,131 (Dec. 8, 1970), H. Stern and J. E. Reardon (to Uni-Tech Chemical Manufacturing Co.).
55. R. L. Reeves, S. A. Harkaway, and A. R. Sochor, *J. Polym. Sci. Polym. Chem. Ed.* **19**, 2427 (1981).
56. J. A. Plaizier-Vercammen and R. E. De Neve, *J. Pharm. Sci.* **70**, 1252 (Nov. 1981).
57. Y. E. Kirsh, T. A. Soos, and T. M. Karaputadze, *Eur. Polym. J.* **15**, 223 (1979).
58. G. A. Digenis, D. A. Wesner, A. Schwartz, and M. Schwartz, *paper presented at the American Pharmaceutical Association Academy of Pharmaceutical Sciences, 31st National Meeting*, Orlando, Fla., Nov. 15–19, 1981.

59. U.S. Pat. 3,376,110 (April 2, 1968), D. A. Shiraeff (to GAF Corporation).
60. U.S. Pat. 3,891,509 (June 24, 1975), D. R. Warren and L. W. Busse (to Clinical Convenience Products, Inc.).
61. H. G. Biedermann, W. Graf, and F. Steininger, *Chem. Ztg.* (3), (1976).
62. S. W. Young, H. H. Muller, and B. Marincek, *Radiology* **138,** 97 (Jan. 1981).
63. R. L. Goldemberg, *J. Soc. Cosmet. Chem.* **28,** 667 (1977).
64. *Ibid.,* **30,** 415 (1979).
65. *Publication No. 9620-003,* GAF Corporation, Wayne, N.J.
66. M. A. Askarov and S. N. Trubitsyna, *Khim i Fizkhm. Prirodn Sintetich Polimerov Akad. Nauk Uz· SSR Inst. Khim Polimerov* (2), 118 (1964); *Chem. Abstr.* **62,** 640 (1965).
67. U.S. Pat. 2,938,017 (May 24, 1960, F. Grosser (to GAF Corporation).
68. J. W. Copenhaver and M. H. Bigelow, *Acetylene and Carbon Monoxide Chemistry,* Reinhold Publishing Corp., New York, 1969, p. 67–74.
69. V. A. Agasandyan, E. A. Trosman, K. S. Bagdasaryan, A. D. Litmanovich, and V. Y. Shtern, *Vysokomol. Soedin.* **8,** 1580 (1966).
70. E. Senogles and R. Thomas, *J. Polym. Sci. Symp.* **49,** 203 (1975).
71. E. Senogles and R. A. Thomas, *J. Polym. Sci. Polym. Lett. Ed.* **16,** 555 (1978).
72. K. Plochocka, *J. Makromol. Sci. Rev. Macromol. Chem.* **20**(1), 67 (1981).
73. S. N. Oshakov and co-workers, *Vysokomolekul. Soedin. Ser. A* **9,** 1807 (1967).
74. U.S. Pat. 3,244,657 (April 1966), F. Grosser and M. R. Leibowitz (to GAF Corporation).
75. U.S. Pat. 1,244,659 (April 1966), F. Grosser and M. R. Leibowitz (to GAF Corporation).
76. Ger. Pat. 1,156,238 (Oct. 24, 1963), F. Grosser and M. R. Leibowitz (to GAF Corporation).
77. Brit. Pat. 1,076,543, A. Merijan, F. Grosser, and E. V. Hort (to GAF Corporation).
78. U.S. Pat. 3,563,968 (Feb. 16, 1971), A. Merijan, E. S. Barabas, and M. M. Fein (to GAF Corporation).
79. U.S. Pat. 2,495,918 (Dec. 31, 1950), E. K. Bolton (to E. I. du Pont de Nemours & Co., Inc.).
80. E. Kirgensons, *Makromol. Chem.* **6,** 30 (1951).
81. U.S. Pat. 3,552,969 (Jan. 5, 1971), R. W. Henn, N. H. King, and J. J. Surash (to Eastman Kodak).
82. K. S. Lyalikov, L. M. Zaitseva, and Z. A. Govorkova, *Tr. Leningr. Inst. Kinoinzh.* (19), 1972.
83. U.S. Pat. 3,359,225 (Dec. 19, 1967), C. F. Weisand.
84. U.S. Pat. 4,258,790 (March 31, 1981), B. W. Hale (to Western Co.).
85. U.S. Pat. 3,749,169 (July 31, 1973), J. Tate (to Texaco, Inc.).
86. U.S. Pat. 3,768,561 (Oct. 30, 1973), J. Tate (to Texaco, Inc.).
87. U.S. Pat. 3,791,446 (Feb. 12, 1974), J. Tate (to Texaco, Inc.).
88. U.S. Pat. 3,924,684 (Dec. 9, 1975), J. F. Tate (to Texaco, Inc.).
89. U.S. Pat. 3,927,717 (Dec. 23, 1975), J. F. Tate (to Texaco, Inc.).
90. U.S. Pat. 4,079,011 (March 14, 1978), J. F. Tate (to Texaco, Inc.).
91. U.S. Pat. 4,219,429 (Aug. 26, 1980), J. C. Allen and J. F. Tate (to Texaco, Inc.).
92. U.S. Pat. 3,500,925 (Dec. 31, 1970), J. P. Beiswanger (to GAF Corporation).
93. U.S. Pat. 4,045,357 (Aug. 30, 1977), M. G. Reed (to Chevron Research Co.).
94. U.S. Pat. 4,207,946 (June 17, 1980), W. C. Haltmar and E. S. Lacey (to Texaco, Inc.).
95. Brit. Pat. 849,063 (Sept. 21, 1960), (to The Dow Chemical Co.).
96. U.S. Pat. 3,296,741 (Jan. 13, 1967), H. Brian and P. Hurley (to The Dow Chemical Co.).
97. Swiss Pat. 382,388 (Nov. 30, 1964), H. Kirner (to Rohner A. G. Pratteln).
98. U.S. Pat. 2,882,255 (April 14, 1959), J. R. Caldwell (to Eastman Kodak).
99. U.S. Pat. 3,105,732 (Oct. 1, 1963), H. Artheil (to Burlington Industries).
100. Fr. Pat. 1,448,353 (Aug. 5, 1966), E. Mazat (to E. I. du Pont de Nemours & Co., Inc.).
101. R. L. Davidson and M. Shig, ed., *Water Soluble Resins,* 2nd ed., Van Nostrand Reinhold Co., New York, 1968, pp. 148–150.
102. U.S. Pat. 3,689,435 (Sept. 5, 1972), R. P. Berni and R. A. Grifo (to GAF Corporation).
103. U.S. Pat. 3,318,816 (May 9, 1967), J. Trowbridge (to Colgate-Palmolive Co.).
104. E. C. Hansen, C. A. Bergman, and D. B. Witwer, *Am. Dyest. Rep.* **43,** 72 (1954).
105. U.S. Pat. 2,820,741 (Jan. 21, 1958), C. J. Endicott and co-workers (to Abbott Laboratories).
106. U.S. Pat. 3,287,441 (Nov. 22, 1966), E. E. Mazat (to E. I. du Pont de Nemours & Co., Inc.).
107. U.S. Pat. 3,278,639 (Oct. 11, 1966), O. J. Matray (to E. I. du Pont de Nemours & Co., Inc.).
108. B. M. Milwillsky, *Soap Chem. Spec.* **39**(4), 53 (1963).
109. U.S. Pat. 3,839,214 (Oct. 1, 1974), J. W. Kiene (to U.S. Borax and Chemical Co.).

110. U.S. Pat. 3,554,759 (Jan. 12, 1971), H. Buschke, H. Reinhardt, and K. Achenbach (to Deutsche Gold-und Silber-Schneidernstalt vormals Roessler).
111. U.S. Pat. 4,166,141 (Aug. 28, 1979), D. H. Westermann and N. J. Huige (to Joseph Schlitz Brewing Co.).
112. U.S. Pat. 2,939,791 (June 7, 1960), W. D. McFarlane (to Canadian Breweries, Ltd.).
113. W. D. McFarlane, E. Wye, and H. L. Grant, *Papers European Brewing Convention 5th Congress*, Elsevier Publishing Co., New York, 1955, pp. 298–310.
114. R. A. Clemens and A. J. Martinelli, *Wines Vines* **39**, 55 (1958).
115. U.S. Pat. 3,222,180 (Dec. 7, 1965), R. Sucietto (to National Distiller and Chemicals Co.).
116. U.S. Pat. 3,557,280 (Jan. 19, 1971), H. A. Weber and A. P. Molenaar (to Koninklijke Nederlandsche Gist-en Spiritusfabrick, N.V.).
117. F. J. Prescott, F. B. Lane, and E. Hahnel, *Tex. J. Pharm.* **4**, 300 (1963).
118. I. Sugimoto and co-workers, *Drug Develop. Ind. Pharm.* **6**(2), 137 (1980).
119. S.-C. Shin, *Arch. Pharm. Res.* **2**(1), 35 (1975).
120. *Ibid.*, 49 (1979).
121. H. Sekikawa and co-workers, *Chem. Pharm. Bull.* **26**, 3033 (1978).
122. G. H. Svoboda and co-workers, *J. Pharm. Sci.* **62**, 333 (1971).
123. E. I. Stupak and T. R. Bates, *J. Pharm. Sci.* **62**, 1806 (1973).
124. *Ibid.*, **61**, 400 (1972).
125. S. A. Said and S. F. Saad, *Aust. J. Pharm. Sci.* **4**(4), 121 (1975).
126. R. F. Shangraw, J. W. Wallace, and F. M. Bowers, *Pharm. Technol.*, 44 (Oct. 1981).
127. *Bus. Week*, 122 (Jan. 11, 1982).

EUGENE V. HORT
B. H. WAXMAN
GAF Corporation

VINYL POLYMERS, POLY(VINYL FLUORIDE).   See Fluorine compounds, organic.

VINYLTOLUENE.   See Styrene.

VIRAL INFECTIONS, CHEMOTHERAPY.   See Chemotherapeutics, antiviral.

VISCOMETRY.   See Rheological measurements.

VISCOSE.   See Rayon.

VISCOSITY.   See Rheological measurements.

VISCOSITY BREAKING, VISBREAKING.   See Petroleum, refinery processes.

VITAMIN K.   See Blood, coagulants, and anticoagulants; Prostaglandins (Supplement Volume).